Mass Spectrometry *in* Cancer Research

John Roboz, Ph.D.
Professor
The Mount Sinai School of Medicine
New York, New York

CRC PRESS

Boca Raton London New York Washington, D.C.

Library of Congress Cataloging-in-Publication Data

Roboz, John, 1931-
Mass spectrometry in cancer research / John Roboz.
p. ; cm.
Includes bibliographical references and index.
ISBN 0-8493-0167-X (alk. paper)
1. Cancer—Research—Methodology. 2. Mass spectrometry. I. Title.
[DNLM: 1. Neoplasms—chemistry. 2. Antineoplastic Agents—metabolism. 3. Biological Markers—chemistry. 4. Carcinogens—metabolism. 5. Spectrum Analysis, Mass—methods. QZ 200 R666m 2001]
RC267 R634 2001
616.99′4′0072—dc21 2001043046

Direct all inquiries to CRC Press LLC, 2000 N.W. Corporate Blvd., Boca Raton, Florida 33431.

Visit the CRC Press Web site at www.crcpress.com

International Standard Book Number 0-8493-0167-X
Library of Congress Card Number 2001043046
Printed in the United States of America 1 2 3 4 5 6 7 8 9 0
Printed on acid-free paper

Dedication

For my daughters Gail and Joanne,
with love and respect

Table of Contents

About the Author xiii
Foreword xv
Mass Spectrometry: An Oncologist's Viewpoint xvii
Preface xix
Acknowledgments xxi

Terminology and Abbreviations 1

Chapter 1
Overview and Scope of Applications 5
1.1 The Essentials 5
1.2 Functions and Types of Instrument Components 7
1.2.1 Sample Introduction Systems 7
1.2.2 Ion Sources 9
1.2.3 Mass Analyzers 12
1.2.4 Ion Detectors 14
1.2.5 Vacuum Systems 15
1.2.6 Data Systems 15
1.3 Measures of Performance 15
1.3.1 Resolution and Resolving Power 15
1.3.2 Mass Range 16
1.3.3 Mass Accuracy 16
1.3.4 Scan Speed 16
1.3.5 Specificity, Sensitivity, and Limit of Detection 17
1.4 Information from Mass Spectra and Scope of Applications 17

Chapter 2
Instrumentation and Techniques of Mass Spectrometry 23
2.1 Ion Sources 23
2.1.1 Electron Ionization 23
2.1.2 Chemical Ionization 24
2.1.3 Atmospheric Pressure Chemical Ionization 25
2.1.4 Electrospray Ionization 26
2.1.4.1 Nano-Electrospray 29
2.1.4.2 Microfabricated, Silicone Chip-Based Electrospray Ion Sources 29
2.1.4.3 Multiplexed Electrospray 29
2.1.5 Fast Atom and Ion Bombardment 30
2.1.6 Matrix-Assisted Laser Desorption Ionization 31
2.1.6.1 Surface-Enhanced Laser Desorption Ionization (SELDI) 33
2.1.7 Photoionization 33
2.1.7.1 Direct Laser Desorption Ionization 33
2.1.7.2 Resonance-Enhanced Multiphoton Ionization 33

2.1.8 Inductively Coupled Plasma....34
2.1.9 Miscellaneous Ionization Techniques....34
2.1.9.1 Secondary Ion Mass Spectrometry....34
2.1.9.2 Field Ionization and Field Desorption....34
2.1.9.3 Californium-252 Plasma Desorption....35
2.1.9.4 Glow Discharge....35
2.1.9.5 Spark Source....35
2.1.9.6 Thermal Ionization....36
2.2 Mass Analyzers....36
2.2.1 Time-of-Flight Analyzers....36
2.2.1.1 Orthogonal Acceleration TOF....38
2.2.1.2 Reflectrons (Ion Mirrors)....38
2.2.1.3 Delayed Extraction....39
2.2.2 Quadrupole Analyzers....39
2.2.2.1 Quadrupoles as Ion Guides....41
2.2.2.2 Hexapoles....41
2.2.3 Quadrupole Ion Traps....41
2.2.4 Magnetic Analyzers....42
2.2.4.1 Magnetic Sector and Electrostatic Analyzers....42
2.2.5 Fourier Transform Ion-Cyclotron Resonance Analyzers....44
2.2.6 Tandem Analyzers....45
2.2.6.1 Magnetic Analyzers as Components of Tandem Systems....46
2.2.6.2 Quadrupole-Time-of-Flight Hybrid Analyzer....47
2.2.6.3 Hexapole-TOF Hybrid Analyzer....47
2.2.6.4 Accelerator Mass Spectrometry....48
2.3 Ion Current Detectors....49
2.3.1 Single Point Ion Collectors....50
2.3.1.1 Faraday Cup Detectors....50
2.3.1.2 Secondary Electron Multipliers....50
2.3.1.3 Scintillation Counters....50
2.3.2 Multipoint (Array) Detectors....51
2.4 Interfaces for Sample Introduction....52
2.4.1 Gases and Volatile Liquids....53
2.4.1.1 Reservoir and Batch Inlets....53
2.4.1.2 Membrane Inlets....54
2.4.2 Static Direct Inlet Probes....54
2.4.3 Interfaces for Coupling with Gas Chromatographs....54
2.4.4 Interfaces for Coupling with Liquid Chromatographs....55
2.4.5 Interfaces for Coupling with Capillary Electrophoresis Systems....56
2.4.6 High Throughput Inlets and Microfabricated Fluidic Systems....58
2.5 Mass Spectrometry/Mass Spectrometry (MS/MS)....59
2.5.1 Ion Activation Methods....59
2.5.1.1 Metastable Ions....60
2.5.1.2 Collision-Induced Dissociation....60
2.5.1.3 Electron-Capture Dissociation....60
2.5.1.4 Photodissociation and Surface-Induced Dissociation....60
2.5.1.5 In-Source Fragmentation....61
2.5.2 Tandem-in-Space (Transmission) Techniques....61

2.5.2.1 Product-Ion Scanning Mode 62
2.5.2.2 Precursor Scanning Mode 62
2.5.2.3 Constant Neutral Loss Scanning Mode 62
2.5.2.4 Selected Reaction Monitoring 62
2.5.2.5 Post-Source Decay 62
2.5.3 Tandem-in-Time (Trapped-Ion) Techniques 63
2.6 Identification and Quantification 63
2.6.1 Mass Spectral Libraries and Bioinformatics 63
2.6.2 Accurate Mass Measurement 64
2.6.3 Selected Ion and Reaction Monitoring 65
2.6.3.1 Selected Ion Monitoring 65
2.6.3.2 Selected Reaction Monitoring 67
2.6.4 Stable Isotope Dilution 67
2.7 Resources 69

Chapter 3
Relevant Concepts of Cancer Medicine and Biology 81
3.1 Classification and Epidemiology 81
3.1.1 Solid Tumors 81
3.1.2 Hematologic Malignancies 82
3.1.3 Incidence and Distribution of Cancers 82
3.1.4 Risk Factors 83
3.2 Cellular Proliferation and Communication 85
3.2.1 Growth Patterns 85
3.2.2 Cell Cycle 86
3.2.3 Cellular Communication and Signal Transduction 89
3.3 Genetic Abnormalities in Cancer 90
3.3.1 Cancer as a Genetic Disorder of Somatic Cells 90
3.3.2 Mutation 90
3.3.3 Oncogenes 91
3.3.4 Tumor Suppressor Genes 94
3.3.5 Carcinogenesis Requires an Accumulation of Mutations 94
3.4 Tumor Immunology 95
3.4.1 Hematopoiesis 95
3.4.2 Antigens and Antibodies 96
3.4.3 Types of Immune Response 97
3.5 Diagnosis and Treatment 98
3.5.1 Clinical Presentation 98
3.5.2 Diagnostic Workup 98
3.5.3 Treatment Modalities 100
3.5.4 Functional Status, Complications, and Causes of Death 101
3.6 Some Pertinent Experimental Methods 101
3.6.1 DNA and RNA Analysis 101
3.6.2 Polymerase Chain Reaction 102
3.6.3 Determination of Clonality 102
3.6.4 Cell Cultures 103
3.6.5 Experimental Mouse Models 104
Recommended Books 105

Chapter 4

Metabolism and Biomarkers of Carcinogens 107
4.1 Classification, Mechanism of Action, and Biomarkers 107
4.1.1 Classification and Risk Assessment 107
4.1.1.1 Risk Assessment 108
4.1.2 Metabolism, Enzymatic Activation, and Mechanism of Action 109
4.1.2.1 Absorption, Distribution, and Storage 109
4.1.2.2 Key Steps in Chemical Carcinogenesis 111
4.1.2.3 Functions of Phase I and II Enzymes 112
4.1.3 Methodologies for Biomarkers 113
4.1.3.1 DNA Adducts 113
4.1.3.2 Hemoglobin Adducts 116
4.2 Occupational Carcinogens 117
4.2.1 Aromatic Hydrocarbons 117
4.2.1.1 Benzene 117
4.2.1.2 Risk Assessment of Polycyclic Aromatic Hydrocarbons in Coke Workers 122
4.2.1.3 Gasoline and Diesel Exhaust 125
4.2.2 Aromatic Amines 125
4.2.2.1 Benzidine 125
4.2.2.2 Miscellaneous 127
4.2.3 Plastic Monomers 128
4.2.3.1 Acrylonitrile 128
4.2.3.2 Ethylene and Propylene Oxides 129
4.2.3.3 Vinyl Chloride 131
4.2.3.4 Butadiene 132
4.2.3.5 Styrene and Styrene Oxide 133
4.2.4 Exposure by Health Care Personnel to Cytotoxic Drugs 134
4.3 Environmental Carcinogens 137
4.3.1 Tobacco 137
4.3.1.1 Quantification of Nicotine and Its Metabolites 137
4.3.1.2 Benzene and Polycyclic Aromatic Hydrocarbons 142
4.3.1.3 4-Aminobiphenyl 143
4.3.1.4 Tobacco-Specific N-Nitrosamines 146
4.3.2 Air 152
4.3.2.1 Gasoline and Diesel Fuels 152
4.3.2.2 Polycyclic Aromatic Hydrocarbons 155
4.3.2.3 Monitoring Miscellaneous Complex Mixtures 156
4.3.3 Food 158
4.3.3.1 Aflatoxins 158
4.3.3.2 Heterocyclic Aromatic Amines 161
4.3.3.3 Nitrosamines 168
4.3.3.4 Polycyclic Aromatic Hydrocarbons 170
4.3.3.5 Dioxins and Pesticides 170
4.3.4 Water and Soil 171
4.3.4.1 Membrane Inlet Mass Spectrometry 171
4.3.4.2 Carcinogen Identification in Well Water by High-Resolution MS 172
4.3.4.3 Vinyl Chloride Monomer in Drinking Water 173

4.3.4.4 Phenolic Xenoestrogens in Water 174
4.3.4.5 Mutagenic Heterocyclic Amines in the Danube River 174
4.3.4.6 Polycyclic Aromatic Hydrocarbons in Soils 174
4.3.4.7 Composted Municipal Sludge and Compost-Amended Soil 176
4.3.5 *Helicobacter Pylori* 176
4.3.6 Elements 177
4.3.6.1 Arsenic 177
4.3.6.2 Selenium 178
4.3.6.3 Chromium 179
4.4 DNA Damage by Radiation and Oxidation 179
References 180

Chapter 5
Mechanism of Action and Metabolism of Antineoplastic and Chemopreventive Agents 201
5.1 Cytotoxic Therapy 201
5.1.1 Basic Principles 201
5.1.1.1 Cell Kill Hypothesis, Classification, and Mechanisms of Action 201
5.1.1.2 Toxicity, Resistance, and Combination Chemotherapy 204
5.1.1.3 Drug Discovery and Development, Clinical Trials 206
5.1.2 Alkylating Agents 211
5.1.2.1 Mechanism of Action 211
5.1.2.2 Nitrogen Mustards 211
5.1.2.3 Alkyl Sulfonates 224
5.1.2.4 Nitrosoureas 227
5.1.2.5 Nonclassic Alkylating Agents 228
5.1.2.6 Platinum Compounds 229
5.1.3 Topoisomerase Inhibitors 237
5.1.3.1 Topoisomerase I Inhibitors 238
5.1.3.2 Topoisomerase II Inhibitors 243
5.1.4 Antimicrotubule Agents 254
5.1.4.1 Microtubules and Tubulins 254
5.1.4.2 Vinca Alkaloids 256
5.1.4.3 Taxanes 258
5.1.4.4 Miscellaneous 263
5.1.5 Antimetabolites 265
5.1.5.1 Purine and Pyrimidine Analogs 265
5.1.5.2 Antifolates 268
5.1.6 Opportunistic Fungal Infections 271
5.2 Endocrine Therapy 271
5.2.1 Mechanism of Action 271
5.2.2 Aromatase Inhibitors 273
5.2.2.1 Structure and Function 273
5.2.2.2 Nonsteroidal Inhibitors 273
5.2.2.3 Steroidal Inhibitors 275
5.2.2.4 Aromatase Inhibitory Activity of Androgens 277
5.2.3 Antiestrogens 278
5.2.3.1 Tamoxifen 278
5.2.3.2 Toremifene 288

5.3 Targeted Drug Delivery and Therapy 289
5.3.1 Drugs Conjugated to Proteins, Antibodies, or Lipids....289
5.3.2 Angiogenesis Inhibitors....290
5.3.3 Antisense Oligonucleotides....299
5.3.4 *Ras* Oncoprotein Inhibitors....301
5.3.5 Antibody-Directed Enzyme Prodrug Therapy....302
5.3.6 Miscellaneous Strategies for Targeting....302
5.3.6.1 Photodynamic Therapy....302
5.3.6.2 Differentiating Agents....304
5.3.6.3 Arsenic Trioxide....304
5.4 Nutritional and Chemical Prevention....305
5.4.1 Objectives and Classifications....305
5.4.2 Phytoestrogens....307
5.4.2.1 Multicomponent Analysis in Food....308
5.4.2.2 Multicomponent Analysis in Body Fluids....312
5.4.2.3 Metabolism and Antineoplastic Effects....316
5.4.3 Flavan-3-ols (Catechins)....321
5.4.3.1 Tea....321
5.4.3.2 Wine....321
5.4.4 Terpenes and Ginsenosides....322
5.4.4.1 Metabolism of *d*-Limonene....322
5.4.4.2 Ginsenosides....325
5.4.4.3 Carotenoids....325
5.4.5 Sulfur-Containing Compounds....327
5.4.5.1 Isothiocyanates....327
5.4.5.2 Garlic....329
5.4.5.3 Oltipraz and N-acetylcysteine....330
5.4.6 Retinoids, Vitamins, and Selenium....331
5.4.6.1 Retinoids....331
5.4.6.2 Vitamin E....332
5.4.6.3 Selenium....332
References....333

Chapter 6
Strategies, Techniques, and Applications in Cancer Biochemistry and Biology....361
6.1 Proteins and Peptides....361
6.1.1 Proteome Technology....361
6.1.1.1 Proteomics and Strategies of Proteome Analysis....361
6.1.1.2 Protein Isolation, Digestion, and Delivery into the Ion Source, Instrumentation....365
6.1.1.3 Sequencing by Edman Degradation....369
6.1.1.4 Mass Spectrometric Peptide Mapping....370
6.1.1.5 Peptide Ladder Sequencing....371
6.1.1.6 Sequencing by Tandem Mass Spectrometry....372
6.1.1.7 Protein Identification by Database Searches....377
6.1.1.8 Mass Measurement of Intact Proteins....381
6.1.1.9 Quantification....383

6.1.2 Posttranslational Modifications 385
6.1.2.1 Phosphorylation 387
6.1.2.2 Selected Other Posttranslational Modifications 395
6.1.3 Miscellaneous Proteins 398
6.1.3.1 Enzymes 398
6.1.3.2 Antigens 402
6.1.3.3 Non-Glycosylated Cytokines 403
6.1.4 Tumor-Associated Proteins 407
6.1.4.1 Tumor Proteomics 407
6.1.4.2 Hematologic Malignancies 409
6.1.4.3 Gynecologic Malignancies 418
6.1.4.4 Genitourinary Malignancies 424
6.1.4.5 Gastrointestinal Malignancies 428
6.1.4.6 Other Cancers 434
6.1.5 Peptides and Class I Major Histocompatibility Complexes 437
6.1.5.1 Class I MHC Molecules 437
6.1.5.2 Strategies and Techniques 440
6.1.5.3 Peptide Antigens in Malignancies 442
6.1.6 Noncovalent Interactions and Protein Folding 449
6.1.6.1 Protein-Ligand Interactions 449
6.1.6.2 Protein-Protein Interactions 450
6.1.6.3 Protein-DNA Interactions 453
6.1.6.4 Protein Conformation and Folding 455
6.2 Lipids 460
6.2.1 Ceramides 462
6.2.2 Eicosanoids 468
6.2.3 Lysophospholipids 470
6.2.4 Platelet-Activating Factor 471
6.3 Oligonucleotides and Nucleic Acids 472
6.3.1 Mass Spectra 472
6.3.1.1 Nomenclature of Fragmentation 472
6.3.1.2 Electrospray Ionization 473
6.3.1.3 MALDI-TOFMS 474
6.3.1.4 High-Resolution MALDI-FTICRMS 474
6.3.1.5 Analysis of Large Nucleic Acids 475
6.3.2 Techniques and Strategies 475
6.3.2.1 Polymerase Chain Reaction 475
6.3.2.2 Differential Sequencing 476
6.3.2.3 Mass Spectrometric Methods for Genotyping 476
6.3.2.4 Detection of Mutation 480
6.3.3 DNA Sequencing and Genetic Diagnosis 480
6.3.3.1 Sequencing Exons of the p53 Gene 480
6.3.3.2 Identifying Microsatellite Alleles 481
6.3.3.3 Familial Adenomatous Polyposis 482
6.3.3.4 Multiple Endocrine Neoplasia Type 2A 483
6.3.3.5 Breast Cancer Susceptibility Gene BRCA1 485
6.4 Carbohydrates and Glycoconjugates 486
6.4.1 Carbohydrates 486

6.4.1.1 Nomenclature and Techniques for Collision-Induced Dissociation.........486
6.4.1.2 Sialic Acids 488
6.4.1.3 Polysaccharides and Glycosaminoglycans 488
6.4.2 Glycosylation........489
6.4.3 Glycoproteins........491
6.4.3.1 Mucins........492
6.4.3.2 Antigens 499
6.4.3.3 Miscellaneous Glycoproteins........503
6.4.4 Glycolipids (Gangliosides) in Various Malignancies 509
Recommended Books 518
References........518

Index........545

About the Author

John Roboz is a Professor in the Department of Medicine (Division of Medical Oncology) of the Mount Sinai School of Medicine, New York. He holds a B.S. degree (1955) from Eötvös University, Budapest, Hungary, and M.S. (1960) and Ph.D. (1962) degrees in Physical Chemistry from New York University.

After immigrating to the U.S. from Hungary in 1957, he worked as a Senior Research Chemist for General Telephone & Electronics Research Laboratories (Bayside, NY) and then at the Central Research Laboratories of the Air Reduction Company (Murray Hill, NJ). In 1969, he joined the Mount Sinai School of Medicine as an Associate Professor in the Department of Pathology and in 1974 also joined the faculty of the Biomedical Sciences Doctoral Program. In 1980, he became Professor of Neoplastic Diseases in the Department of Neoplastic Diseases, which was later incorporated within the Department of Medicine.

His primary research interests have been in the development of mass spectrometric techniques in analytical pharmacology and biochemistry and the application of these methods in many collaborative studies with basic scientists developing new antineoplastic agents and clinicians conducting Phase I clinical trials. He has also conducted research on the occurrence of hyaluronic acid in mesothelioma, and on the role of D-arabinitol as a marker for the early diagnosis and monitoring of candidiasis in cancer patients. Current areas of interest include the use of surface-enhanced laser desorption time-of-flight and electrospray mass spectrometry for the detection and identification of unique proteins as potential markers for several malignancies.

Dr. Roboz has more than 130 publications in mass spectrometry, including several review chapters. One of these was the first review on mass spectrometry in cancer research (1978). In 1968 (reprinted 1979) he published a textbook entitled *Introduction to Mass Spectrometry: Instrumentation and Techniques* (Wiley-Interscience, New York). It was subsequently reprinted in 2000 by the American Society for Mass Spectrometry as Volume 3 in the *Classic Works in Mass Spectrometry* series.

Foreword

Over the past decade mass spectrometry (MS) has experienced a remarkable growth, spurred primarily by the introduction of ionization techniques such as electrospray and matrix-assisted laser desorption, and the development of relatively user-friendly instrumentation. Nowhere is this growth more evident than in the biological sciences where MS-related techniques (e.g., LC/MS, GC/MS, MALDI) are considered an essential analytical arsenal and are used routinely.

While the focus of this book is on mass spectrometry in cancer research, there is material in it for almost everyone. For the biological scientist who is a novice in mass spectrometry, the first chapter provides a comprehensive summary of the principles of the technique with emphasis on their applicability to the analysis of biomolecules. For the "purist" in mass spectrometry who is becoming involved in analytical biochemistry or cancer research, many of the essential fundamentals in the biological sciences are introduced with sufficient detail such that they allow one to see the broader picture and gain an appreciation of how valuable the technique can be. Finally, current researchers in the field will benefit from the generally extensive list of citations, which should enable them to further expand their research endeavors.

The reader is treated to a wide spectrum of MS applications to cancer research. These applications range from topics related to the use of MS techniques for the detection of biomarkers arising from environmental exposure to the elucidation of the mechanism of action of chemopreventive agents, both synthetic drugs and naturally occurring compounds. The coverage of the different topics is accompanied by presentation of well-selected examples with a clear discussion of the subject matter and figures illustrating the finer points of the analysis. The final chapter, which discusses techniques and strategies in cancer biochemistry and biology, provides a broad perspective of the field and introduces a variety of thought-provoking concepts that can serve as a guide in the design of future research projects.

Paul Vouros
Professor of Chemistry
Northeastern University
Boston

Mass Spectrometry: An Oncologist's Viewpoint

Yesterday's oncologist might question the role of mass spectrometry in cancer research. The early trial and error methods, the indiscriminate screening of every compound on the shelf, and acceptance of partial tumor inhibition in the mouse, no longer suffice to initiate a clinical trial hoping for eventual cures or prevention. The explosive expansion of knowledge about DNA, proteins, and intracellular chemistry have provided a new calculus for understanding the cancer process. Understanding the normal pathways of signal transduction, from ligand to receptor to the cascade of enzymes that lead to gene expression or repression, is necessary to illuminate the relationships of cells in a multicellular organism. Aberrations in these complex processes exist in cancer. Although the altered patterns may be mechanistically operative in bringing about cancer cell survival and growth, they may well be epiphenomena. Altered DNA function brought about by loss, mutation, or viral insertion or usurpation appear with today's knowledge to underlie the cancer process.

Every one of the involved molecules must be identified. To understand the kinetics of reactions they must be quantified. Many analytic techniques provide qualitative or even semi-quantitative indications of the compounds involved. Precision analysis is required for certainty and this book elegantly sets forth the contributions that mass spectrometry makes to that precision and certainty.

Cancer therapies still deal primarily with surgery and radiotherapy. Both these modalities are concerned predominantly with local and regional disease. The control of disseminated cancer cells relies on systemic therapies, chemical, and immunologic. Precise measurements of all aspects of chemotherapeutic compounds during synthesis and studies of pharmacokinetics and pharmacodynamics are essential. Metabolic products of the administered drug, both anabolic and catabolic, must be identified and quantified to arrive at optimal dosing regimens. Individuals may vary widely in their disposition of drugs. Mass spectrometry is unsurpassed as the gold standard in these areas.

Cancers secrete large and small molecules of many known and countless unknown structures. Enzymes that allow cancers to invade and metastasize and surface molecules and compounds of unknown function often serve as critical parameters of cancer behavior. Discovery of trace compounds of unique nature that could indicate the presence of early cancer is still theoretically possible, and still hoped for. Identification of such a compound in extremely small quantities in biologic fluids containing hundreds of other compounds is a classic undertaking for mass spectrometry. Coordinated immunologic assay, isotopic, spectroscopic, nuclear-magnetic resonance, and mass spectrometric analyses of such putative markers could advance our diagnostic acumen so that we might recognize pre-cancerous changes, or cancers so early in their course that a cure could be readily achieved.

Cancer is a vast collection of diseases with some common characteristics that usually represent abnormalities in somatic cell genetics, not germ line genetics. Most causes of cancer are environmental, not genetic. A treasure trove of possibilities in cancer prevention and detection awaits in the field of molecular epidemiology. Identification of changes in DNA that arise from interaction with individual environmental agents is of profound interest. Contributions of mass spectrometry to this infant field are largely untapped.

This volume displays the rich understanding of cancer that a brilliant mass spectrometrist has absorbed by proximity to and collaboration with physicians and biological scientists working on cancer. The advances in mass spectrometry have kept pace over the past 25 years with the advances in molecular biology and clinical cancer research. It is a safe bet that continuing advances in all three areas are imminent, and that mass spectrometry can contribute to the future of both laboratory and bedside cancer investigations. This book provides an admirable foundation for understanding the field and for perceiving the challenges that lie ahead.

James F. Holland, MD
Distinguished Professor of Medicine
Mount Sinai School of Medicine
New York

Preface

Modern cancer research is inevitably and increasingly becoming multidisciplinary with the realization of the interwoven complexity of the biochemical, structural, therapeutic, and clinical questions that must be answered. Such problems require integrated, synergystic approaches that employ an assortment of biochemical or immunological manipulations, chromatographic or electrophoretic separations, sequencing strategies and … more and more, mass spectrometry.

The diversity of the disciplines in which mass spectrometry is now a critical component can be seen by inspecting the ever increasing number of publications dedicated to "real world" applications as opposed to those dealing with methodology development. In these applications the expertise of the co-authors often cover multiple disciplines including biochemistry, molecular biology, immunology, pharmacology, microbiology, and even reach areas such as surgery and molecular modeling. Accordingly, it is particularly relevant that a book on mass spectrometry in cancer research gives an overview of how mass spectrometry provides an analytical link between all these fields.

This book is intended for: (a) mass spectrometrists involved in cancer research or in providing core services; (b) "customers," i.e., researchers in the biological, medical, pharmaceutical, or environmental sciences who use mass spectrometry in their work; (c) potential clients contemplating how to solve their intractable analytical problems; and (d) academic and industrial managers anxious to understand the approaches and results of the mass spectrometry being used in the projects they direct or administer.

The Overview (Chapter 1) is intended for those who just want to obtain a quick understanding of how mass spectrometers work, what the differences are between the various ion sources and analyzers, and what are the potential applications for the available analytical techniques in cancer research and in support of therapy. Chapter 2 is intended to provide the scientist and physician, not expert in the field, with an understanding of mass spectrometric instrumentation and pertinent methodologies. In contrast, Chapter 3 presents relevant concepts of cancer medicine and biology. They are intended for mass spectrometrists who are, *horribile dictu*, not also experts in both medicine and molecular biology. The intent of Chapters 4, 5, and 6 on occupational and environmental carcinogens, antineoplastic and chemopreventive agents, and proteins, lipids, nucleic acids, and glycoconjugates, is to provide a broad rather than exhaustive examination of current strategies and techniques illustrated with relevant applications. In a book designed to provide an overview, it is likely that experienced practitioners may be disappointed by the coverage of their specialty. If so, I ask for their indulgence for my naive and superficial, but, hopefully, not erroneous, approach.

It is emphasized that the length of a section does not automatically indicate the importance of the subject. In fact, it is often the case for new and developing methods and areas of application that they may be potentially of major importance but are covered only briefly. This is because there were few publications, albeit often pioneering, available at the time this manuscript was completed. In addition, it often felt as if new and relevant publications, especially in areas such as tumor proteomics and targeted therapies, were appearing every week. Similarly, the relative importance of specific mass spectrometric applications keeps changing, particularly with respect to the clinical uses of chemotherapeutic agents. An arbitrary halt had to be imposed to the search for new and relevant articles and therefore the literature was covered through August 2001. The selection of one reference over another is inevitably subjective. Those references selected should be of value to the reader in that they describe original research, are particularly instructive, or provide detailed reviews of a particular subject. Intentionally there are no references in the Overview. In addition, only a brief list of general texts is provided at the end of Chapter 3. My indebtedness to the quoted references is obvious. Sometimes the original text was so succinct and well-written that it was difficult to recompose. In such cases I can do no more than honor the original authors.

Mass spectrometry grew almost unbelievably in the final quarter of the 20th century. According to industrial sources, instrument sales reached the $2 billion level in 2000. During the last decade there was an astonishingly rapid development of technology and methodology for the determination of the molecular masses, sequences, and higher structures of large, hitherto untouchable molecules, including proteins, nucleic acids, and glycoconjugates. Publications describing and/or proposing new and novel strategies, approaches, schemes, and techniques abound. If one has to add a caveat it should be said that many of these reports are "proof of concept" studies in which the applicability of the new technique is demonstrated in "model" systems, i.e., in carefully selected situations where the solutions are known or easily predicted. Conclusions drawn usually include a catalogue of true and/or perceived advantages, often exaggerated, with few, if any, limitations listed. These reports play an essential role in the transition from innovative ideas to practical applications. Therefore, when such papers are included in this text, it is done without critical evaluation, using only the criterion that the technique has potential applicability to problems in cancer research.

The reader will find some annoying inconsistencies in the naming of compounds and their abbreviations, in concentration units, spelling, etc. When confronted with widely differing designations in the copyrighted figures and tables, I found that, regretfully, I had to give up my feeble attempts at uniformity when describing work on the same subject by different authors.

It has been said that predictions are dangerous, particularly when they concern the future. After several attempts, I abandoned the inclusion of a section on general prospects and personal prognostications. The sage suggestion of the great New York Yankee catcher and philosopher/poet Yogi Berra applies directly to the apparently endless series of breathtaking breakthroughs in the technology of mass spectrometry: "The future ain't what it used to be."

Acknowledgments

Throughout the preparation of the manuscript, I have profited from the intellectual input of many of my colleagues in the scientific and medical community. Foremost, my heartfelt thanks go out to Dr. John Greaves, Director, Mass Spectrometry Facility, University of California at Irvine, for the time he has invested, and for the care with which he has reviewed almost every sentence of this manuscript. Dr. Greaves has shown an extraordinary talent in reading the text from the point of view of an informed mass spectrometrist who intends to use this book as background material in preparing to apply a particular mass spectrometric technique to research not yet pursued by others. It was a pleasure, and provided satisfaction, to observe the text becoming visibly improved as I made modifications based on his suggestions, often by changing just a word or the structure of a sentence. Equally frequently, his challenging questions helped materially in the illumination of the strategy, technique, or result being discussed.

I am strongly indebted to Dr. Lawrence Phillips, who read several chapters of the manuscript and offered numerous suggestions for improvements, and to Dr. Jozsef Lango, for his useful criticism of the sections on basic principles and instrumentation. Acknowledgments and recognition are due to Drs. Ramu Avner, Stephen Carmella, Steven Dikman, Stephen Hecht, Joel Graber, James Holland, James Maggs, Mark McKeage, James-Gilmour Morrison, Kevin Park, Gail Roboz, Paul Vouros, Rong Wang, Martin Winkler, and Ralph Zimmermann, who read various parts of the manuscript and contributed in significant ways. I wish to thank the many individuals with whom I have conversed over the years and whose ideas and thoughts can be found in many corners of the book. As I did not always heed the good advice offered, I am alone responsible for the shortcomings of this book.

I am grateful to the many authors who provided copies of selected illustrations from their papers and reprints of their work. Thanks are due to Micromass Co. and Agilent Technologies Inc. for permission to include illustrations from their publications. The courtesy of Dr. George Wright, Jr., is acknowledged for providing a figure that shows unpublished results of his research on biomarkers of prostate cancer. Among members of my staff, I would like to thank Longhua Ma, for his superb help in making new illustrations, Demetra Silides, for spending long hours entering references into a database, and Lin Deng, for helping with many tasks.

Special thanks are due to James F. Holland, M.D., Distinguished Professor of Medicine at Mount Sinai School of Medicine, New York, for his interest and steadfast support of my research for almost 30 years. He has been an incomparable boss, collaborator, and friend. I wish to express my indebtedness to the T.J. Martell Foundation for Leukemia, Cancer, and AIDS Research, and to Mr. Derald Ruttenberg for their generous support of my research over several decades.

I would like to acknowledge the contributions of the staff of CRC Press. Special thanks are due to Barbara Norwitz, for her remarkable patience and understanding of my myriad excuses for not meeting promised deadlines. Michele Berman provided expert editorial assistance and also exhibited a much appreciated calmness as I bombarded her with revision upon revision of the text. I would also like to thank Robert Caltagirone for performing small-scale miracles with some of the poor quality illustrations. Helena Redshaw provided expert help in a variety of administrative issues. Other members of the staff of CRC Press, including Fequiere Vilsaint, who helped in a variety of areas.

Finally, I would like to express very special thanks to my wife, Julia, for her understanding and encouragement, and all too often, for just tolerating me.

John Roboz

Terminology and Abbreviations

The first nomenclature of mass spectrometry was published some 25 years ago. Since that time, the Committee of Measurements and Standards of the American Society for Mass Spectrometry, ASMS, has been soliciting debates about old terms as well as new definitions prompted by advances in technology. Lists of accepted and recommended definitions of terms and symbols have been published.[1–3] Glossaries are posted by the ASMS (http://www.asms.org), and two Internet resources (http://base-peak.wiley-com/news/glossary.html and http://www.micromass.co.uk/basics). There is a most useful, but occasionally personal and controversial, desk reference on the language of mass spectrometry.[4] Of special interest are the concepts of *m/z* scales,[5] the various means of calculating masses and their usage,[6] and proposed (but not widely used) pictograms for experimental parameters.[7]

As in all fields experiencing rapid progress, mass spectrometry also has its contradictory and confusing definitions, e.g., the definitions of "resolution" and "resolving power" by the Current IUPAC Recommendations and the definitions on the ASMS Web site, or the spelling of the adjectives describing the number of charges on an ion, such as "double-charge" or "doubly charged" ion. In this text, relevant terms are defined or described pragmatically (not rigorously or critically) in the text or tables when used for the first time. Consulting the available published lists is strongly recommended.

Please note: (a) A number of abbreviations/acronyms, often names of compounds or reagents, are used only within a section on particular subjects. These are defined at their first occurrence and are not listed below. (b) The following abbreviations/acronyms are used throughout.

MASS SPECTROMETRY

Current conventions and suggestions (not always logical or consistently followed) concerning instrumentation, techniques, and ionic species include:

- A single type of instrument or technique is abbreviated without hyphens, e.g., ESIMS, MALDIMS, TOFMS, FTMS.
- When two or more different analytical techniques are coupled in tandem, a slash (solidus) is placed between the abbreviations, e.g., LC/MS, LC/ESIMS.
- A hyphen is used to emphasize or highlight a particular technique or instrument component, e.g., EI-MS/MS, ESI-FTICRMS.
- Ionic species are enclosed in square brackets and the number of charges and polarity given as a superscript, e.g., $[M+2H]^{2+}$, $[M–H]^{-}$.
- The interaction of two or more species is indicated by a slash, e.g., ion/molecule reaction, while a hypen indicates the dual nature a species, e.g., ion-radical.

Ion Sources, Ionization

APCI Atmospheric pressure chemical ionization
API Atmospheric pressure ionization
CI Chemical ionization (positive ion CI)
CID Collision-induced dissociation
CF-FAB Continuous-flow fast atom bombardment
DCI Desorption chemical ionization
EI Electron ionization
ESI Electrospray ionization

FAB Fast atom bombardment (also refers to fast ion bombardment)
FD Field desorption
LD Laser desorption
LSI Liquid secondary ion
[$M^{+\cdot}$] Positively charged molecular ion radical
$[M+H]^+$ Protonated molecule (also $[MH]^+$)
MALDI Matrix-assisted laser desorption ionization
nano-ESI Nanoflow electrospray ionization
NCI Negative ion chemical ionization
SELDI Surface enhanced laser desorption ionization
TSP Thermospray

Mass, Mass Analyzers

B Magnetic sector - also used in instrument configuration
Da Dalton, mass of an ion, calculated using the ^{12}C scale
E Electrostatic sector — also used in instrument configuration
ICR Ion cyclotron resonance
FTICR Fourier transform ion cyclotron resonance
FWHM Full width at half-maximum intensity
IT Ion trap
kDa Kilodalton
oa Orthogonal acceleration
QqQ Triple quadrupole mass spectrometer (second stage rf only)
TOF Time-of-flight

Miscellaneous, Mass Spectrometry Related

e^- Electron
eV Electron volt
GC/MS Gas chromatography-mass spectrometry; gas chromatograph-mass spectrometer; gas chromatographic-mass spectrometric
i.s. Internal standard
keV Kiloelectron volt
LC/MS Liquid chromatography-mass spectrometry; liquid chromatograph-mass spectrometer; liquid chromatographic-mass spectrometric
LOD Limit of detection
LOQ Limit of quantification
MRM Multiple-reaction monitoring
MS Mass spectrometry, mass spectrometer, mass spectrometric
MS/MS Mass spectrometry-mass spectrometry
PSD Post-source decay
SIM Selected ion monitoring
SRM Selected reaction monitoring
TIC Total ion current

Chromatography and Electrophoresis

1-D One-dimensional
2-D Two-dimensional

1-DE One-dimensional gel electrophoresis
2-DE Two-dimensional gel electrophoresis
CE Capillary electrophoresis
GC Gas chromatography; gas chromatograph; gas chromatographic
HPLC High-performance liquid chromatography; high performance liquid chromatograph; high-performance liquid chromatographic
LC Liquid chromatography or liquid chromatograph; liquid chromatographic
PAGE Polyacrylamide gel electrophoresis
SDS Sodium dodecyl sulfonate
TLC Thin layer chromatography

Chemical Derivatization and Biochemistry

ACTH Adrenocorticotropic hormone
ATP Adenosine triphosphate
AUC Area under the curve
BSA Bovine serum albumin
cDNA Complementary DNA
CTL Cytotoxic T-cell lymphocyte
CYP450 Cytochrome P450
dNTP Deoxyribonucleoside triphosphate
DTT Dithiothreitol
EDTA Ethylenediaminetetraacetic acid
GSH Glutathione
GST Glutathione-S-transferase
GTP Guanosine 5′-triphosphate
IR Infrared
Hb Hemoglobin
HSA Human serum albumin
MHC Major histocompatibility complex
NMR Nuclear magnetic resonance
RIA Radioimmunoassay
TFA Trifluoroacetic acid
TMS Trimethylsilyl

Miscellaneous Organizations

ASMS American Society for Mass Spectrometry
EPA Environmental Protection Agency, USA
FDA Food and Drug Administration, USA
IARC International Agency for Research on Cancer
NCI National Cancer Institute, USA
NTP National Toxicology Program, USA
OSHA Occupational Safety and Health Administration, USA
WHO World Health Organization

REFERENCES

1. Todd, J. F. J., Recommendations for nomenclature and symbolism for mass spectroscopy (including an appendix of terms used in vacuum technology), *Pure Appl. Chem.,* 63, 1541–1566, 1991.
2. Price, P., Standard definitions of terms relating to mass spectrometry, *J. Am. Soc. Mass Spectrom.,* 2, 336–348, 1991.
3. Voyksner, R., Price, P., Bartmess, J., Little, J., and Iverson, D.W., ASMS terms and definitions, ASMS conference poster, http://www.asms.org, 1997.
4. Sparkman, O.D., *Mass Spectrometry Desk Reference,* Global View Publishing, Pittsburgh, 2000.
5. Dougherty, R., Eyler, J., Richardson, D., and Smalley, R., The mass-to-charge ratio scale, *J. Am. Soc. Mass Spectrom.,* 5, 120–123, 1994.
6. Carr, S.A. and Burlingame, A.L., The meaning and usage of the terms monoisotopic mass, average mass, mass resolution, and mass accuracy for measurements of biomolecules. Appendix XI, in *Mass Spectrometry in the Biological Sciences,* Burlingame, A.L. and Carr, S.A., Eds., Humana Press, Totowa, NJ, 546–553, 1996.
7. Lehmann, W.D., Pictograms for experimental parameters in mass spectrometry, *J. Am. Soc. Mass Spectrom.,* 8, 756–759, 1997.

1 Overview and Scope of Applications

1.1 THE ESSENTIALS

- **Mass spectrometers.** Ion optical devices that produce a beam of gas-phase ions from samples, mass spectrometers sort the resulting mixture of ions according to their mass-to-charge ratios or a derived property, and provide analog or digital output signals (peaks) from which the mass-to-charge ratio and intensity (abundance) of each detected ionic species may be determined. The analyte molecules *must* be ionized and the ions *must* be in the gas phase. There are several mechanisms of ionization (Table 1.1) leading to the formation of a variety of ion types (Table 1.2).

- **Mass-to-charge ratio (*m/z*).** Masses are not measured directly. Mass spectrometers are *m/z* analyzers. The mass-to-charge ratio of an ion is obtained by dividing the mass of the ion (*m*), by the number of charges (*z*) that were acquired during the process of ionization. The mass of a particle is the sum of the atomic masses (in daltons) of all the atoms of the elements of which it is composed. It is noted that although the "official" symbol for a mass unit is *u* (unified atomic mass unit), representing 1/12 the mass of the most abundant naturally occurring stable isotope of carbon (^{12}C), the term *dalton*, Da, is now used in both biochemistry and mass spectrometry.

- **Mass spectra.** The mass spectrum of a compound provides, in a graphical or tabular form, the intensities of all or a selected number of the acquired *m/z* values from the ionic species formed. Mass spectral peaks are observed in analog form (each peak with a height and a width) or digital form (each peak a simple line). Ion intensities are usually obtained, in arbitrary units, as heights or areas of the mass peaks and may be presented as such. However, it is more common to present intensities *normalized* with respect to the "base peak," i.e., the peak of highest intensity (in the selected mass range) taken as 100%.

- **Singly and multiply charged ions.** In most cases, there is only one charge on the ions, thus the measured *m/z* is equivalent to the mass. However, the presence of a second charge will move the spectral peak to almost half the value of the singly charged ion. For example, the addition of a single proton to a molecule of 399 produces an ion of *m/z* 400; the addition of a second proton yields a doubly charged ion ($z = 2$) with *m/z* $401/2 = 201.5$. *Electrospray ionization*, which is one of the most important current techniques for the analysis of large, complex biomolecules, produces an envelope of multiply charged ions from a single analyte, with *z* up to 75 or more. Special deconvolution algorithms are used to calculate the masses of the analyte ions.

- **Mass.** In conventional instruments under routine conditions, masses are usually determined with an accuracy of ±0.5 Da. High resolution (Section 1.3) permits the discrete determination of the mass of an ion often with an accuracy of ~0.001 Da. Because the isotopes of the constituent elements of the analyte ions have masses that are mass positive

TABLE 1.1
Mechanisms of Ionization of Analytes

Process	Result
Neutral Analyte in the Gas Phase	
• Ejection of an electron	Produces an ion with one positive charge
• Capture of an electron	Produces an ion with one negative charge
• Protonation	One positive charge for every proton added
• Depronotation	One negative charge for every proton ejected
• Cationization/Anionization	Noncovalent association of ions with neutral analyte molecule (e.g., Na^{+}, K^{+}, Cl^{-}, $HCOO^{-}$)
Ionic Analyte in the Gas Phase	
• Collisionally induced dissociation (CID)	Structurally significant fragment ions formed in controlled collisions between selected and accelerated analyte ions and neutral gas molecules introduced into the collision cell
Ionic Analyte in Condensed Phase	
• ESI and MALDI	Transfer of analyte ions to gas phase

TABLE 1.2
Major Types of Ions

Ion, positive: Atom, molecule, molecular moiety and/or radical that has lost one or more electrons (or negatively charged particles) or acquired one or more positively charged particles (e.g., H^{+}, Na^{+}).

Ion, negative: Same entities that have acquired (captured) one or more negatively charged particles (e.g., electrons, Cl^{-}, $HCOO^{-}$) or lost positively charged particles (e.g., H^{+}, Na^{+}).

Molecular ion: Positive or negative ion of the analyte without fragmentation; the mass of molecular ions corresponds to the sum of the masses of the most abundant naturally occurring isotopes of the constituent atoms.

Isotopic ion: Contains one or more of the less abundant naturally occurring isotopes of the constitutive elements (e.g., ^{13}C, ^{37}Cl, ^{34}S).

Fragment ion: A charged dissociation product (X^{+}, X^{-}, $X^{+\bullet}$, $X^{-\bullet}$) of an ionic fragmentation resulting from bond cleavages in the analyte ion. Fragment ions may dissociate further into additional fragment ions of successively lower mass.

Singly, doubly, etc., charged ion: Atom, molecule, or molecular moiety that has gained or lost one or more charged particles. These ions have correspondingly reduced *m/z* values. When the number of charge is not designated: *multiply charged* ion.

Protonated molecule (not "protonated molecular ion"): Even electron, positive ions resulting from the acquisition of a proton in a reaction with another ion in the gas or liquid phase.

Adduct ion: Ion formed by noncovalent addition of an ion to a molecule, e.g., cation, anion, or solvent (water and/or organic), often within the ion source.

Dimeric or polymeric ion: Ions in which one or more neutral molecules of the same analyte are attached to the molecular ion; the attachment may take place in the ion source, or the analyte may exist as a dimer, trimer, etc., in the gas phase.

Precursor (parent) ion: A molecular or fragment ion, formed in a primary ionization, that is selected for secondary fragmentation (dissociation, decomposition) to form one or more charged and one or more neutral species in an MS/MS experiment.

Product (daughter) ion: An ion produced by the fragmentation of a mass selected precursor ion is excited and made to collide with a neutral target gas in a collision cell.

Progeny fragment ion: Generic term for subsequent generations of product ions formed from product ions, i.e., second, third, etc., order (generation) product ions.

or negative (e.g., H = 1.0078 Da and O = 15.9949 Da), the resulting *measured accurate masses* of the analyte ions and radicals can be used to determine the elemental composition and the empirical formula of the analyte ion (Section 1.4). The *nominal mass* of an ion is calculated by summing the nominal masses of the elements (i.e., the integer masses of their most abundant naturally occurring stable isotope, usually equal to the mass number) in its empirical formula. The integer value of the nominal mass of an ion is not the same as the *monoisotopic mass* of the ion. The monoisotopic mass (calculated exact mass) of an ion is the sum of the monoisotopic masses of the elements in their formula. The term *molecular mass* refers to the determined mass of the molecule or molecular ion, while *relative molecular mass* (M_r) is based on the atomic masses of the elements. Neither term should be used with reference to fragment or adduct ions, or radicals.

- **Separation of ions.** Ion separation is accomplished *in space and/or in time* with the aid of electromagnetic and/or electric fields and may also involve field-free regions within the mass analyzers. Mass spectrometers must operate in high vacuum to reduce deleterious effects from collisions of analyte ions with neutral molecules and to ensure proper operating conditions for the ion optical system.
- **Mass spectrometry/mass spectrometry (MS/MS).** An important current technique, *MS/MS* is based on controlled collisions between selected ions (*precursors*) and neutral gas molecules in pressurized *collision cells* placed in specific regions of the instrument. The initial formation/fragmentation of ions from a sample and the subsequent collision-induced dissociation of selected ions is carried out in the ion sources of instruments with "tandem in space" or "tandem in-time" *m/z* analyzers. The collision-induced dissociation (CID) of the selected ions results in *product* (previously called *daughter*), precursor (*parent*), or *neutral loss* types of spectra that provide vital structural information needed for the identification of complex biomolecules (Section 1.4). *Selected reaction monitoring,* another mode of MS/MS operation, provides highly specific and sensitive quantification of target analytes.
- **Diversity of mass spectra.** Mass spectrometry does not deal with a well-defined property of molecules, thus there is no such thing as the *correct* mass spectrum of a compound. The type and abundance of ions in a mass spectrum depend greatly on the mode of ionization (type of ion source) and on experimental conditions used to obtain the spectrum. The fact that multiple kinds of mass spectra may be obtained for an analyte (Section 1.4) is a major advantage because judicious selection of the ionization technique and experimental conditions greatly expands applicability and the structural information obtained.

1.2 FUNCTIONS AND TYPES OF INSTRUMENT COMPONENTS

It is emphasized that mass spectrometers are *systems,* or integrated assemblies of interacting elements, designed to carry out cooperatively two functions, i.e., to determine the *m/z* ratios and relative intensities of the various ionic species generated from the sample. Mass spectrometer systems consist of six functional elements: sample inlet (interface), ion source, mass analyzer, ion detector, vacuum system, and a dedicated computer (Figure 1.1). The functions and major types of these elements are summarized in Table 1.3, outlined below, and discussed in the respective sections of the next chapter.

1.2.1 SAMPLE INTRODUCTION SYSTEMS

The functions of the sample introduction systems are to produce vapors from the samples (or reduce the pressure of gaseous samples) and to introduce a sufficient quantity of the sample into the ion

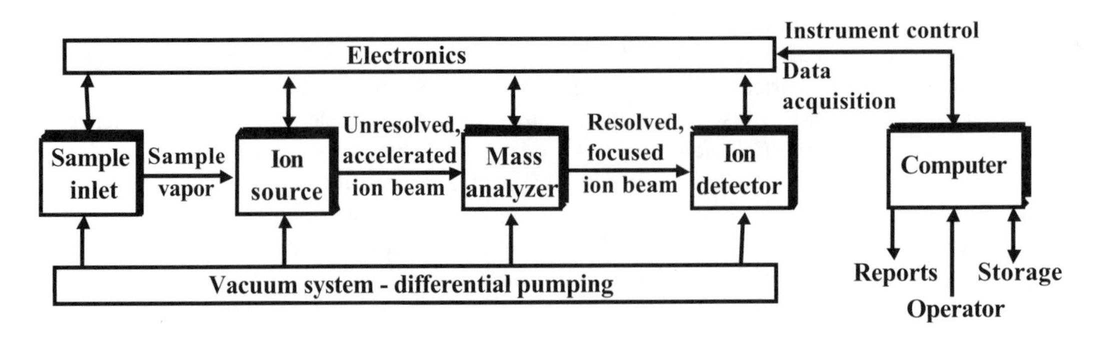

FIGURE 1.1 Major components of mass spectrometer systems.

TABLE 1.3
Elements of Mass Spectrometer Systems, Their Functions, and Major Types

Element	Function	Type
Sample introduction	Reduce pressure of gaseous samples Produce vapors from solid or liquid samples Introduce effluents on-line from GC, HPLC, CE	Batch inlet: gases Direct probe: liquids, solids Direct interface to ion source ESI in various formats
Ion source	Produce analyte ions Form, shape, and eject a suitable ion beam into analyzer Produce fragment ions in special collision cells used in MS/MS operations	*Gaseous phase:* Electron ionization, EI Chemical ionization, CI Atmospheric pressure CI, APCI *Condensed phase*: Electrospray, ESI Matrix-assisted laser desorption, MALDI Fast atom bombardment, FAB Liquid secondary ion, LSI
m/z analyzer	Resolve ions of the same *m/z* from all other ions present Focus the individual ion beams onto the detector Resolve ions obtained by collision-induced dissociation (CID) in MS/MS operation	Quadrupole, Q Ion trap, IT Time-of-flight, TOF Ion cyclotron resonance, ICR (FT) Magnetic, B
Ion detector	Measure the abundance ions in each separated ion beam	Electron/photo multiplier Multipoint array detectors
Vacuum system	Prevent adverse effects due to collisions with neutral particles Prevent gas discharges by high voltages	Mechanical, turbomolecular, diffusion, cryogenic pumps
Computer system	Control instrument operations; e.g., tuning, scanning Data acquisition, data processing, backup, hardcopies Library and Internet searches for identification and interpretation	Modern instruments controlled by PCs

source in such a way that its composition represents that of the original sample. Sample inlets include batch inlets for gases, and direct heated probe inlets to evaporate solids, liquids, and collected chromatographic effluents. There are also interfaces with various levels of sophistication for the on-line introduction of effluents from gas and liquid chromatographs and electrophoresis instruments. It is noted that in the two most important current ion sources, electrospray and matrix-assisted laser desorption, sample introduction and ion formation are accomplished simultaneously.

1.2.2 Ion Sources

All ion sources have two main functions: (1) to effectively produce a beam of ions representative of the sample and (2) to form, shape, and accelerate the ion beam into the mass analyzer. There are more than a dozen distinct and significantly different methods of ionization, and the relative importance of these techniques has been changing continually as new methods are developed and old ones revitalized. The first step in the ionization process is the acquisition of energy by all neutral molecules present, including the analyte(s), other constituents if the sample is a mixture, and background air and other residual molecules. Currently used *agents* capable of bringing about ionization (transferring charges) include electrons, ions (ionic products of ion-molecule reactions), rapidly moving energetic atoms and ions, electric fields with spraying, photons, and inductively coupled plasma (for element analysis).

As discussed and illustrated repeatedly below and throughout the text, the major considerations in selecting an ion source concern the *volatility* and *thermal stability* of the analyte and also the kind of information sought, such as sensitivity, confirmation or identification, and bulk vs. surface analysis. Other important properties of the sample include the type of matrix (pure analyte, biological fluid, tissue) and the quantity available. Techniques of ion production may be divided into two major groups: ionization in the gas phase or some form of desorption from the condensed phase (Table 1.4). The advantages and limitations of currently used ion sources for volatile and nonvolatile analytes are summarized in Tables 1.5 and 1.6, respectively. Three types of ion sources

TABLE 1.4
Available Ionization Techniques as a Function of Analyte Volatility

1. When analyte or chemical derivative is volatile:

Vaporize under vacuum and form gas-phase ions by interacting with:

• Electrons	Electron ionization	EI
• Gaseous ions	Chemical ionization	CI
• Photons (UV, laser)	Photoionization	PI

When analyte in liquid phase has some volatility:

Vaporize under atmospheric pressure (AP) and form gas-phase ions ny interacting with:

• Corona discharge	Chemical ionization	APCI

2. When analyte is not volatile:

→ Ionization in solid or liquid phase with subsequent desorption into gas phase
→ Ionization in solid or liquid phase with simultaneous desorption into gas phase
→ Rapid desorption of neutrals followed by ionization in gas phase

a. Sudden energy burst:

Impacting particle			**Sample in**
• Energetic atoms	Fast atom bombardment	FAB	Liquid matrix
• Energetic ions	Secondary ion MS	SIMS	Liquid or solid matrix
• Photons	Matrix-assisted laser desorption	MALDI	Crystalline matrix
• Photons	Laser desorption	LAMMA	Microprobe: solid matrix
• ^{252}Cf fission fragment	Plasma desorption	PD	Solid matrix

b. Ionization within liquid phase

• Electric field	Electrospray	ESI	Liquid matrix
• Electric field	Thermospray	TSP	Liquid matrix

TABLE 1.5
Advantages and Limitations of Ion Sources for Volatile Analytes

Advantages	Disadvantages
Electron ionization	
• Well understood • Applicable to most volatile compounds • Reproducible mass spectra • Structural information from fragmentation • Large libraries for manual or computerized identification ("fingerprints") • High sensitivity (high femtomole) • Possibility for high resolution	• Sample must be volatile and thermally stable (low mass range, ~1.2 kDa) • Often, lack of molecular ion
Chemical ionization	
• Molecular mass information often when not shown by EI • Simple spectra, little or no fragmentation • High sensitivity (high femtomole) • Possibility for high resolution	• Sample must be volatile and thermally stable (low mass range, ~1.2 kDa) • Strong dependence on nature of sample type and pressure of reagent gas • Less fragmentation than EI, limited or no availability of spectral libraries
Negative chemical ionization	
• Efficient ionization and very high sensitivity • Very high selectivity for certain environmentally and biologically important analytes	• Strong dependence on nature of sample
Atmospheric pressure chemical ionization	
• Excellent interfacing with HPLC • High sensitivity and wide application of analytes to medium polarity	• Upper mass limit ~1.5 kDa

have been used for the majority of applications reviewed in this text: electron ionization, electrospray ionization, and matrix-assisted laser desorption ionization.

In *electron ionization* (EI, electron impact), the neutral molecules of the analytes in the gas phase are bombarded with energetic (70 eV) electrons. The primary product of EI is an excited odd-electron (radical) molecular ion, $[M]^{+}$. EI is a "hard" ionization method in which the acquired energy is considerably more than that required for ionization. The excess energy is distributed inside the ions, and when the energy concentrated in a particular bond becomes equal to the dissociation energy of that bond, radicals or neutral molecules are ejected and fragmentation occurs, yielding ions of lower mass with lower internal energy. The fragmentation of large polyatomic molecules is a complex process, determined by the chemical nature of the analyte and the prevailing energy conditions. Mass spectral *fragmentation patterns* obtained by EI provide a great deal of information about the structure of the analyte. However, volatility requirements impose a practical upper mass limit of ~1.2 kDa for analytes including analytes derivatized to increase volatility.

In contrast to EI, *electrospray ionization* (ESI) is a very "soft" ionization method. Here ions that are in solution (usually in presence of electrolytes) are transferred into the gas phase by spraying from a needle (0.1–0.3 mm i.d.) held at 3–5 kV. Evaporation from the charged micro droplets starts at atmospheric pressure and continues in the heated interface with subsequent volumes maintained at increasingly higher vacuum until the size of the shrinking droplets reach a point of charge density on their surfaces where the integrity of the droplets can no longer be maintained and the ions are released into the gas phase. Spraying is enhanced by a supporting flow of a liquid or a nebulizing gas; a countercurrent drying gas also increases efficiency.

TABLE 1.6
Advantages and Limitations of Ion Sources for Nonvolative Analytes

Advantages	Disadvantages
Electrospray	
• Upper mass limit: ~100 kDa • Arguably the softest ionization method; excellent for native noncovalent interactions • Excellent LC/MS compatibility for flow rates up to 100 μL/min • Relatively little matrix interference • Excellent for all types of MS/MS	• Possible problems with mixtures due to overlapping sets of multiple charged ions (may be overcome by high resolution, e.g., FTICRMS) • Salts and other buffer components interfere
Nano-ESI	
• Low detection limits in terms of both molar concentration and absolute amount of analyte consumed • 1–2 μL samples last for 30–120 min, provides adequate time to optimize experimental conditions in MS/MS mode for each eluted analyte ion in turn	• Inconvenient: analyses must be set up one at a time • New capillary needed for each analysis
MALDI	
• Upper mass limit: > 350 kDa • High sensitivity: low femtomole • Rapid, high throughput, and simple operation • Mixtures may be analyzed	• Interference by matrix • Difficult to quantify • Not an on-line technique • Ion suppression by salts (some tolerance)
FAB	
• Upper mass limit: ~10 kDa • Wide availability	• Problems in LC/MS: matrix required; flow rates limited to <2 μL/min • Difficult to quantify • Ion suppression by salts • Low sensitivity compared with MALDI and ESI

The ions generated by ESI are characterized by the adduction of a proton. A unique feature of ESI is that multiple charging may occur to produce spectra with a Gaussian-type envelope representing a series of multiply charged ions with consecutive ions differing by a single charge. This feature permits the analysis of large biomolecules with *m/z* analyzers of low mass range. Algorithms are available to calculate the molecular mass of the analyte (Figure 1.2). *Nano*-ESI is a very low flow rate (<25 nL/min) variation of ESI that has the advantages of reduced sample quantity requirement, significantly increased sensitivity, and sustained sample flow for MS/MS analysis.

In *matrix-assisted laser desorption ionization* (MALDI), the analyte is co-crystallized with an excess of a *matrix* (e.g., sinapinic acid) that has a functional group or molecular portion that absorbs the photons of the laser. Upon a sudden input of energy from the laser, the evaporating matrix forms ionized species that in turn ionize the analytes. The neutrals that also evaporate have a beneficial cooling effect during their dissipation from the plume under vacuum, thus contributing to the stability of the analyte ions formed. The selection of the matrix, which is often an aromatic acid, is critical because different compound classes exhibit remarkable, matrix-dependent differences in ionization efficiency. Combinations of MALDI with time-of-flight analyzers (see below) represent a major technological advance in the routine, rapid, and accurate analysis of biologically important analytes with molecular masses up to 350 kDa.

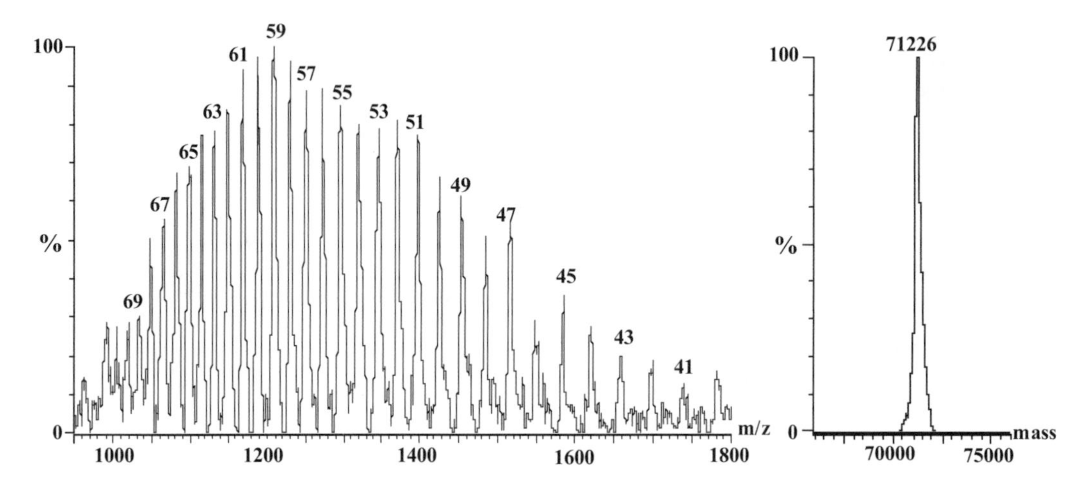

FIGURE 1.2 Electrospray ionization (ESI) mass spectrum of a heat shock protein. The left side shows the typical set of multiply charged ions characteristic of ESI. The numbers represent the number of positive charges. The right side shows the molecular mass of the analyte, ~71 kDa, obtained with a transformation algorithm. One of the advantages of the multiply charged ions is the ability to use analyzers of low mass range to analyze analytes of high mass.

1.2.3 Mass Analyzers

The heart of any mass spectrometer is the mass selective analyzer. The main functions of all mass analyzers are to resolve ions of the same *m/z* from all other ions present and to focus, sequentially or sometimes simultaneously, the individual ion beams of discrete mass onto a detector or, in the case of tandem analyzers, into a collision cell or second ionization chamber. There are five types of analyzers currently in use. Three of these accomplish ion separation in a linear, *in-space*, mode: quadrupole, magnetic, and time-of-flight. The other two accomplish separation *in-time*: ion trap and ion cyclotron resonance (Fourier transform). Because each type of analyzer is based on significantly different principles, each has unique advantages and limitations (Table 1.7). The selection of the proper analyzer is just as important as that of the ion source.

With respect to the mechanism of ion separation, the *time-of-flight* (TOF) analyzers are certainly the simplest. Pulses of ions are accelerated out of the source with the same kinetic energy, provided by an electric field, and travel through an evacuated drift tube (typically 1 m long) to a detector. The ions will separate according to their *m/z* ratios; lighter ions drift more quickly, heavier ions drift more slowly. The resolution and mass measuring accuracy of TOF analyzers has been improved significantly with the development of the *delayed extraction* technique, which reduces the number and energy of collisions in the expanding plume from MALDI, and with the introduction of *ion mirrors* (*reflectrons*) of varying sophistication. Together with other improvements, including advances in sample preparation and ion introduction (*orthogonal acceleration*) techniques, it is now possible to routinely detect peptides at low attomole level with mass accuracies in the low ppm range. Further developments, particularly the commercial realization of *hybrid* analyzers and the *post source decay* technique (PSD), have significantly increased the applicability of the MALDI-TOF approach to obtain sequence or other structural information from proteins and oligonucleotides.

A *transmission-quadrupole* mass analyzer consists of four cylindrical rods (each with a theoretically hyperbolic cross-section) that are ~0.2 m long, ~6 mm diameter, and arranged precisely in a square. To one set of diagonally paired rods, a positive d.c. voltage and a superimposed radiofrequence (rf) are applied, while the other pair of rods receives a negative d.c. voltage and an rf voltage 180° out of phase. Under a given set of d.c. and rf voltages, only ions of a certain *m/z* ratio have stable trajectories and are able to pass toward the detector, while those with unstable trajectories

TABLE 1.7
Comparison of Mass Analyzers

	In-space			In-time	
	Quadrupole	**Magnetic**	**TOF**	**Ion trap**	**FTICR**
Ion beam	Continuous	Continuous	Pulsed	Pulsed	Pulsed
Mechanism	Band pass filter	Momentum	Flight time	Band pass filter	Frequency separation
	Scanning	Scanning	Scanning (time)	Eject analyte ions	Circular path
m/z[a]	$k(V/\omega^2r^2)$	$(B^2r^2)/2V$	$(2t^2V)/L^2$	$k(V/\omega^2r^2)$	$B/2\pi\omega$
Variations	Single quad	Single focus	Linear	Zoom scan	Single/dual cells
	Triple quad	Double focus	Reflectron	Internal source	Internal source
	Ion pipe	Hybrid	Hybrid	External source	External source
	Hexa-, octapole		Delayed extraction		
			Postsource decay		
Detection	Ion current	Ion current	Ion current	Ion current	Image current
Pressure, torr	10^{-6}	10^{-6}	10^{-6}	10^{-4}	10^{-10}
Upper limit, kDa	4; 100 with ESI	8	>350	4	15
Resolution	$>10^3$	$>10^4$	$>10^3$ linear $>10^4$ reflectron	$>10^3$	$>10^6$
Benefits	Excellent for ESI	Capable of:	Highest mass range	Simple, small	Highest resolution
	Versatile	exact mass	Fast scan	MS^n (n = 2 – 6)	MS^4
	Easy ± switching	high resolution	Excellent MALDI	Inexpensive	
Limitations	4 kDa mass limit	Slow scanning	Needs pulsed source	Limited mass range	High vacuum
	No MALDI	High voltages			High field magnet
		Cumbersome			Cumbersome

[a] Derived formulas must include factors to account for constants and conversion to appropriate units
V = accelarating voltage, B = magnetic field strength, r = radius, t = time, L = length of drift tube, ω = frequency

hit the rods and are lost. Data acquisition (scanning) is accomplished by holding the ratio of rf amplitude to that of a d.c. voltage constant, while systematically increasing both (according to complex relationships). The monitoring of the intensities of selected ions is also easy with this analyzer. In addition to obtaining mass spectra (narrow band operation), quadrupoles are also used as *ion guides* with an rf-only operation (wide band mode), permitting controlled passage of ions within a broad mass range. An application of the latter mode is in the use of the triple quadrupole analyzers for MS/MS experiments, where the second, rf-only quadrupole guides the selected ions within the collision cells.

In contrast to transmission quadrupoles, the *quadrupole ion traps* prevent the passage of ions with stable trajectories; the ions of interest can only be detected when their trajectories are made unstable. These analyzers consist of a ring electrode (0.5 to 1.5 cm diameter) and two electrically connected end-cap electrodes (usually grounded). A three-dimensional quadrupole electric field is created by applying superimposed rf (~10^6 Hz frequency) and d.c. (small voltages) potentials on the ring electrode. Ions are created in external ion sources and, with the aid of electrostatic lenses, are introduced into the trap where they assume stable trajectories near the center of the trap in a predominantly sinusoidal motion. With the appropriate selection of the frequencies, all injected ions, regardless of their *m/z* values, may be trapped in stable trajectories. To measure an ion with a particular *m/z* value, the frequencies are changed so that the ion trajectory of the analyte ions becomes unstable resulting in their ejection from the trap volume (through an end-cap) onto a detector. Scanning a mass range (or obtaining a full mass spectrum) of the analyte is accomplished by the sequential changing of frequencies, ejecting ions with different *m/z* values one after the other. In an alternative method of operation, potentials can be chosen such that only ions with one selected *m/z* have a stable trajectory, i.e., remain trapped, while all others with different *m/z* values assume unstable trajectories and are ejected from the trap. The remaining trapped ions of a given *m/z* may be exposed to multiple stages of dissociation to obtain structural information. Ion traps are excellent for the generation and analysis of multiple generation CID products, up to MS^6.

The classical *magnetic sector* mass analyzers are, in fact, momentum analyzers. Ions leaving the source are accelerated to high velocity and enter a magnetic sector in which the magnetic field is applied perpendicularly to the direction of the motion of the ions. The resulting ion motion is circular, and the radius and angle of the arc depend on the ion optical design. To compensate for the loss of resolution due to the fact that the ions of a particular *m/z* formed in the source do not all have exactly the same energy, and thus the same velocity, after acceleration, *electric sectors* may be used to focus ions according to their kinetic energies. Although there are a number of realized optical design variations, the use of magnetic type analyzers has been declining in recent years, except for special applications requiring high resolution, e.g., the analysis of dioxins.

The fact that ions move in a circular path in a magnetic field is used in a different way in the *ion cyclotron resonance* (ICR) analyzers. The frequency of the rotation of the ions in their orbits depends on their *m/z* ratio and the strength of the magnetic field. A cubic ICR cell (~5 cm) consists of three pair of parallel plates—excitation, receiver, and trapping plates—centered in a strong, homogenous magnetic field generated by a superconducting magnet. The ions, which are moving in circular orbits perpendicular to the magnetic field (*xy* plane), are trapped parallel to the magnetic field (*z* axis) with the aid of an electrostatic potential applied to two plates. The trapped ions may be stored for hours, as long as high vacuum is maintained to prevent destabilizing collisions with residual neutral molecules. When an rf field is superimposed perpendicular to the direction of the magnetic field, any ion can be made to move in a spiral path by adjusting the rf frequency to be equal to the natural cyclotron frequency of the ion. The selected ion will absorb energy only at a particular ratio of the irradiation frequency and the magnetic field strength. The operation of ICR is governed by a series of discrete events, known as the *experimental pulse sequence,* and each step is under computer control. Ions are eventually detected simultaneously by an *image current* they induce in an amplifier connected to the receiver plates. The nondestructive image current signal is converted to a voltage, amplified, digitized, and decoded by *Fourier transformation* (FT) to yield a frequency spectrum that, in turn, can be converted into a mass spectrum. Since frequencies can be measured with high precision, ion masses can be determined to up to one part in 10^9 (ppb). Because of the inverse relationship between *m/z* and frequency, high-mass ions (low-frequency) are more difficult to resolve than low-mass (high-frequency) ions. Still, recent advances in instrumentation as well as in methodologies have established FTICRMS as the technique that provides higher resolution for a given mass than any of the other analyzer types. These and other significant advantages have made FTICR instruments the fastest growing segment of the current market for mass spectrometers.

1.2.4 Ion Detectors

The ions in each separated ion beam of different *m/z* eventually arrive at a collector where they are "counted" to yield analog signals which, in turn, are usually "converted" into an electron current that can be amplified to provide ion intensities that are representative of the number of collected ions of particular *m/z* values. There are single- and multi-point (array) detectors to serve different analyzers. FTICR analyzers produce unique image current signals from which mass spectra are derived using FT.

1.2.5 Vacuum Systems

All mass spectrometers must operate under high vacuum to eliminate unwanted collisions among the ions and between the ions and neutral molecules of air and/or molecules other than those of the sample (background). High vacuum is also needed to prevent gas discharges by the high voltages (up to kV levels) often used to accelerate and move gas-phase ions from one region to another and to operate certain ion detectors. Yet another important function of the vacuum system is to assure the production of ion beams representative of the sample under investigation by minimizing background and cross contamination between successive samples (memory effect), particularly when separated components are introduced in rapid succession (10 s or less) in chromatographic or electrophoretic effluents.

Although the official unit of pressure is the Pascal (Pa) vacuum in mass spectrometers is still commonly expressed in *torr* units (1 torr = 133.322 Pa = 1.3 mbar; 1 atm = 760 torr = 1.013×10^5 Pa). Various parts of mass spectrometers require different levels of vacuum to prevent ion scattering. The mean free path, i.e., the average distance that a particle travels between successive collisions with other particles in the gas phase, is inversely proportional to the pressure. The best vacuum is needed within the analyzers, where ~10^{-7} torr (3×10^9 molecules/cm^3) is routinely maintained. At this pressure, the mean free path is ~500 m which is more than adequate considering that the flight length of ions is ~0.1 m in a quadrupole analyzer and ~1 m in sector magnetic and TOF analyzers. In contrast, the path lengths of ions in most ion sources is <1 cm, thus the mean free path is long enough even at 10^{-3} torr pressure, to prevent scattering.

Often radical pressure differences are required between close segments even within single instrument components; e.g., in electrospray ion sources at least two orders of pressure difference must be maintained between the area of the entrance of the liquid spray and the entrance of the ion beam into the mass analyzer. The difficult task of *differential pumping* is accomplished with independent pumping systems that maintain a pressure differential between separate regions by connecting them via apertures of low conductance. In this way the high vacuum in the analyzer section can be maintained regardless of the pressure fluctuations in the ion source. The vacuum systems of mass spectrometers consist of different types of pumps (mechanical, diffusion, turbo, molecular cryogenic), baffles, valves, pressure gauges, tubes, and slits. Some components are in series; others are in parallel configuration.

1.2.6 Data Systems

Computer systems are now integral components of mass spectrometers. Dedicated computers control instrument operation on a variety of levels, ranging from relatively simple functions such as automatic tuning, mass calibration, and saving data to the highly sophisticated automatic switching from one mode of operation to another, such as from full mass spectra acquisition to MS/MS mode, when triggered by the presence of ions of particular m/z values above a specified threshold in a scan. Large quantities of acquired data are often transferred for storage to off-line, central facilities. In the new area of computerized data evaluation, truly spectacular advances have occurred in using databases available on the Internet, e.g., the computer-based identification of proteins directly from limited amino acid sequence information provided by mass spectral analyses.

1.3 MEASURES OF PERFORMANCE

1.3.1 Resolution and Resolving Power

Resolution refers to the ability of a mass analyzer to separate ions of slightly different m/z values. Although resolution can be calculated from ion optics, it is usually determined experimentally using either doublet peaks or a single peak. The "valley" definition expresses resolution ($m/\Delta m$) in terms

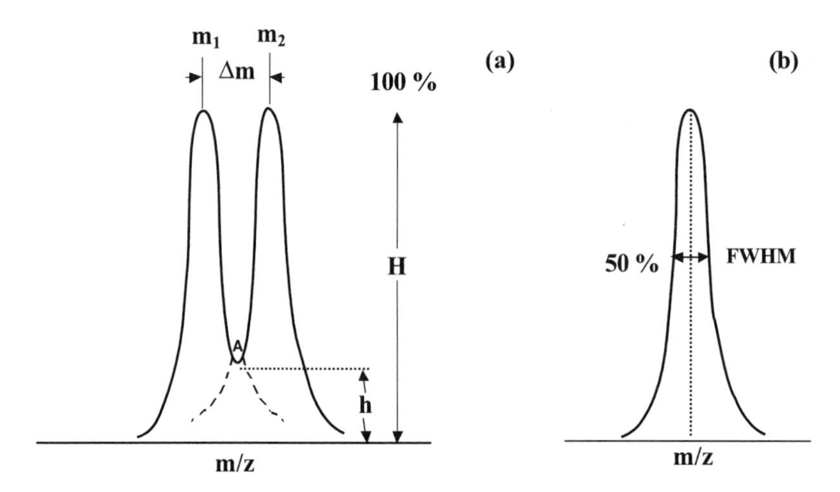

FIGURE 1.3 Definitions of resolution. (a) Valley definition using two adjacent peaks; resolution = $m/\Delta m$; (b) Using a single peak, where FWHM means "full width at half maximum"; resolution = m/FWHM.

of the highest mass at which two adjacent peaks of equal height exhibit a valley between the peaks not greater than a certain percentage, such as 10% (magnetic analyzers) or 50% (quadrupole analyzers) of the peak height (Figure 1.3a). When the resolution is a function of m, $m/\Delta m$ values should be given for several m values. When only an isolated peak of mass m (should be from a singly charged ion) is available for measurement, the "peak width" definition of resolution is used: Δm is taken as the full width of the peak at half its maximum height, FWHM (Figure 1.3b). While resolution is usually a large number, *resolving power*, the inverse of resolution, $\Delta m/m$, is a small number, often expressed in terms of parts per million (ppm). Low-resolution (<2000) and high-resolution (>2000) analyzers provide significantly different information (Section 1.4). It is noted that some published lists of terms define resolution and resolving power opposite to the way defined above.

1.3.2 Mass Range

The maximum measurable m/z is obviously a critical consideration in the selection of an instrument for a particular project. The upper mass limits of the various sources and analyzers are included in Tables 1.6 and 1.7, respectively. The mass range of an instrument is not directly related to its resolution, thus poor resolution could severely reduce the useful portion of the mass range.

1.3.3 Mass Accuracy

The difference, $\Delta m_{accuracy}$ between the true mass (m_{true}) and the measured mass ($m_{measured}$) of an ion may be expressed in terms of millimass units (0.001 Da) or in ppm, which is calculated as $(\Delta m_{accuracy}/m_{measured})10^6$. The minimum accuracy of mass measurements should be to the nearest mass over the entire mass range of interest with ~0.1 Da drift over the time period of the measurement. High mass accuracy or exact mass measurements, i.e., within a few mDa, which are usually associated with high resolution, have significant applications. It is noted that relatively high mass measuring accuracy may also be achieved at low resolution as long as there are no interfering adjacent peaks present.

1.3.4 Scan Speed

Scan speed is the rate at which mass spectra are acquired, is measured in mass units/unit time. Two of the in-space type analyzers, quadrupole and magnetic, are of the scanning type: certain

TABLE 1.8
Specificity, Sensitivity, and Limit of Detection

Instrumentation and Analytical Method	
• **Specificity**	
Mass spectrometer	Unique response to analyte; may require more than one characteristic, e.g., presence of structurally significant species in proper intensity ratios.
Analytical method	Ability to assay only the analyte without interaction with, or interference from, other components in the medium or method.
• **Sensitivity**	
Mass spectrometer	Instrument response to ions of an analyte at an arbitrary *m/z* value, i.e., ratio of ion current change to sample flux change in the ion source.
Analytical method	Ability to detect a particular analyte or a small change in its quantity (incremental sensitivity) in a specified sample matrix.
• **Limit of Detection**	
Mass spectrometer	Smallest sample quantity (or lowest partial pressure) providing a signal that can be distinguished from the background noise in the instrument, e.g., signal/noise ratio of 3; experimental conditions should be provided.
Analytical method	Smallest quantity of the analyte detectable using a specified technique (including sample preparation) in a defined matrix (e.g., serum, urine, tissue). Often expressed as a concentration, e.g., μg/mL serum. The required sample quantity should also be specified.
Clinical Diagnosis	
• **Specificity**	Proportion of negative test results obtained when the test is applied to patients known to be *free* of the disease.
• **Sensitivity**	Proportion of positive test results obtained when the test is applied to patients known to *have* the disease.

instrumental parameters are swept (scanned) continuously in order to focus ions with successively higher or lower *m/z* values onto a detector where their intensities are recorded as a function of time. With respect to overall performance, scanning analyzers have the inherent disadvantage that only a small fraction of all the ions is recorded at any given moment. In contrast, nonscanning analyzers such as ion traps can accumulate ions selectively and thus are well suited for multiple stages of tandem mass spectrometry.

1.3.5 Specificity, Sensitivity, and Limit of Detection

The meanings of specificity and sensitivity are significantly different from the points of view of analytical methodology and clinical diagnosis (Table 1.8). Even within the technical area, it is important to distinguish between the basic, mass spectrometry-related definitions of specificity and sensitivity, which are vital to the mass spectrometrist but not to the "customers" or collaborators, and the overall, method-related definitions which specify the usefulness of a particular technique for the problem at hand. Of major practical importance are incremental sensitivity (the minimum difference between concentrations that a technique can distinguish reliably) and the limit of detection LOD, the smallest detectable quantity of the analyte by the method using a given quantity of serum, urine, tissue, etc. The limit of quantification is often taken to be about three times that of LOD.

1.4 INFORMATION FROM MASS SPECTRA AND SCOPE OF APPLICATIONS

Mass spectrometers answer the basic questions of WHAT is present and HOW MUCH is present by determining ionic masses and intensities. Accordingly, the major areas of applications of mass spectrometry have been qualitative analysis and quantification (Table 1.9).

TABLE 1.9
Global Scope of Mass Spectrometry

• **Qualitative analysis**	
Confirmation of identities	Compare to spectra of standards
	Compare to spectra in libraries
Identification of unknowns	Molecular mass
	Exact mass for empirical formula
	Fragmentation and isotope pattern for structure determination
	Comparison, using libraries and databanks (proteins, nucleic acids, carbohydrates, lipids)
• **Quantification**	Selected constituents (from major to trace)
	Dynamic monitoring
	Isotopic ratios
• **Ionic reactions (gas phase)**	Mechanics and energy (physical chemistry)
• **Industrial process monitoring**	
• **Preparative mass spectrometry**	

As mentioned already, the type and abundance of ions in a mass spectrum depend greatly on the kind of ion source and the experimental conditions used for obtaining the spectrum. For example, when medroxyprogesterone acetate is analyzed in an EI source, the mass spectrum reveals extensive fragmentation with a number of diagnostically useful fragment ions; however, the molecular ion is barely detectable (Figure 1.4a). In contrast, when the same compound is analyzed using chemical ionization (CI), a much softer ionization method than EI, an abundant $[M+H]^+$ ion appears, representing the protonated molecule, and there is little fragmentation (Figure 1.4b). As an example of the importance of experimental conditions, consider the ESI spectra of this compound obtained using different cone voltage conditions in the ion source (nozzle-skimmer fragmentation). The spectrum obtained at lower cone voltage (Figure 1.4c) is useful for the determination of the molecular mass and provides limited structural information. At higher cone voltage, there is intense fragmentation and considerable structural information is revealed (Figure 1.4d). This example underscores the importance of the judicious selection of a suitable ion source as the most important step toward a successful analysis. Indeed, mass spectrometric techniques are often described with reference to the ion source employed. It is emphasized that mass spectra are usually reproducible when experimental parameters can be duplicated. It is concluded that for the full utilization of the diversity of mass spectra, it is more important in mass spectrometry than in other areas of spectroscopy to be familiar with the available alternative instrument designs, analytical techniques, and operational parameters.

Low-resolution analyzers provide *unit* mass resolution separating masses that differ by 0.5 to 1.0 Da. Quadrupole, ion trap, single-focusing magnetic, and TOF (without reflectron) analyzers provide resolutions in the 2000 to 3000 range. These instruments are widely used for the quantification of targeted analytes, usually in an on-line coupled mode with GC and HPLC to separate isobaric (same nominal mass) and other constituents present. Another major area of application is rapid confirmation and/or identification of analytes using the large EI data libraries available.

High-resolution analyzers permit the separation of isobaric ions, thereby significantly increasing selectivity. For example, the resolution needed to separate the isobaric ions of hexachlorobiphenyl ($C_{12}H_4Cl_6$) at m/z 357.8444 and isotopic pentachloro-dibenzo-p-dioxin ($C_{12}H_3{}^{35}Cl_3{}^{37}Cl_2O_2$) at m/z 357.8519 is $m/\Delta m$ = 357.8519/0.0075 or ~48,000, a difficult but doable task for double-focusing magnetic analyzers. However, isomers cannot be separated by high-resolution analyzers, and chromatography is usually required to provide the needed selectivity.

The measurement of ion masses with an accuracy to three or four digits beyond the decimal point may be utilized to calculate the elemental composition of ions. The basis of this is the fact

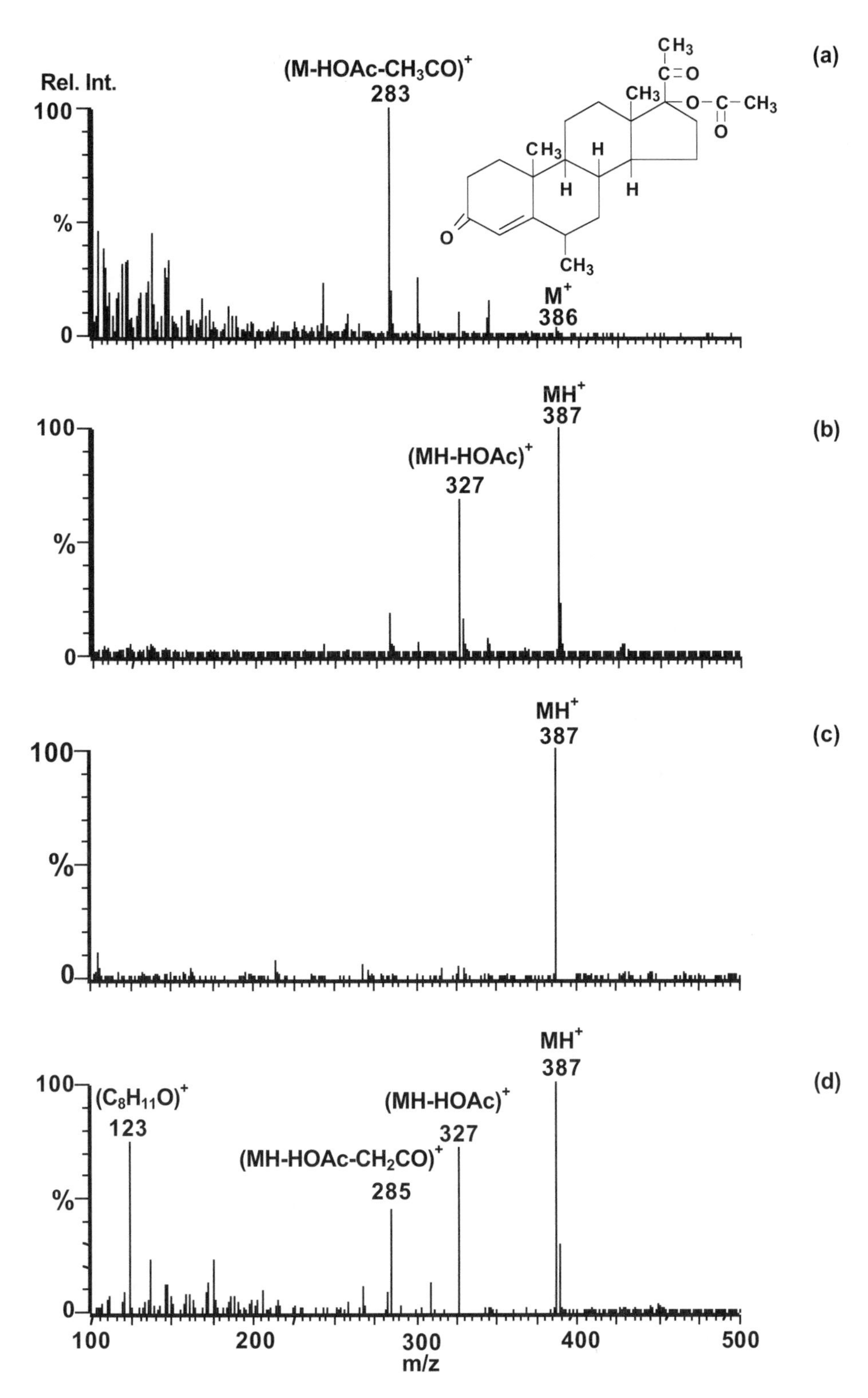

FIGURE 1.4 Mass spectra of medroxyprogesterone acetate obtained using different conditions: (a) electron ionization; (b) chemical ionization; (c) electrospray ionization using low cone (nozzle-skimmer) voltage; and (d) electrospray ionization at higher cone voltage.

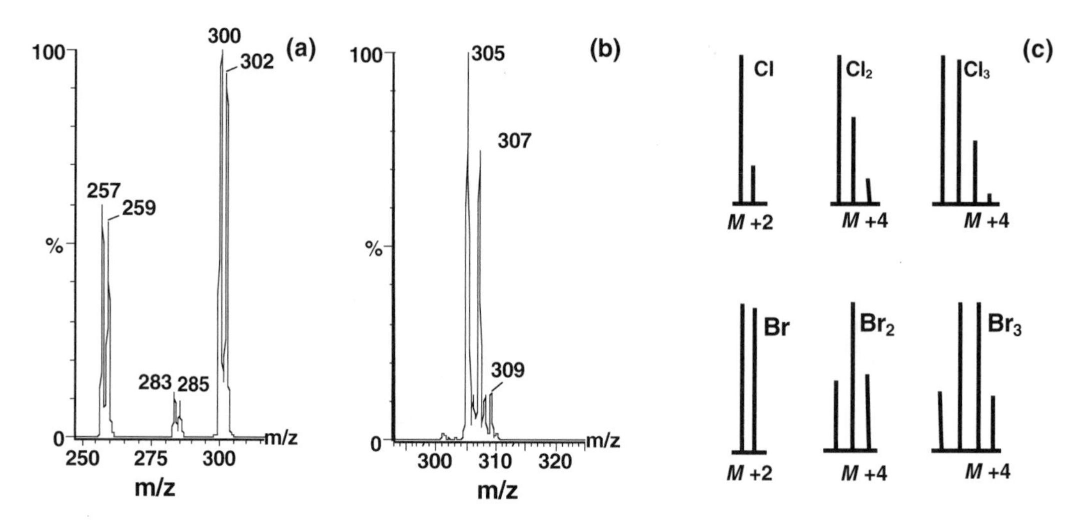

FIGURE 1.5 Isotope patterns. Electrospray mass spectra of: (a) 3-(bromoacetamido)-benzoylurea which contains one bromine; (b) L-m-sarcolysin, L-3-[bis(2-chloroethyl)amino]-L-phenylalanine which contains 2 chlorines; and (c) characteristic isotope patterns for one-, two-, and three-halogen-containing compounds; M refers to the mass of the compound with the lowest mass isotope.

that when the $^{12}C = 12.0000$ scale is used, other elements have fractional masses, thus ionic masses are usually not integer values. An important application of high-resolution MS has been the identification of unknown analytes based on the determination of their elemental compositions. However, the required resolution increases rapidly as the mass increases, thus this approach is restricted to analytes of <1 kDa. Technological advances in TOFMS and FTICRMS have opened a rapidly expanding application area of combined high-resolution and accurate mass determination for the identification and characterization of biomolecules of high mass.

The fact that all organic compounds have a distribution of masses resulting from the natural abundance of atomic isotopes, e.g., 1.1% for ^{13}C, results in the appearance of *isotope peaks*. In certain instances, isotope peaks may appear as doublets, e.g., in compounds containing one bromine atom (Figure 1.5a) or two chlorine atoms (Figure 1.5b). Characteristic patterns appear when there are more halogen atoms (Figure 1.5c). Isotopic peaks may appear in more complex *isotopic clusters,* such as when the compound contains platinum. There are simple equations to calculate the statistical distribution of the isotopes, and thus the expected relative abundances of the mass peaks. When analytes >10 kDa are analyzed using FTICRMS operated at high resolution individual peaks representing species with one or more ^{13}C atoms appear, replacing the more abundant ^{12}C atoms. This approach permits elemental composition analysis.

Structure elucidation of low molecular mass analytes is based on the interpretation of mass spectral fragmentation patterns obtained by EI. Decades of experience have resulted in textbooks and monographs providing data and methodologies for a wide variety of organic compound classes, ranging from simple hydrocarbons to complex alkaloids. Important structural information may also be obtained when selected product ions of the primary fragmentation of the analyte are made to undergo secondary or higher order fragmentation using CID. There are many types of ions (Table 1.2) and, as amply illustrated throughout the text, this is one of the reasons for the unparalleled versatility of mass spectrometry. Still, it is rarely possible to determine the structure of complex organic or biological materials by MS alone, without information from NMR, UV, and IR spectroscopy and other analytical methods.

While the "softness" of ionization by ESI and MALDI is a prerequisite for obtaining molecular mass information of large and thermally labile biomolecules, this is at the expense of the formation of structurally significant fragment ions. Fortunately and conveniently, the technique of CID provides fragment ions with a great deal of structural information. In CID, fragment ions are formed when ions preselected from the mass spectrum of the analyte are accelerated and collide with molecules of an inert gas in a separate collision cell placed between sequential analyzers of tandem mass spectrometers or inside the ion storage area of time sequence-based mass analyzers. The distinct types of possible CID experiments include:

1. *Precursor-product* mode provides fragmentation mass spectra of a selected ion from the mass spectrum of an analyte (product ion MS/MS spectrum). A common use is the identification of targeted compounds present in a complex mixture.
2. *Precursor (parent) ion* mode obtains all of the precursor ions that dissociate to the mass-selected product ion. Useful for the characterization of structurally related compounds that produce a particular ion indicative of some common structural feature.
3. *Constant neutral loss* mode where the mass difference between the precursor and product ions is specified so that the spectrum displays all the reactant ions that dissociate by loss of neutral species of the selected mass. Typical uses include the recognition of functional groups in complex mixtures or the identification of components in mixtures derivatized to yield a characteristic neutral loss.

Quantification of selected constituents present in trace quantities is possible with the techniques of:

1. *Selected ion* monitoring makes the MS a specific ion detector by recording only one or more preselected m/z values, excluding all other ions present.
2. *Selected reaction* monitoring is yet another mode of MS/MS operation where a selected precursor ion is allowed to undergo CID but only the product of a characteristic transformation is detected.

These techniques are applicable to the quantification of a wide variety of compounds, including hitherto inaccessible biological molecules, in almost any matrix, including body fluids and tissues, with minimal sample preparation. Dynamic monitoring of selected constituents is now employed routinely in diverse areas ranging from respiratory and blood gas analysis during surgery to process control in the biotechnology industry.

Table 1.10 summarizes the major areas of applications of mass spectrometry in cancer research. The description and discussion of the type of information provided, together with representative applications, are the subject of Chapters 4, 5, and 6.

TABLE 1.10
Examples of the Scope of Mass Spectrometry in Cancer Research Illustrating the Range of Applications from Small Molecules to Biopolymers

Carcinogens

- Identification, quantification, metabolism
- Compound speciation and quantification of carcinogenic elements and components
- Unique biomarkers and biomarker profiles
- Monitoring subjects at risk (dosimetry)

Pharmacology

- Drug development
  - Structure and purity of natural and synthetic agents
  - Distribution, excretion, metabolism
  - Conjugation, protein binding, loading values
  - Pharmacodynamics
- Chemotherapy
  - Toxic concentrations of drugs and/or metabolites in blood, urine, and tissues
  - Pharmacokinetics; concentration X time curves
  - Pharmacogenetics
  - Global analysis of protein expression in the development of chemoresistance (*in vitro* and *in vivo*)
  - Diagnosis and monitoring of opportunistic infections: circulating microbial metabolites

Biology

- Elucidation, at the molecular level, of cellular or structural changes leading to oncogenesis
  - Nature of relevant mutations and time of their occurrence
  - Identification and quantification of epitopes
  - Function-critical post-translational modifications
  - Changes in cellular proteins in apoptosis
  - Changes in cellular proteins in progression of tumors
  - Identification of critical protein-protein associations
- Diagnosis
  - Detection of small genetic alterations in the background of normal genes
  - Differentially expressed proteins: upregulated, downregulated, or unique

2 Instrumentation and Techniques of Mass Spectrometry

2.1 ION SOURCES

2.1.1 ELECTRON IONIZATION

Historically, electron ionization (EI, electron impact, electron bombardment) has been the most commonly used method to ionize relatively small (<1.2 kDa) molecules which can be vaporized without thermal decomposition (often after derivatization). A typical EI source (Figure 2.1) is a small box (~1 mL volume, stainless steel, maintained at ~200°C to prevent sample condensation) with an inlet tube for the introduction of gaseous samples or vapors from liquid or solid samples, and another inlet for a direct probe tip (Section 2.4.2). Electrons (20–300 μA current), produced by thermionic emission from a rhenium or tungsten filament heated to 2000°C, enter the source through a small slit and are accelerated by a potential drop between the filament and an electron trap at the opposite side of the box in such a way that the electrons attain 70 eV energy in the middle of the ion source where they encounter the analyte molecules. A small permanent magnet (50–300 gauss) forces the electrons to move in a narrow helical path (collimation). This leads not only to a better-regulated electron beam but also provides a smaller volume within which the ionization takes place. The end results are a reduction in the initial energy spread of the ions and an increased ionization cross section. The source slit is usually grounded and all other source potentials are positive with respect to it. The ion repeller electrode is usually at a positive potential with respect to the source to aid in pushing the positive ions in the box toward the exit slit. The ion beam current (10^{-7} to 10^{-14} A) is drawn out of the ionization chamber perpendicular to the electron beam, and a set of extracting, focusing, and accelerating lenses form and shape the exiting ion beam prior to its entering the mass analyzer. Sample vapor pressure in the ion source must be in the 10^{-7} to 10^{-2} torr range. The ion source must be differentially pumped to keep the immediate region of ion formation almost vacuum tight to optimize the residence time of the analytes (maximize sensitivity), while the rest of the source should be exposed to fast pumping.[1]

Ionization occurs, and molecular ions form, when the energy absorbed by the neutral molecules from the bombarding electrons is equal to or more than their ionization potential, resulting in the ejection of an electron to yield a net positive charge. The bombarding electrons are usually accelerated to 70 eV energy to maximize the efficiency of ionization; still, only 0.01 to 0.5% of the neutral molecules are ionized. Because the ionization potential of most organic compounds is ~10 eV, the molecular ions often have a considerable excess of internal energy that provides the energy needs (activation energy) for the formation of a variety of fragment ions. The resulting mass spectral fragmentation patterns depend on the chemical composition and molecular structure of the parent compound. Fragmentation patterns result from a series of competitive and consecutive unimolecular reactions that are explained by the quasiequilibrium and other theories.[2] The rules governing fragmentation of small ions by EI, and their application in the interpretation of observed spectra, are well documented.[3] The structure dependency of fragmentation patterns is an apparent advantage because similar structures give comparable spectra. Disadvantages include that sensitivities

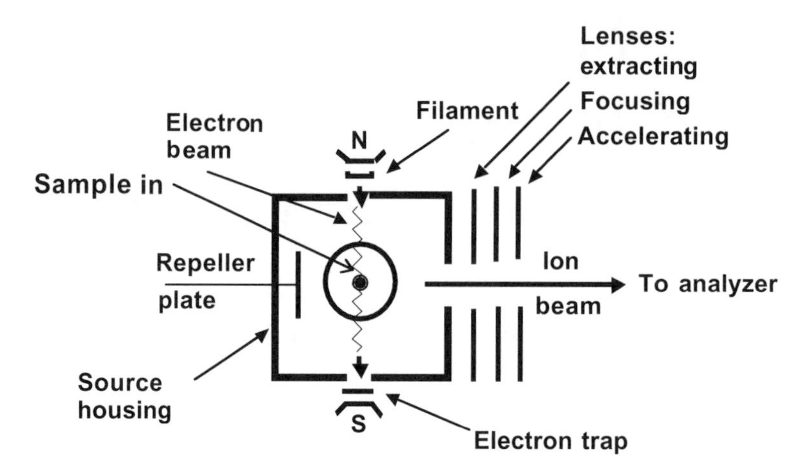

FIGURE 2.1 Electron ionization source.

of different compounds may vary significantly (significant fragmentation results in the dilution of the ion current among many fragments) and that molecular ions often do not appear at all.

The basis for the widespread use of EI is that the mass spectra of individual compounds show remarkable constancy as long as experimental conditions are constant. A vast amount of highly reproducible data is available on the fragmentation patterns of >160,000 organic compounds; computer searchable commercial databases can be used to identify unknowns or to search for common substructures (Section 2.6.1). Despite disadvantages (Table 1.5), EI is the robust method that remains the technique of choice in many applications involving anticancer drugs, metabolites, and environmental carcinogens.

2.1.2 Chemical Ionization

In chemical ionization (CI), ion-molecule reactions are used to produce ions from the analytes.[4,5] First, electron bombardment is used to produce a large number of molecular and fragment ions from a reagent gas which is introduced into a modified EI source at a pressure of 0.2–2 torr (compared to a typical 10^{-6} torr in EI). Because of the high pressure, these primary ions frequently collide (multiple collisions) with unreacted neutral molecules of the reagent gas. The result of such secondary ion-molecule reactions is a high abundance of ions derived from the reagent gas. These ions are efficient proton donors or can form adducts. For instance, with methane, which is the most common reagent gas, a large number of $CH_4^{+\bullet}$, CH_3^+, CH_2^+, CH^+, H_2^+, and $H^{+\bullet}$ ions are formed in the primary EI reaction, followed by the formation of an abundance of CH_5^+, $C_2H_5^+$, and $C_3H_7^+$ ions in the secondary ion-molecule reactions.

When a small amount of sample vapor is introduced (reagent gas/analyte ratio: 1000:1), the analyte molecules are ionized in ion-molecule collisions by the secondary ions formed from the reagent gas. If the analyte molecules are good proton acceptors, the reagent ions act as Brönsted acids and protonate the sample molecules, e.g., $CH_5^+ + BH \rightarrow BH_2^+ + CH_4$. Here the reaction involves the transfer of a proton. The CH_5^+ ion is the most powerful donor known in gas-phase reactions. The mass of the product ions (tertiary ions) are one Da higher than that of the intact molecule, i.e., $[M+H]^+$. The additional energy acquired by the neutral analyte molecules in the process of CI is frequently less than that required to break bonds, thus the $[M+H]^+$ are often the only ions of appreciable abundance present in the mass spectrum. However, significant fragmentation does occur occasionally in CI. For example, an $[M+H]^+$ ion may undergo cleavage when a powerful donor is involved such as with molecules containing trimethylsilyl (TMS) ether groups, yielding $[MH\text{-}Si(CH_3)_3]^+$ ions (loss of 73 Da) in considerable abundance. When fragmentation occurs, the pathways usually relate to those observed in EI.

Other frequently used reagent gases include ammonia (selective ionization of alcohols and amines) and isobutane (minimizes fragmentation). Hydrogen, methanol, nitric oxide, other reagent gases and mixtures of reagent gases have been explored as reagents for special purposes.[6] Depending on the nature of the reagent gas, the mechanism of ionization may be different from that of the proton-transfer mechanism observed with methane. For example, with ammonia as the reagent gas, analytes that are more basic than NH_3 yield abundant $[M+H]^+$ ions by proton exchange while less basic analytes yield $[M+NH_4]^+$ ions by attachment. In the latter case, the most abundant ion in the spectrum is 18 Da higher than M. Beneficial results may be obtained by exploring a variety of ion-molecule reactions through consideration of the functionalities of the analyte and judicious selection of the reagent gas.

Negative ion CI (NCI) is, in most instances, a misnomer because it does not involve an ion-molecule reaction but occurs when an analyte with several electronegative atoms (e.g., polychlorodioxins) captures low-energy electrons present in the CI plasma.[7] When applicable, NCI provides extremely high sensitivity concurrent with high specificity, e.g., quantification of dioxin[8,9] or analysis of toulenediamines in urine (Section 4.2.2.B). True negative chemical ionization involving a chemical reaction is rarely used but can be achieved using gases that generate OH^- ions.

There are significant design differences between EI and CI sources. To sustain the required higher pressure, the ionization chamber of the CI source should be almost gas-tight and the pumping speed in the ion source region must be high (up to 1000 L/s) so that the excess reagent gas can be removed quickly. Differential pumping is essential to maintain the required low pressure in the analyzer. The energy of the bombarding electrons is increased to 250–500 eV (typically 70 eV in EI) to ensure penetration of the electrons into the dense gas plasma for efficient primary ionization of the reagent gas. The electron beam is sprayed into the reagent gas. The collimating magnets are not used in a CI source.

Mass spectra from EI and CI are different (Figure 1.4a and 1.4b) but they complement each other. The major advantages of CI are: (a) it often provides molecular mass information, via $[M+H]^+$ ions while EI spectra may yield no (or low abundance) molecular ions; and (b) the simple mass spectra and the abundance of structurally significant ions permits easy quantification by using the SIM technique (Section 2.6.3). A significant limitation is the need for the analytes to be reasonably volatile and thermally stable (same as in EI), thus chemical derivatization is usually required. The mass range is limited to no more than 1.2 kDa, and because CI is much less reproducible than EI, there is a lack of extensive libraries of spectra.

Following its inception in 1966, the growth of CI had been rapid. Today it is a mature technique with thousands of applications. An important area of application of CI is the analysis of isomers in environmental and biological samples where differentiation by EI is usually unattainable. In contrast, the reactivities of isomers toward the CI reagent gases are often significantly different, e.g., thiocyanates and isothiocyanates. A review[10] details the theoretical and practical aspects of both positive and negative CI.

2.1.3 Atmospheric Pressure Chemical Ionization

In atmospheric pressure chemical ionization (APCI), sample ionization occurs outside the vacuum system at atmospheric pressure.[11] Samples are introduced into the ion source as GC or HPLC effluents, directly injected solutions, or vapors that have been evaporated from a metal (often Pt) tip. A makeup or carrier gas (often a mixture) moves the samples into the ionization area, acts as the CI reagent gas, and also sweeps the source volume clean (Figure 2.2). The primary ionization of the CI reagent gas (e.g., the volatilized HPLC solvent) is usually accomplished by a corona discharge (well suited for HPLC effluents) or, less frequently, with the aid of high-energy radiation emitted from a ^{63}Ni foil. The resulting ions are introduced into the mass analyzer through a 25 to 50 μm orifice. Because of the high pressures, a highly efficient differential pumping system is essential. Quadrupole analyzers are often used with APCI because of their tolerance for high pressures.

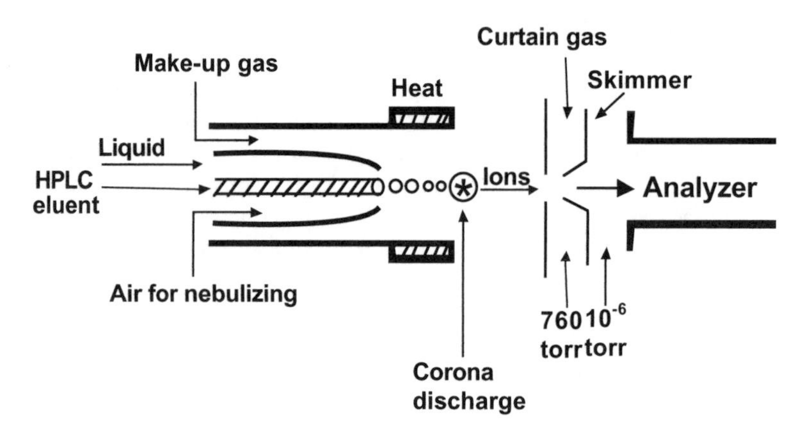

FIGURE 2.2 Atmospheric pressure chemical ionization (APCI) source.

In APCI, ions are formed in a series of complex ion-molecule reactions involving proton transfer or electron capture (both associative and dissociative), charge transfers and exchanges, clusterings, and substitutions. The judicious choice of the reagent gas, similar to CI, permits manipulation of the ion-molecule reactions to lead in the direction best suited for the particular analytical problem. The major difference between APCI and CI is that in the former the ionization takes place at a pressure about 10^3-fold higher than in CI. In APCI there is a high density of neutral molecules, thus the analyte ions attain thermal equilibrium after only a few collisions with the carrier gas molecules. Ionization is not only more efficient in APCI than in CI but there is usually even less fragmentation than in CI.

2.1.4 Electrospray Ionization

Because of its current major importance, electrospray ionization (ESI) has been reviewed extensively, including overviews,[12–16] theoretical and technical aspects, such as the mechanisms of ion formation,[17–20] source design,[21] interfacing with LC[22] and capillary electrophoresis (CE).[23]

In ESI, the analytes, dissolved in an appropriate solution, are electrostatically nebulized by spraying them, at atmospheric pressure, through the tip of a stainless steel nozzle that is maintained at ~4 kV relative to the walls of the ionization chamber and a counter electrode (Figure 2.3a). Alternatively, a fused silica tube with a liquid junction to allow for electrical conductivity can be used. The highly charged droplets are next exposed to either heat or a drying gas (usually nitrogen), or both, as they are passing toward the mass spectrometer inlet which is kept at ground potential. The density of the electric field on the surface of the droplets increases during desolvation, i.e., as their size decreases upon the evaporation of the solvent. When the repulsive Coulombic forces between the like charges on the surface of the droplets exceed the forces of surface tension ("Rayleigh instability limit") the droplets disintegrate into smaller units ("Coulombic or jet fission") but remain charged. Eventually, the electric field, due to the surface charge density, begins to support direct ion evaporation, and ions begin to leave the droplets in "Taylor cone" shapes (Figure 2.3b). The ions thus formed enter the mass analyzer, while neutral molecules are pumped out in chambers maintained at progressively higher vacuum (Figure 2.3a).

The required flow rate in conventional ESI sources is in the 1 to 5 μL/min range, which readily permits direct flow injection of samples. Commercial instruments offer modifications to tolerate flow rates up 1 mL/min to facilitate LC-MS using microbore (down to 100 μ i.d.), narrow bore (0.5 mm i.d.), as well as conventional (4.6. mm i.d.) HPLC columns. Low-flow sheathless electrospray interfaces have also been developed.[24]

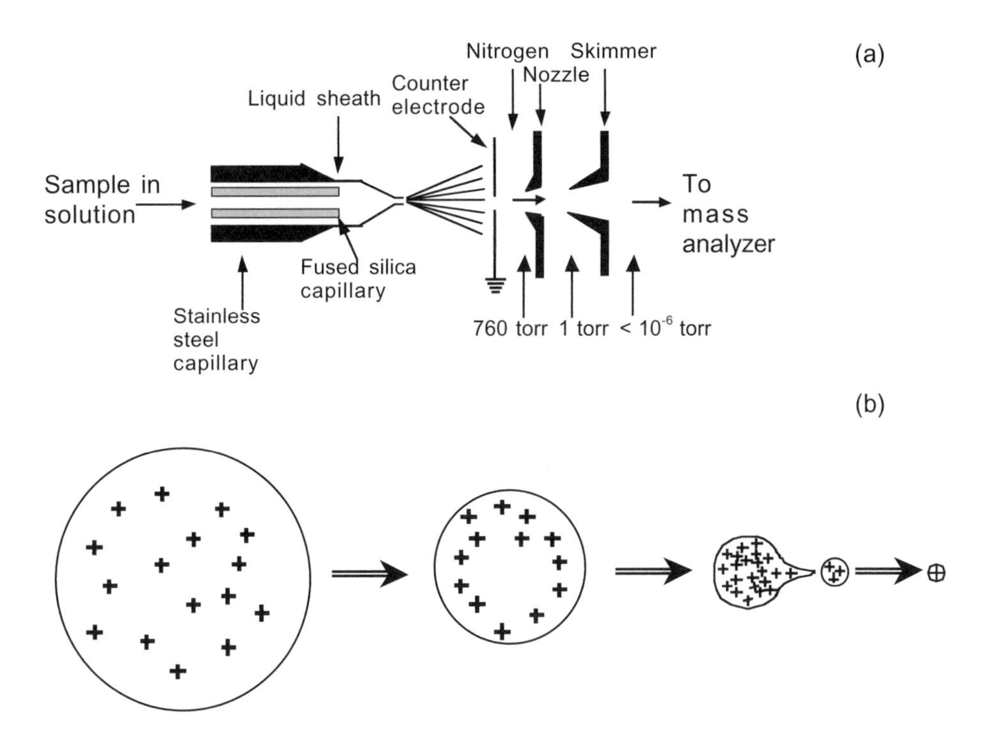

FIGURE 2.3 Electrospray ionization: (a) schematic of one of several available configurations; (b) release of ions from charged droplets.

Positive ions are formed from neutral analytes via protonation or cation attachment. Deprotonation, dissociation of other cations or anion attachment occur in the negative ion mode. The effectiveness of the methanol-water system as an ESI carrier may be explained by its high ionizing power in comparison to other solvent systems.[25–27] One important feature of ESI is the abundant formation of multiply charged ions resulting in Gaussian-shaped clusters of peaks representing an "envelope" of multiply protonated ions, $[M+nH]^{n+}$, with neighboring ions usually differing by one charge (Figure 1.2). Because mass analyzers determine the values of m/z rather than m, multiple charging leads to ions with reduced m/z values that make it possible to analyze molecules of very high molecular mass (routinely up to 100 kDa) with simple and inexpensive mass analyzers, e.g., quadrupoles and ion traps which normally have a mass range of 1 to 4 kDa. Commercial software is available for the transformation of the measured raw data on multicharged ions into the average mass, M_r, of the analyte by solving a set of simultaneous equations.[28,29] Techniques are also available for the automated reduction and interpretation of high resolution ESI spectra of large molecules.[30] The combination of ESI with FTICR analyzers has special advantages for the analysis of macromolecules.[31]

Down-time of ESI sources for cleaning can be significantly reduced when the source is modified so that ions do not enter the analyzer in a line-of-sight mode. In one design, the commercially available Z-spray ESI source, the ions are deflected twice with the aid of an electric field gradient so that they flow in a somewhat flattened Z-shape through two skimmers (Figure 2.4). The remaining solvent and other neutral molecules continue to move in the initial line-of-site direction. The efficient separation of the ions from the neutrals significantly reduces the build-up of residues on the narrow second skimmer entrance into the analyzer region so that sensitivity may be maintained for extended periods of time. The same effects may be achieved by alternative commercial designs where the ions are electrosprayed orthogonally (aided by pneumatic nebulization) to a capillary orifice (Figure 2.5).

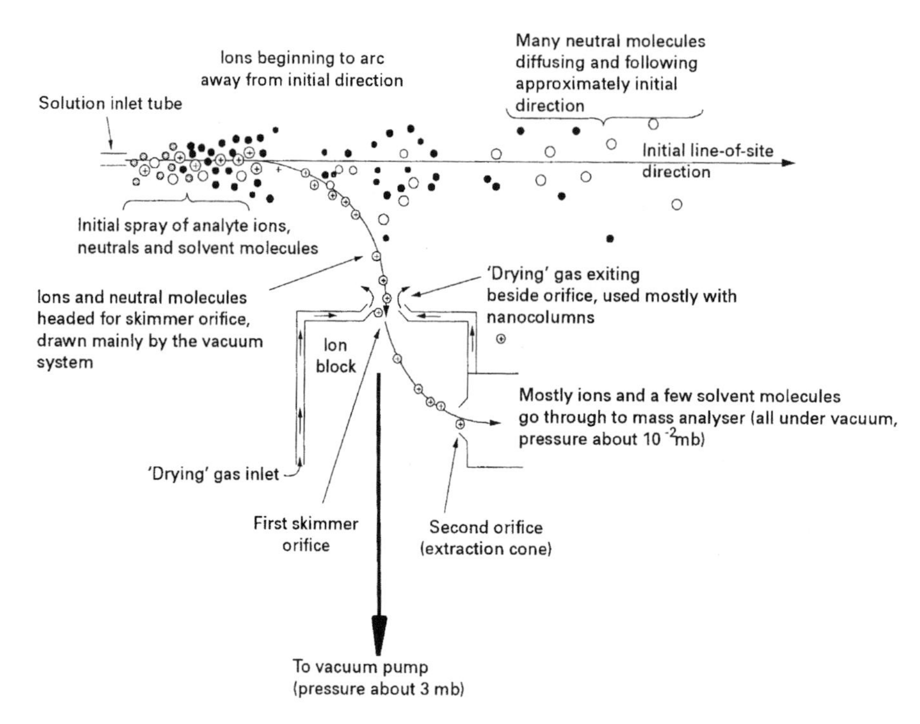

FIGURE 2.4 Electrospray ion source with "Z-spray" type introduction of ions into the analyzer. (Reprinted courtesy of Micromass Co. With permission.)

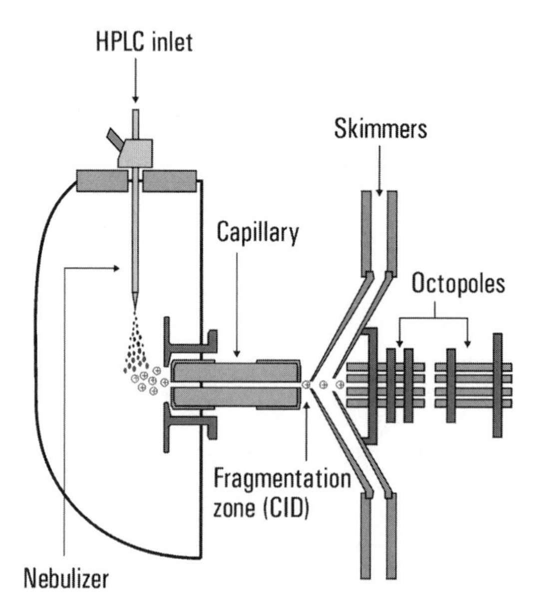

FIGURE 2.5 Sample spraying orthogonally to the capillary orifice of ESI (or APCI) source ensures efficient ion sampling and reduced contamination; also shown is the the area where "in-source" fragmentation (a version of collision-induced dissociation, Section 2.5) may be carried out. An octopole lens system guides the ion beam into an ion trap analyzer (shown in Figure 2.10). (Reprinted courtesy of Agilent Technologies. With permission.)

2.1.4.1 Nano-Electrospray

Because the signal intensity in ESI is concentration dependent, i.e., almost independent of the flow rate employed, absolute sensitivity can be increased up to a 1000-fold when the residence time of the analytes in the source is extended by reducing the flow rate to 10 to 50 nL/min. A simple nano-ESI source consists of a metallized glass capillary needle with a spraying tip of 1 μm i.d.[32] The volume of the initial droplets thus provided are ~10 to 100-fold smaller than those in conventional ESI sources. Various nano-ESI sources are available commercially. Several analytical characteristics of nano-ESI, e.g., tolerance for relatively high salt concentrations, may be explained by a model based on different "predominant fission pathways" which depend on the size of the initial droplets.[33] A 1 to 2 μL sample will flow for 30 to 120 min providing more than adequate time to optimize critical experimental conditions, e.g., collision energy in the MS/MS analysis of selected individual peptide ions of a mixture.[34] Several relevant applications will be described in subsequent chapters.

2.1.4.2 Microfabricated, Silicone Chip-Based Electrospray Ion Sources

Aiming to further increase sensitivity with concurrent reduction of cross-contamination of analytes, multiple ESI sources were made as parallel etched channels on microchips. In a device in which electrospraying was performed directly from the chip, using flow rates of 100 to 200 nL/min, the sensitivity limit for standard proteins (e.g., angiotensin) was 60 fmol/μL.[35] Technological improvements in fabricating chip-based ESI sources have made it possible to accomplish highly efficient on-chip electrophoretic separations.[36,37]

In an alternative design, etched sample and buffer reservoirs were connected by way of channels to a microsprayer, similar to the approach used for CE.[38] The analytes, placed in different sample reservoirs, were sequentially mobilized with the aid of a high-voltage power supply and moved into the microsprayer by electroosmotic pumping. Analyte ions were subsequently injected into an ion trap (IT) analyzer. The average LOD was 2 fmol/μL. A disadvantage was the possibility of cross-contamination because only one spraying capillary was used.[39] The chips were used to deliver not only single samples but also to infuse tryptic digests containing several analytes.[40] When effusions from a capillary/chip interface were analyzed using fast data acquisition in a time-of-flight (TOF) analyzer, subattomole quantities of peptides could be detected.[41]

A microfabricated microfluidic solvent delivery system was designed for gradient delivery at nL/min flow rates into microcolumns and, subsequently, to an ESI source connected to an IT. The system operated by computer controlled differential electroosmotic pumping of the required mobile phases. The performance of the system was impressive: peptides present at 100 amol/μL were detected. Anticipated future advances using this mixing technology include the monitoring of reactions by MS/MS using microvolumes of reactants.[42] Microfabricated devices with integrated liquid junctions have also been developed for using CE as the inlet for ESI sources.[36,37]

2.1.4.3 Multiplexed Electrospray

A novel LC interface has been devised for the parallel analysis of four or eight LC streams (single components or mixtures) with negligible crosstalk.[43] The technique was utilized to obtain accurate masses using an orthogonal acceleration (oa) TOF analyzer[44] (Section 2.2.1.1). In another recently developed interface, nine protein samples on a single chip were analyzed in a continuous flow operation when sequential voltages were applied to each sample reservoir, causing the samples to flow into and through the effusion channel and out through a single ESI needle.[40]

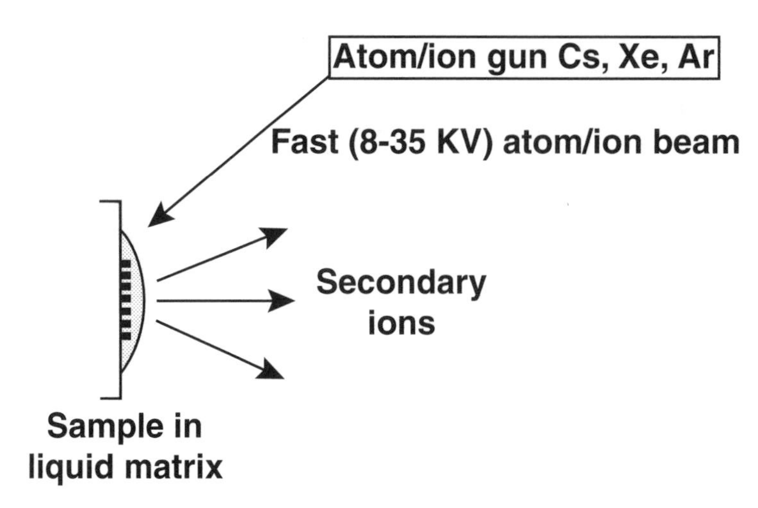

FIGURE 2.6 Fast atom or ion bombardment source.

2.1.5 Fast Atom and Ion Bombardment

In the fast atom bombardment (FAB) ion source, ionization is accomplished by bombarding the sample, which is dissolved in a special liquid matrix (see below), with a high-energy atomic beam[45,46] (reviewed in Reference 47). This results in molecular sputtering leading to the desorption of secondary ions of the sample, from the liquid-vacuum interface, into the gas phase. The ion beam is extracted from the ion source and mass analyzed, usually by quadrupole or magnetic sector analyzers in single or multistage combinations. The related fast ion bombardment technique, also known as *liquid secondary-ion mass spectrometry* (LSIMS), employs a stream of energetic (~30 keV) heavy ions (usually cesium) to accomplish the ionization.

FAB sources may be static or dynamic (continuous flow). In the static mode the beam of atoms, obtained in a special gun, bombards a conventional direct inlet probe modified to have a tip (ribbon of metallic or nonmetallic material) upon which μL size drops of the dissolved sample have been placed (Figure 2.6). Effluents from HPLC can be analyzed by FAB with the *continuous-flow* FAB technique where the eluate is continually mixed with the matrix and flows to the FAB target by passing through a needle where the tip is bombarded in a pulsed or continuous manner.[48]

The beam of neutral argon atoms to be used for bombarding the target is obtained in special FAB guns. An argon ion beam (obtained by EI) is focused into a collision chamber where the energetic argon ions (~5 to 8 kV translational energy) become rapid neutral atoms upon exchanging their charges with neutral argon atoms. The remaining argon ions are deflected from the beam by a potential placed on a set of electrodes, while the momentum of the neutral argon atoms propels them to the probe. Xenon may be used in a similar manner to produce a more energetic bombarding species.

Because FAB only ejects ions into the gas phase that are already present in ionic form in the solution (or ionized by proton transfer), the selection of the proper liquid matrix is critical.[49,50] Obvious requirements are that the sample must be soluble in the matrix and have a high diffusion rate through the matrix toward the surface. Other chemical and physical considerations include acid/base characteristics, purity, vapor pressure, viscosity, and surface tension. Most FAB work has been done with glycerol, thioglycerol, and *m*-nitrobenzyl alcohol as the matrix because they dissolve many polar compounds and have low volatility. When sample-matrix interaction occurs, alternative matrix materials have to be explored.

The ions formed in FAB include protonated molecules and/or cationized species; there is very little or no fragmentation. The negative mode is characterized by loss of proton or adduction of Cl^-, Br^-, or CH_3COO^- ions. For suitable polar analytes, up to 10 kDa, FAB efficiently produces long-lasting

(~10 min), steady ion beams, permitting the setting of experimental conditions for subsequent MS/MS experiments. There is a potential problem in the analysis of compounds with high molecular mass by FAB as well as by the other soft ionization techniques: the most abundant ion observed may not correspond to the monoisotopic mass of the analyte. This results from the distribution of ions in the molecular ion region caused by higher mass isotopes, e.g., the ^{13}C isotope. As discussed in connection with several applications in Chapter 6, this phenomenon assumes major significance in the analysis of compounds containing >100 carbon atoms.

Although largely replaced by ESI and MALDI in recent years, a substantial number of pioneering papers have been published using FAB and LSIMS ionization techniques in most areas of cancer research; several are reviewed in the chapters on applications.

2.1.6 Matrix-Assisted Laser Desorption Ionization

Matrix-assisted laser desorption ionization (MALDI) is currently the most important technique for rapidly depositing enough energy into involatile analytes in the condensed phase to facilitate their ionization and desorption into the gas phase in one continuous step.[51–53] MALDI differs from FAB in four respects: (1) the analytes are dissolved in a matrix and co-crystallized into solid phase rather than being dissolved and analyzed in a liquid matrix; (2) the analytes are exposed to an intense laser light rather than to a beam of energetic atoms or ions; (3) the laser light is applied in pulses of short duration (primarily to accommodate the commonly used TOF analyzers) in contrast to exposure to a continuous beam of energetic atoms in FAB; and (4) the analyte is ionized by energy transfer from the matrix rather than being "ripped" or "sputtered" from a liquid matrix. In a typical MALDI analysis, a 10 μM solution is made of the analyte in water or other appropriate solvent, and a 1 μL aliquot (~10 pmol) is added to an ~10^4-fold excess of a matrix solution (see later) that has already been placed onto a target. The solution is dried and allowed to crystallize so that it is comprised of microcrystals of the matrix into which the analytes are embedded. The target is placed, through a vacuum interlock, into the path of the laser beam in the ion source (Figure 2.7). An important practical advantage is the possibility of placing many samples in individual wells on a single target allowing for rapid analysis of multiple samples. Technological developments include the coupling of MALDI to liquid separation,[54] using HPLC with continuous-flow MALDI[55] and atmospheric pressure MALDI.[56,57]

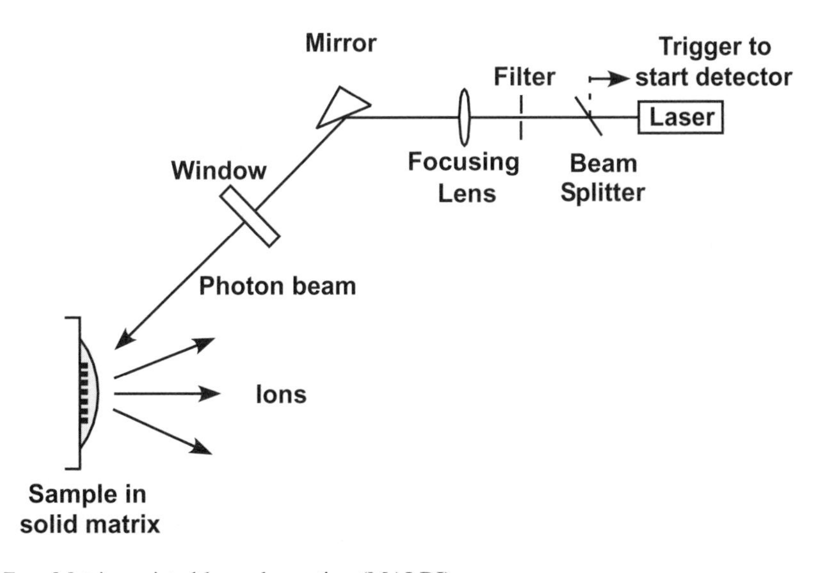

FIGURE 2.7 Matrix-assisted laser desorption (MALDI) source.

In commercial MALDI ion sources, the pulsed laser light of <10 nsec duration is often produced by a nitrogen laser at 337 nm or a Q-switched Nd:YAG laser (with tripled or quadrupled fundamental wavelength, λ at 354 nm or 266 nm). The laser light is focused onto the solid sample that contains both the matrix and the analyte. The exposure usually consists of 50 to 200 single shot pulses. Each pulse leaves a small crater in the target region. As the laser fluence (irradiance) is increased, a sharp threshold is reached at which analyte ions suddenly appear, e.g., 10 mJ/cm^2 (10 ns pulse width) for proteins. Laser fluence and/or sample position are adjusted manually or under computer control during data acquisition to maintain a fresh sample surface and a laser irradiance at ~20% above the threshold. These conditions are typical for optimal overall performance based on resolution and signal-to-noise criteria. A fuzzy logic feedback control system has been suggested to regulate laser fluence to maximize ion intensity while maintaining high resolution.[58]

The mechanism of MALDI is complex.[59,60] According to the hydrodynamic model, the intense heat generated by the rapid absorption of the laser energy by the excess matrix results in a plume of ejected material which undergoes supersonic expansion and subsequent cooling. It has been theorized that matrix molecules are ionized while still in the solid phase, and that these matrix ions will ionize the analytes via which are entrained in the matrix plume, via gas phase ion/molecule charge transfer reactions.

Initially, the matrix for MALDI was nicotinic or cinnamic acid or a simple derivative of these compounds; however, experimentation revealed that the matrix material had major effects on the efficiency of ionization.[61] In addition to being a good absorber of energy at the wavelength of the laser, the matrix has several distinct roles: (a) absorption of most of the laser light, thus minimizing sample damage by the laser irradiation; (b) vaporization when irradiated, thus causing some of the co-crystallized analytes to vaporize by transferring energy in a controlled manner (ablation step); (c) participation in the ionization of the vaporized analytes in the gas phase; (d) serving as a co-crystallization agent for the analyte; and (e) reducing intermolecular forces and minimizing the aggregation of the analyte molecules by being present in a vast excess over the sample. Matrices, including those mentioned above, are typically low molecular weight organic acids, e.g., 4-methoxy cinnamic acid or 3-hydroxypicolinic acid. The formulation of the matrix solution and the matrix:analyte ratio can have significant effects on analytical performance.[61–63] Examples of matrix materials for various type of analytes (reviewed in Reference 64) are given together with particular applications in Chapters 4, 5, and 6.

In TOF mass spectrometers, the gas phase ions produced in MALDI ion sources are subjected to controlled extraction and acceleration to a fixed energy by a strong electrostatic field (25 to 30 KeV). The ion packets thus generated are focused by plates at high voltages (Einzel lenses) into the field-free region of the flight tube of a TOF analyzer for subsequent determination of the *m/z* values of the separated ions. TOF analyzers are well suited to handle the ion packets produced by the discontinuous MALDI process. Ion traps or FTICR analyzers have also been employed.

There are at least six advantages of MALDI ionization with TOF analysis: (1) very high mass range, to 1 MDa or higher; (2) high sensitivity with LOD routinely at low pmoles and often to amole levels; (3) suitability for mixture analysis because the technique is not especially subject to suppression and discrimination effects or to formation of multiple charge states; (4) ability to do analyses in minutes (sample preparation takes longer); (5) little or no analyte fragmentation; and (6) reduced need for involved sample purification because of a reasonable tolerance for salts and other common biochemical additives, often to mM concentrations. On the negative side, background ions from the matrix may be a problem for analytes with molecular weight <1000 Da,[65] limited multiply charged ion clusters may occur complicating spectra (although this is not as prevalent a problem as it is in ESI), and there is the possibility of photodegradation. During the last decade, MALDI-TOFMS has become a primary tool for the study of large, complex, nonvolatile biomolecules, particularly proteins.[66] Dozens of applications are illustrated in subsequent chapters.

The recent ability to conduct MALDI at atmospheric pressure ionization (AP-MALDI) involves the pneumatically assisted transfer of ions from the region of atmospheric pressure to high vacuum by a stream of nitrogen. Sample preparation for AP-MALDI (including matrix selection) is similar to that with conventional MALDI. When used with an oa-TOF analyzer (Section 2.2.1.1), the advantages of AP-MALDI include the possibility of using the same instrument for both ESI and MALDI analyses, linearity of mass calibration, and less need to identify "sweet spots" on the target surface. The increased sample consumption with respect to vacuum MALDI is a disadvantage.[57]

2.1.6.1 Surface-Enhanced Laser Desorption Ionization (SELDI)

In MALDI, the role of the probe is to present the sample to the mass spectrometer. The surface of the sample probe does not have an active role in the analytical process other than to hold the sample. The concept of *SELDI* is based on surface-enhanced affinity capture where the probe surface plays an active role in a number of aspects of the processing of the analytes, functioning as a solid phase extraction technique involving extraction, structural modification, amplification, and presentation.[67–69] In the commercial *ProteinChip*™ *SELDI* technology[67] a number of arrays with chemical or specific biochemical surfaces have been used in the detection and assay of protein biomarkers for various cancers (Section 6.1.4.).

2.1.7 Photoionization

2.1.7.1 Direct Laser Desorption Ionization

Ionization by lasers has been studied for decades (reviewed in Reference 70). In this technique, analyte molecules are desorbed as intact molecules with the aid of laser beams of 1 to 200 ns duration pulses, focused on 10 to 300 μm o.d. spots at a variety of incident angles (30° to 75°). Ion sources have been designed with a variety of lasers, including UV lasers, such as nitrogen (337 nm), excimer (at various wavelengths in the 193 to 351 nm range), frequency-doubled excimer-pumped dye (220 to 300 nm), Q-switched, frequency-quadrupled Nd:YAG (266 nm), and IR emitting lasers.[51] The pulsed nature of the ionization mode requires a TOF or FTICR mass analyzer.

Instruments providing *direct* laser desorption, such as the laser microprobe analyzers (e.g., LAMMA), have been used to study the composition of thin (0.1 mm) biological sections, prepared in the same way as for transmission electron microscopy, or particulate samples embedded into thinly cut transparent layers of mylar or colloidion. The samples are mounted on a movable x-ray stage, viewed with a binocular microscope, and the area of analysis is selected by illuminating a spot with a low-powered (2 mW) He-Ne pilot laser, operated in the visible range. Next, the main laser provides 15 ns pulses which are focused on 0.5 mm diameter spots. A minute amount of analyte is evaporated and a microplasma is formed consisting of neutral fragments, molecular ions, fragment ions, and alkali-cationized ions. The ions are then accelerated and analyzed by TOF. Microregions of both organic and inorganic samples may be analyzed and ions may be obtained from areas as little as 1 mm in diameter. The LOD for nearly all elements is 10^{-18} to 10^{-20} g. Sensitivity for organic compounds varies widely. In the absence of a matrix, the technique is restricted to samples of <2000 Da. Sample preparation is difficult and reproducible focusing is problematic. In addition, sample homogeneity does not exist on a microscale. Unique applications include the detection of platinum in the kidney cells of dogs after cisplatin treatment (Section 5.1.2.6).

2.1.7.2 Resonance-Enhanced Multiphoton Ionization (REMPI)

In this technique the energy for ionization is obtained by the simultaneous absorption of two or more UV laser photons. The two-photon energy must exceed the ionization energy of the analyte. The efficiency of the two-photon absorption, i.e., the yield of ionization, may be increased by several orders of magnitude if the photon energy (laser wavelength) is in resonance with the analyte's UV-spectroscopic transition.[71] In addition to providing high sensitivity, REMPI is also highly selective because only those compounds can be analyzed that exhibit the required specific UV-transition at the laser wavelength applied. REMPI is a soft ionization technique, hence there is virtually no fragmentation. Since REMPI is a pulsed type method of ionization, TOF is an ideal mass analyzer. Several sample inlet systems are available for different applications.[72] REMPI is a two-dimensional technique, combining both optical and mass selectivity. Advantages are high selectivity, sensitivity, and speed. Because it takes only ~20 to 40 μs to obtain a spectrum, the method is well suited for time-resolved analyses of target compounds or specific compound classes in complex samples. A relevant application is the real-time on-line monitoring of aromatic pollutants in waste incinerator flue gases (Section 4.3.2.3).

2.1.8 Inductively Coupled Plasma

This technique is used for the simultaneous quantification of dissolved elements. A spray (gas or aerosol) of the sample solution is introducced into a flame of ~10,000°K where it is rapidly desolvated, atomized, and ionized with nearly 100% efficiency. The predominantly singly charged ions of the analyte elements are usually analyzed by quadrupole analyzers, although Q analyzers (using CID to disrupt clusters) and magnetic sector instruments are also used.

The main components of a typical inductively coupled plasma (ICP) ion source consist of three concentric tubes surrounded by an rf coil (reviewed in Reference 73). A self-sustaining plasma is established by the plasma gas (usually argon) that is introduced through the two outer tubes through which it flows at different rates. The gas is made initially conductive using Tesla sparks. The torch flame is established by the electromagnetic field set up by energizing the rf coil. The sample is sprayed through the inner tube. Sample introduction into the ICP is a critical step. Ultrasonic or concentric nebulizers are used to increase the efficiency of nebulization which, in turn, improves detection limits. There are enormous differences between the detection limits for individual elements, ranging from 0.001 μg/L (e.g., Na, Cr, Cu) to >10 μg/L (e.g., K, P, S). For the elemental speciation of mixtures, it is necessary to combine ICP with a separation technique. On-line HPLC is used most commonly[74,75] while the utilization of the high resolving power of CE is a viable, but not yet fully developed, alternative.[76]

2.1.9 Miscellaneous Ionization Techniques

2.1.9.1 Secondary Ion Mass Spectrometry (SIMS)

Originally developed for the analysis of inorganic materials and polymeric surfaces, SIMS is based on the generation of sputtered secondary ions of the analytes from solid surfaces by bombardment with energetic primary ions (e.g., cesium). The secondary ions are formed at or just above the surface of the metal on which the substance to be analyzed is coated.[77] The mechanism of ionization involves charge exchange processes and ion-molecule reactions. In addition to abundant $[M]^+$ and $[M+H]^+$ ions, $[M+metal]^+$ and fragment ions are also formed. *Dynamic* SIMS uses a high primary ion beam flux ($>10^{-6}$ A/cm^2) for the generation of high sputtering rates. Applications center on elemental "depth profiling" of material surfaces and microanalysis, primarily in the materials science. *Static* SIMS uses low primary ion beam flux ($<10^{-9}$ A/cm^2) which generate reduced sputtering rates, permitting the analysis of monolayers on surfaces. Increasing the analysis area partially compensates for reduced sensitivity. The matrix-assisted SIMS approach has evolved as a tool for obtaining ions of nonvolatile and thermally labile biopolymers and other large molecules.[78]

2.1.9.2 Field Ionization (FI) and Field Desorption (FD)

When a molecule in the gas phase is subjected to an electric field force of 10^8 V/cm, established in the vicinity of a metal electrode (traditional FI and FD emitters were carbon dendrites covering a metal wire), it is possible for an electron to leave its molecular orbital by quantum mechanical tunneling and enter a vacant orbital in the metal. The necessary electric field is established by placing a voltage difference of 2 to 20 kV between an array of fine points on a grid, a thin wire, or a blade anode and a cathode 0.5 to 2 mm apart; the cathode also serves as the exit slit from the source. FI was the first *soft* ionization technique. In FI, the analyte must be admitted as a vapor, i.e., it must be thermally stable. Sensitivity is an order of magnitude lower than that of EI.

The *field desorption* technique utilizes the principle of FI but does not require a volatile sample. Approximately 0.2 μL of the sample, in a suitable solution, is placed on an *emitter* consisting of a tungsten wire, ~10 μm diameter, on which carbon micro-needles (whiskers) are grown and subsequently activated. When the electric field is applied, ionization takes place by the tunneling of electrons from the analyte into the carbon filament. The resulting cations are rapidly desorbed and repelled from the positively charged emitter into the mass analyzer. Until the development of FAB, FD was the method of choice for determining the molecular masses of involatile and thermally unstable molecules. The major problem associated with FD has been the reproducible making and conditioning of the emitters.

2.1.9.3 Californium-252 Plasma Desorption

Another technique for rapidly heating samples to prevent thermal degradation takes advantage of the fact that the pairs of fragments that form during the spontaneous fission of ^{252}Cf nuclei (e.g., $^{142}Ba^{18}$ and $^{106}Tc^{22}$ or ^{144}C and ^{108}Tc) carry kinetic energy of 100 to 200 MeV. When the fission fragments pass through a thin nickel foil on which analytes are coated, local hot spots (~10,000°K) are created, rapid sample volatilization occurs, and both positive and negative ions are formed through ion-pair formation and ion-molecule reactions. Molecular ions with or without proton transfer are formed in abundance and there is little fragmentation. The desorbed ions are analyzed in a TOF analyzer. Because ions are formed as a result of individual fission events, mass spectra are obtained by monitoring the arrival of ions over periods of minutes to hours. To assign masses the time of flight must be obtained by measuring the exact times of both ion formation and detection. The former is determined by detecting those nuclear fission fragments that move off in the opposite direction. The required instrumentation is complex.[79] The upper mass range of the technique is ~50 kDa, resolution is ~1000. This novel ionization technique has been superseded by ESI and MALDI.

2.1.9.4 Glow Discharge

In this technique, a gas discharge (usually in argon) is produced between the sample, which acts as a cathode, and a metal anode (reviewed in Reference 80). When accelerated argon ions from the plasma hit the surface upon which the sample is plated, sputtered neutral atoms of the sample are lifted into the gas phase. Here they are ionized by EI, CI, and Penning ionization in a region close to the ion exit (often an aperture in the anode) and subsequently introduced into the mass analyzer.

2.1.9.5 Spark Source

In the radiofrequency (rf) spark source, pulses of an ac potential of 20 to 100 kV are generated for a few microseconds between two electrodes made of the sample material. The repetition rate of the spark is variable, ranging from single pulses (sparks) to 10^4 pulses/s. The solid samples are evaporated in the electrical discharge and an abundance of positive ions is formed. The complex mechanism of ion formation in the spark is not well understood. Because the ions are formed with a high spread of kinetic energy (100 to 10^3 eV), a double-focusing magnetic mass analyzer with an efficient energy filter is necessary. Ions are detected by photoplates, and quantification is by computerized densitometry. Since the efficiency of ionization is nearly the same for all elements and there is little matrix effect, almost all elements in the periodic table can be analyzed. LOD are in the 0.1 to 0.001 ppm range for most elements irrespective of matrix. Coverage of most elements in the periodic table is accomplished in a single analysis, response is linear with concentration, and μg size samples can be analyzed. The technique is used primarily for the determination of trace quantities of elements in solids.

2.1.9.6 Thermal Ionization

The basis for the thermal ionization (thermal emission, surface emission) source is the fact that when neutral atoms or molecules are heated on, or impinge upon, a hot metallic surface, there is a probability that ions as well as neutral particles will evaporate. The technique is applicable to inorganic compounds which have low ionization potentials (3 to 6 eV); applications to organic compounds (ionization potentials in the 7 to 15 eV range) are limited.

2.2 MASS ANALYZERS

The primary function of all analyzers is to resolve an ion beam of mass m from another beam of $m+\Delta m$; this is a dispersive, or prism, action. The secondary function is to maximize the resolved ion intensities; this is a focusing, or lens, action. Upon their formation, all ions have slightly different kinetic energies due to the Boltzmann kinetic energy distribution and field inhomogeneities in the ion sources. These energy differences remain after the ions are accelerated out of the sources into the mass analyzers. *Velocity-focusing* of some kind is needed to compensate for the ion energy inhomogeneity, e.g., by the electrostatic field in *double-focusing* magnetic analyzers or by *delayed extraction* in TOF analyzers (see below).

The efficiency of a mass analyzer might be characterized by its degree of transmission and duty cycle. In beam-type, scanning mass spectrometers (magnetic sector and quadrupole analyzers), only a very small fraction of the ions issued from the ion source reaches the detector. The degree of transmission refers to the fraction of ions that fall within the nominal m/z window of stable trajectories. During mass scanning the conditions are set in such a manner that only ions having an m/z within a selected window can reach the detector, followed by a sequential changing of conditions so that all m/z values of interest will eventually be focused on the detector, one m/z value after another. The duty cycle is the ratio of the width of the transmitted m/z window to the total width of the m/z range of interest (meaning the fraction of all ions detected vs. all ions traversing the analyzer). Hence, during scanning the duty cycle is usually <1%, i.e., most of the ions formed are lost. When there is no scanning, the duty cycle is nearly 100%, e.g., in single ion monitoring (Section 2.6.3) or with array detectors (Section 2.3.2). An inherent disadvantage of all scanning analyzers is that only a small fraction of all the ions is recorded at a given moment. In contrast, most analyzers in the nonscanning mode accumulate ions selectively; this is an advantage in multiple stage MS/MS applications.

2.2.1 Time-of-Flight Analyzers

In time-of-flight (TOF) analyzers, ions are separated according to their velocities. Ions leaving the ion source are accelerated to 25 to 50 kV by a voltage applied to the final accelerating grid in the source and enter a 1 to 2 m long, field-free drift tube where they travel with various velocities toward an ion detector (Figure 2.8a). The kinetic energy of all extracted ions is nearly constant, regardless of their masses. The velocity of the drifting ions is inversely proportional to the square root of their m/z values. Mass separation is achieved because the higher the mass, the lower the velocity of the individual ions, and conversely, the lower the ion's mass, the greater its velocity. The flight time to the detector depends directly on the length of the flight tube and inversely on

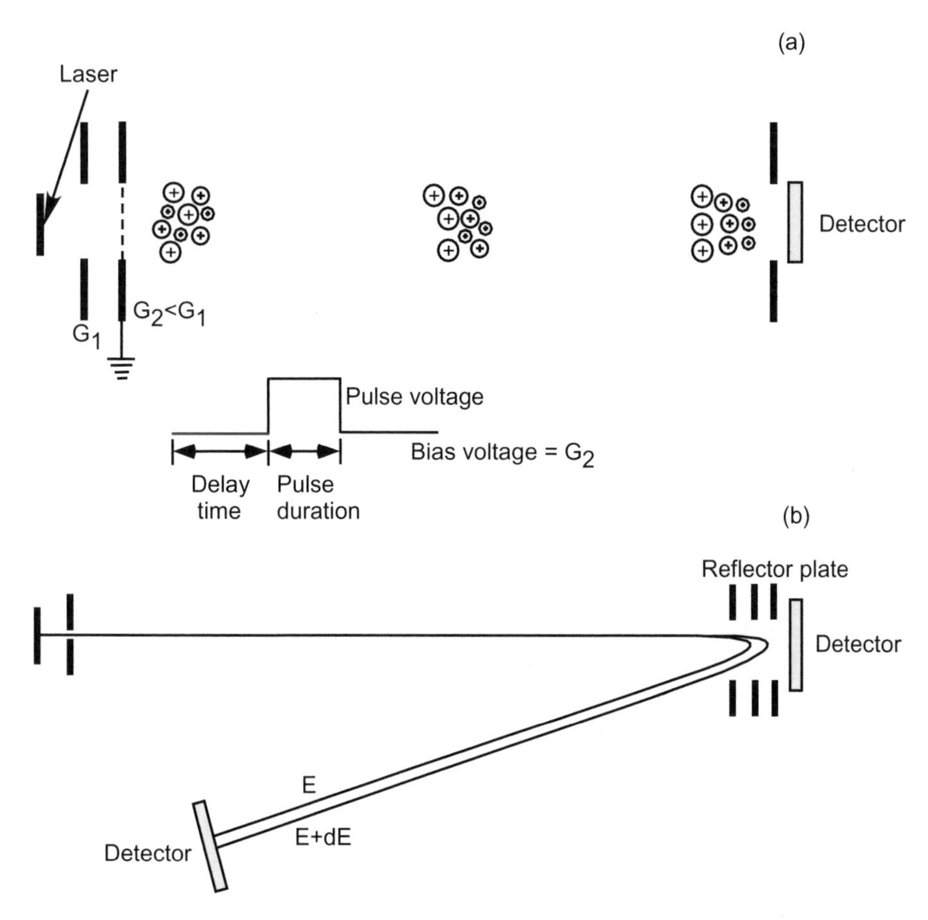

FIGURE 2.8 Time-of-flight (TOF) analyzer: (a) ion separation in a linear analyzer; also shown is the voltage arrangement in *delayed ion extraction* to compensate for the initial velocity distribution of ions; (b) *reflectron* analyzer which provides energy focusing of ions and significantly improves resolution.

ion velocity. The smallest ions arrive at the detector in a few μs, followed by the larger ions at intervals a few fractional μs later. The *m/z* of each ion may be calculated from the time interval between the time it enters the flight tube and the time it reaches the detector. A mass scale is established by calibration with standards of known *m/z* values. The high repetition rate (thousands of spectra/s) produces a virtually continuous spectrum on the instrument display of the ion beam arriving at the detector.

Because the ions must enter the drift tubes in discrete packages, MALDI sources, which produce ions in rapid, ns pulses, are ideal for TOF analyzers. Ion beams emanating from sources that produce ions continuously, e.g., ESI, must be extracted in packets (pulses) by applying a voltage pulse to repeller electrode for a few ns or, preferably, by injecting the ions *orthogonally* into the flight tube by "pusher" electrodes (see below). The range of *m/z* values leaving the source or arriving at the detector may be controlled with the aid of voltage pulses applied to *ion-gates* that are placed between the source and the flight tube or at the end of the flight tube prior to the detector.

A major advantage of TOF analyzers is their theoretically unlimited mass range; the practical upper mass limit is currently ~350 kDa. The sensitivity of TOF analyzers is high because almost all ions produced are detected from each entering ion burst. Another advantage is the possibility for high-speed analysis (e.g., complete mass spectra in 50 μs), permitting unique applications such as reaction kinetic studies of explosives and combustion processes.[81] Major improvements in resolution and sensitivity have resulted from the development of orthogonal injection, delayed extraction, use of reflectrons, and hybridization of instruments.[82]

2.2.1.1 Orthogonal Acceleration (oa) TOF

A continuous ion beam emanating from an ion source is usually accelerated through an electric field resulting in a stream of ions having momenta proportional to the applied accelerating voltage and the *m/z* of the individual ions, but without any temporal separation. When the ion beam reaches the "orthogonal zone," a pulsed electric field gives the ions a second velocity component that is at a right angle to the original velocity. The pulsed ion beam then enters a TOF analyzer that provides temporal separation so that ions of differing *m/z* values arrive at the detector at different times. The pulsing rate may be as high as 30 kHz, which means that ~30,000 spectra/sec can be collected and summed. High performance data acquisition systems have been developed for combined CE/TOF systems which provide a spectral storage rate of 80 spectra/sec and LOD in the 10 to 25 amol range for continuous sample infusions.[83]

Orthogonal acceleration permits the use of "continuous" ion sources, such as ESI, with TOF (reviewed in References 84 and 85). The high sampling efficiency, e.g., 30%, means a much higher duty cycle than in scanning instruments, resulting in substantial increases in sensitivity for obtaining full spectra. TOF instruments with oa provide significant advantages in MS/MS operations (Section 2.5.2). There has been a flurry of recent development of instrumentation involving TOF analyzers, particularly as components of hybrid instruments with oa sample introduction (Section 2.2.6).

2.2.1.2 Reflectrons (Ion Mirrors)

The resolution of TOF analyzers is limited for several reasons including: (a) as the mass of the analyte increases, the difference in arrival times of ions at the detector becomes smaller and smaller, and more difficult to differentiate; and (b) there is a spread (aberration) in the temporal, spatial, and kinetic energy distribution of the groups of ions of the same mass as they are pulsed out of the source into the flight tube. This means that two ions of the same mass formed at different times, or at different locations in the source, will arrive at the detector at different times. The same is true when two ions of the same mass form that have the same kinetic energy but with initial velocities in different directions. The end result of these distribution problems is reduced resolution due to overlapping of ions of neighboring masses within ion pulses. The resolution of linear TOF analyzers of <2 m length is ~700 to 1000.

The reflectron is a homogeneous electrostatic ion mirror consisting of electrodes typically located in front of the detector used if the instrument is in the linear mode operation. The polarity of the voltages on the electrodes is the same as that of the ions. When the electrode voltages are adjusted to potentials slightly higher than that used to accelerate ions out of the ion source, ions arriving at the end of the drift tube experience a retarding potential, eventually come to a stop, and then accelerate in the opposite direction. The reflected ions are usually made to travel a second length of drift tube set at a small angle to the first one (V-shape), to be collected at a second detector (Figure 2.8b). In coaxial reflectrons, the reflected ions move backward on the same axis and are detected with an annular plate detector close to the ion source. Some mirror designs are composed of grids (wire meshes) through which the ions pass, but these are prone to ion losses through scattering. A preferential design, which minimizes such losses, is a series of rings, with varying potential, that create an electric field into which the ions penetrate according to their energies before having their paths reversed. Single-stage ion mirrors consist of two parallel electrodes with a homogenous electric field between them. A variety of more sophisticated, multistage mirrors with nonlinear fields have been constructed to achieve second or higher orders of focusing.[86]

Reflectrons achieve energy focusing of the ions as follows: when two ions of the same mass but with different initial energies (arising from the ionization process and the initial kinetic energy distribution) arrive at the reflecting field, the ions with higher energy penetrate the field to a greater extent, spend more time inside the reflectron, and exit later than the ions with lower kinetic energy. Thus, the lower energy ions can "catch up" (refocused) with those of higher energy so all ions of the same *m/z* reach a common plane at the final detector and are detected at the same time (Figure 2.8b). The result is reduced peak broadening and therefore higher resolution. A second advantage of reflectrons is that they provide increased spatial separation of ions with different *m/z* values because their flight times are effectively doubled. Resolving power may be as high as 20,000, a 20-fold increase from the linear mode of operation. However, the price of excellent resolution is reduced mass range and sensitivity. The latter is due to the loss of ions by collisions and dispersion in both the reflectron and the second drift tube. Accordingly, despite their improved resolution, reflectrons are often not used for analytes of very high mass, particularly when they are present in trace quantities. It is noted that fragment and metastable ions are also reflected, but are not energy focused. Neutral species are unaffected by the reflecting field. Fragment ions formed inside the ion source are reflected; however, metastable ions formed inside the flight tube are reflected only under special postsource decay conditions (Section 2.5.2.5).

2.2.1.3 Delayed Extraction (DE)

While the ion mirror is a postsource compensation method for energy inhomogeneity, delayed extraction (time-lag focusing) is an in-source method of energy compensation. Delayed extraction involves imposing a time delay between the ionization pulse and the application of the voltage to move the ions from the source into the flight tube (Figure 2.8a). First-order velocity focusing is achieved and resolution is improved because DE compensates for the initial velocity distribution of the MALDI generated ion packet.[87] This approach, which is similar but not identical to the *time-lag focusing* technique,[88,89] involves: (a) the expansion of the ions formed in MALDI ion sources into a field-free region between a repeller and the first extraction grid; (b) a certain time lag or delay; and (c) a pulsed change in the relative voltages of the repeller and grid to create a potential difference and therefore extract the ions into the flight tube.[90] Resolving power to 2,500 may be achieved with DE alone. The narrower peaks also improve sensitivity on account of improved signal-to-noise ratio. DE loses effectiveness at masses >20 kDa.[91]

2.2.2 Quadrupole Analyzers

A quadrupole mass analyzer consists of four parallel cylindrical (theoretically hyperbolic) rods, typically 0.1 to 0.3 m long and ~6 mm diameter, arranged precisely in a square (Figure 2.9a). To one pair of diagonally opposite rods a positive direct current (d.c.) voltage is applied while the other pair receives a negative d.c. voltage. Superimposed radiofrequency (rf) voltages are also applied to each pair of rods so that the opposite pairs are 180° out of phase (Figure 2.9b). Mass scanning is accomplished by varying, at constant rf frequency, either the rf or d.c. voltages while keeping a constant ratio between the two pairs of rods. Mass separation (*filtering*) is based on the fact that ions entering the two-dimensional quadrupole field created by these voltages undergo oscillation. The ion trajectories are described by the complex Mathieu equations (reviewed in Reference 92). At a particular combination of the d.c. and rf fields, only ions of a specific *m/z* can move along the axis, i.e., maintain a stable trajectory (bounded oscillation), to be collected at the end of the analyzer. All other ions experience unbounded oscillation and move in unstable trajectories with oscillations perpendicular to the flight path. The amplitudes of the ion oscillations eventually become greater than the diameter of the quadrupole array, and they are deposited on one of the rods and lose their charges. The range of ions of different *m/z* capable of passing through the quadrupole mass analyzer is determined by their position on the stability diagram that shows the values of parameters a and q for which the oscillations performed by the ions are stable (Figure 2.9c). The parameters a and q are related to the rf amplitude and d.c. potential, respectively.

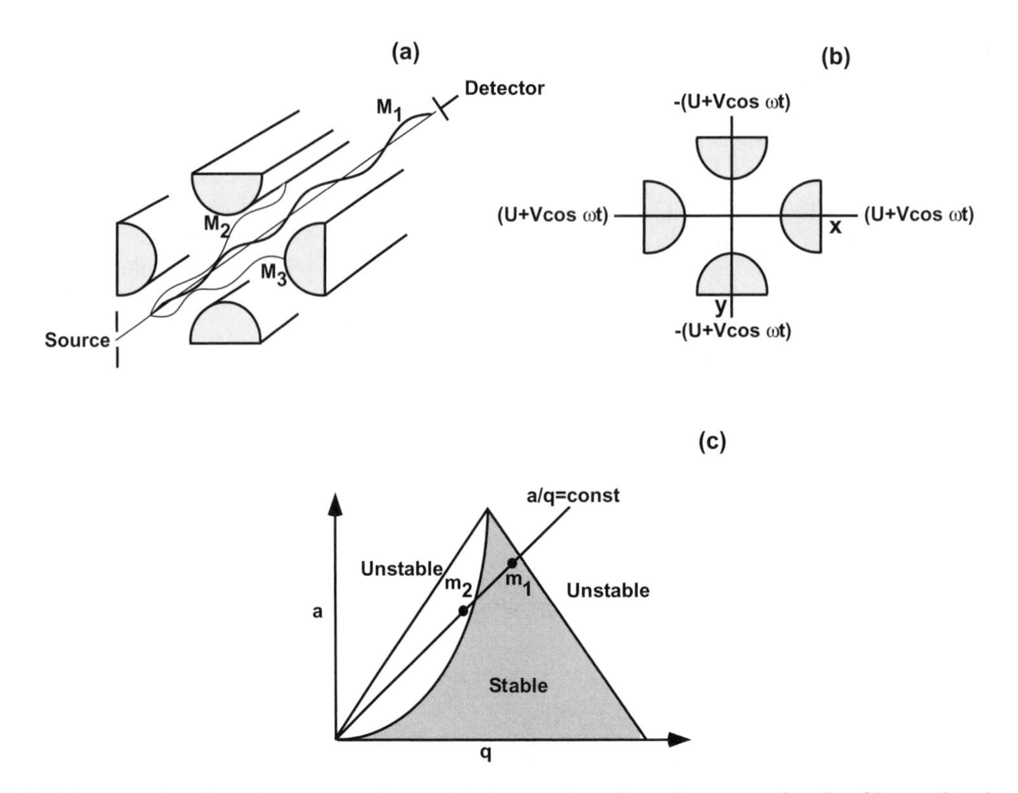

FIGURE 2.9 Quadrupole mass analyzer: (a) the quadrupole rod array and path of ions; (b) electrical connections to the rods; U = magnitude of the d.c. potential, $V \cos \omega t$ is the time dependent rf voltage in which V is the rf amplitude and ω is the rf frequency; (c) stability (Matthieu) diagram showing values of parameters a and q for which the oscillations of the ions are stable; the a/q is the operating scan line; resolution is established by fixing the ratio of the rf and d.c. potentials.

An important advantage of the quadrupole analyzers is that the mass scale is linear. At the same time, however, resolution decreases with increasing mass — unlike in magnetic analyzers (see later) which show constant resolution at all masses. Accordingly, in quadrupole mass analyzers, resolution is given by the definition based on the determination of full peak width at half-maximum height (FWHM) at a selected mass (Figure 1.3). In typical analyses, resolution is often set at a constant 0.5 Da throughout the mass range although the variable resolution may be utilized to optimize sensitivity at low masses. One limitation of quadrupoles is that decreasing sensitivity (mass discrimination) sets the upper limit of the useful mass range at 4 kDa. A contributing factor to the relatively low sensitivity is the fact that the accelerating voltage in quadrupole analyzers is <100 eV (in contrast to the several thousand volts in magnetic analyzers) thus the "slow" ions arriving at the detector do not have enough energy to overcome the work function of the first dynode of conventional electron multiplier detectors.

Other advantages of quadrupole analyzers include the possibility of: (1) fast scanning (0.1 s/mass decade); (2) continuous scanning with almost no reset time; (3) rapidly scanning narrow mass ranges; and most importantly, (4) monitoring one or several selected masses anywhere within the mass range by rapidly jumping from mass to mass. These advantages are fully utilized in combined GC/MS, LC/MS, and CE/MS applications both for qualitative analysis and for high sensitivity quantification using the selected ion monitoring (SIM) technique (Section 2.6.3). A particular advantage in on-line interfacing is that the acceleration voltage used to expel ions from the source is only 5 to 10 eV (~10^3 times less than in magnetic sector instruments), thus discharges and other problems associated with high voltages are eliminated. Other benefits include ease of use, small footprint, and lower cost.

2.2.2.1 Quadrupoles as Ion Guides

When a quadrupole is used as a mass analyzer, as described above, it is said to operate in a narrow band pass mode, acting as an electronic gate. In a second mode of operation, called the wide band pass mode, the same rf voltage is applied to all four quadrupole rods but no d.c. field is applied. This means that the *a* ordinate on the stability diagram is zero, i.e., the operating line lies horizontally along the abscissa (Figure 2.9c), indicating that all ions emerging from the source are transmitted regardless of their *m/z* values. In this mode, the quadrupole assembly acts as an ion guide. In multi-analyzer instruments the ability to rapidly switch between the two modes of operation permits experiments in which ions can be passed through one analyzer to a second with or without specific selection of masses in the first analyzer. Several applications are described in later sections.

2.2.2.2 Hexapoles

In hexapole assemblies there are six parallel rods spaced evenly around a central axis. Unlike quadrupoles, hexapoles are not used to separate ions of different *m/z* values and are used in the wide pass mode with only rf voltages being applied, i.e., all ions in the beam pass through regardless of their *m/z* values. Ion beams leaving combined inlet/ionization sources are often spread out due to collisions with residual solvent and/or gas molecules and mutual ion repulsion. It is the function of hexapoles to refocus such beams so that they leave the hexapole assembly as a constrained, narrow beam. Concurrently, pumping of the hexapole section is used to reduce the background pressure. Frequently, two hexapole units are used in tandem, separated by a very small orifice to restrict entry of neutral gas molecules; differential pumping is used to maintain a pressure differential of two orders of magnitude between the two hexapoles. An important application of hexapoles is to act as "bridges" between API sources and TOF analyzers. The latter are usually operated in an orthogonal fashion.

2.2.3 Quadrupole Ion Traps

In the quadrupole ion traps (IT), molecular (or atomic) cations or anions in the gas phase are confined within the trap for extended periods of time with the aid of electric fields. Although the principles of operation are similar to those of quadrupole analyzers, there is a major functional difference. In a quadrupole analyzer, only ions with stable trajectories are transmitted to the detector while those with unstable trajectories hit the rods and are lost. The reverse is true for quadrupole ion traps: in this case ions with stable trajectories remain trapped and the ions of interest can be detected only when their trajectories are made unstable.

A quadrupole ion trap consists of three cylindrically symmetrical electrodes, accurately machined to provide hyperbolic inner surfaces. There are two end caps and a toroidal ring electrode, with an ~1 cm internal radius (Figure 2.10). In earlier designs, ions were generated *in situ,* usually by EI. In recent designs, ions created in external ion sources, e.g., ESI or APCI, are introduced into the trap with the aid of electrostatic lenses[93] or ion guides, such as shown in Figure 2.10.

When a three-dimensional quadrupole electric field is created by applying superimposed rf (~10^6 Hz frequency) and d.c. potentials (small voltages) on the ring electrode and the end caps are grounded, all ions are constrained into stable trajectories near the center of the trap in a predominantly sinusoidal motion. With the appropriate selection of the frequencies, all injected ions, i.e., all *m/z* values, may be trapped with stable trajectories. To measure an ion with a particular *m/z* value, the rf and/or d.c. component of the field is changed so that the ion trajectory (along the *z* axis) becomes unstable resulting in the ejection of the specific *m/z* from the trap volume, through holes in one of the end caps onto a detector (e.g., electron multiplier). A mass spectrum of the

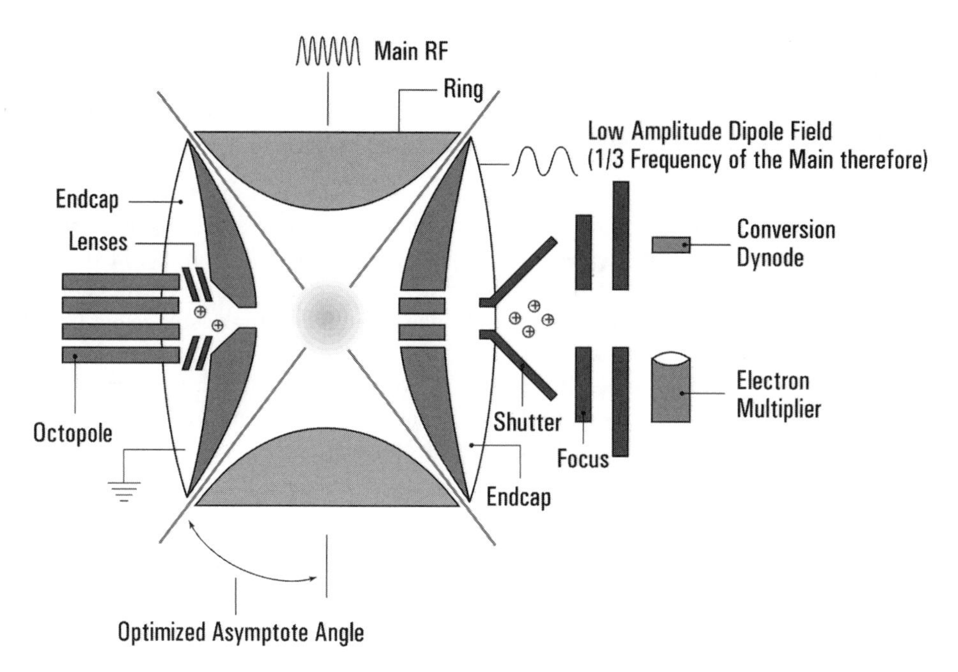

FIGURE 2.10 Ion trap analyzer. Ions from an ESI or APCI source are introduced with the aid of an octapole ion guide. (Reprinted courtesy of Agilent Technologies. With permission.)

analyte, i.e., a scan of a selected mass range, is obtained by the sequencial changing of frequencies, ejecting one m/z at a time.

In an alternative method of operation, potentials can be chosen such that only ions with a selected m/z have a stable trajectory, i.e., remain trapped, while all others with different m/z values assume unstable trajectories and are ejected from the trap. The remaining trapped ions of the selected m/z may be exposed to multiple collisions with an injected buffer gas to produce fragmentation. Indeed, one of the major advantages of ion traps is the possibility to perform CID experiments (up to MS^6) without having complex and expensive multiple analyzers.

Quadrupole ion traps are remarkable devices. They function as mass spectrometers of impressive performance and also as ion stores in which both positive and negative ions can be confined in the gas phase for a period of time. All this is accomplished in notably simple designs that have small footprints, are easy to operate, and cost considerably less than other mass analyzers. Other advantages are similar to those of quadrupole analyzers, including fast scanning, efficient ion transmission (no slits), and easy switching from positive to negative ion mode. A disadvantage (of current commercial models) is low resolving power, typically only to ~1500. Cost, convenience, and ability to do MS/MS experiments easily have resulted in the selling of thousands of ion traps during the last few years. Various aspects of ion traps have been reviewed in books as well as in papers, ranging from tutorials and general aspects[94] through the complex ion optics[95] to applications in protein and peptide sequencing.[96]

2.2.4 Magnetic Analyzers

2.2.4.1 Magnetic Sector and Electrostatic Analyzers

When a positive ion with a charge z ($z = ne$, n is the integral number of units of charge, e is the fundamental unit of charge, 1.602×10^{-19} Coulomb) and mass m is accelerated out of an ion source through a potential difference (accelerating potential/voltage) of V (volts), it will acquire a kinetic energy of $1/2\ (m/z)v^2$, where v is velocity (meters/second). Upon entering a magnetic field B (in

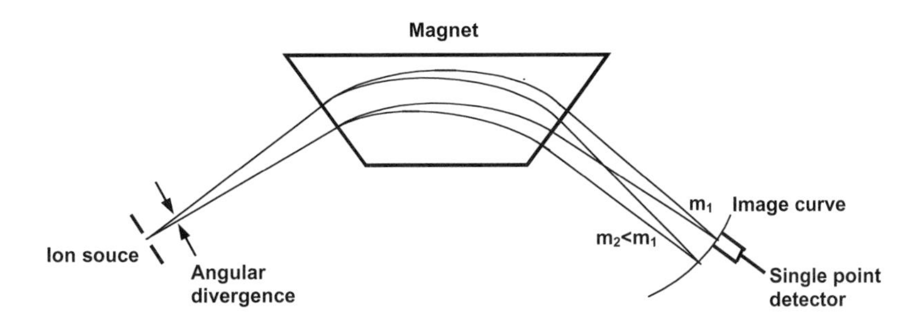

FIGURE 2.11 Magnetic sector analyzer.

Tesla) which is directed perpendicular to the plane (direction) of the flight, the ion will experience a force that is perpendicular to both the magnetic field and its direction of travel ("left-hand rule"). The radius of curvature, *r*, of the path imposed on the ion will be determined by the requirement that the centripetal force, mv^2/r be balanced by the magnetic coercion Bzv. It is seen from the equation $mv^2/r = Bzr$ that magnetic mass spectrometers are, in fact, momentum analyzers, $mv = Bzr$. However, magnetic sectors can be used to separate ions according to their m/z ratios, provided all ions are of equal kinetic energy. When the velocity term v is eliminated, the basic equation of magnetic mass spectrometers is obtained, $m/z = B^2r^2/2V$.

The most common way to separate ions in a magnetic analyzer according to their m/z ratios is to have the ions leave the ion source exposed to the same potential (voltage) drop (V = constant) and change the strength of the magnetic field (B), in a continuous manner (magnetic scanning) so that ions of increasing or decreasing m/z are successively focused onto a single point ion detector placed at a fixed position, i.e., r = constant (Figure 2.11). Alternatively, when both V and B are kept constant, the different masses describe different paths and may be detected simultaneously with a multipoint array (or photoplate) detector placed along the image curve (Figure 2.11). The third possibility for m/z separation, scanning of the acceleration voltage, is rarely used.

The deflection angle is usually 60° or 90°, and the radius normally ranges from 10 to 30 cm. (In the 180° configuration, the ions describe a half circle and both source and detector are immersed inside the magnetic field). Magnetic sectors are *single-focusing*, i.e., the focusing only compensates for the inhomogeneity in the direction of the ions (angular divergence) leaving the source (Figure 2.11). The resolving power is determined by a combination of the magnetic field applied and source and/or collector slits, of adjustable width, that reduce the angular dispersion of the beam. Sector type single-focusing magnetic analyzers, which have a high dynamic range, were used almost exclusively for decades, until the commercial appearance and subsequent proliferation of quadrupole analyzers in the 1970s.

Although all ions are accelerated to the same final voltage (2 to 8 kV) upon leaving the ion source, the energy inhomogeneity within the ion beam may be as high as a few volts and this significantly limits the resolution of single-focusing analyzers. Energy-focusing is accomplished using an *electrostatic* (electric) analyzer (E) which consists of two parallel curved plates describing an arc on the circumferences of two concentric cylinders with a potential difference of a few hundred volts.[97] Ions passing through such a radial electrostatic field move in a circular trajectory. The electrostatic force exerted upon the ions entering the analyzer is balanced against the ion's centrifugal force, so that only ions with a predetermined energy can move along a circular path and leave the energy filter. Ions of higher and lower velocity are deflected and hit the plates. Ions of the same m/z which possess slightly higher or lower energies are "energy refocused" and emerge from the electric sector to enter the magnetic sector for subsequent mass separation (in "forward geometry" type double-focusing sector instruments). Electrostatic analyzers do not analyze masses,

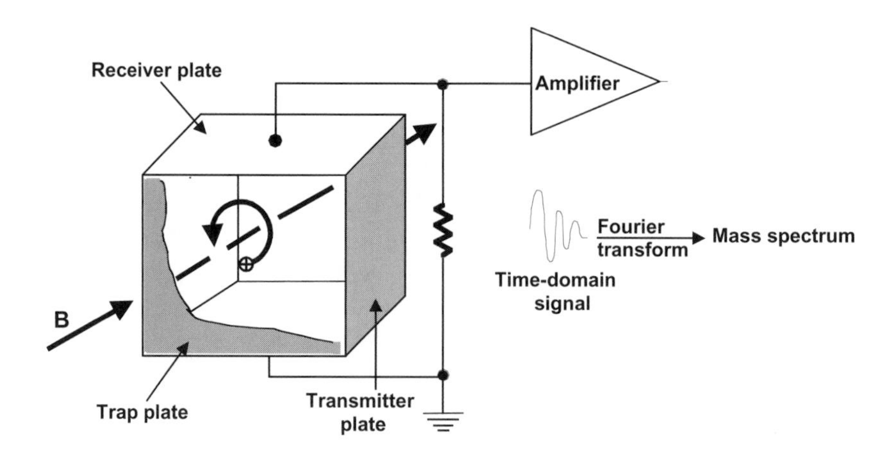

FIGURE 2.12 Fourier transform ion-cyclotron resonance analyzer.

m does not appear in the equation, $R = 2V/E$, where R = radius, V = acceleration voltage, and E = electrostatic field strength. Double-focusing mass spectrometers are discussed in Section 2.2.6.

2.2.5 Fourier Transform Ion-Cyclotron Resonance Analyzers

The force exerted on ions placed in an area exposed to constant and uniform magnetic fields moves those ions in a circular orbit that is perpendicular to the direction of the magnetic field.[98] An ion-cyclotron resonance (ICR) analyzer is a cubic cell (~2 cm each edge), consisting of two opposite trapping plates, two opposite excitation plates, and two opposite receiver plates. The cell is placed between the poles of an electromagnet or superconducting solenoid magnet with typical field strengths in the 3 to 7 T range (Figure 2.12). Ions, which may be formed by almost any currently used ion source, become trapped upon introduction into the cell and begin to "cyclotron," or move in circular orbits.

The frequency of the circular motion (natural cyclotron frequency, f_c) is directly proportional to the strength of the magnetic field (B) and inversely proportional to the *m/z* ratio of the ions, according to the equation $v_c = 2\pi f_c = v/r = kzB/m$, where f_c is frequency, v is ion velocity, r is the radius of the orbiting circle, and k is a proportionality constant. When an rf field (a frequency-swept "chirp" signal) is superimposed perpendicular to the direction of the magnetic field, those ions with cyclotron frequencies equal to the excitation frequency will absorb energy from the rf field and move into orbits of larger radii. These ions are translationally excited and move coherently, i.e., in phase with the exciting field, between the receiver plates. When an external conducting network is attached to the receiving plates, the ions transmit a complex rf signal that contains frequency components related to their *m/z* values. When a group of coherently moving positive ions approaches one of the receiver plates the ions attract electrons thus creating a current; as they continue moving on their orbit and approach the other plate, they again attract electrons. (Negative ions move in the opposite direction and repel electrons as they approach the receiving plates.) The image current signal begins to decay as the coherency is disturbed over time. The complex time-domain image currents thus produced can be transformed into frequency-domain signals by Fourier transform analysis (FT) to yield the component frequencies of the different ions from which mass spectra can be obtained using the equation given above.[98]

The operation of FTICR is governed by a series of computer-controlled discrete events, known as *experimental pulse sequence.* There are three categories of individual events. First: application of d.c. or rf excitation to increase the amplitude of the cyclotron motion, in order to: (a) establish coherent high-amplitude ion packets for detection; (b) eject ions from the cell; (c) facilitate CID.

Second: altering trap plate potentials, in order to: (a) inject or eject ions along magnetic field lines or (b) manipulate the *z*-amplitude of the ions. Third: delaying events, that occur by not making changes in the electric fields and allowing time for: (a) relaxation (homogeneous and inhomogeneous); (b) collisions (dissociative and reactive); and (c) detection of the image current.

There are four steps common to all FTICR experiments: (1) establish a large electric field gradient between the trap plates; during this *quench period* all trapped charged particles are removed from the cell; (2) inject ions from an external ion source or ionize sample within the cell; (3) excite the ions; and (4) detect ions by image current. Mass scanning can be accomplished by varying the rf pulses (frequency of irradiation) at a fixed magnetic field.[98,99]

There is a trend toward using higher magnetic fields, the main advantage of which is improved detection of high-mass ions. This is because of the increased cyclotron frequency associated with these magnets, which makes the slower moving high mass ions easier to detect by avoiding the environmental noise found in the low frequency region.[100,101] Other advantages include improvements in resolution and detection efficiency[102] and routine mass measurements to part-per-million accuracy.[103]

Because of the low drift velocity of the ions and the long cycloidal paths, the actual time it takes for the ions to traverse the analyzer is 5 to 10 ms in contrast to the few μs flight times in other analyzer types. The low drift velocity necessitates operation in very low vacuum, typically in the 10^{-10} torr region. The long flight times of ions mean that a gas pulse (the type of gas may be varied) that raises the pressure to 10^{-6} torr is typical for CID experiments. This amount of gas is readily pumped out of the cell, making these instruments particularly suited for MS^n experiments (Section 2.5).

FTICRMS has several unique advantages: (a) extremely high mass resolution, up to 3 million, may be achieved because *m/z* values are calculated from cyclotron frequency determinations that can be made to nine significant figures; (b) several masses may be detected simultaneously; (c) MS/MS experiments up to MS^4 can be conducted; and (d) change from one operational mode to another, e.g., from high to low resolution or from full mass spectra to multiple ion detection, can be made by changing only electronic parameters. Limitations include a drop in resolution with increasing mass (although the mass range is up to 10 kDa or higher), moderate dynamic range, and difficulties in quantification. Despite the relatively simple mechanical structure of the ICR cells, the need for high intensity magnetic fields, ultrahigh vacuum, and sophisticated computer techniques make the instrumentation expensive.

FTICR is perhaps the most promising of current MS techniques. Advantages include high-accuracy (±0.001%) molecular mass determinations of biopolymers when in combination with ESI or MALDI.[100,104] Improved ion transmission efficiency from ESI sources with an electrodynamic ion funnel, together with some other technological advances,[105] resulted in LOD of ~30 zmol (~18,000 molecules) for proteins in the 8 to 20 kDa range.[106] The field has been reviewed extensively both with respect to principles[99,107] and applications.[108–110]

2.2.6 Tandem Analyzers

Combined magnetic and electrostatic sector instruments have been used for many years to improve resolution by correcting various ion optical aberrations. However, the term *tandem mass analyzers* is more commonly applied to instruments where the analytical stages can be operated independently of each other in order that ions selected in the first stage can be manipulated (e.g., by collisions with neutral gases) in an area between the stages and analyzed in a subsequent stage. The *triple quadrupole* (QqQ) analyzer is one of the most common tandem analyzers and is often used in MS/MS experiments (Section 2.5.2). Many other hybrid analyzer configurations have been constructed to combine the specific features of analyzers for synergistic effects. For example, the combination of a Qq quadrupole analyzer with a TOF analyzer provides for high sensitivity detection of product ion spectra. This format of instrument has been used in multiple applications including

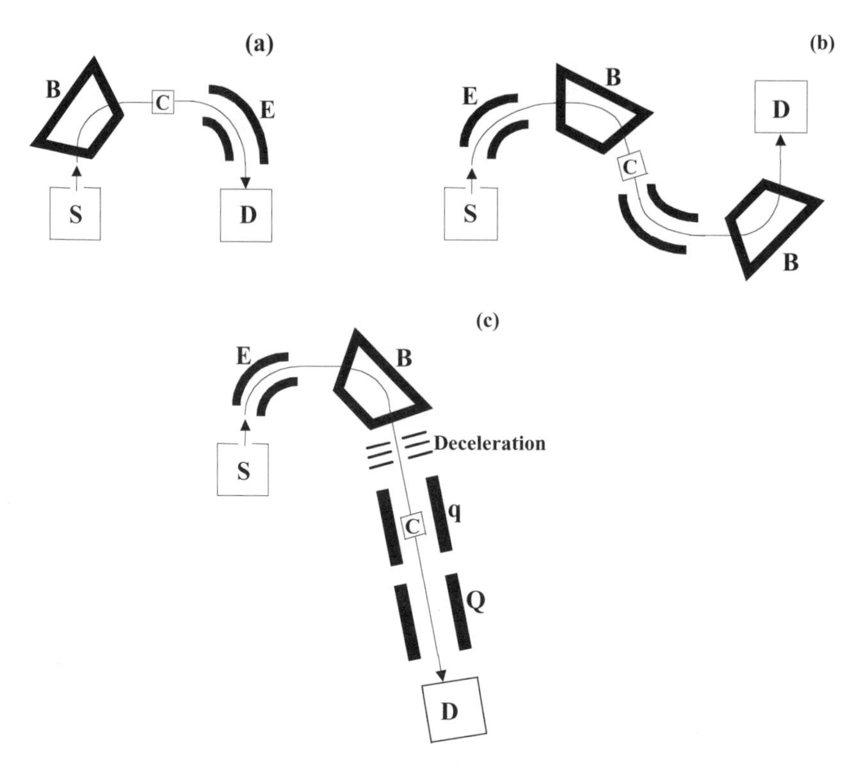

FIGURE 2.13 Various tandem configurations involving double-focusing magnetic analyzers: (a) combination of magnetic (B) and electrostatic (E) sectors in "reverse geometry" mode; (b) four-stage EB-EB magnetic-electric sector combination; (c) four-stage hybdrid configuration of EB combination followed by ion deceleration and analysis in a two-stage quadrupole system. S = source, D = detector, C = collision cell.

the investigation of proteins separated by electrophoresis and ionized by MALDI.[111] The collision energies available in a TOF-TOF tandem instrument are advantageous for improved CID studies of peptides[112] (see Section 2.5.2). A few widely used tandem analyzers are described below to illustrate salient features referred to in the description of numerous relevant applications involving sophisticated structural determinations, as well as quantification, at high sensitivity with concurrent specificity.

2.2.6.1 Magnetic Analyzers as Components of Tandem Systems

In *double-focusing* magnetic analyzers an electrostatic analyzer provides energy focusing (kinetic energy filter) and a magnetic analyzer provides both direction focusing and the required mass dispersion.[113] In the *Nier-Johnson* geometry, the two fields deflect the ions in the same direction (C-shaped) and the detector is placed at the point where double-focusing is reached, i.e., the focal point where the velocity- and direction-focusing curves intersect. Masses with different m/z values are brought onto a single-point focus detector by changing (scanning) the magnetic field strength. In the *forward,* or *EB,* geometry (E = electrostatic field, B = magnetic field), a 90° electric analyzer is followed by a 60° magnetic analyzer; the opposite is the case in the more popular *reverse* geometry, *BE* (Figure 2.13a). When a collision cell (C) is placed at the intermediate focal point, CID experiments may be carried out. In an alternative double-focusing geometry (*Mattauch-Herzog* design), ions are deflected in opposite directions (S shaped). Double-focusing is achieved for all mass-resolved ions simultaneously in the same plane, at the exit boundary of the magnet, where a focal plane detector such as a multichannel array detector or photoplate is placed. Large, four-sector, double-focusing, high-resolution instruments (Figure 2.13b) have unique advantages in some applications such as in high-energy CID (up to 10 kV) where they provide more extensive fragmentation, higher mass range, and better reproducibility than the low energy CID spectra obtained with triple quadrupoles (Section 2.5.2). Multisector hybrid instruments have also been constructed, such as configurations where ion beams separated and focused in an *EB* or *BE* configuration are decelerated and enter a quadrupole system for subsequent analysis, usually involving CID (Figure 2.13c) or coupled with an oa-TOF.

The resolution of large, double-focusing instruments may be as high as 10^5 and masses can be determined with an accuracy of <1 ppm or better.[114] Because of their high price, large size, heavy weight, and inconvenient operation, the popularity of double-focusing magnetic analyzers has significantly declined in recent years in favor of the other types of mass analyzers.

Single-focusing sector magnetic analyzers have been incorporated into a number of hybrid configurations including IT and TOF analyzers with both inline and orthogonal ion injection systems of various complexity.

2.2.6.2 Quadrupole-Time-of-Flight Hybrid Analyzer (Q-TOF)

In this versatile configuration, a quadrupole ion guide serves to manipulate the ion beam while the TOF section carries out the mass analysis (Figure 2.14a). The major analytical advantage of this type of hybrid instrument is the ability to obtain information on both the unequivocal relative molecular mass of the intact analyte as well as structural information using controlled fragmentation. The inherent sensitivity, resolution, and rapid data acquisition of the TOF analyzer are particular assets to this combination of analyzers. These advantages have recently been further extended by the development of a rapidly switchable MALDI-ESI ion source combination.[115]

The Q-TOF instrument has two basic modes of operation. The first is the MS only format when the quadrupole analyzer is operated in the wide band pass mode and passes all ions into the TOF analyzer. The second is the MS/MS format, and in this case the quadrupole is operated in the narrow band pass mode and is set (under computer control) to select designated ions that are then subjected to CID and the products analyzed in the TOF stage.

Ions are formed in the source (ESI is the type most commonly used with Q-TOF), with thermal energies corresponding to their ground state at room temperature. However, gas expansion accelerates them to supersonic speed as they pass the skimmer. The relatively high gas pressure in the first rf-only hexapole stage slows the passing ions back to their original thermal energies. In the MS-only operation all ions traverse through the quadrupole which is operated in the wide band pass mode and then are further collimated by two hexapole lenses. An additional lens then accelerates the collimated beam and a pusher electrode directs pulses of the ions (1 kV acceleration, 3 ms duration) orthogonally into a reflectron-type TOF analyzer where a full mass spectrum is obtained.

Alternatively, the system (Figure 2.14b) may be used for CID experiments. In this case, after the initial cooling and focusing in the hexapole ion-bridge, the desired masses are selected in the quadrupole by appropriate adjustments of the superimposed d.c. and rf voltages. The transmitted ions then enter the first of the post quadrupole hexapoles, which is enclosed in a pressured gas cell, and collide with neutral gas molecules gaining sufficient internal energy to fragment via CID processes. Both product and unchanged precursor ions are subsequently introduced orthogonally into the TOF analyzer to produce product-type MS/MS spectra.

2.2.6.3 Hexapole-TOF Hybrid Analyzer

This configuration is designed to provide clean, well-collimated ion beams from atmospheric pressure-type sources, such as APCI or ESI with the sample originating from static solution, direct injection (called *flow injection*), or as an on-line HPLC effluent. Because there is no quadrupole section, ions of particular masses cannot be preselected, thus ions of all masses pass into the TOF

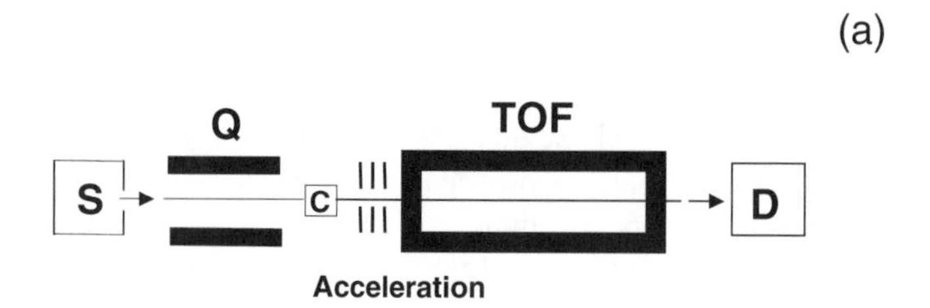

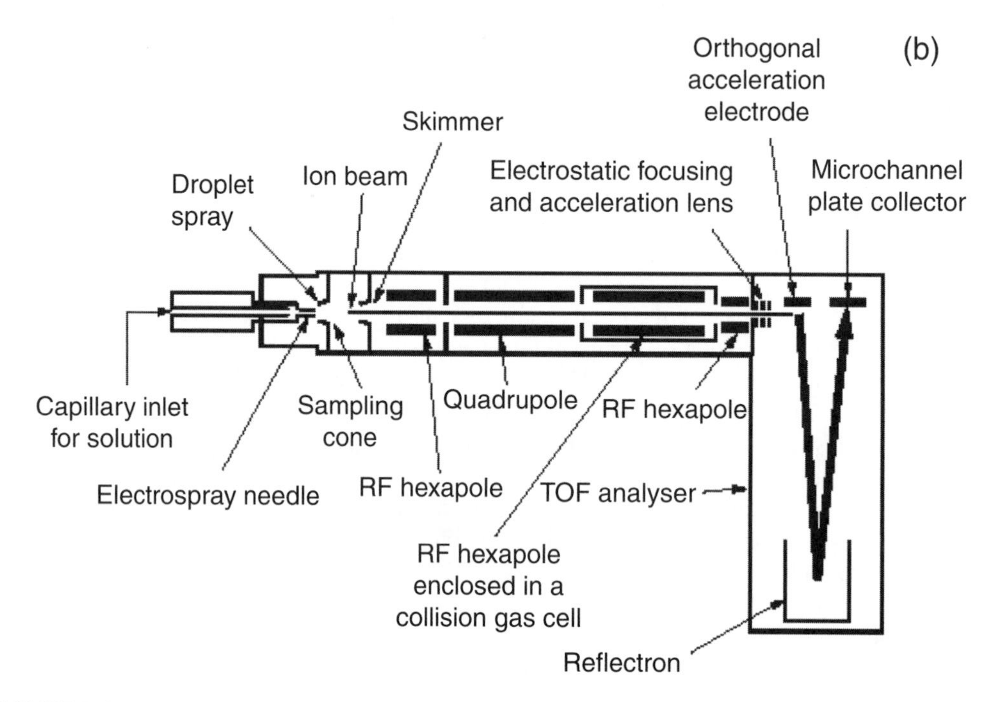

FIGURE 2.14 Hybrid quadrupole-time-of-flight (Q-TOF) analyzer: (a) schematic of the configuration; (b) schematic of a versatile commercial version, including an rf hexapole enclosed in a collision gas cell for CID experiments and orthogonal acceleration of ions into a reflectron TOF. (Reprinted courtesy of Micromass Co. With permission.)

analyzer. The application of differential pumping of the two hexapole sections, with the aid of a very small diameter orifice between the hexapoles, facilitates the passing of ions of the analytes by successively eliminating contributions from solvents and residual gases. The pressure reduction across the hexapole bridge is about 100-fold. The well defined (collimated) and mostly interference-free ion beam leaving the hexapole bridge is delivered orthogonally to the TOF analyzer by the pusher electrode in the form of detached pulses of about 3 ms duration. The TOF analyzer provides full mass spectra of all ions emanating from the source, yielding molecular mass information on each component.

2.2.6.4 Accelerator Mass Spectrometry

An accelerator mass spectrometer (AMS) consists of a magnetic sector mass spectrometer for negative ions (operated at ~20 to 100 keV/ion acceleration energy) connected to an electrostatic

(Van de Graaf) accelerator which, in turn, is linked to a high energy (5 to 150 MeV/ion) momentum/charge spectrometer system for positive ions. Negative ions are produced from solid samples by bombardment with 3 to 10 keV cesium ions in a cesium-sputtering ion source (similar to FAB). After separation by the low-energy mass spectrometer, the selected negative ions are accelerated to high energy (~5 to 10 meV) inside the Van de Graaf accelerator. On the way toward the positive terminal of the accelerator, the negative ions pass through a carbon foil that removes electrons from the ions, making them positively charged. The positive ions are next accelerated back to their ground potential, and preselected isotopes with the desired charge state are focused onto the entrance of the second dipole magnetic MS for subsequent detection by a multi-anode gas ionization detector.[116]

Although used primarily to detect ^{14}C, AMS may be used to detect any isotope with a half-life >10 yr. The technique measures isotope ratios rather than absolute quantities. Quantification is accomplished from the measured $^{14}C/^{13}C$ ratios with the aid of the total carbon content of the analyte (separately determined). The sensitivity of AMS is 10^6-fold larger than that of conventional decay counting techniques; LOD is 0.01 dpm per mL or g sample. This makes it possible to administer only nCurie quantities of radioactivity to humans for the study of the biotransformation of carcinogens, mutagens, or drugs. A relevant application involved the quantification of carcinogenic heterocyclic amines in normal and tumor tissues (Section 4.3.2.2).

2.3 ION CURRENT DETECTORS

Ion currents in most mass spectrometers are in the 10^{-9} to 10^{-16} A range. The arrival of one ion per second corresponds to 1.6×10^{-19} A. The accuracy of ion current measurements obviously depends on the total number of ions arriving at the detector. When the ion current is very low, integration may be used as long as ions of a particular mass continue to arrive at the detector. When it is possible to scan a spectrum repeatedly (sometimes up to 100 times), *multichannel averaging* can be used to integrate individual signals and to average noise. Therefore, signal-to-noise ratios increase with averaging, and both detection limits and precision improve. Statistical considerations impose severe limitations on the accuracy of ion current determination when fast scanning is used, particularly at high resolution or in the pulsed ion extraction systems of time-of-flight instruments, because there may be only a few hundred ions striking the collector during the observation period.

There are two classes of detectors to determine the abundance of separated ions of different *m/z* values: point and array ion detectors. *Point* ion detectors are used with analyzers that sequentially focus ions with different *m/z* values at one point, i.e., ions are detected in the *time domain,* e.g., quadrupole analyzers. Magnetic sector analyzers disperse ions in space, thus the arrival of separated ions may be detected either sequentially at a point detector (by scanning the strength of the magnetic field) or simultaneously in *space* by placing an *array* (or photographic plate) detector at the focal plane of the analyzer. *Cryogenic detectors,* a new class of detectors under development, are energy-sensitive calorimetric detectors operating at low temperatures. In contrast to microchannel plates whose efficiency decreases considerably as the mass of the arriving ions increases, cryogenic detectors may have near 100% efficiency for very large, slow-moving molecules.[117]

Although TOF analyzers can use point ion detectors, microchannel plate detectors are used almost exclusively, particularly with the orthogonal-TOF designs. Although a microchannel plate detector is not a point detector as such, in practicality it is, because single *m/z* are detected and recorded given that different *m/z* are in fact separated in the time domain as a result of their different drift speeds.[113]

Ion trap analyzers may be used with either point or array detectors or even without a conventional detector using different electric field frequencies in flight according to *m/z* values. The unique frequency array method of detection of FTICRMS (Section 2.2.5) permits the simultaneous detection of all *m/z* present in the ICR cell.

2.3.1 Single Point Ion Collectors

2.3.1.1 Faraday Cup Detectors

The time-honored *Faraday cup* detector consists of an open-ended metal cage (cup) having a metal collector plate that is grounded through a high-ohmic resistor (10^{10} to 10^{12} Ω). Positive ions arriving at the collector plate are neutralized by accepting electrons from the grounded metal plate. Negative ions donate electrons. The depletion or acquisition of electrons creates a potential drop across the resistor that is proportional to the abundance (number) of the arriving ions. The flow of electrons resulting from the potential drop constitutes a small electric current that can be amplified and converted to a voltage for subsequent display and/or recording. The shape of the detector improves accuracy by preventing the escape of reflected ions and ejected secondary electrons.

Faraday cups are frequently used for ratio measurements of specific species of ions. In such relative abundance measurements, the ratio of the intensities of the two arriving beams must be measured simultaneously and accurately. Therefore, to avoid errors due to momentary instrument instabilities, double collectors (each with its own amplifier) are used and the ion current ratios are determined by feedback circuitry. Simultaneous detection of up to five ion beams with different *m/z* values is possible with multicup arrangements. An example of the use of these detectors is for monitoring the gaseous constituents of breath and blood using EI.

2.3.1.2 Secondary Electron Multipliers

In these detectors (reviewed in Reference 118), positive ions leaving the analyzer are accelerated into the multiplier by a –2 to 5 kV potential applied to the front of the multiplier. The output of the multiplier is referenced to ground (the reverse is true for negative ions). The accelerated ions impinge on a *conversion dynode plate* (made of Cu-Be alloy) from which two secondary electrons are ejected for every ion arriving (Figure 2.15a). The released electrons are then accelerated and strike a second dynode from which about two electrons are ejected for every arriving electron. This cascading continues through a series of stages. In a 12-stage design there is a ~10^6-fold gain, yielding an output current of several μA. This current is passed outside the vacuum system into a low-noise preamplifier for subsequent amplification by a conventional amplifier. The gain of all types of secondary electron amplifiers decreases with age as the impinging ions "burn" into the surface of the conversion dynodes.

Even higher gain (10^8-fold) is available in the *channel* type multipliers where the cascading of the secondary electrons takes place along the inside wall of a curved leaded glass tube coated with a resistive (10^9 Ω) material. This system acts as a continuous dynode when a potential difference is placed between the ends of the tube (Figure 2.15b). The cascading of the secondary electrons is repeated approximately 20 times. The gain is a function of the length-to-diameter ratio. These multipliers may be used in the analog or pulse counting modes. Channel electron multipliers are often used with quadrupole instruments where the kinetic energy of the ions in the analyzer is only a few electron volts.

The efficiency of this type of secondary electron multipliers is a function of the energy with which ions hit the conversion dynode and therefore, because of their slower speed (actually, lower energy per square area), large ions produce fewer secondary electrons upon hitting the conversion dynodes than the faster moving smaller ions. Accordingly, all electron multipliers, including array detectors, exhibit mass discrimination, i.e., the higher the mass, the lower the detection efficiency (gain) for ions with constant energy.

2.3.1.3 Scintillation Counters

Ions leaving the analyzer strike a dynode, as occurs with the electron multipliers. The released secondary electrons then strike a phosphorous screen from which photons are released that, in turn,

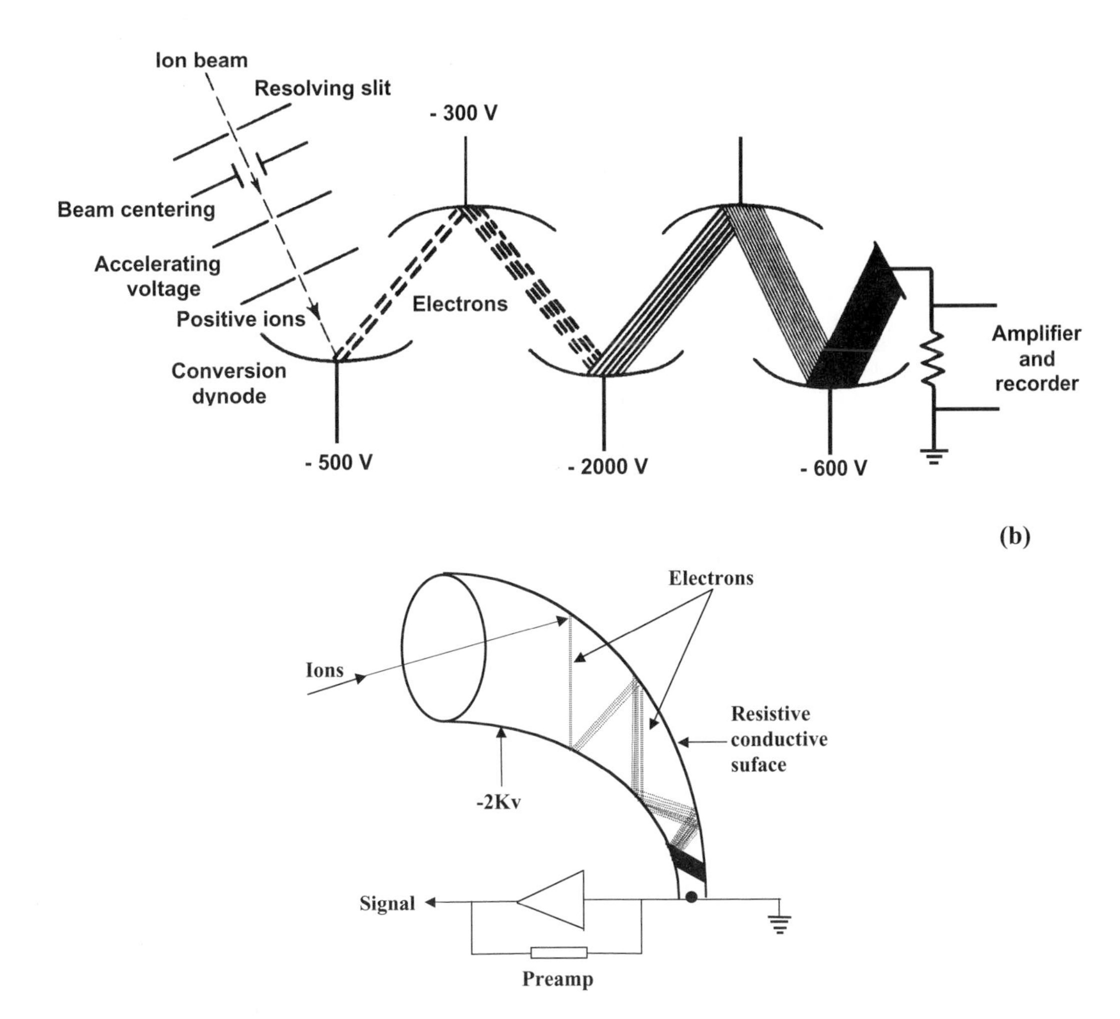

FIGURE 2.15 Principles of (a) secondary electron multiplier; (b) channel type multiplier.

are detected by a sensitive photomultiplier (placed behind the scintillator screen) which converts the photon energy into an electric current for subsequent amplification. The arrival of even a single ion may be observed with these detectors.

2.3.2 Multipoint (Array) Detectors

In the Mattauch-Herzog type double-focusing analyzers, all resolved ion beams are focused simultaneously along a plane, permitting the use of array detectors. The first of these was the *ion-sensitive photoplate* detector which consisted of AgBr crystals suspended in gelatin spread over a glass plate. When energetic ions impinge, a latent image is formed and a chemical amplification occurs during development. Masses and their abundances are determined from the position and the intensity (blackness) of the lines. Photoplates, no longer popular due to inconvenience (e.g., having to break vacuum, need for a darkroom) and poor sensitivity, have been replaced by array detectors that act as electronic photoplates.

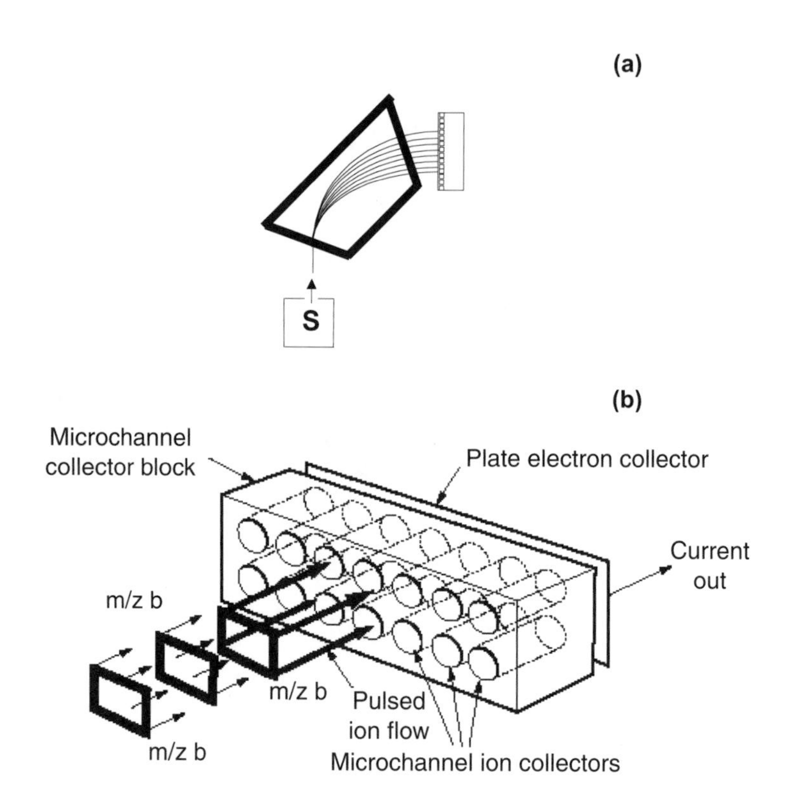

FIGURE 2.16 Multipoint detectors: (a) array detector with magnetic analyzer; (b) multichannel detector for TOF analyzers. (Reprinted courtesy of Micromass Co. With permission.)

In the various array detectors, a portion of the mass scale being scanned is detected simultaneously. The spatial array detectors are electro-optical ion detectors based on a number of very small microchannel electron multipliers (*channeltrons*) arranged in a line and/or one above the other in a very small space (Figure 2.16a). Because of space and cost considerations, only a limited number of array elements, such as 100, are used, thus only a limited section of the mass range can be covered at a given time. The resolution may be increased by scanning a narrower mass range.

In current usage, the term *array* refers to "an assemblage of small single point detectors, which remain as individual ion monitoring elements" while *microchannel plate* is used to describe "an assemblage of small single point detectors, all of which are connected so as to act as a single large monitoring element" (reviewed in "Back to Basics," see http://www.micromass.co.uk/basics/). The most important use of array collectors is with rapid, nanosecond scale, pulsed ion configurations, such as MALDI, where the abundance of a range of ions produced at the same instant must be detected in a short interval of time. Array detectors also have advantages in trace analysis, even when compared to SIM with point ion detectors, because of higher specificity on account of the wider mass range covered. TOF analyzers use microchannel plate detectors (Figure 2.16b) almost exclusively,[119] particularly with the oa-TOF configuration, as they are capable of generating almost instantaneous (~30 μs) spectra. In addition, the oa-TOF-microchannel plate combination can sum ~30,000 spectra per second giving very high signal/noise ratios.

2.4 INTERFACES FOR SAMPLE INTRODUCTION

The main function of inlets is to introduce a sufficient quantity of the sample into the ion source in such a way that its composition represents that of the original sample as accurately as possible.

TABLE 2.1
Comparison of Capillary-GC, HPLC, Capillary-HPLC, and Capillary Electrophoresis

Parameter	c-GC	HPLC	c-HPLC	CE
Solvent flow rate	1–3 mL/min	1–3 mL/min	1–10 μL/min	0.5–10 μL/min
Injection volume	1–100 μL	1 μL to 1 mL	1–500 nL	1 pL to 100 nL
Plates (theoretical)	100–200 K	30–100 K	50–400 K	2.7 million
Molecular mass limit	1.2 kDa	5000 kDa[a]	5000 kDa[a]	2×10^5 kDa
Detection limit	1 pg to 1 ng	1 pg to 1 ng	1 pg to 1 ng	1 ag to 1 ng
Pre-equilibration[b]	No	Yes	Yes	Yes
Applications				
Volatiles	Yes	No	No	No
Semivolatiles	Yes	Yes	Yes	Yes
Polar, nonvolatiles	No[c]	Yes	Yes	Yes

[a] With size exclusion columns: 15×10^7 kDa
[b] Equilibration time is ~5 min for CE, several hours for HPLC and c-HPLC
[c] Yes if derivatization is possible

In the "direct" inlets this means the reduction of the pressure of gases or the production of vapors from solid or liquid samples. In some "static" probes the steps of sample vaporization and ionization take place simultaneously, e.g., in FAB and MALDI. The advent of on-line coupling of gas and liquid chromatographs and various forms of electrophoresis with mass spectrometry necessitated the development of interfaces, of various levels of sophistication, aimed at reducing the atmospheric pressure of gas or for liquid eluates (consisting of analyte[s], other sample components, mobile phase, dissolved stationary phase) by removing most of the eluant (i.e., the chromatographic mobile phase) prior to the introduction of the analyte(s) into the ion source.

There are major differences in the performance parameters of the different chromatographic and electrophoretic techniques (Table 2.1). Judicious selection of the separation technique and the sophistication of the interface technology are essential for obtaining the optimal analytical technique for the problem at hand.

2.4.1 Gases and Volatile Liquids

These inlets must meet the following requirements: (1) the flow rate of the gas or vapor should remain constant during analysis; (2) the chemical composition should not change with time (e.g., dehydration); and (3), in case of mixtures, the partial pressures of individual components should be independent of other components present.

2.4.1.1 Reservoir and Batch Inlets

Gases and volatile liquids are introduced into EI or CI sources from heatable (to 350°C) glass or metal reservoirs of known volume (e.g., 2 L) by way of gold or steel foils (0.25 mm thick) with several pinholes (0.005 to 0.02 mm diameter). Volatile liquids may be frozen with dry ice or liquid nitrogen at the bottom of a small tube connected to a pumping manifold. The vapor pressure of the analyte(s) must be at least 0.2 torr at <300°C without decomposition. Batch systems allow the sample to leak into the ion source at a controlled rate over a long (min to h) period of time. Sample utilization is very poor (0.01%) and most of the sample is wasted. Batch systems are used for the routine quantitative analysis of inorganic gases, volatile and nonvolatile liquids and solids, including waxes and tars (accuracy 0.1 to 1.0 mol %). The dual viscous inlets, used for isotope ratio measurements, include automatic switching for the rapid alternate introduction of the sample gas and the standard gas. Batch inlets are used routinely to introduce mass calibration compounds at a constant flow into a variety of ion sources.

2.4.1.2 Membrane Inlets

Some semipermeable polymers, e.g., silicone membranes, have significantly different permeability for volatile organic compounds relative to water or the main constituents of air. Thus, certain compounds present in liquid or gaseous streams may be introduced selectively into the mass spectrometer ion source. The most relevant properties of the analyte are its vapor pressure and its solubility and diffusivity in the membrane material. There is no need for any sample preparation in most cases, but the sample must not have any constituent that would dissolve the membrane. Analyte concentrations may be monitored in the parts-per-hundred to parts-per-trillion range, and analysis times range from a few seconds to minutes. Membrane inlet mass spectrometry has been employed increasingly in diverse areas ranging from on-site, on-line monitoring of industrial processes and waste effluents, to physiological and metabolic monitoring. The theory of pervaporation, available membrane materials and technologies, classes of analytes, and diverse applications have been reviewed.[120]

2.4.2 Static Direct Inlet Probes

In the direct insertion probes of EI and CI sources, pulverized solids or samples dissolved in a volatile solvent are placed in a small, one-end-closed quartz, glass, special surface-treated metal (aluminum) or inert (platinum or gold) capillary, which is introduced through a vacuum lock into the ion source. Samples are evaporated into the ionization region by heating the capillary up to 400°C with a small electrical coil. Because samples need to be heated only until their vapor pressure reaches 10^{-6} torr, adequate vapor pressure is often reached at temperatures below the point of decomposition of thermally sensitive compounds. It is possible to use temperature programming for the microdistillation of crude mixtures with relatively little overlap of the constituents. Cooling of the insertion probe directly with liquid nitrogen or indirectly with the aid of copper conductors prevents sample evaporation in the warm ion source before desired operational conditions are attained. Efficiency is good because only the very small volume of the ionization chamber must be filled with sample vapor and samples are volatilized only a few mm away from the ionizing electrons or ions derived from of the CI reagent gas. Adequate mass spectra can be obtained from 100 ng of material. As little as a 5 ng sample will often yield usable information. Probes may also be used for the off-line analyses of fractions collected from GC or HPLC or scrapings from TLC plates.

In the probes of FAB and LSIMS sources, droplets of the dissolved samples are placed, one at a time, on the sample stage, inserted into the source through a vacuum lock, and bombarded by energetic atoms or ions. In MALDI sources, mechanically sophisticated sample stages permit the introduction of as many as 100 samples, in solid, crystalline form, for subsequent exposure to laser pulses.

2.4.3 Interfaces for Coupling with Gas Chromatographs

The enormous synergystic analytical potential of combining the separation power of GC with the identification and quantification capabilities of MS was first realized in 1957. In GC/MS the components of samples, even complex mixtures with dozens of constituents, are first separated inside the heated GC columns and then eluted directly into the ion source of the MS. Using EI and CI sources, GC-MS became, and remains, a widely used analytical technique for the identification and quantification of a large variety of volatile or derivatized chemicals, drugs, carcinogens, metabolic products, and environmental contaminants with molecular masses to ~1.2 kDa. Thousands of applications have been reported.

A GC and an MS can be combined without major modifications to either instrument as long as the approximately atmospheric pressure gas flow requirement of the GC and the vacuum requirement of the MS are reconciled. Packed GC columns, used for decades, required a molecular separator (enricher) to reduce the pressure from atmospheric to 10^{-5} to 10^{-6} torr. In the jet separator, which is still in occasional use, the GC effluent passes through a 50 to 100 μm diameter orifice. In the expanding supersonic jet stream, the light carrier gas molecules diffuse into the area around the orifice and are pumped away, while ~50% of the heavier organic molecules continue on a straight line into the ion source.

The currently used highly efficient capillary GC columns, often with on-column injection, require no interface as long as ion source pumps of appropriate capacity and speed are provided. The decreased sample capacity of capillary columns is partly compensated for by the efficiency of the columns that result in peaks a few seconds wide and therefore the arrival of an increased concentration of analytes into the ion source. When a CI source is used, the same gas, e.g., methane or isobutane, may be used as both the GC carrier gas and CI reagent gas.

Although GC/MS instruments have been realized using most types of analyzers,[121] most commercial instruments are based on quadrupole analyzers. Computer-controlled scanning of the voltages and rf fields in quadrupoles routinely meets the scan rate requirements of capillary columns, 0.2 to 0.5 s/scan, e.g., 50 to 500 Da. There are no scanning considerations in GC/FTMS.

GC/MS systems have been described as combinations in which either the MS serves as a sophisticated chromatographic detector or the GC serves as an inlet system of the MS. There is more than a semantic distinction between these points of view, e.g., the combination may serve for the routine quantification of known analytes present in ultratrace quantities in mixtures, i.e., using the MS as a detector, or as a research tool for the identification of one or more unknown substances in complex mixtures. Recent technological developments include high-speed GC/MS.[122] Several current aspects of instrumentation and techniques as well as applications have been reviewed extensively.[121,123]

2.4.4 Interfaces for Coupling with Liquid Chromatographs

Although the role of the LC mobile phase has changed with the development of atmospheric pressure ionization techniques, i.e., the constituents of the mobile phase have become essential in analyte ionization, the need for appropriate pressure reduction has remained. A major advantage of ESI over MALDI is that ESI is a direct liquid introduction technique, thus interfacing ESI to LC (and also to CE) is straightforward. Apart from some necessary compromises in the selection of more volatile buffers, virtually all types of LC separation techniques can be coupled to MS detectors without making major trade-offs (reviewed in References 124–126). A comparison of conventional, narrow-bore, and capillary columns connected on-line to different ion sources revealed their respective advantages and pitfalls.[127] A multipurpose interface has been developed that incorporates different API type sources, including ESI and APCI.[128]

In MALDI, samples must be cocrystallized with the matrix and dried on a solid surface prior to insertion into the source. Despite this obvious limitation, several attempts have been made to couple MALDI to LC, both off-line and on-line. Off-line coupling simply requires collecting the appropriate HPLC fraction for subsequent MALDI analysis. On-line coupling methods to deliver the separated effluents directly into the MALDI source include continuous-flow, e.g., the insertion of the probe through a hole in the ring of an ion trap with laser irradiation through another hole on the opposite side,[129] and aerosol techniques (reviewed in Reference 54). A compromise between off- and on-line can be achieved by depositing the eluate from a low flow rate HPLC directly onto a MALDI plate that has been precoated with matrix. Upon introduction of the plate into the source, computer controlled rastering by the laser generates the spectra and therefore the HPLC profile.

2.4.5 Interfaces for Coupling with Capillary Electrophoresis Systems

Capillary electrophoresis (CE) involves the migration (mobility) of ions in solution in an electric field, i.e., the movement in narrow bands of ions of the appropriate charge in fused silica capillaries (<100 μm diameter) toward a cathode or anode between which there is a high potential difference. The charged analyte species migrate with different speeds in the buffer solutions resulting in a plug-like profile due to the electroosmotic flow created by the electric field. The obvious advantages of CE compared to GC and HPLC include high efficiency separation, small sample size, and low solvent consumption (Table 2.1); the latter is of significance in nanoscale applications. Although CE has been coupled with MALDI,[130] the ion source of choice is ESI, just like with LC interfacing, because of the direct transfer of samples from the liquid to the gaseous phase. As there are no restrictions with respect to the type of analyzers for CE, applications have been reported using QqQ, FTICR, IT, as well as TOF analyzers.

Currently used interfaces for ESI include sheathless, liquid junction, and coaxial liquid sheath interfaces (Figure 2.17) (reviewed in References 131–132). Sheathless interfaces of various levels of sophistication have been constructed, each using only a single capillary.[133–135] These are particularly useful for coupling to nano-ESI (Section 2.1.4.1) because of compatible flow rates and no dilution of the analytes. In contrast, liquid-liquid junction interfaces are partially disconnected, both physically and electrically from the ESI emitter. A fused silica capillary is inserted into a T-section and connection to the ESI emitter is through the electrolytes flowing from the buffer reservoir.[136] The frequently used coaxial sheath-flow interfaces consist of two concentric stainless-steel capillary tubes. The sheath liquid and the CE buffer mix at the tip of the capillary.[137] The MS behaves as a concentration-sensitive detector at high flow rates (>180 nL/min) and as a mass-flux sensitive detector at lower flow rates. Disadvantages include sensitivity loss due to dilution and increase in the chemical background originating from the sheath liquid.[132] The current status of realized CE/MS interfacing is illustrated by the design of a sheathless CE probe connected to an orthogonal type ESI for both conventional and MS/MS modes of operation.[135]

In addition to the different types of interfaces, there are significantly different forms of CE, each with its own set of advantages and limitations. Simplest is the conventional *capillary zone* electrophoresis (CZE) which can be optimized easily with respect to efficiency, selectivity, and analysis time by changing such operational parameters as the type, ionic strength, and pH of the buffer solution, the dimensions of the separating capillary, and the operating voltage. Disadvantages include the need for extensive sample preparation to minimize matrix interferences and limited loading capacity (nL sample size) which, however, may be improved by various *stacking* techniques.[132] CZE is used routinely for the analysis of small (<1 kDa), charged molecules often with MS/MS to obtain structural information. Relevant applications, discussed in later sections, include the analysis of nucleotide adducts and proteolytically digested glycoconjugates.

In *capillary isotachophoresis* (CITP), the sample is introduced between two different buffers, one with higher and the other with lower electrophoretic mobility than the constituents of the sample. The buffer solutions bracket the bands of analytes during the electrophoretic separation. Components of the sample with different electrophoretic mobilities yield sharp bands, the length of which are proportional to the respective quantities. The preconcentrating ability of CITP is often used to analyze dilute solutions. Novel CE techniques include *capillary isoelectric focusing* (c-IEF) which is used for the separation of proteins with different isoelectric points, *capillary gel electrophoresis* (CGE) which is based on separation by size exclusion, and various forms of CE, in which ionic *micelles* are used as stationary phases, to permit the separation of both ionic and neutral analytes. An interesting hybrid technique, *capillary electrochromatography* (CEC) combines CE and HPLC with electrically driven mobile phase for the hydrophobic separation of neutral analytes without the use of surfactants; sub-fmole to amole sensitivities have been predicted (reviewed in Reference 138).

The advent of chip-based technologies (see below) is opening hitherto unimagined areas of application of CE.[139,140] A major area of application is the rapid and sensitive separation of trace

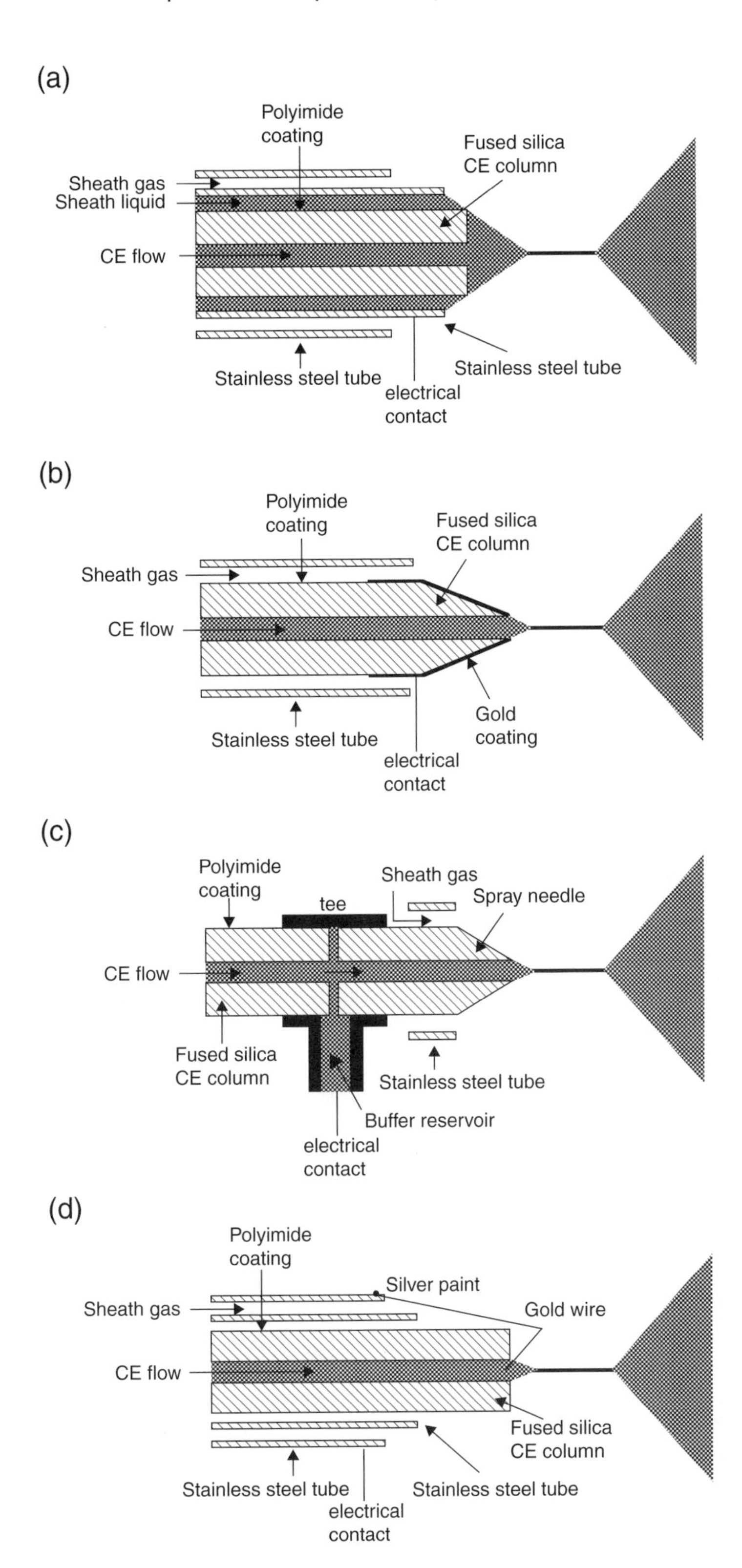

FIGURE 2.17 Various types of interfaces for capillary electrophoresis. Reprinted from Banks, J.F., *Electrophoresis,* 18, 2255–2266, 1997. With permission.

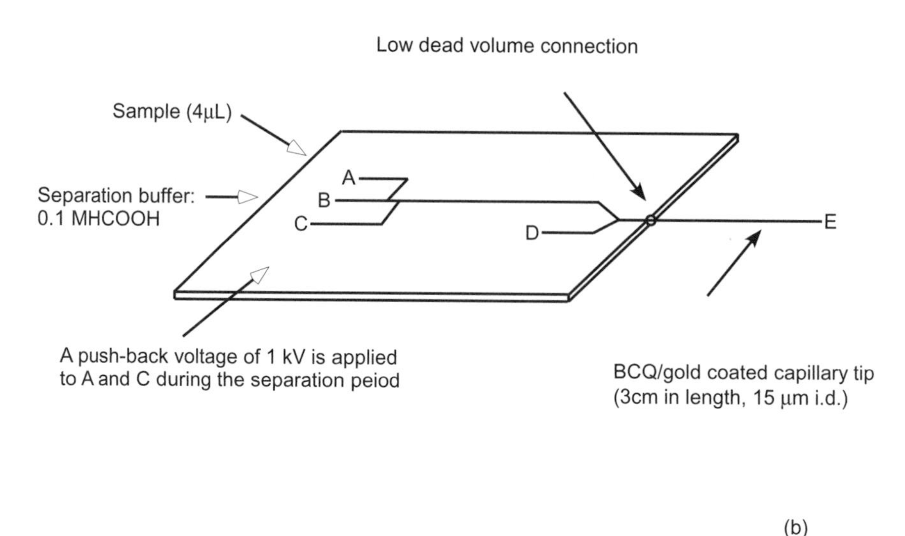

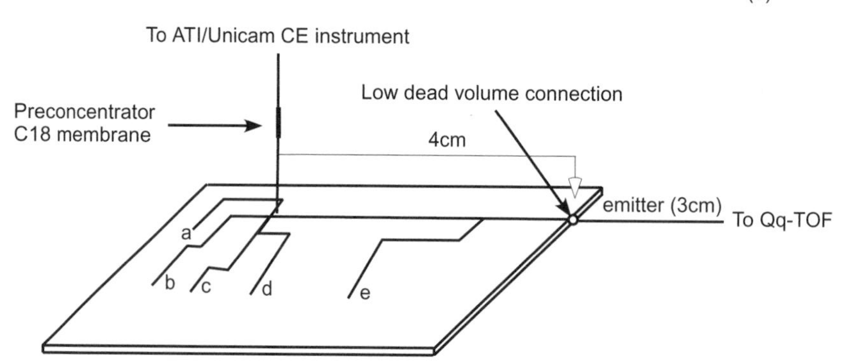

FIGURE 2.18 Schematic representation of the chip capillary electrophoresis configuration using (a) a disposable nano-ESI emitter; (b) a solid-phase extraction (SPE) preconcentrator. In all cases, the microfabricated device was coated with the reagent [(acryloyamino)propyl]trimethylammonium chloride (BCQ) and coupled to a 3 cm BCQ/gold-coated nano-ESI emitter (15 μm ID). Reprinted from Lee, J. et al., *Electrophoresis,* 21, 198–210, 2000. With permission.

levels of protein digests for which chip CE configurations have been developed for interfacing with Q-TOF hydrid analyzers (Figure 2.18).[141]

2.4.6 High Throughput Inlets and Microfabricated Fluidic Systems

Efforts to eliminate sample preparation (extraction) as the rate-limiting step in high throughput analyses include the approach of injecting biological samples directly into LC/ESI-MS/MS systems using special columns and switching valves. In one application, serum was directly injected onto a large particle size column which permitted the rapid passage to waste of serum proteins and other large biomolecules using aqueous mobile phase, but retained small analytes for subsequent elution into the ion source with an organic mobile phase; analysis time was <5 min/sample.[142]

Another approach to improve throughput combined the 96-well format and solid phase extraction with an automated robotic liquid handling sample processor; 96 samples could be prepared unattended in 1 to 1.5 hr.[143,144] The 96-well automation approach has been extended to include liquid-liquid extraction of plasma samples[145] and protein precipitation using acetonitrile[146] followed by the quantification of analytes by LC/ESI-MS/MS using selected reaction monitoring (SRM, Section 2.6.3), employing transitions of representative precursor ions to previously identified product ions. In an application to quantify methotrexate and its major metabolites in human plasma, 384 samples were prepared in ~90 min by one person, and some 820 samples were analyzed by LC/MS with SRM in 24 hr.[147] In another technological development, a 96-channel programmable liquid handling workstation was used with rapid gradient LC separation to obtain clean biological extracts for MS/MS analysis. The method was applied to analyze antineoplastic protease inhibitors in plasma.[148]

The concept of "lab-on-a-chip" is rapidly becoming a trend in instrumentation (reviewed in Reference 149). The current status of advanced multiplexed and integrated MS systems is represented by the coupling of a multifaceted, enhanced microfluidic chip to an ESI-TOF mass spectrometer and applying it to the analysis of protein digests under various analytical conditions.[150]

Microarray systems are arrays of μm or tens of μm sized spots of material immobilized upon solid supports (reviewed in Reference 151). Microfluidic systems are devices where liquids and gases are transported within miniature channels having cross-sectional dimensions on the order of 10 to 100 μm. The manipulation of the fluids is achieved with either hydrostatic displacement pumps (e.g., syringe pump) or electroosmotic flow; the latter offers the possibility of CE separations. The common techniques for the fabrication of microfluidic systems include photolithography and wet etching, radiation-induced etching, replica molding, and laser-based technologies (reviewed in References 152–154). There are current efforts aimed at developing MS-interfaced devices using materials other than glass[155–158] and reducing solvent compatibility problems.[150]

Rapid technological developments in nanotechnology (reviewed in Reference 159) have led to the exploration of microfluidic devices both for sample preparation and as sample introduction devices for mass spectrometric analyses. There is an increasing need for large-scale, high-throughput analysis of exceedingly small sample quantities in several research areas,[160] most notably in proteomics (Section 6.1.1). Despite the huge differences in physical dimensions, microdevices can be conveniently interfaced with ESI sources because of similar flow rate requirements (reviewed in References 160,161). Sample preparation protocols, such as cleanup by dialysis,[157] enzymatic digestion,[162] preconcentration,[141] and separation of proteins and their digested products by electrokinetic or chromatographic approaches have been accomplished on single microchips.[35,39,163,164] In most cases, microfabricated and chip-based interfaces for CE-MS have used ESI for ionization.[165,166] A 100-element microfabricated array was used in a pilot study to interface chips with MALDI ionization for DNA analysis.[167]

2.5 MASS SPECTROMETRY/MASS SPECTROMETRY

2.5.1 ION ACTIVATION METHODS

The fundamental process involved in MS/MS is $m^+_{precursor} \rightarrow m^+_{product} + N$, i.e., fragmentation (decomposition) of a *precursor* ion into a smaller *product* ion, accompanied by the loss of a neutral fragment, *N*. The terminology is the same for negative ions. The precursor ion is selected in the first analyzer, reacted, and the product ions are measured in a second analyzer, hence the term *tandem mass spectrometry.* As will be discussed later, precursor and product ions can be separated *in space* in subsequent analyzers of the same type or in hybrid configurations, or *in time* with the different steps carried out sequentially in the same analyzer. The current availability of hybrid instruments provide hitherto unattainable advantages for MS/MS studies.

There are several ways to provide energy to afford fragmentation of the ions between the source and the analyzer. The "amount" of energy acquired and the way it is distributed within an activated ion significantly influences the type and quantity of the product ions: small internal energy distribution favors simple bond cleavages while broad internal energy distribution leads to increased, information-rich fragmentation.

2.5.1.1 Metastable Ions

When an analyte is ionized, the molecular ion formed either remains as such or decomposes immediately (life time $<10^{-7}$ s) into fragment ions while still in the ion source. Most ions leaving the ion source travel unchanged through the mass analyzer to be detected by the ion collector. It is possible, however, for the traveling ions to undergo unimolecular decomposition 1 to 100 μs after leaving the ion source. Ions undergoing such transitions are called *metastable ions.* In fact, metastable ions do not result from ion activation, though it may be difficult to distinguish them from ions formed in collisions with the background gas in the ion source.

In single-focusing magnetic sector instruments, metastable ions may form in the "field-free" region between the ion source and the magnetic field. In double-focusing magnetic instruments there are several field-free areas for metastable ion formation to occur. The availability of various analyzer configurations has provided a fertile field for the study of the chemistry and physics of gaseous ions. Metastable ions are detected as wide, Gaussian-shaped peaks, usually of low abundance, appearing at nonintegral masses as part of the normal mass spectrum. Metastable transitions can be detected in reflector TOF instruments when a retarding potential is applied at the end of the drift tube.

2.5.1.2 Collision-Induced Dissociation

When gaseous ions with relatively high translational energy are permitted to collide with neutral atoms with a high ionization potential (usually argon) or molecules (nitrogen) maintained at relatively high pressure as an immobile target in a special collision cell, a portion of the ions' kinetic energy (acquired during initial acceleration) is converted into excess internal vibrational and/or electronic energy; the process is called *collisional activation.* If the excess energy is adequate to break bonds, these ions undergo *collision-induced dissociation* (CID) also called *collisionally activated decomposition* (CAD) (reviewed in References 168,169).

Low-energy collisions, with kinetic energies of ions in the range of 1 to 100 eV, primarily excite the vibrational states of an ion, leading to narrow internal energy distribution, and production of a limited variety and quantity of product ions. The nature and degree of fragmentation depend strongly on the collision energy employed and, to a lesser degree, on the type and pressure of the collision gas. Low-energy collisions can be carried out conveniently in triple quadrupole or hybrid instruments.

When the precursor ions are accelerated to ~1 kV, the *high-energy collisions* (often with helium) excite the electronic state of the analyte and produce a broad distribution of acquired internal (vibrational) energies, resulting in significant fragmentation and therefore a great deal of structural information. In addition, nonenergy-related collision conditions, such as the type, pressure, and temperature of the collision gas, have little influence, making the spectra more reproducible than those obtained by low-energy CID. However, high-energy collisions can be carried out only in multisector magnetic or hybrid instruments.

2.5.1.3 Electron-Capture Dissociation

It has been observed in FTICR mass spectrometers that trapped positive ions may fragment upon encountering a high-current beam of near-zero energy electrons, i.e., ion activation may result from ion-electron collisions.[170,171] When multiply charged cations from ESI undergo electron capture dissociation, there is a unique fragmentation of the N-C_α bonds in peptides and proteins which provides information complementary to CID and photodissociation, enabling *de novo* sequencing in some cases.[172]

2.5.1.4 Photodissociation and Surface-Induced Dissociation

Ions that have a chromophore that absorbs light at a given wavelength may undergo electronic excitation upon irradiation with photons with well-defined energy, usually UV laser beams. Photodissociation is

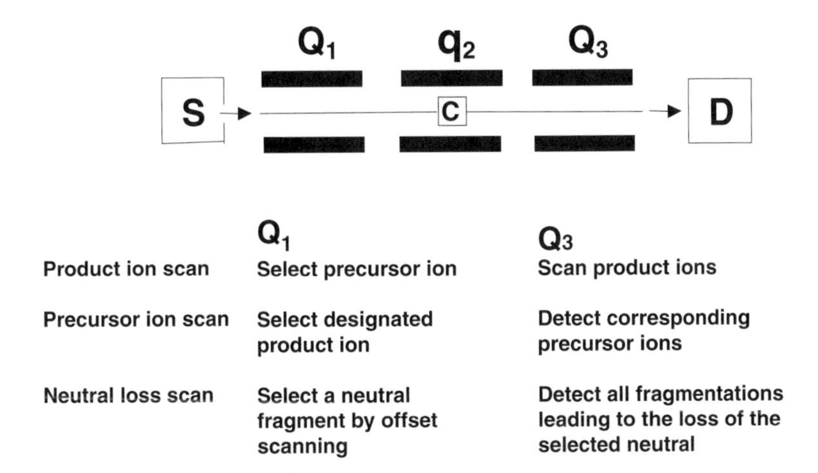

FIGURE 2.19 Collision-induced dissociation in triple-quadrupole ($Q_1q_2Q_3$) analyzer. S = ion source; D = detector; C = collision cell.

highly selective because only those ions are activated that absorb at the wavelength of the irradiating laser. Trapped ion instruments provide a desirable environment for such experiments because ions can be trapped for prolonged periods of time in regions irradiated by the light (reviewed in Reference 70). A tandem TOF instrument was developed in which a linear TOF analyzer was used for primary mass separation and precursor ion selection. The selected ions were irradiated with a high energy laser to induce photodissociation, and the product ions were analyzed with a second, oa-reflectron-TOF analyzer.[173]

Ions may gain internal energy, and subsequently decompose, upon colliding with a target surface placed perpendicular to, or at an angle with, their paths, or even when passing through very narrow channels in which they bounce off the channel surfaces. Major variables include the physical nature and composition of the surface. Surface-induced dissociation (SID) is efficient because even low collision energies leads to high internal energies with narrow energy distributions. The structure, energetics, and fragmentation mechanism of protonated peptides have been studied with SID (reviewed in Reference 174).

2.5.1.5 In-Source Fragmentation

Increasing the "cone voltage" in ESI sources often results in significant fragmentation of the analytes, providing both protonated molecules and intense fragment ions (with the proper isotopic clusters) and a reduction of the charge state for multiply charged species. Two important advantages of this method are the possibility to obtain MS/MS-like spectra without a tandem analyzer, e.g., with a single quadrupole or oa-TOF instruments, and to provide intense fragments for subsequent second-generation CID in tandem instruments. Other advantages include simple operation (often programmable), high sensitivity, and no loss of resolution. However, the technique is much less specific than CID and is not readily applicable to mixtures.

2.5.2 Tandem-in-Space (Transmission) Techniques

The four major operating modes of MS/MS are described below for the triple quadrupole, QqQ configuration (Figure 2.19) that is used frequently because of its relative simplicity and versatility (reviewed in Reference 169). A series of novel pictograms proposed for the visual documentation of tandem mass analyzers and MS/MS scan modes has not yet been generally accepted.[175] Relevant aspects of the preparation of biological samples for coupled LC/MS/MS analysis have been investigated.[176]

2.5.2.1 Product-Ion Scanning Mode

During customary operation to obtain conventional mass spectra in a QqQ instrument, the first quadrupole is scanned and the other two quadrupoles are operated in the wide band mode to pass all masses. In the *product (daughter) mode,* a selected ion is focused through the first quadrupole and reacted with the neutral atoms or molecules of the target gas in the second quadrupole that serves as the collision cell. The ions formed by CID are separated according to their *m/z* in the third quadrupole. The resulting mass spectrum shows the product ions derived from the selected precursor ion. This is the most common type of MS/MS experiment, primarily used to learn about the fragmentation and structure of the selected precursor ion.

2.5.2.2 Precursor Scanning Mode

The *precursor (parent) mode* of operation is the opposite of the product scanning mode. All ions of the analyte are mass separated in the first quadrupole and are allowed to undergo CID in the second quadrupole followed by the passing of only one preselected mass in the third quadrupole. The result of this is a spectrum of parent ions which dissociate to give a specific product ion. An application is that, in a complex mixture of unknowns, particular product ions characteristic of a class of compounds can be used to identify components in the mixture that belong to the same, preselected compound class.

2.5.2.3 Constant Neutral Loss Scanning Mode

In this mode, both the first and third quadrupoles are scanned at the same rate but with a chosen mass offset. The results give information on specific neutral mass losses that occur during the CID process in the collision cell of the second quadrupole. When a particular neutral loss is known to occur in a class of compounds, this type of MS/MS experiment will identify components in mixtures belonging to that compound class.

2.5.2.4 Selected Reaction Monitoring

In the selected reaction monitoring (SRM) mode, a preselected precursor ion is introduced into the collision area followed by the selection of one (or more) specific product ion for detection. The other products of the CID process (and also the baseline noise) are rejected. This MS/MS acquisition mode is analogous to selected ion monitoring in single analyzer instruments (Section 2.6.3). The use of SRM, when both precursor and product masses are known to be characteristic of the analyte, permits quantification of target compounds with high sensitivity and concurrent high specificity.

2.5.2.5 Post-Source Decay

Although MALDI is considered to be a "soft" ionization technique that mainly produces intact protonated molecules, accelerated ions often fragment during their travel through the field-free region of a TOF analyzer prior to detection. For example, both neutral molecule losses (water, ammonia) and random cleavages at peptide bonds occur routinely in the metastable ion fragmentation of peptides and proteins. The fragmentation mechanisms include not only metastable ion decay but also both low- and high-energy collision-induced dissociation. Products of post-source metastable fragmentation cannot be distinguished with linear TOF analyzers because both precursor and product ions move with the same velocity and, because of the rules concerning conservation of momentum, arrive simultaneously at the detector. In contrast, reflector TOF analyzers separate precursor and post-source decay (PSD) product ions by their difference in kinetic energy in the ion mirror. Despite the time and practice needed for the interpretation of PSD spectra, MALDI-TOF with PSD has become a useful complementary approach for obtaining primary structural information on peptides (reviewed in Reference 177).

An alternative to PSD is to obtain mass spectra of *ion-source decay products*. Although prompt ion fragmentation does not occur in MALDI sources to any significant degree, a short time (<100 ns) delay may be applied between ion formation and ion extraction, permitting fragmentation of the analytes into smaller ions and neutrals to occur in the ion source. The extracted ions may be analyzed using a linear or reflectron TOF. There are structurally significant differences between in-source decay and PSD spectra. Useful data on unknowns can often be obtained in a few minutes.[178]

The high sensitivity associated with MALDI has been combined with the comprehensive fragmentation provided by high-energy (keV) CID using a TOF-TOF mass spectrometer. The first mass analyzer is used as a timed ion selector for the isolation of a precursor. The second stage sector includes the collision cell and ion optics for the transfer of ions between the two analyzers. The final TOF analyzer has a reflectron to provide high-resolution spectra of the fragment ions. The versatility of the system has been illustrated by determining peptide sequences using different ionization, collisional cell, and other operational conditions.[112]

2.5.3 Tandem-in-Time (Trapped-Ion) Techniques

Significant advantages of trapped-ion analyzers in MS/MS experiments include the ability to readily conduct MS^n experiments, the simplicity of ion traps, and the high resolution and other unique features of FTICRMS. However, a major limitation is that tandem-in-time instruments can be operated only in the product ion mode.

In quadrupole ion traps (IT), MS/MS is carried out in time rather than in space. There are four basic steps: (1) selection of the mass of the precursor (parent) ion; (2) application of the excitiation voltage to the end cap electrodes of the IT and injection of a collision gas; (3) dissociation of the energized ions to form product ions; and (4) mass analysis of the product ions. In IT, the ions momentarily accelerated to higher kinetic energies in the presence of the collision gas are decelerated by the excitation waveform. This acceleration/deceleration process, which lasts for ~1 ms, leads to multiple collisions and other low-energy processes that generally produce spectra that contain only a limited number of fragment-ion types. This may necessitate the carrying out of additional CID experiments to generate a second generation of product ions from the initial set of fragments, e.g., MS^3. Fortunately, such experiments are relatively easy tasks to accomplish with IT analyzers. As mentioned already, a significant limitation is that this is the only type of MS/MS experiment that can be done with IT analyzers. Also, ion intensities may be lower than those obtainable with the QqQ and Q-TOF geometries.

In FTICR instruments, precursors are selected by ejecting all other ions from the cell by applying an rf pulse. The isolated precursors are next accelerated to a higher kinetic energy with another rf pulse and are trapped for several milliseconds to allow them to collide with the target gas. The resulting CID takes place under low-energy conditions and involves multiple collisions. The product ions are next excited into coherent motion with a further rf pulse and finally the ion image currents are detected for subsequent FT. The advantages include capability to provide ion cooling followed by reacceleration and redetection, to analyze product ions at very high resolution, and combination with other methods of product ion formation such as photodissociation.

2.6 IDENTIFICATION AND QUANTIFICATION

2.6.1 Mass Spectral Libraries and Bioinformatics

Unknown analytes of molecular mass <2000 Da can often be identified by searching one of several available mass spectral libraries.[179] The 1998 release of the NIST/EPA/NIH mass spectral database for Windows™ (National Institute of Standards and Technology, Gaithersburg, MD) contains ~130,000 spectra of ~108,000 compounds and several mass spectral deconvolution and interpretation tools. An alternative commercial library contains ~150,000 spectra of ~112,000 compounds for which there is a Peak Index of ~400,000 entries.[180] The "Eight Peak Index of Mass Spectra" (Royal Society of Chemistry, Cambridge, 1991) of ~80,000 spectra of ~65,000 compounds is classified according to several keywords, e.g., most intense mass and second most intense mass. In addition, there are several specialized libraries, e.g., steroids, pollutants. All the above libraries are composed of only EI spectra for both historical reasons and the reproducibility of EI spectra from instrument to instrument. The obvious drawback of such libraries is the limited number of useful spectra of labile, biologically important, compounds.

There are a number of commercial front-end programs to link libraries with data files from mass spectrometer data systems. Most instrument manufacturers offer software of varying levels of sophistication designed for their products. There are several approaches to searching databases, e.g., forward vs. reverse searching.[181–183] In critical evaluations of mass spectral library search results[184] and of methods to build comprehensive reference libraries, the expert evaluation of submitted spectra has been listed as a particularly challenging task.[184]

A new multidisciplinary field, bioinformatics, is now being developed rapidly for the identification of large, polar, biologically active molecules, including biopolymers. Analytical information, including *inter alia* mass spectral, chromatographic, electrophoretic, etc., information must be integrated with other information in a wide variety of available molecular biological databases. Of particular importance is the generation of protein sequence databases against which peptide sequences can be searched to determine the identity of an isolated protein. Relatively short sequences are required to provide unequivocal identifications. Several relevant approaches are summarized in Section 6.1.1.

2.6.2 Accurate Mass Measurement

It was demonstrated in 1954 that the measurement of the exact mass of ions can be used to determine their elemental compositions within defined margins of error and, therefore, establish a set of probable empirical formulae for the analyte. The approach is based on permutation and matching of the unknown with the known noninteger atomic mass values of the constituent elements.[185] Accurate mass determinations may be used both to confirm the molecular composition of a tentatively identified analyte and to identify unknown substances.

Accurate mass measurement using magnetic sector analyzers at high resolution has been used routinely for the identification of low molecular mass analytes for decades. The measurements are based on the simultaneous introduction of the analyte and a calibration compound with known mass; the mass defect must be considered in mass scale calibrations.[186] Perfluorinated hydrocarbons (PFK) have been the most frequently used calibration compounds because of both the fragmentation patterns they generate and, especially because their masses are slightly lower than integer values, while the masses of most organic compounds are greater than integral values. This difference in mass means that in a magnetic sector instrument, operating at high resolution, interference between the calibrant and analyte (and instrument background ions) is rare. Both manual techniques ("peak matching") and computer-controlled techniques have been developed for the comparison of the masses of standards with analytes.

Another area where knowledge of the exact mass of an ion is important is in selected ion monitoring (Section 2.6.3), using double-focusing magnetic sector analyzers at high resolution. The precision afforded by the technique has been used in, most notably, the quantification of the carcinogenic dioxin and related compounds in environmental samples (Section 4.3.3.5) and in other applications.[187]

Although there are limits to the various approaches for accurate mass characterization of large biomolecules,[188] there have been developments of techniques for accurate mass measurement (<10 ppm) using other types of analyzers (see below). These have led to expanding applications in drug metabolite and carcinogen studies[189] and are currently acquiring significance, particularly in combination with database searching, in protein identification strategies[190–192] (Section 6.1.1).

Accurate mass determination is possible in certain cases with analyzers of low resolution, such as single quadrupoles with continuous flow-FAB, e.g., vinca alkaloids;[193] ESI, e.g., polar molecules;[194] and MALDI-linear TOF.[195] However, the low resolution of these instruments means that there is a need for extensive purification of samples to remove possible isobaric interferences thereby limiting the utility of the method.

The linearity of the mass scale, over a wide dynamic range, and the routinely available 5,000 to 15,000 resolution in oa-TOF analyzers has led to their use in the determination of accurate masses. In one example, accurate mass determinations were made of CID product ions for the structural elucidation of drug metabolites using a hybrid analyzer configuration in which the third quadrupole of an ESI-QqQ instrument was replaced by an oa-TOF analyzer. External mass calibration in the 130 to 830 Da range was made with ions from CsI and an octapeptide. Masses of CID product ions could be determined with an accuracy of ~5 ppm even in directly infused crude plasma or bile extract, provided the precursor ion was free of isobaric interferences.[196] An alternative approach uses the previously determined mass of the parent ion as a lock-mass (internal calibrant) for the determination of the accurate masses of CID product ions. Although this method compounds errors, it is useful in peptide sequence determinations where the limited number of possible combinations of amino acids allows for some relaxation of mass accuracy while maintaining the ability to determine sequence.

In another approach, a Z-spray™-coupled oa-TOF analyzer was used for the accurate mass determination of samples eluting from HPLC by taking advantage of a previously developed multiplexed ESI interface[43] which permitted the use of two separate electrosprays for introducing the reference and analyte independently. This technique eliminated the post-column introduction of the calibration standards and also provided lock masses in nozzle-skimmer or cone voltage-based in-source CID operations (leucine enkephalin and terfenadine were used for positive ESI and a small peptide for negative ESI). In proof of concept experiments, masses of protonated and deprotonated molecules and fragment ions of several known and experimental drugs (to ~400 Da) and eluting impurities were determined. Mass deviations were within a few mDa (<5 ppm) from the theoretical values.[44]

In FTICRMS, the determination of accurate masses depends on the precise measurement of ion orbital frequencies. Recent improvements in technology to (a) maintain the homogeneity of the static magnetic field, (b) improve the symmetry of the trapping potentials within the cell, and (c) reduce the pressure in the ICR cell have resulted in improved sustained coherent ion orbits. These have afforded extended times (seconds) for accurate frequency measurements for groups of only 100 to 1000 ions stored in the cell. In addition to the extremely high resolution attainable with FTICRMS, up to 10^6, an additional major advantage is the ability to provide exact masses for ions produced in several stages of CID. This unique feature of FTICRMS, not available with any other type of analyzer, opens up new possibilities in MS^{pm} (*pro re nata,* latin for "as needed") as an alternative to MS^n type experiments for the study of the structure of ions produced by CID, including low-intensity and/or isotope-resolved product ions.[197] In a proof of principle demonstration, molecular masses of peptides and proteins up to 5700 Da were determined to 0.001% (10 ppm) without the need of internal calibration standards.[104]

2.6.3 Selected Ion and Reaction Monitoring

2.6.3.1 Selected Ion Monitoring

To obtain the "full" mass spectrum of a compound, the abundance of every *m/z* value is determined within the selected mass range. This is accomplished in most mass spectrometers by focusing one mass after another onto the detector using a process called *scanning*. Most of the scan time, however, is spent recording the intensities of masses containing little or no useful information. For example, if the mass range 100 to 600 is scanned linearly in 5 seconds, only 0.01 second is spent recording data for each mass. Selected ion monitoring (SIM) (terms like *multiple ion detection* and *selected ion recording* are obsolete) is a technique that makes the MS a specific ion detector by recording, in rapid repetitive succession (e.g., 10 to 20 times/s), only one or more preselected *m/z* values; all other ions present are excluded. Detection limits are increased by a factor of 10^2 to 10^3 compared to those for the same ions in the full-scanning mode.

The mechanics of SIM operation are simple in quadrupole instruments because the low voltages applied to the rods can be switched rapidly to focus the selected ions onto the detector. Any number of masses may be monitored in any desired order. The *dwell time,* i.e., the period of time spent to integrate the ion current of the selected mass, is easily adjustable and may be set individually depending whether the accuracy of ion current measurement (requiring long dwelling times) or chromatographic profile representation (>10 data points) for a chromatographic peak is to be optimized for a given mass. Dwell times are typically in the 20 to 100 ms range but may be as high as 1 s.

In magnetic instruments SIM operation may be based on changing either the accelerating voltage or the magnetic field strength. Laminated magnets permit scanning the magnetic field at constant accelerating voltage, thus the entire mass range can be covered. Disadvantages include hysteresis effects preventing repetitive reestablishment of a selected field strength, drifts in mass scale calibration, and available switching rates significantly slower than in the electric fields of the quadrupole instruments.

SIM profiles are similar in appearance to chromatograms, except there are several traces, each corresponding to the monitoring of a preselected mass; thus, an ion profile may be considered as an "ion chromatogram." The identity of the compound analyzed is confirmed by comparing the time taken for the appearance of the peaks with the chromatographic retention time of the authentic compound and also by checking the ratios of the intensities of the selected fragments against those of an authentic sample.

Although SIM may be used to confirm the presence of suspected analytes at trace quantities, quantitative analysis of known constituents in mixtures is its most important application. Components present in fmole amounts may be quantified by such measurements.

It is obvious that the greater the proportion of the total ion current of the analyte that is associated with the ion being monitored, the lower will be the detection limit. Thus, CI, which results in high abundance of the $[M+H]^+$ ions or other ions in the molecular mass area, is especially useful in SIM. In addition, CI is highly compatible with GC, the technique most frequently used for the initial separation of the analytes in mixtures. It is fortunate that the derivatization techniques often necessary to optimize GC separation frequently yield compounds with CI spectra containing ions of high abundance as well as high mass. The latter increases specificity by reducing the likelihood of encountering background ions in the MS or from sources such as column bleed. In addition, the reduced fragmentation in CI lessens the variety of ionic species present and therefore the possibility of isobaric interference. SIM of ions formed by negative CI may provide exceptional sensitivity and specificity for analytes with high electron affinity. There is a wide selection of derivatives for most compound classes suitable for generating appropriate ions for both positive and negative SIM. Stable isotope-labeled analogs of the desired analyte are nearly ideal internal standards for SIM because of their similar extraction and chromatographic properties.

In routine work only two ions are usually monitored, one for the sample and one for the internal standard. The disadvantage that SIM is intrinsically less specific than full scanning can be minimized by the judicious selection of both the chosen *m/z* values to be monitored and the number of ions to be monitored. It is essential to establish whether or not the ions selected for monitoring are also present in the matrix (often happens when body fluids are analyzed). Increasing the number of ions increases specificity but reduces sensitivity. Specificity may also be improved by checking appropriate ion ratios. Isomeric compounds can be analyzed only if they are separated chromatographically. SIM in the high-resolution mode greatly improves the specificity of the technique compared with low-resolution operation. The monitoring of only those ions of a selected exact mass, that correspond to a particular elemental composition, is of importance in environmental toxicology (particularly in litigated cases) where chromatographic separation of closely related interfering compounds may be difficult, such as with some dioxins. SIM has been commonly associated with

GC-MS,[198] but the technique is applicable to most means of sample introduction and ionization. It is one of the most frequently used techniques in mass spectrometry.

Quantification using SIM does have pitfalls.[199] Internal standards (i.s.) are used in most applications to compensate for a variety of error sources during both sample preparation, e.g., recovery losses, and the measurement process. Internal standards should be added at as early a stage in the assay as possible. The selection of an appropriate i.s. is of considerable importance.[200] The structure, chemical, and physical properties of the i.s. should be as close to those of the analyte as possible. Structural homologs, isomers, and compounds from the same chemical class as the analyte are widely used.

In analytical mass spectrometry, stable isotopes have been used increasingly as i.s. Stable-isotope enriched compounds, preferably containing three to four heavy atoms, to reduce required corrections from naturally occurring heavy isotopes, are nearly ideal i.s. for the detection and elimination of procedural errors. Such i.s. are structurally the same as the target analytes and behave identically during all phases of sample preparation including derivatization. The mass spectrum of a stable isotope analog is usually identical to that of the analyte but shifted by the respective number of isotopic atoms. Stable isotope enriched standards have the same partition coefficients, rate constants, and chromatographic properties.

An interesting additional advantage is the "carrier effect." Because procedural losses tend to be a constant amount, stable isotope labeled species can also be added in a large excess to act as a carrier; e.g., when the procedural loss is 50 ng, the addition of 900 ng labeled compound to 100 ng analyte will reduce the loss from 50% to 5%. It has been claimed that when a 100- or 1000-fold excess of a labeled i.s. is added, it will act as a carrier to pass small quantities of unlabeled analytes through the system, thereby reducing losses of trace quantities of analytes due to adsorption. However, in this circumstance, a different i.s. of comparable concentration to the analyte is preferable because a large excess of an i.s. used as a carrier compound makes quantification unreliable. The techniques of quantitative MS have been reviewed extensively.[200] As described in subsequent chapters, stable isotope labeled i.s. have been used in dozens of methods for the quantification of carcinogens as well as chemotherapeutic agents in biological fluids and tissues.

2.6.3.2 Selected Reaction Monitoring

This data acquisition mode is similar to that of SIM, however it is carried out in the MS/MS mode in tandem instruments. A selected precursor ion is allowed to undergo CID, but only the product of a characteristic, selected transformation, or reaction is detected. The reaction to be monitored must be selected from the established product ion spectrum of the precursor. It is possible to monitor multiple products from the same precursor ion.

The significantly increased selectivity of SRM, with respect to SIM, is due to the facts that the *m/z* of *both* the precursor and product ions are selected and, most importantly, that the product ions can only arise from the selected precursor ion. The second major advantage of SRM is that there is a significant gain in sensitivity because the increased signal-to-noise ratio inherent in tandem instruments more than compensates for the losses occurring during ion transmission. Quantification to attomole levels has been achieved[201] and the boundaries of quantitative LC-MS/MS techniques have been explored recently.[202] With the proliferation of tandem mass spectrometers, the relaxed sample preparation requirements of SRM are utilized increasingly for the specific, sensitive, and rapid monitoring of target analytes in complex mixtures. A number of applications are described and cited in Chapters 4, 5, and 6.

2.6.4 Stable Isotope Dilution

The fact that many elements in their natural states contain isotopes results in the appearance of characteristic *isotope patterns* in mass spectra.[203] The determination of minute changes in the natural isotope content of substances has applications in widely diverse areas, ranging from the atomic weapons industry, geochronology, the study of the mechanisms of chemical and biochemical reactions caused by isotope kinetic effects to direct applications in clinical chemistry, pharmacokinetics, and toxicology (reviewed in Reference 204).

The principle of the stable isotope dilution technique is simple: the concentration of the analyte is determined from the change produced in its natural isotopic composition by the addition of a known quantity of the same compound, the isotopic composition of which has been artificially altered by incorporation of stable isotopes. There are only a few experimental steps. First, a known quantity of the sample, which includes the analyte, is equilibrated with a measured quantity of the isotopically enriched analyte, the *spike*. Next, the analyte and spike are removed from the matrix by an appropriate extraction procedure, derivatized if necessary, and the altered isotope ratio is determined by MS. Precision is significantly increased by measuring only the intensities of the relevant ions at the expense of the advantages of scanning the entire mass spectrum. In contrast to most quantitative analytical techniques that make relative measurements, i.e., calibrating response against standards, isotope dilution is an absolute method, in the sense that only ratios within a sample need to be determined and not concentrations. This means that extraction need not be quantitative and irreproducible losses during sample preparation, e.g., inadvertent and possibly unobserved spilling can be tolerated.

Because of the abundant and inexpensive availability of highly enriched nuclides of deuterium, carbon, nitrogen, and oxygen, there is an increasing number of D-, ^{13}C-, ^{15}N-, and ^{18}O-enriched synthesized organic and biochemical compounds available. Although usually more expensive than their deuterium counterparts, ^{13}C-labeled compounds are preferable because they are intrinsically stable to a wide range of chemical conditions and undergo isotopic exchange less frequently than D- or ^{18}O-enriched compounds. Large kinetic effects may occur when deuterium is substituted for protium because the rate of cleavage of C-^{1}H bonds is ~15 times slower than that of C-D bonds. Because of the low natural abundances of the isotopes, labeled molecules are identified by the number and position(s) of the incorporated heavy isotopes. *Positional isotopomers* have identical global isotopic composition except for the positions of the heavy atoms in the molecule, e.g., [1-^{13}C]glucose and [2-^{13}C]glucose. *Mass isotopomers* differ by the number of heavy atoms in their molecules, e.g., [$^{14}N_2$]urea, [$^{15}N_2$]urea, and [^{14}N-^{15}N]urea; these compounds have different molecular masses. Molecules with different heavy atoms may have the same isotopomeric mass, e.g., [^{13}C]urea and [^{15}N]urea. Mass measuring accuracy of 0.1 Da is adequate for most applications, however, some pairs require much higher resolution, e.g., [^{13}C-^{15}N]urea and [$^{15}N_2$]urea. Mass spectra provide information on the distribution of the mass isotopomers and the enrichment (number of isotope atoms/mole, analogous to specific activity in radioisotope determinations). Various aspects of the mass isotopomer distribution analysis (MIDA) have been reviewed with reference to the study of fatty acid and cholesterol biosynthesis.[205]

Single-focusing magnetic sector instruments have been used predominantly for accurate isotope ratio determinations, preferably with Faraday-type collectors to obtain *flat top* peaks that provide stable and linear responses over a wide dynamic range with a precision of 0.001%. Disadvantages are the large amount of sample required (up to several mg) and the fact that the positional identity of the incorporated stable isotope is lost because only total isotopic enrichment is measured unless one can identify a specific fragment ion. Compound-specific isotope analysis of GC effluents may be accomplished indirectly, after the controlled combustion of the analytes to H_2, CO_2, or N_2, followed by the determination of isotopic ratios in the gaseous form (reviewed in References 206, 207).

Provided that the level of precision described above is not essential, isotopic ratios in intact molecular as well as fragment ions can be determined at low concentrations in mixtures of compounds separated by GC-MS or LC-MS. High chromatographic resolution is preferred to avoid affecting the isotopic ratios with co-eluting unknowns (e.g., from urine or tissue extracts) that have ions in the same mass selected for SIM. The isotopic abundances of the ions of interest are measured by SIM of the relevant isotope ions as the analyte elutes from the chromatograph. Using this method it is possible to measure isotope ratios with an accuracy of 0.1% or better using only fractional ng sample quantities; in addition, positional identity of the isotope is retained. When derivatization is needed prior to GC-MS analysis, some common derivatizing groups, e.g., those including Si or B atoms, should be avoided to prevent interferences caused by their isotopes. Also, fully ^{13}C-labeled or ^{13}C-depleted derivatization groups should be used if possible. A variety of methodology-related aspects of mass isotopomer analysis has been reviewed in connection with applications in nutrition research[208] and studies of metabolic kinetics.[209]

Stable isotope techniques have been used extensively in metabolism studies, both for exogenous compounds (e.g., drugs and pollutants) and endogenous constituents (e.g., steroids). Stable isotopes can normally be used safely in clinical studies even involving infants and pregnant women as long as the toxicity, purity, and sterility requirements associated with the particular compound or drug in its original form are met. The substance under study is labeled with a stable isotope at a particular site in the molecule which does not affect and is not affected by the metabolic process under investigation. The eventual products of the labeled compound after equilibration and metabolism will have the same isotopes in the same ratio as the substrate, thereby providing unequivocal evidence of origin. In the drug metabolism area, stable isotopes have been used to study bioavailability, chronic administration, and the effects of drug enantiomers.

A variety of techniques have been developed for the estimation of error propagations and isotope enrichments, including elements analyzed by ICPMS[210] and as tracers of the same compound, using isotopomer distributions of fragment ions for *in vivo* human studies.[211] Recent advances in the analysis of large biomolecules revealed the expected complexities of their isotopic distributions, i.e., substantial broadening of the isotopic clusters of CHNO-containing ions, necessitating the development of new algorithms for practical calculations. Several polynomial methods have been developed (briefly reviewed in Reference 212). A fast and accurate new algorithm, which uses FT methods to calculate molecular isotope distributions starting from molecular formulas, was tested by calculating the isotope distribution of a >123 kDa, 400 bases long, single-stranded DNA oligomer, $C_{3900}H_{4902}N_{1500}O_{2401}P_{400}$. Calculations were completed in 1 s with standard deviation of 0.15 ppm.[212]

2.7 RESOURCES

Particulars on virtually all material presented in this chapter can be found in one or more of the recommended books or journals listed below.

There are five monthly journals (in English) devoted solely to mass spectrometry: *Journal of the American Society for Mass Spectrometry* (the official journal of the American Society for Mass Spectrometry), *Journal of Mass Spectrometry, Rapid Communications in Mass Spectrometry, The International Journal of Mass Spectrometry,* and *European Journal of Mass Spectrometry.* Original papers on analytical techniques appear in virtually every issue of *Analytical Chemistry.* The *Journal of Chromatography* (both sections A and B) frequently publishes methodologically oriented papers involving combined GC/MS and LC/MS techniques. With the upsurge of interest in using CE for the separation of complex mixtures prior to MS analysis, *Electrophoresis* has been devoting an increasing amount of space to relevant technological aspects.

The quarterly journal, *Mass Spectrometry Reviews*, publishes critical reviews of current research topics. A comprehensive list of reference titles is available monthly in *The Mass Spectrometry Bulletin* and up-to-date references are provided by *CA Selects Mass Spectrometry.* The biennial reviews of mass spectrometry in *Analytical Chemistry* are excellent sources of information on advances in instrumentation and applications. A directory of companies that supply mass spectrometers, components, and services is published annually by *Rapid Communications in Mass Spectrometry.*[213] As expected, the Internet has become a major resource for all aspects of mass spectrometry.[214,215]

Of the more than 20 societies devoted to mass spectrometry worldwide, The American Society for Mass Spectrometry (ASMS) is the largest and most active. ASMS sponsors a popular and well-attended yearly meeting on virtually all aspects of mass spectrometry. In addition, conferences and workshops are frequently presented on specific subjects. Regularly scheduled international meetings are an excellent source of major reviews of selected subjects both for technological advances and applications.

2.7.1 Recommended Books

Note: A very long list of books on virtually all aspects of mass spectrometry is available on the Internet: http://www.amazon.com. This site also lists books about to be published.

Back to Basics, Micromass Co., http://www.micromass.co.uk/basics. A useful review of basic principles related to commercial instruments.

Busch, K.L., Glish, G.L., and McLuckey, S.A., *Mass Spectrometry/Mass Spectrometry,* VHC, New York, 1988.

Cotter, R.J., *Time-of-Flight Mass Spectrometry,* American Chemical Society, Washington, D.C., 1997.

De Hoffmann, E., Charette, J., and Stroobant, V., *Mass Spectrometry: Principles and Applications,* Wiley, Chichester, UK, 1996.

Johnstone, R.A.W. and Rose, M.E., *Mass Spectrometry for Chemists and Biochemists,* 2nd ed., Cambridge University Press, Cambridge, 1996.

McCloskey, J.A., Ed., *Mass Spectrometry,* in *Methods in Enzymology,* vol. 193, Academic Press, New York, 1990.

Niessen, W.M.A., *Liquid Chromatography-Mass Spectrometry,* Dekker, New York, 1998.

Watson, J.T., *Introduction to Mass Spectrometry,* 3rd ed., Lippincott-Raven, Philadelphia, 1997.

Willoughby, R., Sheehan, E., and Mitrovich, S., *A Global View of LC/MS,* Global View Publishing, Pittsburgh, 1998.

The ASMS Education Committee has been publishing a series of "Classic Works in Mass Spectrometry." Three volumes have been published to date:

Vol. 1. Klaus Biemann, *"Mass Spectrometry: Organic Chemical Applications."*

Vol. 2. John Beynon, *"Mass Spectrometry and Its Application to Organic Chemistry."*

Vol. 3. John Roboz, *"Introduction to Mass Spectrometry: Instrumentation and Techniques."*

REFERENCES

1. Busch, K.L., Electron ionization, up close and personal, *Spectroscopy,* 10, 39–42, 1995.
2. Lango, J., Szepes, I., Csaszar, P., and Innorta, G., Studies on unimolecular decomposition processes of organometallic ions, *J. Organomet. Chem.,* 269, 133–145, 1984.
3. McLafferty, F.W., *Interpretation of Mass Spectra.,* 3rd ed., University Science Books, Mill Valley, CA, 1980.
4. Busch, K., Chemical ionization mass spectrometry, Part I, *Spectroscopy,* 11, 40–42, 1996.
5. Busch, K.L., Chemical ionization mass spectrometry, Part II, *Spectroscopy,* 11, 28–30, 1996.
6. Srinivas, R., Devi, A.R., and Rao, G.K.V., Tetramethylsilane CI of aromatic compounds, *Rapid. Commun. Mass Spectrom.,* 10, 12–15, 1996.
7. Budzikiewicz, H., NCI of organic compounds, *Mass Spectrom. Rev.,* 5, 345–380, 1986.
8. Laramee, J.A., Arbogast, B.C., and Deinzer, M.L., ECNI of 1,2,3,4-tetrachlorodibenzo-p-dioxin, *Anal. Chem.,* 58, 2907–2912, 1990.
9. Wong, B. and Castellanos, M., Enantioselective measurement of the Candida metabolite D-arabinitol in human serum using multidimensional gas chromatography and a new chiral phase, *J. Chromatogr.,* 495, 21–30, 1989.
10. Harrison, A.G., *Chemical ionization mass spectrometry,* 2nd ed., CRC Press, Boca Raton, FL, 1992.
11. Bruins, A.P., MS with ion sources operating at atmospheric pressure, *Mass Spectrom. Rev.,* 10, 53–78, 1991.
12. Metzger, J., Stevanovic, S., Brunjes, J., Wiesmuller, K., and Jung, G., Electrospray mass spectrometry and multiple sequence analysis of synthetic peptide libraries, *Methods,* 6, 425–431, 1994.

13. Banks, J.F. and Whitehouse, C.M., Electrospray ionization mass spectrometry, *Methods Enzymol.*, 270, 486–519, 1996.
14. Smyth, W.F., The use of electrospray mass spectrometry in the detection and determination of molecules of biological significance, *Trends Anal. Chem.*, 18, 335–345, 1999.
15. Gaskell, S.J., Electrospray: principles and practice, *J. Mass Spectrom.*, 32, 677–688, 1997.
16. Dass, C., Recent developments and applications of high-performance chromatography electrospray ionization mass spectrometry, *Current Organic. Chem.*, 3, 193–209, 1999.
17. Fenn, J., Ion formation from charged droplets: roles of geometry, energy, and time, *J. Am. Soc. Mass Spectrom.*, 4, 524–535, 1993.
18. Kebarle, P., Ho, Y., and Cole, R.B., Eds., 1. On the mechanism of electrospray mass spectrometry, in *Electrospray Ionization Mass Spectrometry: Fundamentals Instrumentation & Applications*, John Wiley & Sons, New York, 3–63, 1997.
19. Bruins, A.P., Mechanistic aspects of electrospray ionization, *J. Chromatogr. A*, 794, 345–357, 1998.
20. Kebarle, P., A brief overview of the present status of the mechanisms involved in electrospray mass spectrometry, *J. Mass Spectrom.*, 804–817, 2000.
21. Bruins, A.P., ESI source design and dynamic range considerations, in *Electrospray Ionization Mass Spectrometry*, Cole, R.B., Ed., John Wiley & Sons, New York, 107–136, 1997.
22. Voyksner, R.D., Combining liquid chromatography with electrospray mass spectrometry, in *Electrospray Ionization Mass Spectrometry*, Cole, R.B., Ed., John Wiley & Sons, New York, 323–342, 1997.
23. Severs, J.C. and Smith, R.D., Capillary electrophoresis electrospray ionization mass spectrometry, in *Electrospray Ionization Mass Spectrometry*, Cole, R.B., Ed., John Wiley & Sons, New York, 343–382, 1997.
24. Barnidge, D.R., Nilsson, S., and Markides, K.E., A design for low-flow sheatless electrospray emitters, *Anal. Chem.*, 71, 4115–4118, 1999.
25. Wang, G. and Cole, R., Effect of solution ionic strength on analyte charge state distributions in positive and negative ion electrospray mass spectrometry, *Anal. Chem.*, 661, 3702–3708, 1994.
26. Van Berkel, G. and Zhou, F., Electrospray as a controlled-current electrolytic cell: electrochemical ionization of neutral analytes for detection by electrospray mass spectometry, *Anal. Chem.*, 67, 3958–3964, 1995.
27. Berkel, G., The electrolytic nature of electrospray, in *Electrospray Ionization Mass Spectrometry*, Cole, R.B., Ed., John Wiley & Sons, New York, Chap. 2, 65–103, 1997.
28. Mann, M., Meng, C., and Fenn, J., Interpreting mass spectra of multiply charged ions, *Anal. Chem.*, 61, 1702–1708, 1989.
29. Hagen, J. and Monnig, C., Method for estimating molecular mass from electrospray spectra, *Anal. Chem.*, 66, 1877–1883, 1994.
30. Horn, D.M., Zubarev, R.A., and McLafferty, F.W., Automated reduction and interpretation of high resolution electrospray mass spectra of large molecules, *J. Am. Soc. Mass Spectrom.*, 11, 320–332, 2000.
31. Lorenz, S.A., Maziarz, E.P., and Wood, T.D., Electrospray ionization Fourier transform mass spectrometry of macromolecules: the first decade, *Applied Spectroscopy*, 53, 18A–36A, 1999.
32. Wilm, M. and Mann, M., Analytical properties of the nanoelectrospray ion source, *Anal. Chem.*, 68, 1–8, 1996.
33. Juraschek, R., Dulcks, T., and Karas, M., Nanoelectrospray–more than just a minimized-flow electrospray ionization source, *J. Am. Soc. Mass Spectrom.*, 10, 300–308, 1999.
34. Wilm, M., Shevchenko, A., Houthaeve, T., Brelt, S., Schwelgerer, L., Fotsls, T., and Mann, M., Femtomole sequencing of proteins from polyacrylamide gels by nano-electrospray mass spectrometry, *Nature*, 379, 466–469, 1996.
35. Xue, G., Foret, F., Dunayevskiy, Y.M., Zavracky, P.M., McGruer, N.E., and Karger, B.L., Multichannel microchip electrospray mass spectrometry, *Anal. Chem.*, 69, 426–430, 1997.
36. Zhang, B., Liu, H., Karger, B.L., and Foret, F., Microfabricated devices for capillary electrophoresis-electrospray mass spectrometry, *Anal. Chem.*, 71, 3258–3264, 1999.
37. Zhang, B., Foret, F., and Karger, B.L., A microdevice with integrated liquid junction for facile peptide and protein analysis by capillary electrophoresis/electrospray mass spectrometry, *Anal. Chem.*, 72, 1015–1022, 2000.
38. Cai, J. and Henion, J., Capillary electrophoresis-mass spectrometry, *J. Chromatogr. A*, 703, 667–692, 1995.

39. Figeys, D., Ning, Y., and Aebersold, R., A microfabricated device for rapid protein identification by microelectrospray ion trap mass spectrometry, *Anal. Chem.*, 69, 3152–3160, 1997.
40. Figeys, D., Gygi, S.P., McKinnon, G., and Aebersold, R., An intergrated microfluidics-tandem mass spectrometry system for automated protein analysis, *Anal. Chem.*, 70, 3728–3734, 1998.
41. Lazar, J.M., Ramsey, R.S., Sundberg, S., and Ramsey, J.M., Subattomole-sensitivity microchip nanoelectrospray source with time-of-flight mass spectrometry detection, *Anal. Chem.*, 71, 3627–3631, 1999.
42. Figeys, D. and Aebersold, R., Nanoflow solvent gradient delivery from a microfabricated device for protein identifications by electrospray ionization mass spectrometry, *Anal. Chem.*, 70, 3721–3727, 1998.
43. de Biasi, V., Haskins, N., Organ, A., Bateman, R., Giles, K., and Jarvis, S., High throughput liquid chromatography/mass spectrometric analyses using a novel multiplexed electrospray interface, *Rapid Commun. Mass Spectrom.*, 13, 1165–1168, 1999.
44. Eckers, C., Wolff, J., Haskins, N.J., Sage, A.B., Giles, K., and Bateman, R., Accurate mass liquid chromatography/mass spectrometry on orthogonal acceleration time-of-flight mass analyzers using switching between separate sample and reference sprays. 1. Proof of concept, *Anal. Chem.*, 72, 3683–3688, 2000.
45. Barber, M., Bordoli, R., Sedgwick, R., and Tyler, A., Fast atom bombardment of solids (FAB): a new ion source for mass spectrometry, *J. Chem. Soc. Chem. Commun.*, 81, 325–327, 1981.
46. Barber, M., Bordoli, R., Elliott, G., Sedgwick, R., and Tyler, A., Fast atom bombardment mass spectrometry, *Anal. Chem.*, 54, 645–657, 1982,.
47. Seifert, W.E. and Caprioli, R.M., Fast atom bombardment mass spectrometry, *Methods Enzymol.*, 270, 453–486, 1996.
48. Caprioli, R., Continuous-flow fast atom bombardment mass spectrometry, *Anal. Chem.*, 62, 477–485, 1990.
49. Cook, K., Todd, P., and Friar, D., Physical properties of matrices used for fast atom bombardment, *Biomed. Mass Spectrom.*, 12, 492–497, 1985.
50. De Pauw, E., Agnello, A., and Derwa, F., Liquid matrices for liquid secondary ion mass spectrometry-fast atom bombardment: an update, *Mass Spectrom. Reviews*, 10, 283–301, 1991.
51. Hillenkamp, F., Karas, M., Beavis, R., and Chait, B.T., Matrix-assisted laser desorption/ionization mass spectrometry of biopolymers, *Anal. Chem.*, 63, 1193A–1203A, 1991.
52. Karas, M., Bahr, U., and Giessman, U., Matrix-assisted laser desorption ionization mass spectrometry, *Mass Spectrom. Reviews*, 10, 335–357, 1991.
53. Karas, M., Gluckmann, M., and Schafer, J., Ionization in matrix-assisted laser desorption/ionization: singly charged molecular ions are the lucky survivors, *J. Mass Spectrom.*, 35, 1–12, 2000.
54. Murray, K.K., Coupling matrix-assisted laser desorption/ionization to liquid separation, *Mass Spectrom. Reviews*, 16, 283–299, 1997.
55. Whittal, R.M., Russon, L.M., and Li, L., Development of liquid chromatography-mass spectrometry using continuous-flow matrix-assisted laser desorption ionization time-of-flight mass spectrometry, *J. Chromatogr. A*, 794, 367–375, 1998.
56. Erickson, B., Atmospheric pressure MALDI, *Anal. Chem.*, 72, 186A–186A, 2000.
57. Laiko, V.V., Baldwin, M.A., and Burlingame, A.L., Atmospheric pressure matrix-assisted laser desorption/ionization mass spectrometry, *Anal. Chem.*, *72,* 652–657, 2000.
58. Jensen, O.N., Mortensen, P., Vorm, O., and Mann, M., Automation of matrix-assisted laser desorption/ionization mass spectrometry using fuzzy logic feedback control, *Anal. Chem.*, 69, 1706–1714, 1997.
59. Bencsura, A. and Vertes, A., Dynamics of hydrogen bonding and energy transfer in matrix-assisted laser desorption, *Chem. Physics. Lett.*, 247, 142–148, 1995.
60. Busch, K.L., Mechanisms of MALDI, *Spectroscopy,* 14, 14–19, 1999.
61. Cohen, S. and Chalt, B., Influence of matrix solution conditions on the MALDI-MS analysis of peptides and proteins, *Anal. Chem.*, 68, 31–37, 1996.
62. Sze, E.T., Chan, T.W., and Wang, G., Formulation of matrix solutions for use in matrix-assisted laser desorption/ionization of biomolecules, *J. Am. Soc. Mass Spectrom.*, 9, 166–174, 1998.
63. Yao, J., Scott, J.R., Young, M.K., and Wilkins, C.L., Importance of matrix:analyte ratio for buffer tolerance using 2, 5-dihydroxybenzoic acid as a matrix in matrix-assisted laser desorption/ionization-Fourier transform mass spectrometry and matrix-assisted laser desorption/ionization-time-of-flight, *J. Am. Soc. Mass Spectrom.*, 9, 805–813, 1998.

64. Busch, K.L., Matrices for MALDI, *Spectroscopy,* 14, 14–16, 1999.
65. Hanson, C. and Just, C., Selective background suppression in MALDI-TOF mass spectrometry, *Anal. Chem.*, 66, 3676–3680, 1994.
66. Stults, J., Matrix-assisted laser desorption/ionization mass spectrometry, *Current Opin. Struct. Biol.,* 5, 691–698, 1995.
67. Merchant, M. and Weinberger, S.R., Recent advancements in surface-enhanced laser desorption/ionization-time-of-flight-mass spectrometry, *Electrophoresis*, 21, 1164–1174, 2000.
68. Hutchens, T.W. and Yip, T.T., New desorption strategies for the mass spectrometric analysis of macromolecules, *Rapid Commun. Mass Spectrom.*, 7, 576–580, 1993.
69. Hutchens, T.W. and Yip, T.T., Affinity mass spectrometry, *Protein Sci.*, 2, 92, 1993.
70. Lubman, M., *Lasers in Mass Spectrometry,* Oxford University Press, New York, 1990.
71. Zimmermann, R., Heger, H.J., Kettrup, A., and Boesl, U., A mobile resonance-enhanced multiphoton ionization time-of-flight mass spectrometry device for on-line analysis of aromatic pollutants in waste incinerator flue gasses: first results, *Rapid Commun. Mass Spectrom.*, 11, 1095–1102, 1997.
72. Zimmermann, R., Boesl, U., Heger, H.J., Rohwer, E.R., Ortner, E., Schlag, E.W., and Kettrup, A., Hyphenation of gas chromatography and resonance-enhanced laser mass spectrometry (REMPI-TOFMS): a multidimensional analytical technique, *J. High Resolut. Chromatogr.,* 20, 461–470, 1997.
73. B'Hymer, C., Brisbin, J.A., Sutton, K.L., and Caruso, J.A., New approaches for elemental speciation using plasma mass spectrometry, *Am. Lab.,* 32, 17–39, 2000.
74. Owen, L.M., Rauscher, A.M., Fairweather Tait, S.J., and Crews, H.M., Use of HPLC with inductively coupled plasma mass spectrometry (ICP-MS) for trace element speciation studies in biological materials, *Biochem. Soc. Trans.,* 24, 947–952, 1996.
75. Donais, M.K., How to interface a liquid chromatograph to an inductively coupled plasma-mass spectrometer for elemental speciation studies, *Spectroscopy,* 13, 30–35, 1998.
76. Michalke, B. and Schramel, P., Hyphenation of capillary electrophoresis to inductively coupled plasma mass spectrometry as an element-specific detection method for metal speciation, *J. Chromatogr. A.,* 750, 51–62, 1996.
77. Fragu, P., Clerc, J., Briancon, C., Fourre, C., Jeusset, J., and Halpern, S., Recent developments in medical applications of SIMS microscopy, *Micron., 25,* 361–370, 1994.
78. Wu, K.J. and Odom, R., Matrix-enhanced secondary ion mass spectrometry: a method for molecular analysis of solid surfaces, *Anal. Chem.*, 68, 873–882, 1996.
79. Macfarlane, R., Californium-252 plasma desorption mass spectrometry. Large molecules software and the essence of time, *Anal. Chem.*, 55, 1247–1264, 1983.
80. King, F.L. and Steiner, R.E., Glow discharge mass spectrometry: trace element determinations in solid samples, *J. Mass Spectrom.*, 30, 1061–1075, 1995.
81. Weickhardt, C., Moritz, F., and Grotemeyer, J., Time- of- flight mass spectrometry: state-of-the-art in chemical analysis and molecular science, *Mass Spectrom. Reviews*, 15, 139–162, 1996.
82. Cotter, R.J., The new time-of-flight mass spectrometry, *Anal. Chem.,* 71, 445A–451A, 1999.
83. Lazar, I.M., Rockwood, A.L., Lee, E.D., Sin, J.C.H., and Lee, M.L., High-speed TOFMS detection for capillary electrophoresis, *Anal. Chem.,* 71, 2578–2581, 1999.
84. Guilhaus, M., Selby, D., and Mlynski, V., Orthogonal acceleration time-of-flight mass spectrometry, *Mass Spectrom. Reviews*, 19, 65–107, 2000.
85. Chernushevich, I.V., Ens, W., and Standing, K.G., Orthogonal-injection TOFMS for analyzing biomolecules, *Anal. Chem.*, 71, 452A–461A, 1999.
86. Zhang, J., Gardner, B.D., and Enke, C.G., Simple geometry gridless ion mirror, *J. Am. Soc. Mass Spectrom.*, 11, 765–769, 2000.
87. Vestal, M., Juhasz, P., and Martin, S., Delayed extraction matrix-assisted laser desorption time-of-flight mass spectrometry, *Rapid Commun. Mass Spectrom.*, 9, 1044–1050, 1995.
88. Cotter, R.J., Time-of-flight mass spectrometry for the structural analysis of biological molecules, *Anal. Chem.,* 64, 1027 A–1039 A, 1992.
89. Whittal, R.M., Russon, L.M., Weinberger, S.R., and Li, L., Functional wave time-lag focusing matrix-assisted laser desorption/ionization in a linear time-of-flight mass spectrometer: improved mass accuracy, *Anal. Chem.*, 69, 2147–2153, 1997.
90. Whittal, R.M. and Li, L., Time-lag focusing MALDI-TOF mass spectrometry, *Amer. Lab.,* 24, 30–36, 1997.
91. Guilhaus, M., Mlynski, V., and Selby, D., Perfect timing: time-of-flight mass spectrometry, *Rapid Commun. Mass Spectrom.*, 11, 951–962, 1997.

92. Dawson, P.H., *Quadrupole Mass Spectrometry and its Applications,* Elsevier Science, New York, 1976.
93. McLuckey, S.A., Van Berkel, G., Goeringer, D., and Glish, G.L., Ion trap mass spectrometry of externally generated ions, *Anal. Chem.*, 66, 689A–696A, 1994.
94. March, R.E., An introduction to quadrupole ion trap mass spectrometry, *J. Mass Spectrom.,* 32, 351–369, 1997.
95. March, R.E., Quadrupole ion trap mass spectrometry: theory, simulation, recent developments and application, *Rapid Commun. Mass Spectrom.*, 12, 1543–1554, 1998.
96. Schwartz, J.C. and Jardine, I., Quadrupole ion trap mass spectrometry, *Meth. Enzymol.,* 270, 552–586, 1966.
97. Burgoyne, T.W. and Hieftje, G.M., An introduction to ion optics for the mass spectrograph, *Mass Spectrom. Reviews*, 15, 241–259, 1996.
98. Amster, I.J., Fourier transform mass spectrometry, *J. Mass Spectrom.*, 31, 1325–1327, 1996.
99. Marshall, A.G., Hendrickson, C.L., and Jackson, G.S., Fourier transform ion cyclotron resonance mass spectrometry: a primer, *Mass Spectrom. Reviews*, 17, 1–35, 1998.
100. Buchanan, M.V. and Hettich, R.L., Fourier transform mass spectrometry of high-mass biomolecules, *Anal. Chem.,* 65, 245A–259A, 1993.
101. Solouki, T., Gillig, K., and Russell, D., Detection of high-mass biomolecules in Fourier transform ion cyclotron resonance mass spectrometry: theoretical and experimental investigations, *Anal. Chem.*, 66, 1583–1587, 1994.
102. Guan, S., Marshall, A., and Wahl, M., MS/MS with high detection efficiency and mass resolving power for product ions in Fourier transform ion cyclotron resonance mass spectrometry, *Anal. Chem.,* 66, 1363–1367, 1994.
103. Easterling, M.L., Mize, T.H., and Amster, I.J., Routine part-per-million mass accuracy for high-mass ions: space-charge effects in MALDI-FT-ICR, *Anal. Chem.*, 71, 624–632, 1999.
104. Li, Y. and McIver, R., High-accuracy molecular mass determination for peptides and proteins by Fourier transform mass spectrometry, *Anal. Chem.,* 66, 3077–2083, 1994.
105. Marshall, A.G. and Guan, S.H., Advantages of high magnetic field for Fourier transform ion cyclotron resonance mass spectrometry, *Rapid Commun. Mass Spectrom.,* 10, 1819–1823, 1996.
106. Belov, M.E., Gorshkov, M.V., Udseth, H.R., Anderson, G.A., and Smith, R.D., Zeptamole-sensitivity electrospray ionization-Fourier transform ion cyclotron resonance mass spectrometry of proteins, *Anal. Chem.,* 72, 2271–2279, 2000.
107. Dienes, T., Pastor, S.J., Schurch, S., Scott, J.L., Yao, J., Cui, S., and Wilkins, C., Fourier transform mass spectrometry — advancing years (1992–mid. 1996), *Mass Spectrom. Reviews*, 15, 163–211, 1996.
108. Henry, K.D., Williams, E.R., Wang, B.H., McLafferty, F.W., Shabanowitz, J., and Hunt, D.F., Fourier-transform mass spectrometry of large molecules by electrospray ionization, *Proc. Natl. Acad. Sci. USA,* 86, 9075–9078, 1989.
109. Pyrek, J.S., Fourier transform-ion cyclotron resonance-mass spectrometer as a new tool for organic chemists, *Synlett*, 2, 249–266, 1999.
110. Sannes-Lowery, K.A., Drader, J.J., Griffey, R.H., and Hofstadler, S.A., Fourier transform ion cyclotron resonance mass spectrometry as a high throughput affinity screen to identify RNA binding ligands, *Trends Anal. Chem.,*19, 481–490, 2000.
111. Shevchenko, A., Loboda, A., Shevchenko, A.E., Ens, W., and Standing, K.G., MALDI quadrupole time-of-flight mass spectrometry: a powerful tool for proteomic research, *Anal. Chem.,* 72, 2132–2141, 2000.
112. Medzihradszky, K.F., Campbell, J.M., Baldwin, M.A., Falick, A.M., Juhasz, P., Vestal, M.L., and Burlingame, A.L., The characteristics of peptidic collision-induced dissociation using a high-performance MALDI-TOF/TOF tandem mass spectrometer, *Anal. Chem.,* 72, 552–558, 2000.
113. Matsuda, H., High-resolution high-sensitivity mass spectrometers, *Mass Spectrom. Reviews*, 2, 299–325, 1983.
114. Matsuo, T., High performance sector mass spectrometers: past and present, *Mass Spectrom. Reviews,* 8, 203–236, 1989.
115. Krutchinsky, A.N., Zhang, W., and Chait, B.T., Rapidly switchable matrix-assisted laser desorption/ionization and electrospray quadrupole-time-of-flight mass spectrometry for protein identification, *J. Am. Soc. Mass Spectrom.,* 11, 493–504, 2000.

116. Vogel, J.S., Turteltaub, K.W., Finkel, R., and Nelson, D.E., Accelerator mass spectrometry, *Anal. Chem.*, 67, 353A–359A, 1995.
117. Frank, M., Labov, S.E., Westmacott, G., and Benner, W.H., Energy-sensitive cryogenic detectors for high-mass biomolecule mass spectrometry, *Mass Spectrom. Reviews*, 18, 155–186, 1999.
118. Busch, K.L., The electron multiplier, *Spectroscopy*, 15, 28–33, 2000.
119. Laprade, B. and Labich, R., Microchannel plate-based detectors in mass spectrometry, *Spectroscopy*, 9, 26–27, 1994.
120. Johnson, R.C., Cooks, R.G., Allen, T.M., Cisper, M.E., and Hemberger, P.H., Membrane introduction mass spectrometry: trends and applications, *Mass Spectrom. Reviews*, 19, 1–37, 2000.
121. Ragunathan, N., Krock, K.A., Klawun, C., Sasaki, T.A., and Wilkins, C.L., Gas chromatography with spectroscopic detectors, *J. Chromatogr. A*, 856, 349–397, 1999.
122. Leclercq, P.A. and Cramers, C.A., High-speed GC-MS, *Mass Spectrom. Reviews*, 17, 37–49, 1998.
123. Henry, C.M., GC/MS not the same old combination, *Anal. Chem.*, 71, 401A–406A, 1999.
124. Niessen, W. and Tinke, A., Liquid chromatography-mass spectrometry. General principles and instrumentation, *J. Chromatogr. A*, 703, 37–57, 1995.
125. Chen, Y., Jin, X., Misek, D., Hinderer, R., Hanash, S.M., and Lubman, D.M., Identification of proteins from two-dimensional gel electrophoresis of human erythroleukemia cells using capillary high performance liquid chromatography/electrospray-ion trap-reflectron time-of-flight mass spectrometry with two-dimensional topographic map analysis of in-gel tryptic digest products, *Rapid Commun. Mass Spectrom.*, 13, 1907–1916, 1999.
126. Niessen, W.M.A., State-of-the-art in liquid chromatography-mass spectrometry, *J. Chromatogr. A*, 856, 179–197, 1999.
127. Mann, M. and Talbo, G., Developments in matrix-assisted laser desorption/ionization peptide mass spectrometry, *Current Opin. Biotechnol.*, 7, 11–19, 1996.
128. Sakairi, M. and Kato, Y., Multi-atmospheric pressure ionization interface for liquid chromatography-mass spectrometry, *J. Chromatogr. A*, 794, 391–406, 1998.
129. He, L., Liang, L., and Lubman, D., Continuous-flow MALDI mass spectrometry using an ion-trap/reflectron time-of-flight detector, *Anal. Chem.*, 67, 4127–4132, 1995.
130. Choudhary, G., Chakel, J., Hancock, W., Torres Duarte, A., McMahon, G., and Wainer, I., Investigation of the potential of capillary electrophoresis with off-line matrix-assisted laser desorption/ionization time-of-flight mass spectrometry for clinical analysis: examination of a glycoprotein factor associated with cancer cachexia, *Anal. Chem.Anal. Chem.*, 71, 855–859, 1999.
131. Banks, J.F., Recent advances in capillary electrophoresis/electrospray/mass spectrometry, *Electrophoresis*, 18, 2255–2266, 1997.
132. Ding, J. and Vouros, P., Advances in CE/MS, *Anal. Chem.*, 71, 378A–385A,1999.
133. Olivares, J., Nguyen, N., Yonker, C., and Smith, R., On-line mass spectrometric detection for capillary zone electrophoresis, *Anal. Chem.*, 59, 1230–1232, 1987.
134. Chang, Y.Z. and Her, G.R., Sheathless capillary electrophoresis/electrospray mass spectrometry using a carbon-coated fused-silica capillary, *Anal. Chem.*,72, 626–630, 2000.
135. McComb, M.E. and Perreault, H., Design of a sheathless capillary electrophoresis-mass spectrometry probe for operation with a Z-Spray ionization source, *Electrophoresis*, 21, 1354–1362, 2000.
136. Lee, E., Muck, W., Henion, J., and Covey, T., Liquid junction coupling for capillary zone electrophoresis/ion spray mass spectrometry, *Biomed. Environ. Mass Spectrom.*, 18, 844–850, 1989.
137. Smith, R., Olivares, J., Nguyen, N., and Udseth, H., Capillary zone electrophoresis—mass spectrometry using an electrospray ionization interface, *Anal. Chem.*, 60, 436–441, 1988.
138. Warriner, R.N., Craze, A.S., Games, D.E., and Lane, S.J., Capillary electrochromatography/mass spectrometry—a comparison of the sensitivity of nanospray and microspray ionization techniques. *Rapid Commun. Mass Spectrom.*, 12, 1143–1149, 1998.
139. Li, J., Kelly, J.F., Chernushevich, I., Harrison, D.J., and Thibault, P., Separation and identification of peptides from gel-isolated membrane proteins using a microfabricated device for combined capillary electrophoresis/nanoelectrospray mass spectrometry, *Anal. Chem.*, 72, 599–609, 2000.
140. Shultz-Lockyear, L.L., Colyer, C.L., Fan, Z.H., Roy, K.I., and Harrison, D.J., Effects of injector geometry and sample matrix on injection and sample loading in integrated capillary electrophoresis devices, *Electrophoresis*, 20, 529–538, 1999.

141. Li, J., Wang, C., Kelly, J.F., Harrison, D.J., and Thibault, P., Rapid and sensitive separation of trace level protein digests using microfabricated devices coupled to a quadrupole-time-of-flight mass spectrometer, *Electrophoresis,* 21, 198–210, 2000.
142. Jemal, M., Yuan-Qing, and Whigan, D.B., The use of high-flow high performance liquid chromatography coupled with positive and negative ion electrospray tandem mass spectrometry for the quantitative bioanalysis via direct injection of the plasma/serum samples, *Rapid Commun. Mass Spectrom.,* 12, 1389–1399, 1998.
143. Allanson, J.P., Biddlecombe, R.A., Jones, A.E., and Pleasance, S., The use of automated solid phase extraction in the 96-well format for high throughput bionalaysis using liquid chromatography coupled to tandem mass spectrometry, *Rapid Commun. Mass Spectrom.,* 10, 811–816, 1996.
144. Simpson, H., Berthemy, A., Buhrman, D., Burton, R., Newton, J., Kealy, M., Wells, D. and., and Wu, D., High throughput liquid chromatography/mass spectrometry bionanalysis using 96-well disk solid phase extraction plate for the sample preparation, *Rapid Commun. Mass Spectrom.,* 12, 75–82, 1998.
145. Jemal, M., Teitz, D., Ouyang, Z., and Khan, S., Comparison of plasma sample purification by manual liquid-liquid extraction, automated 96-well liquid-liquid extraction and automated 96-well solid-phase extraction for analysis by high-performance liquid chromatography with tandem mass spectrometry, *J. Chromatogr. B,* 732, 501–508, 1999.
146. Biddlecombe, R.A. and Pleasance, S., Automated protein precipitation by filtration in the 96-well format, *J. Chromatogr. B,* 734, 257–265. 1999.
147. Steinborner, S. and Henion, J., Liquid-liquid extraction in the 96-well plate format with SRM LC/MS quantitative determination of methotrexate and its major metabolites in human plasma, *Anal. Chem.,* 71, 2340–2345, 1999.
148. Peng, S.X., King, S.L., Bornes, D.M., Foltz, D.J., Baker, T.R., and Natchus, M.G., Automated 96-well SPE and LC-MS-MS for determination of protease inhibitors in plasma and cartilage tissues, *Anal. Chem.,* 72, 1913–1917, 2000.
149. Figeys, D. and Pinto, D., Lab-on-a-chip: a revolution in biological and medical sciences. *Anal. Chem.,* 72, 330A–335A, 2000.
150. Pinto, D.M., Ning, Y., and Figeys, D, An enhanced microfluidic chip coupled to an electrospray Qstar mass spectrometer for protein identification, *Electrophoresis,* 21, 181–190, 2000.
151. Sanders, G.H. and Manz, A., Chip-based microsystems for genomic and proteomic analysis, *Trends Anal. Chem.,* 19, 364–378, 2000.
152. McCreedy, T., Fabrication techniques and materials commonly used for the production of microreactors and micro total analytical systems, *Trends Anal. Chem.,* 19, 396–401, 2000.
153. Becker, H. and Gartner, C., Polymer microfabrication methods for microfluidic analytical applications. *Electrophoresis,* 21, 12–26, 2000.
154. McDonald, J.C., Duffy, D.C., Anderson, J.R., Chiu, D.T., Wu, H., Schueller, O.J., and Whitesides, G.M., Fabrication of microfluidic systems in poly(dimethylsiloxane), *Electrophoresis,* 21, 27–40, 2000.
155. Mitchell, P.G., Magna, H.A., Reeves, L.M., Lopresti Morrow, L.L., Yocum, S.A., Rosner, P.J., Geoghegan, K.F., and Hambor, J.E., Cloning, expression, and type II collagenolytic activity of matrix metalloproteinase-13 from human osteoarthritic cartilage, *J. Clin. Invest.*, 97, 761–768, 1996.
156. Xiang, F., Lin, Y., Wen, J., Matson, D.W. , and Smith, R.D., An integrated microfabricated device for dual microdialysis and on-line-ESI-ion trap mass spectrometry for analysis of complex biological samples, *Anal. Chem.,* 71, 1485–1490, 1999.
157. Xu, N., Lin, Y., Hofstadler, S.A., Matson, D., Call, C.J., and Smith, R.D., A microfabricated dialysis device for sample cleanup in electrospray ionization mass spectrometry, *Anal. Chem.,* 70, 3553–3556, 1998.
158. Chan, J.H., Timperman, A.T., Qin, D., and Aebersold, R., Microfabricated polymer devices for automated sample delivery of peptides for analysis by electrospray ionization tandem mass spectrometry, *Anal. Chem.,* 71, 4437–4444, 1999.
159. Guetens, G., Van Cauwenberghe, K., De Boeck, G., Maes, R., Tjaden, U.R., Van Der Greef, J., Highley, M., Van Oosterom, A.T., and De Bruijn, E.A., Nanotechnology in bio/clinical analysis, *J. Chromatogr. B,* 739, 139–150, 2000.
160. Henry, C., Micro meets macro: interfacing microchips and mass spectrometers, *Anal. Chem.*, 69, 359 A–361 A, 1997.

161. Oleschuk, R.D. and Harrison, D.J., Analytical microdevices for mass spectrometry, *Trends Anal. Chem.,* 19, 379–388, 2000.
162. Ekstrom, S., Onnerfjord, P., Nilsson, J., Bengtsson, M., Laurell, T., and Marko-Varga, G., Integrated microanalytical technology enabling rapid and automated protein identification, *Anal. Chem.,* 72, 286–293, 2000.
163. Coyler, C.L., Tang, T., Chiem, N., and Harrison, D.J., Clinical potential of microchip capillary electrophoresis systems, *Electrophoresis,* 18, 1733–1741, 1997.
164. Kutter, J.P. Current developments in electrophoretic and chromatographic separation methods on microfabricated devices, *Trends Anal. Chem.,* 19, 352–363, 2000.
165. Wen, J., Lin, Y., Xiang, F., Matson, D.W., Udseth, H.R., and Smith, R.D., Microfabricated isoelectric focusing device for direct electrospray ionization-mass spectrometry, *Electrophoresis,* 21, 191–197, 2000.
166. Foret, F., Zhou, H., Gangl, E., and Karger, B.L., Subatmospheric electrospray interface for coupling of microcolumn separations with mass spectrometry, *Electrophoresis,* 21, 1363–1371, 2000.
167. Little, D.P., Cornish, T.J., O'Donnell, M.J., Braun, A., Cotter, R.J., and Köster, H., MALDI on a chip: analysis of arrays of low-femtomole to subfemtomole quantities of synthetic oligonucleotides and DNA diagnostic products dispensed by piezoelectric pipet, *Anal. Chem.,* 69, 4540–4546, 1997.
168. McLuckey, S.A., Principles of collisional activation in analytical mass spectrometry, *J. Am. Soc. Mass Spectrom.,* 3, 599–614, 1992.
169. de Hoffmann, E., Tandem mass spectrometry: a primer, *J. Biol. Mass Spectrom.,* 31, 129–137, 1996.
170. Kelleher, N.L., From primary structure to function: biological insights from large-molecule mass spectra, *Chem. Bio.,* 7, R37–R45, 2000.
171. Zubarev, R.A., Kruger, N.A., Fridriksson, E.K., Lewis, M.A., Horn, D.M., Carpenter, B.K., and McLafferty, F.W., Electron capture dissociation of gaseous multiply-charged proteins is favored at disulfide bonds and other sites of high hydrogen atom affinity, *J. Am. Chem. Soc.,* 121, 2857–2862, 1999.
172. Zubarev, R.A., Kelleher, N.L., and McLafferty, F.W., Electron capture dissociation of multiply-charged protein cations. A nonergodic process, *J. Am. Chem. Soc.*, 120, 3265–3266, 1998.
173. Quiniou, M.L., Yates, A.J., and Langridge-Smith, P.R., Laser photo-induced dissociation using tandem time-of-flight mass spectrometry, *Rapid Commun. Mass Spectrom.,* 14, 361–367, 2000.
174. Dongre, A.R., Somogyi, A., and Wysocki, V.H., Surface-induced dissociation: an effective tool to probe structure, energetics and fragmentation mechanisms of protonated peptides, *J. Mass Spectrom.,* 31, 339–350, 1996.
175. Lehmann, W.D., Pictograms for experimental parameters in mass spectrometry, *Am. Soc. Mass. Spectrom.,* 8, 756–759, 1997.
176. Henion, J., Brewer, E., and Rule, G., Sample preparation for LC/MS/MS, *Anal. Chem.,* 70, 650A–656A, 1998.
177. Spengler, B., Post-source decay analysis in matrix-assisted laser desorption/ionization mass spectrometry of biomolecules, *J. Mass. Spectrom.,* 32, 1019–1036, 1997.
178. Brown, R.S. and Lennon, J.J., Sequence-specific fragmentation of matrix-assisted laser-desorbed protein/peptide ions, *Anal. Chem.,* 67, 3990–3999, 1995.
179. Busch, K.L., Mass spectral libraries, *Spectroscopy,* 9(9), 24–26, 1994.
180. McLafferty, F.W. and Stauffer, D.B., *The Wiley/NBS Registry of Mass Spectral Data,* Palisade Corp., New York, 1989.
181. Stauffer, D., McLafferty, F., Ellis, R., and Peterson, D., Adding forward searching capabilites to a reverse search algorithm for unknown mass spectra, *Anal. Chem.,* 57, 771–773, 1985.
182. Stein, S. and Scott, D., Optimization and testing of mass spectral library search algorithms for compound identification, *J. Am. Soc. Mass Spectrom.,* 5, 859–866, 1994.
183. Fink, S., Thompson, W., and Slayback, J., A fully automated system for confirmational mass spectrometric analysis, *Spectroscopy,* 11, 26–32, 1996.
184. Sparkman, O.D., Evaluating electron ionization mass spectral library search results, *Am. Soc. Mass Spectrom.,* 7, 313–318, 1996.
185. Beynon, J.H. and Williams, A.E., *Mass and Abundance Tables for Use in Mass Spectrometry,* Elsevier Science, New York, 1963.
186. Busch, K.L., Mass-scale calibration and the case of the mass defect, *Spectroscopy,* 9(5), 15–17, 1994.

187. Allan, A. and Roboz, J., Small interval multiple channel selected ion monitoring at high resolution, *Rapid Commun. Mass Spectrom.*, 2, 1988.
188. Zubarev, R.A., Demirev, P.A., Hakansson, P., and Sundqvist, B.U.R., Approaches and limits for accurate mass characterization of large biomolecules, *Anal. Chem.*, 67, 3793–3798, 1995.
189. Busch, K.L., Using exact mass measurements, *Spectroscopy*, 9(7), 21–22, 1994.
190. Clauser, K.R., Baker, P., and Burlingame, A.L., Role of accurate mass measurement in protein identification strategies employing MS or MS/MS and database searching, *Anal. Chem.*, 71, 2871–2882, 1999.
191. Green, M.K., Vestling, M.M., Johnston, M.V., and Larsen, B.S., Distinguishing small molecular mass differences of proteins by mass spectrometry. *Anal. Biochem.*, 260, 204–211, 1998.
192. Russell, D.H. and Edmondson, R.D., High-resolution mass spectrometry and accurate mass measurements with emphasis on the characterization of peptides and proteins by matrix-assisted laser desorption/ionization time-of-flight mass spectrometry, *J. Mass Spectrom.*, 32, 263–276, 1997.
193. Roboz, J., McDowall, M., Hillmer, M., and Holland, J.F., Accurate mass measurement in continuous flow fast atom bombardment quadrupole mass spectrometry, *Rapid Commun. Mass Spectrom.*, 2, 64–66, 1988.
194. Tyler, A.N., Clayton, E., and Green, B.N., Exact mass measurement of polar organic molecules at low resolution using electrospray ionization and a quadrupole mass spectrometer, *Anal. Chem.*, 68, 3561–3569, 1996.
195. Wu, J., Fannin, S., Franklin, F., Mollinski, T., and Lebrilla, C., Exact mass determination for elemental analysis of ions produced by matrix-assisted laser desorption, *Anal. Chem.*, 67, 3788–3792, 1995.
196. Hoptgartner, G., Chernushevich, I.V., Covey, T., Plomley, J.B., and Bonner, R., Exact mass measurement of product ions for the structural elucidation of drug metabolites with a tandem quadrupole orthogonal-acceleration time-of-flight mass spectrometer, *J. Am. Soc. Mass Spectrom.*, 10, 1305–1314, 1999.
197. Busch, K.L., The resurgence of exact mass measurement with FTMS, *Spectroscopy*, 15, 22–27, 2000.
198. Busch, K.L. and Chen, C.H., Selected ion monitoring, *Spectroscopy*, 12, 30–35, 1997.
199. Claeys, M., Markey, S.P., and Maenhaut, W., Variance analysis of errors in SIM. A practical study case, *Biomed. Mass Spectrom.*, 4, 122–128, 1977.
200. Millard, B.J., *Quantitative Mass Spectrometry*, Wiley-Heyden, London, 1978.
201. Chatman, K., Hollenbeck, T., Hagey, L., Vallee, M., Purdy, R., Weiss, F., and Siuzdak, G., Nanoelectrospray mass spectrometry and precursor ion monitoring for quantitative steroid analysis and attomole sensitivity, *Anal. Chem.*, 71, 2358–2363, 1999.
202. Lagerwerf, F.M., van Dongen, W.D., Steenvoorden, R.J.J.M., Honing, M., and Jonkman, J.H.G., Exploring the boundaries of bioanalytical quantitative LC-MS-MS, *Trends Anal. Chem.*, 19, 418–427, 2000.
203. Busch, K.L., Isotopes and mass spectrometry, *Spectroscopy*, 12, 24–26, 1997.
204. Leenheer, A. and Thienport, L,. Applications of isotope dilution-mass spectrometry in clinical chemistry, pharmacokinetics, and toxicology, *Mass Spectrom. Reviews*, 11, 249–307, 1992.
205. Lee, P.W.N., Stable isotopes and mass isotopomer study of fatty acid and cholesterol synthesis, *Adv. Exp. Med. Biol.*, 399, 95–114, 1996.
206. Meier-Augenstein, W., Applied gas chromatography coupled to isotope ratio mass spectrometry, *J. Chromatogr. A*, 842, 351–371, 1999.
207. Brand, W.A., High precision isotope ratio monitoring techniques in mass spectrometry, *J. Mass Spectrom.*, 31, 225–235, 1996.
208. Brunengraber, H., Kelleher, J.K., and DesRosiers, C., Applications of mass isotopomer analysis to nutrition research, *Ann. Rev. Nutr.*, 17, 559–596, 1997.
209. Patterson, B.W., Use of stable isotopically labeled tracers for studies of metabolic kinetics: an overview, *Metabolism*, 46, 322–329, 1997.
210. Patterson, K., Velllon, C., and O'Haver, T., Error propagation in isotope dilution analysis as determined by Monte Carlo simulation, *Anal. Chem.*, 66, 2829–2834, 1994.
211. Vogt, J.A., Chapman, T.E., Wagner, D.A., Young, V.R., and Burke, J.F., Determination of the isotope enrichment of one or a mixture of two stable labelled tracers of the same compound using the complete isotopomer distribution of an ion fragment; theory and application to *in vivo* human tracer studies, *Biol. Mass Spectrom.*, 22, 600–612, 1993.

212. Rockwood, A., Van Orden, S., and Smith, R., Rapid calculation of isotope distributions, *Anal. Chem.*, 67, 2699–2704, 1995.
213. Lammert, S.A., 2000 Directory of mass spectrometry manufacturers and suppliers, *Rapid Commun. Mass Spectrom.*, 14, 725–739, 2000.
214. Busch, K.L., Electronic resources for mass spectrometrists. MS-related internet sites and how to use them, *Spectroscopy*, 11, 32–34, 1996.
215. Murray, K.K. Internet resources for mass spectrometry, *J. Mass Spectrom.*, 34, 1–9, 1999.

3 Relevant Concepts of Cancer Medicine and Biology

3.1 CLASSIFICATION AND EPIDEMIOLOGY

The words *tumor* (Latin, tumere = to swell) and *neoplasm* (neo = new, plasma = formation) are used synonymously to describe abnormal tissues characterized by unregulated cellular growth to form a mass without a defined structure. The pathologic process that results in the formation and growth of neoplasms is called *neoplasia. Cancer* (Latin, for crab) is a general term used to describe any of various types of *malignant* neoplasms causing over 200 diseases of multicellular organisms. Cancers share several characteristics, including that they originate from mutant, genetically dysfunctional cells; they escape normal growth controls; and they invade and colonize normal tissues. On its way to becoming a full-fledged malignant tumor, a nascent neoplastic clone must prevail through a remarkable series of events and must undertake a number of exceptional maneuvers. These include bypassing programmed cell death, circumventing growth restraining signals, growing without growth factors from other cells, evading immunological surveillance, appropriating a vascular supply, burrowing into surrounding tissues, traversing blood or lymph vessels to travel to distant sites, recognizing an appropriate environment for crossing out of the vessels, and finally establishing secondary tumors. Only about 0.1% of the breakaway cells survive after entering the circulation, and most of those are captured in the first capillary bed encountered. In the specific case of cells in the lymph system, those cells arrested in the subcapsular sinuses of the lymph nodes may begin to grow. Other lymph borne-malignant cells may enter the blood vascular system via the lymphatic interconnections in the venous system.

3.1.1 Solid Tumors

All solid neoplasms have a parenchyma that comprises the neoplastic proliferating cells and a stroma that consists of the supporting connective tissue and blood supply required for growth. Based on the biological behavior of the parenchyma, solid tumors may be classified into three main groups. At one extreme are *benign* tumors that may arise in any tissue, grow only locally and slowly, and cause damage (which may be significant) by local pressure or obstruction. Patient survival rates are generally high after surgical removal of benign tumors. The second group contains *in-situ* tumors that develop in the epithelium and have the morphological appearance of cancer cells. However, these cells remain in the epithelial layer without invading the basement membrane and supporting mesenchyme. When they acquire the ability to penetrate the basement membrane they become invasive *malignant* tumors that may arise in any tissue, grow rapidly, invade, and destroy the underlying local mesenchyme. They often migrate to local lymph nodes and distant organs where they produce secondary tumors.

Histology is the microscopic study of organs, tissues, and cells. Solid tumors are characterized according to the tissue from which they originate. Benign tumors are described by a prefix designating a specific tissue followed by the suffix *oma,* e.g., adenomas are tumors of glandular tissue. (Exceptions are melanoma and hepatoma, which are malignant). Malignant tumors also use the suffix *oma*; however, this is modified by the roots *carcin-* for tumors of epithelial origin or *sarc-* for those of connective tissue origin. There may also be a prefix indicating the type of epithelial tissue from which the carcinoma or sarcoma originates, e.g., osteosarcoma.

Approximately 90% of human malignancies are epithelial *carcinomas,* i.e., they derive from the layers of cells that cover the body's surface or line internal organs and various glands. Further classification is site-specific, e.g., lung, ovarian, or breast carcinoma. This level of classification does not, however, determine the malignancy, prognosis, or outcome of the neoplasia given that most carcinomas are heterogeneous in both their biological and clinical behavior. The less common *sarcomas* originate from mesodermal tissues which are the supporting (or connective) tissues of the body, such as bones, muscles, fat fibrous tissues, and blood vessels. *Melanomas* start in pigment cells (melanocytes) located among the epithelial cells of the skin.

3.1.2 Hematologic Malignancies

There are three types of these *diffuse* tumors: leukemias, malignant lymphomas, and multiple myeloma. *Leukemias* are characterized by the abnormal proliferation of leukocytes (white blood cells) that infiltrate the bone marrow, peripheral blood, and other organs. Leukemias are classified according to their cell of origin, lymphoid or myeloid (from bone marrow), and the general pace of the disease. The maturity of the predominant cells is indicated by the suffixes *-blast* for immature and *-cyt* for mature. There are four broad groups: acute lymphoblastic leukemia (ALL), acute myelogenous leukemia (AML), chronic lymphocytic leukemia (CLL), and chronic myelogenous leukemia (CML). These diseases are clinically heterogeneous and require significantly different treatment based on assignment into additional subclasses.

Malignant *lymphomas* originate from lymphoid tissues, including the lymph nodes, thymus, and spleen. These modified connective tissues serve as part of a circulatory network to filter impurities and to generate immune responses. Malignant lymphomas are grouped into two broad categories, Hodgkin's disease and the non-Hodgkin's lymphomas. The latter group includes many different disease entities derived mostly from B cell precursors. The lymphomas require significantly different treatment modalities depending on the cell type, degree of differentiation, and growth pattern. *Multiple myeloma* is characterized by a proliferation of malignant plasma cells (mature B lymphocytes). These cells frequently elaborate quantities of immunoglobulin which can be identified in the serum. Despite the presence of excessive numbers of plasma cells and immunoglobulin, myeloma patients, like many patients with hematologic malignancies, are immunocompromised and susceptible to infection.

Neoplasms of cells of uncertain or unknown origin are sometimes named by the person who first described them. Eponymous neoplasms include Ewing's sarcoma (primitive neuroepithelial cell), Hodgkin's lymphoma (early lymphoid cell?), Burkitt's lymphoma (B lymphocyte), Kaposi's sarcoma (vascular endothelial cell), and Wilm's tumor (nephroblastoma).

3.1.3 Incidence and Distribution of Cancers

Cancer epidemiology is the study of the *incidence* (number of newly diagnosed cases), *prevalence* (number of existing cases), and *mortality* (number of deaths) of specific diseases in defined populations during specific periods of time. The ultimate aim is the determination of *risk factors* for individual cancers with respect to geographic, demographic, socioeconomic, behavioral, and genetic patterns. *Descriptive* epidemiology studies are used to determine a possible cause for a particular cancer and may be refined by inclusion of the *time factor,* such as cross-sectional studies (present time), case-control studies of past exposure (retrospective studies), and attempts to establish if selected exposed populations develop a disease in the future (prospective studies). The discipline of *molecular epidemiology* incorporates molecular, cellular, and other biological measurements into epidemiological research.

According to the World Health Organization, in 1997, ~10 million people were diagnosed with cancer and ~6 million died from the disease worldwide. In the U.S., cancer is the second most common cause of death among adults (~24% of total deaths) following heart disease (~33%). About 1500 people die of cancer every day. In both females and males, more than one-half of cancer deaths are attributable to four malignancies. For females they are lung (25%), breast (17%), and colon (10%), while for males the distribution is lung (32%), prostate (14%), and colon (9%).

The incidence of specific cancers varies significantly across the world, possibly due to different environmental factors. For example, the incidence of melanoma is ~150-fold greater in Australia than in Japan, prostate cancer occurs ~70 times more frequently in the U.S. than in China, and liver cancer is ~50 times more prevalent in China than in Canada. At the same time, the incidences of certain malignancies such as leukemia, ovarian cancer, and breast cancer are similar in most countries. Even within these general differences, the use of *cancer maps* has, in certain instances, disclosed some dramatic variations that are likely to be related to local socioeconomic or environmental factors.

The lifetime probability of developing a malignancy is ~1 in 2 for males and ~1 in 3 for females. These numbers increase steadily with age. For example, for males in the 40 to 59-year age bracket there is a 1 in 78 probability for occurrence of prostate cancer which increases to 1 in 6 in the 60 to 79-year-old group. Similarly, for females, the probability for breast cancer in the same age brackets increases from 1 in 26 to 1 in 14.

Survival analysis provides the link between incidence and mortality. Although it has no particular biological significance, the 5-year survival rate has become the standard outcome measure in the clinical literature. A hundred years ago, cancer was invariably fatal. Today, in the U.S. the overall 5-year survival rate for all cancers is ~55% (by race 58% white, 42% black). There have been spectacular advances in the early detection and treatment of some uncommon cancers, such as childhood leukemia and testicular cancer. The 5-year survival for advanced stages of common malignancies has also improved by 2 to 10% during the past decade.

3.1.4 Risk Factors

The exact cause of most cancers remains unknown. Potential etiologies may be divided into those within the body and those related to the environment, with the latter defined broadly to include anything that interacts with humans (Table 3.1).

The common, *sporadic* cancers appear at an age normally expected for the site. These cancers seldom occur in first- or second-degree relatives or in subsequent generations, are rarely bilateral and only infrequently present with precursor states. In contrast, *hereditary* cancers typically appear 10 to 15 years earlier than expected for a particular site and often occur among first- or second-degree relatives with transmission across three or more generations. Precursor states appear frequently, e.g., multiple polyps in familial adenomatous polyposis and the dysplastic nevus syndrome in malignant melanoma. The neoplasms often appear in both paired organs. Hereditary cancers result from mutations within germ cells (egg or sperm) that are then inherited by subsequent generations through *Mendelian* genetics. In *autosomal recessive* inheritance, the trait is expressed only if both gene loci are abnormal, i.e., both parents must have the abnormal gene. Thus only 1 in 4 progeny expect to be affected, although 1 in 2 might be carriers. The absence of normal genes may result in a failure to produce factors necessary for control of cell growth. Transmission occurs most commonly via *autosomal dominant* inheritance, i.e., when only one allele of the pair is abnormal and the offspring has a 50% chance of expressing the abnormal trait. Here, the abnormal gene may express a molecule that directly contributes to neoplasia. Because multiple mutations are needed for the development of cancer (see later), the inheritance of a mutated gene confers only a *predisposition* to developing a cancer. Examples of neoplasias associated with autosomal dominant genes include the rare childrens' cancers retinoblastoma and Wilms's tumor and the important familial adenomatous polyposis syndrome involving the *adenomatous polyposis coli* (APC) gene. Some mass spectrometric techniques to facilitate diagnosis of hereditary cancers are described in Section 6.3.3.

About 15% of human cancers are associated with viruses, approximately 80% of which result from infection by one of three types of DNA viruses: (1) hepatitis B and C (hepatocellular

TABLE 3.1
Etiologic Factors Associated with Carcinogenic Risk (Industrialized Nations)

Carcinogenic Risk Factor	Associated Neoplasm(s)	Probable % of Cases
Within the Body		
Sporadic genetic mutations	Any	10
Inherited genes (familial)	Breast, colon	5
Reproductive history		
Late first pregnancy	Breast	
Zero or low parity	Ovary, breast	
Sexual promiscuity	Cervix	
Occupational		5
Asbestos	Lung, mesothelioma	
Aniline dye	Bladder	
Benzene	Leukemia	
Vinyl chloride	Liver	
Chromium, cadmium, nickel	Lung	
Environment		15
Viral infections	Leukemia, lymphoma	
Pollutants		
Radiation		
Ionizing	Leukemia, breast, thyroid	
Ultraviolet	Skin, melanoma	
Radon	Lung	
Medical Treatments		1
Alkylating agents	Leukemia, bladder	
Diethylstilbestrol	Vaginal (in offspring of exposed woman)	
Estrogens	Endometrium	
Tamoxifen	Endometrium	
Radiation	Skin, lung, breast	
Lifestyle		
Smoking	Lung, bladder, mouth, pharynx, lip	30
Alcohol	Esophagus, liver, larynx	5
Diet	Colon, breast, gall bladder	30 (?)
Food additives (salt)	Stomach	1
Aflatoxin	Liver	
Sedentary lifestyle	??	3

carcinoma, highly prevalent in Southeast Asia and sub-Saharan Africa); (2) papillomaviruses (cervical and anal cancers); and (3) herpesviruses, e.g., Epstein-Barr (Hodgkin's disease and Burkitt's lymphoma). Other cancers may be sequelae of infections by members of the oncoviridae subfamily of RNA viruses, e.g., HTLV-1 and HTLV-2 (human T-cell leukemia) and the mammalian type C tumor viruses. Interestingly, the study of the life cycle and leukemogenesis of the RNA retroviruses has played a major role in the discovery and characterization of oncogenes during the last 30 years. The study of pathogenic human viruses has acquired special emphasis in the age of AIDS and AIDS-related neoplasms, e.g., Herpes virus 8 in Kaposi's sarcoma.

Some medical treatments may cause DNA damage leading to eventual cancer. Secondary leukemias and solid tumors may develop after the primary tumors have been treated with alkylating agents or topoisomerase II inhibitors, as well as after radiation therapy regimens.

TABLE 3.2
Abnormalities of Cell Growth, Differentiation, and Maturation

	Abnormalities of Growth (quantitative)
Hypertrophy	Increase in size of cells
Atrophy	Decrease in size of cells
Hyperplasia	Increase in number of cells
Hypoplasia	Decrease in number of cells
Aplasia	Failure of cell production
	Abnormalities in Differentiation and Maturation (qualitative)
Metaplasia[a]	Replacement of mature cells with cells of another type Organization of tissue maintained Reversible
Dysplasia	Somewhat increased rate of cell multiplication Partial loss of control and organization Cytologic abnormalities Partially reversible or may progress to neoplasia
Neoplasia	Significant increase in number of cells Complete loss of control, variable loss of organization Cytologic abnormalities Irreversible
Anaplasia	No morphologic resemblance to normal tissue (a hallmark of highly malignant neoplasia)

[a] Results from abnormal differentiation of stem cells. It commonly involves epithelium and typically occurs following chronic physical or chemical irritation.

It is now generally believed that most cancers associated with occupational and environmental exposures, as well as with lifestyle (Table 3.1), may possibly be avoided or the risks reduced significantly by limiting exposure to the offending agents. Numerous relevant studies involving mass spectrometry are reviewed in Chapters 4 and 6.

3.2 CELLULAR PROLIFERATION AND COMMUNICATION

3.2.1 Growth Patterns

Although there may be quantitative abnormalities in nonneoplastic growth patterns (Table 3.2), normal cells typically proliferate only until there are enough cells to satisfy current physiological needs. Several feedback mechanisms exist to inhibit further replication. In contrast, malignant neoplasms are characterized by major changes that are qualitative, quantitative, and often irreversible. Normal cells are *differentiated,* i.e., they have developed a specific morphology and function. In cancer, differentiation refers to the extent to which cancer cells resemble comparable normal cells. In fact, one of the hallmarks of neoplastic growth is *anaplasia,* i.e., the absence of morphologic resemblance to normal tissue and the inability to perform the physiological functions of their mature tissue of origin.

There are significant differences between normal and cancer cells in cell culture. Normal cells grow only in the presence of growth factors that are usually provided in the form of serum or other cells which secrete them, purified growth factors, or certain nonphysiological mitogens. Dividing cells also need a solid support because of their *anchorage dependency.* Cell growth is limited by *contact inhibition,* attained when a continuous monolayer has been formed. The lifespan of cultured normal cells is limited, e.g., fibroblasts stop growing and die after 30 to 50 divisions. Cells from malignant tumors are capable of an infinite number of population doublings (immortality) as long as nutrients are present, even in suboptimal proportions. The growth of cancer cells does not require anchorage and is not limited by contact inhibition or adhesiveness. Cancer cells are also characterized by loss of control at checkpoints in the cell cycle (see later).

As normal cells divide and age, there is a progressive shortening of the *telomeres* at the ends of their chromosomes. *Telomerase* is an enzyme that catalyses the synthesis of new telomeric units, thus preventing the shortening of telomeres. Telomerase activity is absent in most normal somatic tissues. In contrast, about 85% of malignant tumors overexpress telomerase to such an extent that in some cases there is a correlation with the stage and severity of the malignancy.

An important concept in the classical model of tumor growth is *tumor volume doubling time* (DT), i.e., the time needed for a tumor mass to double its volume. This is always slower than the generation time of tumor cells, since some cells die and some do not divide. The average DT of primary solid tumors is 2 to 3 mo (range 1 to 7 mo). The smallest tumor capable of physical diagnosis (palpation) is about 1 g in mass, 1 cm^3 in size, and has already progressed through almost 30 doublings (Figure 3.1). The tumor burden after 40 doublings (1 kg size) is usually considered lethal. Another concept of tumor growth is *growth fraction* (GF), i.e., the ratio of the total number of cells present to the number of proliferating cells. The GF decreases as tumor volume increases. In the latter stages of tumor growth, cell multiplication occurs only at the periphery of the tumor; the center becomes increasingly dormant and eventually turns necrotic. *Gompertzian growth* kinetics describes an initial nonlinear, rapid growth phase, followed by a phase of exponential growth, leading eventually to a progressive decline in GF at the plateau phase (Figure 3.1).

All cells within a given tissue are in one of three possible states: quiescent, proliferating, or dying. Normally, most cells are in a resting, quiescent state that may last for months or even years. When there is a need to maintain or replace tissue, cells enter the cell cycle by responding to specific extracellular stimuli. Cell division stops when the necessary growth is complete. Cells may die due to necrosis or apoptosis. *Necrosis* is a process in which cells undergo irregular swelling followed by dissolution. In contrast, cells undergoing *apoptosis* (programmed cellular suicide) shrink, and the nuclear chromatin condenses and fragments. Apoptotic cells segment into membrane-bound "apoptotic bodies" that are eventually phagocytosed by macrophages or other surrounding cells. Proliferation, quiescence, and dying are normal physiological events that are kept in balance by regulatory mechanisms characterized by multiple levels of redundancy, i.e., there are several regulatory paths leading to a given end state. In neoplasia, the homeostatic balance is disturbed and the rate of any of the three processes may increase or decrease individually or in combination. The abnormal gene expressions that significantly affect cell proliferation (see later) may also lead to the impairment of the cell's ability to undergo apoptosis. The survival of such cells that contain genetic abnormalities may also contribute to the refractoriness of the cancer to treatment. The development of resistance to chemotherapy and radiotherapy is affected by any inability of cells to undergo apoptosis. This is because most treatment modalities kill cancer cells by stimulating apoptosis as a result of the damage that they induce.

3.2.2 Cell Cycle

All proliferating cells, both normal and malignant, pass through a cell cycle of four distinct phases (Figure 3.2). Normally, cells are in a quiescent, nonproliferating state (G_o). When there is a need to maintain or replace tissue, cells are stimulated to enter the cell cycle. In the first gap phase (G_1), cells manufacture the enzymes necessary for DNA synthesis, which takes place in the second synthetic phase (S). RNA and proteins needed for the mitotic phase are synthesized in the third phase of the cycle (second gap, G_2). Proteins synthesized include the topoisomerase (Topo)

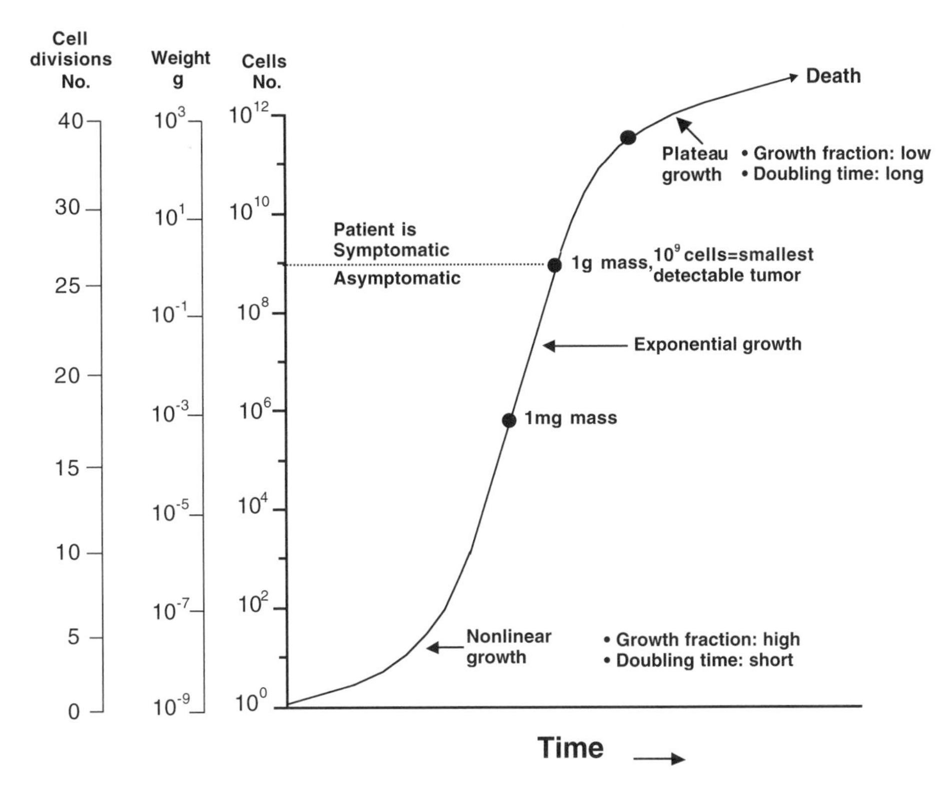

FIGURE 3.1 This Gompertzian growth curve shows that the initial nonlinear growth is followed by an exponential growth phase that eventually becomes a plateau phase with long doubling time and low growth fraction.

enzymes, which eventually make the required cuts in DNA (Topo I: single stranded cuts; Topo II: double stranded cuts).

Cell division or mitosis takes place in the last phase of the cell cycle (M). Although mitosis is a continuous process, it takes place through four phases: prophase, anaphase, metaphase, and telophase. Processes that occur during *prophase* include: (a) increased separation of the two centrosomes (cellular organelles that each contain a pair of centrioles) inside the cytoplasm; (b) formation of the *mitotic apparatus* in which *microtubules* are grown around the centrioles resulting in the formation of the *spindle* between the two separating centrioles; (c) transferring chromatin from the nucleus into pairs of chromosomes each of which contains two filaments (chromatids) that are complete copies of the genetic material of that chromosome. During *metaphase,* the paired chromosomes line up between the centrioles along the equatorial plate. Metaphase is followed by *anaphase* in which each chromosome (containing maternal and paternal chromatids) is assembled from the chromosomal pair. The 92 chromosomes are pulled toward the centriole by the microtubules resulting in the movement of a full set of 46 chromosomes to each of what will become the daughter cells. Next, during the first part of *telophase,* the polarized, rodlike chromosomes return to being threadlike structures and a nuclear membrane is formed around each set of chromosomes. The actual division of the cells occurs during the second part of telophase when the cytoplasm splits into two new daughter cells, each being an exact replica of the parent cell. The newly generated daughter cells may continue to cycle, enter the G_0 resting phase, become permanently differentiated, or die by apoptosis. Differentiation, senescence, and apoptosis are negative controls on the cell cycle progression with significant potential effects on proliferation.

Transitions within the cell cycle depend on separate positive and negative regulatory controls. These intrinsic regulatory pathways are responsible for the proper ordering of the events in the cycle. There are also extrinsic regulatory pathways, and when differences between normal and

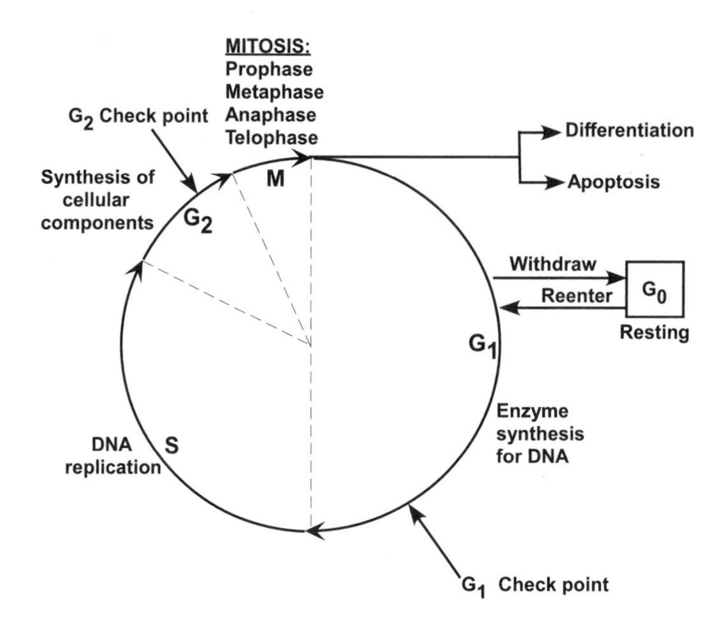

Phase	G_0	G_1	S	G_2	M
Function	Resting	RNA and protein synthesis	DNA synthesis	RNA protein synthesis	Mitosis
DNA		2n	2-4n	4n	
Duration, h	Indefinite	Variable	6-8	3-4	0.5-1

FIGURE 3.2 The cell cycle.

neoplastic cells are considered, it is control by extrinsic regulation that is usually more relevant. In normal cells these pathways respond to detected changes in environmental conditions. Cancer cells are usually less responsive to environmental stresses than are normal cells. The cell cycle proceeds via a number of "checkpoints" at which the correct completion of a process is evaluated and a decision is made whether to go on to the next stage. A major checkpoint occurs at the end of G_1, just before the initiation of DNA replication. If conditions are not acceptable, e.g., there is a lack of growth factors, cells will exit G_1 and return to G_0. A second major checkpoint is located before the cells enter the mitosis phase.

Cell cycle progression is tightly controlled by *cyclins,* a family of proteins with molecular masses in the 30 to 50 kDa range. The G_1 cyclins function by activating their cognate protein-serine-threonine kinases (cdks). These G_1-cyclin-dependent proteins function by phosphorylating the proteins that initiate the S phase. After cells pass this checkpoint, the activating G_1 cyclins are destroyed by proteases and the activity of the G_1-cyclin-dependent kinases declines. The G_2-M checkpoint is controlled by an activated M-cyclin-dependent protein that initiates the mitotic phase by catalyzing the phosphorylation of acceptor proteins. The initial activation of this M-cyclin-dependent protein is complex and involves a cascade of phosphorylation and dephosphorylation processes, including the activation and deactivation of several protein kinases as well as a number of substrates each of which have different functions. As discussed in Section 3.3, specific genes participate in the cell cycle regulatory scheme (and repair damaged pathways within the cell cycle) and aberrations in these genes can have significant regulatory consequences.

3.2.3 Cellular Communication and Signal Transduction

Because information must constantly pass across cell membranes as well as between locations within the cell, multiple complex mechanisms have evolved to provide the necessary communication. Signal transduction is a part of these pathways and is the process that occurs whenever information is converted from one form into another. One area of particular concern is the ability to communicate across the hydrophobic cell membrane between the extra- and intracellular environments, both of which are hydrophilic.

Because most extracellular signal molecules (*ligands*) are water-soluble, they have to deliver their messages at the cell surface for transmittal of the information across the membrane and subsequently to the relevant location within the cell. This is accomplished through the agency of integral plasma membrane proteins, called *receptors*, which are integral components of the membranes. They are glycoproteins that have three functional domains: the extracellular domain to which ligands can bind, one or several transmembrane helices that penetrate the lipid bilayer of the cell surface, and the intracellular domain that initiates the internal signaling process. For these proteins, signal transduction is the process that occurs when the ligand binding on the extracellular fraction of the protein activates the relevant internal domain on the cytoplasmic side of the membrane. The intracellular signaling occurs through a further set of *effector* mechanisms that initiate a cascade of protein-protein interactions to relay the messages into the interior of the cell. Signal transduction occurs at any point along this cascade where the information is converted from one form to another. Signal transduction also involves signal amplification because the cytosolic signal is significantly greater than the original extracellular signal. Signal transduction processes are mediated by growth factors, cytokines, and hormones. These factors may influence cell proliferation in both positive or negative ways depending on the responses induced in the target cells.

There are multiple pathways for the transmission of signals on both extra- and intra-cellular levels. These include small molecules, protein kinases, G-proteins, and enzyme-linked mechanisms. One mechanism used for intercellular communication is *remote signaling* by secreted chemicals called *chemical messengers*. The extracellular messenger (ligand) serves as the "first messenger" that binds to the exterior domain of a receptor protein on the target cell. Subsequently, the induced cytosolic signal increases the synthesis of a small molecule, the "second messenger" inside the cell. Second messengers function by activating protein kinases which, in turn, complete the cascade by phosphorylating amino acid residues (mainly serine, threonine, and tyrosine) on the target proteins. The small molecule mechanism often involves G-proteins (see below) leading to the production of second messengers such as cyclic AMP. Intracellular communication by second messengers is involved in many processes, both normal and abnormal.

Another mechanism of signal transduction involves the activation of a protein kinase within the cytosolic domain of the receptor resulting in *autophosphorylation* of the receptor itself. The activated receptor protein then generates a new intracellular signal that initiates a subsequent signaling cascade within the cell, culminating in the activation of a final receptor protein at the destination site. The last protein of the cascade carries out the intended function of the initial signal, e.g., switching on the expression of a gene. This gene product may activate or repress other genes. The intracellular proteins involved in this overall mechanism behave as molecular switches, turning from inactive to active states upon receiving the signal, and then back to inactive by the action of another mechanism as the signal is passed to the next component of the cascade. An indispensable requirement of proteins to act as *molecular switches* is the ability to be phosphorylated by a protein kinase and dephosphorylated by a protein phosphatase. The transmission of a signal by a phosphorylation cascade usually involves a series of such on and off steps.

Another class of proteins that can be switched on and off are the guanosine triphosphate (GTP)-binding proteins, the G proteins. There are hundreds of G-protein-linked cell-surface receptors. These regulatory proteins are active when bound to GTP and inactive when bound to guanosine diphosphate (GDP). Activated G-proteins hydrolyze back to the GDP following the discharge of their regulatory effect, and thus become available for reactivation. G-proteins occur in monomeric and trimeric states. The most important function of the G-proteins in their small, monomeric form is their involvement in cell growth, e.g., *ras* protein. The trimeric form is involved in the complex processes of coupling receptor and catalytic enzymes, including either stimulatory or inhibitory effects on adenyl cyclase thus affecting, in turn, the intracytoplasmic concentration of cAMP which controls the interconversion of the important active and inactive states of the protein kinases. Mutations of the genes coding for various subunits of G-proteins are implicated in several neoplasms.

The *enzyme-linked* type of signal receptors are another protein-based class of messengers that are of major importance in the regulation of cell growth, proliferation, and differentiation. Signal molecules that act as local mediators controlling these enzyme-linked receptors include epidermal growth factor (EGF), platelet-derived growth factor (PDGF), and colony stimulating factor (CSF). Although a 10^{-9} to 10^{-11} M concentration of these molecules is adequate to initiate action, a relatively large number of intracellular transduction steps are required to precipitate changes in gene functions. Like G-proteins, an enzyme-linked receptor is also a transmembrane protein. Its cytosplasmic domain often functions as a tyrosine kinase capable of phosphorylating tyrosine side-chains on particular intracellular proteins, thus triggering an elaborate complex of intracellular signals which may involve as many as 10 to 20 different molecules. This may include the activation of the *ras* protein, the GTP-binding protein that, in turn, initiates a cascade of protein kinases which enables the transduction of signals from the plasma membrane to the nucleus. Intracellular tyrosine kinases are believed to stimulate the transcription of growth regulatory genes, coding for a variety of growth factors, making the existence and control of this pathway of major importance in the role of oncogenes in abnormal growth.

3.3 GENETIC ABNORMALITIES IN CANCER

3.3.1 Cancer as a Genetic Disorder of Somatic Cells

A gene is a segment of DNA that encodes, in the nucleotide base sequence, a message that defines the nature and sequence of the amino acids in a protein. The gene probably also contains the instructions to control the expression of this information. The region of DNA represented by a gene comprises discrete hereditary characteristics of an organism and as such it is the fundamental unit of inheritance. The DNA of normal human cells contains 50,000 to 100,000 genes, however, only a very small fraction of these are expressed, i.e., are in use, in a particular cell at any given time.

Genes exert their effects by transcription of a specific section of the DNA into a messenger RNA (mRNA) molecule which is, in turn, translated into a protein that is the final effector of the gene's action. Proteins are not encoded by continuous segments of DNA. Most genes contain several noncoding, intervening regions (*introns*) whose sequences are not represented in the translated product. The relevant protein sequences are dictated by the coding sequences, the *exons*. By controlling the synthesis of proteins, genes may be the ultimate determinants of phenotypic expression. Cancer is often a consequence of pathological changes in the information carried by DNA. Although it is a simplified view, transformation of normal cells into malignant cells is likely to occur when genes whose products control signal transduction, the cell cycle, genomic stability, and cellular senescence and apoptosis are altered. The consequence of the alteration is that there is a failure of one or more of these regulatory functions.

3.3.2 Mutation

A mutation is defined as a stable, heritable change in the chemical composition (primary nucleotide sequence) of a gene. Mutations may be spontaneous or arise from the interaction between the genetic material and certain external agents. Mutations may be beneficial since, in combination with natural selection, they control evolution. However, the products of mutation may also be harmful. Most (but not all) mutagens are also carcinogens. Gene mutations result from the addition or deletion of one or more bases or from the substitution into the DNA molecule of an incorrect base. Alterations may involve a single nucleotide or millions of base pairs in the structure of chromosomes. For example, in *point* mutations, only a single nucleotide is involved: a nucleotide may be lost, altered, or substituted by another, or an additional nucleotide may be inserted.

The genetic code is read in codons, composed of three-nucleotide frames which code for specific amino acids or instructions. In a mutagenic event, when the number of base pairs added or deleted is not a multiple of three, the downstream codons will be offset by one or two bases and the coded protein will have an altered amino acid sequence distal to the addition or deletion (a *frame-shift* mutation). Almost invariably this leads to changes in the biological properties of the protein. Substitution type mutations consist of *transitions* and *transversions*. The former involves the replacement of one purine by another (adenine or guanine) or one pyrimidine by another (cytosine or thymidine). A transversion occurs when a purine is replaced by a pyrimidine, or vice versa. Even change in a single codon can change amino acids in the protein, which can alter the folding of the protein. The consequences of these changes in protein function range from minor to major.

When a mutagen/carcinogen is involved, another type of change in the DNA molecule may occur. This is the covalent incorporation of a typically electrophilic chemical, or part of it, into the DNA to form *adducts*. As with the changes described above, the formation of DNA adducts results in significant alterations in subsequently coded proteins. These include substitution with a different amino acid, changes in the sequence of the amino acids, or shortened proteins (because of misreading of the DNA as if it were the termination codon). Such effects almost invariably lead to proteins with altered biological properties.

Both normal and mutated genes produce proteins. However, malignant transformation is often associated with the production of normal cellular proteins in inappropriate amounts, i.e., overexpression or underexpression (quantitative change) or in structurally altered forms (qualitative change). This implies that all proteins involved in signal transduction processes are potentially oncogenic. The major types of proteins involved in malignant transformation are growth factors, growth factor receptors, membrane-associated binding proteins, cytoplasmic kinases, nuclear proteins, and related transcription factors.

3.3.3 Oncogenes

Normal cell proliferation is governed by two opposing systems: a growth stimulatory system regulated by cellular *proto-oncogenes* and a growth inhibitory system controlled by *tumor suppressor genes*. Because these genes are present in each chromosome, progression of tumors to the deregulated growth of full malignancy may involve changes in both of these control systems, i.e., activation of oncogenes and inactivation of suppressor genes. There are significant differences between the characteristics of these two types of genes (Figure 3.3 and Table 3.3). For example, mutations in proto-oncogenes are restricted to somatic cells while mutations of tumor suppressor cells may affect either the somatic or germ lines, leading to sporadic or hereditary malignancies, respectively.

Proto-oncogenes are present in all cells and are part of the normal regulatory system of the cell. It is only when they are subject to mutation or adduction that their contribution to neoplasia ensues. Given that they are part of the normal regulation of cell function, proto-oncogenes are passed through generations. The typical protein products coded for by proto-oncogenes are growth factors, receptors for growth factors, and the enzymes involved in various phases of the signal transduction processes that eventually activate DNA polymerase and result in transcription of DNA for the production of growth regulatory proteins. Mutations that disrupt the normal functioning of

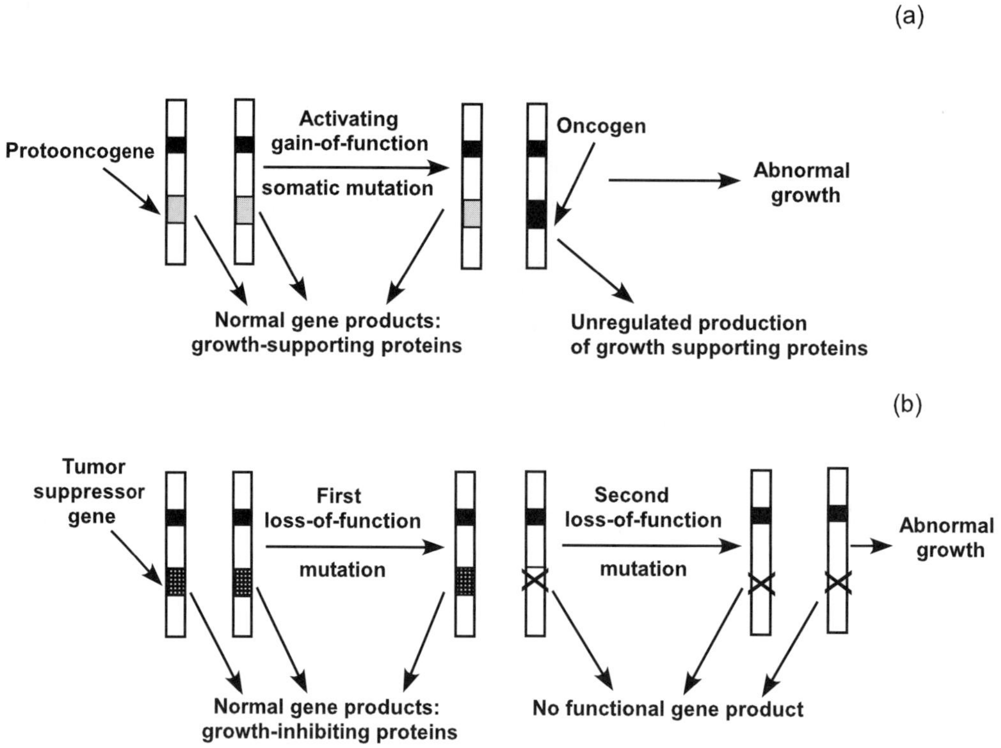

FIGURE 3.3 (a) A single, "gain-of-function" type activating mutation of a cellular proto-oncogene is adequate to produce proteins that abnormally stimulate growth. (b) Two "loss-of-function" mutation of tumor suppresor genes are necessary for uninhibited synthesis of proteins that elicit abnormal cell growth.

TABLE 3.3
Comparison of the Characteristics of Oncogenes and Tumor Suppressor Genes (Examples are shown in Tables 3.4 and 3.5)

Characteristics	Oncogenes	Tumor Suppressor Genes
Normal function	Activation of cell proliferation	Inhibition of growth promoting signals; negative regulation
Mutation	Somatic tissue	Present in germ or somatic cell
Germline inheritance	No	Yes
Number of needed mutational events	One of the two alleles is sufficient for activity	Mutations in both alleles or mutation in one followed by loss/reduction of homozygosity in the second
Types of genetic change	Overexpression Point mutations Gene rearrangements Amplification	Deletions Point mutations Frame shift mutations
Effect of mutation on protein function	Gain of function (dominant)	Loss of function (recessive)
Tissue specificity	Some	Extensive

the proto-oncogenes create *oncogenes* which, in turn, produce oncoproteins that propel the growth of tumor cells. The translocation or amplification of proto-oncogenes in chromosomes can significantly alter regulation leading to malignant transformation. Table 3.4 summarizes the main classes of oncogenes, their protein products and representative neoplasms resulting from their uncontrolled

TABLE 3.4
Examples of Uncontrolled Oncogene Expression

Type	Function	Tumor Association
	Secreted Proteins	
Growth Factors		
c-cis	PDGF family	Osteogenic sarcomas
int-2	FGF family	Breast
Receptor Tyrosine Kinases		
erb BI	EGF receptor	Glioblastoma
neu	EGF	Breast
	Transmembrane Proteins	
GTP-Binding Proteins		
H-ras	Signal transduction	Breast, skin
K-ras	Signal transduction	Colon, pancreatic
	Cytoplasmic	
Nonreceptor Tyrosine Kinases		
c-abl	DNA binding and integrin signaling	Chronic myelocytic leukemia
c-src	Same	Sarcoma
	Nuclear	
Transcription Factors		
c-myc	Nuclear regulatory	Breast, lung, Burkitt
n-myc	DNA binding	Neuroblastoma

TABLE 3.5
Examples of Tumor Suppression Genes; the Loss of Function or Deletion of These Recessive Genes Has Been Associated with Cellular Transformation

Gene	Chromosomal Location[a]	Encoded Protein (Biochemical Function)	Spontaneous Tumor
p53	17p	Nuclear phosphoprotein (Transcription factor)	~50% of all cancers
Rb	13q	Nuclear phosphoprotein (Transcription modulator)	Retinoblastoma
WT-1	11p13	Transcription factor (Nuclear protein)	Wilm's tumor
APC	5q31	(Cell adhesion)	Adenomatous polyposis
hMSH2	2p16	(DNA repair)	Nonpolyposis colon cancer

[a] Note: p = short arm, q = long arm of chromosome

expression. The most widespread cellular oncogene is the *ras* oncogene (first described in a *ra*t *s*arcoma) which occurs in about 25% of all human neoplasms (90% in pancreatic carcinomas). The *ras* proto-oncogene is apparently converted into the *ras* oncogene by a single missense mutation, the substitution of valine for glycine at codon 12. The consequence of this mutation is a reduction in GTPase activity. This results in the affected protein being maintained in its active state, thereby influencing cell proliferation. Given the importance and widespread occurrence of the *ras*-protein, it is a target of cancer chemotherapy via the inhibition of farnesylation at the carboxy terminus of the protein (Section 6.1.2.2).

The framework for the human genome is provided by packaging the DNA into 22 homologous pairs of chromosomes (autosomes) and a pair of sex chromosomes (XX in females and XY in males). Each chromosome is composed of a single duplex DNA molecule that is complexed with the nuclear protein chromatin. Chromosomal abnormalities can be observed by light microscopy when the chromosomes are spread out by cytogenetic techniques. *Numerical* abnormalities include any number other than the diploid number of 46. *Structural* abnormalities may involve chromosome breakage, followed by reconstitution in abnormal combinations, various translocations, formation of ring chromosomes and interstitial or terminal deletions. Chromosomal abnormalities have played an important role in the characterization of genetic alterations in hematologic malignancies because cells are readily accessible for analysis. For example, a characteristic cytogenetic abnormality occurs in chronic myelogenous leukemia: the *abl* proto-oncogene becomes translocated from its normal position on chromosome 9 to chromosome 22 where it fuses with the *bcr* gene. The result is a combination of the tyrosine kinase activity of *abl* with the abnormal regulation and cellular localization activity of *bcr*. The altered chromosome (*Philadelphia* chromosome) can be identified and it has diagnostic and prognostic significance for the disease. A second example is the translocation of the *myc* proto-oncogene and its malignant consequences in *Burkitt's lymphoma*. There are at least a dozen malignancies in which the information provided by cytogenetic analysis is of importance for diagnosis, prognosis, and selection of appropriate therapy.

3.3.4 Tumor Suppressor Genes

In contrast to the growth-promoting activity of oncogenes, tumor suppressor genes constitute a growth inhibitory system. When these genes are deleted, altered, or mutated, the growth-constraining action is removed and cells may become neoplastic and exhibit uncontrolled growth. Inactivation of tumor suppressor genes is frequently observed in both malignant transformations and tumor progression. Some tumor suppressor genes are associated with the regulation of signal transduction while others are of major importance in the regulation of the cell cycle (Table 4.5). Loss (by deletion of chromosome 17) or mutation of the p53 tumor suppressor gene results in the elimination or production of a protein (53 kDa molecular mass) that is structurally altered (one amino acid replaced by another in a missense mutation) in more than half of all human malignancies. The p53 protein may also be inactivated (degraded) by its interaction with some other protein, e.g., an oncoprotein from the human papillomavirus. The p53 gene has been described as both the "guardian of the genome" and "guardian of tissues" because it has a dual role in the processes that prevent the replication of damaged DNA until repair can be effected (during the G_1 phase of the cell cycle) and that promote apoptosis when repair is not possible.

3.3.5 Carcinogenesis Requires an Accumulation of Mutations

Mutations may have a variety of consequences for the cells in which they occur. It is likely that in many instances the effects are either lethal or neutral and as such are impossible to detect. The third possible result is that the mutated cell acquires a competitive advantage over its normal neighbors, and multiplies. It is now accepted that transformation into malignancy, and the establishment of the full cancer phenotype, requires at least two, but often five or more, independent mutations. There is direct evidence for the serial accumulation of independent mutations during carcinogenesis. Mutant proto-oncogenes, tumor-suppressor genes, and mismatch-repair genes have been found by histopathological techniques at different stages during the progression of cells from normal to transformed malignant cells. The number of required mutations varies from one cancer type to another. For example, multiple gene alterations in breast carcinoma involve the *c-myc* and *erbB2/neu* oncogenes and both the *RBI* and *p53* tumor suppressor genes. In contrast, colorectal carcinoma appears to involve only one oncogene (*K*-ras) but three tumor suppressor genes (*p53, MCC,* and *DCC*). These mutations typically take place over long periods of time, even decades. Thus, once a mutant cell survives and gives rise to a small similar population, there is the opportunity for another chance mutation to occur in an already altered cell. This further alteration in the genome, combined with natural selection, may then favor those cells with multiple mutations, and eventually, probably after further mutagenic steps, the neoplastic phenotype appears.

3.4 TUMOR IMMUNOLOGY

3.4.1 Hematopoiesis

About 8% of body weight is circulating blood (~5.5 L in a 70 kg person) which is a form of connective tissue. Blood consists of liquid *plasma* (~55%) in which solid *formed elements,* erythrocytes (red cells), leukocytes (white cells), and thrombocytes (platelets) are suspended. The composition of blood is affected by tumors and by treatment with chemotherapy or radiation. Blood is also the major avenue for the metastasis of cancer cells, as well as for the delivery of anticancer agents to tumors.

The main function of the hematopoietic (blood-making) system is to differentiate and proliferate specific stem cell progenitors that subsequently produce various mature cells. All cellular elements of blood derive from uncommitted, undifferentiated, pluripotent stem cells which migrate during early fetal development from the embryonic yolk sac into fetal liver, spleen, and bone marrow where the specific cells of the immune system are produced by subsequent differentiation. Hematopoietic precursors differentiate into myeloid and lyphoid lineages, which mature in the bone marrow or lymphoid tissues, respectively. In hematologic malignancies, rapidly proliferating immature cells crowd out developing normal cells and spill over into circulating blood.

Erythrocytes and Platelets

Adult erythrocytes have no nucleus, are biconcave in shape, and contain hemoglobin (~15 mg/mL blood), the iron-containing red pigment that is responsible for the transport of oxygen. Erythrocytes live for approximately four months after which they are destroyed by the spleen. Late stage cancers are often accompanied by anemia ($<4 \times 10^6$ cells/μL blood) because of a complex set of factors including decreased production of red cells and/or iron deficiency. Platelets which have a life span of ~8 days and a normal count of 150,000 to 350,000 per μL assist in the blood clotting cascade. A marked increase in the number of platelets (thrombocytosis) is associated with the early stages of certain leukemias. A marked decrease of the platelet count (thrombocytopenia, <10,000 platelets/μL blood) may occur from chemotherapy or radiation therapy resulting in bleeding in certain hematologic malignancies.

Granulocytes

The largest fraction of the circulating leukocytes consists of *granulocytes* (~5,000/μL of the total of ~8,000/μL), the characteristic granules of which can be visualized with staining. About 70% of the granulocytes are *neutrophils* (lifetime 6 to 8 h) which are the body's first line of defense against infection since they ingest (*phagocytize*) and lyse bacteria and kill other pathogens. A significant side effect of cytotoxic chemotherapy can be an abnormal decrease of circulating granulocytes (*neutropenia*) increasing the likelihood of life-threatening opportunistic infections (Section 5.1.6).

A major advance has been the manufacture of naturally occurring cytokines (cellular products that act on other cells) for the stimulation of neutrophil production, e.g., granulocyte and granulocyte/monocyte stimulating factors (G-CSF, and GM-CSF). Another subgroup of granulocytes are the *basophils* (1% of total). These are granulocytes that participate in immediate hypersensitivity and inflammatory reactions; significant increases in the numbers of these cells are associated with chronic granulocytic leukemia. *Eosinophils,* a third subgroup of granulocytes, also participate in inflammatory reactions by releasing toxic granules next to pathogens.

Macrophages

The mononuclear leukocytes include lymphocytes and *monocytes*. Monocytes have oval, notched, or horseshoe-shaped nuclei and migrate from blood to certain tissues such as the lungs, where they mature, with the aid of GM-CSF, into *macrophages.* These cells have multiple functions including fighting infection by ingesting pathogens and "processing" antigens which they then present on their surfaces thereby facilitating immunological reactions with T-cell lymphocytes.

Lymphocytes

Lymphocytes, which have few if any granules in their cytoplasm, are also derived from the pluripotent stem cells in the bone marrow. A large portion of the progeny of these cells migrate to, and differentiate in, the thymus becoming mature thymus-derived or *T-lymphocytes.* The T-cell system controls the cell-mediated response of the immune system (see later). Representing 70 to 80% of all lymphocytes, T-lymphocytes reside and circulate in the lymphatic system, where they can live for months or years and from which they can enter the bloodstream. There are four types of immunoreactive T-lymphocytes: cytotoxic T-lymphocytes (CTL), helper/inducer T-cells, suppressor T-cells, and T-cells responsible for delayed type hypersensitivity (DTH). These lymphocytes can be distinguished by their specific *cell determinant (CD)* surface antigens, e.g., CD8 for CTLs and CD4 for helper T-cells. The most important of the T-cells are the CTL and their subclasses which include *natural killer (NK)* and *lymphokine activated killer (LAK)* cells. The function of these T-cells is to react to foreign proteins including foreign cells or cells with new surface antigens, which sometimes can be killed. Target cells in the latter category are tumors, virally infected cells, and allografts.

A second set of the progenitor lymphoid cells remains in the bone marrow and differentiates to become bone marrow-derived or *B-lymphocytes.* They constitute 20 to 30% of circulating lymphocytes. The B-cell system controls the humoral (antibody) branch of the immune system (see later). The T- and B-lymphocytes, together with helper and suppressor lymphocytes, which act to regulate the activity of other lymphocytes, are of major importance in tumor immunology.

3.4.2 Antigens and Antibodies

Antibody and antigen are defined in terms of each other. An *antibody* is an immunoglobulin, with a specific amino acid sequence, which is produced by plasma cells (terminally differentiated B-cells) in response to specific recognition of a foreign, nonself molecular configuration, the *antigen* (immunogen). Each antibody has a specific affinity for the antigen that stimulated its synthesis. The body has a nearly unlimited ability to produce different antibodies against specific antigens. However, the specific affinity of an antibody is not for the entire macromolecular antigen, but only for a particular site on the antigenic molecule, called the *epitope* or *antigenic determinant,* that selects and defines the binding site of the antibody to which it is complementary (Section 6.1.5.1). Thus, the specificity in the antigen-antibody reaction is of a uniquely structured antibody against a unique epitope. Some cross-reactivity can occur if a particular molecule mimics an epitope too closely.

Monoclonal antibodies (MoAbs) are specific antibodies that are chemically and immunologically homogeneous. They derive from a single clone of cells in certain neoplasms, of plasma cell or B-cell origin, or by using *hybridoma* techniques that involve fusing, in cell cultures, specific antibody-producing lymphocytes of limited life span with immortal tumor plasma cells. Because of their high specificity in binding to target antigens, MoAbs from hybridomas are now used in diagnosis, as immunological probes in a variety of assays, and in therapy, to deliver radiation (yoked to radionucleotides) or other therapy directly to cancerous tissues.

3.4.3 Types of Immune Response

Immunity, which may be nonspecific or acquired, is a mechanism for the protection of the integrity of the body against foreign substances or agents. The theory of *nonspecific* immune response (which is innate) postulates a continuous immune surveillance by macrophages and natural killer cells that recognize as foreign cancer cells, which arise continually during lifetime, and destroy them. In contrast, *acquired immunity* takes weeks to mature, and its level of functioning declines with advancing age (a possible cause for the increased incidence of cancers in the elderly). Acquired immunity has two mechanisms of action, humoral and cell-mediated. In both types lymphocytes are the effector cells. *Humoral immunity* is elicited when an antigen triggers the immune system to rapidly produce B-cells which further differentiate into plasma cells (potentiated by T-helper cells and inhibited by T-suppressor cells) which, in turn, produce antibodies against the offending antigen. *Cell-mediated* immunity is more important with respect to tumor cells. The cascade starts with a reaction of macrophages to the tumor antigen. These activated antigen-presenting cells trigger a proliferation of T-cells which, in turn, induce a series of immunologic reactions, eventually producing CTLs to attack and kill the foreign cells.

To launch a protective response against something foreign invading the body, the immune system must be able to distinguish *self* from *nonself.* The immune system responds to some tumors in a manner similar to that for transplanted foreign tissues. The antigens expressed on cell surfaces, called *histocompatibility* or *transplantation antigens,* determine the compatibility or incompatibility of a transplanted tissue. The most important histocompatibility antigens are found in high concentrations on lymphocytes and other white blood cells. These are the *human leukocyte antigens (HLAs)* which are the products of the genes that make up of the *major histocompatibility complex (MHC)* (see Section 6.1.5).

As a malignant tumor develops, the host is interacting with *tumor-associated antigens* (TAAs), the nature of which may change continuously. Examples of antigens associated with human malignancies include: (a) oncofetal antigens, e.g., carcinoembryonic antigen (CEA) in colorectal and pancreatic carcinoma, and α-fetoprotein in primary liver and some testicular and ovarian cancers; (b) viral antigens, e.g., hepatitis B in primary liver cancer, Epstein-Barr in Burkitt's lymphoma, and human papilloma viruses in cervical carcinoma; and (c) other antigens, e.g., prostate-specific antigen and prostatic acid phosphatase in prostatic cancer, and certain glycoproteins in solid tumors. Mass spectrometric aspects of relevant antigens are discussed in Chapter 6.

The new antigens presented by tumor cells are recognized by the immune system as foreign. In response, specific antibodies, natural killer cells, and activated macrophages are mobilized as the body attempts to eliminate the tumor cells. As neoplastic cells are constantly developing over a lifetime, it must be the case that the immunogenic response is usually successful. However, there are some tumors that are simply not immunogenic, and even for tumors that are immunogenic, there are several potential mechanisms to escape the immune surveillance of the host. These mechanisms include the secretion of immunosuppressive molecules; the production of tumor-growth-enhancing antibodies which block acess to the tumor by killer lymphocytes; the loss of adhesion molecules thus preventing CTLs from attaching to tumor cell membranes; the modulation of antigens on the surface so they are no longer available to target the killing of lymphocytes that make contact with tumor cells; and the production of soluble antigens that neutralize the host's protective responses. Other factors that can favor tumor growth include reduced immune response in the elderly and immunologic deficiency which may be hereditary or induced by infection (e.g., HIV) or immunosuppressive drugs.

3.5 DIAGNOSIS AND TREATMENT

3.5.1 Clinical Presentation

The signs and symptoms of cancer vary widely, depending on the type and site of the tumor, as well as the stage of its development. Tumors of the skin and breast often present as lumps. Hollow tubes in the body, such as trachea, bronchi, ureter, intestine, and bile duct can be partially occluded by tumors, leading to symptoms from compromise of their function. Cancer can ulcerate and bleed into any of these hollow pathways. Pain occurs when hollow organs attempt to overcome the obstruction by contractions or when the tumor cells press on nerve fibers. Cancers can lead to weight loss by competing for energy supplies and diminishing appetite. Because the symptoms of cancer often imitate those of a broad category of other diseases, physicans must first eliminate possible alternative causes of the symptoms. Cancer diagnosis has enormous importance, but a false diagnosis is worse. Cancer sometimes presents with nonspecific constitutional complaints, such as anorexia, weight loss, fatigue, generalized malaise, or fever. Also, it is not uncommon for cancers to be identified incidentally during routine diagnostic or screening procedures, without any symptom at all.

A wide variety of additional symptoms may become apparent when there is spread of the malignancy to other organs. Pulmonary metastases are common because the lung is often the first organ to act as a filter and trap for malignant cells. Metastasis to the liver may originate via portal venous circulation from the gastrointestinal tract from cancers of colon, rectum, pancreas, and stomach. Metastasis to liver, lung, bone, or brain is generally associated with a grim prognosis.

3.5.2 Diagnostic Workup

The diagnostic workup of cancer, depending on the tissue and systems involved, may consist of several components aimed at locating tumors: (a) detailed medical history and physical examination; (b) laboratory evaluation, including complete blood count, serum chemistries, liver function tests, and special chemical tests for particular organ system tumors; (c) radiographic studies, including x-ray, mammogram, barium study (GI tract), intravenous pyelogram (IVP, urinary tract), computerized tomography (CT, multiple cross sectional images of internal structures), magnetic resonance imaging (MRI), ultrasonography, positron emission tomography (PET), and nuclear medicine studies (bone scan, thyroid scan, gallium scan); (d) endoscopy (e.g., bronchoscopy, colonoscopy, gastroscopy for visualization and to obtain tumor samples for pathology); (e) laparoscopic laparotomy; and (f) bone marrow aspiration and biopsy.

Diagnostic Cytology, Histology, and Cytogenetics

Cytology involves the microscopic examination of cells derived from a neoplasm. Observable morphological signs of malignancy include changes of the nucleus and nucleolus, e.g., variable shape and size of nuclei (*pleomorphism*) and the pronounced presence of nuclear chromatin (*hyperchromatism*). The microscopic examination of tissues by a pathologist is still the most reliable method to decide: (a) whether the structure of the candidate cells is significantly different from that of normal cells; (b) whether or not the tumor is malignant; (c) the type of malignancy; (d) the degree (grade) of malignancy; and (e) the extent to which a tumor has spread. Tissue samples for histological examination may be obtained using a variety of techniques, e.g., from the tumor removed by surgery (excisional biopsy) or from small samples removed by incisional biopsy. Cytology techniques look at cells scraped from the surface of tumors, washed from the surfaces (as in sputum or bronchial lavage), or aspirated from solid tumors such as lymph nodes, liver, or lung masses. Cytological techniques study cellular detail rather than tissue architecture. In addition to the standard microscopic techniques, there are powerful immunohistochemical methods that use antibodies (both polyclonal and monoclonal) against tumor associated- or tumor-specific antigens, e.g., cell surface proteins, mucins, and glycolipids. After applying the antibody solution to cells or tissue sections, the product of the antibody-protein reaction is detected by a second antibody against the foreign protein of the first antibody, marked by a fluorescent probe or one of several available enzymatic systems such as horseradish peroxidase, the use of which generates a colored product upon reaction with a substrate. Other diagnostic techniques of importance include electron microscopy, flow cytometry, and *in-situ hybridization* methods using labeled probes to detect specific nucleic acid segments that are characteristic of gene rearrangements associated with particular neoplasms.

As mentioned already, both hematologic neoplasms and solid tumors are often characterized by chromosomal abnormalities. Cytogenetic techniques look at "spreads" of individual chromosomes. The procedures are complex and labor intensive and require culturing of the cells from a buffy coat of peripheral blood, inspection to identify neoplastic cells, followed by the analysis of their chromosomes to yield a *karyotype*. The karyotype obtained is matched against the normal human pattern of 23 pairs of chromosomes aligned together and in order. Karyotypes may reveal deletions, duplications, and/or rearrangements of chromosomes that may help to localize the region of DNA abnormally regulated in a particular neoplasm. The information provided by cytogenetic analysis of malignant cells is often important for diagnosis, selection of appropriate therapy, and also as a prognostic indicator. For example, the Philadelphia chromosome (translocation between chromosomes 9 and 22) is present in >90% of patients with chronic myelogenous leukemia.

Tumor Markers

Markers should ideally fulfill a number of criteria, including specificity to the tumor, a quantitative relationship with tumor mass, and detectability in microscopic disease burden. Ideally, markers should become positive with a reasonably long lead time to permit the initiation, or a change of treatment, before the disease becomes advanced. Tissue tumor markers may be divided into *diagnostic* (e.g., providing differentiation between cell types), *prognostic* (e.g., for angiogenesis), or *predictive* (e.g., for estrogen receptor-positive breast tumors). In April 2000, the Early Detection Research Network was established by the NCI to develop and validate effective cancer markers. Evolving mass spectrometric approaches based on proteomic technology, such as surface-enhanced laser desorption ionization (SELDI, Section 2.1.6) have already yielded promising results in correlating cellular patterns of protein expression with early detection of several cancers (Section 6.1.4).

An important function of serial laboratory tests for markers is monitoring for progression or regression of disease. For example, the serial measurment of carcinoembryonic antigen (CEA) in colon cancer and prostate specific antigen (PSA) in prostate cancer may lead to additional surgical interventions or changes in the course of chemotherapy. Mass spectrometric techniques for the characterization of several tumor markers, e.g., PSA, *ras*, and mucins, are illustrated in Chapter 6.

Grading and Staging

Grading of the growth stage of a neoplastic tumor is based on the degree of cellular anaplasia that can be observed microscopically. There are also a variety of relevant morphological and other criteria, such as necrosis, that vary from tumor to tumor. Less differentiated tumor cells are generally more malignant or virulent. Grade I neoplasms are well differentiated and closely resemble the normal parent cells. Grade II and Grade III cells are moderately and poorly differentiated, respectively, i.e., there is progressively less resemblance of the tumor cells to those of the tissue of origin with increasing variation in the size and shape of the cells and greatly increased frequency of mitoses. Although the prognostic value of grading is controversial in some tumors, progression to higher grades does generally correlate with increased aggressiveness of the lesion. Specific grading systems that often include distinct subpatterns have been developed for individual cancers.

By combining the results of all available diagnostic tests, clinicians are able to establish the stage of the disease, i.e., the extent of the spread of the malignancy in a patient. In the TNM system (tumor-nodes-metastasis), T describes the size and degree of local extension of the primary tumor, N discloses clinical findings in regional lymph nodes, and M indicates the presence or absence of metastasis. Each category is further qualified by a number indicating the extent of involvement according to distinct clinical criteria. In general, Stage 1 refers to cases where the tumor mass is limited to the organ of origin, and the lesion is resectable. The chance of survival is usually good. Stage II indicates local spread into surrounding tissues and/or first-station lymph nodes. The completeness of surgical removal of the entire tumor becomes uncertain, and the chance of survival is often reduced. Stage III refers to extensive local invasion by the primary tumor with malignant invasion of lymph nodes. Surgery can often be performed only for debulking purposes and the probability for long-term survival may be significantly reduced. When distant metastasis is evident, the patient has a Stage IV tumor which is generally inoperable and often not amenable to curative treatment.

3.5.3 Treatment Modalities

There are four treatment modalities for cancer: surgery, radiotherapy, chemotherapy, and biotherapy. In *adjuvant* therapy, one treatment modality is used to supplement another. The ongoing integration of molecular medicine and clinical practice has resulted in profound changes in therapeutic regimens during the last decade.

Surgical resection, which is the oldest and most frequently used modality, aims at reducing the tumor burden to such a degree that the remaining malignant cells can be destroyed by the immune system or by systemic antineoplastic chemotherapy. In *radical surgery*, both the tumor and nearby tissues and lymph glands are removed. *Debulking surgery* is used when the malignancy cannot be removed totally by resection, but a significant decrease of the tumor burden is expected to enhance the efficacy of other treatment modalities. *Palliative surgery* is used to relieve symptoms, provide comfort, and reduce the development of new symptoms. *Preventive surgery*, particularly where there are strong hereditary factors, is gaining support.

Radiation therapy is used when the tumor is unsuitable for surgery, and in combination with limited surgery, e.g., lumpectomy in breast cancer, or in conjunction with chemotherapy. The intent may be curative, e.g., for oral cancers or Hodgkin's lymphoma, or palliative, e.g., for treatment of localized bony metastases. The objective is to destroy the tumor by damaging its DNA content. The radiosensitivity of tumors varies widely. *External beam radiation* treatment entails the irradiation of the tumor by x-rays or gamma rays. *Internal radiation* therapy (*brachytherapy*) involves the placing of a radioactive substance, like radioactive iridium in a sealed container, directly into the tumor, or a soluble radioisotope such as radioactive iodine into the body by ingestion or injections.

Chemotherapy is used alone, or in combination with surgery and/or radiation, to treat a wide variety of cancers. Initially, the approaches to chemotherapy emphasized mechanistic targets and DNA damage. As knowledge of the biology of cancer has increased, however, so has the opportunity to tailor chemotherapeutic regimens against specific molecular targets, e.g., to interfere with cell cycle control or induce apoptosis. The basic principles and strategies of chemotherapy are summarized in Chapter 5. This is followed by a discussion of the metabolism, mechanism of action, and pharmacokinetics of the various classes of antineoplastic and chemopreventive drugs and phytochemicals, with particular reference to the application of mass spectrometric techniques for the quantification and study of the metabolism of individual agents.

Biotherapy, which attempts to use immunology to destroy cancer cells, has a long history but was, until recently, rarely the treatment of choice. It is divided into *active* biotherapy in which the patient's immune system is stimulated to fight the tumor, or *passive* biotherapy in which external factors, e.g., monoclonal antibodies, are used to treat the tumor. In active immunotherapy, the *biological response modifiers* (BRMs) that have been used for *nonspecific* stimulation of the immune system include: (a) macrophage-activating agents, e.g., Bacille Calmette-Guérin vaccine (an attenuated strain of *Mycobacterium tuberculosis*); (b) synthetic molecules that induce interferon production, e.g., polyinosinic-polycytidylic acid; (c) cytokines that act on lymphocytes, e.g., interleukin-2; and (d) hormones that modulate T-cell function, e.g., thymosin. Passive immunotherapies include monoclonal antibodies, e.g., against CD20 for the treatment of lymphomas, against CD33 for the treatment of acute leukemias, and against Her/2 for breast cancer. Monoclonal antibodies have been used alone or coupled to drugs, pro-drugs, toxins, cytokines, or radioactive isotopes. Several cytokines have also been used for immunotherapy, e.g., interferon-α (prolonged remissions of hairy-cell leukemia), interleukin-2 (remissions of renal cancer and melanoma), and tumor necrosis factor-α (reduced malignant ascites). Several examples are discussed in Chapter 6.

A new area of biotherapy is addressing genetic defects by *gene targeting,* i.e., correction of the faulty gene, or *gene augmentation,* i.e., the addition of a functional or therapeutic gene. The methods use vectors that carry a homologous but genetically altered segment of DNA that is to be introduced at a specific gene locus. The therapy is based on the ability of cells to carry out *homologous recombination,* i.e., DNA molecules of similar sequence can interact and undergo either reciprocal exchange or unidirectional transfer of genetic information. Gene targeting has been successful in generating strains of mice with defined genetic alterations. Despite numerous ongoing clinical trials, gene therapy is still experimental; the prospects are exciting, however.

3.5.4 Functional Status, Complications, and Causes of Death

The functional (performance) status of patients is a major prognostic factor during both diagnosis and treatment. There are two scales that are used to evaluate the patient's ability to perform the normal activities of daily living. The "Karnofsky scale" ranges in steps of 10%, from 100% (asymptomatic) to 0% (death). The system of the Eastern Cooperative Oncology Group ranges in four steps from asymptomatic, through symptomatic, in bed <50% of the day, to completely bedridden. These functional status assessments do not, however, adequately assess *quality of life* issues, which are a combination of both subjective factors such as pain and objective factors such as weight loss.

Complications resulting from advanced neoplastic disease are frequent and may be local or systemic in nature. The development of *malignant effusions* in the pleural, peritoneal, and pericardial spaces is common with many solid tumors and may cause urgent therapeutic problems. Aspirated fluids may be routinely examined for cell count and differentiation, cytological study, and protein content. They may also serve as a source of material for analytical biochemical studies aimed at aiding the diagnosis or monitoring of disease progression, e.g., hyaluronan in malignant mesothelioma (Section 6.4.1.4). Systemic complications include *opportunistic infections* that occur as a result of impaired host defense mechanisms or the myelosuppressive and immunosuppressive effects of chemotherapy. Mass spectrometric techniques for the rapid diagnosis and monitoring of systemic candidiasis are described in Section 5.1.6.

Major causes of death from cancer include infection, organ failure, carcinomatosis, and hemorrhage. Infections may be the common causes of pneumonia or caused by gram negative bacilli such as *Escherichia coli, Klebsiella* species, and *Pseudomonas aeruginosa,* and fungi such as *Aspergillus* or *Candida* species. Dysfunction of the respiratory, cardiac, hepatic, renal, or central nervous system leads to organ failure. Thromboses in the lung, heart, mesenteric arteries, or brain may also cause death. Carcinomatosis associated with advanced metastatic disease can be clinically undetected, but autopsy may reveal extensive invasion of vital organs. Hemorrhage, secondary to thrombocytopenia (abnormally low platelet count), or invasion of major blood vessels occur in the GI tract, lungs, and brain. As a result of multiple combined complications and metabolic derangements in patients with advanced metastatic disease, it is often not possible to establish an immediate single cause of death.

3.6 SOME PERTINENT EXPERIMENTAL METHODS

3.6.1 DNA and RNA Analysis

Because the DNA molecules in human chromosomes are enormous (~10^8 base pairs), they must be broken into smaller fragments for subsequent isolation, probing, cloning, or sequencing. The key to the reduction of the DNA into more manageable fragments is the *restriction endonucleases.* These are bacterial enzymes that cleave double-stranded DNA selectively into large double-stranded restriction fragments of 1 to 30 kb sizes.

Southern Blot Analysis

The technique is used for the separation and study of restriction fragments. The first step is the extraction of high molecular mass DNA from cells in the sample of interest, e.g., blood, bone marrow, surgical biopsy specimen, or fine needle aspirate, using organic solvents, proteases, and detergents. Next, the extracted DNA is digested with restriction endonucleases and the fragments generated are separated by size using electrophoretic techniques on cross-linked agarose or polyacrylamide gels. In the case of large fragments, pulsed field gel electrophoresis is used to achieve the separation. In this method the direction of the electric current is alternated between several different sets of electrodes. Because the restriction fragments are double-stranded and the *molecular probes* used for recognition (see below) can detect only single-stranded DNA, the separated DNA fragments must be denatured with NaOH into single chains. This is followed by transfer of the fragments onto a nitrocellulose membrane support by blotting (capillary action), vacuum suction, or electrophoretic transfer. The blots thus obtained are assayed for specific base pair sequences by hybridization with *molecular probes.* These probes are single-stranded synthetic oligonucleotides (≥17 nucleotides long) that are complementary to a sequence on the restriction fragment. To facilitate final recognition, the probes are labeled with a radionuclide (tritium or ^{32}P), a fluorescent group, or moieties for specific chemical reactions. Southern blotting requires ~10 μg DNA (~1 mL blood or ~10 mg biopsy tissue). The primary use of this technique is to isolate and identify size differences in the restriction fragments obtained from both normal and abnormal cells. Differences may reflect variations between normal individuals, clonal rearrangements or other pathological changes. Another technique, *Northern blotting,* is used, in a fashion similar to Southern blotting, for the recognition of alterations in the quantity or nature of transcribed RNA with known base sequences.

3.6.2 Polymerase Chain Reaction

This technique, also called PCR (more details in Section 6.3.2.1) amplifies small target segments of DNA (up to 1 kbp) by multiple repetitions of three basic steps: (1) denaturation of double-stranded DNA at ~90°C; (2) annealing of primers complementary to both strands of DNA flanking the target segment at ~60°C; and (3) chain elongation by a heat-stable DNA polymerase (usually Taq, derived from a thermophilic bacterium) at ~72°C. The sensitivity of PCR is up to 10^4-fold higher than that of Southern blotting but PCR is limited to approximately 1 kbp. The most important application of PCR is the production of large quantities of homogenous DNA from ng quantities of DNA for subsequent cloning and sequencing. Other applications include the detection of point mutations by amplifying gene segments and the recognition of chromosomal breakpoints clustered in narrow regions (up to 1 kbp). PCR has also been extended to the amplification of RNA by adding a reverse transcriptase step to produce cDNA for use as a substrate in the RNA amplification process. Relevant applications are described in Section 6.3.3.

3.6.3 Determination of Clonality

Significant clonal proliferative advantage develops when abnormal cells constitute the majority of cells present. The analysis of clonality provides valuable insight into the pathogenesis, molecular genesis, and often the clinical prognosis of several neoplasms. Approaches to determine clonality include cytogenic analysis, fluorescence in-situ hybridization (FISH), microsatellite analysis to determine loss of heterozygosity (LOH), X-inactivation, and characterization of point mutations.

In FISH, solubilized, fixed metaphase cells are hybridized with probes consisting of specific cDNA probes into which biotin-containing nucleotides, digoxigenin, or fluorescence dyes have been incorporated. Counterstaining with fluorescein isothiocyanate-labeled avidin or other detector reagents provides the number of copies of specific chromosomes or genes depending on the nature of the cDNA probe used. In addition, chromosomal location of specific genes may also be obtained by mapping the fluorescence spots.

Because tumor suppressor genes are recessive, it requires inactivation of both alleles before their activities are lost. This may occur by accumulation of inactivating mutations in both alleles of the gene or by a mutation in one allele and the excision of the second allele by formation of a microsatellite or other chromosomal change that renders that locus haploid instead of diploid. This LOH may be assayed by PCR amplification of polymorphic microsatellite markers for a particular gene or locus. The loss of genetic material from one allele at a specific genetic locus is often associated with the loss of function of tumor suppression genes. The occurrence of LOH at a given locus may identify a cell population with a clonal proliferative advantage. A disadvantage of LOH analysis is that the identity of the gene to be analyzed must be known.

The basis of the X-inactivation clonality assay is that the pattern of random X-activation (Lyonization) that occurs in females during embryogenesis is reproduced in all progeny, including tumor cells. Active and inactive X-chromosomes can be differentiated, at a number of X-linked polymorphic sites, based on differential methylation and expression. The paternal and maternal copies can be differentiated simultaneously. X-inactivation assays assess the presence or absence of clonally derived cells based on the state of expression of the X-chromosome. In contrast to LOH, the assay does not require prior knowledge of the nature of the mutation that gave rise to the clonal proliferation of the cells. Similar to LOH, this test cannot measure residual disease because it fails to detect less than a level of 10% clonally derived cells in a background of polyclonal cells.

Both germ line and acquired point mutations are implicated in the pathogenesis of cancer (e.g., p53 and *ras* genes), and as such may be used as clonal markers of disease when their frequency is high in a given neoplasm.

3.6.4 Cell Cultures

Permanent cell cultures from vertebrate organisms are of two kinds, those derived from nontumorigenic tissues and those that are neoplastic in origin. Normal vertebrate cells grow in culture only for a limited number of divisions (~12 generations for mouse, ~40 for human), after which there is a poorly understood period of *crisis* during which most of the cells die. Those cells that survive the crisis become *established* and can be perpetuated indefinitely. Although several specific properties of such established nontumorigenic lines differ significantly from those of the primary line (e.g., chromosome complement), the general characteristics of normal cells are retained, such as their cytoskeletal organization, growth-controlling properties including dependence on anchorage and nutrients (serum or growth factors), and density-dependent proliferation inhibition. These features are in contrast with those of cells cultured from tumors that have undergone neoplastic *transformation* resulting in an ability to grow in a much less constricted fashion than normal cells. Transformation resulting in lack of proliferative inhibition and immortalization are considered the hallmarks of tumors in cell culture.

An example of the many types of applications of cell cultures is the comparison of normal and transformed cells in efforts to identify the genetic changes involved in the process of transformation. Other application areas include the decades-long use of tumor cell cultures for the investigation of the cytotoxic properties of candidate antineoplastic agents, and the study of the efficacy of drugs against slowly growing solid tumors. A wide variety of animal and human types of cell cultures have been developed for the controlled study of cellular phenomena and processes that cannot be observed or followed in humans. The American Type Culture Collection has some 950 cancer cell

TABLE 3.6
Major Types of Mouse Models Available for Research

Type	Definition	Characteristics
Inbred	Sibling mated for >20 generations	Genetically alike, homozygous at all loci
Hybrid	Crossing of 2 different inbred strains	Heterozygous at all loci at which parents have different alleles
Spontaneous mutation	Genetic change that occurs rarely (includes mutations induced by radiation)	Wildtype phenotype becomes aberrant
Induced mutation	(a) Transgenes* (b) Targeted ("knockout")* (c) Retroviral (d) Chemical	Specific, designed for distinct applications
Congenic	Mating strain with a gene of interest to an inbred strain. Progeny still carrying the gene are mated with the same inbred strain. Selected mating repeated for each generation.	Identical at all loci except for the transferred locus and a linked segment of chromosome

* See text for description.

lines, including ~700 human cancer lines, descriptions of which are available on the Internet at www.attc.org.

3.6.5 Experimental Mouse Models

There is a rapidly expanding supply of mouse models, bred or "designed" for medical research, ranging from dermatology to toxicology. Searchable databases are available on the Internet, e.g., www.jax.org/jaxmice. In cancer research, mouse models that display an increased incidence for a wide variety of tumors are available, e.g., the well-known BALB strain for mammary gland tumors. In addition, dozens of models carrying specific mutations are available for specialized biology research. These include oncogens, tumor suppressor genes, growth factors, receptors, cytokines, as well as strains suited to toxicological research. The major model types are listed in Table 3.6.

Transgenic Animals and Gene Targeting

Cloned DNA can be altered *in vitro* to create mutant genes. In *transgenic* animals, one or more new genes have been incorporated. The inclusion of these genes may result in the expression of their own characteristic products. Alternatively, they may influence the normal genes of the mouse through suppression, overexpression, or mutation through insertion at that point in the genome of the innate mouse gene of interest. The method involves the microinjection of ~100 copies of a cloned gene (the transgene) that has the desired regulatory elements into the male pronucleus of a fertilized oocyte and implanting it into a pseudopregnant mouse. A small fraction of the embryos (<1 in 10^6) will have one or more copies of the exogenous gene incorporated at random positions into any of the chromosomes before the first cell division, thus making the animals heterozygous for the transgene. The heterozygous transgenic progeny are inbred to create a transgenic line that is homozygous for the desired altered genes. The presence of the modified gene may be confirmed by Southern blotting.

The unique integration site of the transgenes in the host chromosomes may alter endogenous gene expression producing altered phenotypes. There are at least three scenarios for the influence and activity of the transgene on the normal cells. These are: (a) gene replacement where only the mutant gene is active; (b) gene "knockout" in which neither gene is active; and (c) gene addition, where both the endogenous and the added mutant gene are active. The properties or behavior of a particular molecule, expressed by a gene in specific cells, or the function of the whole organism may be studied using transgenic animals.

An alternative approach to the introduction of modified cloned genes at random positions is *gene targeting* where a generated or cloned gene (e.g., a drug resistance gene) is transferred to an animal to interact or recombine with the endogenous gene for a particular protein. This renders the targeted gene nonfunctional by deletion (*knockout mouse*) or point mutation. The altered gene is next injected into a pluripotent embryonic stem cell where it recombines with the endogenous gene, followed by injection into a blastocyst and transfer to the uterus of pseudopregnant females. The chimeric mice, which carry the targeted genes in their germ line, are then used to establish the desired specific model. Gene targeting by this "homologous recombination" process makes it possible to produce a wide variety of mutants for many genes.

RECOMMENDED BOOKS

Alberts, B., Bray, D., Johnson, A., Lewis, J., Raff, M., Roberts, K., and Walters, P., Eds., *Essential Cell Biology*, Garland Publishing, New York, 1998.

Bast, R.C., Kufe, D.W., Pollock, R.E., Weichselbaum, R.R., Holland, J.F., and Frei E., Eds., *Holland-Frei Cancer Medicine*, 5th ed., Decker, Hamilton, Ontario, 2000.

DeVita, V.T., Hellman, S., and Rosenberg, S.A., *Cancer Principles and Practice of Oncology*, 6th ed., Lippincott, Philadelphia, 2001.

Lewin, B., *Genes VII*, Oxford University Press, Oxford, 2000.

Mendelsohn, J., Howley, P., Israel, M., and Lioppa, L., Eds., *Molecular Basis of Cancer*, 2nd ed., Faunders, New York, 2001.

Ross, D.W., *Introduction to Oncogenes and Molecular Cancer Medicine*, Springer, New York, 1998.

Ruddon, R.W., *Cancer Biology*, 3rd ed., Oxford University Press, Oxford, 1995.

Tannock, I.F. and Hill, R.P., Eds., *The Basic Science of Oncology*, 3rd ed., McGraw-Hill, New York, 1998.

Vogelstein, B. and Kinzler, K.W., Eds., *The Genetic Basis of Human Cancer*, McGraw-Hill, New York, 1998.

4 Metabolism and Biomarkers of Carcinogens

4.1 CLASSIFICATION, MECHANISM OF ACTION, AND BIOMARKERS

4.1.1 Classification and Risk Assessment

There are more than 6 million chemical compounds registered with the Chemical Abstracts Service. About 10% of these compounds are used commercially, and fewer than 1000 have been thoroughly evaluated with respect to their potential for cancer causation. In a large, ongoing study (>100 books published), the International Agency for Research on Cancer (IARC) has been evaluating carcinogenicity in humans and laboratory animals based on the following groupings: chemicals, groups of chemicals, complex mixtures, occupational exposures, behavioral and life-style exposures, biological agents, and physical agents. The IARC classifies the *evidence of carcinogenicity* into four categories.[1] Group 1: members are carcinogenic to humans, based on sufficient evidence that a causal relationship has been established between exposure and cancer (Table 4.1). Group 2A: members are probably carcinogenic to humans, and Group 2B: possibly carcinogenic to humans (Table 4.2). Group 3 classification means that the available information is inadequate with respect to quality, consistency, or statistical power. Demonstrated lack of human carcinogenicity results in classification into Group 4. There are other evaluation systems, such as those by the NTP, ACGIH, and NIOSH. There are no major variations with respect to the classification of the most important carcinogens.[2] Yet another classification is based on potency grading, defining *potency* as the magnitude of carcinogenic activity of a given dose of a chemical in individual species. Ten criteria are listed that affect allocation to high-, medium-, and low-potency groups.[3]

Classifications are revised continually, based on published reports and committee recommendations.[4] For example, in 1997 dioxin was designated as a known human carcinogen, 1,3-butadiene (occupational exposure occurs in the production of synthetic rubber) was upgraded from "reasonably anticipated" to "known" human carcinogen, and tobacco was named as the proven cause of a variety of cancers. Several other compounds, such as chloramphenicol and 5-nitro-o-anisidine, were delisted because of insufficient evidence of human carcinogenicity. The 8th Report on Carcinogens by the NTP presents profiles for agents, substances, mixtures, or exposure circumstances known and reasonably expected to be human carcinogens (http://ehis.niehs.nih.gov/roc/toc8.html).

Based on their mode of action, carcinogens are classified as epigenetic (nongenotoxic) or genotoxic. *Epigenetic* carcinogens do not react with DNA but promote the growth of transformed cells by increasing the rate of cell multiplication.[5] In contrast, *genotoxic* carcinogens or their enzymatically activated metabolites initiate carcinogenesis by producing DNA damage, including mutations, and attacking other cellular macromolecules (Section 4.1.2.2). Almost all known human carcinogens are genotoxic. Those chemicals that are mutagenic based on the bacterial Ames test[6] and/or on the liver cell-based Williams test[7] are always genotoxic. Genotoxic carcinogens may be subdivided based on the type of residues transferred to DNA into groups of alkylating, arylaminating, and aralkylating agents (Figure 4.1). Alkylating agents include mustards, aliphatic epoxides, nitrosamines, triazines, nitrosoureas, haloalkanes, sulfones, lactones, and aflatoxins. Arylaminating agents include aromatic amines and amides, nitroaromatics, aminoazo dyes, and heterocyclic aromatic amines. Aralkylating agents include the polycyclic aromatic hydrocarbons, pyrrolizidine alkaloids, and alkenyl benzenes. This system eliminates the division of carcinogens into those

TABLE 4.1
Known Human Carcinogens (Group 1)

Individual Compounds	Cancer Site	Use or Occurrence
Aflatoxins	Liver	Toxins from some *Aspergillus* species
4-Aminobiphenyl	Bladder	Color additive, rubber antioxidant
Benzene	Leukemia, Hodgkins	Principal component: light oil, solvent
Benzidine	Bladder	Dye manufacture
Bis (chloromethyl) ether	Lung	Solvent
1,3-Butadiene		Production of synthetic rubber
Chloromethyl methyl ether	Lung	Solvent
2,3,7,8-Tetrachlorodibenzo-*p*-dioxin (TCDD)		Environmental contaminant
Diethylstilbestrol	Testis, vagina	Growth promoter: cattle, sheep
Ethylene oxide	Leukemia	Ripening: fruits; fumigant: food, textiles
Mustard gas	Lung	
2-Naphthylamine	Bladder	
Vinyl chloride	Angiosarcoma, liver	

Elements and Their Compounds
Arsenic (lung, skin, hemangiosarcoma), beryllium (lung), cadmium (prostate), chromium (lung), nickel (nasal, lung).

Mixtures
Alcoholic beverages, coal tars and pitches, untreated mineral oils, salted fish (Chinese style), shale oils, tobacco (all products, smoke, smokeless betel quid).

Drugs and Medicinals
Azathioprine, busulfan, chlorambucil, chlornaphazine, cyclosporin, cyclophosphamide, 8-methxypsoralen (with UV radiation), combined chemotherapy with alkylating agents, melphalan, methyl-CCNU, estrogens (replacement therapy, nonsteroidal), oral contraceptives, thiotepa, tresulfan.

Miscellaneous Other
Asbestos, talc and erionite (lung, mesothelioma); Helicobacter pylori (stomach), chronic hepatitis B or C virus (liver); radon and decay products (lung), solar radiation (skin).

Based on various IARC publications.

intrinsically reactive to DNA, e.g., alkylating agents, and those requiring metabolic activation, e.g., nitrosamines.[8]

4.1.1.1 Risk Assessment

There have been numerous attempts to estimate the risk factor for specific cancers and determine the responsible carcinogen. For example, nitrosamines and polyaromatic hydrocarbons are the most likely responsible agents (*causal* role) in ~75 to 90% of cancers of the lung, bronchus, larynx, oral cavity, and esophagus in Western countries; 4-aminobiphenyl and 2-naphtylamine are estimated to be the *contributory* cause in ~37 to 45% of all bladder cancers.[9–13] Possible carcinogenic hazards from common human exposures to rodent carcinogens, including naturally occurring chemicals in the diet, have been ranked based on the human exposure/rodent potency (HERP) index (http://potency.berkeley.edu/herp.html). It was suggested that animal tests, which are often carried out at the maximum tolerated dose, may be misinterpreted to mean that low doses of synthetic chemicals and industrial pollutants are relevant to human cancer. It has been argued that "there are more rodent carcinogens in a single cup of coffee than potentially carcinogenic pesticide residues

TABLE 4.2
Probable and Possible Human Carcinogens

Probable	Possible
Acrylonitrile	Acetaldehyde
Benz[a]anthracene	Benzo[b]- [j]- and [k]fluoranthrene
Benzo[a]pyrene	Carbon tetrachloride
Cadmium and Cd compounds	Chloroform
Dibenz[a,h]anthacene	Chlorophenols
Ethylene dibromide	Dichlorodiphenyltrichloroethane (DDT)
Formaldehyde	Dibenz[a,h]- and [a,j]acridine
Polychlorinated biphenyls	Dibenzo[c,g]carbazole
Propylene oxide (propylene)	Dibenzo[a,e]-, [g,h]-, [a,i]-, and [a,l]pyrene
Styrene oxide	*p*-Dichlorobenzene
	1,2-Dichloroethane
	Dichloromethane (methylene chloride)
	Di(2-ethylhexyl)phthalate
	Hexachlorobenzene
	Hexachlorocyclohexanes
	Lead and lead compounds
	5-methylchrysene
	5-Nitroacenaphthene
	Polybrominated biphenyls
	Styrene
	Tetrachloroethylene
	Toxaphene

Based on various IARC publications.

in the average American diet in a year, and there are still a thousand chemicals left to test in roasted coffee."[14]

Molecular epidemiology is a new and evolving area of research that combines laboratory determinations of biomarkers for exposure and biological response, and biologic effects with a variety of conventional epidemiologic methodologies (reviewed in References 15, 16). Major objectives are to determine: (a) the internal and bioeffective doses of exogenous and endogenous agents;[17–19] (b) early biological effects likely to be predictive of cancer;[18] and (c) variations in individual susceptibility to carcinogens.[20] The monitoring of carcinogen-DNA adducts is becoming an important component of molecular epidemiology study designs.[21] The probable effects of carcinogens in humans are estimated by using mathematical models that first attempt to correlate the results of animal bioassays, in which tumor incidence is determined in animals exposed to high doses, followed by efforts to extrapolate to the consequences of probable levels of human exposure. The measurement of DNA adducts (*molecular dosimetry*) may help establish the relationship between exposure and molecular dose as well as providing possible validation information in the extrapolations from high- to low-doses and the differences between toxicokinetic factors among different species.

4.1.2 Metabolism, Enzymatic Activation, and Mechanism of Action

4.1.2.1 Absorption, Distribution, and Storage

After the absorption of carcinogens (or other toxicants, including antineoplastic agents) through the skin, lung, or gastrointestinal tract, there are several mechanisms for the passage of the

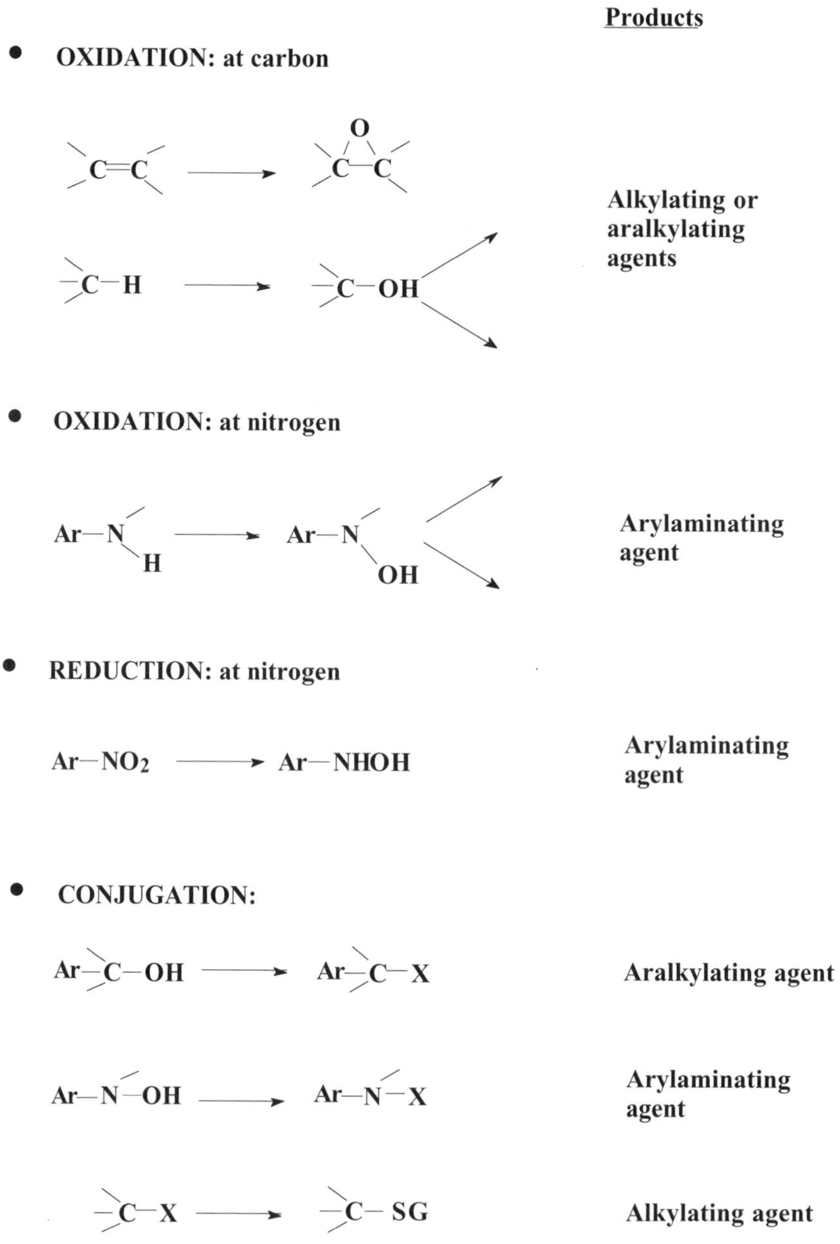

Products

- **OXIDATION: at carbon**

C=C → C–C (epoxide, O bridging)

C–H → C–OH → **Alkylating or aralkylating agents**

- **OXIDATION: at nitrogen**

Ar—N(H) → Ar—N(OH) → **Arylaminating agent**

- **REDUCTION: at nitrogen**

Ar—NO_2 → Ar—NHOH **Arylaminating agent**

- **CONJUGATION:**

Ar–C–OH → Ar–C–X **Aralkylating agent**

Ar—N–OH → Ar—N–X **Arylaminating agent**

C–X → C– SG **Alkylating agent**

FIGURE 4.1 Summary of the types of metabolic reactions that have been associated with the generation of reactive ultimate carcinogens. Reprinted from Dipple, A., *Carcinogenesis,* 16, 437–441, 1995. With permission.

xenobiotics through cell membranes. In *passive diffusion,* the rate is directly related to both the concentration gradient across the membrane and to the lipid solubility of the xenobiotic. This works for neutral compounds, but passage may be hindered for ionized compounds because of poor lipid solubility, thus the pH of the medium is of importance. *Filtration* depends on pore sizes and the molecular weight of the compound. The size of pores in most cells is ~4 nm which limits passage into the intracellular fluid to compounds of <200 Da. In contrast, the size of pores in the membranes of capillaries and glomeruli are ~70 nm, thus the water flowing through such pores (due to osmotic or hydrostatic pressure) may carry compounds with molecular masses up to 60 kDa. *Active transport,* which operates even against a concentration gradient, requires energy, usually provided by ATPase. *Carrier-mediated transport* involves the formation of a complex between the xenobiotic and a macromolecular carrier, diffusion of the complex through the membrane, release of the chemical on the other side, and return of the macromolecule to the original surface. *Facilitated transport* does not move molecules against a concentration gradient, requires no energy, and cannot be inhibited by metabolic poisons. *Endocytosis* is a special transportation process where particles are removed from alveoli and blood by being engulfed within the surrounding cell's reticuloendothelial system. When solids are engulfed it is termed *phagocytosis;* for liquids it is *pinocytosis.*

The rate of distribution of xenobiotics into organs depends not only on the blood flow to particular organs but also on the ability of the chemical to cross the local capillary walls and on the affinity of the chemical for specific components of the cells of that organ. Penentration of the blood-brain barrier is a special case that is particularly dependent on lipid solubility. Binding by liver and kidney tissues is often strong. If covalent binding of the xenobiotic occurs in a tissue, reversibility is limited and based on the repair mechanism of the cell. The half-life of such covalent complexes is long and is strongly associated with toxicity. Noncovalent binding to proteins (usually albumin) is readily reversible and is of importance because it often involves a major portion of the dose. While the protein-bound chemical is not immediately available for distribution to the extravascular space, the protein-xenobiotic complex acts as a rapidly equilibrating reservoir, replenishing the concentration of the unbound chemical in the cytosol. Compared to the protein-xenobiotic reservoir, adipose tissues are important long-term storage depots for lipid-soluble carcinogens, e.g., polychlorinated biphenyls or pesticides, either storing the compounds directly or after their conjugation to fatty acids. The stored chemicals may be released back into the plasma.

4.1.2.2 Key Steps in Chemical Carcinogenesis

The mechanism of action of carcinogens proceeds in several steps (Figure 4.2). The critical step is the reaction of the carcinogen itself (*direct action* or *ultimate* carcinogen) or an *activated, reactive* metabolite of the carcinogen with some key targets in the genome, primarily DNA. The "target dose" refers to the concentration of the ultimate genotoxic agent that evades detoxification by Phase II enzymes that conjugate the carcinogen and allows it to be excreted. Once the carcinogen has reacted with the DNA, there is a separate recovery phase that is the responsibility of the DNA repair mechanism. Persistent DNA damage produces permanent changes, e.g., preventing the proper folding of DNA that in turn may cause problems in the transcription process. Alteration in gene expression and a loss of growth control as a consequence of these events can lead to carcinogenesis. It is noted, however, that there is a substantial period of time for tumors to become evident, thus DNA damage is apparently a necessary but not sufficient event for tumorigenesis.

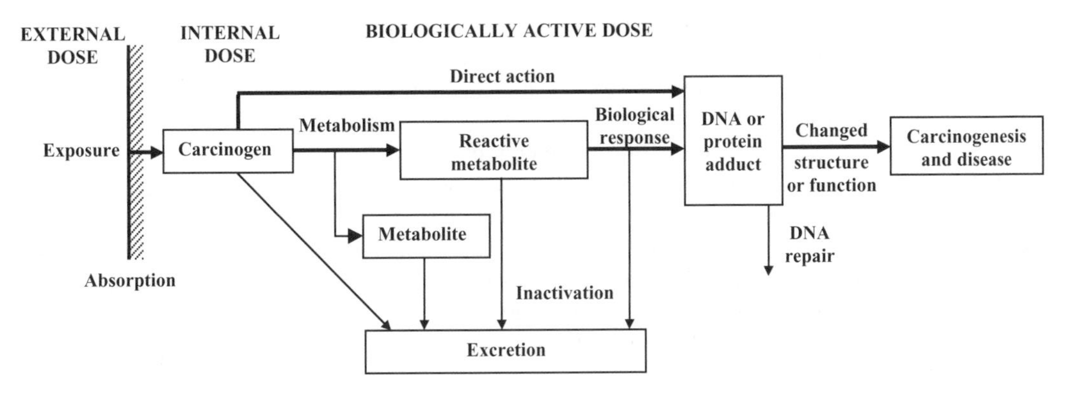

FIGURE 4.2 Keys steps in chemical carcinogenesis. A variety of biomonitoring techniques are available for each step.

Chemical carcinogens or their activated electrophilic metabolites form covalent adducts with nucleophilic centers within nucleic acids. The structure of DNA may change as a result of alkylation, oxidation, dimerization, deamination, or the formation of aromatic adducts. Some electrophilic carcinogens alkylate DNA and other macromolecules directly. Most carcinogens are *procarcinogens* that require cell- or tissue-specific biochemical activation to an electrophilic form that can react with the nucleophilic moieties of cellular macromolecules. Biotransformation with the aid of specific Phase I enzymes yields ultimate carcinogens through an intermediate stage (*proximate carcinogens*). The degree and rate of the activation is counteracted by the simultaneous ongoing process of detoxification.

The Ames test[6] can distinguish between direct-acting mutagens and those requiring metabolic activation and carcinogen-DNA formation, including cases where metabolic activation is required.[8] Carcinogens requiring metabolic activation may still fit well into the proposed classification into alkylating, arylaminating, and aralkylating agents based on the activation reactions (Figure 4.1). The formation of alkylating or aralkylating agents upon the oxidation of double or single carbon-carbon bonds depends on whether the bond is aliphatic or aromatic in origin. Arylaminating agents form upon the oxidation of aromatic amines or amides or the reduction of aromatic nitro compounds. Another process for the formation of reactive genotoxic carcinogens involves lipid peroxidation yielding a variety of substances that may function as both as DNA-reactive genotoxins and as promoters. Examples include species that contain unpaired electrons, e.g., alkoxyl and peroxyl radicals, carbonyl-containing compounds, e.g., malondialdehyde, reactive oxygen radicals, e.g., ·OH radicals and peroxides.[22]

4.1.2.3 Functions of Phase I and II Enzymes

The biotransformation of carcinogens (and antineoplastic agents) is often a complex process of reactions taking place in parallel or in sequence, yielding a variety of metabolites and conjugates depending on host, environmental, and chemical factors. Xenobiotic metabolism usually proceeds in two stages (reviewed in Reference 23). The current paradigm concerning susceptibility to environmental carcinogens assumes that there is an increased level of activity of the Phase I enzymes which produces activated metabolites from *procarcinogens*, combined with decreased conjugating and detoxifying activities of Phase II enzymes, leading to the accumulation of the activated genotoxic metabolites thus increasing the risk of cancers, e.g., certain bladder and lung cancers. The Phase I enzymes which convert carcinogens to electrophilic metabolites by oxidative metabolism are the cytochromes P-450 (CPY). The Phase II conjugating enzymes, which inactivate the electrophilic metabolites include glutathione-S-transferases (GST), glycosyltransferases, and N-acetyltransferases. There are significant correlations between cancer susceptibility and polymorphisms among several of these enzymes.[24]

Phase I biotransformations are degradation reactions involving oxidation, reduction (mostly by intestinal bacteria), and hydrolysis. The oxidation process involves the reduction of one oxygen atom of molecular oxygen into the substrate with reduction of the other oxygen atom to water. Important enzymes catalyzing oxidation are cytochrome P-450 and NADPH cytochrome P-450 reductase. In *in vitro* studies of metabolism, the endoplasmic reticulum, where these enzymes are located, is homogenized into microsomes, and the catalytic enzymes are then known as *microsomal mixed-function oxidases*. The variety of oxidation reactions that occur may include aliphatic oxidation, aromatic hydroxylation, epoxidation, N-, O-, or S-dealkylation, N-oxidation, and sulfoxidation. Another important class of Phase I enzymes is the esterases. Many carcinogens and other toxicants include ester-type bonds that are subject to hydrolysis into the active species by a variety of esterases that are usually located in the soluble fraction of cells. Examples include acyl-, carboxyl-, and acetylesterases.

Phase II biotransformations involve the production of conjugates from the carcinogen, or its metabolite, that are more water soluble and more readily excreted than the parent compound. The most common type of conjugation for alcohols, carboxylic acids, amines, and sulfhydryl compounds is glucuronide formation. This is facilitated by uridine disphosphate glucuronyl transferase and the coenzyme uridine-5′-diphospho-α-D-glucuronic acid. Other Phase II reactions include sulfate conjugation, methylation, acetylation, and amino acid conjugation.

Another important Phase II enzyme is GST. The cofactor, *glutathione,* γ-L-glutamyl-L-cysteinylglycine (GSH), is a ubiquitous endogenous tripeptidic constituent of both animals and plants. Because GSH has a free sulfhydryl group, it can serve both as a nucleophile, by coupling with electrophilic metabolites (in conjunction with GSH-dependent glutathione transferases) and as a reducing agent by forming stable oxidized glutathione conjugates with reactive oxygen species and hydroxyperoxides (in conjunction with glutathione peroxidases). The formation of these polar GSH conjugates leads to facile excretion, thus providing an important mechanism for detoxification and elimination of the reactive metabolites of xenobiotics.

4.1.3 Methodologies for Biomarkers

Environmental monitoring of carcinogens, such as monitoring of concentrations in air or water, is useful but insufficient for the determination of the internal and biologically active doses acquired by individuals. There are major differences between individuals with respect to the degree of absorption and distribution as well as the pharmacokinetics, metabolism, and toxicokinetics of carcinogens.

A biomarker is an observable end point that indicates events in the processes leading to disease. An example of a biomarker of *exposure* is the concentration of a carcinogen or its metabolite in serum or urine. These biomarkers are specific to individual carcinogens. A biomarker of *effect* refers to the biological response to exposure, e.g., somatic cell mutation following exposure to a carcinogen. These biomarkers usually reflect compound classes. Biomarkers of *susceptibility* indicate the biological response of an individual after exposure, e.g., the activity of a specific enzyme controlling the activation or detoxification of a carcinogen. Thus, biomarkers may be used to follow the metabolism and disposition of a toxicant and the biological response including the progression of diseases. The internal dose may be large enough at target tissues to produce altered structures and/or function, eventually leading to clinical disease.[25]

4.1.3.1 DNA Adducts

The reaction of chemical carcinogens with DNA, yielding DNA adducts, is believed to be one of the earliest events in the initiation phase of cancer. These covalent products are the precursor lesions at which subsequent mutation occurs through replication errors during DNA synthesis at the adduct site or in its vicinity. Once formed, DNA adducts may alter the three-dimensional structure (e.g., static bends or flexible hinge joints) and several biophysical properties of the DNA duplex. The DNA damage sites may not be recognized properly by the DNA repair enzymes or transcription factors.

There are at least 18 potential sites for adduct formation in DNA. These include N1, N3, N7, and C8 (within ring) and the N^2 in the amino group at C6 in deoxyadenosine, N3 (within ring), the N^2 in the amino group at C4, and the O at C2 in deoxycytidine, N1, N3, N7, and C8, the N^2 amino group at C2, and the O at C6 in deoxyguanosine, N3 and the O at C2, and C4 in thymidine, and an OH that can be alkylated on the phosphate backbone. The degree of involvement of the different sites depends on the chemical nature of the reactive species, the nucleophilicity of the DNA sites, and steric factors.[26] The site of the DNA reaction can be predicted to some extent by the electrophilicity of the species formed during the bioactivation of the carcinogen, e.g., through the P-450 mono-oxygenase pathway.[8] Deuterium exchange, with tandem MS detection of product ions, has been used to study the mechanism of adduct formation and the release of modified bases in urine.[27] A novel approach to estimate the mutagenic or carcinogenic potential of environmental contaminants has been demonstrated for the effect of benzoyl chloride on adenine using ion-molecule reactions and MS/MS to determine the reaction products of nucleophyl/electrophile reactions.[28]

Because most (but not all) genotoxic xenobiotics interact with the nucleophilic centers of the DNA bases, determination of DNA modifications at the nucleoside or base level may be used for both molecular dosimetry and for advancing the understanding of the carcinogenic process. The occurrence of adducts reveals the presence of genotoxic carcinogens and reflects the effective doses, particularly resulting from chronic exposure. The major advantages of dosimetry based on DNA adducts are: that the matrix is the actual target tissue of the carcinogen, and that the adducts have a half-life of >120 days. There are also disadvantages because, in contrast to the case of protein adducts, DNA adducts are usually determined in a nonsite-specific manner. Another disadvantage is that high sensitivity is required for adduct analysis because of the low concentration of DNA in cells (~1 mg/g wet tissue, ~50 μg/mL blood) and the relatively few adducts (1 in 10^7 or more nucleotides) occurring in the DNA. The usual methodology is that adducted nucleotides or their degradation products are released from carcinogen-modified DNA by enzymatic hydrolysis or chemical reactions. The products of DNA repair and excision, which are often excreted intact in urine, may also be monitored and used as biomarkers.

In addition to MS, there are several other physico-chemical and biochemical techniques for measuring DNA adducts. Most important of these are ^{32}P-postlabeling and various immunoassays (reviewed in References 29, 30). Others, e.g., those based on fluorescence and alkyltransferase (briefly reviewed in Reference 31), are used to determine specific types of human DNA adducts. A large-scale coordinated study on the genotoxicants resulting from petrochemical combustion or processing, such as polycyclic aromatic hydrocarbons and low-molecular weight alkylating agents, has utilized all the major analytical techniques for DNA adduct analysis (also protein adducts) and revealed their respective advantages and shortcomings.[32]

^{32}P-Postlabeling

Developed in 1981 (reviewed in Reference 33), ^{32}P-postlabeling is based on the labeling of adducted nucleoside-3′-monophosphates by the enzymatic transfer of a ^{32}P moiety from [γ-^{32}P]ATP to the 5′ hydroxyl group of nucleotides from fully hydrolyzed DNA. Next, the resulting radiolabeled nucleoside-3′5′diphosphates are separated by multidirectional TLC. The adducted as well as normal nucleotide spots are visualized by radiography, and finally quantified by scintillation counting of excised region(s) from the TLC plate. Postlabeling by ^{32}P is the most sensitive method currently available: the LOD is 0.01 fmol (1 to 10 μg DNA required) and it is now possible to detect 1 DNA adduct per 10^{10} normal bases. Additional advantages are that the structure of the adduct need not be known (the method is generic) and that DNA adducts resulting from complex mixture exposures may be studied. Disadvantages include that the method provides no information on the structure of adducts and it is less useful for carcinogens with low molecular weights. Quantification errors may occur because of differences in the hydrolysis and labeling efficiencies of normal and adducted nucleotides.

In a controlled study a DNA sample modified with 4-aminobiphenyl was synthesized to serve as quantification standard for verification of the ^{32}P-postlabeling method. The structure of the final synthesis product and the degree of modification achieved were determined by LC/ESIMS and then the utility of the standard was tested against DNA from carcinogen-treated mice using both ^{32}P and LC/ESIMS for the quantification of the DNA adduct.[34]

Immunoassays

The first step is to raise monoclonal or polyclonal antibodies against the carcinogen-modified DNA or carcinogen-nucleoside adducts that have been coupled to a protein carrier. The quantification of specific adducts is accomplished by conducting an experiment in which the unknown sample to be assayed is competing with a known amount of added adduct standard for a limited quantity of available antibody. The result is a reduction in the quantity of the added standard (the constant competitor) that binds to the antibody. In *radioimmunoassay* (RIA), the unknown adduct, the radiolabeled constant competitor, and a known quantity of antibody are incubated, the resulting antigen-antibody complex is precipitated with a second antibody (anti-IgG), and the radioactivity of the pellet is determined. Because the antigen-antibody reaction is concentration-dependent, the radioactivity of the antigen-antibody complex is inversely related to the amount of the unknown adduct. An alternative approach is the *enzyme-linked immunosorbent assay* (ELISA). In this case a solution of the unknown adduct and the antibody, labeled with a chromogen, is exposed to the unlabeled constant competitor which is bound to the bottom of a microtiter well. The antibody-antigen complexes that remain in solution are recovered and exposed to a second enzyme-conjugated antibody that can cleave the chromogen, such as *p*-nitrophenylphosphate, the quantity of which can finally be determined spectrophotometrically. ELISA is usually more sensitive than RIA with an LOD of 1 vs. 40 fmol (1 adduct per 10^8 nucleotides by competitive ELISA). The immunological techniques are simple, rapid, and inexpensive, and many samples can be analyzed simultaneously. However, specificity is always suspect because the antibodies used may react with a spectrum of environmental carcinogens (reviewed in Reference 35). The sensitivity of combined immunoaffinity-GC/MS techniques (5 mL size urine samples) is of the order of 0.2 pmol/mL.[36]

Mass Spectrometric Techniques

During the past 25 years, mass spectrometric approaches have, predictably, followed technological advances from early techniques using FAB, FD, or TSP ionization through GC/MS assays, with a variety of derivatization techniques, to recent methods based on LC/ESIMS and CE/ESIMS.[37] Until recently, the method of choice for analysis of specific DNA adducts (both in DNA and urine) has been GC/MS, using EI and CI, stable-isotope labeled analogs as i.s., and SIM for quantification (occasionally in high-resolution SIM mode). A variety of DNA adducts (listed in Reference 30), have been analyzed by GC/MS both in DNA and urine. The usual sensitivity of GC/MS methods has been 0.3 to 1 adduct per 10^8 nucleotides. Highest sensitivities, which approach that of ^{32}P-postlabeling and which have concurrent high specificity, have been achieved by derivatization of the analytes with an electrophile, usually a fluorinated species, and ionization by electron capture negative ion CI (reviewed in Reference 38). An example for heterocyclic amines is given in Reference 39 with other individual applications described elsewhere under specific carcinogens.

LC/ESIMS techniques, particularly in the MS/MS mode, have been used recently to characterize DNA adducts at both the nucleoside and nucleotide levels (reviewed in Reference 40). Sample preparation steps include the hydrolysis and enzymatic digestion, with ribonucleases, of extracted DNA yielding a mixture of modified and unmodified DNA bases. The excess of unmodified bases are then removed using solid phase extraction or HPLC. LODs have been an order of magnitude better than those achieved by GC/MS.[37] MS/MS may also be used for the determination of the structure of individual DNA adducts as well as for short fragments of DNA oligomers to which carcinogens are bound. The possibility of controlled MS/MS, or in-source fragmentation, through adjustment of operation parameters often permits the selective detection of DNA adducts in complex mixtures.

CE/ESIMS is particularly well suited to the analysis of negatively charged species, i.e., nucleotides (reviewed in Reference 41). The availability of the *stacking* techniques to inject increased sample volumes significantly increases sensitivity.[42] A range of methodologies and applications for the detection of modified nucleosides, nucleotides, and oligonucleotides has been listed.[40] The combination of CE techniques with immunoassay and ^{32}P-postlabeling further increases the potential of the various available CE techniques to provide high-performance separation.

Accelerator mass spectrometry (AMS, Section 2.2.6.4) has been used increasingly for the detection of DNA adducts because of its high sensitivity. This is despite the lack of molecular or structural information obtained with AMS. Applications include measurement of DNA adducts in experimental animals that have been exposed to doses of carcinogens comparable to those found in human food (Section 4.3.3).

4.1.3.2 Hemoglobin Adducts

Although carcinogen-albumin adducts have been successfully investigated, e.g., using nano-ESI-MS/MS to study the peptides in adducts of benzo[a]pyrene diol-epoxide and related compounds,[43] most carcinogen-protein adduct dosimetry investigations have been carried out on hemoglobin (Hb). It has been known for some 25 years that Hb adducts may serve as molecular markers of carcinogen dose and act as surrogates for DNA adducts.[44] In certain cases Hb adducts may be used to estimate the risk of disease, e.g., the 4-aminobiphenyl-Hb adduct levels in smokers reflect the number of cigarettes smoked and have been associated with acetylation phenotype.[45] The nitrogens in four terminal amino acids in Hb, one for each popypetide chain, are major nucleophilic sites. Other sites that may be adducted include the sulphydryl in cysteine, the imidazole in histidine, carboxylic groups in aspartic and glutamic acids, and the amino group in the N-terminal valine.[46] Advantages of the determination of Hb adducts include: (a) integration of the total cumulative exposure over the relatively long lifetime of human erythrocytes (~120 d); (b) linear dose dependency because the adducts are covalent; (c) absence of adduct repair; and (d) relative ease of obtaining the required quantities of Hb from blood. A major disadvantage is the lack of direct relevance of Hb adduct formation to the process of carcinogenesis, i.e., the measurement is removed from the event for which it serves as a surrogate. The relationship of Hb adducts to the biologically relevant DNA adducts is not predictable and is usually not well established. The concentration of "background" adducts may affect sensitivity significantly. These background adducts may originate from products of normal physiological processes, metabolic pathways, or inflammations. GC/MS has been used to identify several potentially interfering adducts.[47] Control subjects must always be included in any study design involving protein adducts,[48] e.g., ethylene oxide,[49] a benzene adduct.[50] The sensitivity of Hb adduct analysis by GC/MS has been in the range of 0.05 to 10 pmol/g Hb using 50 to 1000 mg globin samples. About 100 mg of Hb can be isolated from 1 mL blood.

GC/MS is adequate for most carcinogen-protein complexes because the isolated amino acid adducts, or the hydrolyzed carcinogen-derived products, are usually sufficiently volatile or may be derivatized into volatile products for analysis. Although GC/MS with EI or positive CI is satisfactory when analyte concentrations are high, quantification of Hb adducts is usually performed by GC/MS using electron capture NCI with methane as the reagent gas (reviewed in Reference 51). Fluorinated derivatives are particularly applicable to a wide range of compound types with ECNCI yielding high sensitivity concurrently with high selectivity. TMS derivatives are also frequently used and are particularly suited to EI and positive CI. The best internal standards are stable isotope-labeled analogs. Deuterated compounds are often available as i.s., however, the incorporation of several deuterium atoms per molecule is desirable to avoid interferences from isotope peaks, e.g., from ^{13}C, ^{34}S, and ^{37}Cl. Fluorinated or chlorinated analogs may also be used as i.s. A sensitive GC/ECNICI-MS assay for alkylated adducts to the N-terminal valines of Hb is based on a combination of the Edman method with derivatization using pentafluorophenyl isothiocyanate to form the pentafluorophenylhydantoin of the N-alkylated valines (i.s.: globin alkylated with deuterated analogue).[52] Developments in MS/MS-based techniques have increased the resolving power of the approach to the point that even variations of endogenous background levels could be determined reliably.[53] LC-MS/MS has also been used for analytes of inadequate volatility, e.g., the Hb adducts of acetaldehyde.[54]

When the adducted Hb is treated with a weak acid or base to release a carcinogen-derived residue, analyzed by GC/MS, no specific information is obtained as to the site of adduct formation. An alternative approach is the analysis of a carcinogen-modified amino acid that is derived from the polypeptide chain of Hb, thus the structure of the adduct is known with certainty. General procedures and early applications have been reviewed.[55,56] A number of applications have been developed for the N-terminal valine adducts in Hb (reviewed in Reference 48), including those formed with acrylonitrile,[57] epichlorohydrin,[58] ethylene oxide,[59] propylene oxide,[53] 1,3-butadiene epoxide,[60,61] and styrene oxide;[62] several of these are described later.

4.2 OCCUPATIONAL CARCINOGENS

Although only 2 to 8% of human cancers are of occupational origin, the risks are high for specific populations of exposed workers. For example, employment in the rubber industry has been strongly associated with bladder, stomach, and lung cancers and leukemia. The following are manufacturing/production environments in which there is exposure to known human carcinogens (IARC Group I): aluminum production, auramine manufacture, boot and shoe manufacture and repair, coal gasification, coke production, furniture and cabinet making, hematite mining, iron and steel founding, isopropanol and magenta manufacturing, painting, the rubber industry, and heavy exposure to sulfuring acid mists. Probable human carcinogen exposure (Group 2A) occurs for hairdressers and barbers. Possible human carcinogens (Group 2B) affect workers in the carpentry and textile industries. The types of occupation-related epidemiology studies include case reports, sentinel events, ecologic investigations, and analytic epidemiology.[63]

4.2.1 Aromatic Hydrocarbons

4.2.1.1 Benzene

Benzene has been classified as a Group I carcinogen because chronic exposure leads to acute myeloid leukemia. Benzene is a constituent of engine emissions and tobacco smoke and is widely used in the chemical, paint, and dye industries (reviewed in Reference 64). Unleaded gasoline contains ~1% benzene and the main source of occupational exposure is the production, distribution, and handling of gasoline. Nonoccupational benzene exposure occurs primarily through smoking during which the quantities involved are significantly larger than those encountered in virtually any occupation (Section 4.3.1.2). Regulated as a toxic air pollutant in the US under the Clean Air Act, the threshold limit of benzene exposure has been lowered repeatedly over the years and is currently down to 0.1 ppm for 8 h exposure. Benzene, however, is still being used as a solvent in developing countries, resulting in much higher exposures.

Benzene is metabolized in the liver to a variety of hydroxylated and ring-opened products (reviewed in Reference 65). The first step of metabolism is the formation (involving cytochrome P-4502E1) of benzene oxide (epoxide) which is in equilibrium with the unstable oxepin. These reactive intermediates are the subjects of several subsequent metabolic pathways (Figure 4.3). In one pathway, nonenzymatic rearrangements result in the formation of phenol, the main metabolite of benzene (13 to 50% of the total dose), catechol, and some other minor hydroxylated products. These phenolic metabolites, which are genotoxic, can accumulate in the bone marrow.[66] They also undergo Phase II metabolism with the sulfate and glucuronide conjugates formed being excreted in the urine. Most of these conjugated metabolites were identified by FABMS.[67]

In a second pathway, benzene oxide is hydrolyzed by epoxide hydrolase to form benzene dihydrodiol (benzene glycol), which undergoes further enzymatic action either by dehydrogenation to catechol or via ring-opened to trans-trans-muconaldehyde which is in turn oxidized to trans-trans-muconic acid (t,t,-MA, an important biomarker, see below). One direct Phase II metabolic pathway of benzene involves its conjugation with glutathione leading to the urinary metabolite S-phenylmercaptopuric acid (S-PMA, <0.5% of total exposure). Another critical reaction that can occur is that benzene oxide may covalently bind to DNA, RNA, or proteins. Most metabolites and adducts of benzene have been considered as potential biomarkers.[68] A variety of techniques have been developed, examples of which are discussed below, for the determination of nonmetabolized benzene, the urinary metabolites, and the DNA and protein adducts. Methods developed prior to 1994 are reviewed in Reference 69.

Accelerator MS was used to study the tissue distribution and macromolecular binding of extremely low doses of [^{14}C]-benzene in mice. Dosing solutions containing 8 pCi and 254 pCi labeled benzene were administered via i.p. to achieve concentrations in the 700 pg benzene/kg body weight to 500 mg/kg body weight range. After the isolation of DNA and proteins and acid

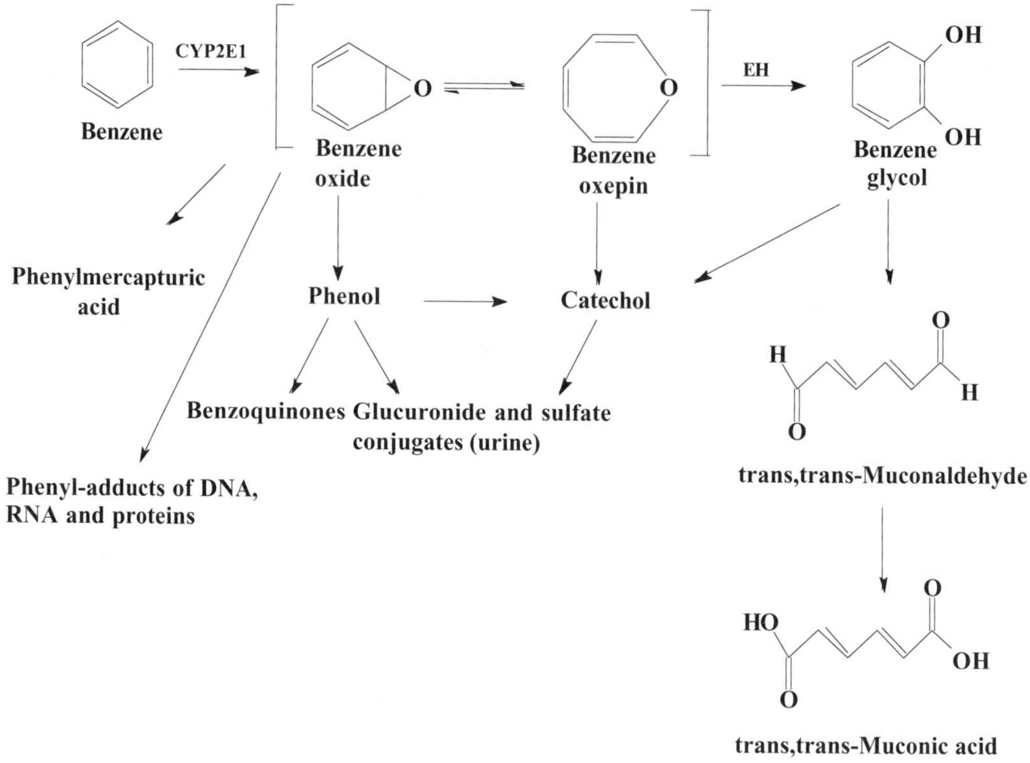

Benzene, CYP2E1, Benzene oxide, Benzene oxepin, EH, Benzene glycol, Phenylmercapturic acid, Phenol, Catechol, Benzoquinones, Glucuronide and sulfate conjugates (urine), Phenyl-adducts of DNA, RNA and proteins, trans,trans-Muconaldehyde, trans,trans-Muconic acid

FIGURE 4.3 Simplified scheme of the metabolism of benzene. CYP, cytochrome P-450, EH, epoxide hydrolase.

precipitation of the macromolecules, samples were oxidized in quartz tubes by heating to 900°C in the presence of copper oxide, and then reduced to filamentous graphite in the presence of titanium hydride and zinc powder. Isotope ratios were determined by measuring ^{14}C concentrations relative to ^{13}C. After normalizing to the $^{14}C/^{12}C$ ratio of a carbon standard, the results were converted into picograms of [^{14}C]-benzene/g tissue and background corrections were applied. Carbon contents were determined by conventional C:N:S analysis. The results showed that [^{14}C]-benzene concentrations were highest in the liver 0.5 h after exposure. DNA adduct levels in liver and bone marrow DNA peaked at 0.5 h and 12 to 24 h, respectively, after exposure. Dose-response curves were obtained over eight orders of magnitude, which encompassed known human exposure levels. The data established that benzene binds to proteins to a much greater degree than to DNA, that the concentrations of the bioactive metabolites are higher in the bone marrow than in the liver, and that protein adducts significantly contribute to the hepatotoxicity of benzene.[70]

Urinary Trans,Trans-Muconic Acid (t,t-MA) as a Dosimeter

An ultimate metabolite of benzene, t,t-MA forms from benzene via muconaldehyde through benzene epoxide and oxepin (Figure 4.3). Interestingly, *t,t*-MA was detected in the urine of leukemia patients >60 yr ago, when treatment with benzene was considered a cure. Although *t,t*-MA is only a minor metabolite (2% of the total) with a half-life of <6 hr, it is considered to be a more suitable biomarker for benzene exposure than phenolic metabolites because of its very low endogenous occurrence and its preferential formation in low-level benzene exposures.[71] However, other factors may significantly modify urinary *t,t*-MA concentrations, including coexposure to toluene, consumption of sorbic acid, genetic susceptibility, and pregnancy.[72]

A GC/EIMS technique for the quantification of t,t-MA in urine includes extraction with ethyl ether, derivatization to form the TMS derivative, and quantification using SIM of the $[M\text{-}CH_3]^+$ ions (i.s.: D_4-t,t-MA, biosynthesized in benzene-treated rats, or ^{14}C-*t,t*-MA).[73] Application of this

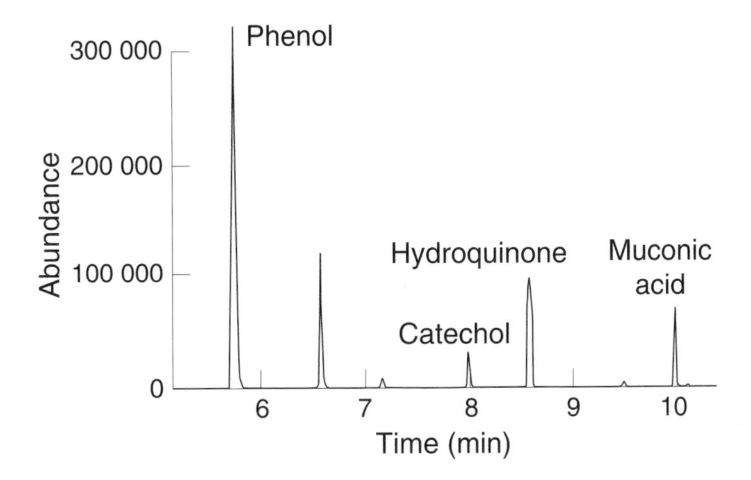

FIGURE 4.4 Selected ion chromatogram for the simultaneous quantification of benzene metabolites phenol, catechol, hydroquinone, and muconic acid in the urine of a worker exposed to >31 ppm time weighted average over an 8-hr day. Reprinted from Rothman, N. et al., *Occup. Environ. Med.*, 55, 705–711, 1998. With permission.

method led to the detection of *t,t*-MA in Chinese workers exposed to as little as 4.4 ppm of benzene in 8 hr.[71] Further expansion of the methodology allowed the monitoring of the urinary excretion of phenol, cathecol, hydroquinone, and *t,t*-Ma in a study of 44 workers exposed to a median of 255 ppm benzene. Monitoring the $[M]^+$ ions of the analytes ($[M\text{-}CH_3]^+$ ions for *t,t*-MA) by SIM all analytes were separated in a single GC/MS analysis (Figure 4.4). Although urinary metabolite concentrations do not reflect their concentrations in target tissues, it was concluded that the risk of adverse health effects has a supralinear rather than linear relation with the external dose.[74] GC/MS has also been used to determine urinary *t,t*-MA concentrations in urban children exposed to benzene in air (Section 4.3.2).

An HPLC-ESIMS/MS technique was used for the determination of urinary *t,t*-MA, S-phenyl-mercaptopuric acid (S-PMA), hydroquinone, catechol, and benzene triol, along with phenol, by GC/MS, in Chinese workers who were exposed to ~30 ppm benzene/day in glue- and shoe-making factories. Correlations were found between the occupational exposure and the concentrations of each of the metabolites. Large differences in metabolite concentrations were found depending on whether the samples were taken before or after work. The half-lives of these metabolites were in the 12 to 16 h range. Because of the high background levels of some of the metabolites, S-PMA and *t,t*-MA were recommended as the most sensitive markers for low level benzene exposure.[75]

Benzene Oxide

To confirm the identity of benzene oxide (3,5-cyclohexadiene-1,2-oxide, BO) as a product of benzene metabolism (Figure 4.3), mouse, rat, and human liver microsomes were incubated with benzene and the metabolic products were extracted with methylene chloride. The putative BO peak, easily separated from phenol (another major metabolite of benzene) by GC, was identified in all three species by GC/EIMS and confirmed using synthetic standards. BO was not detected when the incubation with benzene was carried out using heat-inactivated microsomes.[76]

Although the electrophilic BO binds covalently to cell macromolecules, BO was originally considered to be only a reactive intermediate rather than a direct contributor to benzene carcinogenicity because of its purported inability to diffuse out of the hepatocyte after its formation. A GC/CIMS technique (i.s.: toluene) was used to determine the stability and half-life (~8 min) of BO in blood *in vitro*. Using a GC/EIMS method for an *in vivo* study, six prominent ions of BO

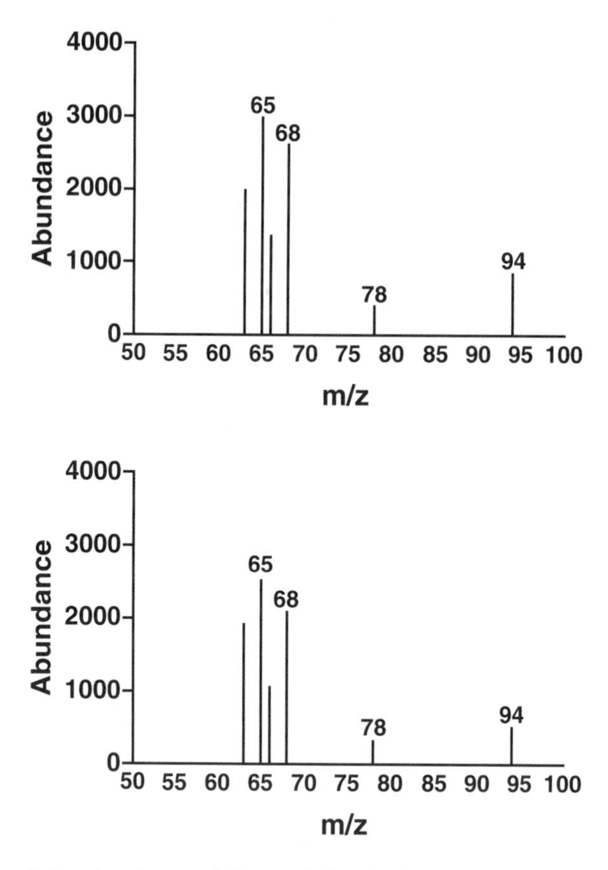

FIGURE 4.5 Upper panel: Ion abundances of diagnostic ions in the EI mass spectrum of 300 pg of authentic benzene oxide. Lower panel: Ion abundances in an extract of blood from a F344 rat, 3 h after dosing with 400 mg benzene/kg body weight. Reprinted from Lindstrom, A. B. et al., *Carcinogenesis,* 18, 1637–1641, 1997. With permission.

were detected in an extract from the blood of a rat treated with a dose of 400 mg benzene/kg (Figure 4.5); no metabolite ions were found in the controls. The maximum BO concentration of ~8.5 ng/mL, measured <1 h after exposure, remained reasonably constant for ~9 h, followed by an exponential decline over the next 12 h until it was no longer detectable 24 h after the dose. It was, therefore, concluded that, contrary to the initial belief that it did not leave the hepatocytes, BO can easily be distributed throughout the body after being formed in the liver and may well contribute to the genotoxicity of benzene.[77]

In contrast to the urinary biomarkers of benzene, such as *t,t*-MA and S-PMA, which reflect recent exposure, the presence of Hb adducts of benzene metabolites reflects a history of exposure because of the 120-day life span of red cells. GC/EIMS techniques were developed for assaying the S-phenylcysteine adducts of Hb and other proteins in experimental animals exposed to benzene.[78–80] Quantification of the cysteinyl BO-Hb adduct is based on derivatization to form phenyltrifluorothioacetate (PTTA) derivative and GC/NCIMS[81] using SIM of PTTA and a [i.s.: $^{2}H_5$]PTTA (Figure 4.6). Exposure of 21 workers to <31 ppm and >31 ppm benzene resulted in median BO-Hb levels of 46.7 and 129 pmol/g globin, respectively. The adduct concentration in 42 normal controls was 32.0 pmol/g globin. There was a significant correlation between the adduct concentrations in exposed workers and the level of benzene exposure. Median BO-albumin adduct levels (351 and 2010 pmol/g Alb) also exhibited a significant correlation with the two levels of benzene

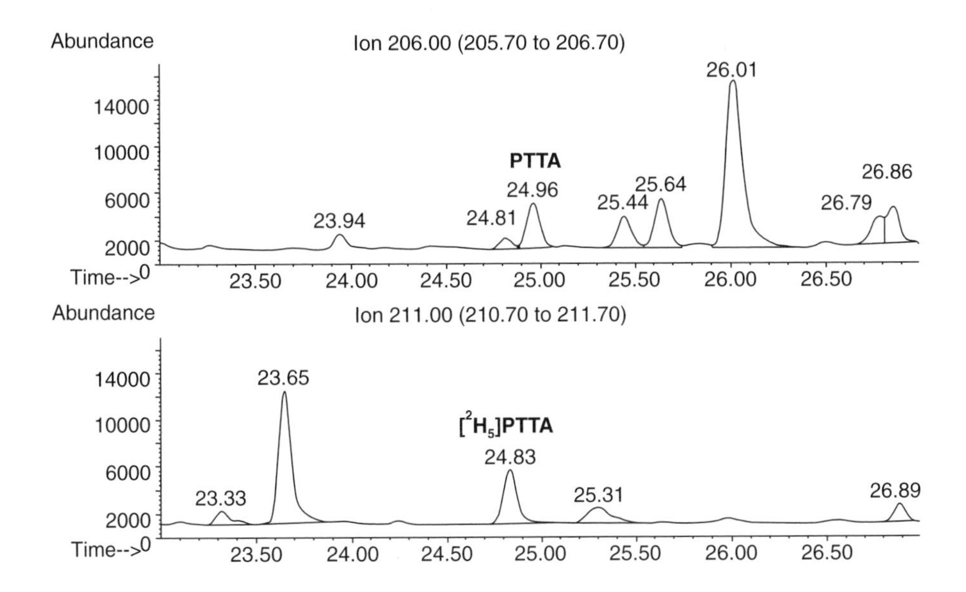

FIGURE 4.6 Typical GC-MS (negative chemical ionization) selected ion monitoring trace obtained following the reaction of albumin (4 mg) from a worker exposed to the median level of benzene (31. ppm, 8 h) with trifluoroacetic anhydride and metanesulfonic acid to yield PTTA. [$^2H_5$]SPC (3 pmol) was added as the internal standard. Reprinted from Yeowell-O'Connell, K. et al., *Carcinogenesis,* 19, 1565–1571, 1998. With permission.

exposure. While portions of the measured background adduct concentration for both Hb and albumin were shown to be of endogenous origin, the rest of the background was proven to represent assay artifacts when recombinant human Hb and Alb were used to form BO adducts.[82]

Benzoquinones

The tissue disposition of three benzene metabolites, BO, 1,2-benzoquinone, and 1,4-benzoquinone, were studied in rats and mice exposed to benzene, with particular focus on the cysteine adducts in Hb and other proteins, as these account for a large portion of the binding of benzene metabolites in the bone marrow. After digestion with protease, the benzoquinones were released from the cysteine residues using a Ni-Al catalyst, followed by derivatization with heptafluorobutyrylimidazole and analysis by GC/NCIMS.[83] Of the cystine adducts, 1,4-benzoquinone predominated in total mouse blood, benzene oxide was the major adduct in Hb, but 1,2-benzoquinone had the highest concentration in the bone marrow. Because there are high background concentrations of both of these benzoquinone adducts, these adduct analytes can only be used as markers when there have been high levels of benzene exposures.[84]

Para-benzoquinone was adducted to a 7-mer oligonucleotide, d(GTTCTTG), to form d(GTT(C-BQ)TTG) where (C-BQ) is the 3-hydroxy-1,N^4-benzoethano-cytidine residue. The fragmentation of the $[M\text{-}3H]^{3-}$ and $[M\text{-}4H]^{4-}$ ions of the unmodified and modified oligonucleotides were compared using ESI-MS/MS at different collision energies. The fragmentation of the modified oligonucleotide followed the patterns that are known for unmodified nucleotides.[85] The full sequence and placement of the adduct were determined from the position of the adduct on the central cytidine base and the assignments of the fragments (Figure 4.7). It was concluded that the observation of site-specific adducts in oligonucleotides should lead to further understanding of the molecular mechanism of carcinogenesis. Advanced instrumentation will provide the required sensitivity for *in vivo* studies.[86]

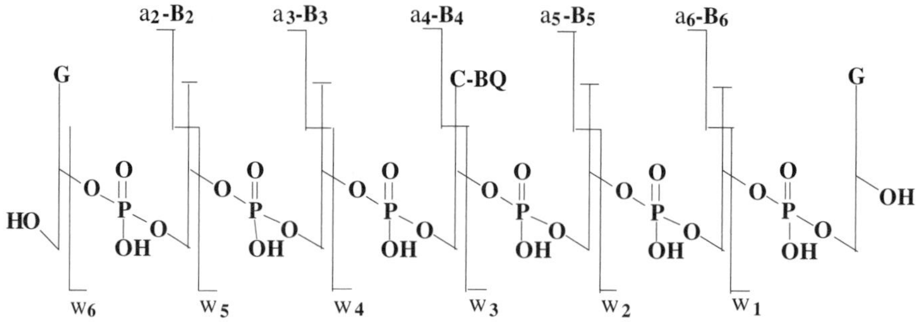

FIGURE 4.7 Schematic diagram of the product ion of the carcinogen-modified oligonucleotide d(GTT(C-BQ)TTG) showing full sequence analysis. Reprinted from Glover, R. P., Lamb, J. H., and Farmer, P. B., *Rapid Commun. Mass Spectrom.*, 12, 368–372, 1998. With permission.

4.2.1.2 Risk Assessment of Polycyclic Aromatic Hydrocarbons in Coke Workers

Polycyclic aromatic hydrocarbons (PAH) are a family of lipophylic, nonpolar chemicals comprising three or more fused benzene rings containing only carbon and hydrogen. The general population has an average daily intake of 2 to 16 μg PAH from food, 0.2 μg from air, and 0.03 μg from water. Of the several hundred known PAH there are only a few that are known to be carcinogenic. Benzo[a]pyrene, one of the most carcinogenic PAH, is used as a marker for PAH exposure. The metabolism of PAH involves conversion to epoxy and hydroxy derivatives by P-450-dependent monooxidases followed by Phase II conjugation to form stable urinary products. However, it is believed that short-lived intermediary metabolites may be responsible for initiating malignant cell growth by altering genetic information through reaction with DNA.

Coke workers are a population whose occupation results in their being continually exposed to aromatic hydrocarbons and PAH. As such they are a readily accessible group that can be assessed for exposure to these compounds, and there have been a number of studies on the presence of the parent compounds and their metabolites, some of which are discussed below. In one study on coke workers, some 25 PAH metabolites, phenols, and dihydrodiols of phenantrane, fluoranthene, pyrene, chrysene, and benzo[a]pyrene were quantified in urine. Samples were first treated with glucuronidase and arylsulfatases to hydrolyze the conjugated metabolites and then extracted with benzene and toluene. This was followed by dividing the extract into two fractions, the first of which was derivatized with diazomethane to convert the phenols into methylesters. The second fraction was treated with acid to convert the dihydrodiols into phenols which were then derivatized. Quantification was carried out using GC/EIMS with SIM of the $[M]^+$, $[M-15]^+$, and $[M-43]^+$ ions (i.s.: indenol [1,2,3-cd]fluoranthene and benzo[c]phenanthrene). The on-column LOD was ~0.01 ng. The metabolic profiles of individual coke workers were fairly constant on days 2, 4, 6, 8, and 10 following tar pitch exposure. There was correlation between inhaled PAH and the quantity of excreted metabolites. Interindividual variability, however, was significant, as illustrated by comparing profiles of the isomeric metabolites of four PAH (Figure 4.8). While the amount of excreted 1-hydroxypyrene was ~20% of the total of all metabolites, the absolute amounts were significantly different, ranging from 1218 to 6574 ng/L. The variations in the amounts of 1-hydroxypyrene excreted, while the proportion of metabolite remained constant, lead to the conclusion that the individual urinary metabolite profiles were proportional to exposure.[87]

The low molecular weight of aromatic hydrocarbons and some PAH mean that coke workers are exposed to these compounds in the breathing zone air. In an environmental evaluation using personal air sampling, GC-EIMS was used to identify 14 PAHs, benzene, toluene, naphtalene, m-xylene, and o-xylene at different workplaces. Urine samples from the same workers were hydrolyzed and o-cresol, 1- and 2-naphthol, methylhippuric acid, and 1-hydroxypyrene were identified.

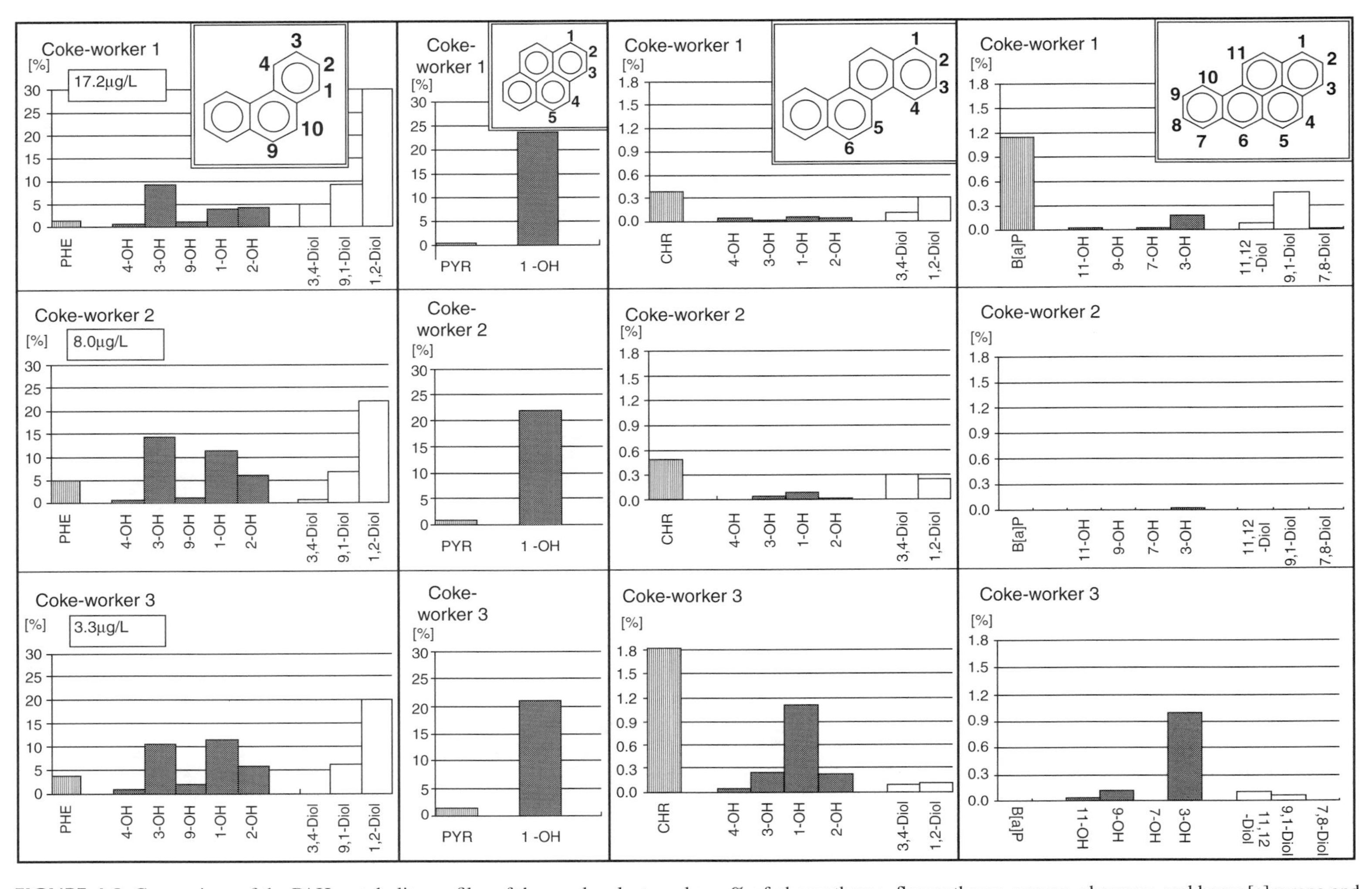

FIGURE 4.8 Comparison of the PAH metabolite profiles of three coke plant workers: % of phenanthrene, fluoranthrene, pyrene, chrysene, and benzo[a]pyrene and all isomeric metabolites thereof. Reprinted from Grimmer, G. et al., *Int. Arch. Occup. Environ. Health*, 69, 231–239, 1997. With permission.

PAH concentrations varied significantly among workers in different departments, and there was correlation between inhaled toluene, naphthalene, xylenol, and the urinary excretion of the respective metabolites. The concentrations of PAH did not exceed permissible levels in the breathing zone, but exposure levels varied significantly among workers in various job categories.[88]

The relationship between exposure to xylenol isomers, determined by personal air sampling, and the excretion of urinary metabolites, obtained after acid hydrolysis and solid-phase extraction, was studied by separating and quantifying the following hydroxylated metabolites in 75 exposed workers and 34 unexposed controls in a coke plant: phenol, three cresol and five xylenol isomers, benzoic and phenylacetic acids, and phenylacetic acid methyl ester. Quantification by GC/MS revealed significant correlations between 3-cresol, 2,4-, 2,5-, 3,4-, and 3,5-xylenol concentrations in urine and the breathing zone. Only minor exposures occured in the high-temperature tar-distillation workshop.[89]

There have been several studies aimed at assessing the significance of various biomarkers of PAH exposure in coke workers. In one study, GC/EIMS was used to quantify the Hb adduct, hydroxyethylvaline as a potential biomarker. After mixing isolated globin with the i.s. (globin treated with D_4-ethylene to form deuterated hydroxyethylvaline), pentafluorophenyl isothiocyanate was used to carry out Edman degradation, followed by purification, TMS derivative formation, and quantification by SIM. While significant differences were found between the concentration of hydroxyethylvaline in Hb among smokers and nonsmokers, the differences between smokers and controls and nonsmokers and controls were not significant. Although 1-hydroxypyrene correlated well with PAH exposure, no increased amount of antibody to benzo[a]pyrene was found.[90]

A series of potential biomarkers, including urinary 1-hydroxypyrene (internal dose), aromatic DNA adducts in leukocytes (biologically effective dose), serum p53 protein (response marker), genetic polymorphism of cytochrome P-4501A1, and glutathione-S-transferase M1 (susceptibility markers), as well as "traditional" PAH exposure, were evaluated in a cross-sectional study of 25 randomly selected male workers (matched for age and smoking status) in three work areas of a coke oven in China. GC/EIMS with SIM was used to quantify some 15 PAH extracted from particulate matter trapped in air filters with benzene, as well as gaseous phase compounds recovered on solid resins (eluted with tetrahydrofuran). The results confirmed the expected correlation between urinary 1-hydroxypyrene levels and PAH exposure. The correlations applied to workplace, stationary and personal sampling, as well as to current and cumulative doses. Positive correlations were also found between aromatic DNA adducts, and urinary 1-hydroxypyrene and between serum p53 levels and cumulated benzo(a)pyrene levels.[91]

The genotoxic potential of the constituents of complex environmental mixtures has usually been determined by the "chemical approach," i.e., individual constituents have been identified first, followed by assessment of their genotoxicity. The alternative "biological approach" has been explored using extracts from coke oven emissions to evaluate DNA adduct formation (by ^{32}P-post-labeling) in exposed multiple mammalian cell culture lines. The LOD of the PAH and nitro-PAH was as low as 0.1 μg/mg extractable organic matter (EOM) in some cell lines. The highest level of DNA adduct formation was observed with hepatocytes which, given their ability to metabolize PAH, suggests that there is a broad variety of genotoxic compounds generated. This data was compared to the "chemical approach" (serving as control) in which GC/EIMS with SIM was used to identify and quantify some 16 PAH and 7 nitro-PAH in crude EOM and several subfractions. The total quantities of these 23 compounds represented only $<5\%$ of the EOM and $<10\%$ of the aromatic fraction, suggesting that the major components of coke extracts were of unknown structure and genotoxic potential and indicating the likely difficulties in the comparison of these two approaches.[92]

A GC/NCIMS technique was developed for the quantification of r-7,t-8,9,c-10-tetrahydroxy-7,8,9,10-tetrahydrobenzo[a]pyrene (*trans-anti*-BaP-tetraol) which is the hydrolysis product of r-7,t-8,9,c-10-dihydroxy-t-9,10-epoxy-7,8,9,10-tetrahydrobenzo[a]pyrene (*anti*-BPDE), the ultimate carcinogen of benzo[a]pyrene. Urines were treated with β-glucuronidase and sulfatase and, after extraction and purification steps, the analytes were converted to tetramethyl ethers which were quantified using SIM (external standard: ^{13}C-labeled benzo[a]pyrene). LOD was at the ppt level. The *trans-anti*-BaP-tetraol metabolite was found in 13/13 exposed coke workers (mean, 4.1 fmol/mL urine) but was virtually undetectable in the urine of 5 control subjects. It was concluded that the method can be part of a phenotyping approach for assessing occupational benzo[a]pyrene exposure.[93]

4.2.1.3 Gasoline and Diesel Exhaust

About 100 million people in the U.S. are exposed to the constituents of gasoline for periods of a few minutes during refueling at gasoline stations; however, the health risks for consumers are generally considered negligible (reviewed in Reference 94). Service station workers who receive longer-term exposures are, however, at an increased risk of developing kidney and nasal cancers.[95] Volatile aliphatic ethers, such as methyl-*tert*-butyl ether (MTBE) and related compounds, have been blended into gasoline to improve octane performance and reduce emissions. A number of assays have been developed for the quantification of these compounds utilizing GC/EIMS or GC/CIMS.[96] The route-to-route (exposure from inhalation to exposure via leaking into groundwater) toxic potency of MTBE has also been investigated.[97]

Diesel exhaust is probably carcinogenic to humans because exposure is associated with an increased lung cancer rate. To compare biomarkers of exposure in bus garages to general air pollution, hydroxyethylvaline-Hb adducts were determined in serum by GC/MS.[98] The adduct concentration was significantly higher in exposed workers, 33.3 (range 25.4 to 58.8) pmol/g Hb vs. 22.1 (range 8.0 to 37.0) pmol/g Hb in controls. PAH from diesel exhausts were the likely source of these genotoxins.[99]

To explore the potential of 1-nitropyrene (a weak carcinogenic nitroarene) as a marker for exposure to diesel exhaust, Hb adducts were estimated by determining 1-aminopyrene (1-AP) in the blood of rats after oral administration of 10 mg 1-nitropyrene. The adducts were extracted from blood with hexane, after alkaline hydrolysis (both mild and strong). After derivatization with heptafluorobutyric anhydride, the products of the hydrolysis were identified by GC/EI-MS/MS. Among several products identified, 1-AP was the only one unique to the exposure with 1-nitropyrene. Quantification (i.s.: D_9-1-AP) was accomplished using a product ion derived from the precursor ion of the analyte in the MS/MS mode of operation of an IT analyzer. The LOD was 0.7 to 1.0 pg 1-AP/mg globin. The peak concentration of 1-AP was 39 pg/mg globin at 3 h after exposure, followed by a decrease to ~6 pg/mg globin 24 h after administration.[100]

4.2.2 Aromatic Amines

4.2.2.1 Benzidine

Benzidine (BZ) and its congeners, which are precursor intermediates in the preparation of azo dyes, are made by reduction of nitrobenzenes. The dyes are subsequently produced by the diazotization of the amino groups on the benzidines followed by azo coupling to other dye intermediates containing reactive aromatic ring systems. Although the association between cancer of the urinary bladder and occupational exposure to dyes and associated chemicals has been known for a century, BZ was classified as a carcinogen only ~25 years ago. An initiative to investigate relevant aspects of 3,3′-dimethylbenzidine, 3,3′-dimethoxibenzidine, and some 25 selected prototypical dyes was implemented only ~10 years ago.[101]

Although BZ and its congeners are not carcinogenic prior to metabolic activation, their presence in urine is significant because it indicates exposure of the urinary bladder. Metabolites implicated in the development of bladder cancer include N′-acetylbenzidine (ABZ) which has high neoplastic activity. Another metabolite, N,N′-diacetylbenzidine (DABZ) is a detoxified, benign product. In a technique for the simultaneous quantification of BZ, ABZ, and DABZ in urine, sample preparation included solid-phase extraction, reduction with $LiAlH_4$ and derivatization with pentafluoropropionic

TABLE 4.3
Benzidine, N-Acetylbenzidine, and N,N′-Diacetylbenzidine Concentrations in Urine from Workers Exposed to Benzidine and Benzidine-Based Dyes and from Unexposed (Control) Workers in India; Unit: ng/mL Urine

Sample	Benzidine	N-Acetybenzidine	N,N′-Diacetylbenzidine
S-1 unexposed	ND[a]	ND	ND
S-2 unexposed	ND	ND	ND
S-3 benzidine-based dye	0.1	5.6	1.3
S-4 benzidine-based dye	0.2	2.9	0.2
S-5 benzidine	17	172	6.3
S-6 benzidine	37	732	17

[a] ND = not detected

Reprinted from Hsu, F. F. et al., *Anal. Biochem.*, 234, 183–189, 1996.

anhydride. The analytes were separated and quantified by GC/NCIMS with monitoring of the base peak, $[M\text{-}HF]^-$, and several other diagnostic ions (i.s.: deuterated analogs of the analytes). LOD were 0.5, 0.8, and 1.5 ppt for BZ, ABZ, and DABZ, respectively. Testing exposed subjects and controls in India revealed that concentrations of the carcinogenic ABZ were many times greater than those of either BZ and DABZ (Table 4.3).[102]

Glucuronidation is one of the most important detoxification processes. It is therefore of particular interest that for this known bladder procarcinogen, the half-life of benzidine-N-glucuronide in urine is only ~3 min. Studies of the behavior of this glucuronide as well as the N-glucuronide of ABZ have been undertaken as part of an investigation to understand the carcinogenicity of azo dyes. N-acetylbenzene-N″-glucuronide was synthesized to be used as a standard for the study of the N″-glucuronidation of ABZ. The structure of the compound and its penta-TMS-derivative were confirmed by high-resolution negative ESIMS. In *in vitro* experiments, the rate of formation of BZ-N-glucuronide formation was found to be 4.3-fold and 1.6-fold higher than that of ABZ-N″-glucuronide in dog and rat livers, respectively, with no difference in human livers. However, the order of the relative amounts of BZ-N-glucuronide formation was human > dog > rat. A proposed model for arylmono- and aryldiamine-induced bladder carcinogenesis suggests that the hepatic detoxification step, i.e., the N-glucuronidation of both BZ and ABZ, is followed by the transportation of protein-bound N-glucuronides into the kidney and passage into the urine. The next particularly relevant step is that because the glucuronides are very acid sensitive, they are hydrolyzed in the urine releasing arylamines which can enter the bladder where they are activated by bladder enzymes followed by binding covalently to DNA, thereby initiating the neoplastic process. This hypothesis of secondary generation of the ultimate carcinogen is consistent with the lack of differences in cancer rates between normal and fast acetylator phenotypes in a population of Chinese factory workers exhibiting a 158-fold increase in risk for bladder cancer.[103]

After the observation of reductive cleavage of the azo bond of benzidine-based azo dyes *in vivo*,[104] it was suggested that the metabolic liberation of 3,3′-dichlorobenzidine (DCB), a known carcinogen, may be hazardous to those individuals exposed to diarylide azo pigments. In a study to determine whether DCB was released from selected pigments fed to rats in their diet or drinking water, the incidence of DCB-Hb adducts was determined by GC/NCIMS. After hydrolyzing dried Hb, the released diarylamines were extracted with toluene, derivatized with heptafluorobutyric anhydride, and four diagnostic ions were monitored for the identification and quantification of DCB and monoacetyl-DCB (i.s.: 2,2′-DCB). No bioavailability of DCB was observed within the LOD of the method.[105]

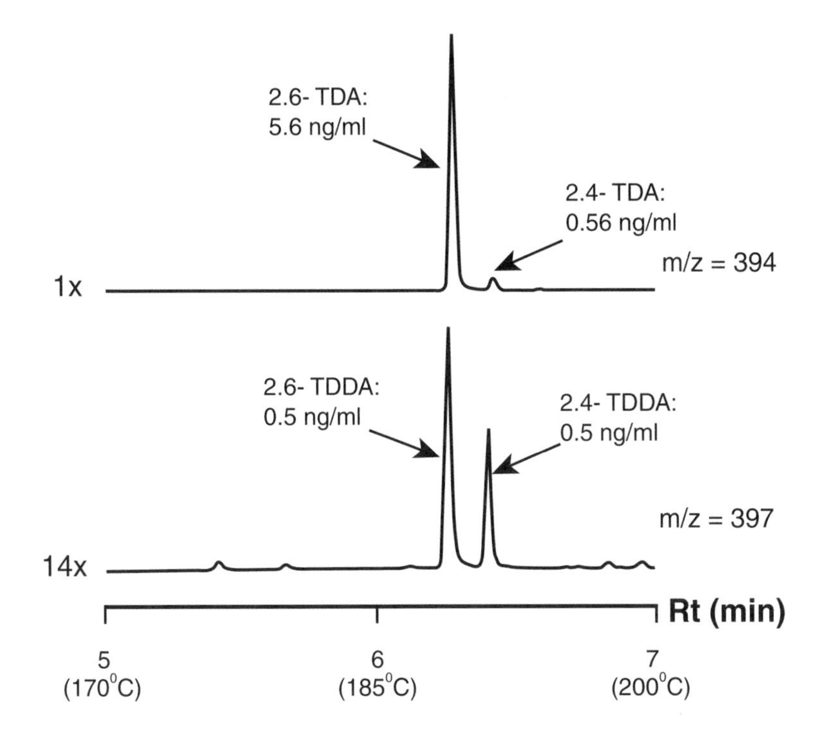

FIGURE 4.9 Determination of toluenediisocyanates 2,6-TDA and 2,4-TDA in hydrolyzed plasma samples from an exposed worker. The pentafluoropropionic anhydride derivatives of the analytes were separated by GC and quantified by negative ion chemical ionization (ammonia reagent gas) using selected ion recording (SIR) of the the $[M\text{-}20]^-$ ions of the analytes at *m/z* 394. The internal standards, the trideutero analogs, 2,6-TDDA, and 2,4-TDDA, were monitored at *m/z* 397. Reprinted from Lind, P. et al., *Occup. Environ. Med.*, 53, 94–99, 1996. With permission.

4.2.2.2 Miscellaneous

Toluenediamines

The aromatic amines 2,4-toluenediamine (2,4-TDA) and 2,6-toluenediamine (2,6-TDA), both of which are potent animal carcinogens, are widely used intermediates in the production of toluene diisocyanate and, in turn, polyurethane foam, coatings, and elastomers. To quantify TDA in hydrolyzed urine and plasma, the five known isomers (2,3-, 3,4-, 2,6-, 2,4-, and 2,5-TDA) were extracted with toluene, derivatized with pentafluoropropionic anhydride, and quantified by GC/NCIMS (i.s.: trideuterated TDA) using SIM of the $[M\text{-}HF]^-$ ions. The LOD in urine or plasma was 1 to 5 fg injected material for each TDA isomer.[106] The technique was employed to assess the toxicokinetics of the 2,4 and 2,6 isomers in blood and urine samples from exposed workers at two polyurethane foam plants. The monitoring revealed 5.6 and 0.5 ng/mL of 2,4-TDA and 2,6-TDA, respectively (Figure 4.9). The metabolic patterns of the two isomers were similar but appeared to be affected by the nature of the exposure as the plasma half-life of the analytes in chronically exposed workers was twice as long as that for workers with short-term exposure. Urinary elimination exhibited a two-phase pattern, the first one being related to metabolism and excretion of material derived from recent exposure, while the second phase is probably linked to the excretion of excised and released toluenediisocyanate adducts.[107]

The metabolism of TDA involves oxidation to the respective arylhydroxylamine by hepatic mixed-function cytochrome P-450 isozymes, followed by further oxidation into the corresponding arylnitrosoarene. In the case of erythrocytes, the aryl-NO derivatives can form covalent conjugates, with the thiol residues of Hb, eventually yielding the sulphinic acid amide Hb adduct, aryl-NH-SO-Hb. The parent amine may be liberated from this complex by acidic or alkaline hydrolysis yielding aryl-NH_2+Hb. This primary amine is the basis of a GC/EIMS technique using SIM of $[M]^+$ and $[M-C_3F_7]^+$ ions of the derivatized analyte. In experiments with TDA-treated rats (0.5 to 250 mg/kg), the Hb-adduct yield was >92%, dose response was linear, the highest adduct concentration was observed 18 to 24 h following administration, and ~90% of the dose was cleared by day 30.[108]

6-nitrochrysene

Although present in the environment at low concentrations, 6-nitrochrysene is one of the most potent lung carcinogens tested. A proposed pathway for activation involves nitroreduction to form 6-aminochrysene. In metabolic studies in rats 6-aminochrysene was found to be a major fecal metabolite, the identification of which was confirmed by MS including a comparison with a synthetic standard. Additional metabolic and DNA adduct studies led to the conclusion that nitroreduction and ring oxidation pathways result in the activation of 6-nitrochrysene as a carcinogen in the rat model.[109]

4.2.3 Plastic Monomers

4.2.3.1 Acrylonitrile

Billions of pounds of acrylonitrile (ACN) are used in the manufacture of elastomers, resins, and acrylic fibers. More than 125,000 workers are exposed to ACN. Cancers identified in exposed workers include those of the prostate, lung, and stomach. ACN is also present in tobacco smoke in a considerable quantity. Classified as a possible human carcinogen, the current exposure standard is 2 ppm for an 8 h period.

ACN-mediated carcinogenesis originates from the alkylation of DNA by the epoxide metabolite, 2-cyanoethylene oxide (CEO), which forms in the liver by P-450-mediated oxidation, followed by transport to the target organs in the blood stream. When ACN is inhaled, pulmonary metabolism may also contribute to the formation of CEO. The latter was investigated by incubation of rat lung cells, as well as lung and liver microsome preparations, with ACN and quantifying the CEO produced, in methylene chloride extracts, using GC/EIMS. The analyte fragment ion, C_2H_3N, was monitored with high resolution SIM (4000 M/ΔM) at *m/z* 41.0265 to distinguish it from ions of the hydrocarbon fragment, C_3H_5, *m/z* 41.0391. It was concluded that lung tissue could metabolize ACN (though liver microsomes were more active) and that CEO formation in the lung was cell type specific.[110]

Both ACN and CEO are electrophilic, thus the formation of Hb adducts may be used as a long-term marker of exposure. In one technique, the product of the interaction of ACN with the N-terminal amino group of Hb, N-(2-cyanoethyl)valine (CEV) was released from the Hb by an N-alkyl-Edman procedure. The pentafluorophenylisothiocyanate (PFPITC) derivative and the resulting thiohydantoin were quantified by GC-NCIMS. One application determined CEV in Hb from rats exposed to ACN in drinking water. A linear relationship was found between adduct levels and water concentrations <10 ppm. At higher levels there was a saturation of the adduct formation process.[111]

Another technique used an isotope-labeled i.s., N-(2-cyanoethyl)-$[D_8]$Val-Leu-Ser, which replicates the CEO adducted N-terminal sequence of Hb. This standard was characterized by NMR, FABMS, ESIMS, and ESI-MS/MS.[112] After addition of the i.s., the alkylated valines were coupled with PFPITC and released from their respective peptides using a modified Edman degradation procedure to produce their pentafluorophenylthiohydantoin (PFPTH) derivatives. The adduct was quantified using GC-EIMS by monitoring the molecular ions of the analyte and the deuterated internal standard in SIM mode (Figure 4.10). LOD was 1 pmol CEV/g globin. The level of CEV in globin was quantified in 11 control subjects (office workers) and 16 nonsmoking workers in a factory where ACN polymerization was carried out. The CEV levels were in the 93.9 to 5746 pmol/g globin range among workers, with no difference between maintenance mechanics and those working

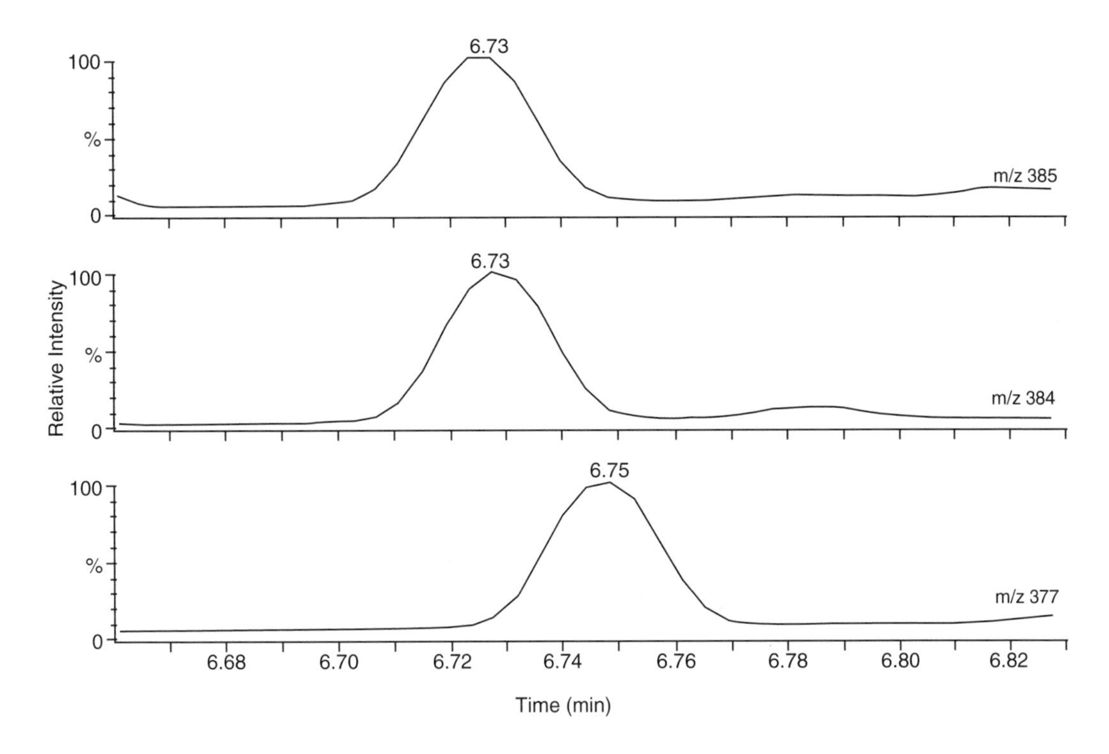

FIGURE 4.10 Representative selective ion monitoring of ions at *m/z* 377, 384 (deuterated internal standard) and 385 from the analysis of N-(2-cyanoethyl)valine (CEV) adduct in the hemoglobin of a worker exposed to acrylonitrile. CEV levels in exposed workers were in the 94 to 5750 pmol/g globin range. The detection limit of the technique was 1 pmol CEV/g globin. Reprinted from Tavares, R. et al., *Carcinogenesis,* 17, 2655–2660, 1996. With permission.

in the continuous polymerization area. These CEV concentrations were significantly higher than those observed in the controls, 8.5 to 70.5 pmol/g globin.[57]

4.2.3.2 Ethylene and Propylene Oxides

Ethylene oxide (EtO) is used both as a chemical intermediate and as a sterilizing agent for medical supplies. Close to 3 million tons of EtO are produced annually in the U.S. and nearly 300,000 workers are exposed to the gas which is classified as a probable human carcinogen and as a hazardous air pollutant. Exposure is associated, in a dose-related manner, to brain and stomach cancers and leukemia.[113] The current short term limit of exposure is 1 ppm.[114] The metabolic precursor of EtO is ethylene. The elimination half-life of EtO is 40 to 55 min and its metabolites are ethylene glycol and its glutathione conjugates.

Hemoglobin Adducts

It has been determined using radioactive EtO that its reaction with Hb leads to the hydroxyethylation of the component nucleophilic amino acids, histidine, cysteine, and the N-terminal valine, in a dose-dependent manner. Assuming that EtO reacts with the imidazole nitrogens of the histidinyl residues of Hb, the degree of alkylation may be estimated by quantification of the N3-(2-hydroxyethyl)histidine in the Hb. In one such assay using GC/EIMS, separated Hb was hydrolyzed, the products separated by ion exchange chromatography, and the O-bis-(heptafluorobutyl) derivatives of the analyte and a deuterated i.s. quantified by SIM. In one application, the concentration of alkylated Hb was significantly greater (0.5 to 13.5 nmol/g globin) in exposed sterilization workers than in controls (0.05 nmol/g globin). This technique has been improved by simplifying the purification stages and increasing sensitivity by monitoring the $[M\text{-}COOCH_3]^+$ ions in a double-focusing instrument. By comparison, point measurements of EtO in work site air were uncertain and therefore inadequate to assess safety.[115]

Quantification of an alternative biomarker for EtO is based on the extension of the previous method to the determination of N-(2-hydroxyethyl)valine as its pentafluorophenylhydantoin derivative after the release of the adducted terminal valine from the Hb using a modified Edman degradation method. The derivatized compound was detected using GC/NCIMS with SIM of representative ions of the analyte and an i.s. (synthetic deuterium-labeled N-terminal valine). This technique has been applied to rats and mice that inhaled EtO in the 0.3 to 300 ppm range for various time periods. Comparing these results with those of simulations, from an *in vitro* model designed to study the formation and removal of Hb adducts, pointed to the discrepancies between exposure estimates based upon Hb adduct and air sample determinations. It was concluded that the current safety standards of EtO exposure are adequate.[116]

DNA Adducts

Upon metabolic activation, EtO acts as a highly reactive epoxide that alkylates in a monofunctional manner. Similar to many alkylating agents, the preferential targets in the DNA are nucleophilic sites, such as the N7 of guanine. Because simple heating of DNA in water releases such DNA adducts as free nucleobases, the determination of the N7-G adduct may be used as a biomarker of exposure. One GC/MS technique yielded background adduct levels of 0.1 to 6 adducts per 10^6 nucleotides in experimental animals but this was inadequate for biomonitoring.[117] A later technique involved a multistep chemical treatment of the precipitated DNA, including reaction with nitrous oxide to form xanthines, derivatization with pentafluorobenzyl bromide of both the NH and OH groups, solid phase extraction, and GC/MS analysis using full mass range scanning with subsequent reconstruction of ion chromatograms. The LOD was 100 pg of adduct spiked into 100 μg DNA. This technique has also been applied to methyl-, phenyl-, and styrenyl- DNA adducts.[118]

In a recent study aiming to correlate DNA adducts and the induction of mutagenic effects in rats exposed to EtO, as a basis for cancer risk assessment, the concentration of the pentafluorobenzyl derivative of N7-(2-hydroxyethyl)guanine (N7-HEG) was determined by GC/NCIMS with SIM [i.s.: N7-(2,3-dihydroxypropyl)guanine]. It was concluded that while EtO at subchronic exposures is only weakly mutagenic in rats, the long-term occupational exposure to airborne EtO at concentrations at or below 1 ppm produces an unacceptable increased risk in humans.[119]

Propylene Oxide

There is considerable occupational exposure to propylene, a monomer/precursor that is widely used in the production of polypropylene, acrylonitrile, isopropyl alcohol, and propylene oxide. Propylene oxide is a biotransformation product of propylene as well as a manufactured chemical. Although propylene is not carcinogenic, propylene oxide is a contact carcinogen in rodents; human data are inadequate. To evaluate the use of N7-(2-hydroxy-propyl)guanine as a dosimeter, DNA samples were analyzed by GC-high-resolution MS in nasal and hepatic tissues of rats after heavy (500 ppm) and long (6 h/day, 5 d/week, for 4 wk) exposure to propylene oxide. The following quantities of adducts were found, immediately after cessation of exposure, expressed as pmol adduct/μmol guanine: 835 ± 80 (respiratory), 397 ± 53 (olfactory), and 35.6 ± 3 (liver). Quantities decreased significantly after 3 d of recovery. No endogenous adduct formation was detected in controls. Concurrent data obtained using ^{32}P-postlabeling were comparable. It was concluded that considerable alkylation of DNA occurred in those tissues, such as respiratory tissues, where contact exposure elicits a carcinogenic response. This is in contrast to liver tissue where levels of alkylation are low and there is no carcinogenic response.[120] In another study, the DNA and Hb adducts and apurinic/apyrimidinic sites were quantified by GC/high-resolution MS in tissues of rats exposed to propylene oxide by inhalation.

The data revealed that the nasal respiratory tissue, which is known to be the target tissue for carcinogenesis, had a much higher degree of alkylation of DNA than non-target tissues.[121]

4.2.3.3 Vinyl Chloride

Vinyl chloride monomer (VCM) is used in the production of polyvinyl chloride (PVC) thermoplastic resins which are fabricated into electrical insulation, pipe, wrapping film, and a variety of other products. In the mid-1970s, >25 deaths were reported among PVC workers due to angiosarcoma of the liver, a rare disease with a latency period of ~20 yr. Animal studies and epidemiological surveys established VCM as the causative agent. VCM is classified as a known human carcinogen. Because >350,000 workers were exposed, the maximum permitted exposure to VCM was set at 1 ppm/8 h (time-weighted average) with a ceiling of 5.0 ppm, averaged over any 15 min period.[122]

Metabolic studies have shown that VCM is converted by a cytochrome P-450-dependent monooxydases to the reactive and carcinogenic chlorethylene oxide (CEO) which may rearrange to 2-chloroacetaldehyde. Using 3,4-dichlorobenzenethiol as a trapping agent of the reactive metabolites formed *in vitro* from reacting VCM with rat liver homogenate, two products were identified by GC/EIMS: 3,4-dichlorophenylthioacetate (in the neutral fraction) and 3,4-dichlorophenylacetic acid methyl ester (in the bicarbonate soluble fraction). It was concluded that the proposed metabolic route to chloroacetaldehyde via CEO oxide does occur as long as other possible routes are not operating.[123]

In studies on rats using ^{14}C-labeled VCM (both inhaled and oral routes), three metabolites were detected in urine. Two were identified using GC/EIMS at both low and high resolution, with masses being determined to within ±0.005 Da, as N-acetyl-S-(2-hydroxyethyl)cysteine (30% of the total radioactivity) and thiodiglycolic acid (25% of the total) while the third remained unidentified.[124] These metabolites were consistent with the proposed mechanism involving the conjugation of CEO and chloroacetaldehyde with glutathione and cysteine leading to the urinary metabolites of thiodiacetic acid and S-(carboxymethyl)cysteine that were identified by GC/EIMS in rats after a 48-hour exposure to 1000 ppm VCM[125] and the determination of thiodiglycholic acid in the urine of VCM exposed workers.[126]

CEO and 2-chloroacetaldehyde are responsible for the formation of DNA adducts. One such cyclic DNA adduct is 1,N^2,3-ethenoguanine (EG). A GC/NCIMS assay for the quantification of EG is based on forming the di-pentafluorobenzyl derivative (i.s.: [$^{13}C_4$]EG). Monitoring the [M-181]$^-$ fragment ions yielded an LOD of 190 fmol/μmol guanine. The EG concentration in liver DNA of rats exposed to 600 ppm VCM from day 10 through day 14 after birth was 1.8 ± 0.3 pmol/μmol guanine.[127]

Other reactions of CEO with DNA, nucleosides, or bases yield, as well as EG, at least four etheno (ε) adducts that are identical to those formed after lipid peroxidation (reviewed in Reference 128). These adducts include 1,N^2-ε-Gua and N^2,3-ε-Gua. These may also be derived from reaction with 2-chloroacetaldehyde, the rearrangement product of CEO. LC/ESI-MS/MS methods were developed for the characterization and quantification of EG and 5,6,7,9-tetrahydro-7-hydroxy-9-oxoimidazole[1,2-α]purine (HO-ethenoGua), the cyclized form of N^2-2(2-oxo-ethyl)Gua], a new DNA adduct. A number of characteristic transitions from the protonated molecules were selected for SRM transitions, e.g., the *m/z* 176→121 transition for the N^2-ε-Gua adduct. The method was used for the measurement of adduct formation in calf thymus DNA after *in vitro* treatment with CEO. The concentration of the adducts was generally proportional to that of CEO, and the ranking of adducts formed was similar to that of 2-halooxiranes with nucleosides. N^2,3-ε-Gua has been reported to be highly miscoding in bacterial systems. The technique afforded very short analysis time compared to ^{32}P-postlabeling; however, it was not sensitive enough in its present form for assaying DNA in unexposed animals and humans.[129]

Several other MS-based techniques have been developed for the quantification of the promutagen etheno adducts of DNA. Using an LC/ESI-MS technique, N^2-3-ε-Gua was quantified (i.s.: $^{13}C_4$ labeled analyte) in human liver DNA at 0.06 ± 0.01 pmol/mg DNA. In CEO-treated rats, the concentration of ε-Gua was inversely related to the time of exposure, suggesting rapid repair of the adduct.[130] An alternative technique utilized extraction by immunoaffinity chromatography, with columns highly specific for ε-Gua, and quantification of the pentafluorobenzyl derivatives using GC/NCIMS operated in high resolution mode. The concentration of ε-Gua in CEO-treated calf thymus DNA was ~45 fmol/mg, regardless of which of three different hydrolysis techniques was used. Results were comparable to those obtained with the LC/ESIMS technique.[131]

The quantification of the VCM-induced adduct ^{1}N6-ethenoadenine (ε-A) in rat urine involved extraction by immunoaffinity chromatography and quantification by LC/ESI-MS with SIM of the analyte (i.s.: ^{15}N-labeled ε-A). The LOD was 30 pmol injected analyte. The endogenous concentration of ε-A was ~20 pmol/mL. Following treatment with CEO, 90% of ε-A was excreted in 24 h and levels returned to normal in ~130 h.[132]

The ethenoadenosine and ethenedeoxycytidine concentrations of crude DNA hydrolyzates from untreated and treated mice and rat livers and human placenta were determined using an LC/ESIMS technique. Using 100 mg size DNA samples, it was possible to analyze one ethenoadenosine adduct in 10^8 normal nucleotides. The sensitivity for 3,N^4-etheno-2′deoxycytidine (ε-dC) was five-fold lower due to interferences.[133] A subsequent method utilized immunoaffinity chromatography and LC-ES-MS/MS for the quantification of ε-dC. The LOD was 5 adducts in 10^8 normal nucleotides using 100 μg samples of DNA. The assay was applied to commercial DNA products and untreated mouse and rat livers.[134]

The availability of these sensitive and rapid techniques for the quantification of VCM biomarkers will permit mechanistic studies of carcinogenesis and DNA repair in animal models as well as epidemiological studies in exposed humans. MS techniques for the analysis of VCM are described in Section 4.3.4.3.

4.2.3.4 Butadiene

1,3-butadiene (BD), a colorless gas with a mildly aromatic odor, is used in the production of synthetic rubber for automobile tires and thermoplastic resins. In the U.S., where >3 billion pounds are made annually, BD is a major air pollutant affecting some 10,000 workers at >200 industrial sites. The allowable workplace exposure is 1 ppm. The carcinogenicity and genotoxicity of BD and its major metabolites, 1,2-butadiene-monoepoxybutane (BMO) and diepoxybutane (DEB), are well established, and BD is classified as a known human carcinogen associated with an increase in leukemia mortality in the butadiene-styrene industry.[135]

Metabolism

When incubated with liver microsomes (mouse, rat, or human), BD is activated by cytochrome P-4502EIi-mediated oxidation to BMO. Further epoxidation by cytochrome P-450 isoenzymes to BDE occurs forming enantiomeric products. The *meso* and (±) diastereomers of DEB have been separated by GC and characterized by EIMS using standards synthesized by oxidation of BMO with 3-chloroperoxybenzoic acid.[136] Both BMO and DEB are direct-acting mutagens *in vitro* and *in vivo*. However, there are striking differences in the biological activities between the epoxide enantiomers, similar to benzo[a]pyrene diol epoxide and styrene oxide. Also, (±)-DEB is much more active than BMO. The inactivation of both of these metabolites to their conjugated derivatives is catalyzed by epoxide hydrolyase and GSH-S-transferase.

In vitro, the rate of metabolism of BD in liver and lung microsomes is mice $\gg$ rats $\cong$ humans. The *in vivo* metabolism is similar, although both qualitative and quantitative differences exist. The distribution and kinetics of elimination of both metabolites were studied in rats dosed with BMO and BDE via i.v. bolus injection. Blood samples taken at various time points were extracted with methylene chloride, and the intact nonconjugated analytes quantified (i.s.: 1-butanol) by GC/EIMS using SIM.[137] Pharmacokinetic data were similar for the two epoxides, including rapid distribution and elimination; half-lives were in ~1.5 to ~2.7 min range.[138]

A second route for the metabolism of BMO is hydrolysis to form 3-butene-1,2,-diol (BDD). Both the plasma clearance and urinary excretion of unchanged and conjugated BDD were studied by GC/EIMS. The glucuronide and sulfate conjugates in urine were incubated with sulfatase and β-glucuronidase, followed by extraction and TMS derivatization. Quantification was accomplished by SIM using the $[M\text{-}CH_2OTMS]^+$ ion (i.s.:1,4-butanediol). The elimination of BDD consisted of metabolism, mediated by both cytochrome P-450 and alcohol dehydrogenase, resulting in the formation of thiol-reactive intermediates.[139]

Other urinary metabolites that are potential biomarkers include 1,2-dihydroxy-4-(N-acetylcysteinyl-S)butane, which is formed by GSH conjugation of butanediol, and 1-hydroxy-2-(N-acetylcyteinyl-S)-3-butene, which is formed by conjugation of GSH with BMO. Both metabolites have been identified by MS. A GC/MS technique for their quantification was applied to find out whether there are statistically different metabolite concentrations in exposed workers vs. controls (reviewed in Reference 140).

The urinary metabolites 1,2-dihydroxybutyl mercapturic acid (DHBMA) and a mixture of monohydroxy-3-butenyl mercapturic acids (MHBMA), and also MHBVal adducts of Hb were explored as possible biomarkers for workers exposed to BD. The urinary metabolites were analyzed as methyl and pentafluorobenzoyl derivatives using GC/NCI-MS/MS with MRM of characteristic transitions. The MHBVal adducts were quantified by another GC/NCI-MS/MS method.[141] There were correlations between the urinary metabolites and 8 h airborne BD levels and the MHBVal adducts and average airborne BD levels over 60 days.[142]

Macromolecular Adducts

Because it is not known which DNA adducts are responsible for the carcinogenicity of BD, efforts have been made to identify (and quantify) these adducts for both BMO and BDE. In an *in vitro* study, purine nucleobases or nucleosides were reacted with epoxybutane and diepoxybutane. The products were separated by HPLC and analyzed by several spectroscopic techniques including LSIMS (cesium ion gun) and negative ESI-MS/MS. A number of guanine and adenine adducts and intermediates were identified and characterized, and it was shown that nucleophilic nitrogens of guanine and adenine first attack one of the epoxy groups of diepoxybutane giving intermediates that hydrolyze to 2′,3′,4′-hydroxybutyl adducts that form cross links with DNA or proteins.[61]

A method has been developed for the quantitative analysis of BD-induced DNA adducts *in vivo* (DNA isolated from BD-exposed animals) and *in vitro* (cultures treated with various BD metabolites) using LC/ESI-MS/MS with SRM of the transition of $[M+H]^+$ ions of the adducts to the corresponding protonated nucleobases (i.s.: ^{13}C- and ^{15}N-labeled analytes). Two regioisomers of N7-EB (EB = epoxybutone) guanine adducts and two N-3-EB-adenine isomers were identified and quantified. This approach is useful for the analysis of BD-DNA adducts that are not amenable to derivatization for GC analysis because of the presence of adjacent polar functional groups.[143]

An investigation comparing the effects of low exposure to ethylene oxide, vinyl chloride, and butadiene on DNA used techniques that can detect DNA adduct concentrations as low as 1 adduct per 10^9 or 10^{10} nucleotides. However, it was concluded that the endogeneous presence of a relatively high number of DNA adducts in all cells prevents the establishment of causal relationships for low level exposures.[144]

4.2.3.5 Styrene and Styrene Oxide

Styrene is an important organic monomer used in high-volumes in the production of plastics, resins, and synthetic rubber. It is classified as probably carcinogenic to humans (Group 2A). Exposure, which may entail gram quantities daily, occurs mainly through inhalation in such occupational settings as the manual layup or spraying of glass-reinforced polyester products, e.g., boats. More than 90% of inhaled styrene is absorbed by humans. Metabolic activation consists of the formation of the intermediate styrene-7,8-oxide (SO), facilitated by microsomal cytochrome P-450. SO is a direct-acting carcinogen in experimental animals. Detoxification involves the hydration of SO to styrene glycol and the subsequent urinary excretion of the final metabolic products, mandelic and phenylglyoxylic acid.

A straightforward GC/EIMS technique with SIM was used to compare blood concentrations of styrene in 76 exposed workers and 81 controls. Significant differences were found at various time points during the workweek. For example, even 16 h after a work-shift, the average blood styrene concentration was ~100 μg/L, compared to ~0.2 μg/L for controls. The half-life of blood styrene was ~4 h.[145] In another study among workers exposed to styrene in factories using fiberglass-reinforced plastics, a correlation coefficient of 0.8 was established between styrene exposure and urinary mandelic acid determined by GC/MS.[62] In another investigation, two alternative GC/MS techniques were used to determine styrene and SO in blood samples from exposed reinforced plastics workers. The analytes were quantified in pentane extracts of blood using positive ion GC/CIMS and SIM (i.s.: D_8-styrene). LOD was 0.05 μg SO/L blood. The other method utilized an initial reaction with valine, followed by derivatization with pentafluorophenyl isothiocyanate, and quantification with GC/NCIMS. LOD for SO was 0.025 μg/L blood. There was moderate to good agreement between the two technique for the determination of SO in exposed workers. The concentrations of styrene in blood were proportional to the corresponding air exposures.[146]

SO reacts with different bases of DNA resulting in N7 (predominant), N^2-, and O^6-guanine adduct formation. MS has played a significant role in the elucidation of structures of these products and the mechanisms of their formation.[147,148] In addition, CE/ESIMS has been used to identify other SO adducts that form at the nucleophilic sites of adenine.[149]

SO also binds to proteins at cysteinyl residues, for example, at the α and β carbon positions of β-C_{93} in Hb. These cysteine adducts can be cleaved using "Raney nickel" to yield the positional isomers 1- and 2-phenylethanol (1-PE and 2-PE). In one technique, isolated globin and albumin were purified by dialysis, digested using protease, reacted with Raney nickel, and derivatized with pentafluorobenzoyl chloride. Quantification was accomplished by GC/NCIMS using SIM to monitor the molecular ions (i.s.: 4-methyl-SO-Hb and 4-methyl-SO-Alb).[150] The technique was applied to the study of the SO adducts in male workers exposed to styrene in fiberglass-reinforced boat factories. After confirming the identity of the background adducts of controls by determining the exact masses of the 1-PE and 2-PE derivatives at a resolving power of 10,000, quantification was carried out as described above. While there was no correlation between SO-Hb and styrene exposure, linear correlations were observed between both 1-PE and 2-PE derived from SO-Alb and exposure to styrene.[151] In a subsequent study, the excretion of the two urinary metabolites, mandelic acid and phenylglyoxylic acid, was correlated with the levels of SO-Alb and SO-Hb adducts. After stratification to exposure (zero, low-level, and high-level) based on the concentrations of the urinary metabolites, further division was made according to smoking or nonsmoking status. Clear correlations between exposure and Alb adduct levels were only found among the categories of workers with high-level exposure and/or smokers. It was concluded that smoking was the major source of the high background SO-Alb adduct observed among the controls not occupationally exposed to styrene. In their present state, these methods are inadequate for the biological monitoring of occupational exposure to styrene because there are significant influences from smoking and other background sources, such as dietary and/or endogeneous, on adduct levels. In addition, urinary metabolites reflect short-term (1-day) exposure, while protein adducts represent longer-term (1 to 3 month) exposure.[152]

4.2.4 Exposure by Health Care Personnel to Cytotoxic Drugs

Several antineoplastic agents have been classified as known human carcinogens (Table 4.1). Possible acute toxic effects (irritation of skin, eyes, mucous membranes, alopecia, vomiting) and long-term health risks (mutagenicity, teratogenicity, and carcinogenicity) to pharmacy technicians and nursing personnel preparing and handling these agents has led to the formulation of guidelines and recommendations by several organizations to reduce occupational exposure and protect employees (listed in Reference 153). Analytical techniques of various levels of sophistication have been developed for the monitoring of individual agents in exposed workers (reviewed in Reference 154). Because any measurable concentration is considered a hazard, analytical sensitivity is a major issue.

A GC/EI-MS/MS technique has been used for the quantification of α-fluoro-β-alanine, the main metabolite of 5-fluorouracil in pharmaceutical plant workers (urinary excretion) as well as 5-fluorouracil in their environment (air, filters, ground, gloves, cleaning procedures, etc.).[155] An LC/ESI-MS/MS method that has been used to quantify methotrexate in urine (i.s.: 7-hydroxymethotrexate) employed solid phase extraction for analyte purification and preconcentration. The LOD (<0.2 ng/L) was adequate to detect trace concentrations in hospital personnel.[156]

Cyclophosphamide

Several studies have been carried out on workplace contamination with cyclophosphamide (CP). In a GC/EIMS technique, using an IT to provide improved sensitivity over a previous method that used a quadrupole analyzer, CP was extracted from urine with diethyl ether, derivatized with TFA, and quantified by SIM of the $[M+H-CH_3Cl]^+$ ions of both CP and the i.s., ifosphamide (IF). The two compounds were separated by GC, there were no interferences, and the blank urine was clean in the area of interest (Figure 4.11). The LOD of CP was 0.25 ng/mL urine.[157] Another GC/EI-ITMS technique quantified both CP and IF (i.s.: trophosphamide) using MRM. Urinary concentrations of CP and IF were determined simultaneously with LOD of 0.1 and 0.5 ng/mL, respectively.[158]

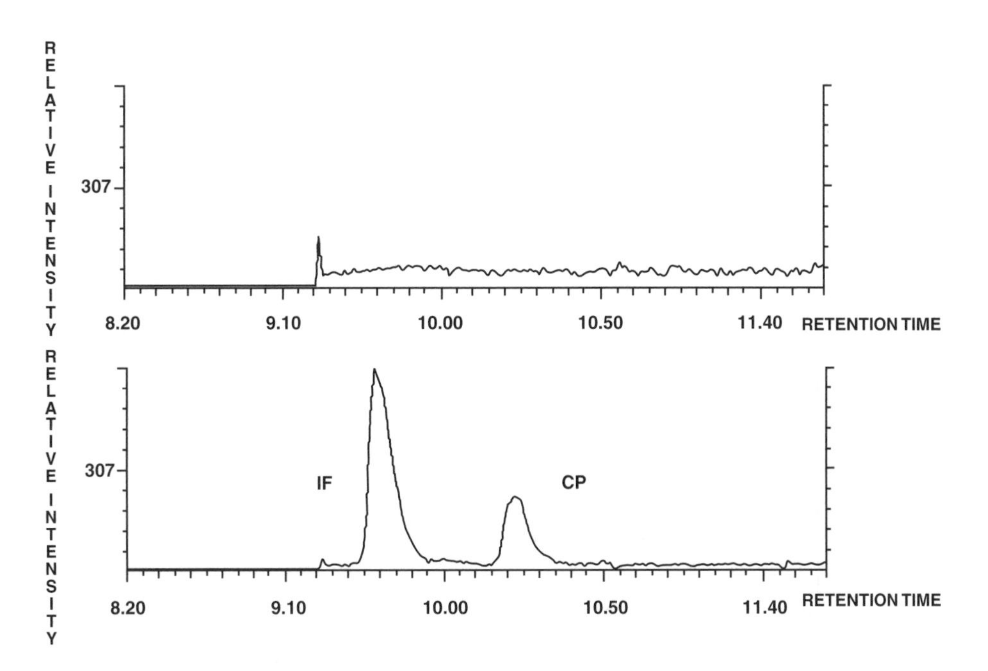

FIGURE 4.11 Total and reconstructed ion chromatograms of *m/z* 307. Upper curve: Blank urine. Lower curve: Urine sample of a pharmacy technician involved in the preparation of antineoplastic agents, including cyclophosphamide (CP). The concentrations of cyclophosphamide and ifosphamide (IF) were 18 ng/mL and 100 ng/mL, respectively. Reprinted from Sessink, P. J. et al., *J. Chromatogr.*, 616, 333–337, 1993. With permission.

A further increase in sensitivity was achieved using LC/ESI-MS/MS (i.s.: IF). The use of HPLC allowed for omission of the derivatization step, therefore eliminating the possibility of thermal decomposition of CP during the formation of the N-TFA derivatives required for the GC-MS method. Specificity and sensitivity were provided by using MRM utilizing the *m/z* 261.2 → 140.2 transition of the $[M+H]^+$ ion. The LOD was 0.05 ng/mL urine, the LOQ was 0.2 ng/mL. CP was in the 0.1 to 1.9 ng/mL range in 12 of 24 hospital workers.[159] In another study in two hospitals, both CP and IF were quantified not only in urine but also in air, wipe samples, gloves, and pads. All 21 air samples taken near or on top of the hoods used for drug preparation were below detection limit (<2 ng/m^3). However, considerable amounts of both compounds were detected in three personal air monitors, 240 ng/m^3 CP, and 20 and 40 ng/m^3 IF. In 49 wipe samples, CP concentration ranged from <0.001 to 383.3 μg/dm^3, and for IF from <0.001 to 141.5 μg/dm^3. The levels were always higher in Hospital B than in Hospital A. Pads placed on arms, legs, and chest were also often contaminated. Both CP and IF were detected on the internal sides of gloves used during drug preparation and administration. In an experiment in which a double pair of vinyl gloves was used during drug preparation, even the internal side of the inner glove contained measurable levels of IF, 0.6 μg/pair vs. 39 μg/pair for the outer glove (Figure 4.12). Urinary CP concentrations were above the detection limit in 12 of 24 workers, ranging from 0.11 to 2.1 μg/L and were also always higher in Hospital B. This correlates with the higher pad and wipe data that were also observed for that hospital. Conclusions included: recommendation for the installation and use of vertical laminar airflow hoods, which were shown to reduce drug levels below detection; a suggestion that dermal uptake was possibly the main exposure route for hospital workers; and the claim that the LC-MS/MS approach was the most sensitive technique available.[153]

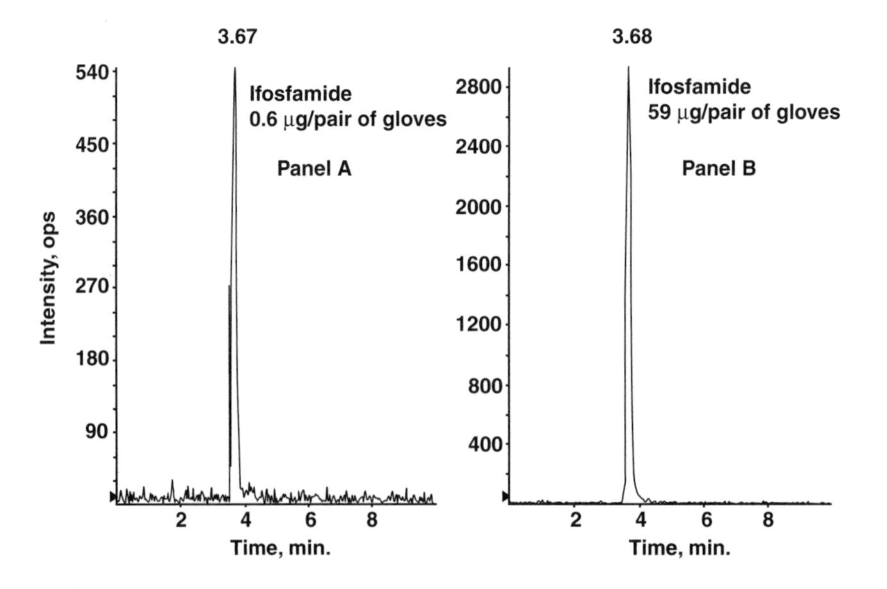

FIGURE 4.12 Selected reaction monitoring (*m/z* 261.2 → 92 transition) curves of (a) the inner and (b) outer pair of gloves worn by a subject involved in the preparation of the drugs. Reprinted from Minoia, C. et al., *Rapid Commun. Mass Spectrom.*, 12, 1485–1493, 1998. With permission.

4.3 ENVIRONMENTAL CARCINOGENS

4.3.1 Tobacco

Approximately 30 to 40% of 10^2- to 10^4-year-old human remains from excavation sites in China were shown to contain 12 to 475 ng/g nicotine. Both nicotine and its metabolite, cotinine were identified by GC/MS, suggesting that the individuals were smokers.[160] Prior to the 20th century, tobacco was mainly used for pipe and cigar smoking among a limited population. During the past 20 years, cigarette production worldwide has been increasing 2.2% annually, in contrast to the 1.7% growth in population. The use of smokeless tobacco (plug, leaf, and snuff) has also been increasing.

Epidemiologic studies have repeatedly suggested that ~30% of cancer deaths in the U.S. are attributable to smoking.[9,10,161] More than 80% of lung cancers and 50% of bladder cancers result from smoking.[162] In the U.S., more people die of lung cancer than of colon, prostate, and breast cancers combined. Five-year survival for lung cancer is only ~13% compared with an average of ~40% for cancers in general. The smoke from pipes and cigars is of higher alkalinity (more irritating) than that from cigarettes, resulting in high rates of cancers of the buccal cavity and pharynx. During "dipping snuff" (placing moist powder between cheek and gum), nicotine and carcinogens are directly absorbed through the oral tissues, frequently leading to oral cancers.[163]

Tobacco use is the greatest single preventable cause of premature mortality in the U.S.[164,165] and the fight against its use has been increasing rapidly, with hardly a month passing without news about lawsuits and state and local legislation involving smoking. Still, the number of smokers remained at ~20% during the 1990s and the number of teenagers who smoke has been increasing steadily to the current rate of one in four high school students. The percentages of smokers among young women in the western world and among all people, both adults and children, in the developing countries continues to increase.

During the burning of a cigarette, some 400 to 500 mg of smoke emerges in the form of an aerosol that contains 40 to 500 gaseous components and up to 10^{10} particles/mL. Some 3500 chemicals have been identified in the particulate phase. At least 43 of these compounds have been identified as carcinogenic in exposed laboratory animals.[9,13,166] For several of the compounds there is sufficient direct evidence, e.g., from occupational exposures, of their carcinogenicity to humans (Table 4.4). As discussed later, several carcinogens can be detected as free compounds and/or DNA-adducts in the urine and/or relevant tissues of smokers.[167]

Passive smoking (second-hand smoking) refers to the inhalation of *environmental tobacco smoke* (ETS) by nonsmokers in proximity of burning tobacco. The smoke exhaled by the smoker (*mainstream smoke*) is only a minor component of ETS. The major component is *sidestream smoke* that is emitted from the burning tip of cigarettes between puffs. In many respects, sidestream smoke is more hazardous than mainstream smoke because it has not been filtered either by the cigarette or by the smoker's lungs and thus contains higher concentrations of nicotine and many carcinogens (Table 4.4). Concern about ETS has been expressed for almost 20 years[168] and exposure to ETS is considered to be an occupational hazard for nonsmoking workers in the railroad and restaurant industries, enclosed offices, etc.[169] Passive smoking has been causally associated with lung cancer,[170] and in 1998 ETS was classified as a known human carcinogen.[11] Salient characteristics of possible biomarkers of ETS have been compared.[171]

4.3.1.1 Quantification of Nicotine and Its Metabolites

Nicotine is a tertiary amine consisting of pyridine and pyrrolidine rings (Figure 4.13a); it is present in tobacco in the levorotatory (S) form. Over the years, the nicotine content of cigarettes has been reduced from ~9 mg to 0.05 to 1.8 mg per cigarette. About 15 to 30% of nicotine is distilled at the tip of the burning cigarette and absorbed onto the particular matter droplets (tar) that are inhaled.

TABLE 4.4
Representative Carcinogens Present in Cigarette Smoke and Its Particulates, ng/per Cigarette

Compound	Mainstream	Sidestream
Polycyclic Aromatic Hydrocarbons		
Benz[a]anthracene	20–70	40–200
Benzo[a]pyrene	20–40	40–70
Chrysene	40–60	
Aza-arenes		
Quinoline	$1–2 \times 10^3$	$15–20 \times 10^3$
Dibenz(a,j)acridine	3–10	
N-nitrosamines		
N-Nitrosopyrrolidine (in vapor)	1.5–110	30–390
N-Nitrosodimethylamine (in vapor)	0.1–180	200–1040
N-nitrosonornicotine, NNN[b]	$0.1–3.7 \times 10^3$	$0.1–1.7 \times 10^3$
NNK[b]	$0.1\text{-}0.8 \times 10^3$	$0.2–1.4 \times 10^3$
Aromatic Amines		
2-Toluidine	30–200	3×10^3
2-Naphtylamine	1–22	70
4-Aminobiphenyl	2–5	14
Aldehydes		
Acetaldehyde	$18–1400 \times 10^6$	
Formaldehyde (in vapor)	$70–100 \times 10^3$	1500×10^3
Misc. Organic Compounds		
Acrylonitrile	$3.2–15 \times 10^3$	
Benzene (in vapor)[a]	$12–48 \times 10^3$	400×10^3
2-Nitropropane	$0.7–1.1 \times 10^3$	
Vinyl chloride	1–16	
Hydrazine (in vapor)	14–51	90
Inorganic Compounds		
Arsenic[a]	40–120	
Chromium[a]	4–70	
Radioactive elements, e.g., polonium-210[a]		

[a] Sufficient evidence of human carcinogenicity; the remainder are based on animal testing.
[b] These are carcinogenic derivatives of nicotine (see Figure 4.18).

Source: Compiled from several sources.

It takes 8 to 10 s for nicotine to travel from the lungs to the brain where it mimics acetylcholine at the cholinergic receptor sites and alters the way the brain processes information. Nicotine is not a carcinogen. Instead, it facilitates addiction by acting on the neurons of the mesolimbic reward system in the midbrain producing an initial transient excitation that is followed by depression or transmission blockade.[172]

The primary metabolite (70 to 80%) of nicotine is cotinine (Figure 4.13a) that is formed via the cytochrome P-450-mediated oxidation of position 5 of the pyrrolidine ring to nicotine-Δ-1′,5′-iminium ion, that is then converted into cotinine by a cytosolic aldehyde oxidase. Several secondary

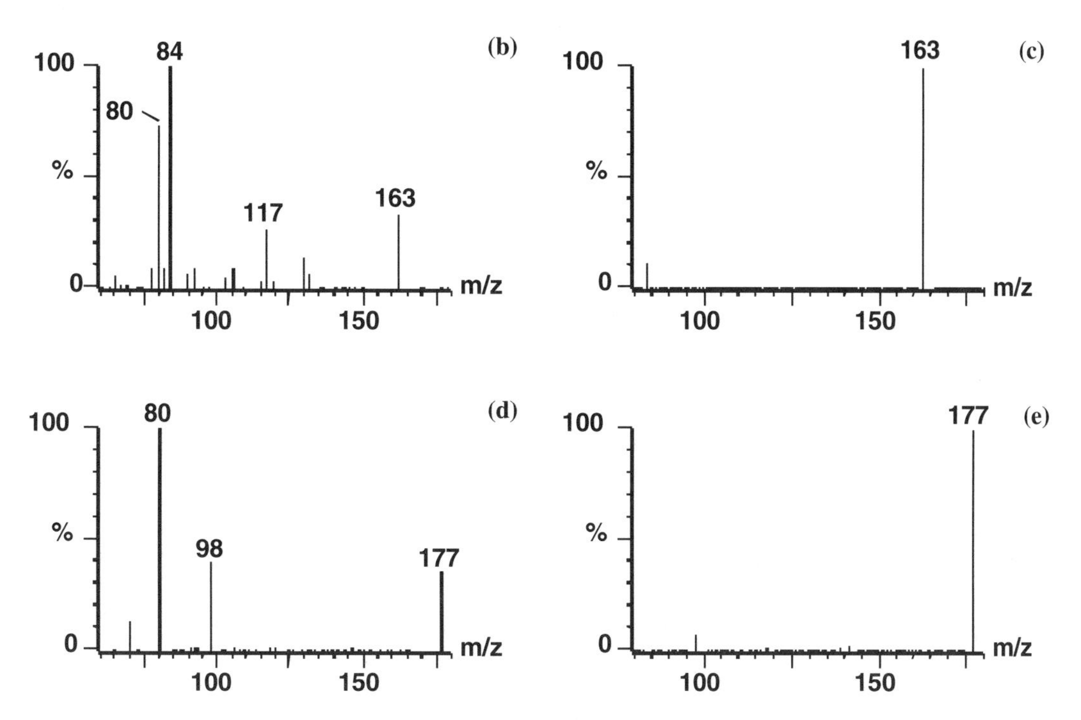

FIGURE 4.13 (a) Cotinine formation from nicotine by enzymatic oxidation. (b) Collision-induced decomposition of the protonated molecule of nicotine. (c) Parent ion spectrum of structurally significant product ion *m/z* 84 from nicotine. (d) Collision-induced decomposition of the protonated molecule of cotinine. (e) Parent ion spectrum of structurally significant product ion *m/z* 98 from cotinine.

metabolites form from cotinine (see later) including 3′-hydroxycotinine, which is the most abundant urinary metabolite of nicotine.[173] In addition to the GC/MS and LC/MS techniques used for the quantification of nicotine and its metabolites in various matrices (reviewed below), a "proof of concept" work has been reported on the use of CE/ESI-MS/MS using a custom-made coaxial sheath-flow interface. Some isobaric metabolites that could not be identified by migration time, e.g., *trans*-3′-hydroxycotinine and 5′-hydroxycotinine, were identified by monitoring specific CID transitions.[174]

Quantification: Serum

In earlier GC/MS techniques, the identity of the metabolites was usually confirmed by EI, as their trifluoroacetyl derivatives,[175] and quantified by SIM of the molecular and abundant fragment ions (Reference 176 and references therein). Using the heptafluorobutyric anhydride derivatives, sensitivities in the femtogram range were obtained with GC/NCIMS.[177] A GC/EIMS technique for both serum and urine quantified extracted nicotine (i.s.: D_3-nicotine) in the 1.25 to 100 ng/mL and cotinine (i.s.: D_3-cotinine) in the 1.25 to 1000 ng/mL range.[178]

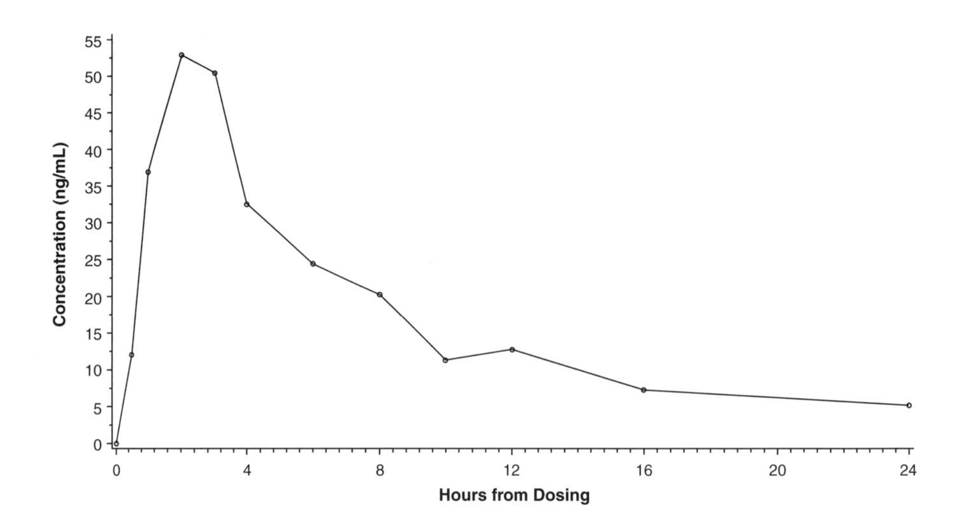

FIGURE 4.14 Pharmacokinetic profile of a subject dosed with a 45 mg nicotine patch. Reprinted from Xu, A. S. et al., *J. Chromatogr. B,* 682, 249–257, 1996. With permission.

The CID spectra of nicotine and cotinine are straightforward (Figure 4.13d) and precursor-product transitions, such as the 177 → 98 transition for cotinine that is due to the loss of the 3-pyridyl group from the molecular ion, may be selected for quantification by SRM. This approach was utilized in an LC/APCI-MS/MS technique where the analytes were extracted from plasma with methylene chloride and quantified by SRM (i.s.: deuterated analytes). Quantification was validated for 1 to 50 ng/mL nicotine and 10 to 500 ng/mL cotinine. The method was applied to obtain time profiles in volunteers treated with 45 mg nicotine (Figure 4.14). Some 400 samples could be analyzed daily.[179] Another paper describes essentially the same technique in more detail and with elaborate validation procedures; throughput was 100 samples/day.[180]

Quantification: Urine

Several LC/TSPMS techniques have been developed for the simultaneous quantification of nicotine and several of its metabolites in urine. In one technique, underivatized urine was injected directly and nicotine, cotinine, *trans*-3′-hydroxycotinine, nicotine-N-oxide, and demethylcotinine were quantified (20 to 200 ng/mL range) using SIM of the protonated molecules.[181] The TSPMS technique was subsequently extended to include the N-glucuronide conjugates of nicotine and (N)-*trans*-3′-hydroxycotinine, both directly[182] and after release of the aglycon with β-glucuronidase. Cotinine-N-glucuronide was an abundant urinary metabolite but there was wide interindividual variability.[183] The major urinary metabolite (35%) was *trans*-3′-hydroxycotinine. The following 8 metabolites were also quantified (in decreasing order of concentration): cotinine-glucuronide, cotinine, nicotine, *trans*-3′-hydroxycotinine-glucuronide, nicotine-N′-oxide, cotinine-N-oxide, nicotine-glucuronide, and demethylcotinine.[184] A larger study involving the quantifications of metabolites in urine concluded that published total nicotine yields for individual brands were useful in predicting nicotine uptake despite individual variability.[185]

There has been considerable interest in urinary cotinine as an index of smoke exposure.[186] One application involved children (4 to 11-yr-old) and nonsmoking adults placed in a bus for 2 h. Smoke was generated by persons smoking some 80 cigarettes, raising the nicotine content to 110 mg/m^3 air. Nicotine in the air and the urinary cotinine of the subjects were quantified (i.s.: D_3-cotinine) by GC/MS using both EI and CI and validated.[187] Urinary cotinine increased for ~6 h and remained constant for 12 h. This was followed by a steady decline. The mean maximum urinary cotinine concentration was 22 mg/L in children and 13 mg/L in adults.[188]

A GC/EIMS technique for the quantification of cotinine and *trans*-3′-hydroxycotinine in subjects receiving transdermal nicotine (11 to 44 mg/day) during smoking cessation therapy involved the hydrolysis of the conjugated metabolites with β-glucuronidase, solvent extraction, TMS derivative formation, and quantification by SIM (i.s.: deuterated analytes). About 17% of the total nicotine intake was converted to cotinine, in about equal amounts of the free and conjugated forms. Hydroxycotinine, the major metabolite (20% of total nicotine intake), was present in the free form.[189]

Because dietary nicotine also produces cotinine, care must be taken when using urinary cotinine as a biomarker for tobacco use. The nicotine content of some foods (e.g., tomatoes, potatoes, and teas) has been determined by GC/EIMS using SIM. The highest nicotine content was found in instant tea, which contained up to 285 ng/g.[190]

Quantification: Hair

Some drugs enter germinal hair cells from the blood by passive transfer where they become tightly bound during keratogenesis, move along the shaft as the hair grows (about 1 cm/mo), and remain for the lifetime of the hair. A GC/EIMS technique involved incubating 50 to 100 mg hair with NaOH, extracting with ethyl ether, and monitoring analyte ions using SIM (i.s.: ketamine). Concentration ranges (ng/mg hair) of nicotine in the hair of smokers and nonsmokers were 0.91 to 38.27 and 0.06 to 1.82, respectively, while for cotinine the ranges were 0.09 to 4.99 and 0.01 to 0.13, respectively. Thus, smokers with nicotine >2 ng/mg hair could be differentiated from nonsmokers.[191] The uptake of nicotine into the hair from smokers and nonsmokers was determined after continuous exposure of individuals to 20, 200, or 2000 μg/m^3 nicotine vapor in a 650 L dynamic exposure chamber for 72 h. Nicotine concentrations were achieved by intermixing fresh air with air that had been saturated with nicotine by being bubbling through a reservoir. Strict protocols were used for cutting and storage of the hair samples. After removing the protein matrix with NaOH, the nicotine was extracted with diethyl ether and quantified by GC/EIMS (i.s.: 2,6-di-t-butyl-4-methylphenol) using SIM of representative ions.[192] The relation of nicotine uptake to applied concentration was second order and it was suggested that the monitoring nicotine in hair is the best technique available for the estimation of ETS exposure.[193]

The time course of cotinine formation has also been measured in beard and saliva after chewing nicotine-containing (4 mg) gum. After digestion with NaOH, analytes were extracted with dichloromethane and quantified by GC/EIMS. While cotinine concentrations in the saliva peaked at 1.5 h and disappeared in 24 h, cotinine only appeared in the beard on day 3, peaked on day 5 (2 ng/mg hair), and disappeared by day 7.[194]

Quantification: Seminal Plasma, Saliva, and Sweat

The presence of nicotine, cotinine, and *trans*-3′-hydroxycotinine was investigated in the seminal plasma of smokers and nonsmokers using LC/particle beam-EIMS with SIM of structurally significant fragment ions. Neither nicotine nor its metabolites were detected in nonsmokers. In smokers, nicotine concentrations were ~2-fold larger in seminal plasma than in serum while the metabolite concentrations were comparable. Cotinine concentrations were significantly correlated with the daily nicotine intake. There was also correlation with the forward motility of spermatozoa, suggesting that the presence of these analytes in seminal plasma may provide a warning of adverse effects of smoking on reproduction.[195]

GC/EIMS with SIM (i.s.: cotinine-D_9) was employed to investigate the relationship between self-reported ETS exposure and salivary cotinine concentrations in nonsmokers. Cotinine was found (0.5 to 7.4 ng/mL) in 83% of a large (n~200) study population and was related to ETS exposure in the household and/or workplace.[196]

Drugs in sweat may be recovered over several-day collection periods by using transcutaneous drug collection patches (acrylate adhesive layer on a polyurethane film with an adsorbent pad).

Nicotine in sweat was collected this way, extracted with methanol and quantified by GC/EIMS with SIM (i.s.: nicotine-D_4). Recovery was ~75% and the test was linear in the 50 to 2500 ng/patch range with an LOD of 10 ng/patch. Nicotine concentrations were 150 to 2498 ng/patch for smokers, 87 to 266 ng/patch for passive smokers, and nondetectable in controls.[197]

4.3.1.2 Benzene and Polycyclic Aromatic Hydrocarbons

Trans,Trans-Muconic Acid (t,t-MA)

The main source of nonoccupational benzene exposure is smoking. The benzene content of mainstream cigarette smoke ranges from ~6 μg/cigarette (low yield) to ~70 μg/cigarette (high yield and nonfilter). Thus, smoking 20 cigarettes results in the inhalation of ~1 mg benzene, which comes close to the occupational threshold limit. The amount of benzene exhaled by smokers is at least fivefold higher than that from nonsmokers. Urinary *t,t*-MA, a ring-opened metabolite of benzene (Figure 4.3) has been considered as a marker of benzene exposure in smokers. Based on the conversion of benzene to *t,t*-MA at a rate of 2 to 25%, the expected daily urinary excretion from smokers is in the 0.0026 to 0.62 mg range.[72] Urinary *t,t*-MA is a more suitable biomarker of low-level benzene exposure than other, more abundant benzene metabolites because its background concentrations are low. However, specificity may be reduced because *t,t*-MA is also a metabolite of sorbic acid, a common food preservative.[198] Urinary *t,t*-MA is often quantified by HPLC. To confirm peak identities, collected HPLC fractions from standards and smokers' urines were derivatized with pentafluorobenzyl bromide and analyzed by GC/NCIMS with methane reagent gas, using the molecular ion and the base peak, which corresponded to the loss of $C_6F_5CH_2$ (Figure 4.15). Urinary *t,t*-MA concentrations were nearly identical among male and female smokers, but significant differences existed between smokers and nonsmokers, including pregnant nonsmokers.[199,200]

In a GC/EIMS technique, urine samples were cleaned up by anion-exchange chromatography followed by derivatization of the *t,t*-MA with methanolic BF_3 to produce the dimethyl ester (the i.s., 2-bromohexanoic acid, was analyzed as methyl ester). Quantification was accomplished by SIM of the molecular ions and structurally significant fragment ions. The LOD of *t,t*-MA was 0.01 mg/L, much better than the 0.05 to 0.1 mg/L values reported for HPLC. SIM also avoided interferences from unrelated constituents often encountered with HPLC. Urinary *t,t*-MA in smokers was significantly higher than that from nonsmokers, 0.09 ± 0.04 vs. 0.05 ± 0.02 mg/g creatinine.[201] In a subsequent study, nonsmokers living in the city tended to have higher background excretion rates of *t,t*-MA than those living in the suburbs. Dietary sorbic acid (6 to 30 mg/day) was a confounding factor, accounting for 10 to 50% of the *t,t*-MA levels in nonsmokers and 5 to 25% in smokers.[198]

Urinary *t,t*-MA and *S*-phenylmercapturic acid (S-PMA) may be quantified simultaneously without derivatization by using gradient LC/ESIMS/MS operated in the negative ion mode. For *t,t*-MA, the product ion resulting from the loss of CO_2 was utilized for SRM (i.s.: $^{13}C_6$-*t,t*-MA) while for S-PMA, the product ions resulting from transitions involving the loss of CO_2 and CH_2=CH-$NHCOCH_3$ were used (i.s.: $^{13}C_6$-S-PMA). Data were obtained for both analytes in urines from smokers and nonsmokers and significant correlations with smoking were obtained for both metabolites.[202]

Polycyclic Aromatic Hydrocarbons (PAH)

The mechanism of the genesis of H-*ras* mutations caused by PAH involves the rapid depurination of adducted bases after the PAH are metabolized to electrophiles that bind to DNA bases and destabilize the N-glycosyl bonds. The depurinated benzo[a]pyrene (BP) adducted bases 7-(benzo[a]pyren-6-yl)guanine and 7-(benzo[a]pyren-6-yl)adenine were quantified by MS/MS. The BP-adducted bases were detected in the urines of 3/7 cigarette smokers (0.1 to 0.6 fmol/mg creatinine equivalent) with BP intake of ~0.8 μg/day, and 3/7 women exposed to coal smoke (60 to 340 fmol/mg) with BP intake of ~23 μg/day, but not in 13 nonexposed controls. The high exposure of coal smoke-exposed women resulted in urinary BP-6-N7-Gua concentrations 20 to 300 times larger than those in cigarette smokers.[203]

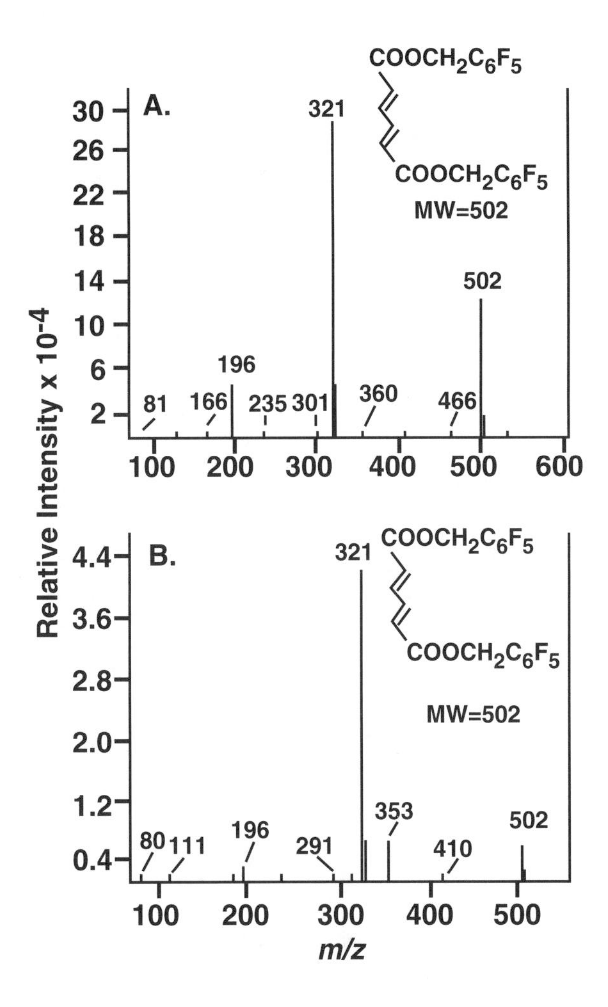

FIGURE 4.15 Negative chemical ionization mass spectra of the perfluorobenzyl derivatives of *trans, trans*-muconic acid. (A) standard; (B) sample from a smoker's urine. Reprinted from Melikian, A., Prahalad, A., and Secker-Walker, R., *Cancer Epidemiol. Biomark. Prev.*, 37, 239–244, 1994. With permission.

4.3.1.3 4-Aminobiphenyl

Hemoglobin Adducts as Biomarkers of Smoke Exposure

The aromatic amine 4-aminobiphenyl (4-ABP, biphenylamine) is a known human bladder carcinogen. There is ~4 ng 4-ABP/cigarette in mainstream smoke and ~145 ng/cigarette in sidestream smoke. Smoke from black (air-cured) tobacco contains more arylamines (including 4-ABP and 2-naphthylamine) than that from blond (flue-cured) tobacco. 4-ABP undergoes hepatic, cytochrome P-4501A2-mediated, N-oxidation to hydroxylamine. The latter reacts covalently with Hb in the circulation, and the complex is transported to the bladder where reaction with urothelial DNA initiates tumorigenesis.[45,204] A competing detoxification and elimination process occurs in the liver and involves N-acetylation, catalyzed by the noninducable N-acetyltransferase, that is under autosomal genetic control (see "acetylators" below). 4-ABP covalently binds to the β-Cys_{93} cysteine of Hb as a stable sulfinic acid amide. The complex is removed *in vivo* at a rate similar to that of Hb.

A technique for the quantification of the adduct was based on the generation of the parent amide by mild basic hydrolysis. Red blood cells were separated from whole blood and lysed with

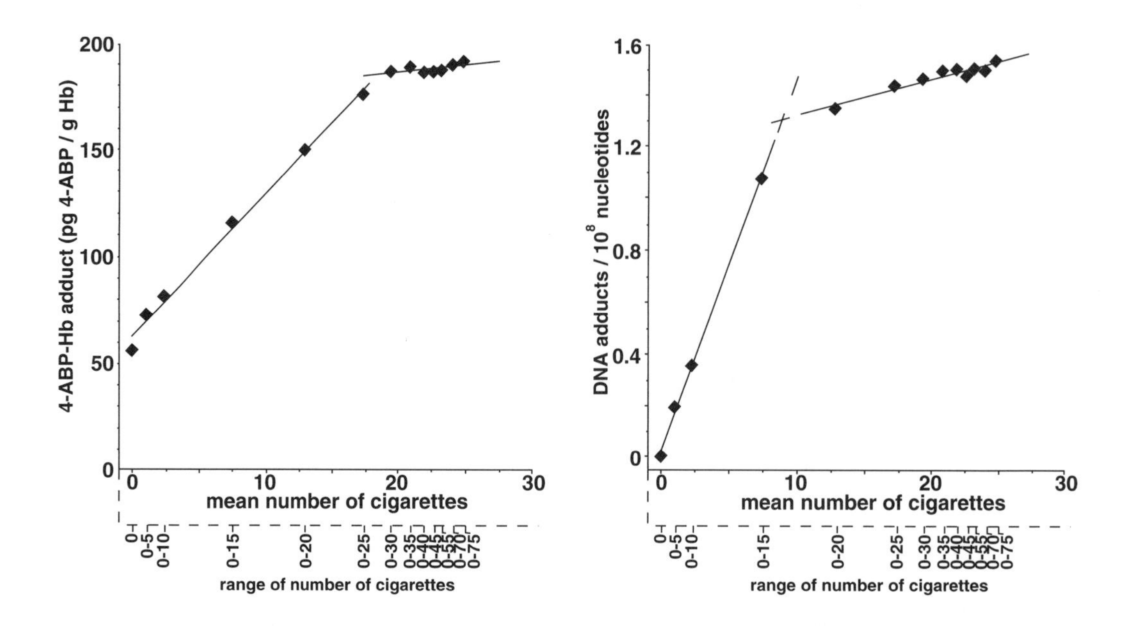

FIGURE 4.16 Relation of hemoglobin-aminobiphenyl adduct formation to mean cigarette consumption. Left: 4-aminobiphenyl-Hb adducts; right: lymphocytic DNA adducts. A saturation effect is seen in both cases. Reprinted from Dallinga, J. W. et al., *Cancer Epidemiol. Biomark. Prev.*, 517–577, 1998. With permission.

water. Additional preparation steps included further lysing, dialysis, and derivatization with pentafluoropropionic anhydride. Quantification was carried out using GC/NCIMS (i.s.: 4′F-4-ABP) using SIM of the $[M-HF]^-$ fragments. The method was linear over six orders of magnitude and the LOD was <10 pg/10 mL blood.[204] The mean 4-ABP was 154 pg/g Hb for smokers (20 to 50 cigarettes/day) and 28 pg/g Hb for nonsmokers, without any overlap of the ranges. Cessation of smoking led to a return to nonsmoker levels in 6 to 8 weeks. The background levels in nonsmokers, which were low, were attributed to the ubiquitous presence of 4-ABP in air, food, and water.[204] LC/ESIMS has been used for the characterization and quantification of a synthetic 4-ABP adduct standard.[34]

In another study, aimed at correlating Hb-4-ABP adduct concentrations with cigarette consumption, significant differences were found in the levels between smokers and nonsmokers, 202 ± 11 pg/g Hb vs. 57 ± 9 pg/g Hb, respectively. Adduct formation correlated linearly with cigarette consumption up to ~20 cigarettes/day at which point there was an apparent saturation (Figure 4.16, left). There were no correlations between adduct level and the age, sex, or duration of smoking. When DNA adducts were examined there was a significant correlation with cigarette consumption and a similar saturation in the DNA-adduct formation to that observed for Hb-adducts, in this case occurring at ~15 cigarettes/day (Figure 4.16, right). It was concluded, based on statistical studies, that the Hb-4-ABP adduct was a better biomarker than the aromatic DNA adduct for monitoring long-term smoking.[205]

After it was indicated that genetically determined "slow acetylators" (homozygous) are at increased risk for bladder cancer, the above GC/NCIMS technique was applied to investigate the role of *acetylator phenotype,* independent of smoking habits. The quantity of Hb-4-ABP adducts depended not only on the quantity and type of tobacco smoked (31 smokers of blonde tobacco and

16 smokers of black tobacco, 50 controls) but also on their classification as "slow" or "rapid" acetylators, as determined from urinary excretion of caffeine metabolites.[206] For example, among smokers of black tobacco, there was ~175 pg adduct/g Hb for the slow acetylators and ~117 pg/gHb for the rapid acetylators.[207] However, a subsequent study revealed that the speed of acetylation was irrelevant at high levels of smoking.[208] Other researchers, using similar methodology, failed to detect any effect in relation to the acetylator genotype.[205]

Hemoglobin Adducts: Fetuses

In an application of the GC/NCIMS technique, adduct levels in both fetal and maternal blood were found to be significantly higher in smokers than in nonsmokers during labor and delivery: 92 ± 54 vs. 17 ± 13 pg 4-ABP/g Hb in fetal blood and 183 ± 108 vs. 22 ± 8 pg/g in maternal blood. The degree of smoke exposure (number of cigarettes smoked, amount of cigarette smoked, depth of inhalation) correlated with 4-ABP concentration in the mothers but not in the fetal blood. Still, it was proven that 4-ABP could induce transplacental carcinogenesis in fetal tissues by damaging DNA.[209] Transplacental crossing of 4-ABP in smoking mothers was confirmed in paired maternal and fetal blood from umbilical veins, taken immediately after delivery. The adduct released from the Hb was quantified by GC/MS using pentafluoropropionic derivatives and SIM, and also separately by HPLC. The identities of the HPLC peaks were confirmed in isolated fractions of maternal and fetal blood by comparisons with authentic 4-ABP using GC/EIMS without analyte derivatization (Figure 4.17). Results of the background adduct levels in paired maternal-fetal samples were 29.6 ± 16.2 pg 4-ABP/g Hb in maternal blood and 14.0 ± 6.5 pg 4-ABP/g Hb in fetal blood. The corresponding concentrations in smokers were 488 ± 174 pg 4-ABP/g Hb in maternal blood and 244 ± 91 pg 4-ABP/g Hb in fetal blood. It was confirmed that the background levels did not originate from the analytical procedure. Taking the maternal 4-ABP adduct as the independent variable and the fetal adduct as the dependent variable, linear regression analyses gave a significant correlation.[210]

Hemoglobin Adducts: Bladder Carcinoma

A similar GC/NCIMS technique (i.s.: $^{2}H_{9}$-4-ABP) was used to compare patients with histologically confirmed transitional bladder carcinoma with smoking controls. The modest difference in Hb-4-ABP adduct content, 103 ± 47 vs. 65 ± 44 pg adduct/g Hb, was statistically significant for samples paired for urinary cotinine levels.[211] A GC/EIMS technique developed for the quantification of Hb-4-ABP adducts in smokers and nonsmokers[212] was applied to determine the adducts in erythrocytes to evaluate the role of 4-ABP in the development of bladder cancer in dogs. It was concluded that the Hb adduct formation was a direct consequence of the hepatic N-oxidation of 4-ABP to N-OH-ABP.[213]

DNA Adducts: Lung and Bladder

The predominant DNA adduct of 4-ABP is N-(deoxyguanosin-8-yl)-ABP. Because the parent molecule can be released by alkaline hydrolysis, free 4-ABP can be determined by GC/NCIMS. The technique, validated against a ^{32}P postlabeling method, was linear in the 1 adduct/10^8 to 1/10^4 nucleotide range. Adduct levels were in the <0.3 to 50 adducts/10^8 nucleotide range in 10/11 human lungs and in the <0.3 to 4 adducts/10^8 nucleotide range in 5/8 urinary bladders. However, there was no correlation with the number of cigarettes smoked.[214]

In another study, alkaline hydrolysis of DNA from biopsy samples, followed by quantification using GC/NCIMS, established that adduct concentrations were higher not only in current smokers compared to ex-smokers, but also in patients with tumors of advanced histological grades. However, there was no correlation with the number of cigarettes smoked. There was no evidence of modulation of the adduct formation by polymorphic N-acetyltransferase (speed of acetylation) or glutathione-S-transferase. Also, there was no pattern for *p*53 mutations, as determined by PCR amplification and sequencing.[215]

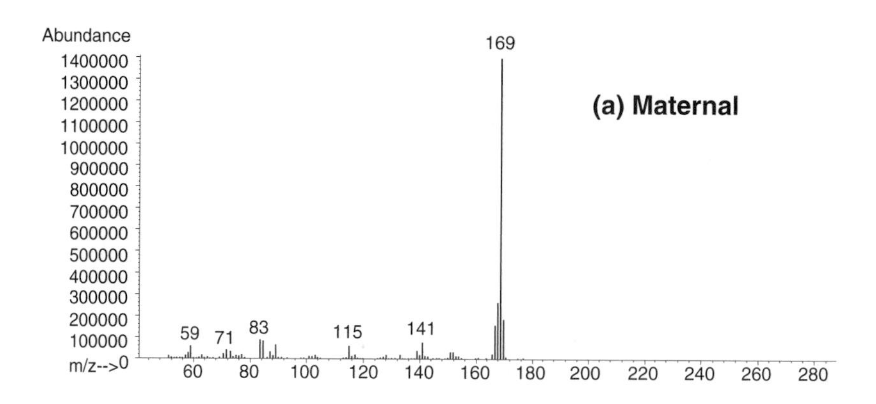

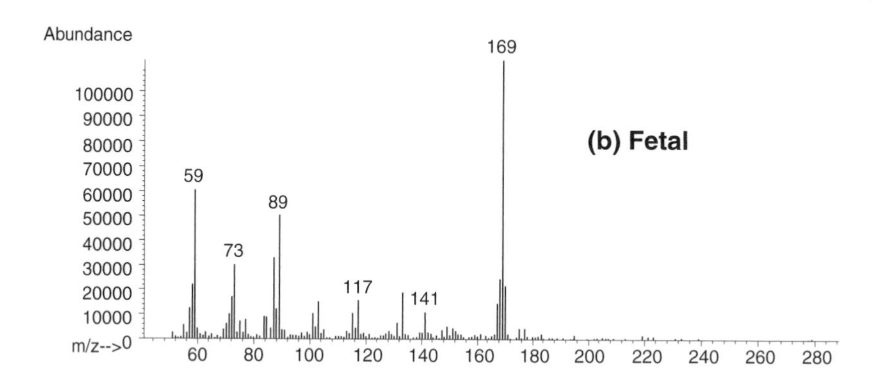

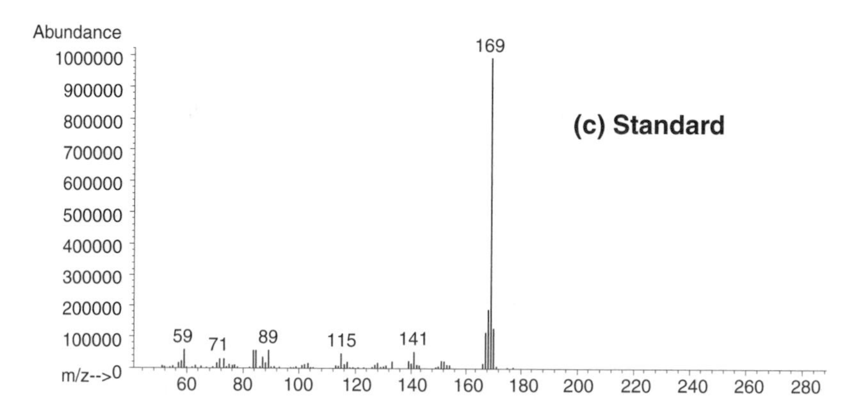

FIGURE 4.17 Full electron ionization mass spectra of underivatized 4-aminobiphenyl isolated from: (**a**) maternal and (**b**) fetal blood, compared to (**c**) authentic standard. Reprinted from Pinorini-Godly, M. and Myers, S. R., *Toxicology,* 107, 209–217, 1996. With permission.

4.3.1.4 Tobacco-Specific N-Nitrosamines

Metabolism and Carcinogenic Action

More than 100 papers have appeared in the past 10 years on the formation and metabolism of three important tobacco-specific N-nitrosamines (TSNA), 4-(methylnitrosoamino)-1-(3-pyridyl)-1-butanone (NNK), NNN, and NNA (see abbreviations in Figure 4.18). NNK is a strong carcinogen

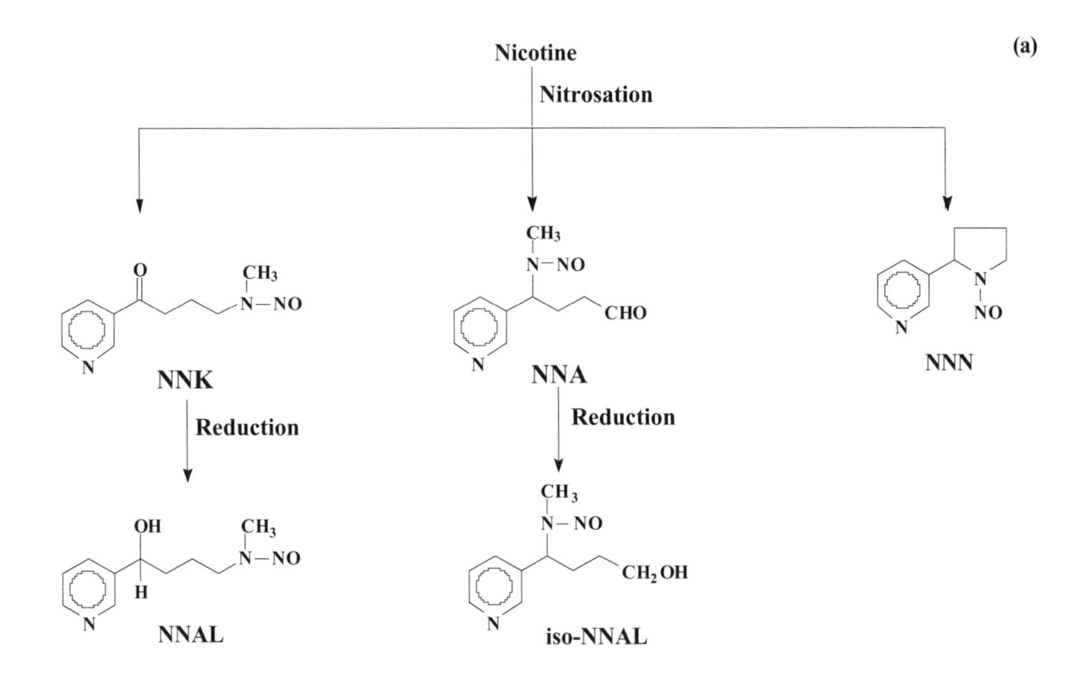

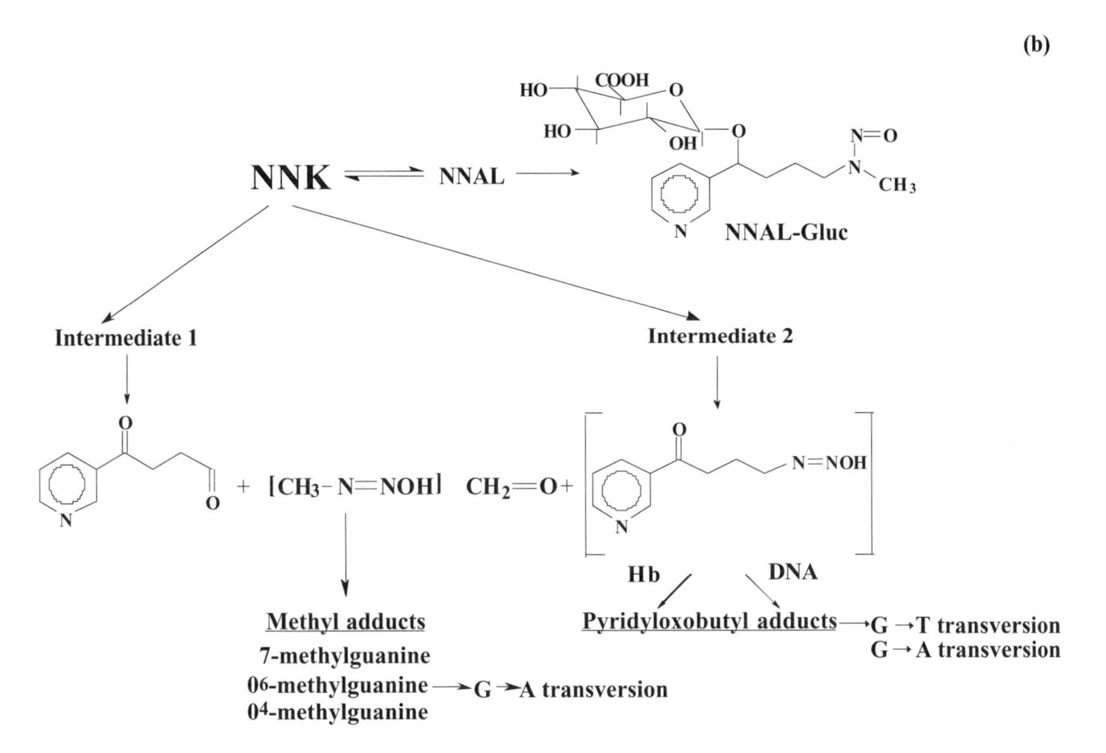

FIGURE 4.18 (a) Overview of the formation tobacco-specific nitrosamines and their metabolism. (b) Formation of the major urinary metabolite of NNK and possible pathways to form carcinogenic adducts. NNK, 4-(methylnitrosamino)-1-(3-pyridyl)-1-butanone; NNA, 4-(methylnitrosamino)-4-(3-pyridyl)butanal; NNN, N′-nitrosonornicotine; NNAL, 4-(methylnitrosamino)-1-(3-pyridyl)-1-butanol; NNAL-Gluc, [4-(methylnitrosamino)-1-(3-pyridyl)but-1-yl]-β-O-D glucosiduronic acid.

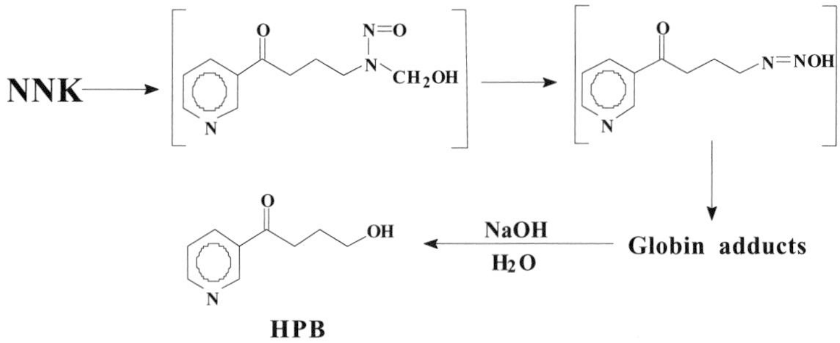

FIGURE 4.19 The formation of hemoglobin adducts of NNK and the hydrolysis into HPB that can be quantified. NNK, 4-(methylnitrosamino)-1-(3-pyridyl)-1-butanone. HPB, 4-hydroxy-1-(3-pyridyl)-1-butanone.

that causes tumors (independently of the route of administration) of the lung, pancreas, oral cavity, and liver in rats, mice, and hamsters. NNK is also involved in the initiation of these neoplasms in smokers and in snuff-dippers.[216,217] The main metabolic pathway of NNK (reviewed in References 218, 219) leads to the N-nitroso alcohol (NNAL) that is then conjugated to the glucuronide NNAL-Gluc (Figure 4.18b). Urinary NNAL and NNAL-Gluc have been detected in both smokers and nonsmokers exposed to ETS.

Possible metabolic activation pathways of NNK involve α-methylene hydroxylation, yielding methane diazohydroxide that methylates Hb and DNA, and a process via several intermediates leading to the pyridyloxibutylation of purine bases in DNA (reviewed in Reference 220), thus initiating carcinogenesis in extrahepatic tissues (Figure 4.18b). Possible approaches to reduce NNK-induced lung tumorigenesis include the reduction of the delivery of NNK by enhancing hepatic cytochrome P-450 enzyme activity by enzyme inducers or, perhaps better, by directly inhibiting the P-450 enzymes responsible for NNK activation in the lung, e.g., with isothiocyanates. Alternative mechanisms may operate for the chemopreventive actions of tea polyphenols and selenocyanates, e.g., inhibition of NNK-induced DNA methylation.

Hemoglobin Adducts of NNK as Dosimeters

Although there are several MS techniques for the determination of cotinine in plasma or urine (Section 4.3.1.1), cotinine concentrations provide no information on the metabolic activation of TSNA. An alternative approach is based on the observation that the α-methyl hydroxylation of NNK produces an intermediate which spontaneously decomposes into the electrophilic pyridyloxobutanediazohydroxide that, in turn, alkylates hemoglobin (reviewed in Reference 221). When globin, isolated from Hb, is treated with aqueous NaOH, 4-hydroxy-1-(3-pyridyl)-1-butanone (HPB) a keto alcohol is liberated (Figure 4.19). HPB can also be obtained from the globin adducts of NNN although these adducts are formed by a different mechanism. The process of HPB formation is linear with dose over four orders of magnitude, making it possible to use HPB as a dosimeter for TSNA. Quantification was accomplished by derivatizing HPB with pentafluorobenzoylchloride to form the pentafluorobenzoate derivative (i.s.: 4,4′-dideutero analyte) followed by GC/NCIMS and SIM of the molecular ions of the analyte (Figure 4.20a). The LOD was 1 fmol/injection. The concentrations of NNK-derived Hb adducts found in snuff-dippers were significantly higher than those in smokers that, in turn, were significantly greater than those in nonsmokers (Figure 4.20b). The snuff used by the participants contained 0.7 μg NNK/g snuff. Thus, using 1 g of snuff five times per day resulted in an exposure to 3.3 μg of the carcinogen NNK.[222a] Details of sample preparation and MS analysis using a validated method have been described.[222b]

NNK in Cervical Mucus

Cervical cancer is the most common cancer among women in developing countries, and the third and sixth most common among Hispanic and Caucasian women, respectively, in the U.S. It is believed that human papilloma virus is involved in >90% of cervical cancers. The risk of cervical

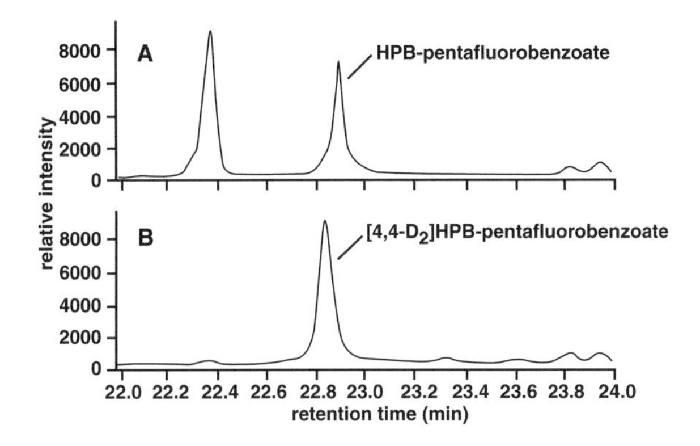

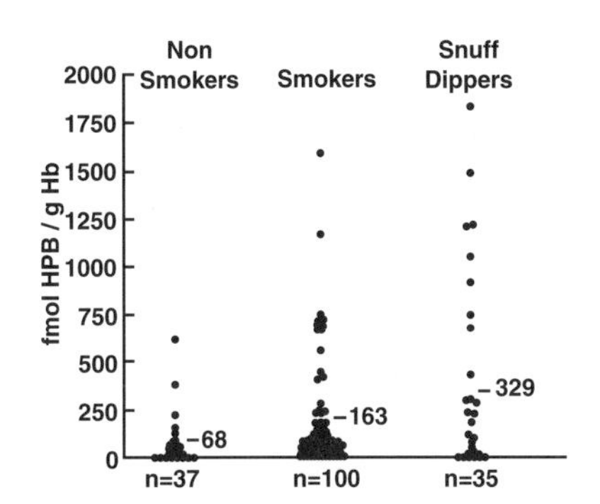

FIGURE 4.20 GC-NCIMS-SIM traces of top figure: (**A**) HPB-pentafluorobenzoate and (**B**) internal standard, [4,4-D_2]HPB-pentafluorobenzoate, obtained from a smoker's hemoglobin. HPB, 4-hydroxy-1-(3-pyridyl)-1-butanone. Bottom figure: comparison of nonsmokers, smokers, and snuff dippers. Reprinted from Hecht, S. S. et al., *Cancer Res.*, 54, 1912S–1917S, 1994. With permission.

cancer is some fourfold larger among smokers than nonsmokers. To investigate the presence of NNK in the cervical mucus of smokers, specimens were collected from the cervical canals of women during the preovulatory phase of their menstrual period. Sample preparation included addition of the i.s. (CD_3-NNK), supercritical fluid extraction with methanolic CO_2, and collection in *n*-hexane. Quantification was accomplished by GC/ESI-MS/MS in the SRM mode, monitoring the 177 $\rightarrow$ 146 transition (Figure 4.21). The base peak was the ion at *m/z* 177, corresponding to the loss of a nitroso group from the molecular ion. The mean concentration of NNK among 15 smokers was 46.9 ± 32.5 ng/g mucus (range: 11 to 115 ng/g) which was significantly higher than the mean of 13.0 ± 9.3 ng/g mucus (range: 4 to 31 ng/g) found in nonsmokers ($n = 9$). This was the first time a TSNA was implicated in the development of cancer of the uterine cervix among smokers. The NNK in the nonsmoking subjects is likely to originate from ETS.[223]

Urinary NNK Metabolites

The main metabolic pathway of NNK (Figure 4.18a) includes its carbonyl reduction product, 4-(methylnitrosamino)-1-(3-pyridyl)-1-butanol (NNAL), that is glucuronated to [4-methyl-nitrosamino)-1-(3-pyridyl)but-1-yl]-β-O-D-glucosiduronic acid (NNAL-Gluc). The (S)-NNAL enantiomer is more tumorigenic (in mice lungs) than the (R) enantiomer, and the same is true for the corresponding diastereomeric glucuronides. The critical step in cancer induction is the activation of both NNK and the NNAL enantiomers by α-hydroxylation. Quantification of these metabolites in smokers' urine has been carried out using chiral stationary phase-GC with nitrosamine-selective

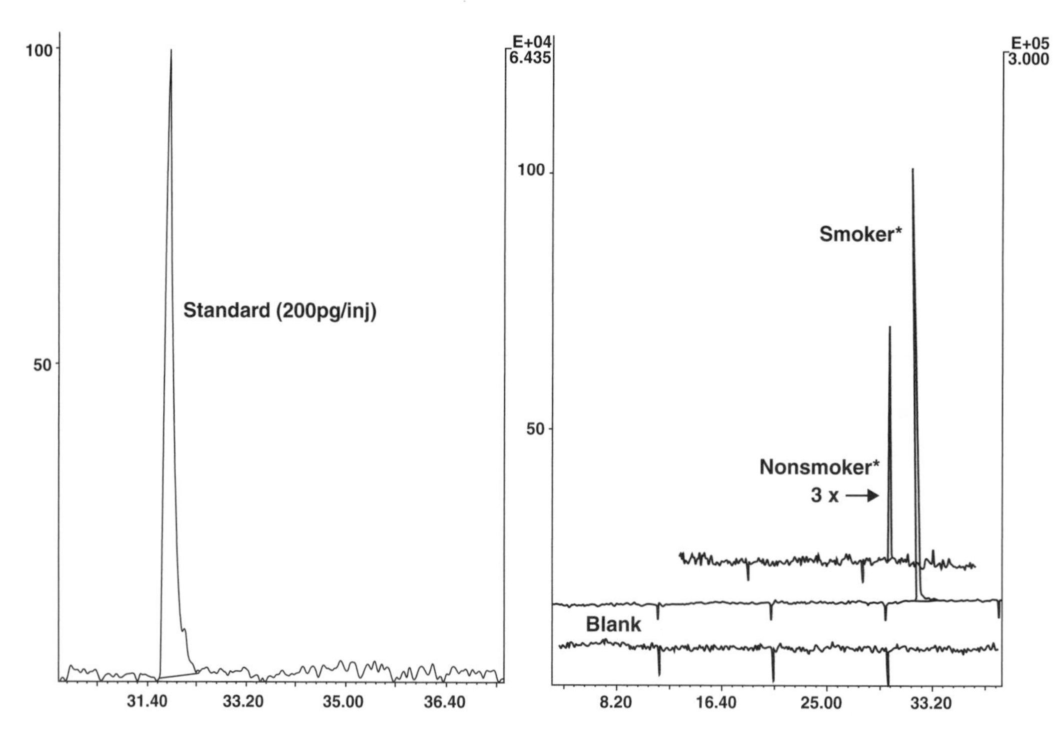

FIGURE 4.21 Capillary gas chromatography-selected reaction monitoring-tandem mass spectrometry of human cervical mucus for 4-(methylnitrosamino)-1-(3-pyridyl)-1-butanone (NNK). The peaks observed result from monitoring the major product ion of *m/z* 146. * Self-reported smoking status. NNK peak in nonsmoker mucus has an identical retention time with that in smoker mucus but has been shifted for illustration clarity; its intensity has been increased three times (3×). X axis, time, min; Y axis, relative ion intensity. Reprinted from Prokopczyk, B. et al., *J. Natl. Cancer Inst.,* 89, 868–873, 1997. With permission.

detection and confirmation by LC-APCI-MS/MS. Urines were separated into fractions containing free NNAL and NNAL-Gluc, followed by hydrolyzing with β-glucuronidase.[224] For the LC/MS analysis, the silylation step used for the GC analysis was replaced by treatment with (R)-(+)-α-methylbenzyl isocyanate (MBIC) to produce the diastereomeric carbamates. The transition of the *m/z* 357 ion ($[M+H]^+$) to a product ion (*m/z* 162) was monitored by SRM (i.s.: CD_3-NNAL). Because of the restricted rotation around the N–N=O bond, each diastereomer produces both the (E)- and (Z)- rotamers, however, two of these coeluted and this had to be corrected by calculation. The LC-MS/MS chromatograms of the NNAL fraction of smokers' urine exhibited the [R-(E)]-NNAL-(R)-MBIC and corresponding (R) peaks at approximately the same intensity (Figure 4.22a), while the (S) enantiomer was about twice as intense as the R enantiomer in the NNAL-Gluc fraction (Figure 4.22b). Quantitative results of the GC and LC methods agreed well. In smokers' urine the enantiomeric distribution of NNAL and NNAL-Gluc was 54% and 68% (S) and 46% and 32% (R), respectively.[225]

Nonsmokers

It has been suggested that NNK and polycyclic aromatic hydrocarbons are not only the most likely cause of lung cancer in smokers, but these carcinogens may also be implicated in the development of lung cancer in nonsmokers.[226] In fact, the two major metabolites of NNK, NNAL, and NNAL-gluc were found in the urine of nonsmokers exposed to relatively high levels of ETS in a room with limited ventilation.[227] In a separate study, NNAL-gluc was analyzed in the urine of nonsmoking

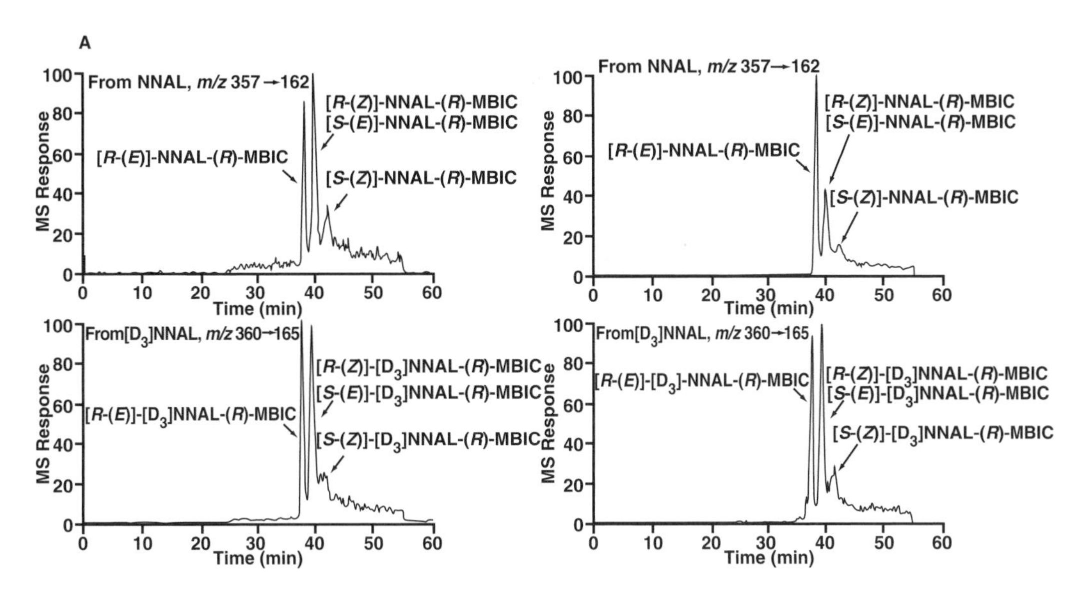

FIGURE 4.22 LC/MSMS chromatograms of NNAL fraction (**A**) and NNAL-Glu fraction (**B**) in a smokers' urine. Top: Selected reaction monitoring for the analyte at transition *m/z* 37 → 162. Bottom: Corresponding analysis of the deuterated internal standard. Reprinted from Carmella, S. G. et al., *Cancer Res.*, 59, 3602–3605, 1999. With permission. (Note: All R and S values were switched inadvertently in the original publication).

hospital workers exposed to ETS under field conditions, e.g., duties in inadequately ventilated smoking lounges. The identity of the analyte, released from NNAL-gluc with β-glucuronidase and trimethylsilylated, was confirmed by GC/CIMS in the MS/MS mode by monitoring the product ions of the $[M+H]^+$ ions by transitions of *m/z* 282 → 162 ($[MH\text{-}HOTMS\text{-}NO]^+$) for NNAL-TMS and *m/z* 282 → 132 ($[MH\text{-}HOTMS\text{-}NO\text{-}CH_3NH]^+$) for NNAL-TMS, used as i.s. (Figure 4.23). The mean urinary concentration of NNAL-Gluc was ~0.06 pmol/mL urine (range 0.005 to 0.11 pmol/mL), significantly more than that found in controls, 0.012 pmol/mL urine. The levels found in the exposed subjects correlated well with urinary cotinine concentrations.[228]

Smoking Cessation

Aiming to investigate the decrease of concentrations in urinary NNAL and NNAL-Gluc after smoking cessation, the analytes were quantified by GC/thermal energy analyzer[224] and their identities confirmed by GC/CIMS/MS.[228] Nicotine and cotinine were quantified by GC/EIMS using SIM to monitor molecular ions and structurally significant fragment ions. Unexpectedly, ~35% of the combined NNK metabolites were still present one week after smoking cessation, whereas the corresponding values for nicotine and cotinine were ~1% and ~0.5%, respectively. About 8% of the NNK metabolites were detectable even 6 weeks after cessation, lasting up to 281 d in some subjects. Parallel treatment of rats, acutely or chronically with NNK, in drinking water, revealed that NNAL has a large volume of distribution. Studies with nicotine patches confirmed that NNK is not formed endogenously from nicotine. It was concluded that NNK and its metabolites are retained (or sequestered) in a high-affinity compartment from which their release is slow.[229]

Newborns

NNAL-Gluc and NNAL were both detected by GC/EIMS/MS[228] in 22/31 and 4/31 urines from newborns and mothers who smoked during pregnancy, respectively. Neither compound was detected in newborns of nonsmoking mothers.[230]

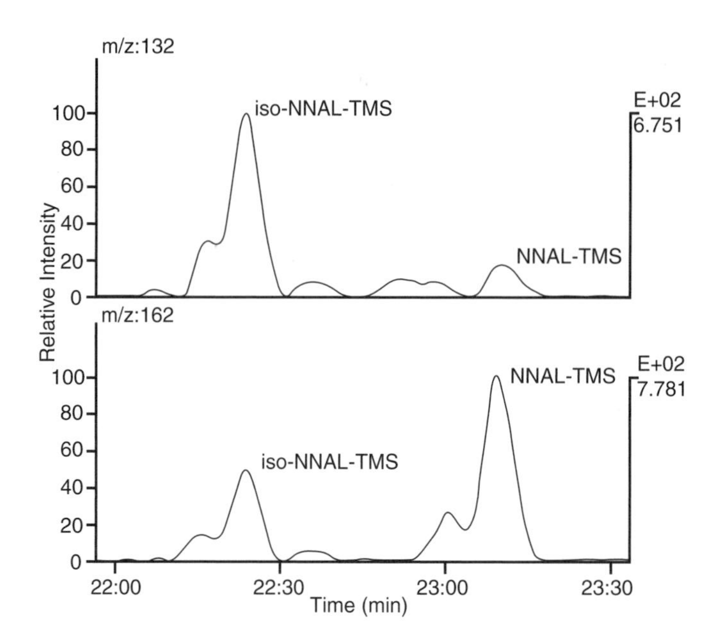

FIGURE 4.23 GC-MS/MS analysis of NNAL-Gluc in the urine of a hospital worker (cleaner) who has stopped smoking 10 yr before. Iso-NNAL-TMS is the derivatized internal standard and NNAL-TMS is the derivatized aglycone of NNAL-Gluc. Top: selected reaction monitoring of *m/z* 282 → 132 transition. Bottom: The *m/z* 282 → 162 transition. NNAL-Gluc concentration: 0.053 pmol/mL. Reprinted from Parsons, W.D. et al., *Cancer Epidemiol. Biomark. Prev.*, 7, 257–260, 1998. With permission.

Other Methodologies

Methodologies based on LC/ESIMS/MS have been developed for the confirmation of the presence of NNK-N-oxide and NNAL-N-oxide in urine. It was concluded that pyridine-N-oxidation is a relatively minor detoxification pathway of NNK and NNAL in humans.[231] The methodological aspects of supercritical fluid extraction as a sample preparation tool for subsequent quantification and analyte confirmation by GC/MS have been described.[232]

4.3.2 Air

4.3.2.1 Gasoline and Diesel Fuels

Gasoline Composition and Vehicle Emissions

Despite the reduction of emissions from gasoline-powered cars since the 1960s, release of hydrocarbons, carbon monoxide, and nitrogen oxides still contribute significantly to air pollution in both cities and countryside. An APCI-MS/MS technique was developed for the study of the transient behavior of regulated and nonregulated exhaust emissions under various engine operating conditions. The method has a real-time response time of 20 ms and LOD were ~1 ppb.[233] Variable operation parameters for the engines included addition of unburned fuels, adjustment of the fuel/air ratio, as well as alteration of the composition and volatility of the fuel. In one study, the patterns of emission of benzene, toluene, and xylene were followed under various experimental conditions, with emphasis on the level of benzene emissions as a function of aromatic content of the fuel. Generally, highest emission concentrations occurred during the first 80 s of operation, i.e., before the catalyst in the emission control system reached operating temperature. Toluene was detected in emissions even when an isooctane/n-heptane fuel (9:1) contained no added aromatic compounds

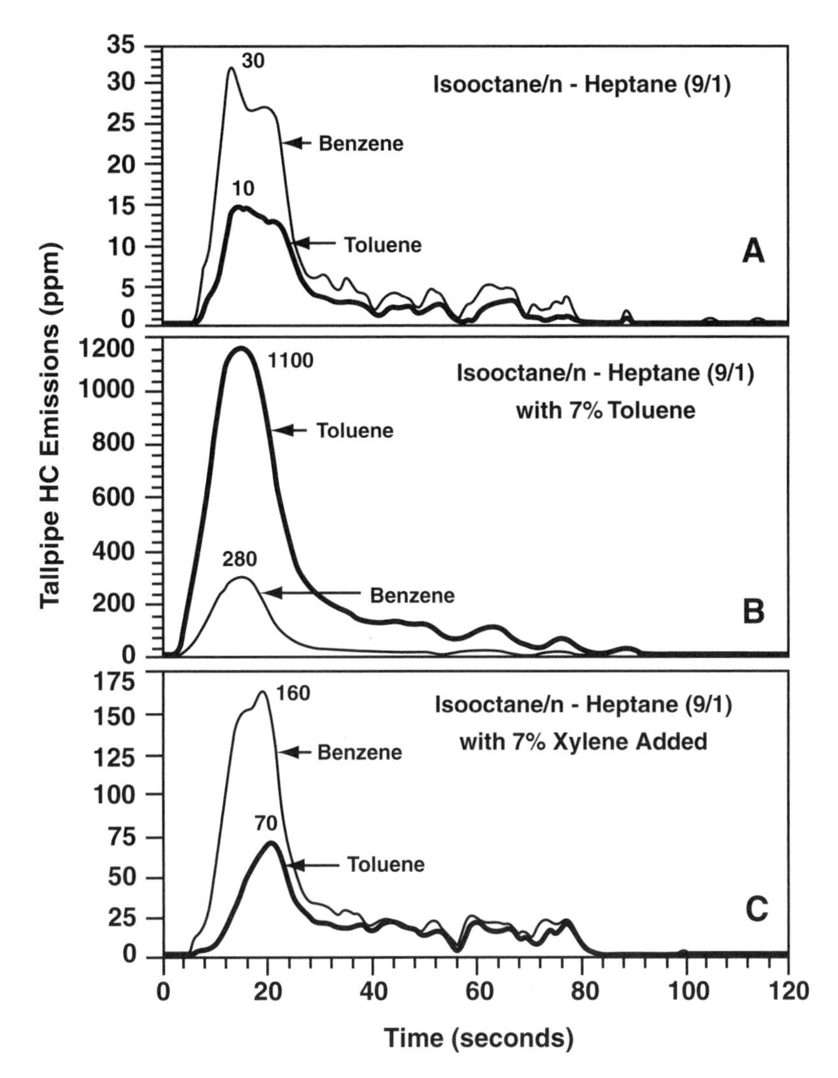

FIGURE 4.24 Real-time atmospheric pressure ionization/mass spectrometric (API/MS) analysis of tailpipe benzene emission as a function of added toluene and xylene during cold-start operation of a 1990 3.0 L Ford Taurus. Reprinted from Schuetzle, D. et al., *Environ. Health Perspect.,* 102 (Suppl. 4), 4–12, 1994. With permission.

(Figure 4.24A) indicating that *de novo* synthesis of aromatic hydrocarbons can occur. When the composition of the fuel was modified by the addition of toluene and xylene, the concentrations of benzene and toluene in tailpipe emissions changed significantly as a function of time during the vehicle's cold-start operation. The fact that the toluene concentration increased significantly when toluene was added to the fuel indicated that unburned fuel escaped the emission control system during cold-start operation (Figure 4.24B). The total quantity of benzene emitted increased significantly when either toluene or xylene was added to the fuel (compare the ordinates in Figure 4.24A, B, and C), probably due to dealkylation of toluene or xylene either during combustion or when it crossed the emission control catalyst. Additional real-time analyses have contributed to the understanding of the complex relationships between the fuel, engine, and emission control system.[234]

Diesel Fuel

To investigate the composition of an A2 grade diesel sample, some 85 constituents were separated on a capillary GC column. Individual mass spectra, obtained by ITMS, were compared to those in NBS/EPA data libraries. The main constituents (~30%) were alkylnaphthalenes (C_1-C_5); the rest were distributed among related compound classes.[235]

Urinary Muconic Acid (t,t-MA) in Children Exposed to Benzene in Air

Urinary *t,t*-MA concentrations (Section 4.2.1.1) were studied by GC/MS[71] in 79 African-American urban children exposed to a variety of benzene sources, including garages, time spent on buses, living near gas stations, and time spent playing on the street. Elevated *t,t*-MA concentrations were confirmed by GC/MS (Figure 4.25). Mean *t,t*-MA was 176 ± 341 ng/mg creatinine. Correlations were found with respect to the time of day when the samples were taken (highest concentrations were in the afternoon) and the time spent playing on the streets. There was no association with the known increased lead concentrations (~25 µg/dL) of the children.[236]

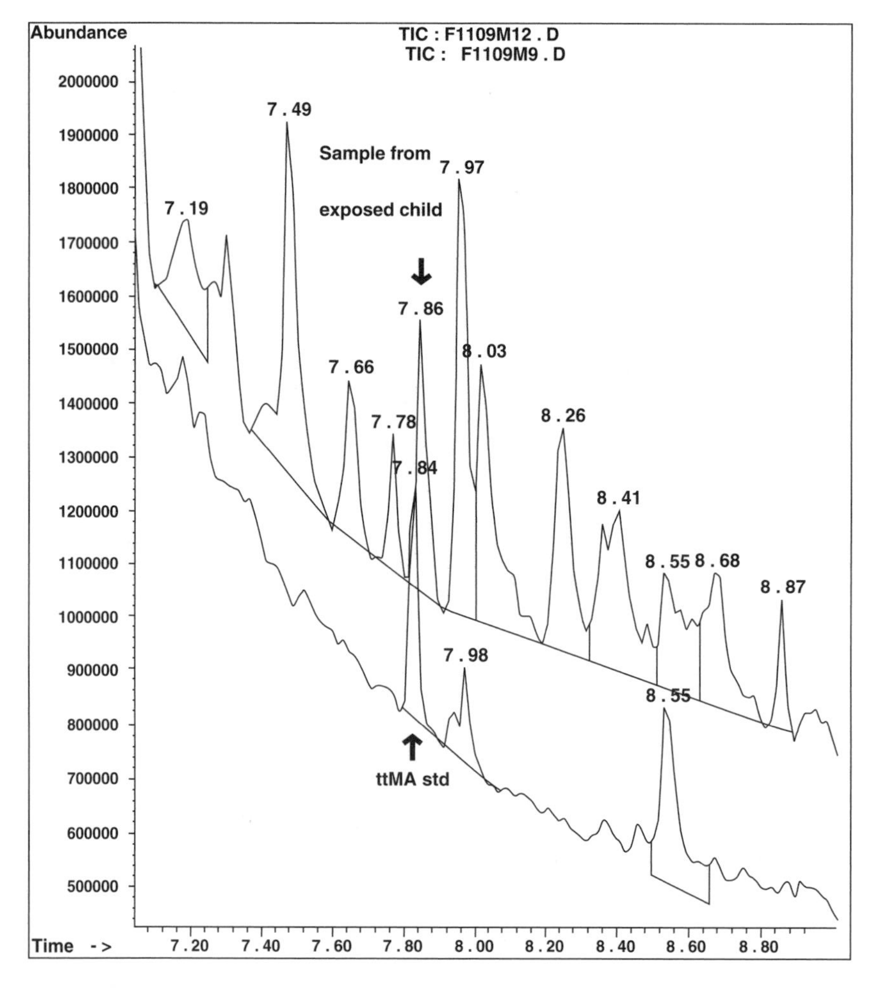

FIGURE 4.25 GC/MS chromatogram of a urine sample from a child with elevated *trans, trans*-muconic acid (MA) level. The lower trace shows the injection of an MA standard. Reprinted from Weaver, V. et al., *Environ. Health Perspect.*, 104, 318–323, 1996. With permission.

4.3.2.2 Polycyclic Aromatic Hydrocarbons

Formation of Atmospheric Mutagens from PAH

The lifetimes of 2- to 4-ring PAH in the atmosphere are less than one day due to reactions initiated by hydroxyl and NO_x radicals that lead to the formation of mutagenic nitro-PAH isomers and other nitropolycyclic aromatic compounds. These atmospherically derived compounds are distinct from similar compounds originating as direct emissions from combustion sources. For example, GC/MS with SIM monitoring of the molecular ions (all at *m/z* 247) of the nitrofluoranthenes (NF) and nitropyrenes (NP) in extracts of diesel exhaust particles and in an extract of ambient air samples collected on filters revealed significant differences (Figure 4.26). The diesel sample contained only one component, 1-nitropyrene, while the additional 2-nitrofluoranthene and nitropyrenes found in the ambient sample were the type expected from gas-phase OH-radical-initiated reactions. Based on additional studies at other locations, where atmospheric nitro-PAH lactones and methylnitrobenzopyranones were also identified by GC/MS, it was concluded that the atmospheric transformation of combustion emissions into potentially mutagenic products should be considered as a component in health risk assessments.[237]

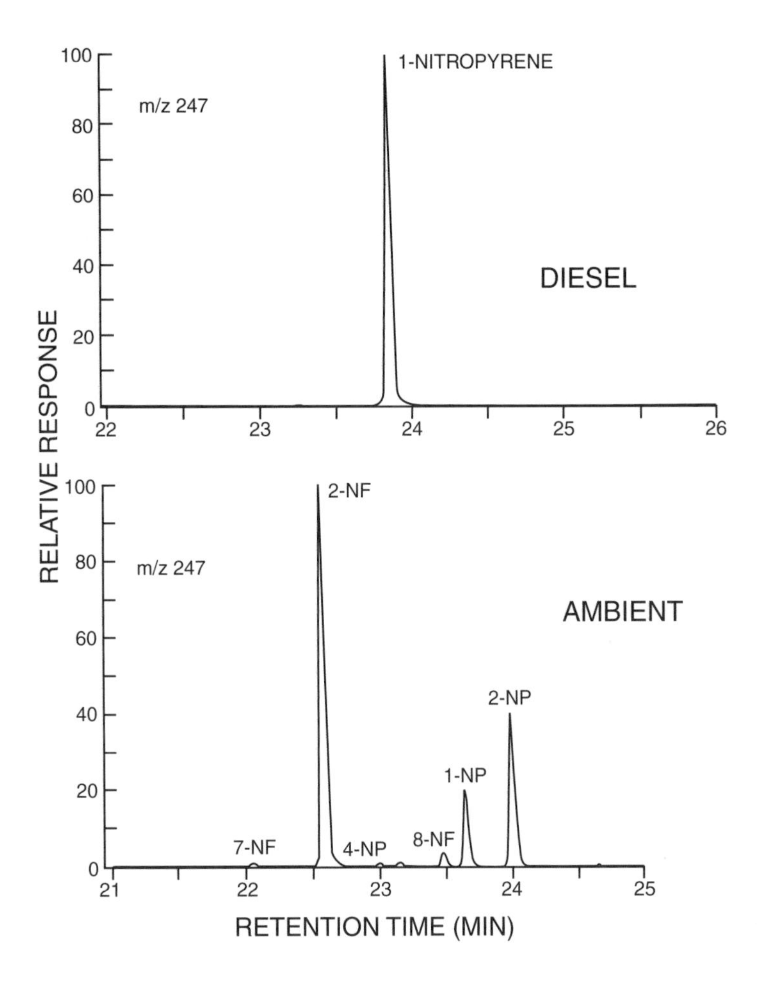

FIGURE 4.26 GC/MS SIM traces for the molecular ions of the nitrofluoranthrenes (NF) and nitropyrenes in extracts from diesel particulates and ambient particles collected on filters in Torrance, California. Reprinted from Atkinson, R. and Arey, J., *Environ. Health Perspect.*, 102 (Suppl. 4), 117–126, 1994. With permission.

Coal Fly Ash

The PAH vapors that form during the combustion of coal rapidly adsorb onto fly ash particles and are transported this way around the plants where they are generated as well as into the environment. Some 21 PAH, including most known carcinogenic PAH, were determined in a methylene chloride extract of coal fly ash using a GC/NCIMS technique. The masses of the base peaks corresponded to either the $[M]^-$ ions for PAH with high electron affinity, or $[M-H]^-$ ions for PAH with electron affinities <0.5 eV. Anthracene-related compounds yielded both types of ions in high abundance. Also observed were $[M+15]^-$ ions, the identity of which was shown (using perdeuterated PAH) to be $[M+O-H]^-$ rather than $[M+CH_3]^-$ as previously assumed. There was a strong correlation between the NCI response factor and the carcinogenicity of several of the identified PAH. For example, the highest total ion current ratios (negative/positive ion intensities) were observed for the highly carcinogenic benzo[b]- and benzo[k]fluoranthene, benzo[a]pyrene and indeno[1,2,3-cd]pyrene, each composed of a five-membered ring and exhibiting a high degree of carcinogenicity. This approach may be used to pinpoint compounds with high electron capturing ability having concurrent carcinogenic activity in the complex NCI profiles of coal fly ash with a plethora of unidentified compounds.[238]

4.3.2.3 Monitoring Miscellaneous Complex Mixtures

Urban Air

Methods for monitoring the concentrations of pollutants include the collection of particulates from the air followed by extraction and analysis of the mixtures obtained. Highly volatile contaminants are cryotrapped from the air followed by GC/MS analysis. An example involved on-line cryofocusing GC/MS for the monitoring of benzene and toluene that are considered to be representative of the volatile aromatic organic compounds of urban pollutants. Air samples were obtained with a membrane pump and a known quantity of the i.s., benzene-d_6, was continuously introduced through a permeable device. The spiked air was then cryofocused in a thermal desorption cold trap and the samples introduced into a GC/EIMS in a 5 to 14 min cycle. Analyses were carried out by SIM of the *m/z* 78, 84, and 91 ions of benzene, i.s., and toluene, respectively. A five-day monitoring of air in front of the research institute conducting the study revealed that the highest concentrations of both analytes occurred during rush hours (Figure 4.27). The high concentrations observed on the second day were due to a strong thermal inversion that increased air pollution in the city (Milan) in January.[239]

Aromatic Pollutants in Waste Incinerator Flue Gases

The analytical potential of REMPI-TOF (Section 2.1.7.2) for aromatic molecules, including a number of carcinogens, has been demonstrated under various experimental conditions, e.g., specific ionization of a class of molecules introduced into a cooled supersonic beam.[240] To explore the advantages of REMPI-TOF for time-resolved on-line analysis of target compounds or compound classes in complex samples, a new, mobile instrument was used for the real-time analysis of flue gases from a pilot plant.[241] An example of the application of this setup used contaminated wood as a feedstock. Gases from the end of the post-combustion chamber were sampled continuously while the gate was operated under gasificiation conditions (i.e., understochiometric O_2 supply). Using one-color REMPI (laser wavelength: 248 nm), a time resolution of 0.5 s (derived from laser repetition rate of 50 Hz and the averaging of 25 primary spectra) a number of monocyclic and polycyclic aromatic compounds and their alkylated derivatives were identified in the 10 to 100 ppb range (Figure 4.28). The benzene, naphtalene, methylated derivatives, anthacene, and pyrene observed were typical products of incomplete combustion while the phenol was produced during wood gasification. Because the REMPI was conducted at a fixed wavelength, isomers of larger PAH could not be resolved. Changing the conditions of the combustion process resulted in significant changes in the time-intensity profiles of the products.[242] Subsequent work has extended the areas of application, e.g., to the measurement of chlorobenzene as a surrogate for halogenated

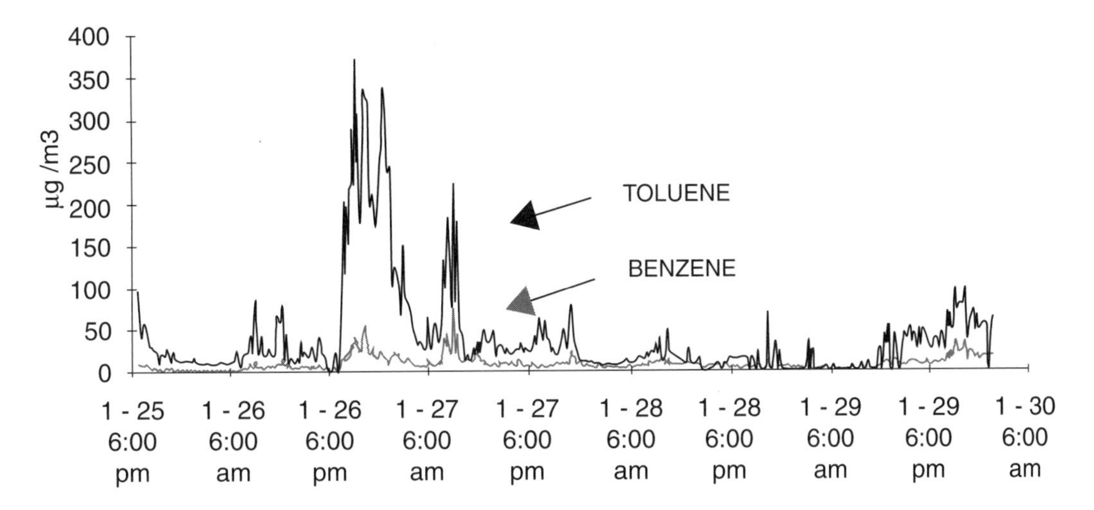

FIGURE 4.27 Results of a 5-day monitoring of benzene and toluene in urban air. Full analytical cycle was 14 min. The highest concentrations each day were observed during the rush hour. Differences on different days are probably due to weather conditions. Reprinted from Davoli, E. et al., *J. Am. Soc. Mass Spectrom.*, 5, 1001–1007, 1994. With permission.

dibenzo-*p*-dioxins/furans.[243] There have also been improvements in sensitivity to the ppt (by volume) level, e.g., for naphthalene.[244]

Sugar Cane Soot

The introduction of ethanol as an automotive fuel has necessitated an increase in the production of sugar cane. To make harvesting easier, and also to increase the sugar content by weight of shipped material, the leaves of the sugar cane are removed by burning before the cane is harvested. This leaf burning is an incomplete process in which numerous carcinogenic/mutagenic compounds are released into the atmosphere. Field workers and the local population (primarily in Brazil) are exposed to PAH as a result of inhaling these combustion-generated aerosols that include submicron-size particles onto which the PAH have been adsorbed. The presence of these compounds was determined in collected fly soot samples that were sieved, extracted with solvents, and separated into three factions with Sep-Pak cartridges. A GC/EIMS technique was used to identify (full scanning) and quantify (SIM) 31 PAH and 7 thiophenes, many of them known carcinogens.[245]

Tricholoethylene Combustion-Generated Aerosols

In the course of the incomplete combustion of trichloroethylene and the pyrolysis of plastics, there are transient discharges (e.g., 20 s long) called *puffs,* which include large quantities of toxic and potentially carcinogenic compounds. A variety of biological and biochemical assays have been carried out that have detected and confirmed chloro-dioxin-like toxic effects. Some 250 compounds were detected in the total ion chromatograms (GC/EIMS) of soot extracts derived from the combustion of these materials (Figure 4.29). A diverse set of stuctures was encountered, most of which were heavily chlorinated (four to six chlorine atoms) or even perchlorinated. Compounds included chlorinated derivatives of mono- and polyunsaturated aliphatics, cyclic polyenes, 1-, 2-, and 3-ring aromatics, phenols, and fulvanes (structural isomers of benzene). Although the LOD for both 2,3,7,8-tetrachlorodibenzo-p-dioxin and 2,3,7,8-tetrachlorodibenzofuran was 1 pmol, neither of the target compounds were detected. However, it was still concluded that the large number of heavily chlorinated compounds identified assisted in the explanation of the wide array of toxic effects associated with exposure to these aerosols.[246]

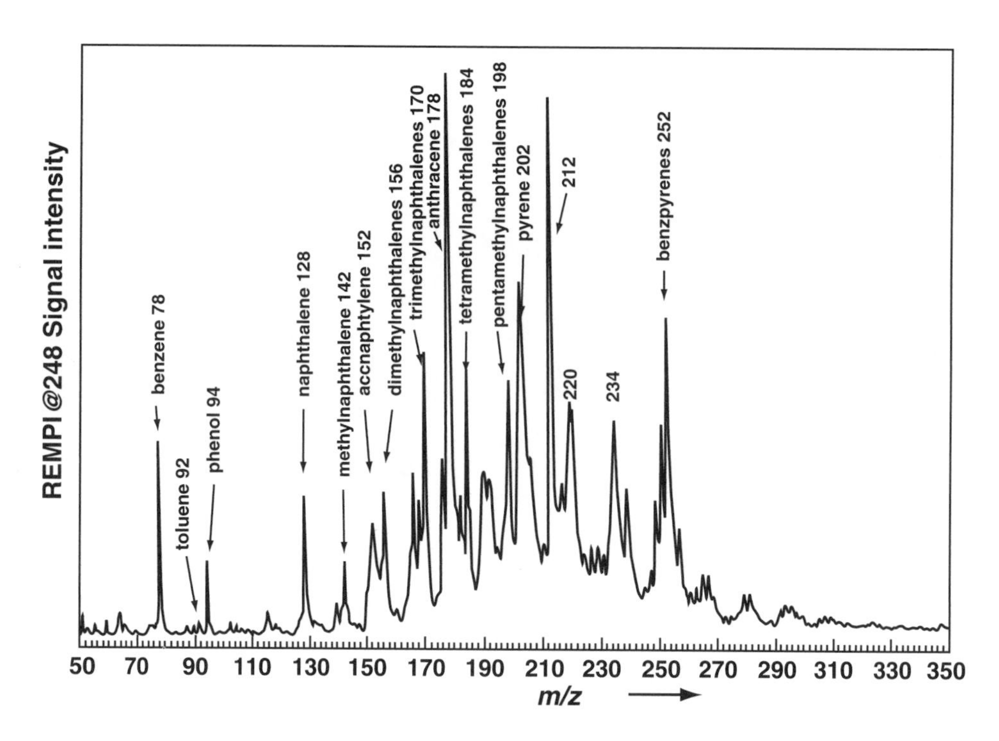

FIGURE 4.28 Resonance-enhanced multiphoton ionization TOF mass spectrum obtained at the second sampling point, at the end of post-combustion chamber, during disturbed postcombustion conditions. The grate was operated under gasification conditions (understochiometric oxygen supply), with contaminated wood as feedstock. The selectivity of REMPI (detection of PAH and methylated PAH) allows the assignment of several peaks. Benzene, naphthalene, its methylated derivatives, anthracene, and pyrene are typical products of incomplete combustion. The spectrum also shows phenol, a typical product of wood gasification. Isomeric ensembles could not be resolved, since REMPI with a fixed laser wavelength has been used. Therefore, in the assignment, especially of the larger PAH, only one of the possible isomers is mentioned. Reprinted from Zimmermann, R. et al., *Rapid Commun. Mass Spectrom.,* 11, 1095–1102, 1997. With permission.

4.3.3 Food

4.3.3.1 Aflatoxins

The aflatoxins, which belong to the mycotoxin group, are highly substituted coumarins containing a fused dihydrofurofuran moiety. There are two major groups of aflatoxins, B (blue fluorescence) and G (green fluorescence), of which aflatoxin B_1 (AFB_1, Figure 4.30) is both the most abundant and carcinogenic. Aflatoxin-induced hepatic toxicity and liver cancer occur frequently in certain regions of Africa and Asia where grains and other foodstuffs, e.g., corn, peanuts, copra, and rice, are exposed to high heat and humidity during growth, harvest, or storage and become contaminated with the fungal mold strains *Aspergillus flavus* and *A. parasiticus.* The aflatoxins are metabolites produced by these fungi. The concentrations of these contaminants in foodstuff range from 1 ppb to 1 ppm. It is interesting that the initial distribution of these toxins is uneven, e.g., while only 1 peanut in 10,000 may contain the toxin, that peanut may contain several hundred μg aflatoxin, leading to the contamination of an entire batch during processing. Permissible total aflatoxin levels in the U.S. are 20 ppb in agricultural commodities and 0.5 ppb in milk.

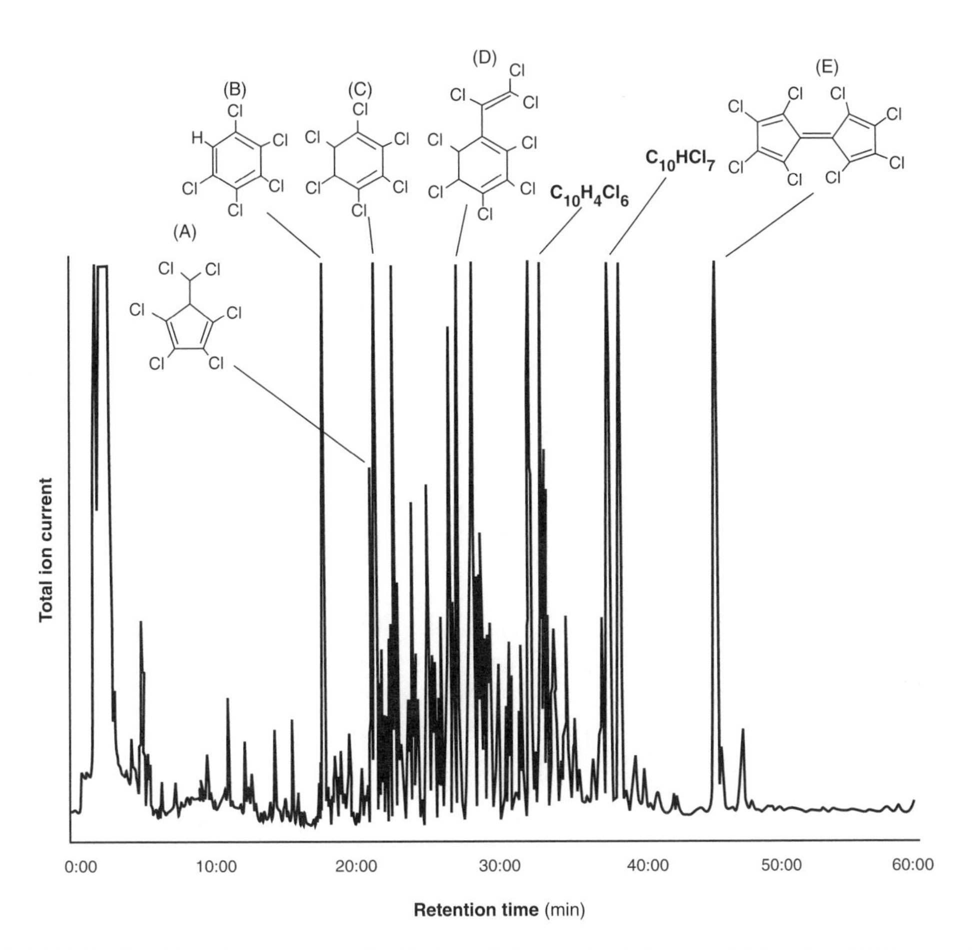

FIGURE 4.29 Total ion chromatogram of tricholoroethylene soot whole extract. More than 250 incomplete combustion byproducts were formed during pyrolysis. **A**: hexachlorofulvene; **B**: pentachlorobenzene; **C**: hexachlorobenzene; **D**: octachlorostyrene; **E**: octachlorofulvalene. Reprinted from Villalobos, S. A. et al., *Environ. Health Perspect.*, 104, 734–743, 1996. With permission.

CYP450
DNA

Aflatoxin B_1
AFB_1

AFB_1-8,9-exo epoxide

AFB_1-N7-guanine DNA adduct

FIGURE 4.30 The 8,9-exo epoxide metabolite of aflatoxin B_1 forms DNA adducts predominantly at the N7 position of guanine.

The high degree of carcinogenicity of the aflatoxins has led to extensive investigation of the metabolism of these compounds. The mutagenic DNA adducts are produced from the exo-8,9-AFB_1 epoxide metabolite which binds predominantly at the N7 position of guanine (Figure 4.30). The competing process that detoxifies the epoxide is mediated by members of the GST enzyme family. The glutathione conjugates formed undergo further metabolic processing to yield a mercapturic acid that appears in urine and may serve as an intermediate end point in chemoprotection trials. MS has been used to confirm the identity and structure of both synthetically- and *in vivo*-produced exo-AFB_1 mercapturate as well as the conjugate diastereoisomers.[247] A second group of enzymes that metabolize aflatoxins include members of the aldo-keto reductase superfamily (AKR7A3). The metabolic route used by these enzymes generates AFB_1 dihydrodiol as confirmed by the action of recombinant human AKR7A3. This structure was confirmed by LC/ESIMS/MS. In addition to detecting this *in vitro* metabolic product at the expected mass of the dihydrodiol by full mass scanning, CID revealed two characteristic fragmentation products corresponding to sequential losses of water from the precursor ion. It was concluded that the presence of this second group of AFB_1-metabolizing enzymes (GST being the first group) may provide a significantly beneficial competing pathway for *in vivo* aflatoxin detoxification.[248]

An enzyme induced by cancer chemopreventive agents is AFB_1 aldehyde reductase (AFAR) which catalyzes the NADPH-dependent reduction of the dialdehyde to a dialcohol. Two monoalcohol products of human AFAR activity were identified by MS and their positions in the metabolic schemes of AFAR as it protects against AFB_1 toxicity were established.[249]

Three AFB_1 adducts derived from AFB_1 epoxide treatment of the symmetric oligonucleotide, 5′-CCGGAGGCC, were separated by HPLC and analyzed by ESI-ITMS. The MS/MS fragmentation pattern (Figure 4.31) is consistent with the known mechanism according to which the loss of a nucleobase promotes fragmentation of the 3′-C-O bond of the sugar to which the base was attached.[250] The predominant backbone cleavages yielded the *a* and *w* ions, as explained by the nomenclature of oligonucleotide fragmentation.[85] The presence of abundant doubly and triply

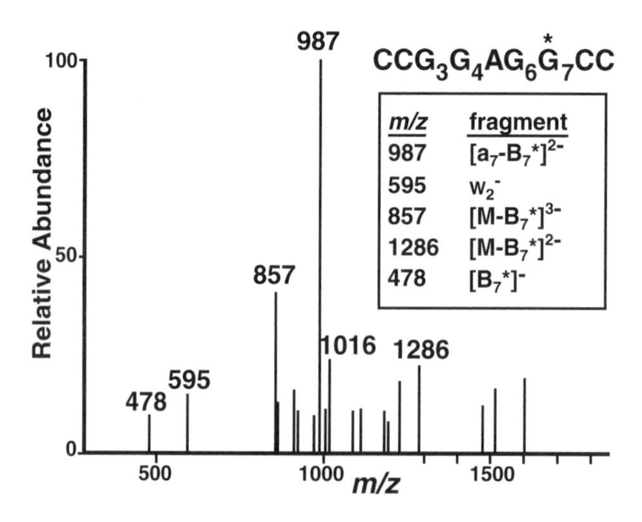

FIGURE 4.31 MS/MS spectrum of the $[M-3H]^{3-}$ ion at *m/z* 1016 of AFB_1-5′-CCGGAGG*CC adduct peak. Product ions reveal the site of base modification to be the seventh nucleotide from the 5′ end of the oligonucleotide. The modified base is designated with an asterisk. Reprinted from Marzilli, L.A. et al., *J. Am. Soc. Mass Spectrom.*, 9, 676–682, 1998. With permission.

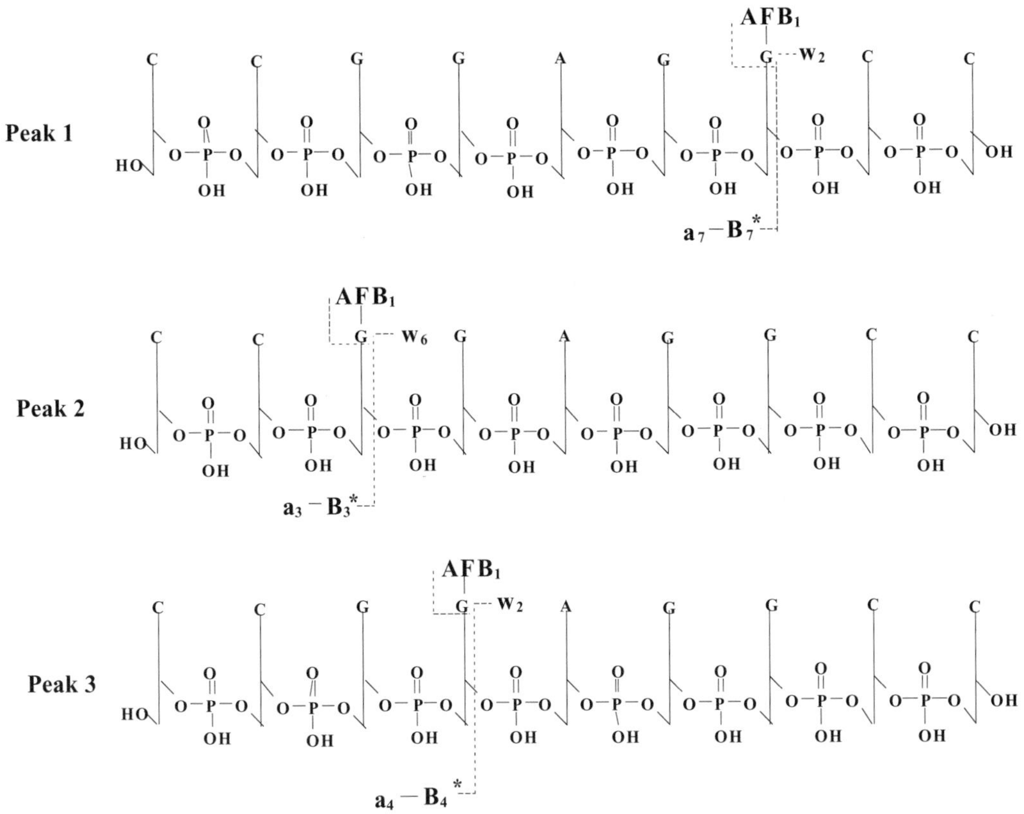

FIGURE 4.32 Summary of MS/MS fragmentation of three AFB_1-modified oligonucleotide isomers. Reprinted from Marzilli, L. A., Wang, D., Kobertz, W. R., Essigmann, J. M., and Vouros, P., Mass spectral identification and positional mapping of aflatoxin B_1-guanine adducts in oligonucleotides, *J. Am. Soc. Mass Spectrom.*, 9, 676–682, 1998. With permission.

charged ions at *m/z* 1286 and 857 was considered a confirmation of the ease with which the AFB_1 modified guanine directs fragmentation during CID. This promotion occurs because of the depurination of AFB_1-N7-Gua that reflects the formation of a positive charge on the imidazole ring potentiated by the aflatoxin adducted to the guanine. Additional MS/MS experiments, using the MS^3 capability of ITMS, have confirmed that the covalent modification to the oligonucleotide, 5′-CCGGAGGCC, consisted of the addition of an intact AFB_1 moiety at one of the structurally unique guanines (Figure 4.32). The site of the modification was confirmed by comparison with the distinctive fragmentation patterns observed for each of three synthetic structural isomers.[251]

The availability of techniques to detect and quantify the concentrations of covalent adducts of aflatoxins with cellular DNA and blood proteins may be utilized in chemoprevention preclinical and clinical trials by considering these adducts as intermediate markers of the predisposition for mutation and neoplasia in exposed individuals (reviewed in Reference 252).

4.3.3.2 Heterocyclic Aromatic Amines

It has been known for about 20 years that cooked (charred) fish and beef exhibit high mutagenic activity as assessed by the Ames/Salmonella test.[253,254] The mutagens were identified as heterocyclic amines (HAA) that form during the pyrolysis of proteins, amino acids, creatine, creatinine, and sugars during cooking.[255–257] At least ten HAA identified in cooked food have been shown to be

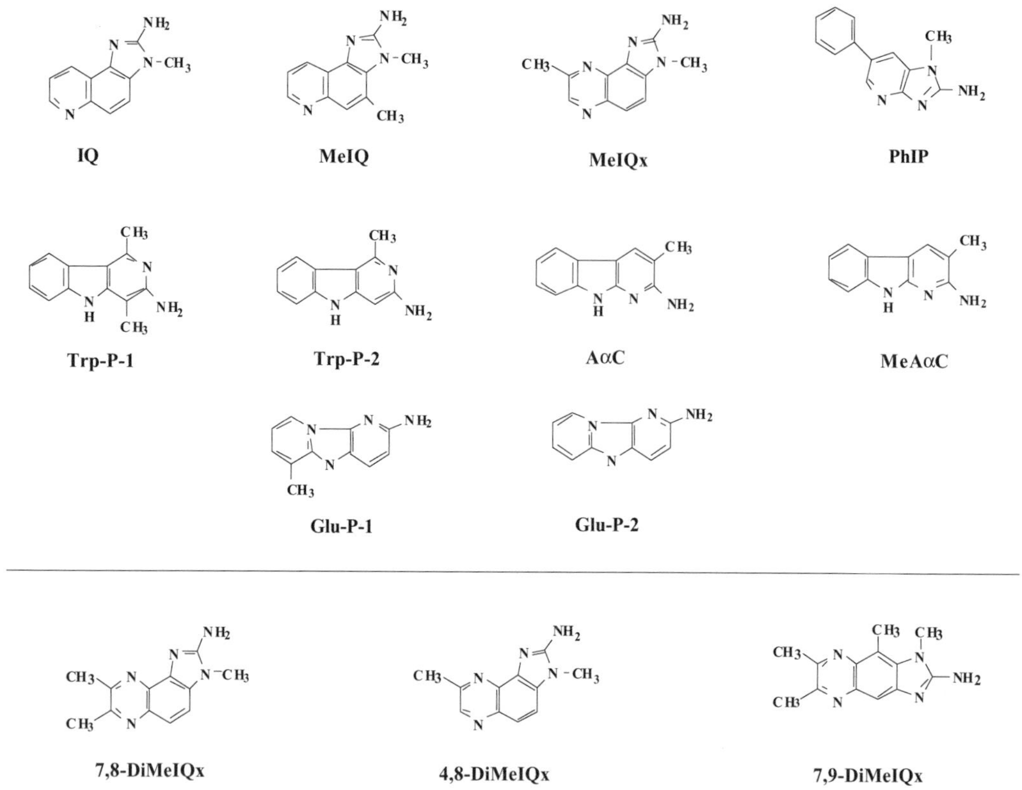

IQ MeIQ MeIQx PhIP

Trp-P-1 Trp-P-2 AαC MeAαC

Glu-P-1 Glu-P-2

7,8-DiMeIQx 4,8-DiMeIQx 7,9-DiMeIQx

FIGURE 4.33 Chemical structure of representative heterocyclic aromatic amines (HAA): **IQ,** 2-amino-3-methyl-imidazo[4,5-f]quinoline; **MeIQ**, 2-amino-3,4-dimethyl-imidazo[4,5-f]quinoline; **MeIQx**, 2-amino-3,8-dimethyl-imidazo[4,5-f]quinoxaline; **PhIP**, 2-amino-1-methyl-6-phenylimidazo[4,5-b]pyridine; **Trp-P-1**, 3-amino-1,4-dimethyl-5H-pyrido[4,3-b]indole; **Trp-P-2**, 3-amino-1-methyl-5H-pyrido[4,3-b]indole; **AαC,** 2-amino-9H-pyrido[2,3-b]indole; **MeAαC**, 2-amino-3-methyl-9H-pyrido[2,3-b]indole; **Glu-P-1**, 2-amino-6-methyl-dipyrido[1,2-α3′2′-d]imidazole; **Glu-P-2**, 2-aminodipyrido[1,2-α3′,2′-d]imidazole; **7,8-DiMeIQx**, 2-amino-3,7,8-trimethylimidazo[4,5-f]quinoxaline, **4,8-DiMeIQx**, 2-amino-3,4,8-trimethylimidazo[4,5-f]quinoxaline; **7,9-DiMeIQx**, 2-amino-3,7,9-trimethylimidazo[4,5-f]quinoxaline.

capable of inducing, often in a synergistic manner, organ tumors in mice, rats, and monkeys. Figure 4.33 shows the structures, chemical names, and abbreviations of these ten HAA. IQ is classified by the IARC as a probable (Group 2A) human carcinogen, while the others are listed as possible carcinogens. Three additional compounds, the ubiquitous 7,8-DiMeIQx and its two isomers, are strongly mutagenic but probably not carcinogenic. Chemically, these compounds are either aminoimidazoazaarenes or carbolines. Those in the former group (IQ group) have a 2-amino-imidazo group fused to a quinoline (IQ and MeIQ), a quinoxaline (MeIQx and DiMeIQx) or a pyridine (PhIP) ring. The carboline group includes the aminopyridoindoles (Trp-P-1, Trp-P-2, AαC, and MeAαC) and the aminopyridoimidazoles (Glu-P-1 and Glu-P-2).[258]

The most frequently occurring carcinogenic HAA in cooked beef are IQ, MeIQ, MeIQx, and PhIP all of which occur at ppb levels. Considerable differences have been reported in dietary exposure to the individual compounds, averaging from the low ng/d to a few μg/d range. As reviewed below, several aspects of the HAA have been explored, including animal experimentation on carcinogenicity, elucidation of metabolism and DNA adduct formation, quantification in meat and other foods, and the correlation between cooked food intake and urinary concentrations of individual HAA. To date, the only human population-based study reported involved determination of the usual dietary intake of HAA in Sweden. Using previously validated techniques for the quantification of

IQ, MeIQ, MeIQx, DiMeIQx, and PhIP,[259] the median daily intake of 77 ng HAA (range 66 to 96 ng) for controls (n = 553) was similar to those of patients with cancer of the colon (n = 352), rectum (n = 249), bladder (n = 273), and kidney (n = 138). Significantly elevated HAA, in the 1.9 to 6.8 μg range, were only found in four colon, two bladder, and one kidney cancer cases, and not in any of the controls. It was suggested that previously reported positive associations between meat intake and risk of colorectal (and perhaps other) cancer might not be due to HAA but rather to other carcinogenic substances, e.g., PAH or nitrosamines.[260]

Methodologies

A number of different techniques have been developed for the analysis of HAA, including HPLC, CZE, immunoassay, GC, GC/MS, LC/TSPMS, and LC/ESIMS (reviewed in Reference 261). A GC/NCIMS technique was developed for the quantification of the di-3,5-bis-trifluoromethylbenzyl derivatives of unconjugated MeIQx and PhIP (i.s.: stable isotope labeled analogue of MeIQx) in urine. The method was used to study the metabolism of MeIQx[262] (see below) and the intra- and interindividual variability in the systemic exposure of humans to these compounds based on meat consumption.[263] Other GC/NCIMS techniques have been applied to the quantification of individual HAA in beef and other foodstuffs cooked under various conditions.[264–266]

Earlier LC/MS techniques using TSP[267,268] have been replaced by LC/ESIMS methods.[269] The on-column LOD for seven HAA for an LC/ESIMS technique, on a QqQ instrument, were 0.12 to 2.2 ng and the 5.4 to 44 pg ranges for the full scanning and SRM modes, respectively.[270] For meat extracts, the LOD were in the 0.9 to 11.2 ng/g range.[271] LC/APCIMS techniques have also been developed using QqQ analyzers both for quantification[272] and identification using in-source fragmentation.[273] Methodological aspects of CE/ESIMS have also been investigated, particularly with respect to the CE separation of HAA having closely related structures and to the optimization of the ESI conditions for on-line operation. Although the LOD was ~1 pbb for most analytes, it was concluded that the method was inferior to LC/ESI-MS/MS because of low sample loading capacity and the lack of robustness of CE.[274]

Sample preparation investigations have included comparisons of the recoveries of the HAA by solid-phase extraction cartridges packed with different adsorbents[275] and an investigation of whether some shortcomings of a simplified extraction procedure could be compensated for by using an IT analyzer. A solid-phase extraction technique[276] has been automated using a robotic workstation, thus allowing unattended sample preparation in which the HAA were separated into two groups, a polar extract (IQ, MeIQ, MeIQx, 4,8-DiMeIQx, 7,8-MeIQx, PhIP, Glu-P-1, and Glu-P-2) and a nonpolar extract containing other HAA.[277]

The accelerator MS has gained an increasingly important role in biological applications, such as HAA metabolism studies, because of its attomole sensitivities. These limits of detection enable the safe administration to humans of nmole quantities of ^{14}C-labeled carcinogens both in single doses and in low chronic doses, the latter attempting to mimic chronic dietary exposure (see later).

Quantification in Food

To quantify MeIQx and DiMeIQx in fried beef, lean minced beef patties were cooked until well done, the i.s. ($^{13}C,^{15}N_2$MeIQx) was added, and the analytes were isolated in a procedure that included homogenization, removal of fats and oils, and extraction of the amines. The di-bis-trifluoromethylbenzyl derivatives then were formed, separated on GC, and quantified by NCIMS using SIM of structurally significant high-intensity fragment ions. Neither compound was detected in raw, uncooked beef. In two cooked patties, the quantities were 2.4 and 1.2 ng/g meat for MeIQx and 1.2 and 0.5 ng/g for DiMeIQx (Figure 4.34). It was concluded that, even in conventional diets, hundreds of nanograms of these compounds may be ingested daily.[278]

To increase both sensitivity and specificity, an LC/APCIMS/MS technique was developed. After investigating the CID products of the protonated molecules of the analytes under both low and higher collision energy conditions, a detailed fragmentation pathway was determined for MeIQx.

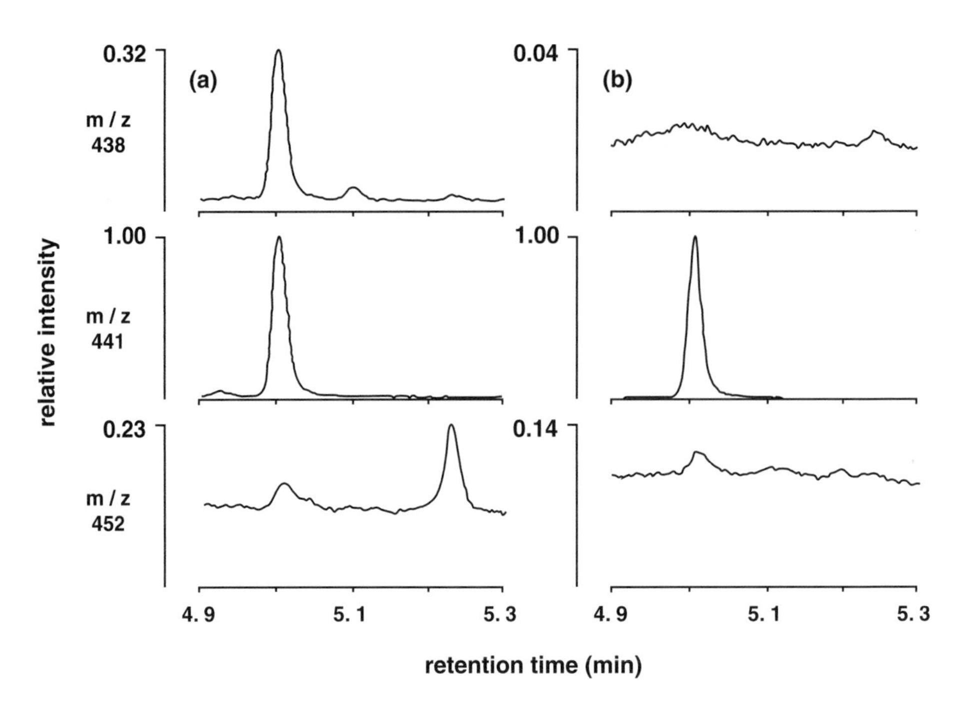

FIGURE 4.34 Selected ion monitoring traces for the analysis of 2-amino-3,8-dimethylimidazo[4,5-f]quinoxaline (MeIQx, *m/z* 438) and 2-amino-3,4,8-trimethylimidazo[4,5-f]quinoxaline (DiMeIQx, *m/z* 452) in (**a**) fried beef and (**b**) raw, uncooked beef. I.s. = *m/z* 441. Reprinted from Murray, S. et al., *Carcinogenesis,* 9, 321–325, 1988. With permission.

For quantification, the cleavages of the N-CH_3 and N-CD_3 (i.s.) bonds were monitored in the SRM mode, e.g., the *m/z* 214.1 → 199.1 transition for MeIQx. The limits of detection and quantification approached 0.015 and 0.045 μg/kg (ppb), respectively, using only 4 g of meat. In a meat-based bouillon sample that contained <1 ppb total HAA, IQ, MeIQx and DiMeIQx could readily be identified and quantified (i.s.: deuterated analytes), however, PhIP was not detectable (Figure 4.35). The method was also used for the quantification of five HAA in meat extracts, as well as grilled and pan-fried bacon. The range of HAA concentrations was wide, 45 to 45,500 ng/kg. MeIQx predominaned and 7,8-DiMeIQx was the least abundant. There were large differences in the HAA concentrations found in grilled bacon, with fivefold higher concentrations observed when the heating surface was smooth compared to heating on grills. This confirmed that fat drippings, presumably because they acted as an extraction solvent, contained 10- to 100-fold higher levels of HAA than cooked meats. Use of the constant neutral loss mode of MS/MS permitted the identification of a rarely observed HAA, 2-amino-1,7,9-trimethylimidazo[4,5-g]quinoxaline.[279]

These and similar techniques have been applied to determine the concentrations of individual HAA in a variety of other food products. Examples include microwaved and vacuum-dried meat extracts, grilled beef, merguez sausage, chicken-flavored paste and peanut butter,[277] process flavors and their ingredients, bouillon concentrates and pan residues,[280] barbecue sauce and barbecued beef,[281] and wine.[282]

In another application, it was determined by LC/ESIMS that ~270 ng of MeIQx per frying fish or other Chinese food could be detected in the aerosol above the cooking surface in a few minutes. Because Chinese woman spend ~1 h per day preparing meat, it was suggested that the heavy exposure to this carcinogen in cooking fumes may be responsible for the high incidence of adenocarcinoma of the lung among the Chinese nonsmoking female population.[283]

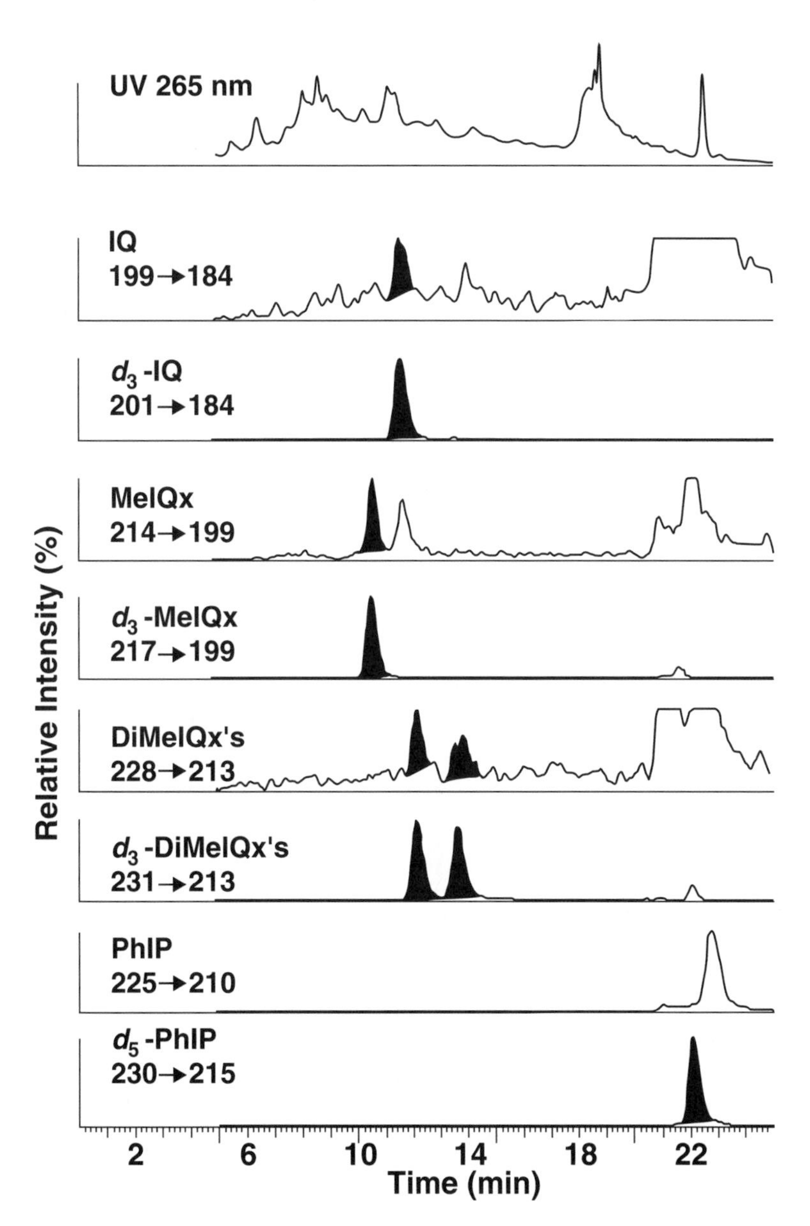

FIGURE 4.35 LC/APCI-MS/MS analysis of a meat-based bouillon sample containing a low amount (<1 ppb) of total heterocyclic aromatic amines (HAA) obtained in selected reaction monitoring (SRM) mode. The UV trace on top was recorded at 265 nm and corresponds to a meat extract sample contaminated with high levels of total HAA. See Figure 4.33 for the chemical structures of the HAA. Reprinted from Guy, P. A. et al., *J. Chromatogr. A*, 883, 89–102, 2000. With permission.

Quantification in Urine

The GC/NCIMS technique described above[278] was used for the quantification of MeIQx in urine. The LOD was 5 pg/mL. When fried beef was given to six subjects, MeIQx was detected in all the post consumption 12 h urine collections but not in later samples. The total amount of unchanged MeIQx in the urine ranged from 9.1 to 15.5 ng. Comparison of these quantities of the amine with those determined in the cooked meal revealed that only 2 to 5% of the total MeIQx was excreted unchanged. Additional quantitative *in vitro* studies with human microsomal fractions has led to the suggestion that dietary MeIQx may not only be absorbed efficiently but also biotransformed extensively, thus increasing the carcinogenic risk.[284] It was also determined that the percentage of unchanged MeIQx excreted, although variable from person to person, remained constant for each subject irrespective of the ingested dose.[263] In another study, a negative correlation was found between unmetabolized urinary MeIQx and the activity of the metabolizing enzyme, CYP1A2 (see below).[285]

Both low-resolution GC/NCIMS and high-resolution (resolution 8000) GC/EIMS techniques have been used to quantify MeIQx and PhIP in urine for up to 24 hr following a meal of fried meat. Derivatization for NCI involved acylation with heptafluorobutyric acid anhydride followed by methylation of the resulting amide and also any phenolic hydroxyl groups, meaning that 4′-OH-PhIP could also be detected. Post-meal HAA were measured both in untreated and acid-hydrolyzed urine, the latter providing higher concentrations of HAA. The quantities of the parent compounds relative to hydrolysable metabolites varied considerably among individual subjects, probably due to differences in their capacity for glucuronidation and sulfation.[286]

The contribution of N-oxidation to the metabolism of MeIQx was investigated by measuring the degree of N^2-glucuronide conjugation in urine following a diet of cooked meat that contained known amounts of MeIQx. After solid-phase extraction and immunoaffinity separation of the glucuronide, a deaminated MeIQx derivative was relesed from the conjugate with acetic acid and then derivatized to form the 3,5-bis-trifluoromethylbenzyl ether that was quantified by GC/NCIMS in the SIM mode. The LOD was 80 pg/8 mL urine. The quantity of the N-OH-MeIQx-N^2-glucuronide recovered 0 to 12 h after the test meal was ~2 to 17% of the ingested dose of MeIQx.[287] A related study, using the same methodology, concerned the relationship between the activating and detoxification enzymes (cytochrome P-4501A2 and N-acetyltransferase) and of N-OH-MeIQx-N^2-glucuronide formation among healthy subjects fed a controlled meal cooked at high temperature containing a known quantity of MeIQx. There was no association between the concentration of the urinary products and the activity of either enzyme, confirming that the biotransformation of MeIQx by CYP1A2 oxidation to the N-hydroxylamine followed by N^2-glucuronidation is not a rate limiting set of reactions but is instead a general pathway for the removal of these compounds in humans.[288]

Other approaches to the quantification of free and conjugated PhIP in urine samples involved purification by immunoaffinity chromatography and quantification by both LC/ESIMS using SIM and LC/ESIMS/MS using SRM. The LOD was the same for both procedures, 4 pg/mL urine. For reasons not yet known, the mean concentrations of PhIP were ~3-fold higher in Asian- and African-Americans than in whites (in Los Angeles County). Comparison of the PhIP levels observed with those found for MeIQx, in the same and comparable samples, led to the conclusion that quantification of single HAA provides only an approximate measure for estimating the overall exposure to these compounds.[289]

A solid-phase extraction LC/ESI-MS/MS method has been developed for the quantification of four metabolites of PhIP in human urine after a controlled meal of well-done chicken containing 9 to 21 μg of PhIP and also after a conventional meal of chicken in a restaurant. The four metabolites assayed were: N^2-OH-PhIP-N^2-glucuronide (major metabolite), PhIP-N^2-glucuronide, PhIP-4′-sulfate, and N^2-OH-PhIP-N3-glucuronide. There was an eightfold variation in the total amount of these metabolites and 20-fold variation in the relative amounts of individual metabolites.[290]

Metabolic Activation and DNA Adducts

Being promutagens/pro-carcinogens, HAA require metabolic activation before DNA adduct formation can occur (reviewed in Reference 291). For example, the major metabolic pathway of MeIQx involves Phase I N-oxidation to the N-hydroxy metabolite[262] that is mediated by the hepatic cytochrome P-4501A2 (CYP1A2). This is followed by a Phase II esterification to form a reactive O-acetoxy derivative by N-acetyltransferase (NAT2) that can eventually covalently modify DNA.

A study of healthy subjects on a controlled diet of lean beef cooked at low and high temperatures established that an inverse relationship exists between CYP1A2 activity and urinary MeIQx concentrations, the latter being determined by GC/NCIMS.[262] These results suggest that interindividual variations in the activity of this enzyme may be relevant to the occurrence of cancers associated with these HAA. There was no relationship between diet and the level of NAT2 activity.[285] The same GC/NCIMS method was used to study the effects of furafylline, a potent and selective inhibitor of CYP1A2, both in human liver microsomes and in healthy volunteers. Administration of the inhibitor before ingestion of fried beef resulted in 14-fold and fourfold increases in the urinary excretion of unchanged MeIQx and PhIP, respectively. There was no effect when a placebo was administered. It was concluded that these HAA are oxidized to mutagenic species by CYP1A2 in humans but not in experimental animals where deactivation through C-oxidation also occurs.[292]

One of the HAA, IQ, is a potent hepatocarcinogen in nonhuman primates. The major DNA adducts of IQ are formed at the C8 and N^2 positions of guanine to give N-(deoxyguanosine-8-yl)-2-amino-3-methylimidazo[4,5-f]quinoline (dG-C8-IQ) and 5-(deoxyguanosine-N^2-yl)-2-amino-3-methylimidazo[4,5-f]quinoline (dG-N^2-IQ). An LC/ESI-MS/MS technique (with a QqQ instrument) using both nozzle-skimmer and CID fragmentation has been applied to elucidate and quantify the DNA adduct formation by IQ both *in vitro* and in monkeys. The LOD were 1 adduct in 10^4 unmodified bases using the QqQ analyzer in constant neutral loss mode (for compound class analysis) and 1 adduct in 10^7 unmodified bases using SRM (for sensitive target analysis) from 300 μg DNA samples. An application of the method involved the analysis of DNA isolated from kidney tissues of monkeys exposed to both a single dose (20 mg/kg) of IQ and chronic administration (5 doses/wk of 10 or 20 mg/kg for 3.6 yr). Adduct levels after the single dose were below the detection limit (Figure 4.36, curve B). The chromatograms of samples from the chronically exposed animals revealed both the dG-C8-IQ and dG-N^2-IQ adducts (Figure 4.36, curve C). The technique permitted the detection of 1 modification in 10^7 nucleotides in a kidney DNA sample. The ratio of the two types of adducts was ~4:2 which was in agreement with results obtained by ^{32}P-postlabeling. It was concluded that the advantages of the LC/MS method included the ability

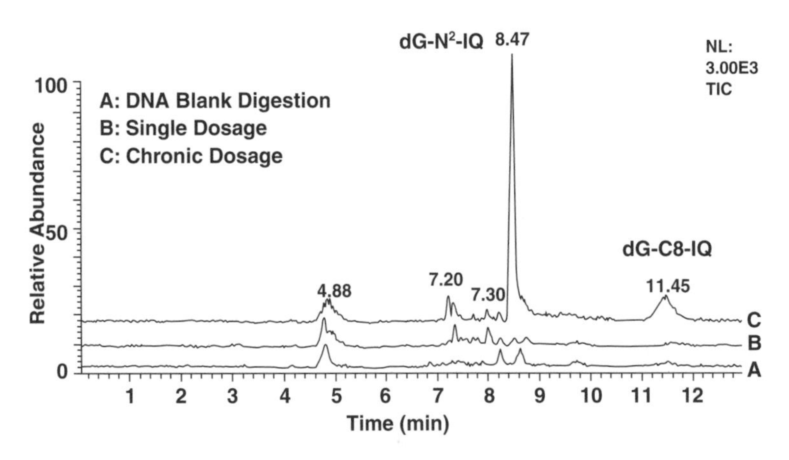

FIGURE 4.36 Capillary LC/ESI-MS/MS using selected reaction monitoring (SRM) scanning of the *m/z* 464 → 348 transition. (**A**) Procedural blank digest of calf thymus DNA. (**B**) Digest of the IQ-DNA adduct from kidney tissue of nonhuman primate 24 h after a single oral dose of IQ (20 mg/kg). (**C**) Digest of DNA from kidney tissue of nonhuman primate undergoing carcinogen bioassay. IQ, 2-amino-3-methylimidazo[4,5-f]quinoline. Reprinted from Gangl, E. T., Turesky, R. J., and Vouros, P., *Chem. Res. Toxicol.*, 12, 1019–1027, 1999. With permission.

to confirm adduct structures and recognize new adducts at a sensitivity approaching that of ^{32}P-postlabeling, thus providing two parallel techniques for the *in vivo* detection and quantification of IQ adducts of DNA.[293] In subsequent work, the methodology was modified such that analyte detection was improved significantly, to 1 adduct in 10^9 unmodified base using ~500 μg of DNA. In an illustration, the DNA adducts DG-C8-IQ and DG-N^2-IQ were detected in monkey pancreas tissues 24 h after a single administration of 10 mg/kg IQ.[294] The same methodology was applied to the quantification of dG-C8-IQ in rat livers. In a dose-response study, 0.05, 0.50, 1.0, and 10 mg/kg of body weight quantities of IQ were administered to rats. The DNA adduct formation was not a linear function of IQ dose because IQ must first undergo metabolic N-oxidation by cytochrome P-4501A2 in the liver before it can bind to DNA, and at high IQ concentrations the enzyme is saturated by the substrate and some of the dose is eliminated as unmetabolized IQ.[295]

Applications Using Accelerator Mass Spectrometry (AMS)

AMS has been used with increasing frequency for studies on the metabolism and macromolecular adduct formation of HAAs in humans and rodents at very low doses. It was demonstrated some ten years ago, that MeIQx adducts (derived mainly from guanine) could be determined at the 1 adduct per 10^{11} nucleotide level with a reproducibility of ±2%.[296] The concentration of the adducts was linear with respect to the administration of doses from 5 mg/kg body weight down to 500 ng/kg body weight.[297] When [2-^{14}C]PhIP was administered to mice at a dose equivalent to the consumption of two 100 g beef patties (41 ng/kg), peak tissue concentration in the GI tract was achieved within 3 h. In other experiments, linear adduct-dose relationships were established in a number of organ tissues.[298] Dose-response curves from animal studies were compared to those obtained from colon cancer patients taking low doses (228 μg) of [^{14}C]MeIQx prior to surgical resection of their tumors. Findings revealed that human colon DNA adduct levels were ~tenfold greater than those in rodents at the same dose and time point after exposure. It was also established in animal studies that adduct formation begins to plateau at high chronic doses, suggesting that low-dose exposure may lead to higher levels of DNA damage than extrapolations from high-dose studies might suggest.[299] Rodent models apparently do not represent accurately the human response to HAA exposure because there is more N-hydroxylation, i.e., bioactivation, of both MeIQx and PhIP and less ring oxidation, i.e., detoxification, in humans.[300]

In one study, 20 μg of 2-^{14}C-MeIQx was given orally to patients prior to colorectal surgery. DNA samples taken from normal and tumor tissues were extracted and purified, the adducts isolated and then converted to graphite for AMS analysis. MeIQx-DNA adducts were quantified and found to be in the 20 to 42 adducts/10^{12} base range in both normal and tumor tissues.[31] The same approach was used to explore the metabolism and adduct formation of MeIQx in rat liver DNA using low, human equivalent, doses (1×10^{-6} to 3×10^{-2} mg/kg day) in both acute (24 hr) and chronic (7 d and 42 d) exposures. The relationship between administered dose and both hepatic MeIQx content (~100 pg/g liver) and the number of MeIQx-DNA adducts (~10^3 adducts/μg DNA) was linear, with the DNA adducts persisting up to 14 d after exposure.[301]

The aim of another study using AMS was to examine the ability of PhIP and B[a]P to form DNA adducts in human breast tissues. Prior to undergoing breast surgery, patients were orally administered 20 μg of ^{14}C PhIP (182 kBq) or 5 μg of ^{14}C B[a]P (36 kBq). Adduct levels in resected breast tissues ranged from 26 to 473 and 6 to 208 adducts/10^{12} nucleotides for the ^{14}C PhIP and ^{14}C B[a]P, respectively. There was no difference between the levels of these adducts in normal or tumor tissues. The adduct concentrations were comparable to those previously found in resected colon tissues.[302]

4.3.3.3 Nitrosamines

The nitrosamines are potent carcinogens that are important contributors to the adverse effects of smoking (Section 4.3.1). A positive association has been established between nitrosodimethylamine

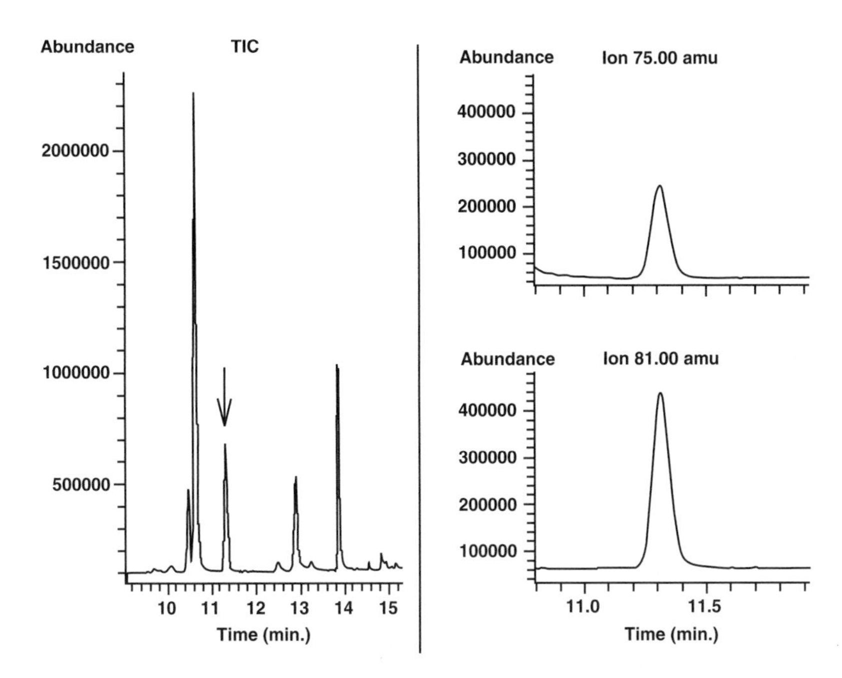

FIGURE 4.37 Total ion chromatogram (TIC) and selected ion monitoring (SIM) traces relative to the protonated molecule of N-nitrosodimethylamine (NDMA, *m/z* 75.0) and [$^2H_6$]NDMA (*m/z* 81.0) from a beer sample containing 1.36 μg/kg NDMA (arrow in the TIC). Reprinted from Longo, M., Lionette, C., and Cavallaro, A., *J. Chromatogr. A*, 708, 303–307, 1995. With permission.

(NDMA) from smoked and salted fish (but not cured meat and other fish) and subsequent occurrence of colorectal cancer (reviewed in Reference 303). NDMA is also present in beer and other malt beverages where it occurs as a result of the reaction of alkaloids with nitrogen oxide during the hop drying process.

A GC/CIMS technique for NDMA in beer started with 200 g samples. The analyte was extracted with dichloromethane and quantified by SIM of the protonated analyte (i.s.: deuterated NDMA). Although several unrelated compounds present had the same mass as the analyte, there were no interferences at the respective retention times in some 40 beer samples analyzed (Figure 4.37). The LOD was 0.04 μg/kg, with remarkable linearity (r = 0.9999) in the 0.10 to 4.00 μg/kg range.[304] Using a different GC/MS method, a study of Chinese beers revealed 0.1 to 6 pbb NDMA in 146 of 176 samples.[305]

Eight volatile nitrosamines were quantified in gastric juice samples from patients with various upper abdominal complaints. Aliquots of the dichloromethane extracts of the gastric juice were analyzed directly with GC/EIMS double-focusing using instruments operated at a resolution of 1000, and using SIM. The main nitrosamines found were N-nitrosodiethylamine (~3 nmol/L), NDMA (~0.9 nmol/L), and N-nitrosopyrrolidine (~0.4 nmol/L). The LOD were ~0.05 nmol/L gastric juice (0.001 pmol injected). A correlation was observed between the total volatile nitrosamine content and the intragastric pH values.[306]

An LC/APCI-MS/MS method was used for the identification of N-nitrosodiethylamine, N-nitrosopyrrolidine, and N-nitrosopiperidine in dry sausages. Quantification was carried out using SIM. It was concluded that nitroso compounds may form during ripening by reaction between residual nitrite and amines originating from reactions from inadvertent fermentation.[307]

N-nitrosodiethanolamine (NDELA), a potent liver carcinogen in animals, is a common trace contaminant in personal care and tobacco products, metalworking fluids, and pesticides. The carcinogenic activation of the compound (and similar β-hydroxynitrosamines) involves α-hydroxylation leading to the production of α-nitrosaminoaldehydes that can transfer their nitroso groups to other amines (transnitrosation) that, after further metabolic steps, can result in the deamination of primary amines. If this process occurs to the primary amino groups of DNA bases it may result in deamination at that location in the base and the eventual formation of carbonyls with the potential for transcription errors as a consequence. An investigation of the deamination of DNA bases in oligonucleotides and calf thymus DNA by three α-nitrosaminoaldehydes, including N-nitroso-2-hydroxymorpholine (NHMOR), a metabolite of NDELA, was undertaken using GC/EIMS. SIM of the silylated analytes was used to quantify the extent of deamination of guanine, adenine, and cytosine. It was determined that NHMOR, studied at concentrations close to those to which humans are exposed, produced significant deamination of DNA in a short period of time, suggesting an active role for this metabolite in the mechanism of NDELA genotoxicity.[308]

4.3.3.4 Polycyclic Aromatic Hydrocarbons

Polycyclic aromatic hydrocarbon (PAH) exposure is both occupational and environmental. The DNA damage in exposed subjects has been reviewed.[309] The occurrence of PAH in edible fats and oils and available analytical methods have also been reviewed.[310] Most GC/MS methods for the analysis of PAH are straightforward, using capillary columns for separation and SIM for quantification.[311] The approach is often toward group-type analysis because, despite the high resolving power of GC, there are often large numbers of unresolved components of similar structure. PAH are often expressed as total benzo[a]pyrene equivalents. Supercritical fluid extraction (SFE) is increasingly used instead of solvent-based extraction. The "SFE plus C_{18}" extraction method is based on the addition of absorbent beads to the initial sample slurry, and placing the dried mixture including the adsorbent into the SFE chamber where lipids are preferentially retained by the beads. The recovery of PAH is nearly complete concurrent with the removal of ~85% of the lipids.[312]

Data gathering type applications have included the elucidation, in processed or restaurant food, of the formation of PAH (up to 1 ng/g) and HAA (0.1 to 14 ng/g) with up to 38 ng/g being found in hamburgers fried in laboratories.[313] Large intraindividual variations of DNA adduct levels were also found to depend on the location of sampling.[314]

A number of reported applications of GC/MS for the analysis of PAH in food have included the identification and quantification of some 34 PAH in liquid smoke flavorings,[315] determination of 12 PAH in smoked meats at levels of 3 to 52 ng/g wet weight,[316] and establishment of shellfish harvest reopening criteria following an oil spill.[317] To explore the possibility of cooking-released PAH contributing to the high incidence of lung cancer among nonsmoking Chinese women, benzo[a]pyrene, dibenz[a,h]anthracene, benzo[b]fluoranthene, and benzo[a]anthracene were identified in fumes from safflower, vegetable, and corn oil and the correlation of measured quantities with the operation of fume extractors was established.[318]

4.3.3.5 Dioxins and Pesticides

Dioxins

In 1997 the NIEHS voted to list 2,3,7,8-tetrachlorodibenzo-*p*-dioxin (TCDD, often called *dioxin,* Figure 4.38) as a known human carcinogen.[319] TCDD has been called "the most toxic man-made chemical." However, there is a wide range of sensitivities among animal species, e.g., 10^4-fold higher in guinea pigs than in hamsters. Scientific, public, and political debates about the dioxin "dilemma" have been raging for decades.[320]

There have been numerous reviews of the available analytical methodologies, including a historical review of congener-specific human tissue measurements.[321] High-resolution GC, combined with high-resolution, isotope-dilution MS, is still the most reliable method for the analysis of dibenzo-*p*-dioxins and -furans at the ppt level.[322] Details of the sample preparation, analytical

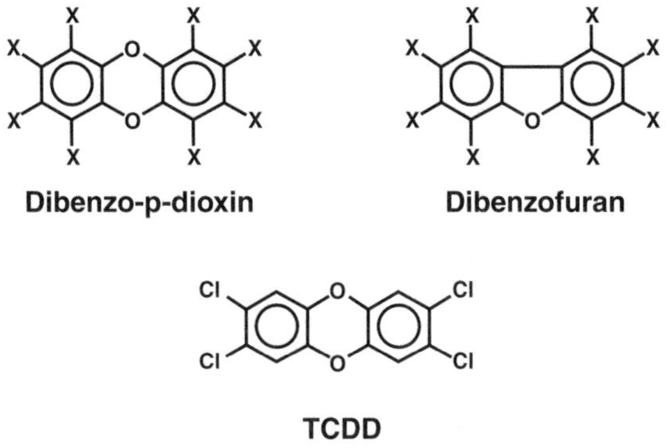

Dibenzo-p-dioxin Dibenzofuran

TCDD

FIGURE 4.38 Chemical structure of dibenzo-p-dioxin, dibenzofuran, 2,3,7,8-tetrachlorodibenzo-*p*-dioxin, TCDD.

techniques, and quality control procedures are described in US EPA Method 1613. A representative example of the application of this method is the quantification of these analytes in 65 samples from the back fat of an animal population in the U.S. selected to include representative geographical regions. The method LOD (whole weight basis) were 0.05 ppt for TCDD, 0.1 ppt for TCDF, 0.5 ppt for selected penta isomers, and 3.0 ppt for the octa-CDD and -CDF. Statistically evaluated replicate data were published for 17 analytes.[323]

Different methodological approaches include GC/MS (reviewed in Reference 324) tandem-in-space MS,[325] and tandem-in-time MS.[326] Applications, using various methodologies, include the analysis of breast milk to assess exposure in agricultural villages in Kazakstan,[327] human blood to correlate with the consumption of crabs in Norway,[328] and muscle and roe of fishes from the Great Lakes.[329]

Pesticide Residues

The introduction of the insecticide dichlorodiphenyl-trichloroethane (DDT) in the late 1940s heralded the use of a range of halogenated pesticides. Although DDT is not toxic to humans and is carcinogenic only in mice, the compound bioaccumulates and is highly persistent in the environment. There have been disparities in cancer risk estimates for pesticide residues in food (reviewed in Reference 330). Many methods have been developed for the assay of pesticides. One example is the application of supercritical fluid extraction, GC separation, and identification and quantification using an ion trap, all in a technique designed to analyze 46 pesticides in potatoes, carrots, broccoli, and grapes. Starting with 3 g sample sizes, Hydromatrix was used to absorb water effectively, and chrysene-D_{12} or phenanthrene-D_{10} was added as i.s. The extracted pesticides were collected on octadecylsilane-coated silica gel and recovered by elution with acetonitrile. More than 20 pesticides detected in a potato extract were present at >5g/g level.[331,332]

4.3.4 Water and Soil

4.3.4.1 Membrane Inlet Mass Spectrometry

Flow-through and other membrane inlets have been used with most types of ion sources and analyzers and applied to a wide variety of environmentally important volatile analytes (e.g., benzene, chloroform, carbon tetrachloride) in water. The LOD for these compounds are of the order of 0.1 μg/L with assay linearity being 0.1 to 1000 μg/L.[333,334] On-line monitoring of multiple target compounds is often used in wastewater stream management, e.g., for oil refineries. The technique benefits from the fact that measurements may be made frequently, even every hour if indicated, or

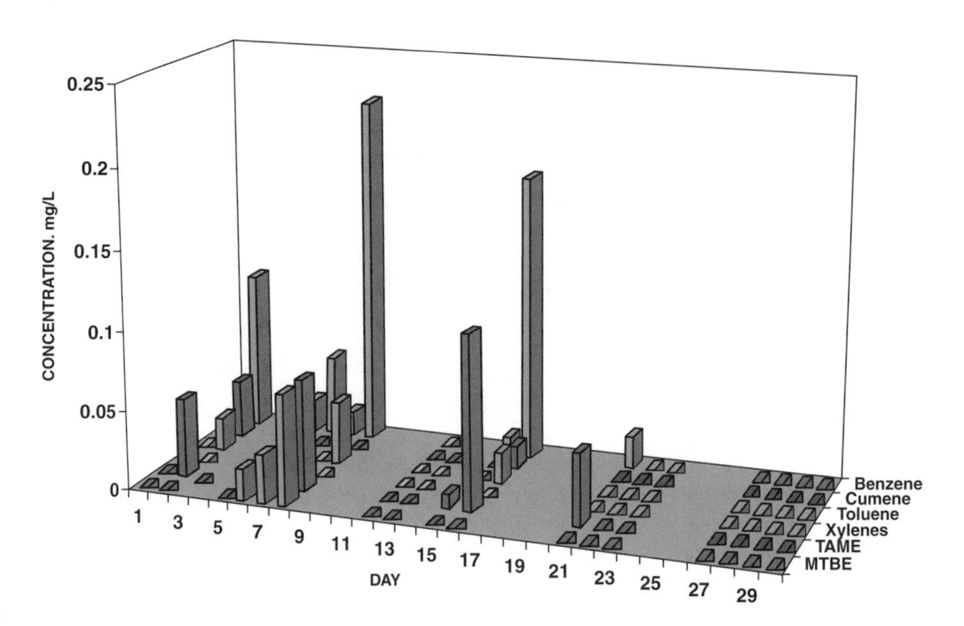

FIGURE 4.39 Using membrane-inlet mass spectrometry for the monitoring of day-to-day changes in the concentration of six contaminants in the wastewater of an oil refinery over a one-month period. Courtesy of Dr. T. Kotiaho. Figure based on Figure 5 in Kotiaho, T. et al., *Am. Environ. Lab.*, 1997, March, 19–22.

daily as shown in Figure 4.39. Membrane inlet MS (MIMS) is subject to the limitation that there is no chromatographic separation of the analytes prior to MS analysis. Other limitations and advantages of the technique as it applies to water analysis have been reviewed.[335]

4.3.4.2 Carcinogen Identification in Well Water by High-Resolution MS

Several techniques have been employed in a still ongoing attempt to identify the cause of the elevated incidence of childhood cancers in a small town located near an industrial waste dump in New Jersey. Analyzing extracts from the municipal well water by standard GC/MS methods yielded four peaks in the total ion chromatograms that could not be matched to any entry in mass spectral libraries. Effluents from the GC were then analyzed by EIMS in a double-focusing magnetic sector instrument at 20,000 resolution, using a special technique, *mass peak profiling from selected-ion-recording data* (MPPSIRD) that provided a 100-fold increase in sensitivity, and 6-fold improvement in scan cycle time with respect to conventional scanning techniques.[336] A *profile generation model* was developed for data acquisiton and interpetation.[337] First, accurate masses were obtained at various resolutions of the presumed molecular ions of the four unidentified GC peaks, followed by the tabulation of possible compositions based on C, H, N, O, P, S, Si, and F atoms at various error limits. Chlorine, bromine, and some other possible elements were excluded based on the observed isotopic abundance patterns. Interpretation of the high-resolution spectra of one of the analytes of nominal mass 210 led to a potential elemental composition of $C_{14}H_{14}N_2$. Although there had not been an exact library match previously, it was now possible to return to the libraries to seek assistance in the interpretation of the fragmentation process. Knowledge of the isotope formula reduced possible library matches from ~1600 (based on nominal mass) to <50. Subsequent hypotheses based on the observed fragmentation patterns as well as further library searches suggested that the structures of the observed isomers contained α-cyanoethyl and/or β-cyanoethyl groups attached to a tetralin (Figure 4.40). It was concluded that the unknown compounds were products originating from an industrial styrene-acrylonitrile polymerization process that was no longer in use. However, three of six isomers of the initial four unidentified analytes in the water extract were

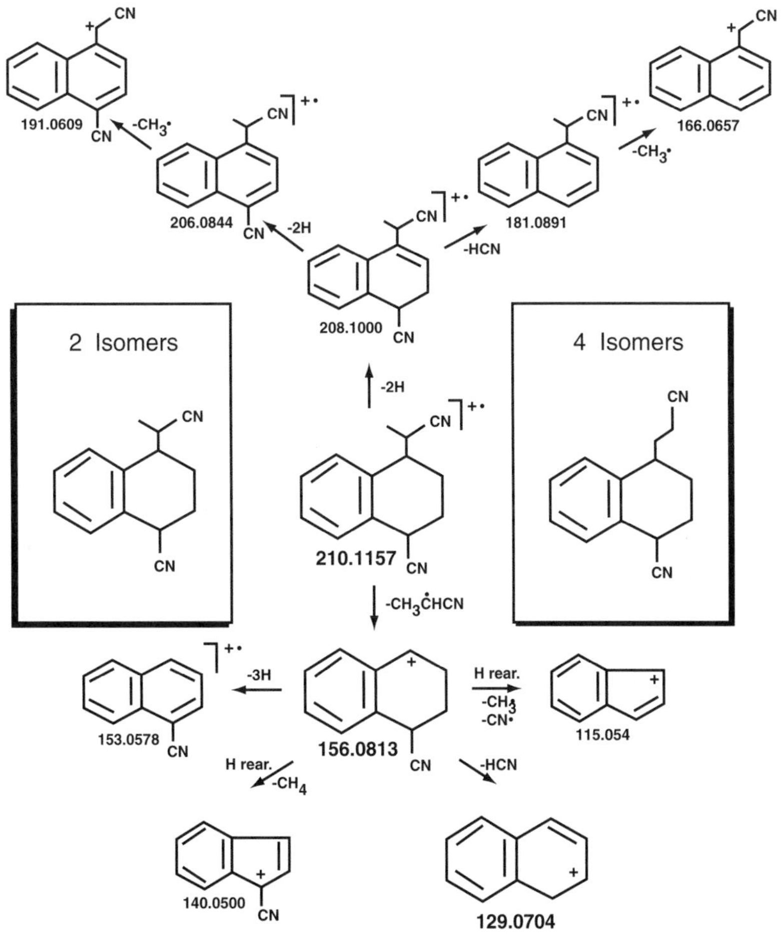

FIGURE 4.40 In the boxes, structures for the isomers with α-cyanoethyl and β-cyanoethyl groups, and a plausible fragmentation scheme for an isomer with an α-cyanoethyl group for the 10 fragment ions studied. Isomers result from the asymetric C atoms; the ring attachment positions are the same for each isomer. Reprinted from Grange, A. H. et al., *Rapid Commun. Mass Spectrom.,* 12, 1161–1169, 1998. With permission.

shown to be present in other industrial processes still in use. Isolation of adequate quantities of the putative carcinogens for toxicity/carcinogenicity studies has been in progress.[338]

4.3.4.3 Vinyl Chloride Monomer in Drinking Water

Monitoring vinyl chloride monomer (VCM) concentrations in PVC-bottled drinking water by GC/MS (i.s.: trideuterated VCM) revealed a linear increase in the VCM concentration of ~1 ng/L/day. It was concluded that individual daily intake of the monomer may exceed 100 ng.[339]

Because of the potential exposure to VCM, water bottles are no longer made of PVC. The VCM content of waste water from a plastics manufacturing plant was in the 0.2 to 9.3 mg/liter range. With the direct injection of 1 mL water (using modified injectors), 0.1 ppb of VCM could be detected.[340]

4.3.4.4 Phenolic Xenoestrogens in Water

Some 40 nonsteroidal anthropogenic substances have been shown to mimic the effects of 17β-estradiol, a natural estrogen. Many of these xenoestrogens are phenolic compounds, e.g., 4-nonyl- and 4-octyl-bisphenol, and 4-hydroxybiphenyl, which end up in the aquatic environment via sewage and sewage treatment plants. A GC/MS technique has been developed for the identification and quantification of a number of phenolic xenoestrogens in surface water and sewage. The analytes were recovered by solid phase extraction, methylated with phenyl-trimethylammoniumhydroxide (i.s.: biphenyl), and quantified by GC/EIMS with SIM of the molecular ions of the methylethers. Nine analytes were studied: bisphenol, 2-OH-biphenyl, 4-OH-biphenol, 4-Cl-2- and 4-Cl-3-methylphenol, 3-*t*-butyl-4-OH-anisole, 4-*t*-octylphenol, 4-nonylphenol, and 2-t-butyl-4-methylphenol. The LOD for these analytes in water were ~0.01 ng/L or less, with the LOQ some 2 to 5 times higher.[341] This technique has been applied to the input/output balance of a large municipal sewage plant in Germany. The contribution of these phenolic xenoestrogens to the total estrogenic activity in the sewage was 0.7 to 4.3%.[342]

4.3.4.5 Mutagenic Heterocyclic Amines in the Danube River

Three HAA, IQ, Trp-P-1, and AαC (Figure 4.33), were extracted from the river (in Vienna, Austria), identified by GC/EIMS as their N-dimethylaminomethylene derivatives, with quantification of the individual analytes using SIM of the respective molecular ions (i.s.: TriMeIQx). The total quantity of these three compounds accounted for ~26% of the total mutagenicity of the water at various sampling sites. The sources of these HAA were food processing (e.g., smoked sausage) and wood burning.[343]

4.3.4.6 Polycyclic Aromatic Hydrocarbons in Soils

There are 16 PAH (many of them carcinogenic) in the list of priority pollutants established by the U.S. Environmental Protection Agency (EPA). Sample preparation of soils and sediments is an obvious problem, particularly when the PAH concentrations are low. Supercritical fluid extraction and accelerated solvent extraction have been compared as sample preparation methods. Comparisons have also been made of GC/ITMS in the full scan mode and GC/MS in the SIM mode. The LOD for individual PAH was 1 μg/kg soil.[344]

Although most PAH of environmental interest have been those with molecular mass <300, there are carcinogenic PAH of higher molecular mass, e.g., dibenzopyrenes and dibenzo(*cd,lm*)perylene. Conventional HPLC methods are inadequate for the separation of most high-molecular weight PAH. In a study with a particle beam LC/EI instrument using a polymeric C_{18} HPLC column to separate the targeted analytes, 16 PAH were selected in the 300 to 450 Da range. Sample preparation consisted of an initial extraction from the soil samples with chlorobenzene, the removal of the low-molecular-weight PAH with solid phase extraction cartridges followed by extraction of the analytes with chlorobenzene. Analyses were carried out in both full scanning and SIM modes. The LOD were in the 10 to 54 ng range for the full scan mode and from 0.1 to 1.3 ng for SIM. The mass spectra of these analytes were simple, with the molecular ions and doubly charged molecular ions predominating. An advantage of the particle beam inlet is that PAH <228 Da do not efficiently pass through the interface, thus improving the sensitivity of the method for the higher molecular weight species.

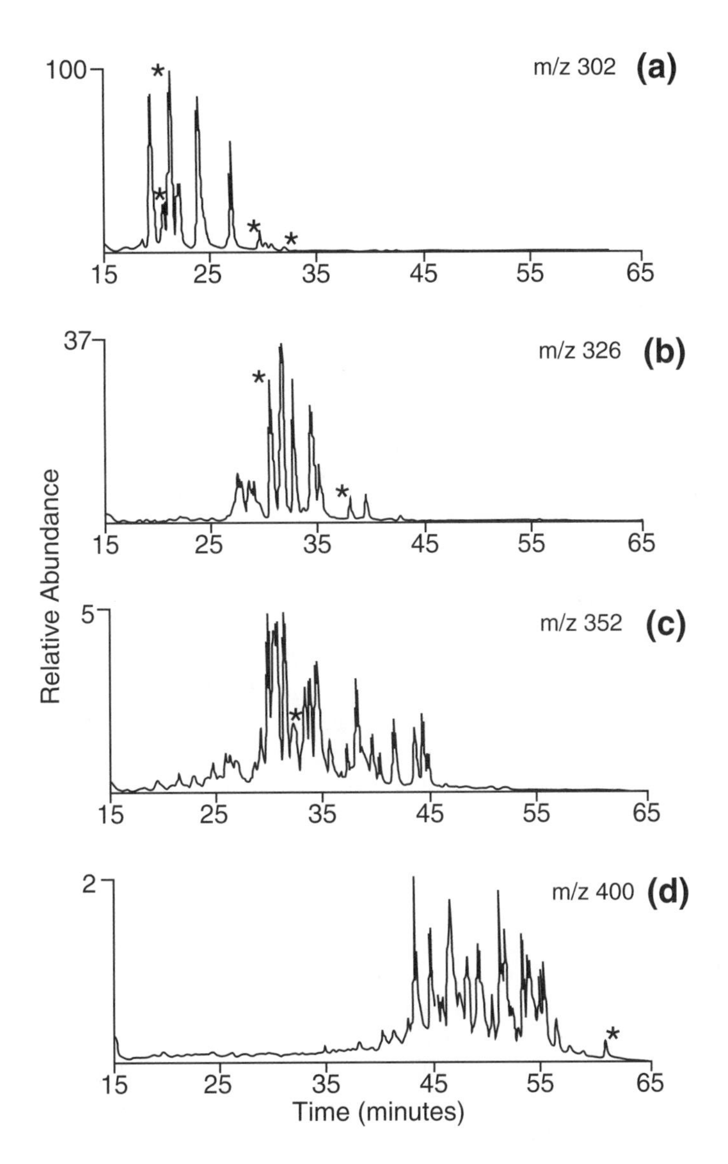

FIGURE 4.41 Selected ion monitoring chromatograms of (a) *m/z* 302, (b) *m/z* 326, (c) *m/z* 352, and (d) *m/z* 400 from the Pacific Northwest polycyclic aromatic hydrocarbons (PAH)-contaminated soil extract. Targeted PAH are labeled with asterisks. The *y*-axis scalings are relative to the *m/z* 302 abundance. Reprinted from Pace, C. M., and Betowski, L. D., *J. Am. Soc. Mass Spectrom.*, 6, 597–607, 1995. With permission.

The ion chromatograms of four selected ions monitored for high molecular weight PAH in a contaminated soil sample from the Pacific Northwest revealed how an increasing number of potential isomers occur and was observed as the molecular weight increased. Targeted PAH are labeled with asterisks in Figure 4.41. After using separate analysis of known compounds with increasing molecular weights to determine response factors, full mass scans were obtained on the extracted soils. Some 20 components were characterized and 14 were identified and quantified, with molecular weights ranging from 284 to 424. It was suggested that particle beam LC-MS can be a valuable tool not only for identification of PAH but also possibly for regional pattern recognition of sample locations. However, it is essential that the PAH standards selected for calibration and sample preparation schemes should be for PAH that are environmentally relevant.[345]

4.3.4.7 Composted Municipal Sludge and Compost-Amended Soil

The benefits of applying recycled municipal wastes to increase soil fertility has been tempered by evidence that carcinogenic polycyclic aromatic compounds (PAC) from the composts and sludges may enter the food chain. In one study, the composition of three types of samples were compared: compost derived from municipal refuse and two different anaerobic sewage soil samples taken over a 3 yr period. After lyophilization, PAC were extracted with methylene chloride, fractionated with neutral alumina Sep-Pak cartridges, and the constituents of the eluted fractions separated on a fused-silica capillary column. Components were identified by EIMS by comparison with standards. Recovery using a 7-component standard mixture was 78 to 100%. About 50 PAC were identified including the carcinogenic mono-, di-, and trimethyl derivatives of naphthalene, phenanthrane, fluorene, dibenthiophene, and naphthothiophene. A wide range of concentrations was detected; however, there were no indications of PAC accumulation over time.[346]

4.3.5 *Helicobacter Pylori*

The stomachs of about half of the people in the world are colonized (within the gastric mucus) by *H. pylori,* a gram-negative organism. While most colonized individuals are asymptomatic, a subpopulation of 10 to 20% develops peptic ulcers that may in turn evolve into adenocarcinoma. It has been estimated that 30 to 90% of gastric cancers are tied to this microorganism.[347] Because *H. pylori* has been assigned as a Class 1 carcinogen,[348] there are significant clinical and economic aspects of screening for, and diagnosis of, this infection.[349]

The standard method of diagnosis is the ^{13}C-urea breath test which exploits the observation that while humans have no endogenous urease activity in the stomach, *Helicobacter* species have high urease activity (a number of other bacteria, such as *Staphylococcus epidermedis* and *Mycobacterium scrofulaceum* make urease, but these species do not inhabit the stomach). When orally administered, ^{13}C-enriched urea is hydrolyzed in the stomach to $^{13}CO_2$ and NH_3 by *H. pylori* and the $^{13}CO_2/^{12}CO_2$ ratio in the expired air is diagnostic. Results obtained this way were in complete agreement with those obtained by histological examination.[350] Being a stable, nonradioactive isotope, ^{13}C can be administered safely to children and pregnant women. The excess ^{13}C in exhaled breath can be determined accurately with dual-inlet gas isotope-ratio mass spectrometers using instruments of various degrees of sophistication that have been available for >20 years. The precision of the $^{13}CO_2/^{12}CO_2$ ratio influences the diagnostic certainty of the method, e.g., a precision of 0.3% corresponds to 3 parts excess ^{13}C per 10^6 parts CO_2.[351] While minor procedural modifications have been described,[352,353] the isotope-ratio mass spectrometric method, validated by the histological comparison, has remained a reference standard for the evaluation of alternative methodologies such as a ^{13}C-urea blood test[354,355] and nondispersive isotope-selective infrared spectroscopy.[356] Benchtop GC/MS instruments (SIM mode) have also been evaluated for $^{13}CO_2/^{12}CO_2$ measurements, and both sensitivity and specificity were in the 96 to 98% range using histology as the "gold standard."[357]

Mass spectrometric techniques have also been applied to the search for specific biomarkers of the bacterium. In a methodology-oriented study, lysates and extracts from six different *H. pylori* strains were analyzed by MALDI-TOFMS with the aim of investigating the effects of different solubilizing detergents and MALDI matrices. It was concluded that the strain-specific biomarkers identified might be used in a fingerprinting technique for strain typing.[358] In another MALDI-TOFMS investigation, a potential biomarker of 58,268 Da was adequate to distinguish the *H. pylori* from *H. mustalae* and *Campylobacter* species. However, when applied to intact *H. pylori* cells, the technique was unable to distinguish between three stains of the species. Additional, nonspecific ions common to the three strains studied were located in the 61.4 to 61.5 and 44.1 to 44.3 kDa ranges. It was concluded that the technique is adequate for the rapid detection of these organisms in foods, beverages, or manufactured products.[359] MALDI-TOFMS was also used to search for unique proteins in proteolytic digests of *H. pylori* lysates from blood samples of infected patients. Some 20 candidate proteins were identified with the aid of predictive information from the *H. pylori* genome. It was concluded that this approach has potential for vaccine development.[360]

It has been suggested that *H. pylori* binds to a series of specific leukocyte-related gangliosides. Using FABMS and GC/MS to characterize the ceramide portions isolated from a complex series of gangliosides, it was established that the bacterium only recognizes sialic acids linked through the 3 position. The fact that the *in vitro* binding of *H. pylori* to gastric cells occurs almost exclusively in a sialic acid-dependent manner may have clinical significance.[361]

4.3.6 Elements

There have been a number of investigations attempting to associate the concentrations of elements in body fluids and tissues with various cancers, aimed at developing diagnostic applications, as well as exploring the use of elements in the diet as chemopreventive agents. For example, Br, Rb, Br/Rb ratio, K, Fe, and Se concentrations have been found at significantly lower than normal levels in the plasma of patients with colorectal cancer.[362] The concentrations of Zn, Cu, Fe, Mn, Mg, and Sr varied significantly in the blood of patients with thyroid cancer.[363] Elevated concentrations of Cd have been found in tissues from breast cancer patients[364] and variations in tissue concentrations of K, Se, and Fe have been associated with the progression of gastric cancers.[365] However, in many other studies, no significant correlations were found between the concentrations of elements in body fluids or tissues and cancer, e.g., a prospective study failed to establish a correlation between toenail trace element levels and breast cancer.[366]

The adverse effects of an element depend not only on the quantity present but also on the chemical form in which it occurs. Elemental *speciation* refers to the identification of the chemical form in which the element exists in the sample. Speciation includes the oxidation state of an inorganic form, e.g., the pentavalent and trivalent forms of arsenic (arsenate and arsenite, respectively). Speciation also refers to the determination of the type and number of substituents of organometallic compounds, e.g., methylarsonic acid and dimethylarsinic acid.

The method of choice for elemental analysis has been inductively coupled plasma-mass spectrometry (ICP/MS, Section 2.1.8), commonly using quadrupole analyzers. Advantages of ICP/MS include nearly 100% ionization efficiency for most elements, dynamic range of five orders of magnitude, relatively simple spectra, and very low background levels. Response is linear, covering the fg/mL to ng/mL range. Assay precision is good when isotope dilution is employed. Limitations include mass overlaps and interferences from chlorine (see later), molecular ions of acids used in sample preparation and signal suppression due to large quantities of alkali salts, e.g., in sea water. Sample preparation and instrument operation techniques have been developed that reduce or eliminate some of these limitiations.[367] For speciation, an efficient separation procedure, such as ion-exchange, ion-pairing, or reversed-phase HPLC is commonly interfaced with ICP/MS. The unique nature of ICP ionization poses special requirements for interfacing chromatographs.[368]

4.3.6.1 Arsenic

Arsenic compounds have been known for millenia. The normal metabolic role of arsenic in humans, if any, is unknown. In mammals, inorganic arsenic exists as both arsenate and arsenite, and both forms undergo enzymatic mono- and dimethylation. Several arsenic compounds are toxic and can cause acute and chronic poisoning. The criminal use of arsenic trioxide (no taste in food and dissolves in drinks) has equated “arsenic” with “poison.” There are many inorganic and organic arsenic compounds. Trivalent arsenic is considered to be the most toxic and carcinogenic valence state. A proven association of arsenic-induced malignant transformation with DNA hypomethylation and aberrant gene expression supports the hypothesis that arsenic carcinogenicity has an epigenetic mechanism[369] and probably involves the modulation of signaling pathways that control cell growth.[370]

Environmental exposure to arsenic, particularly in drinking water, is associated with increased risk of skin, urinary, bladder, and respiratory tract cancers. For example, water that contains >50 μg/L arsenic presents an unacceptable risk of 1 in 1000 for development of bladder cancer. Very high levels of arsenic in drinking water have recently raised an alarm regarding serious future health problems for India and Bangladesh. In an ongoing epidemiological study in New Hampshire, in the U.S., ICP/MS measurements revealed arsenic concentrations >10 μg/L in >10% of private wells and >50 μg/L in 2.5% of the wells.[371]

A limitation of the analysis of arsenic by ICP/MS is an interference of $[^{40}Ar^{35}Cl]^+$ at 74.930 Da with the monoisotopic ^{75}As isotope at 74.921 Da. A resolution of >8000 is required for baseline separation. This interference can be eliminated by coupling the ICP source with a high-resolution magnetic sector analyzer operated at m/Δm ~8,000; however, sensitivity is reduced significantly. In contrast, hydride generation (HG) ICP/MS, using tetrahydrodiborate as the reducing reagent, provided accurate results for arsenic concentrations in the 0.01 to 1 μg/L range in drinking water.[372] An HG-magnetic sector combination instrument has been used to correlate the arsenic content of tap water in the households of normal subjects with the arsenic content of their toenails for the quantification of low-level arsenic exposure.[373]

After >1500 papers have been published on various aspects of the implications and behavior of "total" arsenic in the environment, the current availability of coupled LC/ICP techniques now permits the elucidation of the role of specific arsenic compounds in the biological environment. Several microwave-induced plasma (MIP) sources (combined with quadrupole analyzers) have been developed that eliminate the $[^{40}Ar^{35}Cl]^+$ interference while retaining the ability to speciate with on-line HPLC. A sophisticated MIP system has been described that allows for the speciation of monomethylarsonate (MMA), dimethylarsinate (DMA), arsenobetaine (AsB), and several other organic arsenic compounds in urine. The LOD were tenfold lower than obtained with conventional LC/ICPMS.[374] Speciation of arsenic oxides may also be accomplished using single-step laser desorption ionization, combined with a TOF analyzer that yields high-mass cluster ions (in the negative mode) that are unique to the oxidation state of each oxide. Certain advantages of the technique have been demonstrated by analyzing an air filter from a model incinerator stack.[375] Organic arsenic compounds may also be analyzed by positive ESI-MS/MS while inorganic arsenic species can be identified using negative ion ESIMS.[376]

The LC/ICP analysis of the arsenic species excreted in urine revealed that maximum arsenical excretion occurred five days after an acute arsenic exposure. Excreted species included MMA, DMA, AsB, and trivalent inorganic arsenic. The data confirmed the existence of two successive enzymatically mediated methylation steps.[377] The same technique was employed to study the composition of air in six art glass factories that use arsenic as a fining agent. Arsenic concentration was often significantly above the mandated threshold limit of 10 μg/m^3 air.[378] The concentration order of excreted arsenicals ranged from AsB (34%) through DMA and MMA to total inorganic arsenicals (12%). A good correlation was found between arsenic in the air and total excreted inorganic trivalent arsenic. It was concluded that the concentration of trivalent arsenic in urine was the best indicator of potential carcinogenic exposure.[379]

4.3.6.2 Selenium

As a constituent of selenoproteins, selenium has several vital structural and enzymatic roles. Increased intake of selenium-enriched food has been shown to yield direct, inverse, or null associations with cancer risk.[380,381] The most important selenium-containing organic compounds are selenomethionine (present in plants) and selenocysteine (present in animal proteins). A coupled CE/HG-ICPMS technique was developed for the speciation of two selenium (and four arsenic) species. Selenate (Se-VI) was reduced to selenite (Se-IV) by mixing the CE effluent with concentrated HCl. Injection LOD for the two species studied in drinking water were 24 and 10 pg, respectively.[382]

Parameters of sample preparation and analytical methodology using ion-pair HPLC/ICPMS were investigated, and optimal separations of more than 20 selenium compounds obtained. The LOD were in the 20 to 80 ng/mL sample range.[383] The predominant selenium species in both garlic (296 μg/g Se) and yeast (1922 μg/g Se) were γ-glutamyl-methylselenocysteine and selenomethionine. In rats, selenium from garlic was significantly more effective than selenium from yeast in suppressing the development of premalignant lesions and the formation of breast adenocarcinomas.[384]

4.3.6.3 Chromium

It has been suggested that inhaled chromium may promote, directly or indirectly, the generation of reactive oxygen species that can activate or inactivate transcription factors leading to epigenetic carcinogenesis (reviewed in Reference 385). However, chromium may also act as a direct carcinogen through the formation of DNA adducts. Assuming that the human small airway epithelial (HSAE) cells are the target for chromium(IV)-induced carcinogenesis in the lung, the dose-dependent formation of Cr-DNA adducts was investigated in lead chromate-treated HSAE cells using ICPMS.[386] The background levels of Cr-DNA and Pb-DNA were 17.7 and 1.1 adducts/10^4 bp, respectively. Treatment of HSAE cells with lead chromate yielded a dose-dependent increase in both Cr-DNA and Pb-DNA, with significant increases of both adducts. There appeared to be two competing processes of DNA adduct formation, both of which could induce carcinogenesis.[387]

4.4 DNA DAMAGE BY RADIATION AND OXIDATION

It has been demonstrated repeatedly that ionizing radiation damages the DNA in cells, often leading to both chromosomal aberrations and mutations. The direct effects of ionizing radiation include breaks to single or both DNA strands, base modifications, and damage to the carbohydrate components associated with the DNA. The most important indirect effect of radiation is the production of free radicals from cellular water, which react with the DNA (reviewed in Reference 388). Tumors may develop when the cellular DNA repair mechanisms fail to detect and correct these changes or are themselves compromised.

Concomitant damage to both the base and sugar moieties of the same nucleotide subunit has been studied using sample preparation techniques that release the sugars and use acidic or enzymatic hydrolyses to free the bases. The sugars have been identified and quantified by GC/MS after reduction with $NaBD_4$ to the corresponding polyalcohols, followed by silylation. The methodology has also been used to detect the 8,5′-cyclopurine nucleosides that are diagnostic of radiation damage. Products of damaged DNA could be quantified even at radiation doses as low 0.1 Gy (Gray, unit of absorbed dose of ionizing radiation, 1 Gy = 100 rad).[389]

Mass spectrometric methodologies have been developed for the study of oxidative DNA damage produced by free radicals, particularly by the highly reactive hydroxyl radical (·OH), as well as for the investigation of enzymatic repair of the DNA (reviewed in Reference 390). Enzymatic hydrolysis, with deoxyribonuclease and/or exonucleases, produced nucleosides from DNA. This was followed by acidic hydrolysis, often with formic acid, to separate the intact and modified bases by cleavage of the glycosidic bonds between the base and sugar moieties. The hydrolysis products were silylated and analyzed by GC/MS. Full scanning EI provided adequate structural details for compound identification, while SIM was used for quantification. The four intact DNA bases, ~20 modified DNA bases, ~12 modified DNA 2′-deoxynucleosides, and DNA base-amino acid cross-linked products could be identified and quantified in a single analysis. Analogs labeled with ^{13}C and/or deuterium were used as i.s. in studies to assess the presence and position of carbonyl and deoxy groups in modified/damaged sugars. Using conventional GC/MS, LOD were of the order of 1 fmol/analyte or 1 to 3 modified residues in 10^6 bases.[391] Another isotope dilution GC/MS technique has been used to identify and quantify nine modified DNA bases in children with acute lymphoblastic leukemia and in controls. More than threefold differences were found between the patients and controls in the cases of 5-hydroxy-5-methylhydantoin and 2,6-diamino-4-hydroxy-5-formamidoguanine.[392]

Base modifications arising from ·OH radical-induced hydroxylation and cleavage reactions have been studied in DNA that was extracted from ductal carcinoma tissues excised from patients with invasive breast cancer. After hydrolyzing the DNA and forming TMS derivatives, GC/EIMS was used to quantify the modified bases. The concentrations of 8-hydroxyguanine, 2,6-diamino-4-hydroxy-5-formamidopyrimidine, and 8-hydroxyadenine were up to ninefold higher than those of DNA isolated from calf-thymus. There were no differences in the profiles of the unmodified bases from the two tissues. It has been suggested that the ·OH attack on DNA may be associated with escalations in H_2O_2 concentrations that bring about "automutagenic" progressive alterations in DNA.[393] Care should be taken in the interpretation of results from GC/MS techniques because these methods often produce derivatization-related artifacts that may be present in considerable quantities (reviewed in Reference 394).

Another application of the above methodologies has been the investigation of genomic base damage in the lymphocytes of lung cancer patients undergoing radiation therapy. Because the concentrations of 8-OH-guanine and some other modified bases have been shown to differ significantly among individuals,[395] this research was based on lymphocyte chromatin samples from cancer patients using their own preradiation samples as controls. Observed DNA modifications that indicated damage by ·OH radicals included several purine- and pyrimidine-derived modified bases. The quantitative differences were credited to the intensity of applied radiation (dictated by the nature of the neoplasm), individual radiation sensitivity, and the capability of the cells from different individuals to repair DNA damage. These modifications in the DNA, caused by the radiation, were in qualitative agreement with those resulting from the treatment of experimental animals with carcinogens.[396] It was concluded that DNA damage by therapeutic radiation may contribute significantly to the development of secondary malignancies, such as leukemias.[397]

REFERENCES

1. Hopkins, J., The role of cancer mechanism in IARC carcinogen classification, *Food. Chem. Toxicol.,* 32, 193–198, 1994.
2. Sanner, T., Dybing, E., Kroese, D., Roelfzema, H., and Hardeng, S., Carcinogen classification systems: similarities and differences, *Regul. Toxicol. Pharmacol.*, 23, 128–138, 1996.
3. Sanner, T., Dybing, E., Kroese, D., Roelfzema, H., and Hardeng, S., Potency grading in carcinogen classification, *Mol. Carcinog.*, 20, 280–287, 1997.
4. Neumann, H., Thielmann, H.W., Filser, J.G., Gelbke, H.-P., Greim, H., Kappus, H., Norpoth, K.H., Reuter, U., Vanvakas, S., Wardenbach, P., and Wichmann, H.-E., Proposed changes in the classification of carcinogenic chemicals in the work area, *Regulatory. Toxicol. Pharmacol.*, 26, 288–295, 1997.
5. Cohen, S. and Ellwein, L., Cell proliferation in carcinogenesis, *Science*, 249, 1007–1011, 1990.
6. Ames, B.N., McCann, J., and Yamasaki, E., Methods for detecting carcinogens and mutagens with the Salmonella/mammalian microsome mutagenicity test, *Mutation. Res.*, 31, 347–363, 1995.
7. Williams, G.M., Mori, H., and McQueen, C.A., Structure-activity relationships in the rat hepatocyte DNA—repair test for 300 chemicals, *Mutation. Res.*, 221, 263–286, 1989.
8. Dipple, A., DNA adducts of chemical carcinogens, *Carcinogenesis,* 16, 437–441, 1995.
9. Surgeon General, Reducing the Health Consequences of Smoking. 25 Years of Progress, U.S. Dept Health & Human Services, 1989.
10. Peto, R., Smoking and death in the past 40 years and the next 40, *Brit. Med. J.,* 309, 937–939, 1994.

11. Gritz, E.R., Fiore, M.C., and Henningfield, J.E., Smoking and cancer, in *Clinical Oncology,* American Cancer Society, New York, 1998.
12. Engstrom, P.F., Rosvold, E.A., Boyd, N.R., and Orleans, C.T., Prevention of tobacco-related cancers, in *Cancer Medicine,* Williams & Wilkins, Baltimore, 447–463, 1997.
13. Hoffman, D. and Hecht, S.S., Advances in tobacco carcinogenesis, in *Chemical Carcinogenesis and Mutagenesis,* Springer-Verlag, Berlin, 63–102, 1990.
14. Ames, B.N. and Gold, L.S., The causes and prevention of cancer: the role of environment, *Biotherapy,* 11, 205–220, 1998.
15. Sram, R.J. and Binkova, B., Molecular epidemiology studies on occupational and environmental exposure to mutagens and carcinogens, 1997–1999, *Environ. Health Perspect.,* 108, 57–70, 2000.
16. Perera, F.P. and Weinstein, I.B., Molecular epidemiology: recent advances and future directions, *Carcinogenesis,* 21, 517–524, 2000.
17. Wogan, G.N., Molecular epidemiology in cancer risk assessment and prevention: recent progress and avenues for future research, *Environ. Health Perspect.,* 98, 167–178, 1992.
18. Perera, F.P., Molecular epidemiology: insights into cancer susceptibility, risk assessment, and prevention, *J. Natl. Cancer. Inst.,* 88, 496–509, 1996.
19. Perera, F.P. and Dickey, C., Molecular epidemiology and occupational health, *Ann. N.Y. Acad. Sci.,* 837, 353–359, 1997.
20. Perera, F.P., Environment and cancer: who are susceptible? *Science,* 278, 1068–1073, 1997.
21. Poirier, M.C., Santella, R.M., and Weston, A., Carcinogen macromolecular adducts and their measurement, *Carcinogenesis,* 21, 353–359, 2000.
22. Burcham, P.C., Genotoxic lipid peroxidation products: their DNA damaging properties and role in formation of endogenous adducts, *Mutagenesis,* 13, 287–305, 1998.
23. Guengerich, F.P., Metabolism of chemical carcinogens, *Carcinogenesis,* 21, 345–351, 2000.
24. Raunio, H., Husgafvel-Pursiainen, K., Anttila, S., Hieyanen, E., and Hirvonen, A., Diagnosis of polymorphisms in carcinogen-activating and inactivating enzymes and cancer susceptibility—a review, *Gene,* 159, 113–121, 1995.
25. Selzer, R.R. and Elfarra, A.A., Characterization of N^1-and N^6-adenosine adducts and N^1-inosine adducts formed by the reaction of butadiene monoxide with adenosine: evidence for the N^1-adenosine adducts as major initial products, *Chem. Res. Toxicol.,* 9, 875–881, 1996.
26. La, D.K. and Swenberg, J.A., DNA adducts: biological markers of exposure and potential applications to risk assessment, *Mutat. Res.,* 365, 129–146, 1996.
27. Cushnir, J., Naylor, S., Lamb, J.A., and Farmer, P., Deuterium exchange studies in the identification of alkylated DNA bases found in urine, by tandem mass spectrometry, *Rapid Commun. Mass Spectrom.,* 4, 426–431, 1990.
28. Freeman, J., Johnson, J., Hail, M., Yost, R., and Kuehl, D., Estimation of mutagenic/carcinogenic potential of environmental contaminants by ion-molecule reactions and tandem mass spectrometry, *J. Am. Soc. Mass Spectrom.,* 1, 110–115, 1990.
29. Strickland, P.T., Routledge, M.N., and Dipple, A., Methodologies for measuring carcinogen adducts in humans, *Cancer Epidemiol. Biomark. Prev.,* 2, 607–619, 1993.
30. Kristensen, D.B., Imamura, K., Miyamoto, Y., and Yoshizato, K., Mass spectrometric approaches for the characterization of proteins on a hybrid quadrupole time-of-flight (Q-TOF) mass spectrometer, *Electrophoresis,* 21, 430–439, 2000.
31. Garner, R.C., The role of DNA adducts in chemical carcinogensis, *Mutat. Res.,* 402, 67–75, 1998.
32. Farmer, P.B., Sepai, O., Lawrence, R., Autrup, H., Nielsen, P.S., Vestegard, A.B., Waters, R., Leurati, C., Jones, N.J., Stone, J., Baan, R.A., van Delft, J.H., Steenwinkel, M.J., Kyrtopoulos, S.A., and Souliotis, V.L., Biomonitoring human exposure to environmental carcinogenic chemicals, *Mutagenesis,* 11, 363–381, 1996.
33. Keith, G. and Dirheimer, G., Postlabeling: a sensitive method for studying DNA adducts and their role in carcinogenesis, *Current Opin. Biotechnol.,* 6, 3–11, 1995.
34. Beland, F.A., Doerge, D.R., Churchwell, M.I., Poirier, M.C., Schoket, B., and Marques, M.M., Synthesis, characterization, and quantitation of a 4-aminobiphenyl-DNA adduct standard, *Chem. Res. Toxicol.,* 12, 68–77, 1999.
35. Santella, R.M., Immunological methods for detection of carcinogen-DNA damage in humans, *Cancer, Epidemiol. Biomark. Prev.,* 8, 733–739, 1999.

36. Shuker, D.E. and Bartsch, H., Detection of human exposure to carcinogens by measurement of alkyl-DNA adducts using immunoaffinity clean-up in combination with gas chromatography-mass spectrometry and other methods of quantitation, *Mutat. Res.*, 313, 263–268, 1994.
37. Leclercq, L., Laurent, C., and DePauw, E., High performance liquid chromatography/electrospray mass spectrometry for the analysis of modified bases in DNA:7-(2-hydroxylethyl)guanine, the major ethylene oxide-DNA adduct, *Anal. Chem.*, 69, 1952–1955, 1997.
38. Giese, R.W., Saha, M., Abdel-Baky, S., and Allam, K., Measuring DNA adducts by gas chromatography-electron capture-mass spectrometry: trace organic analysis, *Methods Enzymol.*, 271, 504–522, 1996.
39. Friesen, M.D., Kaderlik, K., Lin, D., Garren, L., Bartsch, H., Lang, N.P., and Kadlubar, F.F., Analysis of DNA adducts of 2-amino-1-methyl-6-phenylimidazol[4,5-b]pyridine in rat and human tissues by alkaline hydrolysis and gas chromatography-electron capture mass spectrometry: validation by comparison with ^{32}P-postlabeling, *Chem. Res. Toxicol.*, 7, 733–739, 1994.
40. Andrews, C.L., Vouros, P., and Harsch, A., Analysis of DNA adducts using high-performance separation techniques coupled to electrospray ionization mass spectrometry, *J. Chromatogr. A*, 856, 515–526, 1999.
41. Apruzzese, W.A. and Vouros, P., Analysis of DNA adducts by capillary methods coupled to mass spectrometry: a perspective, *J. Chromatogr. A*, 794, 97–108, 1998.
42. Wolf, S.M. and Vouros, P., Incorporation of sample stacking techniques into the capillary electrophoresis CF-FAB mass spectrometric analysis of DNA adducts, *Anal. Chem.*, 67, 891–900, 1995.
43. Harriman, S.P., Hill, J.A., and Tannenbaum, S.R., Detection and identification of carcinogen-peptide adducts by nanoelectrospray tandem mass spectrometry, *J. Am. Soc. Mass Spectrom.*, 9, 202–207, 1998.
44. Meyer, M.J. and Bechtold, W.E., Protein adduct biomarkers: state of the art, *Environ. Health Perspect.*, 104(5), 879–882, 1996.
45. Vineis, P., Caporaso, N., Tannenbaum, S.R., Skipper, P.L., Glogowski, J., Bartsch, H., Coda, M., Talaska, G., and Kadlubar, F., Acetylation phenotype, carcinogen-hemoglobin adducts, and cigarette smoking, *Cancer Res.*, 50, 3002–3004, 1990.
46. Farmer, P., Carcinogen adducts: use in diagnosis and risk assessment, *Clin. Chem.*, 40, 1438–1443, 1994.
47. Farmer, P.B. and Shuker, D.E.G., What is the significance of increases in background levels of carcinogen-derived protein and DNA adducts? Some considerations for incremental risk assessment, *Mutat. Res. Fundamen. Mol. Mech. Mutagenesis*, 424, 275–286, 1999.
48. Farmer, P.B., Studies using specific biomarkers for human exposure assessment to exogenous and endogenous chemical agents, *Mutat. Res.*, 428, 69–81, 1999.
49. Filser, J.G., Denk, B., Törnqvist, M.A., Kessler, W., and Ehrenberg, L., Pharmacokinetics of ethylene in man; body burden with ethylene oxide and hydroxyethylation of hemoglobin due to endogenous and environmental ethylene, *Arch. Toxicol.*, 66, 157–163, 1992.
50. Carbone, P.P., Douglas, J.A., Larson, P.O., Verma, A.K., Blair, I.A., Pomplun, M., and TutschK, D., Phase 1 chemoprevention study of piroxicam and α-difluoromethylornithine, *Cancer Epidemiol. Biomark. Prev.*, 7, 907–912, 1998.
51. Wishnok, J.S., Quantitative mass spectrometry of hemoglobin adducts, *Methods Enzymol.*, 231, 632–643, 1994.
52. Törnqvist, M.A., Mowrer, J., Jensen, S., and Ehrenberg, L., Monitoring of environmental cancer initiators through hemoglobin adducts by a modified Edman degradation method, *Anal. Biochem.*, 154, 255–266, 1986.
53. Törnqvist, M.A., Kautiainen, A., Gatz, R., and Ehrenberg, L., Hemoglobin adducts in animals exposed to gasoline and diesel exhausts. 1. Alkenes, *J. Appl. Toxicol.*, 8, 159–170, 1988.
54. Conduah-Birt, J.E., Shuker, D.E., and Farmer, P.B., Stable acetaldehyde-protein adducts as biomarkers of alcohol exposure, *Chem. Res. Toxicol.*, 11, 136–142, 1998.
55. Farmer, P., Analytical approaches for the determination of protein-carcinogen adducts using mass spectrometry, in *Molecular Dosimetry and Human Cancer: Analytical, Epidemiological, and Social Considerations*, CRC Press, Boca Raton, FL, 189–210, 1991.
56. Garner, R.C., Farmer, P.B., Steel, G.T., and Wright, A.S., The N-alkyl Edman method for hemoglobin adduct measurement: updating and applications to humans, in *Human Carcinogen Exposure*, IRL Press, Oxford, 411–419, 1991.

57. Tavares, R., Borba, H., Monteiro, M., Proenca, M.J., Lynce, N., Rueff, J., Bailey, E., Sweetman, G.M., Lawrence, R.M., and Farmer, P.B., Monitoring of exposure to acrylonitrile by determination of N-(2-cyanoethyl)valine at the N-terminal position of haemoglobin, *Carcinogenesis,* 17, 2655–2660, 1996.
58. Landin, H.H., Osterman-Golkar, S.M., Zorcec, V., and Törnqvist, M.A., Biomonitoring of epichlorohydrin by hemoglobin adducts, *Anal. Biochem.*, 240, 1–6, 1996.
59. Farmer, P.B., Cordero, R., and Autrup, H., Monitoring human exposure to 2-hydroxyethylating carcinogens, *Environ. Health Perspect.,* 104, 449–452, 1996.
60. Rydberg, P., Magnusson, A.L., Zorcec, V., Granath, M., and Törnqvist, M.A., Adducts to N-terminal valines in hemoglobin from butadiene metabolites, *Chem. Biol. Interact.,* 101, 193–205, 1996.
61. Tretyakova, N.Y., Lin, Y., Upton, P.B., Sangaiah, R., and Swenberg, J.A., Macromolecular adducts of butadiene, *Toxicology,* 113, 70–76, 1996.
62. Severi, M., Pauwels, W., Van Hummelen, P., Roosels, D., Kirsch Volders, M., and Veulemans, H., Urinary mandelic acid and hemoglobin adducts in fiberglass-reinforced plastics workers exposed to styrene, *Scand. J. Work. Environ. Health,* 20, 451–458, 1994.
63. Stellman, J.M. and Stellman, S.D., Cancer and the workplace, *Cancer J. Clin.*, 46, 70–92, 1996.
64. Duarte-Davidson, R., Courage, C., Rushton, L., and Levy, L., Benzene in the environment: an assessment of the potential risks to the health of the population, *Occup. Environ. Med.*, 58, 2–13, 2001.
65. Snyder, R. and Hedli, C.C., An overview of benzene metabolism, *Environ. Health Perspect.,* 104, Suppl. 6, 1165–1171, 1996.
66. Smith, M.T. Mechanistic Studies of Benzene Toxicity—Implications for Risk Assessment, in *Biological Reactive Intermediates,* Plenum Press, New York, 259–296, 1996.
67. Schrenk, D., Orzechowski, A., Schwartz, L.R., Snyder, R., Burchell, B., Ingelman-Sundberg, M., and Bock, K.W., Phase II metabolism of benzene, *Environ. Health Perspect.,* 104, Suppl. 6, 1183–1188, 1996.
68. Medeiros, A.M. and Bird, M.G., Potential biomarkers of benzene exposure, *J. Toxicol. Environ. Health,* 51, 519–539, 1997.
69. Ong, C. and Lee, B., Determination of benzene and its metabolites: application in biological monitoring of environmental and occupational exposure to benzene, *J. Chromatogr. B,* 660, 1–22, 1994.
70. Creek, M.R., Mani, C., Vogel, J.S., and Turteltaub, K.W., Tissue distribution and macromolecular binding of extremely low doses of [^{14}C]-benzene in BCF mice, *Carcinogenesis,* 18, 2421–2427, 1997.
71. Bechtold, W.E., Lucier, G., Birnbaum, L.S., Yin, S.N., Li, G.L., and Henderson, R.F., Muconic acid determinations in urine as a biological exposure index for workers occupationally exposed to benzene, *Am. Ind. Hyg. Assoc. J.,* 52, 473–478, 1991.
72. Scherer, G., Renner, T., and Meger, M., Analysis and evaluation of trans,trans-muconic acid as biomarker for benzene exposure, *J. Chromatogr. B,* 717, 179–199, 1998.
73. Bartczak, A., Kline, S., Yu, R., Weisel, C., Goldstein, B., and Witz, G., Evaluation of assays for the identification and quantitation of muconic acid, a benzene metabolite in human urine, *J. Toxicol. Environ. Health,* 42, 245–258, 1994.
74. Rothman, N., Bechtold, W.E., Yin, S.N., Dosemeci, M., Li, G.L., Wang, Y.Z., Griffith, W.C., Smith, M.T., and Hayes, R.B., Urinary excretion of phenol, catechol, hydroquinone, and muconic acid by workers occupationally exposed to benzene, *Occup. Environ. Med.*, 55, 705–711, 1998.
75. Qu, Q., Melikian, A.A., Li, G., Shore, R., Chen, L., Cohen, B., Yin, S., Kagan, M.R., Li, H., Meng, M., Jin, X., Winnik, W., Li, Y., Mu, R., and Li, K., Validation of biomarkers in humans exposed to benzene: urine metabolites, *Am. J. Industr. Med.,* 37, 522–531, 2000.
76. Lovern, M.R., Turner, M.J., Meyer, M., Kedderis, G.L., Bechtold, W.E., and Schlosser, P.M., Identification of benzene oxide as a product of benzene metabolism by mouse, rat, and human microsomes, *Carcinogenesis,* 18, 1695–1700, 1997.
77. Lindstrom, A.B., Yeowell, O.-K., Waidyanatha, S., Golding, B.T., Tornero Velez, R., and Rappaport, S.M., Measurement of benzene oxide in the blood of rats following administration of benzene, *Carcinogenesis,* 18, 1637–1641, 1997.
78. Bechtold, W.E., Willis, J.K., Sun, J.D., Griffith, W.C., and Reddy, T.V., Biological markers of exposure to benzene: *S*-phenylcysteine in albumin. *Carcinogenesis*, 13, 1217–1220, 1992.
79. McDonald, T.A., Yeowell-O'Connell, K., and Rappaport, S.M., Comparison of protein adducts of benzene oxide and benzoquinone in the blood and bone marrow of rats and mice exposed to [^{14}C/^{13}C$_6$]benzene, *Cancer Res.,* 54, 4907–4914, 1994.

80. Melikian, A.A., Prahalad, A.K., and Coleman, S., Isolation and characterization of two benzene-derived hemoglobin adducts in vivo in rats, *Cancer Epidemiol. Biomark. Prev.,* 1, 307–313, 1992.
81. Yeowell-O'Connell, K., McDonald, T.A., and Rappaport, S.M., Analysis of hemoglobin adducts of benzene oxide by gas chromatography-mass spectrometry, *Anal. Biochem.*, 237, 49–55, 1996.
82. Yeowell-O'Connell, K., Rothman, N., Smith, M.T., Hayes, R.B., Li, G., Waidyanatha, S., Dosemeci, M., Zhang, L., Yin, S., Titenko-Holland, N., and Rappaport, S.M., Hemoglobin and albumin adducts of benzene oxide among workers exposed to high levels of benzene, *Carcinogenesis,* 19, 1565–1571, 1998.
83. McDonald, T.A., Waidyanatha, S., and Rappaport, S.M., Production of benzoquinone adducts with hemoglobin and bone-marrow proteins following administration of $[^{13}C_6]$benzene, *Carcinogenesis,* 14, 1921–1925, 1993.
84. Rappaport, S.M., McDonald, T.A., and O'Connell, K.Y., The use of protein adducts to investigate the disposition of reactive metabolites of benzene, *Environ. Health Perspect.,* 104, Suppl. 6, 1235–1237, 1996.
85. McLuckey, S.A., Van Berkel, G.J., and Glish, G.L., Tandem mass spectrometry of small, multiply charged oligonucleotides, *J. Am. Soc. Mass Spectrom.,* 3, 60–70, 1992.
86. Glover, R.P., Lamb, J.H., and Farmer, P.B., Tandem mass spectrometry studies of a carcinogen modified oligodeoxynucleotide, *Rapid Commun. Mass Spectrom.,* 12, 368–372, 1998.
87. Grimmer, G., Jacob, J., Dettbarn, G., and Naujack, K.W., Determination of urinary metabolites of polycyclic aromatic hydrocarbons (PAH) for the risk assessment of PAH-exposed workers, *Int. Arch. Occup. Environ. Health,* 69, 231–239, 1997.
88. Bieniek, G., Aromatic and polycyclic hydrocarbons in air and their urinary metabolites in coke plant workers, *Am. J. Ind. Med.,* 34, 445–454, 1998.
89. Bieniek, G., Urinary excretion of phenols as an indicator of occupational exposure in the coke-plant industry, *Int. Arch. Occup. Environ. Health,* 70, 334–340, 1997.
90. Ovrebo, S., Haugen, A., Farmer, P.B., and Anderson, D., Evaluation of biomarkers in plasma, blood and urine samples from coke oven workers: significance of exposure to polycyclic aroatic hydrocarbons, *Occup. Environ. Med.,* 52, 750–756, 1995.
91. Pan, G., Hanaoka, T., Yamano, Y., Hara, K., Ichiba, M., Wang, Y., Zhang, J., Feng, Y., Shujuan, Z., Guan, D., Gao, G., Liu, N., and Takahashi, K., A study of multiple biomarkers in coke oven workers — a cross-sectional study in China, *Carcinogenesis,* 19, 1963–1968, 1998.
92. Topinka, J., Schwarz, L.R., Kiefer, F., Wiebel, F.J., Gajdos, O., Vidova, P., Dobias, L., Fried, M., Sram, R.J., and Wolff, T., DNA adduct formation in mammalian cell cultures by polycyclic aromatic hydrocarbons (PAH) and nitro-PAH in coke oven emission extract, *Mutat. Res.,* 419, 91–105, 1998.
93. Simpson, C.D., Wu, M., Christiani, D.C., Santella, R.M., Carmella, S.G., and Hecht, S.S., Determination of r-7,t-9,0,c-10-tetrahydroxy-7,8,9,10-tetrahydrobenzo[a]pyrene in human urine by gas chromatography-negative ion chemical ionization/mass spectrometry, *Chem. Res. Toxicol.,* 13, 271–280, 2000.
94. Caprino, L. and Togna, G.I., Potential health effects of gasoline and its constituents: a review of current literature (1990–1997) on toxicological data, *Environ. Health Perspect.,* 106, 115–125, 1998.
95. Lynge, E., Andersen, A., Nilsson, R., Barlow, L., Pukkala, E., Nordlinder, R., Boffetta, P., Grandjean, P., Heikkila, P., Horte, L.G., Jakobsson, R., Lundberg, I., Moen, B., Partanen, T., and Riise, T., Risk of cancer and exposure to gasoline vapors, *Am. J. Epidemiol.,* 145, 449–458, 1997.
96. Kanal, H., Inouye, V., Goo, R., Chow, Yazawa, L., and Maka, J., GC/MS analysis of MYBE, ETBE, and TAME in gasolines, *Anal. Chem.,* 66, 924–927, 1994.
97. Dourson, M.L. and Felter, S.P., Route-to-route extrapolation of the toxic potency of MTBE, *Risk Anal.,* 17, 717–725, 1997.
98. Bailey, E., Farmer, P.B., Tang, Y.S., Vangikar, H., Gray, A., Slee, D., Ings, R.M., Campbell, D.B., McVie, J.G., and Dubbelman, R., Hydroxyethylation of hemoglobin by 1-(2-chloroethyl)-1-nitrosoureas, *Chem. Res. Toxicol.*, 4, 462–466, 1991.
99. Nielsen, P.S., Andreassen, A., Farmer, P.B., Ovrebo, S., and Autrup, H., Biomonitoring of diesel exhaust-exposed workers. DNA and hemoglobin adducts and urinary 1-hydroxypyrene as markers of exposure, *Toxicol. Lett.,* 86, 27–37, 1996.

100. van Bekkum, Y.M., Scheepers, P.T.J., van den Broek, P.H.H., Velders, D.D., Noodhoek, J., and Bos, R.P., Determination of hemoglobin adducts following oral administration of 1-nitropyrene to rats using gas chromatography-tandem mass spectrometry, *J. Chromatogr. B,* 701, 19–28, 1997.
101. Morgan, D.L., Dunnick, J.K., Goehl, T., Jokinen, M.P., Matthews, H.B., Zeiger, E., and Mennear, J.H., Summary of the national toxicology program benzidine dye initiative, *Environ. Health Perspect.,* 102, 63–67, 1994.
102. Hsu, F.F., Lakshmi, V.M., Rothman, B., Bhatnager, V., Hayes, R., Kashyap, R., Parikh, D., Kashyap, S., Turk, J., Zenser, T.V., and Davis, B, Determination of benzidine, N-acetylbenzidine, and N,N′-diacetylbenzidine in human urine by capillary gas chromatography/negative chemical ionization mass spectrometry, *Anal. Biochem.,* 234, 183–189, 1996.
103. Babu, S., Lakshmi, V.M., Hsu, F.F., Kane, R., Zenser, T.V., and Davis, B., N-acetylbenzidine-N′-glucuronidation by human, dog and rat liver, *Carcinogenesis,* 14, 2605–2611, 1993.
104. Rinde, E.A.W. Metabolic reduction of benzidine-containing azo dyes to benzidine in the rhesus monkey, *J. Natl. Cancer. Inst.,* 55, 181–182, 1975.
105. Sagelsdorff, P., Haenggi, R., Heuberger, B., Joppich-Kuhn, R., Jung, R., Weideli, H., and Joppich, M., Lack of bioavailability of dichlorobenzdine from darylid azo pigments: molecular dosimetry for hemoglobin and DNA adducts, *Carcinogenesis,* 17, 507–514, 1996.
106. Skarping, G., Dalene, M., and Lind, P., Determination of toluenediamine isomers by capillary gas chromatography and chemical ionization mass spectrometry with special reference to the biological monitoring of 2,4- and 2,6-toluene diisocyanate, *J. Chromatogr. A,* 663, 199–210, 1994.
107. Lind, P., Dalene, M., Skarping, G., and Hagmar, L., Toxicokinetics of 2,4- and 2,6-toluenediamine in hydrolysed urine and plasma after occupational exposure to 2,4- and 2,6-toluene diisocyanate, *Occup. Environ. Med.,* 53, 94–99, 1996.
108. Wilson, P., Hee, S., and Froines, J., Determination of hemoglobin adduct levels of the carcinogen 2,4-diaminotoluene using gas chromatography-electron impact positive-ion mass spectrometry, *J. Chromatogr. B,* 667, 166–172, 1995.
109. Chae, Y.-H., Delclos, K.B., Blaydes, B., and el-Bayoumy, K., Metabolism and DNA binding of the envirommental colon carcinogen 6-nitrochrysene in rats, *Cancer Res.,* 56, 2052–2058, 1996.
110. Roberts, A., Lacy, S., Pilon, D., Turner, M., and Rickert, D., Metabolism of acrylonitrile to 2-cyanoethylene oxide in F-344 rat liver microsomes, lung microsomes, and lung cells, *Drug Metab. Dispos.,* 17, 481–485, 1989.
111. Osterman-Golkar, S.M., MacNeela, J.P., Turner, M.J., Walker, V.E., Swenberg, J.A., Sumner, S.J., Youtsey, N., and Fennell, T.R., Monitoring exposure to acrylonitrile using adducts with N-terminal valine in hemoglobin, *Carcinogenesis,* 15, 2701–2707, 1994.
112. Lawrence, R.M., Sweetman, G.M., Tavares, R., and Farmer, P.B., Synthesis and characterization of peptide adducts for use in monitoring human exposure to acrylonitrile and ethylene oxide, *Teratog.Carcinog. Mutag.,* 16, 139–148, 1996.
113. Steenland, K., Stayner, L., Greife, A., Halperin, W., Hayes, R., Hornung, R., and Nowlin, S., Mortality among workers exposed to ethylene oxide, *New Eng. J. Med.,* 424, 1402–1407, 1991.
114. Dellarco, V. and Generoso, W.M., Review of the mutagenicity of ethylene oxide, *Environ. Molec. Mutagen.,* 16, 85–103, 1990.
115. Calleman, C., Ehrenberg, L., Jansson, B., Osterman-Golkar, S.M., Segerback, D., Svensson, K., and Wachtmeister, C., Monitoring and risk assessment by means of alkyl groups in hemoglobin in persons occupationally exposed to ethylene oxide, *J. Environ. Pathol. Toxicol.,* 2, 427–442, 1978.
116. Walker, V.E., MacNeela, J.P., Swenberg, J.A., Turner, M.J.J., and Fennell, T.R., Molecular dosimetry of ethylene oxide: formation and persistence of N-(2-hydroxyethyl)valine in hemoglobin following repeated exposures of rats and mice, *Cancer Res.,* 52, 4320–4327, 1992.
117. Fost, U., Mrarczynski, B., Kasemann, R., and Peter, H., Determination of 7-(2-hydroxyethyl)guanin with gas chromatography/mass spectrometry as a parameter for genotoxicity of ethylene oxide, *Arch. Toxicol.,* 13 (Suppl.), 250–253, 1989.
118. Saha, M., Abushamaa, A., and Giese, R.W., General method for determining ethylene oxide and related N7-guanine DNA adducts by gas chromatography-electron capture mass spectrometry, *J. Chromatogr. A,* 712, 345–354, 1995.

119. van Sittert, N.J., Boogaard, P.J., Natarajan, A.T., Tates, A.D., Ehrenberg, L.G., and Törnqvist, M.A., Formation of DNA adducts and induction of mutagenic effects in rats following 4 weeks inhalation exposure to ethylene oxide as a basis for cancer risk assessment, *Mutat. Res.,* 447, 27–48, 2000.
120. Boogaard, P.J., Rocchi, P.S., and Sittert, N.J., Biomonitoring of exposure to ethylene oxide and propylene oxide by determination of hemoglobin adducts: correlation between airborne exposure and adduct levels, *Int. Arch. Occup. Environ. Health,* 72, 142–150, 1999.
121. Rios-Blanco, M.N., Plna, K., Faller, T., Kessler, W., Hakansson, K., Kreuzer, P.E., Ranasinghe, A., Filser, J.G., Segerback, D., and Swenberg, J.A., Propylene oxide: mutagenesis, carcinogenesis and molecular dose, *Mutat. Res.,* 380, 179–197, 1997.
122. Levine, S.P., Hebel, K.G., Bolton, J., and Kugel, R.E., Industrial analytical chemists and OSHA regulations for vinyl chloride, *Anal. Chem.,* 47, 1075A–1080A, 1975.
123. Göthe, R., Calleman, C.J., Ehrenberg, L., and Wachtmesiter, C.A., Trapping with 3,4-dichlorobenzenethiol of reactive metabolites formed in vitro from the carcinogen vinyl chloride, *Ambio,* 3, 224–226, 1974.
124. Watanabe, P., McGowan, G., Madrid, E., and Gehring, P., Fate of ^{14}C-vinyl chloride following inhalation exposure in rats, *Toxicol. Appl. Pharmacol.,* 37, 49-59, 1976.
125. Müller, G., Norpoth, K., and Eckard, R., Identification of two urine metabolites of vinyl chloride by GC-MS-investigations, *Int. Arch. Occup. Environ. Health,* 38, 69–75, 1976.
126. Müller, G., Norpoth, K., and Wickramasinghe, R.H., An analytical method, using GC-MS, for the quantitative determination of urinary thiodiglycolic acid, *Int. Arch. Occup. Environ. Health,* 44, 185–191, 1979.
127. Fedtke, N., Boucheron, J.A., Turner, M.J., and Swenberg, J.A., Vinyl chloride-induced DNA adducts, I: Quantitative determination of N-2,3-ethenoguanine based on electrophore labeling, *Carcinogenesis,* 11, 1279–1285, 1990.
128. Swenberg, J.A., Bogdanffy, M.S., Ham, A., Holt, S., Kim, A., Morinello, E.J., Ranasinghe, A., Scheller, N., and Upton, P.B., Formation and repair of DNA adducts in vinyl chloride- and vinyl fluoride-induced carcinogenesis, *IARC Sci. Publ.,* 150, 29–43, 1999.
129. Muller, M., Belas, F.J., Blair, I.A., and Guengerich, F.P., Analysis of 1,N^2-ethenoguanine and 5,6,7,9-tetrahydro-7-hydroxy-9-oxoimidazo[1,2-α]purine in DNA treated with 2-chlorooxirane by HPLC/electrospray MS and comparison of amounts to other DNA adducts, *Chem. Res. Toxicol.,* 10, 242–247, 1997.
130. Yen, T.Y., Christova-Gueoguieva, N.I., Scheller, N., Holt, S., Swenberg, J.A., and Charles, M.J., Quantitative analysis of the DNA adduct N2,3-ethenoguanine using liquid chromatography/electrospray ionization mass spectrometry, *J. Mass Spectrom.,* 31, 1271–1276, 1996.
131. Ham, A.J., Ranasinghe, A., Morinello, E.J., Nakamura, J., Upton, P.B., Johnson, F., and Swenberg, J.A., Immunoaffinity/gas chromatography/high-resolution mass spectrometry method for the detection of N(2),3-ethenoguanine, *Chem. Res. Toxicol.,* 12, 1240–1246, 1999.
132. Yen, T.Y., Holt, S., Sangaiah, R., Gold, A., and Swenberg, J.A., Quantitation of 1,N6-ethenoadenine in rat urine by immunoaffinity extraction combined with liquid chromatography/electrospray ionization mass spectrometry, *Chem. Res. Toxicol.,* 11, 810–815, 1998.
133. Doerge, D.R., Churchwell, M.I., Fang, J.L., and Beland, F.A., Quantification of etheno-DNA adducts using liquid chromatography, on-line sample processing, and electrospray tandem mass spectrometry, *Chem. Res. Toxicol.,* 13, 1259–1264, 2000.
134. Roberts, D.W., Churchwell, M.I., Beland, F.A., Fang, J., and Doerge, D.R., Quantitative analysis of etheno-2′-deoxycytidine DNA adducts using on-line immunoaffinity chromatography coupled with LC/ES-MS/MS detection, *Anal. Chem.,* 73, 303–309, 2001.
135. Divine, B.J. and Hartman, C.M., Mortality update of butadiene production workers, *Toxicology,* 113, 169–181, 1996.
136. Krause, R.J. and Elfarra, A.A., Oxidation of butadiene monoxide to meso- and (+/-)-diepoxybutane by cDNA-expressed human cytochrome P450s and by mouse, rat, and human liver microsomes: evidence for preferential hydration of meso-diepoxybutane in rat and human liver microsomes, *Arch. Biochem. Biophys.,* 337, 176–184, 1997.
137. Himmelstein, M., Turner, M., Asgharian, B., and Bond, J., Comparison of blood concetrations of 1,3-butadiene and butadiene epoxides in mice and rats exposed to 1,3-butadiene by inhalation, *Carcinogenesis,* 15, 1479–1486, 1994.

138. Valentine, J.L., Boogaard, P.J., Sweeney, L.M., Turner, M.J., Bond, J.A., and Medinsky, M.A., Disposition of butadiene epoxides in Sprague-Dawley rats, *Chem. Biol. Interact.,* 104, 103–115, 1997.
139. Kemper, R.A., Elfarra, A.A., and Myers, S.R., Metabolism of 3-butene-1,2-diol in B6C3F1 mice, *Drug Metab. Dispos.,* 26, 914–920, 1998.
140. Osterman-Golkar, S.M., and Bond, J.A., Biomonitoring of 1,3-butadiene and related compounds, *Environ. Health Perspect.,* 104, 907–915, 1996.
141. Richardson, K.A., Megens, H.J., Webb, J.D., and van Sittert, N.J., Biological monitoring of butadiene exposure by measurement of hemoglobin adducts, *Toxicology,* 113, 112–118, 1996.
142. van Sittert, N.J., Megens, H.J., Watson, W.P., and Boogaard, P.J., Biomarkers of exposure to 1,3-butadiene as a basis of cancer risk assessment, *Toxicol. Sci.,* 56, 189–202, 2000.
143. Tretyakova, N.Y., Chiang, S.-Y., Walker, V.E., and Swenberg, J.A., Quantitative analysis of 1,3-butadiene-induced DNA adducts in *vivo* and in *vitro* using liquid chromatography electrospray ionization tandem mass spectrometry, *J. Mass Spectrom.,* 33, 363–376, 1998.
144. Swenberg, J.A., Ham, A., Koc, H., Morinello, E., Ranasinghe, A., Tretyakova, N., Upton, P.B., and Wu, K.Y., DNA adducts: effects of low exposure to ethylene oxide, vinyl chloride and butadiene, *Mutat. Res.,* 464, 77–86, 2000.
145. Brugnone, F., Perbellini, L., Wang, G.Z., Maranelli, G., Raineri, E., De Rosa, E., Saletti, C., Soave, C., and Romeo, L., Blood styrene concentrations in a "normal" population and in exposed workers 16 hours after the end of the workshift, *Int. Arch. Occup. Environ. Health,* 65, 125–130, 1993.
146. Tornero-Velez, R., Waidyanatha, S., Perez, H.L., Osterman-Golkar, S., Echeverria, D., and Rappaport, S.M., Determination of styrene and styrene-7,8-oxide in human blood by gas chromatography-mass spectrometry, *J. Chromatogr. B,* 59–68, 2001.
147. Pongracz, K., Burlingame, A.L., and Bodell, W., Identification of N^2-substituted 2′-deoxygaunosine-3′-phosphate adducts detected by ^{32}P-postlabeling of styrene-oxide-treated DNA, *Carcinogenesis,* 13, 315–319, 1992.
148. Kaur, S., Pongracz, K., Bodell, W., and Burlingame, A.L., Bis(hydroxyphenylethyl)deoxygauanosine adducts identified by ^{32}P-postlabeling and four-sector tandem mass spectrometry: unanticipated adducts formed upon treatment of DNA with styrene 7, 8-oxide, *Chem. Res. Toxicol.*, 6, 125–132, 1993.
149. Schrader, W. and Linscheid, M., Styrene oxide DNA adducts: *in vitro* reaction and sensitive detection of modified oligonucleotides using capillary zone electrophoresis interfaced to electrospray mass spectrometry, *Arch. Toxicol.,* 71, 588–595, 1997.
150. Rappaport, S.M., Ting, D., Jin, Z., Yeowell-O'Connell, K., Waaidyanatha, S., and McDonald, T., Application of Raney nickel to measure adducts of styrene oxide with hemoglobin and albumin, *Chem. Res. Toxicol.,* 6, 238–244, 1993.
151. Yeowell-O'Connell, K., Jin, Z., and Rappaport, S.M., Determination of albumin and hemoglobin adducts in workers exposed to styrene and styrene oxide, *Cancer Epidemiol. Biomark. Prev.,* 5, 205–215, 1996.
152. Fustinoni, S., Colosio, A., Colombi, A., Yeowell-O'Connell, K., Lastrucci, L., and Rappaport, S.M., Albumin and hemoglobulin adducts as biomarkers of exposure to styrene in fiberglass-reinforced-plastics workers, *Int. Arch. Occup. Environ. Health,* 71, 35–41, 1998.
153. Minoia, C., Turci, R., Sottani, C., Schiavi, A., Perbellini, L., Angeleri, S., Draicchio, F., and Apostoli, P., Application of high performance liquid chromatography/tandem mass spectrometry in the environmental and biological monitoring of health care personnel occupationally exposed to cyclophosphamide and ifosfamide, *Rapid Commun. Mass Spectrom.*, 12, 1485–1493, 1998.
154. Sessink, P.J. and Bos, R.P., Drugs hazardous to healthcare workers, *Drug. Safety,* 20, 347–359, 1999.
155. Sessink, P.J., Timmersmans, J.L., Anzion, R.B., and Bos, R.P., Assessment of occupational exposure of pharmaceutical plant workers to 5-fluorouracil. Determination of alpha-fluoro-beta-alanine in urine, *J. Occup. Med.,* 36, 79–83, 1994.
156. Turci, R., Fiorentino, M.L., Sottani, C., and Minoia, C., Determination of methotrexate in human urine at trace levels by solid phase extraction and high-performance liquid chromatography/tandem mass spectrometry, *Rapid Commun. Mass Spectrom.*, 14, 173–179, 2000.
157. Sessink, P.J., Scholtes, M., Anzion, R., and Bos, R, Determination of cyclophosphamide in urine by gas chromatography-mass spectrometry, *J. Chromatogr.*, 616, 333–337, 1993.
158. Sannolo, N., Miraglia, N., Biglietto, M., Acampora, A., and Malorni, A., Determination of cyclophosphamide and ifosphamide in urine at trace levels by gas chromatography/tandem mass spectrometry, *J. Mass Spectrom.*, 34, 845–849, 1999.

159. Sottani, C., Turci, R., Perbellini, L., and Minoia, C., Liquid-liquid extraction procedure for trace determination of cyclophosphamide in human urine by high-performance liquid chromatography tandem mass spectrometry, *Rapid Commun. Mass Spectrom.*, 12, 1063–1068, 1998.
160. Balavanova, S., Wei, B., Rosing, F., Buhler, G., Scherer, G., Mayerhofer, C., Chen, Z., Zhang, W., and Rosenthal, J., Detection of nicotine in prehistorical skeletal remains of South China, *Anthropol. Anz.*, 54, 341–353, 1996.
161. Crawford, W., On the effects of environmental tobacco smoke, *Arch. Environ. Health*, 43, 34–37, 1988.
162. Wynder, E. and Hoffmann, D., Smoking and lung cancer: scientific challenges and opportunities, *Cancer Res.*, 54, 5284–5295, 1994.
163. Brunnemann, K., Rivenson, A., Adams, J., Hecht, S., and Hoffman, D., A study of snuff carcinogenesis, in *The Relevance of N-Nitroso Compounds to Human Cancer, IARC Scientific Publication No. 84,* Lyon, 456–459, 1987.
164. Cinciripini, P.M., Hecht, S.S., Henningfield, J.E., Manley, M.W., and Kramer, B.S., Tobacco addiction: implications for treatment and cancer prevention, *J. Natl. Cancer Inst.*, 89, 1852–1867, 1997.
165. Hecht, S.S., Tobacco and cancer: approaches using carcinogen biomarkers and chemoprevention, *Ann. N.Y. Acad. Sci.*, 833, 91–111, 1997.
166. Lofroth, G., Environmental tobacco smoke: overview of chemical composition and genotoxic components, *Mutat. Res.*, 222, 73–80, 1989.
167. Phillips, D.H., DNA adducts in human tissues: biomarkers of exposure to carcinogens in tobacco smoke, *Environ. Health Perspect.*, 104 Suppl. 3, 453–458, 1996.
168. Lefooe, N., Ashley, M., Pederson, L., and Keays, J., The health risks of passive smoking. The growing case for control measures in enclosed environments, *Chest*, 84, 90–95, 1983.
169. Hammond, S.K., Exposure of U.S. workers to environmental tobacco smoke, *Environ. Health Perspect.*, 107, 329–340, 1999.
170. Environmental Protection Agency, *Respiratory Health Effects of Passive Smoking: Lung Cancer and Other Disorders,* Washington, D.C. Office of Health Env. Assessment, 1–4, 1992.
171. Benowitz, N.L., Biomarkers of environmental tobacco smoke exposure, *Environ. Health Perspect.*, 107, 349–355, 1999.
172. Ember, L.R., The nicotine connection, *C&EN*, 8–18, 1994.
173. Zevin, S., Jacob, P., III, and Benowitz, N., Cotinine effects on nicotine metabolism, *Clin. Pharm. Therap.*, 61, 649–654, 1997.
174. Palmer, M.E., Smith, R.F., Chambers, K., and Tetler, L.W., Separation of nicotine metabolites by capillary zone electrophoresis and capillary zone electrophoresis/mass spectrometry, *Rapid Commun. Mass Spectrom.*, 15, 224–231, 2001.
175. Neurath, G. and Pein, G., Gas chromatographic determination of trans-3′-hydroxycotinine, a major metabolite of nicotine in smokers, *J. Chromatogr.*, 415, 400–406, 1987.
176. McAdams, S.A. and Cordeiro, M.L., Simple selected ion monitoring capillary gas chromatographic-mass spectrometric method for the determination of cotinine in serum, urine and oral samples, *J. Chromatogr.*, 615, 148–153, 1993.
177. Cooper, D. and Moore, J., Femtogram on-column detection of nicotine by isotope dilution gas chromatography/negative ion detection mass spectrometry, *Biol. Mass Spectrom.*, 22, 590–594, 1993.
178. James, H., Tizabi, Y., and Taylor, R., Rapid method for the simultaneous measurement of nicotine and cotinine in urine and serum by gas chromatography-mass spectrometry, *J. Chromatogr. B*, 708, 87–93, 1998.
179. Xu, A.S., Peng, L., Havel, J.A., Petersen, M.E., Fiene, J.A., and Hulse, J.D., Determination of nicotine and cotinine in human plasma by liquid chromatography-tandem mass spectrometry with atmospheric-pressure chemical ionization interface, *J. Chromatogr. B*, 682, 249–257, 1996.
180. Bernert, J.T., Turner, W.E., Pirkle, J.L., Sosnoff, C.S., Akins, J.R., Waldrep, M.K., Ann, Q., Covey, T.R., Whitfield, W.E., Gunter, E.W., Miller, B.B., Patterson, D.G.J., Needham, L.L., Hannon, W.H., and Sampson, E.J., Development and validation of sensitive method for determination of serum cotinine in smokers and nonsmokers by liquid chromatography/atmospheric pressure ionization tandem mass spectrometry, *Clin. Chem.*, 43, 2281–2291, 1997.
181. McManus, K., deBethizy, J., Garteiz, D., Kyerematen, G., and Vesell, E., A new quantitative thermospray LC-MS method for nicotine and its metabolites in biological fluids, *J. Chromatogr. Sci.*, 28, 510–516, 1990.

182. Byrd, G., Uhrig, M., deBethizy, J., Caldwell, W., Crooks, P., Ravard, A., and Riggs, R., Direct determination of cotinine-N-glucuronide in urine using thermospray liquid chromatography/mass spectrometry, *Biol. Mass Spectrom.*, 23, 103–107, 1994.
183. Byrd, G., Chang, K., Greene, J., and deBethizy, J., Determination of nicotine and its metabolites in urine by thermospray liquid chromatography/mass spectrometry, in *The Biology of Nicotine: Current Research Issues*, Raven Press, Ltd., New York, 71–83, 1992.
184. Byrd, G., Chang, K.-M., Greene, J., and deBethizy, J., Evidence for urinary excretion of glucuronide conjugates of nicotine, cotinine, and trans-3′-hydroxycotinine in smokers, *Drug Metab. Dispos.*, 20, 192–197, 1992.
185. Byrd, G.D., Robinson, J.H., Caldwell, W.S., and deBethizy, J.D., Comparison of measured and FTC-predicted nicotine uptake in smokers, *Psychopharmacology*, 122, 95–103, 1995.
186. Haufroid, V. and Lison, D., Urinary cotinine as a tobacco-smoke exposure index: a minireview, *Int. Arch. Occup. Environ. Health*, 71, 162–168, 1998.
187. Skarping, G., Willers, S., and Dalene, M., Determination of cotinine in urine using capillary gas chromatography and selective detection, with special reference to the biological monitoring of passive smoking, *J. Chromatogr.*, 454, 293–301, 1988.
188. Willers, S., Skapping, G., Dalene, M., and Skerfving, S., Urinary cotinine in children and adults during and after semiexperimental exposure to environmental tobacco smoke, *Arch. Environ. Health*, 50, 130–138, 1995.
189. Allena, J., Lawson, G.M., Anderson, R., Dale, L.C., Croghan, I.T., and Hurt, R.D., A new gas chromatography-mass spectrometry method for simultaneous determination of total and free *trans*-3′-hydroxycotinine and cotinine in the urine of subjects receiving transdermal nicotine, *Clin. Chem.*, 45, 85–91, 1999.
190. Davis, R., Stiles, M., deBethizy, J., and Reynolds, J., Dietary nicotine: a source of urinary cotinine, *Food Chem. Toxic.*, 29, 821–827, 1991.
191. Kintz, P., Gas chromatographic analysis of nicotine and cotinine in hair, *J. Chromatogr.*, 580, 347–353, 1992.
192. Zahlsen, K. and Nilsen, O.G., Nicotine in hair of smokers and nonsmokers: sampling procedure and gas chromatographic/mass spectrometric analysis, *Pharmacol. Toxicol.*, 75, 143–149, 1994.
193. Zahlsen, K., Nilsen, T., and Nilsen, O.G., Interindividual differences in hair uptake of air nicotine and significance of cigarette counting for estimation of environmental tobacco smoke exposure, *Pharmacol. Toxicol.*, 79, 183–190, 1996.
194. Gwent, S.H., Wilson, J.F., Tsanaclis, L.M., and Wicks, J.F., Time course of appearance of cotinine in human beard hair after a single dose of nicotine, *Ther. Drug Monit.*, 17, 195–198, 1995.
195. Pacifici, R., Altieri, I., Gandini, L., Lenzi, A., Pishini, S., Rosa, M., Zuccaro, P., and Dondero, F., Nicotine, cotinine, and trans-3-hydroxycotinine levels in seminal plasma of smokers: effects on sperm parameters, *Ther. Drug Monit.*, 15, 358–363, 1993.
196. Emmons, K., Abrams, D., Marshall, R., Marcus, B., Kane, M., Novotny, T., and Etzel, R., An evaluation of the relationship between self-report and biochemical measures of environmental tobacco smoke exposure, *Prevent. Med.*, 23, 35–39, 1994.
197. Kintz, P., Henrich, A., Cirimele, V., and Ludes, B., Nicotine monitoring in sweat with a sweat patch, *J. Chromatogr. B*, 705, 357–361, 1998.
198. Ruppert, T., Scherer, G., Tricker, A.R., and Adlkofer, F., Trans,trans-muconic acid as a biomarker of non-occupational environmental exposure to benzene, *Int. Arch. Occup. Environ. Health*, 69, 247–251, 1997.
199. Melikian, A., Prahalad, A., and Hoffman, D., Urinary trans,trans-muconic acid as an indicator of exposure to benzene in cigarette smokers, *Cancer Epidemiol. Biomark. Prev.*, 2, 47–51, 1993.
200. Melikian, A., Prahalad, A., and Secker-Walker, R., Comparison of the levels of the urinary benzene metabolite trans,trans-muconic acid in smokers and nonsmokers, and the effects of pregnancy, *Cancer Epidemiol. Biomark. Prev.*, 37, 239–244, 1994.
201. Ruppert, T., Scherer, G., Tricker, A., Rauscher, D., and Adlkofer, F., Determination of urinary *trans,trans*-muconic acid by gas chromatography-mass spectrometry, *J. Chromatogr. B*, 666, 71–76, 1995.
202. Melikian, A.A., O'Connor, R., Prahalad, A.K., Hu, P.F., Li, H.Y., Kagan, M., and Thompson, S., Determination of the urinary benzene metabolites S-phenylmercapturic acid and trans,trans-muconic acid by liquid chromatography-tandem mass spectrometry, *Carcinogenesis*, 20, 719–726, 1999.

203. Casale, G.P., Singhal, M., Bhattacharya, S., Ramanathan, R., Roberts, K.P., Barbacci, D.C., Zhao, J., Jankowiak, R., Gross, M.L., Cavalieri, E.L., Small, G.J., Rennard, S.I., Mumford, J.L., and Shen, M., Detection and quantification of depurinated benzo[a]pyrene-adducted DNA bases in the urine of cigarette smokers and women exposed to household coal smoke, *Chem. Res. Toxicol.*, 14, 192–201, 2001.
204. Bryant, M.S., Skipper, P.L., Tannenbaum, S.R., and Maclure, M,. Hemoglobin adducts of 4-aminobiphenyl in smokers and nonsmokers, *Cancer Res.*, 47, 602–608, 1987.
205. Dallinga, J.W., Pachen, D.M., Wijnhoven, S.W., Breedijk, A., van't Veer, L., Wigbout, V., van Zandwijk, N., Maas, L.M., van Agen, E., Kleinjans, J.C., and van Shooten, F.J., The use of 4-aminobiphenyl hemoglobin adducts and aromatic DNA adducts in lymphocytes of smokers as biomarkers of exposure, *Cancer Epidemiol. Biomark. Prev.*, 571–577, 1998.
206. Grant, D.M., Tang, B.K., and Kalow, W., A simple test for acetylator phenotype using caffeine, *Br. J. Clin. Pharmacol.*, 17, 459–464, 1984.
207. Cadet, J., Douki, T., and Ravanat, J.L., Artifacts associated with the measurement of oxidized DNA bases, *Environ. Health Perspect.*, 105, 1034–1039, 1997.
208. Landi, M.T., Zocchetti, C., Bernucci, H., Kadlubar, F.F., Tannenbaum, S., Skipper, P., Bartsch, H., Malaveille, C., Shields, P., Caporaso, N.E., and Vineis, P., Cytochrome P45401A2: enzyme induction and genetic control in determining 4-aminobiphenyl-hemoglobin adduct levels, *Cancer Epidemiol. Biomark. Prev.*, 5, 693–698, 1996.
209. Coghlin, J., Gann, P., Hammond, S., Skipper, P., Taghizadeh, K., Paul, M., and Tannenbaum, S., 4-Aminobiphenyl hemoglobin adducts in fetuses exposed to the tobacco smoke carcinogen in utero, *J. Natl. Cancer Inst.*, 83, 274–280, 1991.
210. Pinorini-Godly, M., and Myers, S.R., HPLC and GC/MS determination of 4-aminobiphenyl haemoglobin adducts in fetuses exposed to the tobacco smoke carcinogen in utero, *Toxicology*, 107, 209–217, 1996.
211. Del Santo, P., Moneti, G., Salvadori, M., Saltutti, C., Rose, A., and Dolara, P., Levels of the adducts of 4-aminobiphenyl to hemoglobin in control subjects and bladder carcinoma patients, *Cancer Letters*, 60, 245–251, 1991.
212. Bryant, M., Skipper, P., Tannenbaum, S., and Maclure, M., Hemoglobin adducts of 4-aminobiphenyl in smokers and nonsmokers, *Cancer Res.*, 47, 612–618, 1987.
213. Kadlubar, F.F., Dooley, K.L., Teitel, C.H., Roberts, D.W., Benson, R.W., Butler, M.A., Bailey, J.R., Young, J.F., Skipper, P.W., and Tannenbaum, S.R., Frequency of urination and its effects on metabolism, pharmacokinetics, blood hemoglobin adduct formation, and liver and urinary bladder DNA adduct levels in beagle dogs given the carcinogen 4-aminobiphenyl, *Cancer Res.*, 51, 4371–4377, 1991.
214. Lin, D., Lay, J.O.J., Bryant, M.S., Malaveille, C., Friesen, M., Bartsch, H., Lang, N.P., and Kadlubar, F.F., Analysis of 4-aminobiphenyl-DNA adducts in human urinary bladder and lung by alkaline hydrolysis and negative ion gas chromatography-mass spectrometry, *Environ. Health Perspect.*, 102(Suppl 6), 11–16, 1994.
215. Martone, T., Airoldi, L., Magagnotti, C., Coda, R., Randone, D., Malaveille, C., Avanzi, G., Merletti, F., Hautefeuille, A., and Vineis, P., 4-aminobiphenyl-DNA adducts and p53 mutations in bladder cancer, *Int. J. Cancer*, 75, 512–516, 1998.
216. Preston-Martin, S., N-Nitroso compounds as a cause of human cancer, *IARC Sci. Publ.*, 477–484, 1987.
217. Hecht, S.S., Biochemistry, biology, and carcinogenicity of tobacco-specific N-nitrosamines, *Chem. Res. Toxicol.*, 11, 559–603, 1998.
218. Brunnemann, K.D., Prokopczyk, B., Djordjevic, M.V., and Hoffman, D., Formation and analysis of tobacco-specific N-nitrosamines, *Crit. Rev. Toxicol.*, 26, 121–137, 1996.
219. Hecht, S.S., Recent studies on mechanism of bioactivation and detoxification of 4-(methylnitrosamino)-1-(pyridyl)-1-butanone (NNK), a tobacco-specific lung carcinogen, *Crit. Rev. Toxicol.*, 26, 163–181, 1996.
220. Hecht, S.S., DNA adduct formation from tobacco-specific N-nitrosamines, *Mutat. Res.*, 424, 127–142, 1999.
221. Hecht, S.S., Carmella, S.G., and Murphy, S.E., Tobacco-specific nitrosamine-hemoglobin adducts, *Methods Enzymol.*, 231, 657–667, 1994.
222a. Hecht, S.S., Carmella, S.G., Foiles, P.G., and Murphy, S.E., Biomarkers for human uptake and metabolic activation of tobacco-specific nitrosamines, *Cancer Res.*, 54, 1912S–1917S, 1994.
222b. Atawodi, S.E., Lea, S., Nyberg, F., Mukeria, A., Constantinescu, V., Aherns, W., Brueske-Hohlfeld, I., Fortes, C., Boffetta, P., and Friesen, M., 4-Hydroxy-1-(3-pyridyl)-1-butanone-hemoglobin adducts as biomarkers of exposure to tobacco smoke: validation of a method to be used in multicenter studies, *Cancer Epidemiol. Biomark. Prev.*, 7, 817–821, 1998.

223. Prokopczyk, B., Cox, J.E., Hoffman, D., and Waggoner, S.E., Identification of tobacco-specific carcinogen in the cervical mucus of smokers and nonsmokers, *J. Natl. Cancer Inst.*, 89, 868–873, 1997.
224. Carmella, S.G., Akerkar, S., Richie, J.P., and Hecht, S.S., Intraindividual and interindividual differences in metabolites of the tobacco-specific lung carcinogen 4-(methylnitrosamino)1-(3-pyridyl)-1-butanone (NNK) in smokers' urine, *Cancer Epidemiol. Biomark. Prev.*, 4, 635–642, 1995.
225. Carmella, S.G., Ye, M., Upadhyaya, P., and Hecht, S.S., Stereochemistry of metabolites of a tobacco-specific lung carcinogen in smokers' urine, *Cancer Res.*, 59, 3602–3605, 1999.
226. Pongracz, K., Kaur, S., Burlingame, A.L., and Bodell, W.J., O6-substituted-2′-deoxyguanosine-3′-phosphate adducts detected by ^{32}P post-labeling of styrene oxide treated DNA, *Carcinogenesis*, 10, 1009–1013, 1989.
227. Hecht, S., Carmella, S., Murphy, S., Akerkar, S., Klaus, M., Brunnemann, K., and Hoffmann, D., A tobacco-specific lung carcinogen in the urine of men exposed to cigarette smoke, *New. Eng. J. Med.*, 329, 1543–1546, 1993.
228. Parsons, W.D., Carmella, S.G., Akerkar, S., Bonilla, L.E., and Hecht, S.S., A metabolite for the tobacco-specific lung carcinogen 4-(methylnitrosamino)-1-(r-pyridyl)-1-butanone in the urine of hospital workers exposed to environmental tobacco smoke, *Cancer Epidemiol. Biomark. Prev.*, 7, 257–260, 1998.
229. Hecht, S.S., Carmella, S.G., Chen, M., Koch, J.F., Miller, A.T., Murphy, S.E., Jensen, J.A., Zimmerman, C.L., and Hatsukami, D.K., Quantitation of urinary metabolites of tobacco-specific lung carcinogen after smoking cessation, *Cancer Res.*, 59, 590–596, 1999.
230. Lackmann, G.M., Salzberger, U., Tollner, U., Chen, M., Carmella, S.G., and Hecht, S.S., Metabolites of a tobacco-specific carcinogen in urine from newborns, *J. Natl. Cancer Inst.*, 9, 459–465, 1999.
231. Carmella, S.G., Borukhova, A., Akerkar, S.A., and Hecht, S.S., Analysis of human urine for pyridine-N-oxide metabolites of 4-(methylnitrosamino)-1-(3-pyridyl)-1-butanone, a tobacco-specific lung carcinogen, *Cancer Epidemiol. Biomark. Prev.*, 6, 113–120, 1997.
232. Song, S. and Ashley, D.L., Supercritical fluid extraction and gas chromatography/mass spectrometry for the analysis of tobacco-specific nitrosamines in cigarettes, *Anal. Chem.*, 71, 1303–1308, 1999.
233. Dearth, M.A., Gierczak, C.A., and Sieglt, W.O., On-line measurement of benzene and toluene in dilute vehicle exhaust by mass spectrometry, *Environ. Sci. Technol.*, 26, 1573–1580, 1992.
234. Schuetzle, D., Siegl, W.O., Jensen, T.E., Dearth, M.A., Kaiser, E.W., Gorse, R., Kreucher, W., and Kulik, E., The relationship between gasoline composition and vehicle hydrocarbon emissions: a review of current studies and future research needs, *Environ. Health Perspect.*, 102 (Suppl. 4), 4–12, 1994.
235. Williams, P. and Andrews, P.R., Analysis of the polycyclic aromatic compounds of diesel fuel by gas chromatography with ion-trap detection, *Biomed. Environ. Mass Spectrom.*, 15, 517–519, 1988.
236. Weaver, V., Davoli, C., Heller, P., Fitzwilliam, A., Peters, H., Sunyer, J., Murphy, S., Goldstein, G., and Groopman, J., Benzene exposure assessed by urinary trans,trans-muconic acid, in urban children with elevated blood lead levels, *Environ. Health Perspect.*, 104, 318–323, 1996.
237. Atkinson, R. and Arey, J., Atmospheric chemistry of gas-phase polycyclic aromatic hydrocarbons: formation of atmospheric mutagens, *Environ. Health Perspect.*, 102 (Suppl. 4), 117–126, 1994.
238. Low, G., Bately, G., Lidgard, R., and Duffield, A., Determinations of polycyclic aromatic hydrocarbons in coal fly ash using gas chromatography/negative ion chemical ionization mass spectometry, *Biomed. Environ. Mass Spectrom.*, 13, 95–104, 1986.
239. Davoli, E., Cappellini, L., Moggi, M., and Fanelli, R., Automated high-speed analysis of selected organic compounds in urban air by on-line isotopic dilution cryofocusing gas chromatography/mass spectrometry, *J. Am. Soc. Mass Spectrom.*, 5, 1001–1007, 1994.
240. Lubman, D.M. and Kronick, M.N., Mass spectrometry of aromatic molecules with resonance-enhanced multiphoton ionization, *Anal. Chem.*, 54, 660–665, 1982.
241. Zimmermann, R., Heger, H., Dorfner, R., Boesl, U., Blumenstock, M., Lenoir, D., and Kettrup, A., A mobile laser mass spectrometer (REMPI-TOFMS) for continuous monitoring of toxic combustion byproducts: real-time-on-line analysis of PAH in waste incineration flue gases, *Combust. Sci. Tech.*, 134, 87–101, 1998.
242. Zimmermann, R., Heger, H.J., Kettrup, A., and Boesl, U., A mobile resonance-enhanced multiphoton ionization time-of-flight mass spectrometry device for on-line analysis of aromatic pollutants in waste incinerator flue gasses: first results, *Rapid Commun. Mass Spectrom.*, 11, 1095–1102, 1997.

243. Zimmermann, R., Heger, H., Blumenstock, M., Dorfner, R., Schramm, K.W., Boesl, U., and Kettrup, A., On-line measurement of chlorobenzene in waste incineration flue gas as a surrogate for the emission of polychlorinated dibenzo-p-dioxins/furans (I-TEQ) using mobile resonance laser ionization time-of-flight mass spectrometry, *Rapid Commun. Mass Spectrom.*, 13, 307–314, 1999.
244. Heger, H., Zimmermann, R., Dorfner, R., Beckmann, M., Griebel, H., Kettrup, A., and Boesl, U., On-line emission analysis of polycyclic aromatic hydrocarbons down to ppt concentration levels in the flue gas of an incineration pilot plant with a mobile resonance-enhanced multiphoton ionization time-of-flight mass specrometer, *Anal. Chem.*, 71, 46–57, 1999.
245. Zamperlini, G.C., Silva, M.R.S., and Vilegas, W., Identification of polycyclic aromatic hydrocarbons in sugar cane soot by gas chromatography-mass spectrometry, *Chromatographia*, 46, 655–663, 1997.
246. Villalobos, S.A., Anderson, M.J., Denison, M.S., Hinton, D.E., Tullis, K., Kennedy, I.M., Jones, A.D., Chang, D.P., Yang, G., and Kelly, P., Dioxinlike properties of a trichloroethylene combustion-generated aerosol, *Environ. Health Perspect.*, 104, 734–743, 1996.
247. Scholl, P.F., Musser, S.M., and Groopman, J.D., Synthesis and characterization of aflatoxin B1 mercapturic acids and their identification in rat urine, *Chem. Res. Toxicol.*, 10, 1144–1151, 1997.
248. Knight, L.P., Primiano, T., Groopman, J.D., Kensler, T.W., and Sutter, T.R., cDNA cloning, expression and activity of a second human aflatoxin B1-metabolizing member of the aldi-keto reductase superfamily, AKR7A3, *Carciogenesis*, 20, 1215–1223, 1999.
249. Guengerich, F.P., Cai, H., McMahon, M., Hayes, J.D., Sutter, T.R., Groopman, J.D., Deng, Z., and Harris, T.M., Reduction of aflatoxin B1 dialdehyde by rat and human aldo-keto reductases, *Chem. Res. Toxicol.*, 14, 727–737, 2001.
250. McLuckey, S.A. and Habibi-Goudarzi, S., Ion trap tandem mass spectrometry applied to small multiply charged oligonucleotides with a modified base, *J. Am. Soc. Mass Spectrom.*, 5, 740–747, 1994.
251. Marzilli, L.A., Wang, D., Kobertz, W.R., Essigmann, J.M., and Vouros, P., Mass spectral identification and positional mapping of aflatoxin B_1-guanine adducts in oligonucleotides, *J. Am. Soc. Mass Spectrom.*, 9, 676–682, 1998.
252. Kensler, T.W., Groopman, J.D., and Roebuck, B.D., Use of aflatoxin adducts as intermediate endpoints to assess the efficacy of chemopreventive interventions in animals and man, *Mutat. Res.*, 402, 165–172, 1998.
253. Sugimura, T., Overview of carcinogenic hetrocyclic amines, *Mutat. Res.*, 376, 211–219, 1997.
254. Robbana-Barnat, S., Rabache, M., Rialland, E., and Fradin, J., Heterocyclic amines: occurrence and prevention in cooked food, *Environ. Health Perspect.*, 104, 280–288, 1996.
255. Jagerstad, M., Olaaon, K., Grivas, S., Negishi, C., Wakabayashi, K., Tsuda, M., Sato, S., and Sugimura, T., Formation of 2-amino-3,8-dimetylimidazo[4,5-f]quinoxaline in a model system by heating creatinine, glycine, and glucose, *Mutat. Res.*, 126, 239–244, 1984.
256. Grivas, S. and Nyhammar, T., Isolation and identification of the food mutagens IQ and MEIQx from a heated model system of creatinine, glycine and fructose, *Food Chem.*, 20, 127–136, 1986.
257. Shioya, M. and Wakabayashi, K., Formation of a mutagen 2-amino-1-methyl-6-phenylimidazo[4,5-b]-pyridine (PhIP) in cooked beef, by heating a mixture containing creatinine, phenylalanine and glucose, *Mutat. Res.*, 191, 133–138, 1987.
258. Felton, J. and Knize, M., Heterocyclic-amine mutagens/carcinogens in foods, in *Chemical Carcinogenesis and Mutagenesis I*, Spinger-Verlag, Berlin, New York, 471–502, 1990.
259. Skog, K., Augustsson, K., Steineck, G., Stenberg, M., and Jagerstad, M., Polar and non-polar heterocyclic amines in cooked fish and meat products and their corresponding pan residues, *Food Chem. Toxicol.*, 35, 555–565, 1997.
260. Augustsson, K., Skog, K., Jagerstad, M., Dickman, P.W., and Steineck, G., Dietary heterocyclic amines and cancer of the colon, rectum, bladder, and kidney: a population-based study, *Lancet*, 353, 703–707, 1999.
261. Kataoka, H., Methods for the determination of mutagenic heterocyclic amines and their applications in environmental analysis, *J. Chromatogr. A*, 774, 121–142, 1997.
262. Boobis, A.R., Lynch, A.M., Murray, S., de la Torre, R., Solans, A., Farre, M., Segura, J., Gooderham, N.J., and Davies, D.S., CYP1A2-catalyzed conversion of dietary heterocyclic amines to their proximate carcinogens is their major route of metabolism in humans, *Cancer Res.*, 54, 89–94, 1994.
263. Lynch, A.M., Knize, M.G., Boobis, A.R., Gooderham, N.J., Davies, D.S., and Murray, S., Intra- and interindividual variability in systemic exposure in humans to 2-amino-3,8-dimethylimidazo[4,5-f]quinoxaline and 2-amino-1-methyl-6-phenylimidazo[4,5-b]pyridine, carcinogens present in cooked beef, *Cancer Res.*, 52, 6216–6223, 1992.

264. Murray, S., Lynch, A.M., Knize, M.G., and Gooderham, M.J., Quantification of the carcinogens 2-amino-3,8-dimethyl- and 2-amino-3,4,8-trimethylimidazo[4,5-f]quinoxaline and 2-amino-1-methyl-6-phenylimidazo[4,5-b]pyridine in food using a combined assay based on gas chromatography-negative ion mass spectrometry, *J. Chromatogr.*, 616, 211–219, 1993.
265. Friesen, M., Garren, L., Bereziat, J.-C., Kadlubar, F., and Lin, D., Gas chromatography-mass spectrometry analysis of 2-amino-1-methyl-6-phenyl-imidazo[4,5-b]pyridine in urine and feces, *Environ. Health Perspect.*, 99, 179–181, 1993.
266. Tikkanen, L., Sauri, T., and Latva-Kala, K., Screening of heat-processed Finnish foods for the mutagens 2-amino-3,8-dimethylimidazo[4,5-f]-quinoxaline,2-amino-3,4,8-tri-methylimidazo[4,5-f]-quinoxaine and 2-amino-1-methyl-6- phenylimidazo[4,5-b]pyridine, *Food Chem. Toxic.*, 31, 717–721, 1993.
267. Yamaizumi, Z., Kasai, H., Nishimura, S., Edmonds, C., and McCloskey, J., Stable isotope dilution quantification of mutagens in cooked foods by combined liquid chromatography-thermospray mass spectrometry, *Mutat. Res.*, 173, 1–7, 1986.
268. Turesky, R., Bur, H., Huynh-Ba, T., Aeschbacher, H., and Milon, H., Analysis of mutagenic heterocyclic amines in cooked beef products by high performance liquid chromatography in combination with mass spectrometry, *Food Chem. Toxicol.*, 26, 501–509, 1988.
269. Richling, E., Haring, D., Herderich, M., and Schreier, P., Determination of heterocyclic aromatic amines(HAA) in commercially available meat products and fish by high performance liquid chromatography-electrospray tandem mass spectrometry (HPLC- ESI-MS-MS), *Chromatographia*, 48, 258–262, 1998.
270. Galceran, M.T., Moyano, E., Puignou, L., and Pais, P., Determination of hetrocyclic amines by pneumatically assisted electrospray liquid chromatography-mass spectrometry, *J. Chromatogr. A*, 730, 185–194, 1996.
271. Toribio, F., Moyano, E., Puignou, L., and Galceran, M.T., Determination of heterocyclic aromatic amines in meat extracts by liquid chromatography-ion-trap atmospheric pressure chemical ionization mass spectrometry, *J. Chromatogr. A*, 869, 307–317, 2000.
272. Pais, P., Moyano, E., Puignou, L., and Galceran, M.T., Liquid chromatography-atmospheric-pressure chemical ionization mass spectrometry as a routine method for the analysis of mutagenic amines in beef extracts, *J. Chromatogr. A*, 778, 207–218, 1997.
273. Pais, P., Moyano, E., Puignou, L., and Galceran, M.T., Liquid chromatography-electrospray mass spectrometry with in-source fragmentation for the identification and quantification of fourteen mutagenic amines in beef extract, *J. Chromatogr. A*, 775, 125–136, 1997.
274. Zhao, Y., Schelfaut, M., Sandra, P., and Banks, F., Capillary electrophoresis and capillary electrophoresis-electrospray-mass spectroscopy for the analysis of heterocyclic amines, *Electrophoresis*, 19, 2213–2219, 1998.
275. Toribio, F., Moyano, E., Puignou, L., and Galceran, M.T., Comparison of different commercial solid-phase extraction cartridges used to extract heterocyclic amines from a lyophilised meat extract, *J. Chromatogr. A*, 880, 101–112, 2000.
276. Johansson, M.A.E., Fay, L.B., Gross, G.A., Olsson, K., and Jagerstad, M., Influence of amino acids on the formation of mutagenic/carcinogenic hetrocyclic amines in a model system, *Carcinogenesis*, 16, 2553–2560, 1995.
277. Fay, L.B., Ali, S., and Gross, G.A., Determination of heterocyclic aromatic amines in food products: automation of the sample preparation method prior to HPLC and HPLC-MS quantification, *Mutat. Res.*, 376, 29–35, 1997.
278. Murray, S., Gooderham, N., Boobis, A., and Davies, D., Measurement of MeIQx and DiMeIQx in fried beef by capillary column gas chromatography electron capture negative ion chemical ionization mass spectrometry, *Carcinogenesis*, 9, 321–325, 1988.
279. Guy, P.A., Gremaud, E., Richoz, J., and Turesky, R.J., Quantitative analysis of mutagenic heterocyclic aromatic amines in cooked meat using liquid chromatography-atmospheric pressure chemical ionization tandem mass spectrometry, *J. Chromatogr. A*, 883, 89–102, 2000.
280. Solyakov, A., Skog, K., and Jagerstad, M., Heterocyclic amines in process flavours, process flavour ingredients, bouillon concentrates and a pan residue, *Food Chem. Toxicol.*, 37, 1–11, 1999.
281. Nerurkar, P.V., Le Marchand, L., and Cooney, R.V., Effects of marinating with Asian marinades or western barbecue sauce in PhIP and MeIQx formation in barbecued beef, *Nutr. Cancer*, 34, 147–152, 1999.

282. Richling, E., Decker, C., Haring, D., Herderich, M., and Schreier, P., Analysis of heterocyclic aromatic amines in wine by high-performance liquid chromatography-electrospray tandem mass spectrometry, *J. Chromatogr. A*, 791, 71–77, 1997.
283. Yang, C.-C., Jeng, S.N., and Lee, H., Characterization of the carcinogen 2-amino-3,8-dimethylimidazol[4,5-f]quinoxaline in cooking aerosols under domestic conditions, *Carcinogenesis*, 19, 359–363, 1998.
284. Murray, S., Gooderham, N.J., Boobis, A.R., and Davies, D.S., Detection and measurement of MeIQx in human urine after ingestion of a cooked meat meal, *Carcinogenesis*, 10, 763–765, 1989.
285. Sinha, R., Rothman, N., Mark, S.D., Murray, S., Brown, E.D., Levander, O.A., Davies, D.S., Lang, N.P., Kadlubar, F.F., and Hoover, R.N., Lower levels of urinary 2-amino-3,8-dimethylimidazo[4,5-f]-quinoxaline (MeIQx) in humans with higher CYPIA2 activity, *Carcinogenesis*, 16, 2859–2861, 1995.
286. Reistad, R., Rossland, O.J., Latva-Kala, K.J., Rasmussen, T., Vikse, R., Becher, G., and Alexander, J., Heterocyclic aromatic amines in human urine following a fried meat meal, *Food Chem. Toxicol.*, 35, 945–955, 1997.
287. Stillwell, W.G., Turesky, R.J., Sinha, R., Skipper, P.L., and Tannenbaum, S.R., Biomonitoring of heterocyclic aromatic amine metabolites in human urine, *Cancer Lett.*, 143, 145–148, 1999.
288. Stillwell, W.G., Turesky, R.J., Sinha, R., and Tannenbaum, S.R., N-oxidative metabolism of 2-amino-3,8-dimethylimidazo[4,5-f]quinoxaline (MeIQx) in humans: excretion of the N2-glucuronide conjugate of 2-hydroxyamino-MeIQx in urine, *Cancer Res.*, 59, 5154–5194, 1999.
289. Kidd, L.C.R., Stillwell, W.G., Yu, M.C., Wishnok, J.S., Skipper, P.L., Ross, R.K., Henderson, B.E., and Tannenbaum, S.R., Urinary excretion of 2-amino-1-methyl-6-phenylimidazo[4,5-b]pyridine (PhIP) in white, African-American, and Asian-American men in Los Angeles county, *Cancer Epidemiol. Biomarkers Prev.*, 8, 439–445, 1999.
290. Knize, M.G., Kulp, K.S., Malfatti, M.A., Salmon, C.P., and Felton, J.S., Liquid chromatography-tandem mass spectrometry method of urine analysis for determining human variation of carcinogen metabolism, *J. Chromatogr. A*, 914, 95–103, 2001.
291. Schut, H.A. and Snyderwine, E.G., DNA adducts of theterocyclic amine food mutagens: implications for mutagenesis and carcinogenesis, *Carcinogenesis*, 20, 353–368, 1999.
292. Gooderham, N.J., Murray, S., Lynch, A.M., Yadollahi-Farsani, M., Zhoa, K., Rich, K., Boobis, A.R., and Davies, D.S., Assessing human risk to hetrocyclic amines, *Mutat. Res.*, 376, 53–60, 1997.
293. Gangl, E.T., Turesky, R.J., and Vouros, P., Determination of *in vitro*- and *in vivo*-formed DNA adducts of 2-amino-3-methylimidazo[4,5-f]quinoline by capillary liquid chromatography/microelectrospray mass spectrometry, *Chem. Res. Toxicol.*, 12, 1019–1027, 1999.
294. Gangl, E.T., Turesky, R.J., and Vouros, P., Detection of *in vivo* formed DNA adducts at the part-per-billion level by capillary liquid chromatography/microelectrospray mass spectrometry, *Anal. Chem.*, 73, 2397–2404, 2001.
295. Soglia, J.R., Turesky, R.J., Paehler, A., and Vouros, P., Quantification of the heterocyclic aromatic amine DNA adduct N-(deoxyguanosin-8-yl)-2-amino-3-methylimidazole[4,5-f]quinoline in livers of rats using capillary liquid chromatography/microelectrospray mass spectrometry: a dose-response study, *Anal. Chem.*, 73, 2819–2827, 2001.
296. Turteltaub, K.W., Felton, J., Gledhill, B., Vogel, J., Southton, J., Caffee, M., Finkel, R., Nelson, D., Proctor, L., and Davis, J., Accelerator mass spectrometry in biomedical dosimetry: relationship between low-level exposure and covalent binding of heterocyclic amine carcinogens to DNA, *Proc. Natl. Acad. Sci. USA*, 87, 5288–5292, 1990.
297. Felton, J., Knize, M., Roper, M., Fultz, E., Shen, N., and Turteltaub, K.W., Chemical analysis, prevention, and low level dosimetry of heterocyclic amine from cooked food, *Cancer Res.*, 52, 2103s–2107s, 1992.
298. Turteltaub, K.W., Vogel, J., Frantz, C., Buonarati, M., and Felton, J. Low-level biological dosimetry of heterocyclic amine carcinogens isolated from cooked food, *Environ. Health Perspect.*, 1993, 99, 183–186.
299. Turteltaub, K.W., Mauthe, R.J., Dingley, K.H., Vogel, J.S., Frantz, C.E., Garner, R.C., and Shen, N., MeIQx-DNA adduct formation in rodent and human tissues at low doses, *Mutat. Res.*, 376, 243–252, 1997.
300. Turteltaub, K.W., Dingley, K.H., Curtis, K.D., Malfatti, M.A., Turesky, R.J., Garner, R.C., Felton, J.S., and Lang, N.P., Macromolecular adduct formation and metabolism of heterocyclic amines in humans and rodents at low doses, *Cancer Lett.*, 143, 149–155, 1999.

301. Frantz, C., Bangerter, C., Fultz, E., Mayer, K., Vogel, J., and Turteltaub, K.W., Dose-response studies of MelQx in rat liver and liver DNA at low doses, *Carcinogenesis*, 16, 367–373, 1995.
302. Lightfoot, T.J., Coxhead, J.M., Cupid, B.C., Nicholson, S., and Garner, R.C., Analysis of DNA adducts by accelerator mass spectrometry in human breast tissue after administration of 2-amino-1-methyl-6-phenylimidazole[4,5-b]pyridine and benzo[a]pyrene, *Mutation Res.*, 472, 119–127, 2000.
303. Knekt, P., Jarvinen, R., Dich, J., and Hakulinen, T., Risk of colorectal and other gastro-intestinal cancers after exposure to nitrate, nitrite and N-nitroso compounds: a follow-up study, *Int. J. Cancer*, 80, 852–856, 1999.
304. Longo, M., Lionette, C., and Cavallaro, A., Determination of N-nitrosodimethylamine in beer gas chromatography-stable isotope dilution chemical ionization mass spectrometry, *J. Chromatogr. A*, 708, 303–307, 1995.
305. Song, P.J. and Hu, J.F., N-nitrosamines in Chinese foods, *Food Chem. Toxicol.*, 26, 205–208, 1988.
306. Dallinga, J.W., Pachen, D.M.F.A., Lousberg, A.H.P., van Geel, J.A.M., Houben, G.M.P., Stockbrugger, R.W., vanMaanen, J.M.S., and Kleinjans, J.C.S., Volatile N-nitrosamines in gastric juice of patients with various conditions of the gastrointestinal tract determined by gas chromatography-mass spectrometry and related to intragastric pH and nitrate and nitrite levels, *Cancer Lett.*, 124, 119–125, 1998.
307. Eerola, S., Otegui, E., Saari, L., and Rizzo, A., Application of liquid chromatography-atmospheric pressure chemical ionization mass spectrometry and tandem mass spectrometry to the determination of volatile nitrosamines in dry sausages, *Food Addit. Contam.*, 15, 270–279, 1998.
308. Park, M. and Loeppky, R.N., *In vitro* DNA deamination by α-nitrosaminoaldehydes determined by GC/MS-SIM quantitation, *Chem. Res. Toxicol.*, 13, 72–81, 2000.
309. Schoket, B., DNA damage in humans exposed to enviromental and dietary polycyclic aromatic hydrocarbons, *Mutat. Res.*, 424, 43–153, 1999.
310. Moret, S. and Conte, L.S., Polycyclic aromatic hydrocarbons in edible fats and oils: occurrence and analytical methods. *J. Chromatogr. A*, 882, 245–253, 2000.
311. Gmeiner, G., Krassing, C., Schmid, E., and Tausch, H., Fast screening method for the profile analysis of polycyclic aromatic hydrocarbon metabolites in urine using derivatization — solid-phase microextraction, *J. Chromatogr. B*, 705, 132–138, 1998.
312. Ali, M.Y. and Cole, R.B., SFE plus C18 lipid cleanup method for selective extraction and GC/MS quantitation of polycyclic aromatic hydrocarbons in biological tissues, *Anal. Chem.*, 70, 3242–3248, 1998.
313. Knize, M.G., Salmon, C.P., Pais, P., and Felton, J.S., Food heating and the formation of heterocyclic aromatic amine and polycyclic aromatic hydrocarbon mutagens/carcinogens, *Adv. Exp. Med. Biol.*, 459, 179–193, 1999.
314. Eder, E., Intraindividual variations of DNA adduct levels in humans, *Mutat. Res.*, 424, 249–261, 1999.
315. Guillen, M.D., Sopelana, P., and Partearroyo, M.A., Determination of polycyclic aromatic hydrocarbons in commercial liquid smoke flavorings of different compositions by gas chromatography-mass spectrometry, *J. Agric. Food Chem.*, 48, 126–131, 2000.
316. Wang, G., Lee, A.S., Lewis, M., Kamath, B., and Archer, R.K., Accelerated solvent extraction and gas chromatography/mass spectrometry for determination of polycyclic aromatic hydrocarbons in smoked food samples, *J. Agric. Food Chem.*, 47, 1062–1066, 1999.
317. Gilroy, D.J. Derivation of shellfish harvest reopening criteria following the New Carissa oil spill in Coos Bay, Oregon, *J. Toxicol. Environ. Health. A.*, 60, 317–329, 2000.
318. Chiang, T.A., Wu, P.F., and Ko, Y.C., Identification of carcinogens in cooking oil fumes, *Environ. Res. Section. A*, 81, 18–22, 1999.
319. International Agency for Research on Cancer, IARC Working Group on the Evaluation of Carcinogenic Risks to Humans: Polychlorinated Dibenzo-para-dioxins and Polychlorinated Digenzofurans, 1997.
320. Hoover, R.N., Dioxin dilemmas, *J. Natl. Cancer Inst.*, 91, 745–746, 1999.
321. Schecter, A., A selective historical review of congener-specific-human tissue measurements as sensitive and specific biomarkers of exposure to dioxins and related compounds, *Environ. Health Perspect.*, 106, 737–742, 1998.
322. Nygren, M., Hansson, M., Sjostrom, M., Rappe, C., Kahn, P.C., and Gochfeld, M., Development and validation of a method for determination of PCDDs and PCDFs in human blood plasma, *Chemosphere*, 17, 163–193, 1988.

323. Ferrario, J., Byme, C., McDaniel, D., and Dupuy, J., Determination of 2,3,7,8-chlorine-substituted dibenzo-p-dioxins and -furans at the part per trillion level in United States beef fat using high-resolution gas chromatograph/high resolution mass spectrometry, *Anal. Chem.*, 68, 647–652, 1996.
324. Clement, R. and Tosine, H., The gas chromatography/mass spectrometry determination of chloro-dibenzo-p-dioxins and dibenzofurans, *Mass Spectrom. Reviews*, 7, 593–636, 1988.
325. Huang, L., Eitzer, B., Moore, C., McGown, S., and Tomer, K., The application of hybrid mass spectrometry/mass spectrometry and high-resolution mass spectrometry to the analysis of fish samples for polychlorinated dibenzo-p-dioxins and dibenzofurans, *Biol. Mass Spectrom.*, 20, 161–168, 1991.
326. Hayward, D.G., Hooper, K., and Andrzejewski, D., Tandem-in-time mass spectrometry method for the sub-parts-per-trillion determination of 2,3,7,8-chlorine substitututed dibenzo-p-dioxins and -furans in high fat food, *Anal. Chem.*, 71, 212–220, 1999.
327. Hooper, K., Petreas, M.X., Chuvakova, T., Kazbekova, G., Druz, N., Seminova, G., Sharmanov, T., Hayward, D., She, J., Visita, P., Winkler, J., McKinney, M., Wade, T.J., Grassman, J., and Stephens, R.D., Analysis of breast milk to assess exposure to chlorinated contaminants in Kazakstan: high levels of 2,3,7,8-tetrachlorodibenzo-*p*-dioxin (TCDD) in agricultural villages of southern Kazakstan, *Environ. Health Perspect.,* 106, 797–806, 1998.
328. Johansen, H.R., Alexander, J., Rossland, O.J., Planting, S., Lovik, M., Gaarder, P.I., Gdynia, W., Bjerve, K.S., and Becher, G., PCDDs, PCDFs, and PCBs in human blood in relation to consumption of crabs from a contaminated fjord area in Norway, *Environ. Health Perspect.*, 104, 756–764, 1996.
329. Giesy, J.P., Kannan, K., Kubitz, J.A., Williams, L.L., and Zabik, M.J., Polychlorinated dibenzo-p-dioxins (PCDDs) and dibenzofurans (PCDFs) in muscle and eggs of salmonid fishes from the Great Lakes, *Arch. Environ. Contam. Toxicol.*, 36, 432–446, 1999.
330. Gold, L.S., Stern, B.R., Slone, T.H., Brown, J.P., Manley, N.B., and Ames, B.N., Pesticide residues in food: Investigation of disparities in cancer risk estimates, *Cancer Lett.*, 117, 195–207, 1997.
331. Lehotay, S.J. and Ibrahim, M.A., Supercritical fluid extraction and gas chromatography/ion trap mass spectrometry of pentachloronitrobenzene pesticides in vegetables, *J. AOAC Int.*, 78, 445–452, 1995.
332. Lehotay, S.J. and Valverde Garcia, A., Evaluation of different solid-phase traps for automated collection and clean-up in the analysis of multiple pesticides in fruits and vegetables after supercritical fluid extraction, *J. Chromatogr. A*, 765, 69–84, 1997.
333. Virkki, V.T., Ketola, R.A., Ojala, M., Kotiaho, T., Komppa, V., Grove, A., and Facchetti, S., On-site environmental analysis by membrane inlet mass spectrometry, *Anal. Chem.*, 67, 1421–1425, 1995.
334. Kotiaho, T., Ketola, R.A., Ojala, M., Mansikka, T., and Kostiainen, R., Membrane inlet mass spectrometry in environmental analysis, *Am. Environ. Lab.*, March, 19–22, 1997.
335. Lopez-Avilla, V., Benedicto, J., Prest, H., and Bauer, S., Automated MIMS for direct analysis of organic compounds in water, *Amer. Lab.*, 31, 32–37, 1999.
336. Grange, A.H., Donnelly, J.R., Sovocool, G.W., and Brumley, W.C., Determination of elemental compositions from mass peak profiles of the molecular ion (M) and the M+1 and M+2 ions, *Anal. Chem.*, 68, 553–560, 1996.
337. Grange, A.H. and Brumley, W.C., A mass peak profile generation model to facilitate determination of elemental compositions of ions based on exact masses and isotopic abundances, *J. Am. Soc. Mass Spectrom.*, 8, 170–182, 1997.
338. Grange, A.H., Sovocool, G.W., Donnelly, J.R., Genicola, F.A., and Gurka, D.F., Identification of pollutants in a municipal well using high resolution mass spectrometry, *Rapid Commun. Mass Spectrom.*, 12, 1161–1169, 1998.
339. Benfenati, E., Natangelo, M., Davoli, E., and Fanelli, R., Migration of vinyl chloride into PVC-bottled drinking-water assessed by gas chromatography-mass spectrometry, *Food Chem. Toxic.*, 29, 131–134, 1991.
340. Fujii, T., Trace determination of vinyl chloride in water by direct aquenous injection gas chromatography-mass spectrometry, *Anal. Chem.,* 1977, 49, 1985–1987.
341. Bolz, U., Korner, W., and Hagenmaier, H., Development and validation of a GC/MS method for determination of phenolic xenoestrogens in aquatic samples, *Chemosphere*, 40, 929–935, 2000.
342. Körner, R., Bolz, U., Sussmuth, W., Hiller, G., Schuller, W., Hanf, V., and Hagenmaier, H., Input/output balance of estogenic active compounds in a major municipal sewage plant in Germany, *Chemosphere*, 40, 1131–1142, 2000.

343. Kataoka, H., Hayatsu, T., Hietsch, G., Steinkellner, H., Nishioka, S., Narimatsu, S., Knasmuller, S., and Hayatsu, H., Identification of mutagenic heterocyclic amines (IQ, Trp-P-1 and A α C) in water of the Danube River, *Mutat. Res.*, 466, 27–35, 2000.
344. Berset, J.D., Ejem, M., Holzer, R., and Lischer, P., Comparison of different drying, extraction and detection techniques for the determination of priority polycyclic aromatic hydrocarbons in background contaminated soil samples, *Anal. Chim. Acta*, 383, 263–275, 1999.
345. Pace, C.M. and Betowski, L.D., Measurement of high-molecular-weight polycyclic aromatic hydrocarbons in soils by particle beam high-performance liquid chromatography-mass spectrometry, *J. Am. Soc. Mass Spectrom.*, 6, 597-607, 1995.
346. Gonzalez-Vila, F., and Lopez, G.L., Determination of polynuclear aromatic compounds in composted municipal refuse and compost-amended soils by a simple clean-up procedure, *Biomed. Environ. Mass Spectrom.*, 16, 423–425, 1988.
347. Segal, E.D., Consequences of attachment of *Helicobacter pylori* to gastric cells, *Biomed. Pharmacother.*, 51, 5–12, 1997.
348. Forman, D., *Helicobacter pylori* and gastric cancer, *Scand. J. Gastroenterol. Suppl.*, 215, 48–51, 1996.
349. Fendrick, A.M., Chernew, M.E., Hirth, R.A., Bloom, B.S., Bandekar, R.R., and Scheiman, J.M., Clinical and economic effects of population-based *Helicobacter pylori* screening to prevent gastric cancer, *Arch. Intern. Med.*, 159, 142–148, 1999.
350. Graham, D.Y., Klein, P.D., Evans, D.J., Evans, D.G., Alpert, L.C., Openkun, A.R., and Boutton, T.W., *Campylobacter pylori* detected noninvasively by the ^{13}C-urea breath test, *Lancet*, 1, 1174–1177, 1987.
351. Schoeller, D.A. and Klein, P.D., A microprocessor controlled mass spectrometer for the fully automated purification and isotopic analysis of breath carbon dioxide, *Biomed. Mass Spectrom.*, 6, 350–355, 1979.
352. Lotterer, E., Ludtke, F.E., Tegeler, R., Lepsien, G., and Bauer, F.E., The ^{13}C-urea breath test — detection of *Helicobacter pylori* infection in patients with partial gastrectomy, *Z. Gastroenterol.*, 31, 115–119, 1993.
353. Braden, B., Duan, L.P., Caspary, W.F., and Lembcke, B., More convenient ^{13}C-urea breath test modifications still meet the criteria for valid diagnosis of *Helicobacter pylori* infection, *Z. Gastroenterol.*, 32, 198–202, 1994.
354. Cutler, A.F. and Toskes, P., Comparison of ^{13}C-urea blood test to ^{13}C-urea breath test for the diagnosis of *Helicobacter pylori*, *Am. J. Gastroenterol.*, 94, 959–961, 1999.
355. Chey, W.D., Murthy, U., Toskes, P., Carpenter, S., and Laine, L., The ^{13}C-urea blood test accurately detects active Helicobacter pylori infection: a United States, multicenter trial, *Am. J. Gastroenterol.*, 94, 1522–1524, 1999.
356. Savarino, V., Mela, G.S., Zentilin, P., Bisso, G., Pivari, M., Mansi, C., Mele, M.R., Bilardi, C., Vigneri, S., and Celle, G., Comparison of isotope ratio mass spectrometry and nondispersive isotope-selective infrared spectroscopy for ^{13}C-urea breath test, *Am. J. Gastroenterol.*, 94, 1203–1208, 1999.
357. Lee, H.S., Gwee, K.A., Teng, L.Y., Kang, J.Y., Yeoh, K.G., Wee, A., and Chua, B.C., Validation of ^{13}C-urea breath test for *Helicobacter pylori* using a simple gas chromatograph-mass selective detector, *Eur. J. Gastroenterol. Hepatol.*, 10, 569–572, 1998.
358. Nilsson, C.L., Fingerprinting of *Helicobacter pylori* strains by matrix-assisted laser desorption/ionization mass spectrometric analysis, *Rapid Commun. Mass Spectrom.*, 13, 1067–1071, 1999.
359. Winkler, M.A., Uher, J., and Cepa, S., Direct analysis and identification of *Helicobacter* and *Campylobacter* species by MALDI-TOF mass spectrometry, *Anal. Chem.*, 71, 3416–3419, 1999.
360. McAtee, C.P., Fry, K.E., and Berg, D.E., Identification of potential diagnostic and vaccine candidates of *Helicobacter pylori* by "proteome" technologies, *Helicobacter*, 3, 163–169, 1998.
361. Johansson, L. and Miller-Podraza, H., Analysis of 3- and 6-linked sialic acids in mixtures of gangliosides using blotting to polyvinylidene difluoride membranes, binding assays, and various mass spectrometry techniques with application to recognition by *Helicobacter pylori*, *Anal. Biochem.*, 265, 260–268, 1998.
362. Shenberg, C., Feldstein, H., Cornelis, R., Mees, L., Versieck, J., Vanballenberghe, L., Cafmeyer, J., and Maenhaut, W., Br, Rb, Zn, Fe, Se, and K in blood of colorectal patients by INAA and PIXE, *J. Trace Elem. Med. Biol.*, 9, 193–199, 1995.
363. Leung, P.L. and Li, X.L., Multielement analysis in serum of thyroid cancer patients before and after a surgical operation, *Biol. Trace Elem. Res.*, 51, 259–266, 1996.

364. Antila, E., Mussalo-Rauhamaa, H., Kantola, M., Atroshi, F., and Westermarck, T., Association of cadmium with human breast cancer, *Sci. Total. Environ.*, 186, 251–256, 1996.
365. Wu, C.W., Wei, Y.Y., Chi, C.W., Lui, W.Y., P'Eng, F.K., and Chung, C., Tissue potassium, selenium and iron levels associated with gastric cancer progression, *Dig. Dis. Sci.*, 41, 119–125, 1996.
366. Garland, M., Morris, J.S., Colditz, G.A., Stampfer, M.J., Spate, V.L., Baskett, C.K., Rosner, B., Speizer, F.E., Willett, W.C., and Hunter, D.J., Toenail trace element levels and breast cancer: a prospective study, *Am. J. Epidemiol.*, 144, 653–660, 1996.
367. Hsiung, C.-S., Andrade, J.D., Costa, R., and Ash, K.O., Minimizing interferences in the quantitative multielement analysis of trace elements in biological fluids by inductively coupled plasma mass spectrometry, *Clin. Chem.*, 43, 2303–2311, 1997.
368. Donais, M.K., How to interface a liquid chromatograph to an inductively coupled plasma-mass spectrometer for elemental speciation studies, *Spectroscopy*, 13, 30–35, 1998.
369. Zhoa, C.O., Young, M.R., Diwan, B.A., Coogan, T.P., and Waalkes, M.P,. Association of arsenic-induced malignant transformation with DNA hypomethylation and aberrant gene expression, *Proc. Natl. Acad. Sci. USA*, 94, 10907–10912, 1997.
370. Simeonova, P.P. and Luster, M.I., Mechanisms of arsenic carcinogenicity: genetic or epigenetic mechanisms? *J. Environ. Pathol. Toxicol. Oncol.*, 19, 281–286, 2001.
371. Karagas, M.R., Tosteson, T.D., Blum, J., Morris, J.S., Baron, J.A., and Klaue, B., Design of an epidemiologic study of drinking water arsenic exposure and skin and bladder cancer risk in a U.S. population, *Environ. Health Perspect.*, 106, 1047–1050, 1998.
372. Klaue, B. and Blum, J.D., Trace analyses of arsenic in drinking water by inductively coupled plasma mass spectrometry: high resolution versus hydride generation, *Anal. Chem.*, 71, 1408–1414, 1999.
373. Karagas, M.R., Tosteson, T.D., Blum, J., Klaue, B., Weiss, J.E., Stannard, V., Spate, V., and Morris, J.S., Measurement of low levels of arsenic exposure: a comparison of water and toenail concentrations, *Am. J. Epidemiol.*, 152, 84–90, 2001.
374. Chatterjee, A., Shibata, Y., Yoshinaga, J., and Morita, M., Determination of arsenic compounds by high-performance liquid chromatography-ultrasonic nebulizer-high power nitrogen-microwave-induced plasma mass spectrometry: an accepted coupling, *Anal. Chem.*, 72, 4402–4412, 2000.
375. Allen, T.M., Bezabeth, D.Z., Smith, C.H., McCauley, E.M., Jones, A.D., Chang, D.P.Y., Kennedy, I.M., and Kelly, P.B., Speciation of arsenic oxides using laser desorption/ionization time-of-flight mass spectrometry, *Anal. Chem.*, 68, 4052–4059, 1996.
376. Florencio, M.H., Duarte, M.F., de Bettencourt, A.M.M., Gomes, M.L., and Vilas Boas, L.F., Electrospray mass spectra of arsenic compounds, *Rapid Commun. Mass Spectrom.*, 11, 469-473, 1997.
377. Apostoli, P., Alessio, L., Romeo, L., Buchet, J.P., and Leone, R., Metabolism of arsenic after acute occupational arsine intoxication, *J. Toxicol. Environ. Health*, 52, 331–342, 1997.
378. Apostoli, P., Giusti, S., Bartoli, D., Perico, A., Bavazzano, P., and Alessio, L., Multiple exposure to arsenic, antimony, and other elements in art glass manufacturing, *Am. J. Ind. Med.*, 34, 65–72, 1998.
379. Apostoli, P., Bartoli, D., Alessio, L., and Buchet, J.P., Biological monitoring of occupational exposure to inorganic arsenic, *Occup. Environ. Med.*, 56, 825–832, 1999.
380. Rayman, M.P., The importance of selenium to human health, *Lancet*, 356, 233–241, 2001.
381. Alaejos, M.S. and Diaz Romero, F.J., Selenium and cancer: some nutritional aspects, *Nutrition*, 16, 376–383, 2001.
382. Magnuson, M.L., Creed, J.T., and Brockhoff, C.A., Speciation of selenium and arsenic compounds by capillary electrophoresis with hydrodynamically modified electroosmotic flow and on-line reduction of selenium(VI) to selenium(IV) with hydride generation inductively coupled plasma mass spectrometric detection, *Analyst*, 122, 1057–1061, 1997.
383. Kotrebai, M., Tyson, J.F., Block, E., and Uden, P.C., High-performance liquid chromatography of selenium compounds utilizing perfluorinated carboxylic acid ion-pairing agents and inductively coupled plasma and electrospray ionization mass spectrometric detection, *J. Chromatogr. A.*, 866, 51–63, 2000.
384. Ip, C., Birringer, M., Block, E., Kotrebai, M., Tyson, J.F., Uden, P.C., and Lisk, D.J., Chemical speciation influences comparative activity of selenium-enriched garlic and yeast in mammary cancer prevention, *J. Agric. Food Chem.*, 48, 2062–2070, 2001.
385. Ding, M., Shi, X., Castranova, V., and Vallyathan, V., Predisposing factors in occupational lung cancer: inorganic minerals and chromium, *J. Environ. Pathol. Toxicol. Oncol.*, 19, 129–138, 2001.

386. Mclean, J.A., Zhang, H., and Montaser, A., A direct injection high efficiency nebulizer for inductively coupled plasma mass spectrometry, *Anal. Chem.*, 70, 1012–1020, 1998.
387. Singh, J., Pritchard, D.E., Carlisle, D.L., Mclean, J.A., Montaser, A., Orenstein, J.M., and Patierno, S.R., Internalization of carcinogenic lead chromate particles by cultured normal human lung epithelial cells: formation of intracellular lead-inclusion bodies and induction of apoptosis, *Toxicol. Appl. Pharmacol.*, 161, 240–248, 1999.
388. Halliwell, B., Oxygen and nitrogen are pro-carcinogens. Damage to DNA by reactive oxygen, chlorine and nitrogen species: measurement, mechanism and the effects of nutrition, *Mutat. Res.*, 443, 37–52, 1999.
389. Dizdaroglu, M., Chemical characterization of ionizing radiation-induced damage to DNA, *BioTech.*, 4, 536–546, 1986.
390. Cadet, J., Delatour, T., Douki, T., Gasparutto, D., Pouget, J.P., Ravanat, J.L., and Sauvaigo, S., Hydroxyl radicals and DNA base damage, *Mutat. Res.*, 424, 9–21, 1999.
391. Dizdaroglu, M., Chemical determination of oxidative DNA damages by gas chromatography-mass spectrometry, *Methods Enzymol.*, 234, 3–16, 1994.
392. Senturker, S., Karahalil, B., Inal, M., Yilmaz, H., Muslumanoglu, H., Gedikoglu, G., and Dizdaroglu, M., Oxidative DNA base damage and antioxidant enzyme level in childhood acute lymphoblastic leukemia, *FEBS Lett.*, 416, 286–290, 1997.
393. Malins, D. and Haimanot, R., Major alterations in the nucleotide structure of DNA in cancer of the female breast, *Cancer Res.*, 51, 5430–5432,1991.
394. Cadet, J., D'Ham, C., Douki, T., Pouget, J.P., Ravanat, J.L., and Sauvaigo, S., Facts and artifacts in the measurement of oxidative base damage to DNA, *Free Radic. Res.*, 29, 541–550, 1998.
395. Jaruga, P., Zastawny, T.H., Skokowski, J., Dizdaroglu, M., and Olinski, R., Oxidative DNA base damage and antioxidant enzyme activities in human lung cancer, *FEBS Lett.*, 341, 59–64, 1994.
396. Kasprzak, K.S., Jaruga, P., Zastawny, T.H., North, S.L., Riggs, C.W., Olinski, R., and Dizdaroglu, M., Oxidative DNA base damage and its repair in kidneys and livers of nickel(III)-treated male F344 rats, *Carcinogenesis*, 18(2), 271–277, 1997.
397. Olinski, R., Zastawny, T.H., Foksinski, M., Windorbska, W., Jaruga, P., and Dizdaroglu, M., DNA base damage in lymphocytes of cancer patients undergoing radiation therapy., *Cancer Lett.*, 106, 207–215, 1996.

5 Mechanism of Action and Metabolism of Antineoplastic and Chemopreventive Agents

5.1 CYTOTOXIC THERAPY

5.1.1 BASIC PRINCIPLES

5.1.1.1 Cell Kill Hypothesis, Classification, and Mechanisms of Action

Every tumor cell must be killed to cure a patient because even a single cancer cell is capable of eventually killing the host. The *fractional cell kill hypothesis* states that the efficacy of cytotoxic drugs obeys first order kinetics, i.e., a given drug concentration applied for a specific period of time will kill a constant fraction of the cell population, regardless of the absolute number of cells present. Each subsequent treatment cycle will kill the same fraction of the remaining cells. For example, if a tumor has 1 million cells and a particular treatment regimen has a 90% cell kill rate, the first treatment will leave 100,000 cells. Even if regrowth of the cancer cells doubles the number of cancer cells present by the time the next chemotherapy course is given, it is theoretically possible to eventually reduce the tumor cell population to only one cell which, hopefully, will be killed by the immune system (Figure 5.1a). In practice, the major bulk of the tumor is often removed by surgery prior to chemotherapy (Figure 5.1b). The fractional cell kill hypothesis does not consider the fact that cancer cells can mutate over time and, therefore, that an increasingly large proportion of the cells may become resistant to chemotherapy (Figure 5.1c). Also not considered is the fact that heterogeneous solid tumors grow slowly and are, therefore, less susceptible to cytotoxic agents.

To cause a lethal cytotoxic lesion within a cell that can, in turn, arrest the tumor progression, the attack of cytotoxic chemotherapeutic agents is directed against specific targets essential to cell replication (Figure 5.2, Table 5.1). For example, many clinically useful cytotoxic drugs are targeted against enzyme actions, such as the inhibition of enzymes involved in purine or pyrimidine biosynthesis and the operation of DNA polymerases. Other targets include the corruption of the nucleic acids (DNA rather than RNA) by alkylation or intercalation and the disruption of tubulin-facilitated processes by the inhibition either of their assembly (e.g., by vinca alkaloids) or disassembly (e.g., by taxanes). Recently, approaches to chemotherapy have been focusing on more clearly defined targets with the aim of assaulting the tumor cells themselves by specifically affecting unique pathways operating in the particular neoplastic cells being treated. Examples include protein receptor kinases, protein phosphorylation sites, membrane-bound and membrane-associated targets such as G-protein-coupled receptors, and intracellular targets such as nuclear hormone receptors. Representative examples of currently active targeted approaches are described in Section 5.3.

The number of tumor cells present at a given time is a function of the fraction of the cell population proliferating at that time and the rate of cell loss for any reason, while the rate of growth of that tumor is a function of the cell cycle time, i.e., the time required for individual cells to divide. In chemotherapy, the types of drugs used and their sequence of administration are influenced by both factors. For instance, the neoplastic cells that divide ceaselessly are susceptible to drugs targeted at the cell-cycle.[1] More effective chemotherapy may be induced by using the concept of *synchronization,* a process that increases the percentage of tumor cells in a specific phase of the

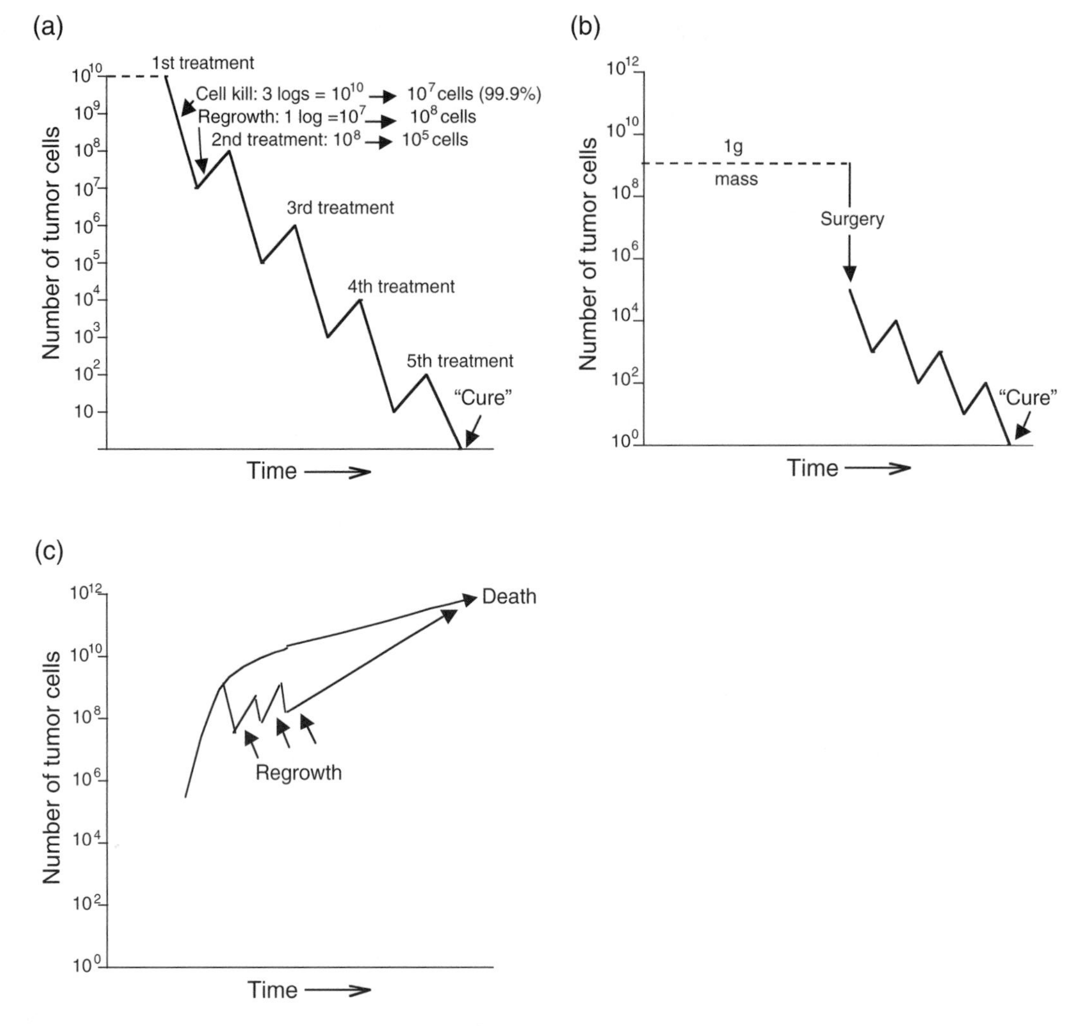

FIGURE 5.1 **(a)** Reduction and regrowth of cells during chemotherapy. **(b)** In practice, the bulk of the tumor is often removed by surgery prior to chemotherapy. **(c)** Cells can mutate over time; an increasingly larger proportion of the cells may become resistant to chemotherapy.

cell cycle and may therefore be susceptible to specific drugs. This may be accomplished by administering *cytostatic* agents which block or retard cell development at a specific phase of the cell cycle, e.g., cytosine arabinoside arrests cells at the G_1/S boundary. However, one consequence of using phase-sensitive *cytotoxic* agents that kill cells in a particular phase is an increase in the percentage of cells present in the insensitive phase, which will act as a reservoir for the disease.

Drugs that are not cell phase-specific, e.g., alkylating agents and antitumor antibiotics, will kill any proliferating cell in any phase of the cell cycle. These agents are usually active against large tumors and are often given as a single-bolus injection. *Cell cycle phase-specific* drugs kill cells only while they are at a specific phase of the cell cycle, e.g., the vinca alkaloids and taxol exert their effects in the M phase, and etoposide is active during the G_2 phase during which its target enzyme, topoisomerase II, is formed (see later). These agents are most effective against actively growing tumors. Phase-specific drugs are administered at minimal effective concentrations via continuous dosing. A few drugs are self-limiting in their toxicity to tumors, i.e., while cytotoxic in one phase, they inhibit cell growth in another phase, thereby limiting the transition of other cells

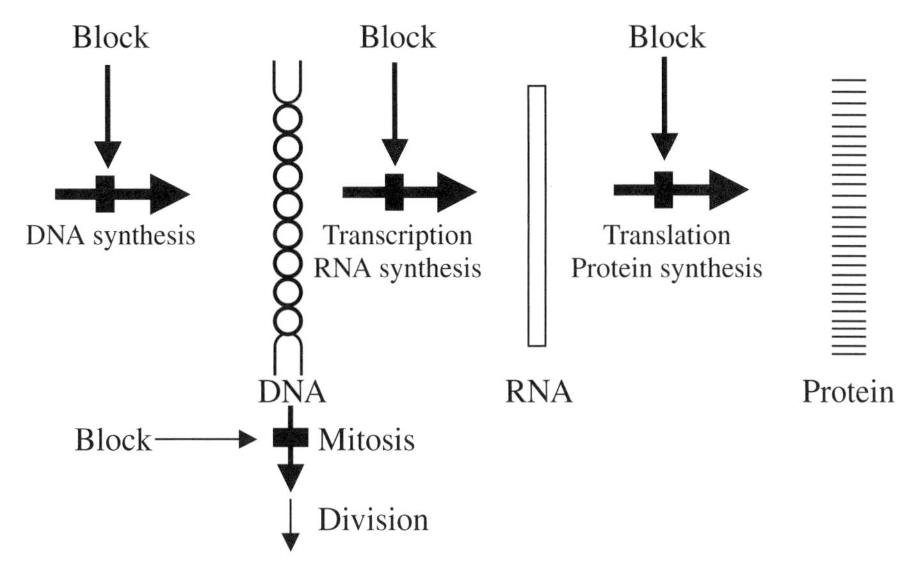

FIGURE 5.2 The attack of cytotoxic chemotherapeutic agents is directed against specific targets essential to cell replication. See also Table 5.1.

TABLE 5.1
Classification of Cytotoxic Agents by Mechanism of Action

Mechanism or Site of Action	Example
Cause Direct DNA damage [a]	
• Alkylation — break DNA helix	
Nitrogen mustards	Cyclophosphamide, ifosfamide, melphalan, chlorambucil
Nitrosoureas	Carmustine (BCNU), lomustine (CCNU)
Others	Thiotepa, busulfan
• Platinum coordination/cross linking	Cisplatin, carboplatin
• Double-strand cleavage via topoisomerase II antibiotics	
Podophyllotoxins[d]	Etoposide, teniposide
• Single-strand cleavage via topoisomerase I	Irinotecan, topotecan
• Intercalation	Doxorubicin, daunomycin
Microtubulin Spindle Poisons [b]	Vincristine, vinblastine, vinorelbin Paclitaxel, docetaxel
Antimetabolites [c]	
• Thymidylate synthase	5-Fluorouracil, 5-fluoro-2′-deoxyuridine
• Dihydrofolate reductase	Methotrexate
• DNA polymerase	Cytosine arabinoside
• Ribonucleotide reductase	Hydroxyurea
• Phosphoribosylpyrophosphate aminotransferase	6-mercaptopurine, 6-thioguanine

[a] Cell cycle nonspecific
[b] Cell cycle specific for M phase
[c] Cell cycle specific for S phase
[d] Cell cycle specific for G2 phase

into the phase where they are cytotoxic. An example is methotrexate: it kills cells in the S-phase but inhibits RNA synthesis in both G phases, thus restricting the number of cells entering the S phase.

Many cytotoxic drugs are not cytotoxic per se but exert their effects only after intracellular activation by enzymatic or chemical reactions in normal and/or tumor tissues. Activation reactions include microsomal oxidation (cyclophoshamide) or reduction (bleomycin), phosphorylation (cytosine arabinoside), polyglutamylation (methotrexate), phosphoribosylation (5-fluorouracil), hydrolysis (irinotecan), and aquation (cisplatin). These reactions are discussed in the sections on individual drugs (below). The rate of formation of the active species is limited by competition for the particular enzyme from its naturally occurring substrates, and also by factors related to the development of drug resistance, such as the rates of transmembrane influx and efflux of the drug.

5.1.1.2 Toxicity, Resistance, and Combination Chemotherapy

Most cytotoxic agents are preferentially toxic to frequently dividing cells, neoplastic as well as normal, with a slight selectivity for tumor cells. Normal cells that are significantly affected include those of the bone marrow, the mucosal lining of the gastrointestinal tract, skin, hair follicles, and germinal cells (sperm and ova). The side effects of chemotherapy resulting from the effects on these rapidly dividing normal cells include immunosuppression, gastrointestinal problems (nausea, vomiting, bleeding, ulcers), alopecia, nephrotoxocity, abortion, and teratogenesis. Treatment schedules have been designed to allow time for the recovery of affected normal cells, particularly the bone marrow. The bone marrow can supply mature cells to the peripheral blood for 8 to 10 days after being damaged by cytotoxic drugs. Nadir blood counts occur between days 14 and 18; recovery begins around 21 days after treatment.

An area of active research is the development of drugs and biological isolates that selectively protect normal cells from the toxicity of the chemotherapeutic agents. For example, a breakthrough has been achieved recently by identifying, purifying, and cloning several hematopoietic growth factors that regulate the proliferation and maturation of hematopoetic stem cells and control hematopoiesis at both the basal and emergency levels.[2] Drug-specific *cytoprotector* agents have been designed to minimize treatment-limiting toxicities to normal organs while preserving the cytotoxic effects of the drug,[3] e.g., Na-2-mercaptoethanesulfonate (MESNA) that protects against the uroepithelial toxicity induced by ifosfamide and high-dose cyclophosphamide.[4]

Pharmacogenomics is the integration of pharmacogenetics with the new technologies of genomics. The goal of pharmacogenomics is to maximize drug response while minimizing adverse effects for individuals by tailoring drug doses according to the patient's genotype.[5,6] The benefit of applying pharmacogenetics in a clinical setting has been demonstrated in the case of 5-fluorouracil. Mutations in the dihydropyrimidine dehydrogenase gene, which codes for the rate-limiting enzyme in the catabolism of the drug, lead to severe toxicity. It is recommended that patients be tested for these mutations prior to treatment.[7] There is an impressive number of genetic polymorphisms that affect drug efficacy/metabolism. MALDI-TOFMS techniques have been developed for analyzing single nucleotide polymorphisms or locating, identifying, and characterizing the function of specific genes[8] (see also Section 6.3.2.3), and applications in pharmacogenomics are likely to follow. Available mass spectrometric techniques for therapeutic drug monitoring (see later) are also suitable for investigating the often large differences in optimum dose requirements among individuals.

There are several types of chemotherapy, depending on the type and nature of the tumor and the clinical status of the patient (Table 5.2). The response to a particular chemotherapeutic treatment greatly depends on the sensitivity of the tumor to the drugs used. At one extreme, there are potentially complete cures for a few intrinsically drug-sensitive tumors, e.g., childhood acute lymphoblastic leukemia, some non-Hodgkin's lymphomas, and testicular cancer. At the other extreme, some tumors have considerable intrinsic resistance to most antineoplastic agents and, therefore, their response to conventional chemotherapy is limited to a minority of cases, e.g., non-small cell lung cancer, colon cancer, and malignant melanoma. However, drug resistance

TABLE 5.2
Types of Chemotherapy

Induction	Primary treatment of advanced cancer for which no alternative treatment exists.
Adjuvant	Systemic treatment of patients with no overt evidence of residual cancer following the elimination or destruction of the primary tumor with surgery or radiotherapy. Intended to treat micrometastases.
Primary	Treatment of localized cancer before (or in place of) surgery or (*neoadjuvant*) radiotherapy. Often allows the sparing of vital organs (e.g., bladder) as the size of the primary tumor is reduced and rendered easier to deal with by local therapies. Good response identifies patients likely to benefit from further treatment.
Combination	Administration of two or more agents. Intended to effect a synergistic response.

eventually develops in most patients and prohibits cure by preventing the eradication of all the tumor cells. *Acquired* resistance is the unresponsiveness of the tumors in individual patients after a period of successful treatment with a particular drug. Unfortunately, most tumors that are highly responsive to initial treatments eventually become refractory to further therapy, developing resistance not only to the antineoplastic agents used in the initial therapy, but even for drugs to which the patient was never exposed. Examples include breast and ovarian carcinomas and small cell lung cancers.

The cellular and biochemical mechanisms proposed to explain the development of drug resistance, both in general and as applied to individual drugs, include decreased drug accumulation resulting from decreased drug influx, increased drug efflux, altered intracellular trafficking, increased repair of or tolerance to drug-induced damage of DNA, proteins and membranes, and a variety of altered gene expressions (reviewed in the recommended books on cancer medicine, see Chapter 3).

The most important type of resistance is pleiotropic *multidrug resistance* (MDR), referring to the phenotype that expresses resistance not only to the particular agent being used for the treatment but also for a number of other drugs, often including drugs with unrelated chemical composition. Cross-resistance is limited to certain drug classes. For example, MDR occurs among the naturally derived vinca alkaloids, taxanes, and DNA intercalators, but does not occur for alkylating agents or antimetabolites. Although the processes that induce MDR are multifactorial, a major role is played by the P-glycoprotein (p170), a membrane protein encoded by the MDR-1 gene. The p170 protein that has been found at high concentrations in drug-resistant cells, elicits an overexpression of the membrane efflux pump. This leads to the enhanced removal of cytotoxic drugs from the interior of the cells, which, in turn, leads to a major decrease of intracellular drug concentration, presumably below therapeutic levels.

Another proposed mechanism of drug resistance is based on *kinetic* reasons: when tumors are in the plateau growth phase, their growth factor is small, thus only a small number of cells is susceptible to drug attack. Strategies to overcome this kind of resistance include the reduction of tumor load by surgery or radiotherapy, the use of combination therapy to affect cells in the resting stage of the cell cycle, and the use of therapy schedules that synchronize cell populations to points in the cell cycle where phase specific drugs can be used. *Pharmacological* reasons for drug resistance include inadequate blood concentrations of the drug(s) due to erratic absorption, excretion, or catabolism. This type of patient-dependent resistance may be diagnosed by therapeutic drug monitoring and reduced by dose-modification schemes. Another pharmacological reason for resistance is inadequate transportation of agents to the tumor site, such as for tumors originating in or metastasizing to, the central nervous system.

Elimination or reduction of cross-resistance have been the major rationale for *combination* chemotherapy. In many tumors, judicious combinations of drugs are significantly more effective in producing responses than when the same drugs are used sequentially. The reasons for the effectiveness of drug combinations include the possibility for simultaneously attacking cells at different stages of the cell cycle; the use of cycle-nonspecific drugs to reduce the tumor population and thereby induce a response that recruits cells into a more active dividing state where they can be destroyed by cycle-specific drugs; the possibility of biochemical enhancements, such as reducing metabolic inactivation of the therapeutic agent, e.g., using tetrahydrouridine to inhibit the cytidine deaminase inactivation of cytosine arabinoside; prevention or reduction of the probability of forming drug resistant clones; and rescuing the host from toxic effects, e.g., leucovorin therapy following treatment with high-dose methotrexate. Considerations for the expedient selection of drugs for combination chemotherapy to a specific cancer include drug effectiveness (drugs that are active individually should be selected), the biochemical or pharmacological rationale (based on animal models), and differences in dose-limiting toxicities. Perhaps the most dramatic results obtainable with combination chemotherapy have been demonstrated in acute lymphocytic leukemia. For this disease combinations of as many as six or nine agents in highly complex regimens, that maximize the synergistic effects of toxicologically acceptable doses ("summation" dose intensities), have induced complete response in >95% of the cases and cures have approached 75 to 80%. Other successful multidrug regimens include MOPP (mechlorethamine + oncovine + procarbazine + prednisone) for Hodgkin's disease; CAF (cyclophosphamide + doxorubicin + 5-fluorouracil) for metastatic breast cancer; MVAC (methotrexate + vinblastine + adriamycin + cisplatin) for bladder cancer; paclitaxel + carboplatin for ovarian carcinoma, and mitoxantrone + prednisone for prostate cancer.

5.1.1.3 Drug Discovery and Development, Clinical Trials

Historically, anticancer drug discovery has involved the screening of a large number of compounds for antineoplastic activity against a panel of rapidly proliferating murine tumors, primarily the L1210 and P388 murine leukemias. Many of these agents originated from plant, microbial, or marine animal sources, while others were synthetic compounds. After finding a "lead" compound, *rational* drug design has been aimed at synthesizing structurally related compounds that have greater therapeutic activity, fewer side effects, and desirable chemical and physical properties. During 1955 to 1985, some 40,000 new chemicals were tested, resulting in about 60 antineoplastic drugs that were eventually approved and used, mostly against hematologic malignancies. Animal solid tumor and human tumor xenografts were subsequently added as a secondary *in vivo* tumor panel against which drug candidates were tested. In 1985, a new primary screening system was established, based on the automated *in vitro* initial screening of compounds against a panel of 60 cell lines, including 8 solid tumor types.[9] Current drug discovery includes efforts to screen compounds for inhibitory activity against specific *molecular targets*, such as enzymes, growth factor receptors, and oncogene products.[10] The capacity of the current screening program of the NCI is 10,000 compounds per year. About 5% of these compounds progress to evaluation in *in vivo* screens or biochemical-molecular assays as well as comparative analyses using information available in databases. Pharmaceutical companies often have their own propriety drug discovery programs.

It is a thought-provoking fact, particularly as the rate of loss of ecological diversity increases, that about one-third of the effective antineoplastic agents currently used are from natural sources, e.g., microbial antibiotics and vinca alkaloids. More than 1000 papers have been published in which MS (and NMR) have played critical roles in the characterization of the complex structures (including chiral centers) of active principles, isolated from natural sources, as well as the intermediates and final products produced during efforts to synthesize the active compounds and numerous analogs. Representative applications include the structure characterization of antileukemic triterpene glycosides,[11] cephalostatins from marine worms,[12] novel cytotoxic diterpenes from corals,[13] the apoptosis-inducing pierisin isolated from the cabbage butterfly,[14] ecteinascidin 743 isolated from the Caribbean sea squirt that exhibits remarkable activity against soft-tissue sarcomas,[15] cytotoxic constituents from leaves of *Aglaia elliptifolia*,[16] and bioactive constituents from gum guggul.[17]

Combinatorial Chemistry

Combinatorial chemistry has shifted drug design from the traditional one-compound-at-a-time approach (synthesize, purify, characterize, test) to high-throughput, automated syntheses in which thousands of chemical variations derived from a given skeleton are produced.[18] The resulting "chemical diversity" significantly increases the chances of finding compounds that will react usefully with selected molecular targets. Techniques based on solid-phase synthesis have resulted in libraries often containing millions of compounds. This has led to problems in identifying the active components. Current approaches produce collections of thousands of discrete molecules of a designed structure, in parallel syntheses, using resin beads (one bead — one compound) or plastic pins. One of many advantages of parallel synthesis in the microtiter plate format is the possibility of direct biological screening.

The first objective of combinatorial chemistry in drug development is the discovery of "lead" compounds or "hits" within the library. Lead compounds are then synthesized for the second objective, "lead optimization," which includes testing based on a variety of considerations ranging from physical properties to desired pharmacokinetic and pharmacodynamic behavior. An example has been the use of combinatorial biocatalysis and enzymatic acylation to synthesize hundreds of paclitaxel derivatives that may solve problems with the drug caused by its poor solubility in water.[19] Although already extensively used in basic cancer research, there have been only a few reports on the application of combinatorial chemistry to anticancer drug development, a situation that is likely to change significantly in the immediate future, particularly in the area of therapeutic targeting of signaling pathways.[20,21]

During the past few years, mass spectrometric techniques have acquired a major role in drug discovery projects by providing organic synthesis support and bioanalytical support (reviewed in References 22–25). The roles of mass spectrometry in supporting organic synthesis include characterization of the compounds synthesized (confirmation of structure), estimation of the purity of the products and, more recently, "mass-directed" preparative purification of libraries.[24] Bioanalytical support includes absorption, distribution, metabolism, and excretion (ADME, see below) analyses for early drug discovery and target compound analysis in support of pre-clinical and clinical studies. The current buzzword is "high-throughput" because mass spectrometric analyses are inserted between rapid, automated processes, e.g., synthesis and screening. However, the quest of increasing throughput by increasing the number of mass spectra acquired per unit time should not be at the expense of the information content of the spectra, such as mass accuracy and resolving power.

An obvious objective of product characterization is to search for discrepancies between theoretical and actual composition. For example, differences between expected and observed masses in peptide libraries can often be attributed to under- or over-coupling of monomer/reagent at particular stages of the synthesis. Another example is the well-known 80 Da mass difference between unphosphorylated and phosphorylated peptides. Sensitivity must be high because the loading of growing structures may be as low as 100 picomoles per resin bead. Isobaric molecules, which are represented by a single ion in the mass spectrum, must be chromatographically separated before detection by SIM or analyzed directly using a mass spectrometric *ladder sequence* technique (reviewed in Reference 26). High-throughput characterization of the products in combinatorial libraries may be carried out using flow-injection analysis (FIA), a technique that can be automated easily. Techniques have been developed to use FIA in serial or parallel fashion, providing spectacular throughput results by needing only 10 to 30 sec for the characterization of individual samples.[24] A multichannel device with an array of ESI tips combined with FIA permitted the analysis of 720 samples per hour.[27] Sophisticated automatic data handling and reporting approaches have been developed to improve throughput.[28] Examples of approaches to high-throughput analysis are illustrated in References 29–32. Instrumentation and techniques, including the use of supercritical fluid chromatography which often allows better resolution than HPLC, have also been developed for the high-throughput characterization of libraries in a parallel fashion.[22,24]

There is a growing awareness of the key roles that pharmacokinetics and drug metabolism play as determinants of *in vivo* drug action. Accordingly, one aim of bioanalytical support is the *in vitro*, parallel determination of such relevant properties of lead compounds as absorption, distribution, metabolism, and excretion (ADME) at an early stage of the drug development.[33] A commonly performed LC/MS screening assay quantifies permeability using Caco-2 cells. One technique utilized LC/MS/MS and used simple peak area ratios between the apical and basal sides to calculate concentrations. APCI rather than ESI ion sources were used to reduce compound-dependent ion suppression.[34] Another important area of bioanalytical support is the assaying of metabolism using liver microsomes, hepatocytes, or related cytochrome P-450 isoforms. In one application, LC/MS/MS was used to identify and quantify substrate isoform products. Analysis time was 0.5 min and the total cycle time was 1.5 min.[35] In another application, LC/TOFMS was used to obtain both qualitative metabolic and quantitative pharmacokinetic information from cassette-dosed plasma samples.[36]

An instructive illustration of the use of MS for the identification of lead compounds in combinatorial libraries is shown by a set of experiments designed to recognize and isolate biologically active components based on their competitive interaction with a receptor. The ESI spectrum of the ~600 compounds that comprised the library was highly complex, revealing numerous singly charged molecular ions, many of them isobaric, in the 350 to 800 Da mass range (Figure 5.3a). All compounds in the library were then exposed to a receptor so that any active component could form a protein/ligand complex that could be separated from the nonreactive species by HPLC. The protein/ligand complexes were then disrupted and the isolated ligands analyzed by ESIMS. Two

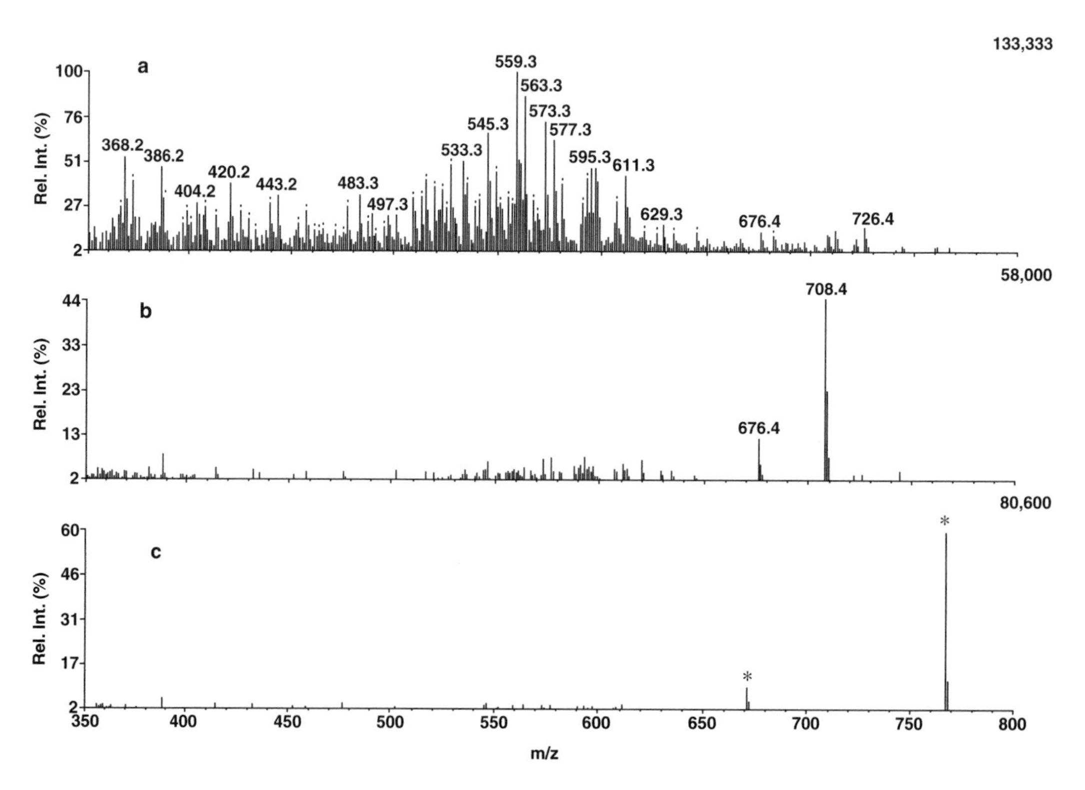

FIGURE 5.3 ESI spectra showing (**a**) combinatorial library containing ~600 compounds; (**b**) compounds selected by affinity selection in the absence of a competitor; (**c**) compounds selected by affinity selection in the presence of a competitor. Asterisks indicate multiply-charged ions corresponding to the competitor. Reprinted from Kaur, S. et al., *J. Protein Chem.*, 16, 505–511, 1997. With permission.

ions of potential interest were observed (Figure 5.3b). These ions did not appear in parallel control experiments in which a known competitor to the protein was also included in the incubation (Figure 5.3c), i.e., they did not bind at all in the presence of the competitor. Further ESIMS/MS experiments were used to determine the structure of the compounds. These compounds were then synthesized, purified, and their competitive binding properties confirmed. By keying in on the structures thus obtained and using a smaller, less diverse library, additional compounds were found that could bind to the receptor in the absence of the competitor. Even relative binding affinities could be obtained from the relative abundance of appropriate ions.[37]

There is also a need for efficient MS techniques to screen for biological activity, e.g., assessing enzyme substrate specificity[38] or conducting cytochrome P-450 enzyme inhibition studies.[35] One technique that improves sample throughput is based on replacing the chromatographic steps by combining pulsed ultrafiltration with ESIMS; after incubating the mixture to be screened with a receptor having the desired biological activity, the mixture is forced under pressure to cross a filtration membrane, thus removing unbound small molecules. The bound ligands are then released from the antibody/receptor (e.g., with methanol) and these analytes then pass through the membrane into the ESIMS system for subsequent identification.[39,40]

Drug Development

Candidate compounds that pass the primary screening are submitted to the arduous, time-consuming, and expensive process of drug development. This sequence often begins with a critical *in vivo* efficacy study in which the human tumor cell line that exhibited the highest *in vitro* sensitivity to the candidate drug is used as a xenograft in a subcutaneous implant in nude mouse models. Subsequently, preclinical studies establish interspecies variations in the candidate drug's behavior with mice, rats, and dogs. The next step is the collection of pharmacokinetic data in animals with the aim of establishing the all-important initial safe starting doses and escalation schedules for human clinical trials. This often involves at least two species to investigate qualitative and quantitative organ toxicities and their reversibilities. Additional preclinical objectives include the exploration of formulation possibilities, e.g., liposomal encapsulation.

Pharmacokinetics and pharmacodynamics reflect the interactions between the drug and the host (Figure 5.4).[41] *Pharmacokinetics* deals with what the body does with drugs over time, i.e., the kinetics of the absorption, distribution, metabolism (biotransformation), and excretion that regulate the fate of a drug in the body as well as in cell cultures or isolated organs, e.g., perfused liver.

The efficiency of absorption and the rate of distribution are strongly affected by the chemical and physical properties of the drugs as well as by the route of administration. For example, while

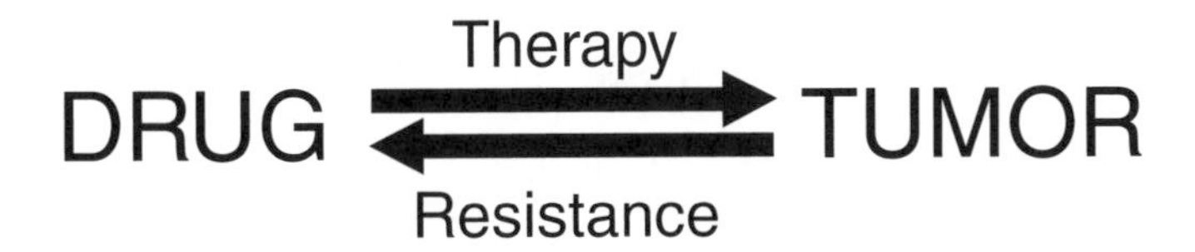

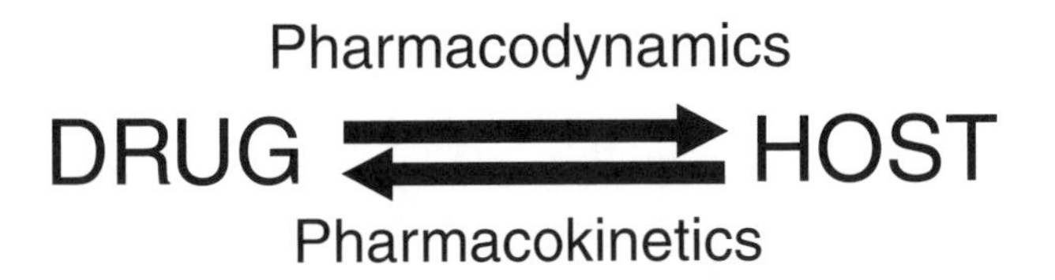

FIGURE 5.4 Pharmacokinetics and pharmacodynamics reflect the interactions between the drug and the host.

the slow absorption after oral administration usually results in relatively low blood concentrations and the likelihood of a low degree of toxic side-effects, the efficacy of the drug may be compromised because some drugs undergo extensive hepatic metabolism (first pass effect) before reaching the desired site of action. In contrast, parenteral administration, which bypasses the alimentary tract, quickly delivers large quantities of the drug to the site of action, and usually in a well-controlled manner; however, the rapid absorption is likely to increase toxicity. Drug absorption after intravenous administration obeys zero-order kinetics, i.e., a constant amount of drug is absorbed. All other routes of administration follow first order or exponential kinetics, i.e., a constant fraction of the drug is absorbed.

In contrast to pharmacokinetics, *pharmacodynamics* studies the relationship between the dose of a drug and the response in the host over time. Examples include the determination of the response of cells *in vitro* to varying doses of a new agent, or the determination of the dose-limiting toxicities in patients during Phase I clinical trials.

Therapeutic Drug Monitoring

Although drug dosages are usually calculated on the basis of body surface area, e.g., mg/m^2,[42] dose individualization based on pharmacokinetics has obvious advantages.[43,44] It is generally assumed that a threshold serum (plasma) concentration of a drug is needed at the receptor site to initiate and maintain a pharmacological response. Area under the concentration-time curve (AUC) information is generally accepted to be the best pharmacokinetic parameter for correlations with the antineoplastic efficacy, as well as toxicity, of a therapeutic agent. Reliable AUC determinations require serum concentration measurements at 8 to 12 time points, an inconvenient and costly process. Accordingly, *limited sampling models* have been developed to provide a reliable estimate of AUC values.[45,46]

Clinical Trials

The primary objective of *Phase I* trials of a drug is to determine the maximum tolerated dose in humans, with the aim of determining a safe initial therapeutic dose for subsequent efficacy testing. The typical safe starting dose is one-tenth of the equivalent dose that is lethal to 10% of nontumor-bearing animals. Secondary objectives are to obtain quantitative and temporal pharmacokinetic data as well as pharmacodynamic information on the drug. Study subjects are volunteers who have failed available conventional treatments for their disease. Although not much is expected in terms of response, even minimal signs of response are monitored. The purpose of *Phase II* trials is to determine the spectrum of the activity of the agent in patients with any of several designated signal tumor types. For solid tumors the extremes of response criteria are *complete response* (no demonstrable disease), *no response* (<50% tumor reduction at any site), or *progression* (continued increase in tumor size). *Partial response* is usually defined as >50% reduction in the product of the two longest diameters of the tumor.

Phase III trials are large, multi-institutional studies aimed at comparing the new treatment with standard treatments. These are double-blind, randomized studies in which potential biases are eliminated by arbitrarily assigning patients to treatment groups with neither patient nor physician knowing which group receives which drug. In *stratified randomization,* patients are divided into risk groups for factors known to be associated with expected response, followed by random assignment to one of the treatment arms. *Phase IV* studies are intended to evaluate new regimens in large scale general populations.

Protocols for initiation of a clinical trial include therapeutic objectives, criteria for patient eligibility, measures of response and toxicity, methods for reporting data and adverse reactions, and the often elaborate, informed consent forms. Major ethical aspects of clinical trials, such as autonomy, beneficence, and justice, have been extensively debated at all levels of sophistication, ranging from theological theories to tabloid journalism.

5.1.2 Alkylating Agents

5.1.2.1 Mechanism of Action

The mechanism of the cytotoxicity of the polyfunctional alkylating agents is based on their producing highly reactive carbonium ions. Alkylation occurs when these positively charged, highly reactive, electrophiles form covalent bonds with electron-rich nucleophilic groups, such as amino, carboxyl, phosphate, or sulfhydryl moieties of cell constituents. The most important such interaction is the alkylation of the highly nucleophilic N7 in the purine base guanine of DNA. When the structural configuration of guanine is altered, a miscoding occurs with thymidine leading to abnormal base pairing. When the action is bifunctional, i.e., the same reaction also occurs on an adjacent guanine, two strands of DNA become cross-linked by covalent bonding of two guanines. This is illustrated by the alkylating action of methchlorethamine, the simplest alkylating agent (Figure 5.5a). As a result of crosslinkages between chains and a depurination, strand breaks occur after the development of nicks in the DNA chain. This leads to an inability of DNA to replicate. At the same time, the synthesis of other cellular constituents, such as RNA and protein, continues. The resulting imbalance in cell growth eventually causes the cell to die. Although these agents do not discriminate between cycling and resting cells, maximum toxicity is exerted upon cells in the rapid division phase. Despite the common mechanism of action, alkylating agents differ in effectiveness against various neoplasms, and there are significant differences among their metabolism, pharmacokinetics, and toxicity. Of current interest is the use of high-dose induction therapy with alkylating agents followed by autologous bone marrow transplantation.

Alkylating agents are grouped into four broad classes: nitrogen mustards, alkyl sulphonates, nonclassic alkylating agents, and platinum-based drugs. The structures and chemical names of representative agents, except for platinum compounds (see later), are shown in Figures 5.5b and 5.6. During the past 25 years, mass spectrometry has played a major role in studies of the mechanism of action of these agents, as well as in the quantification of the parent drugs and their metabolites in body fluids, to support clinical investigations. For example, close to 100 publications have appeared on cyclophosphamide, more than on any other antineoplastic agent.

5.1.2.2 Nitrogen Mustards

All therapeutically active nitrogen mustards contain two chloroethylamino chains. In the simplest nitrogen mustard, mechlorethamine, a single nitrogen atom bears both chloroethyl substituents (Figure 5.5b). Mechlorethamine was developed for use in chemical warfare during World War I. It was suggested that the mustard may be used to treat leukemia after observation that it produced profound lymphocytopenia. The drug is currently used as part of the MOPP combination regimen (mechlorethamine, oncovine, prednisone, and procarbazine) in the treatment of Hodgkin's disease.

Cyclophosphamide

Cyclophosphamide (CP) (Figure 5.5b) is widely used as a component of multi-agent chemotherapy regimens, as a single agent for the mobilization of hematopoietic stem cells and in conditioning regimens for bone marrow transplantation. Unlike most alkylating agents, CP is preferentially administered orally. CP and its structural analog ifosfamide (IF) [(Figure 5.5b) see later] are chiral molecules because they contain an asymmetrically substituted phosphorus atom. Thus both drugs and several of their metabolites exist in two forms, (+)(R) and (-)(S). Both drugs are used clinically in a 50:50 (racemic) mixture of the two stereoisomers. There has been evidence that differences exist between the enantiomers with respect to metabolism, disposition, efficacy, and toxicity.

Metabolism

There are two relevant reviews on CP, a personal one covering the early research on metabolism[47] and one on the chemistry of the metabolites.[48] CP itself is not cytotoxic; it is a prodrug. The CP

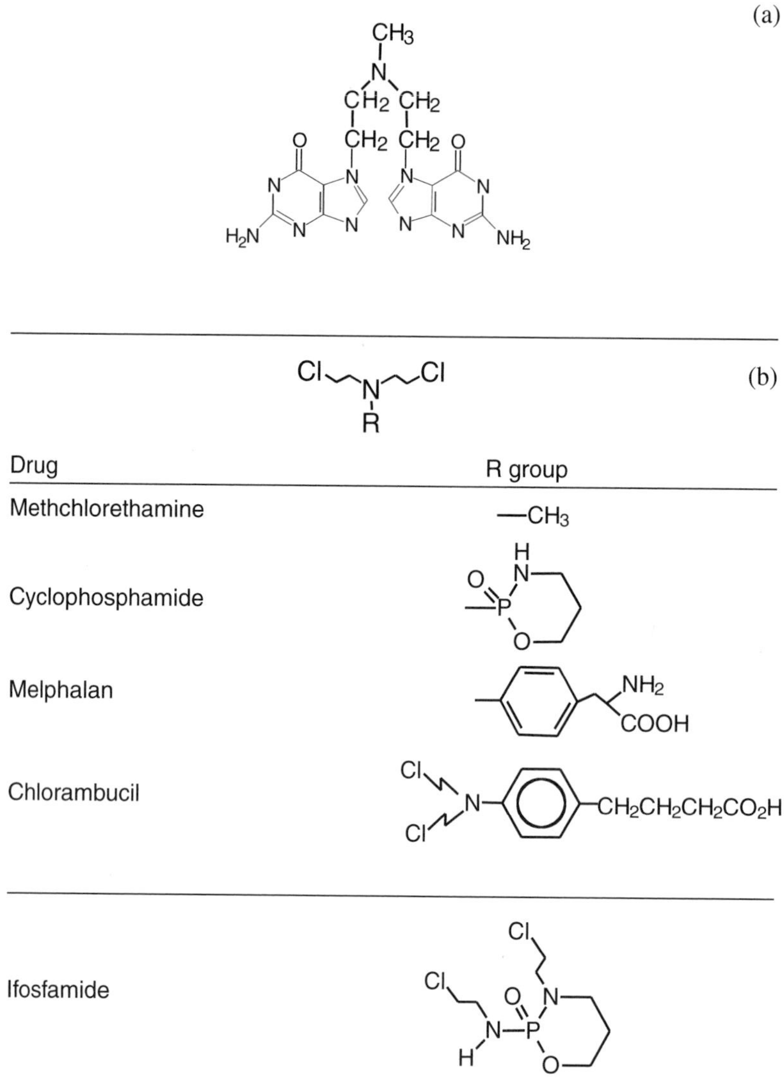

(a)

(b)

Drug	R group
Methchlorethamine	—CH_3
Cyclophosphamide	
Melphalan	
Chlorambucil	
Ifosfamide	

FIGURE 5.5 Mechanism of alkylation of a bifunctional-alkylating drug causing both interstrand and intrastrand cross-linking as it binds to DNA such as with (**a**) methclorethamine; (**b**) formulas of representative analogs of methclorethamine: cyclophosphamide (cytoxan), 2-[bis(2-chloroethyl)-amino]tetrahydro-2H-1,3,2-oxaza-phosphonine 2-oxide monohydrate. Ifosfamide (Ifosphamide), 3-(2-chloroethyl)-2-(chloroethylamino)-2H-tetrahydro-1,3,2-oxazaphosphorine-2-oxide. Melphalan (L-PAM), phenylalanine mustard, bis(2-chloroethyl)amine. Chlorambucil (Leukeran), *p*-(di-2-chlorethyl)amino-γ-phenylbutyric acid.

activation pathway starts by biotransformation in the liver to a hydroxylated intermediate, 4-hydroxycyclophosphamide (4-OH-CP), by the hepatic microsomal cytochromes P-450 2C9 and 3A4 (Figure 5.7). This precursor is transported by the systemic circulation into the target tissues. The hydroxylated intermediate, 4-OH-CP, is in equilibrium with the open-ring aldehyde, aldophosphamide (AP), which decomposes spontaneously to the active cytotoxic agent phosphoramide mustard (PM) and the inactive acrolein. The cytotoxic action is achieved by the interaction of PM with DNA. A degradation pathway also exists for CP. Further oxidation at the 4-position of the primary metabolites yields inactive metabolites that are excreted in the urine (Figure 5.7). The

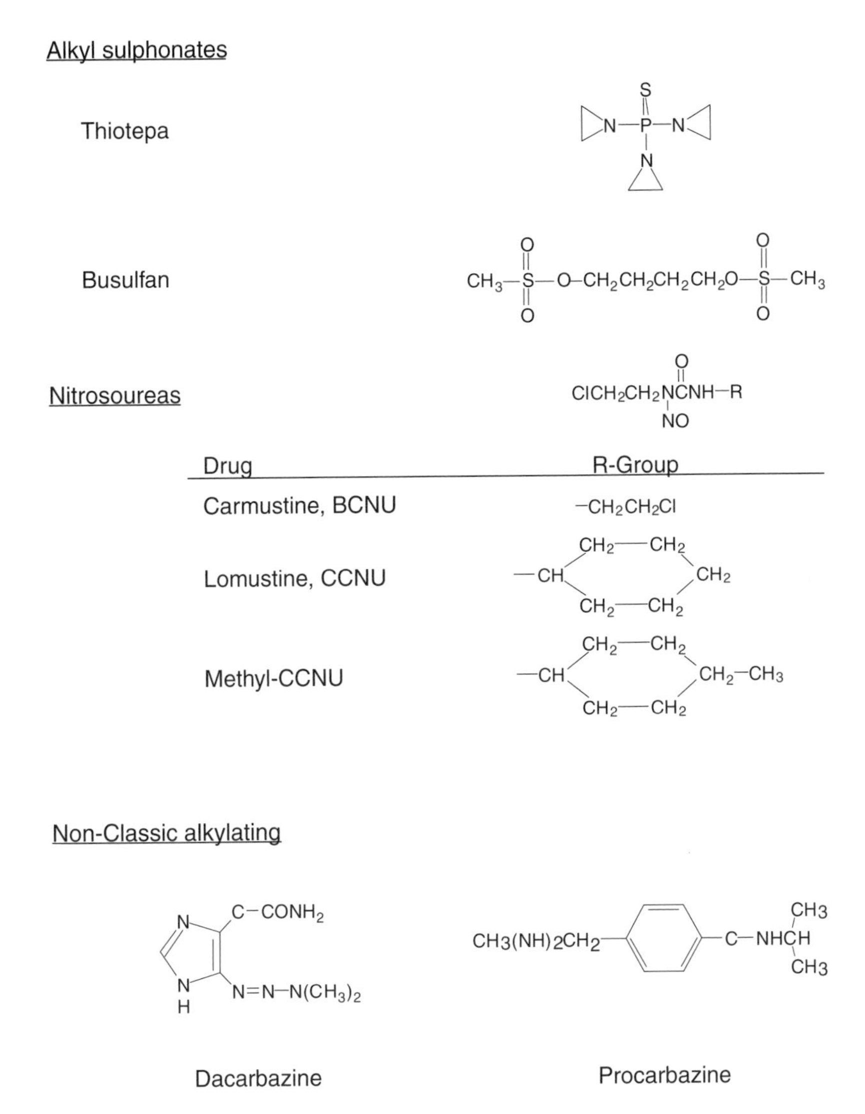

FIGURE 5.6 Formulas of alkyl sulphonates, nitrosoureas, and nonclassic alkylating agents. Thiotepa, N,N′N″-triethylenethiophosphoramide or 1,1′,1″-phosphino-thioylidenetris-aziridine. BCNU (Carmustine), bischloroethylnitrosourea. CCNU (Lomustin), cyclo-hexylchloroethylnitrosourea. Methyl-CCNU, methylcyclohexylchloroethylnitroso-urea. Busulfan (Myleran), 1,4-butanediol dimethanesulphonate. Dacarbazine, 5-(3,3-dimethyl-1-triazeno)-imidazole-4-carboxamide. Procarbazine, N-isopropyl-(2-methyl-hydrazino)-*p*-toluamide.

dechlorethylation step of the degradation pathway yields chloroacetaldehyde, which is neurotoxic. This step is minor in the case of CP but of considerable importance in IF (see later).

The reason for CP treatment failure is the development of resistance to the drug. Similar to other cytotoxic drugs, resistance is the result of several interrelated factors (Section 5.1.1.2) including the DNA repair process (reviewed in Reference 49). Some isozymes of aldehyde dehydrogenase (ALDH) catalyze the oxidation of AP leading to the formation of inert carboxyphosphamide. It has been observed that tumors resistant to oxazaphosphorines, such as CP, overexpress ALDH. It was also demonstrated that resistance can be induced in human leukemia cells by transfecting them with the gene for the human ALDH isozyme 3 (hALDH3). TOFMS has played

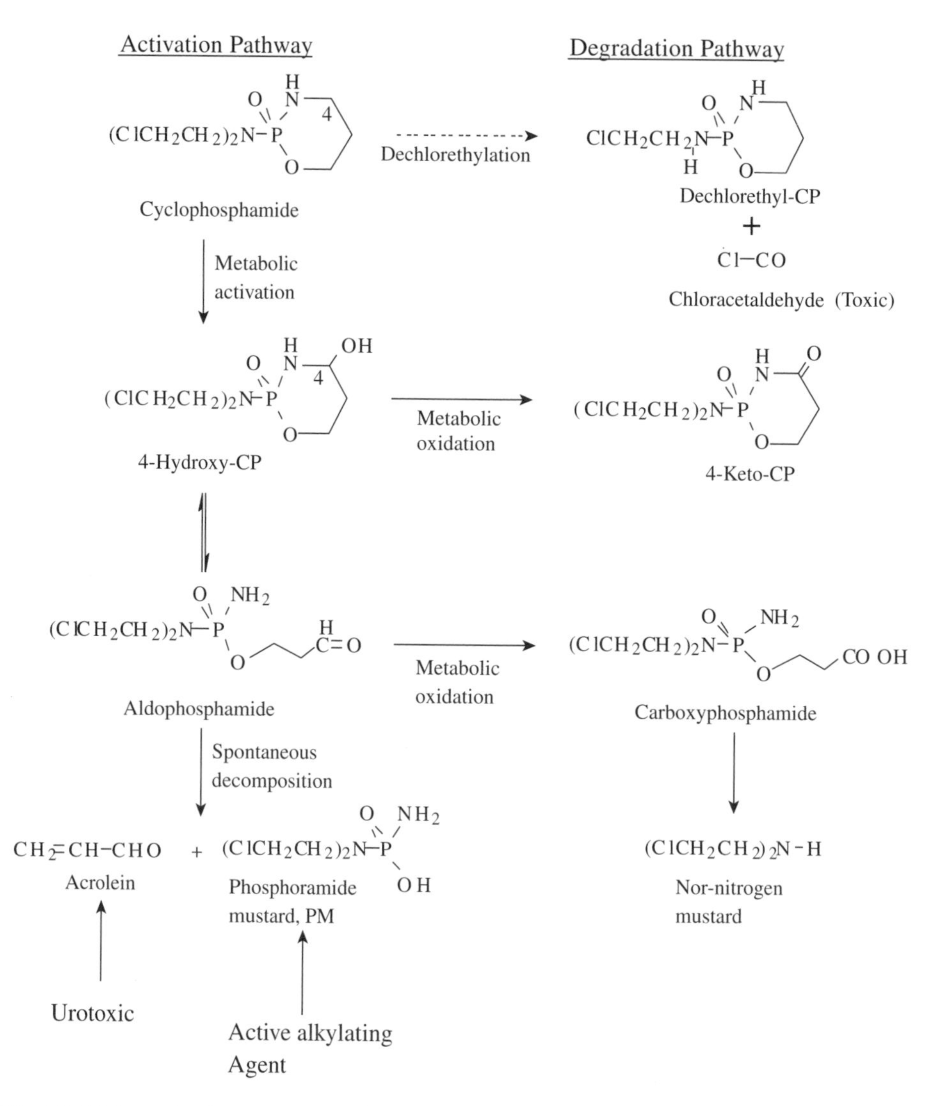

FIGURE 5.7 Major activation and degradation pathways of cyclophosphamide.

an important role in these experiments by confirming the molecular mass of recombinant hALDH3 as 50,280 ± 150 Da, in close agreement with the value predicted from cDNA, 50,247 Da.[50]

Metallothionein (MT) is a family of metal binding (Zn and Cd), high cysteine-containing proteins that occurs ubiquitously in a variety of cell types. It has been suggested that an overexpression of MT contributes to the development of resistance to alkylating agents. Mass spectrometric techniques have been used to investigate the covalent bonding between MT and alkylating agents (see Reference 51 for mechlorethamine and later in the sections on melphalan and chlorambucil). It was established using ESIMS that although the main adduct with MT was formed with nornitrogen mustard (NNM), the alkylation of MT was carried out by the PM metabolite. The molecular masses of MT-NNM and MT-PM were 6230.0 Da (calculated: 6230.8 Da) and 6308.1 Da (calculated: 6309.8 Da), respectively. MSMS revealed that the principal site of alkylation was at Cys-48 of MT. Based on these and additional MS results, it was concluded that intracellular MT may sequester and thereby reduce the concentration of active PM during chemotherapy.[52]

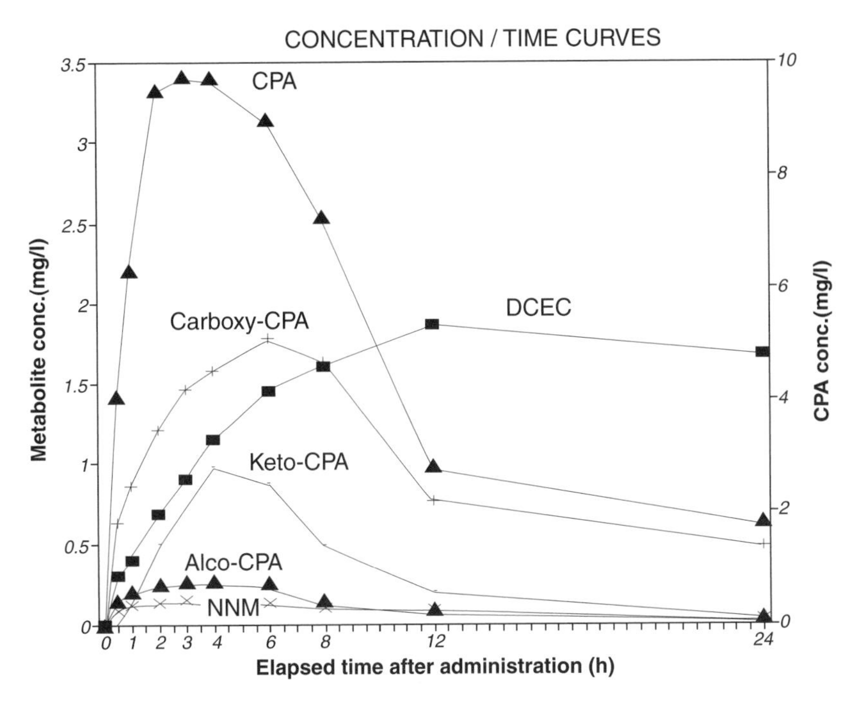

FIGURE 5.8 Example of a concentration vs. time profile for cyclophosphamide (CPA) metabolites in blood plasma of a breast cancer patient. DCEC, N-dechloroethyl-CPA; Alco-CPA, alcophosphamide; NNM, nor-nitrogen mustard, $HN(CH_2CH_2Cl)_2$. Reprinted from Momerency, G. et al., *Biol. Mass Spectrom.*, 23, 149–158, 1994. With permission.

Quantification and Pharmacokinetics

The analytical consequences of the decomposition of PM, 4-keto-PM and carboxyphosphoramide during sample storage, extraction, and derivatization and the problems of incomplete derivatization were presented >20 yr ago in a pioneering paper on the quantification of PM and metabolites by GC/CIMS.[53] A review of the analysis of CP and its metabolites covers a number of techniques, including NMR, TLC, HPLC, GC, methods for the separation of the enantiomers as well as GC/MS and LC/MS techniques.[54] Because the intermediary metabolites are transported to tumor tissues by blood circulation, the concentrations of circulating 4-OH-CP and AP, the immediate precursors of PM, reflect the cytotoxic effects of CP better than that of the parent drug. A technique for the quantification of CP and six metabolites in blood plasma utilized solid–phase extraction, derivatization to form stable methyltrifluoroacetyl derivatives and quantification (i.s.: IF) by SIM using GC/NCIMS. The technique was applied to obtain concentration vs. time profiles in breast cancer patients after the oral administration of 150 mg CP (Figure 5.8).[55] The paper also compares the advantages and limitations of the technique by contrasting it to a GC/CIMS technique based on forming stable cyanohydrin derivatives that were silylated prior to quantification using SIM.[56]

A technique for the routine monitoring of 4-OH-CP and its acyclic tautomer AP in whole blood utilized O-(pentafluorobenzyl)hydroxylamine-HCl to rapidly form a stable aldophosphamide oxime derivative (PBOX). Whole blood samples were added to previously prepared reaction tubes containing the derivatization reagent and the i.s. (D_4-PBOX).[57] The analyte products were stable up to 8 d after derivatization. The PBOX products were next isolated and silylated. It was determined by separate LSIMS and GC/EIMS experiments using synthetic compounds that both 4-OH-CP and AP ultimately produced the same *t*-butyldimethylsilyl-PBOX compound. The ions used for quantification by GC/EIMS with SIM were breakdown products of the analyte formed in the injection

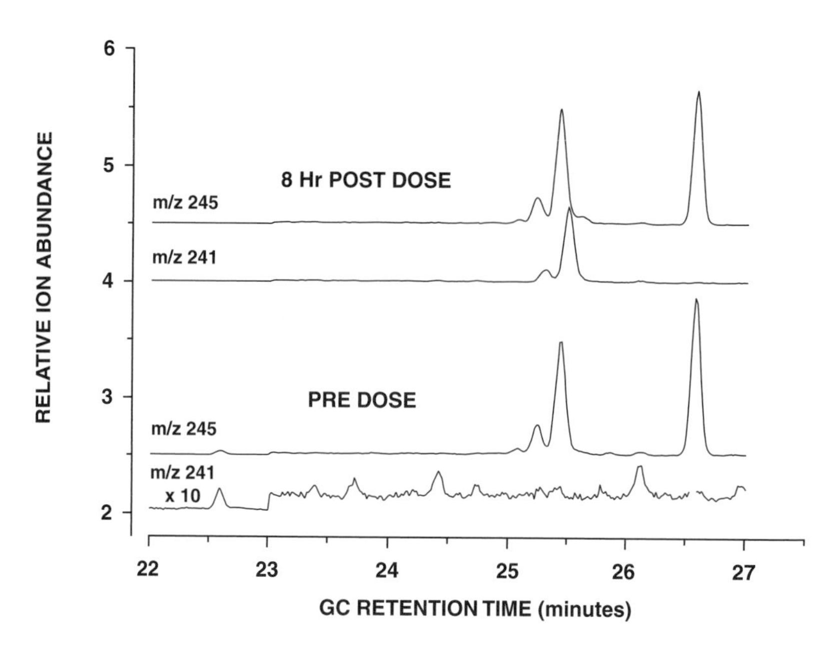

FIGURE 5.9 Reconstructed ion chromatograms of patient samples before and 8 h after administration of cyclophosphamide; *m/z* 245 is the ion monitored for the i.s., and *m/z* 241 is the ion monitored for the analyte from patient samples. Reprinted from Anderson, L. et al., *J. Chromatogr. B,* 667, 247–257, 1995. With permission.

port of the GC. Additional ions were also monitored to confirm peak identities. They were linear in the 0.085 μM (25 ng/mL) to 34 μM (10 μg/mL) range. The technique was applied to analyze blood from patients receiving 4 g/m^2 CP by i.v. over a 90 min period as part of induction therapy prior to bone marrow transplantation. There was no interference in the predose monitoring and the 4-OH-CP/AP concentration was 6 μM in the sample taken 8 h post-dose (Figure 5.9). Analyte concentrations monitored during and after drug infusion revealed that maximum concentrations, in the 10 to 14 μM range, were reached in 1.5 to 3.5 h, followed by a steady decrease to 0.5 to 0.7 μM in 24 to 28 h post-infusion (Figure 5.10).[58]

Subsequently the technique was modified to include the quantification of CP in whole blood (i.s.: D_4-CP) and used to obtain pharmacokinetic data in women with metastatic breast cancer receiving two different regimens of CP in combination with high-dose thiotepa prior to autologous bone marrow transplantation. Whole blood and plasma CP concentrations were compared in three patients during and following a 1.5 h i.v. infusion of 4 g/m^2 CP and a 96 h i.v. infusion of 6 g/m^2 CP (Figure 5.11). Despite some differences between individual whole-blood and plasma concentrations among the paired patients, differential partitioning of CP was not significant. In contrast, there was statistically significant nonproportional differential partitioning between the whole-blood exposures of both the parent drug and the metabolite. The dose-normalized AUC_{CP} values, in μM × h/g/m^2, were 1112 vs. 1579 for the first and second courses, respectively, i.e., there was a 1.4-fold increase for the second course. The corresponding $AUC_{4\text{-}OHCP}$ values were 25 and 21 for the two courses, revealing a 22% decrease from the first to the second course. These results are consistent with earlier suggestions, and concurrent *in vitro* human liver microsome experiments, concerning drug interactions between CP and thiotepa.[59] Noting the importance of potential drug interactions in combination chemotherapy, the methodologies described above were used in a follow-up clinical study on breast cancer patients treated with similar CP-thiotepa regimens. The data obtained permitted the establishment of nonlinear pharmacokinetics and computer simulations aiming to explore the complex relationship between CP and its active metabolite.[60]

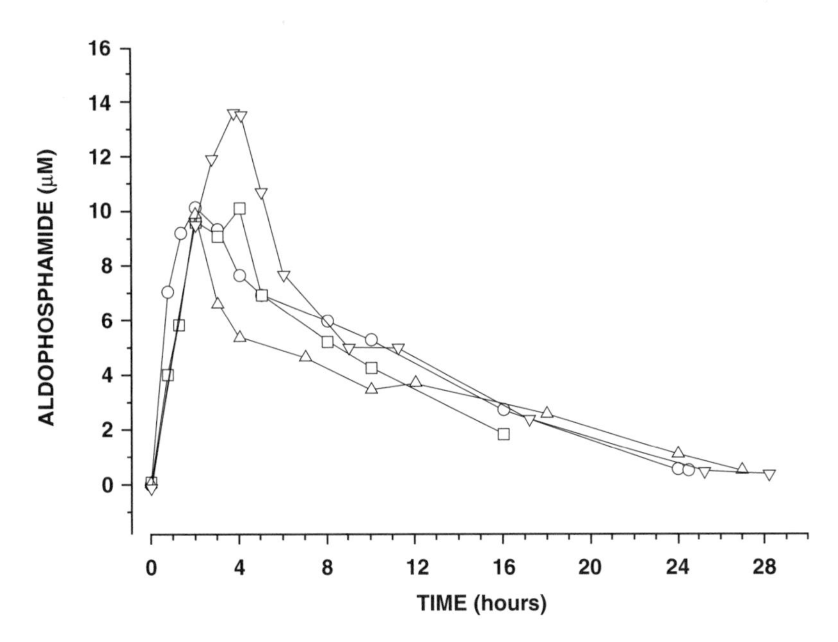

FIGURE 5.10 Concetrations of 4-hydroxy-cyclophosphamide (4-OH-CP) and its tautomer, aldophosphamide (AP) in the plasma of 4 patients during and after a 90 min intravenous infusion of 4 g/m^2 cyclophophosphamide. Reprinted from Anderson, L. et al., *J. Chromatogr. B,* 667, 247–257, 1995. With permission.

A technique developed for the quantification of CP and five metabolites in plasma used a sample preparation scheme to separate nonpolar compounds into an organic layer with the polar compounds remaining in aqueous solution. Nonpolar analytes were quantified using GC with a nitrogen-phosphorus detector. The parent drug and two metabolites, carboxyethylphosphoramide mustard and 3-hydropropylphosphoramide mustard (a potential "transport intermediate"), were quantified by LC/ESIMS (i.s.: deuterated analytes). LOD for all analytes were in the low to sub-micromolar range using 0.5 mL size samples. Application to samples from patients receiving CP infusion (60 mg/kg, once a day for 2 days) yielded CP and metabolite concentrations in the 1 to 15 μM range during 50 h of monitoring.[61] An alternative LC/ESIMS approach extracted CP and several metabolites from plasma with C_{18} solid phase extraction and used methylhydroxylamine to convert 4-OH-CP to the stable methyloxime form (i.s.: ifosfamide). LOD were 15 ng/mL for CP, 3-dechloro-ethyliphosphamide, 4-keto-CP, and 30 ng/mL for 4-OH-CP and carboxyphosphamide. To illustrate the utility of the technique, kinetic elimination data were obtained for CP and 4 metabolites in the plasma of 2 patients at several time points up to 24 h after i.v. administration of 1100 mg CP.[62] Three MS/MS-based techniques have been developed for the quantification of CP and IF in urine;[63–65] these are discussed in Section 4.2.4.1 on occupational hazards.

The pharmacokinetics of the (R) and (S) enantiomers of CP and its dechloroethylated metabolites were studied in cancer patients using a chiral GC/MS methodology developed for IF (see later and in Reference 66). Using an enantiospecific compartmental pharmacokinetic model that can evaluate the proportion of CP dose transformed to active or toxic metabolites, twofold differences were found between the formation and clearance of (R)- and (S)-CP and their metabolites, but not for several other pharmacokinetic parameters. Differences were also observed between the pharmacokinetic and metabolic profiles of adult and pediatric patients.[67]

Although *acrolein* is a noncytotoxic end product of CP metabolism (Figure 5.7), the compound significantly contributes to urotoxicity when high doses of CP and IF are used. In a GC/MS technique developed for the monitoring of acrolein toxicity, sample preparation consisted of heating

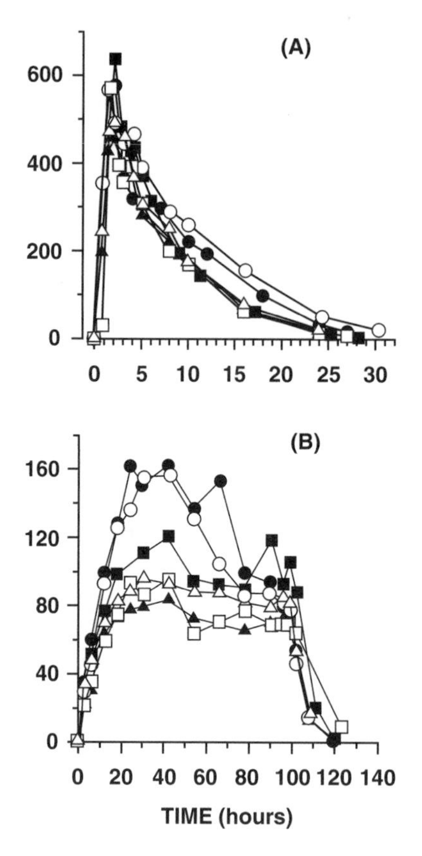

FIGURE 5.11 Comparison of whole-blood (open circle, square, and triangle) and plasma (filled symbols) cyclophosphamide (CP) concentrations for three patients. **(A)** during and following an 1.5 h intravenous infusion of 4 g/m^2 dose of cyclophosphamide; **(B)** during and following a 96 h intravenous infusion of 6 g/m^2 cyclophosphamide. Reprinted from Anderson, L. W. et al., *Clin. Cancer Res.,* 2, 1481–1487, 1996. With permission.

urine at 80°C for 5 min. After injecting an aliquot into the headspace volume of a GC/MS instrument, the analyte was quantified using SIM (i.s.: propionaldehyde). A separate calibration curve was needed with each sample to compensate for the wide variation of evaporation rates in individual urines. Maximum acrolein excretion (16.5 nmole using 280 mg/2 hr infusion of CP) occurred at the end of the transfusions.[68]

DNA Adducts

The formation of adducts between PM, the ultimate cytotoxic metabolite of CP, and DNA were investigated *in vitro* using FABMS. Three expected adducts were synthesized: two mono-DNA adducts, N-(2-chlorethyl)-N[2-(7-guaninyl)ethyl]amine (NOR-G), N-(2-hydroxyethyl)-N[2-(7-guaninyl)ethyl]amine (NOR-G-OH), and the cross-linked product N,N-bis[2-(7-guaninyl)ethylyl]amine (G-NOR-G). The FAB mass spectra of these products were compared to those obtained by incubating purified PM with calf-thymus DNA. The mass spectra of the alkylated DNA exhibited the same characteristic ions for NOR-G and NOR-G-OH as a mixture of synthetic compounds, but not the expected ions corresponding to [G-NOR-G]$^+$. However, the increased sensitivity and specificity of precursor-ion scans in MS/MS permitted the detection of G-NOR-G in the alkylated DNA.[69]

Ifosfamide

Ifosfamide (IF, also spelled ifosphamide) is an oxazaposphorine nitrogen mustard, a structural isomer of CP, involving a shift of a 2-chlorethyl group from the nitrogen mustard side-chain to the nitrogen of the heterocyclic ring (Figure 5.5). The drug has important applications in the salvage therapy of testicular cancer and recurrent lymphoma and, as a single agent or in combination therapy, for osteosarcoma and soft tissue sarcoma. Like CP, IF itself is inactive. The mechanism of action of the drug is similar to that of CP (Figure 5.7), including the need for metabolic activation, especially 4-hydroxylation, that leads to the active metabolite 4-hydroxyifosphoramide (4-OH-IF) which differentially partitions among cell types.[70] 4-OH-IF is in an equilibrium with the open-ring tautomer, aldoifosphamide (AIF). The ultimate intracellular alkylating metabolite, ifosfamide mustard (IPM), is generated from AIF upon the β-elimination of acrolein.

The degradation pathway of the 2-chloroethyl chains leads to the production of 2- or 3-dechloroethylifosfamide (2-DCEI and 3-DCEI) and equimolar amounts of chloroacetaldehyde. A major difference in the metabolism of IF and CP is that the side-chain biotransformation of IF is much greater (~50% of the dose) than that of CP (<10% of the dose). In addition, the dose of IF has to be threefold larger than those of CP to achieve equipotent alkylation. Thus, the concentration of the side-chain products of IF may be up to 100-fold larger than those of CP. Of the three metabolic products, the two DCEI isomers are not nephrotoxic; however, chloroacetaldehyde is an unstable, highly reactive compound that has been implicated in the nephropathy associated with IF treatment. According to one hypothesis, the toxic chloroacetaldehyde is produced locally within the renal tubular cells where IF is metabolized by cytochrome P-450 enzymes. Both DCEIs were detected and identified by GC/EIMS when human kidney microsomes were incubated with IF. The biotransformation of IF into these metabolites was time- and concentration-dependent. The chloroacetaldehyde metabolite could not be identified due to its instability. The presence of the DCEIs implied that the human kidney is able to produce the nephrotoxic chloroacetaldehyde.[71]

A comprehensive sample preparation approach has been developed for the analysis of IF and four metabolites associated with both the activation and degradation pathways in plasma (Figure 5.12). In one plasma aliquot the labile 4-OH-IF/AIF metabolites were first converted to the stable cyanohydrin adducts with KCN, followed by the extraction and silylation of the analytes, IF, and its metabolites. The cytoactive IPM was extracted with a C_{18} resin from another plasma aliquot, followed by silylation. Deuterium-labeled analogues were synthesized for each analyte to serve as i.s. Quantification was accomplished using GC/CIITMS and SIM of the dehydrochlorinated ions of each analyte. The LOD were in the 0.1 to 0.5 μg/mL range using 100 μL plasma samples. In an illustrative application, the concentrations of the parent drug and the 4 metabolites were monitored in plasma taken from treated rats for 8 h after drug administration.[72]

An alternative technique utilized solid phase extraction for the isolation of four acidic and one neutral metabolites (2-DCEI and 3-DCEI, 4-ketoifosfamide, carboxyifosfamide, and the weakly acidic IPM), followed by a two-step derivatization procedure involving methylation and trifluoroacetylation to form stable methyltrifluoroacetyl derivatives. Two previously undetected metabolites, chloroethylamine and 1,3-oxazolidine-2-one, were extracted prior to derivatization. Quantification of each analyte was accomplished using GC/EIMS and GC/NCIMS with SIM of specific ions. The method was linear in the 3 ng to 20 μg/mL plasma range. Typical concentration-time profiles of IF and metabolites in plasma samples from a patient who received an oral dose of 500 mg IF are shown in Figure 5.13.[73] This technique is similar to the one developed for CP and its metabolites (Figure 5.9).[55] The technique was applied to determine IF concentrations in a study that revealed that the pharmacokinetics of IF change in a sequence-dependent manner when IF is used in combination with docetaxel (Section 5.1.4.3), resulting in increased clinical toxicity.[74]

A GC/EIMS technique was developed for the quantification of enantiomers of IF and its 2-DCEI and 3-DCEI metabolites in plasma and urine. The enantiomers were separated by GC using a chiral stationary phase based on heptakis(2,6-di-O-methyl-3-pentyl)-β-cyclodextrin. LOD were

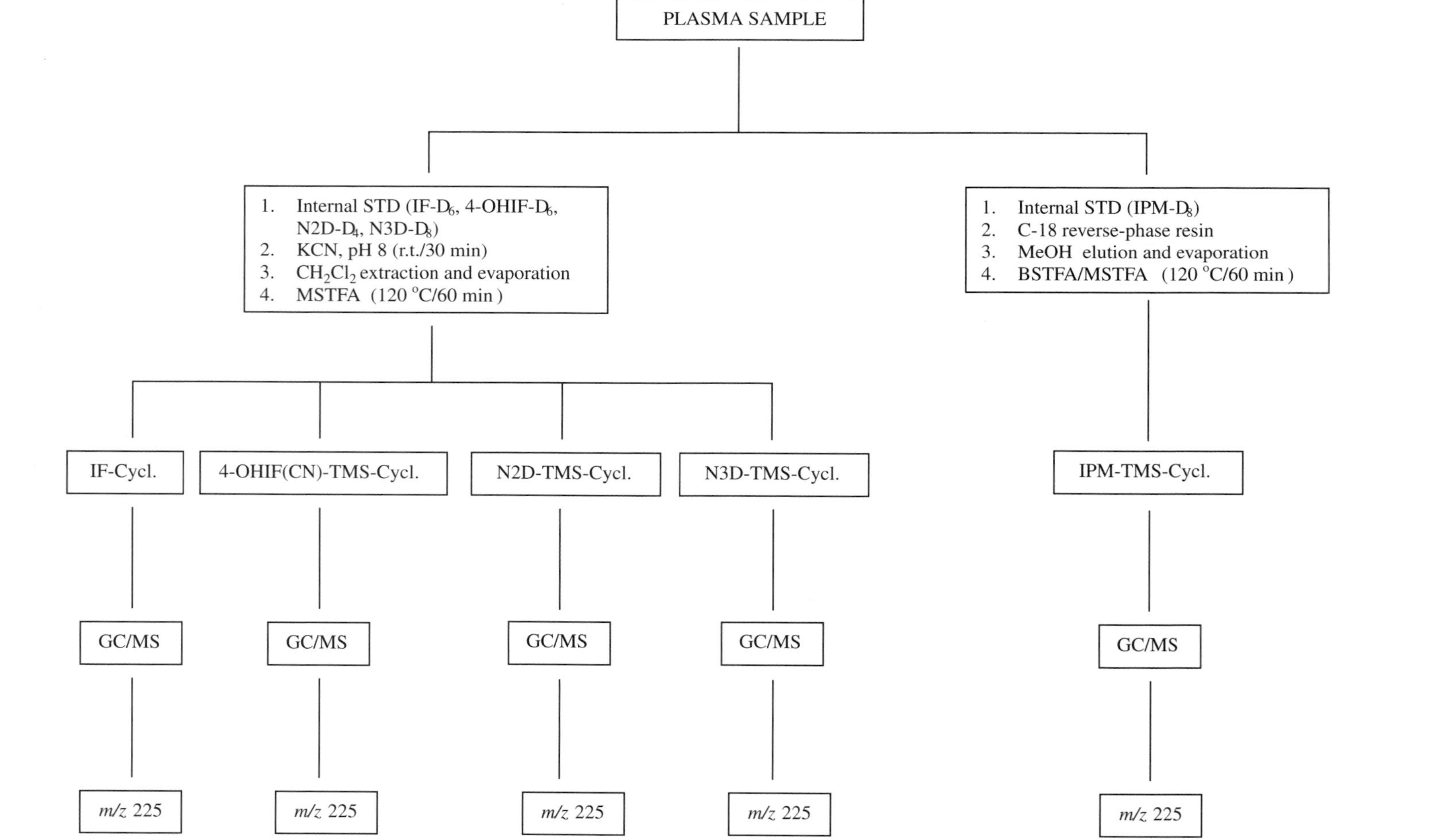

FIGURE 5.12 Comprehensive analytical scheme for the analysis of ifosfamide and major metabolites in plasma. STD, standard; IF-Cycl, cyclic dehydrochlorinated ifosfamide; 4-OHIF(CN)-TMS-Cycl, cyclic silylated dehydrochlorinated 4-hydroxy-ifosfamide cyanohydrin; N2D-TMS-Cycl, cyclic silylated dehydrochlorinated N2-dechloro-ethylifosfamide; N3D-TMS-Cycl, cyclic silylated dehydrochlorinated N3-dechloro-ethylifosfamide; IPM-TMS-Cycl, cyclic silylated dehydrochlorinated ifosfamide mustard. Redrawn from Wang, J. and Chan, K. K., *J. Chromatogr. B,* 674, 205–217, 1995. With permission.

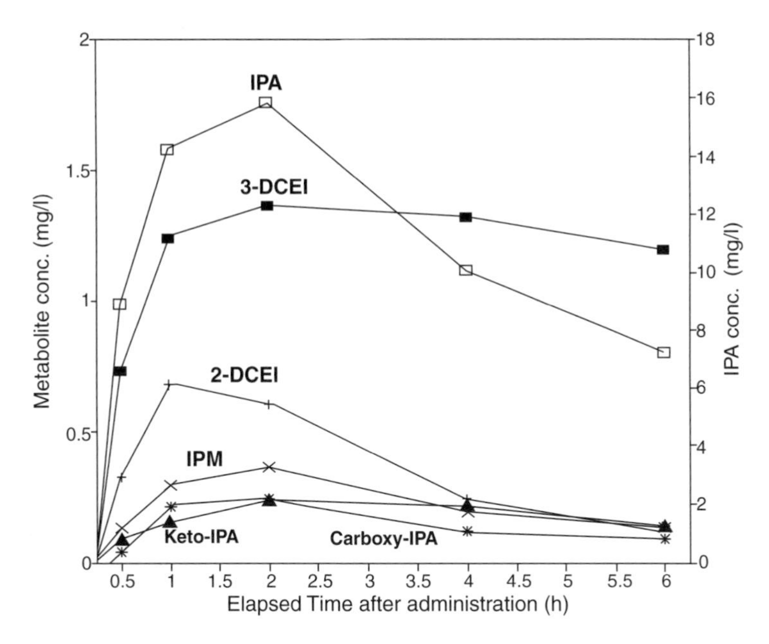

FIGURE 5.13 Example of a profile of concentration against time for IPA metabolites in the blood plasma of a cancer patient who received 500 mg ifosfamide. IPA, ifosfamide; 2-DCEI, 2-dechlorethylifosfamide; 3-DCEI, 3-dechloroethylifosfamide; keto-IPA, ketoifosfamide; carboxy-IPA, carboxyifosfamide methylester; IPM, methylifosfamide mustard. All analytes were monitored as stable methyl and/or trifluoroacetyl derivatives. Reprinted from Momerency, G. et al., *J. High Resol. Chrom.,* 17, 655–661, 1994. With permission.

~500 ng/mL in both plasma and urine. The utility of the technique was illustrated by the separation of the enantiomers of IF and the metabolites in plasma and urine from patients with cervical cancer who were administered 3 g/m^2 IF in 170 min i.v. infusions.[75] In a study aimed to confirm that the DCEI metabolites of the (S)-enantiomer lead to higher neurotoxicity, purified enantiomers were used in an *in vivo* rat mammary carcinoma model to study various aspects of toxicity and their relationships to the time course of plasma concentrations of the pure enantiomers. The drug extracted from plasma was monitored underivatized (i.s.: hexobarbital) using GC/EIMS with SIM. There were significant differences in the AUCs between racemic mixtures vs. the (S) enantiomer, and the (S) vs. the (R) forms, but not between the racemic mixture and the (R) enantiomer. Although there was no difference between the antitumor efficacy of the enantiomers, the (R) form had higher lethality and bone marrow suppression than the (S) form. This agreed with the pharmacokinetic observation that the (R) form was metabolized to a greater extent and cleared from the plasma faster than the (S) form. It was suggested that purified (R)-ifosfamide might yield higher concentrations of the desirable cytotoxic metabolite at the expense of the inactive but highly neurotoxic metabolite formed in the degradation pathway.[66]

Transportation of Activated Metabolites by Erythrocytes

Blood samples taken from patients receiving an i.v. infusion of IF were separated into plasma and erythrocyte fractions, and the concentrations of the parent drug and seven metabolites were determined by the GC/MS method described above.[73] In the erythrocytes the concentrations of all compounds were higher or equal to the concentrations in plasma. The red cell fraction contained significant quantities of the cytotoxic IPM (as much as 77% of the total blood concentration) and carboxyifosfamide. It was concluded that erythrocytes transport the major fraction of activated IF from the site of its formation to the tissues.[76]

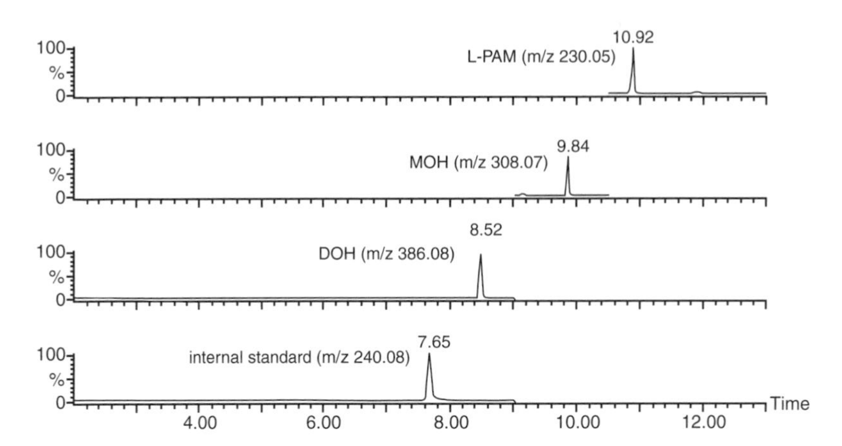

FIGURE 5.14 Selected ion monitoring of *p*-[bis(2-chloroethyl)amino]-phenylacetyl acid methyl ester (internal standard), *p*-[bis-N-(2-hydroxyethyl)amino]phenylalanine (DOH), *p*-[(2-hydroxyethyl)(2-chloroethyl)amino]-phenylalanine (MOH), and phenyl alanine mustard (L-PAM) from a leg perfusate plasma sample, recorded 90 min after administration of L-PAM. Reprinted from De Boeck, G. et al., *J. High Resol. Chrom.*, 20, 697–700, 1997. With permission.

Melphalan

Melphalan (L-PAM, L-phenylalanine mustard, L-sarcolysin), 4-[bis(2-chloroethyl)amino]-L-phenylalanine (Figure 5.5) has been used clinically for >30 years for the treatment of breast and ovarian carcinomas, malignant melanoma, and multiple myeloma. Early efforts using MS concentrated on the interpretation of the EI spectra of the compound, derivatives for GC and hydrolytic decomposition products. The first attempt at quantification involved the direct probe introduction of diazomethylated melphalan and SIM of the $[M\text{-}CH_2NH_2CO_2Me]^+$ ion.[77] A more convenient GC/CIMS technique (i.s.: deuterated analyte) using SIM of the protonated molecules was applied to the analysis of plasma samples from treated patients.[78] A GC/EIMS technique utilized trifluoroacetic anhydride to derivatize the amino and hydroxyl functions, and diazomethane to derivatize the carboxyl function, and SIM of diagnostic ions (i.s.: dichloroethylamino phenylacetic acid methyl ester). The technique was applied to determine melphalan and its monohydroxy- and dihydroxy-ethylamino metabolites (MOH and DOH) in spiked plasma and in samples taken 30 and 90 min after the infusion of 100 mg melphalan to a cancer patient. Plasma taken 30 min after infusion contained 16.4 mg/L melphalan, 0.3 mg/L MOH, and no DOH. The 90-min sample contained 9.5 mg/L melphalan, 1.3 mg/L MOH, and 0.6 mg/L DOH (Figure 5.14).[79]

The suggestion that reduced glutathione (GSH) conjugation of melphalan may contribute to the development of cross-resistance against nitrogen mustards has been based upon the observation of increased cellular concentrations of both GSH and glutathione-S-transferase (GST) in various resistant tumors. In one study, 4-(glutathionyl)phenylalanine was identified by FABMS when melphalan was incubated with glutathione and immobilized microsomal GST. The MS data also led to an exploration of the mechanism of the formation of the GSH adducts of alkylating agents.[80] Subsequently, two major GSH conjugates were identified in biliary excretions by FABMS: mono-hydroxy-mono-GSH-L-PAM and di-GSH-L-PAM. However, quantification in bile or liver perfusate samples, from both rats and patients treated with melphalan, led to the conclusion that hepatic GSH conjugation of melphalan is unlikely to be of importance in the development of cross-resistance against alkylating agents.[81]

Another study investigated the role of the human multidrug resistance protein (MRP1) as a mediator of the transport of melphalan (and also chlorambucil) conjugated to glutathione. ESIMS was used to identify three different glutathione-S-conjugates: monoglutathionyl-S-conjugate, monohydroxy-glutathionyl-S-conjugate, and the bisglutathionyl-S-conjugate. It was concluded that MRP1, although active, was not the rate-limiting step in the resistance exhibited to melphalan and chlorambucil.[82]

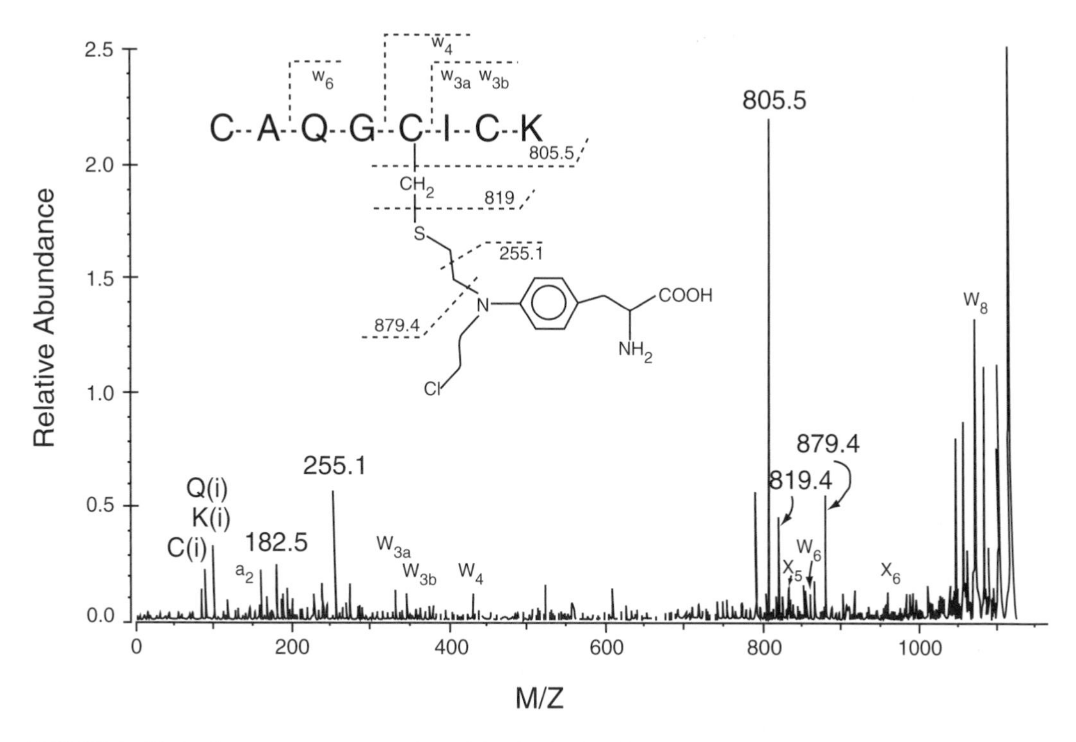

FIGURE 5.15 Tandem mass spectrum of peptide [CAQGCICK]-melphalan, in which -SH groups have been converted to $-SCH_3$ groups. Major fragments are illustrated in the inset, which indicate alkylation at Cys-48. Reprinted from Yu, X., Wu, Z., and Fenselau, C. C., *Biochemistry,* 34, 3377–3385, 1995. With permission.

As already mentioned in connection with CP, it has been suggested that an overexpression of metallothionein (MT) may contribute to the development of resistance against alkylating agents. When commercial MT was incubated with melphalan, new HPLC peaks appeared, and ESIMS determined that each peak contained protein species with molecular masses 268.3 Da higher than that of the previously determined mass of the major subisoform of MT, 6125 Da.[83] This was consistent with the addition of one melphalan moiety. When the melphalan-alkylated MT samples were methylated at free cysteine thiols and hydrolyzed with trypsin, sequence analyses by FABMS and ESIMS established that one of three collected HPLC fractions contained a component corresponding to the adduction of one melphalan to the peptide CAQGCICK, at residues 44 to 51. The other two HPLC peaks represented melphalan-containing peptides at residues 32 to 43 and 1 to 22. The exact site of alkylation was determined by MS/MS in samples further hydrolyzed by endoproteinase Asp-N. The reaction site in peptide CAQGCICK was at Cys-48 (Figure 5.15). The sites of alkylation were identified by MS/MS in the other two melphalan-containing peptides and it was established that two cysteines in the α-domain, Cys-48 and Cys-33, accounted for ~90% of the total alkylation. The direct covalent sequestration of melphalan was shown to occur rapidly and efficiently *in vitro* at near-physiological conditions. The sulfhydryl concentration of MT (~20 sulfhydryl groups per molecule) was comparable to that provided by glutathione. Thus, activated Ha-*ras*-induced synthesis of MT might lead to intracellular MT concentrations that, in turn, would induce covalent sequestration of melphalan. This case supports the hypothesis about the role of MT in the development of cross-resistance against alkylating agents.[84]

Chlorambucil

Chlorambucil, an aromatic derivative of mechlorethamine (Figure 5.5), is a bifunctional nitrogen mustard alkylating agent, which has been widely used in the treatment of a variety of malignancies.

Chlorambucil has a cytotoxic metabolite, phenylacetic mustard, 2-[4-N,N-bis(2-chloroethyl)aminophenyl] acid (PAM), that is formed at the butyric acid side-chain by β-oxidation.

A GC/EIMS technique was developed for the quantification of chlorambucil and its metabolite in plasma and urine. After precipitating proteins with perchloric acid, the analytes were extracted and the TMS derivatives were analyzed using SIM (i.s.: D_8-chlorambucil). Recovery was ~94%.[85] In another approach, the carboxyl group was derivatized by extractive alkylation with allyl bromide. Increased chemical stability was achieved when the nitrogen mustard group was converted into a tetrahydrothiazane derivative with sodium sulfide prior to alkylation. Quantification was linear in the 10 to 640 ng/ml range using GC/EIMS with SIM. The LOD of the metabolite was 2 ng/mL.[86] A study of the pharmacokinetics of chlorambucil and the metabolite revealed that the absorption of chlorambucil was slower in the presence of food than when fasting; however, the AUCs were unaffected.[87]

A GC/EIMS technique developed for the simultaneous quantification of chlorambucil and prednimustine (the C21-chlorambucil ester of prednisolone) in plasma extracts is based on the transesterification of chlorambucil in the aqueous phase and prednimustine in the organic phase (i.s.: D_8-chlorambucil). LOD was 8 ng/mL.[88] Prednimustine rapidly metabolizes in plasma to its components, prednisolone, chlorambucil, and the metabolite PAM. Treatments with prednimustine and prednisolone plus chlorambucil were compared. The appearance, elimination times, and quantities of prednisolone, chlorambucil, and PAM revealed significant differences between the two regimens. It was suggested that prednimustine expresses the classic alkylating action of chlorambucil combined with the corticosteroid action of prednisolone.[89]

In a more recent approach to the quantification of chlorambucil and its β-oxidation metabolite in human serum and plasma, automated solid-phase extraction of the analytes was followed by quantification by LC/ESI-MS/MS with SRM using representative transitions for the parent drug and the metabolite (i.s.: ^{13}C,D-labeled analogs). The method was validated in the 4 to 800 ng/mL serum or plasma range. Using 200 μL samples, a batch of 96 samples (74 study samples, 14 standards, 6 quality controls, 2 blanks) could be analyzed in 50 min. The method has been applied to provide comparative pharmacokinetic data following single-dose oral administration of three different formulations of chlorambucil.[90]

The hypothesis about the possible role of metallothionein (MT) in acquired drug resistance, described for melphalan in the previous section, was also investigated for chlorambucil. The sites of modification induced by chlorambucil in MT were the sulfur atoms of cysteines 33 and 48, which co-chelate the same metal atom in native MT. In one experiment aiming to monitor the stability of the MT metal clusters after alkylation by chlorambucil, the reactions between rabbit MT with various cadmium/zinc metal ion contents and chlorambucil were monitored directly using ESIMS, which was scanned continuously as the reaction mixtures were infused. The values of the observed masses of the metallated MT species were within 0.1% of those calculated. At the beginning of the experiment (t = 3 min), the ratios of the 4+ charge states of Cd_4Zn_3-MT, Cd_5Zn_2-MT, Cd_6Zn_1-MT, and Cd_7-MT were 1:2:2:1 (Figure 5.16, upper part). Sixty minutes later the measured masses of MT had shifted by the amount expected from the modification caused by chlorambucil, however, the metal compositions did not change (Figure 5.16, bottom part). Unmodified MT disappeared as the reaction proceeded. The fact that chlorambucil, melphalan, and CP react predominantly with Cys-48 and Cys-33 was utilized in molecular dynamics calculations aiming to establish the structural requirements for the sequestration of these drugs by MT.[91]

5.1.2.3 Alkyl Sulfonates

Thiotepa

Thiotepa (Figure 5.6) is an analog of the ring-closed intermediates of the nitrogen mustards. The drug has a broad spectrum of antineoplastic activity and has been used in cancer therapy for >40 years. The drug is now used in high-dose combination regimens against solid tumors. Thiotepa is stable in dilute aqueous solutions. Acid- and salt-induced hydrolysis products found by FIMS

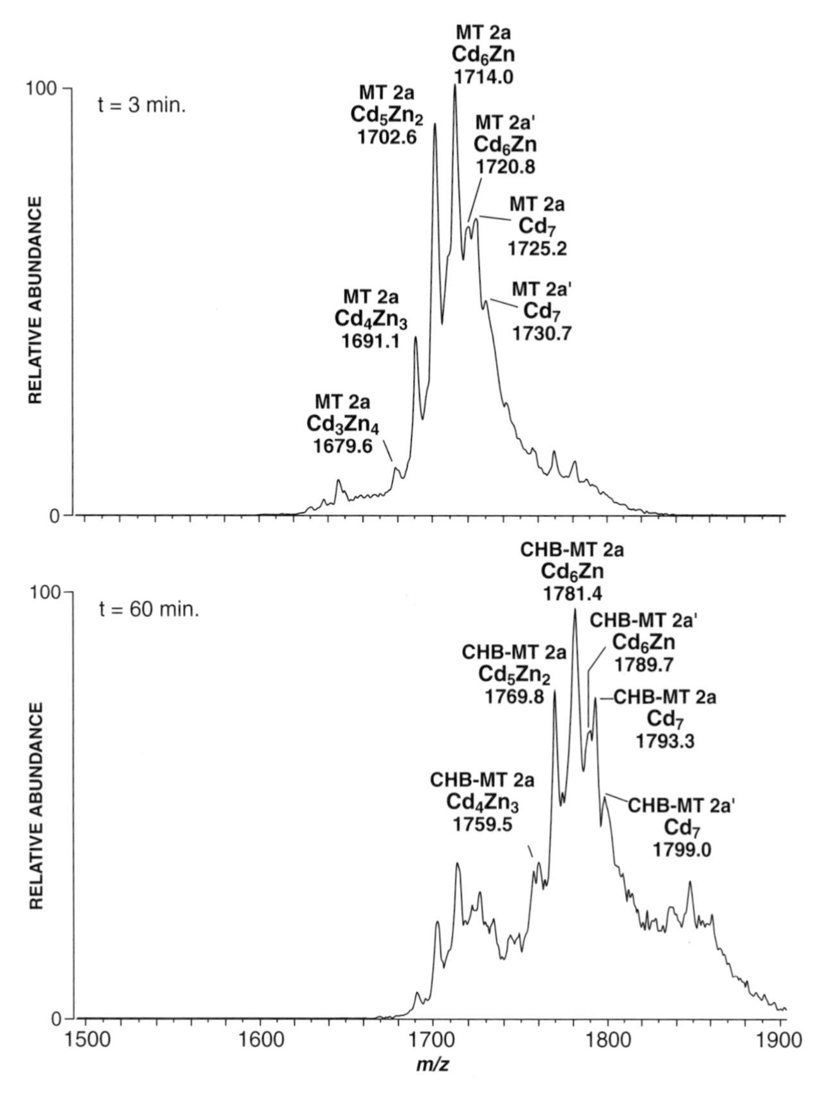

FIGURE 5.16 ESIMS spectra obtained by monitoring the reaction between metallothienin (MT) and chlorambucil (CHB) on-line: (**a**), t = 3 min; (**b**) = 60 min. All ions are in the 4+ charge state and correspond to rabbit metallothionein 2a or 2a′ with various cadmium/zinc metal ion compositions. Reprinted from Zaia, J. et al., *Biochemistry,* 35, 2830–2835, 1997. With permission.

included mono and dichloro derivatives, products formed by the cleavage between P and N and release of aziridine.[92] Two metabolites detected in acidic urine samples from treated patients were identified by EIMS as the mono- and dichloro derivatives, resulting from the opening of the aziridine rings and conversion to 2-chlorethyl moieties.[93] An important metabolite, N,N′N″-triethylenephosphoramide (TEPA) is formed in the liver after oxidative desulfuration, mediated by cytochrome P-450 enzymes; TEPA also exhibits alkylating activity. In a method developed for the simultaneous quantification of thiotepa and TEPA, the analytes were extracted from plasma or urine (i.s.: diphenylamine), the solvents were evaporated, and aliquots from the reconstituted residues were analyzed by GC/CIMS. Quantification was possible in the 0.1 to 25 μg/mL plasma or urine range. The utility of the technique was demonstrated by obtaining concentration-time profiles for both thiotepa and TEPA using plasma samples from patients treated with 40 mg/m^2 thiotepa. The half-life of TEPA was found to be 2 to 7 times longer than that of the parent drug.[94]

Because the alkylating activity of urines from treated patients exceeded that expected for the total of thiotepa and TEPA content, a search was made for additional urinary metabolites. A new metabolite was identified by GC/CIMS as N,N′-diethylene-N″-2-chloroethylphosphoramide (monochloroTEPA) that formed from the conversion of an aziridinyl function of TEPA into a β-chloroethyl moiety. Another metabolite was identified by FABMS as thiotepa-mercapturate resulting from glutathione conjugation followed by the loss of two amino acid residues from the glutathionine adduct. The CID product ions indicated the loss of one and two aziridine molecules. The concentrations of these metabolites in urine exceeded that of TEPA which used to be considered the major metabolite of thiotepa.[95] An LC/ESIMS technique was subsequently developed for the quantification of thiotepa-mercapturate in urine (i.s.: sulphadiazine). The therapeutically relevant concentration range was 1 to 25 μg/mL. An application to monitor urines from patients treated with thiotepa revealed that the new metabolite was present in urines at larger concentrations and for longer times than TEPA. The clinical importance of thioTEPA-mercapturate is not known.[96]

The interaction between thiotepa and all four DNA bases were explored using FI after incubating the drug with bases methylated at various positions. Alkylation was confirmed by obtaining complexes with thiotepa binding at N3-Thy, M1-Cyt, and N9-Ade sites.[97] The alkylation adducts of deoxyguanosine (dGuo) and DNA, involving both one and two dGuo moieties, were confirmed by FABMS. Decomposition led to hydrolysis products of the glycosidic bond via the opening of the imidazole ring. The alkylation of DNA was also studied after detaching the alkylated purines by heating. The site of the monothiotepa alkylation of guanosine in both dGuo and DNA was at N7.[98] Thiotepa-DNA adducts could be depurinated from DNA and identified by LC/TSPMS as the modified bases without the ribose moiety attached.[99]

It has been suggested that increased glutathione (GSH) and/or glutathione-S-transferase (GST) activity might lead to the deactivation of thiotepa, by conjugating this electrophilic compound with the GSH, and thus may be a major factor in acquired cross-resistance to alkylating agents.[100] The major product of the reaction between GSH and thiotepa was determined by ESIMS to be monoglutathionyl-thiotepa. It was shown that allelic variants of human PI class GST (hGSTP1-1), which differ in their primary structures at amino acids in two positions (104 and/or 113), have different efficiencies in catalyzing the formation of this conjugate. Accordingly, subjects homozygous for this allele are at a greater risk of developing GST-mediated resistance to thiotepa than heterozygotes or homozygotes with a different amino acid at position 104.[101]

It was shown by ESIMS that thiotepa is connected covalently to the β-chain subunit of human hemoglobin upon *in vitro* incubation. The location of the connection was determined to be at the conformationally accessible Cys-93 residue. The stochiometry of the complex formation was 1:1.[102]

Busulfan

Busulfan (Figure 5.6) is a straight chain bifunctional alkylating agent. After using low doses for the treatment of chronic myelogenous leukemia since the 1950s, there has been a renewed interest in the profound effects of high-dose regimens of this drug (in combination with CP or etoposide) on the granulocyte stem cells when used as a preparatory regimen for bone marrow transplantation in the treatment of leukemias.

A study with radiolabeled drug led to the identification of three urinary metabolites by GC/EIMS in the rat: 3-hydroxysulfolane, tetrahydrothiophene and sulfolane. As these metabolites are not cytotoxic, it was suggested that the *in vivo* antineoplastic effect of busulfan is mediated by the parent compound.[103] The biotransformation of busulfan may be followed by quantifying tetrahydrothiophene (THT) which forms from a positively charged sulfonium ion, an intermediate product, either enzymatically (e.g., by β-lyases) or chemically. A GC/EIMS technique using SIM was developed for the quantification of THT in human liver cytosols incubated with busulfan (i.s.: 2-ethylthiophene). The LOD was 2 ng/mL. The technique was tested in enzyme kinetic studies in human liver cytosol.[104]

In a technique developed for the quantification of busulfan extracted from plasma, cerebrospinal fluid (CSF), or pleural effusions, sensitivity was increased by treatment with NaI to convert the analyte first to 4-iodo-1-butanol methanesulphonate and then to 1,4-diiodobutane.[105] Quantification was accomplished by GC/EIMS (i.s.: D_4- and D_8-busulfan)[106] using SIM of the $[M-127]^+$ ions (loss of one iodine). The LOD was 0.5 ng/mL. In one study of children receiving high-dose busulfan, the CSF:plasma busulfan concentration ratios reflected the dose-dependent neurotoxocity of the drug (that could be efficiently prevented by administering clonazepam).[107]

Based on the straightforward EI fragmentation pattern of underivatized busulfan, an LC/EIMS technique was developed for its quantification in serum and CSF. After extraction with dichloromethane from serum or spinal fluid, the HPLC eluent was sprayed into the desolvation chamber of a particle beam interface and a fragment ion monitored. Despite the relatively poor LOD, 100 ng/mL, both serum and CSF concentrations could be determined in children receiving high busulfan doses prior to bone marrow transplantation.[108]

In a GC/MS technique developed for the quantification of busulfan in plasma, CIMS was used to confirm the identity of the products of derivatization with 2,3,4,6-tetrafluorothiophenol. Quantification of the solvent-extracted busulfan was accomplished by EIMS (i.s.: D_8-busulfan) in the therapeutically important 10 to 2000 ng/mL range. The technique was applied to the determination of the clearance and volume of distribution of busulfan in children treated with 1 mg/kg busulfan.[109]

When busulfan is administered at high doses to patients undergoing bone marrow transplantation, there is a high incidence (20 to 40%) of a hepatic veno-occlusive disease (HVOD) that is fatal in ~50% of cases. Although increased plasma concentrations of the drug have been reported occasionally in patients who developed HVOD, no causal relationship has been established between HVOD and busulfan. In one application, individual plasma levels were monitored by GC/MS in bone marrow-transplanted children (<3 yr old) receiving 600 mg/m^2 doses of busulfan.[110] Pharmacokinetic evaluation of the plasma concentrations revealed significant changes in elimination half-life, volume of distribution, and clearance in the presence of lysosomal storage disease, suggesting a need for dosage changes to achieve optimal myeloablation before transplantation.[111] A subsequent pharmacokinetic investigation of children receiving busulfan doses ranging from 16 to 600 mg/m^2 confirmed the already demonstrated wide interpatient variability in the disposition of busulfan. Despite the high incidence of HVOD among children receiving the highest doses of busulfan, a study of the pharmacodynamic relationship between busulfan disposition and HVOD in combination chemotherapy revealed that there were no alterations in the previously observed interpatient variability of busulfan disposition due to the presence of HVOD.[112] A chronopharmacology study in bone marrow-transplanted children revealed a significant circadian rhythm; however, the presence of HVOD did not alter the acrophase (maximum) plasma levels occurring at about 6 am, or other observed mean and amplitude values, suggesting that no dosage compromises were necessary in the presence of the disease.[113]

5.1.2.4 Nitrosoureas

The nitrosoureas (formulae in Figure 5.6) are highly lipophilic compounds and as such they are able to penetrate the blood-brain barrier. Thus, nitrosoureas are important in the treatment of brain tumors and meningeal leukemias. Clinical applications have been limited by the instability of these drugs, severe and prolonged myelosuppression, and pulmonary toxicity. The therapeutic and toxic effects of the nitrosoureas are based on the ability of their active metabolites to alkylate DNA and other nucleophiles. The active metabolites are also carbamoylate proteins. The rapid decomposition of nitrosoureas in aqueous environment yields two major products: chloroethyl (or methyl) carbonium ions that form adducts at N7 and O^6-guanines, and a substituted isocyanate, which carbamoylates the lysine residues of proteins. Cytotoxicity is expressed only in dividing cells, thus nondividing cells escape the action of nitrosoureas.

In a study of the metabolism of CCNU and methyl-CCNU, the major product of their nonenzymatic degradation in physiologic buffers was identified by GC/EIMS as 2-chloroethanol, which resulted from the deprotonation of N3. The 2-chloroethyl moiety was trapped with Cl^-, Br^-, and I^- ions. Degradation products included acetaldehyde, vinyl chloride, ethylene, and cyclohexylamine.[114] The *in vivo* metabolism of CCNU in rats was studied using ^{14}C-labeled drug with GC/EIMS for identification. One metabolite was identified by GC/EIMS as a thiodiacetic acid. The rapid microsomal hydroxylation of the cyclohexyl ring yielded five additional metabolites, *cis* or *trans*-2-hydroxy-, *trans*-3-hydroxy-, *cis*-3-hydroxy-, *cis*-4-hydroxy-, and *trans*-4-hydroxy-CCNU. In another *in vitro* study of microsomal metabolism EIMS was used for identification. The products of CCNU and methyl-CCNU were ring-hydroxylated derivatives. The main metabolic product of BCNU was identified as 1,3-bis(2-chloroethyl)urea. Quantification of the metabolites led to the conclusion that microsomal metabolism took place before chemical decomposition.[115]

The rapid decomposition of nitrosoureas has presented problems in quantification. In one technique, sample extracts from rat-liver microsomal incubations and rat and human plasma were introduced by a direct insertion probe into a CI source. The protonated molecules and abundant fragment ions (loss of chlorines) were used for quantification (i.s.: BCNU-$^2D_8$). LOD was to 50 ng/mL.[116] Employing the technique on patients revealed first-order kinetics with an elimination half-life of ~15 min.[117] In a subsequent study, several ether-extracted lipophilic chloroethylnitrosoureas were converted to O-methylcarbamate derivatives with anhydrous methanol and quantified by GC/CIMS using SIM of the $[MH\text{-}HOCH_3]^+$ ions of the analytes. LOD was 100 ng/mL. Clearance curves and pharmacokinetic data were obtained from treated rats.[118] In an alternative approach, BCNU, CCNU, and Me-CCNU were converted to their 1,3-diacyl-1,3-dialkylurea derivatives prior to quantification by GC/EIMS (i.s.: chloropropyl homologs) using SIM. Clearance curves were obtained from treated dogs.[119]

Fotemustine (FM), diethyl-1-[3-(2-chloroethyl)-3-nitrosoureido]ethylposphonate, is a nitrosourea with encouraging clinical responses for brain metastases of melanoma. As determined by FABMS, FM decomposed rapidly into a reactive diethyl ethylphosphonate, in the absence of GSH, in both buffers and rat liver fractions. Further decomposition yielded diethyl 1-aminoethyl-phosphonate; however, this step could be blocked by the addition of GSH.[120] The decomposition profile of FM, obtained by ESIMS during a 5 to 100 min incubation period, revealed several products relevant to DNA alkylation, including diethyl[1-(2-hydroxyethyl)ureido]ethylphosphonate, 2-chloroethyldiazohydroxide, and diethyl(1-aminoethyl)phosphonate. It was concluded that the initial activity of FM is caused by 2-chloroethyldiazohydroxide which rapidly generates O^6-guanine lesions. A long-lived iminol tautomer is responsible for both the alkylation of the remaining guanine and the toxicity of fotemustine.[121]

The electrophylic reactivity of the metabolites of nitrosoureas is also displayed toward nucleophilic sites of proteins, suggesting the possible use of hemoglobin adducts as biomarkers in treated patients or exposed subjects. In an *in vitro* study, BCNU was incubated with several model peptides, the products were separated using HPLC, and collected fractions were analyzed by FABMS. The main reaction product with angiotensin was the 2-chloroethylcarbamoyl derivative as evidenced by a gain of 105 Da in mass and the appearance of the characteristic chlorine isotope pattern. The site of modification within the angiotensin molecule was at the N-terminus, determined by FAB-MS/MS, which revealed an abundant product ion at *m/z* 1401, an "a" series fragment, and other structurally significant products. Analyses of other, minor reaction products revealed modifications of the N-terminus-Arg-2 site of angiotensin.[122]

5.1.2.5 Nonclassic Alkylating Agents

Although the "nonclassic"-alkylating agents do not contain the two chloroethyl groups of the classic alkylating agents needed for bifunctional alkylation of DNA, there is at least one common structural feature, namely the presence of an N-methyl group that facilitates monofunctional covalent binding to macromolecules. Nonclassic alkylating agents are *prodrugs*, i.e., their biologic action is expressed by active intermediates formed in a series of metabolic transformations.

Dacarbazine

Dacarbazine (DTIC) (Figure 5.6) is used as a single agent and in combination therapy in the treatment of metastatic malignant melanoma, malignant neuroendocrine tumors, and in the ABVD regimen (adriamycin, bleomycin, vinblastine, dacarbazine) for Hodgkin's disease. The drug undergoes both photolytic and hydrolytic degradation in aqueous solution. The main degradation product was identified as 2-azahypoxanthine.[123] Numerous dacarbazine analogs have been developed, e.g., 3-(3,3-dimethyl-1-triazenyl)-5-methyl-4,5-dihydroisoxazole, with selective antimetastatic properties. The structures of these compounds have been verified by EIMS.[124]

Procarbazine

Procarbazine (Figure 5.6), a methylhydrazine derivative, is primarily used as part of the MOPP (mustargen, oncovin, procarbazine, prednisone) combination regimen for the treatment of Hodgkin's disease. The drug is also active in malignant lymphomas, leukemias, and brain tumors. A GC/CIMS technique (i.s.: N-ethyl-*p*-toluamide) developed for the quantification of procarbazine and seven metabolites in plasma used derivatization with acetic acid anhydride to yield conversion of the parent drug to its triacetyl form. The LOD was 1 ng/mL. In the rat model, ~70% of the total drug and metabolites were present as aldehyde, rather than azo, metabolites 1 h after an oral dose.[125]

5.1.2.6 Platinum Compounds

Structures and Clinical Indications

The platinum-based therapeutic agents are inorganic coordination complexes with a planar configuration. The spatial configuration of platinum atoms (fixed bond angles) depends on their oxidation states. The simplest platinum compound with antineoplastic activity is *cis*-dichlorodiammineplatinum (cisplatin) in which the platinum has an oxidation state of +2 [Pt(II)], thus, there are four ligands (Figure 5.17). The *trans* isomer of this complex has no antineoplastic activity. When the platinum is in the +4 oxidation state [Pt(IV)], the configuration of the element is hexahedral, i.e.,

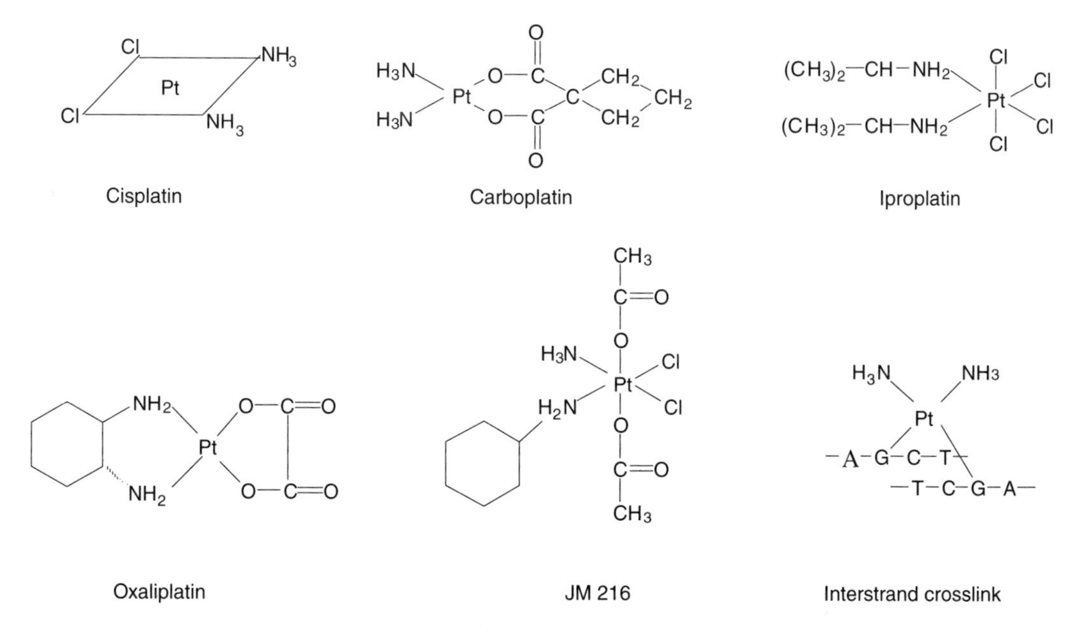

FIGURE 5.17 Structures and formulas for some platinum complexes of current clinical interest. Cisplatin, *cis*-dichlorodiammineplatinum. Carboplatin (paraplatin), 1-cyclobutane-dicarboxylatodiammine-platinum(II). Iproplatin, *cis*-dichloro-*trans*-dihydroxy-bis-isopropylamine-platinum(IV). Oxaliplatin, *trans-l-(R,R0*-1,2-diamino-cyclo-hexane oxalatoplatium. JM 216, bis(acetato)ammine-dichloro-(cyclohexylamine)-platinum(IV).

there are four bonds in a planar square, as with Pt(II), plus one above and one below the platinum atom, and there are six ligands, e.g., iproplatin (Figure 5.17). The antineoplastic properties of cisplatin were discovered serendipitously in 1965, during experiments examining the effects of electricity on the growth of *E. coli*. Since that time, hundreds of Pt(II) and Pt(IV) platinum coordination complexes have been synthesized, and their structures characterized by NMR and MS, in a search for compounds with enhanced antineoplastic properties as well as reduced nephrotoxicity and neurotoxicity.

The structures of some platinum drugs of current clinical interest are shown in Figure 5.17. Cisplatin and carboplatin are among the most widely used and broadly active cytotoxic anticancer drugs.[126] Advanced testicular cancer can now often be cured using platinum-based combination chemotherapy with vinblastine and bleomycin. Cisplatin and carboplatin are also components of approved standard treatment regimens for several types of cancer. Oxaliplatin is a third generation platinum complex of the 1,2-diaminocyclohexane platinum (DACH-Pt) family. It is a water soluble compound that does not show cross-resistance with cisplatin or carboplatin. The first orally administrable platinum drug, JM216 (satraplatin), possesses antitumor activity comparable to that obtainable with parenterally administered cisplatin or carboplatin and produces relatively mild toxicity.[127]

The platinum compounds are phase-nonspecific agents. Their mechanism of action is similar to that of bifunctional alkylating agents, namely the formation of a macrochelate with two adjacent guanines in a DNA chain, as suggested by FABMS.[128,129] Both MALDI-TOFMS and ESIMS, together with NMR, have been used to characterize a variety of representative complexes of platinum compounds and hybrid molecules that contain sulfur and nitrogen ligands after digestion with proteases and reaction with hydrogen peroxide. It was suggested that a considerable portion of the administered platinum drug doses is likely to be lost as indicated by the poor efficiency with which the "therapeutic complex" (i.e., the GG-cisplatin complex) is formed.[130]

Mass Spectra

EI and FD spectra were typically used for the characterization of synthesized complexes.[131] Both positive and negative FAB spectra have been obtained for cisplatin analogs including complexes with dibasic bidentate ligands.[132,133] The CID spectra of the $[M+H]^+$ and $[M-H]^-$ precursor ions of cisplatin analogs and bis(nucleobase)-diammineplatinum(II) complexes revealed several structurally significant fragment ions.[128]

The ESI spectra of cisplatin, iproplatin, and paraplatin exhibited the expected characteristic clusters of ions derived from the isotopes of both chlorine and platinum. The CID spectra obtained, either in the collision cell in the second quadrupole analyzer or in the ESI source in the high pressure region between the glass capillary tube and skimmer, were straightforward, e.g., for iproplatin, the abundant product ions at *m/z* 58 corresponds to the $[(CH_3)_2CHNH]^+$ ion derived from the side-chain.[134]

In square-planar platinum complexes the *trans effect* refers to the differential ESI/CID fragmentation of the *cis* and *trans* isomers of the dinuclear platinum complexes, $[cis\text{-}Pt(NH_3)_2Cl)_2\mu\text{-}(NH_2(CH_2)_nNH_2)](NO_3)_2$ (with the diamine being 1,4-butanediamine or 1,6-hexanediamine) presented a problem because the precursor ions of the isomers have the same *m/z* values. Experiments, using *surface-induced* dissociation where the precursor ions collide (at low energy) with a surface inside the ion source, revealed that the cleavage of the Pt-N bond *trans* to the chloride was the most favorable decomposition pathway of the complexes.[135]

Distribution

Determination in Kidney by Laser Microprobe

Cisplatin is widely distributed in the tissues, with high concentrations occurring in the kidneys, liver, and prostate. Demonstration of the localization of platinum in the renal proximal tubular cells of dogs treated with cisplatin was accomplished by laser microprobe-TOFMS (Section 2.1.7.1).

Sections of 0.35 μm thick renal biopsy samples, taken before and 60 min after treatment, were cut with an ultramicrotome, mounted on copper-rhenium grids, and introduced into the laser source. The platinum isotope pattern was detected in the renal proximal tubular cells but not in the extra-cellular, interstitial volumes. The platinum concentrations in the sampled spot areas (6 to 7 μm^2) were estimated to correspond to 14 to 60 μg Pt/g tissue. These numbers were in reasonable agreement with values obtained by x-ray fluorescence analysis.[136]

Platinum in the Brain, Peripheral Nerves, and Intervertebral Disks

It has been suggested that the neurotoxicity of cisplatin is limited to peripheral damage because the chemical nature of the drug prevents it from crossing the blood-brain barrier. This hypothesis was tested, using ICPMS, to determine the occurrence of platinum in seven regions of the brains of mice after the administration of 3 mg/kg cisplatin. The degree and retention of accumulated platinum in certain regions of the brain were found to increase significantly when short-term hypoxia was induced, by varying the oxygen content of the air to which the animals were exposed, during the administration of the drug. It was concluded that the initial hypothesis was incorrect and that platinum can readily cross the blood-brain barrier and accumulates in all areas of the brain. Retention in the cerebral cortex was significantly longer than in the kidney and blood.[137]

Using ICPMS, platinum concentrations were 1.1 to 10.3 μg Pt/g dry tissue in the intervertebral discs (obtained postmortem) of cisplatin-treated patients with ovarian tumors. These levels were about four-fold higher than those found in the vertebrae. No platinum was detected in control subjects.[138]

Separation and Characterization of Hydrated Complexes

It has been assumed that the monohydrated complex of cisplatin is responsible for cytotoxicity upon binding to DNA inside the target cells. Hydrated complexes are probably also involved in renal toxicity. To study the hydrated complexes of cisplatin and its *trans* isomer, the compounds were hydrolyzed with NaOH in the dark for 24 h. The conditions used favored the presence of the hydrated complexes in their less reactive, uncharged form, i.e., only $<1\%$ of the monohydrated complexes were converted to dihydrated complexes. The ESI spectrum of unreacted cisplatin showed the sodiated adduct ion (Figure 5.18A) with the isotope pattern expected for an ion containing one platinum and two chlorine substituents. The spectrum of the monohydrated species contained ions corresponding to the $[M]^+$, $[M\text{-}H_2O]^+$, and $[M\text{-}H_2O\text{+}MeCN]^+$ clusters (Figure 5.18B). The structure of the MeCN-containing ion of the monohydrated complex (Figure 5.18B) was confirmed by the product ions in the CID spectrum of the *m/z* 304 ion (Figure 5.18C). The protonated molecule (with an isotope cluster corresponding to one platinum and no chlorine) appeared when the same sample was diluted with methanol instead of acidified acetonitrile (Figure 5.18D). No sodiated adduct was observed for transplatin (Figure 5.18E). This work has demonstrated that the closely related hydrated species of cisplatin (and also transplatin) can be separated by judicially selected chromatographic conditions, and characterized utilizing the ability of ESI to produce alkali metal adducts without proteolytic properties.[139]

Quantification in Plasma and Urine

All platinum drugs bind irreversibly to plasma proteins following i.v. administration, at which point the cytotoxic properties are lost, e.g., up to 90% of cisplatin becomes protein-bound. Only the free drugs and/or their metabolites have therapeutic effects through their reaction with DNA. It is generally accepted that drug concentrations in urine and plasma ultrafiltrates represent non-protein bound or free drug, while concentrations in whole blood or plasma provide information on the fraction of protein-bound drug.[140]

Isotope Dilution GC/MS

Urine samples were spiked with ^{192}Pt isotope (50 atom % enriched), partially purified, derivatized with lithium bis(trifluoroethyl)dithiocarbamate (Li-FDEDTC), and the platinum chelate formed extracted and analyzed using GC/EIMS with SIM of the cluster of ions corresponding to

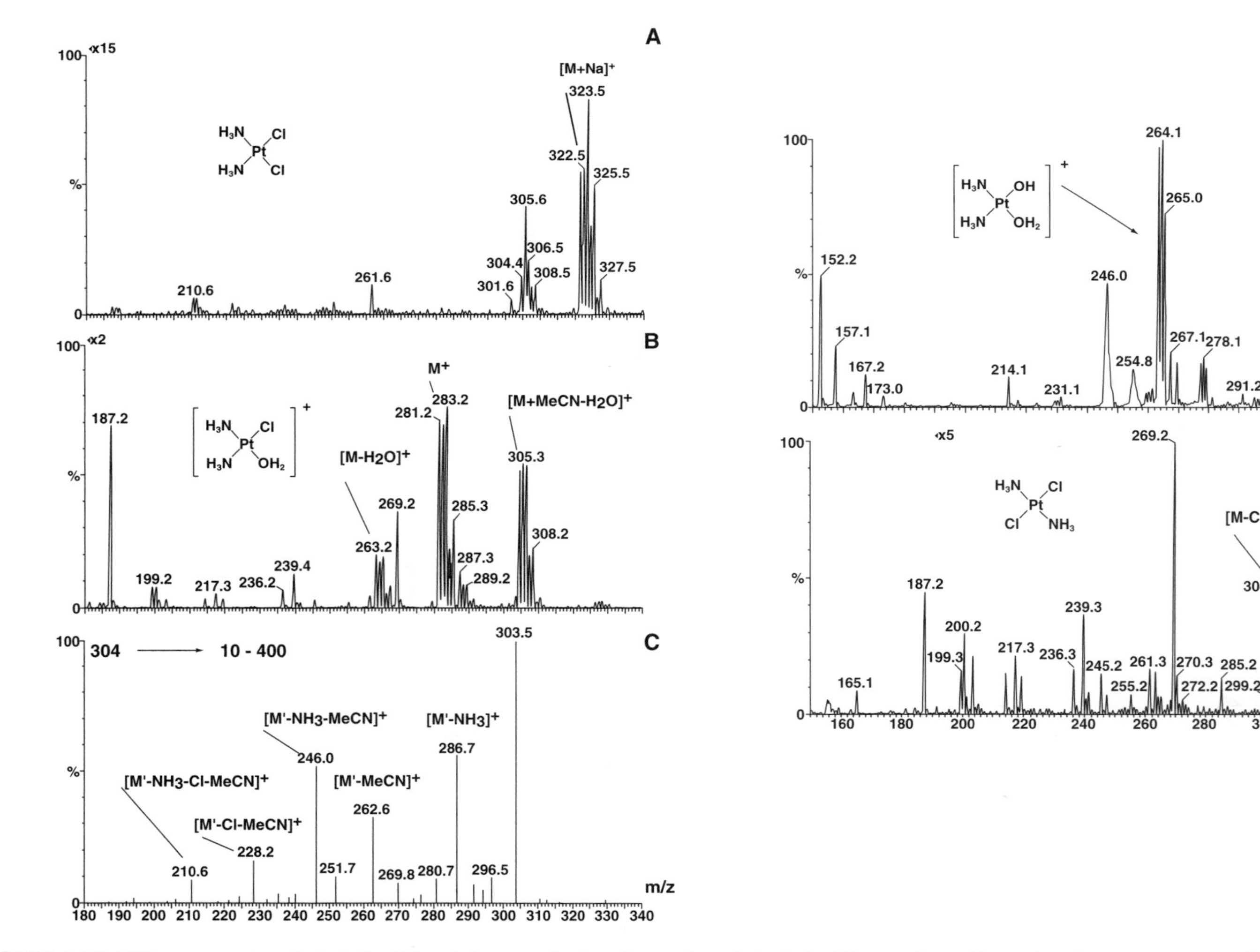

FIGURE 5.18 ESI mass spectra of cisplatin (**A**) and the monohydrated complex of cisplatin (**B**), together with product ion mass spectra of *m/z* 304 produced from the monohydrated complex of cisplatin (**C**), the dihydrated complex of cisplatin (**D**), and transplatin (**E**). Spectra A, B, and E were obtained using water-acetontrile-acetic acid; spectrum D was obtained using water-methanol-sodium hydroxide. Reprinted from Ehrsson, H. et al., *Anal. Chem.*, 67, 3608–3611, 1995. With permission.

$[Pt(FDEDTC)_2]^+$ which contains the six platinum isotope peaks. (The base peak of the spectrum was the $(CF_3CH_2)_2NCS^+$ ion which does not contain platinum). The LOD was 3 μg Pt/L urine and precision was ±1%. The technique was extended to the analysis of whole blood, plasma, and its ultrafiltrate.[141]

Inductively Coupled Plasma Mass Spectrometry (ICPMS)

The use of ICPMS techniques for the quantification of platinum in blood plasma is expanding because the experimental procedures are simple and sensitivities are higher than those of any other technique.[142] Ongoing technological developments in ICPMS include efforts to improve sample preparation using solid-phase extraction[143] and other sample-handling techniques.[144] In one technique, the sample preparation for total plasma platinum determination consisted of a 20-fold dilution of plasma with water containing 1% nitric acid (i.s.: europium), while for free platinum the plasma was ultracentrifuged and the supernatant diluted 7-fold with the same diluent. Ions of the ^{195}Pt isotope and the i.s. were monitored using ICPMS. Recovery was ~100%. The LOD was 0.05 μg Pt/L. In patients, the total plasma concentrations were in the 100 to 1000 μg Pt/L range, while the corresponding free platinum concentrations were 3 to 7% of the total. There was a positive correlation between the total platinum concentrations and the percentage of free platinum.[145]

Cisplatin and Carboplatin

The therapeutic monitoring of platinum in the serum of individual patients is advantageous because of the putative correlation between renal toxicity and peak dose concentrations. The combination of flow-injection of μL serum samples that have been diluted tenfold, offers the opportunity for high sampling rates and reduced build-up of solids on the injector torch. In one application, LOD was ~0.015 μg Pt/L serum. This was more than adequate for clinical studies as the total platinum concentrations observed in patients receiving therapeutic doses of cisplatin or carboplatin were in the 150 to 300 μg Pt/L range. Even though there is only a small percentage that is free platinum, the LOD is still well below the expected concentrations of free platinum determined by re-analyzing the samples after ultrafiltration.[146]

To investigate the long-term pharmacokinetic behavior of platinum, total and ultrafiltrable concentrations were monitored by ICPMS in patients with head and neck cancers who received three courses of 25 mg/m^2 cisplatin by i.v. infusion (1 h) daily for 4 days every 3 weeks. LOD was 0.05 μg Pt/L. Based on the analytical results the patients could be divided into two groups. Patients in Group I exhibited low initial levels followed by a progressive increase in platinum concentrations over the three sets of treatments. This group also had plasma concentrations of platinum that displayed long terminal-half lives. In Group II the platinum concentrations were high from the beginning and did not show a progressive increase but instead the maximal concentrations remained the same after each course of treatment (Figure 5.19, upper part). Comparison of the concentrations of free (i.e., nonbound drug and all biotransformation products) and protein-bound platinum, exhibited proportionality throughout the treatment period (Figure 5.19, lower part) with the free platinum concentrations being approximately 5% of the total platinum.[147]

After treatment with cisplatin or carboplatin, platinum has been detected in plasma for up to 2 yr (decreasing by an inverse square function). Platinum has also been detected in urine for up to 5 yr. This indicates there must be a reservoir within the body from which there is slow removal, possibly at concentrations that are below the LOD of plasma. Analysis of necropsy samples has revealed retention of platinum in the liver amounting to up to 2% of the initial platinum dose. These results may be relevant to the explanation of long-term toxicity.[148]

Oxaliplatin

ICPMS methods have been developed and validated for oxaliplatin-derived platinum in different blood compartments. In one method, ultrasonic nebulization dramatically affected sensitivity. The LOD for plasma ultrafiltrate was 0.001 μg Pt/mL compared with conventional nebulizing where the LOD for whole blood and plasma was 0.1 μg Pt/mL. The sample size was typically 100 μL.

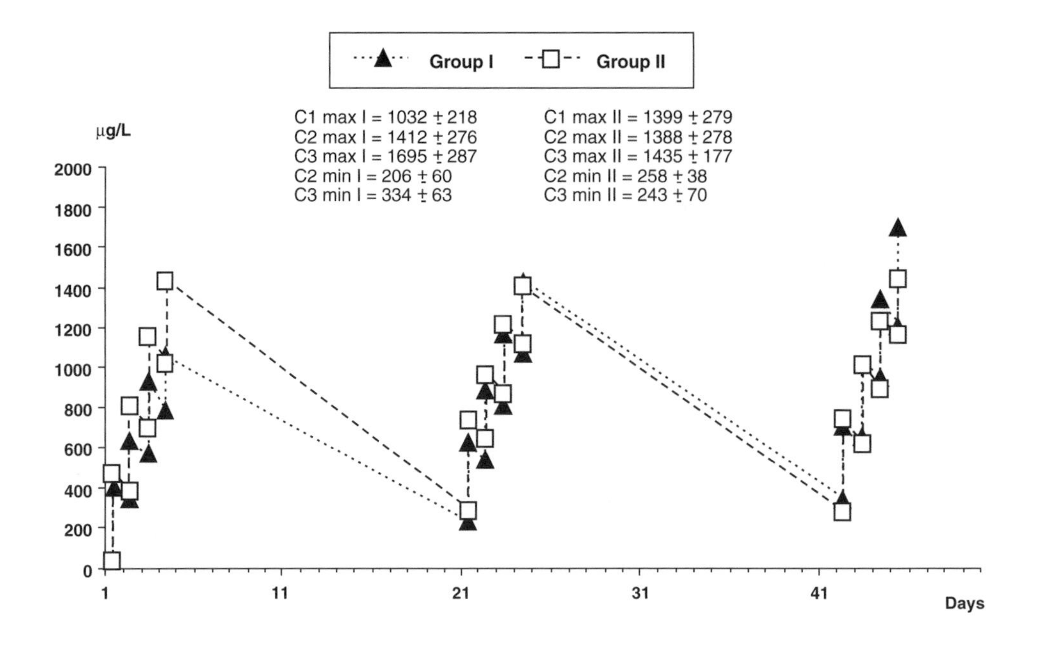

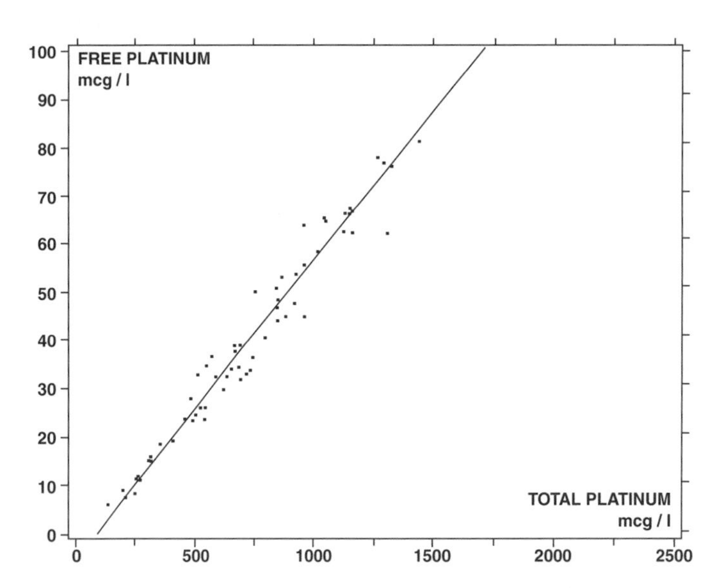

FIGURE 5.19 Upper panel: Mean total platinum levels measured during three courses of cisplatin treatment in two groups (I and II). The profiles of the platinum curves differed greatly between the two groups of patients. In group II, plasma platinum levels reached the steady state as early as during the first course. In group I they increased continually throughout the three treatment courses. Lower panel: Ratio of free and total platinum levels measured during three courses of cisplatin therapy. The highly sensitive ICPMS method revealed that free and total platinum levels were strictly proportional, and that free platinum persisted for the duration of the treatment period. Reprinted from Gamelin, E. et al., *Cancer Chemother. Pharmacol.,* 37, 97–102, 1995. With permission.

An application of this method in a clinical pharmacokinetic study revealed that platinum declined in a triphasic manner in all blood compartments assayed. Platinum was detected in the plasma ultrafiltrate for up to 3 weeks after administration.[149] Platinum has also been quantified by ICPMS in a pharmacokinetic study of the behavior of oxaliplatin (combined with 5-fluorouracil) in patients with colorectal cancer. In contrast to cisplatin, which is rapidly distributed into its free and bound forms in the plasma, oxaliplatin was found to preferentially accumulate in erythrocytes over the repeated cycles of administration at the recommended doses.[150] The 3-compartmental model of the pharmacokinetics of oxaliplatin was confirmed in another study using ICPMS. This study also confirmed that, because of its lipophilicity, this drug rapidly crosses the cell membrane. About 40% of the total platinum administered was found in the erythrocytes by the end of a 2 h infusion. Determination of the terminal elimination half-life has shown major differences with a 26 h half-life being the result obtained using a flameless atomic absorption assay and a 270 h half-life measured by ICPMS.[151] The latter technique has a >tenfold lower LOD suggesting perhaps that these two very different half-lives reflect different phases in the elimination of the drug, with the longer half-life observed with ICPMS reflecting the ability of the technique to observe a slow, low-level release of platinum into the blood-stream from a reservoir in the body. No relationship between pharmacokinetic and pharmacodynamic data has been established (reviewed in Reference 152).

JM216

The pharmacokinetics of this drug, based on the total and ultrafiltrable platinum in plasma as well as the platinum in urine, was studied on days 1 and 14 of the first cycle of treatment in the course of a Phase I study of JM216 given daily for 14 days every 4 to 5 weeks to patients with solid tumors refractory to conventional treatments. The LOD of the ICPMS technique used was 0.001 μg Pt/L. Platinum was detectable in both the plasma and urine as early as 30 min after drug administration with peak levels being observed at ~3 h. Elimination of platinum from the plasma was very slow and the levels of platinum in the urine were variable, with platinum still being present in both at the time of subsequent dosing 4 weeks later. The ultrafiltrate/plasma ratio of platinum concentrations were ~7. This was higher than for other related platinum drugs but still of similar magnitude. The 24 h urinary excretion varied from 1.5% to 18% of the dose. Only C_{max} on day 1 was related to the dose. Later C_{max} and AUC data of both total and ultrafiltrable platinum were highly variable among patients. There was no relationship between AUC values and either myelotoxicity or gastrointestinal toxicity.[153] The variability of the pharmacokinetic data may be related to the complex metabolism of JM216 as discussed below.

An LC/ICPMS technique was developed and validated for the quantification of JM216 and its three methanol-extractable metabolites. The HPLC eluent was introduced into the plasma of the ICP source without using any desolvation device. The LOD were in the 1 to 2 μg Pt/L range.[154]

Another application of ICPMS involved the concurrent determination of thallium and platinum in non small cell lung carcinoma cells. It was concluded that the cellular thallium accumulation subsequent to ^{201}Th scintigraphy may be used as an indicator of cytotoxicity and developing resistance during chemotherapy with platinum.[155]

Biotransformation

Oxaliplatin

Oxaliplatin is converted into monochloro, dichloro, or diaquo DACH platinum species that, in turn, may react with sulfur-containing compounds such as cysteine, reduced glutathione, methionine, and other substances. Several biotransformation products separated by HPLC, e.g., Pt(DACH)(Met), were identified by LC/ESIMS in combination with double isotope labeling experiments.[156]

JM216

The first orally administrable platinum drug, JM216 (Figure 5.18), has a complex metabolism, with at least six metabolites (but not the parent drug) observed by analyzing for platinum-containing compounds remaining in the ultrafiltrates of plasma from treated patients and also in plasma ultrafiltrates incubated with the drug. Several platinum-containing metabolites were characterized by accurate mass determinations using an ESIMS with a magnetic analyzer at a resolution of 4000, and additional structural information obtained with an ESI-MS/MS system. The major metabolite was *cis*-dichlorodiammine(cyclohexylamine) platinum(II) (JM118), resulting from the reduction of the parent compound via the loss of the two axial carboxylate groups. The cytotoxicity of this metabolite was fivefold higher than that of cisplatin. The mass spectrum of the second metabolite, bis(acetato)ammine(cyclohexylamine) dihydroxyplatinum(IV), revealed that the two chlorine atoms of JM216 had been displaced. The structures of the other minor metabolites were also established by ESIMS. The quantitative profiles of the major metabolites revealed differences between ultrafiltrates from patients and *in vitro* incubation experiments. It was concluded that the Pt(II) and Pt(IV) metabolites from this oral drug were significantly different from those platinum drugs that are administered systemically.[157,158]

Adducts

Similarly to cisplatin, oxaliplatin binds to all DNA bases following intracellular hydrolysis or aquation. Reactions occur preferentially with the N7 positions of guanine and adenine. Crosslinks are produced by reaction of the platinum with two bases either on an interstrand basis between opposite strands (Figure 5.17) or within the same strand (intrastrand) (reviewed in Reference 152). It has been generally accepted that platinum(IV) complexes are reduced in the body to platinum(II) compounds that subsequently react with DNA, the major target of platinum-based chemotherapy. An alternative mechanism that might involve the passive diffusion of platinum(IV) complexes into the cells for subsequent direct reaction with the nucleobases of DNA has been explored by investigating the binding of a synthetic (ethane-1,2-diamine)platinum tetraacetate [ethdmnPt$(Ac)_4$] complex to 5′-GMP. Three major compounds were identified by ESIMS operated in the negative ion mode: the starting analytes, free 5′-GMP, and ethdmnPt$(Ac)_4$ as well as the monoadduct, ethdmnPt$(Ac)_3$5′-GMP. The adduct was stable for weeks under physiological conditions. All species could be reduced by adding sodium ascorbate resulting in a final Pt(II) adduct, ethdmnPt$(5'\text{-GMP})_2$, observed within 7 days of adding the reducing agent.[159] This result implies that the Pt(II) valence state is the active species, as only after reduction did the di-adduct form. This is the model for the therapeutic cross-linking of DNA by platinum drugs in tumor cells.

Because platinum compounds are known to have an affinity for nucleophilic sulfur ligands, GSH was proposed as a protective agent against cisplatin-induced toxicity. *In vitro* experiments were conducted to search for cisplatin-GSH adducts that might have protective properties. Two such compounds were identified by LC-APCIMS, with structure confirmation by CID, as *cis*-$[Pt(NH_3)_2Cl(SG)]$ and *cis*-$[Pt(NH_3)_2Cl]_2(\mu\text{-SG})$, the latter being an adduct, the existence of which had only been postulated in previous studies. However, these two adducts could not be detected in rats or patients, and it was concluded that GSH appears to offer no protection against the side effects of cisplatin.[160]

The structure of complexes formed between cisplatin and four oligonucleotide tetramers, dGCGC, dGGCC, dTGAT, and dTGCT, were studied by both positive and negative ion FABMS. Platination of the oligonucleotides took 3 days in the dark at 3°C with a tenfold molar excess of cisplatin. Positive ion FAB yielded abundant $[M+H]^+$ ions and fragments that corresponded to Pt-base, Pt-nucleoside, or Pt-nucleotide complexes. The presence of both singly and doubly charged complexes indicated the reduction from Pt(II) to Pt(I) and/or Pt(0) during ionization. Progeny fragments from $[M+H]^+$ precursors included abundant platinated ions. Platination decreased the abundance of specific, sequence-dependent ions compared with those obtained from the nonplatinated oligonucleotides. The most prominent platinum containing fragment ions were platinated bases and, in particular, those derived from guanine. The platinum-directed fragmentation enabled the FAB technique (negative ion mode) to reveal the location of the platinum complex for each oligonucleotide sequence even though the platinum itself was not observed in the spectra.[161] Similar experiments on the reaction of cisplatin with dGG, dAG, and dCTAG were also evaluated by negative-ion FAB. The $[M\text{-}H]^-$ ion corresponded to the adducts *cis*-$Pt(NH_3)_2$ dinucleotide, while the fragment ions from the platinated tetradeoxynucleotide gave information on the sequence and the location of the platination.[162]

The reaction products of a deoxyribonucleotide 18-mer, dAACGGTTAACCGTTAATT with $[Pt(NH_3)_3(H_2O)]^{2+}$ and *cis*-$[Pt(NH_3)_2(H_2O)_2]^{2+}$ were studied using MALDI. The triamino-monoaqua complex gave peaks corresponding to monofunctional adducts, 18-mer+n$[Pt(NH_3)_3]$ with n values ranging from 1 to 4. The complex with the diamino-diaqua platinum compound yielded a chelate (18-mer+$[Pt(NH_3)_2]$ involving the two adjacent guanine residues.[163] In another study of this deoxyribonucleotide and of a single-stranded octamer, the products of experiments were separated by HPLC and the fractions treated enzymatically to release the platinated bases. The mass spectra revealed that the monoadduct formed with the GG sequence was located at the 5′ position of the guanine for both platinum complexes.[164] Similar methodology has also been employed to confirm the structures of the complexes of diaqua cisplatin with a 14-base oligonucleotide in a study on the selectivity of platination occurring at the 5′G and 3′G sites of guanine.[165]

Platinum-DNA adducts have been determined using ICPMS in two studies on platinum-resistant cell lines. In one study, both elemental and DNA-bound platinum were found to be lower in an ovarian carcinoma cell line resistant to cisplatin and flavopiridol when compared with a nonresistant cell line after both had been incubated with cisplatin. The addition of flavopiridol further reduced platinum concentrations in the resistant cell line.[166] In another study, an estrogen receptor negative human ovarian carcinoma cell line from which a cisplatin-resistant subline has been generated was used. After the incubation of the two cell lines with cisplatin, the platinum content of separated, purified, and enzymatically cleaved DNA was determined by ICPMS.[145] The resistant cell line had a reduced level of platinum-DNA adducts. When the experiments were repeated in the presence of an experimental steroidal antiestrogen, it was found that the drug potentiated the cell growth reduction produced by cisplatin alone, and the magnitude of this activity was greater in resistant cells.[167]

A CE/ESIMS technique has been used to study the formation of the adducts of cisplatin with various DNA nucleotides. Monochloro, monoaqua, and bifunctional adduct species were identified and the conversion rates and their time dependencies among these species were determined.[168]

Complexes with Lipids

Platinum drugs can also form complexes with lipids. When cisplatin was incubated in model membranes with phosphatidylserine (PS, 1,2-dioleoyl-*sn*-glycero-3-phosphoserine), the resulting platinated complex was extracted with chloroform and identified with positive-ion FAB as $C_{42}H_{83}N_3O_{10}PPt$ using exact mass measurement. The interaction of cisplatin with PS led to the loss of two chloride ions. CID spectra revealed that the structure of the complex coordinated the platinum with the carboxylate and amine group of the serine moiety. The CID spectra also confirmed that the platinum was associated with the hydrophilic headgroup of the compound and not with the fatty acyl chains attached to the glycerol backbone. The cisplatin-PS complex formation was also demonstrated in the plasma membrane of human erythrocytes.[169]

5.1.3 Topoisomerase Inhibitors

Topoisomers are circular molecules of DNA with the same sequence of bases but differing in the number of their twists, i.e., the number of turns made by one strand around the other. *Topoisomerases* (topos) are cellular enzymes that catalyze the interconversion of DNA topoisomers. This is accomplished by introducing or eliminating supertwists in DNA double helices thereby modifying the number of interlacings. Accordingly, topos are intricately involved in the maintenance of the topographic structure of circular DNA which, in turn, affects the processes of translation, transcription, and mitosis (reviewed in Reference 170). Type I topoisomerases bring about the transitory cutting and re-fusion of a single strand of the duplex; they also mediate the relaxation of superhelical DNA. Leukemias, lymphomas, and colon cancer express high levels of topoisomerase I. Type II topoisomerases produce a transient strand break in the duplex and also convert relaxed, closed circular DNA to a superhelical form. The direct action of topoisomerase inhibitors on the structure of DNA during replication justifies the increasing interest in developing new agents of this type.[171] Figure 5.20 shows the formulas and lists the chemical composition of important topoisomerase inhibitors.

5.1.3.1 Topoisomerase I Inhibitors

Camptothecin

During the large-scale screening of plant products for antitumor activity in the late 1950s, the active component of a crude extract of *Camptotheca accuminata* was isolated and identified as camptothecin (CPT) by several techniques, including MS.[172] Clinical development of the water-soluble sodium carboxylate form for i.v. administration was halted because of poor efficacy and adverse side effects. Interest in CPT analogues was rekindled after the demonstration that the target of CPT was topoisomerase I and that an intact lactone ring was necessary for biological activity.

The metabolism of CPT has been studied in an isolated perfused rat liver model. Addition of CPT to the perfusion medium resulted in an increase in bile flow, indicating a concentrative excretion of the drug into bile. Three metabolites, separated by HPLC, could not be hydrolyzed with β-glucuronidase or sulfatase, thus Phase II conjugations with glucuronic acid or sulfate were ruled out. The metabolites were identified by LC/APCIMS, in the negative ion mode, as two dihydroxy-CPT products and monohydroxy-CPT originating from the CPT core. Comparisons of the spectra to that of CPT confirmed the loss of a carbon dioxide group and the hydrolysis of the lactone to the hydroxy acid, in a fashion similar to that described for irinotecan (see below). The pharmacological or toxic effects of these CPT metabolites are not yet known.[173]

The CPT analog 9-nitro-20(S)-camptothecin (9NC), that could be administered orally or intramuscularly, has significantly reduced toxicity concurrent with increased antitumor activity. It was confirmed by both conventional and tandem MS (magnetic, high resolution) that 9NC converts to 9-aminocamptothecin in the liver after oral administration. The molecular ion of the metabolite obtained in blood plasma, $[C_{20}H_{17}N_3O_4]^+$ (at *m/z* 363) and the EI fragmentation pattern (Figure 5.21) in plasma samples from treated patients and dogs, and from dog liver incubation experiments were identical to that from authentic compound.[174]

Irinotecan

Irinotecan (CPT-11), a semisynthetic, water-soluble derivative of CPT, differs from CPT by a 10-1,4′-bipiperidine-1′-carbonyloxy side-chain and a 7-ethyl group (Figure 5.20). The drug is a potent topoisomerase I poison with activity against pretreated colorectal, cervical, and several other cancers. CPT-11 is a prodrug. The cytotoxic activity is expressed by the major metabolite, 7-ethyl-10-hydroxycamptothecin (SN-38, Figure 5.22A) that results from the hydrolysis of the parent drug by carboxylesterases. *In vitro*, SN-38 is a 200-fold more potent inhibitor of topoisomerase I than CPT. At least 16 metabolites of CPT-11 were detected in the bile of a patient with liver metastasis of colon carcinoma. Eight of these metabolites were also detected in urine. After determining their molecular masses using LC/ESIMS, structural information was obtained by CID of representative precursors. The dominant metabolic pathway consists of oxidation of the terminal piperidine ring of the side-chain of the parent drug leading, eventually, to the formation of a primary amine. Other metabolites were shown to originate from the oxidation of the camptothecin nucleus or the 7-ethyl group, and decarboxylation of the carboxylate form of the parent drug. It was concluded that carboxylesterases, UDP-glucuronosyl transferases and cytochrome P-450 enzymes are involved in the complex metabolism of CPT-11.[175]

A polar metabolite was also detected by LC/ESIMS in the plasma of patients receiving CPT-11. The $[M+H]^+$ ion of this metabolite (*m/z* 619) was 32 Da higher than that of the parent compound (Figure 5.22B). The fragment ions at *m/z* 393 and 227 (Figure 5.22C) suggested that the addition occurred on the bipiperidino moiety. The ion at *m/z* 502 indicated that the distal piperidine was the

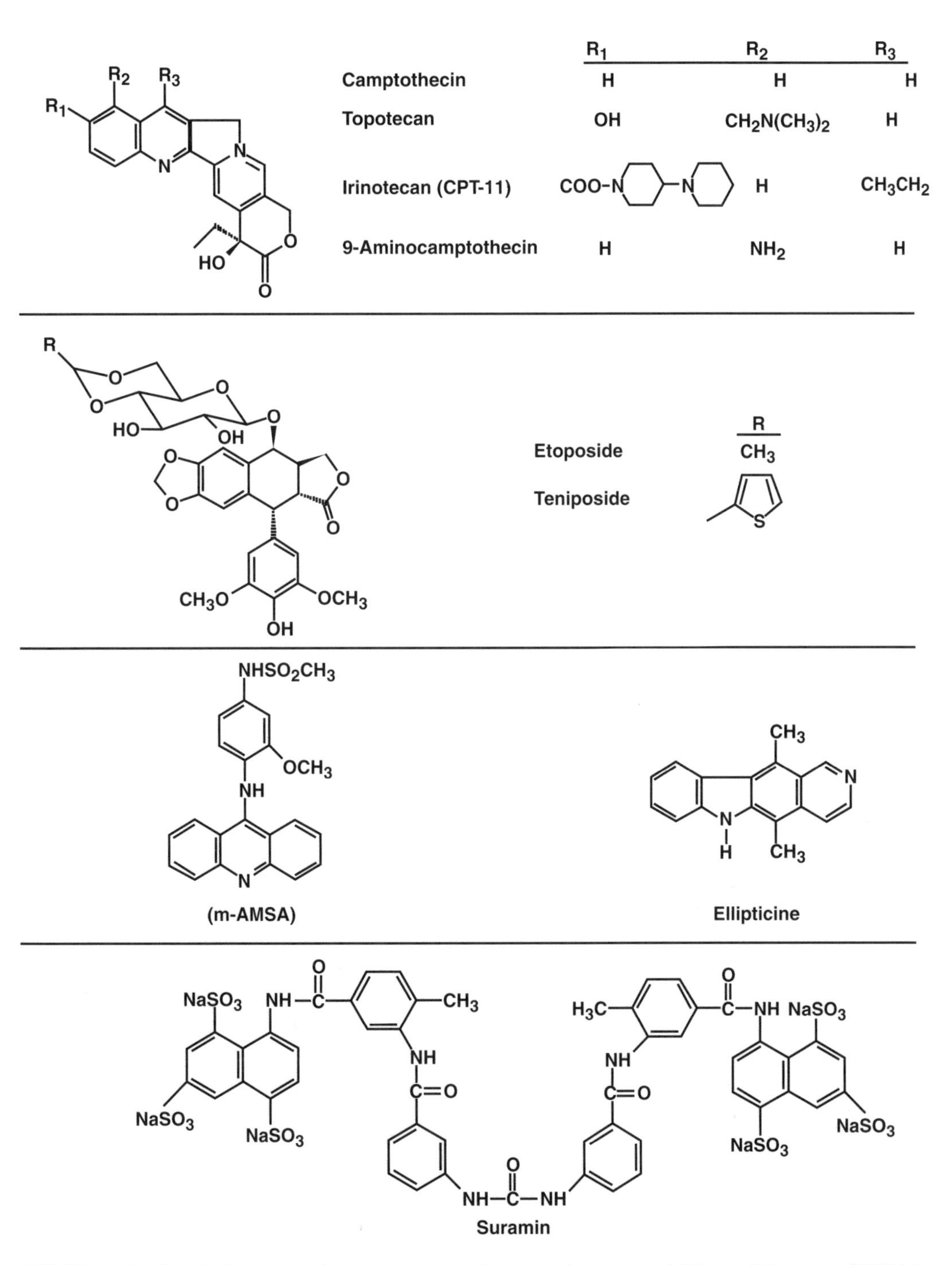

FIGURE 5.20 Chemical composition of representative toposiomerase inhibitors. Irinotecan (CPT-11), 7-ethyl-10[4-(1-piperidino)-1-piperidino]carbonyl-oxycamptothecin. Etoposide (VP-16), 4′-demethylepipodophyllotoxinethylidene-β-D-glucopyranoside. Tenoposide (VM-26), 4′-demethylepipodophyllotoxinthenylidene-β-D-glucopyranoside. Amsacrine (m-AMSA), 4′-(9-acridinylamino)methanesulfone-*m*-anisidide. Ellipticine, 2-methyl-9-hydroxyellipticinium acetate. Suramin, hexasodium salt of 8,8′-[[carbonyl-bis[imino-3,1-phenylenecarbonylimino-(4-methyl-3,1-phenylene)carbonyl-imino]]bis-1,3,5-naphthalenetrisulfonic acid.

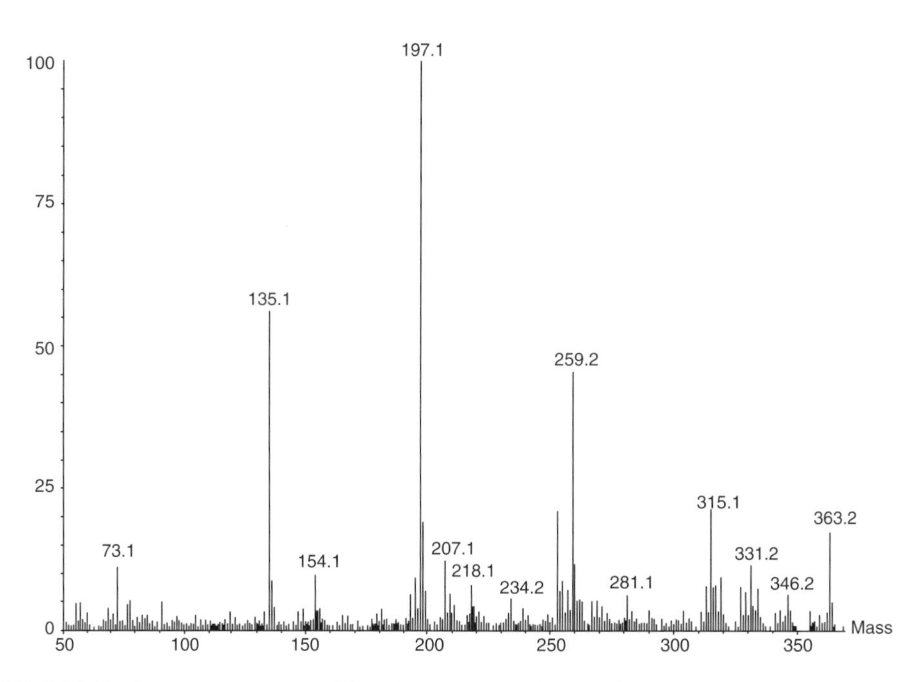

FIGURE 5.21 Tandem mass spectrum of 9-aminocamptothecin identified in a biologically active fraction of a dog liver incubate. *m/z* 363 represents the $[C_{20}H_{17}N_3O_4]^+$ ion. Reprinted from Hinz, H. et al., *Cancer Res.*, 54, 3096–3100, 1994. With permission.

site of the ring-opening oxidation. The metabolite was identified as 7-ethyl-10-[4-N-(aminopentanoic acid)-1-piperidino]carbonyloxycamptothecin (APC, Figure 5.22C). The high plasma concentration of APC suggests that the oxidation of the bipiperidino side-chain may be an important metabolic pathway for CPT-11.[176]

Another polar metabolite, detected in human liver microsomes incubated with CPT-11, was identified by LC/APCIMS as the dealkylated product of APC, 7-ethyl-10-(4-amino-1piperidino)carbonyloxycamptothecin (NPC). The compound was also identified in the urine (Figure 5.23) and plasma of patients treated with CPT-11. As shown in the insets, the CID spectra of NPC and a standard were identical.[177]

It has been suggested that 4-piperidinopiperidine (4PP), which is released during the esterolysis of CPT-11 to SN-38, may be involved in the putative anti-angiogenic properties of the drug. An LC/ESI-MS/MS technique was developed for the quantification of 4PP in plasma from patients with advanced colorectal adenocarcinoma (refractive to 5-fluorouracil) receiving CPT-11 in various regimens. Quantification was accomplished by MRM of the predominant product ions from the $[M+H]^+$ ions of the analyte and the i.s. (1-piperidinepropionitrile). The LOD was 14.8 nM. The AUC correlated with the drug dose, but not with the concentration-time profile of SN-38. The half-life of 4PP was ~30 h. 4PP appeared unlikely to contribute to the antitumor activity of CPT-11, as the submicromolar plasma concentrations were too low for the induction of apoptosis.[178]

An LC/ESIMS technique was developed for the quantification of CPT-11 and SN-38 in human serum. The parent drug could be analyzed in serum directly after protein precipitation. SN-38 required liquid-liquid extraction. The analytes were quantified (i.s.: camptothecin) using MRM of precursor ions obtained by in-source CID. The LOD were 2.5 ng/mL for CPT-11 and 0.25 ng/mL for SN-38.[179]

The pharmacokinetics, metabolism, and excretion of CPT-11 were studied in patients with solid tumors following the infusion of ^{14}C-labeled drug. Of 19 metabolites detected in excreta, 14 were fully or partially characterized by LC/ESI-MS/MS using MRM of $[M+H]^+$ ions obtained by CID

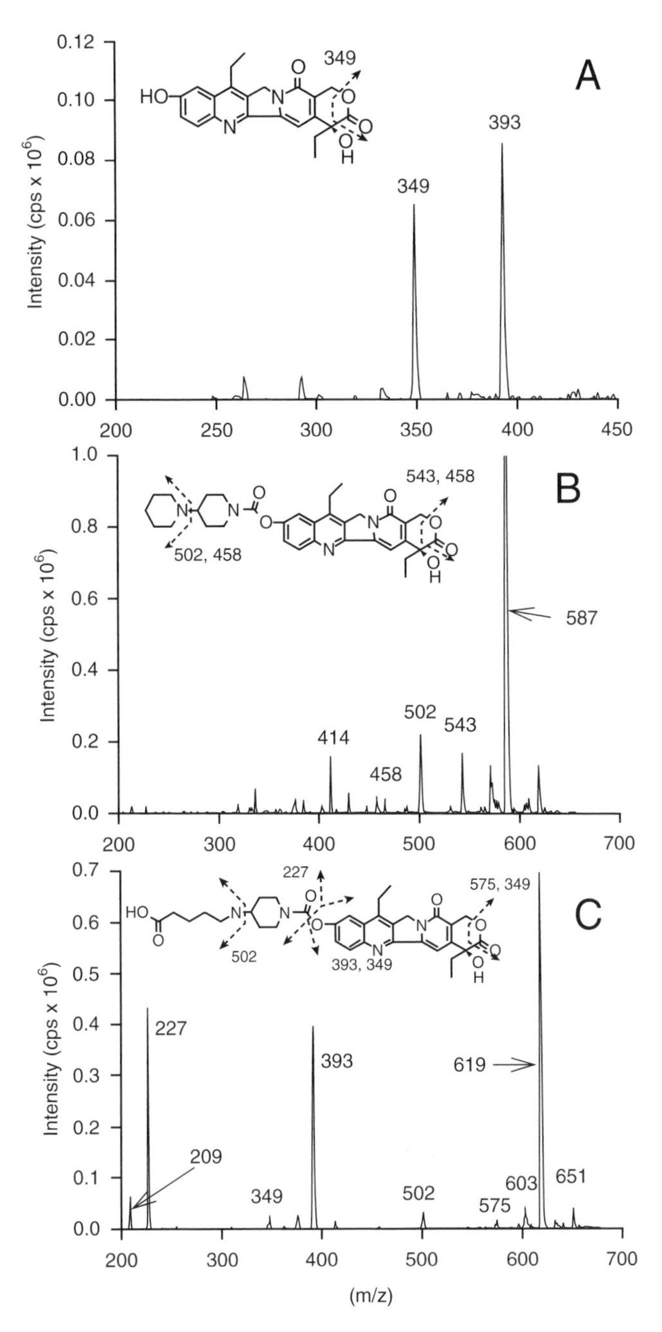

FIGURE 5.22 Mass spectra collected from the peaks corresponding to (**A**) 7-ethyl-10-hydroxycamptothecin (SN-38); (**B**) 7-ethyl-10-[4-(1-piperidino)-1-piperidino]carbonyloxycamptothecin (CPT-11, NPC); (**C**) 7-ethyl-10-[4-N-(5-aminopentanoic acid)-1-piperidino]-carbonyloxycamptothecin (APC). The formulas show proposed fragmentation. Reprinted from Rivory, L. P. et al., *Cancer Res.*, 56, 3689–3694, 1996. With permission.

in the ion source region. Identifications were based on published interpretations of the mass spectra of major and minor metabolites, and degradation products of CPT-11.[175,177,179,180] Some minor metabolites were tentatively identified, such as the quinoline-N-oxide of SN-38. The three major metabolites were SN-38, SN-38-glucuronide, and APC, and their relative quantities were determined in urine, bile, and feces. The major elimination pathway was fecal.[181]

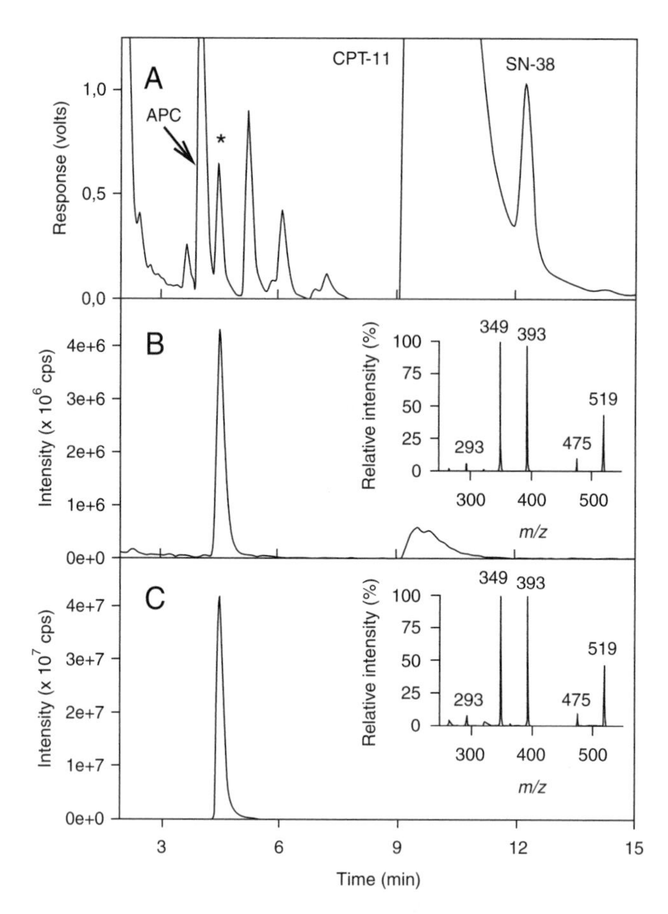

FIGURE 5.23 Chromatograms of extracted urine samples: (**A**) Trace from fluorescence detector; (**B**) Selected reaction monitoring of the fluorescent peak of interest at its characteristic *m/z* value of 519. (**C**) Chromatogram of a pure standard of NPC (see Fig. 5.22 for composition) at the same *m/z*. Insets: Collision-induced dissociation spectra of the samples. Reprinted from Dodds, H.M. et al., *J. Pharmacol. Exper. Therapy,* 286, 578–583, 1998. With permission.

CPT-11 undergoes photodegradation in aqueous solutions exposed to laboratory light. The structure of five products, identified by ESIMS, resulted from extensive modifications of the lactone ring.[182] Subsequently, the photodegradation products, were identified by LC/APCIMS in clinical drug solutions exposed to light during infusion. It was concluded that the photodegradation products were already present in the drug ampoules and were, therefore, the likely source of these compounds found in the plasma and urine of patients.[183]

Topotecan

Topotecan, another semisynthetic analog of CPT (Figure 5.20), is known to undergo a pH-dependent, reversible hydrolytic dissociation of its lactone moiety into the hydroxylated form. In the course of analyzing plasma and urine samples from both patients and dogs by HPLC, an unknown peak was identified by LC/ESIMS as N-desmethyl topotecan, based on the loss of 14 mass units from topotecan and comparison with an authentic standard.[184] Another unknown peak detected in clinical urine samples was shown to contain two components, identified by LC/ESIMS as topotecan-O-glucuronide and N-desmethyl topotecan-O-glucuronide. The concentrations of these metabolites in urine were 10% and 3.5%, respectively, of the concentration of the parent drug.[185]

5.1.3.2 Topoisomerase II Inhibitors

Nonintercalators: Etoposide and Teniposide

Etoposide (VP-16) and teniposide (VM-26) are semisynthetic derivatives of the plant lignan podophyllotoxin (Figure 5.20). These agents are active against refractory non-Hodgkin's lymphoma, lung cancer, and ovarian cancer.[186] The mechanism of action of the parent compound, which is not a clinically useful drug, is the inhibition of microtubule polymerization. In contrast, the cytotoxicity of the glucoside derivatives, etoposide and teniposide, is due to the damage to DNA achieved by interfering with the scission-reunion reaction of topoisomerase by stabilizing the cleavable complex.[187] As an alternative mechanism, it was shown by LC/ESIMS that the tyrosinase-catalyzed oxidation of etoposide resulted in oxidation to its *o*′-quinone and aromatized derivative via intermediate formation of the phenoxyl radical. The oxidation could be prevented by ascorbate and thiols, both independently and additively.[188]

Etoposide is a chiral drug and the stereochemistry of the trans-lactone is essential for cytotoxicity. An important metabolic step is the formation of the hydroxy acid derivative after hydrolysis of the lactone ring. A human urinary metabolite was identified by EIMS as the 4′-demethylepipodophyllic acid glycopyranoside congener. In other experiments, three radioactive peaks were isolated by HPLC from the bile of rats and rabbits whose livers were infused with ^{3}H-etoposide. The first two peaks were identified as etoposide glucuronides based on the $[M-H]^-$ and $[M+Na-2H]^-$ ions in their negative ion FAB spectra (Figure 5.24a) and the $[M+Na_2-H]^+$ and $[M+Na_3-2H]^+$

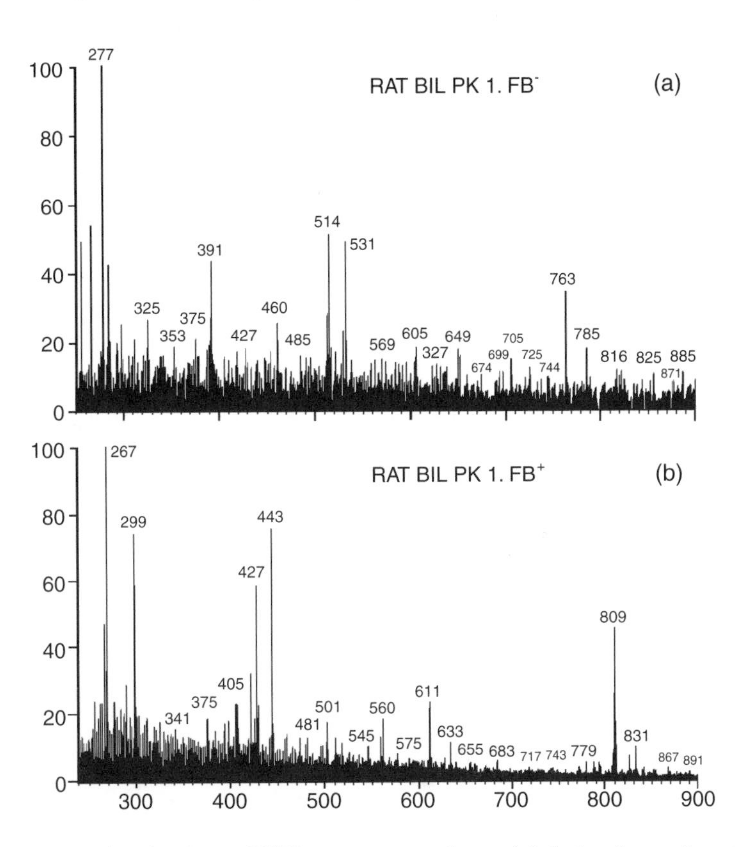

FIGURE 5.24 Fast atom bombardment (FAB) mass spectra of material eluting from relevant peaks separated by HPLC. (**a**) Negative FAB spectrum; (**b**) Positive FAB spectrum. Reprinted from Hande, K. et al., *Cancer Res.*, 48, 1829–1834, 1988. With permission.

peaks in their positive ion FAB spectra (Figure 5.24b). The third peak was intact etoposide. The glucuronides accounted for 30 to 40% of drug disposition. Possible chemical reasons were offered for the presence of two separate peaks for the glucuronides.[189]

A technique for the quantification of etoposide in plasma involved sample extraction and separation by HPLC or TLC followed by the measurement of the intensities of the protonated analyte (i.s.: teniposide) using ^{252}Cf-plasma desorption.[190] The LOD, 50 ng/mL, was much lower than the quantities found during 24 h kinetic studies.[191] The method was applied to obtain pharmacokinetic data in patients of different ages and individuals with impaired renal and liver functions, using several dosage, infusion time, and polychemotherapy protocols.[192] A limited-sampling model, developed using a training set of 15 patients, utilized AUC data from only two concentration points, 4 h and 8 h after the end of the infusion.[193]

An ESIMS technique was developed to measure the molecular masses of intact complexes of etoposide (588.6 Da) with albumin (Alb) and hemoglobin (Hb). The objectives were to determine the number of bound etoposide molecules under different molar ratios and incubation times in models, to analyze serum spiked to mimic therapeutic concentrations, and to test the approach using blood from patients under treatment. Complexes present in trace concentrations were identified in the presence of large protein excesses by mass transformation scan-by-scan along the total ion current curve and by monitoring preselected multiply charged ions. Masses were determined within 0.2% of calculated values. With up to a tenfold excess of Alb, the complex contained only 1 etoposide/Alb at all incubation times. Increasing the relative quantity of etoposide led to complexes containing up to 10 etoposides. Hemoglobin complexes with up to 8 etoposide/Hb were detected using 1:1 to 1:50 molar ratios, e.g., 15,750 Da (+1 etoposide) and 18,189 Da (+5 etoposides). A plasma sample from a patient receiving 80 mg/m^2 etoposide contained albumin complexes with 1 and 3 attached etoposide (67,256 Da and 68,254 Da) and ~25 μg/mL free etoposide. A complex of Hb with one etoposide molecule attached was found in the erythrocyte fraction (Figure 5.25). Knowledge of quantity and half-life of the complexes may lead to a better understanding of the demonstrated influences of protein binding on the therapeutically important free drug fraction (e.g., adults vs. children) and to the monitoring of complexes for toxicity.[194]

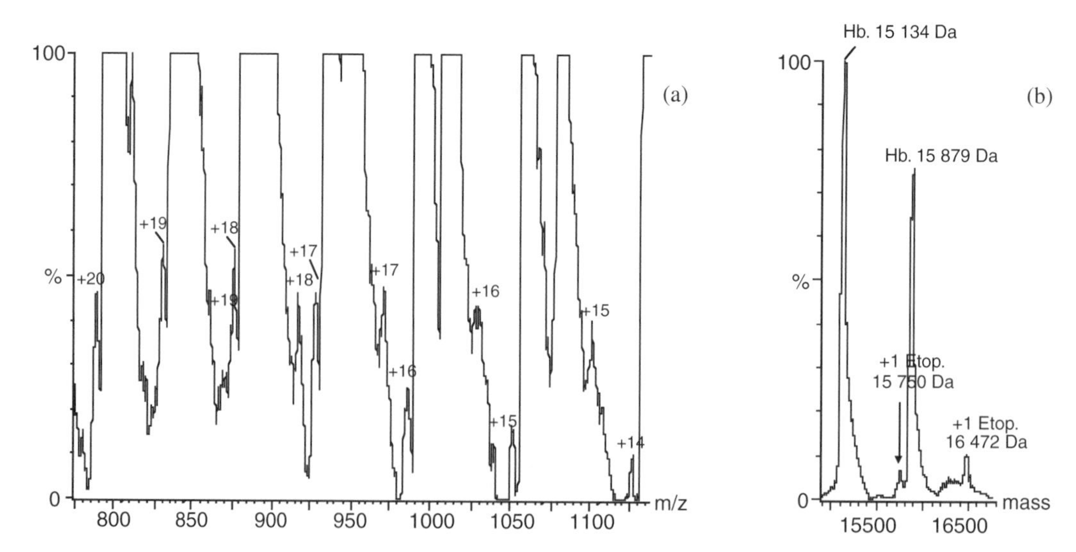

FIGURE 5.25 Complex of hemoglobin (Hb) + 1 etoposide molecule in patient serum. (a) ESI spectrum showing a segment of multiply charged ions with Hb ions off scale. The numerical values indicate the number of positive charges on the Hb-etoposide complex attached to each of the Hb chains; (b) Molecular masses after transformation. Based on Roboz, J., Deng, L., and Ma, L., *Proc. 47th ASMS Conf. Mass Spectrom.,* 1999.

Amsacrine

Amsacrine (m-AMSA) (Figure 5.20) is a rationally synthesized acridine derivative. It is active against hematologic malignancies but not against solid tumors. In the rat, the major route of elimination is via the bile. The conjugation of m-AMSA to glutathione in the bile involves nucleophilic displacement (thiolysis) at the C9-position. The conjugate was identified by FDMS as the 9-acridinyl thioether of glutathione. This product could also be obtained *in vitro* by incubating m-AMSA with glutathione, cysteine, and N-acetylcysteine. The thiolytic cleavage pathway product, 4-amino-3-methoxymethanesulfonalidine, was identified in rat serum, as well as in all *in vitro* incubation experiments.[195] As confirmed using FDMS in studies with synthetic glutathione conjugation products, and other thiol derivatives related to the oxidative metabolism of m-AMSA, the pathways of spontaneous thiolysis involved both the acridine rings and aniline ring of the parent drug.[196]

The major *in vitro* metabolite of amsacrine in rat liver was identified as N′1-methane-sulphonyl-N′4-(9-acridinyl)-3′-methoxy-2′-5′-cyclohexadiene-1′,4′-diimine which yields 5′-amsacrine-GSH upon conjugation with GSH at the 5′-position of the aniline ring. Identical mass spectra were obtained from a second biliary GSH conjugate, isolated from the bile ducts of rats, and gall bladders, of mice treated with m-AMSA. Accurate mass determinations of the molecular ions suggested isomeric GSH conjugates of m-AMSA. Comparison of CID spectra of the two metabolites with those from authentic 5′- and 6′-amsacrine-GSH conjugates revealed that the product ions in the high-mass region were different for the two conjugates. Both metabolites were detected in approximately equal amounts in m-AMSA-treated mice, while the 6′-GSH conjugate was dominant in rat bile. The possibility that the formation of 6′-conjugate of m-AMSA may be catalyzed by GSH transferase (which may have implications in hepatotoxicity) is suggested by a study of *acridine carboxamide*, N-[2′-(dimethylamino)ethyl]acridine-4-carboxamide, a third-generation derivative of amsacrine that has much higher activity against solid tumors than m-AMSA. The cytosol-mediated *in vitro* metabolism of the drug revealed only one metabolite, identified by high-resolution LSIMS as the corresponding 9(10H)-acridone-carboxamide.[197] A study of acridine carboxamide in mouse urine, bile, and feces by FABMS led to the identification of a number of major and minor metabolites.[198]

Ellipticine

Ellipticine (Figure 5.20) has been used in the treatment of bone metastasis from breast cancer, soft-tissue sarcomas, and renal cancer. The metabolism of the drug was investigated in rat urine and bile. Components of urine separated by HPLC were sprayed onto a moving belt interface leading into a double-focusing magnetic MS. One metabolite was identified as the N-acetylcysteine adduct of the parent drug. A second sulphydryl metabolite, a glutathionyl adduct, was identified in bile by LC/MS analysis.[199] A metabolite isolated by HPLC from the urine of a patient was identified by FAB as a conjugate with glutathione.[200]

Intercalators and Alkylators; Daunorubicin and Doxorubicin

Daunorubicin (DRB, daunomycin) and doxorubicin (DOX, adriamycin) are anthracycline antibiotics consisting of a planar, hydrophobic tetracyclic quinoid aglycone with a glycosidic linkage to the aminosugar daunosamine (Figure 5.26). Although DRB and DOX differ only by a single hydroxyl at position C14, their antitumor properties differ significantly. DOX is one of the most widely used drugs used for the treatment of acute lymphocytic leukemia, breast and lung cancer, Hodgkin's disease, and several other malignancies. In contrast, DRB is mostly used in the treatment of acute myelocytic leukemia.

The action of anthracyclines involves the stabilizing of the topoisomerase-DNA complex in the cleaved configuration, thereby maintaining the single-strand breaks. Intercalation with DNA is achieved between adjacent base pairs and binding to the sugar-phosphate backbone of DNA. The selective effects of DOX on rapidly dividing tumors may be due to the rapid rise of the concentration

(a)

O OH R OH OCH3 O O O O CH3 OH NH2

	R
Doxorubicin	—C(=O)—CH_2OH
Doxorubicinol	—CH(OH)—CH_2OH
Daunorubicin	—C(=O)—CH_3
Daunorubicinol	—CH(OH)—CH_3

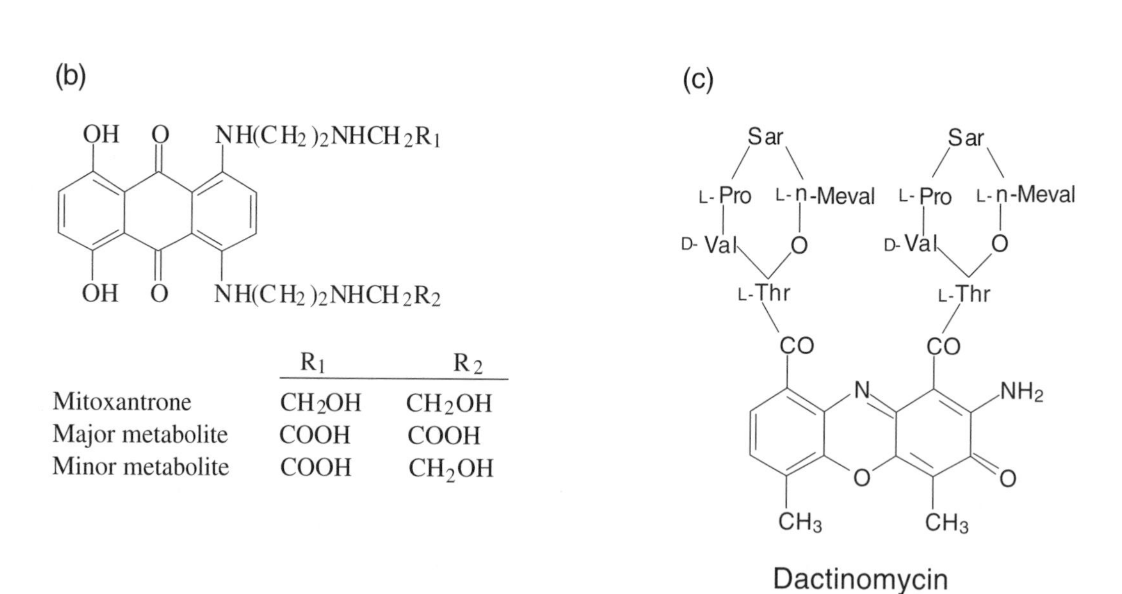

	R_1	R_2
Mitoxantrone	CH_2OH	CH_2OH
Major metabolite	COOH	COOH
Minor metabolite	COOH	CH_2OH

FIGURE 5.26 Chemical structures of some intercalators: (**a**) anthracyclines and metabolites; (**b**) mitoxantrone and metabolites; (**c**) dactinomycin.

of topoisomerase II in vulnerable cells during mitosis. Anthracyclines may also be considered as alkylating agents (see below). The dose-limiting and cumulative cardiac toxicity of anthracyclines is likely to result from their ability to bind ferric (Fe^{+3}) iron within the cytosol, leading to the generation of highly reactive hydroxyl radical species by a one-electron reduction of the hydroxyquinone structure. The drug-iron complex causes oxidative cell damage by binding to cell membranes. Cardiac tissues lack the superoxide dismutase or glutathione needed for the repair of the oxidative damage.

The major metabolites of DRB and DOX are daunorubicinol (13-hydrodaunorubicin) and doxorubicinol, respectively (Figure 5.26), which display reduced cytotoxic activity. Further reduction by deglycosylation leads to their respective 7-deoxyaglycone metabolites that do not possess cytotoxic activity. All of these compounds were isolated from urines of patients with acute lymphocytic leukemia and identified by EIMS during the 1970s.[201–203] Other metabolic processes involve carbonyl reduction, O-demethylation, O-sulfation, and O-glucuronidation. The metabolism of DRB,[204] DOX,[205] and 4′-epidoxorubicin[206] are qualitatively similar although there are significant quantitative differences. For example, for a given dose, daunorubicinol is present in urine in larger quantities than doxorubicinol, indicating a preference for DRB as a substrate for the cytoplasmic aldo-keto-reductase and microsomal glycosidases.

The monitoring of serum levels of the anthracycline drugs during treatment is important because significant interpatient heterogeneity has been reported for the same dose. Four anthracyclines of current clinical interest, DRB, DOX, epirubicin, and idarubicin, have the same single sugar moiety and their structures are similar. A technique developed for the simultaneous quantification of these four drugs and their respective 13-dihydro metabolites in human serum involved automated solid-phase extraction followed by LC/ESIMS analysis with SIM of the $[M+H]^+$ or $[M+NH4]^+$ ions (i.s.: aclarubicin, an analog with a chain of three sugars). The LOD were in the 0.5 to 2.5 ng/mL range. The verified linearity, up to 2000 ng/mL for the parent drugs and 200 ng/mL for the metabolites, permit single run analyses from the end of infusions to trough levels for routine therapeutic monitoring.[207]

The ongoing search for anthracyclines with reduced toxic effects led to the development of a variety of analogs. GC/MS techniques using both EI and CI have been used for the characterization of the TMS derivatives of compounds with alterations of the glycosidic portions.[208] The lead of a new class of related drugs with minimal cardiotoxicity is 4-demethoxy-3′-deamino-3′-aziridinyl-4′-methylsulphodaunorubicin (PNU-159548). A technique for the determination of the parent drug and its 13-hydroxy metabolite involved direct introduction of plasma onto a clean-up LC column, to remove proteins and polar endogenous constituents, and backflushing the analytes onto an analytical column. The separated analytes were quantified using ESIMS with MRM of characteristic precursor-to-product ion transitions. Linearity, in the 0.05 to 10.3 ng/mL range for the parent drug and 0.1 to 10.4 ng/mL for the metabolite, was adequate for Phase I clinical studies.[209]

DOX loses cytotoxicity upon exposure to illumination with 365 nm light. The process is strongly mediated by light-excited riboflavin, which is known to have the capability of oxidizing a variety of organic compounds. It was shown that photooxidation proceeded after DOX molecules had formed complexes with riboflavin. Mass spectra revealed that the mechanism of the photooxidation included the opening of the central ring of DOX; 3-methylsalicilic acid was one of the resulting products.[210]

The accepted mechanism of action of the anthracyclines is DNA intercalation (reviewed in Reference 211). ESIMS and ESI-MS/MS techniques have been developed for the direct demonstration of abundant duplex ions resulting from the formation of complexes with double-stranded DNA. Self-complementary deoxyoligonucleotide models, such as 5′-dGCGCGCGC-3′ or 5′-dGGCTAGCC-3′, were annealed under various conditions in the presence of DRB or nogalamycin at differing molar excesses. Nogalamycin, which is not important clinically and has a much more complex hydrophobic sugar moiety than DRB, was selected because the double helix must "breathe" to allow entry of this dumbbell-shaped molecule during intercalation. Using 5:1 molar drug excess with respect to (duplex) oligonucleotide, complexes containing up to six DRB and four nogalamycin were observed. According to the "neighbor exclusion" principle, intercalation can occur between every other base pair. For example, there are three or four binding sites (*) for an 8mer nucleotide: NN*NN*NN*NN or N*NN*NN*NN*N. The ESI results of binding from the DRB and nogalomycin to the two oligonucleotides studied were consistent with this principle. It has been suggested that ESIMS may be used to detect differences in sequence selectivity.[212]

An additional mechanism of action of the anthracyclines on DNA involves alkylation. An important consequence of oxidative stress induced by the anthracyclines is the production of formaldehyde which was measured, in cell lysates of breast cancer cells treated with anthracyclines, using selected ion flow tube-CIMS.[213] ESI techniques have been used to investigate the nature of the alkylation and cross-linking of DNA by anthracyclines using a low molecular weight, self-complementary deoxyoligonucleotide model, 5′-dGCGCGCGC-3′ [$(GC)_4$], a sequence that provides multiple GC sites for covalent bonding (reviewed in Reference 214). The negative ion ESI spectrum of DRB with $(GC)_4$ showed peaks of double-stranded DNA with each strand bound to one molecule of DRB via a methylene group (Figure 5.27). DNA-drug adduct formation involves a cascade of reactions including the reduction of the drug, catalytic production of reactive oxygen species, and the oxidative synthesis of formaldehyde. The last of these acts as a mediator in the alkylation of DNA (Figure 5.28). Alternative sources for the formation of formaldehyde include spermine, a polyamine associated with DNA *in vivo*.[215,216]

To explore the hypothesis about DNA alkylation via formaldehyde, two drug-conjugates of formaldehyde, named *doxoform* and *daunoform*, were synthesized. As confirmed by ESI, these conjugates consist of two molecules of the parent drug bound with three methylene groups, two forming oxazolidine rings and one binding the oxazolidines at their 3′-amino nitrogens.[217] When doxoform and daunoform were reacted with the model $(GC)_4$, they formed drug-DNA adducts faster than DOX or DRB. Doxoform was significantly more toxic than DOX to both sensitive and resistant breast cancer cell lines.[214] A hydrolytically more stable anthracycline-formaldehyde complex, *epidoxoform*, was shown by ESI to consist of two molecules of epidoxorubicin (the 4′-epimer of DOX) bound with three methylene groups in a 1,6-diaza-4,9-dioxybicyclo[4.4.1]undecane ring system. Epidoxoform reacted with $(GC)_4$ similar to doxoform and daunoform, and was much more toxic than epidoxorubicin against DOX-resistant breast cancer cells.[218,219]

Mitoxantrone

Mitoxantrone, a DOX analogue, is a planar, tetracyclic compound with two symmetrical amino acid side arms but no glycosidic substituent (Figure 5.20). It has less cardiac toxicity and fewer untoward side effects than DOX.

Mass spectral data for mitoxantrone have been obtained by EI, CI, FD, FAB, and TSP.[220] In one study, two polar human urinary metabolites were isolated by preparative HPLC and evaporated from a gold wire in the solids probe into a NCI source. The structure of both metabolites involves the oxidation of the terminal hydroxy groups of the side-chains of mitoxantrone, to dicarboxylic acid in the major metabolite, and to monocarboxylic acids in the minor metabolite (Figures 5.20 and 5.29). These metabolites are detoxification products as proven by the inactivity of synthesized samples against P388 mouse leukemia.[221]

In a study of electrophilic intermediates that may be responsible for the P-450-induced cytotoxicity of mitoxantrone, ESIMS was used for the identification of metabolites separated by HPLC. The human metabolites described above were detected in rabbit and human hepatocytes but not in rat hepatocytes. The latter produced glutathione, cysteinylglycine, and cysteine derivatives of the parent drug. The only metabolite in human HepG2 hepatoma cells was 2-(L-cyteine-S-yl)mitoxantrone. The *in vivo* biliary metabolites from minipigs included the same thioether derivatives seen in rat hepatocytes and several glucuronic acid derivatives. The site of reaction with glutathione was the hydroquinone moiety of the parent drug. The cytotoxic activity of mitoxantrone could be

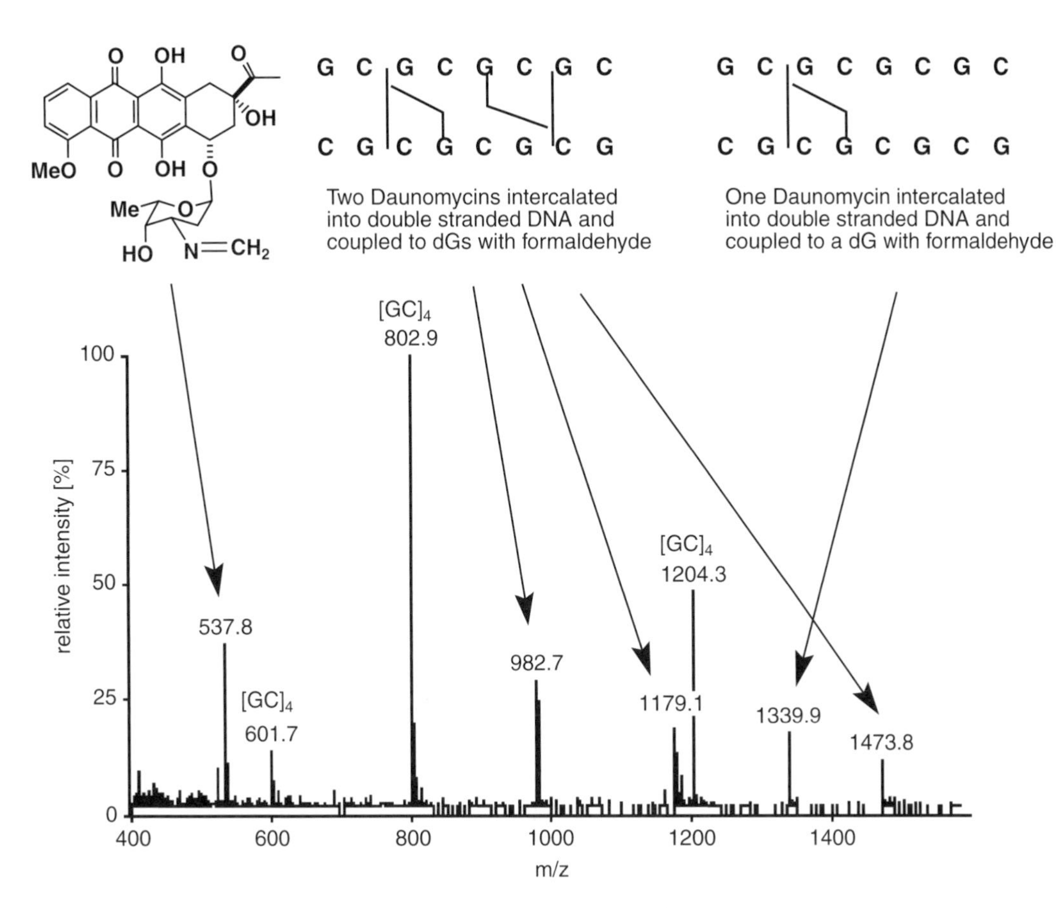

FIGURE 5.27 Negative ion electrospray mass spectrum of the major adduct formed by reaction of daunorubicin with the self-complementary DNA oligonucleotide $(GC)_4$. The adduct is proposed to have two molecules of daunorubicin (DRB), each intercalated at a CpG site in the double-stranded DNA and each linked from its 3′-amino to the 2-amino of the neighboring G base, consistent with the appearance of the ions at *m/z* 982.7 (calcd for $[M–6H]^{6–}$, 982.5 Da), 1179.1 (calcd for ($[M–5H]^{5–}$, 1179.2 Da), and 1473.8 (calcd for $[M–4H]^{4–}$, 1473.3 Da). Significant intensity at *m/z* 982.7 and 1473.8 also results from single-stranded DNA covalently bound to one DRB as $[M–3H]^{3–}$ and $[M–2H]^{2–}$, respectively. The adduct decomposes in the mass spectrometer to yield double-stranded DNA linked to a single DRB, *m/z* 1339.9 ($[M–4H]^{4–}$]), single-stranded DNA, *m/z* 601.7, 802.9, and 1204.3; and DRB linked to a methylene (likely the Schiff base as shown), *m/z* 537.8 ($[M–1H]^{1–}$). Reprinted from Taatjes, D. J. et al., *J. Med. Chem.*, 39, 4135–4138, 1996. With permission.

eliminated when the formation of the thioether conjugates was prevented by metyrapone which inhibited the cytochrome P-450 isoenzymes.[222]

Suramin

Suramin is an organic polyanion with two trisulfonated naphthalene moieties (Figure 5.20). Used for the treatment of certain parasitic diseases since the 1920s, the drug has been shown to exhibit activity against hormone-refractory prostate cancer. Suramin is a controversial drug because positive clinical results have been tempered by severe neurotoxicity associated with serum concentrations >350 µg/mL. Suramin, being a highly acidic compound that is difficult to ionize, MALDI spectra could only be obtained when noncovalent complexes were formed with peptides or small proteins rich in arginine that served as basic components of the complexes. The utility of this approach was also demonstrated for several polysulfated, polysulfonated, and polyphosphorylated compounds of biological interest. Molecular masses could be determined with an accuracy of ± 0.1%.[222a]

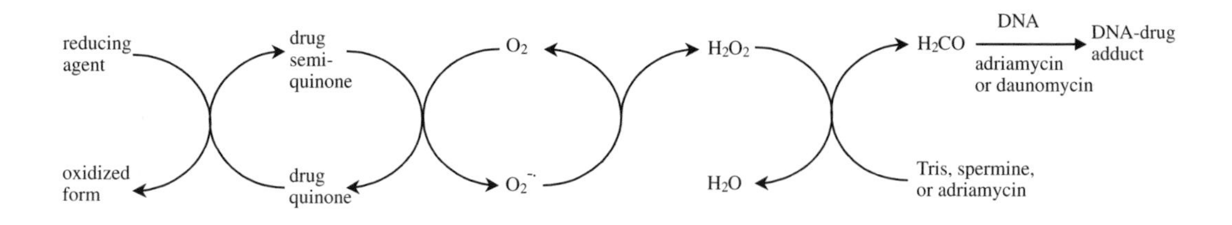

FIGURE 5.28 Scheme proposed for the pathway from daunorubicin reduction to DNA-drug adduct formation. Reprinted from Taatjes, D. J. et al., *J. Med. Chem.,* 39, 4135–4138, 1996. With permission.

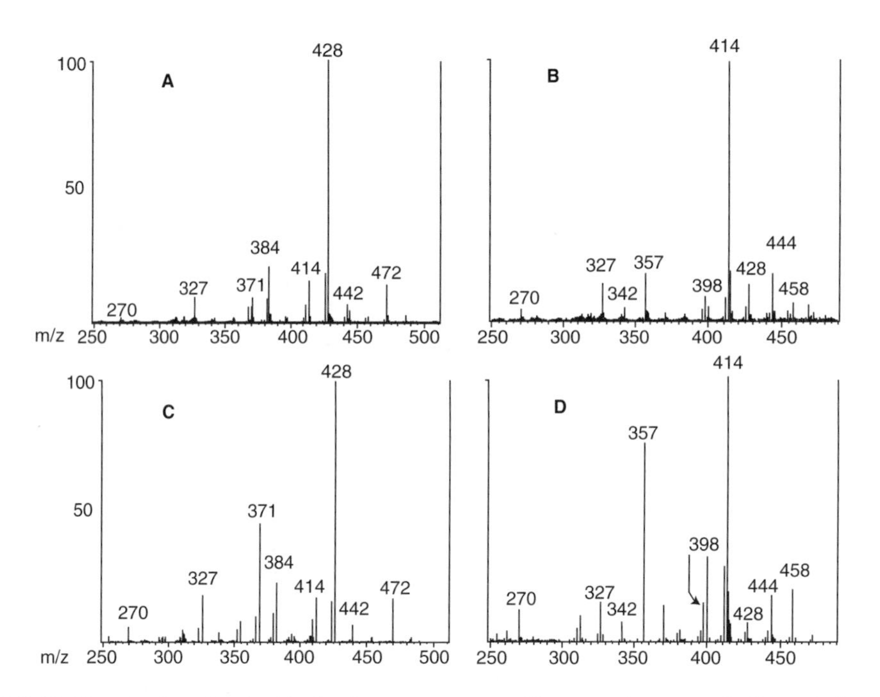

FIGURE 5.29 Negative ion chemical ionization mass spectra of human urinary metabolites of mitoxantrone. (**A**) and (**B**), major and minor isolated metabolites, respectively; (**C**) and (**D**), the corresponding synthetic compounds. The two metabolites are the di- and mono-carboxylic acids resulting from oxidation of the terminal hydroxyl groups of the side-chain(s). Reprinted from Chiccarelli, F. et al., *Cancer Res.,* 46, 4858–4861, 1986. With permission.

Suramin binds strongly to albumin, up to 90% in serum, and this may contribute to the persistence of drug in patients for weeks following a single intravenous administration. An ESIMS technique has been developed to measure the molecular masses of the complexes of albumin and suramin in a model system and in human serum at concentrations associated with toxicity. The normally bell-shaped and smooth profile of multiply charged ions in the ESI spectrum of albumin became ragged when an aliquot of a albumin-suramin mixture (1:1 molar ratio, incubated for 1 h at 37°C) was flow-injected (Figure 5.30a). Deconvolution revealed two molecular masses, unchanged albumin at 66,539 Da and a complex of albumin+1 suramin at 67,881 Da (Figure 5.30b). Incubation with excess albumin also yielded a complex with one attached suramin molecule. When excess suramin was used, the number of bound suramin molecules increased significantly. The

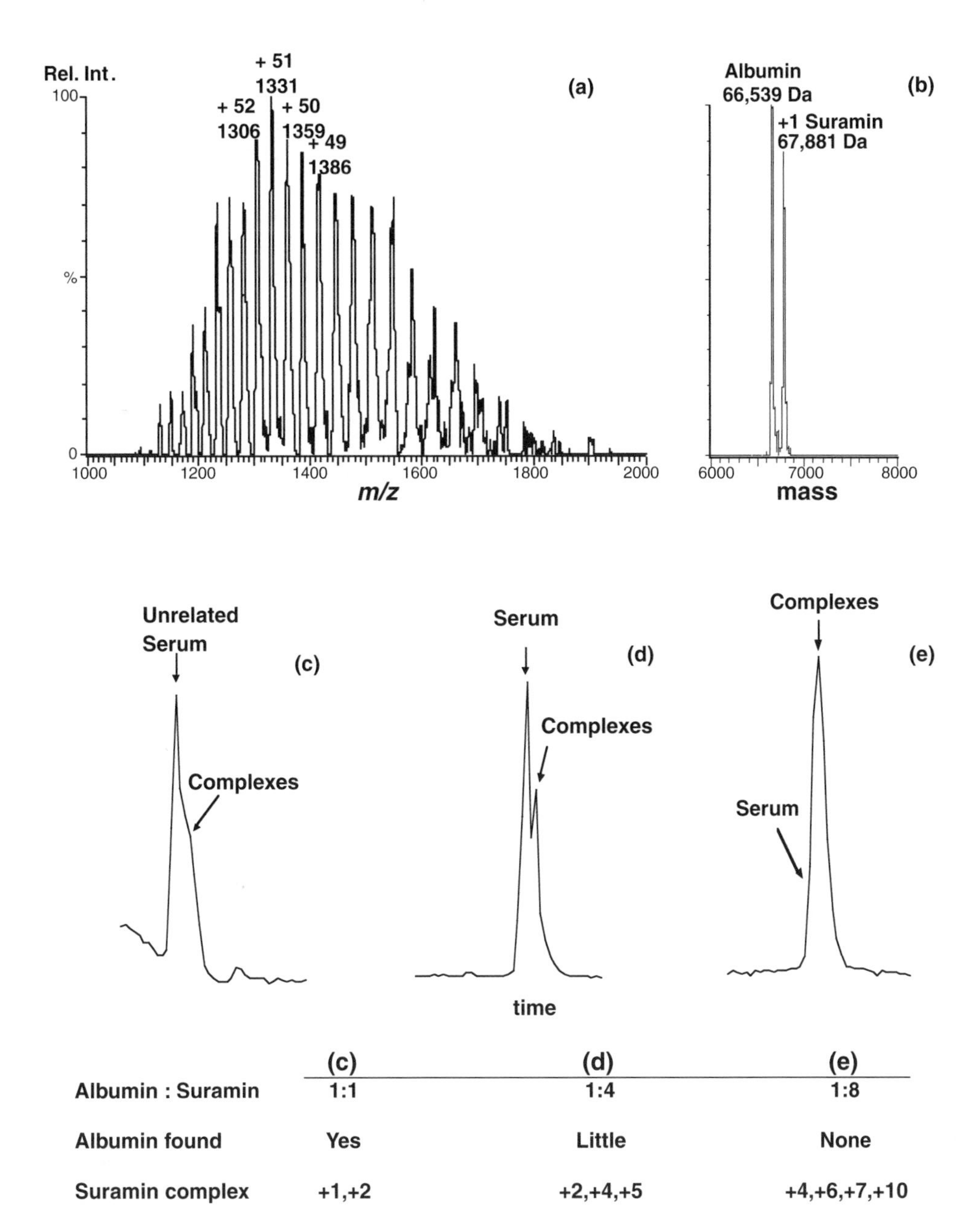

FIGURE 5.30 **(a)** The normally bell-shaped and smooth profile of multiply charged ions in the ESI spectrum of albumin (not shown) became ragged when an aliquot of an albumin-suramin mixture (1:1 molar ratio, incubated for 1 h at 37°C) was flow-injected. **(b)** Deconvolution revealed two molecular masses, unchanged albumin and a complex of albumin+1 suramin. **(c)** Total ion monitoring profile of a serum sample spiked with suramin to provide a 1:1 molar ratio (incubated for 10 min at 37°C). The arrow shows the point where transformation revealed unreacted albumin and two albumin-suramin complexes, one with 1, the other with 2 bound suramins. Spectra taken at other points along the profile revealed only unrelated serum constituents. **(d)** Recommended suramin therapeutic doses yielded serum concentration levels of 280 to 300 μg/mL. This corresponds to a suramin:albumin molar ratio in the 3:1 to 4:1 range. Under these conditions complexes with +2, +4, and +5 suramins were observed. **(e)** When the suramin concentration was increased to molar ratios corresponding to significantly toxic conditions, complexes with up to 10 suramin molecules were found and unreacted albumin diminished. Based on data reported in Roboz, J. et al., *Rapid Commun. Mass Spectrom.*, 12, 1319–1322, 1998.

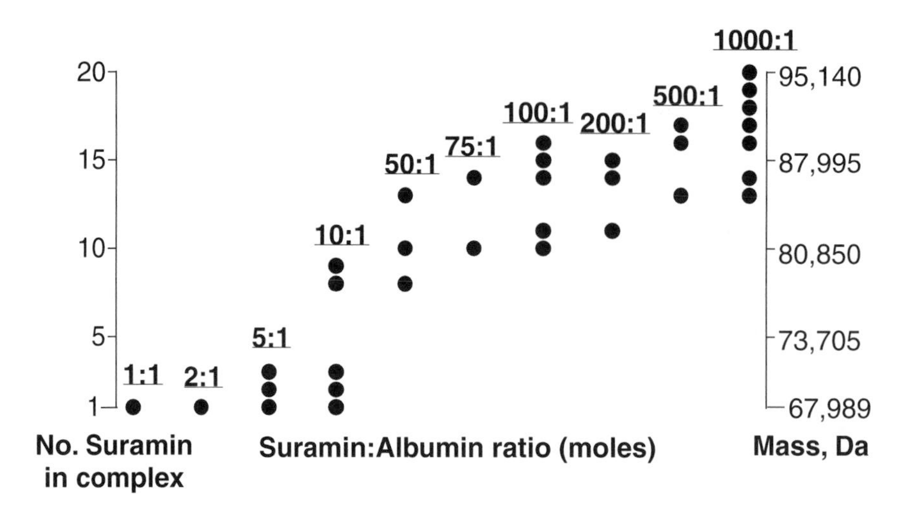

FIGURE 5.31 The number of suramin molecules bound to albumin and the molecular masses of the complexes at various suramin:albumin molar ratios.

highest number of bound suramin molecules observed was 20 suramins/albumin at 95,140 Da. The number of bound suramin molecules observed at various suramin:albumin molar ratios is shown in Figure 5.31.

When human serum was incubated with suramin, the nature and composition of the complexes found were similar to those observed in the model. Figure 5.30c shows the total ion monitoring profile of a serum sample spiked with suramin to provide a 1:1 molar ratio (incubated for 10 min at 37°C). The arrow shows the point where transformation revealed the presence of unreacted albumin and two albumin-suramin complexes, one with 1, the other with 2 bound suramins. Spectra taken at other points of the total ion monitoring profile revealed unrelated serum constituents but no albumin-suramin complexes. Recommended suramin therapeutic doses yield serum concentration levels of 280 to 300 μg/mL. This corresponds to a suramin:albumin molar ratio in the 3:1 to 4:1 range. Under these conditions complexes with +2, +4, and +5 suramins were observed (Figure 5.30d). When the suramin concentration was increased to molar ratios corresponding to significantly toxic conditions, complexes with up to 10 suramin molecules were found and unreacted albumin diminished (Figure 5.30e). The binding of this many suramins can only be explained by assuming that there are changes in the secondary and/or tertiary structure of albumin during complex formation. This technique could be a useful tool for the development of algorithms to maximize loading regimens for the regulation of the quantity of circulating free drug and also for monitoring serum to determine the onset of toxicity.[223]

Hedamycin

Hedamycin (Figure 5.32) is a naturally occurring antitumor antibiotic. In addition to being an intercalator of DNA, hedamycin also alkylates with a preference for guanines located in 5′-CGT-3′ and 5′-CGG-3′ sequences, suggesting that the N7 of guanine is the nucleophilic site on DNA. It was shown by ESIMS that hedamycin binds covalently to the self-complementary hexadeoxyribonucleotide 5′-CACGTG-3′ and other similar oligonucleotides, increases duplex stability, and forms base-paired duplexes in the gas phase.[224,225] Subsequently, it has been shown that the sequence selectivity of base alkylation by hedamycin can be determined unequivocally by ESI-MS/MS. While the CID spectra of the $[M-2H]^{2-}$ ions of 5′-CACGTG-3′ contain ~30 fragment ions with no dominant pathway of fragmentation, the CID of the same ions from the single-stranded complex of hedamycin with the same oligonucleotide is notably simple (Figure 5.32). The five fragments originate from a single fragmentation pathway that involves the loss of the alkylated guanine, by

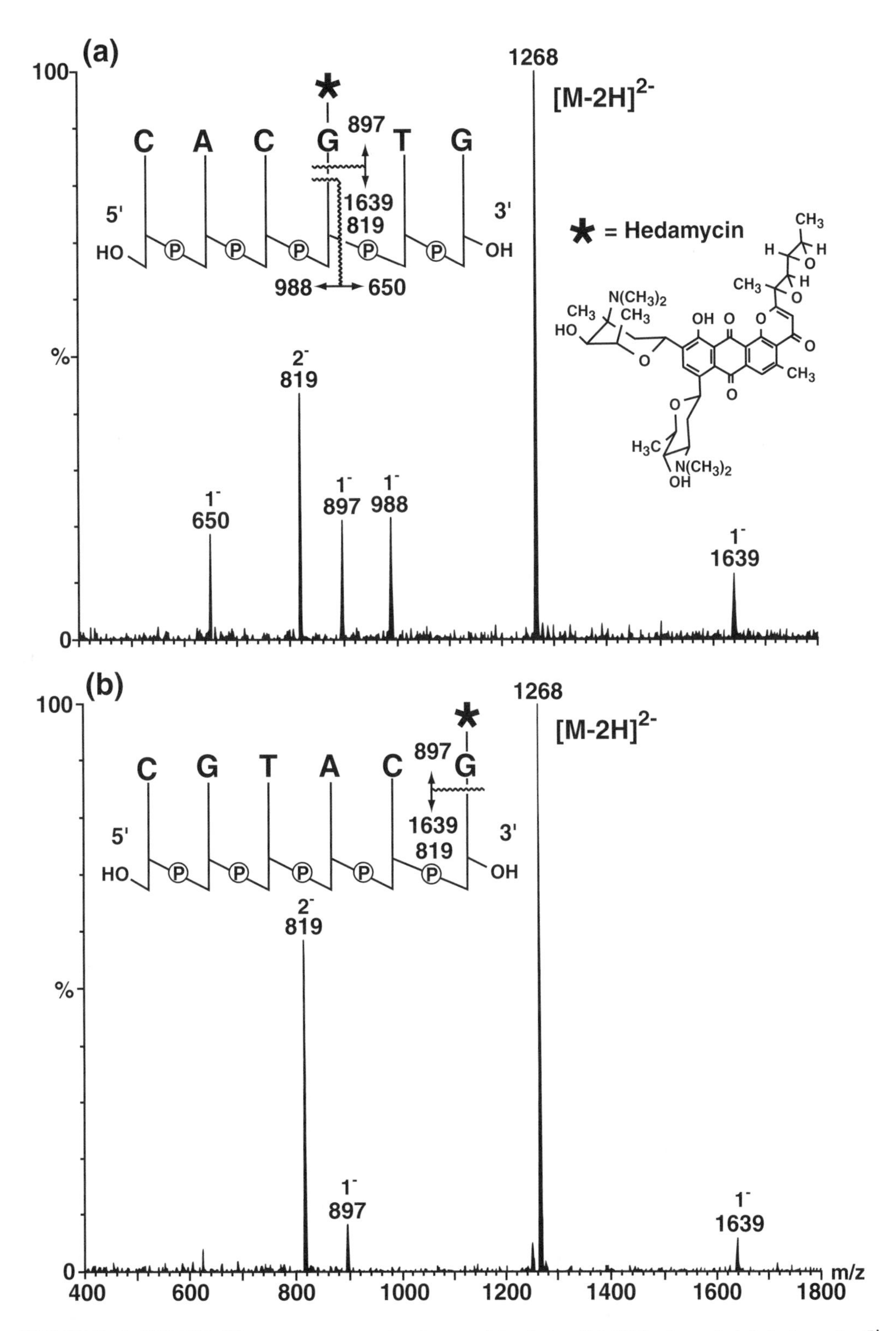

FIGURE 5.32 ESI-MS/MS spectrum (negative ion mode) of the $[M{-}2H]^{2-}$ ion of: (**a**) hedamycin–5′-CACGTG-3′ single-stranded adduct; (**b**) hedamycin–5′-CGTACG-3′ single-stranded adduct. Reprinted from Iannitti, P., Sheil, M. M., and Wickham, G., *J. Am. Chem. Soc.*, 119, 1490–1491, 1997. With permission.

N-glycosidic bond cleavage, yielding either the doubly charged ion of the depurinated oligonucleotide (*m/z* 819) or the corresponding singly charged ion (*m/z* 1639). The product ion at *m/z* 897 is the singly charged hedamycin-guanine adduct. The remaining fragments originate from the alkylated guanine. Similar CID fragmentation was observed for other complexes of hedamycin with comparable oligonucleotides. It was concluded that the technique is applicable to the structural analysis of alkylating ligand-DNA adducts.[226]

Dactinomycin

The actinomycins, a large group of antitumor antibiotics, are derivatives of phenoxazone in which cyclic polypeptides are attached through the 1 and 9 carbonyl groups. Dactinomycin (DACT, actinomycin D, Figure 5.26) was introduced as an antineoplastic agent almost 50 years ago and it is still being used, often in combination chemotherapy, against several neoplasms. DACT has been described as a dual topoisomerase poison with mixed groove binding/intercalation as the mode of its DNA binding.[170] According to one model, DNA–DACT intercalation occurs by the phenoxazone ring intercalating between base pairs with the peptide lactone rings lying in the minor groove. There are no known metabolites of DACT.

The CID of DACT and several analogs has been studied by FAB-MS/MS. Pathways for the structurally significant product ions of CID were rationalized by assuming an initial McLafferty rearrangement at the ester linkage between the methylvaline and the threonine of both rings to form linear pentapaptides. The subsequent fragmentation of either pentapeptide led to typical sequence ions and cleavages from both rings.[227]

The intact complexes of DACT with albumin and hemoglobin have been investigated using the same ESIMS technique described above for etoposide and suramin. With albumin, the maximum number of attached DACT molecules was 3, observed at 70,332 Da, in samples having a 1:1 mole ratio, incubated for 48 h.

The formation of hemoglobin (Hb)–DACT complexes is illustrated in Figure 5.33 where sections from the mass spectra of incubated samples representing the +19 and +20 charge states are enlarged. The control samples (Figures 5.33a and 5.33d) show the lower mass side of the off-scale size peak representing unreacted Hb and the arrows show the position where the complex should appear. Figures 5.33b and 5.33e indicate the appearance of the +19 and +20 charge states of a complex when a 1:1 molar ratio mixture was incubated for 3 h at 37°C. The molecular mass of the complex was 16,382 Da, corresponding to the uptake of 0.99 molecule of DACT. Figures 5.33c and 5.33f show similar data for a mixture of 1:50 molar ratio, incubated for 1 h. The molecular mass was 16,348 Da indicating the uptake of 0.98 DACT molecule. Comparing the relative quantities of the Hb–DACT complex with one incorporated DACT molecule, incubation for 0.5 and 1.0 h reveal significant differences (Figure 5.33g). Note that at a 1:1 molar ratio the 0.5 h incubation yielded no detectable complex, while incubation for 3 h resulted in easily detectable quantities (Figures 5.33b and 5.33e).

5.1.4 Antimicrotubule Agents

5.1.4.1 Microtubules and Tubulins

The cytoskeleton of eukaryotic cells contains an extensive intracellular network of filaments of different diameters that perform multiple cellular mechanical functions, ranging from morphogenesis to intracellular transport. Microtubules are cylinders of 13 longitudinally arranged protofilaments, each consisting of noncovalently bonded heterodimers of α- and β-tubulin. Tubulins are globular proteins of ~50 kDa (~450 amino acids) with considerable homology among different species. Microtubules grow out from *centrosomes,* near the center of cells. In the presence of microtubule-associated ~200 kDa size proteins and GTP (one tightly bound, another interchangeable with unbound GTP), the tubulin dimers assemble in a head-to-tail, alternating manner, into long protein fibers called *protofilaments*. As assembly continues for several minutes, C-shaped sheets

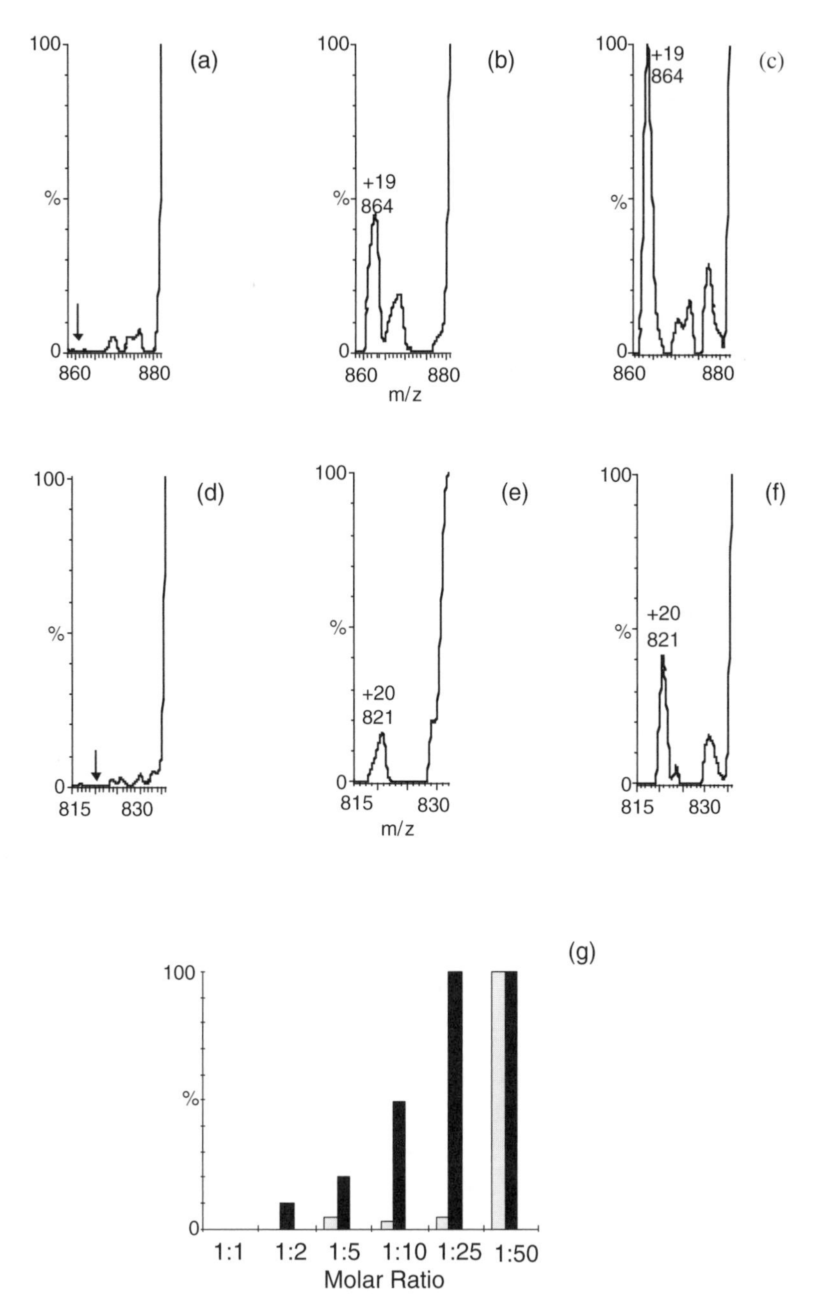

FIGURE 5.33 The formation of hemoglobin (Hb)–dactinomycin (DACT) complexes. Enlarged sections are shown from the ESI mass spectra of incubated samples representing the +19 and +20 charge states. **(a)** and **(d)**, Lower mass side of the off-scale size peak representing unreacted control Hb; the arrows show the position where the Hb-DACT complex should appear. **(b)** and **(e)**, the +19 and +20 charge states of a complex appeared when a 1:1 molar ratio mixture was incubated for 3 h at 37°C. The molecular mass of the complex was 16,382 Da, corresponding to the uptake of 0.99 molecule of DACT. **(c)** and **(f)**, Similar data for a mixture of 1:50 molar ratio, incubated for 1 h. The observed molecular mass was 16,348 Da corresponding to the uptake of 0.98 DACT molecule. **(g)** Comparison of the relative quantities of the Hb-DACT complex with 1 incorporated DACT molecule; incubation for 0.5 and 1.0 h reveal significant differences. Note: at a 1:1 molar ratio the 0.5 h incubation yielded no detectable complex, while the 3 h incubation resulted in readily detectable quantities, **(b)** and **(e)**. Shaded: 0.5 h and solid: 1.0 h.

form that finally curl around to become pipe-like microtubules of 15 nm internal and 25 nm external diameter. A vital function of microtubules is the formation of the mitotic spindle. Other functions of tubulins include serving as an internal scaffold and participating in intracellular transport. The rate at which cytoplasmic microtubules disassemble and reassemble increases during mitosis. This assists the chromosomes to capture the growing microtubules, leading to the formation of intricate *mitotic spindles* that are vital in the segregation of chromosomes prior to cell division.

Microtubules can disassemble at one location and reassemble in another, and at any given moment, cells contain a mixture of free tubulin subunits and micotubules. Microtubules are in a state of continuous dynamic instability with the direction of the dynamic equilibrium being promoted by GTP and inhibited by Ca^{2+}. The spindles are functional only when the microtubules are able to both assemble and disassemble. This mechanism provides a target for antimitotic cytotoxic agents. Dividing cells may be killed by agents which destabilize the microtubules or by agents that stabilize them (reviewed in Reference 228). For example, the vinca alkaloids bind to a site on β-tubulin, thereby destabilizing the process of polymerization. The blocking of the region where heterodimer attachment occurs prevents the growth of the microtubule polymer. In contrast, the taxenes act by blocking microtubule depolymerization when they bind tightly to the microtubules. Thus, the microtubules can grow but cannot shrink.

5.1.4.2 Vinca Alkaloids

The vinca alkaloids are dimeric compounds in which indole and dihydroindole are joined together with other complex ring systems (Figure 5.34). Modifications have been made on both the velbamine (catharantine) and vindoline moieties. Note that vincristine (VCR) and vinblastine (VBL) differ only in the presence of a formyl or methyl group, respectively, on the vindoline moiety. Despite this modest difference in structure, which does not alter the mechanism of action of binding to tubulin, there are considerable differences in the clinical spectrum of the efficacy and toxicity of these drugs. Vinorelbine differs from the other vinca alkaloids: while the vindoline moiety of vinorelbine is the same as that of VBL, the catharantine moiety has been changed, with an 8-membered ring replacing the hydroxyl and 9-membered ring. As a consequence, the lipophilicity of this compound is significantly increased. The vinca alkaloids are relatively hydrophobic molecules that partition into lipid bilayers in the unchanged state, altering the structure and function of membranes.

The antineoplastic activity of vinca alkaloids is attributed to their ability to disrupt the reversible binding of tubulin, the subunit protein of microtubules. This causes dissolution of mitotic spindles and arrest of dividing cells in metaphase. The disruption of microtubules also leads to toxicity in

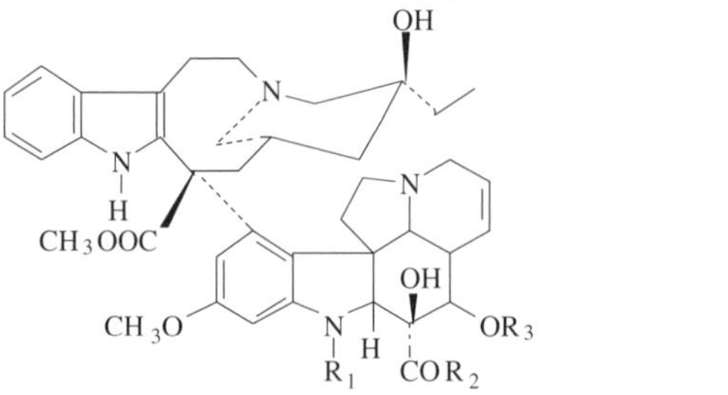

	R_1	R_2	R_3
Vinblastine	$-CH_3$	$-OCH_3$	$-COCH_3$
Vincristine	$-CHO$	$-OCH_3$	$-COCH_3$

FIGURE 5.34 Chemical structures of some vinca alkaloids.

non-mitotic neoplastic cells. The vinca alkaloids are classified as mitotic inhibitors, however, their antineoplastic activity arises not only from perturbation of several microtubule-dependendent processes but also from the disruption of the cell cycle and from induction of apoptosis.

The ESIMS and CID spectra of pure VCR and VBL have been obtained and interpreted using standards.[229] These techniques were then applied to monitoring the production yields of these drugs in *Catharanthus roseus* obtained from somatic embryogenesis. An *in vitro* cell culture was first established from a mother plant with pink flowers. Next, two phenotypes were obtained by somatic embyrogenesis. These plants produced mainly white flowers. Both flowers and leaves were extracted with water/0.1% methanolic HCl and filtered. Acidic extracts were flow-injected (FIA) into the ESI source. In addition to the detection of such previously observed components as the monoglycosides and 3-O(6-O-*p*-coumaroyl)glycosides of petunidin, hirsutidin and malvidine, ESI-MS/MS confirmed the presence of catharantine and vindoline as well as VCR and VBL. Unexpectedly, the vinca analytes were present in the flowers as well as in the leaves. Although involved chromatographic separation was needed for the full separation of all constituents, FIA permitted rapid and easy semiquantitative analysis. The LOD was 2 ng for both analytes. For cultivated *C. roseus* the VBL and VCR contents were 5.8 and 1.0 μg/g fresh weight, respectively, for leaves, and 1.9 and 0.4 μg/g for pink flowers. The corresponding results from the plants regenerated by somatic embryogenesis were 14.5 and 2.2 μg VBL and VCR per g of leaves, and 12.5 and 1.5 μg/g for pink flowers. The vinca alkaloid content of the leaves from the regenerated plants with white flowers were significantly different, 5.1 μg/g and 0.8 μg/g of VBL and VCR, respectively. The vinca contents of the white flowers were comparable to those from pink flowers. These data show that while leaves are the preferential sites of accumulation of VBL and VCR in cultivated plants, the white flowers of plants obtained by somatic embryogenesis contained a larger amount of alkaloids than the leaves of the same plant, and the leaves and flowers of the pink phenotype produced comparable quantities. In all cases, the quantity of VBL was much greater than that of VCR. It was concluded that the semiquantitative ESI technique was a simple and fast approach to screening *C. roseus* for highly productive cell lines or crops.[230]

Vinca alkaloids are usually quantified in human plasma by HPLC. The LOD of an LC/APCI-MS/MS technique for the quantification of VCR (i.s.: VBL) was 0.1 ng/mL.[231] This technique was applied to investigate the effects of the antiepileptic agents carbamazepine and phenytoin on the pharmacokinetics of VCR in patients with brain cancer receiving VCR as a component of combination chemotherapy. These antiepileptic agents that are used to prevent seizures, are potent inducers of CYP3A-4, a cytochrome P-450 enzyme that is known to partially mediate the biotransformation of VCR in the liver. There was a ~40% reduction of the total AUC and ~60% increase in the systemic clearance of VCR in the patients receiving the antiepileptics. The clinical significance of the faster drug elimination, with respect to the efficacy of VCR, has not yet been established.[232]

An investigation of the CID decomposition pattern of vinca alkaloids using FAB has led to the selection of strategic ions that may be utilized for searching for metabolites in urine, serum, or tissues. The FAB spectra of VCR could be divided into two parts (Figure 5.35a). The substituent-dependent ions, appearing at their respective masses, were almost identical for VCR and VBL. The $[M+H]^+$ ions were the base peaks, while other substituent-dependent ions, including those associated with the loss of water, methanol, the R_1 substituents, and the loss of the catharantine moiety were present at low intensities in all vincas. In the substituent-independent groups, at masses <500 Da, a number of structurally significant peaks appeared, e.g., at *m/z* 154 and 122. The peak at *m/z* 555, although of low intensity, is a marker of the presence of vindoline. As evidenced by the precursor mode spectrum (Figure 5.35b), the peak is genetically related to the precursor molecule. The CID spectrum of the low-intensity ion at *m/z* 353 (not shown) may be used to explain the appearance of several structurally important peaks at lower masses. An important step in a strategy for metabolite detection is a search for the ion at *m/z* 555. The absence of this ion proves that there are no metabolic changes involving R and C18 substituents. The presence of the ion at *m/z* 353 indicates that the metabolic process involved only the vindoline parts which contain the R substituent. In

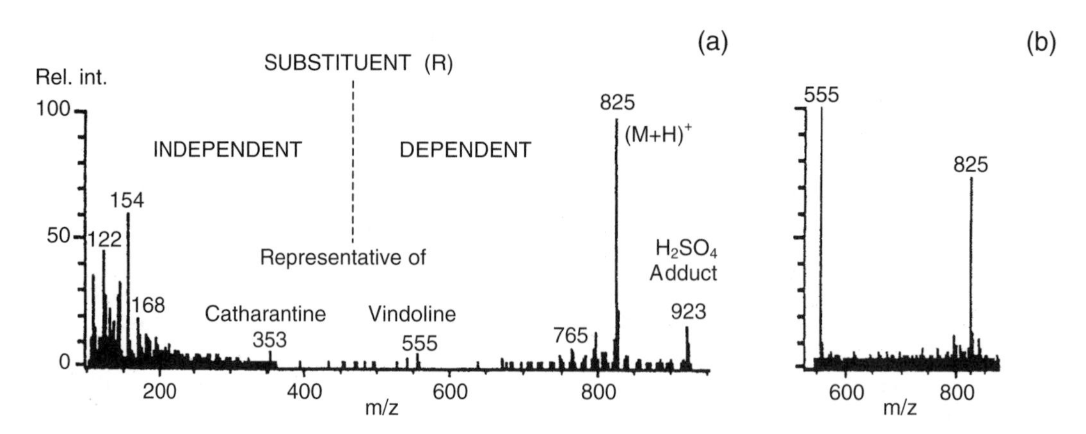

FIGURE 5.35 Collision-induced dissociation pattern of vincristine using FABMS, leading to the selection of strategic ions that may be utilized for searching for metabolites in urine, serum, or tissues. (a) The spectrum may be divided into two parts: substituent-dependent and substituent-independent groups. The peak at *m/z* 555, although of low intensity, is a marker for the presence of vindoline. (b) The precursor ion mode spectrum confirms that the *m/z* 555 ions is genetically related to the parent molecule. The CID spectrum of the low-intensity ion at *m/z* 353 (not shown) may be used to explain the appearance of several structurally important peaks at lower masses.

the absence of the *m/z* 353 ion, the presence of characteristic ions at lower masses indicate metabolic changes on substituents on the carbons at the C16 or C18 positions. The precursor ion mode spectra of selected ions may help to detect and identify metabolites in biological materials.

5.1.4.3 Taxanes

Paclitaxel (taxol) is a complex diterpenoid ester consisting of an oxetan ring attached to a derivative of taxane (Figure 5.36). Isolated from the stem bark of the Pacific yew, *Taxus brevifolia*, in 1971, taxol was found to have antitumor and antileukemic activity against murine tumors (reviewed in References 233, 234). Paclitaxel is FDA-approved for metastatic ovarian cancer, metastatic or relapsed breast cancer, and AIDS-related Kaposi's sarcoma. Under investigation are combinations with 5FU, doxorubicin, and platinum drugs for several malignancies. Docetaxel (taxotere), an analog that is more potent than taxol and is also more water soluble, is approved for locally advanced or metastatic breast cancer that has failed anthracycline therapy. The semisynthetic 10-deacetyl baccatin III, a precursor of taxol, and several related diterpenoids have also been investigated.[235]

The mechanism of action of the taxanes involves the stabilization of microtubules during cell division. The high-affinity binding to polymerized microtubules (1:1 molar ratio) prevents depolymerization. As a result, cells are arrested in the G_2 and M phases of the cell cycle and exhibit disorganized arrays of microtubules. Resistance to taxol is multifaceted, including alterations or mutations in the α- and β-tubulin chains, and overexpression of P-glycoprotein and the MDR phenotype.

Mass Spectra

Interpretation of some 60 fragment ions observed by EI, CI, and FAB was facilitated by sorting the ions into three categories representative of the major portions of the molecule. The M series included the molecular ion, pronated molecule, ion-molecular species (e.g., the ammonium adduct), and all fragments formed by losses from either the taxane ring or the C13 side-chain, with the O-C13 bond remaining intact. The taxane ring or T series included fragments formed by the cleavage of the O-C13 bond with the charge residing on the taxane ring, and fragmentations involving the

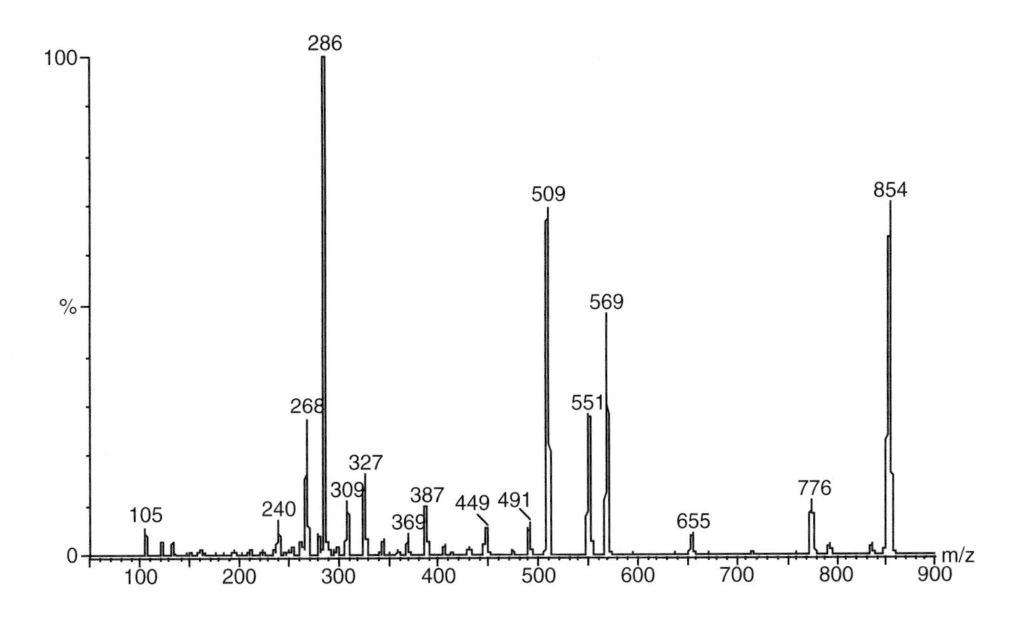

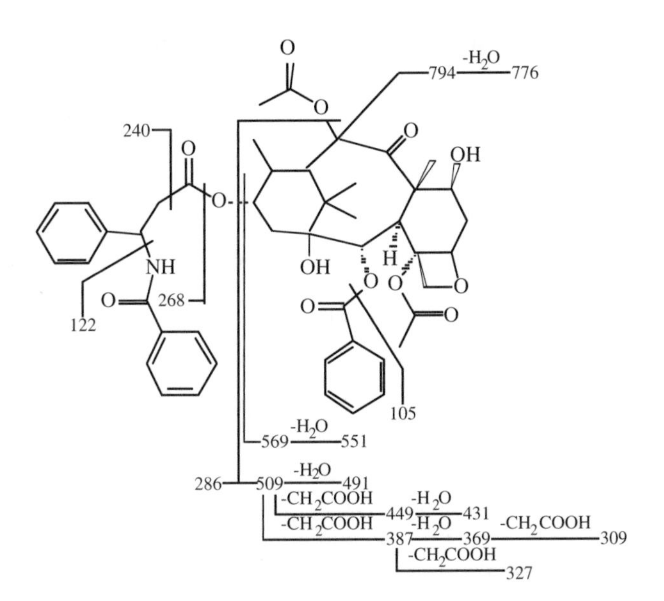

FIGURE 5.36 Taxol, tax-11-en-9-one,5β,20-epoxy-1,2α,4,7ß,10ß,13α-hexahydroxy-4,10-diacetate-2-benzoate-13-α-phenylhippurate. Upper: collision-induced decomposition mass spectrum of the protonated molecule obtained using ESIMS. Lower: explanation of the formation of important fragments.

loss of functionalities attached to the taxane ring and cleavage of the taxane ring. The C13 ester side-chain or S series ions included the even-electron ion side-chain and associated fragments that can be used to identify the attached functional groups. Additional CID experiments and accurate mass determinations using high-resolution EI led to the establishment of fragmentation pathways for the three ion series, thereby establishing a database for subsequent studies of analogs and metabolites.[236] Figure 5.36 shows some of the diagnostic products in the CID spectrum of paclitaxel that can be used for identification.

Screening for Taxanes

A potentially important observation has been that taxol and related compounds may be obtained as fungal products. For example, when *Taxomyces andreanae*, a fungal endophyte of the inner bark of the Pacific yew, was grown in semi-synthetic liquid medium, mass spectral analyses played a major role in identifying the fungal products.[237] Bioprospecting for taxol has also been carried out in a variety of plants; e.g., taxol was detected in the needles of Korean native yews[238] and filbert trees by TSPMS and confirmed by LC/ESIMS.[239]

Analysis in Cell Cultures

In an LC/ESIMS technique, sample preparation consisted of homogenization of biomass from cell culture, evaporation, partitioning between water and dichloromethane, purification using solid phase extraction, and reconstitution of the evaporated eluates for subsequent HPLC separation of the constituents. The LOD of taxanes was 5 ng using full mass range scanning. SIM of the $[M+H]^+$ ions permitted the analysis of 100 pg quantities of taxol, baccatin III, cephalomannin, and 10-deacetylcephalomannin, even in mixtures. The taxane ring is conserved within different taxanes, thus the simultaneous presence of selected T-series fragments served to locate and identify taxanes in mixtures, e.g., taxol and cephalomannin in the methanol extract of biomass from cell cultures.[240]

An LC/ESI-MS/MS technique with MRM has been developed for the quantification of paclitaxel in cell cultures. The LOD was 0.1 μM. The technique was applied to study the accumulation of paclitaxel and an analog in normal and multidrug resistant cell lines using incubations at therapeutic concentrations. It was shown that the analog was less subject to MDR than the parent drug, suggesting that the increased therapeutic effect may be due to differences in uptake and efflux at the cellular level.[241]

Quantification in Plasma/Serum

Sample preparation consisted of automated solid phase extraction (i.s.: 2′-methylpaclitaxel).[242] Quantification was accomplished by ESI-MS/MS using MRM of a characteristic transition. The LOQ was 5 ng/mL. In an application, the variation of the plasma levels of paclitaxel was monitored in a patient receiving a dose of 50 mg/m^2/day by continuous infusion. A steady state concentration of ~0.3 μM was maintained for ~80 h (Figure 5.37). The half-life of paclitaxel was 23 h.[243]

A high-throughput on-line solid phase extraction sample preparation system, with the extraction cartridges also functioning as short analytical columns, has been combined with APCI-MS/MS (negative ion mode) detection. The cycle time was 80 sec for the entire assay, permitting the analysis of 116 samples in 3 h. The linear dynamic range of the method was from 1 to 1000 ng/mL.[244]

Metabolism

After undergoing successive hydroxylation reactions, paclitaxel is eliminated by biliary and intestinal secretion of both the parent drug and its metabolites.[245] The major metabolic products of paclitaxel, identified by LC/APCIMS, are two monohydroxylated products, 6α-hydroxypaclitaxel and 3′-*p*-hydroxypaclitaxel, and a dihydroxylated compound, 6α,3′-*p*-dihydroxypaclitaxel.[246–248]

Of nine metabolites detected in the bile of paclitaxel-treated rats by HPLC, several were identified by FABMS. A major metabolite (13% of injected taxol) was formed by hydroxylation on the phenyl group at C3′ on the side-chain of C13 as evidenced by an increase of 16 mass units with respect to the protonated taxol ion. Another hydroxylation (5% of injected taxol) occurred in the *m*-position on the benzoate of the side-chain at C2. While both of these metabolites exhibited biological activity similar to that of taxol, another metabolite, baccatin III, showed no activity, probably because of the loss of the side-chain at C13.[249] The FABMS technique has been replaced by an LC/ESI-MS/MS technique that has permitted the detection of paclitaxel metabolites in rat bile without isolation by HPLC, with structural characterization using CID fragmentation. The nine metabolites were identified as four monohydroxytaxols, three dihydrotaxols, deacetyltaxol, and a

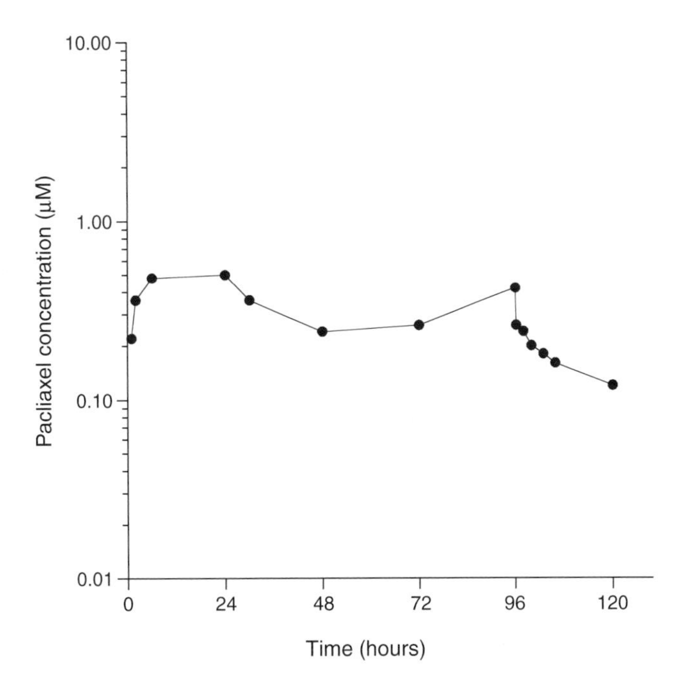

FIGURE 5.37 Plasma levels of paclitaxel in a leukemic patient during a continuous infusion of 50 mg/m²/day of the drug. Reprinted from Sottani, C. et al., *Rapid Commun. Mass Spectrom.*, 12, 251–255, 1998. With permission.

metabolite containing the taxane ring.[250] Deacetyltaxol was also identified by FAB-MS/MS in a study of the dexamethasone-induced metabolism of paclitaxel in the rat.[251]

The major metabolite of taxol in human liver microsomes was identified by FABMS as 6α-hydroxytaxol. The CID spectra were used to explain structural differences between the hydroxylated metabolites obtained from the rat and in human liver microsomes.[252] 6α-hydroxytaxol was also the principal metabolite in bile and liver from treated patients but not from rats. The negative-ion LSIMS spectrum revealed an intense 3-nitrobenzyl alcohol adduct ion. This and other fragment ions supported the conclusion, confirmed by NMR, that the hydroxy group is placed stereospecifically *trans* to the hydroxy group at position C7.[246]

Infusing tritium-labeled (100 μCi) taxol into patients, it was determined that 70% of the drug was excreted in feces, mostly as the 6α-hydroxytaxol metabolite. Unchanged taxol and five metabolites were isolated by HPLC from fecal extracts and urine, freeze-dried, redissolved in glycerol and methanol, and analyzed by FAB. All fecal metabolites gave intense $[M+Na]^+$ ions. The main CID fragments of paclitaxel represented the side-chain and the taxane ring system, with additional fragments resulting from the loss of acetic acid from the taxane and from the formation of phenyl ketene at the C2 site. Another metabolite, with a hydroxy group in the *para* position of the 3′-phenyl group, was also found in the rat model.[249] The presence of the non-hydroxylated phenyl ketene fragment, $[C_6H_5CO]^+$, in all human metabolites indicated that the aromatic ring in the 2-position was not a site for metabolic oxygenation. In contrast, this site was found to be a prominent site of metabolism in the rat model.[253]

The major metabolites in human plasma, 6α-hydroxypaclitaxel, 3′-*p*-hydroxypaclitaxel, and 6α,3′-*p*-dihydroxypaclitaxel, were extracted in mg quantities from human feces and identified by FAB, based on fragments from the taxane nucleus, the side-chain attached to the C13 position of the taxane nucleus, and the hydroxylated C13 side-chain. The biological activity of these metabolites was much less than that of the parent drug.[247]

6α-hydroxylation is catalyzed by cytochrome CYP2C8,[254,255] whereas the formation of 3′-*p*-hydroxypaclitaxel depends on CYP3A4.[256] The reaction products in these studies were determined by both positive and negative ion FABMS with CID of the protonated molecules.[246] Rats lack the human-specific CYP2C8-mediated pathway to 6α-hydroxytaxol and a reinterpretation of observed mass spectra has suggested that rats form 2-*m*-hydroxytaxol.[257] In a study of the biliary disposition of paclitaxel metabolites, dihydroxypaclitaxel was the major metabolite. The structure of the metabolite was confirmed by LC/APCIMS.[248] It was shown that the hydroxylation at the C6 position of the taxene ring was mediated by cytochrome CYP2C8, while the hydroxylation at the phenyl C3′ position on the C13 side-chain was catalyzed by CYP3A4. Pretreatment with corticoids modified the metabolism of paclitaxel by inducing the CYP3A4 isoform.[258]

Despite the similarity of the chemical structures of paclitaxel and docetaxel, there are significant differences in their metabolism. The major metabolite of paclitaxel results from hydroxylation at the 6α-position of the taxene nucleus. This metabolite has not been found for docetaxel. Also, hydroxylation at the para-position on the phenyl group at C5, which is a major human and murine metabolic pathway of paclitaxel, barely occurs for docetaxel. In contrast, as proven by the previously mentioned LC/APCIMS technique,[248] hydroxylation in docetaxel occurs at the butyl group of the C13 side-chain. In human liver microsomes, docetaxel hydroxylation, correlated with the CYP3A4 content of the microsome. Metabolite formation was significantly reduced by CYP3A4 inhibitors, such as ketoconazole. When microsomes were incubated with both paclitaxel and docetaxel, there was significant mutual inhibition of those metabolites mediated by CYP3A4 but not with the 6α-hydroxylation product.[259]

In a study of the disposition of ^{3}H-labeled paclitaxel in combination with the formulation vehicle chemophor EL in a patient with recurrent ovarian cancer, who had severely impaired renal function due to prior treatment with cisplatin, five metabolites were identified using LC/APCIMS. Cremophor EL significantly altered the kinetics of the parent drug and metabolites, particularly species lacking an intact ring fragment.[260]

Detection in Intact Tissues

Aiming to establish a microprobe technique that can confirm whether a drug has reached the intended site of action, a custom MALDI-ITMS was constructed. The external MALDI source provided the spatial resolution for the determination of the location of the drug within the intact tissue sample, and the unique MS/MS capabilities of the IT analyzer were utilized for the resolution of the target compound from the background tissue matrix. The performance of the instrument was demonstrated by detecting paclitaxel in a thin section of liver tissue that had been incubated in a solution of the drug, and in a human ovarian tumor implanted into a nude mouse. Liver and tumor slices of 0.5 mm thickness were exposed to an average of 30 laser shots. Abundant of $[M + H]^+$ and $[M + Na]^+$ ions of paclitaxel were evident in the tumor sample (Figure 5.38A). After moving the laser beam to new locations in the sample, reproducible product ion spectra were obtained from the sodiated ion (Figure 5.38B). The fragmentation pattern of the analyte was almost identical to that from a paclitaxel standard (Figure 5.38C). The holes blasted by the laser beam were ~100 μm diameter and had burned completely through the tissue. Based on the reported concentration of paclitaxel in the tissue (50 mg/kg), the amount detected in the tissue was ~290 pg. The thick coating layer on the top of the tissue slice, and the inability to visually inspect the tissue during analysis, prevented the correlation of individual spectra with specific regions of the tissue. A combination of laser desorption (no matrix needed) with CI was suggested as an alternative technique for the visualization of the surfaces of tissue samples during microsampling.[261]

Modifications to Increase Solubility

The poor solubility of paclitaxel makes it difficult to prepare clinical formulations and requires special techniques for administration. Paclitaxel has been modified by introducing a phosphate

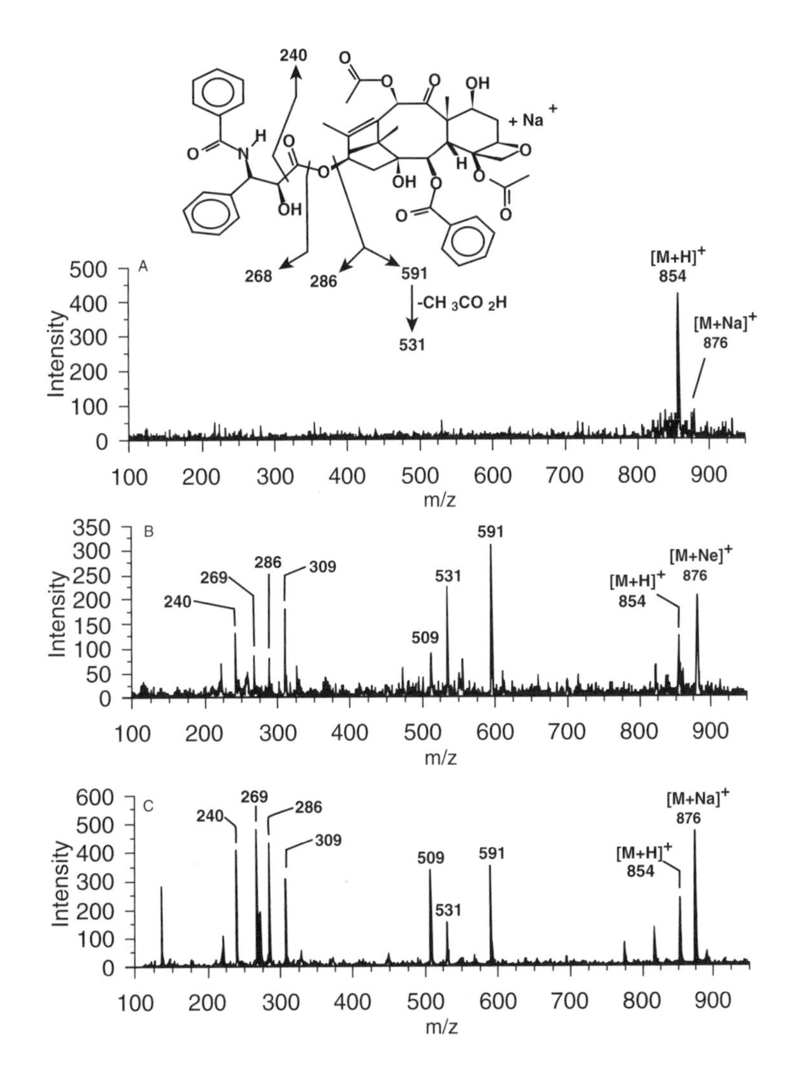

FIGURE 5.38 (**A**) MALDI mass spectrum of paclitaxel in an ovarian cancer tumor $[M+H]^+$ ion at *m/z* 854. (**B**) Product ion spectrum of the $[M+Na]^+$ ion from a different location on the same sample. (**C**) Product ion spectrum of the $[M+Na]^+$ ion of a paclitaxel standard. Reprinted from Troendle, F. J., Reddick, C. D., and Yost, R. A., *J. Am. Soc. Mass Spectrom.*, 10, 1315–1321, 1999. With permission.

moiety, phosphonoxyphenylpropionate ester, at positions C2′ or C7, to yield a water-soluble prodrug that, in turn, generate paclitaxel upon enzymatic activation. An *in vitro* system, coupling a metabolic activation system using alkaline phosphatase with a tubulin polymerization assay, has been used to test two synthetic prodrugs. The presence of paclitaxel formed upon enzymatic activation of the prodrugs was confirmed by TSPMS.[262]

5.1.4.4 Miscellaneous

Dolastatin-10 is a linear, lipophylic pentapeptide composed of dolavaline (an α-amino acid), and the γ-amino acids dolaisoleucine and dolaproline which are linked to the primary amine dolaphenine at its carboxyl terminus (Figure 5.39a). These unusual amino acids are unique to *Dolabella auricularia*, a sea hare from which this compound and a series of analogs were isolated. Like the vinca alkaloids, dolastatin inhibits microtubule assembly and tubulin-dependent GTP binding. Exposed cells are arrested in the metaphase. Dolastatin-10 is a very potent agent: the inhibitory concentration,

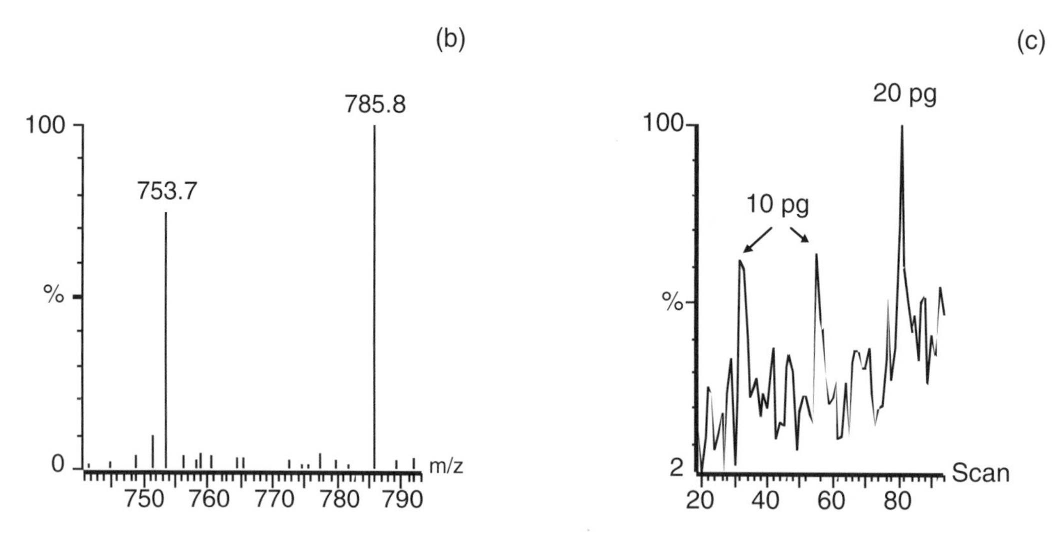

FIGURE 5.39 **(a)** Dolastatin-10 is a linear, lipophylic pentapeptide composed of dolavaline (an α-amino acid), and the γ-amino acids dolaisoleucine and dolaproline which are linked to the primary amine dolaphenine at its carboxyl terminus. **(b)** The product ion spectrum of the protonated molecule, obtained by ESI-MS/MS, includes an intense fragment ion that provides a transition for selected reaction monitoring analysis. **(c)** Flow-injection analyses to detect dolastatin-10 down to 10 pg.

IC_{50}, for P388 murine leukemia cells is 5×10^{-11} M. Strong antitumor activity has been demonstrated in cell cultures against a variety of human leukemias, ovarian carcinoma, melanoma, and several lung cancer xenograft models (reviewed in References 263, 264).

The product ion spectrum of the protonated molecule of dolastatin-10 obtained by ESIMS, includes an intense fragment ion providing a transition for MRM analysis (Figure 5.39b). Convenient flow-injection analyses may be used to detect dolastatin-10 down to 10 pg (Figure 5.39c). Two *in vitro* metabolites were detected upon incubation of dolastatin with an activated rat liver preparation and identified by ESI-MS/MS. The major metabolite was N-demethyl-dolastatin-10; the minor metabolite was hydroxy-dolastatin-10. Only the parent compound and the N-demethylated metabolite were detected in patients. Sample preparation steps in a technique developed for quantification in plasma included extraction with *n*-butyl chloride (i.s.: dolastatin-15, an analogue), evaporation of the organic phase, reconstitution with the mobile phase, and analysis by LC/ESIMS. The method was linear in the 0.005 to 50 ng/mL range. Pharmacokinetic data obtained from treated patients suggested a three-compartment model for the distribution. It was noted that the i.s. might also be considered as a cytotoxic drug, in which case dolastatin-10 would serve as the i.s.[265]

5.1.5 ANTIMETABOLITES

5.1.5.1 Purine and Pyrimidine Analogs

6-Mercaptopurine

6-Mercaptopurine (6MP), the thiol analog of the purine hypoxanthine, is used in the maintenance of remission for childhood acute lymphoblastic leukemia. 6MP is a prodrug; it has no intrinsic cytotoxic activity. It is converted into cytotoxic and inactive metabolites. The complex metabolic pathway[266] begins by intracellular conversion into 6-thioinosine-5′-monophosphate, catalyzed by hypoxanthine phosphoribosyltransferase. The active metabolites are 6-thioguanosine nucleotides (mono-, di-, and triphosphates) that cause cytotoxicity by incorporation into DNA and RNA. The catabolic end product is 6-mercapto-2,8-dihydroxypurine which is excreted by the kidneys.

Another metabolite (20% of the parent drug), detected in plasma during high-dose 6MP infusion therapy, was identified by GC/EIMS as 6-methylmercapto-8-hydroxypurine.[266] Although a GC/EIMS technique has been described for the determination of 6-mercaptopurine in plasma,[267] there are adequate HPLC methods available for quantification in pharmacokinetic studies.

Fluorouracil

5-Fluorouracil (5-FU) is a pyrimidine analog with a stable fluorine atom at position 5 of the uracil ring (Figure 5.40). Alone and in combination with cisplatin, 5-FU has been shown to be an efficient drug in the treatment of advanced gastrointestinal cancers[268] and head and neck cancer. The role of the fluorine is to prevent the conversion of deoxyuridylic acid to thymidylic acid, thus depriving the cells of one of the essential precursors of DNA synthesis; this leads to imbalanced cell growth and cell death. 5-FU is not cytotoxic per se, and has to be converted by uridine phosphorylase and uridine kinase to a deoxynucleotide, 5-fluoro-deoxyuridine monophosphate (5-FdUMP) which competes with deoxyuridine monophosphate (dUMP) for the enzyme thymidylate synthetase (reviewed in Reference 269).

The time-related variations of the plasma concentrations 5-FU and the end product of its catabolism, α-fluoro-β-alanine (FBAL), have been studied in lung cancer patients treated with continuously infused 5-FU. After precipitating plasma proteins and evaporating the liquid phase, the residue was derivatized with *t*-butyl-dimethyl-silyl-trifluoroacetamide and the ions of 5-FU and FBAL monitored by SIM (i.s.: chlorouracil). LOD was 2 ng/mL for 5-FU and 100 ng/mL for FBAL. Mean plasma concentrations of both analytes increased from the first day to the fourth day, from 0.4 to 0.7 μg/mL for 5-FU and from 1.2 ug/mL to 1.8 μg/mL for FBAL. Despite the constant-rate infusion of 5-FU, plasma 5-FU concentrations were significantly lower during daytime than nighttime. These, and other results of this pharmacokinetic study, led to speculations concerning 5-FU clearance, drug interactions, and toxicity.[270]

Several techniques have been developed during the last 25 yr for the quantification of 5-FU, related pro-drugs and their metabolites in plasma, to support clinical pharmacokinetic studies. Most of these are based on GC/EIMS or GC/CIMS. Representative techniques are described in References 271–277. These techniques can be modified for special requirements, such as in the case where GC/NCIMS was used for the quantification of an agent consisting of four related drugs.[278] Often, HPLC techniques are adequate.[279]

A study of the clinical pharmacological effects of administering thymidine (TdR) prior to 5-FU using various dosage protocols involved the monitoring of 5-FU, 5-fluoro-deoxyuridine (FUDR), TdR, and thymine (T), the major metabolite of thymidine, by GC/CIMS. Isobutane served as both the carrier and reagent gas (i.s.: 5-chlorodeoxyuridine and 5-chlorouracil). LOD was 25 ng/mL serum for all analytes. The patients' own sera served to correct for endogenous TdR, $\sim 2 \times 10^{-7}$ M. The monitoring of these analytes in samples from a particular protocol is shown in Figure 5.41. When a bolus of 200 mg/m^2 5-FU was administered alone, the elimination curve of the drug was

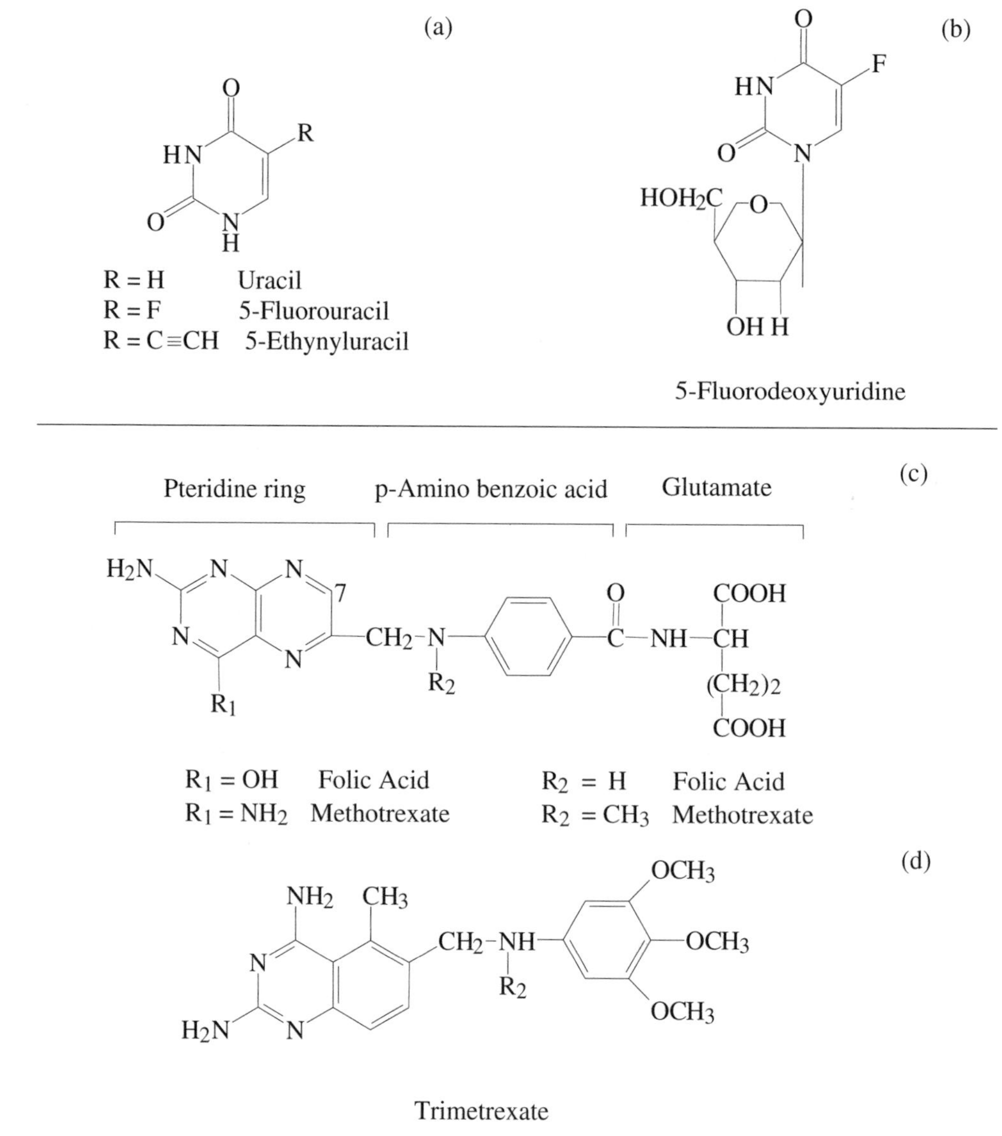

(a)

R = H Uracil
R = F 5-Fluorouracil
R = C≡CH 5-Ethynyluracil

(b)

5-Fluorodeoxyuridine

(c)

Pteridine ring p-Amino benzoic acid Glutamate

R_1 = OH Folic Acid
R_1 = NH_2 Methotrexate

R_2 = H Folic Acid
R_2 = CH_3 Methotrexate

(d)

Trimetrexate

FIGURE 5.40 Structures of pyridines, **(a)** 5-fluorouracil and related compounds; **(b)** 5-fluorodeoxyuridine; and folates; **(c)** folic acid, methotrexate, and **(d)** trimetrexate.

as expected and the 5-FU levels decreased to below detection level in ~4 h. When a 24-h infusion of 8 g/m^2 thymidine preceded the 5-FU bolus, the blood levels of 5-FU increased significantly and a considerable quantity of the drug was still present up to 24 h. The blood levels of FUDR, TdR, and T reasonably paralleled that of 5-FU. The protracted decay of 5-FU appeared to account for the increased toxicity of the combination therapy.[280]

The degree of 5-FU incorporation into RNA was studied by incubating RNA, isolated from tumors, with RNase, alkaline phosphatase, and uridine phosphorylase. This resulted in the complete degradation of RNA and liberation of 5-FU which, in turn, was derivatized with pentafluorobenzyl-bromide and quantified (i.s.: ^{15}N-labeled 5-FU) by GC/NCIMS. The 5-FU concentration in isolated RNA was ~1.5 pmol/µg 24 hr after treatment with 500 mg/m^2 5-FU. Comparable results were obtained using murine tumors.[281]

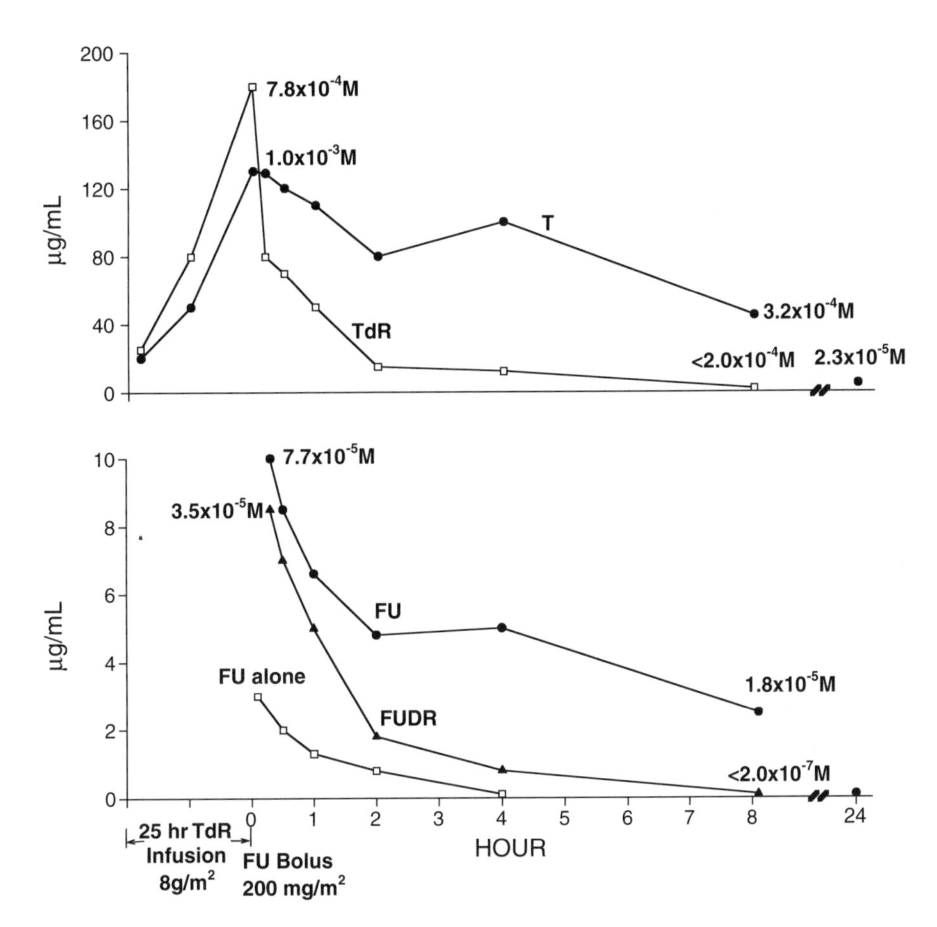

FIGURE 5.41 Monitoring by GC/CIMS of 5-fluorouracil (FU), 5-fluorodeoxyuridine (FUDR), thymidine (TdR), and thymine (T) to study the clinical pharmacologic effects of administering thymidine prior to 5-FU. When a bolus of 200 mg/m^2 5-FU was administered alone, FU levels decreased to below detection level in ~4 h (FU alone). When a 24 h infusion of 8 g/m^2 thymidine preceded the FU bolus, the blood levels of FU increased significantly and a considerable quantity of the drug was still present at 24 h. Blood levels of FUDR, TdR, and T paralleled that of FU. The protracted decay of FU appeared to account for the increased toxicity observed using the combination therapy.

Enzymes involved in the degradation of pyrimidines in humans include dihydropyrimidine dehydrogenase (DPD) and dihydropyrimidinase (DHP). These enzymes degrade 70 to 80% of administered 5-FU to fluorinated α-fluoro-β-alanine. Patients with deficiencies of these enzymes develop severe neurotoxicity upon 5-FU administration. A GC/EIMS technique has been developed for the simultaneous quantification of thymine, uracil, dihydrothymine, dihydrouracil, orotate, and creatinine in urine (i.s.: stable isotope labeled analytes). Abnormal quantities of uracil and thymine have been shown to be indicative of DPD deficiency, while large concentrations of 5,6-dihydrothymine and 5,6-dihydrouracil, and moderate amounts of uracil and thymine suggest DHP deficiency. It was suggested that this simple technique should be applied for the differential diagnosis of homozygous and heterozygous patients in the pyrimidine degradation pathway prior to treatment with 5-FU or other pyrimidine analogues.[282]

Eniluracil

The major determinant of the rate of clearance of 5-FU is dihydropyrimidine dehydrogenase (DPD) which is the rate-limiting enzyme in the catabolism of endogenous pyrimidine bases such as uracil and thymine. Inhibition of DPD has evolved as an attractive target for 5-FU modulation for several reasons, including reduced toxicity due to decreased levels of toxic metabolites and enhanced efficacy by suppressing catabolism in tumor tissues. Eniluracil (5-ethynyluracil, Figure 5.40) binds covalently to DPD thereby inhibiting the catabolism of 5-FU both systemically and in tumor tissues. In Phase I clinical and pharmacologic studies of eniluracil and 5-FU the drug concentrations have been measured by validated GC/EIMS techniques with an LOQ of 1 ng/mL. Several of the expected advantages of the combination therapy, including manageable oral administration, were demonstrated. However, it became obvious that the maximum-tolerated doses of 5-FU decreased in the presence of eniluracil because of marked alteration in 5-FU pharmacokinetics brought about by the DPD inhibitor.[283,284]

Because the covalent binding of eniluracil to DPD is irreversible, lethal drug interactions may occur with inappropriate co-admission of drugs that are normally metabolized by DPD. Uracil is the natural substrate for DPD. The concentration of uracil in plasma may increase 100-fold when DPD is inhibited. A GC/EIMS technique has been developed for the determination of plasma uracil concentrations during treatment with eniluracil. The dynamic range of the technique is wide enough to cover concentrations from basal endogenous levels to the high micromolar levels that occur during treatment so it may serve as an indicator of the extent and duration of the enzyme inhibition. Sample preparation consisted of protein removal with acetonitrile, solid-phase extraction, evaporation, resuspension, and quantification using SIM (i.s.: 5-chlorouracil). Baseline plasma uracil concentrations were ~0.2 μM. Uracil concentrations increased to 15 to 20 μM after eniluracil treatment, and it took weeks to return to normal after the termination of treatment.[285]

5.1.5.2 Antifolates

Methotrexate

It was almost 50 years ago that aminopterin, pteroylglutamic acid, a 4-amino analog of folic acid, was shown to produce remissions in childhood acute leukemia. Methotrexate (MTX, Figure 5.40), is another analog of folic acid. The antifolates, particularly MTX, have been proven to be efficacious, both as single agents and in combination therapy, for the treatment of a wide variety of neoplasms. The mechanism of action of these drugs involves the inhibition of the intracellular enzyme dihydrofolate reductase, causing a deficit of reduced folates which are donors of one-carbon units in the *de novo* biosynthesis of nucleic acids and some amino acids. The blocking of the synthesis of DNA and RNA leads to the death of tumor cells.

At conventional doses, MTX is excreted unchanged in the urine. At high doses MTX is metabolized in the liver in a dose-dependent manner to 7-hydroxymethotrexate (7-OH-MTX) which is excreted in the urine of both humans and primates. Because this metabolite is much less water-soluble than the parent compound, recrystallization in the tubules causes renal damage. High-dose MTX therapy (>50 mg/kg) offers an efficacious treatment for osteogenic and some other sarcomas and lymphomas. The potentially lethal toxicity of MTX levels reaching $>10^{-5}$ M at 24 hr is minimized by the administration of leucovorin (citrovorum factor, folinic acid) as a rescue agent. Because the sensitivity of available MS techniques is inadequate, commercial RIA kits or HPLC techniques have been used for the monitoring of therapeutic MTX doses.[286]

Methotrexate and its analogs are not amenable to EI analysis directly or as TMS derivatives. Methylation with diazomethane has yielded interpretable EI spectra that were used to identify impurities in early parenteral dosage forms.[287] Positive ion CI spectra yield intense $[M+H]^+$ ions with several structurally significant fragments.[288]

A series of MTX-γ-glutamyl-glutathione conjugates have been synthesized aiming to develop cytostatic antifolates with improved cellular uptake and transport properties. FAB techniques have yielded abundant molecular ions and structurally significant fragments at the peroyl amide and peptide bonds.[289] In studies of the multiple folate transport systems in L1210 leukemia cells, FABMS was used to establish the authenticity of fluorescein-methotrexate complexes. In one version, the fluorophore was linked, by a diaminopentane spacer, to one of the carboxyl groups of the glutamate moiety of MTX.[290] In a variant, the spacer group was lengthened by a dissociable disulfide bond and the fluorecein was replaced by biotin. The derivative was used to label the folate transporter proteins in intact cells.[291]

The 2,4-diamino-N^{10}-methylpteroic acid (DAMPA) is normally a minor urinary metabolite of MTX (<5% of total dose). However, a proposed MTX rescue agent, carboxypeptidase-G_2 ($CPDG_2$), converts almost all MTX in the plasma to DAMPA and glutamate, thus the enzyme may be used as a rescue agent in renal failure induced by high-dose MTX therapy. Three suspected metabolites obtained from DAMPA-treated primates and MTX-treated patients, after the administration of the rescue agent were separated by HPLC and identified by LC/ESI-MS/MS product scans as hydroxy-DAMPA, DAMPA-glucuronide, and hydroxy-DAMPA-glucuronide (Figure 5.42). A proposed metabolic pathway of MTX in the presence of aldehyde oxidase (AO) and $CPDG_2$ involves the conversion of MTX to 7-OH-MTX by AO and hydrolysis to DAMPA by $CPDG_2$. Glucuronyl transferase converts DAMPA and OH-DAMPA to glucuronidated products.[292]

Available HPLC techniques for the quantification of MTX and its major metabolite, 7-OH-MTX, in plasma, are usually adequate for providing analytical supports of pharmacokinetic studies. Significant progress in sample preparation has been achieved by the introduction of liquid-liquid extraction using the 96-well plate format. In an application to MTX analysis in plasma, 384 samples were extracted and prepared for MS analysis in 90 min by one operator. The speed of the LC/ESIMS detection with SRM was equally impressive: sample extracts were analyzed within 1.2 min for the two analytes (i.s.: deuterated analog of each analyte). LOD for MTX and 7-OH-MTX were 0.05 and 0.1 ng/mL, respectively.[293]

A technique has been developed for the analysis of MTX in urine samples with solid-phase extraction and quantification of the analyte (i.s.: 7-OH-MTX) using LC/ESI-MS/MS. The LOD was 0.2 μg/L urine. The purpose of the assay was to monitor hospital personnel occupationally exposed to MTX.

Severe disturbances in folate metabolism, characterized by the total folate content in erythrocytes ("folate status"), often accompany treatments with MTX. A GC/MS technique developed for the quantification of folate in whole blood samples used initial acid hydrolysis to release folates from red cells down to their *common* stable moiety, *p*-aminobenzoic acid (*p*ABA). The *p*ABA isotopomers were ethyl-esterified, trifluoroacetylated, and silylated, followed by quantification using GC/EIMS with SIM of the $[M-57]^+$ fragment (i.s.: $^{13}C_6$-*p*ABA). The presence of MTX did not interfere with this assay. Results agreed with the those obtained by the current reference assay based on *Lactobacillus casei*.[293a]

An ultrafiltration technique (Section 5.1.1.3) developed for the screening of combinatorial libraries for inhibitors of dihydrofolate reductase utilized a continuous-flow, hydrophobic, non-specific receptor-binding separation system for the separation of low-affinity (free) and high-affinity ligands and ESI for identification. Twenty-two compound libraries were searched using 6 known inhibitors of the enzyme and 16 compounds without affinity. Two library systems were identified with high affinity for dihydrofolate reductase, aminopterin and MTX.[293b]

Trimetrexate

The structure of trimetrexate (Figure 5.40), the mechanism of its entry into cells, and its intracellular metabolism, are considerably different from those of dihydrofolic acid, folic acid, and MTX.[294] Still, trimetrexate is considered to be a folate analogue ("nonclassical") because of its high potency in inhibiting dihydrofolate reductase and its broader spectrum of cytotoxic activity than MTX, partially because of its enhanced lipophilicity.[295]

The study of the metabolites of trimetrexate is of importance as they contribute to toxicity. The protonated molecule of the glucuronate conjugate of trimetrexate was detected in both human and mouse urine using FABMS. The aglycone and two prominent fragmentation peaks were shown by ESIMS to represent the quinoline portion and the substituted aniline. In the rat, one urinary

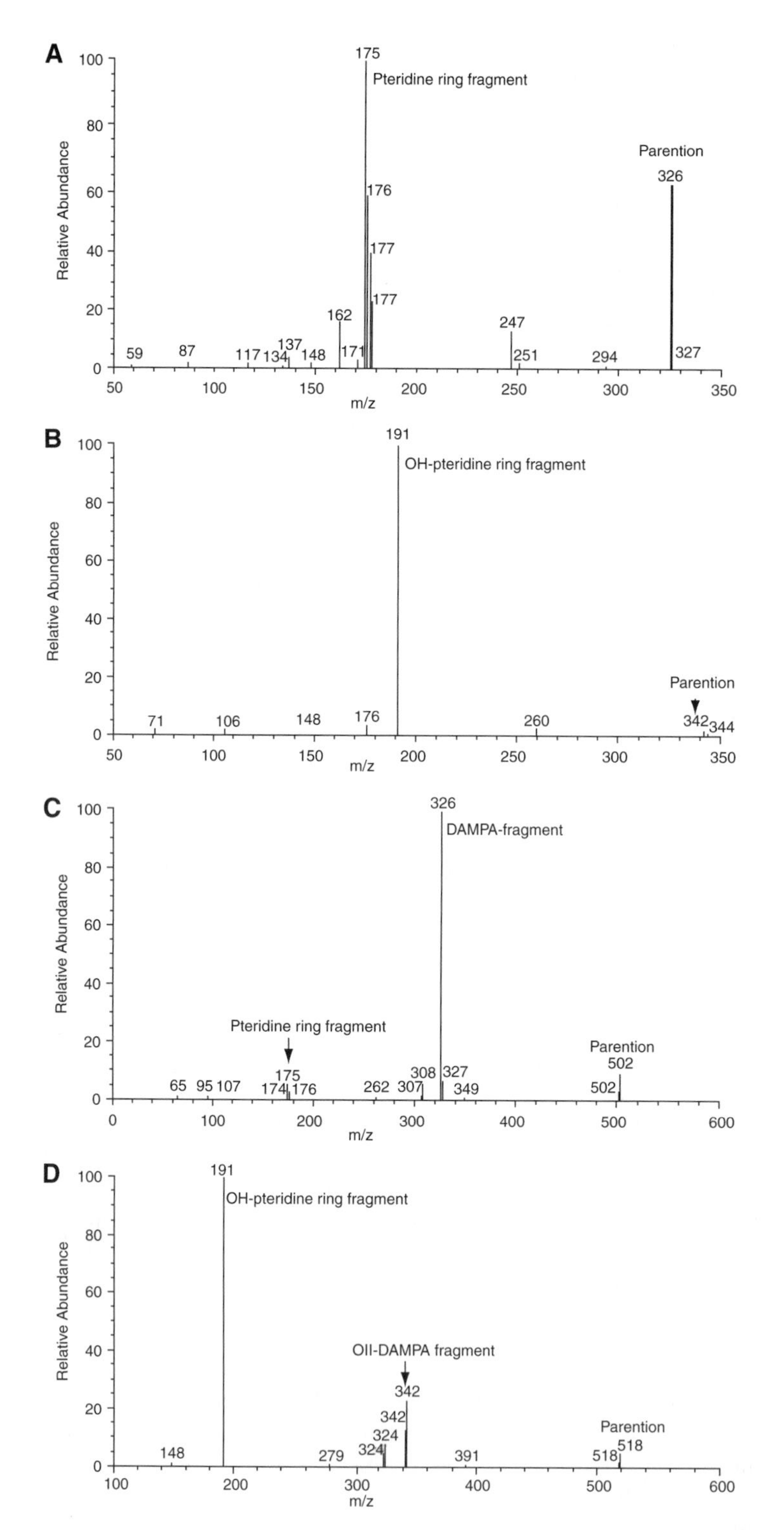

FIGURE 5.42 A proposed MTX rescue agent, carboxypeptidase-G_2 ($CPDG_2$) converts almost all MTX in the plasma to DAMPA and glutamate. Product ion scans of 2,4-diamino-N^{10}-methylpteroic acid (DAMPA). **(A)** Precursor (parent) ion and major pteridine fragment; **(B)** OH-DAMPA; **(C)** DAMPA-glucuronide; **(D)** OH-DAMPA-glucuronide. Reprinted from Widemann, B. C., et al., *J. Pharmacol. Exper. Therapy,* 294, 894–901, 2000. With permission.

metabolite was identified by LC/TSP-MS/MS as 2,4-diamino-5-methyl-6-quinazoline carboxylic acid. The mass spectrum of a second metabolite, identified as the glucuronide of O-desmethyl trimetrexate, included a fragment ion corresponding to the intact quinazoline ring system. Co-administration of unlabeled drug and ^{35}S-sulfate revealed a third urinary metabolite, identified as trimetrexate-4′-O-desmethyl sulfate. The same metabolites were also found in similarly treated dogs.[296]

A GC/EIMS technique developed for the quantification of trimetrexate in plasma is based on the SIM of the abundant fragment ions corresponding to the loss of -NHC_6H_4Cl from the TMS derivatives of the analyte (i.s.: 2,4-diamino-5-methyl-6[(4-chloroanilino)methyl]quinazoline). The LOD was 2 ng/ml plasma.[297]

5.1.6 Opportunistic Fungal Infections

Disseminated fungal infections, such as candidiasis and aspergillosis, continue to be the cause of high morbidity and mortality in immunocompromised patients. These diseases are often present without circulating organisms, thus the rate of false negatives in blood cultures may reach 50%. There has been a dearth of reliable biochemical or immunological diagnostic techniques. Because these diseases are usually not diagnosed early enough for effective treatment, prophylactic antifungal treatments, often at toxic doses, are frequently applied to persistently febrile neutropenic patients without diagnosis. The hypotheses in the search for biochemical markers for disseminated candidiasis have been: (a) there are characteristic and specific circulating metabolic products of fungal species in the serum, urine, respiratory secretions, or sputum of infected patients; (b) the concentrations of these metabolites are proportional to the extent of the disease; and (c) successful treatment results in reduced metabolite concentrations.

The D optical isomer of the 5-carbon sugar alcohol arabinitol is a unique metabolic product of most pathogenic *Candida* species (Figure 5.43a). Initial GC techniques, and even a GC/CIMS method with SIM,[298] were inadequate because of poor specificity in the presence of renal dysfunction.[299,300] This problem has been solved by a methodology that separates the D and L enantiomers of arabinitol with the aid of chiral GC columns. The use of D/L arabinitol ratios, instead of concentrations, has led to considerable improvements in both diagnostic specificity and sensitivity because there was no need to know the volume of the samples. Also, time-consuming calibrations curves could be omitted.[301]

The technique has been extended from serum to plasma, whole blood, and urine, and the introduction of filter paper sampling requires only an unmeasured drop of blood or urine on mailable filter paper.[302] The possibility of making parallel serum and urine analyses significantly increases the reliability of the technique. The endogenous D/L arabinitol ratios in serum and urine have been remarkably close, particularly when considering that the absolute concentration of total arabinitol is some 50-fold higher in normal urine than in normal serum. With the determination of the means and upper limits of endogenous D-arabinitol in serum and urine, observed D/L arabinitol ratios can be used directly for diagnosis. For example, in a culture-confirmed candidiasis case, the D/L ratios in both serum and urine were significantly higher in the patient (Figure 5.43c) than in the control (Figure 5.43b). Potential uses of the arabinitol technique include: diagnosis or confirmation of diagnosis of disseminated candidiasis; serial monitoring of patients at high risk of developing the disease, such as patients receiving bone marrow transplantation; and the monitoring of the progression or regression of the disease during antifungal chemotherapy.

5.2 ENDOCRINE THERAPY

5.2.1 Mechanism of Action

Several human tumors originate from tissues that are sensitive to hormonal growth control. Tumors from these tissues that, in their differentiated form, retain hormone receptors may be treated by

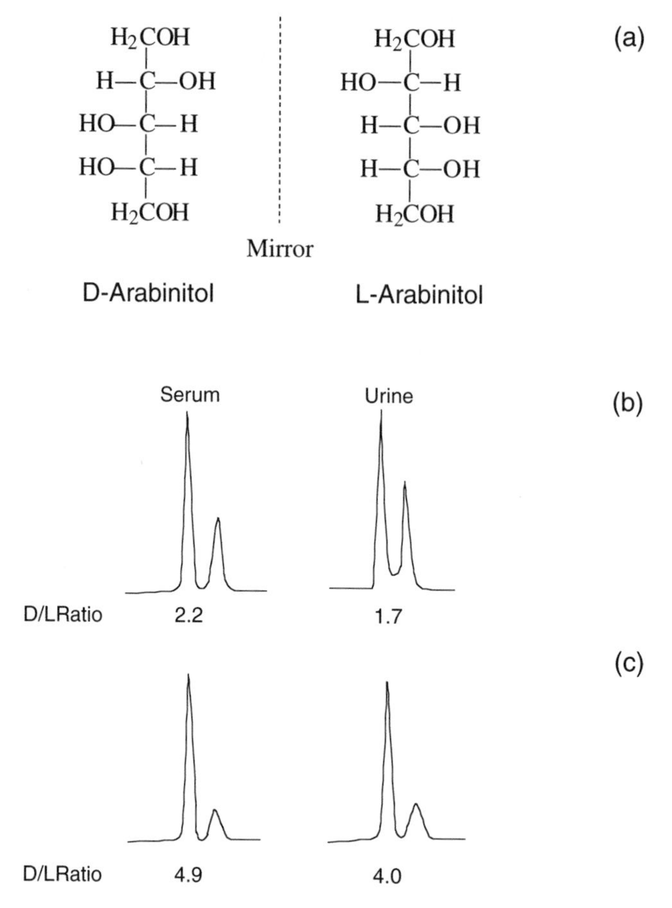

FIGURE 5.43 **(a)** The D and L optical isomers of arabinitol; **(b)** D/L arabinitol ratios in serum and urine samples from a control subject; and **(c)** Significantly increased D/L ratios in both serum and urine indicate disseminated candidiasis in a patient (confirmed by blood culturing).

interruption of this growth control mechanism. These methods include *endocrine ablation*, i.e., removal of those tissues that stimulate tumor growth as a result of their hormone secretions, or *endocrine therapy* using natural or synthetic hormonal substances capable of downregulating tumor growth.

The normal mechanism of hormone activity involves the formation of a stereospecific, non-covalent steroid-receptor complex in the cytoplasm which, after transformation, is translocated to the nuclear chromatin where it binds to a segment of DNA called a hormone response element. This binding process results in the stimulation of the transcription of specific genes that finally, via the synthesis of specific proteins, affect the control processes of cellular growth.[303] Hormonal therapy provides synthetic or natural competitive antagonists that block the receptor proteins, for instance, tamoxifen in breast cancer and estrogens in prostate cancer. The larger the number of available steroid receptors, the more efficient the therapy. Thus it is important to determine the receptor content (in breast cancer both estrogen and progesterone receptors should be assayed) in tumor biopsy specimens.

Although hormonal agents exert probably the most specifically tumor-directed actions of all antineoplastic agents, they are not cytotoxic and therefore proffer no curative potential. They are important because they offer prolonged palliation with relatively little toxicity. Some response occurs in 40 to 80% of cases in steroid receptor-positive breast cancer, prostate cancer, meningioma, and melanoma. In steroid receptor-negative tumors the response rate is $<10\%$. Resistance develops when the hormonally dependent growth characteristics of the tumor are lost. The modalities of endocrine therapy include approaches that may be *ablative*, e.g., ovariectomy, *inhibitive*, e.g., aromatase inhibitors, *competitive*, e.g., antiestrogens, antiandrogens, and antiprogestins, or *additive*, e.g., estrogens, progestins, and androgens.

5.2.2 Aromatase Inhibitors

5.2.2.1 Structure and Function

The ultimate steps in the biosynthesis of estrogens are the conversions of the C19 androgenic precursors, androst-4-ene-3,17-dione, into estrone, and testosterone into estradiol, by the enzyme estrogen synthetase, also called aromatase (Figure 5.44a). Aromatase is a complex protein that consists of a specific cytochrome P-450 heme protein in association with a cytochrome P-450 reductase flavoprotein. During HPLC analysis this complex is denatured and dissociates into the heme unit, its associated apoprotein and the flavoprotein. Each of these units can be detected by ESIMS (Figure 5.45). The heme group and its concentration dependent cluster ions are observed at *m/z* 615, 1230, and 1845 in the upper part of the figure while the reconstructed spectrum for the apoprotein is shown in the upper insert. This apoprotein has a molecular mass of 6150 Da that is within one Da of the expected mass. The spectrum of the much larger flavoprotein is shown in the lower part of the figure, revealing the expected complex pattern of highly charged species. Deconvolution yielded a molecular mass of 59,001 Da which was more than 1200 Da higher than expected and it was suggested that the difference could be attributed to protein microheterogeneity as well as glycosylation.[304]

Suppression of estrogen production by nonsteroidal aromatase inhibitors ("medical adrenalectomy") has been effective in the chemotherapy of breast cancer in postmenopausal women and in prostate cancer.[305] The chemical structures of aromatase inhibitors of current interest are given in Figure 5.44b. The current generation of aromatase inhibitors are mostly used in second-line protocols, against tumors that have become resistant to first-line protocols such as those including the antiestrogen tamoxifen. Drugs like formestan and fadrozol are more specific than the original glutethimide in inhibiting hormone synthesis (reviewed in Reference 306). Some aromatase inhibitors have been considered as potential cancer chemopreventives.[307]

To investigate aromatase activity in breast tissues, samples histologically classified into normal, adenoma, and carcinoma were incubated with androstenedione and testosterone. The levels of estradiol and estrone produced, as well as the remaining androstenedione, were quantified by GC/EIMS. Of the carcinoma tissues 48% had aromatase activity compared to 56% for the adenomas and 80% of the normal tissues. When present, there were no differences in absolute aromatase activities between the three types of tissue, and it was concluded that aromatase inhibition therapy is justified without testing for the status of aromatase activity.[308]

5.2.2.2 Nonsteroidal Inhibitors

Aminoglutethimide (AG, Figure 5.44b), the first aromatase inhibitor to be used clinically, is FDA-approved for second- or third-line therapy for advanced breast cancer in postmenopausal women. Both off-line TLC/EIMS[309] and on-line TSPMS/MS[310–312] techniques have been used for the study of AG and its metabolites. Hydroxylaminoglutethimide, 3-ethyl-3-(4-hydroxylaminophenyl)-2,6-piperidinedione has been identified by EIMS as a major urinary metabolite in treated patients but not in rats. Because the metabolite appeared only after chronic treatment, it was concluded that activation of this metabolic route was responsible for the diminished half-life of the parent drug on prolonged administration.[309] In another study, four additional minor metabolites were identified, by EIMS using their N-acetylated derivatives, in the rat but not in human samples.[313]

Another aromatase inhibitor, *rogletimid,* a pyridylglutarimide (PG), is an analogue of AG (Figure 5.44b). The EI and CI analyses of PG and its N- and C-*n*-octyl analogues provided spectra that illustrated the structural correlations of this compound class.[314] There was extensive biotransformation of PG in post-menopausal patients with breast carcinoma. Ten metabolites were identified

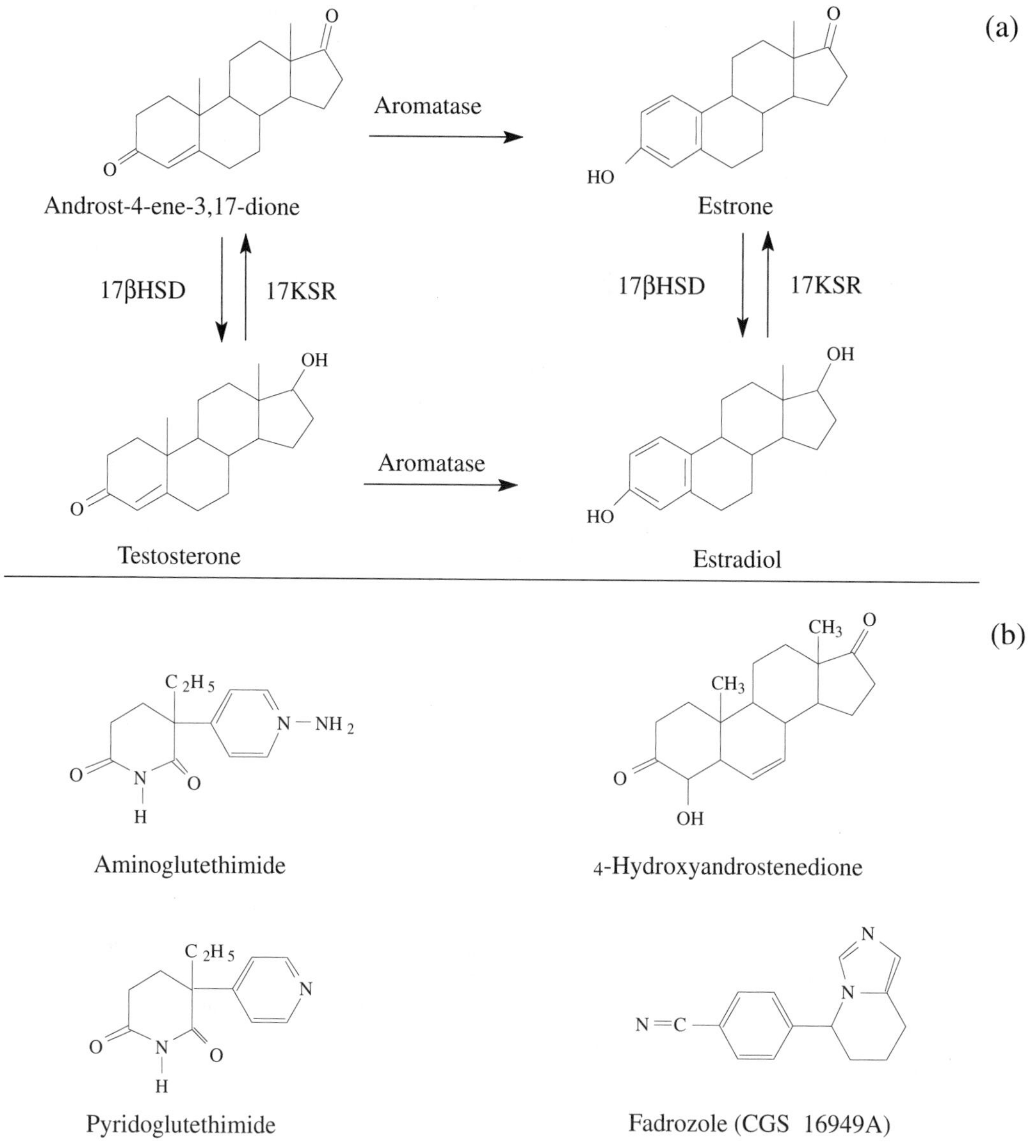

FIGURE 5.44 **(a)** The transformation of androst-4-ene-3,17-dione into estrone and testosterone into estradiol by aromatase (estrogen synthetase); and **(b)** Structure of several aromatase inhibitors: Aminoglutethimide (AG, cytadren), 3-(4-aminophenyl)-3-ethyl-2,6-piperidinedione; formestane, 4-hydroxyandrost-4-ene-3,17-dione (4-OHA, HAD); rogletimid (pyridoglutethimide), 3-ethyl-3-(4-pyridyl)piperidine-2,6-dione; fadrozole, 4-(5,6,7,8-tetrahydroimidazo[1,5-α]pyridin-5-yl)benzonitrile monohydrochloride.

using a combination of LC/TSPMS followed by EIMS for accurate mass determination. Included were: N-oxide of PG, the major metabolite; γ-butyrolactone, formed by terminal hydroxylation of the ethyl side-chain followed by intramolecular cyclization; two hydroxylation products, at the 4 and 5 positions of the glutarimide ring or ethyl side-chain; and 2-ethyl-2-(4-pyridyl)-5-carboxypentanamide along with two of its isomers that were all formed by reductive opening (hydrolytic cleavage) of the glutarimide ring of the PG. Comparison of the metabolism of PG with that of AG revealed several structure-related similarities and differences.[310]

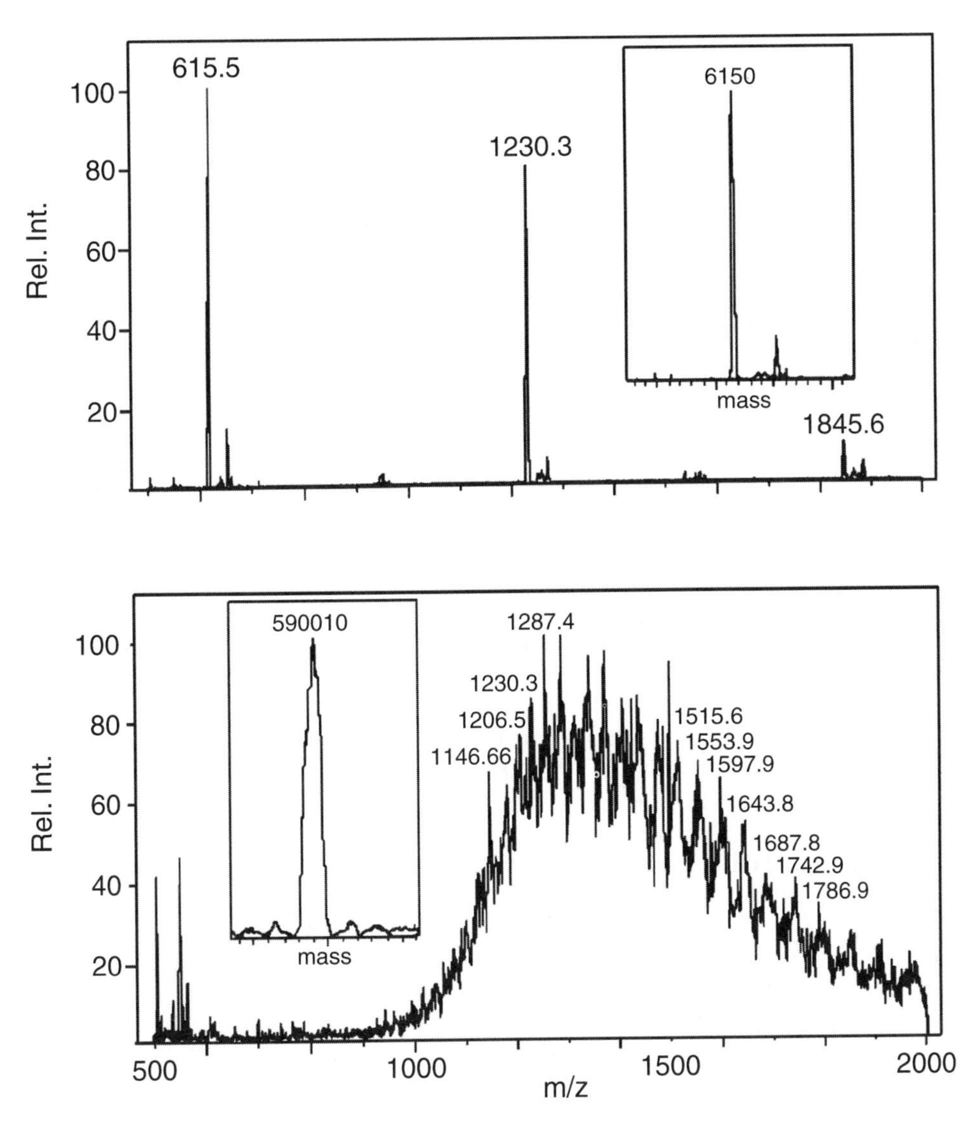

FIGURE 5.45 Electrospray mass spectrum of the aromatase apoprotein. Inset: deconvoluted spectrum showing the determination of the molecular mass. Reprinted from Chen, S. et al., *J. Steroid Biochem. Molec. Biol.*, 44, 347–356, 1993. With permission.

The sample preparation technique for the quantification of *fadrozole* (Figure 5.44b)[303] in plasma and urine required several solvent extraction steps in order to remove interfering endogenous compounds. Analysis was by GC/EIMS using SIM of the molecular ions of the analyte and i.s (deuterated analogue). The LOD was 1 nmol/L in both plasma and urine with the LOQ 5 times the LOD. The drug was absorbed rapidly (t_{max} = 2 h) and was eliminated from the plasma in 24 h. Biotransformation was extensive with only 1% of unchanged drug appearing in the urine.[315] Comparable results were also obtained using an LC/MS technique.[316]

5.2.2.3 Steroidal Inhibitors

4-Hydroxyandrost-4-ene-3,17-dione (4-OHA, Figure 5.44b) was the first structurally designed sterodial aromatase inhibitor. The search for the active species led to the investigation of the metabolism of 4-OHA during which several Phase I metabolites, formed in rat hepatocytes, were identified by EIMS and CIMS. These included products of hydroxylation, of hydroxylation and hydration, and of reduction. Of the compounds identified, only the 2-hydroxy-4-OHA metabolite exhibited aromatase inhibitory activity.[317]

Considerable quantities of 2-hydroxy-4-OH glucuronide were found in human urine and rat bile. In the method used the pre- and post-treatment patient urines were extracted with ethyl acetate, purified, permethylated, and subjected to preparative TLC. A major UV-absorbing band from the post-treatment samples was recovered from the TLC plate and identified by CIMS as the tetra-O-methyl derivative of the 2-hydroxy-4-OHA glucuronide. The identification was based on the presence of the protonated molecule, the ammonium adduct ion, and structurally significant, abundant fragment ions. When the permethylation was carried out with trideuteromethyl iodide, the 12 Da increases in the masses of the corresponding ions confirmed the incorporation of four methyl groups. After enzymatic cleavage of the glucuronide the released 4-OHA was quantified using GC/CIMS and SIM of the perfluorotolyl ether. The 2-hydroxy-4-OHA glucuronide in human urine accounted for 20 to 45% of the administered drug. The same compound was also the major metabolite identified in rat bile, but not in rat urine. No free 4-OHA was found in rat plasma, suggesting extensive first-pass metabolism in the liver.[318]

An isotope dilution GC/MS technique was developed for quantification of 4-OHA in human plasma (i.s.: synthetic 7,7-[D_2]-4-OHA). After solid-phase purification and hexane elution steps, the $[M\text{-}CH_3]^+$ ions of the TMS-ether derivatives were monitored by GC/MS. The LOD was 0.65 pg injected analyte. Quantification, with 3 to 7% reproducibility, was achieved in the 0.5 to 5 ng/mL range.[319]

The metabolism of 4-OHA was also studied in breast cancer patients using LC/TSPMS. One-half mL urine aliquots (compared with 50 mL in Reference 318) were extracted, hydrolyzed with β-glucuronidase, extracted again, evaporated, and redissolved in methanol. The TSP source was operated in the "discharge on" mode to increase fragmentation. Metabolites were quantified by SIM. Seven new metabolites present as their Phase II conjugates were identified by TSPMS with SIM in urine that had been hydrolyzed with β-glucuronidase: 4-OHA-4,6-diene, 3α-hydroxy-5β-androstane-4,17-dione, 3β,4β-dihydroxy-5α-androstane-17-one, 4-hydroxy-testosterone, the two isomers of 3,17-dihydroxyandrostane-4-one, and 4β,5α-dihydroxyandrostane-3,17-dione. The routes of metabolism to form these metabolites included dehydrogenation, reduction of the ketone functional groups, reduction at the C4-C5 double bond, and hydroxylation at the C5 position. Three metabolites exhibited some aromatase inhibitory activity.[311] The same technique was also applied to analyze the urine from patients with prostatic cancer being treated with 4-OHA and three of the same metabolites were identified, 3α-hydroxy-5β-androstane-4,17-dione, 3,17-dihydroxyandrostane-4-one, and 4β,5α-dihydroxyandrostane-3,17-dione.[312] ESI and in-source fragmentation spectra of these and related compounds were obtained with as little as 150 fmol consumption using a 2 pmol/μL sample infusion.[320]

Conjugated 4-OHA metabolites were analyzed using LC/ESI-MS/MS (negative ion mode) in urines from breast cancer patients (single oral dose of 800 mg 4-OHA) and prostate cancer patients (single intramuscular dose of 500 mg). The two main urinary metabolites in breast cancer patients were identified as 4-OHA-glucuronide and 4-OHA-sulfate using full scan mass spectra, CID fragmentation, and comparison with authentic standards. Several other metabolites, present at lower concentrations, were also identified using selected ion chromatograms (Figure 5.46). In contrast, the 4-OHA-sulfate conjugate was not present in the urine of patients with prostate cancer, and there were also other, less pronounced, differences between the levels of conjugated metabolites excreted by the two sets of patients. It was suggested that the conjugated metabolites may serve as storage for the free components, and also that the pathway of 4-OHA metabolism resembles those of testosterone and androstenedione.[321]

Exemestane, 6-methylene-androst-1,4-diene-3,17-dione, is an orally active steroidal aromatase inhibitor that is used for the therapy of metastatic postmenopausal breast cancer. An LC/APCI-MS/MS

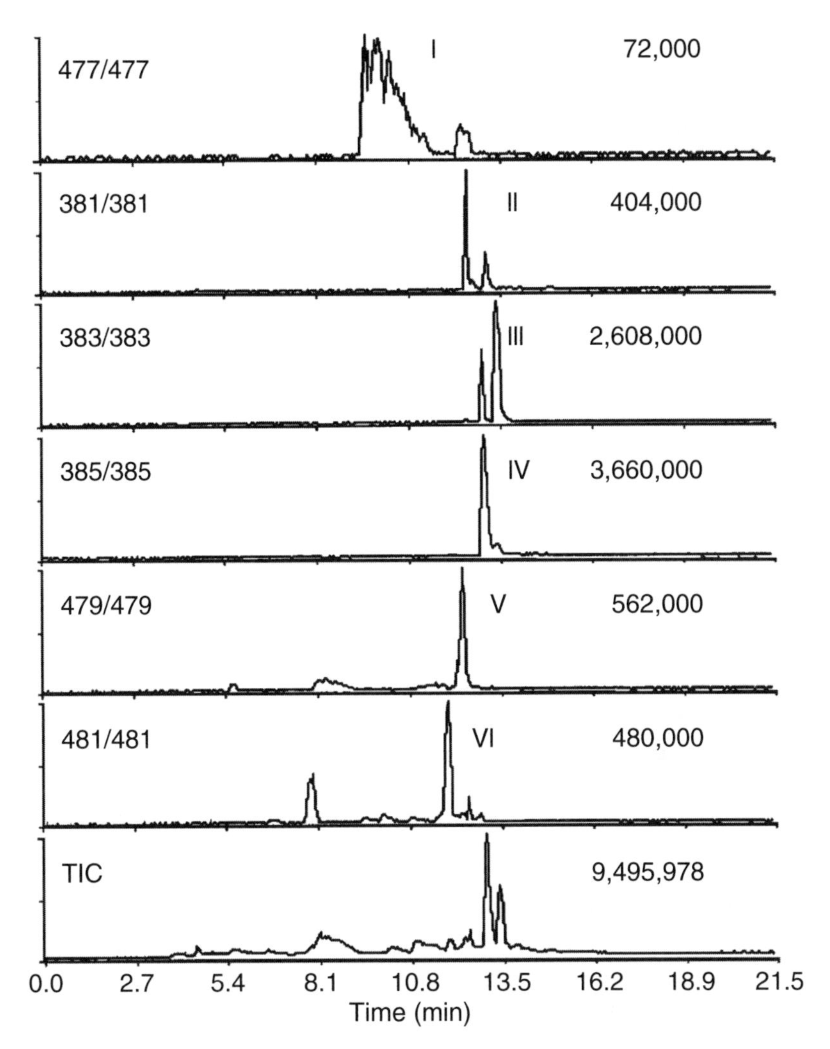

FIGURE 5.46 Selected ion chromatograms and total ion chromatogram (TIC) of extracted urine sample from female patients treated with 4-hydroxyandrost-4ene-3,17-dione (4OHA). **I**, $[M-H]^-$ ion of 4-OHA-glucuronide; **II**, $[M-H]^-$ ion of 4-OHA-sulfate; **III**, either the 4-OHT-sulfate or 3α-OHA-sulfate; **IV**, 3α,17-OHA-monosulfate; **V**, 3α-OHA-glucuronide; **VI**, 3,17-OHA-monoglucuronide (tentative). Reprinted from Poon, G. K. et al., *Drug Metab. Dispos.*, 20, 941–947, 1992. With permission.

technique for the quantification of the drug in human plasma involved solid phase extraction in a 96-well plate format, followed by MRM of characteristic transitions of the analyte (i.s.: ^{13}C-labeled analyte). The method was validated in the 0.05 to 25 ng/mL concentration range.[322]

5.2.2.4 Aromatase Inhibitory Activity of Androgens

The aromatization of a series of 6α- and 6β-alkyl-substituted androst-4-ene-3,17-diones and their androsta-1,4-diene-3,17-dione derivatives by human placental aromatase was studied using GC/MS. The steroids with methyl substituants at the 6 position were aromatized most efficiently in both the α and β series. The rate of aromatization decreased proportionally with increasing length of the 6-alkyl side-chains. The α-substituted compounds were more efficient substrates than their β-substituted counterparts. The significant difference between the ability of the individual compounds in the 6-alkyl androgen series to act either as an inhibitor or a substrate suggests therapeutic implications.[323]

In a study of the clinical pharmacokinetics and metabolism of *formestane* (4-hydroxy-androst-4-ene-3,17-dione) in breast cancer patients, three urinary metabolites have been characterized by FABMS, EIMS, LC/ESIMS, and NMR: 4-O-glucuronide and the epimeric 3-O-sulfuric acid esters of two exocons, 3β,4β-dihydroxy-5α-androstane-17-one and 3α,4β-dihydroxy-5α-androstane-17-one. The exocons were shown to form by stereoselective 3-keto reduction, accompanied by reduction of the 4,5-enol function. Both exocons are devoid of placental aromatase inhibiting properties *in vitro*.[324]

5.2.3 Antiestrogens

5.2.3.1 Tamoxifen

Clinical Indications, Mechanism of Action

Following its introduction in the 1970s, tamoxifen, a nonsteroidal, triphenylethylene type antiestrogen (Figure 5.47), has become the FDA-approved agent of choice for the medical management of hormone-dependent advanced breast cancer in post-menopausal women, for advanced male breast cancer, and for the prevention of contralateral breast cancer in patients who have already developed a primary tumor. The success (~45% of subjects) of large-scale chemopreventive clinical trials using tamoxifen as a prophylactic agent has led to FDA approval for the use of the drug in healthy subjects at high risk of breast cancer. There are, however, ongoing controversies concerning the use of tamoxifen, particularly as a chemopreventive agent. These have occurred because the drug has been shown to be a potent hepatocarcinogen in rats through p53 gene mutations, but not in mice. In addition, there have been reports of the appearance of malignant endometrial cancer and uterine abnormalities, but not liver cancer, in patients treated with tamoxifen.[325–327] Accordingly, there is particular interest in the attempts to determine the clinical relevance of laboratory and animal data,[328] with emphasis on DNA adduct formation[329] (see later). New antiestrogen compounds, called selective estrogen receptor modifiers (SERMs) have been developed, e.g., toremifene, raloxifene, droloxifene, and idoxifene (Figure 5.47). For example, idoxifene, a halogenated tamoxifen

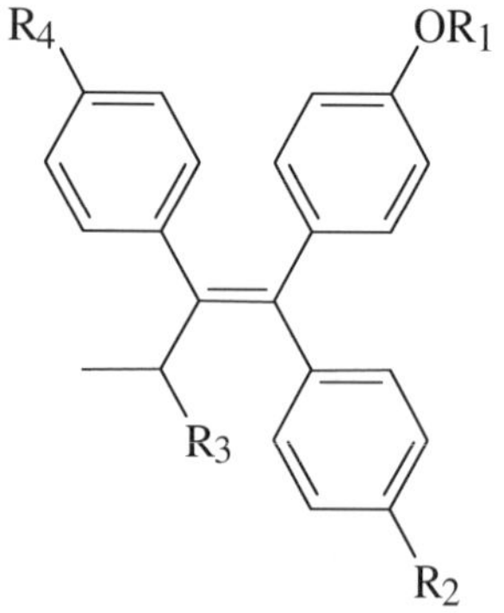

	R_1	R_2	R_3	R_4
(1)	$-CH_2CH_2NMe_2$	-H	-H	- H
(2)	$-CH_2CH_2NMe_2$	-OH	-H	- H
(3)	$-CH_2CH_2NHMe$	-H	-H	- H
(4)	$-CH_2CH_2NH_2$	-H	-H	- H
(5)	$-CH_2CH_2OH$	-H	-H	- H
(6)	$-CH_2CH_2N(\rightarrow O)Me_2$	-H	-H	- H
(7)	$-CH_2CH_2NMe_2$	-H	-H	-OH
(8)	$-CH_2CH_2N(\rightarrow O)Me_2$	-H	-OH	- H

FIGURE 5.47 Structures of tamoxifen and several of its metabolites. Tamoxifen, (Z)-1-[4-(2-dimethylaminoethoxy)phenyl)-1,1-diphenyl-1-butene; α-hydroxytamoxifen, (E)-4-[4-[2-(dimethylamino)-ethoxy]phenyl]-3,4-diphenyl-3-buten-2-ol. (**1**) tamoxifen; (**2**) 4-hydroxytamoxifen; (**3**) N-desmethyltamoxifen; (**4**) N,N-didesmeyl-tamoxifen; (**5**) primary alcohol; (**6**) tamoxifen-N-oxide; (**7**) 4′-hydroxytamoxifen; (**8**) α-hydroxytamoxifen N-oxide. Other antiestrogenic drugs used for the treatment of estrogen-dependent breast cancer include: *Toremifene*, (Z)-4-chloro-1-[4-2(N,N-dimethylamino)-ethoxy-[phenyl]-1-butene; *Raloxifene*, I6-hydroxy-2-(4-hydroxyphenyl)-benzo[b]thien-3-yl][4-[2-(1-piperidinyl)-ethoxy]phenyl]methanone; *Droloxifene*, 3-hydroxytamoxifen; *Idoxifene*, 4-iodopyrrolidinotamoxifen.

derivative has been shown to have a greater affinity for the estrogen receptor than tamoxifen concurrent with reduced uterotrophic effect. Some SERMs, like raloxifene, do not seem to promote the development of endometrial cancer (reviewed in Reference 330). The role of toremifene is controversial (see later).

Tamoxifen inhibits tumor growth by competitively antagonizing the binding of estradiol to the estrogen receptors, resulting in the inhibition of important estrogen-dependent metabolic pathways. However, depending on the species, target organs, and assessed end-points, tamoxifen may act in several ways including acting as a pure agonist (frank estrogen), a partial agonist or as an antagonist. Human metabolism of tamoxifen is complex and, as reviewed below, at least nine metabolites have been identified by MS. Major metabolites resulting from N-oxidation, 4-hydroxylation, or N-demethylation include tamoxifen-N-oxide, N-desmethyltamoxifen, and 4-hydroxytamoxifen (Figure 5.47). Some of the metabolites may be responsible for the therapeutic effects of the drug, and it has been suggested that this may include metabolites that are more potent than the parent drug. Mass spectrometric techniques have been used since the 1970s for the study of tamoxifen metabolism by microsomes, in isolated hepatocytes, experimental animals, and treated patients, as well as for the detection of adducts with DNA and proteins.

Microsomal and Cellular Metabolism

There have been a large number of microsomal and cellular studies on the biology of tamoxifen. Tamoxifen is extensively metabolized by cytochrome P-450 microsomal liver enzymes. The N-oxide metabolite, formed upon incubation of tamoxifen with rat liver microsomes, was first identified by EIMS[331] and confirmed by CIMS.[332] Subsequently, several additional metabolites in hepatocytes were identified, including 4-hydroxytamoxifen-glucuronide.[333,334] A dimeric metabolite of 4-hydroxytamoxifen was also identified in rat liver microsomes using LC/ESIMS. It has been proposed that free radical reactions resulted in the formation of reactive intermediates that may lead either to the formation of the dimer or to reactions with DNA and the creation of adducts.[335]

The rates of formation of four metabolites, tamoxifen N-oxide, N-desmethyltamoxifen, 4-hydroxytamoxifen, and 4′-hydroxytamoxifen, revealed major interspecies differences. Mouse produced the largest quantities and had the fastest rates of formation for all four metabolites. For example, the rate of formation of tamoxifen N-oxide, the predominant metabolite, was ~2900 pmol/min/mg protein which was 3-fold and 10-fold higher than in rats and humans, respectively.[336]

A comparative study has been conducted in rat, mouse, and human microsomes to search for epoxide metabolites that might be expected to form on the aromatic substructures of the molecule and to be reactive intermediates that can bind covalently to liver microsomal proteins and therefore by inference, to DNA.[137] Potential tamoxifen metabolites of these incubates were identified by LC/ESIMS and quantified by SIM. Because the anticipated metabolites were all products of the addition of a single oxygen it was possible in the initial phase of identification to monitor just the $[M+H]^+$ peak at *m/z* 388. This revealed five potential metabolites from which full spectra were obtained, all of which were identical. Subsequently these compounds were identified as 4-hydroxytamoxifen, 4′-hydroxytamoxifen, tamoxifen-N-oxide, 3,4-epoxytamoxifen, and 3′4′-epoxytamoxifen (Figure 5.48). The assignments were confirmed using ESIMS to obtain spectra of the corresponding dihydrodiols that were obtained by the action of epoxide hydrolase, and also by the disappearance of the epoxide peaks upon treatment of the samples with acid. Significantly, 3,4-epoxytamoxifen was present in the microsomal incubates of all three species, with the largest quantities occurring in the rat, but 3′,4′-epoxytamoxifen was only detected in the rat microsomes.[336]

A comparison of the metabolic profiles obtained for human liver homogenate and human Hep G2 cells, each incubated with tamoxifen, with plasma from patients treated with the drug, revealed five metabolites to be present in all samples. An additional five minor metabolites were observed in the patient samples, both 3 h after a single oral dose (60 mg) and after multiple doses (30 mg daily for >6 mo). The latter, long-term regimen resulted in three more metabolites being detected in the plasma of three patients (Figure 5.49). The quantities of the important metabolites II, VII,

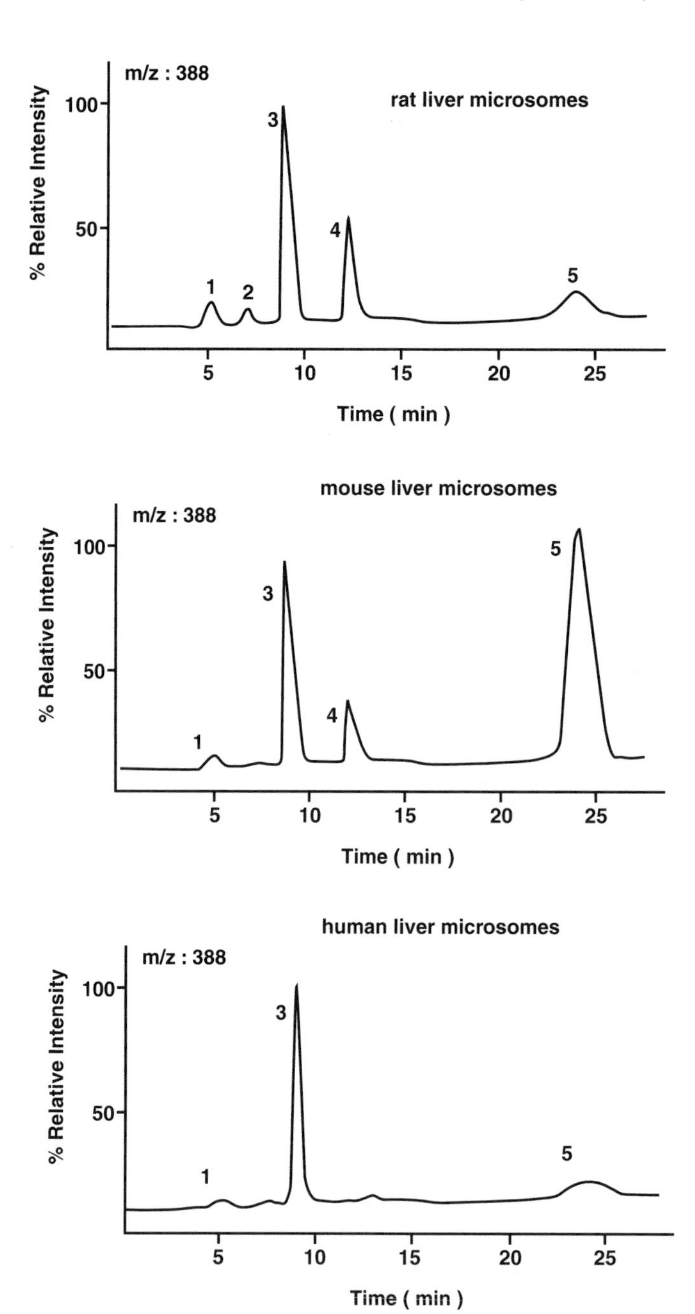

FIGURE 5.48 Selected ion chromatograms of tamoxifen metabolites with $[M+H]^+$ ion at *m/z* 388 formed by rat, mouse, and human liver microsomes. Peaks: 1, 3,4-epoxytamoxifen; 2, 3′,4′-epoxytamoxifen; 3, 4-hydroxytamoxifen; 4, 4′-hydroxytamoxifen; 5, tamoxifen N-oxide. Reprinted from Lim, C. et al., *Carcinogenesis,* 15, 589–593, 1994. With permission.

and X were approximately 75, 50, and 200 pg/mL, respectively, in an individual on the long-term dosing schedule. Because there was no evidence of epoxide formation, which had previously been suggested as being the active species in DNA adduct formation, it was concluded that the α-hydroxy metabolites (II and VI in Figure 5.49c) were likely to be involved in the formation of DNA adducts.[338]

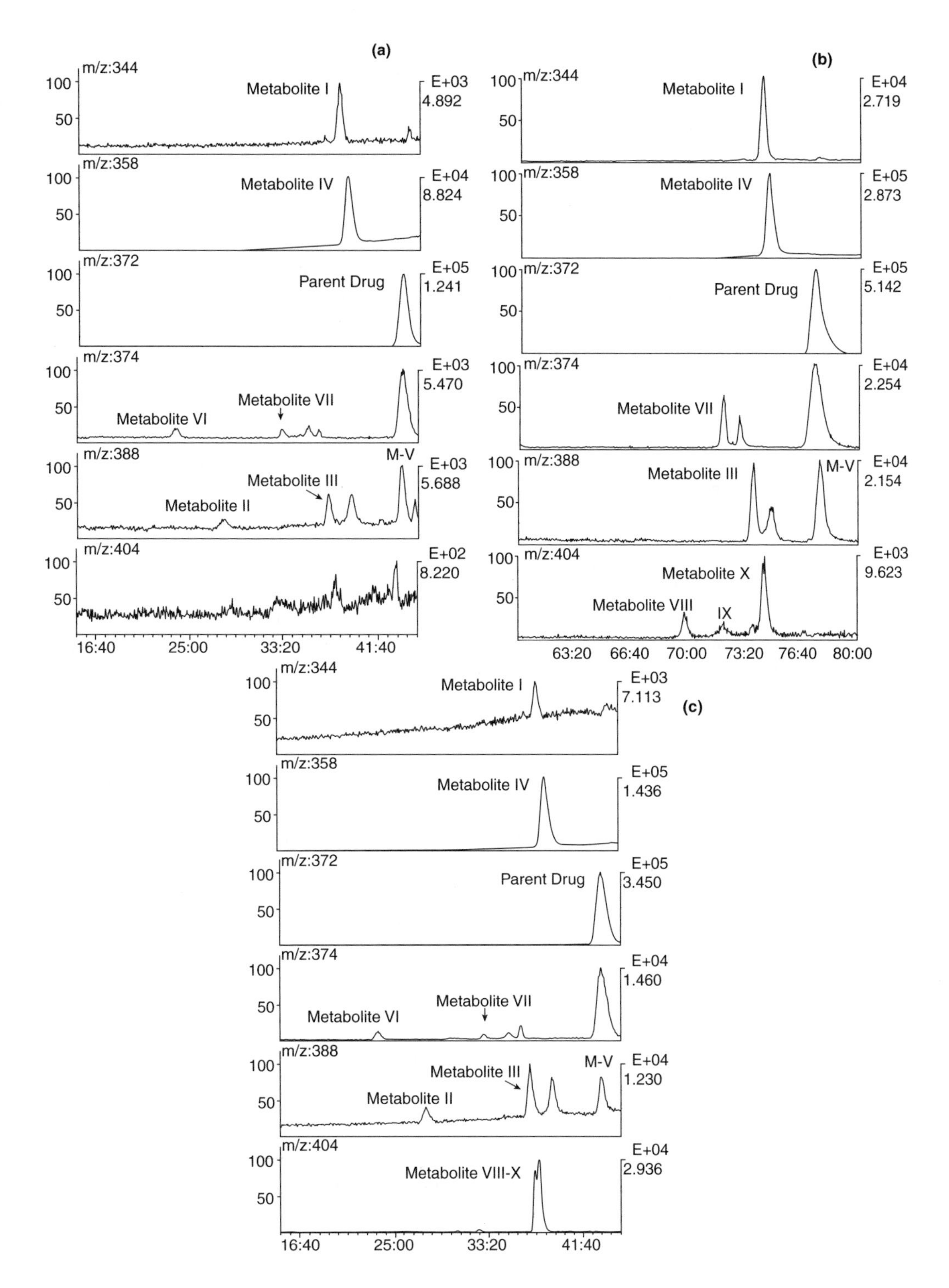

FIGURE 5.49 SIM chromatograms from LC/MS analysis of tamoxifen metabolites in the plasma of a patient: (a) 3 h after receiving a single dose of tamoxifen; (b) 3 h after receiving multiple doses of tamoxifen; (c) after tamoxifen treatment for >6 months. Metabolite I, N-didesmethyltamoxifen; metabolite II, α-OH-tamoxifen; metabolite III, 4-OH-tamoxifen; metabolite IV, N-desmethyltamoxifen; metabolite V, tamoxifen N-oxide; metabolite VI, α-OH-N-desmethyltamoxifen; metabolite VII, 4-OH-N-desmethyltamoxifen; metabolite VIII, 4-OH-tamoxifen N-oxide; metabolites IX and X were dihydroxylated compounds. Reprinted from Poon, G. K. et al., *Drug. Metab. Dispos.*, 23, 377–382, 1995. With permission.

A technique that combines receptor-binding with multiple-stage MS has been developed for the rapid profiling of microsomal metabolites with the aim of differentiating between biologically active and inactive metabolites. This methodology has been applied to tamoxifen that was incubated with rat liver microsomes to which estrogen receptors had been added. After separation by centrifugal ultrafiltration, the receptor-bound microsomal metabolites were released and analyzed by ESI-ITMS in the MS/MS mode. Reconstructed ion chromatograms indicated that the estrogen receptors had captured tamoxifen, N-desmethyltamoxifen, and hydroxytamoxifen (Figure 5.50) but showed that the binding of the dihydroxytamoxifen metabolites was negligible. The technique was also applied to *raloxifene*. Bound ligands identified from the microsomal incubate of this drug included the parent compound, hydroxyraloxifene, and dihydroxyraloxifene. It was concluded that screening for bioactive metabolites by this receptor-binding technique early in drug discovery can help to generate better quality leads or improved second- and third-generation drugs.[339]

CE/ESIMS has been investigated as an alternative methodology for the analysis of tamoxifen and its metabolites in mouse hepatocyte incubates. Using nonaqueous media containing surfactants led to improved solubility and homogeneity in the electrophoretic stream, and higher yields of protonated analytes, than the use of aqueous solutions. The surfactant, sodium dodecyl sulfate also improved the electrophoretic resolution, e.g., the separation of α- and 4-hydroxytamoxifen.[340]

Metabolism: Experimental Animals

Underivatized tamoxifen was quantified in rat plasma more than 20 yr ago by monitoring the molecular ion at *m/z* 371.2249, employing GC/EIMS with a double-focusing sector analyzer at a resolution of 8500 to eliminate endogenous interferences.[341] The TMS derivatives of tamoxifen and desmethyltamoxifen, and the heptaflurobutyl derivative of the 4-hydroxy metabolite have also been quantified in immature rat uterine cytoplasm by a GC/EIMS technique that was initially developed for plasma analysis.[342] Based on additional biochemical experiments, it was concluded that the 4-hydroxytamoxifen metabolite was an important mediator of the antiestrogenic action of the parent drug.[343]

The same two metabolites were also found by GC/MS in tissues from dimethylbenzanthracene-induced mammary tumors in the rat, a model that is frequently used because its growth behavior upon endocrine treatment is similar to that found in patients. The major metabolite, N-desmethyltamoxifen, was analyzed as the heptafluorobutyl derivative. The other metabolite, analyzed as the TMS ether, was identified as 4-hydroxytamoxifen. Administration of the individual metabolites resulted in tumor regression. Tumor volume and the concentration of metabolites in the cytosol fractions were found to be related.[344]

Metabolites identified by LC/TSPMS in extracted brain, lung, liver, kidney, fat, testes and heart tissues of rats included, in order of abundance, N-desmethyltamoxifen, 4-hydroxytamoxifen, 4-hydroxy-N-desmethyltamoxifen, and N-didesmethyltamoxifen. The concentrations of the parent drug and metabolites were ~tenfold higher in tissues than in plasma, with the highest concentrations occurring in the lung and liver.[345]

Metabolism: Human Tumors

The major Phase I metabolite in patient plasma was N-desmethyltamoxifen. Small quantities of the N-oxide metabolite were also identified by ESIMS with CID by using comparisons with synthetic standards. There was virtually no parent drug present in 24 h urine samples, however, four Phase II metabolites were identified as the intact glucuronides of 4-hydroxy-, 4-hydroxy-N-desmethyl-, dihydroxy-, and monohydroxy-N-desmethyltamoxifen. The neutral loss diagnostic of most glucuronides is that of 176 Da (dehydroglucuronic acid) rather than 175 Da as noted in the reference quoted. The diagnostic side-chain fragments of the aglycones and glucuronides were the generally abundant ions at *m/z* 72 and 58, representing the (di)methylaminoethyl moiety. The diagnostic minor ion at *m/z* 300, that was due to the neutral loss of the entire (di)methylaminoethoxy side-chain, was obtained only from the aglycones.[346]

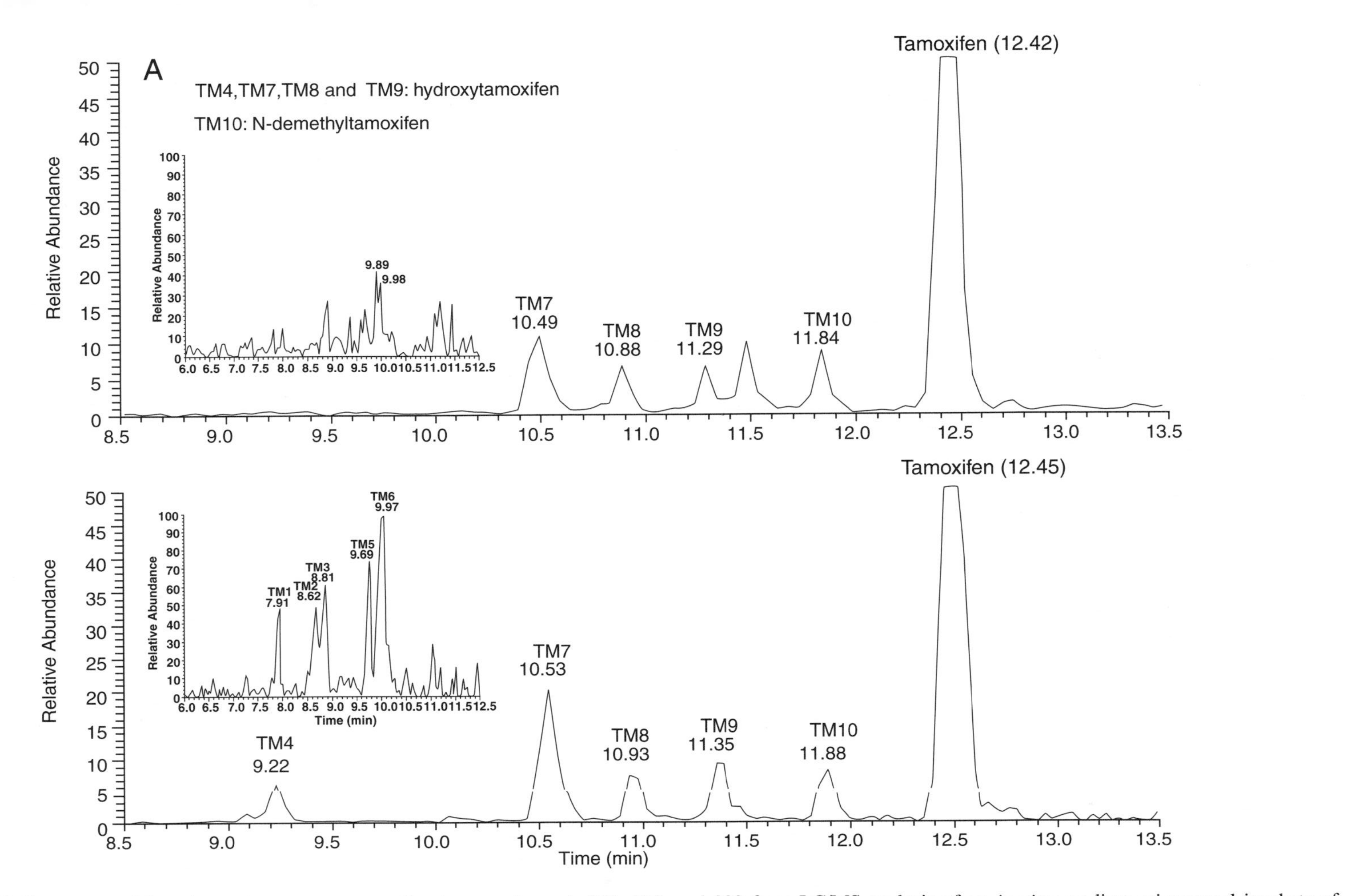

FIGURE 5.50 Reconstructed ion chromatograms corresponding to summing *m/z* 358, 372, and 388 from LC/MS analysis of an *in vitro* rat liver microsomal incubate of tamoxifen after incubation with α-estrogen receptors (A). The lower reconstructed ion chromatogram (B) is from LC/MS analysis of an *in vitro* microsomal incubate of tamoxifen at the same dilution as that used in the binding study. The inserts correspond to the reconstructed ion chromatograms of the dihydroxytamoxifen metabolites. The binding study was carried out at equimolar concentrations of drug and α-estrogen receptor. Reprinted from Lim, H. K. et al., *J. Chromatogr. A,* 831, 227–241, 1999. With permission.

Extracts of tumors from tamoxifen-treated patients were analyzed using a GC/EIMS technique that had been developed for plasma.[342] Desmethyltamoxifen was the most abundant metabolite, but 4-hydroxy-tamoxifen was also present. Based on additional biochemical observations, it was concluded that both metabolites contribute to the antiestrogenic effects of tamoxifen.[347] The LC/TSPMS method used to study rat tissues[345] was also employed on tumor tissues from patients. Parent drug and metabolite concentrations were 10- to 60-fold larger in most tumor tissues, such as from brain tumors, than in serum. Considerable differences were found in the concentrations of metabolites between rat and human, e.g., high levels of the parent drug and the 4-hydroxytamoxifen in rat tissues vs. the occurrence of 4-hydroxy-N-desmethyltamoxifen and concentrations of N-desmethyltamoxifen that were higher than the parent drug in human samples. Such differences in metabolism may explain differences in the pharmacodynamics found between different species.[345]

4-hydroxytamoxifen was detected and quantified, using GC/NCIMS, in the endometrial tissues of a breast cancer patient a month after the termination of treatment.[348] In another study, tamoxifen, N-desmethyltamoxifen, N-didesmethyl-tamoxifen, and 4-hydroxytamoxifen were quantified in endometrial tissues and serum of women treated for >2 mo with tamoxifen. Two-mg tissue samples were used, and for reporting purposes 1 g tissue was considered to be equivalent to 1 mL serum. Both tamoxifen and its metabolites were significantly more concentrated in the endometrium than in serum. The concentrations of tamoxifen were ~100 ng/mL serum vs. ~1900 ng/g tissue while for N-desmethyltamoxifen they were ~320 ng/mL serum vs. ~4200 ng/g tissue. The serum concentrations of the other two metabolites were very small, <30 ng/mL, but their tissue concentrations were 100- to 400-fold higher. A significant correlation was found between the tamoxifen concentrations in hyperplastic and nonhyperplastic endometria.[349]

Estrogenic Metabolites in Tamoxifen-Resistant Breast Tumors

On the assumption that tumor-cell growth depends on the net agonist/antagonist activity ratio of the intracellular ligands for the estrogen receptor, a search was made for estrogenic metabolites both in MCF-7 breast tumor cells with acquired tamoxifen resistance and patients for whom tamoxifen therapy had been unsuccessful. The major metabolite found by ESIMS in the resistant tumor cells from both sources was tamoxifen with a hydroxyl group in place of the dimethylaminoethyl side-chain. The equivalent biphenolic metabolite of 4-hydroxytamoxifen was also identified. Both these metabolites have estrogenic activity *in vitro* and *in vivo*. Resistance to tamoxifen is known to be accompanied by reduced tamoxifen concentrations and increased *cis*- to *trans*-4-hydroxytamoxifen ratios in tissues. It was concluded that the accumulation of estrogenic metabolites may in fact stimulate rather than inhibit tumor growth.[350]

Quantification in Plasma, Urine, and Tumor Tissues

Initial techniques for the quantification of tamoxifen and its active metabolites in plasma and tumor tissues were made by GC/EIMS and applied to breast cancer patients receiving various doses for both brief and extended periods. Typically, metabolite concentrations were in the 0.5 to 100 ng/mL range.[342,351,352]

Sample preparation for the quantification of 4-hydroxytamoxifen in plasma consisted of solid-phase extraction, elution with ether-hexane, and derivatization with pentafluorobenzyl chloride. Sample preparation of tumor tissues consisted of homogenization in ethyl alcohol, evaporation, dissolution into sulfuric acid, removal of the lipid fraction, and solid-phase extraction and derivatization similar to plasma preparation. Extraction efficiency was 95% for plasma and 88% for tumors. Quantification was accomplished by high-resolution GC-NCIMS (i.s.: deuterated 4-hydroxytamoxifen) using SIM of the protonated molecules of the monosubstituted derivatives. The LOD were 20 pg/mL for plasma and 100 pg/g for tumor tissue.[352]

Recent developments for the quantification of SERMs in human plasma have concentrated on high-throughput methodologies employing liquid-liquid extraction in the 96-well format for rapid sample preparation and ESIMS with SRM for specific and sensitive detection (i.s.: stable isotope labeled analytes). The approach was illustrated by the quantification of tamoxifen, 4-hydroxytamoxifen, raloxifene, nafoxidine, and idoxifene in control human plasma with a linear dynamic range of 3 orders of magnitude. LOQ were 5 and 50 ng/mL for tamoxifen and idoxifene, respectively. To demonstrate the speed of the technique, 2112 and 2304 samples were analyzed on 2 separate days.[353] Subsequent work concentrated on idoxifene, one of the more recent SERM compounds. The drug and its pyrrolidinone metabolite were quantified in >600 clinical plasma samples using the above robotic high-throughput injection LC/MS-SRM method. The LOQ was 10 ng/mL for idoxifene and 30 ng/mL for the metabolite. The average run time was 23 s/sample or over 3700 samples/day.[354] In another study, TOF and triple quadrupole analyzers were compared with respect to speed, dynamic range, limit of quantification, and precision for the quantification of idoxifene in human plasma. Although the LOQ of the method using the triple quadrupole instrument was a tenfold improvement over that of the TOF instrument, the latter may still be a viable alternative with several advantages for high-throughput analysis of analytes present in the low ng/mL range.[355]

A straightforward GC/CIMS technique has been developed for the quantification of tamoxifen and its main metabolites in human urine from male patients.[356] Recent techniques for the analysis of tamoxifen and its metabolites in urine and synthetic gastric fluid have been developed based on LC/ESI with either a QqQ or a QTOF analyzer. In the synthetic gastric fluid, tamoxifen, transformed to the (E)-isomer, and at least one known metabolite were detected consistently. In contrast, there was considerable patient-to-patient variation in both the quality and quantity of metabolites in the urines of patients with breast cancer.[357]

In an attempt to develop a nonaqueous CE/ESIMS method for the analysis of tamoxifen, conditions were established to separate α-hydroxytamoxifen and 4-hydroxytamoxifen. Using non-aqueous separation media was shown to provide high ionization efficiency because of the lower surface tension of the electrospray droplet and more rapid evaporation of the methanolic carrier fluid.[340] Another technique has utilized a non-aqueous (without surfactants) CE/ESIMS method for analysis of tamoxifen and seven pseudo-metabolites derived from *in vitro* acid-mediated hydrolysis, designed to mimic the acidic conditions in the stomach.[358]

DNA Adducts

As mentioned already, possible side effects of the use of tamoxifen include endometrial cancer, indicating the induction of mutations presumably through the formation of DNA adducts. Tamoxifen requires metabolic activation before binding to DNA. Although the reactive metabolite responsible for the DNA damage *in vivo* that leads to hepatocarcinogenesis in rats has not yet been identified, several potential routes have been proposed for the activation of tamoxifen to a reactive electrophile that eventually leads to the liver tumors.

α-Hydroxytamoxifen

The most extensively elaborated hypothesis for the formation of a reactive electrophile is that the process starts with an initial α-hydroxylation of the ethyl group, followed by the reversible O-sulphonation of the hydroxy moiety of α-hydroxytamoxifen, resulting in a highly reactive carbocation. In one study, rat liver microsomes were incubated with equimolar quantities of tamoxifen and its D_5-ethyl analogue. ESIMS revealed a large (3:1) deuterium isotope effect on the metabolic rate. In addition the loss of one deuterium atom in the course of the metabolic process permitted the detection of two more metabolites, α-hydroxytamoxifen-N-oxide and α-hydroxy-N-desmethyltamoxifen. The fact that these metabolites were α-hydroxy metabolites was proven by comparing these mass spectra with those obtained for the products of microsomal incubations with both α- and β-D_2-ethyl tamoxifens. Three other tamoxifen metabolites that did not involve α-hydroxylation (tamoxifen-N-oxide, N-desmethyltamoxifen, and 4-hydroxytamoxifen) did not, as expected, exhibit isotopic depletion.[359] In a parallel study, the identity of the α-hydroxy metabolite in extracts from the cell culture media from tamoxifen-treated hepatocytes was confirmed by LC/ESIMS. In addition to the presence of the $[M+H]^+$ ion, 16 Da higher than that of tamoxifen, CID revealed a fragment that has previously been shown by both EI[360,361] and ESI with CID[320] to represent the dimethylaminoethyl side-chain in tamoxifen and several analogs.

The extent of DNA adduct formation was higher by factors of 25 to 50 (determined by ^{32}P-postlabeling) in α-hydroxytamoxifen-treated primary rat hepatocytes than in those treated with the parent compound.[362] The degree of activation of tamoxifen and α-hydroxytamoxifen to form DNA-binding products was compared in rat, mouse, and human hepatocytes using ^{32}P-postlabeling and LC/ESIMS. Incubation with tamoxifen led to readily detectable quantities of DNA adducts in the rat and mouse hepatocytes but not in those of human origin. Incubation of the human cells with 1 and 10 μM concentrations of α-hydroxytamoxifen yielded on the order of 2 and 19 adducts per 10^8 nucleotides, respectively. However, these numbers were ~300-fold lower than the levels observed for rat hepatocytes.[363] The same methodology has been used to directly establish the genotoxicity of tamoxifen in cell cultures from the endometria of hysterectomy patients. A dose-dependent presence of α-hydroxytamoxifen was established with LC/ESIMS; however, the quantities present were apparently inadequate to produce detectable levels of DNA adducts.[364] In one study, using ^{32}P-postlabeling, no DNA adducts were detected in endometrial DNA from patients under treatment with tamoxifen. In another study, using ^{32}P-postlabeling in combination with HPLC, DNA adducts were identified in 8 of 16 human endometrial samples. Thus, the results of studies using the rat model may not apply directly to the human endometrium.

The rates of formation of α-hydroxytamoxifen in incubations of tamoxifen with liver microsomal preparations from women, mice, and rats were determined using LC/ESIMS with SIM of the protonated molecules of tamoxifen, α-hydroxytamoxifen, dihydroxytamoxifen, desmethyltamoxifen, and the i.s., toremifene. In addition, analysis of the glucuronidase hydrolysates of rat bile resulted in the identification of the aglycones of 4-monohydroxytamoxifen, 4-dihydroxytamoxifen and 4-methoxymonohydroxytamoxifen. Three peaks were resolved in the ion current chromatogram of the $[M+H]^+$ ion (*m/z* 388) of the monohydroxytamoxifens in hydrolyzed bile. One peak identified as 4-hydroxytamoxifen was the major metabolite and represented ~10% of the dose. The second peak was not identified while the third peak, the smallest one at 0.1% of the dose, was identified as α-hydroxytamoxifen. Isobaric interferences were excluded under the experimental conditions used and the identification of α-hydroxytamoxifen was corroborated with CID experiments. Although 4-hydroxytamoxifen does not form DNA adducts in either rat tissues or rat hepatocytes, N-desmethyltamoxifen yields hepatic DNA adducts in rats, and α-hydroxy-N-desmethyltamoxifen forms adducts *in vitro* and *in vivo* (see later). Thus, α-hydroxylation is the most likely pathway to DNA forming metabolites (see other possible pathways below). The fact that the rate of formation of α-hydroxytamoxifen was 3-fold higher in the rat than in humans and 10-fold higher than in the mouse might explain why tamoxifen is carcinogenic in the rat but not in the mouse nor hepatocarcinogenic in humans. It was concluded that the glucuronidation of α-hydroxytamoxifen might be a detoxification, mechanism whereas the genotoxic intermediates result from sulfonation.[365]

A subsequent study investigated both the glucuronidation and sulfonation of α-hydroxytamoxifen in the liver, the latter being mediated by hepatic sulfotransferase (SULT). *Trans*-α-hydroxytamoxifen was found to be a better substrate for both rat and human SULT than its more polar *cis* epimer in microsomal preparations. The CID spectrum of the glucuronide of α-hydroxytamoxifen formed by human liver microsomes revealed that the protonated molecule fragmented by loosing a glucuronic acid upon the breaking of the α-carbon-oxygen bond. The ion corresponding to the loss of dehydroglucuronic acid was also observed, confirming earlier MS studies.[346] It was suggested that the relative rates of α-hydroxylation, O-sulfonation, and glucuronidation are determinants of the bioactivation of tamoxifen to form DNA adducts.[366]

1,2-epoxitamoxifen

Both tamoxifen and toremifene (see later) have been shown to be metabolized *in vitro* by direct incubation with horseradish peroxidase and H_2O_2 to their respective N-desmethyl, N-oxide, and epoxides, but not to the 4-hydroxy-metabolites. These products were identified by LC/ESIMS.

Their ability to cause damage to DNA or covalently bind to bovine serum albumin has been determined by ^{32}P-postlabeling. The epoxide metabolites of tamoxifen were 1,2-epoxytamoxifen and two arene oxides, 3,4-epoxytamoxifen and 3′,4′-epoxytamoxifen. The 1,2-epoxy product could not be detected when the activating enzyme was a cytochrome P-450 system.[336] Toremifene yielded the corresponding epoxides and behaved similarly to tamoxifen in most respects except that the chromatographic retention values of two of their DNA adducts were different from those of tamoxifen. The N-demethylation and binding processes catalyzed by horseradish peroxidase were inhibited by ascorbate or reduced glutathione.[367] The analytical methodology was subsequently applied to liver microsomes incubated with tamoxifen using both ESIMS and ESI-MS/MS for metabolite identification. It was suggested that the known hydroxylated metabolites are probably detoxification metabolites, while the arene oxides or related intermediates are responsible for DNA adduct formation.[368] This appears in contrast to the reports that α-hydroxytamoxifen is active in DNA adduct formation.

Quinone Methide

Another potential pathway for the activation of tamoxifen to a reactive electrophile involves the oxidation of one of the phenyl rings to yield 4-hydroxytamoxifen which may then be further oxidized to an electrophilic quinone methide that, in turn, can form covalent DNA adducts through Michael-type addition reactions.[369] In one study, adduct preparation involved the synthesis of 4-hydroxytamoxifen quinone methide by oxidizing synthetic *cis/trans*-4-hydroxytamoxifen with manganese dioxide. This compound was incubated with salmon testes DNA. After sequential solvent extraction of nonbound material and enzymatic hydrolysis to yield the nucleosides, HPLC was used to separate the adducts that were then characterized by NMR and positive ion FAB. CID of several relevant precursor ions revealed that the adducts found were a mixture of 4-hydroxy-tamoxifen-deoxyguanosine diastereomers, with substitution occurring at the allylic carbon of 4-hydroxytamoxifen. The major adduct products were characterized as the E and Z isomers resulting from the addition of the exocyclic nitrogen of deoxyguanosine to the quinone methide. These adducts were chemically similar to adducts previously shown to be associated with mutations in the p53 tumor suppression gene.[370] A related study involved the identification of quinone methide-GSH conjugates as the predominant products involving 4-hydroxytamoxifen. It was established by LC/ESI-MS/MS that the reaction between 4-hydroxytamoxifen quinone methide and GSH is reversible. Although detected in microsomal incubations, no GSH conjugate was detected in breast cancer cell lines.[371]

o-Quinone

Tamoxifen carcinogenesis may involve the oxidation of 3,4-dihydroxytamoxifen to the highly reactive 3,4-dihydroxytamoxifen-o-quinone that may alkylate and/or oxidize cellular macromolecules. In one study, a mixture of the E and Z isomers of the quinone were synthesized and two di-GSH and three mono-GSH conjugates were identified in chemical reactions of the o-quinone with GSH. The same conjugates were also detected both in incubates of tamoxifen with GSH in the presence of microsomal P-450 preparations and in MCF-7 human breast cancer cells. Reaction with deoxynucleosides produced only thymidine and guanosine adducts. The identification of the adducts in these incubation reactions led to the suggestion that the o-quinone formation may be involved in the formation of hepatocarcinogenic adducts in rats upon dosing the animals with tamoxifen.[372]

N-desmethyltamoxifen

This major metabolite of tamoxifen may also be involved in the development of hepatocarcinogenicity in rats. Two tamoxifen-induced DNA adducts were detected using HPLC of ^{32}P-labeled samples of hydrolyzed DNA from the liver of tamoxifen-treated rats. For mass spectrometric analysis nucleic acids were recovered from homogenized liver samples and the RNA removed in a series of purification steps. The DNA-adducts were identified by ESI-MS/MS using CID of the protonated analytes. One adduct was N-desmethyltamoxifen-deoxyguanosine in which the α-position of the metabolite N-desmethyltamoxifen was linked covalently to the amino group of deoxyguanosine. The second adduct was the *trans* isomer of α-(N^2-deoxyguanosyl)-N-desmethyltamoxifen.[373] Based on the prior detection by GC/MS of a large quantity of N-desmethyltamoxifen in treated breast cancer patients, it was suggested that this metabolite may be activated to form DNA adducts.[351,374] In another study, α-hydroxy-N-desmethyltamoxifen was synthesized, and the adduct from α-sulfoxy-N-desmethyltamoxifen obtained *in vitro* was characterized by ESIMS and NMR as (E)-α-(deoxyguanosine-N(2)-yl)-N-desmethyltamoxifen. The compound was also confirmed in the liver of rats treated with both tamoxifen and the synthesized α-hydroxy-N-desmethyltamoxifen.[375]

Accelerator mass spectrometry has been used to determine the accumulation of DNA adducts in the livers of rats and three strains of mice chronically exposed to ~40 mg/kg tamoxifen. After 3 mo only ~70 adducts/10^8 nucleotides were detected in the mice as compared with ~2500 adducts/10^8 nucleotides that were found in rats. Also, in the mice the number of adducts decreased over time and were below the LOD in 2 years. In contrast, adduct levels in the rat decreased into an extended plateau phase 6 mo after dosing. In contrast to the rats, the mice did not develop liver tumors and it was concluded that tamoxifen is noncarcinogenic in mice.[374]

Dimeric Metabolite

A new, hydrophobic metabolite ($[M+H]^+$ at *m/z* 773), found in rat liver microsomes incubated with tamoxifen in the presence of NADPH, was identified by LC/ESIMS to be the dimer of tamoxifen with two additional oxygen atoms and removal of two hydrogen atoms. This dimer could possibly originate from the association of two hydroxylated tamoxifen molecules in a free radical reaction. When microsomal incubations were carried out using 4-hydroxytamoxifen instead, two metabolites with $[M+H]^+$ ions at *m/z* 773 with different chromatographic retention times were observed. The retention time and diagnostic fragment ions of the major metabolite of this pair were the same as those of the new metabolite that had been found in the microsomal tamoxifen metabolism experiments. The more polar second dimeric metabolite, that was present at much lower concentration in this 4-hydroxytamoxifen metabolism study, could not be detected among the tamoxifen metabolites either because it was not being formed or because the LOD of the methodology was not sufficient. A pathway has been proposed for the formation of three possible dimeric structures via free radical reactions as well as for subsequent reactions that result in forming DNA adducts.[335]

5.2.3.2 Toremifene

The structure of toremifene differs from that of tamoxifen by only one atom, i.e., a chlorine instead of hydrogen at the α position on the ethyl chain attached to the central ethylene of the molecule (Figure 5.47). Toremifene has been claimed to have mild and transient side effects and less incidence of endometrial cancer than tamoxifen and the drug is FDA-approved for the treatment of advanced breast cancer in steroid-receptor-positive postmenopausal women. Toremifene produces no liver tumors in rats. This has been attributed to the fact that the α-chlorine substituent in toremifene blocks bioactivation via sequential α-hydroxylation and sulphonation, thus preventing the formation of drug-DNA adducts. However, it is not proven that a similar blocking process operates in the human uterus.[376] The formation of uterine drug-DNA adducts in patients has been disputed even for tamoxifen itself. Toremifene, being a partial agonist, promotes the growth of implanted human endometrial tumors, thus it might promote the growth of existing endometrial tumors in breast cancer patients, including any tamoxifen-induced tumors.

Because the chlorine atom is not lost during metabolism, there are no metabolites that are α-hydroxy metabolites that would be common with those of tamoxifen. Parallel metabolites at other sites of the molecule are, however, possible. Of four unconjugated and three glucuronide-conjugated metabolites that were released as the aglycones by treating patient urine with β-glucuronidase, the only one that could be identified by LC/APCIMS was one of these parallel metabolites, 4-hydroxytoremifene glucuronide.[377]

The importance of quinone methide formation in the hepatocarcinogenic behavior of tamoxifen has also been investigated for toremifene. The formation of the di-GSH conjugates involved the loss of chlorine during the reaction of 4-hydroxytoremifene quinone methine with two molecules of GSH. The rate of formation of the di-GSH conjugates from 4-hydroxytoremifene was threefold higher than that for 4-hydroxytamoxifen. However, as with tamoxifen, no GSH conjugates were detected in breast cancer cell incubates. It was concluded that quinone methide formation was not likely to be a major factor in the genotoxic effects of toremifene and tamoxifen.[371]

Because the degree of hydroxylation of toremifene is only about one-fifth that of N-desmethylation, there is much less 4-hydroxytoremifene available for activation via free radical reactions, as described above for tamoxifen. This might explain why virtually no DNA adducts have been observed with toremifene *in vivo*.

The binding of ^{14}C-radiolabeled tamoxifen and toremifene to rat liver DNA and the DNA of other nontarget organs (spleen, lung, kidney, reproductive, and GI tract) has been compared using AMS. HPLC fractions of isolated DNA samples were converted to graphite by combustion first to CO_2 followed by reduction to filamentous graphite on cobalt. The binding of tamoxifen to the rat liver DNA was linearly dose-dependent in the 0.1 to 1 mg/kg dose range (human therapeutic dose is ~20 mg/person/day). More than 80% of the hydrolyzed products were non-polar. In contrast, the DNA binding of toremifene was 10-fold less than for tamoxifen and there were no non-polar adducted nucleotides. For example, numerical values for the covalent binding (pmol tamoxifen or toremifene equivalents per mg DNA) in liver were ~2 vs. ~0.15 for tamoxifen and toremifene, respectively. Other comparative values included ~1 vs. 0.1 in duodenum, ~0.25 vs. ~0.003 in lung and ~0.08 vs. 0.02 in kidney. Assuming that 1 μg DNA is equivalent to 3×10^9 amol nucleotides, a 40 mg/kg tamoxifen dose results in 2137 fmol/mg DNA. A comparison with data from ^{32}P-postlabeling suggests that the latter technique underestimates the total extent of adduct formation by a factor of about three. This could be explained by the fact that the postlabeling technique detects ~12 individual DNA adducts in the liver. In contrast to the animal dosing experiments, *in vitro* experiments using human, rat, and mouse liver microsomal preparations revealed that through appropriate manipulation of the experimental conditions, toremifene can also form reactive intermediates leading to both DNA and protein adducts.[378]

5.3 TARGETED DRUG DELIVERY AND THERAPY

Most conventional drugs reach the site of action by distribution and passive diffusion. The aims of targeted drug delivery or site-specific drug delivery are, in increasing order of sophistication, to place the drug at or near the intended target site (tissue or cell) in which the receptor is located, to deliver the drug only to the tumor cells and not to normal cells, and to release the drug specifically to the intracellular site of the tumor cells, e.g., by endocytosis followed by lysosomal release. Antitumor drugs and prodrugs have been coupled to antibodies, hormones, and dextran, or encapsulated in liposomes. Techniques have been developed for carbohydrate moieties to produce neoglycoconjugates. Several limitations must be overcome for the commercial production as well as for the use of these complex products to provide appropriate bioavailability. When large molecules are involved, changes in tertiary or quaternary structure, such as aggregation, may result in loss of activity. Chemical inactivation may occur due to proteolytic enzymes or to oxidation involving sulfhydryl groups. Major problems may be caused when the drug-carrier complex is recognized as foreign substances by the body and induces immunological reactions.

5.3.1 DRUGS CONJUGATED TO PROTEINS, ANTIBODIES, OR LIPIDS

One approach for attaching anticancer drugs to antibodies that target tumor sites is the "indirect method" when a carrier molecule is used between the drug and antibody. The carrier may be a protein or a synthetic polymer. The term "loading value" is defined as the number of moles of a drug, or a drug linked to a sugar, relative to the number of moles of the carrier. The main advantage of the indirect method is the possibility for higher loading values than in the direct method where the attachment of the drug is accomplished through a small spacer moiety.

In one application, MALDI-TOFMS spectra were obtained to determine the mass shifts when human serum albumin (HSA) was conjugated to aminopterin and 4-desacetylvincristine hemisuccinate (DAVCR) (Figure 5.51). There was a significant shift in the centroids of the $[M]^+$ ions for the HSA conjugates, and also a broadening of the peak widths upon conjugation relative to the unconjugated HSA (Figure 5.52). Mass spectrometric loading values were calculated from the difference in the measured chemically averaged mass values for the conjugated and unconjugated proteins, divided by the calculated change in mass of the protein when conjugated with one drug or sugar molecule. The average relative error for the mass spectrometrically determined loading values was ±10%. An advantage of using MALDI for such determinations is that only singly, doubly, and triply charged species are detected and these are well resolved from one another, thus average masses could be calculated easily from the centroids.[379] This approach was tested on antineoplastic drugs conjugated through various carbohydrates and chelators.[380]

The calicheamins are potent enediyne-containing cytotoxic antibiotics that act by cleaving DNA. A derivative, NAc-calicheamin-ε (~1500 Da) has been conjugated to monoclonal antibodies (MoAbs, ~150 kDa) with loading of 2 to 3 mole of conjugated drug per mole of monoclonal antibody. Both MALDI-TOFMS and ESIMS (QqQ type) were employed to analyze these conjugates. In the MALDI-TOFMS techniques IR instead of UV lasers were used to avoid decomposition due to UV irradiation. Figure 5.53 shows the narrow-scan ESI data from the pure MoAb and molecular masses obtained by conventional transformation and also by the resolution-enhancing technique of maximum entropy. The figure also illustrates similar data for low- and high-loaded samples. The charge distribution between the MoAb peaks results from the presence of conjugate. Higher abundances also appear on the high-mass side of the series of the pure conjugate. The abundances of these peaks were higher in samples with higher loading values (Figure 5.53). The appearance of a new series of peaks, each ~1460 Da apart, corresponds to the mass change due to the covalent addition of NAc-calicheamin-DMA via an amidation reaction.[381]

A number of lipophylic methotrexate (MTX)-lipoamino acid conjugates coupled with amide or ester linkages were synthesized. One objective has been to modify one or more of the carboxyl groups in the glutamate moiety of MTX because of its involvement in the binding of the drug to dihydrofolate reductase and thymidilate synthetase, the target enzymes. The identities of these conjugates were determined using FABMS. When protonated molecules were not observed, NaI methanol solution was added to produce natriated species. Although the conjugates were found to be less potent than MTX, they have advantageous properties in combating cross-resistance.[382]

The usefulness of 1-β-D-arabinofuranosylcytosine (ara-C), a drug often used to treat acute myelogenous leukemia, is limited by its rapid deamination to its biologically inactive metabolite 1-β-D-arabinofuranosyluracil (ara-U). The metabolism of a new liposomal drug, N^4-octadecyl-1-β-D-arabinofuranosylcytosine (NOAC), was studied in mice. Only the unmetabolized parent drug was found in purified liver homogenates by LC/ESIMS. NOAC has two metabolic pathways. Several metabolites were identified in feces including hydroxylated NOAC and its sulfated derivative (Figure 5.54). Additional structural information on these two metabolites were obtained by negative ESI and by obtaining spectra with changes of the cone voltage to obtain in-source fragmentation. Compounds identified in urine included the unmetabolized drug and the hydrophylic molecules ara-C and ara-U. The extent of the possible contribution of the metabolite ara-C to the cytotoxic effect of NOAC is not known.[383]

5.3.2 Angiogenesis Inhibitors

New capillary growth (*angiogenesis, neovascularization*) occurs only under limited circumstances in the normal adult, e.g., ovulation, pregnancy, and wound healing, and is rare even under pathological conditions, e.g., diabetic retinopathy. While avascular tumors do not usually grow beyond 1 to 2 mm

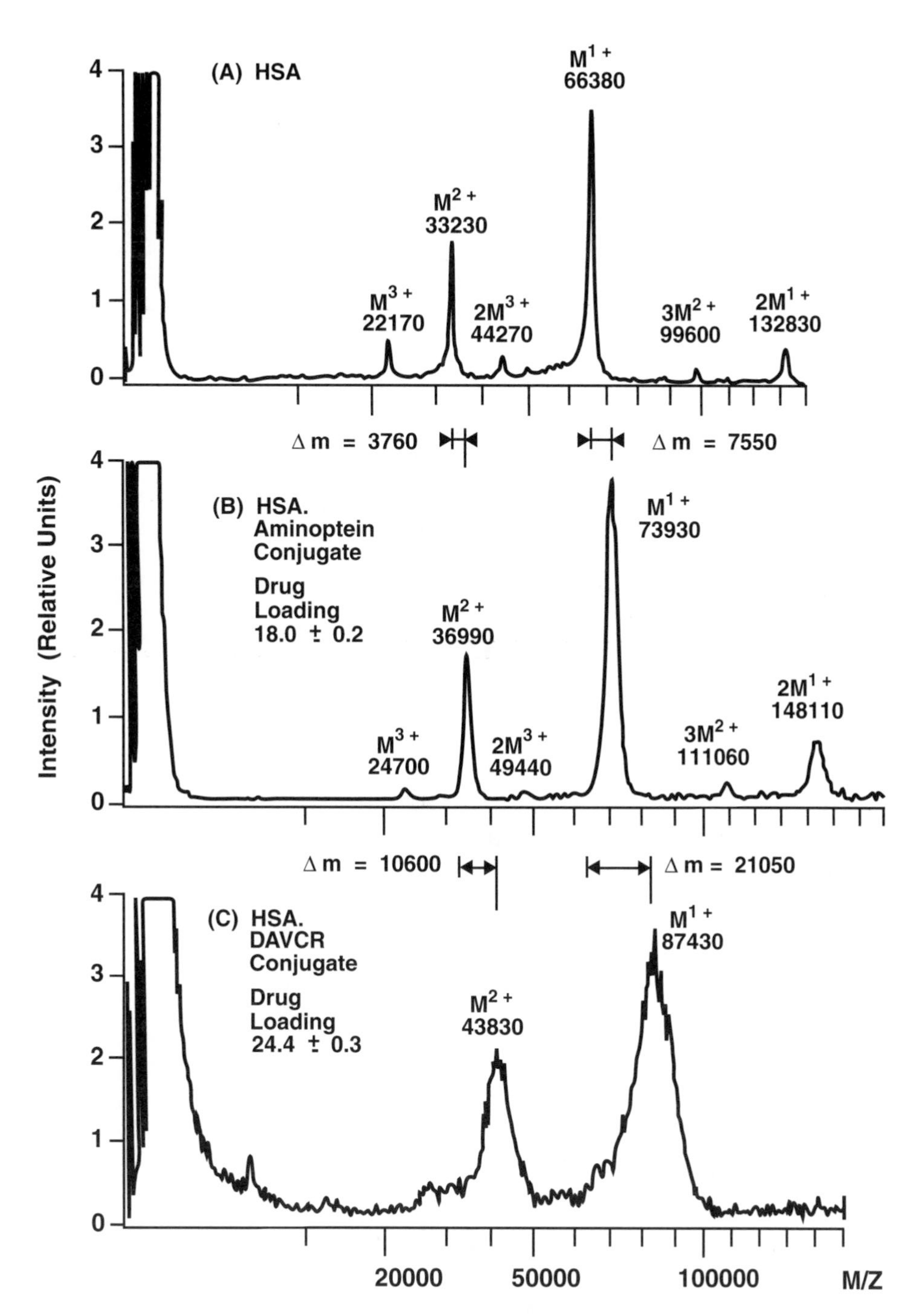

FIGURE 5.51 MALDI mass spectra of: (A) human serum albumin (HSA); (B) HSA–aminopterin conjugate and (C) HSA–4-desacetylvincristine succinate (DAVCR) conjugate. The mass shifts for the $[M]^{+}$ and $[M]^{2+}$ ions for the HSA conjugates relative to HSA are indicated as Δm values. Reprinted from Siegel, M. et al., *Biol. Mass Spectrom.*, 22, 369–376, 1993. With permission.

diameter (~10^6 cells), angiogenesis is essential for the local growth and also the metastasis of the common solid tumors.[384] The newly formed blood vessels supply the tumor tissues with oxygen and nutrients necessary for survival, growth, and spread.[385] At least 20 angiogenesis inhibitors are

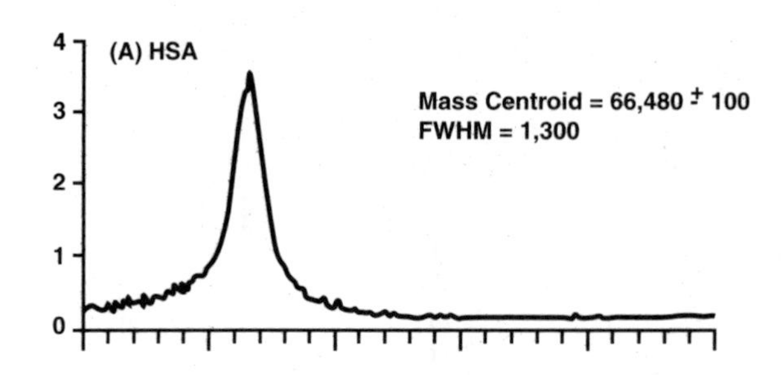

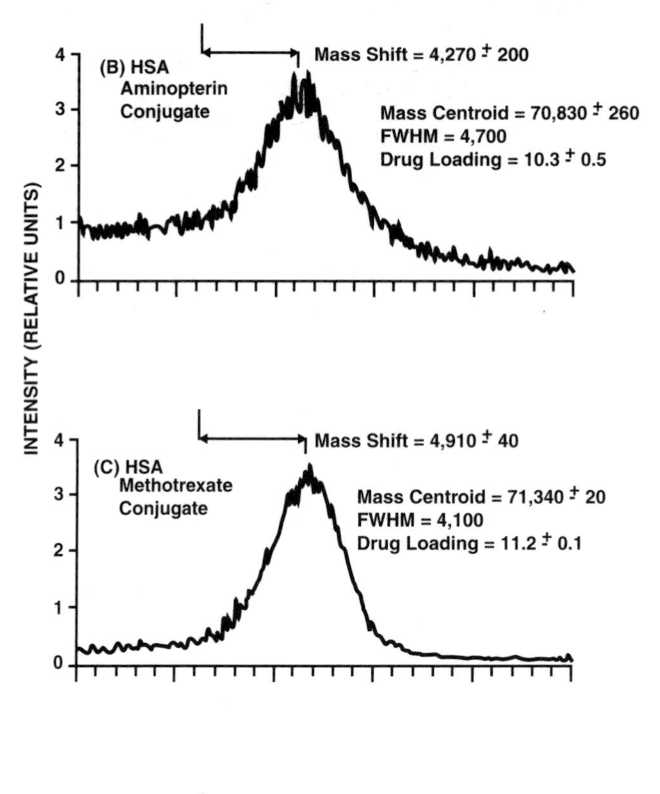

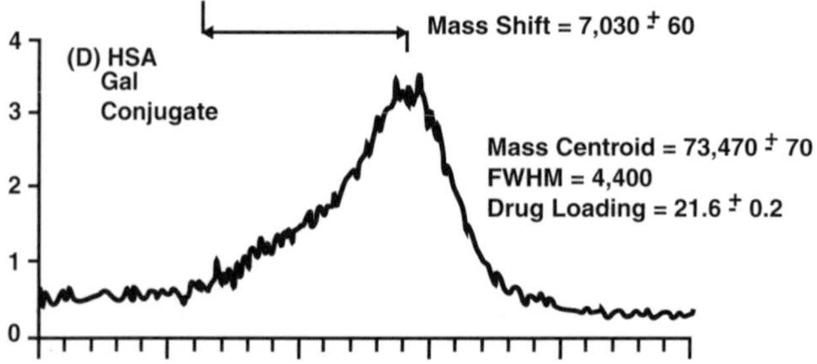

FIGURE 5.52 Exploded views of the molecular ion $[M]^+$ regions of the MALDI mass spectra of human serum albumin (HSA) and HSA conjugates, illustrating the mass shifts of the different conjugates relative to HSA: (A) HSA; (B) HSA–aminopterin conjugate; (C) HSA–methotrexate conjugate; and (D) HSA–lactose (Gal) conjugate. Reprinted from Siegel, M. et al., *Biol. Mass Spectrom.*, 22, 369–376, 1993. With permission.

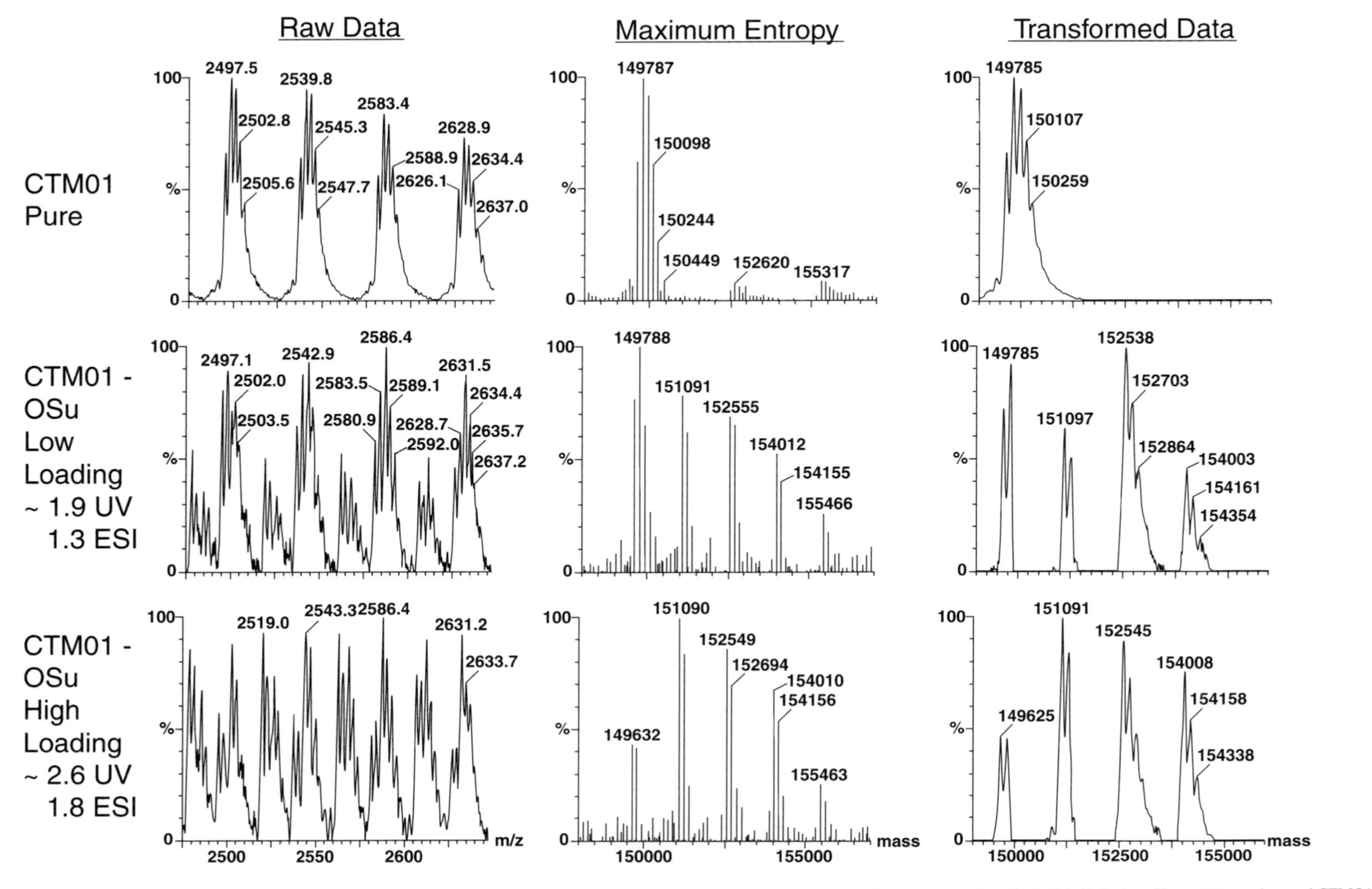

FIGURE 5.53 Narrow-scan ESI mass spectra showing raw data (A-1, B-1, C-1), integrated transformed data (A-3, B-3, C-3), and maximum entropy data (A-2, B-2, C-2) data. Upper (A) panel: pure hCTMO1 monoclonal antibody (MoAb). Middle (B) and lower (C) panels: loaded NAc-calicheamicin-DMA-hVTMO1 MoAb conjugates. Reprinted from Siegel, M. M. et al., *Anal. Chem.*, 69, 2716–2726, 1997. With permission.

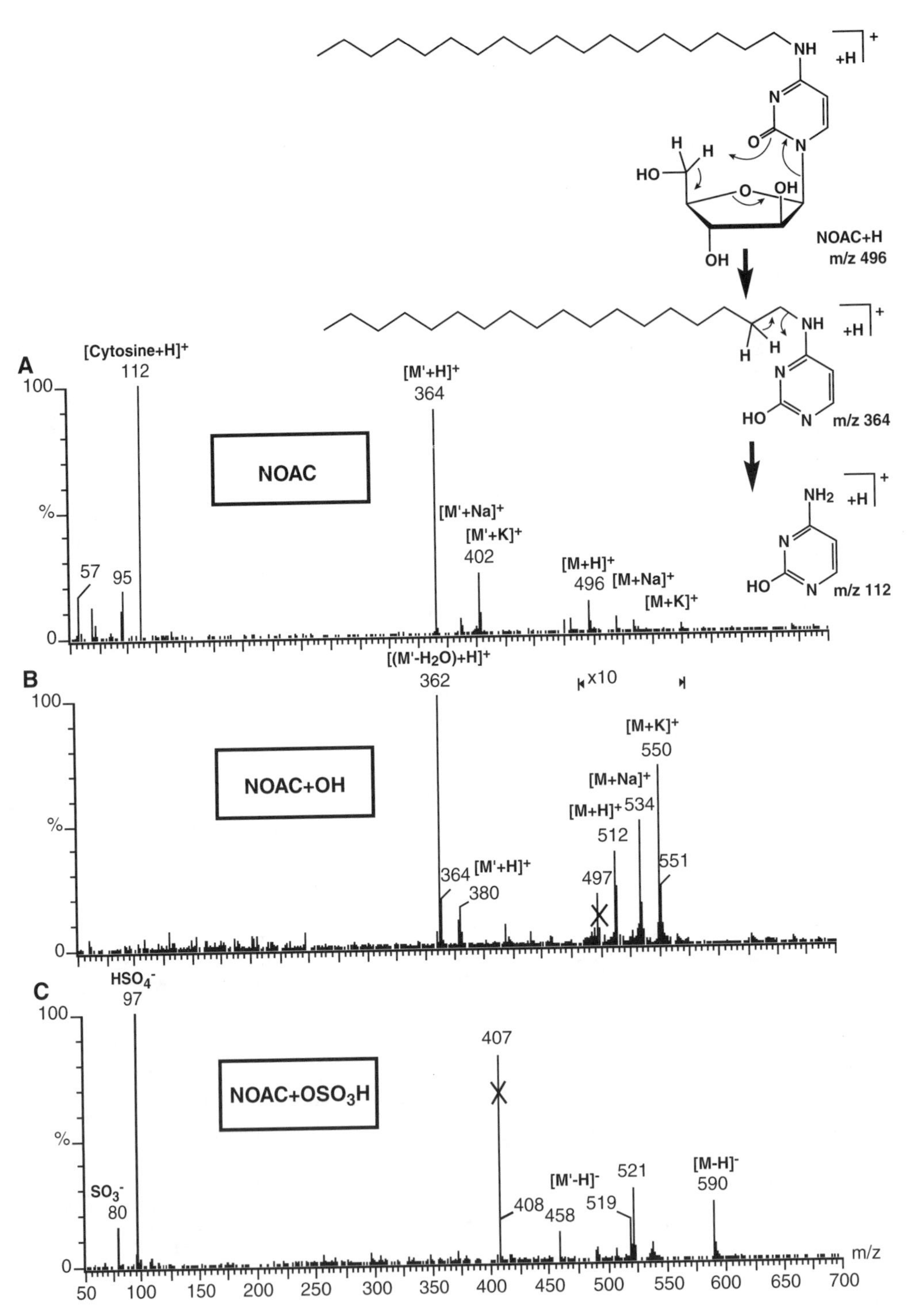

FIGURE 5.54 Compounds detected in feces extracts of N[4]-octadecyl-1-β-D-arabinofuranosylcytosine (NOAC)-treated mice. Rearrangements necessary for fragmentation are represented at the top of the figure. M′ stands for M-132, which is the loss of the arabinose moiety. Peaks in B and C marked with a cross are unidentified compounds. Reprinted from Koller-Lucae, S. K. M. et al., *Drug Metab. Dispos.*, 27, 342–350, 1999. With permission.

currently being tested in human trials, including endostatin, an analog of fumagillol (TNP-470), interferons (α and β), pentosan polysulfate, and platelet factor 4, as well as the dietary, plant-derived genistein (Section 5.4.2).[386]

Endostatin

The 20 kDa endostatin was first isolated from a murine hemangioendothelioma using a rationale similar to that for the isolation of angiostatin, a 38 kDa specific inhibitor of endothelial cell proliferation, another angiogenesis inhibitor.[387] Human circulating endostatin was isolated, purified, and structurally compared to mouse endostatin with demonstrated antiproliferative activity. The MS techniques employed to guide the complex processes of separating ~200 µg of pure protein from 2500 L of human ultrafiltrate, included MALDI with the "dry drop" technique,[388] conventional ESI, and ESI-MS/MS with the "Sherpa" method of interpretation.[389] The ESI spectrum of purified endostatin revealed four major peaks with ionic species of 8 to 11 charges, corresponding to a molecular mass of 18,494 ± 2 Da (Figure 5.55A). This was in accordance with the cDNA-deduced mass of amino acids 514-683 of collagen α1-XVIIIO, which is a class of non-fibrillar proteins of the extracellular matrix. Sixty residues of the N-terminus of endostatin/2 were sequenced. Fragments resulting from the trypsin-treated native and reduced $AcNH_2$ peptides were analyzed by MALDI-TOFMS (Figure 5.55B, solid lines), while the cysteine containing regions were analyzed by ESI-MS/MS after thermolytic digest (Figure 5.55B, bold lines). Figure 5.55C shows ESI-MS/MS analysis of the disulfide bonds of several thermolytically digested (and HPLC separated) fragments of the peptide (boxed in Figure 5.55B) obtained by ESI-MS/MS from the 1,088 Da parent ion.

These and other analyses by MALDI and ESI-MS/MS revealed that the protein sequences of human and mouse endostatins share 86% identity and >90% similarity. A comparison of human endostatin with that from mouse hemangioendothelioma cell medium revealed that human endostatin is 12 N-terminal amino acids shorter than endostatin from mouse tumor (Figure 5.56). It was suggested that circulating endostatin might function in an endocrine fashion. The concentration of circulating plasma endostatin was in the range of 10^{-10} M in subjects without tumors. This is at least an order of magnitude lower than the reported mouse endostatin concentration required for antiproliferative effects. Purified human endostatin exhibited no antiproliferative potency against several endothelial cell types.[390]

A comparative analysis of the disulfide bonds of both circulating endogenous endostatin and the recombinant protein have been made using Edman degradation, MALDI-TOFMS, and ESI-ITMS. Both native and recombinant endostation exhibited a Cys1-Cys4 (Cys^{162}-Cys^{302}) and Cys2-Cys3 (Cys^{264}-Cys^{294}) linkages. The conventional nomenclature for the CID of peptides has been extended to simplify the explanation of data on the more complex disulfide-bridged peptides obtained in MS^n (n up to 5) experiments.[391]

Fumagillol

O-(chloroacetyl-carbamoyl) fumagillol(TNP-470, (3R,4S,5S,6R)-5-methoxy-4-[(2R,3R)-2-methyl-3-(3-methyl-2-utenyl)-oxiranyl]-1-oxaspiro-[2,5]-oct-6-yl-(chloroacetyl) carbamate is a semisynthetic analog of fumagillin, an antitumor antibiotic secreted by the fungus *Aspergillus fumigatus fresenius*. TNP-470 exhibits a 50-fold increase in antiangiogenic activity over fumagillin without the associated toxicity. An HPLC study revealed six metabolites upon incubation with human hepatocytes and microsomal fractions. The two major metabolites yielded easy-to-interpret ESI spectra (Figure 5.57) that agreed with those of authentic synthesized standards. Metabolite M-IV resulted from the parent TNP-470 upon the action of an esterase, while metabolite M-II was derived from M-IV by a microsomal epoxide hydrolase (Figure 5.58). Both processes could be prevented by appropriate inhibitors.[392] An APCIMS technique was also developed for the analysis of these two metabolites in human plasma. LOQ were 0.6 ng/mL for M-IV and 2.5 ng/mL for M-II.[393] Subsequent studies in hepatocytes and microsomes from monkeys, rats, and dogs revealed that M-II was the major metabolite in all species. There were significant quantitative variations in these species.[394]

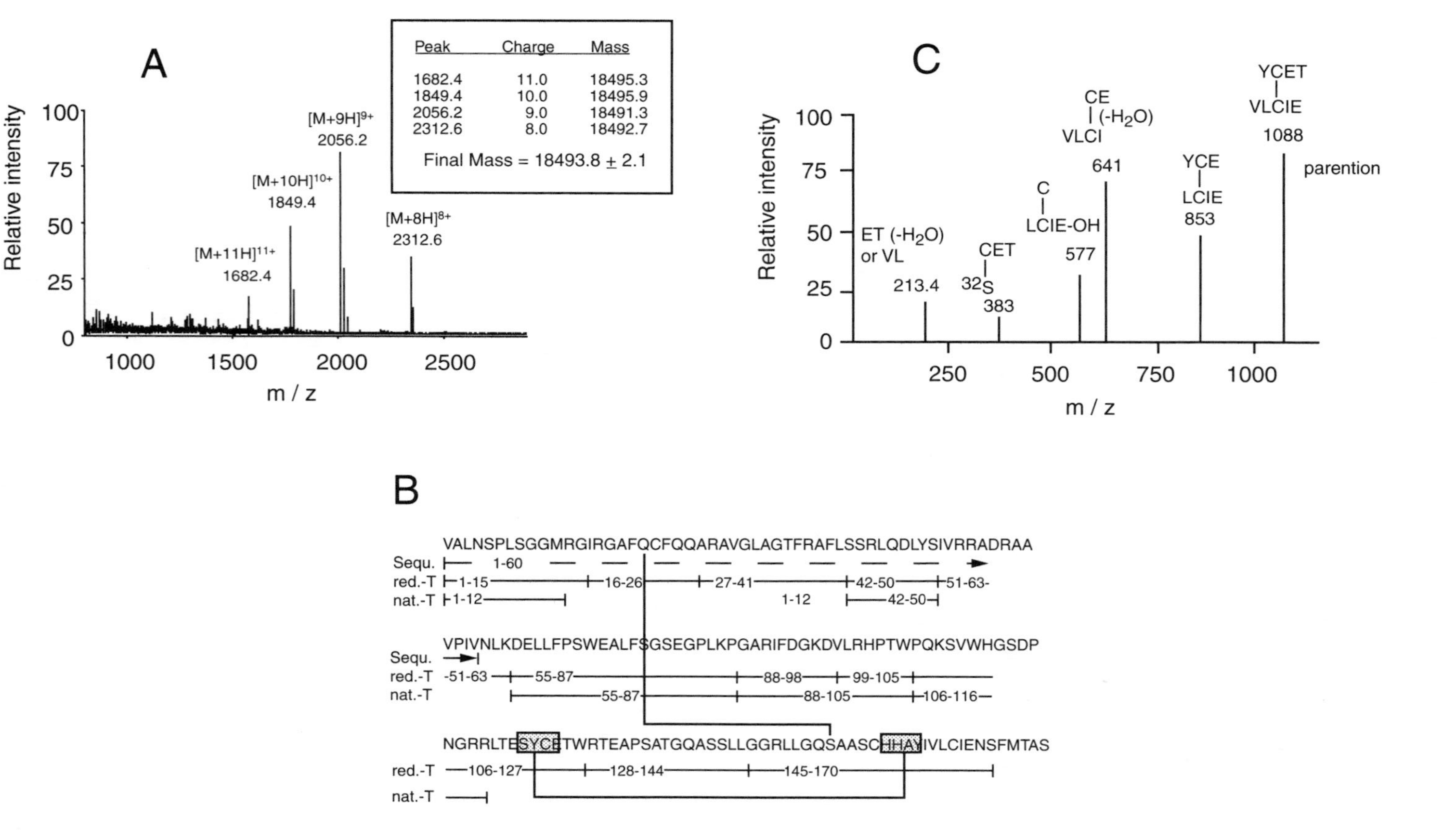

FIGURE 5.55 Identification of human circulating endostatin as collagen α1(XVIII) 514-683. (A) Mass determination of native endostatin using ESI. The mass spectrum consists of four different ions with *m/z* of 1682.4, 1849.4, 2056.2, and 2312.6 representing multiple charged ions of the peptide. The molecular mass of the native peptide was determined to be 18,494 ± 2.1 Da; (B) Sequence analysis and enzymatic digest of human endostatin. Sixty residues of the N-terminus were sequenced (broken line). Native and reduced $AcNH_2$ were cleaved by trypsin (T). The resulting fragments were analyzed by MALDIMS (solid lines). The cysteine-containing regions in the native molecule show high stability against trypsin. Determination of the two disulfide bonds (bold lines) were made by ESI-MS/MS; and (C) Analysis of disulfide bonds of endostatin by ESI-MS/MS. The peptide of 1088 Da (precursor ion, amino acids boxed and shaded in B) and some fragment ion signals are shown with their deduced sequences. The parent ion mass and its fragment pattern were in accordance with the amino acid sequence corresponding to residues 122-125 and 159-163 of human endostatin from hemofiltrate. Reprinted from Standker, L. et al., *FEBS Lett.*, 420, 129–133, 1997. With permission.

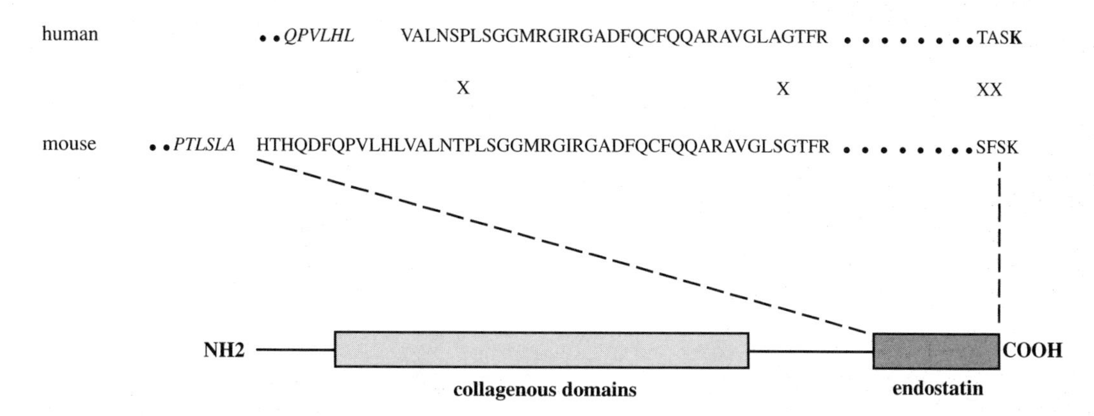

FIGURE 5.56 N-terminal sequences of human circulating endostatin and mouse endostatin. The sequences from the nontriple-helical C-terminal regions of collagen α1(XVIII) (schematic drawing) indicate high homology, mismatches between the mouse and human sequences are denoted by X. The lacking C-terminal lysine (K) in the human sequence is in bold. Note that the human endostatin isolated from hemofiltrate lacks 12 N-terminal amino acids, as compared to the mouse counterpart. Reprinted from Standker, L. et al., *FEBS Lett.*, 420, 129–133, 1997. With permission.

To study the molecular mechanism of action of TNP-470, a radioactive photoaffinity label was attached to the side-chain at the C6 position. A bovine aortic endothelial cell proliferation assay revealed no significant loss of activity and aided the isolation from mouse embryos of a target protein of 67 kDa. After additional biochemical and purification steps, a silver-stained band was excised from PAGE, subjected to in-gel digestion with trypsin, and the tryptic fragments extracted. MALDI-TOF analysis revealed $[M+H]^+$ ions representing 22 distinct peptides (Figure 5.59) of which 5 were related to controls. Searching the EMBL protein database for the remaining 17 peptides revealed that 16 fit those predicted for both rat and human type-2 methionine aminopeptidase, a common bifunctional enzyme that is also known to inhibit the phosphorylation of eukaryotic initiation factor 2α. A comparison of a putative open reading frame, generated by overlying sequences, indicated that all 16 identified peptides of the type-2 methionine aminopeptidase isolated in the angiogenesis inhibitor experiments expressed homology to peptides from the corresponding enzyme in the mouse model (Figure 5.59). This protein is highly conserved among eukaryotes with 55 to 93% identity. Several synthesized analogs of fumagillin with anti-angiogenic potency also inhibited methionine aminopeptidase activity but without affecting the phosphorylation action of the enzyme. It was concluded that this enzyme may be a promising target for antiangiogenic drugs.[395]

In a Phase I study of the pharmacokinetics of dose escalation of TNP-470 and its M-II and M-IV metabolites in AIDS patients with Kaposi's sarcoma, the analytes were quantified in plasma samples collected at numerous time points up to 168 h after the beginning of the infusion. The $[M+NH_4]^+$ adducts of each analyte were monitored (i.s.: ^{13}C-TNPD$_3$) using APCIMS.[393] The predominant metabolite was M-II with $t_{1/2}$ of ~1 h. The average concentration of M-IV, the active metabolite, was ~2% that of M-II, with a $t_{1/2}$ of 0.2 h. To reduced interferences, urine concentrations were determined using MRM of the $[M+NH_4]^+$ adducts. The parent drug was undetectable in urine, metabolite M-II was present in all urine samples (~1% of administered dose) and M-IV appeared only in the urines of patients in the highest dosing cohorts. There was no evident correlation between pharmacokinetic data and clinical pharmacodynamics.[396]

Genistein

This isoflavonoid, a specific tyrosine kinase inhibitor, is a highly potent dietary-derived inhibitor of *in vitro* angiogenesis.[397] The concentration of genistein in the urine of subjects on plant-based diets has been shown to be ~30-fold higher than in subjects on traditional Western diets. Urine

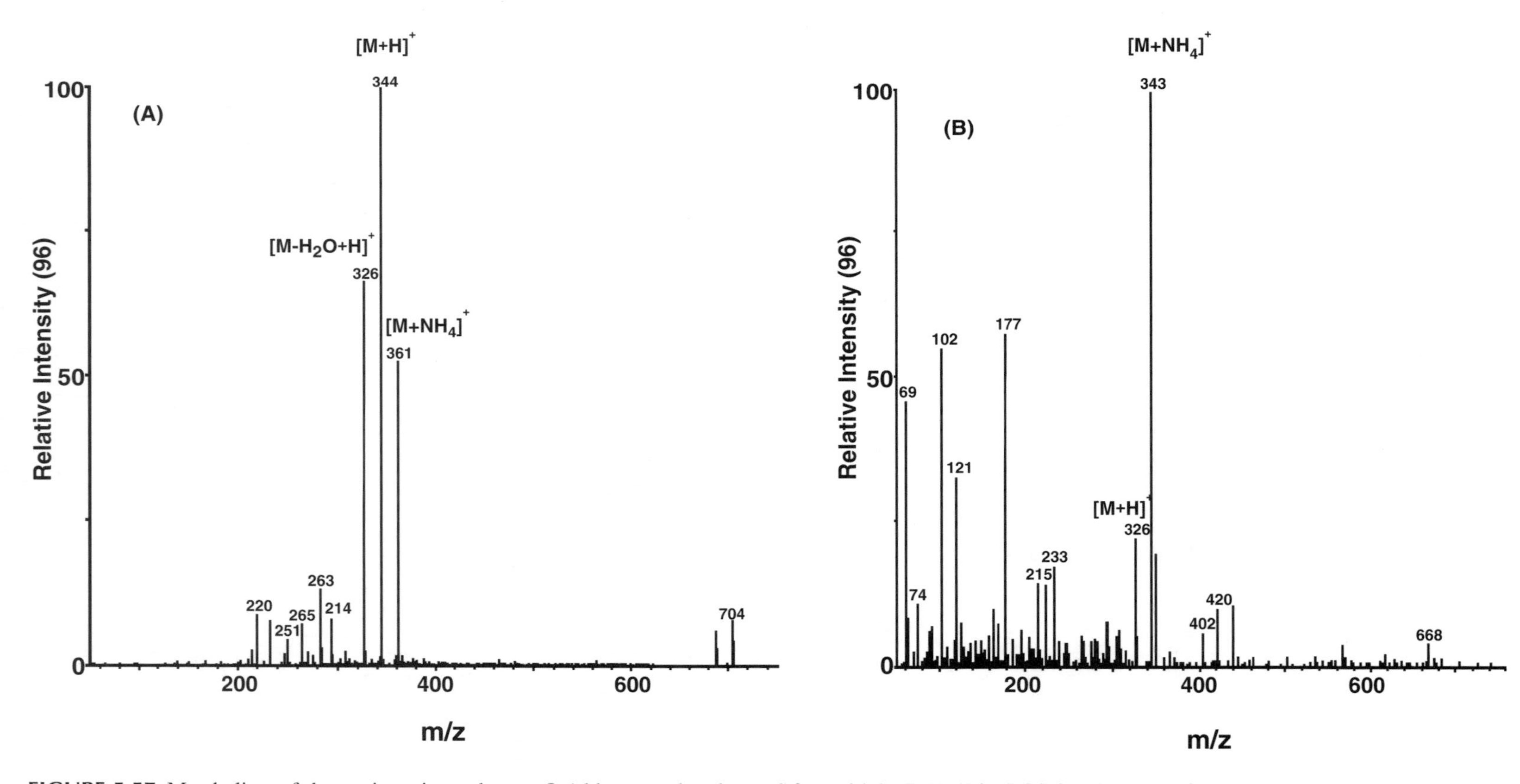

FIGURE 5.57 Metabolites of the angiogenic moderator O-(chloroacetyl-carbamoyl)fumagillol, (TNP-470). LC/ESIMS spectra from two separated peaks, M-II (A) and M-IV; (B) isolated from a primary culture of human hepatocytes. For identities, see Figure 5.58. Reprinted from Placidi, L. et al., *Cancer Res.*, 55, 3036–3042, 1995. With permission.

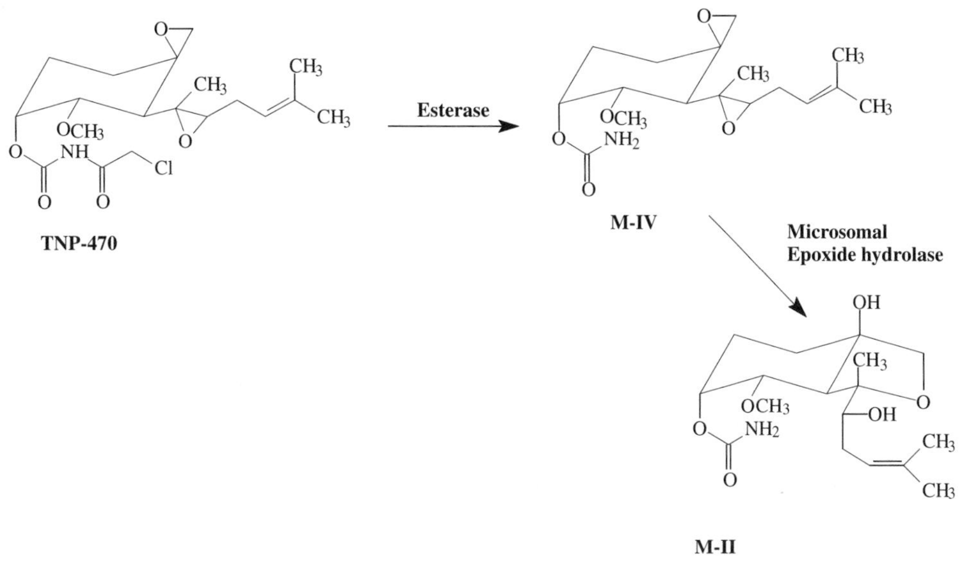

FIGURE 5.58 Tentative metabolic pathway for TNP-470 in liver cells. Also see Figure 5.57. Reprinted from Griffith, E. C. et al., *Chemistry & Biology,* 4, 461–471, 1997. With permission.

samples of healthy subjects on a soy-rich vegetarian diet were separated into six fractions of different compound classes. The fraction exhibiting the highest potency for the inhibition of the proliferation of vascular endothelial cells contained several isoflavonoids, identified by isotope dilution GC/MS (Section 5.4.2.1). These metabolites were synthesized and individually tested in an *in vitro* angiogenesis model. Genistein was 5- to 20-fold more potent than the other isoflavonoids and inhibited the proliferation of cells in several malignant tumor cultures in a dose-dependent manner.[398]

5.3.3 Antisense Oligonucleotides

A strategy for disrupting gene expression aims to disturb the function of mRNA by arresting protein-specific translation. The "antisense" approach relies on the formation of reverse complementary Watson-Crick base pairs between a segment of the mRNA, whose function is to be disrupted, and the targeting antisense oligonucleotides. These constructs, generally 12 to 30 bases long, bind with high specificity to the targeted length of mRNA and the mRNA-DNA complex and suppress the translation of the targeted message into protein. A major advantage of antisense compounds is that they may be rationally designed against any protein as long as the sequence of the corresponding gene is known. Clinical applications include lymphomas[399] and leukemias.[400] Second generation antisense oligonucleotides undergoing preclinical and clinical trials include lipid conjugates, modified nucleobase derivatives, and novel structures involving mixed backbone (modified DNA and RNA) segments. In one study, the pharmacokinetics and tissue disposition of an antisense oligonucleotide (ISIS 2503) were compared with the agent formulated in encapsulated stealth liposomes or unencapsulated in phosphate-buffered saline. Metabolites were identified by ESIMS. In monkeys, the stealth liposomes protected ISIS 2503 from nucleases in blood and tissues and slowed its clearance from systemic circulation.[401]

The characterization of oligonucleotides by ESI has been reviewed.[402] The noncovalent interactions between met- and leu-enkephalins and their antisense peptides have been studied by ESIMS, and it was shown that there is a preferential interaction between the sense and antisense peptides compared with that between the sense and control peptides.[403]

The metabolism of two 2′-deoxyphosphorothioate oligonucleotides, ISIS 11637 (CATCCAAGGCACAGCTTGA) and ISIS 11061 (ATGCATTCTGCCCCAAGGA) have been studied in a rat

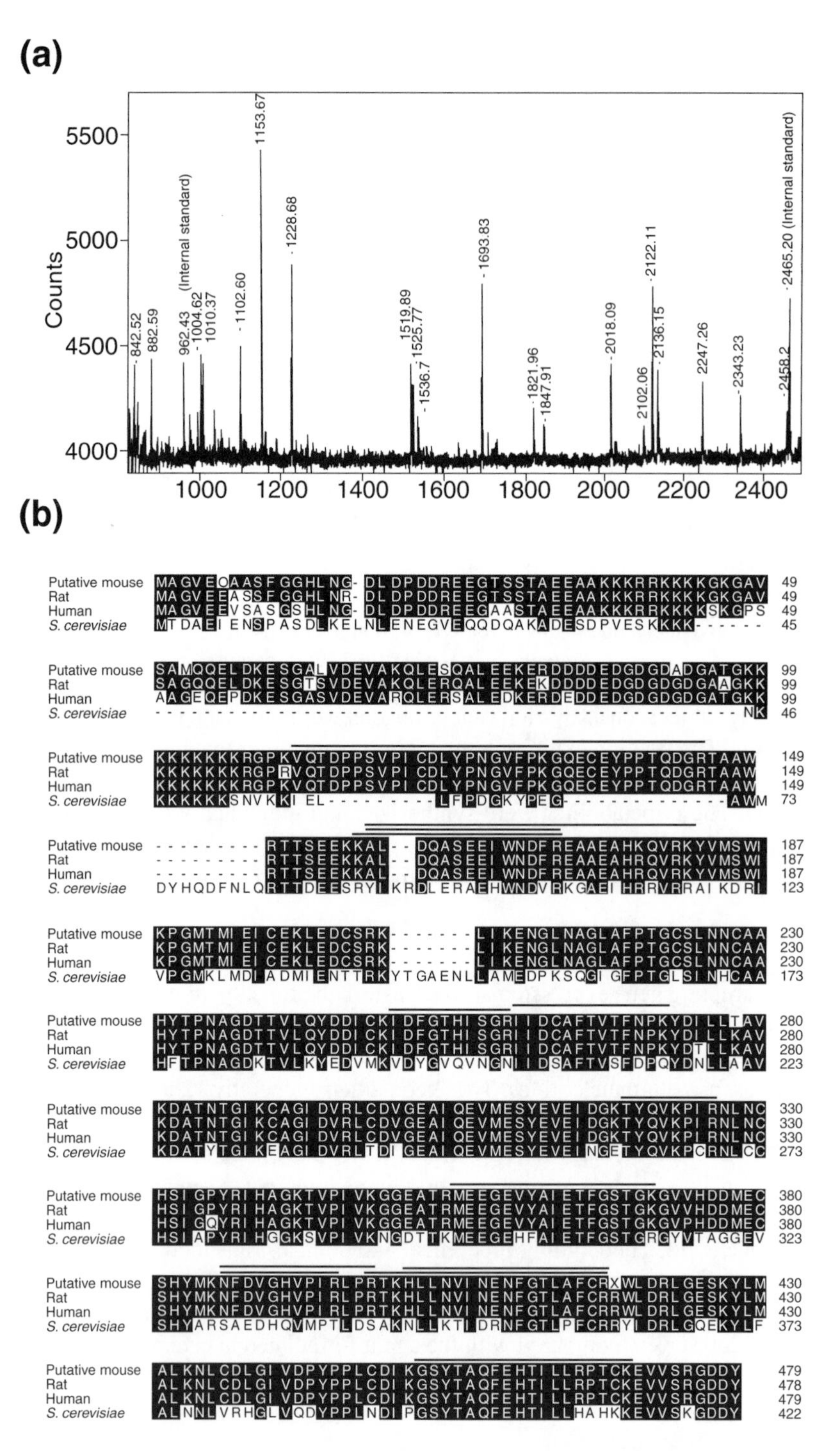

FIGURE 5.59 Isolation and identification of p67 from mouse embryo extracts using biotin-fumagillin and biotin-ovalicin conjugates. (a) MALDITOF spectrum of a tryptic digest of p67. The *m*/*z* values are monoisotopic. Peaks marked with ⁻ result from blank experiment; and (b) MetAP2 is highly conserved among eukaryotes. The putative mouse MetAP2 sequence is compared to those of rat, human, and *Saccharomyces cerevisiae*. Protein sequences were aligned using the program MegAlign. The tryptic fragments identified by mass spectrometry are overlined. Sequence identity is highlighted by shading. Reprinted from Griffith, E. C. et al., *Chemistry & Biology,* 4, 461–471, 1997. With permission.

TABLE 5.3
Observed and Calculated Masses and Sequences of Metabolites of ISIS 11601 from Rat Liver and Kidney Tissue Samples

Oligonucleotide Calcd Mr	Sequence of Observed Oligonucleotide	Observed Mass Liver	Observed Mass Kidney
A, 6367.6	5′ ATG CAT TCT GCC CCC AAG GA-3′	6367.8	6368.6
B,M, 6037.6	5′-ATG CAT TCT GCC CCC AAG G-3′ and 5′-TG CAT TCT GCC CCC AAG GA-3′	6038.9	6038.1
C, 5692.6	5′-ATG CAT TCT GCC CCC AAG -3′	5692.4	5693.1
D, 5347.5	5′-ATG CAT TCT GCC CCC AA-3′	5350.6	5348.3
E, 5018.5	5′-ATG CAT TCT GCC CCC A-3′ (E1) and/or 5′-TG CAT TCT GCC CCC AA-3′ (E2)	5018.8	5018.3
F, 4688.5	5′-ATG CAT TCT GCC CCC -3′ (F1) and/or 5′-TG CAT TCT GCC CCC A-3′ (F2)		4690.1
G, 4383.4	5′-ATG CAT TCT GCC CC -3′		4383.9
H, 4078.4	5′-ATG CAT TCT GCC C -3′		4078.2
I, 3773.4	5′-ATG CAT TCT GCC -3′		3773.9
K, 3468.4	5′-ATG CAT TCT GC -3′		3471.0
N, 5717.6	5′-G CAT TCT GCC CCC AAG GA-3′	5717.8	5717.5
O, 5372.5	5′-CAT TCT GCC CCC AAG GA-3′	5374.1	5375.3
P, 5067.5	5′-AT TCT GCC CCC AAG GA-3′	5069.0	5067.3
Q, 4738.5	5′-T TCT GCC CCC AAG GA-3′ (Q1) and/or 5′ AT TCT GCC CCC AAG G-3′ (Q2)	4740.8	4741.1
W, 5708.6	5′-TG CAT TCT GCC CCC AAG G-3′	5709.8	5709.8
X, 5363.5	5′-TG CAT TCT GCC CCC AAG -3′	5364.1	
Y, 5043.5	5′-CAT TCT GCC CCC AAG G-3′ (Y1) and/or 5′-G CAT TCT GCC CCC AAG-3′ (Y2)	5045.0	5042.5
Z, 6232.7	5′-ATG CAT TCT GCC CCC AAG G′-3′ (A-adenine)	6234.1	

Reprinted from Gaus, H. J. et al., *Anal. Chem.*, 69, 313–319, 1997. With permission.

model using LC/ESIMS. Metabolites in the plasma were consistent with 3′-exonuclease activity. Calculated and observed masses of metabolites obtained from liver and kidney tissues from animals treated with ISIS 11601 agreed well, and a series of sequences were confirmed (Table 5.3). The peaks of highest abundance, corresponding to metabolites B and M, represent the n – 1 species. These species are indicative of either 3′ or 5′ degradation as ISIS 11061 is terminated by the same dA nucleotide at both termini. Metabolites C-K represent n – 2 to n – 9 species, all truncated from the 3′ terminus. Metabolites having masses consistent with 3′ cleavage alone, 5′ cleavage alone, and both 3′ and 5′ cleavages were observed, consistent with 3′- and 5′-exonuclease activity. Sequence Z, which is consistent with the loss of adenine from the parent molecule, was observed only in the liver. The origin of all other sequences shown in the table could be explained rationally.[404]

5.3.4 *Ras* Oncoprotein Inhibitors

The fact that some 90% of pancreatic, 50% of colon, and 40% of lung cancers are linked to the mutation of the *ras* oncogene justifies efforts to develop inhibitors to this oncoprotein. The objective of one approach has been to develop compounds that would deactivate *ras* by preventing the interchange of the inactive *ras*-GTP complex with the active *ras*-GDP complex (reviewed in References 405, 406). More than 50 non-nucleotide type-compounds were synthesized and the drug-protein structures determined by NMR, MS, and molecular modeling.[407] Techniques developed for probing protein-protein noncovalent interactions by ESIMS[408] were extended to study the ternary complex of *ras*-GDP with the experimental drug SCH 54292. Information about the binding site of the drug was obtained by analyzing the products after covalent modification of the protein upon treatment with succinic anhydride. This acts specifically with the ε-amino groups of exposed lysine residues on the surface of the protein, followed by peptide mapping after digestion with endoproteinase Lys-C that does not cleave succinylated Lys residues. Accordingly, the mass spectra exhibited two adjacent peptides plus a succinate group, e.g., SL6,7 refers to succinylated fragment L6 plus L7 due to noncleavage at the succinylated Lys-104 (Figure 5.60A). The enzyme cleaved proteins at the C-terminal end of a lysine (Figure 5.60B). The differences between the two spectra revealed the lysine positions that were protected from succinylation, leading to the conclusion that the binding of the compound to *ras*-GDP occurred at sites near Lys-101 which is located on the underside of a particular cleft as suggested by NMR and modeling. As expected, the sites of binding were different from those for the GDP/GTP nucleotides.[409]

5.3.5 Antibody-Directed Enzyme Prodrug Therapy

Antibody-directed enzyme prodrug therapy (ADEPT) is based upon three steps: attachment of a prodrug-activating exogenous enzyme to a tumor-selective antibody; systemic dosing with the enzyme-antibody conjugate to localize the tumor; and systemic dosing with the nontoxic prodrug. When the targeted enzyme converts the prodrug to the active drug, the chemotherapeutic agent is synthesized selectively at the tumor site. Repeated administration is limited by the development of an immune response as most enzymes that are capable of performing the required drug conversion are foreign to the human body. Alternatives for the selective activation of products in tumor tissues include therapeutic radiation, use of tumor-specific endogenous enzymes, and the application of DNA constructs containing the corresponding gene (GDEPT).

In a Phase I clinical trial in 10 patients with nonresectable colorectal carcinoma, the prodrug was 4-[(2-chloroethyl)(2-mesyloxyethyl) amino]benzoyl-L-glutamic acid, in which the chemically active ionized carboxyl group was masked by linking the L-glutamic acid to the benzoic acid function through an amide group. A bifunctional alkylating drug, 4-[2-chloroethyl)(2-mesyloxyethyl)amino benzoic acid, was released from this prodrug by the action of a tumor-localized bacterial carboxypeptidase G2 enzyme, which was coupled to a murine monoclonal antibody. The presence of the prodrug (*m/z* 450) and drug (*m/z* 321) in plasma extracts was confirmed by the detectin of their $[M+H]^+$ ions using LC/ESIMS. Two additional HPLC peaks in plasma extracts were identified as the monohydrolyzed compounds of both prodrug and active drug.[410] The presence and disappearance of the active drug was also monitored by LC/ESIMS in additional *in vitro* and *in vivo* experimentation.[411] Other metabolites formed by the hydrolysis of the mesyl group were also identified, providing a means of following the fate of both the prodrug and the active drug during the course of the treatment.[412]

5.3.6 Miscellaneous Strategies for Targeting

5.3.6.1 Photodynamic Therapy

The first step in photodynamic therapy is the injection of a photosensitizing drug, such as a light-activated dye that is taken up by cancer cells. The cancerous lesions are next imaged and localized with low light doses, followed 24 to 28 h later by an exposure to a laser light of high fluence to induce fluorescence of the dye which, in turn, generates cytotoxic agents. Repeated use of photodynamic therapy does lead to the development of resistance. Malignancies treated with varying success include basal and squamous cell tumors, malignant melanoma, head and neck cancer, Kaposi's sarcoma, and bladder cancer.

Photosensitizers include hematoporphyrin derivatives and dihematoporphyrin ethers and esters. Photodynamic agents often are mixtures of photoactive compounds, e.g., the widely used *photofrin*

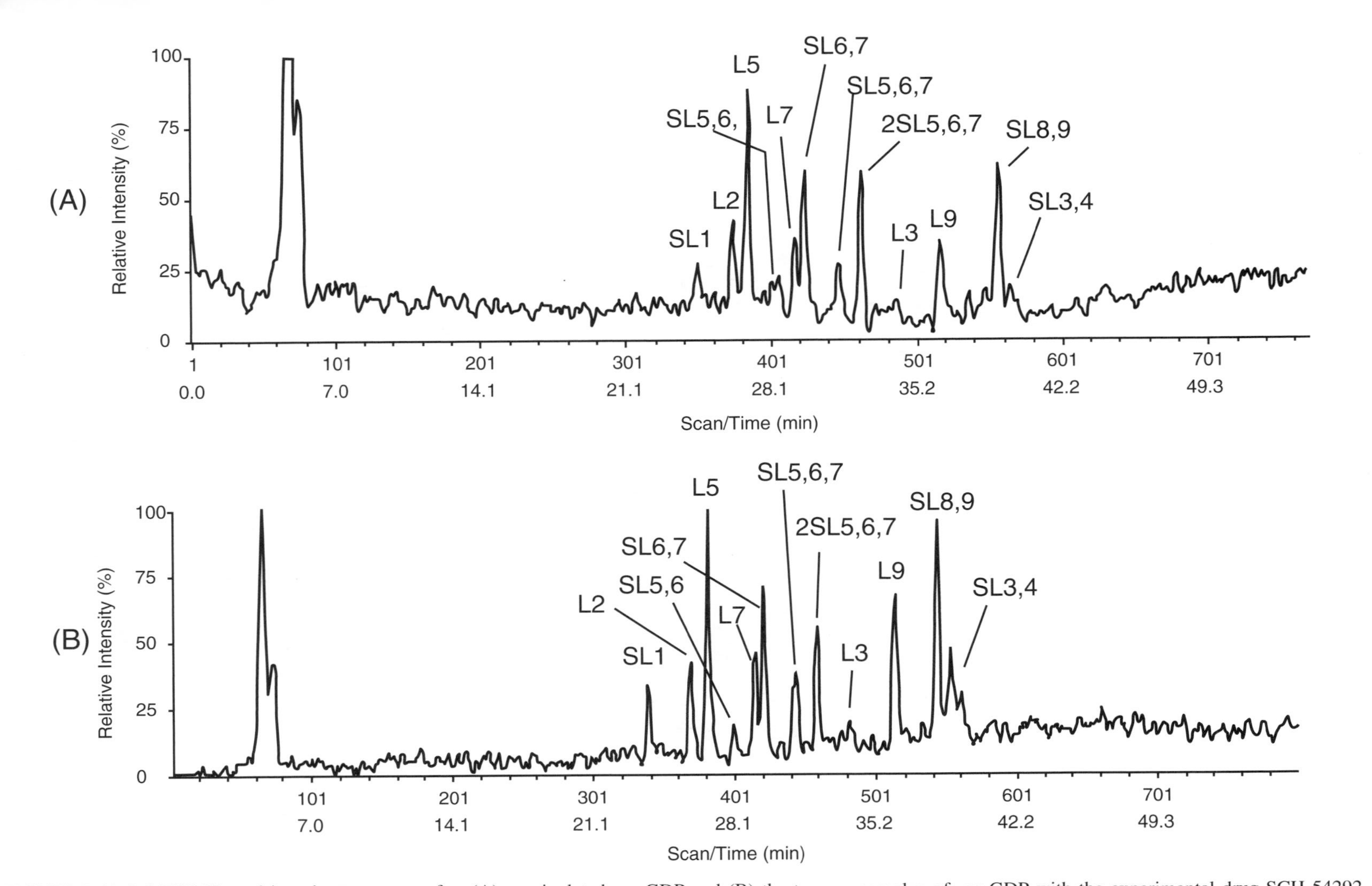

FIGURE 5.60 LC-ESIMS total ion chromatograms for: (A) succinylated *ras*-GDP and (B) the ternary complex of *ras*-GDP with the experimental drug SCH 54292, following digestion using endoproteinase Lys-C. Reprinted from Ganguly, A. K. et al., *Bioorg. Med. Chem.*, 5, 817–820, 1997. With permission.

(porfimer sodium), which is a complex mixture of monomeric and oligomeric porphyrins. *Temoporfin* [5,10,15,20-tetra(*m*-hydroxy-phenyl)chlorine] is another potent single compound photosensitiser.

Experiments on the structure of photosensitizers using EI, FD, CI, and FAB revealed that dihematoporphyrin ether was a tumor-localizing component, suggesting that clinically useful mixtures should be enriched in this component (reviewed in Reference 413). A two-step laser-MS technique established that photofrin does not contain covalently bound porphyrin oligomers.[414] A comparative analysis of photofrin by FAB, UV-MALDI, IR-MALDI, ESI, and laser desorption/jet-cooling photoionization, resulted in similar oligomer distributions with an average oligomer length of 2.7 ± 0.1 porphyrin units.[415] Mass spectral characterization of the conjugate of temoporfin with poly(ethylene glycol) 2000 (PEG 2000) was accomplished with HPLC/ESIMS, which resolved the complex mixture resulting from the presence of four PEG 2000 side-chains, permitting the subsequent determination of the molecular mass distribution of the oligomers.[416]

To investigate the possible advantages of using porphyrin mixtures rather than single agents, a combinatorial library of 666 tetraphenylporphyrins was constructed, that included substituents, such as Br, CN, OAc, and Ph, and the masking of polar groups with lipohylic protecting groups. Structures were determined using quantitative, isotopically-resolved, fragmentation-free LDTOF. Membrane incorporation and possible "mixture effects" were elucidated using spectrometrically monitored "selection" experiments.[417] Peaks indicative of higher molecular weight components unaccounted for by theoretically assigned LDTOF mass spectra of the library, were identified by a custom-designed software. Interfering pentaphorphyrins and short oligopyrroles were removed from the library by tailor-made HPLC phases. It was shown that the total porphyrin incorporation into the photosensitiser was higher for libraries than for single compounds of similar polarity at constant lipid:porphyrin ratio. It was concluded that complex mixtures of porphyrins perform better clinically than individual compounds because of membrane solubility incorporation effects.[418]

Experiments with LC/ESIMS revealed that the single agent, temoporfin, was excreted unchanged into the feces. No *in vitro* metabolites were detected, even after induction with cytochrome P-450 isoenzyme inducers. Previously reported "metabolites" were shown to be impurities in the original drug or artifacts generated by photochemical oxidation during sample preparation or extraction.[419]

5.3.6.2 Differentiating Agents

In contrast to cytotoxic agents whose objective is tumor cell destruction, differentiating agents are supposed to make tumor cells lose their ability to proliferate, i.e., to convert already existing malignant tumors into benign ones. Clinical trials with such agents, e.g., N-methyl-formamide or dimethyl sulfoxide, have been disappointing because the required plasma concentrations indicated by *in vitro* experiments could not be established *in vivo*. After encouraging Phase I clinical trials with *hexamethylene bisacetamide* (HMBA),[420] six acidic and basic metabolites were isolated from both plasma and urine and identified by comparing their EI and CI spectra with those of synthetic standards. Two urinary metabolites of HMBA were identified by GC/EIMS as the ethoxycarbamate-urethane derivative of 6-acetamidohexanoic acid and the *n*-butyl ester derivative of N-acetyl-1,6-diaminohexane.[421] The *in vitro* cell differentiation activity against human leukemia of 6-acetamidohexanal, an aldehydic metabolite, was higher than that of the parent compound.[422]

5.3.6.3 Arsenic Trioxide

Considering the strongly carcinogenic nature of arsenic compounds (Section 4.3.6.1), it is ironic that arsenic trioxide has recently been shown to induce complete remission in a number of patients with acute promyelocytic leukemia (APL) who had relapsed after conventional chemotherapy and treatment with all-trans retinoic acid.[423,424] Arsenic trioxide has been shown to induce both apoptosis and differentiation.[425] In an *in vitro* study on how arsenic trioxide may exert an antileukemic effect via inhibition of angiogenesis by inducing dose- and time-dependent apoptosis of endothelium, ICPMS was used to quantify nanomolar quantities of arsenic in a highly sensitive cell culture system.[426] ICPMS was also used to determine free arsenic in biological fluids in a clinical study of an organic arsenical, melarsoprol, in patients with advanced leukemia. The mean plasma AUC concentration of elemental arsenic was ~6 ng·h/mL, with a gradual increase over the 3-week treatment course.[427] The ongoing investigations of arsenic trioxide and organic arsenicals in cell cultures,[428] as well as the increasing number of arsenic-based clinical trials that are being initiated are likely to expand the use of ICPMS both for the assay of total arsenic and in speciation studies.

5.4 NUTRITIONAL AND CHEMICAL PREVENTION

5.4.1 OBJECTIVES AND CLASSIFICATIONS

The objective of *primary prevention* is the prevention of the occurrence of disease and is therefore directed at apparently healthy subjects. *Secondary prevention* consists of early diagnosis at a preclinical stage, to be followed by intervention aimed at reversing, halting, or at least retarding the progress of the disease. Secondary prevention is best employed when the tumor mass reaches about 1 million cells (~1 g). *Tertiary prevention* is aimed at preventing recurrences or further progression in subjects who have been previously "cured." The objective of cancer chemoprevention is to prevent the development of cancer with the aid of micronutrients or noncytotoxic chemical compounds.[429,430] The basic premise is that intervention is possible during the multiple genetic and phenotypic alterations that occur in the long process leading to invasive cancer, i.e., that there are agents that can suppress or reverse the events that initiate and/or promote carcinogenesis. A basic concept is *field carcinogenesis* which assumes that there are extensive, multifocal, genetically distinct premalignant and malignant lesions that occur *within the entire carcinogen-exposed region* in the subject at risk.[431] An obvious example is the exposure of the lung and upper aerodigestive tract to the carcinogenic effects of tobacco. The objective of chemoprevention for these tissues is to block the occurrence, promotion, and progression of lesions. Thus, cancer chemoprevention is a part of primary prevention. Chemoprevention may also be considered secondary prevention, aiming at reversing carcinogenesis or preventing second primaries which are latent during the intervention.

The public media has been saturated with reports and advertisements, all too often based only on testimonials, about the magical properties of micronutrients and dietary supplements that prevent cancer and delay aging. There has also been a rapid expansion of the scientific literature, particularly on the chemopreventive benefits of certain drugs, e.g., tamoxifen, oltipraz, flutamide,[432,433] as well as other areas including micronutrients (reviewed in *J. Nutrition,* vol. 129, Suppl. 1999).

It was shown about a decade ago that several chemically induced cancers in laboratory animals could be inhibited by a wide range of foods and naturally occurring compounds. There are over 1000 minor or trace compounds in plant-derived foods (*phytochemicals*), most without nutritive value, that that are believed to have cancer preventive properties. Although more than 200 epidemiological studies have indicated that diets high in fruits and vegetables can reduce cancer risk,[434] such studies are difficult because of the myriad micronutrients packed into an everyday diet. In contrast, standardized dosing regimens with individual supplements that would provide needed blood concentrations for efficacy evaluation should be easy, however, most nutritionists have a near-dogmatic resistance to the use of nutritional supplements.

In contrast to Phase I chemotherapy trials which are concerned with the establishment of tolerable doses in patients with refractory disease, the objective of Phase I chemoprevention trials is the establishment of safe doses in healthy subjects. When the general population is targeted, the chemopreventive agents must have minimal toxicity. Agents with some toxicity may be acceptable in specific populations where there is an increased likelihood of developing malignancy, e.g., a genetic predisposition to colonic polyposis or breast cancer. The difficulties associated with conducting chemopreventive trials, in particular with respect to evaluation of the efficacy and the time

TABLE 5.4
Classifications of Chemopreventives with Representative Examples

Based on Chemical Composition	
Plant Phenolic Compounds	
Phytoestrogens	Flavonols: quercetin
	Isoflavonoids: genistein, daidzein, equol
	Lignans: matairesinol, enterodiol, enterolactone
Flavan-3-ols	Epigallocatechin gallate and other catechins
Terpenes and ginsenosides	Mono: limonene; tetraterpenoid: carotenoids
	Triterpene saponins: ginseng extracts
Sulfur-containing compounds	Allyl sulfides: alkylarylisothiocyanates, sulforaphene; oltipraz, N-acetylcysteine
Omega-3 polyunsaturated fatty acids	Eicosapentonic and docosahexenoic acids
Vitamins	Vitamin A, retinoids; vitamins D, E, B12/folate
Selenium	Inorganic: selenite, selenate
	Organic: selenomethionate
Based on Mechanism of Action	
Inhibitors of carcinogen formation	Ascorbic, caffeic, and gallic acids, N-acetylcysteine
Inhibition of cytochrome P-450	Isothiocyanates, diallyl sulfide
Induction of Phase II enzymes	
Glutathione S-transferase	Isothiocyanates, allyl sulfides
UDP-glucoronyltransferase	Polyphenols
Glutathione peroxidase	Selenium
Electrophile scavenging	N-acetylcysteine
Free radicals scavenging	Vitamin E
Suppressing agents	
Induce apoptosis	Selenium, genistein
Inhibit oncogene activity	Genistein, monoterpenes
Restore immune response	Selenium, vitamin E
Inhibit polyamine metabolism	Polyphenols, substituted putrescine

Based on References 423, 570, 571.

spans involved, mean that there is a need for analytical techniques that recognize, identify and statistically validate surrogate intermediates and end point biomarkers that would both reduce the required number of subjects as well as the length of large-scale trials conducted on the general population.[435]

The two main categories of chemopreventive agents are drugs and micronutrient phytochemicals, including vitamins (often also available as supplements). Drugs include: hormone activity suppressors, e.g., tamoxifen and its analogs and the aromatase inhibitors; nonsteroidal antiinflammatory drugs, e.g., aspirin; miscellaneous other drugs, e.g., oltipraz (see later). Phytochemicals may be classified according their chemical classes or by their mechanism of action (Table 5.4). Classification into inhibitors of carcinogen formation or as blocking and suppressing agents is unsatisfactory because most chemopreventive agents act by multiple mechanisms.[432] For example, the isoflavones both enhance and suppress estrogenic activity but can also act in a repressive manner against tyrosine kinase and topoisomerase activity. Other approaches to the classification of chemopreventive agents include: their natural source, e.g., soybeans as a major source of isoflavones; the type of cancer under investigation, e.g., lung cancer prevention with β-carotene, retinoids, vitamin E, oltipraz, or aspirin; or the molecular targets involved, e.g., selective binding to the retinoid receptors.[436,437]

Analytical problems encountered in the analysis of micronutrients derived from plants and fruits include the complex nature of the matrices, the presence of positional isomers, the need to identify low levels of analytes that are chromatographically and mass spectrometrically adjacent to disproportionally high concentrations of other constituents, and difficulties resulting from regional variations in plant growth patterns. For example, an LC/EIMS method successfully utilized a particle beam interface to identify a small quantity of daidzein in the presence of an overwhelmingly large neighboring peak attributed to an unidentified glycoside. In another case, seven flavone-derived compounds were compared in a single species of plants grown in two geographically separated regions.[438] A recurring question in reconciling the high content in natural dietary products of certain constituents of desirable properties and failures of clinical trials to confirm expected benefits, is whether the compounds of interest are actually absorbed sufficiently to reach biologically meaningful concentrations in blood and tissues. A GC/EIMS technique with SIM has been developed for the quantification of three polyphenols (catechin, quercetin and resveratrol) and their conjugates in whole blood, serum, plasma, and urine. The compounds were administered orally and intragastrically to humans and rats. LOD and LOQ were 0.01 μg/L and 0.1 μg/L, respectively, for all compounds. These values are an order of magnitude lower than those of previously available techniques. The concentrations of the conjugates (glucuronides and sulfates) of these compounds were significantly higher than those of their aglycones in both plasma and urine after oral administration, with peak conjugate values occurring within 0.5 to 1.0 h in plasma and within 8 h in urine.[439]

5.4.2 Phytoestrogens

Phytoestrogens are estrogen mimics produced by plants. The observation that sheep became infertile upon eating clover led to the discovery of the estrogenic effect of flavonoids. The structural similarity between the 2-phenylnaphthalane-type (polyphenolic) ring structure of isoflavones and the steroid nucleus of 17β-estradiol suggests functional similarities. Indeed, phytoestrogens and endogenous native estrogens compete in binding to the estrogen receptor. Certain flavonols and lignans, e.g., quercetin and enterodiol, respectively, compete for the catalytic site of estrogen synthase. This leads to the inhibition of this enzyme, which is the catalyst for the conversion of androgens to estrogens. Because these compounds exhibit weak estrogenic as well as antiestrogenic activity, it has been hypothesized that the development and/or progression of the hormone-dependent breast and prostatic carcinomas may be influenced beneficially with diets high in isoflavones and lignan precursors,[440] particularly diets high in soybeans.[441]

Flavonoids are naturally occurring polyphenols that share a similar chemical structure, i.e., two benzene rings linked through a heterocyclic pyran ring (Figure 5.61). *Flavonols* have a double bond at the C2-3 position and an hydroxyl group at C3. The most common flavonols are quercetin and its two glycosylated analogs, quercetrin and rutin (quercetin-3-O-β-rutinoside). These are present in some vegetables and fruits at up to 400 mg/kg concentrations. The *isoflavonoids* have the second benzene ring attached to the C3 position instead of C2. These compounds occur in plants as glycosides that, when consumed, are converted into biologically active products by bacterial action in the gut. Important isoflavonoids include genistein (4′,5,7-trihydroxyisoflavone), biochanin A (the 4-methyl ether derivative of genistein), daidzein (4′,7-dihydroxyisoflavone), and its bacterial decomposition product equol (7-hydroxy-3-[4′-hydroxyphenyl]chroman) (Figure 5.61). Soybean and soybean products contain 100 to 1000 mg/kg isoflavonoids. Genistein is important because it reaches high plasma concentrations and has a relatively long half-life of several hours.[441]

Lignans are another group of plant phenolic phytoestrogens. There are two important plant lignans, secoisolariciresinol, [R-(R*R*)]-2,3-bis[(4-hydroxy-3-methoxyphenyl)-methyl]-1,4-butanediol, and matairesinol, [3R-*trans*-dihydro-3,4,-bis[(4-hydroxy-3-methoxyphenyl)-methyl]-2(3H)-furanone. These are ingested in their glycosidic forms, at which point intestinal metabolism releases the lignan aglycones that are the biologically active species in mammalian systems. For example,

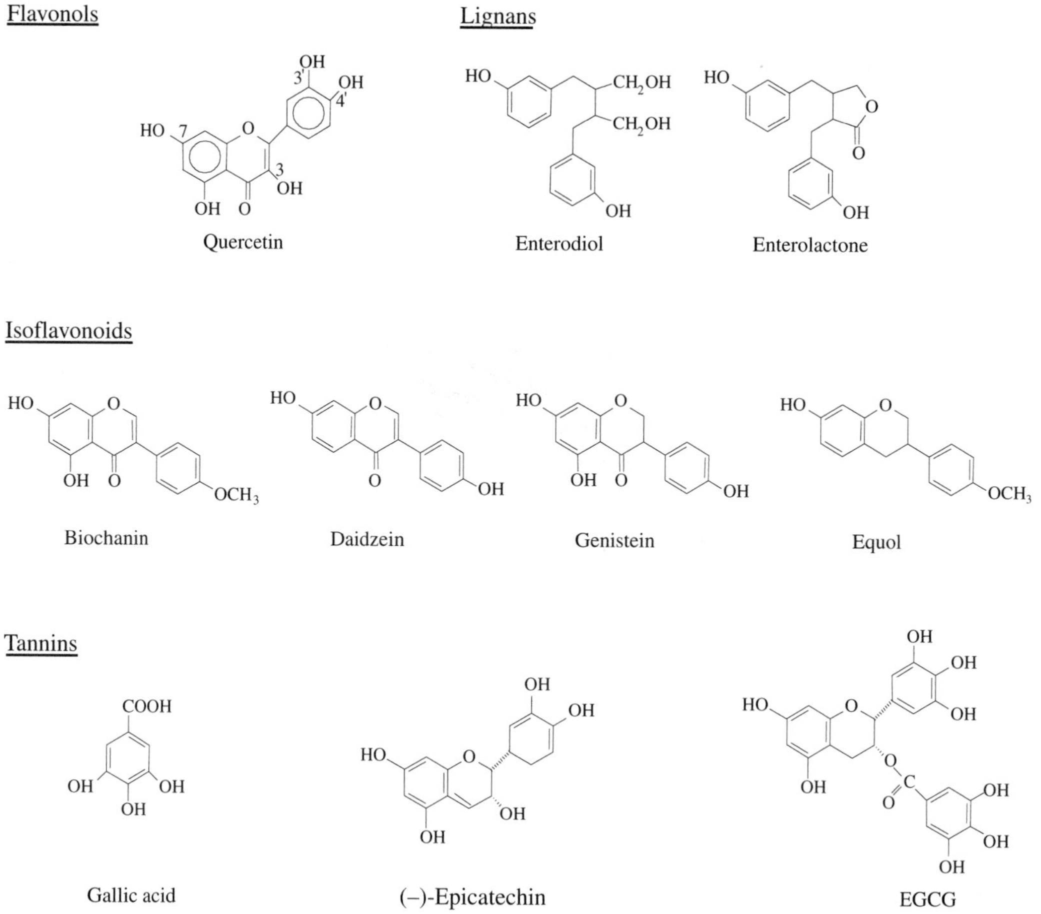

Flavonols

Quercetin

Lignans

Enterodiol

Enterolactone

Isoflavonoids

Biochanin

Daidzein

Genistein

Equol

Tannins

Gallic acid

(–)-Epicatechin

EGCG

FIGURE 5.61 Chemical structures of representative flavonols, lignans, isoflavonoids, and tannins. EGCG = (–) epigallocatechin gallate.

secoisolariciresinol diglucoside is the precursor of enterolactone (*trans*-2,3-bis[3-hydrobenzyl]-γ-butyrolactone) and enterodiol [2,3,bis[3-hydroxybenzyl]-1,4-butandiol) (Figure 5.61). Flaxseed, whole grains, legumes, and vegetables are rich sources of dietary lignans of plant origin. Although the chemopreventive properties of tea have been mainly attributed to catechins (Section 5.4.3), secoisolariciresinol and matairesinol have also been found in significant quantities in unbrewed and brewed teas.[442]

5.4.2.1 Multicomponent Analysis in Food

Mass Spectra

A review of several aspects of the mass spectra of flavonoid glycosides has been presented.[443] CID of ESIMS spectra of flavonoids obtained in the negative ion mode provides a means to distinguish methoxylated flavonoids with identical molecular mass.[444] A technique has been developed to enhance the detection of flavonoids by metal complexation with a neutral auxiliary ligand, 2,2′-bipyridine. The resulting ternary complexes have up to two orders of magnitude larger intensities than the corresponding protonated or deprotonated flavonoids. The formation of ternary complexes with several divalent transition metals was investigated and Cu^{2+} yielded the most abundant complexes and simplest mass spectra.[445]

Sample Preparation and Isotope Dilution-GC/MS-SIM

Aiming to develop a multicomponent screening tool, a general sample preparation technique was developed to separate isoflavonoids and coumestrol from lignans. A three-step hydrolysis procedure converted diphenolic glycosides into their respective aglycones (Figure 5.62).[446] The fractions were silylated, the constituents separated by GC, and quantified by SIM of selected pairs of ions from the analytes and deuterated i.s.[398,447] The LOD were 2 to 3 μg/100 g, thus 50 mg samples were adequate. The technique was used to simultaneously quantify seven constituents in more than 200 food items. For example, the biochanin A, daidzein, and genistein content of a granola candy bar were, in μg/100 g units, 2.2, 51.9, and 78.3, respectively. The values for the same constituents in soy flour were 74.4, 67,369, and 96,914, respectively. In contrast, these three components were found only in trace quantities in sunflower seeds.[446] A subsequent application concentrated on soy products (including tofu) with high isoflavone concentrations, soy drinks containing ~tenfold less and soy-based infant formulas that were devoid of isoflavones.[448] This technique has also been used in a number of other applications as described below.

Soy Isoflavone Content and Metabolism in Human and Cow Milk and Infant Formulas

Arguments concerning the role of soy isoflavonoids and cancer prevention have been intensified by the discovery of phytoestrogens in breast milk.[449] A GC/EIMS method capable of detecting the TMS ether derivatives of eight isoflavones has revealed that only daidzein (Figure 5.63) and genistein are present in human milk after soy consumption.[450] Further investigations using this GC/MS methodology have shown that these compounds also occurred as their glycosides.[451]

In a study aimed at comparing the isoflavone content of soy and cow milk based infant formulas, and human breast milk, as well as investigating phytoestrogen metabolism in early life, analytes were extracted, enzymatically hydrolyzed, and then isolated by gel chromatography. The recovered isoflavone aglycones and their metabolites (i.s.: dihydroflavone) were converted into *tert*-butyldimethylsilyl derivatives and the $[M-C_4H_9]^+$ ions monitored by GC/EIMS (Figure 5.64). Plasma from infants that were four months old and had been fed exclusively soy-based infant formula contained considerable quantities of genistein and daidzein but only trace quantities of equol. The low equol concentrations, and also the inability to detect other expected bacterial metabolites, e.g., dihydrodaidzein, may be explained by reduced biotransformation because of the lack of a fully developed intestinal microflora or limited enzyme activity in the infants. Cow-milk-based formulas contained up to twofold higher quantities of equol.[452]

The total plasma isoflavone concentration in infants who were fed soy-based formulas was 13 to 22 thousand times higher than that of estradiol. The significant cumulative estrogenic activity observed even at the low absolute activity of phytoestrogens supports the alleged long-term benefits of soy-based formulas against hormone-dependent malignancies in China and Japan.[453] An alternative LC/APCIMS technique, developed for the analysis of comminuted baby foods and soy flour, had LOD of 0.2 mg/kg for daidzein and 0.7 mg/kg for genistein.[454]

Effects of Cooking and Processing of Soyfoods

Although isoflavones are present in soyfoods mainly as 6′-O-malonyl-β-glucoside (6OMalGlc), β-glucosides, and 6′-O-acetyl-β-glucoside forms are also present. After exposure to different cooking conditions or commercial processing of soyfood and milk, the identities of extracted glucosides of daidzein, genistein, and glycitein were established by LC/APCIMS from the $[M+H]^+$ ions and aglycone fragments. Grinding to make flour and extraction with hexane to remove fats did not affect 6OMalGlc conjugates. However, wet heat treatments induced decarboxylation of the malonate group to acetate resulting in conversion to acetyl-β-glucoside conjugates. Fermentation to produce miso and tempeh resulted in hydrolysis of glucosides to form the aglycones. Because the bioavalability and pharmacokinetics of the glucoside conjugates may differ significantly, clinical trials should consider the type of cooking/processing treatments used on soyfoods.[455]

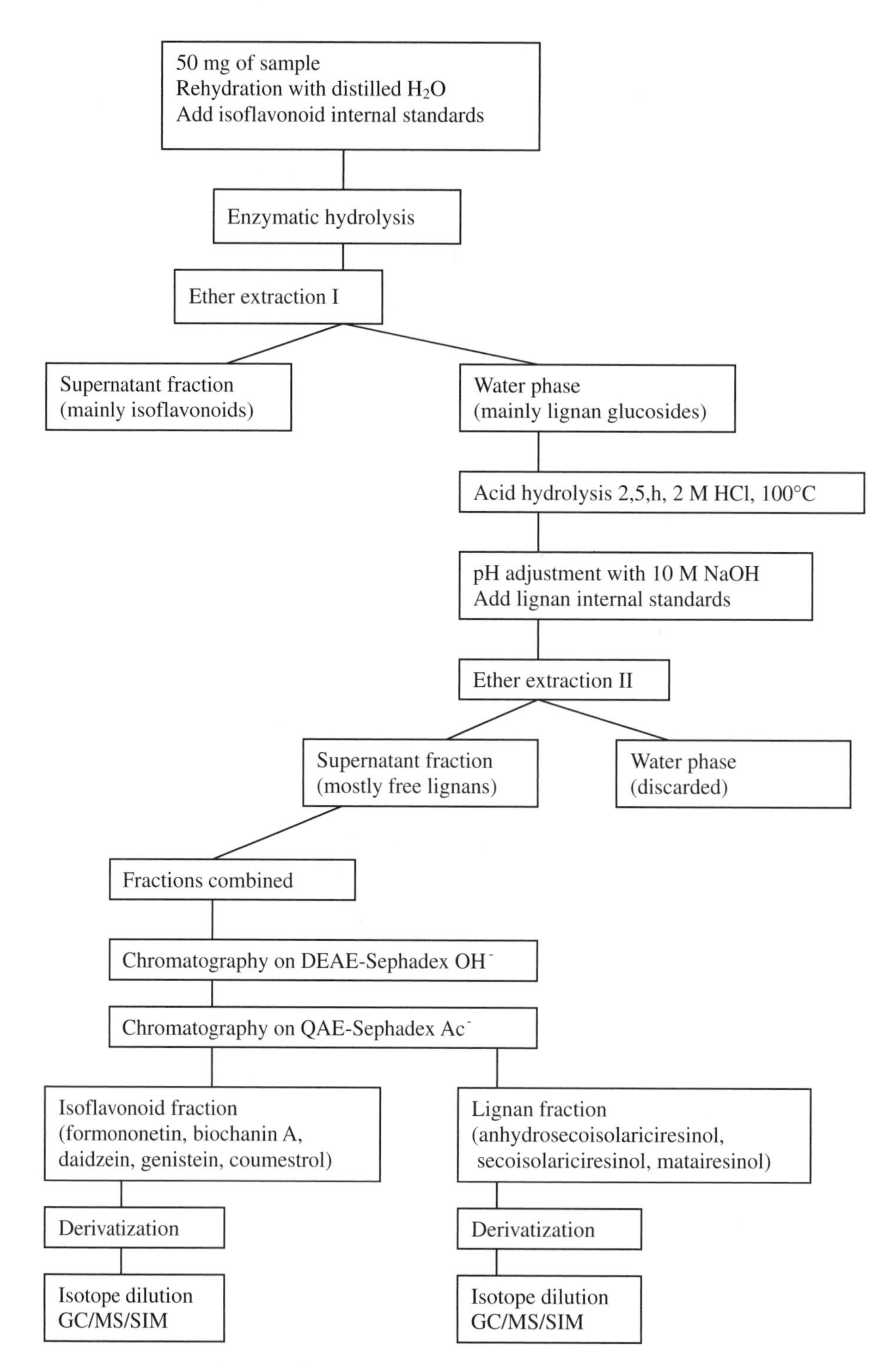

FIGURE 5.62 Flow diagram of the method for the determination of lignans and isoflavonoids in food. Reprinted from Mazur, W. et al., *Anal. Biochem.*, 233, 169–180, 1996. With permission.

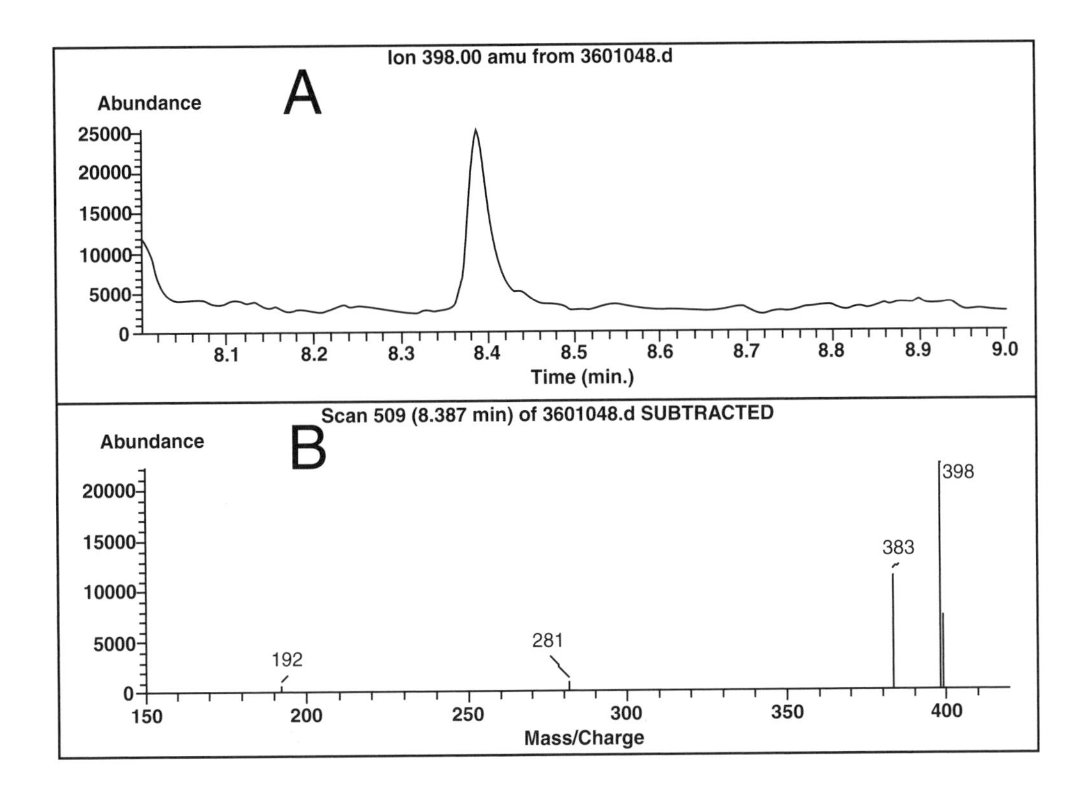

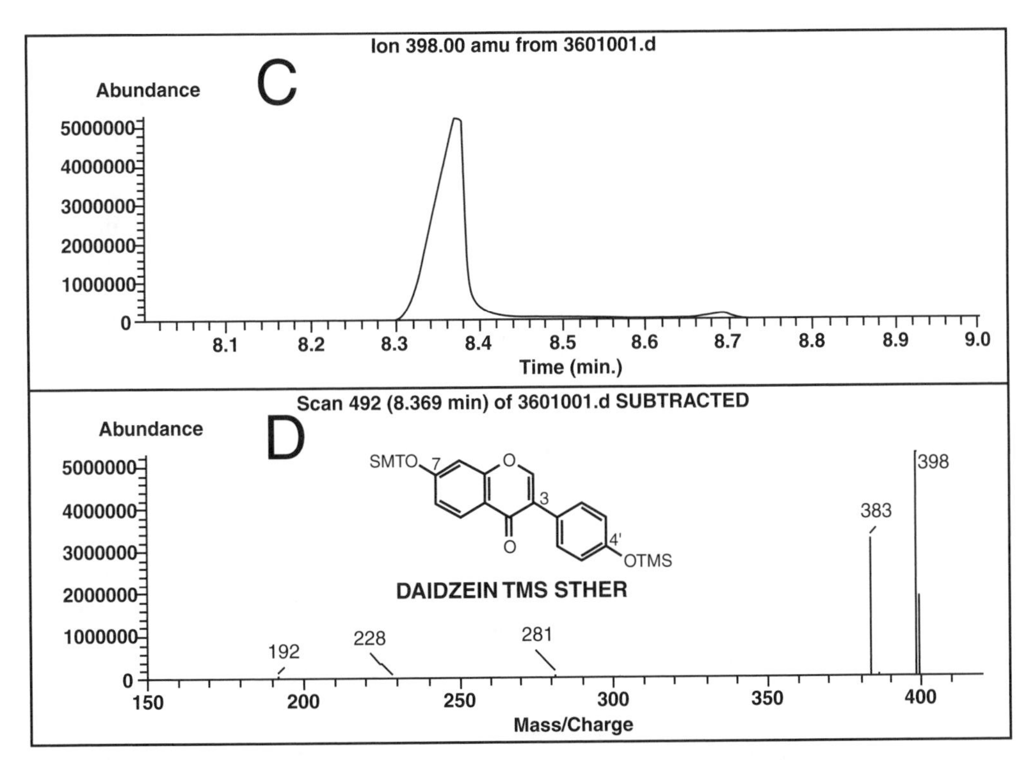

FIGURE. 5.63 (A, C) Gas chromatograms and (B, D) MS/MS fragmentation patterns of trimethylsilylated milk extracts (A, B) and authentic daidzein (C, D). Reprinted from Franke, A. A. and Custer, L. J., *Clin. Chem.*, 42, 955–964, 1996. With permission.

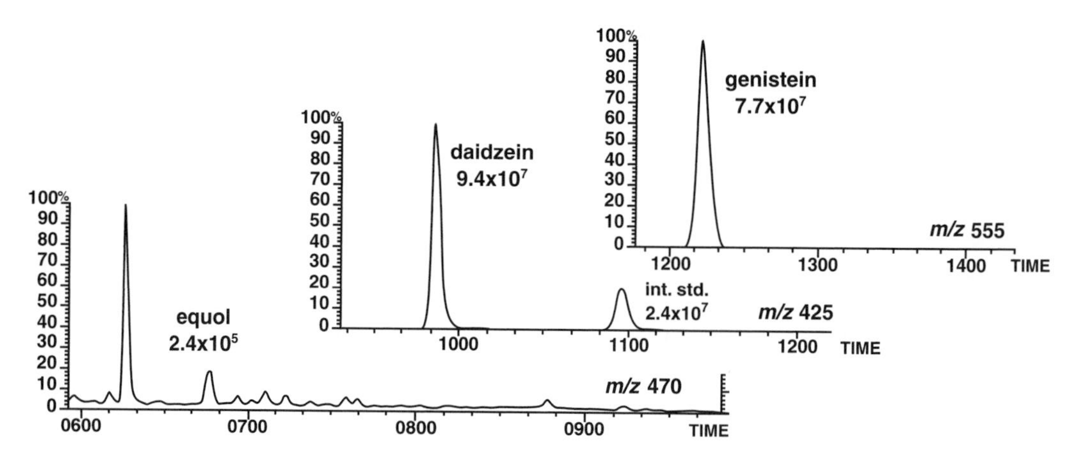

FIGURE 5.64 Selected ion current monitoring from the GC-MS analysis of the *tert*-butyldimethylsilyl ether derivatives of isoflavones isolated from the plasma of a 4 mo-old infant fed exclusively soy-based infant formula. The ions at *m/z* 470, 425, and 555 were monitored for the specific detection of equol, daidzein and the internal standard (dihydroflavone), and genistein, respectively. The integrated ion current for each compound is indicated. The channel recording of equol is amplified 100-fold with respect to the other channels. Reprinted from Setchell, K. et al., *Am. J. Nutr.,* 68 (Suppl.), 1453S–1461S, 1998. With permission.

Berries and Herbs

An LC/ESIMS technique developed for the identification of flavonol aglycones and glycosides utilized four operational modes: (a) full scanning to obtain molecular masses; (b) source-induced dissociation to screen for glycosides; (c) MS-MS for production of structurally important fragment ions from the $[M+H]^+$ ions; and (d) MS^3 for structural analysis using fragments derived from abundant product ions initially obtained from the $[M+H]^+$ precursor ions. In an application to analyze Finnish berries, the aglycones found included quercetin, myricetin, and kaempferol. The identities of the carbohydrate components of the flavonal glycosides, which differed depending on the species of berry, were also determined by GC/EIMS after their isolation, hydrolysis, and silylation.[456] Isotope dilution-GC/MS with SIM[446] has also been used to determine isoflavonoids and lignans in nine edible berries. The lignan secoisolariciresinol was the major constituent (1.4 to 37 mg/kg), while the concentration of matairesinol was <0.8 mg/kg. No isoflavones were found.[457]

The flavonoid composition of common fresh herbs has been analyzed by LC/APCI-MS/MS operated in the negative ion mode. The CID patterns provided information about the glycosyl substituents and the aglycones.[458]

5.4.2.2 Multicomponent Analysis in Body Fluids

Methodologies

A column-switching LC/APCIMS technique has been developed for the identification and quantification of 12 dietary flavonoid glycosides and aglycons in human urine. The first step of the column switching was used for sample clean-up, followed by heart-cutting the flavonoid fraction onto a second column for the separation of the components. Detection was accomplished by SIM in the negative ion mode. The method was linear in the dynamic range of 2.5 to 1000 ng/mL.[459]

Enzymatic deconjugation and centrifugation were the only sample preparation steps required in a technique combining solid-phase extraction and restricted-access media for the analysis of soy isoflavonoids in rat serum. Quantification was accomplished with LC/ESIMS using SIM (i.s.: several deuterated isoflavones). LOD was 0.02 μM. In an application, statistically significant sex differences were found in the serum concentrations of genistein.[460]

Changes in Urine Excretion Profiles with Different Diets

The isotope dilution GC/MS with SIM methodology noted above[398] has been applied to obtain data from experimental populations subjected to controlled diets. The excretion of isoflavonoids was in general 10 to 30 times higher among Japanese men and women consuming a traditional diet (low fat and high in rice, soy, fish, and vegetables), than was observed for subjects living in Boston and Helsinki. There was a correlation between the quantities of isoflavonoids excreted and the amount of soybean consumed. However, the excretion of the lignans, enterodiol and enterolactone, was in the low to normal range for all the subject populations.[461]

Excretion of enterodiol was higher for a high vegetable and fruit diets compared with either the baseline control or legume and allium diets. The reverse was true in the major isoflavonoids for the legume and allium diet with the excretion of equol being lowest in the high vegetable and fruit diet. It was concluded that isoflavonoid excretion may be used to reflect legume consumption if there is no soy consumption.[462] In a subsequent study of 30 men and 30 women who consumed a soy protein supplement product, daily equol excretion could be used to divide the participants into two groups. The "excreters" voided from ~2,000 to 20,000 nmole/day while the "non-excreters" produced <200 nmole/day. Among women, the quantity of fiber ingested appeared to influence the bacterial transformation of daidzein to equol.[463] The same methodology has also been applied to soy-complemented and conventional diets revealing increased isoflavonoid and decreased lignan excretion profiles in the former.[464]

In another application, the urinary excretion of the lignans enterodiol and enterolactone increased when a basal, vegetable free diet was replaced by a series of diets: one with high content of carotenoidal (carrot and spinach) vegetables; one with cruciferous (broccoli and cauliflower) vegetables; and a third with soy (tofu) products. The soy diet led to increased isoflavonoid excretion. While there was no gender effect on the excretion of isoflavonoids, lignan excretion was different with men releasing more enterolactone and less enterodiol than women.[465]

Ingestion of 10 g of flaxseed powder supplement led to rapid and large (3- to 285-fold) increases in the excretion of the lignans enterodiol and enterolactone in premenopausal women. There was no correlation with the phase (follicular or luteal) of the menstrual cycle, and no effect on the excretion of isoflavonoids.[466]

Urinary Excretion of Phytoestrogens in Finnish Omnivorous Men

Free and conjugated analytes were extracted, hydrolysed, and separated into two fractions: one contained equol, enterolactone, enterodiol, matairesinol, and all estrogens, the other contained O-desmethylangolesin, daidzein, and genistein. Following silylation, quantification was carried out by monitoring the $[M]^+$ or $[M\text{-}CH_3]^+$ ions (i.s.: deuterated analytes).[467] The technique was also extended to differentiate and quantify the different types of conjugants formed, including mono- and disulfate conjugates, and mono-, di- and sulfo-glucuronides. The monoglucuronides comprised >60% of all the analytes; with some 20% of the genistein being excreted as the diglucuronide. An important conclusion was that the pattern of the conjugated phytoestrogens excreted in the urine was similar to that of endogenous estrogens.[468,469]

Endogenous Urinary Phytoestrogens in Women from a Stratified Multiethnic Population

Seven phytoestrogens were quantified in 24-h urine samples by LC/APCI-MS/MS using MRM of characteristic precursor/product ion pairs of the analytes and deuterated i.s. The lignans, enterolactone and enterodiol, were highest in white women and lowest among Latina and African Americans, while genistein levels were highest among Latina. The concentrations of the other phytoestrogens, including daidzein and two of its metabolites, did not show ethnic variations.[470]

Quercetin and Its Urinary Metabolites

Confirmed biological activities of quercetin (Figure 5.61), a compound that is found in the outer layers of apples, as well as onions and tea, have included the inhibition of cell proliferation in cultures and prevention of the occurrence of experimental tumors that are normally produced by exposure to specific chemicals. The following metabolites of quercetin were identified by GC/EIMS in the silylated extracts of urine obtained from normal subjects: 3,4-dihydroxyphenylacetic acid, *m*-hydroxyphenylacetic acid, and 4-hydroxy-3-methoxyphenylacetic acid.[471] Flavonoid metabolites were also studied in the blood, urine, and feces of rats exposed to *Gingko biloba* extracts. Seven phenylalkyl acid metabolites, including homovanillic acid, were identified by LC/ESIMS.[472]

Another technique for the quantification of quercetin and kaempferol (another dietary flavonoid) in urine, and also in *Ginkgo biloba* tablets,[473] utilized solid-phase extraction, enzymatic hydrolysis with sulphatase or β-glucuronidase, and TMS derivatization prior to SIM of ions obtained by GC/NCIMS (i.s.: deuterated analytes). The fact that significant differences were observed in the quantities of the analytes before and after enzyme treatment suggested that the majority of the metabolites observed were present as their conjugated products.[474]

Quantification of Lignans and Isoflavonoids in Plasma by Isotope Dilution GC/MS

Unconjugated (free) and sulfated constituents were separated from glucuronides and sulfoglucuronides by DEAE-Sephadex chromatography, followed by hydrolysis of the glucuronides with β-glucuronidase. The separated compounds were determined by monitoring characteristic ions of their TMS derivatives. The free, mono- and disulfate compounds, all of which are considered to be biologically active, included matairesinol, enterolactone, enterodiol, and equol, while the glucuronide group included daidzein, genistein, and O-desmethylangolensin. Requiring 4-mL plasma samples, the technique was capable of determining 0.2 to 1.0 nmole/L concentrations of the diphenols. Good correlation was found between plasma and urine levels.[447] High concentrations of lignans were found in Finnish omnivores and vegetarians; in contrast, the plasma of Japanese subjects contained high levels of isoflavonoids.[475] The reliability coefficient of single measurements of enterolactone over a 2-year period was adequate for prospective epidemiological studies.[476] In a dietary study in which plasma concentrations of genistein and daidzein were determined for individuals on either a soy protein isolate supplement or a casein supplement, isotope dilution GC/MS revealed that the soy-based diet produced dramatic, ~120- and ~150-fold, increases in plasma genistein to ~900 nmole/L and daidzein to ~500 nmole/L, respectively.[477]

A similar sample preparation and GC/MS technique (magnetic analyzer) was employed to determine plasma levels of enterodiol, enterolactone, daidzein, equol, and genistein in post-menopausal Australian women receiving controlled diets. When compared to basal concentrations, large increases were observed in the combined concentrations of enterolactone and enterodiol when the diets were supplemented with linseed. Supplementation with soy flour or clover sprouts led to increases in the concentrations of the phytoestrogens.[478] Observations of this nature support the hypotheses on the beneficial effects of dietary lignans and isoflavonoids in the prevention and/or control of hormone-dependent neoplasms.

Quantification of Isoflavones in Plasma by LC/MS

The mass spectra of several isoflavones, including daidzein and genistein, have been obtained using LC/APCIMS in the negative ion mode. At low extraction cone voltages (–30V) intense $[M-H]^-$ ions predominated. When the extraction voltage was increased (to –70V) structurally diagnostic fragments appeared. The number of exchangeable hydrogen atoms was also determined from the peak shifts induced by deuterium oxide.[479]

An LC/APCIMS technique has been used to quantify four isoflavones as their aglycones or as their glucuronide and sulfate conjugates in spiked human plasma and in plasma from subjects who consumed a soy-based drink. Several i.s. were added to determine the efficiency of the enzymatic hydrolysis of the conjugates with β-glucuronidase and sulfatase. Full scanning, SIM, and SRM spectra were obtained in the negative ion mode. Appropriate precursor-product ion pairs, e.g., *m/z* 269 → 133, were selected for quantification. Plasma samples from subjects taking isolated soy proteins (containing 21 mg genistein and 13.5 mg daidzein) twice in a 24-h period contained 496 to

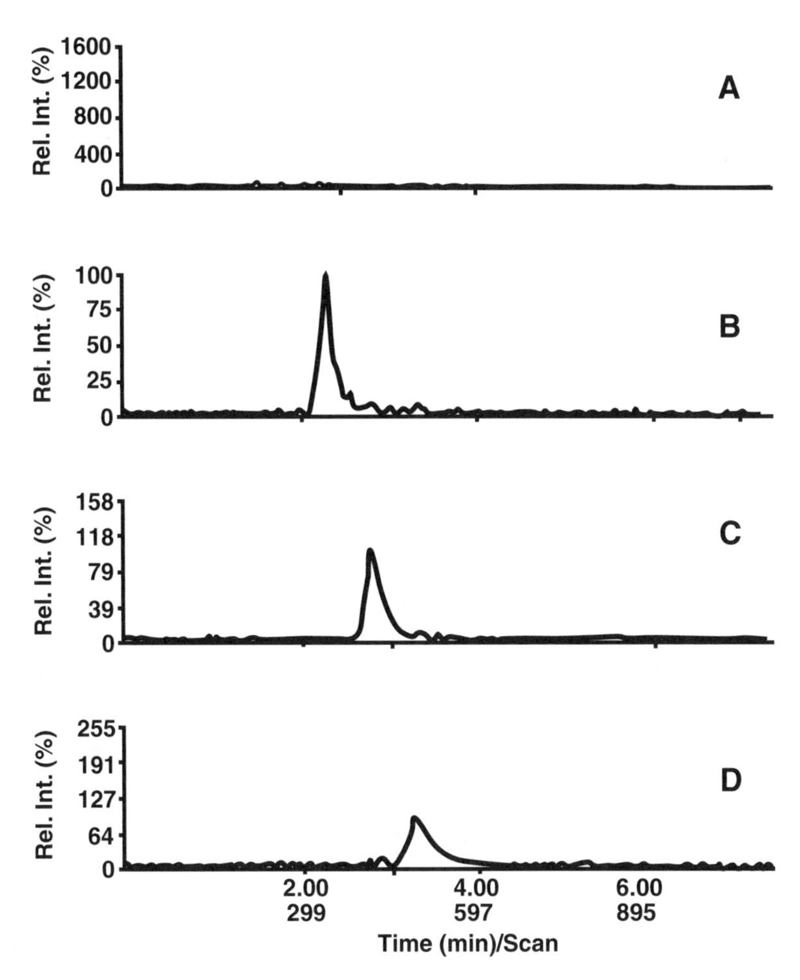

FIGURE 5.65 LC/MS/MS MRM chromatograms for (A) dihydrodaidzein DHD (255/135 transition); (B) daidzein (253/132); (C) genistein (269/133); and (D) O-desmethylangolesin (257/108) of a plasma sample from a non-equol producer who had consumed the soy beverage for 3 days. The ion chromatograms were normalized to the same ion current sensitivity. Reprinted from Coward, L. et al., *Clin. Chim. Acta,* 247, 121–142, 1996. With permission.

644 nM genistein and 289 to 424 nM daidzein. This is about the same mole ratio as the intake. The daidzein metabolite, O-desmethylangolensin, appeared in all samples, while dihydrodaidzein, a precursor of the isoflavone equol, was found in only 25% of the subjects studied (Figure 5.65). The observed genistein concentrations approximated the reported IC_{50} values for the inhibition of epidermal growth factor-induced proliferation of human epithelial cells in tests to evaluate the activities of genistein and genistein-sulfate.[480]

Prostatic Tissue and Fluid

The inhibitory effects of two dietary lignans and six isoflavonoids on 5-α-reductase and 17-β-hydroxysteroid dehydrogenase were studied in genital skin fibroblast monolayers and homogenates, and in prostate tissue. Biochanin A and equol both inhibited 80% of 5-α-reductase activity in skin fibroblasts; however, the biochanin A effect was reversible. All of the compounds inhibited 17-β-hydroxysteroid dehydrogenase activity in the skin fibroblast monolayers.[481] Using the isotope-dilution GC/MS technique described above,[478] lignan and isoflavonoid concentrations were compared in prostatic fluid and plasma from men in Portugal, Hong Kong and England. Enterolactone was very high (>600 ng/mL) in prostatic fluid from Portuguese men. Prostatic fluid and plasma from Hong Kong males both contained considerably higher concentrations (5- to 20-fold differences) of daidzein and equol when compared with those from Portugal and England.[482] A subsequent study confirmed the higher levels of daidzein and equol in the prostatic fluid of Asian men.[483] Because the early molecular events of prostatic cancer are hormone-dependent, dietary factors may influence the pathogenesis of prostate cancer.[484]

Other Body Fluids and Fecal Excretion

The isotope dilution GC/MS technique has been applied to obtain data on several phytoestrogens in human saliva, aspirated breast and cyst fluid, and prostatic fluid, obtained from a variety of sources, for "before" and "after" comparisons on the effects of the ingestion of selected foodstuffs.[485] The similar GC/MS technique previously developed for determining urinary phytoestrogen excretion[398] was modified for the determination of three lignans and four isoflavonoids in their unconjugated forms in human feces. Applications included comparisons of omnivorous and vegetarian women,[486] as well as premenopausal women who were consuming flaxseed powder as a nutritional supplement. The latter subjects exhibited significant increases in their fecal excretion for the lignans enterodiol (32-fold), enterolactone (16-fold), and matairesinol (1.7-fold), while the excretion of the isoflavonoids remained unchanged.[487]

LC/ESI-MS/MS was used to identify a metabolite of genistein in biliary extraction from a rat, as 7-O-β-glucuronide. The single glucuronide group resulted in the characteristic mass increase of 176 Da; however, MS/MS was needed to distinguish the analyte from endogenous biliary constituents appearing at the same mass. (NMR established the site of conjugation to be at the 7-OH group.) There was no evidence of sulfate conjugates in the bile.[488] When re-infused, the genistein 7-O-β-glucuronide was found to be absorbed well from the gut (reviewed in Reference 489).

Eight degradation products of quercetin, quercetin-3-glucoside, and quercetin-3-rhamnoglucoside, produced by *in vitro* fermentation with human fecal flora, was investigated by negative ion LC/ESI-MS/MS. CID fragmentation provided structural information on each metabolite.[490]

Comparative Pharmacokinetics of Soybean Isoflavones in Plasma, Urine, and Feces

After ingesting 60 g of kinako, a baked soybean product, containing 103 μmol daidzein and 112 μmol genistein, isoflavones and lignans were determined by isotope-dilution GC/MS. The main urinary component was daidzein, reaching 2.4 μmol/L in 8 h. Most of the isoflavones (54.7%) were recovered in feces. It was suggested that the high peak concentrations and long half-life (~6 to 8 h) of genistein in plasma may promote interactions with marcomolecules.[491]

5.4.2.3 Metabolism and Antineoplastic Effects

Identification of the Metabolites of Daidzein and Genistein

Urine samples were taken from volunteers at the beginning and end of a 2-week period after including 3 soy bars (~39 g) daily in their normal diet. The metabolites were analyzed as their TMS derivatives by GC/EIMS. At least 11 compounds were found and identified (Figure 5.66). Based on the urinary metabolites, a metabolic pathway was established for daidzein. Only two metabolites were detected for genistein, indicating a difference in the metabolism of these two isoflavones.[492]

Oxidative Metabolites

In one study, enterodiol and enterolactone were administered intraduodenally to bile duct-canulated rats and the 6 h bile was analyzed by GC/MS. Several metabolites involving monohydroxylation of the aromatic rings were positively identified: three in the enterodiol-dosed animals and six in the enterolactone-dosed animals. The mass spectra of several metabolites presumably arising from aliphatic hydroxylation were similar to those previously reported as occurring in microsomal incubation experiments, but the structures could not be identified unequivocally. In other experiments

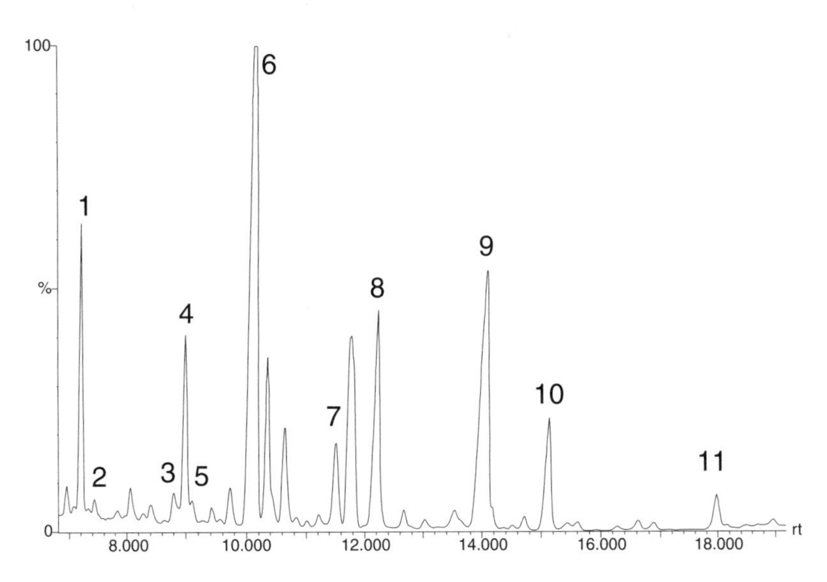

FIGURE 5.66 Total ion current of a urine sample after soy consumption. Compounds isolated and identified are **1**, O-desmethylangolesin (ODMA); **2**, 6′-OH-ODMA; **3**, silylation artifact of dihydrodaidzein, e.g., dehydro-ODMA; **4**, analogous compound of 6′-OH-ODMA tentatively identified as 5′-OH-ODMA; **5**, tentatively identified as 3′-OH-ODMA; **6**, dihydrodaidzein; **7**, dihyrdogenistein; **8**, enterolactone; **9**, daidzein; **10**, genistein; and **11**, glycitein. Reprinted from Heinonen, S. et al., *Anal. Biochem.,* 274, 211–219, 1999. With permission.

in which urine was collected from animals dosed with lignans or fed a flaxseed-containing diet, a number of aromatic and aliphatic metabolites were detected in the urine.[493]

In another study, the oxidative metabolites of the isoflavones daidzein and genistein were generated by cytochrome P-450 activity in rat liver microsomes. The identities of these metabolites were determined by GC/MS and LC/ESIMS. Daidzein was found to be converted into four monohydroxylated, four dihydroxylated, and one trihydroxylated metabolites. The metabolic products of genistein included four monohydroxylated and two dihydroxylated compounds. One of the monohydroxylated metabolites for each of these isoflavones was hydroxylated at the aliphatic position of the C-ring. Apart from this aliphatic OH group it was found that in all other cases the additional OH groups were introduced exclusively into positions *ortho* to existing phenolic hydroxy groups.[494]

Effects of Soy Isoflavones on Urinary Estrogen Metabolism in Pre- and Postmenopausal Women

To test the hypothesis that phytoestrogens exert cancer-preventive effects by modulating estrogen synthesis and metabolism, 10 phytoestrogens and 15 endogenous estrogens and their metabolites were quantified by isotope-dilution GC/MS[398,495] in a randomized, cross-over soy isoflavone study. Increased urinary excretion of isoflavones and lignans was accompanied by decreased excretion of estradiol, estrone, estriol, and total estrogens. In addition, the concentrations of the hypothesized genotoxic estrogenic metabolites 16-α-hydroxyestrone, 4-hydroxyestrone, 4-α-hydroxyestriol,[496] and the genotoxic/total ratio also decreased. These results showed, for the first time, that phytoestrogens may alter estrogen metabolism in the direction of decreased genotoxic and increased inactive metabolites.[497]

A study of phytoestrogens in postmenopausal women consisted of three diet periods during which diets were significantly complemented with soy protein isolates; there were washout periods between the diets. Ten phytoestrogens, 15 endogenous estrogens, and their metabolites were quantified using GC/MS methods similar to those mentioned above. Significant alterations in estrogen metabolism found during the soy-based diet included increased α-(OH)estrone, decreased excretion of 4-(OH)estrone, and decreased ratio of genotoxic to total estrogens.[498]

Effects of Soy and Rye Diets of the Development of Prostate Cancer in Rats

Compared to control populations, fewer and smaller tumors were detected at 14 and 16 weeks on both soy and rye diets. There were significant increases in the urinary excretion of five flavonoids in rats on a soy flour diet, as well as of enterolactone and enterodiol in rats fed rye bran and heat-treated rye bran. None of the diets resulted in differences of testosterone concentrations with respect to the controls.[499]

Inhibition of Cell Proliferation by Genistein

Chromatographically separated fractions of urine from a normal subject on a plant-based diet were assayed for their ability to inhibit the *in vitro* proliferation of vascular endothelial cells and various cancer cells. Several isoflavones were identified by GC/MS in the most potent fractions. Genistein, which was the major isoflavonoid found, inhibited both endothelial cell proliferation, at 5 nmol/L, and *in vitro* angiogenesis, at 150 nmol/L.[500]

Metabolism of Isoflavones in Breast Cancer Cells

To investigate the differences between the sensitivity of normal human mammary epithelial cells (HME line) and transformed human breast cancer cells (MCF-7 line) to growth inhibition by genistein and biochanin A, the cell lines were incubated with ^{14}C-labeled analogs. Metabolites occurring in these cell culture incubates were detected by radio-HPLC, and identified by HPLC-ESIMS (negative ion mode) and NMR. After incubation with the MCF-7 cells, the biochanin A was metabolized to genistein (270 Da) which was, in turn, further degraded to products identical to those obtained when the cells were incubated with genistein itself. One of the HPLC peaks observed for both genistein and biochanin A (designated 2G), detected by scintillation counting, was revealed by ESIMS to consist of two distinct components. The major metabolite had a molecular weight of 350 Da while the minor product was 380 Da (Figure 5.67A and B). Both compounds originated from genistein as indicated by the intensity ratios of the molecular and +2 isotope ions for each of the analytes and the genistein control (Figure 5.67 insert). The major metabolite was 80 Da larger than genistein, corresponding to the uptake of a sulfate group, as verified by CID. NMR confirmed that the sulfate was at the 7 position of genistein, rather than the 4′ or 5′ hydroxyl positions that are also available. The minor metabolite was suggested to be sulfated and hydroxylated genistein. The fact that these metabolites were found in the culture media suggested their immediate release from the cells upon formation. In MCF-7 cell cultures genistein metabolism was correlated with reduced growth inhibition whereas the metabolism of biochanin A was associated with increased growth inhibition presumably on account of its initial metabolism to the bioactive genistein. In contrast, the HME cells did not significantly metabolize either isoflavone.[501]

The same methodology was also used to investigate if the metabolism of genistein and biochanin A determined the level of sensitivity of breast cancer cells to inhibition of their growth by these isoflavones. MCF-7 and T47D cells and the 2- to 4-fold less sensitive ZR-75-1 and BT-20 cells were incubated with the ^{14}C-labeled isoflavones. The metabolites released into the cell culture media were again detected by radio-HPLC and identified by LC/ESIMS (negative ion mode). The major metabolite of genistein in both the T47D and ZR-75-1 lines was genistein-7-sulfate but hydroxylated and methylated metabolites were also produced by the T47D cell line (Figure 5.68a). The metabolic profile of biochanin A was more complex because it led not only to its own direct metabolites, including biochanin A sulfate but also, after demethylation, to genistein and its metabolites (Figure 5.68b). In addition, what was initially assumed to be a considerable quantity of "genistein" derived from biochanin A was shown, by ESIMS, to be biochanin-A-sulfate, $[M\text{-}H]^-$ = 363 Da. The IC_{50} values were only correlated with the occurrence of the hydroxylated and methylated metabolites; however, these metabolites were only produced in the MCF-7 and T47D cells. The ZR-75-1 cells yielded the largest amount of genistein-7-sulfate into the media, much more than the T47D cells, while this metabolite was formed only at low concentration (10%) in BT-20 cells. It was concluded that to overcome intratumor metabolism and Phase II deactivation,

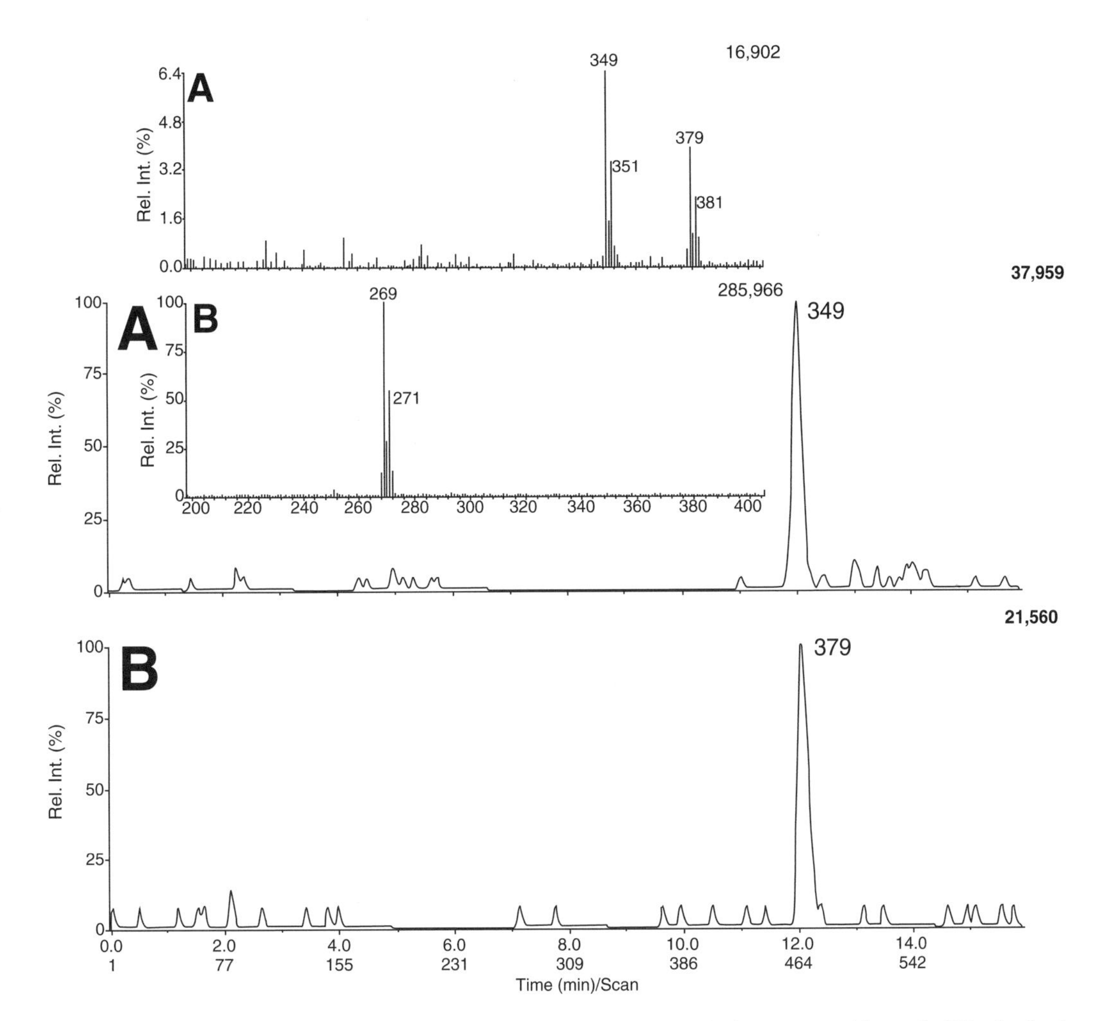

FIGURE 5.67 ESIMS (SIM) analysis of metabolites (from collected chromatographic peak 2G) obtained during the metabolism of genistein by human breast cancer cells (MCF-7). (A) *m/z* 349 and *m/z* 379. Insert: Mass spectra of the peaks identified in selected ion chromatograms show that both metabolites in peak 2G originate from genistein. Note the $[M-H]^-/[M-H+2]^-$ ratios for *m/z* 349 and 379 from peak 2G (A) and *m/z* 269 from $[^{14}C]$genistein (B) are similar. Reprinted from Peterson, T. G. et al., *Carcinogenesis,* 17, 1861–1869, 1996. With permission.

the required dose of genistein, as a drug, would have to be much higher than that which can be provided through dietary sources.[501] In a study of the variations in the growth inhibition of different breast cancer cell lines by genistein and biochanin A, several metabolites were identified by ESIMS, including hydroxylated and methylated form of both isoflavones and a biochanin sulfate.[502]

Isoflavones and Sex Hormone-Binding Globulin Production

One of the many suggested beneficial effects of dietary isoflavonoids is their alleged ability to increase the production of sex hormone-binding globulin (SHBG). This hypothesis was investigated in a study of isoflavonoid metabolism by the human hepatoma cell line HepG2. It was established, using isotope dilution GC/MS, that, in contrast to human urine, the metabolic products were either unconjugated or sulfated compounds. Both daidzein and equol increased SHBG levels in both the intra- and extracellular samples, while treatment with genistein only increased SBGH within the cells, thus resembling the inductive behavior of 17-β-estradiol. It was concluded that the stimulating

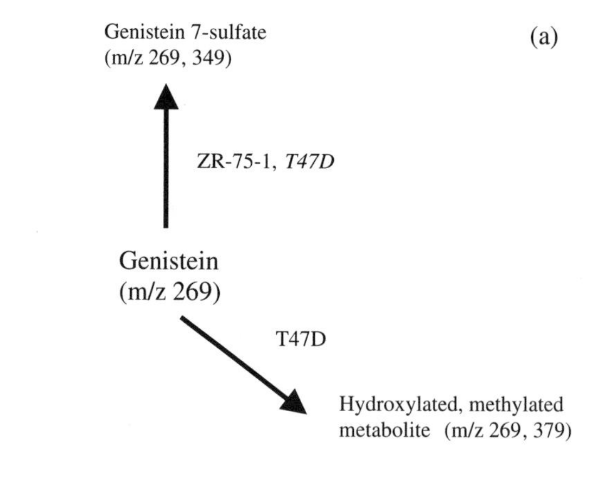

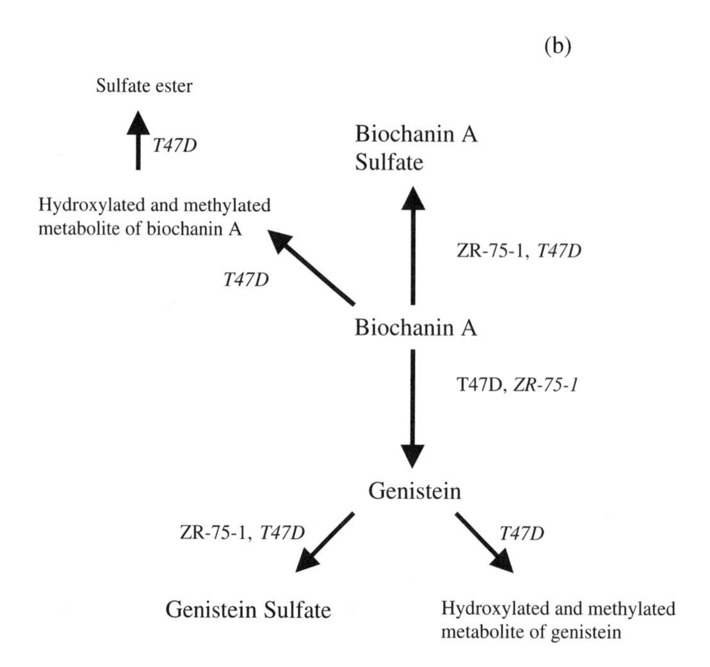

FIGURE 5.68 (a) Summary of the pathways of metabolism of genistein A in human breast cancer lines; and (b) Summary of the pathways of metabolism of biochanin A in human breast cancer cell lines. Cell lines that are italicized indicate cells in which the pathway is minor. Reprinted from Peterson, T. G. et al., *Am. J. Clin. Nutr.,* 68 (Suppl.), 1505S–1511S, 1998. With permission.

effect of the isoflavonoids on SHBG is compound-specific and the regulation of SHBG appears to occur at the posttranslational level.[503]

Fate of Quercetin in Cell Lines

When human hepatocarcinoma cells were incubated with quercetin, a new metabolite was identified by LC/ESIMS: isorhamnetin, an O-methylated metabolite. As evidenced by a 32 Da increase in the mass of the protonated molecule, the initial reaction appeared to involve peroxidation, leading to a dioxetan, with subsequent opening of the C-ring yielding carboxylic acids. The major carboxylic acid was protocatechuic acid. It was postulated that the observed degradative and metabolic changes might contribute to the biological actions of quercetin in cell culture models.[504]

5.4.3 Flavan-3-ols (Catechins)

5.4.3.1 Tea

Flavanols are flavonoids with a *saturated* C2-3 bond and a hydroxyl group at the C3 position. The saturated bond provides (+) and (-) isomerization. The OH groups provide for the possibility of conjugation with gallic acid through ester bond formation. Six biologically active catechins have been identified in green tea: (+)catechin, (-)epicatechin, (-)epigallocatechin, (-)gallocatechin gallate, (-)epicatechin gallate, and (-)epigallocatechin gallate (EGCG) (Figure 5.61). In contrast to green tea, the fermentation of black tea permits polyphenol oxidase, which is activated by the mechanical crushing of tea leaves, to form dark-colored catechin oligomers that are linked by multiple, hydrolysis resistant carbon-carbon crosslinks.

Tea consumption around the world (80% black tea, 20% green tea) ranks second only to water consumption. During the last decade, numerous studies on animal tumor systems, as well as human epidemiological surveys, have concluded that the catechin-type polyphenols present in green tea (also present in black tea but at much lower concentrations) may provide significant protection in apparently all stages of multistage carcinogenesis. The major constituent, EGCG, induces apoptosis and cell cycle arrest, in the G_0-G_1 phase, in several types of human carcinoma cells including prostate.[505] EGCG also prevents angiogenesis.[506]

While both EI and FAB spectra provide useful molecular mass and some structural information, LC/ESIMS is the recommended technique both for the direct analysis of crude tea extracts, including the differentiation of epimers, and for the characterization of synthetic analogues.[507] In one technique, the extraction procedure was designed to simulate conventional brewing conditions by steeping green tea at 80°C for 10 min. C_{18} type HPLC columns were used to resolve six catechins, including diastereomers. SIM of the $[M+H]^+$ ions provided LOD of ~20 ng/component. Using another approach, after crude purification of a tea extract using C_{18} cartridges, catechins were analyzed rapidly (2 to 3 min per sample) by direct flow injection into an ESI source, and monitoring $[M-H]^-$ ions in the negative ion mode. Product ions were also obtained from selected precursor ions using CID. As little as 100 ng of EGCG was detectable in spiked human plasma.[508] It was suggested that this method should be applicable for high-throughput analyses of crude biological extracts.[509]

The metabolism and the time-profiles of (-)epicatechin have been investigated in the urine, plasma, bile, and liver homogenates of rats. In one experiment, extracted urine samples were purified by HPLC. Three major peaks were collected and treated with glucuronidase and sulfatase, rechromatographed, and analyzed by both positive and negative LC/FABMS. One peak was found to be the parent compound, while the other two peaks were the metabolites, 3′-O-methyl-(-)epicatechin and 4′-O-methyl-(-)epicatechin. The occurrence of these metabolites was also confirmed in liver homogenate incubations. In further experiments, two conjugated urinary metabolites were isolated and identified as the 5-O-β-glucuronides of the parent compound and the 3′-O-methyl derivative. When additional tissues were investigated for nonconjugated species, it was found that the free parent (-)epicatechin and the methylated metabolite could be detected in plasma and urine but not in the bile. The distribution of the conjugated forms of the compound was different from that of the free parent with the conjugate of the parent predominating in urine, while conjugates of the methylated metabolites were present in the bile and plasma. Overall, about 8% of the administered drug was excreted in the form of metabolites in 24 h.[510]

5.4.3.2 Wine

The growing interest in the potential anticancer properties of antioxidants in wine, particularly polyphenols, has prompted the development of a GC/EIMS technique for the simultaneous quantification of biologically active nonflavanoids (e.g., hydroxybenzoates), flavonoids, and flavan-3-ols. Wine samples can be kept refrigerated in glass vials up to 5 d but should not be frozen because it results in the precipitation of polyphenols upon thawing. In one approach, sample preparation included extraction with C_8 cartridges and silylation. Identification was based on determining the ratios of the $[M]^+$ and two characteristic fragment ions per analyte (i.s.: 3,3′,4′,7-tetrahydroxyflavone). LOD were established for 15 analytes, including gallic acid 0.05 mg/L, epicatechin 0.32 mg/L, and catechin 0.34 mg/L.[511]

5.4.4 Terpenes and Ginsenosides

5.4.4.1 Metabolism of *d*-Limonene

Formed from two molecules of isopentane, the monoterpene *d*-limonene is a 10-carbon cyclic alkene (Figure 5.69). A major constituent of orange-peel oil, limonene exhibits both a high therapeutic index (>80% regression) in the chemotherapy of chemically induced [by (7,12-dimethylbenz(*a*)anthracene)] rat mammary carcinomas as well as chemopreventive capacity during the initiation and promotive stages in a variety of other chemically induced tumors in rodents. Several Phase I chemoprevention trials with *d*-limonene are in progress. The mechanism of antineoplastic action is the inhibition of the isoprenylation of *ras* and other GTP binding proteins. Limonene is a prodrug because several of its metabolites are more effective inhibitors of protein isoprenylation than is the parent compound.

In a study of the human metabolism of limonene, metabolites were extracted from blood samples of healthy volunteers 4 hr and 24 hr after ingestion of 100 mg/kg d-limonene from orange oil. Analyzed without derivatization, five metabolites were recognized by GC with flame ionization detection and identified by GC/EIMS (Figure 5.70). The formulae and quantities found of these metabolites (i.s.: perillaldehyde) are shown in Figure 5.69. The metabolism was both extensive (>98% of limonene was biotransformed) and rapid, with metabolite concentrations being much higher at 4 h than 24 h after ingestion. The metabolism of limonene in humans parallels that observed in the rat[512] both with respect to the identities of the metabolites and the time course of their formation.[513]

In a study of plasma and urine samples from cancer patients, in a Phase I clinical trial with oral limonene, the presence of the major hydroxylated and carboxylated metabolites in plasma was confirmed by LC/APCIMS. Polar, Phase II metabolites were quantified (i.s.: α-terpene) in both unhydrolyzed and β-glucuronidase hydrolyzed urine using LC/ESIMS. The structures of these urinary metabolites were confirmed by CID as being the glucuronides of the diastereoisomeric isomers of perillic acid, dihydroperillic acid, limonene-8,9-diol, and monohydroxylated limonene (Figure 5.71).[514] Similar methodology was employed to obtain pharmacokinetic data on the parent drug and several metabolites in plasma, and on the glycoconjugates in hydrolyzed urine, from

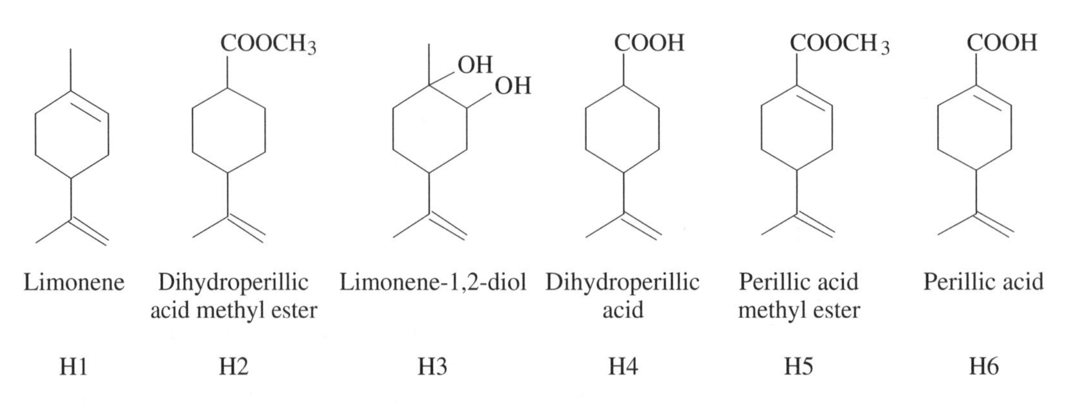

FIGURE 5.69 Chemical structures of limonene and its metabolites. See also Figure 5.70.

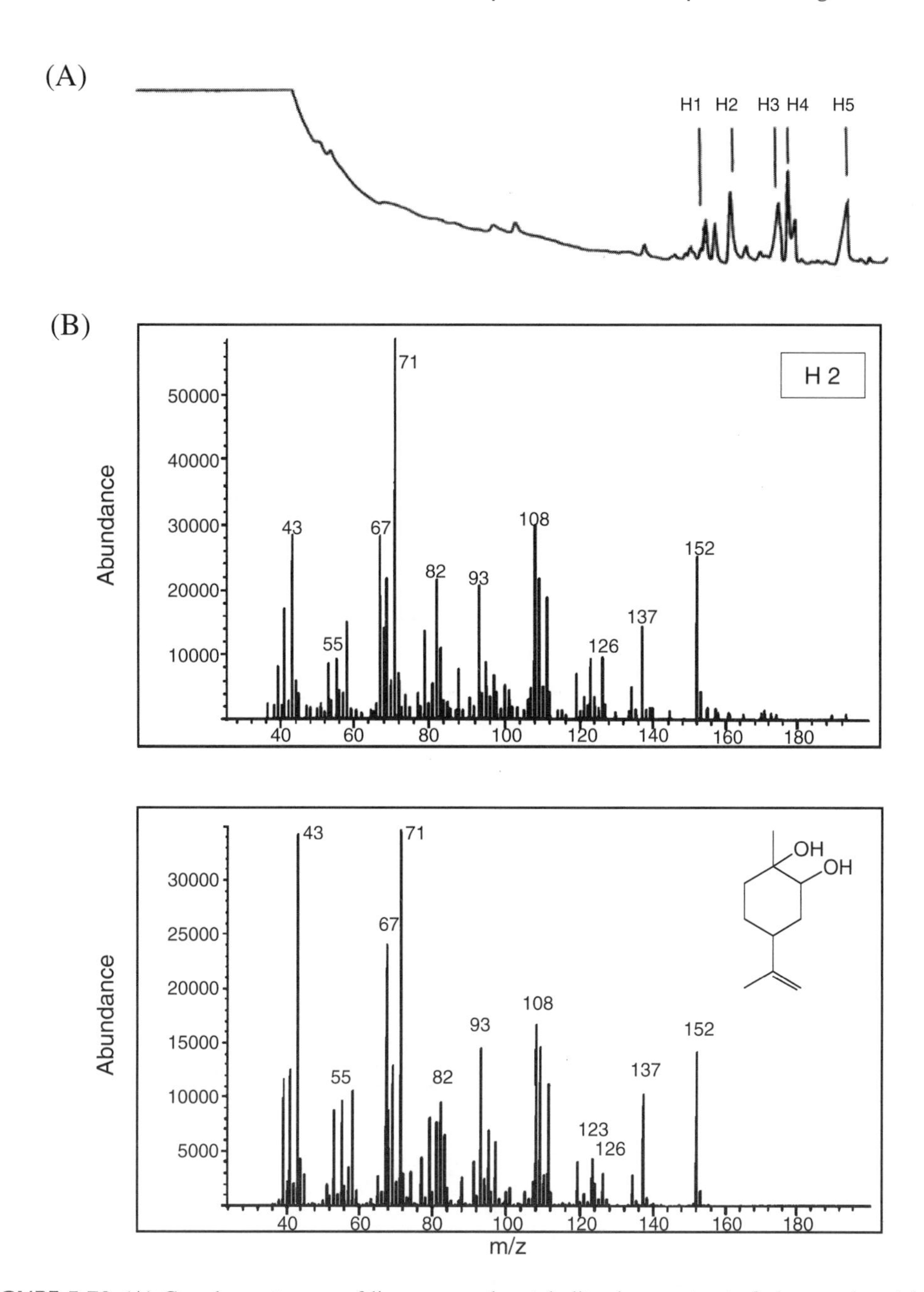

FIGURE 5.70 (**A**) Gas chromatogram of limonene and metabolites in an extract of plasma taken 4 hr after a subject ingested 100 mg/kg *d*-limonene; and (**B**) Identification of metabolite H2 by comparison of its mass spectra with that of an authentic standard of limonene-1,2-diol. See also Figure 5.69. Reprinted from Crowell, P. et al., *Cancer Chemother. Pharmacol.*, 35, 31–37, 1994. With permission.

patients with various solid tumors. It was also determined that the intratumoral concentrations of limonene exceeded corresponding plasma levels.[515]

A previously unrecognized metabolite of limonene, iso-perillic acid, was identified in human plasma by LC/APCIMS. Iso-perillic acid is a structural isomer of perillic acid. The two isomers have the same $[M+H]^+$ as well as the $[M+CH_3OH]^+$ adduct ions. Iso-perillic acid was present in plasma at a concentration of the order of 25% that of perillic acid following the administration of

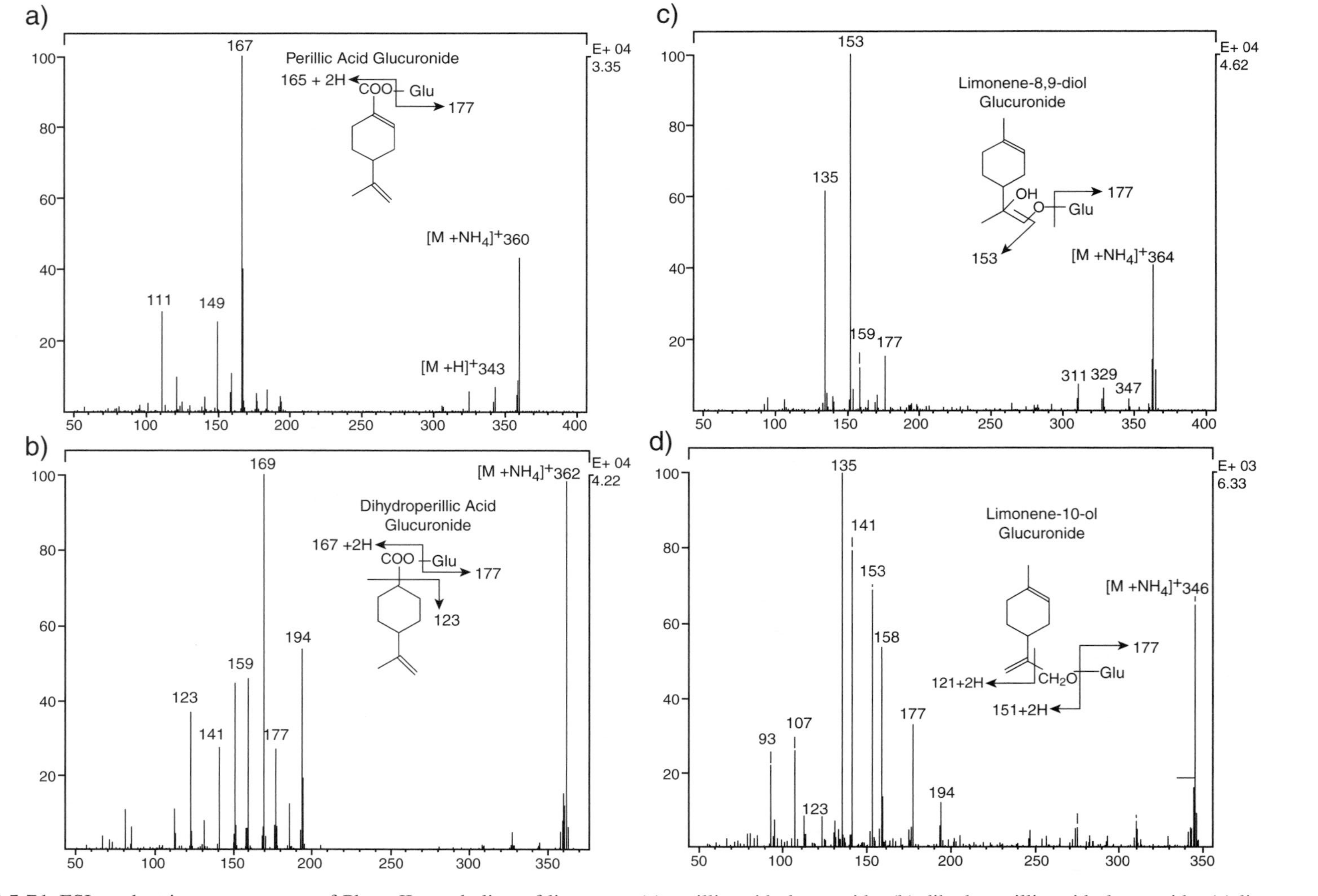

FIGURE 5.71 ESI product ion mass spectra of Phase II metabolites of limonene: (a) perillic acid glucuronide; (b) dihydroperillic acid glucuronide; (c) limonene-8-9-diol glucuronide; (d) limonene-10-ol glucuronide. Reprinted from Poon, G. K. et al., *Drug Metab. Dispos.*, 24, 565–571, 1996. With permission.

several doses of limonene. Iso-perillic acid exhibited similar inhibitory potency against the protein prenylation enzymes farnesyltransferase and geranylgeranyltransferase as perillic acid, thus it may contribute to the *in vivo* antitumor activity of limonene.[516]

5.4.4.2 Ginsenosides

These glycosylated triterpenoids (Figure 5.72), which belong to the class of saponins, may be considered to be biological detergents based on the combination of the hydrophobic aglycones with the hydrophilic sugar moiety.[517] Red ginseng is a traditional Chinese medicine, extracts of which have demonstrated significant cancer chemopreventive activity in several animal tumor models.[518] The structures of complex saponins are usually determined by NMR; however, strategies have been developed for the interpretation of the complex spectra using multiple-stage nano-ESI-ITMS.[519]

Individual ginsenosides in Korean and American ginseng have been detected by LC/ESIMS, as the $[M+Na]^+$ ions. There were significant differences between the relative amounts of identifiable ginsenosides in the ginseng samples from the two countries (Figure 5.72). In addition, three unknown saponins were present only in the Korean ginseng extracts along with one unidentified constituent that was present in both ginsengs.[520] A GC/MS technique has established that there is a linear correlation between ginseng consumed and urinary 20(S)-protopanaxadiol and 20(S)-protopanaxatriol glycoside (i.s.: panaxatriol). Approximately 1% of the dose was recovered over five days.[521] Twenty-five ginsenosides were separated in *Panax ginseng* root. The sugar unit sequences and aglycone moieties were determined by ESI-MS/MS. For example, neutral ginsenosides exhibited a fragmentation pattern corresponding to the successive loss of glycosidic units. Several separated isomers with the same mass spectra were also detected.[522]

5.4.4.3 Carotenoids

The straight chain 40-carbon skeleton carotenoids (tetraterpenoids) include α- and β-carotene (the orange color of carrots), lycopene (the red color of tomatoes), and their hydroxylated derivatives, such as lutein (Figure 5.73). About 10% of the ~600 known carotenoids, including β-carotene, are metabolized enzymatically to vitamin A (retinol) and related compounds, including retinoic acid. The mechanism of the putative cancer chemopreventive action of carotenoids is based on their antioxidant properties that are due to the extensive conjugated double bond systems in the molecules. The use of β-carotene as a dietary supplement has been controversial because of recent clinical results that suggested harmful effects in lung cancer among smokers, possibly due to the pO_2 dependency of antioxidant properties.

In one study, a healthy volunteer swallowed 73 μmol of all *trans*-β-carotene-D_8 to double the endogenous β-carotene concentration in plasma. After solid phase extraction and extensive purification, β-carotene and β-carotene-D_8 were analyzed by GCEI-MS/MS in some 25 blood samples taken during a period of 113 days. The analysis by SRM utilized the 536.4 → 444.4 transition and the corresponding transition for β-carotene-D_8 (Figure 5.74). These transitions represented the loss of C_7H_8 from the molecular ion of the analytes. Quantification revealed that the D_8:H_8-carotene ratio in the plasma reached a maximum at ~7 h, followed by a slow decline over 2 weeks. The serum metabolite, retinol, was quantified by GC/EIMS as its *tert*-butyldimethylsilyl derivative. The retinol-D_4/retinol profile reached a maximum at ~24 h. The technique was adequately stable for pharmacokinetic analysis.[523] In a subsequent study of a plasma from a volunteer ingesting 30 mg of β-carotene enriched with deuterium, the retinol isotopomers were quantified using GC/EI-ITMS, taking advantage of the capability of the ion trap analyzer to provide more accurate isotope ratios than quadrupole analyzers.[524]

Difficulties in data interpretation resulting from the degradation of carotenoids on C_{18} type HPLC columns led to the development of an "engineered" C_{30} type "carotenoid" stationary phase that provided high absolute retention, enhanced shape recognition of structured solutes, and moderate

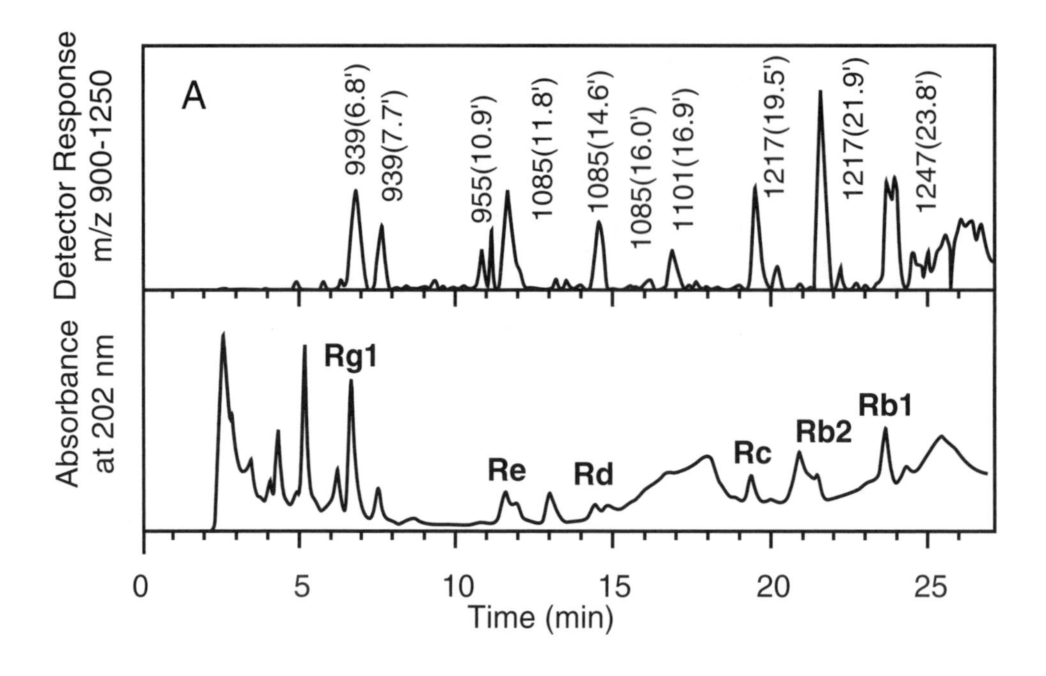

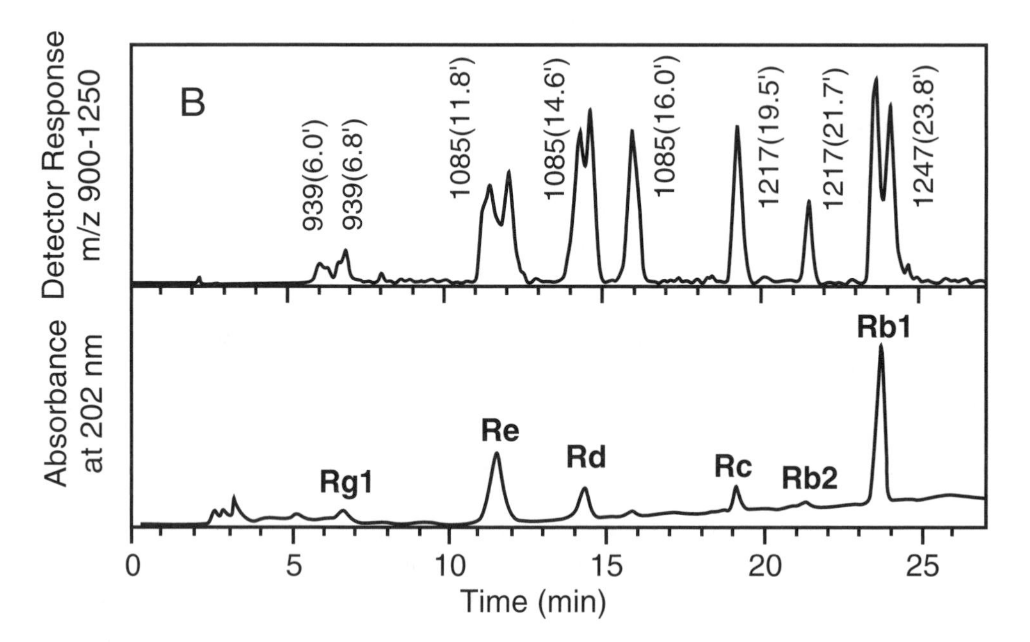

FIGURE 5.72 LC/ESIMS analysis with on-line UV absorbance detection of extracts of (A) Korean ginseng and (B) American ginseng. The upper chromatogram of each set shows the computer-reconstructed mass chromatogram over the range *m/z* 900 to 1250. Each major peak is labeled with the most abundant *m/z* value (corresponding to the $[M+138]^+$ ion), followed by the chromatographic retention time in parenthesis. The detailed structures of the complex triterpane saponins (ginsenosides) Rgl, Re, Rd, etc. are given in Reference 520. Reprinted from Breemen, R. et al., *Anal. Chem.*, 67, 3985–3989, 1995. With permission.

silanol activity.[525] Early MS techniques used with these columns included moving belt, particle beam, CF-FAB, and APCI (reviewed in Reference 526). Subsequently, the *cis* and *trans* isomers were resolved by LC/ESI-MS/MS (Figure 5.75A). LOD were 1 pmol for lutein and 2 pmol for

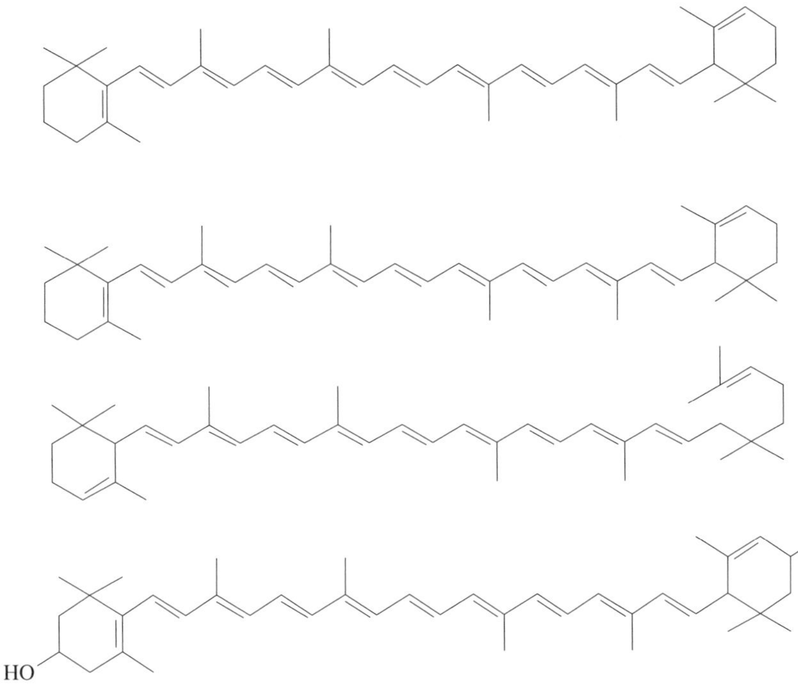

α-Carotene

β-Carotene

Lycopene

Lutein

FIGURE 5.73 Chemical structures of carotene and related compounds.

β-carotene. In an extract from a heat-processed canned potato sample, β-carotene was positively, and lutein and β-cryptoxanthine tentatively, identified (Figure 5.75B and C).[527]

The C_{30} phase has also been employed in the detection and quantification of carotenoids (and tocopherols) by ESIMS. Addition of silver ions, as an $AgClO_4$ solution, resulted in partial oxidation of the analytes to radical cation Ag^+-adducts. This compensated for the lack of sites for protonation or deprotonation in these compounds. These ions were immediately recognizable by using the characteristic (~1:1) $^{107}Ag/^{109}Ag$ isotope ratio. LOD for β-carotene was 300 fmol. Using this method, lycopene and several tocopherols were quantified in tomato and vegetable juices, while tocopherols and high levels of β-carotene were identified in carrot juice. These and other related compounds were also identified in vitamin drinks as well as in infant food.[528]

LC/NCIMS (particle beam interface) was used to identify 34 carotenoids, including 13 geometrical isomers and 8 metabolites, in serum and breast milk from lactating mothers. The presence of carotenoid metabolites in serum has suggested that, in addition to serving as antioxidants, carotenoids may have other immunity- and toxicity-related functions, e.g., stimulation of Phase II enzymes.[529] This reference also includes a useful table of the systematic and trivial names of carotenoids.

Quantification of the geometrical isomers lycopene, α-carotene, and β-carotene in serum has been accomplished with LC/APCIMS using SIM of the separated $[M+H]^+$ ions (i.s.: squalene). Use of a mixture of methanol and acetonitrile as the mobile phase eliminated interferences from serum lipids. The LOD was 3 ng/mL serum.[530]

5.4.5 Sulfur-Containing Compounds

5.4.5.1 Isothiocyanates

The role of the Phase II enzymes in the metabolism of xenobiotics is their detoxification by the conjugation of functionalized metabolic products with endogenous ligands such as glutathione and glucuronic acid. The induction of quinone reductase and glutathione-S-transferase in rodents, fed

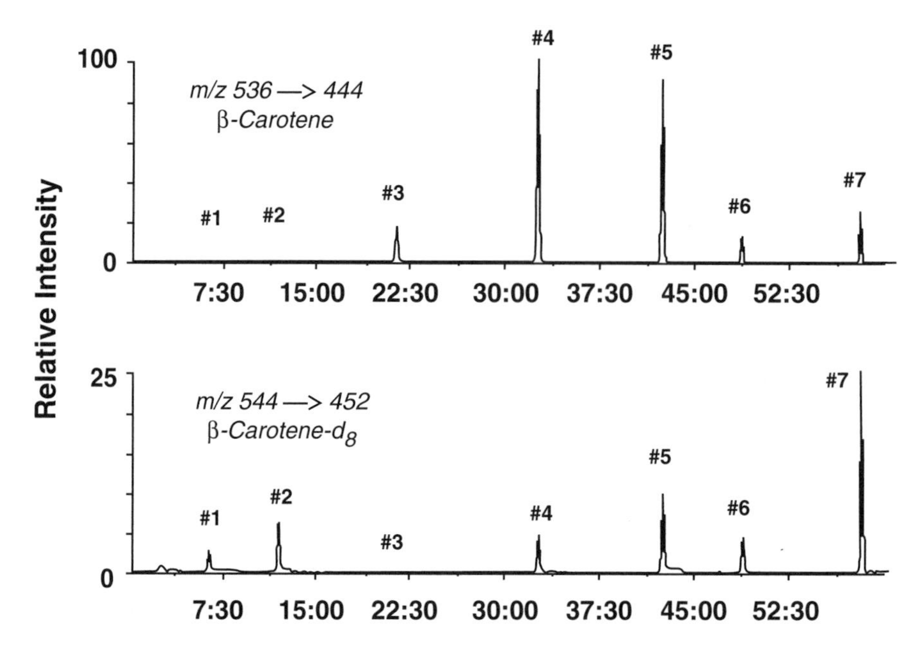

FIGURE 5.74 Ion current profiles measured during selected reaction monitoring of the transititons *m/z* 536.4 → 444.4 (β-carotene) and 544.5 → 452.4 (β-carotene-D_8) for isotopomer standards and analyses of β-carotene in several plasma extracts collected at different times after administration of β-carotene-D_8. The first two injections (#1 and #2) contained only β-carotene-D_8 and showed no peaks in the *m/z* 536.4 → 444.4 channel. The third sample (#3) contained only unlabeled β-carotene and showed no peaks in the *m/z* 544.5 → 452.4 channel. The last four samples (#4 to #7) were extracts of serial plasma samples taken from the volunteer who ingested β-carotene-D_8. All peaks are normalized to peak 4 in the top panel. Variability in the absolute peak height is attributed to variations in the amount of solution loaded onto the direct exposure probe filament. Reprinted from Dueker, S. R. et al., *Anal. Chem.,* 66, 4177–4185, 1994. With permission.

diets of green and yellow vegetables, has been shown to provide protection against chemical carcinogenesis. Isothiocyanates have been identified as inducers of Phase II enzymes. A wide variety of cruciferous vegetables contain glucosinolates, the thioglycosidic conjugates of isothiocyanates. Chewing or other damage to the vegetable cells releases the enzyme myrosinase which, in turn, catalyzes the hydrolysis of glucosinolates to yield the free isothiocyanate. Several of the ~20 known natural and synthetic isothiocyanates have been tested as chemopreventive agents.[531]

After broccoli was shown to exhibit strong Phase II enzyme inducer activity, the most potent component was isolated, biochemically characterized (by monitoring quinone reductase induction in cultured murine hepatoma cells), and identified by NMR, UV, optical rotation, and MS (EI, CI, FAB) as sulforaphane, (-)1-isothiocyanato-(4R)-(methylsulfinyl)butane. The calculated and measured masses for the molecular ion agreed within 0.0005 Da. The fact that a single isothiocyanate may be responsible for most of the Phase II inducer activity of broccoli is important in the search for the active chemopreventive ingredients of other cruciferous vegetables.[532] Fresh (3-day old) broccoli sprouts contained 10 to 100 times more glucoraphanin, the glucosinolate precursor to sulforaphane, than mature broccoli.[533]

Some 35 bifunctional structural analogs of sulforaphane were synthesized and their molecular masses confirmed by high-resolution EIMS.[534] Comparative determination of the potency of these analogs for the induction of quinone reductase revealed that the most potent analogs, assayed with respect to sulforaphane, were those in which the isothiocyanate group was separated from a methylsulfinyl, or an acetyl, group by a three- or four-carbon atom chain.[535]

In a search for urinary metabolites of isothiocyanates, volunteers consumed watercress, which is rich in glucosinolate precursors of isothiocyanates. The N-acetylcysteine conjugate was identified

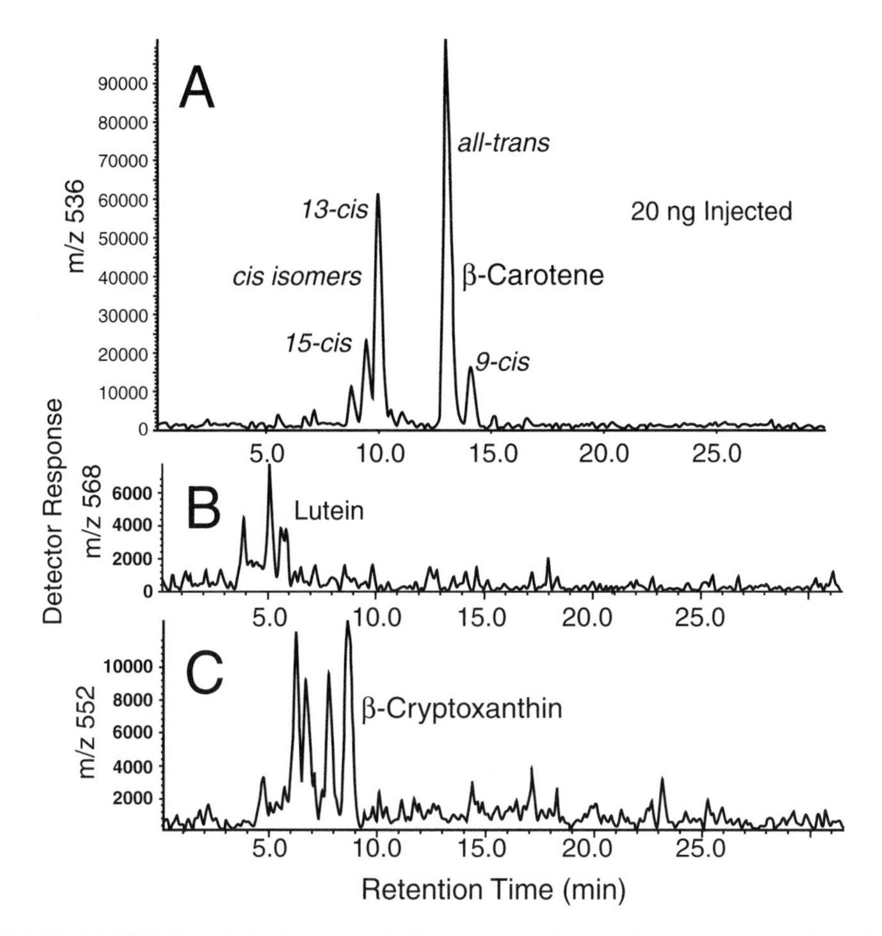

FIGURE 5.75 LC/ESIMS analysis of an extract of heat-processed, canned sweet potatoes using a C_{30} HPLC column with postcolumn addition of heptafluorobutanol. Computer-reconstructed mass chromatograms of: (A) β-carotene molecular ion at *m/z* 536 (~20 ng extract injected); (B) ion at *m/z* 568 corresponding to lutein; and (C) ion at *m/z* 552 showing isomers of β-cryptoxanthine. Reprinted from van Breemen, R. B., *Anal. Chem.*, 67, 2004–2009, 1995. With permission.

as the major metabolite of phenethyl-isothiocyanate by EIMS and NMR. The fact that this metabolite shows dose-dependent excretion suggests its possible use as a marker in epidemiological studies.[536]

5.4.5.2 Garlic

Several investigations have concluded that garlic and its organic allyl sulfur components may provide protection against the initiation of carcinogenesis, but have no efficacy against established tumors.[537] Several thiosulfinates, the predominant flavor principles in garlic, and related compounds have also been identified by GC/MS in onion, leek, scallion, shallot, and chives. Thiosulfinates containing a volatile 2-propenyl (allyl) group could not be analyzed satisfactorily.[538] An alternative approach with LC/TSPMS was also inadequate because of poor chromatographic resolution.[539] Supercritical fluid GC/CIMS identified allicin and showed that it was the predominant thiosulfinate in freshly cut garlic (Figure 5.76). An unidentified peak in a week-old garlic extract was shown to be a decomposition product of the allicin rather than an artifact.[540]

To test the hypothesis that allicin in garlic is an inhibitor of the synthesis of prostaglandins, which are implicated in tumor promotion, endogenous prostaglandin-M (11α-hydroxy-9,15-dioxo-2,3,4,5,20-pentanor-19-carboxy-prostanoic acid), which is an indicator of prostaglandin turnover, was quantified

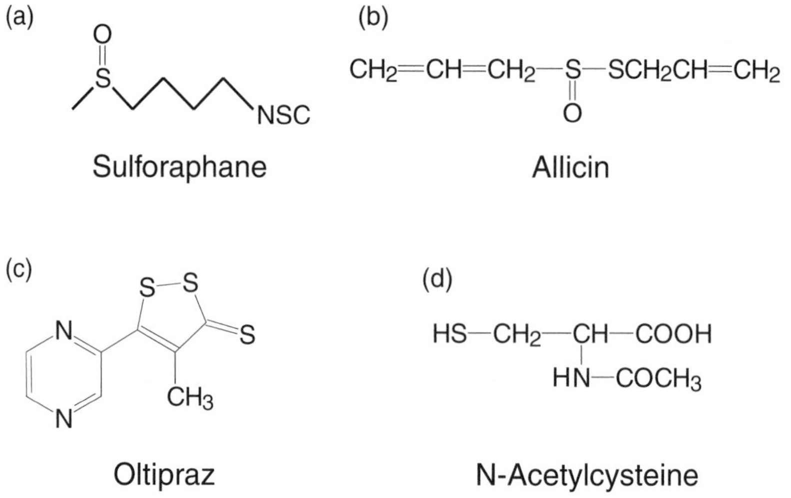

(a) Sulforaphane

(b) $CH_2=CH-CH_2-S(=O)-SCH_2CH=CH_2$

Allicin

(c) Oltipraz

(d) $HS-CH_2-CH(NH-COCH_3)-COOH$

N-Acetylcysteine

FIGURE 5.76 Chemical structures of some sulfur-containing chemopreventives: Sulforaphane, (-)1-isothiocyanato-4(R)-(methylsulfinyl)butane Allicin, 2-propene-1-sulfinothioic acid S-2-propenyl ester. Oltipraz, 5-(2-pyrazinyl)-4-methyl-1,2-dithiole-3-thione.

in subjects who had consumed a garlic extract for 3 mo using GC/MS with SIM of the TMS derivatives. GC/MS studies on garlic have confirmed the presence of allylmercapturic acid (N-acetyl-S-allyl-L-cysteine) in 24-h urine samples from subjects taking garlic extract tablets.[541] This confirmed an earlier finding that garlic consumption was a confounding factor that prevents the use of urinary mercapturic acid as a biomarker of occupational exposure to allyl chloride (3-chloropropane).[542] The potential relevance of garlic in the prevention and treatment of a variety of malignancies has been discussed in a series of publications (*J. Nutr.*, 131, Suppl. 3S, 2001); several of these describe the MS methodologies used.

5.4.5.3 Oltipraz and N-acetylcysteine

Oltipraz is a synthetic dithiolethione (Figure 5.76) that has chemopreventive activity against aflatoxin B_1-induced breast, skin, lung, and bladder cancers in animal models. The proposed mechanism of protection involves a mutagen/carcinogen blocking activity and/or an antioxidant action as a result of the induction of glutathione-S-transferase and other Phase II detoxification enzymes.[543] An LC/MS method for the quantification of oltipraz in serum was based on discharge-assisted TSPMS with SIM of the $[M]^{-}$ ions (i.s.: synthetic 4-(methyl-D_3)-5-(2-pyrazinyl)-1,2-dithiole-3-thione). The LOD was 0.1 ng/mL serum.[544]

In a randomized, placebo-controlled, double-blind Phase II chemoprevention trial, healthy residents in China at high risk of developing hepatocellular carcinoma due to food contaminated with aflatoxins were given a placebo or oral oltipraz at either 125 mg/day or 500 mg/week. It has been assumed that the protective action of the drug is the result of an altered balance between the Phase I hepatic activation of a carcinogen by cytochrome P-450 and its detoxification by glutathione-S-transferase in a Phase II reaction. The molecular masses and structures of the Phase I oxidative metabolite, hydroxylated aflatoxin, and the Phase II metabolite, aflatoxin-B_1-mercapturic acid, were determined by ESIMS and ESI-MS/MS. Compared with those receiving the placebo, the 500 mg/week treatment resulted in a ~50% decrease in the levels of the free hydroxylated metabolite but there was no change in the conjugated metabolite concentrations. The reverse was observed with the 125 mg/day intervention, suggesting that chemoprevention may have occurred by the induction of Phase II enzymes.[545]

The multifaceted chemical properties of the cysteinyl thiol moiety of N-acetylcysteine (NAC, Figure 5.76) include nucleophilicity and redox activity. Clinical chemoprevention trials with NAC have been conducted for a decade.[546] The antioxidant properties of NAC are involved in anticarcinogenic activities both *in vitro* and *in vivo*.[547] NAC is present in plasma and tissues as the free thiol, as well as in various oxidized forms (disulfides) resulting from its reaction with other low-molecular weight thiols, e.g., another NAC, cysteine glutathione, or with high-molecular weight thiols, e.g., proteins. In a GC/MS technique for the quantification of NAC in serum, methyl derivatives were formed with diazomethane. The enantiomers were separated using chiral GC, followed by SIM of selected EI fragments or the protonated molecule obtained by CI (i.s.: N-acetyl-D-cysteine). The LOD for endogenous NAC was 0.5 μmol/L.[548] Another technique utilized a two-step derivatization process for extracted NAC: first, the isopropylic ester of the thiazoline-carboxylic acid was formed with 2-propanol, followed by opening of the thiazole ring with pentafluoropropionic acid. Quantification was accomplished by GC/EIMS with SIM of structurally significant fragments.[549]

The metabolites of NAC include cysteine, cystine, N′N′-diacetylcysteine, and other labile disulfide complexes with plasma and serum proteins. Several of the low molecular weight compounds could be assayed using an LC/APCI-MS/MS technique (i.s.: phenylalanine) with MRM of various precursor-product transitions, e.g., deacetylation for NAC and decarboxylation for L-cysteine and L-cystine. Alternatively, some dimers e.g., N.N′-diacetylcysteine could be determined using LC/ESIMS.[550]

5.4.6 Retinoids, Vitamins, and Selenium

5.4.6.1 Retinoids

Retinoids, including retinol (Vitamin A) and its metabolites, all-*trans*-retinoic acid, 9-*cis*- and 13-*cis*- retinoic acid, as well as synthetic analogs have been used in both the treatment and prevention of cancer. The basis for their antiproliferative and differentiation-inducing action is that they bind to members of the RAR and RXR retinoid receptor families, which are ligand-modulated transcription factors, that regulate the expression of specific genes. There are a number of clinical chemoprevention trials in progress, e.g., a 5-yr Phase III trial on retinol activity against skin cancer.

Earlier LC/MS techniques for the analysis of retinoic acid and its metabolites (listed in Reference 551) required their derivatization to less polar esters. In a subsequent LC/ESIMS technique, four related retinoids were separated by HPLC using a mobile phase to which ammonium acetate and acetic acid had been added to facilitate ion pair-formation. This provided improved separation and promoted deprotonation for enhanced $[M-H]^-$ ion formation in negative ESI. The LOD of retinoic acid was 23 pg. Positive ESI yielded abundant protonated molecules for retinol, while with retinal and retinyl acetate the base peaks corresponded to the elimination of water and acetic acid, respectively. The LOD were in the 0.5 to 10 ng range.[551] Sample preparation for the quantification of all-*trans*-retinoic acid and 13-*cis*-retinoic acid in human plasma consisted of protein precipitation, solid phase extraction, derivatization to form pentafluorobenzyl esters and the collection of relevant fractions after LC separation. The LOD for both compounds were ~0.05 ng/mL plasma using NCIMS and SIM. Endogenous all-*trans*-retinoic acid concentrations were found to be significantly higher in females than in males, ~1.7 ng/mL and ~1.3 ng/mL, respectively, while the 13-*cis*-retinoic acid levels were essentially the same, ~0.9 ng/mL. Although this technique resolved all-*trans*-retinoic acid from 9-*cis*-retinoic acid, the latter being the isomer that binds to the RXR nuclear retinoic acid receptors, no endogenous 9-*cis*-retinoic acid was detected in human samples. The technique was also applicable to the quantification of the 4-oxo-, 4-hydroxy-, and 5,6-epoxyretinoic acid metabolites.[552] Two more metabolites, 9,13-di-*cis*-retinoic acid and 14-hydroxy-4,14-retro-retinol, were identified by GC/CIMS and quantified by HPLC, in the plasma of volunteers who consumed turkey liver. A single meal of this liver contains 10- to 20-fold greater quantities of vitamin A than the level that is known to be teratogenic in animals.[553]

Targretin

Although not a retinoic acid derivative, *targretin*, 4-[1-(3,5,5,8,8-pentamethyl-5,6,7,8-tetrahydro-2-naphthalenic)-1-ethenyl]benzoic acid, is a high-affinity ligand for RXR receptors, with chemopreventive efficacy comparable to that of tamoxifen in carcinogen-induced mammary carcinoma models.[554] Two metabolites of targretin were identified by GC/NCIMS after the carboxylate moiety had been derivatized with perfluorobenzyl bromide. Subsequent derivatization to the methyloxime increased the mass by 29 Da, confirming the presence of the oxo moiety. Additional structural information was obtained from the ESIMS/MS product ion spectra from ^{13}C-D_3-labeled metabolite analogs. This confirmed that the Phase I oxidative metabolism occurred at the C6 and C7 positions. The Phase II metabolism of targretin has been studied by orally dosing rats with an equimolar mixture of unlabeled and ^{13}C-D_3-labeled drugs and analyzing the bile by LC/ESIMS in the negative ion mode. Using an isotope cluster search technique in which all the scans collected were interrogated by the software for doublets separated by 4 Da,[555] 13 clusters with 9 molecular masses were detected simultaneously in a single analysis. The parent drug and both the hydroxy and oxo metabolites were found to yield glucuronide and taurine conjugates. Confirmation of these structures was obtained by MS/MS.[556]

5.4.6.2 Vitamin E

The roles, if any, of vitamins E (α-tocopherol) and C in cancer prevention and therapy are controversial, particularly with respect to their potential as modulators of apoptosis in neoplastic cells.[557,558] Several large-scale randomized intervention trials have been conducted with inconclusive results.[559] Because the antioxidant properties of vitamin E are exerted primarily through the trapping of peroxyl radicals, oxidative products of this compound were quantified in rat liver microsomes at the subpicomole level, using GC/EIMS. The technique involved protein removal, hexane extraction, and the formation of TMS derivatives (i.s.: deuterium-labeled products). SIM was carried out on pairs of ions consisting of the molecular ion and characteristic fragments for each analyte.[560] A GC/EI-ITMS/MS technique for the quantification of vitamin E in plasma also involved protein precipitation, liquid/liquid extraction, and the formation of TMS derivatives (i.s.: 2,2,5,7,8-pentamethyl-6-chromanol), followed by the monitoring of diagnostic product ions formed by CID of the molecular ions. Only ~30 μL of plasma were needed for the determination of endogenous levels in humans, 28 μmol/L. The LOD was 350 fg.[561] The ESIMS technique for carotenoids given in Section 5.5.4.3 may also be employed to analyze tocopherols.[528]

5.4.6.3 Selenium

Selenium deficiency has been linked to several malignancies in both epidemiological and animal model studies,[562] e.g., low selenium status has been suggested as a contributing factor to the risk of developing lung cancer.[563] However, the chemopreventive role of selenium is controversial, particularly with respect to the potential advantages of consuming foods high in selenium, e.g., brasil nuts, as well as food supplements.

Quantification of selenium in whole blood, erythrocytes, and serum is easily carried out by ICPMS with one technique requiring only a 15 to 30-fold sample dilution. The assay had an LOD of 0.02 mmol/L in both serum and red blood cells.[564] The speciation of selenium in urine involves both inorganic selenium species, selenious acid (selenite) and selenic acid (selenate), and organic species such as the selenoamino acids. Selenomethionine appears in plants and is also the main constituent of selenium supplements. Separation and quantification of these compounds has been accomplished by LC/ICPMS. Quantification, based on the major selenium isotope at *m/z* 80 (50%), is problematic because of an interference by $^{40}Ar^{40}Ar$ arising from argon in the plasma source. Of the other 5 selenium isotopes, the ^{78}Se may be the best because of its relatively high abundance (~24%). The LOD of ICPMS with ultrasonic nebulization was in the 0.07 to 0.34 μg/L urine range.[565]

The metabolism of selenium and selenoproteins, and the their roles in cancer prevention have been reviewed.[566] There are three selenium-containing proteins in plasma. Selenomethionine bound to albumin (in a nonspecific manner) is biologically inactive as a source of selenium. Active selenium occurs in the form of selenocysteine that appears in glutathione peroxidase and in selenoprotein P. These selenium-containing proteins have been separated by HPLC with heparin affinity and gel filtration columns used in tandem. Both the *m/z* 77 and 82 peaks had to be monitored by ICPMS to eliminate interferences from $^{40}Ar^{37}Cl$ and $^{81}Br^{1}H$ that originated from the ICP plasma. The distribution of selenium in normal human blood was selenoprotein P 55%, albumin 26%, and glutathione peroxidase 18%. The results also revealed that selenoprotein P preferentially retained selenium among selenium-deficient subjects (human and mouse) and preferentially bonded supplemental selenium in rats.[567]

The retention and distribution of the stable isotopes of selenium consumed as selenite, selenate (^{74}Se), and hydroponically-grown broccoli (^{82}Se) were investigated in healthy young men. Plasma and urine samples were subjected to gel chromatography. The collected fractions were digested with nitric acid and analyzed by ICPMS for total selenium and the quantities of each stable isotope.[568] In addition to providing qualitative and quantitative information about selenium retention after high and low doses, the data revealed retention of selenium from broccoli followed a mechanism that was different from that of inorganic selenium salts.[569]

RECOMMENDED BOOK

Perry, M.C., Ed., *The Chemotherapy Source Book,* 3rd ed., Lippincott, Williams and Wilkins, New York, 2001.

REFERENCES

1. Shapiro, G.I. and Harper, J.W., Anticancer drug targets: cell cycle and checkpoint control, *J. Clin. Invest.*, 104, 1645–1653, 1999.
2. Glaspy, J.A., Hematopoietic growth factors for patients with cancer, *Clin. Oncol. Updates*, 4, 1–13, 2001.
3. Schuchter, L.M., Current role of protective agents in cancer treatment, *Oncology,* 11, 505–517, 1997.
4. Siu, L.L. and Moore, M.J., Use of mesna to prevent ifosfamide-induced urotoxicity, *Support. Care Cancer,* 6, 144–154, 1998.
5. Iyer, L. and Ratain, M.J., Pharmacogenetics and cancer chemotherapy, *Eur. J. Cancer,* 34, 1493–1499, 1998.
6. Destenaves, B. and Thomas, F., New advances in pharmacogenomics, *Curr. Opin. Chem. Biol*, 4, 440–444, 2000.
7. Allegra, C.J., Dihydropyrimidine dehydrogenase activity: prognostic partner of 5-fluorouracil?, *Clin. Cancer Res.*, 5, 1947–1949, 1999.
8. Griffin, T.J., Hall, J.G., Prudent, J.R., and Smith, L.M., Direct genetic analysis by matrix-assisted laser desorption/ionization mass spectrometry, *Proc. Natl. Acad. Sci. USA*, 96, 6301–6306, 1999.
9. Bibby, M.C., Making the most of rodent tumour systems in cancer drug discovery, *British J. Cancer,* 79, 1633–1640, 1999.
10. Gibbs, J.B., Mechanism-based target identification and drug discovery in cancer research, *Science,* 287, 1969–1973, 2000.
11. Kashiwada, Y., Fujioka, T., Mihashi, K., Chen, I.S., Katayama, H., Ikeshiro, Y., and Lee, K.H., Antitumor agents. 180. Chemical studies and cytotoxic evaluation of cumingianosides and cumindysoside A, antileukemic triterpene glucosides with a 14,18-cycloapotirucallane skeleton, *J. Nat. Prod.*, 60, 1105–1114, 1997.
12. Pettit, G.R., Tan, R., Xu, J., Ichihara, Y., Williams, M.D., and Boyd, M.R., Antineoplastic agents. 398. Isolation and structure elucidation of cephalostatins 18 and 19, *J. Nat. Prod.*, 61, 955–958, 1998.
13. Sheu, J.H., Sung, P.J., Cheng, M.C., Liu, H.Y., Fang, L.S., Duh, C.Y., and Chiang, M.Y., Novel cytotoxic diterpenes, excavatolides A-E, isolated from the Formosan gorgonian *Briareum excavatum, J. Nat. Prod.*, 61, 602–608, 1998.
14. Watanabe, M., Kono, T., Koyama, K., Sugimura, T., and Wakabayashi, K., Purification of pierisin, an inducer of apoptosis in human gastric carcinoma cells, from cabbage butterfly, *Pieris rapae, Jpn. J. Cancer Res.,* 89, 556–561, 1998.
15. Rinehart, K.L., Antitumor compounds from tunicates, *Med. Res. Rev.,* 20, 1–27, 2000.

16. Wang, S.K., Cheng, Y.J., and Duh, C.Y., Cytotoxic constituents from leaves of *Aglaia elliptifolia*, *J. Nat. Prod.,* 64, 92–94, 2001.
17. Zhun, N., Raffi, M.M., DiPaola, R.S., Xin, J., Chin, C.K., Badmaev, V., Ghai, G., and Rosen, R.T., Bioactive constituents from gum guggul (*Commiphora wightii*), *Phytochemistry*, 56, 723–727, 2001.
18. Borman, S., Combinatorial chemistry: redefining the scientific method, *C&EN*, 53–65, 2000.
19. Khmelnitsky, Y.L., Budde, C., Arnold, J.M., Usyatinsky, A., Clark, D.S., and Derdick, J.S., Synthesis of water-soluble paclitaxel derivatives by enzymatic acylation, *J. Am. Chem. Soc.*, 119, 11554–11555, 1997.
20. Lam, K.S., Application of combinatorial library methods in cancer research and drug discovery, *Anti-Cancer Drug Design,* 12, 145–167, 1997.
21. Leonard, K.A., Deisseroth, A.B., and Austin, D.J., Combinatorial chemistry in cancer drug development, *The Cancer Journal*, 7, 79–83, 2001.
22. Kyranos, J.N., Cai, H., Wei, D., and Goetzinger, W.K., High-throughput high-performance liquid chromatography/mass spectrometry for modern drug discovery, *Current Opinions Biotech,* 12, 105–111, 2001.
23. Enjalbal, C., Martinez, J., and Aubagnac, J.L., Mass spectrometry in combinatorial chemistry, *Mass Spectrom. Rev.,* 19, 139–161, 2000.
24. Kassel, D.B., Combinatorial chemistry and mass spectrometry in the 21st century drug discovery laboratory, *Chem. Rev.*, 101, 255–267, 2001.
25. Lee, M.S. and Kerns, E.H., LC/MS applications in drug development, *Mass Spectrom. Reviews*, 18, 187–279, 1999.
26. Barnes, C. and Balasubramanian, S., Recent developments in the encoding and deconvolution of combinatorial libraries, *Current Opinions Chem. Biol.*, 4, 346–350, 2000.
27. Liu, H., Felten, Q., Zhang, B., Jedrzejewski, P., Karger, B.L., and Foret, F., Development of multichannel devices with an array of electrospray tips for high-throughput mass spectrometry, *Anal. Chem.*, 72, 3303–3310, 2000.
28. Greig, M., Use of automated HPLC-MS analysis for monitoring and improving the purity of combinatorial libraries, *Am. Lab.*, 31, 28–32, 1999.
29. Blom, K.F., Strategies and data precision requirements for the mass spectrometric determination of structures from combinatorial mixtures. *Anal. Chem*, 69, 4354–4362, 1997.
30. Wang, T., Zeng, L., Strader, T., Burton, L., and Kassel, D.B., A new ultra-high throughput method for characterizing combinatorial libraries incorporating a multiple probe autosampler coupled with flow injection mass spectrometry analysis. *Rapid Commun. Mass Spectrom.*, 12, 1123–1129, 1998.
31. Blom, K.F., Combs, A.P., Rockwell, A.L., Oldenburg, K.R., Zhuang, J., and Chen, T., Direct mass spectrometric determination of bead bound compounds in a combinatorial lead discovery application, *Rapid Commun. Mass Spectrom.*, 12, 1192–1198, 1998.
32. Hsieh, Y., Bryant, M.S., Gruela, G., Brisson, J.M., and Korfmacher, W.A., Direct analysis of plasma samples for drug discovery compounds using mixed-function column liquid chromatography tandem mass spectrometry, *Rapid Commun. Mass Spectrom.*, 14, 1384–1390, 2001.
33. White, R.E., High-throughput screening in drug metabolism and pharmacokinetic support of drug discovery, *Ann. Rev. Pharmacol. Toxicol.*, 40, 133–157, 2000.
34. Wang, Z., Hop, C.E., Leung, K., and Pang, J., Determination of *in vitro* permeability of drug candidates through a caco-2 cell monolayer by liquid chromatography/tandem mass spectrometry, *J. Mass Spectrom.*, 35, 71–76, 2000.
35. Chu, I., Favreau, L., Soares, T., Lin, C.C., and Nomeir, A.A., Validation of higher-throughput high-performance liquid chromatography/atomospheric pressure chemical ionization tandem mass spectrometry assays to conduct cytochrome P 450 CYP2D6 and CYP3A4 enzyme inhibition studies in human liver microsomes, *Rapid Commun. Mass Spectrom.*, 14, 207–214, 2000.
36. Zhang, N., Fountain, S.T., and Rossi, D.T., Quantification and rapid identification in drug discovery using API time-of-flight LS/MS, *Anal. Chem.*, 72, 800–806, 2000.
37. Kaur, S., McGuire, L., Tang, D., Dollinger, G., and Huebner, V., Affinity selection and mass spectrometry-based strategies to identify lead compounds in combinatorial libraries, *J. Protein Chem.,* 16, 505–511, 1997.
38. Wigger, M., Nawrocki, J.P., Watson, C.H., Eyler, J.R., and Benner, S.A., Assessing enzyme substrate specificity using combinatorial libraries and electrospray ionization-Fourier transform ion cyclotron resonance mass spectrometry, *Rapid Commun. Mass Spectrom.,* 11, 1749–1752, 1997.

39. van Breeman, R.B., Huang, C., Nikolic, D., Woodbury, C.P., Zhao, Y., and Venton, D.L., Pulsed ultrafiltration mass spectrometry: a new method for screening combinatorial libraries, *Anal. Chem.*, 69, 2159–2164, 1997.
40. van Breeman, R.B., Nikolic, D., and Bolton, J., Metabolic screening using on-line ultrafiltration mass spectrometry, *Drug Metab. Dispos.*, 26, 85–90, 1998.
41. Ratain, M., Therapeutic relevance of pharmacokinetics and pharmacodynamics, *Sem. Oncol.*, 19 (Suppl. 11), 8–13, 1992.
42. Gurney, H., Dose calculation of anticancer drugs: a review of the current practice and introduction of an alternative, *J. Clin. Oncol.*, 14, 2590–2611, 1996.
43. Masson, E. and Zamboni, W.C., Pharmacokinetic optimisation of cancer chemotherapy, *Clin. Pharmacokin.*, 32, 324–343, 1997.
44. Canal, P., Chatelut, E., and Guichard, S., Practical treatment guide for dose individualisation in cancer chemotheraphy. *Drugs,* 56, 1019–1038, 1998.
45. Minami, H., Beijnen, J.H., Verweij, J., and Ratain, M.J., Limited sampling model for area under the concentration time curve of total topotecan, *Clin. Cancer Res.*, 2, 43–46, 1996.
46. Mick, R., Gupta, E., Vokes, E.E., and Ratain, M.J., Limited-sampling models for irinotecan pharmacokinetics-pharmacodynamics: prediction of biliary index and intestinal toxicity, *J. Clin. Oncol.,* 14, 2012–2019, 1996.
47. Fenselau, C.C. and Morell, S.P., Mass spectrometry in cancer research: 1969–1994, *Adv. Mass Spectrom.,* 13, 71–84, 1995.
48. Ludeman, S.M., The chemistry of the metabolites of cyclophosphamide, *Current Pharm. Design,* 5, 627–643, 1999.
49. Gamcsik, M.P., Dolan, M.E., Andersson, B.S., and Murray, D., Mechanisms of resistance to the toxicity of cyclophosphamide, *Curr. Pharm. Design,* 5, 587–605, 1999.
50. Giorgianni, F., Bridson, P.K., Sorrentino, B.P., Pohl, J., and Blakley, R.L., Inactivation of aldophosphamide by human aldehyde dehydrogenase isozyme 3, *Biochem. Pharmacol.*, 60, 325–338, 2000.
51. Antoine, M., Fabris, D., and Fenselau, C.C., Covalent sequestration of the nitrogen mustard mechlorethamine by metallothionein, *Drug Metab. Dispos.*, 26, 921–926, 1998.
52. Wei, D., Fabris, D., and Fenselau, C.C., Covalent sequestration of phosphoramide mustard by metallothionein—an in vitro study, *Drug Metab. Dispos.,* 27, 786–854, 1999.
53. Jardine, I., Fenselau, C.C., Appler, M., Kan, M., Brundrett, R., and Colvin, M., Quantitation by gas chromatography-chemical ionization mass spectrometry of cyclophosphamide, phosphoramide mustard, and nornitrogen mustard in the plasma and urine of patients receiving cyclophosphamide therapy, *Cancer Res.*, 38, 408–415, 1978.
54. Malet-Martino, M., Gilard, V., and Martino, R., The analysis of cyclophosphamide and its metabolites, *Current Pharm. Design*, 5, 561–586, 1999.
55. Momerency, G., Van Cauwenberghe, K., Slee, P.H., Van Oosterom, A.T., and De Bruijn, E.A., The determination of cyclophosphamide and its metabolites in blood plasma as stable trifluoroacetyl derivatives by electron capture chemical ionization gas chromatography/mass spectrometry, *Biol. Mass Spectrom.,* 23, 149–158, 1994.
56. Hong, P. and Chan, K.K., Analysis of 4-hydroxycyclophosphamide by gas chromatography-mass spectrometry in plasma, *J. Chromatogr.*, 495, 131–138, 1989.
57. Ludeman, S., Shulman-Roskes, E., Wong, K., Han, S., Anderson, L., Strong, J., and Colvin, O.M., Oxime derivatives of the intermediary oncostatic metabolites of cyclophosphamide and ifosfamide: synthesis and deuterium labeling for applications to metabolite quantification, *J. Pharm. Sci.*, 84, 393–398, 1995.
58. Anderson, L., Ludeman, S., Colvin, O., and Grochow, L., Quantitation of 4-hydroxycyclophosphamide/aldophosphamide in whole blood, *J. Chromatogr. B,* 667, 247–257, 1995.
59. Anderson, L.W., Chen, T.-L., Colvin, O.M., Grochow, L.B., Collins, J.M., Kennedy, M.J., and Strong, J.M., Cyclophosphamide and 4-hydroxycyclophosphamide/aldophosphamide kinetics in patients receiving high-dose cyclophosphamide chemotherapy, *Clin. Cancer Res.*, 2, 1481–1487, 1996.
60. Chen, T.L., Kennedy, J., Anderson, L.W., Kiraly, S.B., Black, K.C., Colvin, O.M., and Grochow, L.B., Nonlinear pharmacokinetics of cyclophosphamide and 4-hydroxycyclophosphamide/aldophosphamide in patients with metastatic breast cancer receiving high-dose chemotherapy followed by autologous bone marrow transplantation, *Drug Metab. Dispos.,* 25, 544–551, 1997.

61. Kalhorn, T.F., Ren, S., Howald, W.N., Lawrence, R.F., and Slattery, J.T., Analysis of cyclophosphamide and five metabolites from human plasma using liquid chromatography-mass spectrometry and gas chromatography-nitrogen-phosphorus detection, *J. Chromatogr. B*, 732, 287–298, 1999.
62. Baumann, F., Lorenz, C., Jaehde, U., and Preiss, R., Determination of cyclophosphamide and its metabolites in human plasma by high-performance liquid chromatography-mass spectrometry, *J. Chromatogr. B*, 729, 297–305, 1999.
63. Minoia, C., Turci, R., Sottani, C., Schiavi, A., Perbellini, L., Angeleri, S., Draicchio, F., and Apostoli, P., Application of high performance liquid chromatography/tandem mass spectrometry in the environmental and biological monitoring of health care personnel occupationally exposed to cyclophosphamide and ifosfamide, *Rapid Commun. Mass Spectrom.*, 12, 1485–1493, 1998.
64. Sottani, C., Turci, R., Perbellini, L., and Minoia, C., Liquid-liquid extraction procedure for trace determination of cyclophosphamide in human urine by high-performance liquid chromatography tandem mass spectrometry, *Rapid Commun. Mass Spectrom.*, 12, 1063–1068, 1998.
65. Sannolo, N., Miraglia, N., Biglietto, M., Acampora, A., and Malorni, A., Determination of cyclophosphamide and ifosphamide in urine at trace levels by gas chromatography/tandem mass spectrometry, *J. Mass Spectrom.*, 34, 845–849, 1999.
66. Wainer, I., Granvil, C., Wang, T., and Batist, G., Efficacy and toxicity of ifosamide stereoisomers in an *in vivo* rat mammary carcinoma model, *Cancer Res.*, 54, 4393–4397, 1994.
67. Williams, M.L., Wainer, I.W., Granvil, C.P., Gehrcke, B., Bernstein, M.L., and Durharme, M.P., Pharmacokinetics of (R)- and (S)-cyclophosphamide and their dechloroethylated metabolites in cancer patients, *Chirality*, 11, 301–308, 1999.
68. Sakura, N., Nishimura, S., Fujita, N., Namera, A., Yashiki, M., and Kojima, T., Determination of acrolein in human urine by headspace gas chromatography and mass spectrometry, *J. Chromatogr. B*, 719, 209–212, 1998.
69. Cushnir, J., Naylor, S., Lamb, J., Farmer, P., Brown, N., and Mirkes, P., Identification of phosphoramide mustard/DNA adducts using tandem mass spectrometry, *Rapid Commun. Mass Spectrom.*, 10, 410–415, 1990.
70. Zheng, J.J., Chan, K.K., and Muggia, F., Preclinical pharmacokinetics and stability of isophosphoramide mustard, *Cancer Chemother. Pharmacol.*, 33, 391–398, 1994.
71. Woodland, C., Ito, S., Granvil, C.P., Wainer, I.W., Klein, J., and Koren, G., Evidence of renal metabolism of ifosfamide to nephrotoxic metabolites, *Life Sci.*, 68, 109–117, 2000.
72. Wang, J. and Chan, K.K., Analysis of ifosfamide,4-hydroxyifosfamide, N2-dechloroethylifosfamide, N3-dechloroethylifosfamide and ifosforamide mustard in plasma by gas chromatography mass spectrometry, *J. Chromatogr. B*, 674, 205–217, 1995.
73. Momerency, G., Van Cauwenberghe, K., De Bruijn, E.A., Van Oosterom, A.T., Highley, M.S., and Harper, P.G., Determination of iphoshamide and seven metabolites in blood plasma as stable trifluoroacetyl derivatives, by electron capture chemical ionization GC-MS, *J. High Resol. Chrom.*, 17, 655–661, 1994.
74. Schrijvers, D., Pronk, L., Highley, M., Bruno, R., Locci-Tonelli, D., De Bruijn, E., Van Oosterom, A.T., and Verweij, J., Pharmacokinetics of ifosfamide are changed by combination with docetaxel: results of a phase I pharmacologic study, *Am. J. Clin. Oncol.*, 23, 358–363, 2000.
75. Granville, C., Gehrck, B., Konig, W., and Wainer, I., Determination of the enantiomers of ifosfamide and its 2- and 3-N-dechloroethylated metabolites in plasma and urine using enantioselective gas chromatography with mass spectrometric detection, *J. Chromatogr.*, 622, 21–31, 1993.
76. Highley, M.S., Schrijvers, D., Van Oosterom, A.T., Harper, P.G., Momerency, G., Van Cauwenberghe, K., Maes, R.A., De Bruijn, E.A., and Edelstein, M.B., Activated oxazaphosphorines are transported predominantly by erythrocytes. *Ann. Oncol.*, 8, 1139–1144, 1997.
77. Tattersall, M., Jarman, M., Newlands, E., Holyhead, L., Milstead, R., and Weinberg, A., Pharmacokinetics of melphalan following oral or intravenous administration in patients with malignant disease, *Eur. J. Cancer*, 14, 507–513, 1978.
78. Pallante, S., Fenselau, C.C., Mennel, R., Brundrett, R., Appler, M., Rosensein, N., and Colvin, M., Quantitation by gas chromatography-chemical ionization-mass spectrometry of phenylalanine mustard in plasma of patients, *Cancer Res.*, 40, 2268–2272, 1980.
79. De Boeck, G., Van Cauwenberghe, K., Eggermont, A.M., Van Oosterom, A.T., and De Bruijn, E.A., Determination of melphalan and hydrolysis products in body fluids by GC-MS, *J. High Resol. Chrom.*, 20, 697–700, 1997.

80. Dulik, D. and Fenselau, C.C., Conversion of melphalan to 4-(glutathionyl)phenylalanine, *Drug Metab. Dispos.,* 15, 195–199, 1987.
81. Vahrmeijer, A.L., Snel, C.A.W., Steenvoorden, D.P.T., Beijnen, J.H., Pang, K.S., Schutrups, J., Tirona, R., Keizer, H.J., van Dierendonck, J.H., and van de Velde, C.J.H., Lack of glutathione conjugation of melphalan in the isolated *in situ* liver perfusion in humans, *Cancer Res.,* 56, 4709–4714, 1996.
82. Barnouin, K., Leier, I., Jedlitschky, G., Pourtier-Manzanedo, A., Konig, J., Lehmann, W.D., and Keppler, D., Multidrug resistance protein-mediated transport of chlorambucil and melphalan conjugated to glutathione, *Br. J. Cancer,* 77, 201–209, 1998.
83. Yu, X., Wojciechowski, M., and Fenselau, C.C., Assessment of metals in reconstituted metallothioneins by electrospray mass spectrometry, *Anal. Chem.,* 65, 1355–1359, 1993.
84. Yu, X., Wu, Z., and Fenselau, C.C., Covalent sequestration of melphalan by metallothionein and selective alkylation of cysteines, *Biochemistry,* 34, 3377–3385, 1995.
85. Chang, S., Larcom, B., Alberts, D., Larsen, B., Walson, P., and Sipes, I., Mass spectrometry of chlorambucil, its degradation products, and its metabolite in biological samples, *J. Pharm. Sci.,* 69, 80–84, 1980.
86. Ehrsson, H., Eksbork, S., Wallin, I., Marde, Y., and Joansson, B., Determination of chlorambucil in plasma by GLC with selected ion monitoring, *J. Pharm. Sci.,* 69, 710–712, 1980.
87. Ehrsson, H., Wallin, I., Simonsson, B., Hartvig, P., and Oberg, G., Effect of food on pharmacokinetics of chlorambucil and it main metabolite, phenylacetic acid mustard, *Eur. J. Clin. Pharmacol.,* 27, 111–114, 1984.
88. Jakhammer, T., Olsson, A., and Svensson, L., Mass fragmentographic determination of prednimustine and chlorambucil in plasma, *Acta Pharm. Suec.,* 14, 485–496, 1977.
89. Bastholt, L., Johansson, C., Pfeiffer, P., Svensson, L., Johansson, S., Gunnarsson, P., and Mouridsen, H., A pharmacokinetic study of prednimustine as compared with prednisolone plus chlorambucil in cancer patients, *Cancer Chemother. Pharmacol.,* 28, 205–210, 1991.
90. Davies, I.D., Allanson, J.P., and Causon, R.C., Rapid determination of the anti-cancer drug chlorambucil (Leukeran) and its phenyl acetic acid mustard metabolite in human serum and plasma by automated solid-phase extraction and liquid chromatography-tandem mass spectrometry, *J. Chromatogr. B,* 732, 173–184, 1999.
91. Zaia, J., Jiang, L., Han, M.S., Tabb, J.R., Wu, Z., Fabris, D., and Fenselau, C.C., A binding site for chlorambucil on metallothionein, *Biochemistry,* 35, 2830–2835, 1997.
92. Pyatigorskaya, T., Zhilkova, O., Shelkovsky, V., Arkhangelova, N., Grizodub, A., and Sukhodub, L., Hydrolysis of 1,1′,1″-phosphinothioylidinetrisaziridine (thiotepa) in aqueous solution, *Biomed. Environ. Mass Spectrom.,* 14, 143–148, 1987.
93. Cohen, B., Egorin, M., Nayar, M., and Gutierrez, P., Effects of pH and temperature on the stability and decomposition of N,N′,N″-triethylenethiophosphoramide in urine and buffer, *Cancer Res.,* 44, 4312–4316, 1984.
94. van Maanen, R.J., van Ooijen, R.D., Zwikker, J.W., Huitema, A.D.R., Rodenhuis, S., and Beijnen, J.H., Determination of N,N′,N″-triethylenethiophosphoramide and its active metabolite N,N′,N″-triethylenephosphoramide in plasma and urine using capillary gas chromatography, *J. Chromatogr. B,* 719, 103–112, 1998.
95. van Maanen, M.J., Tijhof, I.M., Damen, J.M., Versluis, C., van den Bosch, J.J., Heck, A.J., Rodenhuis, S., and Beijnen, J.H., A search for new metabolites of N,N′,N″-triethylenethiophosphoramide, *Cancer Res.,* 59, 4720–4724, 1999.
96. van Maanen, M.J. and Beijnen, J.H., Liquid chromatographic-mass spectrometric determination of the novel, recently identified thioTEPA metabolite, thioTEPA- mercapturate, in urine, *J. Chromatogr. B,* 732, 73–79, 1999.
97. Sukhodub, L., Shelkovsky, V., Kosevich, M., Pyatigorskaya, T., and Zhilkova, O., Nucleic acid base complexes with thiotepa as revealed by field ionization mass spectrometry, *Biomed. Mass Spectrom.,* 13, 167–170, 1986.
98. Andrievsky, G., Sukhodub, L., Pyatigorskaya, T., Boryak, O., Limnaskaya, O., and Shelkovsky, S., Direct observation of the alkylation products of deoxyguanosine and DNA by fast atom bombardment mass spectrometry, *Biol. Mass Spectrom.,* 20, 665–668, 1991.
99. Musser, S. and Callery, P., Supercritical fluid chromatography/chemical ionization/mass spectrometry of some anticancer drugs in a thermospray ion source, *Biomed. Environ. Mass Spectrom.,* 19, 348–352, 1990.

100. Dirven, H., Dictus, E., Broeders, N., Van-Ommen, B., and van-Bladeren, P., The role of human glutathione S-tansferase isoenzymes in the formation of glutathione conjugates of the alkylating cytostatic drug thiotepa, *Cancer Res.*, 55, 1701–1706, 1995.
101. Srivastava, S.K., Singhal, S.S., Hu, X., Awasthi, Y.C., Zimniak, P., and Singh, S.V., Differential catalytic efficiency of allelic variants of human glutathione S-transferase glutathione conjugation of thiotepa, *Arch. Biochem. Biophys.*, 366, 89–94, 1999.
102. Bakhtiar, R., *In vitro* exposure of human hemoglobin to the antineoplastic drug thiotepa, *Rapid Commun. Mass Spectrom.*, 14, 534–537, 2000.
103. Hassan, M. and Ehrsson, H., Urinary metabolites of busulfan in the rat, *Drug Metab. Dispos.*, 15, 399–402, 1987.
104. Ritter, C.A., Bohnenstengel, F., Hofmann, U., Kroemer, H.K., and Sperker, B., Determination of tetrahydrothiophene formation as a probe of *in vitro* busulfan metabolism by human glutathione S-transferase AI-I: use of a highly sensitive gas chromatographic-mass spectrometric method, *J. Chromatogr. B*, 730, 25–31, 1999.
105. Ehrsson, H. and Hassan, M., Determination of busulfan in plasma by GC-MS with selected-ion monitoring, *J. Pharm. Sci.*, 72, 1203–1205, 1983.
106. Vassal, G., Re, M., and Gouyette, A., Gas chromatographic-mass spectrometric assay for busulfan in biological fluids using a deuterated internal standard, *J. Chromatogr.*, 428, 357–361, 1988.
107. Vassal, G., Deroussent, A., Hartmann, O., Challine, D., Benhamou, E., Valteau Couanet, D., Brugieres, L., Kalifa, C., Gouyette, A., and Lemerle, J., Dose-dependent neurotoxicity of high-dose busulfan in children: a clinical and pharmacological study, *Cancer Res.*, 50, 6203–6207, 1990.
108. Pichini, S., Altieri, I., Bascosi, A., Di Carlo, S., and Zuccaro, P., High performance liquid chromatographic-mass spectrometric assay of busulfan in serum and cerebrospinal fluid, *J. Chromatogr.*, 581, 143–146, 1992.
109. Quernin, M.-H., Poonkuzhali, B., Montes, C., Krishnamoorthy, R., Dennison, D., Srivastava, A., Vilmer, E., Chandy, M., and Jacqz-Aigrain, E., Quantification of bisulfan in plasma by gas chromatography-mass spectrometry following derivatization with tetrafluorothiophenol, *J. Chromatogr. B*, 709, 47–56, 1998.
110. Vassal, G., Deroussent, A., Challine, D., Hartmann, O., Koscielny, S., Valteau Couanet, D., Lemerle, J., and Gouyette, A., Is 600 mg/m^2 the appropriate dosage of busulfan in children undergoing bone marrow transplantation?, *Blood*, 79, 2475–2479, 1992.
111. Vassal, G., Fischer, A., Challine, D., Boland, I., Ledheist, F., Lemerle, S., Vilmer, E., Rahimy, C., Souillet, G., Gluckman, E. et al., Busulfan disposition below the age of three: alteration in children with lysosomal storage disease, *Blood*, 82, 1030–1034, 1993.
112. Vassal, G., Koscielny, S., Challine, D., Valteau-Couanet, D., Boland, I., Deroussent, A., Lemerle, J., Gouyette, A., and Hartman, O., Busulfan disposition and hepatic veno-occlusive disease in children undergoing bone marrow transplantation, *Cancer Chemother. Pharmacol.*, 37, 247–253, 1996.
113. Vassal, G., Challine, D., Koscielny, S., Hartmann, O., Deroussent, A., Boland, I., Valteau Couanet, D., Lemerle, J., Levi, F., and Gouyette, A., Chronopharmacology of high-dose busulfan in children, *Cancer Res.*, 53, 1534–1537, 1993.
114. Reed, D., May, H., Boose, R., Gregory, K., and Beilstein, M., 2-Chloroethanol formation as evidence for a 2-chloroethyl alkylating intermediate during chemical degradation of 1-(2-chloroethyl)-3-cyclohexyl-1-nitrosourea and 1-(2-chloroethyl)-3-(trans-4-methylcylohexyl)-1-nitrosourea, *Cancer Res.*, 35, 568–576, 1975.
115. Hill, D., Kirk, M., and Struck, R., Microsomal metabolism of nitrosoureas, *Cancer Res.*, 35, 296–301, 1975.
116. Weinkam, R., Wen, J., Furst, D., and Levin, V., Analysis for 1,3-bis(2-chloroethyl)-1-nitrosourea by chemical ionization mass spectrometry, *Clin. Chem.*, 24, 45–49, 1978.
117. Levin, V., Hoffman, W., and Weinkam, R., Pharmacokinetics of BCNU in man: a preliminary study of 20 patients, *Cancer Treatment Report*, 62, 1305–1312, 1978.
118. Weinkam, R. and Liu, T., Quantitation of lipophilic chloroethyl-nitrosourea cancer chemotherapeutic agents, *J. Pharm. Sci.*, 71, 153–157, 1982.
119. Smith, R., Blackstock, S., Cheung, L., and Loo, T., Analysis for nitrosourea antitumor agents by gas chromatography-mass spectrometry, *Anal. Chem.*, 53, 1205–1208, 1981.

120. Brakenhoff, J.P., Commandeur, J.N., de Kanter, F.J., van Baar, B.L., Luijten, W.C., and Vermeulen, N.P., Chemical and glutathione conjugation-related degradation of fotemustine: formation and characterization of a glutathione conjugate of diethyl (1-isocyanatoethyl)phosphonate, a reactive metabolite of fotemustine, *Chem. Res. Toxicol.*, 7, 380–389, 1994.
121. Hayes, M.T., Bartley, J., Parsons, P.G., Eaglesham, G.K., and Prakash, A.S., Mechanism of action of fotemustine, a new chloroethylnitrosourea anticancer agent: evidence for the formation of two DNA-reactive intermediates contributing to cytotoxicity, *Biochemistry,* 36, 10646–10654, 1997.
122. Carbone, V., Salzano, A., Pucci, P., Fiume, I., Pocsfalvi, G., Sannolo, N., DiLanda, G., and Malorini, A., *In vitro* reactivity of antineoplastic drugs carmustin and acrolein with model peptides, *J. Peptide Res.*, 49, 586–593, 1997.
123. Shetty, B., Schowen, R., Slavik, M., and Riley, C., Degradation of carbazine in aqueous solution, *J. Pharm. Biomed. Anal.*, 10, 675–683, 1992.
124. Perissin, L., Facchin, P., Bedini, A., Diamantini, G., Tontini, A., and Tarzia, G., Antimetastatic action of a new analog of dacarbazine in mice bearing Lewis lung carcinoma, *Anticancer Res.*, 20, 1513–1517, 2000.
125. Gorsen, R., Weiss, A., and Menthei, R., Analysis of procarbazine and metabolites by gas chromatography-mass spectrometry, *J. Chromatogr.*, 221, 309–318, 1980.
126. Lebwohl, D. and Canetta, R., Clinical development of platinum complexes in cancer therapy: an historical perpsective and an update, *Eur. J. Cancer,* 34, 1522–1534, 1998.
127. Kelland, L.R., An update on satraplatin: the first orally available platinum anticancer drug, *Expert Opin. Invest. Drugs,* 9, 1373–1382, 2000.
128. Claereboudt, J., De Spiegeleer, B., Lippert, B., De Bruijn, E., and Claeys, M., Fast atom bombardment and tandem mass spectrometry for the structural characterization of cisplatin analogs and bis-nucleobase adducts with cisplatin, *Spectrosc. Int. J.*, 7, 91–112, 1989.
129. Puzo, G., Prome, J., Macquet, J., and Lewis, I., Fast atom bombardment mass spectra of the bis-guanosine adduct with cisplatin, *Biomed. Mass Spectrom.*, 9, 552–556, 1982.
130. Marchan, V., Moreno, V., Pedroso, E., and Grandas, A., Towards a better understanding of the cisplatin mode of action, *Chem. Euro. J.*, 7, 808–815, 2001.
131. Cowens, J., Steview, F., Alderfer, J., Hansen, G., Pendyala, L., and Creaven, P., Synthesis and identification of the derivatives of a platinum containing complex, *J. Mass Spectrom Ion Phys.*, 48, 177–180, 1983.
132. Siegel, M., Bitha, P., Child, R., Hlavka, J., Lin, Y., and Chang, T., Fast atom bombardment mass spectrometry of cisplatin analogs, *Biomed. Mass Spectrom.*, 13, 25–32, 1986.
133. DAlietos, D., Furst, A., Dimitiros, T., and Lee, T., Fast atom bombardment, field desorption, and desorption chemical ionization mass spectrometry of cis-dichloroplatinum amino acid and dipeptide complexes, *Int. J. Mass Spectrom. Ion Proc.*, 61, 141–148, 1984.
134. Poon, G.K., Mistry, P., and Lewis, S., Electrospray ionization mass spectrometry of platinum anticancer agents, *Biol. Mass Spectrom.*, 20, 687–692, 1991.
135. Schaaff, T.G., Qu, Y., Farrell, N., and Wysocki, V.H., Investigation of the *trans* effect in the fragmentation of dinuclear platinum complexes by electrospray ionization surface-induced dissociation tandem mass spectrometry, *J. Mass Spectrom., 33,* 436–443, 1998.
136. Verbueken, A., Van Grieken, R., Paulus, G., Verpooten, G., and De Broe, M., Laser microbe mass spectrometry of platinum in dog kidney after cisplatin administration, *Biomed. Mass Spectrom.*, 11, 159–163, 1984.
137. Minami, T., Ichii, M., and Okazaki, Y., Detection of platinum in the brain of mice treated with cisplatin and subjected to short-term hypoxia, *J. Pharm. Pharmacol.*, 48, 505–509, 1996.
138. Minami, T., Hashii, K., Tateyama, I., Kadota, E., Tohno, Y., Yohno, S., Utsumi, M., Yamada, M., Ichii, M., and Namikawa, K., Acccumulation of platinum in the intervertebral discs and vertebrae of ovarian tumor-bearing patients treated with cisplatin, *Biol. Trace Elem. Res.*, 42, 253–257, 1994.
139. Ehrsson, H., Wallin, I., and Andersson, A., Cisplatin, transplatin, and their hydrated complexes: separation and identification using porous graphitic carbon and electrospray ionization mass spectrometry, *Anal. Chem.*, 67, 3608–3611, 1995.
140. Aggarwal, S., Gemma, N., Kinter, M., Nicholson, J., Shipe, J., and Herold, D., Determination of platinum in urine, ultrafiltrate, and whole plasma by isotope dilution gas chromatography-mass spectrometry compared to electrochemical atomic absorption spectrometry, *Anal. Biochem.*, 210, 113–118, 1993.

141. Aggarwal, S. and Kinter, M., Isotope dilution gas chromatography/mass spectrometry for platinum determination in urine, *J. Am. Soc. Mass Spectrom.*, 2, 85–90, 1991.
142. Minami, T., Ichii, M., and Okazaki, Y., Comparison of three different methods for measurement of tissue platinum level, *Biol. Trace Elem. Res.*, 48, 37–44, 1995.
143. Pyrzynska, K., Recent advances in solid-phase extraction of platinum and palladium, *Talanta*, 47, 841–848, 1998.
144. Johnsson, A., Bjork, H., Schultz, A., and Skarby, T., Sample handling for determination of free platinum in blood after cisplatin exposure, *Cancer Chemother. Pharmacol.*, 41, 248–251, 1998.
145. Allain, P., Berre, S., Mauras, Y., and LeBouil, A., Evaluation of inductively coupled mass spectrometry for the determination of platinum in plasma, *Biol. Mass Spectrom.*, 21, 141–143, 1992.
146. Cairns, W.R.L., McLeod, C.W., and Hancock, B., Determination of platinum in human serum by flow injection inductively coupled plasma-mass spectrometry, *Spectroscopy*, 12, 16–21, 1997.
147. Gamelin, E., Allain, P., Maillart, P., Turcant, A., Delva, R., Lortolary, A., and Larra, F., Long-term pharmacokinetic behavior of platinum after cisplatin administration, *Cancer Chemother. Pharmacol.*, 37, 97–102, 1995.
148. Tothill, P., Klys, H.S., Matheson, L.M., McKay, K., and Smyth, J.F., The long-term retention of platinum in human tissues following the administration of cisplatin or carboplatin for cancer chemotherapy, *Eur. J. Cancer*, 28A, 1358–1361, 1992.
149. Morrison, J.G., White, P., McDougall, S., Firth, J.W., Woolfrey, S.G., Graham, M.A., and Greenslade, D., Validation of a highly sensitive ICP-MS method for the determination of platinum in biofluids: application to clinical pharmacokinetic studies with oxaliplatin, *J. Pharm. Biomed. Anal.*, 24, 1–10, 2000.
150. Gamelin, E., Bouil, A.L., Boisdron Celle, M., Turcant, A., Delva, R., Cailleux, A., Krikorian, A., Brienza, S., Cvitkovic, E., Robert, J., Larra, F., and Allain, P., Cumulative pharmacokinetic study of oxaliplatin, administered every three weeks, combined with 5-fluorouracil in colorectal cancer patients, *Clin. Cancer Res.*, 3, 891–899, 1997.
151. Massari, C., Brienza, S., Rotarski, M., Gastiaburu, J., Misset, J.L., Cupissol, D., Alafaci, E., Dutertre-Catella, H., and Bastian, G., Pharmacokinetics of oxaliplatin in patients with normal versus impaired renal function, *Cancer Chemother. Pharmacol.*, 45, 157–164, 2000.
152. Levi, F., Metzger, G., Massari, C., and Milano, G., Oxaliplatin: pharmacokinetics and chronopharmacological aspects, *Clin. Pharmacokin.*, 38, 1–21, 2000.
153. Sessa, C., Minoia, C., Ronchi, A., Zucchetti, M., Bauer, J., Borner, M., de Jong, J., Pagani, O., Renard, J., Weil, C., and D'Incalci, M., Phase 1 clinical and pharmacokinetic study of the oral platinum analogue JM216 given daily for 14 days, *Annals Oncol.*, 9, 1315–1322, 1998.
154. Galettis, P., Carr, J.L., Paxton, J.W., and McKeage, M.J., Quantitative determination of platinum complexes in human plasma generated from the oral antitumour drug JM216 using directly coupled high-performance liquid chromatography-inductively coupled plasma mass spectrometry without desolvation, *J. Anal. Atomic Spectrom.*, 14, 953–956, 1999.
155. Hanada, T., Isobe, H., Saito, T., Ogura, S., Takekawa, H., Yamazaki, K., Tokuchi, Y., and Kawakami, Y., Intracellular accumulation of thallium as a marker of cisplatin cytotoxicity in nonsmall cell lung carcinoma, *Cancer*, 83, 930–935, 1998.
156. Luo, F.R., Yen, T.Y., Wyrick, S.D., and Chaney, S.G., High-performance liquid chromatographic separation of the biotransformation products of oxaliplatin, *J. Chromatogr. B*, 724, 345–356, 1999.
157. Raynaud, F.I., Mistry, P., Donaghue, A., Poon, G.K., Kelland, L.R., Barnard, C.F., Murrer, B.A., and Harrap, K.R., Biotransformation of the platinum drug JM216 following oral administration to cancer patients, *Cancer Chemother. Pharmacol.*, 38, 155–162, 1996.
158. Poon, G.K., Raynaud, F., Misty, P., Odell, D., Kelland, L., Harrap, K.R., Barnard, C., and Murrer, B., Metabolic studies of an orally active platinum anticancer drug by liquid chromatography-electrospray ionization-mass spectrometry, *J. Chromatogr. A*, 712, 61–66, 1995.
159. Galanski, M. and Keppler, B.K., Is reduction required for antitumour activity of platinum(IV) compounds? Characterisation of platinum(IV)-nucleotide adducts [enPT(OCOCH$_3$)$_3$(5′-GMP)] by NMR spectroscopy and ESI-MS, *Inorganica Chimica Acta*, 300–302, 783–789, 2000.
160. Bernareggi, A., Torti, L., Facino, R., Carini, M., Depta, G., Casetta, B., Farrell, N., Spadacini, S., Ceserani, R., and Tognella, S., Characterization of cisplatin-glutathione adducts by liquid chromatography-mass spectrometry. Evidence of their formation *in vitro* but not *in vivo* after concomitant administration of cisplatin and glutathione to rats and cancer patients, *J. Chromatogr. B*, 669, 247–263, 1995.

161. Martin, L., Schreiner, A., and Breemen, R., Characterization of cisplatin adducts of oligonucleotides by fast atom bombardment mass spectrometry, *Anal. Biochem.*, 193, 6–15, 1991.
162. Sharma, M., Jain, R., and Isac, T., A novel technique to assay adducts of DNA induced by anticancer agent cis-diamminedichloroplatinum(II), *Bioconjug. Chem.*, 2, 403–406, 1991.
163. Guittard, J., Pacifico, C., Blais, J.C., Bolbach, G., Chottard, J.C., and Spassky, A., Matrix-assisted laser desorption ionization time-of-flight mass spectrometry of DNA-Pt(II) complexes, *Rapid Commun. Mass Spectrom.*, 9, 33–36, 1995.
164. Troujman, H. and Chottard, J.-C., Comparison between HPLC and capillary electrophoresis for the separation and identification of the platination products of oligonucleotides with cis-$[Pt(NH_3)_2(H_2O)_2]^{2+}$ and $[Pt(NH_3)_3(H_2O)]^{2+}$, *Anal. Biochem.*, 252, 177–185, 1997.
165. Reeder, F., Guo, Z., Murdoch, P., Corazza, A., Hambley, T.W., Berners-Price, S.J., Chottard, J.-C., and Sadler, P.J., Plantination of a GG site on single-stranded and double-stranded forms of a 14-base oligonucleotide with diaqua cisplatin followed by NMR and HPLC, *Eur. J. Biochem.*, 249, 370–382, 1997.
166. Bible, K.C., Boerner, S.A., Kirkland, K., Anderl, K.L., Bartelt, D., Jr., Svingen, P.A., Kottke, T.J., Lee, Y.K., Eckdahl, S., Stalboerger, P.G., Jenkins, R.B., and Kaufmann, S.H., Characterization of an ovarian carcinoma cell line resistant to cisplatin and flavopiridol, *Clin. Cancer Res.*, 6, 661–670, 2000.
167. Ercoli, A., Battaglia, A., Raspaglio, G., Fattorossi, A., Alimonti, A., Petrucci, F., Caroli, S., Mancuso, S., and Scambia, G., Activity of cisplatin and ICI 182,780 on estrogen receptor negative ovarian cancer cells: cell cycle and cell replication rate perturbation, chromatin texture alteration and apoptosis induction, *Int. J. Cancer*, 85, 98–103, 2000.
168. Warnke, U., Gysler, J., Hofte, B., Tjaden, U.R., Greef, J., Kloft, C., Schunack, W., and Jaehde, U., Separation and identification of platinum adducts with DNA nucleotides by capillary zone electrophoresis and capillary zone electrophoresis coupled to mass spectrometry, *Electrophoresis*, 22, 97–103, 2001.
169. Speelmans, G., Staffhorst, R.W., Versluis, K., Reedijk, J., and de Kruijff, B., Cisplatin complexes with phosphatidylserine in membranes, *Biochemistry*, 36, 10545–10550, 1997.
170. Gatto, B., Capranico, G., and Palumbo, M., Drugs acting on DNA topoisomerases: recent advances and future perspectives, *Current Pharm. Design*, 5, 195–215, 1999.
171. Wang, H.-K., Morris-Natschke, S.L., and Lee, K.-H., Recent advances in the discovery and development of topoisomerase inhibitors as antitumor agents, *Med. Res. Reviews*, 17, 367–425, 1997.
172. Wall, M. and Wani, M., Camptothecin and taxol: discovery to clinic, *Cancer Res.*, 55, 753–760, 1995.
173. Platzer, P., Thalhammer, T., Hamilton, G., Ulsperger, E., Rosenberg, E., Wissiack, R., and Jager, W., Metabolism of camptothecin, a potent topoisomerase I inhibitor, in the isolated perfused rat liver, *Cancer Chemother. Pharmacol.*, 45, 50–54, 2000.
174. Hinz, H., Harris, N., Natelson, E., and Ciovanella, B., Pharmacokinetics of the *in vivo* and *in vitro* conversion of 9-nitro-20(S)-camptothecin to 9-amino-20(S)-camptothecin in humans, dogs, and mice, *Cancer Res.*, 54, 3096–3100, 1994.
175. Lokiec, F., Sorbier, B.M., and Sanderink, G., Irinotecan (CPT-11) metabolites in human bile and urine, *Clin. Cancer Res.*, 2, 1943–1949, 1996.
176. Rivory, L.P., Riou, J.F., Haaz, M.C., Sable, S., Vuilhorgne, M., Commercon, A., Pond, S.M., and Robert, J., Identification and properties of a major plasma metabolite of irinotecan (CPT-11) isolated from the plasma of patients, *Cancer Res.*, 56, 3689–3694, 1996.
177. Dodds, H.M., Haaz, M.C., Riou, J.F., Robert, J., and Rivory, L.P., Identification of a new metabolite of CPT-11 (irinotecan): pharmacological properties and activation to SN-38, *J. Pharmacol. Exper. Therapy*, 286, 578–583, 1998.
178. Dodds, H.M., Clarke, S.J., Findlay, M., Bishop, J.F., Robert, J., and Rivory, L.P., Clinical pharmacokinetics of the irinotecan metabolite 4-piperidinopiperidine and its possible clinical importance, *Cancer Chemother. Pharmacol.*, 45, 9–14, 2000.
179. Ragot, S., Marquet, P., Lachatre, F., Rousseau, A., Lacassie, E., Gaulier, J.M., Dupuy, J.L., and Lachatre, G., Sensitive determination of irinotecan (CPT-11) and its active metabolite SN-38 in human serum using liquid chromatography-electrospray mass spectrometry, *J. Chromatogr. B*, 736, 175–184, 1999.
180. Mick, R., Gupta, E., Vokes, E.E., and Ratain, M. J., Limited-sampling models for irinotecan pharmacokinetics-pharmacodynamics: prediction of biliary index and intestinal toxicity, *J. Clin. Oncol.*, 14, 2012–2019, 1996.

181. Slatter, J.G., Schaaf, L.J., Sams, J.P., Feenstra, K.L., Johnson, M.G., Bombardt, P.A., Cathcart, K.S., Verburg, M.T., Pearson, L.K., Compton, L.D., Miller, L.L., Baker, D.S., Pesheck, C.V., and Lord, R.S., 3rd., Pharmacokinetics, metabolism, and excretion of irinotecan (CPT-11) following I.V. infusion of [(14)C]CPT-11 in cancer patients, *Drug Metab. Dispos.,* 28, 423–433, 2000.
182. Dodds, H.M., Craik, D.J., and Rivory, L.P., Photodegradation of irinotecan (CPT-11) in aqueous solutions: identification of fluorescent products and influence of solution composition, *J. Pharm. Sci.,* 86, 1410–1416, 1997.
183. Dodds, H.M., Robert, J., and Rivory, L.P., The detection of photodegradation products of irinotecan (CPT-11, Camptosar (R)), in clinical studies, using high-performance liquid chromatography atmospheric pressure chemical ionization mass spectrometry, *J. Pharm. Biomed. Anal.*, 17, 785–792, 1998.
184. Rosing, T., Herben, V.M.M., van Gortel-van Zomeren, D.M., Hop, E., Kettenes-van den Bosch, J.J., ten Bokkel Huinink, W.W., and Beijnen, J.H., Isolation and structural confirmation of N-desmethyl topotecan, a metabolite of topotecan, *Cancer Chemother. Pharmacol.,* 39, 498–504, 1997.
185. Rosing, H., van Zomeren, D.M., Doyle, E., Bult, A., and Beijnen, J.H., O-glucuronidation, a newly identified metabolic pathway for topotecan and N-desmethyl topotecan, *Anticancer Drugs,* 9, 587–592, 1998.
186. Hande, K.R., Etoposide: four decades of development of a topoisomerase II inhibitor, *Eur. J. Cancer,* 34, 1514–1521, 1998.
187. van Maanen, J., Retel, J., de Vries, J., and Pinedo, H., Mechanism of action of antitumor drug etoposide: a review, *J. Natl. Cancer Inst.,* 80, 1526–1533, 1988.
188. Kagan, V.E., Yalowich, J.C., Day, B.W., Goldman, R., Gantchev, T.G., and Stoyanovsky, D.A., Ascorbate is the primary reductant of the phenoxyl radical of etoposide in the presence of thiols both in cell homogenates and in model systems, *Biochemistry,* 33, 9651–9660, 1994.
189. Hande, K., Anthony, L., Hamilton, R., Bennett, R., Sweetman, B., and Branch, R., Identification of etoposide glucuronide as a major metabolite of etoposide in the rat and rabbit, *Cancer Res.,* 48, 1829–1834, 1988.
190. Jungclas, H., Schmidt, L., Kohl, P., and Fritsch, H.W., Quantitative matrix assisted plasma desorption mass spectrometry, *Int. J. Mass Spectrom. Ion Proc.,* 126, 157–161, 1993.
191. Kohl, P., Koppler, H., Schmidt, L., Fritsch, H.W., Holz, J., Pfluger, K.H., and Jungclas, H., Pharmacokinetics of high-dose etoposide after short-term infusion, *Cancer Chemother. Pharmacol.,* 29, 316–320, 1992.
192. Pfluger, K.H., Hahn, M., Holz, J.B., Schmidt, L., Kohl, P., Fritsch, H.W., Jungclas, H., and Havemann, K., Pharmacokinetics of etoposide: correlation of pharmacokinetic parameters with clinical conditions, *Cancer Chemother. Pharmacol.,* 31, 350–356, 1993.
193. Holz, J., Koppler, H., Schmidt, L., Fritsch, H., Pfluger, H., and Jungclas, H., Limited stampling models for reliable estimation of etoposide area under the curve, *Eur. J. Cancer,* 31A, 1794–1798, 1995.
194. Roboz, J., Deng, L., and Ma, L., Identification of intact complexes of etoposide with albumin and hemoglobin in models and serum by ESI and LC/ESI [Abstract], *Proc. 47th ASMS Conf. Mass Spectrom.,* 1999.
195. Przybylski, M., Cysyk, R., Shoemaker, D., and Adamson, R., Identification of conjugation and cleavage products in the thiolytic metabolism of the anticancer drug 4′-(9-Acridinylamino) methanesulfon-m-anisidide, *Biomed. Mass Spectrom.*, 8, 485–491, 1981.
196. Gaudich, K. and Przybylski, M., Field desorption mass specrtrometric characterization of thiol conjugates related to the oxidative metabolism of the anticancer drug 4′-(9-acridinylamino)-methanesulfon-m-anisidide, *Biomed. Mass Spectrom.*, 10, 292–299, 1983.
197. Robertson, I., Palmer, B., Officer, M., Siegers, D., Paxton, J., and Shaw, G., Cytosol mediated metabolism of the experimental antitumour agent acridine carboxamide to the 9-acridone derivative, *Biochem. Pharmacol.,* 42, 1879–1884, 1991.
198. Robertson, I., Palmer, B., Paxton, J., and Bland, T., Metabolism of the experimental antitumor agent acridine carboxamide in the mouse, *Drug Metab. Dispos.,* 21, 530–536, 1993.
199. Gouyette, A., Synthesis of deuterium-labelled elliptinium and its use in metabolic studies, *Biomed. Environ. Mass Spectrom.,* 15, 243–247, 1988.
200. Gouyette, A., Voisin, E., Auclair, C., and Paoletti, C., Isolation and characterization of glutathione-elliptinium conjugate in human urine, *Anticancer Res.,* 7, 823–827, 1987.

201. Bachur, N.R., Daunorubicinol, a major metabolite of daunorubicin: isolation from human urine and enzymatic reactions, *J. Pharmacol. Exper. Therapy,* 177, 573–578, 1971.
202. Bullock, F., Bruni, R., and Asbell, A., Identification of new metabolites of daunomycin and adriamycin, *J. Pharmacol. Exper. Therapy,* 182, 70–76, 1972.
203. Cradock, J., Egorin, M., and Bachur, N.R., Daunorubicin biliary excretion and metabolism in the rat, *Arch. Int. Pharmacodyn,* 202, 48–61, 1973.
204. Takanashi, S. and Bachur, N.R., Daunorubicin metabolites in human urine, *J. Pharmacol. Exp. Ther.,* 195, 41–49, 1975.
205. Takanashi, S. and Bachur, N.R., Adriamycin metabolism in man. Evidence from urinary metabolites, *Drug Metab. Dispos.,* 4, 79–87, 1976.
206. van der Vijgh, W., Maessen, P., and Pinedo, H., Comparative metabolism and pharmacokinetics of doxorubicin and 4′-epidoxorubicin in plasma, heart and tumor of tumor-bearing mice, *Cancer Chemother. Pharmacol.,* 26, 9–12, 1990.
207. Lachatre, F., Marquet, P., Ragot, S., Gaulier, J.M., Cardot, P., and Dupuy, J.L., Simultaneous determination of four anthracyclines and three metabolites in human serum by liquid chromatography-electrospray mass spectrometry, *J. Chromatogr. B,* 281–291, 2000.
208. Andrews, P., Callery, P., Chou, F., May, M., and Bachur, N.R., Qualitative analysis of trimethylsilylated daunosamine and N-alkylated analogs by gas chromatography-mass spectrometry, *Anal. Biochem.,* 126, 258–267, 1982.
209. Breda, M., Basileo, G., Fonte, G., Long, J., and James, C.A., Determination of 4-demethoxy-3′-deamino-3′-aziridinyi-4′-methylsulphonyldaunorubicin and its 13-hydroxy metabolite by direct injection of human plasma into a column-switching liquid chromatography system with mass spectrometric detection, *J. Chromatogr. A,* 854, 81–92, 1999.
210. Ramu, A., Mehta, M.M., Liu, J., Turyan, I., and Aleksic, A., The riboflavin-mediated photooxidation of doxorubicin, *Cancer Chemother. Pharmacol.,* 46, 449–458, 2000.
211. Luce, R.A. and Hopkins, P.B., Chemical cross-linking of drugs to DNA, *Methods Enzymol.,* 340, 396–412, 2001.
212. Kapur, A., Beck, J.L., and Sheil, M.M., Observation of daunomycin and nogalamycin complexes with duplex DNA using electrospray ionization mass spectrometry, *Rapid Commun. Mass Spectrom.,* 13, 2489–2497, 1999.
213. Kato, S., Burke, P.J., Fenick, D.J., Taatjes, D.J., Bierbaum, V.M., and Koch, T.H., Mass spectrometric measurement of formaldehyde generated in breast cancer cells upon treatment with anthracycline antitumor drugs, *Chem. Res. Toxicol.,* 13, 509–516, 2000.
214. Taatjes, D.J., Fenick, D.J., Gaudiano, G., and Koch, T.H., A redox pathway leading to the alkylation of nucleic acids by doxorubicin and related anthracyclines: application to the design of antitumor drugs for resistant cancer, *Curr. Pharm. Des.,* 4, 203–218, 1998.
215. Taatjes, D.J., Gaudiano, G., Resing, K., and Koch, T.H., Alkylation of DNA by the anthracycline, antitumor drugs adriamycin and daunomycin, *J. Med. Chem.,* 39, 4135–4138, 1996.
216. Taatjes, D.J., Gaudiano, G., Resing, K., and Koch, T.H., Redox pathway leading to the alkylation of DNA by the anthracycline, antitumor drugs adriamycin and daunomycin, *J. Med. Chem.,* 40, 1276–1286, 1997.
217. Fenick, D.J., Taatjes, D.J., and Koch, T.H., Doxoform and Daunoform: anthracycline-formaldehyde conjugates toxic to resistant tumor cells, *J. Med. Chem.,* 40, 2452–2461, 1997.
218. Taatjes, D.J., Fenick, D.J., and Koch, T.H., Epidoxoform: a hydrolytically more stable anthracycline-formaldehyde conjugate toxic to resistant tumor cells, *J. Med. Chem.,* 41, 1306–1314, 1998.
219. Taatjes, D.J. and Koch, T.H., Nuclear targeting and retention of anthracycline antitumor drugs in sensitive and resistant tumor cells, *Curr. Med. Chem.,* 8, 15–29, 2001.
220. Siegel, M., Hydrogen-deuterium exchange studies utilizing a thermospray mass spectrometer interface, *Anal. Chem.,* 60, 2090–2095, 1988.
221. Chiccarelli, F., Mirrison, J., Cosulich, D., Perkinson, N., Ridge, D., Sum, F., Murdock, K., Woodward, D., and Arnold, E., Identification of human urinary mitoxantrone metabolites, *Cancer Res.,* 46, 4858–4861, 1986.
222. Mewes, K., Blanz, J., Ehninger, G., Gebhardt, R., and Zeller, K.P., Cytochrome P-450-induced cytotoxicity of mitoxantrone by formation of electrophilic intermediates, *Cancer Res.,* 53, 5135–5142, 1993.

222a. Juhasz, P. and Biemann, K., Mass spectrometric molecular-weight determination of highly acidic compounds of biological significance via their complexes with basic polypeptides, *Proc. Natl. Acad. Sci. U.S.A.*, 91, 4333–4337, 1994.
223. Roboz, J., Deng, L., Ma, L., and Holland, J.F., Investigation of suramin-albumin binding by electrospray mass spectrometry, *Rapid Commun. Mass Spectrom.*, 12, 1319–1322, 1998.
224. Wickham, G., Iannitti, P., Boschenok, J., and Sheil, M., The observation of a hedamycin-d(CACGTG) 2 covalent adduct by electrospray mass spectrometry, *FEBS Lett.*, 360, 231–234, 1995.
225. Wickham, G., Iannitti, P., Boschenok, J., and Sheil, M.M., Electrospray ionization mass spectrometry of covalent ligand-oligonucleotide adducts: evidence for specific duplex formation, *J. Mass Spectrom.*, S197–S203, 1995.
226. Iannitti, P., Sheil, M.M., and Wickham, G., High sensitivity and fragmentation specificity in the analysis of drug-DNA adducts by electrospray tandem mass spectrometry, *J. Am. Chem. Soc.*, 119, 1490–1491, 1997.
227. Roboz, J., McCamish, M., Nieves, E., Smith, C., and Holland, J.F., Collisional activation decomposition of actinomycins using tandem mass spectrometry, *Biomed. Environ. Mass Spectrom.*, 16, 67–70, 1988.
228. Jordan, A., Hadfield, J.A., Lawrence, N.J., and McGrown, A.T., Tubulin as a target for anticancer drugs: agents which interact with the mitotic spindle, *Med. Res. Reviews*, 18, 259–296, 1998.
229. Favretto, D., Piovan, A., and Cappelletti, E.M., Electrospray ionization mass spectra and collision induced mass spectra of antitumor *Catharanthus* alkaloids, *Rapid Commun. Mass Spectrom.*, 12, 982–984, 1998.
230. Favretto, D., Piovan, A., Filippini, R., and Caniato, R., Monitoring the production yields of vincristine and vinblastine in *Catharanthus roseus* from somatic embryogenesis. Semiquantitative determination by flow-injection electrospray ionization mass spectrometry, *Rapid Commun. Mass Spectrom.*, 15, 364–369, 2001.
231. Ramirez, J., Ogan, K., and Ratain, M.J., Determination of vinca alkaloids in human plasma by liquid chromatography/atmospheric pressure chemical ionization mass spectrometry, *Cancer Chemother. Pharmacol.*, 39, 286–290. 1997.
232. Villikka, K., Kivisto, K.T., Maenpaa, H., Joensuu, H., and Neuvonen, P.J., Cytochrome P450-inducing antiepileptics increase the clearance of vincristine in patients with brain tumors, *Clin. Pharmacol. Ther.*, 66, 589–593, 1999.
233. Wall, M.E., Camptothecin and taxol: discovery to clinic, *Med. Res. Reviews*, 18, 299–314, 1998.
234. Cragg, G.M., Paclitaxel (taxol): a success story with a valuable lesson for natural product drug discovery and development, *Med. Res. Reviews*, 18, 315–331, 1998.
235. Huizing, M., Misser, V., Pieters, R., Huinink, W., Veenhof, C., Vermorken, J., Pinedo, H., and Beijnen, J., Taxanes: a new class of antitumor agents, *Cancer Investigations*, 13, 381–404, 1995.
236. McClure, T., Schram, K., and Reiner, M., The mass spectrometry of taxol, *J. Am. Soc. Mass Spectrom.*, 3, 672–679, 1992.
237. Stierle, A., Strobel, G., and Stierle, D., Taxol and taxane production by *Taxomyces andreanae*, an endophytic fungus of pacific yew, *Science*, 260, 214–216, 1993.
238. Choi, M., Kwak, S., Liu, J., Park, Y., Lee, M., and An, N., Taxol and related compounds in Korean native yews (*Taxus cuspidata*), *Planta Med.*, 61, 264–266, 1995.
239. Hoffman, A., Khan, W., Worapong, J., Strobel, G., Griffin, D., Arbogast, B., Barofsky, D., Boone, R.B., Ning, L., Zheng, P., and Daley, L., Bioprospecting for taxol in angiosperm plant extracts, *Spectroscopy*, 13, 22–31, 1998.
240. Bitsch, F., Ma, W., Macdonald, F., Nieder, M., and Shackleton, C., Analysis of taxol and related diterpenoids from cell cultures by liquid chromatography-mass spectrometry, *J. Chromatogr.*, 615, 273–288, 1993.
241. Kerns, E.H., Hill, S.E., Detlefsen, D.J., Volk, K.J., Long, B.H., Carboni, J., and Lee, M.S., Cellular uptake profile of paclitaxel using liquid chromatography tandem mass spectrometry, *Rapid Commun. Mass Spectrom.*, 12, 620–624, 1998.
242. Willey, T., Bekos, E., Gaver, R., Duncan, G., Tay, L., Beijnen, J., and Farmen, R., High-performance liquid chromatographic procedure for the quantitative determination of paclitaxel (Taxol) in human plasma, *J. Chromatogr.*, 621, 231–238, 1993.

243. Sottani, C., Minoia, C., D'Incalci, M., Paganini, M., and Zucchetti, M., High-performance liquid chromatography tandem mass spectrometry procedure with automated solid phase extraction sample preparation for the quantitative determination of paclitaxel (taxol) in human plasma, *Rapid Commun. Mass Spectrom.,* 12, 251–255, 1998.
244. Schellen, A., Ooms, B., van Gils, M., Halmingh, O., van der Vlis, E., van de Lagemaat, D., and Verheij, E., High-throughput on-line phase extraction/tandem mass spectrometric determination of paclitaxel in human serum, *Rapid Commun. Mass Spectrom.,* 14, 230–233, 2000.
245. Monsarrat, B., Alvineri, P., Wright, M., Dubois, J., Gueritte-Voegelein, F., Guenard, D., Donehower, R., and Rowinsky, E., Hepatic metabolism and biliary clearance of taxol in rats and humans, *J. Natl. Cancer Inst. Monogr.,* 15, 39–46, 1993.
246. Harris, J., Katki, A., Anderson, L., Chmurny, G., Paukstelis, J., and Collins, J., Isolation, structural determination, and biological activity of 6α-hydroxytaxol, the principal human metabolite of taxol, *J. Med. Chem.*, 37, 706–709, 1994.
247. Sparreboom, A., Huizing, M., Boesen, J., Nooijen, W., Tellingen, O., and Beijnen, J., Isolation, purification, and biological activity of mono- and dihydroxylated paclitaxel metabolites from human feces, *Cancer Chemother. Pharmacol.*, 36, 299–304, 1995.
248. Royer, I., Alvinerie, P., Armand, J., Ho, L., Wright, M., and Monsarrat, B., Paclitaxel metabolites in human plasma and urine: identification of 6-α-hydroxytaxol, 7-epitaxol and taxol hydrolysis products using liquid chromatography/atmospheric pressure chemical ionization mass spectrometry, *Rapid Commun. Mass Spectrom.,* 9, 495–502, 1995.
249. Monsarrat, B., Mariel, E., Cros, S., Gares, M., Guenard, D., Voegelein, F., and Wright, M., Taxol metabolism: isolation and identification of three major metabolites of taxol in rat bile, *Drug Metab. Dispos.*, 18, 895–901, 1990.
250. Sottani, C., Minoia, C., Colombo, A., Zucchetti, M., D'Incalci, M., and Fanelli, R., Structural characterization of mono-and dihydroxylated metabolites of paclitaxel in rat bile using liquid chromatography/ion spray tandem mass spectrometry, *Rapid Commun. Mass Spectrom.,* 11, 1025–1032, 1997.
251. Anderson, C., Wang, J., Kumar, G., McMillan, J., Walle, K., and Walle, T., Dexamethasone induction of taxol metabolism in the rat, *Drug Metab. Dispos.*, 23, 1286–1290, 1995.
252. Kumar, G., Oatis, J., Thornburg, K., Heldrich, F., Hazard, E., and Walle, T., 6α-Hydroxytaxol: isolation and identification of the major metabolite of taxol in human liver microsomes, *Drug Metab. Dispos.*, *22,* 1772–1796, 1994.
253. Walle, T., Walle, K., Kumar, G., and Bhalla, K., Taxol metabolism and disposition in cancer patients, *Drug Metab. Dispos.,* 23, 506–512, 1995.
254. Rahman, A., Korzekwa, K., Grogan, J., Gonzales, F., and Harris, J., Selective biotransformation of taxol to 6α-hydroxytaxol by human cytochrome P450 2C8, *Cancer Res.,* 54, 5543–5546, 1994.
255. Cresteil, T., Monsarrat, B., Alvinerie, P., Treluyer, J., Vieira, I., and Wright, M., Taxol metabolism by human liver microsomes: identification of cytochrome P450 isozymes involved in its biotransformation, *Cancer Res.,* 54, 386–392, 1994.
256. Harris, J., Rahman, A., Kim, B., Guengerich, P., and Collins, J., Metabolism of taxol by human hepatic microsomes and liver slices: participation of cytochrome P450 3A4 and an unknown P450 enzyme, *Cancer Res.,* 54, 4026–4035, 1994.
257. Hauck, C., Structure elucidation of taxol metabolites by liquid chromatography/mass spectrometry and species differences in taxol metabolism, *Rapid Commun. Mass Spectrom.*, 11, 1823, 1997.
258. Monsarrat, B., Chatelut, E., Royer, I., Alvinerie, P., Dubois, J., Dezeuse, A., Roche, H., Cros, S., Wright, M., and Canal, P., Modification of paclitaxel metabolism in a cancer patient by the induction of cytochrome P450 3A4, *Drug Metab. Dispos.,* 26, 229–233, 1998.
259. Royer, I., Monsarrat, B., Sonnier, M., Wright, M., and Cresteil, T., Metabolism of docetaxel by human cytochromes P450: interaction with paclitaxel and other antineoplastic drugs, *Cancer Res.*, 56, 58–65, 1996.
260. Gelderblom, H., Verweij, J., Brouwer, E., Pillay, M., de Bruijn, P., Nooter, K., Stoter, G., and Sparreboom, A., Disposition of [G-(3)H]paclitaxel and cremophor EL in a patient with severely impaired renal function, *Drug Metab. Dispos.,* 27, 1300–1305, 1999.
261. Troendle, F.J., Reddick, C.D., and Yost, R.A., Detection of pharmaceutical compounds in tissue by matrix-assisted laser desorption/ionization and laser desorption/chemical ionization tandem mass spectrometry with a quadrupole ion trap, *J. Am. Soc. Mass Spectrom.,* 10, 1315–1321, 1999.

262. Mamber, S., Mikkilineni, A., Pack, E., Rosser, M., Wong, H., Ueda, Y., and Forenza, S., Tubulin polymerization by paclitaxel (Taxol) phosphate prodrugs after metabolic activation with alkaline phosphatase, *J. Pharmacol. Exper. Therapy,* 274, 877–883, 1995.
263. Poncet, J., The dolastatins, a family of promising antineoplastic agents, *Current Pharm. Design,* 5, 139–162, 1999.
264. Madden, T., Tran, H.T., Beck, D., Huie, R., Newman, R.A., Pusztai, L., Wright, J.J., and Abbruzzese, J.L., Novel marine-derived anticancer agents: a phase I clinical, pharmacological, and pharmacodynamic study of dolastatin 10 (NSC 376128) in patients with advanced solid tumors, *Clin. Cancer Res.,* 6, 1293–1301, 2000.
265. Garteiz, D.A., Madden, T., Beck, D.E., Huie, W.R., McManus, K.T., Abbruzzese, J.L., Chen, W., and Newman, R.A., Quantitation of dolastatin-10 using HPLC/electrospray ionization mass spectrometry: application in a phase I clinical trial. *Cancer Chemother. Pharmacol.,* 41, 299–306, 1998.
266. Keuzenkamp-Jansen, C.W., van-Baal, J.M., De-Abreu, R.A., de-Jong, J.G., Zuiderent, R., and Trijbels, J.M., Detection and identification of 6-mercapto-8-hydroxypurine, a major metabolite of 6-mercaptopurine, in plasma during intravenous administration, *Clin. Chem.,* 42, 380–386, 1996.
267. Rosenfeld, J.M., Taguchi, V.Y., Hillcoat, B.L., and Kawai, M., Determination of 6-mercaptopurine in plasma by mass spectrometry, *Anal. Chem.,* 49, 725–727, 1977.
268. Meropol, N.J., Oral fluoropyrimidines in treatment of colorectal cancer, *Eur. J. Cancer,* 34, 1509–1513, 1998.
269. Tanaka, F., Fukuse, T., Wada, H., and Fukushima, M., The history, mechanism and clinical use of oral 5-fluorouracil derivative chemotherapeutic agents, *Current Pharmaceut. Biotechnol.,* 1, 137–164, 2000.
270. Thiberville, L., Compagnon, P., Moore, N., Bastian, G., Richard, M., Hellot, M., Vincent, C., Kannass, M., Dominique, S., and Thuillez, C., Plasma 5-fluorouracil and alpha-fluoro-beta-alanine accumulation in lung cancer patients treated with continuous infusion of cisplatin and 5-fluorouracil, *Cancer Chemother. Pharmacol,* 35, 64–70, 1994.
271. Aubert, C., Sommadossi, J., Coassolo, P., Cano, J., and Rigault, J., Quantitative analysis of 5-fluorouracil and 5,6-dihydrofluorouracil in plasma by gas chromatography mass spectrometry, *Biomed. Mass Spectrom.,* 9, 336–339, 1982.
272. Odagiri, H., Ichihara, S., Semura, E., Uton, M., Tateishi, M., and Kuruma, I., Determination of 5-fluorouracil in plasma and oral administration of 5′-deoxy-5-fluorouridine using gas chromatography-mass spectrometry, *J. Pharmacobio. Dyn.,* 11, 234–240, 1988.
273. Kubo, M., Sasabe, H., and Shimizu, T., Highly sensitive method for the determination of 5-fluorouracil in biological samples in the presence of 2′-deoxy-5-fluorouridine by gas chromatography-mass spectrometry, *J. Chromatogr.,* 564, 137–145, 1991.
274. Anderson, L., Parker, R., Collins, J., Ahlgren, J., Wilkinson, D., and Strong, J., Gas chromatographic-mass spectrometric method for routine monitoring of 5-fluorouracil in plasma of patients receiving low-level protracted infusions, *J. Chromatogr.,* 581, 195–201, 1992.
275. Zambonin, C.G. and Palmisano, F., Gas chromatography-mass spectrometry identification of a novel N^3-methylated metabolite of 5′-deoxy-5-fluorouridine in plasma of cancer patients undergoing chemotherapy, *J. Pharm. Biomed. Anal.,* 14, 1521–1528, 1996.
276. Wang, K., Nano, M., Mulligan, T., Bush, E.D., and Edom, R.W., Derivatization of 5-fluorouracil with 4-bromomethyl-7-methoxycoumarin for determination by liquid chromatography-mass spectrometry, *J. Am. Soc. Mass Spectrom.,* 9, 970–976, 1998.
277. Zambonin, C.G., Mastrolitti, S., and Palmisano, F., Derivatization reactions for gas chromatograph/mass spectrometry determination of N^3-methyl-5′-deoxy-5-fluorouridine, *Rapid Commun. Mass Spectrom.,* 11, 1529–1535, 1997.
278. Matsushima, E., Yoshida, K., Kitamura, R., and Yoshida, K., Determination of S-1 (combined drug of tegafur, 5-chloro-2,4-dihydroxypyridine and potassium oxonate) and 5-fluorouracil in human plasma and urine using high-performance liquid chromatography and gas chromatography-negative ion chemical ioization mass spectrometry, *J. Chromatogr. B,* 691, 95–104, 1997.
279. Loos, W., de Bruijn, P., van Zuylen, L., Verweij, J., Nooter, K., Stoter, G., and Sparreboom, A., Determination of 5-fluorouracil in microvolumes of human plasma by solvent extraction and high-performance liquid chromatography, *J. Chromatogr. B,* 735, 293–297, 1999.
280. Ohnuma, T., Roboz, J., Waxman, S., Mandel, E., Martin, D., and Holland, J.F., Clinical pharmacological effects of thymidine plus 5-FU, *Cancer Treatment Report,* 64, 1169–1177, 1980.

281. Peters, G.J., Noordhuis, P., Komissarov, A., Holwerda, U., Kok, M., van Laar, J.A.M., van der Wilt, C.L., van Groeningen, C.J., and Pinedo, H.M., Quantification of 5-fluorouracil incorporation into RNA of human and murine tumors as measured with a sensitive gas chromatography-mass spectrometry assay, *Anal. Biochem.,* 231, 157–163, 1995.
282. Kuhara, T., Ohdoi, C., and Ohse, M., Simple gas chromatographic-mass spectrometric procedure for diagnosing pyrimidine degradation defects for prevention of severe anticancer side effects, *J. Chromatogr. B,* 758, 61–74, 2001.
283. Schilsky, R.L., Hohneker, J., Ratain, M.J., Smeltzer, L., Lucas, V.S., Khor, S.P., Diasio, R., Von Hoff, D.D., and Burris, H.A.I., Phase I clinical and pharmacologic study of eniluracil plus fluorouacil in patients with advanced cancer, *J. Clin. Oncol.,* 16, 1450–1457, 1998.
284. Baker, S.D., Diasio, R., O'Reilly, S., Lucas, V.S., Khor, S.P., Sartorius, S.E., Donehower, R.C., Grochow, L.B., Spector, T., Hohneker, J., and Rowinsky, E.K., Phase I and pharmacologic study of oral fluorouracil on a chronic daily schedule in combination with the dihyropyrimidine dehydrogenase inactivator eniluricil, *J. Clin. Onocol.,* 18, 915–926, 2000.
285. Bi, D., Anderson, L.W., Shapiro, J., Shapiro, A., Grem, J.L., and Takimoto, C.H., Measurement of plasma uracil using gas chromatography-mass spectrometry in normal individuals and patients receiving inhibitors of dihydropyrimidine dehydrogenase, *J. Chromatogr. B,* 738, 249–258, 2000.
286. Cociglio, M., Hillaire-Buys, D., and Alric, C., Determination of methotrexate and 7-hydroxymethotrexate by liquid chromatography for routine monitoring of plasma levels, *J. Chromatogr. B,* 674, 101–110, 1995.
287. Hignite, C. and Azarnoff, D., Identification of methotrexate and folic acid analogs by mass spectrometry, *Biomed. Mass Spectrom.,* 5, 161–163, 1978.
288. Cheung, A., Tattam, B., Antonjuk, D., and Boadle, D., Ammonia and methane chemical ionization mass spectra of methotrexate and its amide and ester analogues, *Biomed. Mass Spectrom.,* 12, 11–18, 1985.
289. Kussmann, M., Wiehr, D., Knepper, T., and Przybylski, M., Synthesis, structural and biochemical characterization of cytostatic methotrexate-gammaglutamyl-glutathione conjugates, *Adv. Exp. Med. Biol.,* 338, 453–456, 1993.
290. Fan, J.G., Pope, L.E., Vitols, K.S., and Huennekens, F.M., Affinity labeling of folate transport proteins with the N-hydroxysuccinimide ester of the gamma isomer of fluorescein-methotrexate, *Biochemistry,* 30, 4573–4580, 1991.
291. Fan, J., Vitols, K., and Huennekens, F., Biotin derivatives of methotrexate and folate, *J. Biol. Chem.,* 266, 14862–14865, 1991.
292. Widemann, B.C., Sung, E., Anderson, L., Salzer, W.L., Balis, F.M., Monitjo, K.S., McCully, C., Hawkins, M., and Adamson, P.C. Pharmacokinetics and metabolism of the methotrexate metabolite 2,4,-diamino-N^{10}-methylpteroic acid, *J. Pharmacol. Exper. Therapy,* 294, 894–901, 2000.
293. Steinborner, S. and Henion, J., Liquid-liquid extraction in the 96-well plate format with SRM LC/MS quantitative determination of methotrexate and its major metabolite in human plasma, *Anal. Chem.,* 71, 2340–2345, 1999.
293a. Dueker, S.R., Lin, Y., Jones, A.D., Mercer, R., Fabbro, E., Miller, J.W., Green, R., and Clifford, A.J., Determinatino of blood folate using acid extraction and internally standardized gas chromatography-mass spectrometry detection, *Anal. Biochem.,* 283, 266–275, 2000.
293b. Nikolic, D. and van Breemen, R.B., Screening for inhibitors of dihydrofolate reductase using pulsed ultrafiltration mass spectrometry, *Comb. Chem. High. Throughput Screen.,* 1, 47–55, 1998.
294. Fulton, B., Wagstaff, A., and McTavish, D., Trimetrexate: a review of its pharmacodynamic and pharmacokinetic properties and therapeutic potential in the treatment of *p*neumocystis carinii pneumonia, *Drugs,* 49, 563–576, 1995.
295. Lin, J. and Bertino, J., Update on trimetrexate, a folate antagonist with antineoplastic and antiprotozoal properties, *Cancer Invest.,* 9, 159–172, 1991.
296. Wong, B., Woolf, T., Chang, T., and Whitfield, L., Metabolic disposition of trimetrexate, a nonclassical dihydrofolate reductase inhibitor, in rat and dog, *Drug Metab. Dispos.,* 18, 980–986, 1990.
297. Stetson, P. and Ensminger, M., Determination of plasma trimetrexate levels using GC/MS-SIM, *J. Chromatogr.,* 383, 69–76, 1986.
298. Roboz, J., Suzuki, R., and Holland, J.F., Quantification of arabinitol in serum by selected ion monitoring as diagnostic technique in invasive candidiasis, *J. Clin. Microbiol.,* 12, 594–602, 1980.

299. Roboz, J., Mass spectrometric determination of D- to L-Arabinitol ratios for diagnosis and monitoring of disseminated candidiasis, in *Mass Spectrometry for the Characterization of Microorganisms*, Fenselau, C., Ed., Am. Chem. Soc., Washington, DC, 132–146, 1994.
300. Roboz, J., Mass spectrometry in cancer research, *Adv. Mass Spectrom.*,13, 507–520, 1995.
301. Roboz, J., Diagnosis and monitoring of disseminated candidiasis based on serum/urine D/L-arabinitol ratios, *Chirality,* 6, 51–57, 1994.
302. Roboz, J., Yu, Q., and Holland, J.F., Filter paper sampling of whole blood and urine for the determination of D/L arabinitol ratios by mass spectrometry, *J. Microbiol. Methods,* 15, 207–214, 1992.
303. Lonning, P., Lien, E., Lundgren, S., and Kvinnsland, S., Clinical pharmacokinetics of endocrine agents used in advanced breast cancer, *Clin. Pharmacokin.,* 22, 327–358, 1992.
304. Chen, S., Zhou, D., Swiderek, K., Kadohama, N., Osawa, Y., and Hall, P., Structure-function studies of human aromatase, *J. Steroid Biochem. Molec. Biol.,* 44, 347–356, 1993.
305. Goss, P.E. and Gwyn, K.M., Current perspectives on aromatase inhibitors in breast cancer, *J. Clin. Oncol., 12,* 2460–2470, 1994.
306. Brodowicz, T., Wiltschenke, C., and Zielinski, C., Recent advances in hormonal treatment of breast cancer, *Onkologie,* 21, 429–433, 1998.
307. Kelloff, G.J., Lubet, R.A., Lieberman, R., Eisenhauer, K., Steele, V.E., Crowell, J.A., Hawk, E.T., Booene, C.W., and Sigman, C.C., Aromatase inhibitors as potential cancer chemopreventives, *Cancer Epidemiol. Biomark. Prev.,* 7, 65–78, 1998.
308. Dikkeschei, L.D., Wolthers, B.G., Bos Zuur, I., de la Riviere, G.B., Nagel, G.T., van der Kolk, D.A., and Willemse, P.H., Optimization of a classical aromatase activity assay and application in normal, adenomatous and malignant breast parenchyma, *J. Steroid Biochem. Molec. Biol.,* 59, 305–313, 1996.
309. Jarman, M., Foster, A., Goss, P., Griggs, L., and Howe, I., Metabolism of aminoglutethimide in humans: identification of hydroxylaminoglutethimide as an induced metabolite, *Biomed. Mass Spectrom.,* 10, 620–625, 1983.
310. Poon, G.K., McCague, R., Griggs, L., Jarman, M., and Lewis, I., Characterization of metabolites of the aromatase inhibitor 3-ethyl-3-(4-pyridyl)piperidine-2,6-dione—a potential anti-tumor agent, *J. Chromatogr.,* 572, 143–157, 1991.
311. Poon, G.K., Jarman, M., and Rowlands, M., Determination of 4-hydroxyandrost-4-ene-3,17-dione metabolism in breast cancer patients using high-performance liquid chromatography-mass spectrometry, *J. Chromatogr.,* 565, 75–88, 1991.
312. Poon, G.K., Jarman, M., and McCague, R., Identification of 4-hydroxyandrost-4-ene-3,17-dione metabolites in prostatic cancer patients by liquid chromatography-mass spectrometry, *J. Chromatogr.,* 576, 235–244, 1992.
313. Foster, A., Griggs, L., Howe, I., Jarman, M., Leung, C., Manson, D., and Rowlands, M., Metabolism of aminoglutethimide in humans. Identification of four new urinary metabolites, *Drug Metab. Dispos.,* 12, 511–516, 1984.
314. Seago, A., Baker, M., Houghton, J., Leung, C., and Jarman, M., Metabolism and pharmacokinetics of the N- and C-n-octyl analogues of pyridoglutethimide [3-ethyl-3-(4-pyridyl)piperidine-2,6-dione]: novel inhibitor of aromatase, *Biochem. Pharmacol.,* 36, 573–577, 1987.
315. Ackermann, R. and Kaiser, G., Determination of the new aromatase inhibitor CGS 16 949 in biological fluids by capillary gas chromatography/mass spectrometry, *Biomed. Environ. Mass Spectrom.,* 18, 558–562, 1989.
316. Kochak, G.M., Mangat, S., Mulagha, M.T., Entwistle, E.A., Santen, R.J., Lipton, A., and Demers, L., The pharmacodynamic inhibition of estrogen synthesis by fadrozole, an aromatase inhibitor, and its pharmacokinetic disposition, *J. Clin. Endocrinol. Metab.,* 71, 1349–1355, 1990.
317. Foster, A., Jarman, M., Mann, J., and Parr, I., Metabolism of 4-hydroxyandrost-4-ene-3,17-dione by rat hepatocytes, *J. Steroid Biochem.,* 24, 607–617, 1986.
318. Goss, P., Jarman, M., Wilkinson, J., and Coombes, R., Metabolism of the aromatase inhibitor 4-hydroxyandrostenedione *in vivo*. Identification of the glucuronide as a major urinary metabolite in patients and billary metabolite in the rat, *J. Steroid Biochem.,* 24, 619–622, 1986.
319. Guarna, A., Moneti, G., Prucher, D., Salerno, R., and Serio, M., Quantitative determination of 4-hydroxy-4-androstene-3,17-dione (4-OHA), a potent aromatase inhibitor, in human plasma, using isotope dilution mass spectrometry, *J. Steroid Biochem.,* 32, 699–702, 1989.

320. Poon, G.K., Bisset, G., and Mistry, P., Electrospray ionization mass spectrometry for analysis of low-molecular-weight anticancer drugs and their analogues, *J. Am. Soc. Mass Spectrom.*, 4, 588–595, 1993.
321. Poon, G.K., Chui, C., Jarman, M., Rowlands, M., Kokkonen, P., Niessen, W., and Ven Der Greef, J., Investigation of conjugated metabolites of 4-hydroxyandrost-4-ene-3,17-dione in patient urine by liquid chromatography-atmospheric pressure ionization mass spectrometry, *Drug Metab. Dispos.*, 20, 941–947, 1992.
322. Cenacchi, V., Baratte, S., Cicioni, P., Figerio, E., Long, J., and James, C., LC-MS-MS determination of exemestane in human plasma with heated nebulizer interface following solid-phase extraction in the 96 well plate format, *J. Pharm. Biomed. Anal.*, 22, 451–460, 2000.
323. Numazawa, M., Yoshimura, A., and Oshibe, M., Enzymic aromatization of 6-alkyl-substituted androgens, potent competitive and mechanism-based inhibitors of aromatase, *Biochem. J.*, 329, 151–156, 1998.
324. Lonning, P.E., Geisler, J., Johannessen, D.C., Gschwind, H., Waldmeier, F., Schneider, W., Galli, B., Winkler, T., Blum, W., Kriemler, H., Miller, W.R., and Faigle, J.W., Pharmacokinetics and metabolism of formestane in breast cancer patients, *J. Steroid Biochem. Molec. Biol.*, 77, 39–47, 2001.
325. Wogan, G.N., Review of the toxicology of tamoxifen, *Semin. Oncol.*, 24, S1-87–S1-97, 1997.
326. White, I.N.H., The tamoxifen dilemma, *Carcinogenesis*, 20, 1153–1160, 1999.
327. Smith, L.I., Brown, K., Carthew, P., Lim, C.K., Martin, E.A., Styles, J., and White, I.N., Chemoprevention of breast cancer by tamoxifen: risks and opportunities, *Crit. Rev. Toxicol.*, 571–594, 2000.
328. Svenberg, J.A., Clinical relevance of laboratory and animal data on tamoxifen, *Oncology*, 11, 39–44, 1997.
329. Tannenbaum, S.R., Comparative metabolism of tamoxifen and DNA adduct formation and *in vitro* studies on genotoxicity, *Semin. Oncol.*, 24, S1-81–S1-6, 1997.
330. Lien, E.A. and Lonning, P.E., Selective oestrogen receptor modifiers (SERMs) and breast cancer therapy, *Cancer Treatment Reviews*, 26, 205–227, 2000.
331. Foster, A., Griggs, L., Jarman, M., van Maanen, J., and Schulten, H., Metabolism of tamoxifen by rat liver microsomes: formation of the N-oxide, a new metabolite, *Biochem. Pharmacol.*, 29, 1977–1979, 1980.
332. Bates, D., Foster, A., Griggs, L., Jarman, M., Leclercq, G., and Devleeschouwer, N., Metabolism of tamoxifen by isolated rat hepatocytes: anti-estrogenic activity of tamoxifen N-oxide, *Biochem. Pharmacol.*, 31, 2823–2827, 1982.
333. McCague, R., Parr, I., and Haynes, B., Metabolism of the 4-iodo derivative of tamoxifen by isolated rat hepatocytes, *Biochem. Pharmacol.*, 40, 2277–2283, 1990.
334. McCague, R., Parr, I., Leclercq, G., Leung, O., and Jarman, M., Metabolism of tamoxifen by isolated rat hepatocytes, *Biochem. Pharmacol.*, 39, 1459–1465, 1990.
335. Jones, R.M., Yuan, Z.X., and Lim, C.K., Tamoxifen metabolism in rat liver microsomes: identification of a dimeric metabolite derived from free radical intermediates by liquid chromatography mass spectrometry, *Rapid Commun. Mass Spectrom.*, 13, 211–215, 1999.
336. Lim, C., Yuan, Z., Lamb, J., White, I., De Matteis, F., and Smith, L., A comparative study of tamoxifen metabolism in female rat, mouse and human liver microsomes, *Carcinogenesis*, 15, 589–593, 1994.
337. Lim, C., Lamb, J., Yuan, Z., and Smith, L., Identification of epoxide metabolites of tamoxifen by on-line liquid chromatography-electrospray ionization mass spectrometry, *Biochem. Soc. Trans.*, 22, 165s–166s, 1994.
338. Poon, G.K., Walter, B., Lonning, P., Horton, M., and McCague, R., Identification of tamoxifen metabolites in human Hep G2 cell line, human liver homogenate, and patients on long-term therapy for breast cancer, *Drug Metab. Dispos.*, 23, 377–382, 1995.
339. Lim, H.K., Stellingweif, S., Sisenwine, S., and Chan, K.W., Rapid drug metabolic profiling using fast liquid chromatography, automated multiple-stage mass spectrometry and receptor-binding, *J. Chromatogr. A*, 831, 227–241, 1999.
340. Lu, W., Poon, G.K., Carmichael, P., and Cole, R., Analysis of tamoxifen and its metabolites by on-line capillary electrophoresis-electrospray ionization mass spectrometry employing nonaqueous media containing surfactants, *Anal. Chem.*, 68, 668–674, 1996.
341. Gaskell, S., Daniel, C. and Nicholson, R. Determination of tamoxifen in rat plasma by gas chromatography-mass spectrometry, *J. Endocr.* 1978, 78, 293–294.
342. Murphy, C., Fotsis, T., Pantzar, P., Adlercreutz, H., and Martin, F., Analysis of tamoxifen and its metabolites in human plasma by gas chromatography–mass spectrometry (GC–MS) using selected ion monitoring (SIM), *J. Steroid Biochem.*, 26, 547–555, 1987.

343. Murphy, C., Fotsis, T., Adlercreutz, H., and Martin, F., Analysis of tamoxifen and 4–hydroxytamoxifen levels in immature rat uterine cytoplasm and KCl–nuclear extracts by gas chromatography-mass spectrometry (GC–MS) using selected ion monitoring (SIM), *J. Steroid Biochem.*, 28, 289–299, 1987.
344. Daniel, P., Gaskell, S., and Nicholson, R., The measurement of tamoxifen and metabolites in the rat and relationship to the response of DMBA–induced mammery tumours, *Europ. J. Cancer Clin. Oncol.*, 20, 137–143, 1984.
345. Lien, E.A., Solheim, E., and Ueland, P.M., Distribution of tamoxifen and its metabolites in rat and human tissues during steady-state treatment, *Cancer Res.*, 51, 4837–4844, 1991.
346. Poon, G.K., Chui, Y., McCague, R., Lonning, P., Feng, R., Rowlands, M., and Jarman, M., Analysis of Phase I and Phase II metabolites of tamoxifen in breast cancer patients, *Drug Metab. Dispos.*, 21, 1119–1124, 1993.
347. Murphy, C., Fotsis, T., Pantzar, P., Adlercreutz, H., and Martin, F., Analysis of tamoxifen, N-desmethyltamoxifen and 4-hydroxytamoxifen levels in cytosol and KCl-nuclear extracts of breast tumors from tamoxifen treated patients by gas chromatography-mass spectrometry (GC-MS) using selected ion monitoring (SIM), *J. Steroid Biochem.*, 28, 609–618, 1987.
348. Girault, J., Istin, B., and Fourtillan, J., Quantitative measurement of 4-hydroxy tamoxifen in human plasma and mammary tumors by combined gas chromatoography/negative chemical ionization mass spectrometry, *Biol. Mass Spectrom.*, 22, 395–402, 1993.
349. Giorda, G., Franceschi, L., Crivellari, D., Magri, M.D., Veronesi, A., Scarabelli, C., and Furlanut, M., Determination of tamoxifen and its metabolites in the endometrial tissue of long–term treated women, *Euro. J. Cancer,* S88–S89, 2000.
350. Wiebe, V., Osborne, C., McGuire, W., and DeGregorio, M., Identification of estrogenic tamoxifen metabolite(s) in tamoxifen-resistant human breast tumors, *J. Clin. Oncol.*, 10, 990–994, 1992.
351. Daniel, P., Gaskell, S., Bishop, H., Campbell, C., and Nicholson, R., Determination of tamoxifen and biologically active metabolites in human breast tumors and plasma, *Europ. J. Cancer Clin. Oncol.*, 17, 1183–1189, 1981.
352. Girault, J., Istin, B., and Fourtillan, J., Quantitative measurement of 4-hydroxy tamoxifen in human plasma and mammary tumors by combined gas chromatography/negative chemical ionization mass spectrometry, *Biol. Mass Spectrom.*, 22, 395–402, 1993.
353. Zweigenbaum, J. and Henion, J., Bioanalytical high-throughput selected reaction monitoring-LC/MS determination of selected estrogen receptor modulators in human plasma: 2000 samples/day, *Anal. Chem.*, 72, 2446–2454, 2000.
354. Onorato, J.M., Henion, J.D., Lefebvre, P.M., and Kiplinger, J.P., Selected reaction monitoring LC-MS determination of idoxifene and its pyrrolidinone metabolite in human plasma using robotic high-throughput, sequential sample injection, *Anal. Chem.*, 73, 119–125, 2001.
355. Zhang, H. and Henion, J., Comparison betwen liquid chromatography-time-of-flight mass spectrometry and selected reaction monitoring liquid chromatography-mass spectrometry for quantitative determination of idoxifene in human plasma, *J. Chromatogr. B,* 757, 151–159, 2001.
356. Mihailescu, R., Aboul–Enein, H.Y., and Efstatide, M.D., Identification of tamoxifen and metabolites in human male urine by GC/MS, *Biomed. Chromatogr.*, 14, 180–183, 2000.
357. Li, X.F., Carter, S., Dovichi, N.J., Zhao, J.Y., Kovarik, P., and Sakuma, T., Analysis of tamoxifen and its metabolites in synthetic gastric fluid digests and urine samples using high-performance liquid chromatography with electrospray mass spectrometry, *J. Chromatogr. A,* 914, 5–12, 2001.
358. Li, X.F., Carter, S.J., and Dovichi, N.J., Non-aqueous capillary electrophoresis of tamoxifen and its acid hydrolysis products, *J. Chromatogr. A,* 895, 81–85, 2000.
359. Jarman, M., Poon, G.K., Rowlands, M., Grimshaw, R., Horton, M., Potter, G., and McCague, R., The deuterium isotope effect for the alpha hydroxylation of tamoxifen by rat liver microsomes accounts for the reduced genotoxicity of [D_5-ethyl]tamoxifen, *Carcinogenesis,* 16, 683–688, 1995.
360. Daniel, P., Gaskell, S., Bishop, H., and Nicholson, R., Determination of tamoxifen and an hydroxylated metabolite in plasma from patients with advanced breast cancer using gas chromatography-mass spectrometry, *J. Endocr.*, 83, 401–408, 1979.
361. Parr, I., McCague, R., Leclercq, G., and Stoessel, S., Metabolism of tamoxifen by isolated rat hepatocytes, *Biochem. Pharmacol.*, 36, 1513–1519, 1987.
362. Phillips, D., Carmichael, P., Hewer, A., Cole, K., and Poon, G.K., α-Hydroxytamoxifen, a metabolite of tamoxifen with exceptionally high DNA-binding activity in rat hepatocytes, *Cancer Res.*, 54, 5518–5522, 1994.

363. Phillips, D.H., Carmichael, P.L., Hewer, A., Cole, K.J., Hardcastle, I.R., Poon, G.K., Keogh, A., and Strain, A.J., Activation of tamoxifen and its metabolite α-hydroxytamoxifen to DNA-binding products: comparisons between human, rat and mouse hepatocytes, *Carcinogenesis,* 17, 89–94, 1996.
364. Carmichael, L., Ugwamadu, A., Neven, P., Hewer, A., Poon, G.K., and Phillips, D., Lack of genotoxicity of tamoxifen in human endometrium, *Cancer Res.,* 56, 1475–1479, 1996.
365. Boocock, D.J., Maggs, J.L., White, I.N., and Park, B.K., α-Hydroxytamoxifen, a genotoxic metabolite of tamoxifen in the rat: identification and quantification *in vivo* and *in vitro*, *Carcinogenesis,* 20, 153–160, 1999.
366. Boocock, D.J., Maggs, J.L., Brown, K., White, I.N., and Park, B.K., Major inter-species differences in the rates of O-sulphonation and O-glucuronylation of alpha-hydroxytamoxifen *in vitro*: a metabolic disparity protecting human liver from the formation of tamoxifen-DNA adducts, *Carcinogenesis,* 21, 1851–1858, 2000.
367. Davies, A., Martin, E., Jones, R., Lim, C., Smith, L., and White, I., Peroxidase activation of tamoxifen and toremifene resulting in DNA damage and covalently bound protein adducts, *Carcinogenesis,* 16, 539–545, 1995.
368. Lim, C.K., Yuan, Z.X., Jones, R.M., White, I.N., and Smith, L.L., Identification and mechanism of formation of potentially genotoxic metabolites of tamoxifen: study by LC-MS/MS, *J. Pharm. Biomed. Anal.,* 15, 1335–1342, 1997.
369. Moorthy, B., Sriram, P., Pathak, D.N., Bodell, W.J., and Randerath, K., Tamoxifen metabolic activation: comparison of DNA adducts formed by microsomal and chemical activation of tamoxifen and 4-hydroxytamoxifen with DNA adducts formed *in vivo*, *Cancer Res.,* 56, 53–57, 1996.
370. Marques, M.M. and Beland, F.A., Identification of tamoxifen-DNA adducts formed by 4-hydroxytamoxifen quinone methide, *Carcinogenesis,* 18, 1949–1954, 1997.
371. Fan, P.W., Zhang, F., and Bolton, J.L., 4-Hydroxylated metabolites of the antiestrogens tamoxifen and toremifene are metabolized to unusually stable quinone methides, *Chem. Res. Toxicol.,* 13, 45–52, 2000.
372. Zhang, F., Fan, P.W., Liu, X., Shen, L., van Breeman, R.B., and Bolton, J.L., Synthesis and reactivity of a potential carcinogenic metabolite of tamoxifen: 3,4-dihydroxytamoxifen-o-quinone, *Chem. Res. Toxicol.*, 13, 53–62, 2000.
373. Rajaniemi, H., Rasanen, I., Koivisto, P., Peltonen, K., and Hemminki, K., Identification of the major tamoxifen-DNA adducts in rat liver by mass spectroscopy, *Carcinogenesis,* 20, 305–309, 1999.
374. Martin, E.A., Carthew, P., White, I.N.H., Heydon, R.T., Gaskell, M., Mauthe, J., Turteltaub, K.W., and Smith, L.L., Investigation of the formation and accumulation of liver DNA adducts in mice chronically exposed to tamoxifen, *Carcinogenesis,* 18, 2209–2215, 1997.
375. Gamboa da Costa, G., Hamilton, L.P., Beland, F.A., and Marques, M.M., Characterization of the major DNA adduct formed by alpha-hydroxy-N-desmethyltamoxifen *in vitro* and *in vivo*, *Chem. Res. Toxicol.,* 13, 200–207, 2000.
376. Maenpaa, H., Holli, K., and Pasanen, T., Toremifene: where do we stand?, *Eur. J. Cancer,* 36, S61–S62, 2000.
377. Watanabe, N., Irie, T., Koyama, M., and Tominaga, T., Liquid chromatographic-atmospheric pressure ionization mass spectrometric analysis of toremifene metabolites in human urine, *J. Chromatogr.,* 497, 169–180, 1989.
378. White, I.N., Martin, E.A., Mauthe, R.J., Vogel, J.S., Turteltaub, K.W., and Smith, L.L., Comparisons of the binding of $[^{14}C]$radiolabelled tamoxifen or toremifene to rat DNA using accelerator mass spectrometry, *Chem. Biol. Interact.,* 106, 149–160, 1997.
379. Siegel, M., Tsou, H., Lin, B., Hollander, I., Wissner, A., Karas, M., Ingendoh, A., and Hillenkamp, F., Determination of the loading values for high levels of drugs and sugars conjugated to proteins by matrix-assisted ultraviolet laser desorption/ionization mass spectrometry, *Biol. Mass Spectrom., 22,* 369–376, 1993.
380. Siegel, M., Hollander, I., Hamann, P., James, J., Hinman, L., Smith, B., Farnsworth, A., Phipps, A., King, D., Karas, M., Ingendoh, A., and Hillankamp, F., Matrix-assisted UV-laser desorption/ionization mass spectrometric analysis of monocloncal antibodies for the determination of carbohydrate, conjugate chelator, and conjugate drug content, *Anal. Chem.*, 63, 2470–2481 1991.
381. Siegel, M.M., Tabei, K., Kunz, A., Hollander, I.J., Hamann, P.R., and Bell, D.H., Calicheamicin derivatives conjugated to monoclonal antibodies: Determination of loading values and distributions by infrared and UV matrix-assisted laser desorption/ionization mass spectrometry and electrospray ionization mass spectrometry, *Anal. Chem.,* 69, 2716–2726, 1997.

382. Pignatello, R., Spampinato, G., Sorrenti, V., Di Giacomo, C., Vicari, L., McGuire, J.J., Russell, C.A., Puglisi, G., and Toth, I., Lipophilic methotrexate conjugates with antitumor activity, *Eur. J. Pharmacol. Sci.,* 10, 237–245, 2000.
383. Koller-Lucae, S.K.M., Suter, M.J.F., Rentsch, K.M., Schott, H., and Schwendener, R.A., Metabolism of the new liposomal anticancer drug N^4-octadecyl-1-β-d-arabinofuranoscytosine in mice, *Drug Metab. Dispos.,* 27, 342–350, 1999.
384. Rak, J., St. Crox, B., and Kerbel, R., Consequences of angiogenesis for tumor progression, metastasis and cancer therapy, *Anticancer Drugs,* 6, 3–18, 1995.
385. Folkman, J., Clinical applications of research on angiogenesis, *New Eng. J. Med.,* 333, 1757–1763, 1995.
386. Jones, A. and Harris, A.L., New developments in angiogenesis: a major mechanism for tumor growth and target for therapy, *The Cancer Journal,* 4, 209–217, 1998.
387. O'Reilly, M.S., Boehm, T., Shing, Y., Fukai, N., Vasios, G., Lane, W.S., Flynn, E., Birkhead, J.R., Olsen, B.R., and Folkman, J., Endostatin: an endogenous inhibitor of angiogenesis and tumor growth, *Cell,* 88, 277–285, 1998.
388. Schrader, M., Jurgens, M., Hess, R., Schulz Knappe, P., Raida, M., and Forssmann, W.G., Matrix-assisted laser desorption/ionisation mass spectrometry guided purification of human guanylin from blood ultrafiltrate, *J. Chromatogr. A,* 776, 139–145, 1997.
389. Taylor, J.A., Walsh, K.A., and Johnson, R.S., Sherpa: a Macintosh-based expert system for the interpretation of electrospray ionization LC/MS and MS/MS data from protein digests, *Rapid Commun. Mass Spectrom.,* 10, 679–687, 1996.
390. Standker, L., Schrader, M., Kanse, S.M., Jurgens, M., Forsmann, W.G., and Preissner, K.T., Isolation and characterization of the circulating form of human endostatin, *FEBS Lett.,* 420, 129–133, 1997.
391. John, H. and Forssmann, W.G., Determination of the disulfide bond pattern of the endogenous and recombinant angiogenesis inhibitor endostatin by mass spectrometry, *Rapid Commun. Mass Spectrom.,* 15, 1222–1228, 2001.
392. Placidi, L., Cretton-Scott, E., Sousa, G., Rahmani, R., Placidi, M., and Sommadossi, J.P., Disposition and metabolism of the angiogenic moderator O-(chloroacetylcarbamoyl) fumagillol (TNP-470) in human hepatocytes and tissue microsomes, *Cancer Res.,* 55, 3036–3042, 1995.
393. Moore, J.D., Sommadossi, J.P., and ACTG 215 Study Team, Determination of O-(chloroacetyl-carbamoyl)fumagillol (TP-470) and two metabolites in plasma by high-performance liquid chromatography/mass spectrometry with atmospheric pressure chemical ionization, *J. Mass Spectrom.,* 30, 1707–1715, 1995.
394. Placidi, L., Scott, E.C., DeSousa, G., Rahmani, R., Placidi, M., and Sommadossi, J.-P., Interspecies variability of TNP-470 metabolism, using primary monkey, rat, and dog cultured hepatocytes, *Drug Metab. Dispos.,* 25, 94–99, 1997.
395. Griffith, E.C., Su, Z., Turk, B.E., Chen, S., Chang, Y., Wu, Z., Biemann, K., and Liu, J.O., Methionine aminopeptidase (type 2) is the common target for angiogenesis inhibitors AGM-1470 and ovalicin, *Chemistry & Biology,* 4, 461–471, 1997.
396. Moore, J.D., Dezube, B.J., Gill, P., Zhou, X.J., Acosta, E.P., and Sommadossi, J.P., Phase I dose escalation pharmacokinetics of O-(chloroacetylcarbamoyl)fumagillol (TNP-470) and its metabolites in AIDS patients with Kaposi's sarcoma, *Cancer Chemother. Pharmacol.,* 46, 173–179, 2000.
397. Fotsis, T., Pepper, M., Adlercreutz, H., Hase, T., Montesano, R., and Schweigerer, L., Genistein, a dietary ingested isoflavonoid, inhibits cell proliferation and *in vitro* angiogenesis, *J. Nutr.,* 125, 790S–797S, 1995.
398. Adlercreutz, H., Fotsis, T., Bannwart, C., Wahala, K., Bruow, G., and Hase, T., Isotope dilution gas chromatographic-mass spectrometric method for the determination of lignans and isoflavonoids in human urine, including identification of genistein, *Clinica Chimica Acta,* 199, 263–278, 1991.
399. Cotter, F.E., Antisense therapy for lymphomas, *Hemat. Oncol.,* 15, 3–11, 1997.
400. Gewitz, A.M., Antisense oligonucleotide therapeutics for human leukemia, *Critical Rev. Oncogenesis,* 8, 93–109, 1997.
401. Yu, R.Z., Geary, R.S., Leeds, J.M., Watanabe, T., Fitchett, J.R., Watson, J.E., Mehta, R., Hardee, G.R., Templin, M.V., Huang, K., Newman, M.S., Quinn, Y., Uster, P., Zhu, G., and Levin, A.A., Pharmacokinetics and tissue disposition in monkeys of an antisense oligonucleotide inhibitor of H-ras encapsulated in stealth liposomes, *Pharm. Res.,* 16, 1309–1315, 1999.

402. Crain, P.F., Characterization of oligonucleotides by electrospray mass spectrometry, in *Mass Spectrometry of Biological Materials*, Larson, B.S. and McEwen, C.N., Eds., Marcel Dekker, New York, 389–404, 1998.
403. Madhusudanan, K.P., Katti, S.B., Haq, W., and Misra, P.K., Antisense peptide interactions studied by electrospray ionization mass spectrometry, *J. Mass Spectrom.*, 35, 237–241, 2000.
404. Gaus, H.J., Owens, S.R., Winniman, M., Cooper, S., and Cummins, L.L., On-line HPLC electrospray mass spectrometry of phosphorothioate oligonucleotide metabolites, *Anal. Chem.*, 69, 313–319, 1997.
405. Beaupre, D.M. and Kurzrock, R., *Ras* inhibitors in hematologic cancers: biologic considerations and clinical applications, *Invest. New Drugs*, 17, 137–143, 1999.
406. Scharovsky, O.G., Rozados, V.R., Gervasoni, S.I., and Matar, P., Inhibition of *ras* oncogene: a novel approach to antineoplastic therapy, *J. Biomed. Sci.*, 7, 292–298 2000.
407. Traveras, A.G., Remiszewski, S.W., Doll, R.J., Cesarz, D., Huang, E.C., Kirschmeier, P., Pramanik, B.N., Snow, M.E., Wang, Y.S. et al., *Ras* oncoprotein inhibitors: the discovery of potent, *ras* nucleotide exchange inhibitors and structural determination of a drug-protein complex, *Bioorg. Med. Chem.*, 5, 125–133, 1997.
408. Huang, E., Pramanik, B.N., Tsarbopoulos, A., Reichert, P., Ganguly, A.K., Trotta, P., and Nagabhushan, T., Application of electrospray mass spectrometry in probing protein-protein-ligand noncovalent interactions, *J. Am. Soc. Mass Spectrom.*, 4, 624–630, 1993.
409. Ganguly, A.K., Pramanik, B.N., Huang, E.C., Liberles, S., Heimark, L., Liu, Y.H., Tsarbopoulos, A., Doll, R.J., Taveras, A.G., Remiszewski, S., Snow, M.E., Wang, Y.S., Vibulbhan, B., Cesarz, D., Brown, J.E., del Rosario, J., James, L., Kirschmeier, P., and Girijavallabhan, V., Detection and structural characterization of *ras* oncoprotein-inhibitor complexes by electrospray mass spectrometry, *Bioorg. Med. Chem.*, 5, 817–820, 1997.
410. Springer, C.J., Poon, G.K., Sharma, S., and Bagshawe, K., Identification of prodrug, active drug, and metabolites in an ADEPT clinical study, *Cell Biophys.*, 22, 9–26, 1993.
411. Springer, C.J., Poon, G.K., Sharma, S., and Bragshawe, K., Analysis of antibody-enzyme conjugate clearance by investigation or prodrug and active drug in an ADEPT clinical study, *Cell Biophysics*, 24/25, 193–207, 1994.
412. Martin, J., Stribbling, S.M., Poon, G.K., Begent, R.H., Napier, M., Sharma, S.K., and Springer, C.J., Antibody-directed enzyme prodrug therapy: pharmacokinetics and plasma levels of prodrug and drug in a phase I clinical trial, *Cancer Chemother. Pharmacol.*, 40, 189–201, 1997.
413. Musselman, B., Mass spectrometry of porphyrin photosensitizers, in *Mass Spectrometry of Biological Materials*, McEwen, C. and Larsen, B., Eds., Marcel Dekker, New York, 403–411, 1990.
414. Zhan, Q., Voumard, P., and Zenobi, R., Chemical analysis of cancer therapy photosensitizers by two-step laser mass spectrometry, *Anal. Chem.*, 66, 3259–3266, 1994.
415. Siegel, M.M., Tabei, K., Tsao, R., Pastel, M.J., Pandey, R.K., Berkenkamp, S., Hillenkamp, F., and de Vries, M.S., Comparative mass spectrometric analyses of Photofrin oligomers by fast atom bombardment mass spectrometry, UV and IR matrix-assisted laser desorption/ionization mass spectrometry, electrospray ionization mass spectrometry and laser desorption/jet-cooling photoionization mass spectrometry, *J. Mass Spectrom.*, 34, 661–669, 1999.
416. Lord, G.A., Cai, H., Luo, J.L., and Lim, C.K., HPLC-electrospray mass spectrometry for the analysis of temoporfin-poly(ethylene glycol) conjugates, *Analyst*, 125, 605–608, 2000.
417. Berlin, K., Jain, R.K., Tetzlaff, C., Steinbeck, C., and Richert, C., Spectrometrically monitored selection experiments: quantitative laser desorption mass spectrometry of small chemical libraries, *Chem. Biol.*, 4, 63–77, 1997.
418. Berlin, K., Jain, R.K., and Richert, C., Are porphyrin mixtures favorable photodynamic anticancer drugs? A model study with combinatorial libraries of tetraphenylporphyrins, *Biotechnol. Bioeng.*, 61, 107–118, 1998.
419. Cai, H., Wang, Q., Luo, J., and Lim, C.K., Study of temoporfin metabolism by HPLC and electrospray mass spectrometry, *Biomed. Chromatogr.*, 13, 354–359, 1999.
420. Egorin, M., Snyder, S., Cohen, A., Zuhowski, E., Subramanyam, B., and Callery, P., Metabolism of hexamethylene bisacetamide and its metabolites in leukemic cells, *Cancer Res.*, 48, 1712–1716, 1988.
421. Roth, J., Kelley, J., Chun, H., and Ward, F., Simultaneous measurement of the cell-differentiating agent hexamethylene bisacetamide and its metabolites by gas chromatography, *J. Chromatogr. B*, 652, 149–159, 1994.

422. Subramanyam, B., Callery, P., Egorin, M., Snyder, S., and Conley, B., An active, aldehydic metabolite of the cell-differentiating agent hexamethylene bisacetamide, *Drug Metab. Dispos.*, 17, 398–401, 1989.
423. Shen, Z.-X., Chen, G.-Q., Ni, J.-H., Li, X.-S., Xiong, S.-M., Qiu, Q.-Y., Zhu, J., Tang, W., Sun, G.-L., Yang, K.-K., Chen, Y., Zhou, L., Fang, Z.-W., Wang, Y.-T., Ma, J., Zhang, P., Zhang, T.-D., Chen, S.-J., Chen, Z., and Wang, Z.-Y., Use of arsenic trioxide (As_2O_3) in the treatment of acute promyelocytic leukemia (APL): ll. clinical efficacy and pharmacokinetics in relapsed patients, *Blood*, 89, 3354–3360, 1997.
424. Slack, J.L. and Rusiniak, M.E., Current issues in the management of acute promyelocytic leukemia, *Ann. Hematol.*, 79, 227–238, 2001.
425. Zhu, X.H., Shen, Y.L., Jing, Y., Cai, X., Jia, P.M., Huang, Y., Tang, W., Shi, G.Y., Sun, Y.P., Dai, J., Wang, Z.Y., Chen, S.J., Zhang, T.D., Waxman, S., Chen, Z., and Chen, G.Q., Apoptosis and growth inhibition in malignant lymphocytes after treatment with arsenic trioxide at clinically achievable concentrations, *J. Natl. Cancer Inst.*, 91, 772–778, 1999.
426. Roboz, G.J., Dias, S., Lam, G., Lane, W.J., Soignet, S.L., Warrell, R.P., Jr., and Rafii, S., Arsenic trioxide induces dose- and time-dependent apoptosis of endothelium and may exert an antileukemic effect via inhibition of angiogenesis, *Blood*, 96, 1525–1530, 2000.
427. Soignet, S.L., Tong, W.P., Hirschfeld, S., and Warrell, R.P., Clinical study of an organic arsenical, melarsoprol, in patients with advanced leukemia, *Cancer Chemother. Pharmacol.*, 44, 417–421, 1999.
428. Konig, A., Wrazel, L., Warrell, R.P., Rivi, R., Pandolfi, P.P., Jakubowski, A., and Gabrilove, J., Comparitive activity of melarsoprol and arsenic trioxide in chronic B-cell leukemia lines, *Blood*, 90, 562–570, 1997.
429. Hong, W.K. and Sporn, M.B., Recent advances in chemoprevention of cancer, *Science*, 278, 1073–1077, 1997.
430. Waladkhani, A.R. and Clemens, M.R., Effect of dietary phytochemicals on cancer development, *Int. J. Mol. Med.*, 1, 747–753, 1998.
431. Singh, D.K. and Lippman, S., Cancer chemoprevention part 1: retinoids and carotenoids and other classic antioxidants, *Oncology*, 12, 1643–1660, 1998.
432. Stoner, G.D., Morse, M.A., and Kelloff, G.J., Perspectives in cancer chemoprevention, *Environ. Health Perspect.*, 105, 945–954, 1997.
433. Rui, H., Research and development of cancer chemopreventive agents in China, *J. Cell Biochem. Suppl.*, 27, 7–11, 1997.
434. Ames, B.N., Micronutrients prevent cancer and delay aging, *Toxicol. Lett.*, 103, 5–18, 1998.
435. Boone, C.W. and Kelloff, G.J., Biomarker end-points in cancer chemoprevention trials, *IARC Sci. Publ.*, 273–280, 1997.
436. Sporn, M.B. and Suh, N., Chemoprevention of cancer, *Carcinogensis*, 21, 525–530, 2000.
437. Singh, D.K. and Lippman, S., Cancer chemoprevention part 2: nonclassic antioxidant natural agents, NSAIDs, and other agents, *Oncology*, 12, 1787–1805, 1998.
438. Balogh, M.P., Advancing phytochemical development: characterization of plant extract isoflavonoids using LC-MS, *LC-GC*, 5, 456–466, 1997.
439. Soleas, G.J., Yan, J., and Goldberg, D.M., Ultrasensitive assay for three polyphenols (catechin, quercetin and resveratrol) and their conjugates in biological fluids utilizing gas chromatography with mass selective detection, *J. Chromatogr. B*, 57, 161–172, 2001.
440. Setchell, K.D. and Cassidy, A., Dietary isoflavones: biological effects and relevance to human health, *J. Nutr.*, 129, 758–767, 1999.
441. Adlercreutz, H. and Mazur, W., Phytoestrogens and western diseases, *Ann. Med.*, 29, 95–120, 1997.
442. Mazur, W.M., Wahala, K., Rasku, S., Salakka, A., Hase, T., and Adlercreutz, H., Lignan and isoflavonoid concentrations in tea and coffee, *Br. J. Nutr.*, 79, 37–45, 1998.
443. Stobiecki, M., Application of mass spectrometry for identification and structural studies of flavonoid glycosides. *Phytochem.*, 54, 237–256, 2000.
444. Justesen, U., Collision-induced fragmentation of deprotonated methoxylated flavonoids, obtained by electrospray ionization mass spectrometry, *J. Mass Spectrom.*, 36, 169–178, 2001.
445. Satterfield, M. and Brodbelt, J.S., Enhanced detection of flavonoids by metal complexation and electrospray ionization mass spectrometry, *Anal. Chem.*, 72, 5898–5906, 2000.
446. Mazur, W., Fotsis, T., Wahala, K., Ojala, S., Salakka, A., and Adlercreutz, H., Isotope dilution gas chromatographic-mass spectrometric method for the determination of isoflavonoids, coumestrol, and lignans in food samples, *Anal. Biochem.*, 233, 169–180, 1996.

447. Adlercreutz, H., Fotsis, T., Lampe, J., Wahala, K., Makela, T., Brunow, G., and Hase, T., Quantitative determination of lignans and isoflavonoids in plasma of omnivorous and vegetarian women by isotope dilution gas chromatography-mass spectrometry, *Scand. J. Clin. Lab. Invest.*, 53 (Suppl. 215), 5–18, 1993.
448. Dwyer, J.T., Goldin, B.R., Saul, N., Gualtieri, L., Barakat, S., and Adlercreutz, H., Tofu and soy drinks contain phytoestrogens, *J. Am. Diet. Assoc.*, 94, 739–743, 1994.
449. Slavin, J.L., Phytoestrogens in breast milk—another advantage of breast feeding?, *Clin. Chem.*, 42, 841–842, 1996.
450. Franke, A.A. and Custer, L.J., Daidzein and genistein concentrations in human milk after soy consumption, *Clin. Chem.*, 42, 955–964, 1996.
451. Franke, A.A., Custer, L.J., and Tanaka, Y., Isoflavones in human breast milk and other biological fluids, *Am. J. Clin. Nutr.*, 68, 1466S–1473S, 1998.
452. Setchell, K., Zimmer-Nechemias, L., Cai, J., and Heubi, J.E., Isoflavone content of infant formulas and the metabolic fate of these phytoestrogens in early life, *Am. J. Nutr.*, 68 (Suppl.), 1453S–1461S, 1998.
453. Setchell, K.D., Exposure of infants to phyto-oestrogens from soy-based infant formula, *Lancet*, 350, 23–27, 1997.
454. Barnes, K.A., Smith, R.A., Williams, K., Damant, A.P., and Shepherd, M.J., A microbore high-performance liquid chromatography/electrospray ionization mass spectrometry method for the determination of the phytoestrogens genistein and diadzein in comminuted baby foods and soya flour, *Rapid Commun. Mass Spectrom.*, 12, 130–138, 1998.
455. Coward, L., Smith, M., Kirk, M., and Barnes, S., Chemical modification of isoflavones in soyfoods during cooking and processing, *Am. J. Clin. Nutr.*, 68 (Suppl.), 1486S–1491S, 1998.
456. Hakkinen, S. and Auriola, S., High-performance liquid chromatography with electrospray ionization mass spectrometry and diode array ultraviolet detection in the identification of flavonol aglycones and glycosides in berries, *J. Chromatogr. A*, 829, 91–100, 1998.
457. Mazur, W.M., Uehara, M., Wahala, K., and Adlercreutz, H., Phyto-oestrogen content of berries, and plasma concentrations and urinary excretion of enterolactone after a single strawberry-meal in human subjects, *Br. J. Nutr.*, 83, 381–387, 2000.
458. Justesen, U., Negative atmospheric pressure chemical ionisation low-energy collision activation mass spectrometry for the characterisation of flavonoids in extracts of fresh herbs, *J. Chromatogr. A*, 902, 369–379, 2001.
459. Nielsen, S.E., Freese, R., Cornett, C., and Dragsted, L.O., Identification and quantification of flavonoids in human urine samples by column-switching liquid chromatography coupled to atomospheric pressure chemical ionization mass spectrometry, *Anal. Chem.*, 72, 1503–1509, 2000.
460. Doerge, D.R., Churchwell, M.I., and Delclos, K.B., On-line sample preparation using restricted-access media in the analysis of the soy isoflavones, genistein and daidzein, in rat serum using liquid chromatography electrospray mass spectrometry, *Rapid Commun. Mass Spectrom.*, 14, 673–678, 2001.
461. Adlercreutz, H., Honjo, H., Higashi, A., Fotsis, T., Hamalainen, E., Hasegawa, T., and Okada, H., Urinary excretion of lignans and isoflavonoid phytoestrogens in Japanese men and women consuming a traditional Japanese diet, *Am. J. Clin. Nutr.*, 54, 1093–1100, 1991.
462. Hutchins, A., Lampe, J., Martini, M., Campbell, D., and Salvin, J., Vegetables, fruits, and legumes: effect on urinary isoflavonoid phytoestrogen and lignan excretion, *J. Am. Diet. Assoc.*, 95, 769–774, 1995.
463. Lampe, J.W., Karr, S.C., Hutchins, A.M., and Slavin, J.L., Urinary equol excretion with a soy challenge: influence of habitual diet, *Proc. Soc. Exp. Biol. Med.*, 217, 335–339 1998.
464. Hutchins, A.M., Slavin, J.L., and Lampe, J.W., Urinary isoflavonoid phytoestrogen and lignan excretion after consumption of fermented and unfermented soy products, *J. Am. Diet. Assoc.*, 95, 545–551, 1995.
465. Kirkman, L., Lampe, J., Campbell, D., Martini, M., and Slavin, J., Urinary lignan and isoflavonoid excretion in men and woman consuming vegetable and soy diets, *Nutr. Cancer*, 24, 1–12, 1995.
466. Lampe, J.W., Martini, M.C., Kurzer, M.S., Adlercreutz, H., and Slavin, J.L., Urinary lignan and isoflavonoid excretion in premenopausal women consuming flaxseed powder, *Am. J. Clin. Nutr.*, 60, 122–128, 1994.
467. Wahala, K., Hase, T., and Adlercreutz, H., Synthesis and labeling of isoflavone phytoestrogens, including daidzein and genistein, *Proc. Soc. Exp. Biol. Med.*, 208, 27–32, 1995.
468. Adlercreutz, H., van der Wildt, J., Kinzel, J., Attalla, H., Wahala, K., Makela, T., Hase, T., and Fotsis, T., Lignan and isoflavonid conjugates in human urine, *J. Steroid Biochem. Molec. Biol.*, 52, 97–103, 1995.

469. Adlercreutz, H., Gorbach, S.L., Goldin, B.R., Woods, M.N., Dwyer, J.T., and Hamalainen, E., Estrogen metabolism and excretion in Oriental and Caucasian women, *J. Natl. Cancer Inst.*, 86, 1076–1082, 1994.
470. Horn Ross, P.L., Barnes, S., Kirk, M., Coward, L., Parsonnet, J., and Hiatt, R.A., Urinary phytoestrogen levels in young women from a multiethnic population, *Cancer Epidemiol. Biomark. Prev.*, 6, 339–345, 1997.
471. Gross, M., Pfeiffer, M., Martini, M., Campbell, D., Slavin, J., and Potter, J., The quantitation of metabolites of quercetin flavonols in human urine, *Cancer Epidemiol. Biomark. Prev.*, 5, 711–720, 1996.
472. Pietta, P., Gardana, C., Mauri, P., Maffei-Facino, R., and Carini, M., Identification of flavonoid metabolites after oral administration to rats of a ginkgo bioba extract, *J. Chromatogr. B*, 673, 73–80, 1995.
473. Watson, D.G. and Pitt, A.R., Analysis of flavonoids in tablets and urine by gas chromatography/mass spectrometry and liquid chromatography/mass spectrometry, *Rapid Commun. Mass Spectrom.*, 12, 153–156, 1998.
474. Watson, D.G. and Oliveira, E.J., Solid-phase extraction and gas chromatography-mass spectrometry determination of kaempferol and quercetin in human urine after consumption of *Ginkgo biloba* tablets, *J. Chromatogr. B*, 723, 203–210, 1999.
475. Adlercreutz, H., Fotsis, T., Watanabe, S., Lampe, J., Wahala, K., Makela, T., and Hase, T., Determination of lignans and isoflavonoids in plasma by isotope dilution gas chromatography-mass spectrometry, *Cancer Detect. Prevent.*, 18, 259–271, 1994.
476. Zeleniuch-Jacquotte, A., Adlercreutz, H., Akhmedkhanov, A., and Toniolo, P., Reliability of serum measurements of lignans and isoflavoid phytoestrogens over a two-year period, *Cancer Epidemiol. Biomark. Prev.*, 7, 885–889, 1998.
477. Gooderham, M.H., Adlercreutz, H., Ojala, S.T., Wahala, K., and Holub, B.J., A soy protein isolate rich in genistein and daidzein and its effects on plasma isoflavone concentrations, platelet aggregation, blood lipids and fatty acid composition of plasma phospholipid in normal men, *J. Nutr.*, 126, 2000–2006, 1996.
478. Morton, M.S., Wilcox, G., Wahlgvist, M.L., and Griffths, K., Determination of lignans and isoflavonoids in human female plasma following dietary supplementation, *J. Endocrin.*, 142, 251–259, 1994.
479. Aramendia, M., Garcia, I., Lafont, F., and Marinas, J., Determination of isoflavones using high-performance liquid chromatography with atmospheric-pressure chemical ionization mass spectrometry, *Rapid Commun. Mass Spectrom.*, 9, 503–508, 1995.
480. Coward, L., Kirk, M., Albin, N., and Barnes, S., Analysis of plasma isoflavones by reversed-phase HPLC-multiple reaction ion monitoring-mass spectrometry, *Clin. Chim. Acta*, 247, 121–142, 1996.
481. Evans, B.A., Griffiths, K., and Morton, M.S., Inhibition of 5 alpha-reductase in genital skin fibroblasts and prostate tissue by dietary lignans and isoflavonoids, *J. Endocrinol.*, 147, 295–302, 1995.
482. Morton, M.S., Chan, P.S., Cheng, C., Blacklock, N., Matos Ferreira, A., Abranches Monteiro, L., Correia, R., Lloyd, S., and Griffiths, K., Lignans and isoflavonoids in plasma and prostatic fluid in men: samples from Portugal, Hong Kong, and the United Kingdom, *Prostate*, 32, 122–128, 1997.
483. Morton, M.S., Matos-Ferreira, A., Abranches-Monteiro, L., Correia, R., Blacklock, N., Chan, P.S.F., Cheng, C., Lloyd, S., Chieh-Ping, W., and Griffiths, K., Measurement and metabolism of isoflavonoids and lignans in the human male. *Cancer Lett.*, 114, 145–151, 1997.
484. Griffiths, K., Denis, L., Turkes, A., and Morton, M.S., Possible relationship between dietary factors and pathogenesis of prostate cancer, *Int. J. Urol.*, 5, 195–213, 1998.
485. Finlay, E.M., Wilson, D.W., Adlercreutz, H., and Griffiths, K., The identification and measurement of phyto-oestrogens in human saliva, plasma, breast aspirate or cyst fluid, and prostatic fluid using gas chromatography-mass spectrometry, *J. Endocrin.*, 129 (Suppl.), 49–50, 1991.
486. Adlercreutz, H., Fotsis, T., Kurzer, M.S., Wahala, K., Makela, T., and Hase, T., Isotope dilution gas chromatographic-mass spectrometric method for the determination of unconjugated lignans and isoflavonoids in human feces, with preliminary results in omnivorous and vegetarian women, *Anal. Biochem.*, 225, 101–108 1995.
487. Kurzer, M.S., Lampe, J.W., Martin, M.C., and Adlercreutz, H., Fecal lignan and isoflavonoid excretion in premenopausal women consuming flaxseed powder, *Cancer Epidemiol. Biomark. Prev.*, 4, 353–358, 1995.

488. Sfakianos, J., Coward, L., Kirk, M., and Barnes, S., Intestinal uptake and biliary excretion of the isoflavone genistein in rats, *J. Nutr.*, 127, 1260–1268, 1997.
489. Barnes, S., Sfakianos, J., Coward, L., and Kirk, M., Soy isoflavonoids and cancer prevention, *Adv. Exp. Med. Biol.*, 401, 87–100, 1996.
490. Justesen, U. and Arrigoni, E., Electrospray ionization mass spectrometric study of degradation products of quercetin, quercetin-3-glucoside and quercetin-3-rhamnoglucoside, produced by in vitro fermentation with human fecal flora, *Rapid Commun. Mass Spectrom.*, 15, 477–483, 2001.
491. Watanabe, S., Yamaguchi, M., Sobue, T., Takahashi, T., Miura, T., Arai, Y., Mazur, W., Wahala, K., and Adlercreutz, H., Phamarcokinetics of soybean isoflavones in plasma, urine and feces of men after ingestion of 60 g baked soybean poweder (kinako), *J. Nutr.*, 128, 1710–1715, 1998.
492. Heinonen, S., Wahala, K., and Adlercreutz, H., Identification of isoflavone metabolites dihydrodaidzein, dihydrogenistein, 6′-OH-O-dma, and cis-4-OH-equol in human urine by gas chromatography-mass spectroscopy using authentic reference compounds, *Anal. Biochem.*, 274, 211–219, 1999.
493. Niemeyer, H.B., Honig, D., Lange-Bohmer, A., Jacobs, E., Kulling, S.E., and Metzler, M., Oxidative metabolites of the mammalian lignans enterodiol and enterolactone in rat bile and urine, *J. Agric. Food Chem.*, 48, 2910–2919, 2000.
494. Kulling, S.E., Honig, D.M., Simat, T.J., and Metzler, M., Oxidative *in vitro* metabolism of the soy phytoestrogens daidzein and genistein, *J. Agric. Food Chem.*, 48, 4963–72, 2000.
495. Adlercreutz, H., Fotsis, T., Hockerstedt, K., Hamalainen, E., Bannwart, C., Bloigu, S., Valtonen, A., and Ollus, A., Diet and urinary estrogen profile in premenopausal omnivorous and vegetarian women and in premenopausal women with breast cancer, *J. Steroid Biochem.*, 34, 1–6, 1989.
496. Telang, N.T., Katdare, M., Bradlow, H.L., and Osborne, M.P., Estradiol metabolism: an endocrine biomarker for modulation of human mammary carcinogensis, *Environ. Health Perspect.*, 105, 559–564, 1997.
497. Xu, X., Duncan, A.M., Merz, B.E., and Kurzer, M.S., Effects of soy isoflavones on estrogen and phytoestrogen metabolism in premenopausal women, *Cancer Epidemiol. Biomark. Prev.*, 7, 1101–1108, 1998.
498. Xu, X., Duncan, A.M., Wangen, K.E., and Kurzer, M.S., Soy consumption alters endogenous estrogen metabolism in postmenopausal women, *Cancer Epidemiol. Biomark. Prev.*, 9, 781–786, 2000.
499. Landstrom, M., Zhang, J.X., Hallmans, G., Aman, P., Bergh, A., Damber, J.E., Mazur, W., Wahala, K., and Adlercreutz, H., Inhibitory effects of soy and rye diets on the development of Dunning R3327 prostate adenocarcinoma in rats, *Prostate*, 36, 151–161, 1998.
500. Fotsis, T., Pepper, M., Adlercreutz, H., Hase, T., Montesano, R., and Schweigerer, L., Genistein, a dietary ingested isoflavonoid, inhibits cell proliferation and *in vitro* angiogenesis, *J. Nutr.*, 125, 790S–797S, 1995.
501. Peterson, T.G., Coward, L., Kirk, M., Falany, C.N., and Barnes, S., The role of metabolism in mammary epithelial cell growth inhibition by the isoflavones genistein and biochanin A, *Carcinogenesis*, 17, 1861–1869, 1996.
502. Peterson, T.G., Ji, G.-P., Kirk, M., Coward, L., Falany, C.N., and Barnes, S., Metabolism of the isoflavones genistein and biochanin A in human breast cancer cell lines, *Am. J. Clin. Nutr.*, 68 (Suppl.), 1505S–1511S, 1998.
503. Loukovaara, M., Carson, M., Palotie, A., and Adlercreutz, H., Regulation of sex hormone-binding globulin production by isoflavonids and patterns of isoflavonoid conjugation in HepG2 cell cultures, *Steroids*, 60, 656–661, 1995.
504. Boulton, D.W., Walle, U.K., and Walle, T., Fate of flavonoid quercetin in human cell lines: chemical instability and metabolism, *J. Pharm. Pharmacol.*, 51, 353–359, 1999.
505. Ahmad, N., Feyes, D.K., Nieminen, A.-L., Agarwal, R., and Mukhtar, H., Green tea constituent epigallocatechin-3-gallate and induction of apoptosis and cell cycle arrest in human carcinoma cells, *J. Natl. Cancer Inst.*, 89, 1881–1886, 1997.
506. Cao, Y.H. and Cao, R.H., Angiogenesis inhibited by drinking tea, *Nature*, 398, 381–382, 1999.
507. Miketova, P., Schram, K.H., Whitney, J.L., Kerns, E.H., Valcic, S., Timmermann, B.N., and Volk, K.J., Mass spectrometry of selected components of biological interest in green tea extracts, *J. Nat. Prod.*, 61, 461–467, 1998.
508. Dalluge, J.J., Nelson, B.C., Thomas, J.B., Welch, M.J., and Sander, L.C., Capillary liquid chromatography/electrospray mass spectrometry for the separation and detection of catechins in green tea and human plasma, *Rapid Commun. Mass Spectrom.*, 11, 1753–1756, 1997.

509. Poon, G.K., Analysis of catechins in tea extracts by liquid chromatography-electrospray ionization mass spectrometry, *J. Chromatogr. A,* 794, 63–74, 1998.
510. Okushio, K., Suzuki, M., Matsumoto, N., Nanjo, F., and Hara, Y., Identification of (–)-epicatechin metabolites and their metabolic fate in the rat, *Drug Metab. Dispos.,* 27, 309–316, 1999.
511. Soleas, G.J. and Goldberg, D.M., Analysis of antioxidant wine polyphenols by gas chromatography-mass spectrometry. *Methods Enzymol.,* 299, 137–151, 1999.
512. Crowell, P., Lin, S., Vedejs, E., and Gould, M., Identification of metabolites of the antitumor agent d-limonene capable of inhibiting protein isoprenylation and cell growth, *Cancer Chemother. Pharmacol.,* 31, 205–212, 1992.
513. Crowell, P., Elson, C., Bailey, H., Ekegbede, A., Haag, J., and Gould, M., Human metabolism of the experimental cancer chemotherapeutic agent d-limonene, *Cancer Chemother. Pharmacol.,* 35, 31–37, 1994.
514. Poon, G.K., Vigushin, D., Griggs, L.J., Rowlands, M.G., Coombes, R.C., and Jarman, M., Identification and characterization of limonene metabolites in patients with advanced cancer by liquid chromatography/mass spectrometry, *Drug Metab. Dispos.,* 24, 565–571, 1996.
515. Vigushin, D.M., Poon, G.K., Boddy, A., English, J., Halbert, G.W., Pagonis, C., Jarman, M., and Coombes, R.C., Phase I and pharmacokinetic study of D-limonene in patients with advanced cancer, *Cancer Chemother. Pharmacol.,* 42, 111–117, 1998.
516. Hardcastle, I.R., Rowlands, M.G., Barber, A.M., Grimshaw, R.M., Mohan, M.K., Nutley, B.P., and Jarman, M., Inhibition of protein prenylation by metabolites of limonene, *Biochem. Pharmacol.,* 57, 801–809, 1999.
517. Mahato, S.B. and Garai, S., Triterpenoid saponins, *Fortschr. Chem. Org. Naturst.,* 74, 1–196, 1998.
518. Xiaoguang, C., Hongyan, L., Xiaohong, L., Zhaodi, F., Yan, L., Lihua, T., and Rui, H., Cancer chemopreventive and therapeutic activities of red ginseng, *J. Ethnopharmacol.,* 60, 71–78, 1998.
519. van Setten, D.C., ten Hove, G.J., Wiertz, E.J., Kamerling, J.P., and van de Werken, G., Multiple-stage tandem mass spectrometry for structural characterization of saponins, *Anal. Chem.,* 70, 4401–4409, 1998.
520. Breemen, R., Huang, C., Lu, Z., Rimando, A., Fong, H., and Fitzloff, J., Electrospray liquid chromatography/mass spectrometry of ginsenosides, *Anal. Chem.,* 67, 3985–3989, 1995.
521. Cui, J.F., Bjorkhem, I., and Eneroth, P., Gas chromatographic-mass spectrometric determination of 20(S)-protopanaxadiol and 20(S)-protopanaxatriol for study on human urinary excretion of ginsenosides after ingestion of ginseng preparations, *J. Chromatogr. B,* 689, 349–355, 1997.
522. Fuzzati, N., Gabetta, B., Jayakar, K., Pace, R., and Peterlongo, F., Liquid chromatography-electrospray mass spectrometric identification of ginsenosides in Panax ginseng roots, *J. Chromatogr. A,* 854, 69–79, 1999.
523. Dueker, S.R., Jones, A.D., Smith, G.M., and Clifford, A.J., Stable isotope methods for the study of β-carotene-d_8 metabolism in humans utilizing tandem mass spectrometry and high performance liquid chromatography, *Anal. Chem.,* 66, 4177–4185, 1994.
524. Dueker, S.R., Mercer, R.S., Jones, A.D., and Clifford, A.J., Ion trap mass spectrometry for kinetic studies of stable isotope labeled vitamin A at low enrichments, *Anal. Chem.,* 70, 1369–1374, 1998.
525. Sander, L.C., Sharpless, K.E., Craft, N.E., and Wise, S.A., Development of engineered stationary phases for the separation of carotenoid isomers, *Anal. Chem.,* 66, 1667–1674, 1994.
526. van Breemen, R., Innovations in carotenoid analysis using LC/MS, *Anal. Chem.,* 68, 299A–304A, 1996.
527. van Breemen, R.B., Electrospray liquid chromatography-mass spectrometry of carotenoids, *Anal. Chem.,* 67, 2004–2009, 1995.
528. Rentel, C., Strohschein, S., Albert, K., and Bayer, E., Silver-plated vitamins: a method of detecting tocopherols and carotenoids in LC/ESI-MS coupling, *Anal. Chem.,* 70, 4394–4400, 1998.
529. Khachik, F., Spangler, C.J., Smith, J.C.J., Canfield, L.M., Steck, A., and Pfander, H., Identification, quantification, and relative concentrations of carotenoids and their metabolites in human milk and serum, *Anal. Chem.,* 69, 1873–1881, 1997.
530. Hagiwara, T., Yasuno, T., Funayama, K., and Suzuki, S., Determination of lycopene, α-carotene and β-carotene in serum by liquid chromatography-atmospheric pressure chemical ionization mass spectrometry with selected-ion monitoring, *J. Chromatogr. B,* 708, 67–73, 1998.
531. Hecht, S.S., Chemoprevention by isothiocyanates, *J. Cellular Biochem.*, 22 (Suppl.), 195–209, 1995.

532. Zhang, Y., Talalay, P., Cho, C.-G., and Posner, G., A major inducer of anticarcinogenic protective enzymes from broccoli: isolation and elucidation of structure, *Proc. Natl. Acad. Sci. USA,* 89, 2399–2403, 1992.
533. Fahey, J.W., Zhang, Y., and Talalay, P., Broccoli sprouts: an exceptionally rich source of inducers of enzymes that protect against chemical carcinogens, *Proc. Natl. Acad. Sci. USA,* 94, 10367–10372, 1997.
534. Prochaska, H., Santamaria, A., and Talalay, P., Rapid detection of inducers of enzymes that protect against carcinogens, *Proc. Natl. Acad. Sci. USA,* 89, 2394–2398, 1992.
535. Posner, G., Cho, C.-G., Green, J., Zhang, Y., and Talalay, P., Design and synthesis of bifunctional isothiocyanate analogs of sulforaphane. Correction between structure and potency as inducers of anticarcinogenic detoxication enzymes, *J. Med. Chem.,* 37, 170–176, 1994.
536. Chung, F.L., Morse, M.A., Eklind, K.I., and Lewis, J., Quantitation of human uptake of the anticarcinogen phenethyl isothiocyanate after a watercress meal, *Cancer Epidemiol. Biomark. Prev.,* 1, 383–388, 1992.
537. Milner, J.A., Garlic: its anticarcinogenic and antitumorigenic properties, *Nutr. Rev.,* 54, S82–S86, 1996.
538. Block, E., Naganathan, S., Putman, D., and Zhao, S.H., *Allium* chemistry: HPLC analysis of thiosulfinates from onion, garlic, wild garlic (ramsoms), leek, scallion, shallot, elephant (great-headed) garlic, chive and Chinese chive, *J. Agric. Food Chem.,* 40, 2418–2430, 1992.
539. Block, E., Putman, D., and Zhao, S.H., *Allium* chemistry: GC-MS analysis of thiosulfinates and related compounds from onion, leak, scallion, shallot, chive, and Chinese chive, *J. Agric. Food Chem.,* 40, 2431–2438, 1992.
540. Calvey, E., Roach, J., and Block, E., Supercritial fluid chromatography of garlic (*Allium sativum*) extracts with mass spectrometric identification of allicin, *J. Chromatogr. Sci.,* 32, 93–96, 1994.
541. de Rooij, B.M., Boogaard, P.J., Rijksen, D.A., Commandeur, J.N., and Vermeulen, N.P., Urinary excretion of N-acetyl-S-allyl-L-cysteine upon garlic consumption by human volunteers, *Arch. Toxicol.,* 70, 635–639, 1996.
542. de Rooij, B.M., Boogaard, P.J., Commandeur, J.N., van Sittert, N.J., and Vermeulen, N.P., Allylmercapturic acid as urinary biomarker of human exposure to allyl chloride, *Occup. Environ. Med.,* 54, 653–661, 1997.
543. Primiano, T., Egner, P.A., Sutter, T.R., Kelloff, G.J., Roebuck, B.D., and Kensler, T.W., Intermittent dosing with oltipraz: relationship between chemoprevention of aflatoxin-induced tumorigenesis and induction of glutathione S-transferases, *Cancer Res.,* 55, 4319–4324, 1995.
544. Christensen, R. Determination of oltipraz in serum by high-performance liquid chromatography with optical absorbance and mass spectrometric detection, *J. Chromatogr.,* 584, 207–212, 1992.
545. Wang, J.S., Shen, X., He, X., Zhu, Y.R., Zhang, B.C., Wang, J.B., Qian, G.S., Kuang, S.Y., Zarba, A., Egner, P.A., Jacobson, L.P., Munoz, A., Helzlsouer, K.J., Groopman, J.D., and Kensler, T.W., Protective alterations in phase I and II metabolism of aflatoxin B_1 by oltipraz in residents of Qidong, People's Republic of China, *J. Natl. Cancer Inst.,* 91, 347–35, 1999.
546. De Vries, N. and De Flora, S., N-acetyl-l-cysteine, *J. Cell. Biochem.,* Suppl. 17F, 270–277, 1993.
547. Moldeus, P. and Cotgreave, I.A., N-acetylcysteine, *Methods Enzymol.,* 234, 482–492, 1994.
548. Frank, H., Thiel, D., and Langer, K., Determination of N-acetyl-l-cysteine in biological fluids, *J. Chromatogr.,* 309, 261–267, 1984.
549. Longo, A., Toro, M., Galimberti, C., and Carenzi, A., Determination of N-acetylcysteine in human plasma by gas chromatography-mass spectrometry, *J. Chromatogr.,* 562, 639–645, 1991.
550. Toussaint, B., Ceccato, A., Hubert, P., De Graeve, J., De Pauw, J., and Crommen, J., Determination of L-lysine N-acetylcysteinate and its mono and dimeric related compounds by liquid chromatography mass spectrometry. *J. Chromatogr. A,* 819, 161–170, 1998.
551. van Breemen, R.B. and Huang, C.-R., High-performance liquid chromatography-electrospray mass spectrometry of retinoids, *FASEB J.,* 10, 1098–1101, 1996.
552. Lehman, P.A. and Franz, T.J., A sensitive high pressure liquid chromatography/particle beam/mass spectrometry assay for the determination of all-trans-retonic acid and 13-cis-retonic acid in human plasma, *J. Pharm. Sci.,* 85, 287–290, 1996.
553. Arnhold, T., Tzimas, G., Wittfoht, W., Plonait, S., and Nau, H., Identification of 9-cis-retinoic acid, 9,13-di-cis-retinoic acid, and 14-hydroxy-4,14-retro-retinol in human plasma after liver consumption, *Life Sci.,* 59, 169–177, 1996.

554. Bischoff, E.D., Gottardis, M.M., Moon, T.E., Heyman, R.A., and Lamph, W.W., Beyond tamoxifen: the retinoid X receptor-selective ligand LGD1069 (TARGRETIN) causes complete regression of mammary carcinoma, *Cancer Res.*, 58, 479–484, 1998.
555. Baillie, T.A. and Rettenmeier, A.W., Recent advances in the use of stable isotopes in drug metabolism research, *J. Clin. Pharmacol.,* 26, 481–484, 1986.
556. Shirley, M.A., Wheelan, P., Howell, S.R., and Myrphy, R.C., Oxidative metabolism of a rexinoid and rapid phase II metabolite identification by mass spectrometry, *Drug Metab. Dispos.,* 25, 1144–1149, 1997.
557. Cole, W.C. and Prasad, K.N., Contrasting effects of vitamins as modulators of apoptosis in cancer cells and normal cells: a review, *Nutr. Cancer,* 29, 97–103, 1997.
558. Gey, K.F., Vitamins E plus C and interacting conutrients required for optimal health. A critical and constructive review of epidemiology and supplementation data regarding cardiovascular disease and cancer, *Biofactors,* 7, 113–174, 1998.
559. Blot, W.J., Vitamin/mineral supplementation and cancer risk: international chemoprevention trials, *Proc. Soc. Exp. Biol. Med.,* 216, 291–296, 1997.
560. Liebler, D.C., Burr, J.A., and Ham, A.J., Gas chromatography-mass spectrometry analysis of vitamins E and its oxidation products, *Methods Enzymol.,* 299, 309–318, 1999.
561. VanPelt, C.K., Haggarty, P., and Brenna, J.T., Quantitative subfemtomole analysis of α-tocopherol and deuterated isotopomers in plasma using tabletop GC/MS/MS, *Anal. Chem.,* 70, 4369–4375, 1998.
562. Raich, P.C., Lü, J., Thompson, H.J., and Combs, G.F., Selenium in cancer prevention: clinical issues and implications, *Cancer Invest.,* 19, 540–553, 2001.
563. Knekt, P., Marniemi, J., Teppo, L., Heliovaara, M., and Aromaa, A., Is low selenium status a risk factor for lung cancer?, *Am. J. Epidemiol.,* 148, 975–982, 1998.
564. Sieniawska, C.E., Mensikov, R., and Delves, H.T., Determination of total selenium in serum, whole blood and erythrocytes by ICP-MS, *J. Anal. Atomic Spectrom.*, 14, 109–112, 1999.
565. Li, F., Goessler, W., and Irgolic, K.J., Determination of trimethylselenonium iodide, selenomethionine, selenious acid, and selenic acid using high-performance liquid chromatography with on-line detection by inductively coupled plasma mass spectrometry or flame atomic absorption spectrometry, *J. Chromatogr. A,* 337–344, 1999.
566. Ganther, H.E., Selenium metabolism, selenoproteins and mechanism of cancer prevention: complexities with thioredoxin reductase, *Carcinogenesis,* 1657–1660, 1999.
567. Koyama, H., Omura, K., Ejima, A., Kasanuma, Y., Watanabe, C., and Satoh, H., Separation of selenium-containing proteins in human and mouse plasma using tandem high-performance liquid chromatography columns coupled with inductively coupled plasma-mass spectrometry, *Anal. Biochem.,* 267, 84–91, 1999.
568. Finley, J.W., Vanderpool, R.A., and Korynta, E., Use of stable isotopic selenium as a tracer to follow incorporation of selenium into selenoproteins, *Proc. Soc. Exp. Biol. Med.,* 210, 270–277, 1995.
569. Finley, J., The retention and distribution by healthy young men of stable isotopes of selenium consumed as selenite, selenate or hydroponically-grown broccoli are dependent on the isotopic form, *J. Nutr.,* 129, 865–871, 1999.

6 Strategies, Techniques, and Applications in Cancer Biochemistry and Biology

6.1 PROTEINS AND PEPTIDES

6.1.1 PROTEOME TECHNOLOGY

6.1.1.1 Proteomics and Strategies of Proteome Analysis

A *gene* is a segment of DNA that encodes a message for the synthesis of a protein including the mechanism that controls expression of this information. *Genome* refers to the total genetic information carried by a cell or organism. *Genomics* is the study of the structure and function of genetic material in chromosomes. The spectacular success of genomics has provided entire DNA sequences for several organisms. With the advent of gene chips and microarrays of genes (gene probes), genomics has been moving into a *functional phase* where gene expression is detected through the detection and assay of gene-specific mRNA sequences. However, a full understanding of the molecular processes that lead from health to disease demands a global rather than a reductionist approach. For example, detailed knowledge is needed not only of the DNA and RNA but also of several aspects of the derived proteins ranging from their biosynthesis and post-translational modification to their functions and interactions (Figure 6.1).[1,2] The term *proteome* refers to the total protein complement that is encoded within and expressed by a genome.[3] The concept of *proteomics* refers to the systematic study of the total protein complement expressed by the genome or particular cells or tissues, healthy as well as diseased.[4] *Transcriptomes* are subsets of genes transcribed or expressed in a particular cell or tissue. The expression of these genes provides information about the messages present in a cell at a particular time and, therefore, into the dynamic links between the genome, proteome, and cellular phenotype. An illustration of the integration of MS techniques into obtaining cellular information from the transcriptome has been shown for prostate cancer (Section 6.1.4.4).

The aims of proteomics may, at present, be divided into three areas: (a) large-scale identification of proteins and their post-translational modifications; (b) differential expression comparisons of specific proteins in healthy and diseased states; and (c) studies of protein-protein interaction.[5] Other aspects of proteomics include the participation of proteins in metabolic pathways[6] and in new approaches to drug development.[7]

When there is no known entry point for examination of a biological problem, the approach is *global* proteomics, i.e., analyses of the entire protein content during the course of changes in a cellular system. Such analyses involve the display and quantification of all proteins in that system and the identification of those that express quantitative changes. In contrast, *targeted* proteomics explores the function of individual proteins, including studying their physical relationships with other proteins. There must be a known molecular entry point, i.e., a defined biochemical process or reaction, and there is the assumption that any protein appearing to have an affinity (often noncovalent) with that entry point does have a functional role in the biological process being studied. Purification of the relevant proteins is the most critical step in this approach, and development

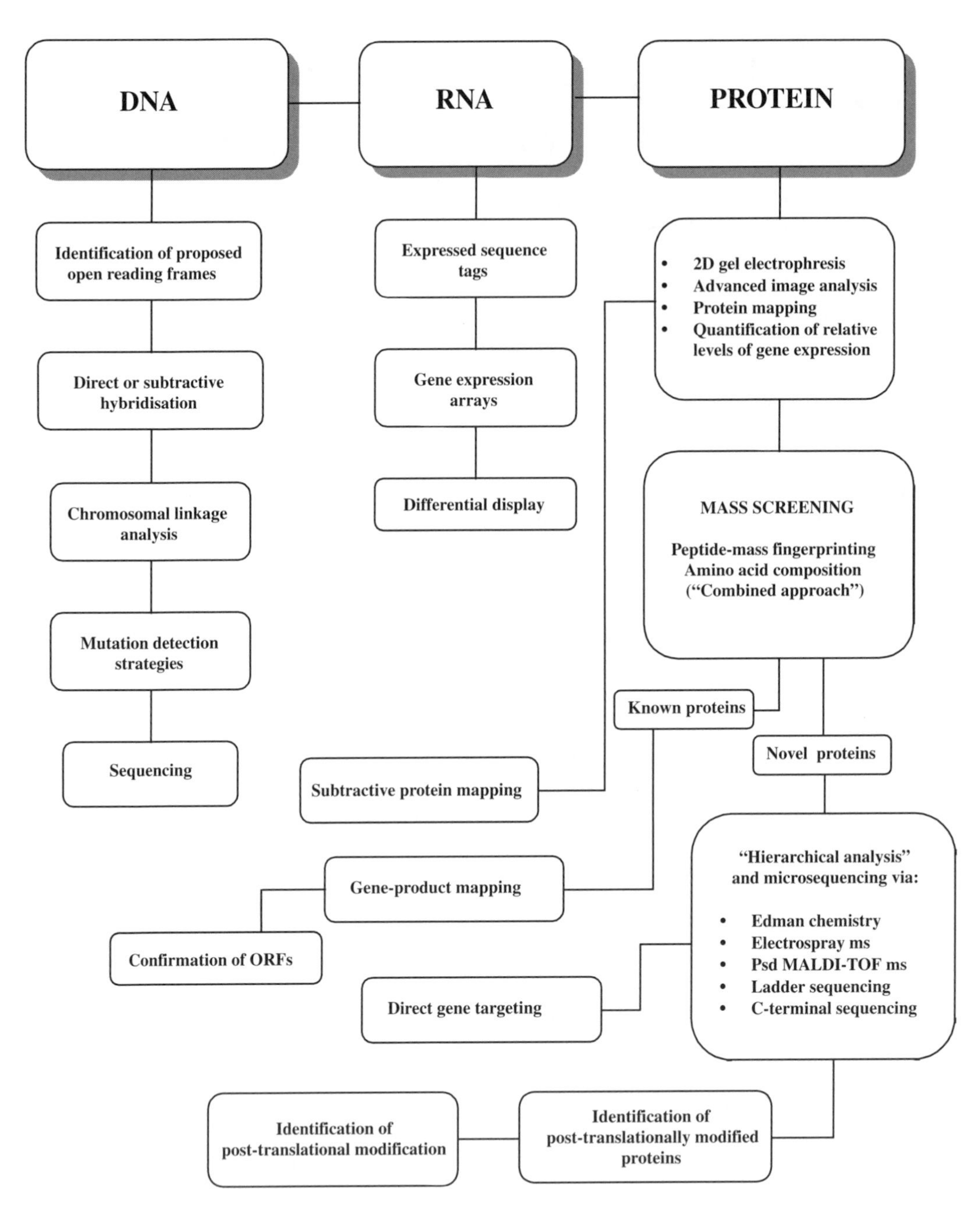

FIGURE 6.1 A schematic overview of genome and proteome analysis. Reprinted from Humphery-Smith, I. and Blackstock, W., *J. Protein Chem.,* 16, 537–544, 1997. With permission.

of appropriate technologies for finding and identifying the partners that interact with a lead protein is currently an active field, e.g., the tandem affinity purification method [TAP][8].

An overview of the biochemistry of protein biosynthesis should help to understand the relationship between genomics and proteomics. The first step, *transcription,* starts with copying a section of DNA to form *heterogenous nuclear RNA* (hnRNA). This process commences with attachment of RNA polymerase, the enzyme that controls transcription, to a specific sequence of bases, called the promoter region. This sequence varies from gene to gene. Next, ribonucleotides are linked together in a sequence that is complementary to the DNA template strand. The synthesis of an hnRNA molecule is completed when the RNA polymerase releases from the template DNA strand at a termination sequence. The second step, *hnRNA processing*, includes five major processes: (a) removal of *introns*, the segments of DNA that do not code for an mRNA product; (b) removal of some nucleotides from the 3′ end of the molecule; (c) addition of a string of adenine nucleotides to help with the subsequent transport of the mRNA into the cytoplasm; (d) addition of a cap to the 5′ end of the mRNA to aid in its attachment to a ribosome during translation; and (e) transferring the resulting mRNA from the nucleus into the cytoplasm of the cell where it associates with the ribosome. This is where the third step, synthesis of the proteins, takes place. The two subunits of ribosomal RNA (rRNA) read the coded instructions on the newly transcribed mRNA molecule and use it as the template for the *translation* process, i.e., the synthesis of proteins. This synthesis is based on the recognition of *codons* as part of the mRNA molecules. A codons consists of a group of three bases. Thus, with 4 bases there are 4^3 or 64 possible codons of which 61 code for amino acids (there is redundancy in this system with multiple codons coding for the same amino acid) while the other 3 are "stop" codons that signal the termination of protein synthesis. The process of protein synthesis is facilitated by a third type of ribonucleic acid, the *transfer RNAs* (tRNAs). These are small RNA molecules, ~80 nucleotides long, the important components of which are the *anticodons*, nucleotide triplets that are complementary to the mRNA codons. They have sites on their surfaces to bind to both codons and the specific amino acids for which they code. Ribosomes read the mRNA one codon at a time, with the appropriate tRNAs transfering the specific amino acid to the growing peptide chain. The biopolymeric end products of the translation often undergo the additional steps of folding (a spontaneous, thermodynamically driven process) and *posttranslational modification* (Section 6.1.2). These last steps convert the protein into the biologically active molecule that is subsequently transported to appropriate cell compartments. Protein structure, and thus function, can be affected at any of these stages from the mutation of the genes through any of the steps of transcription, mRNA processing, posttranslational modification, complex formation, and subcellular localization.

The two attributes of MS that have made it an indispensable tool in proteomics are the ability to determine molecular masses that are "absolute," in the sense that they are based on the intrinsic property of the *m/z* ratio, and the capability of providing partial, and even complete, sequence information. Additionally, technology-oriented advantages of MS include high sensitivity, speed of analysis, capacity to analyze mixtures in on-line combination with HPLC or CE, or even directly, and the prospect for partial or complete automation. Several MS strategies and techniques have been developed recently for applications that range from the identification and characterization of peptides and proteins to the exploration of the biological consequences of changes in the higher order structure of proteins (Table 6.1).

The role of MS in proteomics has been reviewed over a wide variety of aspects ranging from concise summaries[9,10] through general perspectives[11–14] to more specific areas of applications,[12,15–19] including important signaling pathways.[20,21] The analytical challenges of the wide dynamic range of protein expression have been reviewed. For example, for 1000 copies of a protein in each of 10^9 cells there is 1.6 pmole of that protein present. Therefore, for proteins of 25 and 100 kDa size, this translates into 4 and 16 ng, respectively. Current mass spectrometric techniques are adequate to analyze such quantities of proteins. However, when only 10^6 cells are available, the analysis of low-abundance proteins presents major technical challenges both in sample handling and instrument performance.[22]

There are at least five distinct basic strategies for protein identification: (a) use of two-dimensional gel electrophoresis (2-DE) for separation of the proteins followed by attempting to identify the components, by comparison of their isoelectric points and approximate molecular masses, using information available in databases; (b) use of 2-DE for separation, enzymatic digestion of the

TABLE 6.1
When and How to Use MS in Protein Studies from Genome to Function

Problem/Task	Relevant MS Techniques
Is the protein known? Search in genome, protein, or EST databases	Peptide mapping by MALDI or ESI-MS
ESI-MS/MS, MALDI-PSD,* or ESI-MS/MS on intact protein*	
If unknown, produce sufficient sequence information for cloning	ESI-MS/MS, Maldi-PSD* of selected peptides in the peptide map
What does the protein really look like? Secondary modifications, disulfide linkages isoforms (sequence errors) techniques	Molecular mass determination followed by peptide mapping by MALDI or ESI-MS or MS/MS. When appropriate these could be combined with enzymatic digestion or chemical degradation
Higher order structures: folding stability, monomer, or oligomer	Deuterium exchange monitored by ESI-MS, surface labeling monitored by MALDI or ESI*-MS ESI-MS under nondenaturing conditions. Cross-linking monitored by MALDI-MS
With what, where, and how does protein interact?	Affinity-based techniques combined with MALDI-or ESI*-MS. Surface labeling or limited proteolysis monitored by MALDI- or ESI*-MS. Cross-linking or chemical derivatization

* Second choice techniques

Reprinted from Roepstorff, P., *Curr. Opin. Biotech.*, 8, 6–13, 1997. With permission.

protein spots of interest, separation the peptides by HPLC and determination their retention times, obtaining peptide sequences by Edman chemistry, and identification by searching databases; (c) use of the "bottom up" approach which involves the separation of the proteins by 2-DE or some other means, such as protein chip arrays, followed by extensive proteolysis of the protein and identification of the smaller peptides generated (peptide mapping) using MS, and identify proteins with database search; (d) use the "top down" approach, i.e., sequencing the protein directly in the gas phase to produce a complementary set of fragments that eventually add up to the whole protein (the dissociation by MS/MS yields large products and interpretation relies on information obtainable in sophisticated hybrid mass spectrometers); and (e) employ highly accurate mass determinations for direct identification.

Of these approaches, the bottom up is the most commonly used technique.[23] The process includes four discrete, consecutive stages (Figure 6.2): (a) from sample preparation to gel imaging; (b) from imaging to spot digestion; (c) from digestion to database searching; and (d) from search results to data analysis and archiving. Mass spectrometry plays a dominant role in the third stage, including the major steps of peptide mass fingerprinting, accumulation of MS/MS data, and analysis of posttranslational modifications.[24] A commonly used MS strategy starts with the in-gel digestion of the protein samples separated by 2-DE, followed by analysis using MALDI-TOFMS to mass map the peptide sample. Only 2 to 10% of the peptide sample is consumed in the analysis. Highly accurate mass determinations and assignments can be obtained when the TOFMS is operated in the delayed ion extraction and reflectron modes (Figure 6.3). If the database search is negative, a larger portion (~50%) of the peptide mixture may then be analyzed by nano-ESI-MS/MS to acquire sequence tags (see later) to be integrated into the database search. If this approach also fails, *de novo* sequencing is needed (briefly reviewed in Reference 13). As detailed below and in subsequent sections, several variations of this general approach have been developed. There are alternatives that do not involve 2-DE at all (Section 6.1.1.7). The advantages and limitations of the bottom up vs. top down approaches have been compared.[25]

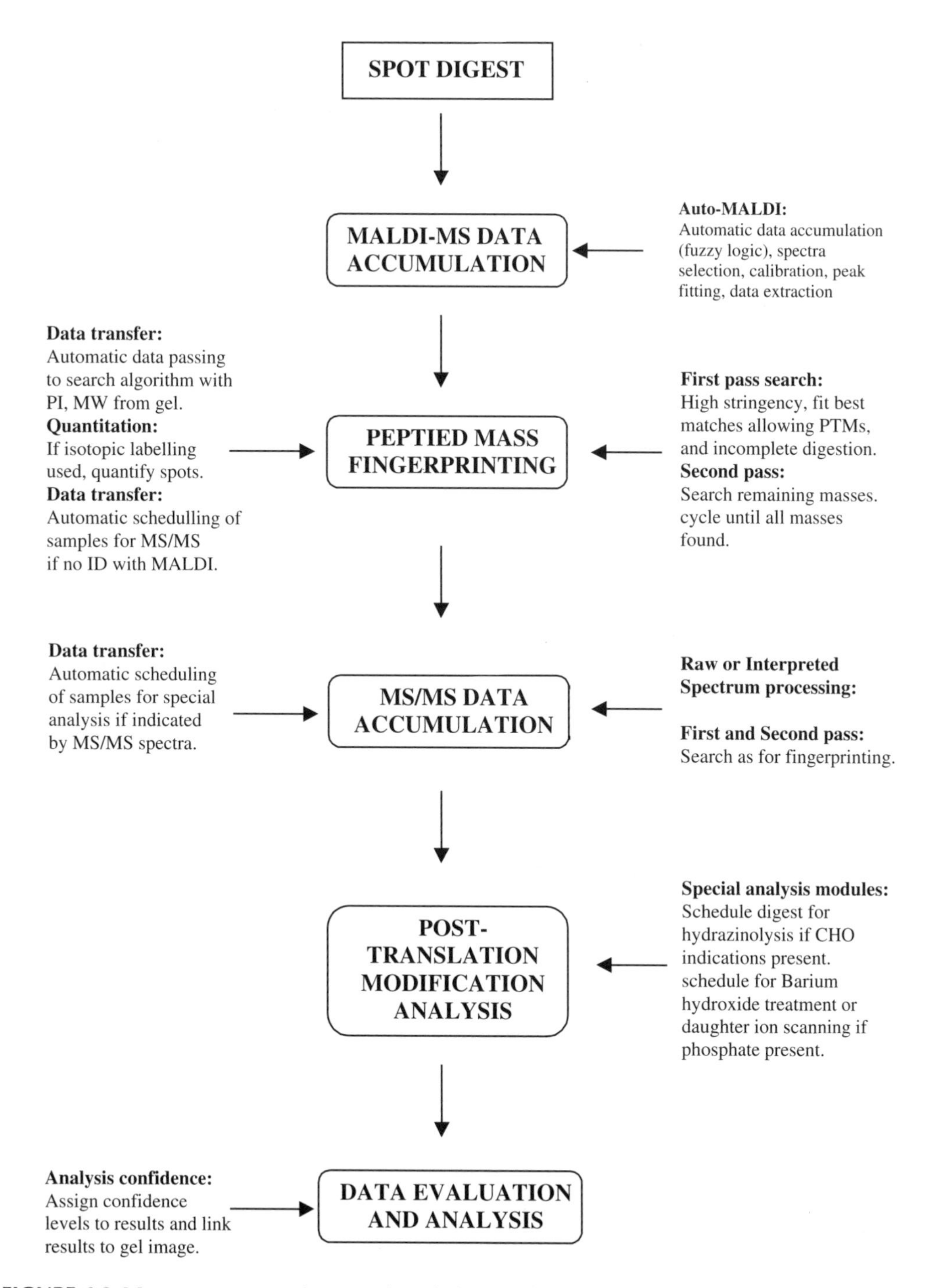

FIGURE 6.2 Mass spectrometry plays a major role in the "digest to data evaluation" stage of the process of comprehensive proteome analysis. Reprinted from Quadroni, M. and James, P., *Electrophoresis,* 20, 664–677, 1999. With permission.

6.1.1.2 Protein Isolation, Digestion, and Delivery into the Ion Source, Instrumentation

The significance of the technological aspects of sample preparation methods, including separation of proteins and concentration of the peptides generated by digestion, has become evident with the increasingly stringent requirements for sample presentation dictated by the ever evolving mass

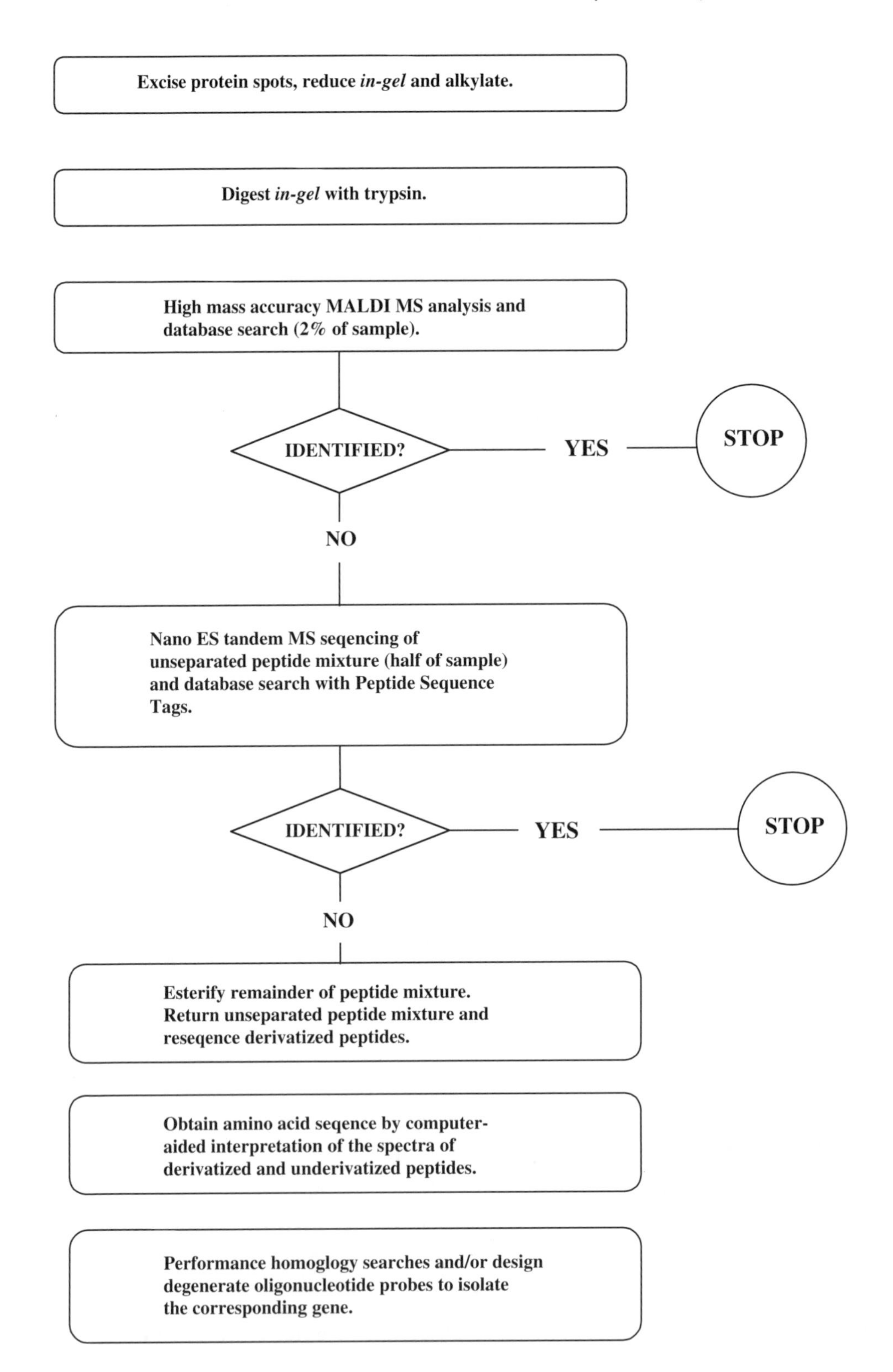

FIGURE 6.3 Strategy to identify or clone proteins using mass spectrometry alone. From Shevchenko, A., Wilm, M., and Mann, M., *J. Protein Chem.,* 16, 481–490, 1997. With permission.

spectrometric techniques. These needs and available preparative techniques have been reviewed from several perspectives[26–29] including approaches to the separation of those proteins that cannot be analyzed satisfactorily by 2-DE, including membrane proteins, highly basic proteins, and very large (>150 kDa) and small (<10 kDa) proteins.[30]

Protein Isolation: One- and Two-Dimensional Gel Electrophoresis (1-DE and 2-DE)

Because the dynamic range of protein concentrations in biological samples may be as high as 10^6, affinity-based purification is often indicated as a first step of purification, particularly when the analyte is present at a low abundance. Protein mixtures thus isolated are usually separated electrophoretically into bands or spots on polyacrylamide gels (PAGE). In simple mixtures, proteins can be identified using the resolving capability of 1-DE followed by LC/ESI-MS/MS of the recovered bands.[31] The advantages of 1-DE separation include the fact that it is usually carried out with sodium dodecyl sulfate (SDS), in which most proteins are soluble, and that both acidic and basic proteins can be separated and easily visualized.

Adding a second dimension of separation *orthogonal* to the first dimension dramatically improves the resolution of components of complex peptide and protein mixtures.[32,33] In the first dimension of 2-DE, separation of the protein mixtures is according to charge. This is accomplished by isoelectric focusing of the components, aided by the presence of high concentrations of urea and nonionic detergents. The second dimension involves further separation with the aid of SDS-PAGE.[34] In SDS-PAGE the protein molecules move as negatively charged SDS-protein complexes in the direction of the anode and are separated in the polyacrylamide matrix according to their molecular mass. The separated proteins may then be visualized by chemical staining, Coomassie staining, or fluorescence dyte staining (reviewed in: http://www.the-scientist.com/yr1998/may/profile1_980511.html). The commonly used silver staining involves fixing the proteins in the gel with trichloroacetic acid and then placing them in a silver nitrate solution to bind silver ions to the proteins. Elemental silver is finally precipitated by reduction. An alternative staining method is based on the triphenylmethane dye, Coomassie blue. Staining intensity is mainly proportional to the amount of protein in the band rather than its amino acid composition. Because of saturation effects, staining is only linear over two orders of magnitude, thus very limited quantitative information is provided. For quantification over a wide dynamic range, separate gels must be prepared with increasing quantities of protein followed by computer-supported image evaluation.

Because of the fragility of polyacrylamide gels, separated proteins are often transferred onto more robust polymer membranes, such as polyvinylidene difluoride (PVDF), nylon, or nitrocellulose by *electroblotting* using an orthogonal electric field. Advantages include the quantitative transfer of proteins free of buffers and SDS, and retention of the relative positions of the separated proteins. A special "protein flume" has been developed for the concentration of proteins from multiple preparative gel plugs for the identification of cytokine-regulated proteins in normal and malignant cells.[35] To identify the separated proteins, the bands or spots are usually excised from the gel directly, digested into peptides with sequence-specific proteases, and the peptides separated by HPLC. The resulting "peptide maps" portray the sequence of the protein. Sequencing may be carried out by Edman degradation or, more practically, by one of several MS techniques to be described.

Technological developments have largely eliminated the problems of interlaboratory reproducibility in the separation patterns obtained, thereby facilitating the worldwide exchange of information and the standardization of protocols.[36] High-resolution 2-DE gels yield hundreds or even thousands of separated proteins. A major limitation of 2-DE technology is its inadequacy in the separation of hydrophobic (low solubility) proteins. This is especially the case for the heterogeneous family of plasma membrane proteins that are known to focus over a wide pH range. Alternative approaches include one that uses SDS-PAGE for separation followed by advanced HPLC/MS techniques.[37] The polymerization of residual enhancers and monomers may cause in-gel modification of the separated proteins, leading to erroneous results.[38] Modification of cysteine residues of proteins does not affect their mobility significantly, thus this is not a problem in 2-D gel separations. However, artifacts may appear because of the addition of 71 Da due to cys-acrylamide formation.

Generation of Proteolytic Peptide Fragments

Direct digestion of the protein in the gel matrix is the preferred technique for generating these fragments because of the small amount of analyte required, ~100 fmol. Several variations have been described for the digestion of proteins in silver-stained gels[39] and for improving the time-consuming desalting step.[40–42] An alternative technique, called "molecular scanning" obviates the need for spot excision, in-gel digestion, and all associated liquid handling. In this technique, sample digestion is carried out, during the process of electroblotting by trypsin that has been immobilized on a membrane placed between the gel and the PVDF membrane that is itself soaked with a MALDI matrix. The membrane onto which the products of the digestion have been transferred is allowed to dry and introduced into the MS for MALDI-TOF analysis. This method retains the spatial-resolution of the original 2-DE separation.[43,44]

Separation and Preconcentration of Peptides for ESI

Both purification and preconcentration may be accomplished by coupled LC/ESIMS.[45,46] Because ESI response is concentration-dependent at flow rates of <100 μL/min, the concentrations of the peptides delivered into the ESI process can be maximized by minimizing the HPLC column elution volume. The advantages and limitations of special techniques, such as "peak parking," stop-flow analysis, and nanoflow infusion have been compared.[47] Solid-phase microextraction may also be used for initial sample clarification.[48]

The separation of peptides by CE is determined by the different migration profiles observed, as derived from their electrophoretic mobility. The flat flow profiles of peptides in CE generate narrow elution bands that, in turn, result in the presentation of the analytes to the ESI source at high concentrations, thus yielding attomole sensitivities.[49] Sample preconcentration of very dilute samples may be achieved with a hydrophobic membrane placed near the head of the column.[50]

Peptide mixtures may be infused directly into an ESI source without separation by either HPLC or CE. One approach has combined on-line sample clean-up on reversed-phase LC packing material followed by batch elution into the ESI source. For small peptides the LOD was 1 fmol.[51] An alternative approach used the same sample cleanup procedure but in an off-line fashion. The use of glass capillaries to reduce the flow to low nL/min rates provides ESI signals that last for tens of minutes. This facilitates conducting multiple MS/MS experiments using only 1 to 2 μL size samples.[52]

Protein Chip Arrays

Rapid profiling of protein extracts from physiological fluids, cells, and tissues has become possible with the commercial introduction of protein chip arrays. These arrays facilitate the differential capture of proteins according to general chemical or specific biochemical properties (reviewed in References 53, 54). After a series of wash protocols, the captured proteins are mixed with MALDI matrices, dried, and released as gaseous ions directly from the arrays by MALDI. Mass determinations can be made in a linear TOFMS or MS/MS studies carried out with a Q-TOF analyzer.

Commercially available chip arrays with SELDI surfaces (reviewed in Reference 53) are available in several *chemically defined* variations including: normal phase silica, surfaces preactivated with reactive carbonyl diimidazole moieties or epoxy groups, hydrophobic surfaces, strong anion exchange, weak cation exchange, and immobilized metal affinity-capture surfaces. Although the use of such surfaces reveals specific chemical properties of the analytes, such as total charge and hydrophobicity, the objective of the approach is still, primarily, to separate entire classes of proteins. In contrast, chips with biochemically modified surfaces have been designed to engender specific *biomolecular interactions* that "bait out" individual proteins from crude biological samples.[55] For example, when a surface contains a specific antibody, only those proteins will be recovered from the sample and remain on the chip that are bound to that antibody. These proteins can then be identified by MALDI-TOFMS and further studied by CID.[56] Relevant applications using chemically defined surfaces are reviewed in Section 6.1.4.

Mass Spectrometric Instrumentation

Peptide mapping (see below), which has been used for the large-scale identification of proteins from species with relatively small, and fully sequenced, genomes, has usually been carried out by MALDI-TOFMS. For *de novo* sequencing using low-energy CID (50 eV, N_2 or Ar gas) the current choices include conventional MALDI-ITMS and the MALDI-QTOF hybrid instrument.[57,58] Performance of commercially available versions of QTOF instruments for both MS and MS/MS analyses include <20 ppm accuracy and >10,000 resolution for both parent and product ions along with <5 fmole sensitivities for peptides. These performance capabilities often obviate the need for cleavage specificity and may permit the identification of a protein in a heterogeneous mixture using only one peptide.

In the recently developed MALDI-TOF/TOF instruments, ions of high velocity, produced by conventional MALDI-TOF, are selected with a timed ion gate and subjected to CID at keV energies. The product ions are further accelerated and analyzed in the second TOF that includes a reflectron. This design eliminates several disadvantages of conventional PSD and adds the capabilities of high energy CID and high speed analysis not available on MALDI or ESI-QTOF instruments.[59]

Recent advances in ESI-FTICRMS instrumentation have increased sensitivities >tenfold.[60,61] The combined capabilities of high dynamic range and ability to determine masses with an accuracy to 1 ppm, mean that FTICRMS can be used for identification and quantification of hundreds of proteins in a single analysis.[62] Technologies are being developed for protein identification by on-line fragmentation using FTICRMS[63] or by "accurate mass tags," thus removing the need of MS/MS (Reference 64, Section 6.1.1.7).

Novel techniques such as "molecular imaging"[65,66] and "molecular scanning"[44] can rapidly analyze a large number of samples and therefore generate a vast quantity of data. Accordingly, automation technology has become essential for sample preparation and introduction, for MS and MS/MS, and for the computerized evaluation of data including database searching. Automation techniques currently available range in sophistication and include robotic systems for gel picking,[67] the use of microchips,[68] and software techniques for exploring the unique advantages of nano-ESI.[69–71]

6.1.1.3 Sequencing by Edman Degradation

The Edman degradation technique for peptide sequencing has been used for some 40 years (reviewed in Reference 72). The basis of this method is a repetitive process in which the order of the amino acids, from the N-terminus only, is determined by the removal and identification of these amino acids one at a time.[73] The first step is a *coupling* reaction, carried out under mildly basic conditions (pH 9), in which the N-terminal amino group is derivatized with phenylisothiocyanate to form the phenylthiocarbamoyl (PTC) derivative. The second step, a *cleavage* reaction, breaks the peptide bond between the first and second amino acids when the PTC derivative is heated in anhydrous TFA. This results in the formation of the phenyl-containing thiazinole derivative of the first amino acid. This is hydrophobic and can, therefore, be extracted into an organic solvent. The third, or *conversion*, step consists of heating the recovered derivative in a strong acid, during which the thiazolinone isomerizes (cyclizes) to form a stable phenylthiohydantoin (PTH) derivative that can, in turn, be identified by its HPLC retention time. The remainder of the sample, i.e., the original peptide minus the amino-terminal amino acid, that is hydrophilic and was partitioned into the aqueous fraction during the extraction process contains a free amino group that can, therefore, be subjected in another degradation cycle. Automated sequencers are available to carry out the sequencing process rapidly and conveniently. A recent improvement in the technique permits amino acid identification using a 20-min cycle.[74] Because Edman degradation is only ~95% efficient, amino acid sequences can be determined only to 10 to 20 residues in length. The sequence eventually becomes heavily staggered as not all peptides in the reaction mixture release the amino acid derivative at each step.

While the amino acid sequences of isolated segments of cleaved proteins can be elucidated with the cleavage+Edman degradation approach, the *order* of the segments can only be determined by obtaining the sequences of *overlapping* peptides. This means that data must be obtained using more than one type of cleavage, e.g., with both trypsin and chymotrypsin individually. If the original protein contains two or more polypeptide chains held together by noncovalent bonds, then the chains must be dissociated, by denaturing with urea or guanidine hydrochloride, and separated prior to cleavage and sequencing. Similarly, to prevent recombination of polypeptide chains normally linked by disulfide bridges, these bonds must be broken, by reduction with β-mercaptoethanol or dithiothreitol, followed by alkylation of the –SH groups with iodoacetate to their –S-carboxymethyl derivatives. N-terminally blocked proteins cannot be directly sequenced by the Edman technique. The blockage is usually caused by posttranslational modifications, most of which are biologically important and added during synthesis; however, they may also be artifacts of sample purification. It may take days to chemically or enzymatically cleave the blocked protein and separate the peptides by HPLC for subsequent sequencing. Even then there is a potential that the total sequence can never be obtained. The chemistry of the Edman degradation method and its implications in sequencing by MS have been reviewed.[75]

6.1.1.4 Mass Spectrometric Peptide Mapping

The two major mass spectrometric strategies of sequencing are (a) mass analysis of relatively short peptides obtained from the analyte proteins by enzymatic or chemical reactions, e.g., ladder sequencing, that can be undertaken on single analyzer instuments, and (b) sequencing by CID analysis of selected ions with known m/z and predictable fragmentation patterns associated with amino acid sequences. The latter experiments are usually carried out using ESI followed by low-energy CID on triple quadrupole, ion trap, or hybrid Q-TOF analyzers. MALDI-TOFMS with PSD to exploit metastable ion fragmentation has also been used.

The concept of peptide mass mapping (*fingerprinting*) as a rapid and reliable method for protein identification, when compared with peptide sequencing by Edman degradation, is based on the fact that the set of masses for the peptides produced by residue-specific enzymatic, or chemical digestion, is unique for any given protein.[21,76] A peptide mass fingerprint is the collective mass spectrum of the peptides derived from a protein upon defined enzymatic or chemical cleavage. Three types of proteolytic enzymes are used all of which cleave peptide bonds on the carboxyl side of specific amino acid residues. The enzymes are trypsin, which cleaves at the basic amino acids (Arg and Lys), chymotrypsin which cleaves at aromatic amino acids (Phe, Tyr, and Trp), and the staphylococcal proteases which cleave at acidic amino acids (Asp and Glu). Fragmentation may also be accomplished with residue-specific chemical reagents such as cyanogen bromide, that acts at the carboxyl side of methionine residues, and hydroxylamine which is specific for Asn-Gly bonds.[77,78] The mass spectra of the resultant mixtures are usually obtained by MALDI-TOF[79] with high mass accuracy being critical for the reduction of false positive identifications.[80–82] The analysis is followed by a database search that consists of looking for a protein that, on digestion with the particular enzyme used, would produce a set of peptides with the same masses as have been obtained experimentally. Because no sequence information is generated, the method is very fast. Appropriate instrumentation and computer capacity mean that the mass determinations and database search can often be completed within one minute. The technique is, therefore, well suited for applications involving the hundreds of proteins that are typically separated with 2-DE. Robotic systems have been developed for all steps of sample preparation, including gel cutting, digestion, and sample introduction.[83,84] Technological improvements of both 2-DE and mass fingerprinting for proteome mapping have been reviewed.[85]

A protein is considered to have been correctly identified when there is a user-defined, statistically significant number of peptide mass matches found between the mass spectrometric results and those predicted for the protein in the database. Mass mapping is most effective in the 20 to 100 kDa molecular mass range. Because there is no sequence information generated by the technique, the method can only be used when the complete sequence of the analyte protein is available in the database. Ambiguous results of database searches for masses of peptides obtained from the digestion of protein mixtures may be compensated for by innovative data search methods.[84] Other limitations of the peptide mass mapping technique and approaches to overcoming them have been reviewed.[86]

6.1.1.5 Peptide Ladder Sequencing

In this strategy, polypeptides are truncated (degraded) progressively by removing single amino acids from the amino terminus by Edman degradation or from either terminus by chemical hydrolysis or biological reagents such as proteolytic exopeptidases (reviewed in Reference 87). The result is a mixture of peptides, called a peptide ladder, in which each member differs from the next by one amino acid residue (Figure 6.4). The molecular masses of all the truncated peptides are then determined in a single MS analysis. The identity of each component amino acid in the ladder is determined from the measured mass differences between adjacent molecular ions, and the sequence is established from the order of the peptides in the ladder. MALDI-TOF has been the instrumentation of choice for the determination of the molecular masses of these peptides because of its high sensitivity, tendency to generate singly charged ions rather than multiply charged ions, and favorable salt tolerance. Advantages of ladder sequencing include the capability for multiple determinations in minutes, sensitivity to fmole levels, and low sample consumption.

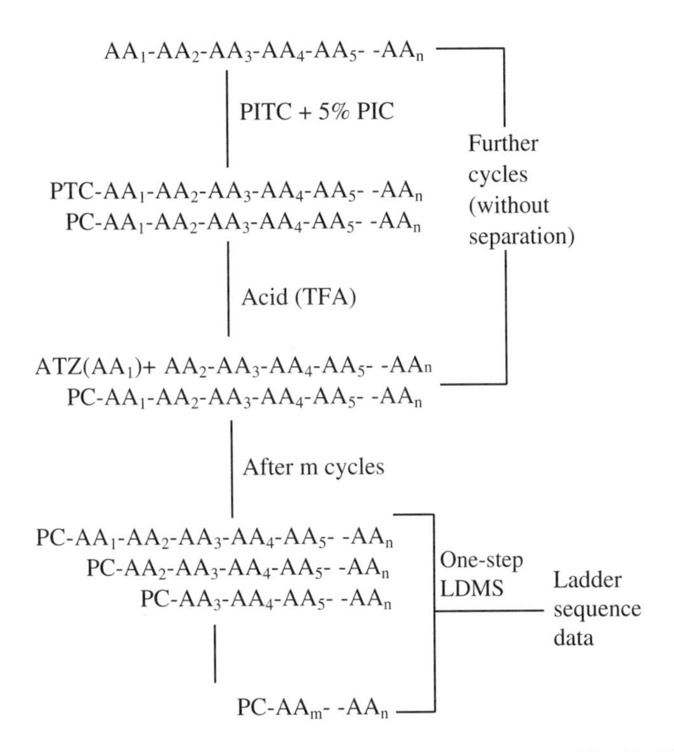

FIGURE 6.4 Peptide ladder sequencing. Based on Chait, B.T. et al., *Science*, 262, 89–92, 1993.

Methods for N-terminal degradation usually employ chemical procedures. In one approach, a peptide ladder was obtained using a modified Edman reaction scheme that substituted phenylisocyanate as the terminating reagent at the N-terminal α-amino group in place of the classical phenylisothiocyanate. A useful consequence of doing such "differential derivatization" is a nested set of peptides differing by 16 Da, since the difference between the two reagents is the mass difference between sulfur and oxygen. Using this approach, mass accuracy with MALDI-MS was 100 ppm or better, the LOD was at the subpicomole level, and peptides with up to 50 residues could be analyzed.[88] An alternative approach used allyl isothiocyanate which offered easy removal of all by-products and several other advantages.[20] Another approach generates multiple small "sequence tags" from N-terminal peptides in unseparated protein digests by stepwise thioacetylation and acid cleavage. Using this method a prototype device has been designed for the parallel identification of proteins in multiple samples.[89]

When C-terminal degradation is required the methods usually employ biological degradation for which there is a variety of enzymes. The carboxypeptidase CPY will penetrate all amino acid sequences but shows a preference for hydrophobic residues. Other carboxypeptidases have their respective advantages and limitations with respect to preferred amino acids. Aminopeptidases that act from the N-terminus are also available. A relevant proof of concept application demonstrating the combined use of a carboxypeptidase and an aminopeptidase with on-plate time-dependent digestion was the analysis of a nine-residue peptide that is a MHC Class I tumor-specific, immunoreactive antigen. The results revealed the first five C-terminal residues followed by three consecutive N-terminal residues.[90] As an alternative to this time-dependent approach, ten parallel microdigestions were carried out on a multiple sample-position stainless steel MALDI target using increasing exopeptidase concentrations. After the reaction volumes (1 μL ea.) had evaporated (~10 min), the matrix was added and all resulting peptides were analyzed by MALDI-TOF. Advantages of this approach included fast optimization of reaction conditions and reduced sample and enzyme requirements.[91]

6.1.1.6 Sequencing by Tandem Mass Spectrometry

Another commonly used sequence-derived approach to protein identification is based on ionizing peptides with ESI followed by CID of selected precursor peptide ions. This results in the production of characteristic amino- and carboxy-terminal-containing fragments. Frequently the peptides in digest mixtures are separated by capillary LC or CE connected on-line to an ESI-MS/MS instrument. An advantage of the MS/MS approach is that sequence information is often obtainable from mixtures with less purification because of the individual precursor ions are "purified" in the first stage of the tandem operation. Sophisticated data system control of instruments allows for acquisition of both MS and MS/MS data in the same analysis through data-dependant switching, and may even allow for multi-collision-energy experiments in fast TOF analyzers. A disadvantage of such data-dependent automatic precursor ion selection is limited dynamic range because usually only the ions of the highest abundance are selected, thus peptides of low intensity are ignored.[92,93] However, data systems are becoming more sophisticated in areas such as charge state recognition and artifact exclusion parameters.

The primary structure of peptides can be established even if several incomplete series of ions are present, provided that there are other series of ions present and as long as there are overlaps. Proteins may be identified from uninterpreted CID data by the automated correlation of the spectra with a sequence database. Alternatively, identifications can be ascertained when complete or partial sequences are derived by interpretation of spectra either manually or with the aid of an algorithm followed by database searches.[94] Frequently, proteins can be identified conclusively from the CID spectrum of a single peptide species.[95]

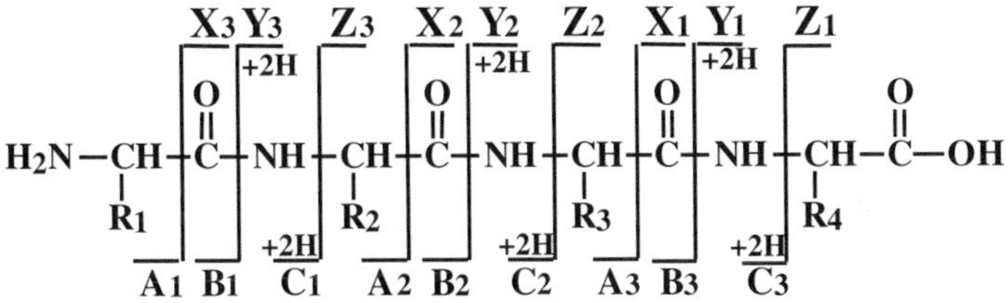

X3 Y3 Z3 X2 Y2 Z2 X1 Y1 Z1

+2H +2H +2H

$H_2N-CH(R_1)-C(=O)-NH-CH(R_2)-C(=O)-NH-CH(R_3)-C(=O)-NH-CH(R_4)-C(=O)-OH$

+2H +2H +2H

A1 B1 C1 A2 B2 C2 A3 B3 C3

Origin	Terminal	Type	Structure
Backbone	N	a_n	$H-(NH-CHR-CO)_{n-1}-\overset{+}{N}H=CHR_n$
Backbone	N	b_n	$H-(NH-CHR-CO)_{n-1}-NH-CHR_n-\overset{+}{C}=O$
Backbone	N	c_n	$H-(NH-CHR-CO)_{n-1}-NH=CHR_n-CO-\overset{+}{N}H_3$
Side-chain	N	d_n	$H-(NH-CHR-CO)_{n-1}-NH-CH(=CR^b_n)$
Backbone	C	x_n	$^{+}CO-NH-CHR_n-CO-(NH-CHR-CO)_{n-1}-OH$
Backbone	C	y_n	$^{+}H_3N-CHR_n-CO-(NH-CHR-CO)_{n-1}-OH$
Backbone	C	z_n	$^{+}CHR_n-CO-(NH-CHR-CO)_{n-1}-OH$
Side-chain	C	v_n	$HN=CH-CO-(NH-CHR-CO)_{n-1}-OH$
Side-chain	C	w_n	$CR^b_n-CH-CO-(NH-CHR-CO)_{n-1}-OH$
Fragments from a double cleavage of the main peptide chain			
Immonium ion			$H_2\overset{+}{N}=CHR$
Internal fragment			$H_2N-CHR_{>1}-(NH-NH-CHR)_{m<n}-\overset{+}{C}=O$

FIGURE 6.5 Nomenclature of peptide fragmentation. Based on the nomenclature proposed in Roepstorff, P. and Fohlman, J., *Biomed. Mass Spectrom.*, 11, 601–604, 1984; and Biemann, K., *Methods Enzymol.*, 193, 886–887, 1990.

Nomenclature and Interpretation of CID Fragmentation of Peptide Ions

The major fragment ions obtained from protonated peptide precursor ions with low-energy CID originate from the cleavage of bonds in the backbone of the peptide. In the commonly used nomenclature,[96–98] ions where there is a charge retention on the N-terminus are designated as a_n, b_n, c_n ions, while C-terminus ions are designated as x_n, y_n, and z_n ions (Figure 6.5). For example, all y-type ions contain the carboxyl terminus of the peptide plus one or more additional residues. The masses of the ions of the same type with adjacent n values differ by the mass of the amino acid residue (Table 6.2). These ions can then be used to assign the sequence of the peptide by calculating the mass differences between the fragment ions and recognizing those that are separated by known amino acid residue masses from the lightest (57 Da) to the heaviest (186 Da). Additional CID product ions may result from losses of water, carbon monoxide, ammonia, or amino acid-specific low mass ions. In instruments with exact mass measurement capability the isobaric (same nominal mass) species Gln and Lys can be differentiated. Further useful information, often obtainable only by high-energy CID, is provided by side-chain-specific fragments that may also be used to differentiate isobaric Gln and Lys residues and particularly peptides containing the isomeric residues Leu and Ile.

In addition to the conventional MS/MS techniques of CID coupled with tandem mass spectrometry[99,100] and the PSD available coupled with MALDI-TOFMS,[101] sequence information of peptides and proteins may also be obtained from in-source decay in MALDI-TOFMS.[102,103] There

TABLE 6.2
Amino Acid Residue Masses

Amino acid	Three letter code	Single letter code	Residue mass Monoisotopic	Residue mass Average
Alanine	Ala	A	71.0371	71.079
Arginine	Arg	R	156.1011	156.188
Asparagine	Asn	N	114.0429	114.104
Aspartic acid	Asp	D	115.0269	115.089
Cysteine	Cys	C	103.0091	103.144
Glutamic acid	Glu	E	129.0426	129.116
Glutamine	Gln	Q	128.0585	128.131
Glycine	Gly	G	57.0214	57.052
Histidine	His	H	137.0589	137.142
Isoleucine	Ile	I	113.0840	113.160
Leucine	Leu	L	113.0840	113.160
Lysine	Lys	K	128.0949	128.174
Methionine	Met	M	131.0404	131.198
Phenylalanine	Phe	F	147.0684	147.177
Proline	Pro	P	97.0527	97.117
Serine	Ser	S	87.0320	87.078
Threonine	Thr	T	101.0476	101.105
Tryptophan	Trp	W	186.0793	186.213
Tyrosine	Tyr	Y	163.0633	163.176
Valine	Val	V	99.0684	99.133

are significant differences between the types of ions produced by these techniques, each with particular advantages and limitations. For example, in-source decay yields relatively unique and specific c-, y-, and/or z-series product ions.[104]

Sequencing with Nano-ESI-MS/MS

A step elution technique that represents a compromise between using micro-flow HPLC and the analysis of excessively complex peptide digest mixtures by standard nanospray ESI has been developed for sequencing proteins isolated by PAGE. The peptide extract obtained after in-gel tryptic cleavage of the analyte protein was passed through a capillary that contained ~100 nL of a perfusion sorbent. After washing, the peptides were step-eluted in an 1 μL volume into the capillary of a nano-ESI source. A new capillary and new resin was used for each analysis to prevent cross-contamination.[105] The technique was demonstrated by determining the sequence of a 45 kDa protein that exhibited antiangiogenic properties. Seven peptides were obtained by enzymatically cleaving a 200 ng sample. These were analyzed by MS/MS in both their native and esterified forms, with the latter providing characteristic mass shifts.[99] A total of 73 amino acids were sequenced correctly with the aid of a database (Table 6.3). Sequence data from the CID spectra of one of the peptides (*m/z* 699.3) (Figure 6.6) and other cleaved peptides revealed significant homology to arginine deiminase. This CID based nano-ESI approach was 10- to 100-fold more sensitive than microsequencing with either in-gel or on-blot digestion followed by Edman degradation. The results were used to generate oligonucleotide primers for cDNA cloning.[52]

"Shotgun" Identification

In this technique gel separation is omitted altogether (reviewed in Reference 11). The proteins are cleaved with a proteolytic enzyme in solution and the resulting peptide mixtures are analyzed using LC-MS/MS. The large number of tandem mass spectra thus obtained is used to identify proteins

TABLE 6.3
Comparison of the Deduced Amino-acid Sequence with the Partial Sequences Obtained by Tandem Mass Spectrometry

Nucleic acid sequence of cloned gene	Deduced amino acid sequence	Amino acid sequence by mass spectrometry
GAA CCT GTT CRR TCA CCT GAA CAC AGA	Glu-Pro-Val-Leu-Ser-Pro-Glu-His-Arg	Glu-Pro-Val-Lxx-Ser-Pro-Glu-His-Arg
GAA TTA ATC CAA TAC ATG ATG GCA GGT ATT ACT AAA	Glu-Leu-Ile-Gln-Tyr-Met-Met-Ala-Gly-Ile-Thr-Lys	Glu-Lxx-Lxx-Gln-Tyr-Met-Met-Ala-Gly-Lxx-Thr-Lys
GAC CCA TTT GCA TCA GTT GGT AAT GGT GTT ACA ATT CAC TAC ATG CGT	Asp-Pro-Phe-Ala-Ser-Val-Gly-Asn-Gly-Val-Thr-Ile-His-Tyr-Met-Arg	Asp-Pro-Phe-Ala-Ser-Val-Gly-Asp-Gly-Val-Thr-Lxx-His-Tyr-Met-Arg
ACT CCA TGA TAC TAT GAC CCA GCA ATG AAA	Thr-Pro-Trp-Tyr-Tyr-Asp-Pro-Ala-Met-Lys	Thr-Pro-Trp-Tyr-Tyr-Asp-Pro-Ala-Met-Lys
ACT GAC CTA GAA ACT ATT ACT TTA TTA GCT AAA	Thr-Asp-Leu-Glu-Thr-Ile-Thr-Leu-Leu-Ala-Lys	Thr-Asp-Lxx-Glu-Thr-Lxx-Thr-Lxx-Lxx-Ala-Lys
ATA GTT GCA ATT AAT GTT CCT AAA	Ile-Val-Ala-Ille-Asn-Val-Pro-Lys	Lxx-Val-Ala-Lxx-Asn-Val-Pro-Lys
TGT ATG TCA ATG CCT CTT TCA CGT	Cys-Met-Ser-Met-Pro-Leu-Ser-Arg	Cys-Met-Ser-Met-Pro-Lxx-Ser-Arg

Note: The only discrepancy observed was the prediction of Asn from the nucleotide sequence of the cloned protein instead of the mass spectrometrically sequenced Asp in one of the peptides (underlined). Because the identity of Asp is absolutely certain owing to the +14 Da mass shift induced on this amino acid by the derivatization procedure in conjunction with the mass spectrometric data on the underivatized peptide, it is concluded that deamidation of the amino acid must have occurred during the purification procedure.

Note: The peptide Cys-Met-Ser-Met-Pro-Lxx-Ser-Arg (Lxx is Leu or Ile), which was at the C-terminal position (as judged by homology) was used to generate an antisera PCR primer. As sense primer, a sequence from the 5′-conserved region of *Mycoplasma arginine* was used to obtain a nearly full length clone using DNA prepared from the contaminated tumor cells as template.

Reprinted from Wilm, M. et al., *Nature,* 379, 466–469, 1996. With permission.

with the aid of databases.[31] As noted above, the CID spectrum of a single peptide is often sufficient to identify a protein so this shotgun method provides the opportunity to recognize many of the protein components of the sample, provided they are present in the databases searched.

Direct Analysis of Large Protein Complexes (DALPC)

This strategy uses the independent physical properties of charge and hydrophobicity to resolve complex peptide mixtures and the predictive powers of whole genome sequences to identify the proteins. This is accomplished without the need for purification to homogeneity of each protein component. The process consists of several steps during which a denatured and reduced protein complex is digested, a fraction of the components of the acidified peptide mixture separated by strong cation exchange chromatography, retained onto an HPLC column, and finally eluted into an ESI-ITMS for automated MS/MS analysis [106] and data-system-based evaluation[107] to infer the amino acid sequences from the detected fragment ions. In an alternative approach, sensitivity was extended into the ng range by improving the separation with an integrated 2-D microcapillary chromatography system and a simplified on-line micro-ESI interface.[108] In one application, a single analysis revealed 749 unique peptides in an extract of the 80S ribosome of *Saccharomyces cerevisiae*, in turn leading to the identification of 189 proteins. It was concluded that the DALPC approach has the potential for the direct and comprehensive analysis of the components of even the largest macromolecular complexes.[93]

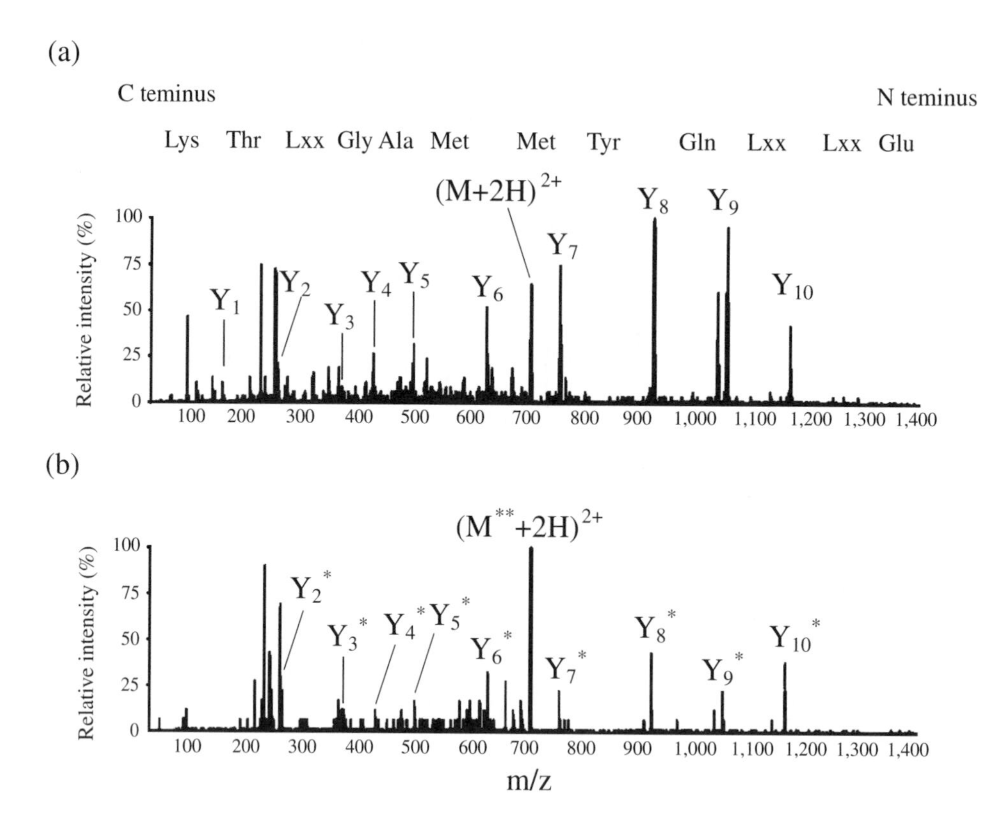

FIGURE 6.6 Sequencing of a peptide from an endothelial cell growth inhibitor. Tandem mass spectra of: (a) a peptide with *m/z* 699.3 from the tryptic digest; (b) same peptide sequenced after esterification of the total digest mixture, resulting in a characteristic mass shift of the C terminus containing fragment ions (Y ions) by 14 Da and an additional 14 Da for each Asp or Glu residue. Asterisks indicate the incorporated methyl groups. Reprinted from Wilm, M. et al., *Nature,* 379, 466–469, 1996. With permission.

De Novo *Sequencing*

It is not unusual to be able to retrieve only a few peptides from the digests of those proteins that are present at low concentrations in the gels. In addition, posttranslational modifications produce tryptic peptides that differ from the theoretical values available in the databases. When the sequence is unknown, derived *de novo* sequencing is required. The manual evaluation of CID spectra, involving identification of ion series and the generation of ladders for each series is not a trivial undertaking. Automatic algorithms have been developed for *de novo* sequencing;[109] however, manual interpretation usually provides lower error rates.

De novo sequencing is often carried out with ESI-MS/MS and low-energy CID to obtain continuous ion series. One approach to simplifying the interpretation of MS/MS spectra is to derivatize the peptides in such a manner that fragmentation is controlled and directed into previously explored pathways. Techniques for charge derivatization[110] and acetylation[111] have been developed and instructive guidelines for the *de novo* sequencing of naphthalenesulfonated peptides have been reported.[112] Another useful technique involves CID fragmentation of peptides before and after methylation with subsequent observation of C-terminal fragments that are characterized throughout the spectrum by a 14 Da/methyl group mass shift.[99,113]

Alternative strategies for *de novo* sequencing use MALDI-TOFMS. In one approach, derivatization with chlorosulfonylacetyl chloride to add a sulfonic acid group to the N-terminus of the analytes resulted in increased fragment ion yields and simple, predictable, PSD fragmentation patterns. Sequencing could be carried out with only 300 fmol of protein. Limitations included the need for non-aqueous solutions for the derivatization step and difficulties in the interpretation of spectra obtained from lysine-terminated peptides.[114]

When proteins are digested in 50% ^{18}O-labeled water, every cleaved peptide will be present as $^{16}O/^{18}O$ isotopic doublets. These doublets can be resolved by many instruments including Q-TOFMS. When these peptides are fragmented, all the y-ions can be readily identified because every C-terminal fragment will contain the $^{16}O/^{18}O$ isotope doublet, whereas N-terminal fragments will retain the normal ^{16}O isotopic distribution with the ^{16}O singlet.[115] An automatic technique has been described for the *de novo* sequencing of proteins using the "differential scanning technique" with partially ^{18}O labeled peptides and nano-ESI-QTOFMS.[116] An alternative technique used IT analyzers for *de novo* sequencing with ^{18}O labeling.[117]

Amino Acid Sequence Determination of Antineoplastic Peptides

The MS/MS approach may also be used for the analysis of simple peptides. The following six synthetic di- and tripeptides (ethyl esters) have the common property of releasing L-*m*-sarcolysin (SL, *m*[di(2-chloroethyl)amino]L-phenylalanine) in the course of their antineoplastic activity: SL-L-Arg(NO_2)-L-Norval, L-Ser-L-*p*-FPhe-SL, L-*p*-FPhe-SL-Asn, L-Pro-SL-L-*p*-FPhe, L-*p*-FPhe-SL-L-Met, and L-*p*-FPhe-Gly-SL-L-Norval. Because these peptides were available in pure form, they could be dissolved in an ESI carrier fluid and injected directly. A comparison of the regular ESI spectrum of L-Pro-SL-L-*p*-FPhe (Figure 6.7a) with the CID spectrum of its $[M+H]^+$ ion (Figure 6.7b) revealed the substructures of the structurally significant ions (Figure 6.7c). The spectra and interpretation of the other five compounds were similar to the one shown in the figure. In-chain and side-chain fragment ions were also identified (Figure 6.7d). Abundant peaks characteristic for SL were present in all compounds and the identity and sequence of the amino acids could be defined unambiguously. In the precursor ion mode, selected ions may be used for the detection and identification of metabolites.

6.1.1.7 Protein Identification by Database Searches

The general role of bioinformatics in protein analysis includes database searching for protein identification through comparisons of peptide maps and peptide structures. The prediction of the structures of peptides is another important aspect but one that is generally carried out on a local level rather than through database searches. These general activities have been reviewed.[118] Whatever analytical approach has been used for the separation and characterization of proteins, the common last step is the use of a computer algorithm for the evaluation of the data obtained.[119] Throughout this section there are references to computer software and Internet-based sources of information. The computer world is a rapidly evolving entity meaning that even a current snapshot must be considered as just that. The onus is on the user to remain aware of current developments.

Several commercial and public domain software packages exist for computerized processing and evaluation of protein spots resolved by 2-DE. The major steps include: (a) characterization by pI and M_r values; (b) quantification by intensities; (c) identification of selected proteins with the aid of the extensive databases available on the Internet; and (d) qualitative and/or quantitative comparison of the images across various sample sets such as control vs. disease or of different disease states. Typical approaches to the computerized analysis of 2-DE images and to spot identification are described in References 120 and 121, respectively.

A rapidly increasing array of software algorithms with various levels of sophistication has been developed for mass spectrometric data interpretation, database searching, and bioinformatics. These include public domain and commercially produced software packages, as well as those available from mass spectrometer manufacturers. Most software offer utilities to search the databases on the Internet. An on-line introductory tutorial to molecular sequence analysis is available at www.sequenceanalysis.com. A concise review of the currently available software tools for database searching to interpret mass spectrometric data also lists a number of original references.[122] Uniform resource locators (URLs) of relevant programs and databases available on the Internet are listed in Reference 15.

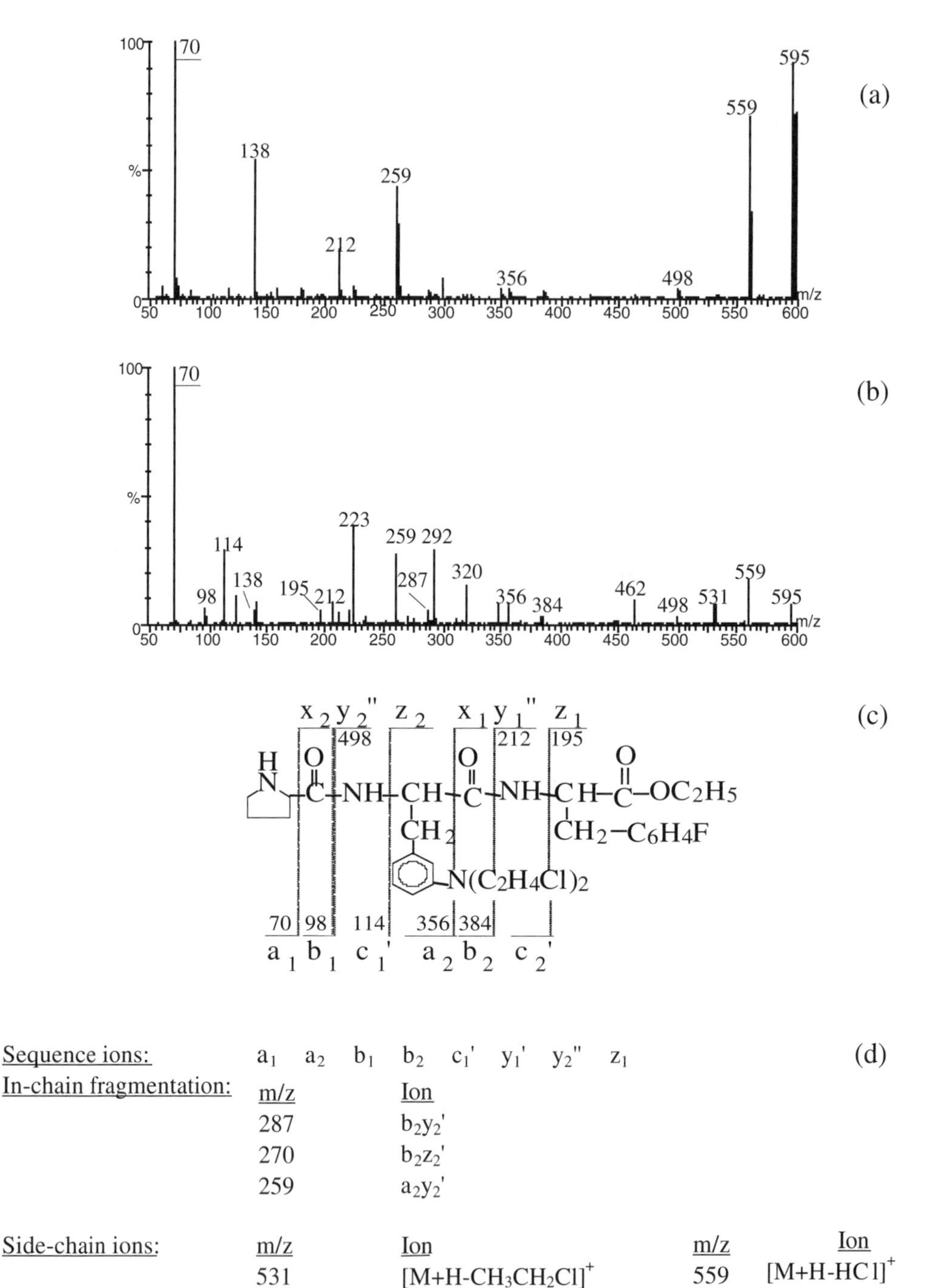

FIGURE 6.7 Sequencing of the antineoplastic sarcolysin-containing peptide Pro-SL-L-*p*-FPhe. (a) ESI spectrum; (b) CID spectrum of the $[M+H]^+$ ion; (c) Substructures of the structurally significant ions; (d) Identified in-chain and side-chain fragment ions.

The comparison of experimentally determined peptide masses with the theoretical peptide masses calculated for each protein in a database results in a list of candidate proteins that are then ordered using a scoring algorithm. There are several scoring algorithms each having the aim of obtaining the highest levels of sensitivity and selectivity, i.e., least number of false negatives and false positives. The sensitivity of a search algorithm refers to its ability to calculate high-ranking scores for distantly related sequences, in essence, its ability to make correct identifications using poor-quality data. Selectivity refers to the ability to calculate low-ranking scores for spurious, random matches that can occur for unrelated sequences. Depending on the sophistication of the algorithm, error windows for a number of parameters must be specified at the time the data are submitted for search. Mass accuracy is a particularly important parameter. When the selected mass window is too large, discrimination is reduced and the number of random matches increases; however, the specification of a window that is too narrow will result in missing valid matches. Use of peptide standards as internal calibrants for the mass scale allows for correction of the measured spectra and permits the use of mass tolerance windows as narrow as 0.0025% (25 ppm) or better. This level of accuracy significantly increases the confidence in any identification obtained. Another important parameter selection involves the choice of permissible fragment ion series in MS/MS-based searches. For example, the inclusion or omission of side-chain ion data may increase random matches or reduce potential matches, respectively.

Software for Protein Identification from Sequence Data

The approach for searching sequence databases is a correlative process that compares experimentally obtained constituent peptide and/or partial sequence information with calculated peptide or fragment ion mass values (theoretical experiments) of candidate proteins in the database. The peptide masses included in the databases are derived from the application of appropriate cleavage rules to the known sequences of individual protein entries. The analyte peptide or protein is identified based on scoring for correspondence between the calculated and observed mass values. If there is no acceptable match, the search is extended to elicit the proteins of closest homology, such as equivalent proteins from related species. Failure of this approach indicates that *de novo* sequencing is probably required.

When high-quality data are available, scoring can be made by counting the number of observed peptide masses that correspond to the calculated masses available for each protein in the database. Software tools for this approach include *PepSea*,[123] *PepIdent/MultiIdent*,[124,125] and *MS-Fit*.[81] The disadvantage of these scoring methods is that they give higher scores to large proteins because the probability of a random match is higher.

More sophisticated software packages optimize selectivity and sensitivity by using additional information available on the proteins in the databases. For example, with the *Mascot* algorithm the scoring is probability-based, i.e., it is assumed that a match between the experimental dataset and each sequence entered in the database is a chance event. Therefore, the match with the lowest probability (P) of being a random occurrence is reported as the best fit. For convenience, scores are reported as $-10\mathrm{Log}_{10}(P)$, i.e., the best match is the one with the highest score. This program can also be used for protein identification from CID fragmentation data.[126] Another algorithm, *ProFound*, ranks the sequences of the protein candidates in the database by their probability of occurrence by using data in addition to the experimental MS data, including the mass range in which the analyte is expected to lie, the accuracy of the mass measurements, the enzyme cleavage chemistry, etc.[127] Additional features offered by some programs include the capability to compare different types of searches, such as by sequence homology, and the possibility of optimizing search parameters through iteration.

Software for Protein Identification from CID Fragmentation Data

In principle, based on probability calculations the sequence information from a single peptide consisting of a number of amino acids should be adequate to identify a protein. Additional mass spectrometric information can radically reduce this number as described below.

Peptide Sequence Tags

The approach is based on the selection of a short known sequence, called the *peptide sequence tag*, obtained from the often-complex MS/MS spectrum of an analyte.[128] The term was inspired by the well-known concept of *expressed sequence tag* (EST) used to represent the short, 200 to 300 base pair long, oligonucleotide sequences randomly collected from genes to characterize the whole gene product in large scale cDNA sequencing projects. The selection of a peptide sequence tag divides that peptide into three regions: a section with known mass (m_1) and sequence and two sections on either side of the tag with unknown sequence but known masses (m_2 and m_3), i.e., $m_1 + m_2 + m_3$ = precursor. When combined with the sequence specificity of the protease used to obtain the digest, the search specificity for a peptide sequence tag (including the m_2 and m_3 values) is ~10^6-fold higher than that of a short sequence alone.[129]

Algorithms have been developed to use this information to provide searching and matching criteria against available sequence databases. Because the search can be made from either direction, assuming either m_2 or m_3 to be the N-terminus, even the direction of the sequence does not need to be known, i.e., whether the ions are b or y type. The candidate peptide suggested by the program must also obey the cleavage conditions that correspond to the proteolytic enzyme used for obtaining the selected sequence tag. A program implementing the sequence tag algorithm, *Peptide Search,* requires the input of only three pieces of information: (a) the total molecular mass of the protein; (b) a set of masses of the products of the proteolytic digestion; and (c) some partial sequence information from the MS/MS analysis. The predicted MS/MS spectrum of candidate sequences is compared with the MS/MS spectra obtained for the analyte. With the inclusion of the other data entered about the protein, identifications can often be made applying CID-derived sequences that consist of only 2-3 amino acids (References 94, 130, see also an application in Section 6.1.3.3).

The popular *SEQUEST* program does not require user intervention to extract sequence information from low-energy (10–50 eV) CID spectra of the candidate peptides. The database is simply searched to identify linear sequences of amino acids that are within a mass tolerance of ±1 Da of the molecular mass of the precursor ion. The scoring of candidate proteins is based on obtaining a cross-correlation function for the experimental data with a protein sequence from the database.[107,131,132] The statistical basis for testing the significance of the scoring obtained with this algorithm has been evaluated and is integrated with the software.[133] An alternative algorithm permits automated *de novo* sequencing of spectra obtained using ESI-QTOF-MS.

The interpretation of CID spectra can be complicated by ions with multiple charge states as well as the influence of isotope patterns. Simplification of the spectra to ease interpretation can be accomplished with a commercial maximum entropy data enhancement program that produces derivative spectra consisting of only monoisotopic peaks in single charge state.[134]

An *in silico* evaluation of the constraints on the mass of proteolytic peptides as a function of mass and the accuracy of mass measurements led to the development of another software tool, *PepFrag*, for protein identification. This program restricts the database search by combining different types of available mass spectrometric information with the experimentally available data from the peptide fragments.[80] The *MS-Tag* program is based on using the possible elemental composition of components in a protein digest as the constraint in the database search. The algorithm calculates possible sequence permutations from the parent mass and immonium ion limited to "near complete amino acid" compositions based on mass measurements obtained to an accuracy of ±10 ppm, such as obtained by MALDI-TOFMS operated with delayed extraction and in the reflectron mode.[81]

The ExPASy Server

This is a multifaceted software package intended to integrate multiple programs into a comprehensive protein identification tool (reviewed in Reference 124). The first objective is to identify the candidate proteins as either already known or as novel. This is accomplished by matching the

TABLE 6.4
Tools for Protein Identification in the ExPASy Server

Compute pI/Mw	Predicts isoelectric points, pI, and molecular weights, M_r
PeptideMass	Theoretically cleaves proteins and calculates the masses of the resulting peptides and known posttranslational modifications
TagIdent	Lists proteins in user-specified pI and M_r regions
	Identifies proteins based on sequence tags up to 6 amino acids long
AACompIdent	Identifies proteins based on their AA* composition, sequence tags, pI, and M_r
AACompSim	Matches theoretical AA compositions against the SWISS-PROT database to find similar proteins
MultIdent types	Combination of the above tools that accepts multiple data

Note: The URL address of the programs is: http://www.expasy.ch/tools.html.

* AA = amino acid.

Based on Wilkins, M.R. et al., *Methods Mol. Biol.,* 112, 531–552, 1999.

observed data against the annotated *SWISS-PROT* protein database and its *TrEMBL* supplement. The server offers tools for several specific purposes (Table 6.4). For example, the *Compute pI/Mw* tool recognizes the approximate region on a 2-DE gel where a particular protein may be found. Another program, the *PeptideMass* tool, is used to interpret peptide mass fingerprinting and other mass spectrometric data. This software also includes the capability for cleaving any protein sequence with a chosen enzyme and computing the masses of the generated peptides.[135] While the *PeptideMass* tool considers discrete posttranslational modifications in peptide mass calculations when the protein data in the *SWISS-PROT* database has the relevant annotations, it will not predict potential posttranslational modifications. However, the *Scanprosite* tool can scan a protein sequence for the occurrence of sites and patterns of posttranslational modifications stored in the *Prosite* database. The *TagIdent* tool can create lists of proteins from organisms, based on user-defined ranges of pI or M_r, and can identify proteins from sequence tags of up to six amino acids that have been derived from the N- or C-termini or internally. It has been reported that tags derived from a C-terminus are more specific than those from an N-terminus tag.[136] The *AACompIdent* tool provides a score that represents the difference between the measured amino acid composition of the analyte protein and corresponding proteins in the database. If the top-ranking protein, i.e., the one with the least difference, does not match certain conditions, unambiguous identification may still be obtained from the list of best matches based on the high specificity of sequence tag data.[137] This tool is also useful for cross-species protein identification.[138]

6.1.1.8 Mass Measurement of Intact Proteins

The measurement of the molecular mass of an intact protein provides rapid confirmation of a suspected or determined sequence, whereas a deviation from this mass indicates an incorrectly determined sequence, the presence of a posttranslational side-chain, or the occurrence of another modification to the protein.[139] Preparative gel techniques have been combined with MS to obtain precise molecular masses.[140] In these methods separated proteins may be cut directly out of polyacrylamide gels or electroblotted onto a membrane. This is followed by placing the dehydrated slices or membranes onto a MALDI stage from which, after the addition of an appropriate matrix solution, the intact protein ions are desorbed by MALDI and their masses determined by TOFMS.[140,141] Alternatively, the analytes may be digested with appropriate enzymes *in situ* on the MALDI plate, after the molecular mass determination, for subsequent CID analysis of the resulting peptides.[142]

The Accurate Mass of a Single Cysteine-Containing Peptide Can Be Used to Identify a Protein

Protein identification from the accurate determination of the mass of a single peptide would obviate the need for either peptide mass mapping or CID analysis. Calculations on the feasibility of this approach indicate that if mass measurements can be made to an accuracy of 0.1 ppm, then 96% of the proteins expressed by certain genomes will be identifiable because they will generate tryptic peptides with unique masses.[64] The stringency of this approach can be modified by the inclusion of additional constraints. In one example, the presence of cysteine in the sequence was introduced as a constraint to reduce the number of potential database hits. Cysteine was selected because of the chemically distinct nature of the sulfhydryl side-chains of cysteine residues. Cysteine-containing peptides were alkylated to form, so-called IDEnT peptides by reaction with 2,4-dichlorobenzyliodoacetamide, an alkylating agent that is specific for cysteine residues, to form their dichlorobenzyl derivatives. Peptide mixtures were then infused directly into an ESI-FTICR instrument capable of determining masses with an accuracy of 1 ppm.[143] Automatic screening of the mass spectra was carried out based on the distinctive natural isotope distribution provided by the chlorinated derivative of the cysteine-containing peptides. Comparison was made between the observed CID spectra of the unknowns and the available data for all possible isobars.[107] Additional database constraints included the specificity of the protease used for protein digestion and the estimated molecular mass of the protein based on 2-DE analyses. The method was illustrated by considering all 6118 possible open reading frames of *Saccharomyces cerevisiae* that yielded 344,855 possible peptide fragments. Applying the cysteine constraint with its added chlorine pattern to 20 peptide isobars, the correct laminin B1 tryptic fragment of the designated test protein, having one cysteine and one tryptic cleavage site was easily identified. When applied to a complex protein mixture obtained from a yeast, the approach permitted the identification of low-abundance proteins that could not be detected by either mass mapping or data-dependent MS/MS analysis. The development of IDEnT reagents that have target specificities other than cysteine, e.g., methionine or tryptophan, would extend the areas of application of the technique.[144]

Resolution of the Fine Isotopic Structure of Proteins

When an analyte molecule contains hundreds of carbon, hydrogen, nitrogen, and oxygen atoms, the binomial distributions of the naturally occurring isotopes of the constituent elements result in a broad and highly complex isotope pattern. For compounds containing the common elements, C, H, N, O, P, S, the mass of the most abundant peak is shifted upward by 1 Da from the monoisotopic peak for every 1.5 kDa of molecular mass. This means that the monoisotopic peak of a protein >15 kDa is no longer detectable. The problem of accurate mass measurement that results from this distribution of isotopes was illustrated using a 107 amino acid mutant, C22A, FK506-binding protein. This protein is an ~11,800 Da species that exhibits peptidyl-prolyl *cis-trans* isomerase activity and has been used as a model in studies of protein folding. In the ESI-FTICR spectrum of this compound there is only one resolved peak that represents a monoisotopic species, $[^{12}C_{527}{}^{1}H_{146}{}^{16}O_{155}{}^{32}S_{3}]^{+}$ (Figure 6.8 top). All other peaks are the result of superposing several unresolved isotopic variants, e.g., the next highest nominal mass peak includes species involving $^{13}C^{12}C_{526}$ and $^{15}N^{14}N_{145}$ as well as other combinations. A novel approach to solving this problem was based on the possibility of obtaining the C22A analyte in a form that was isotopically *depleted* in ^{13}C and ^{15}N. This was achieved by isolating the protein from *E. coli* that had been grown in a nutrient system containing glucose and ammonium sulfate that were significantly (>99.95%) ^{12}C and ^{14}N, enriched. This double isotopic depletion resulted in a significant change in the mass spectrum, namely that the monoisotopic peak, that was present at only 0.65% of the major peak with natural isotope abundances became the most abundant peak of the molecular ion cluster (Figure 6.8 bottom). The resulting monoisotopic molecular mass could be determined unambiguously to be 11,780.01 Da compared with 11,780.07 Da that was computed from the amino acid sequence. It was concluded

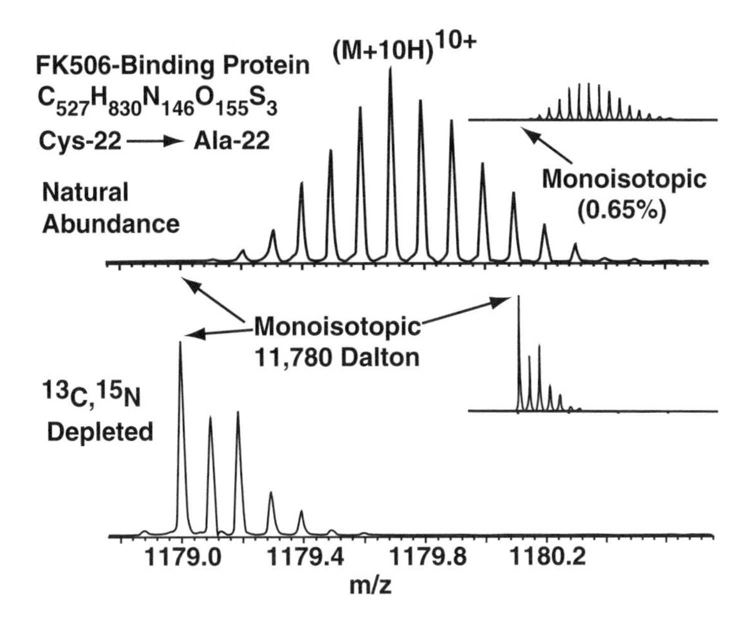

FIGURE 6.8 ESI-FTICR mass spectra of a mutant (C22A) FL506-binding protein. Top: natural-abundance isotopic distribution; bottom: isotopic distribution for the same protein grown on a medium with enriched ^{12}C and ^{14}N. Insets: theoretical isotopic distribution calculated for: (top) natural isotope abundances where the "monoisotopic" peak is barely observable, and each of the other isobaric peaks consists of multiple isotope combinations and (bottom) doubly-depleted protein with highly abundant monoisotopic peak. Reprinted from Marshall, A.G. et al., *J. Am. Chem. Soc.*, 119, 433–434, 1997. With permission.

that this double-depletion method is generally applicable and can be used for a large number of proteins, as well as RNA and DNA for which ^{13}C and ^{15}N enrichment are already available.[145]

In subsequent work, this *isotopic double-depletion* technique was combined with several other techniques[146] to improve the resolution of isotopic fine structure in proteins. The approach was illustrated through the analysis of the p16 tumor suppression protein by ESI-FTICRMS at ultrahigh mass resolution. Comparing the theoretical isotopic distribution for the protein with that obtained from the ^{13}C and ^{15}N doubly depleted analyte revealed a major improvement in the signal-to-noise ratio of the monoisotopic $^{12}C^{14}N^{34}S$ and $^{12}C^{14}N^{18}O$ species, due to reduction in the abundances of the ^{13}C- and/or ^{15}N-containing species (Figure 6.9a). In other words, the double isotopic depletion increased the monoisotopic abundance of the molecular species at mass M and reduced interferences from other ions at M + 2 such as those containing $^{15}N^{33}S$, thereby leaving those containing ^{34}S and ^{18}O as the principal M + 2 species. The resulting isotopic fine structure permitted the determination of the number of sulfur atoms in the doubly depleted analyte, solely by measuring the relative abundance of the M + 2(^{34}S) peak without any prior information about the protein (Figure 6.9b). The number of sulfur atoms found was 5.1 ± 0.3 vs. 5 determined by conventional chemical methods. It was concluded that distinguishing different elemental composition with ultrahigh mass resolution and determining relative abundances within a few percent accuracy are adequate to identify and quantify selected elements in a protein without the need of high mass measurement accuracy.[146]

6.1.1.9 Quantification

One major goal of proteomics is the quantitative measurement, on a global scale, of each protein expressed in healthy and diseased cells or tissues (reviewed in References 147–149). The high degree of variability in the ionization efficiency for different peptides has necessitated the use of

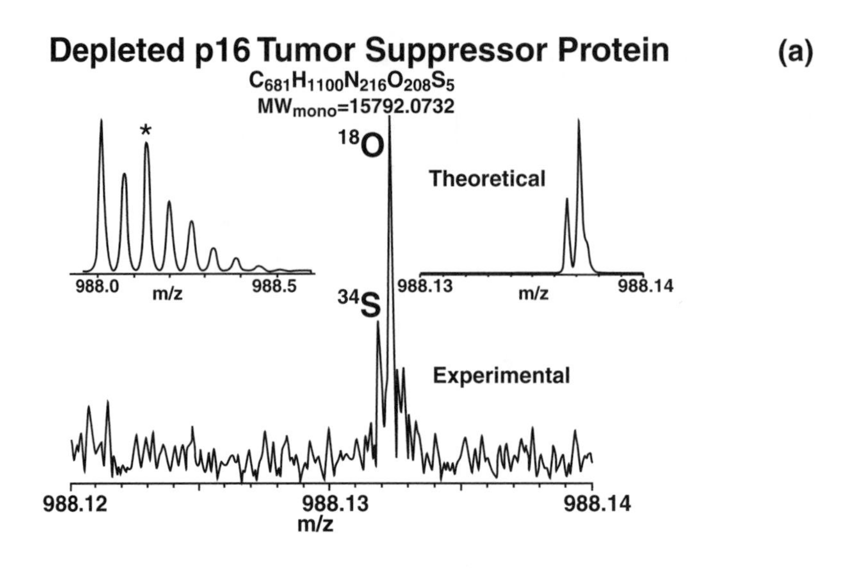

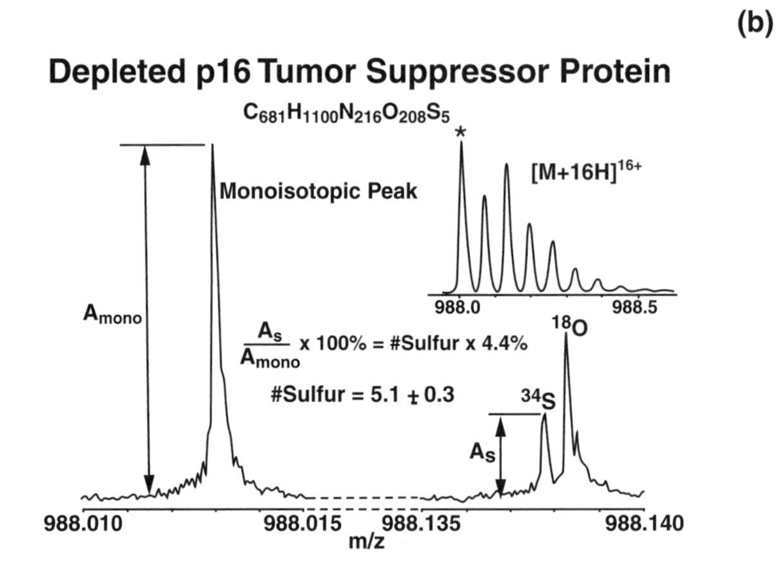

FIGURE 6.9 (a) Mass scale-expanded spectra of one isotopic peak for ^{13}C, ^{15}N doubly depleted p16 tumor suppressor protein, from a single ESIFTICR mass spectrum: (a) Upper left: theoretical isotopic distribution, * denotes the isotopic peak (2 Da above the monoisotopic peak). Upper right: theoretical isotopic fine structure for the starred peak. The signal-to-noise ratio of the monoisotopic ^{34}S and ^{18}O species was significantly increased by reducing the abundances of species containing ^{13}C and/or ^{15}N; (b) Upper right: theoretical isotopic distribution in which * denotes the monoisotopic peak. Left: monoisotopic peak and right: ~2 Da above the monoisotopic mass. The number of sulfur atoms in the protein (5.1 ± 0.3) was determined from the abundance ratio of the resolved ^{34}S ions to the monoisotopic species. Reprinted from Shi, S.D.-H., Hendrickson, C.L., and Marshall, A.G., *Proc. Natl. Acad. Sci. USA*, 95, 11532–11537, 1998. With permission.

stable-isotopically labeled internal standards[150] and isotope dilution (Section 2.6.4) in quantification methodologies. In one approach, control and perturbed yeast cultures were grown in two media; one on a medium enriched in ^{15}N, >95%, and the other in a medium where the ^{15}N was present at its natural abundance, i.e., 0.4%. Following cell lysis, the recovered proteins from the two yeast strains were combined in equal amounts, separated by 2-DE, stained, digested, and analyzed on a

MALDI-TOFMS instrument capable of resolving isotopes. The incorporated ^{15}N resulted in pairs of peptide peaks in which with the masses of the enriched peptides were increased according to the number of nitrogen atoms present. Quantification was based on the ratio of the signal intensities of the unlabeled and labeled peptide pairs. In another application of the method, the expression of 42 proteins was quantified to within ±10% in two *S. cerevisiae* pools with different abilities to express the G1 cyclin CLN2.[151] The approach can be reversed by using media depleted in ^{13}C, ^{15}N, and ^{2}H to generate the internal standards. This was used to quantify some 200 proteins in *E. coli.*[62]

Quantification of Protein Mixtures Using Isotope-Coded Affinity Tags (ICAT)

The functional elements of an ICAT reagent include a group with specific reactivity toward sulfhydryl residues, a normal or heavily deuterated linker, and a biotin affinity tag. The technique assumes that a short (5 to 25 residues) sequence of contiguous amino acids from a protein contains sufficient information to uniquely identify that protein and that the products of peptides treated with the "light" (i.e., no isotopic enrichment) and "heavy" (i.e., isotopically enriched) ICAT reagents are chemically identical. Thus the determination of the relative abundances of the isotopic pairs can be used for quantification because they can serve as mutual internal standards, as in conventional isotope dilution methodology.[150] The technique was tested by comparing protein expression in *S. cerevisiae* when either ethanol or galactose was used as the carbon source. The first step was the derivatization of the cysteinyl residues in reduced protein samples that were representative of cells from the two growth media, with the proteo- and deutero- ICAT reagents, respectively. Next, the derivatized samples were combined and enzymatically cleaved, and the tagged, cysteine-containing peptides were isolated by avidin affinity chromatography. Finally, the isolated peptides were analyzed by LC-MS/MS. An automated, multistage MS operation using an ITMS permitted not only the quantification of tagged peptides that were specific for the media in which the cells were grown using the MS mode, but also the determination of the tagged sequences with the automated MS/MS and therefore the likely protein of origin. Quantification was based on the determination of the relative abundances of peptide ion pairs tagged with the proteo- and deutero-isotopes of the linker. Sequence identification was obtained through data-dependent MS/MS operation followed by automatic correlation with sequence databases. Advantages of this approach include avoidance of 2-D gels, virtually no limitation on sample size, and avoidance of metabolic labeling.[149,152]

Proteolytic ^{18}O Labeling

In this approach, the samples to be compared are differentiated by the inclusion of an ^{18}O label. In the first sample, conventional proteolytic cleavage is used, and in the second sample, two ^{18}O atoms are incorporated universally at the carboxyl termini of the tryptic peptides. This results in the creation of peptide pairs recognized by their 4 Da mass separations. After pooling the mixtures, the components are separated and analyzed by high-resolution ESI-FTICRMS and/or MALDI-QTOFMS. Mass spectrometric identification of short sequences and/or accurate mass measurements are used with the aid of databases to relate the peptides to their precursor proteins, while quantification is accomplished using the relative signal intensities of the isotopically modified peptide pairs. The technique was evaluated by comparing two serotypes of adenovirus. In contrast to the integrated design of the ICAT approach, the proteolytic ^{18}O strategy is considered to be modular.[153]

6.1.2 Posttranslational Modifications

Posttranslational modification refers to the covalent chemical reaction(s) whereby a polypeptide chain, newly synthesized on the ribosome, is converted into a functional protein. Also included are the modifications that regulate the activity of a protein, such as signal transduction pathways. The concept thus includes all changes not coded for by the gene. Posttranslational modifications of eukaryotic proteins, *in vivo*, are common, and >200 of these covalent modifications are known;[154] several are listed in Table 6.5. Almost any part of a protein may be altered by a posttranslational

TABLE 6.5
Representative Posttranslational Modifications

Modification	Mass difference, Da	Example (Reference)
Acetylation	42.04	189, 190
Carboxylation of Asp or Glu	44.01	
Cysteinylation	119.14	186
Farnesylation	204.36	188
Hydroxylation of Pro, Lys, and Asp	16.00	190
Methylation		
N-methylation of Lys, Arg, His, or Gln	14.03	190
O-methylation of Glu or Asp	14.03	190
Myristoylation	210.36	547
Palmitoylation	238.41	548
Phosphorylation of Ser, Thr, or Tyr	79.98	164
Propylation	42.08	
Reduction of disulfide bridge	2.02	
Sulfation of Tyr	80.06	549
Glycosylation of Asn, Ser, and Thr by		
Deoxyhexoses (Fuc)	146.14	547
Hexosamines (GlcN, GalN)	161.16	187
Hexoses (Glc, Gal, Man)	162.14	
Pentoses (Xyl, Ara)	132.12	
Sialic acid (NeuNAc)	291.26	550

Note: More information available at http://www.abrf.org/ABRF/ResearchCommittees/deltamass/deltamass.html.

modification: (a) a specific residue at a terminus, e.g., conversion of an N-terminal glutamine to pyroglutamic acid; (b) a terminus independent of the identity of the residue, e.g., esterification of the C-terminus; and (c) a residue independent of its position in the peptide, e.g., oxidation of methionine. Because these modifications involve the addition or subtraction of mass to the polypeptide chain, mass spectrometry is the technique of choice to detect, identify, and localize these modifications. When N-terminal amino acids have been modified and, therefore, become inert to Edman degradation reactions, mass spectrometry is the only technique for the detection of these modifications and the identification of the products. Mass spectrometric techniques have been developed to provide the following information for peptides up to 30 residues long and at the femtomole level: (a) confirmation of the presence of posttranslational modifications; (b) composition of the product; (c) site(s) of individual amino acids carrying the modifications; and (d) quantities of the modified species.

A software tool called *FindMod* has been developed to predict those amino acids in particular peptides that may carry posttranslational modifications and to examine peptide mass fingerprinting data for mass differences between empirical and theoretical peptides that may indicate the presence of modifications. *FindMod* uses a set of 29 systematic rules that were constructed by examining 5153 documented posttranslational modifications in the *SWISS-PROT* database. The rules determine which amino acids can carry one of the 22 most common posttranslational modifications and whether the modifications are likely to occur on the N- or C-terminus or on a particular internal amino acid. The software was tested by evaluating peptide fragments obtained using MALDI-TOFMS with PSD of proteins from *E. coli* and sheep wool.[155]

6.1.2.1 Phosphorylation

Negatively charged *phosphoproteins* are formed when a phosphate group is added via an oxygen linkage to the hydroxyl group of serine, threonine, or tyrosine. Protein phosphorylation and dephosphorylation are probably the most important and ubiquitous of the posttranslational modification events that occur in eukaryotic organisms: ~30% of intracellular proteins are phosphorylated. In addition to controlling major cellular signaling events, phosphorylation has significant roles in regulating a variety of intracellular functions and metabolic pathways. Phosphorylation is mediated by phosphotransferase enzymes called *protein kinases*. Each protein kinase contains a consensus sequence to catalyze the transfer of the terminal phosphate moiety of ATP or GTP to the nucleophilic hydroxyl groups of serine, threonine, and/or tyrosine of proteins and peptides. The removal of phosphate is controlled by another class of enzymes, the *phosphatases*.

Regulation by multisite phosphorylation has been studied in connection with the activation of the serine/threonine kinase, protein kinase D, via a phorbol ester-dependent pathway. Five *in vivo* phosphorylation sites were identified and their roles investigated by site-directed mutagenesis.[156]

Because several oncogenes have products with tyrosine or serine/threonine kinase activity that are upregulated in a variety of neoplasms, the determination of phosphorylation sites in targeted proteins is particularly relevant in cancer research. Despite extensive ongoing research, several major aspects of phosphorylation remain poorly understood. For example, not all sites of a protein/peptide that could be phosphorylated are actually phosphorylated, and not all copies of a particular protein/peptide are phosphorylated.

Methodologies and Strategies

Because tyrosine has a hydrophobic benzene ring, several of its properties differ significantly from those of serine and threonine. In addition, the β-elimination reaction that removes phosphate from serine and threonine cannot occur for thyrosine-containing phosphoproteins. These and other differences must be considered in designing appropriate hydrolysis and separation strategies. The available MS technologies for the identification of phosphoamino acids have been reviewed.[157-160] A common technique for the identification of phosphorylated amino acid residues, both *in vitro* and *in vivo*, has been radiolabeling with ^{32}P-ATP followed by chemical or enzymatic cleavage into smaller peptides for subsequent amino-terminal sequencing by Edman degradation.

Mass spectrometric techniques are based on the fact that the phosphoryl group ($-PO_3H$) is represented by a mass increment of 80 Da/phosphate, i.e., evidence of phosphorylation may be obtained by observing the loss of the phosphoryl group during fragmentation. Alternatively, when digestion with a suitable endoprotease leads to mismatching peptides, differential peptide mapping is accomplished by undertaking another digestion with both the same endoprotease and alkaline phosphatase, the latter being specific for removing Ser-O- and Thr-O-bound phosphoryl groups. After accounting for the mass loss of $n \times 80$ Da, a peptide match suggests that Ser-O/Thr-O phosphorylation was present. Even the rarely occurring tyrosine phosphorylation may be investigated by MS.[161] ESIMS techniques have the advantage of accommodating on-line chromatographic separation of peptides in proteolytic digests with simultaneous sequencing of several peptides in a single HPLC run. MALDI-TOFMS techniques usually offer greater sensitivity and more rapid analysis than ESIMS, but require off-line chromatographic separation.

Phosphorylated residues may be identified by ESIMS/MS given that there is usually additional fragmentation at the phosphate ester bond during CID.[162,163] These dephosphorylated product ions may be of the b type containing the N-terminus of the molecule, or y type, originating from the C-terminus. General technical aspects and performance of MS techniques have been illustrated using 30 pmol of trypsin-digested bovine β-casein, a well-characterized protein with five phosphorylation sites at serine residues.[164,165] Nano-ESIMS is capable of mapping phosphorylation sites of gel-isolated proteins with a sensitivity of 250 fmol protein applied to the gel.[166] Certain advantages of IT analyzers, e.g., fast scanning and multiple fragmentations (MS^n), have been utilized to identify phosphopeptides, determine their sequences, and localize the phosphorylation sites.[167]

Phosphoserines and phosphothreonines have been identified by MALDI-TOFMS in the PSD mode through recognition of ions that represent the losses of 80 Da ($[M+H-HPO_3]^+$) and 98 Da ($[M+H-H_3PO_4]^+$) from the precursors. The latter ions are absent from phosphotyrosines because elimination of phosphoric acid from the aromatic ring of tyrosine is not favored. The technique has been applied to the study of the dephosphorylation of a phosphotyrosine-containing peptide of the human Tau protein, and to the peptide containing the autophosphorylation site of $pp60^{c\text{-}src}$, a prototype peptide for the *src* family of nonreceptor tyrosine kinases in the Rous sarcoma virus-encoded transforming protein.[168]

Another approach to mapping the phosphorylation sites of proteins used on-line immobilized metal [Fe(III)] affinity chromatography to retain and preconcentrate phosphorylated proteins and peptides. Components of the isolated mixture could then be separated by CE and analyzed by ESI-MS/MS. The use of an IT analyzer to carry out MS^n provided reliable assignments for the locations of the phosphorylated residues. The technique was tested on β-casein for which the stochiometries, i.e., the relative levels of phosphorylation at each modifiable site, were obtained from the mass spectral data.[165]

All current methodologies require purification to near homogeneity of the phosphoprotein of interest before analysis. At least two sophisticated mass spectrometric techniques have been described for probing the phosphoproteome (reviewed in Reference 169). In one technique, six steps were used for the selective isolation of phosphopeptides from a peptide mixture obtained by a proteolytic digestion: (1) peptide amino group protection; (2) condensation reaction; (3) phosphate regeneration; (4) carbodiimide condensation with cystamine and reduction of the internal disulfide of cystamine to produce free sulfhydryl groups for every phosphate group; (5) covalent solid-phase capture of the free sulfhydryl groups with iodoacetic acid immobilized on glass; and (6) phosphopeptide recovery with trifluoroacetic acid. The phosphopeptides were analyzed by automated LC/MS/MS and identified by correlating the CID data with sequence databases. The technique was tested by determining the sites and locations of serine-, threonine-, and tryosine-phosphorylated residues in β-casein and *Saccharomyces cerevisiae*.[170] Another technique started with the removal of all reactive cysteine species in crude cell extracts by oxidation with performic acid. Next, base hydrolysis was used to induce β-elimination of the phosphates from phosphoserine and phosphothreonine. Free sulfhydryls were then formed by adding ethanedithiol to the alkene. The phosphoproteins were purified by coupling the free sulfhydryls to biotinylated moieties, used as affinity handles for immobilized avidin. The biotinylated peptides were further purified with a second avidin binding. MALDI-TOFMS was used to obtain peptide maps. The localization of the biotinylated residue, and hence the site of phosphorylation, was accomplished by ESI-MS/MS. The technique was tested by analyzing the phosphorylation sites of β-casein.[171]

The well-established technique of isotope dilution MS (Section 2.6.4) has been used in a new context to determine changes in the levels of modification at specific sites of phosphorylation. Two cell pools were grown; one was placed on media containing nitrogen with its natural isotope abundance, and the other on media that were isotopically enriched for ^{15}N. After combining the pools, the proteins were purified, digested with trypsin, and analyzed by ESI-ITMS. Peptides undergoing changes in their levels of modification yielded pairs of MS peaks with the $^{14}N/^{15}N$ intensity ratios reflecting the degree of the change. The technique was tested on a specific protein kinase in yeast. Measurement of the intensity ratios of isotopically labeled and unlabeled phosphopeptides revealed the presence of at least four sites of phosphorylation. Both the sites and the degree of phosphorylation were determined. The technique can reliably determine changes >20% using subpicomole amounts of gel-separated mixtures as well as simple protein mixtures, as long as the cell systems can be grown on isotopically enriched media.[151]

An alternative technique, based on the ICAT approach to the quantification of proteins by stable isotope labeling of functional groups (Section 6.1.1.9), was adapted to provide a quantitative comparison of the phosphorylation status of proteins generated under two conditions. The isolated proteins were subjected to β-elimination of the phosphate from serine- and threonine-containing phosphoproteins. This was followed by a Michael addition of either ethanethiol or ethane-D_5-thiol at the resulting vinyl moiety to provide a specific label. Analysis was carried out using an ESI-QTOFMS. MS/MS fragmentation of the three most abundant peptides was undertaken with data-dependent switching. The technique was tested by quantifying mixtures (0 to 100%) of α_{s1}-casein and dephospho-α_{s1}-casein.[172]

An ICPMS technique developed to determine the degree of phosphorylation involves separation of digested peptides by HPLC and determination of the phosphorus (^{31}P) to sulfur (^{32}S) ratio. Although phosphopeptides usually do not contain cysteine or methionine, the intact phosphoproteins customarily have at least one sulfur-containing residue. The degree of phosphorylation can be determined from the measured P/S ratio combined with protein/peptide sequence information. The utility of the technique has been demonstrated using several known phosphopeptides and phosphoproteins.[173]

Phosphorylation of Mitogen-Activated Protein Kinase Kinase (MKK)

A network of signal transduction pathways that leads to cell proliferation and differentiation is initiated by the activation of receptor tyrosine kinases in response to extracellular stimuli. A strongly conserved module in this complex network is a protein kinase cascade involving the phosphorylation of the mitogen-activated protein kinase (MAPK) by MKK. The reverse action, feedback phosphorylation of MKK, has also been demonstrated *in vitro* with the phosphorylation sites being identified by ESIMS and site-directed mutagenesis. Samples were prepared by tryptic digestion or CNBr treatment. The peptide products were then separated by HPLC and analyzed on-line by ESIMS/MS. A large number of both tryptic and CNBr peptides were identified (Figure 6.10). There was agreement between the observed peptide masses and those predicted from the cDNA sequence. Extracting ion-specific chromatograms for selected masses and looking for masses differing by 80 Da (or a multiple of 80) from masses of predicted sequences led to the identification of two sets of phosphopeptides (T33/34 and T40; Figure 6.10). Candidate target sites were at Thr-292 and Thr-386, both of which were within the consensus sequence PXT*P. Replacing these sites with alanine using site-directed mutagenesis significantly reduced phosphate incorporation providing additional information on the structure. Additional biochemical experiments revealed that the phosphorylation of MKK at these sites did not interfere with its *in vitro* catalytic activity.[174]

In a subsequent study, the immunoprecipitated product of the oncogene *v-mos* was used to catalyze the phosphorylation of recombinant human MAP kinase kinase (rMKK) and rabbit skeletal muscle MKK1. The products of digestion with endoproteinase LysC were analyzed by LC/ESIMS and the observed masses compared to those predicted for proteolytic fragments. Peptide sequences were confirmed by MS/MS using b-, y-, and a-type ions. Significant phosphorylation was detected on peptides Tp28 and Tp33a. For example, in Tp33a (residues 292-302: TPGRPLSSYGM), MS/MS of the $[M+2H]^{2+}$ ion revealed cleavage of all peptide bonds except, as would be anticipated, those between GR and PL, as well as n × 80 Da mass shifts corresponding to phosphorylation (Figure 6.11A). Further evaluation of the MS/MS spectrum of the monophosphorylated peptide suggested that there were two isoforms. These two peptides were then separated by gradient HPLC and re-analyzed by MS/MS. One of the spectra showed mass shifts of +80 Da for the b_9 and b_{10} ions, but not for b_8, suggesting that the site of phosphorylation is located at Y-300 (Figure 6.11B). Detailed interpretation of the ions involving the loss of 18 Da in the MS/MS spectrum of the second isoform (Figure 6.11C) was consistent with the phosphorylation of S-298. Diphosphorylated Tp33a was also observed, and its MS/MS spectrum revealed an ion corresponding to the neutral loss of H_3PO_4 from the parent ion (Figure 6.11D). With additional evidence from the MS/MS spectrum, it was concluded that both S-298 and 4-300 were phosphorylated in this peptide. There was no MS evidence for phosphorylation at T-292. Comparison of MS and MS/MS data from rMKK and MKK1 revealed significant differences in the degree of phosphorylation, apparently due to autophosphorylation under physiological conditions.[175]

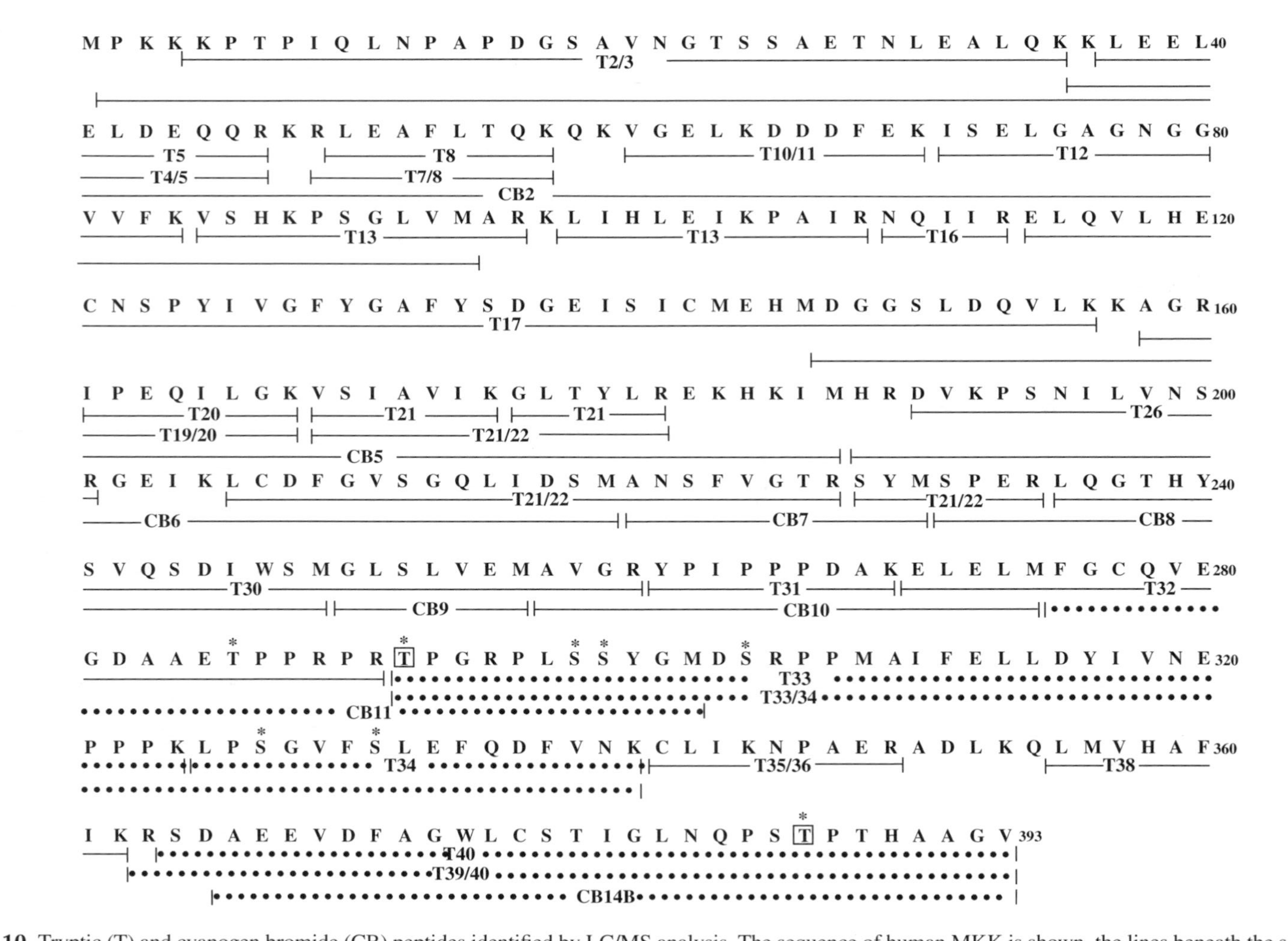

FIGURE 6.10 Tryptic (T) and cyanogen bromide (CB) peptides identified by LC/MS analysis. The sequence of human MKK is shown, the lines beneath the sequence denote the T and CB peptides observed by LC/MS. Overlapping peptides generated by incomplete proteolysis are indicated in some cases. Dotted lines indicate phosphopeptides observed in mitogent-activated protein kinase (MAPK)-phosphorylated mitogen activated protein kinase kinase (MKK) samples. Mutated residues are indicated with asterisks. Those mutations that reduced phosphorylaytion by MAPK are boxed. Reprinted from Resing, K. et al., *Biochemistry,* 34, 2610–2620, 1995. With permission.

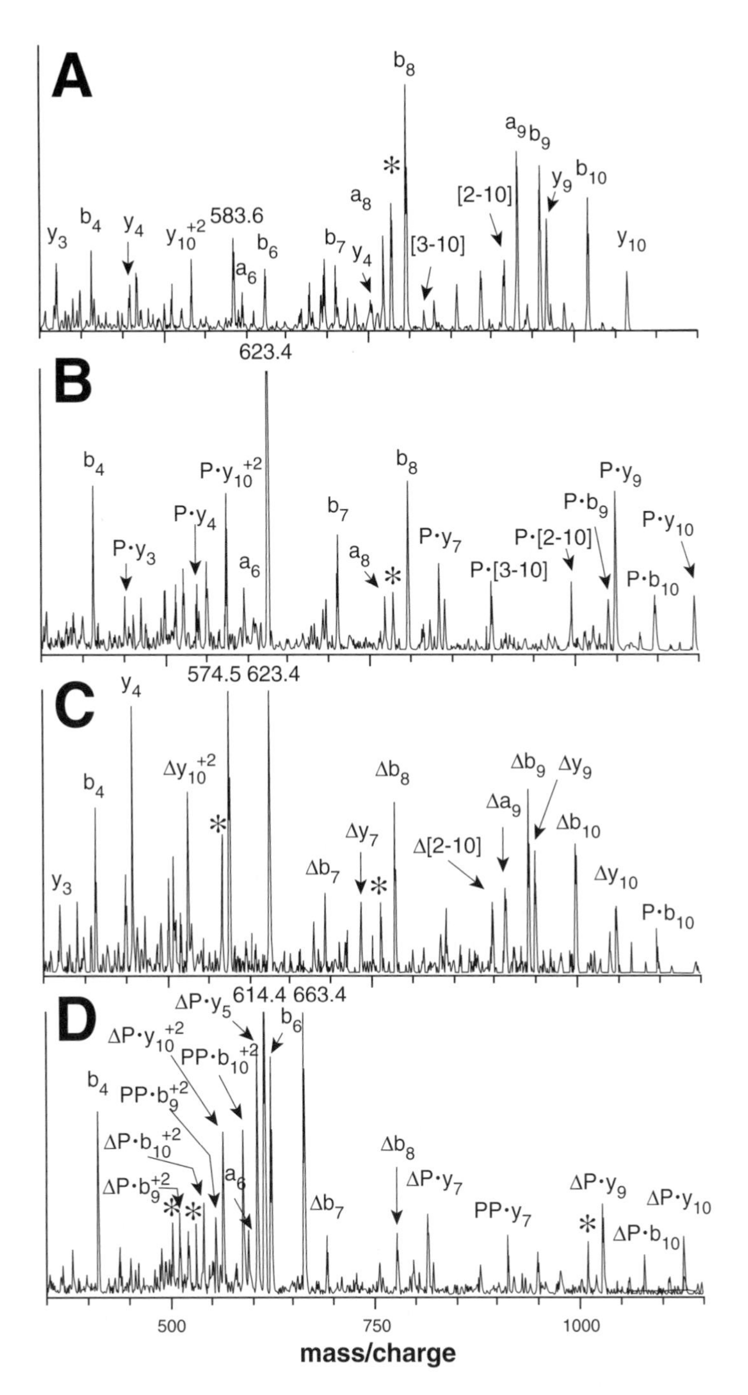

FIGURE 6.11 Identification of the major autophosphorylation sites in recombinant mitogen-activated protein kinase kinase. MS/MS of (A) Tp33a $[M+2H]^{2+}$, m/z = 583.6); (B) Tp33a + 80 Da (one phosphate) early eluting peak $[M+2H]^{2+}$, m/z = 623.4); (C) Tp33a + 80 Da late eluting peak; and (D) Tp33a + 160 Da (two phosphates) $[M+2H]^{2+}$, m/z = 663.4). The m/z of the precursor ion is shown in each spectrum. In (C) and (D), ions at 574.5 and 614.4 Da, respectively, indicate the neutral loss of H_3PO_4 from the precursor ions. The b_6 ion of (B) has the same predicted mass as the precursor ion; its presence is suggested by the observation of the a_6 ion. The a_6 and b_6 ions were not observed in (C). Bracketed numerals indicate internal fragment b ions; e.g., [2-10] corresponds to an internal fragment between residues 2 and 10 of the peptide. Asterisks indicate the loss of H_2O from higher mass ions. Reprinted from Resing, K. et al., *Biochem.*, 34, 2610–2620, 1995. With permission.

Phosphorylation of Stathmin Isoforms

Stathmin (p18, oncoprotein 18, prosolin) is a cytosolic phosphoprotein that is highly expressed in several malignancies including leukemia, lymphomas, myelomas, and melanoma. In various states of phosphorylation, stathmin is apparently involved in several kinase-related transduction pathways and may act as a relay protein integrating signals from several pathways during both differentiation and neoplastic progression.[176] Stathmin is one of the key regulators of microtubule dynamics because it prevents assembly and promotes disassembly of microtubules in a concentration-dependent manner. Mass spectrometry has been playing an important role in the elucidation of the native structure of strathmin and its tubulin interaction domains.[177]

The phosphorylation status of three brain stathmin isoforms was determined by peptide mapping using MALDI-TOFMS and ESI-ITMS. Protein spots separated by 2-DE were excised, digested in-gel with trypsin, the peptides extracted, separated by HPLC, and 0.5 μL aliquots of peptide-containing fractions analyzed by MALDI-TOFMS in the linear mode. One protein was identified as stathmin from mass fingerprinting based on 11/18 matches using the *MS-Fit* algorithm. Each peptide ion was also analyzed by MALDI-TOFMS in the PSD mode in an attempt to determine which of those peptides eluding identification by mass fingerprinting may have been phosphorylated. Only one unidentified peptide was associated with ions corresponding to the expected loss of 80 or 97 Da (HPO_3 or H_2PO_4, respectively). The amino acid sequence was deduced by *MS-Fit* to be SKESVPEFPLSPPKKK, corresponding to residues 28-43 of stathmin, and indicating that it contained one phosphorylated amino acid. Because the MALDI-TOF analysis could not determine which of the three possible sites (Ser-28, Ser-31, and Ser-38) was phosphorylated, the relevant fraction was subjected to LC/ESI-ITMS analysis. Doubly- or triply-charged ions above a preselected abundance threshold in each ESI spectrum were automatically subjected to CID in a subsequent MS/MS scan. The resulting data produced sets of doublets indicative of the presence of a monophosphopeptide and a nonphosphorylated form. One of the pair of doublets revealed an intense doubly-charged ion (Figure 6.12A) that confirmed the presence of a phosphate, and backbone fragment ions that lead to the elucidation of the peptide sequence as SKESVPEFPLpSPPKKK. The CID product spectrum of the triply-charged ion from the second peak of the doublet (Figure 6.12B) was identified as the nonphosphorylated peptide. It was concluded that phosphorylation of this stathmin isoform was at Ser-38. Using the same approach, no phosphorylation was found in the second brain stathmin isoform studied. Partial characterization of the third isoform revealed phosphorylation of Ser-38 but not of Ser-16 and Ser-25.[178]

Protein Kinase C-Mediated Phosphorylation at Specific Sites in the Binding Region of the Microtubule-Associated Protein-2 (MAP-2)

The serine and threonine residues of MAP-2 are extensively phosphorylated. To locate and confirm the critical phosphorylation sites, the effects of phosphorylation of MAP-2 by protein kinase C were studied using taxol-stabilized micotubules. Native MAP-2 was cleaved by thrombin into two domains. One of these, the 22 kDa microtubule-binding region (MTBR), retained the binding affinity for microtubules. Two MTBR constructs were produced by site-directed mutagenesis with methionyl residues being introduced at four locations. One was N123C, where N is the amino-terminal sequence at the thrombin cleavage site and C refers to the sequence from the repeats to the carboxyl terminus. The designation 123 refers to the three nonidentical peptide repeats within the MTBR. The constructs were cleaved with CNBr and shifts in the molecular masses of the products were determined by MALDI-TOFMS. The predicted and observed molecular masses were in good agreement (Figure 6.13). Comparison of HPLC profiles and selected ion traces of unphosphorylated peptides with those that had been phosphorylated over various time intervals with protein kinase C, revealed peaks that differed by multiples of 80 (Figure 6.14). From the overlapping partial cleavage points in the mass spectra of the MTBR 123 region of MAP-2, obtained for incubations from 20 min to 960 min, the phosphoryl groups were localized to the serines at positions 1680, 1703, 1711, 1728, and 1760. The assignments were confirmed by constructing individual

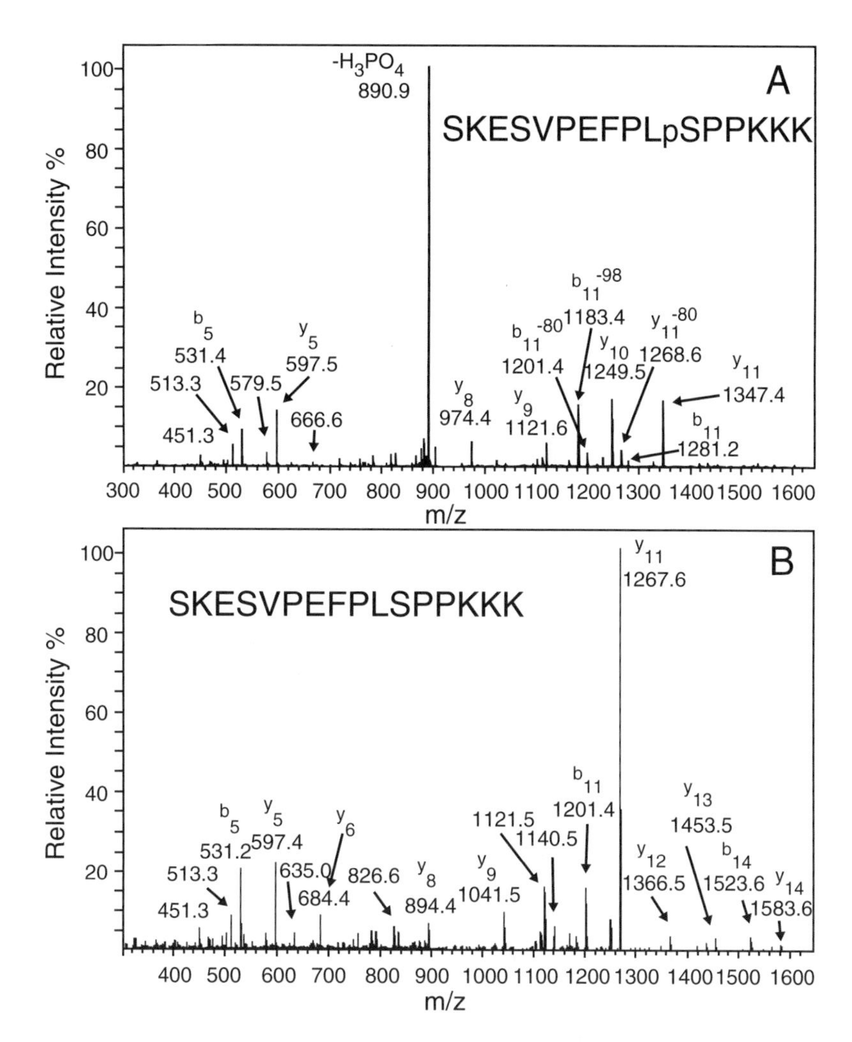

FIGURE 6.12 CID identification of tryptic peptides in one the separated fractions (fraction 15). (A) CID spectrum of peak A identified as the monophosphopeptide SKESVPEFPLpSPPKKK. (B) CID spectrum of the triply-charged ion *m/z* 600.8 that was identified as the nonphosphorylated version of the peptide in peak A. Reprinted from Zugaro, L.M. et al., *Electrophoresis,* 19, 867–876, 1998. With permission.

site-directed mutants where each serine was replaced by an alanine. Mass spectra of these mutant peptides, obtained after phosphorylation, confirmed the expected decrease in phosphoryl incorporation. Additional MS data on the wild-type and mutant N123C, and other single and double mutant constructs, revealed that the Ser-1703 and Ser-1711 positions were the most important in the stimulation of microtubule assembly after phosphorylation by protein kinase C.[179]

Synthesis of the 23/33 Amino Acid C-Terminal Sections of Human p53 in Phosphorylated Forms

The tumor suppressor gene p53 codes for a nuclear phosphoprotein shown to be altered by mutation or deletion in ~50% of human tumors. Phosphorylation sites abound in several domains of this nuclear phosphoprotein, and phosphorylation significantly interferes with transcriptional control. In the course of developing a sensitive and phosphorylation-specific monoclonal antibody, two sets of peptides were synthesized, one 23 amino acids long and the other 33 amino acids long. The peptides were unphosphorylated, contained a single phosphate group on Ser-378 or Ser-392, or were phosphorylated on both serines. The molecular masses of the peptides were in the 2593 to

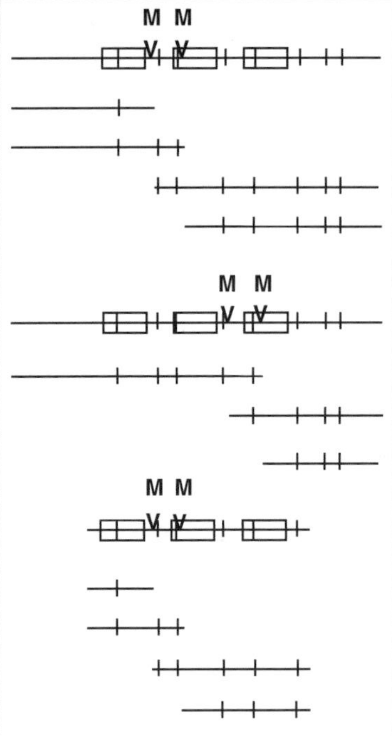

Fragment	Diagrammatic Localization	M/z (predicted)	M/z (observed)
N123C	M M	21,989	22,012
18-1		8,411	8,518
18-2		9,382	9,400
18-3		13,279	13,320
18-4		12,309	12,280
N123C	M M	21,989	22,012
17-1		12,911	13,020
17-3		10,152	10,215
17-4		8,779	8,769
123	M M	10,924	10,945
21-1		3,595	3,538
21-2		4,565	4,571
21-3		7,368	7,380
21-4		6,398	6,313

FIGURE 6.13 Calculated and observed *m/z* ratios for microtubule-binding region (MTBR) phosphopeptides. Overlapping peptides generated upon CNBr cleavage of methionyl-containing MTBR constructs are shown. Consensus sequence-specific protein kinase C phosphorylation sites are indicated with vertical lines. Note that not all of these sites were modified. Each unphosphorylated fragment correlates to a specific molecular weight (predicted) and is comparable with the value observed. The variability in these values (0.1 to 1.5% with $n > 10$) was self-consistent for a particular fragment in both the unphosphorylated and phosphorylated forms, suggesting that certain peptides potentially contained matrix adducts that shifted. Unphosphorylated samples were used as controls. Reprinted from Ainsztein, A. and Purich, D., *J. Biol. Chem.*, 269, 28465–28471, 1994. With permission.

3816 Da range and the MALDI-TOFMS values obtained were within 0.1% of the calculated masses. Biochemical experimentation revealed that one of the monoclonal antibodies generated, mAb p53-18, recognized the p53 protein and its fragments in both human and mouse cells, but only if the C-terminus of the protein was phosphorylated. After purification, the isolated phosphoprotein was subjected to tryptic digestion followed by LC/ESIMS analysis. One of the ions found by selected ion profiling corresponded to the singly charged phosphorylated peptide fragment Val-385-Asp-390 while a second observed mass was consistent with being the doubly charged ion of the phosphorylated peptide fragment Glc-371-Arg-376. Further analysis verified that the phosphorylations were at Ser-389 and Ser-375. Serum samples from cancer patients preferentially recognized the doubly phosphorylated peptide over the mono- or nonphosphorylated analogs.[180]

Human Translational Repression Protein 4E-BP1

This important protein binds to, and thereby deactivates, the subunit of initiation factor 4F that links the mRNA to the ribosomal complex during protein synthesis. Hyperphosphorylation of 4E-PB1 leads to its disassociation from the initiation factor subunit thus allowing it to participate in ribosomal activity and thereby enabling a concomitant increase in protein synthesis. To elucidate the 4E-PB1 phosphorylation sites that occur in human embryonic kidney cells (HEK 293) that had

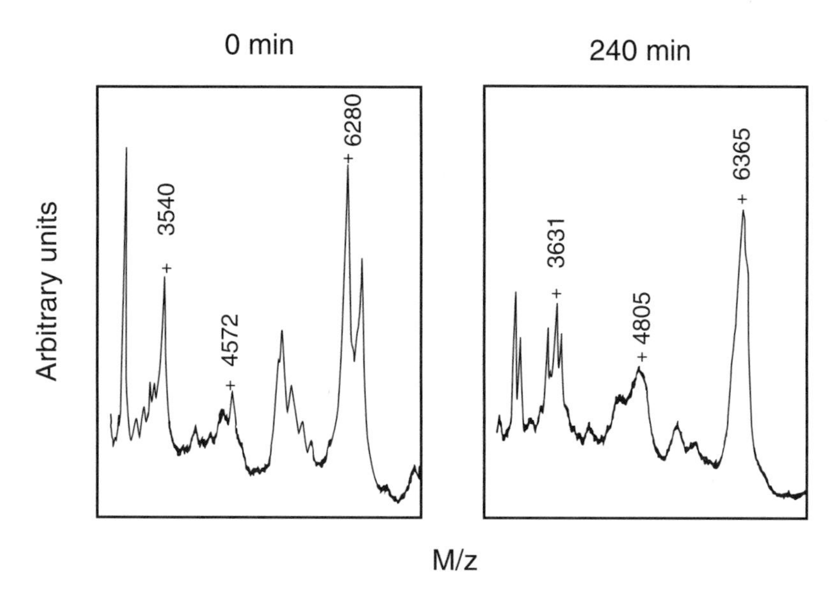

FIGURE 6.14 Representative mass spectra for unphosphorylated and phosphorylated microtubule-binding region peptides (MTBR-123). Shown are two mass spectral tracings for protein kinase C phosphorylation time points, 0 and 240 min. Each peak (marked with +) corresponds to a fragment produced from the MTBR-123 construct (as detailed in Figure 6.13). Unmarked peaks represent doubly charged fragments or peptides shifted by matrix adducts. Reprinted from Ainsztein, A. and Purich, D., *J. Biol. Chem.*, 269, 28465–28471, 1994. With permission.

been stimulated with fetal calf serum, ^{32}P-labeling was used. The labeled cells were lysed, immunoprecipitated, and the proteins were separated by 2-DE. After electroblotting onto a PVDF membrane, the band corresponding to the analyte was identified by radioautography, excised, enzymatically cleaved, and the peptide fragments separated by electrophoresis with detection by their radioactive label. After some additional sample preparation steps, 12 prominent TLC spots were scraped, extracted, and analyzed by LC/ESI-MS/MS. Individual CID spectra were analyzed using the *SEQUEST* program.[107] Proteins were identified unambiguously by the detection of multiple peptides derived from the same protein[46] and the sites of phosphorylation were deduced using an algorithm modified to allow for the 80 Da mass increases.[181] The computer-matched peptides were verified manually. A total of five phosphorylation sites were characterized (Table 6.6). It was noted that spots 4 and 8 had the same sequence, however, spot 8 was a doubly phosphorylated peptide.[182] In subsequent studies, the same MS technique was used to determine phosphorylation sites under different stimulation conditions, leading to the elucidation of the mechanism of the phosphorylation of the carboxy-terminal serum-sensitive sites.[183]

6.1.2.2 Selected Other Posttranslational Modifications

Cysteinylation

LC/ESIMS and T cell epitope reconstitution assays have been used to identify numerous peptide antigens in a variety of tumors (Section 6.1.5). Using a well-established methodology,[184] several of these epitopes were identified as H-Y antigens recognized by two cytotoxic T cell (CTL) clones carrying the HLA-A*0201-restricted human minor histocompatibility antigen H-Y. These peptidic epitopes originate from the Y-chromosomal protein (SMCY protein) and are similar to an 11-residue peptide that defines a human H-Y epitope for an HLA-B*0702-restricted CTL.[185]

TABLE 6.6
***In vivo* Phosphorylation Sites Identified by Capillary LC-MS/MS and Computer Database Searching from 2-DE Phosphopeptide Map of Human Protein 4E-BP1**

Spot no.	Amino acid	Phosphopeptide identified
1	44–51	STT*PGGTR
2	21–42	VVLGDGVQLPPGDYSTT*PGGTLF
3	77–99	DLPTIPGVTS*PSSDEPM$_{OX}$EASQSHLR
4	64–73	NSPVTKT*PPR
5		Ψ n.i.[a]
6		Ψ n.i.
7		Ψ n.i.
8	64–73	NS*PVTKT*PPR
9	20–51	RVVLGDGVQLPPGYSTT*PPGTLFSTT*PGGTR
10	21–51	VVLGDGVQLPPGDYSTT*PGGTLFSTT*PGGTR
11	20–42	RVVLGDGVQLPPGDYSTT*PGGTLF
12		Ψ n.i.

[a] n.i. = not identified
* Phosphorylated amino acid residue

Reprinted from Gygi, S.P. et al., *Electrophoresis,* 20, 310–318, 1999. With permission.

Of some 22 synthesized SMCY-related peptides, a 9-residue peptide, FIDSYICQV was the only one that was active, as displayed by its ability to enable the functionality of the relevant CTL clone. After a series of immunopurification steps intended to recover the peptide from these activated CTL cells, LC/ESIMS was used to search for the $[M+2H]^{2+}$ ion of this peptide in the biologically active fractions. Although several doubly-charged ions at the anticipated *m/z* 544 were observed in these fractions, their CID spectra did not confirm the expected peptide, suggesting that there had been a posttranslational modification of the FIDSYICQV peptide. Assuming the modification to involve the sulfhydryl group of the cysteine residue at position 7, reduction with dithiothreitol (DTT) would be expected to result in the appearance of a new ion at the expected mass in these biologically active fractions. This was shown to be the case, and the FIDSYICQV sequence was confirmed by CID (Figure 6.15a). Additionally, a doubly-charged peptide ion in one of the active fractions disappeared as a result of the treatment with DTT. The CID spectrum confirmed the sequence of this peptide to be FIDSYICQV with a cysteinylated cystine residue at position 7 (Figure 6.15b). This type of posttranslational modification, in which a second cysteine residue has been covalently linked to a cysteine in the peptide sequence via a disulfide bond, was shown to have profound effects on activity. In this case there was a 1000-fold augmentation on the recognition of the peptide by CTL. It was suggested that posttranslational modifications may be commonplace in MHC-associated peptides and, therefore, must be considered in the design of peptide-based immunotherapeutic agents.[186] The same LC/ESI-MS/MS technique was also used to detect another posttranslational modification of Class 1 MHC-associated peptides (Section 6.1.5.1). An asparagine residue was converted to aspartic acid by a combination of N-linked glycosylation and deglycosylation reactions occurring to a peptide derived from tyrosinase.[187]

Farnesylation

Also known as *prenylation*, *farneslyation* is the process by which an isoprenoid group is added to the end of a protein. The extension may consist of either a 15-carbon farnesyl unit or a 20-carbon geranyl-geranyl addition. Such lipophilic extensions contribute to the anchoring of proteins, involved in signaling pathways, to cell membranes. Virtually all members of the *ras* family of proteins have

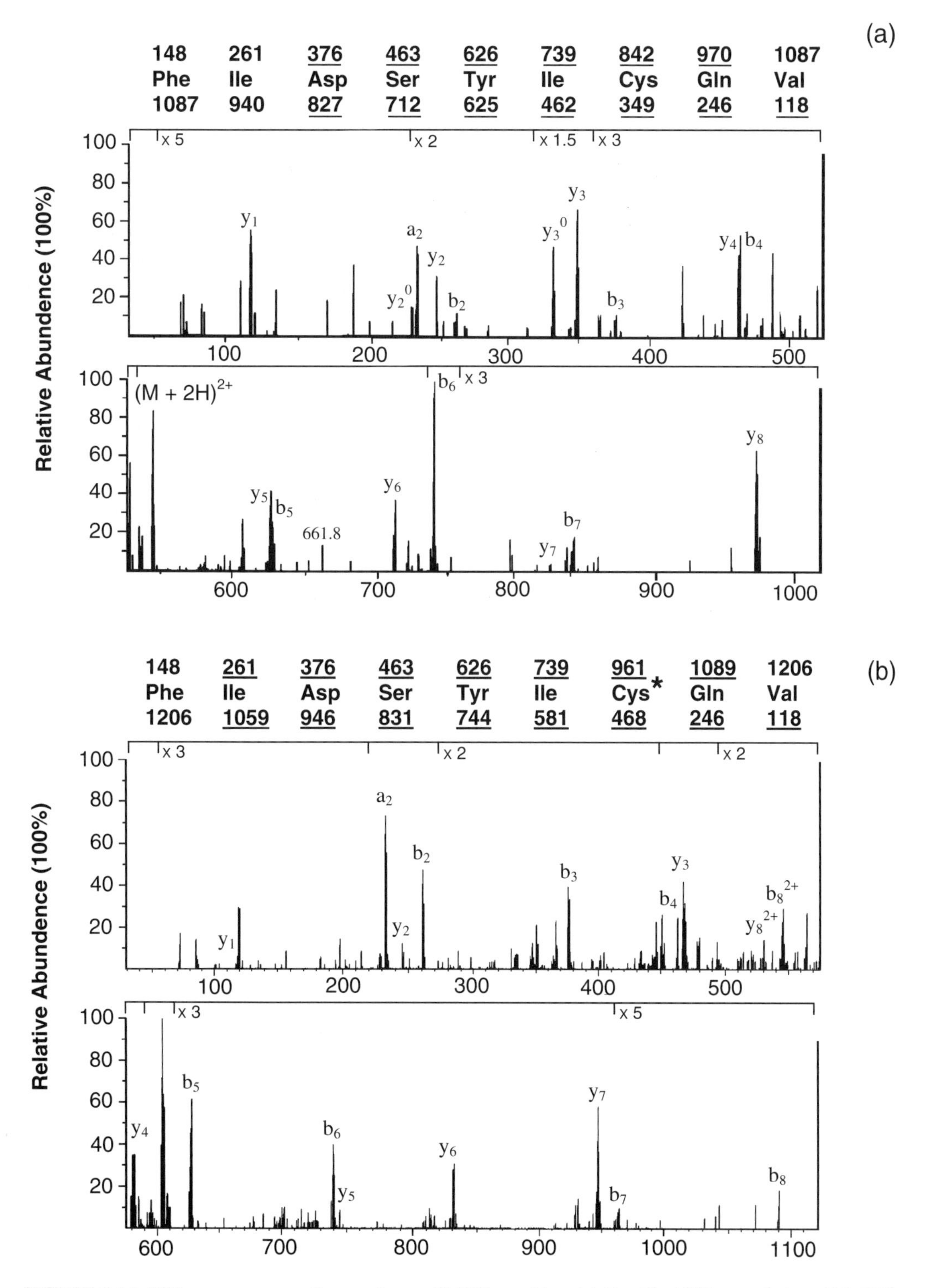

FIGURE 6.15 CID mass spectra of two relevant SMCY peptides. (a) Peptide 1087 recorded on $[M+2H]^{2+}$ ion at *m/z* 544; and (b) peptide 1207 recorded on the $[M+2H]^{2+}$ ion at *m/z* 604 of the microtubule-binding region (MTBR) phosphopeptides.[604] The y- and b-type ions have been labeled. In (b), Cys* microtubule-binding region (MTBR) phosphopeptides represent a cysteinylated cystine residue of 222 mass units. Reprinted from Meadows, L. et al., *Immunity,* 6, 273–281, 1997. With permission.

farnesyl attachments, thus it is obvious that inhibitors of farnesyltransferase, the facilitator of farnesyl attachment, have been suggested as antineoplastic agents. These drugs do not kill normal cells in the course of their action in reducing the growth of cancer cells.[188]

Acetylation

A group of tumor-associated antigens, designated UK101, that were extracted from goat liver, have been shown, in monoclonal antibody experiments, to be reactive with several human tumors. Exploring the hypothesis that similar antigens may also be expressed in neoplastic tissue, a search was conducted for the major antigenic component of UK101, which is a 14 kDa protein designated UK14. This protein was located on the cell membrane of several human tumor cell lines. However, the molecular mass, observed by MALDI-TOFMS, was 14,290 Da. This was 60 Da higher than the expected mass calculated on the basis of the amino acid sequence determined by Edman degradation of the peptides obtained by CNBr and endoproteinase Lys-C treatment of the protein. It was concluded that there was amino-acetylation of the N-terminal methionine.[189]

When 40 S ribosomal proteins from Rat-1 fibroblasts were analyzed by the methodology described above for the study of MKK,[174] observed masses of several proteins did not agree with those expected from known sequences. A number of posttranslational modifications were confirmed, including both N-terminal and internal acetylation as well as internal hydroxylation and methylation.[190]

MALDI-TOFMS and ESI-QTOFMS/MS were used to study posttranslational modifications of the p53 protein in response to ionizing radiation (OCL/AML-3 cell line). Phosphorylation was observed at four sites at the N-terminus and there were several acetylation sites at the C-terminus. It was confirmed that phosphorylation at the N-terminus increased the binding of p53 protein to acetyl transferases that subsequently promoted acetylation of the C-terminus following DNA damage.[190a]

Palmitoylation

Aiming to localize the palmitoylation site known to exist for the Friend murine leukemia virus transmembrane protein, p12E, the protein was extracted from a sample that was comprised of lysed virus particles that had been incubated with palmitic acid. This resulted in the binding of the fatty acid to cysteine residues via a thioester linkage. This was expected to be the modification site in wild-type protein. Of 18 peptide fragments detected in a tryptic digest, 3 contained cysteine residues, and MALDI-TOFMS analyses in the reflectron mode provided support for the position of acylation to the Cys-606 of the p12E protein.[191]

Polyglycylation

In polyglycylation, lateral polyglycine chains are formed in which the first glycine residue of the chain forms an amide bond with the γ-carboxyl group of a glutamate residue in the backbone sequence. Polyglycation is a posttranslational modification that is specific to tubulin. MALDI-TOFMS with PSD and ESI-QTOFMS/MS have revealed that a single molecule of β-tubulin can be glycylated at each of the last four C-terminal glutamate residues, Glu-437, Glu-438, Glu-439, and Glu-441, in the sequence ^{427}DATAEEEGEFEEEGEQ^{442}. Polyglycylation profiles of cytoplasmic and axonemal tubulin have revealed differences in the number and behavior of the glycylated isoforms. The formation of this hitherto unknown negatively charged, three-dimensional, bulky C-terminal domain with multiple isoforms, is likely to significantly affect the structural and regulatory interactions occurring between microtubules and the microtubule-associated proteins.[192,193]

6.1.3 Miscellaneous Proteins

6.1.3.1 Enzymes

Prostatic Acid Phosphatase

Human prostatic acid phosphatase (HPAP) was one of the first tumor markers identified. Significant discrepancies noted in sequences derived from cDNA by different laboratories have justified additional investigations by MS techniques. In one study, multiple cleavages were necessary to reduce the initial HPAP monomer to fragments in the 500 to 2500 Da size for disulfide bond identification by FABMS and for the verification of the full-length amino acid sequence using overlap techniques. Desialylated, purified HPAP was carboxymethylated, followed by proteolysis with trypsin and fractionation by HPLC for off-line analysis by FABMS. While the sequence of the native HPAP agreed with that predicted from independently derived, full length cDNA clones, FAB analyses revealed posttranslational deamidations, the conversion of Gln or Asn to Glu or Asp. Multiple carboxyl residues were also found by FAB, and further confirmed by PD-TOFMS in samples cleaved by CNBr. A number of disulfide-containing peptides were identified, e.g., amino acids 340 to 350 in the mature protein, based on detecting both the S-S fragment ion masses and individual sulfhydryl fragments in the same sample.[194]

Telomerase

Telomeres, which contain repeats of short TTAGGG sequences, are the physical ends of chromosomes and function as protective caps in normal cells. Telomeres act as "molecular clocks" in the aging process; the telomeres shorten each time the cell divides and eventually become so short that the cell can no longer divide and eventually it dies.[195] This is probably a tumor suppressor mechanism. About 95% of all human cancers produce *telomerase*, an enzyme that counteracts the progressive shortening process as it adds simple sequence repeats to the ends of chromosomes by copying the relevant template sequence within the telomerase RNA component. Thus, telomerase blocks the process that limits the lifetime of most cells. By permitting the telomere to maintain its length through unlimited cell divisions, strong telomerase activity allows cancer to spread without restraint. Research interest in telomeres and telomerase (reviewed in Reference 196) has increased to headline-grabbing intensity with the reporting of the extension of the life-span of normal human cells by the introduction of telomerase.[197]

A telomerase protein, purified from *Euplotes aediculatus* (a ciliated protozoan), was partially sequenced by nano-ESI-MS/MS. The active telomerase complex had a molecular mass of ~230 kDa, consisting of two proteins of 123 kDa and ~43 kDa, and a 66 kDa RNA subunit.[198] After separating the polypeptides on SDS-polyacrylamide gels and digesting the 123 kDa band with trypsin, 14 peptides were sequenced *de novo* (Figure 6.16). The use of a 2 Da mass window and 0.2 Da resolution permitted the unambiguous assignment of all amino acid series to identifiable C-terminal ions. More than 150 amino acids were identified and subsequently assigned in the open reading frame. Ten additional peptides were also characterized by comparing partial sequences with gene sequences. Additional molecular biological experiments confirmed the conclusion that the active site of telomerase contains reverse transcriptase motifs essential for chromosome replication. As the available sample quantity was inadequate for traditional sequencing, this was the first instance of the cloning of a protein of major biological significance using MS data alone.[199]

The activity of telomerase may be inhibited by peptide nucleic acids (PNAs) that are complementary to the RNA component of the enzyme. Although PNAs are very potent *in vitro*, they are poorly transported across membranes. A chimeric molecule was constructed by coupling a 13-mer PNA (a potent inhibitor of telomerase in melanoma cells extracts) with a 15 amino acid long, cysteine-containing transport peptide called Antennapedia. The coupling was accomplished via disulfide bonds to the cysteine linker-extended PNA. The purified construct was characterized by ESIMS. Although the construct induced a dose- and time-dependent inhibition of telomerase activity, it failed to induce significant telomere shortening and did not reduce the proliferation of melanoma cells.[200]

FLICE, the Effector Protease in the Apoptotic Death-Inducing Signaling Complex

Members of the caspase family, which are cysteine aspartate proteases, are important *effector* components of the pathway for apoptosis; e.g., transfection of ICE (caspase-1) into cultured cells causes apoptosis. Two cell surface cytokine receptors, CD95 (Fas/APO-1) and TNFR-1, that have

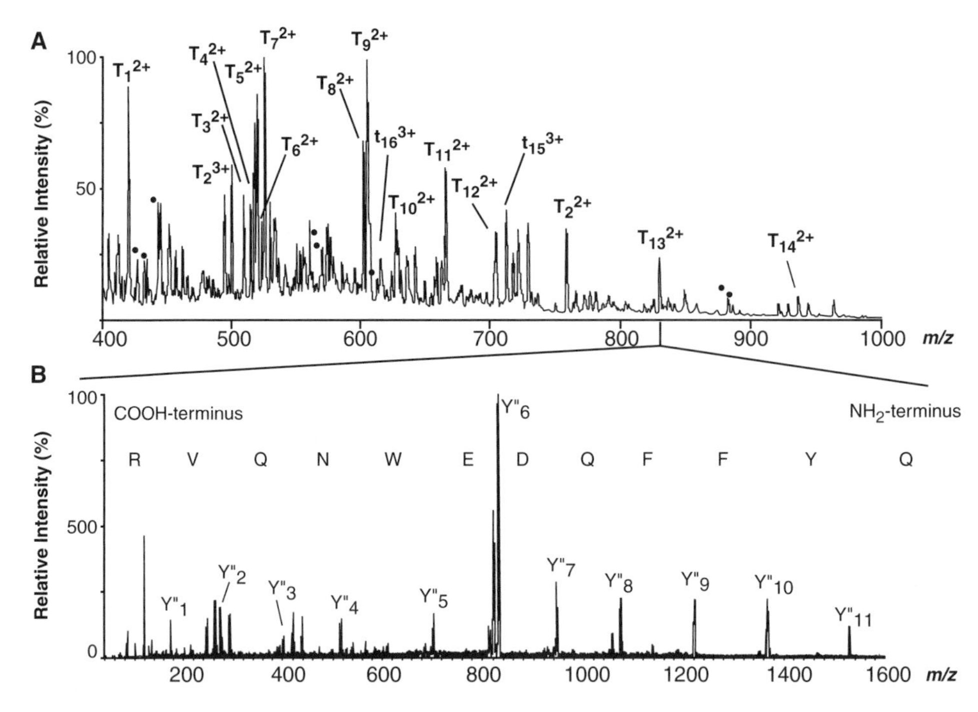

FIGURE 6.16 Sequencing to the p123 subunit of telomerase by nano-ESIMS. (A) mass spectrum of the unseparated peptide mixture. All peptides that were sequenced completely or partially are marked by the letter T or t, respectively. The eight peptide ions from which the sequence tags were generated are marked by filled circles. Most unlabeled peaks correspond to trypsin autolysis products. (B) Tandem mass spectrum of the doubly charged precursor ion at *m/z* 830.4 of (A). Interpretation of the fragment ion mass in (B) and comparison with the esterified form of the peptide allowed the sequence assignment. Reprinted from Lingner, J. et al., *Science,* 276, 561–567, 1997. With permission.

been shown to trigger apoptosis via their natural ligands or specific agonist antibodies, are members of the tumor necrosis factor (TNF)/nerve growth factor receptor family and are known as "death receptors" through their sharing of a region of homology, the "death domain." One of three identified death domain-containing molecules is the Fas-associating protein with death domain (FADD). There are four proteins, designated as CAP (cytotoxicity-dependent APO-1-associated proteins), of which CAP1 and CAP2 are alternate forms of serine-phosphorylated FADD and CAP3 is unidentified. CAP4, a cysteine protease, has been designated FLICE (FADD-like ICE) because it has a prodomain that is homologous to FADD. CAP4 and CAP3 were isolated from cultures of the human full-length CD95 transfectant after a series of biochemical manipulations. The silver-stained spots of the proteins isolated by 2-DE were excised and digested, and the peptides were purified and concentrated. The tryptic peptides were first analyzed by nano-ESIMS (Figure 6.17a). Next, one-half of the peptide mixture was sequenced by MS/MS, yielding the expected product ion series from the fragmentation at the amide bonds, containing the C-terminus (Figure 6.17b). The other half of the tryptic peptide mixture was esterified prior to MS/MS to yield the characteristic 14 Da mass shifts of the C-terminus containing the fragment ions, and the additional 14 Da shift for each Asp and Glu residue[99] (Figure 6.17c). Sequences were obtained for all seven peptides detected (Figure 6.17a), providing a complete characterization of FLICE as a 55 kDa protein. CAP3 was identified as the FLICE prodomain, an N-terminal fragment of FLICE. Overexpression of FLICE has induced apoptosis which could be abrogated by ICE family inhibitors. It was concluded that the death receptor CD95-mediated cascade of apoptosis is initiated by direct physical association of CD95 with FADD, the adaptor molecule, and FLICE, which is the effector protease.[201]

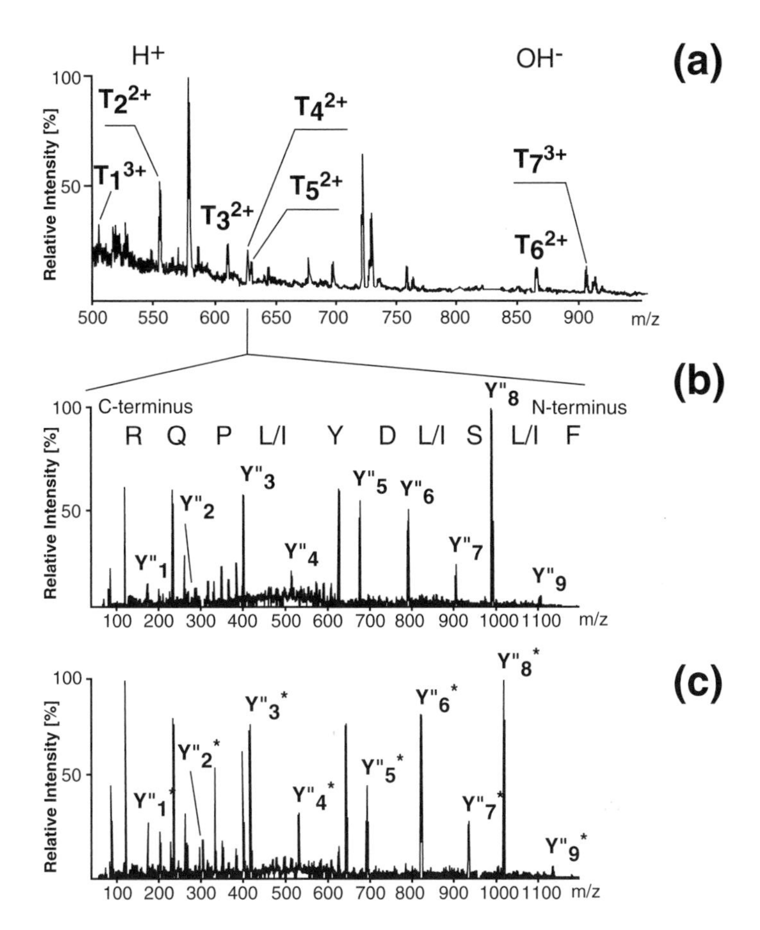

FIGURE 6.17 Nano-ESI-MS/MS sequence of cytotoxicity-dependent APO-1-associated protein, CAP4. (a) Part of the mass spectrum of the peptide mixture extracted after in-gel digestion of the CAP4 spot with trypsin. Peaks of the tryptic peptides (T1-T7) were sequenced by nano-ESI-MS/MS. The unlabeled peaks are trypsin autolysis products. (b) Sequencing of peptide 4 shown in (a) (*m/z* 626.5) using CID. Fragmentation of tryptic peptides at the amide bonds predominantly produced ion series containing the C-terminus (Y_1'', Y_2'', etc.). One-half of the tryptic digest was analyzed in this experiment. (c) CID spectrum of the same peptide after esterification of the second half of the total tryptic digest, resulting in a characteristic 14 Da mass shift of the C-terminus containing the "y"-type fragment ions and an additional shift of 14 Da for each Asp and Glu residue. The following sequences were obtained: T1, LLNDYEEFSKER, T2 FLLQEELSK, T3, QMPQTPFTLR, T4, FLSLDYLPQR, T5, LLNDYEEFSK. The isobaric amino acids isoleucine and leucine could not be distinguished. For peptides T6 and T7, peptide sequence tags were obtained that uniquely identified the peptides VFFIQACQGDNYQK and GIPVETDSEEQPYLEMDLSSPQTR, respectively, within the FLICE sequence. Reprinted from Muzio, M. et al., *Cell,* 85, 817–827, 1996. With permission.

Phospholipid Hydroperoxide Glutathione Peroxidase (PHGPx)

In the course of studying the arachidonate metabolism in human epidermoid carcinoma A431 cells, a putative inhibitor of 12-lipogenase was detected, which masked the biosynthesis of 12(S)-hydroxyeicosatetranoic acid, and also inhibited the activities of 5-lipoxygenase and fatty acid cyclooxygenase. The 22 kDa protein was purified and its structure determined by MALDI-TOFMS techniques (both linear and PSD reflectron modes) developed for highly bridged disulfide-linked peptides.[202] The sequences of the peptides of the analyte protein matched completely those of the known sequence of human PHGPx cloned from a human testis cDNA library. Additional tests for biological activity led to the conclusion that the lipogenase inhibitor in human epidermoid carcinoma cells was a phospholipid hydroperoxide glutathione peroxidase, known to be a selenium-dependent glutathione peroxidase having a unique substrate specificity for peroxidized phospholipids that it reduces to hydroxy compounds.[203]

6.1.3.2 Antigens

Genetic Origin of an Immunodominant Antigen of a Regressor Tumor

The ultraviolet light-induced murine tumor 8101 expresses an antigen that induces cytolytic $CD8^+$ cells. The antigen-positive cells were rejected by naive mice while antigen-negative tumor cells proliferated. After immunoprecipitating the antigen from antigen-positive cells, peptides were acid-eluted and the sensitizing peptide (sensitizing Anti-A cytolytic T cells) analyzed by ESI-MS/MS. Direct CID fragmentation of the $[M+H]^+$ ions specified the sequence of residues 3-8 as FVFAGX, where X is leucine or isoleucine. Residues 1-2 were identified as either SN or NS (Figure 6.18A). To obtain further sequence information, an aliquot of the peptide sample was subjected to a single cycle Edman degradation, followed by LC-ESIMS analysis. The resulting 87 Da shift confirmed

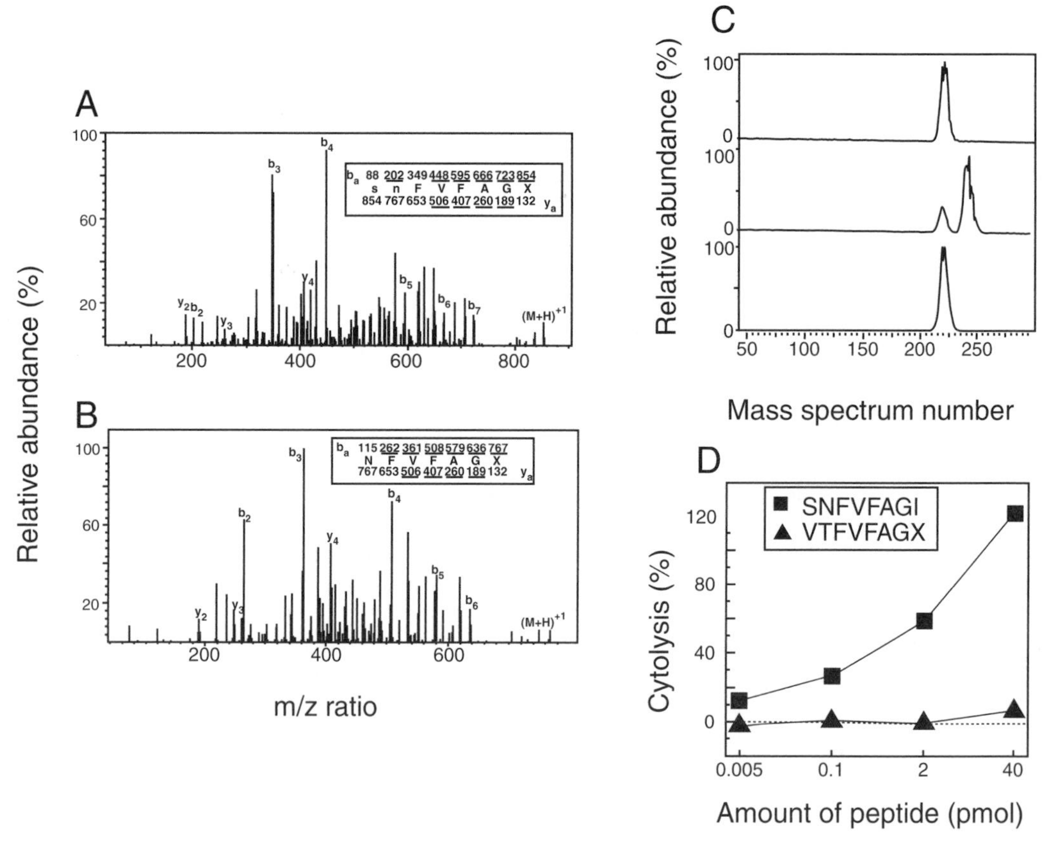

FIGURE 6.18 Structural characterization of the tumor epitope. CID mass spectra of $[M+H]^+$ ions of (A) *m/z* 854 of the tumor antigen; (B) *m/z* 767 from the tumor antigen after a single round of Edman degradation. The ions observed in each spectra are underlined. (C) Results of coelution experiments in which synthetic peptides SNFVTAGL or SNFVTAGI were added to the biologically active subfractions containing the tumor antigen. (D) Synthetic peptides SNFVFAGI and VTFVFAGX (X = L or I) were loaded onto RMA-S cells in the indicated concentrations. The E/T ratio was 5:1. SNFVFAGI is specifically recognized by testing for lysis by the anti-A CTL clone assay. Reprinted from Dubey, P. et al., *J. Exp. Med.*, 185 No. 4, 695–705, 1997. With permission.

the residue to be serine (Figure 6.18B). To differentiate between Leu or Ile at the C-terminal, co-elution experiments were carried out after doping with synthetic peptides SNFVFAGL and SNFVFAGI. Doping with the former peptide revealed two discrete peptide components, confirming isoleucine as the C-terminal residue in the epitope (Figure 6.18C). The final sequence of SNFVFAGI was confirmed by comparison with the synthetic peptide and also by comparison of sensitizing ability with respect to a control peptide VTFVFAGX (Figure 6.18D). Additional sequencing experiments revealed that the identified peptide matched the sequence of the DEAD box protein in the murine p68 RNA helicase except for a substitution of phenylalanine for serine.[204]

Composition of the T-Cell Antigen Receptor (TCR) Complex

TCR is a multimeric transmembrane complex consisting of a covalently linked α,β-dimer, noncovalently associated CD3-subunits γ, δ, and ε, and a $\zeta\zeta$ homodimer. The complex is known to be essential for the activation of T-lymphocytes. After binding of specific ligands to the extracellular domains of TCR, the intracellular domain generates and transduces response signals into the interior of cells.

TCR was immunoprecipitated from unstimulated and antibody-stimulated cells of the mouse T-cell line CD11.3. After resolving the proteins by SDS-PAGE, a set of immunological criteria were used to recognize the proteins that were components of the TCR. The relevant proteins were excised from the gel, digested in-gel, and analyzed by LC/ESI-MS/MS. For example, the TIC chromatogram of the peptides derived from one of the protein bands revealed some 149 components which were subjected to data-dependent MS/MS by automatically switching between the MS and MS/MS modes.[93] Five of these components (P1 to P5, Figure 6.19A) were matched by a *SEQUEST* database search to the TCRζ subunit. Both b-type and y-type ions could be detected in the MS/MS spectra (Figure 6.19B). In a similar fashion, analysis of 26 other relevant immunoprecipitated gel bands led to the identification of a number of proteins from the following additional components of TCR: four isoforms of CD3γ, two isoforms of CD3ε, CD3δ, and TCRα–β. There was an additional protein found associated with the stimulated T-cells, identified as protein tyrosine kinase ZAP-70 and known to be involved in early T-cell signaling. Several other identified proteins have no known association with TCR. Two of these, p110 and p120, were suggested as potential targets for new drugs.[205]

6.1.3.3 Non-Glycosylated Cytokines

Proteins in the Signal Transduction Pathways of Platelet-Derived Growth Factor β Receptor

After stimulating NIH 3T3 fibroblast cells with platelet-derived growth factor (PDGF-BB), cellular proteins were separated using 2-DE, and ~600 phosphorylated proteins were detected with anti-phosphotyrosine and anti-phosphoserine antibodies (~300 each). Mass fingerprints were obtained by MALDI-TOFMS for about 100 in-gel digested proteins that exhibited significant time-dependent changes in phosphorylation. The observed peptide masses of some ten peptides matched data for known proteins in databases. The identities of candidate proteins were confirmed by MS/MS sequencing of selected peptides in the unseparated peptide mixtures using ESI-ITMS.[206] Observed and calculated masses and additional data, including data on the time-dependence of the intensities of phosphorylation, were given for 12 phosphotyrosine and 7 phosphoserine related proteins, several of which are known to be involved in signal transduction, while the appearance of others was reported for the first time. For example, two of the identified proteins in the phosphoserine blots, a 73 kDa heat shock protein and a 58 kDa proto-oncogene tyrosine kinase (FGR) (Figure 6.20), which have been shown to be implicated in signal transduction, are also involved in direct PDGF signaling. CID revealed that the phosphorylation of the proto-oncogen tyrosine kinase FGR occurs at Ser-13.[207]

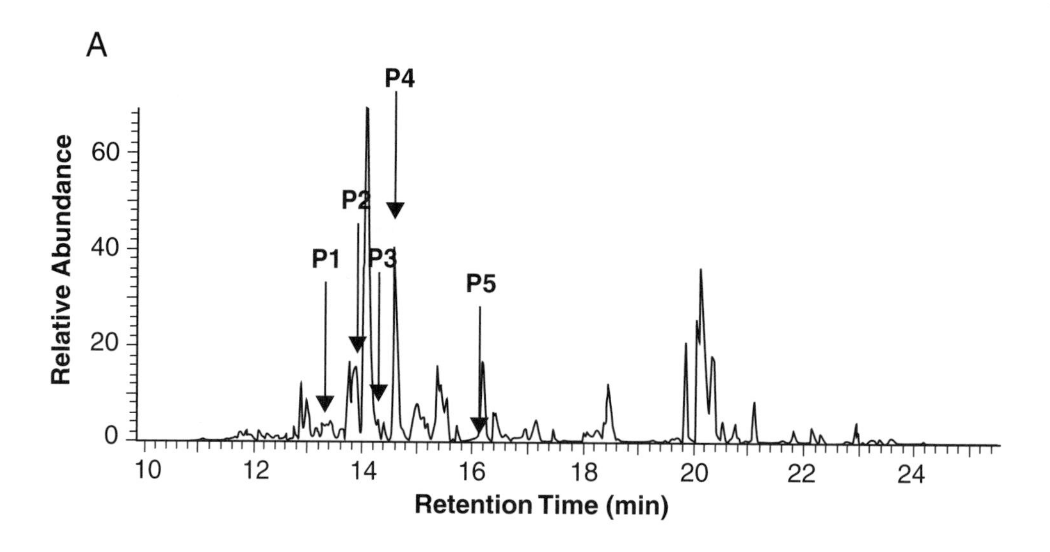

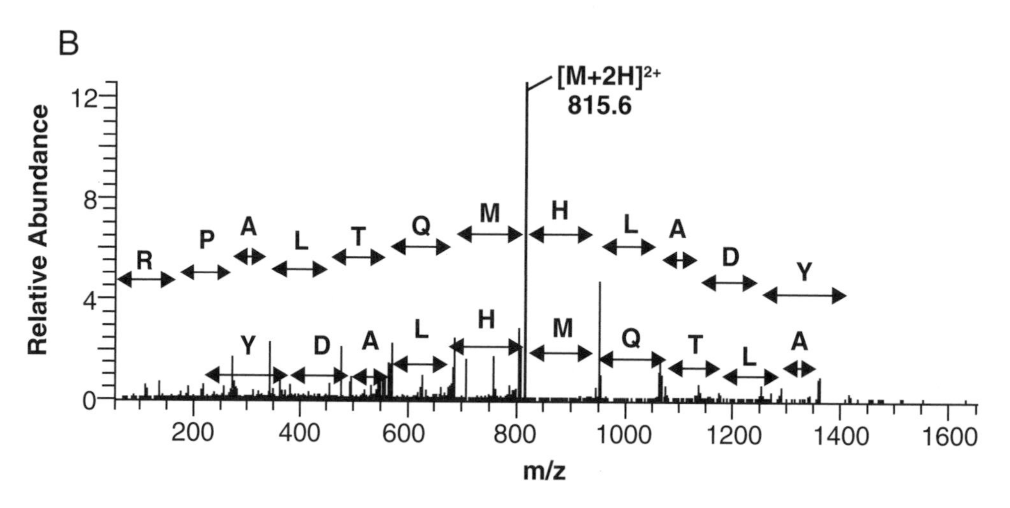

FIGURE 6.19 Automated identification of the p16 protein as TCRζ with LC-ESI-MS/MS analysis. (A) Total ion current chromatogram of the trypsin-digested band excised from SDS-PAGE gel. The masses of five different eluted peptides (marked with arrows) were determined and the MS/MS spectra of these peptides were acquired automatically. Use of the *SEQUEST* program matched the spectra to five tryptic peptides of TCRζ: (K)MAEAVSEIGTH, (R)NPQEGVYNALQK, (K)GHDGLYQGLSTATK, (K)DTYDAL-HMQT-LAPR, and (R)SAETAANLQDPNQLYNELNLGR, respectively. (B) MS/MS spectrum of the P4 peptide from (A). The parent ion $[M+2H]^{2+}$ at *m/z* 815.6 was hardly fragmented (most intense peak in spectra). Nevertheless, almost the entire series of b-ions and y-ions could be detected. This particular MS/MS spectrum resulted in a *SEQUEST* cross-correlation score of 3.81. The other spectra scored between 3.56 (P1) and 5.88 (P5). Reprinted from Heller, M. et al., *Electrophoresis,* 21, 2180–2195, 2000. With permission.

Midkine and Pleiotropin

Midkine (MK), a product of retinoic acid induction, may be a better tumor differentiation agent than retinoic acid, with comparable therapeutic effects but without the teratogenic effects of retinoic acid. *Pleiotropin* (PTN), another retinoic acid-induceable cytokine, has a 65% amino acid sequence identity with MK. Based on a comparison of the expression of MK and PTN in ovarian tumors, an association was suggested between MK and carcinogenesis, and between PTN and neural differentiation.[208] A

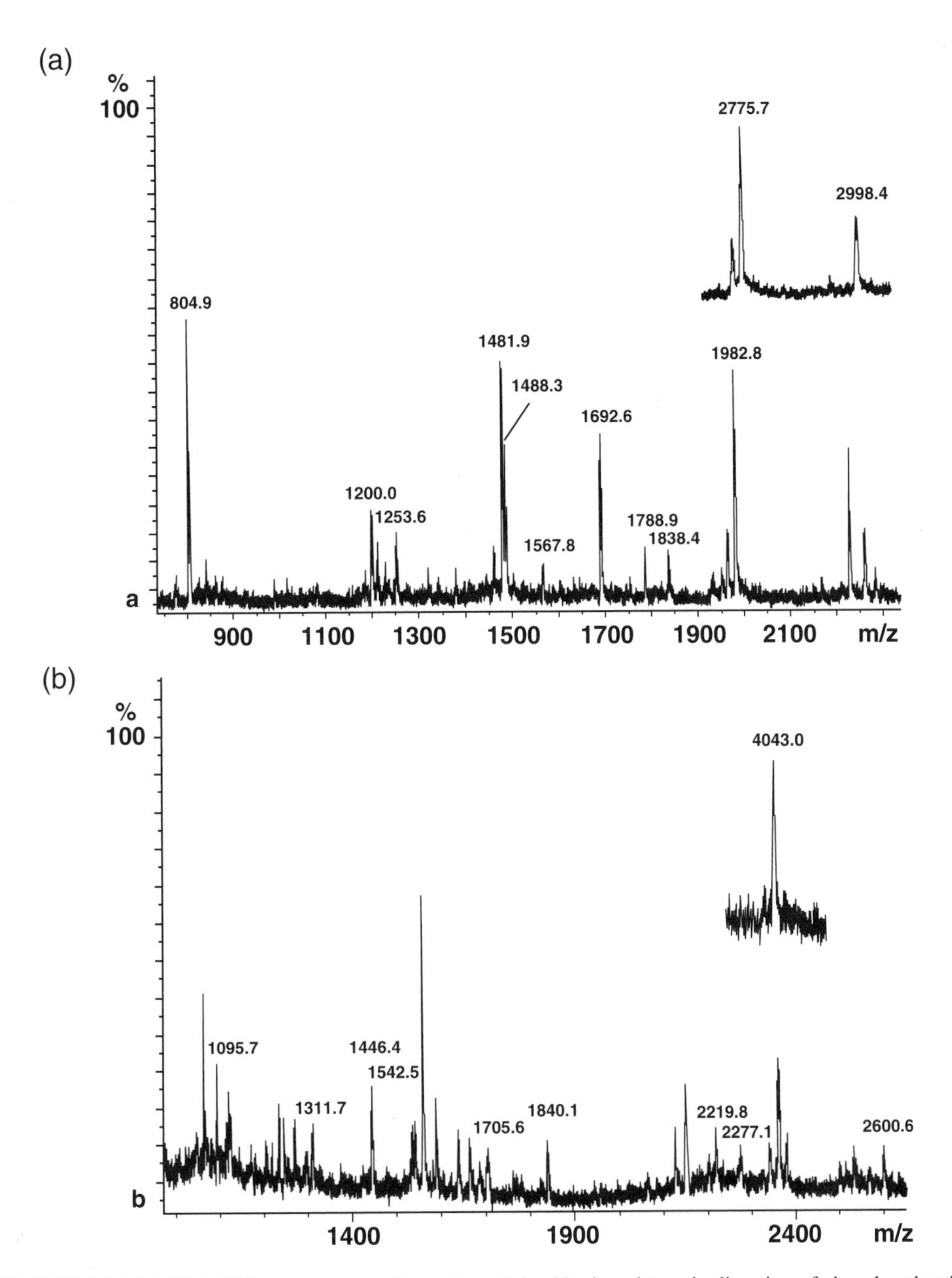

FIGURE 6.20 MALDI-TOF mass spectra of peptides obtained by in-gel trypsin digestion of phosphorylated proteins: (a) s12, heat shock 73 kDa protein; (b) s24, proto-oncogene tyrosine kinase FGR. Reprinted from Soskic, V. et al., *Biochem.*, 38, 1757–1764, 1999. With permission.

difference was observed between the measured molecular mass of purified recombinant MK obtained by SDS-PAGE analysis and the mass calculated from the amino acid sequence, ~19 kDa vs. 13,226 Da. However, MALDI-TOFMS analysis of the intact molecule confirmed the calculated mass within 26 Da. Three disulfide bonds were detected, linking peptide 54-60 to peptide 87-96, peptide 63-76 to

97-104, and peptide 26-32 to 47-52. The cysteines involved in these disulfide bonds were C_{59}-C_{91}, C_{69}-C_{101}, and C_{27}-C_{49}, respectively. Disulfide bond-containing fragments were generated from PTN by digestion with trypsin and further digestion with other proteases when required by the presence of more than one disulfide. The digests were purified by HPLC and the peptides identified by MALDI-TOFMS. Native and recombinant MK and PTN samples revealed strong similarity in disulfide bond assignments. MALDI-TOFMS also detected and characterized several impurities which have confounded the interpretation of N-terminal sequence data.[209]

A Protein with Growth-Inhibitory Activity in a Neuroepithelioma Cell Line

Two distinct components that inhibited the proliferation of human carcinoma and leukemia cells in culture have been detected and fractionated in the conditioned medium of the human peripheral neuroepithelioma cell line A673 (derived from a rhadomyosarcoma). One component was identified as interleukin-1-α. The molecular mass of the second purified component was determined by MALDI-TOFMS. The measured molecular mass, 25,576 ± 4 Da, was nearly identical to the molecular mass of transforming growth factor β1, 25,572 Da, calculated from its known sequence. The bioactive component also had exact protein identity with TGF-β1 over the N-terminal 22 amino acids. The measured amino acid composition was also consistent with TGF-β1.[210]

Cytokine-Regulated Proteins in Human Renal Carcinoma Cell Culture

In a study to investigate the mechanism of the antiproliferative effects of γ-interferon and interleukin-4 (IL-4), 2-DE revealed that treatment of the cell culture with either cytokine resulted in a 30 to 50% decrease in cell proliferation, but without synergistic effect. Five proteins were found by densitometry in >2-fold excess compared with untreated controls. The concentration of all five had increased upon treatment with γ-interferon and two with IL-4. The gel spots were excised and lysed with trypsin, and the peptides, separated by CE or HPLC, were sequenced by MALDI-TOFMS in the PSD mode and MALDI-QTOFMS in the CID mode. Both mass fingerprint and peptide fragment data were evaluated using software and databases available on the Internet. One protein was identified as tropomyosin by MALDIMS using peptide mass fingerprints from the unseparated tryptic digest.[211]

Changes in Protein Synthesis Induced by Fibroblast Growth Factor-2 in Breast Cancer Cells

MALDI-TOF and ESI-MS/MS identified proteins, the synthesis of which are induced and regulated by fibroblast growth factor-2 in human breast cancer cells (MCF-7 cell line). Four proteins were identified (the sequences in bold were obtained by ESI-MS/MS): heat shock proteins HSP90 (87 kDa, ^{386}GVV**DSED**LPLNISR^{399}) and HSP70 (70 kDa, ^{37}TT**PSYVA**FTDTER^{49}), the proliferating cell nuclear antigen (32 kDa, ^{6}**LVQ**GSILK^{13}), and the transcriptionally controlled tumor protein (24 kDa, ^{111}VK**PFM**TGAAEGIK^{123}). The discussion on the roles of these proteins provides new information on the molecular events leading to the growth of breast cancer cells.[212]

Uridine as Growth Factor in Neuroblastoma

It has been postulated that glial cells produce a neuronal growth factor in addition to the already-reported nerve growth factor (NGF) and numerous protein factors collectively known as neutrotropic factors (NTF). After detecting several neuroblastoma growth factors (NBGF) in extracts from neurofibroma cells originating from neurofibromatosis type 1 (NF1, von Recklinghausen neurofibromatosis), the growth-promoting factor from NF1 neurofibromas harvested from six cultivated neuroblastoma cell lines were isolated and purified. The NF1 extract yielded two fractions with the major growth-promoting activity concentrated in the NBGF1 fraction. The active component was identified by GC-EIMS as uridine (TMS derivative). Three other peaks in the extract were identified as adenosine, inosine, and guanosine. It was concluded that dysfunctions in the *de novo* biosynthesis of purine and pyrimidine nucleotides may significantly affect human neuroblastoma cells.[213]

6.1.4 Tumor-Associated Proteins

6.1.4.1 Tumor Proteomics

Differential-Display Proteomics

Important contributions of proteome technology in cancer research include the description of the phenotypes of cancer cells, profiling of signal pathways, and searches for tumor markers (reviewed in Reference 214). Malignant phenotypes may be determined by qualitative and/or quantitative changes in proteins, such as in the levels of expression and in the types and extent of both post-translational modifications and protein-protein interactions. Tumor markers may be used not only for diagnosis and prognosis but also as targets for therapeutic intervention. In differential-display proteomics one would hope to find unique tumor markers, perhaps resulting from differences in the posttranslational modifications of proteins recovered from cancerous cells. This would be particularly desirable for the "stealth," hard-to-detect tumors such those of the pancreas and ovaries. However, it is more common to attempt to recognize those proteins that are significantly upregulated or downregulated in the tumors.

The most important consideration when deciding on the value of a biomarker is its diagnostic specificity. As is illustrated in the applications described below, biomarker discovery must always include comparisons of the protein profiles of the particular tumors to those observed in normal subjects and, if available, in patients with a benign variety of the tumor studied. However, once a candidate has been identified, the specificity of a newly discovered biomarker must be proved through analysis of samples from patients with a variety of different cancers, as well as individuals with a range of different nonmalignant diseases.

Approaches of various levels of sophistication have been developed for the subtractive or relative quantification analyses needed to recognize and quantify proteins that display increased or decreased expression in correlation with the occurrences or extent of a disease. In the case of 2-DE data these evaluations include, at one extreme, differential diagnoses using simple comparisons of 2-DE profiles of tumor and normal tissues, e.g., to distinguish between duodenal and pancreatic cancers. At the other end of the scale is a novel approach that uses data from proteomics for artificial diagnosis and outcome prediction using a learning model based on multivariate analysis. In this case 2-DE protein data from as few as 22 samples were used to classify ovarian tumors into benign, borderline, and malignant groups with attempts to correlate results of the model with biological outcomes.[215] Until recently, mass spectrometry has been playing a minor role in the quantification aspects of proteins separated by 2-DE. A neoteric approach to protein quantification, based on small-molecule isotope-labeling, is applicable in many cases (Reference 152, Section 6.1.1.9).

The recently developed technique of *imaging* mass spectrometry not only joins the methodologies of fluorescence microscopy and immunochemistry for the study of the spatial arrangement of molecules within biological tissues (reviewed in Reference 216) but also provides a novel approach to searching for specific tumor markers and analyzing protein expression. The technique is based on creating molecular images by rastering over the surface of a sample with consecutive laser shots using a modified MALDI-TOFMS. A typical data array consists of 1,000 to 30,000 such spots, and contains the masses and intensities of desorbed ions in the 500 Da to 80 kDa mass range. Hundreds of image maps, each at a discrete molecular mass, can then be evaluated with the aid of special software.[217]

Tumor-Associated Proteins

Protein databases, such as the *SWISS-PROT* database, often provide classifications of proteins based on utility, such as functional and structural proteins, enzymes, chaperones, etc. Annotations in several databases also provide information about the exact function(s) of the proteins, although in many instances the specific function of the identified protein(s) is unknown. When numerous proteins are to be identified on several gels, navigational tools may be useful, e.g., lookup tables in which "master spot numbers" from the experimental data are paired with database accession numbers.[218]

Investigations of proteins in different malignancies, reviewed below, have revealed a rapidly increasing number of identified, and sometimes quantified, proteins that are upregulated or down-regulated. Most publications provide some information or suggestions concerning the role(s) or relevance of the proteins identified. It is beyond the scope of this text to even summarize, or provide references to the biochemistry or molecular biology of the proteins reported, but in certain instances brief comments are included, particularly when newly discovered potential markers are described. There are a few proteins that often appear in reported databases. *α-Enolase* (phosphopyruvate hydratase) is a 46.9 kDa glycolysis enzyme located in the cytoplasm. Enolase is a substrate for cytoplasmic tyrosine kinases, enzymes known to be involved in the differentiation of lymphoma in humans. The presence of phosphorylated enolase may be indicative of increased kinase activity. *Vimentin* (53.6 kDa) is an intermediate filament protein, found in nonepithelial cells, especially mesenchymal cells. Vimentin is present in many tumor cell types but not normally in differentiated epithelial cells. Low levels of expression of vimentin may indicate a more differentiated tumor phenotype. This protein is known to degrade during radiation-induced, Ca^{2+}-mediated apoptosis. *Disulfide isomerase* (PDI) is a 56.6 kDa protein located in the lumen of the endoplasmic reticulum of different types of secretory cells. This is an enzyme that participates in the formation disulfide bonds in newly synthesized proteins. *Prohibitin* (29.8 kDa) is a B-cell-associated protein, a tumor suppressor that inhibits DNA synthesis, and has a role in regulating cell proliferation. The 23 kDa *phosphatidylethanolamine binding protein* (PEBP) is present in many cell lines. The level of expression of PEBP is thought to reflect changes in cell morphology associated with a differentiated phenotype. The cytosolic *Cu/Zn superoxide dismutase* (SOD) has multifaceted functions, including destruction of toxic radicals produced as part of normal cellular activities.

Some current high-throughput techniques, particularly SELDI-TOFMS, provide a great deal of information on hundreds of proteins concomitantly and rapidly from a small number of cells (reviewed in Reference 219). When several potential biomarkers are detected in a single analysis, obtaining individual identifications can be a daunting process. One approach to handling this identification problem is to simply ignore it for the time being. Instead, heuristic pattern recognition algorithms are applied to the protein fingerprints so that detected changes in patterns can be used for diagnosis or prognosis, all without knowing the identities of the components. A related problem is the difficulty in obtaining reliable quantification data with MALDI-MS, particularly when ion signal suppression occurs in the *m/z* regions close to that of the analyte. The problems of using relative intensities for quantification, even with the use of internal standards, restrain the use of these techniques in diagnostic applications. Several aspects of these limitations are illustrated and discussed in the publications reviewed in subsequent sections.

Protein Profiles in Tissues

Collection of protein profiles is a necessary first step for the establishment of comprehensive proteome databases for individual tumor types. Most efforts have used available tumor cell lines or bulk tissue samples as templates. Shortcomings of these approaches include the unknown effects of *in vitro* artifacts, induced by tissue culture, on protein profiles. In the case of bulk tissue samples, there are mixed populations of tumor and normal cells present resulting in a dilution of tumor-specific effects and, therefore, loss of specificity. Problems in these tissues include the unexpected appearance of large concentrations of unrelated constituents, such as an antigen.[220] Interpatient variability is another issue but this can be eliminated by direct intrapatient comparison of proteins expressed by matched, histopathologically confirmed tumor cells and the normal epithelial cells from which they arose. However, it is often difficult to procure relevant normal epithelial cells. Patient-specific analyses do permit marker identifications that would not be possible using single reference samples as representative of tumors, as illustrated by a study on esophageal tumors (Section 6.1.4.5).

Laser capture microdissection (LCM) is a new technique that enables acquisition of very small populations of well-defined normal epithelial and adjacent tumor cells for subsequent analysis.[221,222] In contrast to the ~50,000 LCM-procured cells needed for a 2-DE analysis, SELDI-TOFMS[55] requires only 25 to 100 cells to obtain a usable protein profile. Multiple SELDI-derived profiles have been obtained from 500 cells from individual melanoma and head and neck tumors (Section 6.1.4.6). The simplicity and reproducibility of LCM, combined with the specificity, sensitivity (to attomole levels), and speed of SELDI-TOFMS have made it possible to obtain protein profiles in tissue samples from patients with a wide variety of cancers. The protein profiles of patient-matched normal, premalignant, malignant, and metastatic cell populations obtained with the aid of chemical matrices, with compound class specificity or biochemical baits that display specificity, for a single analyte, are expected to provide hitherto unavailable opportunities to diagnose as well as monitor treatment.[219]

Defining the Urinary Proteome

The protein content of urine has been considered to be qualitatively, but not quantitatively, similar to that of blood. A new approach toward defining the urinary proteome is based on handling data using a technique called "dynamic exclusion" and using automated LC/ESI-QTOF analysis of unfractionated tryptic digests of urine. The mixtures were analyzed by MS/MS using a data-dependent ion selection mode of operation. Analysis of *m/z* 400 to 1795 ions was accomplished by acquiring four sets of data over narrow mass ranges, e.g., *m/z* 560 to 740. Within these acquisitions four precursors were selected for parallel CID analysis. Of the 5768 CID spectra obtained, 1451 were matched by database searches, to 751 amino acid sequences. In turn, this led to identification of 124 species, including both proteins and the peptides that result from translation of expressed sequence tags. The most abundant protein was albumin with 508 matched spectra, accounting for 35% of the total number of identified spectra. Nine other proteins of high abundance accounted for 63% of all spectra. These included serotransferrin, Ig kappa light chain c, Ig gamma heavy chain c, and uromodulin. The rest of the proteins were arbitrarily classified into medium- and low-abundance proteins. The identified proteins were also classified against known or putative functions, e.g., growth factors, plasma proteins, membrane proteins, and secreted enzymes/inhibitors.[223] Analyses such as these are highly complex and various related items were included in a discussion of the instrumental performance and data handling for such mixtures. These included the impact on data fidelity of data-dependent selection of precursor ions, and the consequences of the operational and post-analysis strategies used to optimize and/or reconcile the quality of the data used for the protein identification stage.[224]

6.1.4.2 Hematologic Malignancies

Burkitt's Lymphoma

Protein Profiles of Burkitt's Lymphoma BL60 Cells

These proteins were separated using 2-DE, stained with Coomassie brilliant blue, excised, digested in-gel with trypsin, desalted, and analyzed on a MALDI-QTOFMS without HPLC separation. Thirty-three proteins were identified with the aid of the *SWISS-PROT* protein and nucleotide databases using the *MS-Fit* and *Find/Mod* programs, as well as manual *de novo* sequencing of up to eight peptides per protein. Some identified proteins are associated with apoptotic processes. For example, one spot yielded a series of peptides (Figure 6.21) from which five sequences were determined, leading to the identification of the protein as α-enolase. The relevant peptide sequences were GNPTVEVDLFTSK (15-27), TIAPALVSK (71-79), IGAEVYHNLK (183-192), VISPDQLADLYK (269-280), and IEELGSK (412-419). Several of the other proteins identified were known to be common constituents in eukaryotic cells, e.g., proteins expressed by the Hsp70 gene families. The profiles were to serve as a database for planned subtractive analyses in diagnostic and drug screening

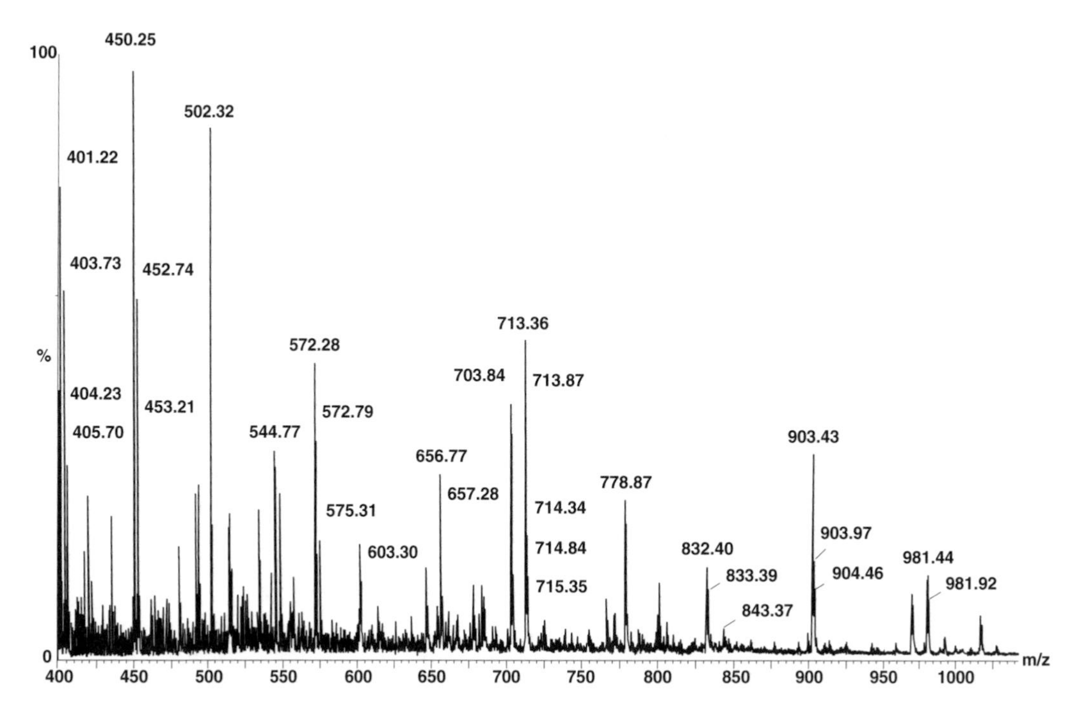

FIGURE 6.21 Nanoflow ESI spectrum of the in-gel tryptic digest of a 2-DE gel spot (B19) from BL60 cells. Twenty percent (1 μL) of the total digest of one spot was used for the nanospray-MS spectrum and five MS/MS spectra derived from the peptides. Reprinted from Muller, E.C. et al., *Electrophoresis,* 20, 320–330, 1999. With permission.

applications.[225] The peptide sequences listed above for α-enolase were obtained on a MALDI-TOFMS. Although there can be applications specific to particular instruments, this is often not the case, with the same results being reported with MALDI-TOFMS and LC/ESI-ITMS.[45]

Proteins Induced by Apoptosis

Apoptosis has been induced in an anti-IgM-sensitive subclone of the human Burkitt's lymphoma cell line BL60-2 by adding anti-IgM $F(ab)_2$. Differential analysis of the dying and vital nondying cells was carried out on populations of cells separated by a method based on the occurrence of phosphatidylserine on apoptotic cells. Of ~3250 proteins separated using 2-DE, the intensities of ~80 spots were significantly altered during apoptosis. After in-gel enzymatic digestion and separation of the resulting peptides by HPLC, peptide mass maps were obtained using MALDI-TOFMS. The spectra were searched using the *Fragmod* program and the sequences were compared with data in the *SWISS-PROT* and *Gen-Bank/EMBL* databases. Twelve proteins and their variants were found to be elevated in apoptotic cells (and their functions discussed). They were actin, heterogeneous nuclear ribonucleoprotein (hnRNP) A_1, hnRNP C1/C2, 60-S acidic ribosomal protein PO (L10E), laminin, nucleolin, FUSE-binding protein, dUTPase, lymphocyte-specific protein LSP1, UV excision repair protein RAD23 homologue B, heterochromatin protein 1 homologue α ($HP1_\alpha$), and neutral calponin. The elevation of the first six of these proteins could be inhibited if the cells were grown in the presence of a selective, irreversible inhibitor of caspase-3. This suggests that these proteins are involved in the control of apoptosis in this cell line and, therefore, probably in B lymphocytes. One of the high-intensity spots in the non-apoptotic cells contained four peptides that matched the sequences found at positions 45-58, 69-83, 103-113, and 114-130 of the 15,367 Da protein dUTPase (Figure 6.22). The intensity of this protein decreased significantly during apoptosis, confirming previous suggestions that the absence of dUTPase results in cell death.[226]

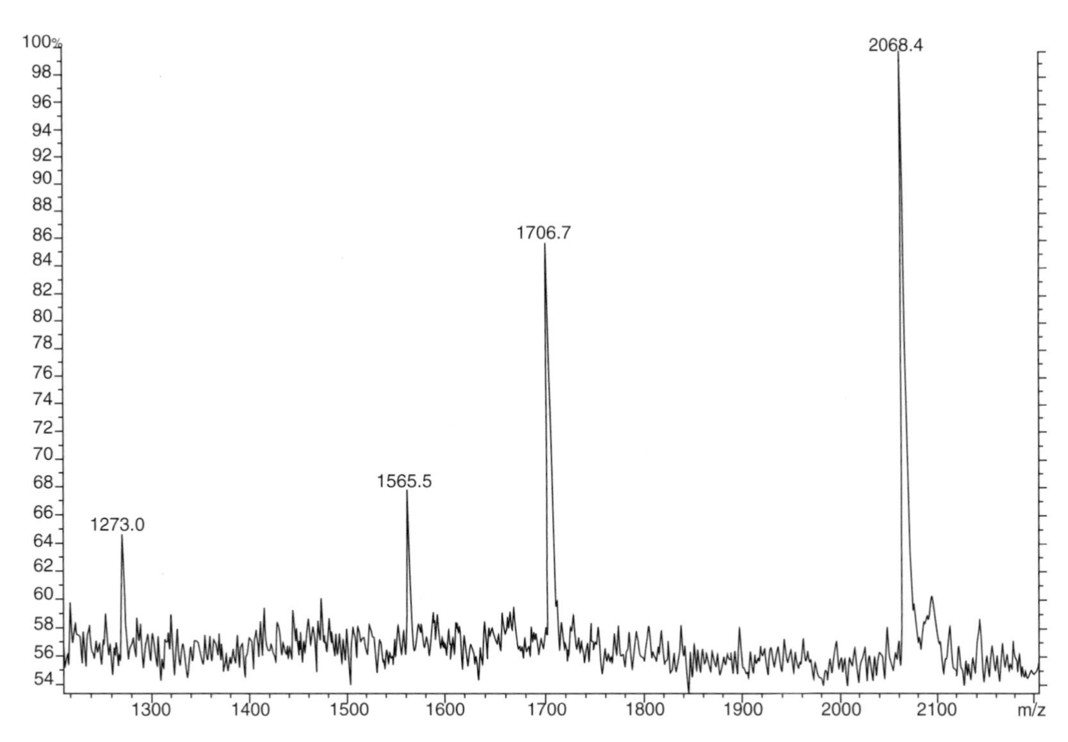

FIGURE 6.22 dUTPase spot identification: MALDIMS of the tryptic digestion mixture. Four peaks matched peptides of the protein dUTPase using the *FRAG-MOD* program. The corresponding peptide sequences for the mass peaks are given in parentheses: 1273.0 Da (103-113), 1565.5 Da (45-58), 1706.7 Da (69-83), and 2068.4 Da (114-130). Reprinted from Brockstedt, E. et al., *J. Biol. Chem.*, 273, 28057–28064, 1998. With permission.

In a related study on the same anti-IgM-sensitive subclone, one protein spot that disappeared after surface IgM cross-linking was identified as D4-GDI. This is a known hematopoetic cell GDP dissociation inhibitor of the *ras*-related Rho family GTPase. The disappearance of the high-intensity spot representing D4-GDI during apoptosis was accompanied by the appearance of a new protein spot. After enzymatic digestion of this protein, five peptides were analyzed by MALDI-TOFMS. The molecular masses, as well as those for the first three and last three amino acids for each of the 5 peptides, were determined. The results indicated that a cleavage protein, formed during apoptosis, had been found and that it corresponded to a 23 kDa fragment of D4-GDI that had been truncated at the Asp-19 position. Neither apoptosis nor cleavage of D4-GDI occurred when the cells were grown in the presence of a caspase-3 inhibitor.[227] The same methodology was also applied to evaluation of other proteins and resulted in the identification of RNA polymerase B transcription factor 3 as a protein that is differentially expressed but present in minor quantities. Another 13 listed potential marker proteins were characterized by Edman microsequencing and peptide mass fingerprinting using MALDI-TOFMS.[228]

Effects of Treatment with Taxol or Epirubicin on the Apoptosis of Burkitt-like BJAB Cells

Treatment of BJAB cells with taxol and epirubicin was followed by 2-DE to search for cleavages of D4-GDI and related species. After in-gel digestion, the proteins present in three spots were identified using ESI-QTOFMS. One of the spots, identified as Rho-GDI 1, a homologous protein, remained unchanged during drug treatment. The second spot, D4-GDI, disappeared in those samples where the drug treatment induced apoptosis and was replaced by the same ~23 kDa spot, truncated at Asp-19, seen with the anti-IgM-sensitive subclone of the Burkitt's lymphoma cell line BL60-2. The cleavage preceded the characteristic morphological and other changes associated with apoptotic cell death. The appearance of the 23 kDa spot occurred simultaneously with an observed activation of caspase-3, suggesting that D4-GDI, and not Rho-GDI 1, is the substrate for caspase-3-mediated (rather than caspase-1-mediated) cleavage during drug-induced proteolytic apoptosis.[229]

Expression of Stathmin Before and After Immortalization of Epstein-Barr Virus (EBV)-Transformed Human Cells

The progression of the immortalization process for three continuously proliferating EBV-infected human B-lymphoblastoid cell lines has been followed by differential display analysis of the transformation-associated proteins using 2-DE. The process appeared to be initiated before the 17th population doubling, with one indicator being the suppression of a 16.3 kDa protein. After digestion with lysilendopeptidase and peptide mass fingerprinting with nano-ESIMS, the protein was identified by the *MS-Fit* database as the phosphoprotein stathmin (Figure 6.23). Amino terminal acetylation had occurred and was detected by CID analysis (see Section 6.1.2.2). Identification of stathmin was confirmed by CID of two doubly-charged ions that yielded the relevant sequences of AIEENNNFSK and PEFPLSPPK (Figure 6.24). The decrease in the expression of stathmin during the process of cellular immortalization appears to be consistent with the numerous proliferation-related functions and interactions reported for this ubiquitous cytosolic phosphoprotein.[230]

Protein Expression of Lymphoblastoid Cells

Lymphoblastoid cell lines immortalized by EBV are the *in vitro* equivalents of B-cell lymphocytes. The protein expression in one of these lines (PRI) has been studied by peptide mass mapping of digested peptides using MALDI-TOFMS (delayed extraction and reflectron modes), followed by protein identification with the aid the *PepIdent* and *ProFound* databases. Some 43 soluble proteins of widely differing types were identified; however, 60% of these were not yet positioned on the 2-DE maps in three of the best-documented databases. A number of posttranslational modifications were also observed and evaluated with the aid of the *PeptideMass* and *ScanProsite* programs. An important example for considering and identifying posttranslational modifications is the case of *L-plastin*, a monomeric actin bundling protein, the expression of which appears to be restricted to lymphoid cells. The *ScanProsite* software lists 24 potential sites of phosphorylation for L-plastin. At least four adjacent protein spots have been attributed to L-plastin based on peptide mass fingerprinting; therefore, it is possible that each of these protein spots could be the result of a different set of posttranslational modifications of L-plastin each providing the protein with an altered functionality. A database devoted to proteins expressed by lymphoblastoid cells has been created.[231]

Erythroleukemia

Conventional 2-DE has been used to separate proteins in whole cell lysates of the HEL human erythroleukemia line. Protein spots cut from the gel were digested by trypsin, extracted with TFA+acetonitrile and aliquots analyzed by both MALDI-TOFMS and LC/ESI-IT-reflectron-TOFMS. A database search resulted in the identification of eight proteins: heat shock protein 60 (61.1 kDa), tropomyosin α-chain (32.8 kDa), stathmin (17.3 kDa), tropomyosin (29.0 kDa), vimentin (41.6 kDa), troponin T (34.6 kDa), matrin 3 (47.0 kDa), and microfibrillar-associated protein 1 (51.9 kDa). The coverage was in the 50 to 90% range for all identified peptides, with the best match obtained for *stathmin* with 19 tryptic peptides detected, providing a 91% coverage (Table 6.7). Although MALDI-TOFMS provided useful information, HPLC/ESI-IT-TOFMS gave better coverage of the sequences.[232]

The HEL cell line noted above has also been used to illustrate the advantages of HPLC, with C_{18} type nonporous silica columns, over gel electrophoresis for the separation of proteins.[233] Separated proteins from the HEL cell line were digested with both CNBr (which generates a few large peptide fragments) and trypsin (to resolve multiple matches found with CNBr), and selected peaks were collected for analysis by MALDI-TOFMS. Of >100 separated proteins detected by UV, 5 were identified with the aid of the *MS-Fit* sequence database, all appearing in the 5.5 to 25 kDa range:

Result Summary

Rank	Mowse Score	# (%) Masses Matched	Protein MW (Da)/pi	Species	SwissProt. R07.09.99 Accession #	Protein name
1	49.3	6/11 (54%)	17302.6/5.76	HUMAN	P16949	STATHMIN (PHOSPHOPROTEIN P 19) (PP 19) (ONCOPROTEIN 18) (OP 18) (LEUKEMIA-ASSOCIATED PHOSPHOPROTEIN P 18) (PP 17) (PROSOLIN) (METABLSATIN) (PR22 PROTEIN)
2	23.7	5/11 (45%)	1727.45/5.76	MOUSE	P54227	STATHMIN (PHOSPHOPROTEIN P 19) (PP 19) (ONCOPROTEIN 18) (OP 18) (LEUKEMIA-ASSOCIATED PHOSPHOPROTEIN P 18) (PP 17) (PROSOLIN) (METABLASTIN) (PR22 PROTEIN)
3	1.42	4/11 (36%)	24802.8/4.64	HUMAN	P17677	NEWUROMODULIN (AXONAL MEMBRANE PROTEIN GAP-43) (PP 46) (B-50) PROTEIN F1) (CALMODULIN-BINDING PROTEIN P-57)

Detail Results

1.6/11 matches (54%). 17302.6 Da, pI = 5.76. Acc. # P16949. HUMAN. STATHMIN (PHOSPHOPROTEIN P 19) (PP 19) (ONCOPROTEIN 18) (OP 18) (LEUKEMIA-ASSOCIATED PHOSPHOPROTEIN P 18) (PP 17) (PROSOLIN) (METABLASTIN) (PR22 PROTEIN).

m/z submitted	MH⁺ Equivalent	MH⁺ matched	Delta ppm	start	end	Peptide Sequence	Modification
445.1000^{+2}	669.1922	669.4631	−304.5790	1	9	(–)ASSDIQVK(E)	Acet N
537.7600^{+2}	1074.5122	1074.5122	− 52.2118	44	52	(K)DLSLEEIQK(K)	
550.8000^{+2}	1100.5922	1101.6016	−916.3666	54	62	(K)LEAAEERRK(S)	
563.2300^{+2}	1165.4522	1165.5489	− 83.0244	86	95	(K)AIEENNNFSK(M)	
663.7800^{+2}	1326.5522	1326.6945	−107.3106	30	41	(K)ESVPEFPLSPPK(K)	
694.7900^{+2}	1388.5722	1388.7273	−111.7031	1	13	(K)ASSDIQVKELEK(R)	Acet N

5 unmatched masses: 567.2800 429.0800^{+2} 480.1900^{+2} 519.0900^{+2} 604.1900^{+2}
The matched peptides cover 35% (53/149 AA's) of the protein.
Coverage Map for this Hlt (MS-Digest index #): 28217

FIGURE 6.23 MS-FIT database search resulted in the assignment of ssp7001 to stathmin in the first rank with the highest score of 49.3. Six masses matched the theoretical products of Lys-C digestion of human stathmin. The results also suggested N-terminal acylation. Reprinted from Toda, T. et al., *Electrophoresis,* 21, 1814–1822, 2000. With permission.

thymosin β 4Y isoform, high mobility group protein 2, putative DNA binding protein, ribonucleoprotein, and calmodulin. Some of the peptide sequences suggested the presence of posttranslational modifications, such as oxidized methionine.[234]

In subsequent work, isoelectric focusing methodology was improved and combined with HPLC to provide a two-dimensional approach. Separation of proteins was therefore accomplished by first

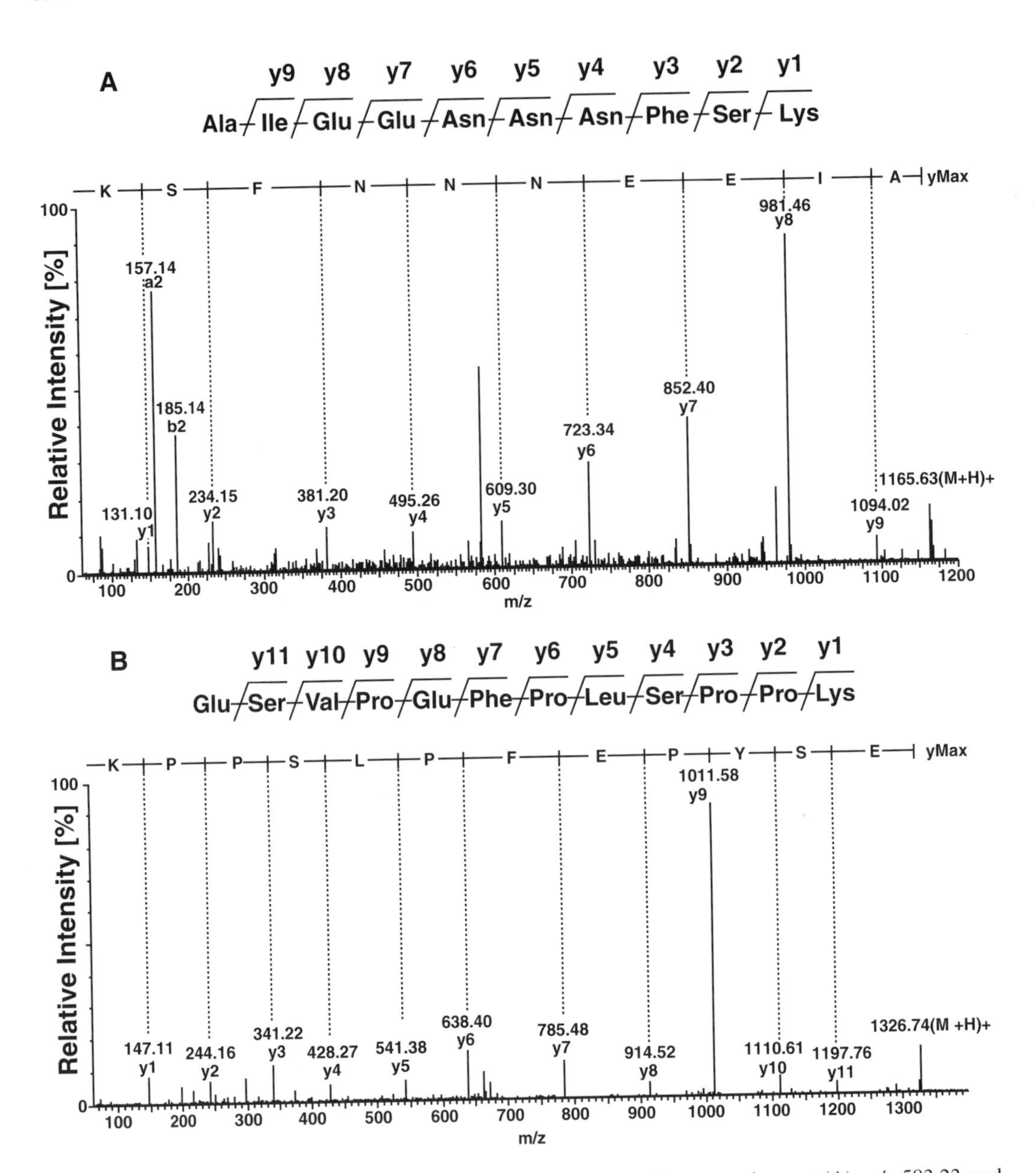

FIGURE 6.24 Results of the *PepSeq* analysis of MS/MS spectra. Precursor ions at (A) *m/z* 583.23 and (B) 663.78 were subjected to CID, and mass data of y-series product ions were analyzed by the *PepSeq* program in the *ProteinLynx* software. The sequences identified were AIEENNFSK and ESVPEFPLSPPK, respectively. Reprinted from Toda, T. et al., *Electrophoresis,* 21, 1814–1822, 2000. With permission.

separating them according to their pI values using isoelectric focusing, followed by a separation based on hydrophobicity by using nonporous reversed-phase HPLC. The two-dimensional patterns thus obtained were similar to the images obtained by 2-DE but with improved quantification provided by the intensities of the peaks eluting from the HPLC. The HPLC peaks were collected, digested with proteolytic enzymes, analyzed by MALDI-TOFMS, and the original proteins eventually identified using *MS-Fit* database searching. A direct comparison of the MALDI-TOFMS spectra of α-enolase isolated by the isoelectric focusing and HPLC approach with those obtained via 2-DE separation revealed some advantages of the former method (Figure 6.25). For example,

TABLE 6.7
Database Search for the Tryptic Digest of Protein Spot #3 from HEL Cell 2-D Gels Analyzed by LC-ESI-IT-Reflectron-TOFMS. Masses Measured in Da. The Protein was Identified as Stathmin

M_{mean}	M_{calc}	ΔM	Residue	Amino acid sequence and modifications
587.5	588.7	–1.2	71–75	(K) QLAEK (R)
674.9	674.8	0.1	10–14	(K) ELEKR (A)
891.2	890.0	1.2	1–9	(–) ASSDIQVK(E) - Acet N
912.2	911.1	1.1	129–135	(K) HIEEVRK (N)
913.3	913.0	0.3	63–70	(K) SHEAEVLK (Q)
943.3	944.1	–0.8	120–126	(K) LERLREK (D)
972.9	974.1	–1.2	54–61	(K) LEAAEERR (K)
1075.1	1075.2	–0.2	44–52	(K) DLSLEEIQK (K)
1153.6	1153.3	0.3	129–137	(K) HIEEVRKNK (E)
1165.5	1164.3	1.2	110–119	(K) ENREAQMAAK (L)
1166.3	1166.2	0.1	86–95	(K) AIEENNNFSK (M)
1201.9	1203.4	–1.5	43–52	(K) KDLSLEEIQK (K)
1327.9	1327.5	0.4	30–41	(K) ESVPEFPLSPPK (K)
1389.6	1389.6	0.0	5–27	(R) ASGQAFELILSPR (S)
1412.3	1411.6	0.7	125–135	(R) EKDKHIEEVRK (N)
1473.3	1478.7	–1.4	1–13	(-) MASSDIQVKELEK (R)
1544.0	1542.8	1.2	28–41	(R) SKESVPEFPLSPPK (K)
1676.1	1677.0	–0.9	96–109	(R) SKESEKLTHKMEANK (E) 1Met-ox
1799.3	1799.1	0.2	28–43	(R) SKESVPEEFPLSPPKKK (D)

Reprinted from Chen, Y. et al., *Rapid Commun. Mass Spectrom.*, 13, 1907–1916, 1999. With permission.

the improved digestion that occurred in the liquid phase reaction enabled a 60% match of the protein sequence to be achieved, vs. 49% match found using the in-gel digest. Other advantages of this liquid 2-D separation method compared with 2-DE included the purification of the analyte proteins in the liquid phase, 50-fold higher separative power, faster analysis, and the potential for automation. Some 38 proteins in the 12 to 75 kDa range were tentatively identified in the HEL line with this approach, in contrast with only 19 identified from 2-DE gels. Three of the identified proteins, α-enolase, glyceraldehyde-3-phosphate dehydrogenase, and heat shock protein HS27, have been linked to various forms of cancer, while two other proteins, NPM and CRKL-D, have been linked specifically to leukemias.[235] These new separative techniques may be applied to the differential diagnosis of the subtypes of acute lymphoblastic leukemia (ALL) and to the differences between ALL and acute myelocytic leukemia (AML) as well as for correlations between protein expression and the stage of the disease. 2-DE without MS identification has already been explored in this area; for instance in the detection of differences in the expression of the 36 kDa proliferating cell nuclear antigen with respect to the subtypes of childhood ALL.[236]

Protein Expression in B-Cell Chronic Lymphocytic Leukemia (B-CLL)

Survival times of patients with B-CLL, the most common form of adult leukemia in the Western world, range from a few months to decades. In recent years, certain observable cytogenic abnormalities have been correlated with disease progression and the development of resistance to therapy. A study was conducted to correlate the protein expression profiles of the B-CLL cells with the patients' karyotypes and survival times. Proteins from mononuclear cells, obtained from the peripheral blood of patients, were separated by 2-DE. The separated protein spots were detected, quantified,

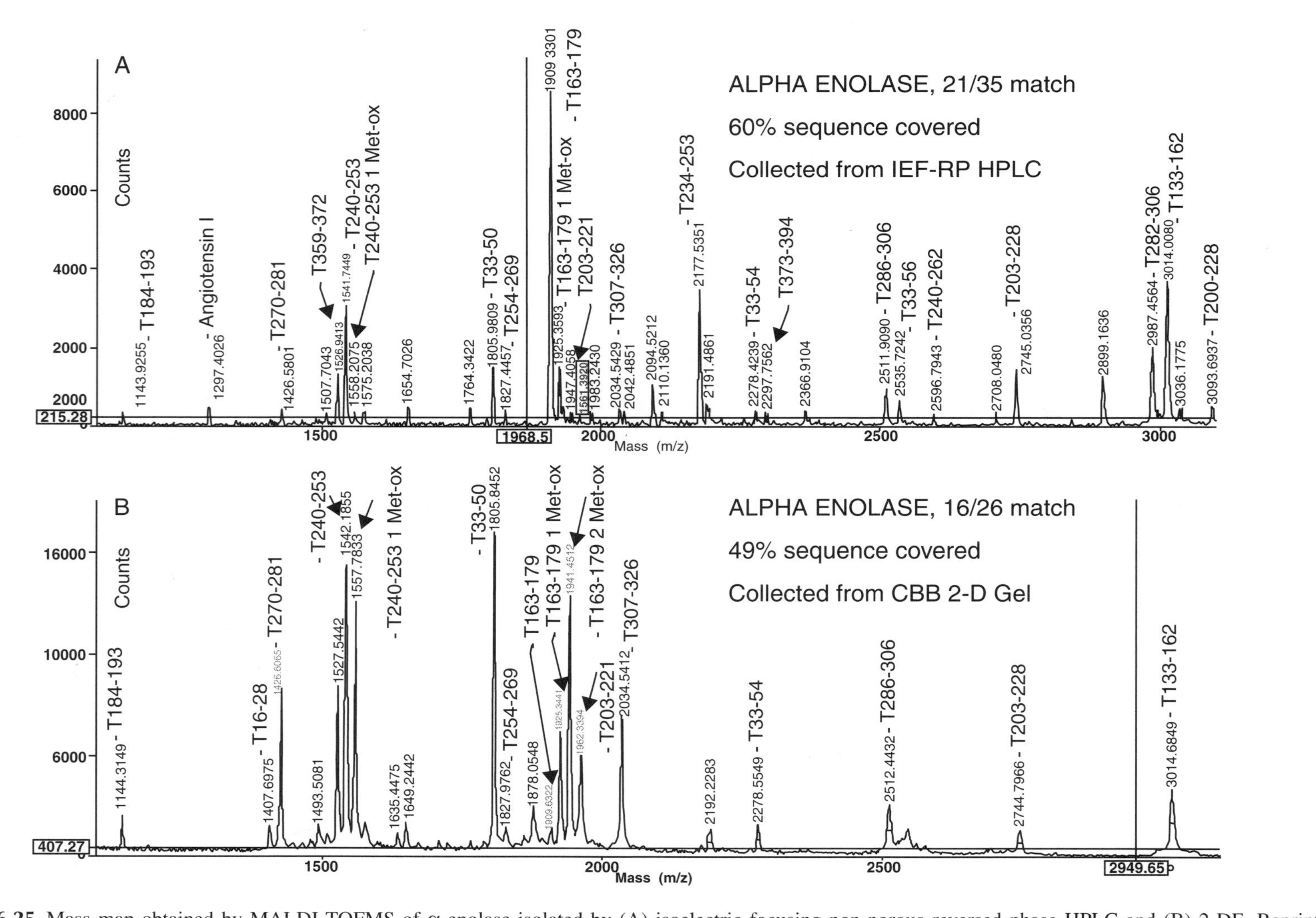

FIGURE 6.25 Mass map obtained by MALDI-TOFMS of α-enolase isolated by (A) isoelectric focusing non-porous reversed phase HPLC and (B) 2-DE. Reprinted from Wall, D.B. et al., *Anal. Chem.*, 72, 1099–1111, 2000. With permission.

and aligned using conventional 2-DE software. The gels with the separated proteins were divided into three groups based on cytogenetically determined deletions in chromosome bands 17p13 ($17p^-$), 11q22.3-11q23.1 ($11q^-$), and 13q14 ($13q^-$). Patients in the first two groups exhibited short survival times while the outcome of patients in the third group was more favorable. An algorithm was developed to correlate the intensity of protein spots with survival rate. Initially, 36 spots could be classified, based on preselected intensity thresholds, as absent or present within the three population groups. After a rigorous statistical analysis for the low- and high-expression groups, 12 spots could be correlated with survival behavior.

Of the protein spots excised, digested, and analyzed by MALDI-TOFMS, 26 could be annotated using the peptide mass fingerprint data as input into the *SWISS-PROT* and *NCBI* protein databases for mass map matches. Relevant parameters for the search included peptide mass tolerance = 50 ppm, minimum number of peptides required per protein = 4, molecular mass range = 1 to 100 kDa, and cysteine modification = amidomethylated. Based on information in the literature, several of the identified proteins appeared to have relevance to B-CLL. For example, glutathione-S-transferase (a p53-response gene) was preferentially absent in the $17p^-$ group; levels of thioredoxin peroxidase (an inhibitor of apoptosis) were low in patients with short survival times; protein disulfide isomerase (associated with membrane proteins with high -SH content) expression was high in patients with prolonged survival; increased levels of heat shock protein P27 correlated with short survival times. The correlation of specific proteins with chromosomal and survival data should become useful both for prognostic evaluations and also for the opportunity to recognize those patients with altered status of their redox-related enzymes that may, in the future, be a point for therapeutic intervention.[237]

Prohibitin Expression in Phorbol-Ester Treated Leukemic B-Lymphocytes

Purified B-CLL cells have the capacity to mature into a plasmacytoid phenotype when treated *in vitro* with tetradecanoyl phorbol acetate (TPA). These TPA-treated B-CLL lymphocytes also exhibit a two-fold increase in the expression of prohibitin. This protein is a ubiquitous 30 kDa inner mitochondrial membrane component that possesses antiproliferative properties through its ability to inhibit progression through the cell cycle at the G_1/S boundary. Using ESI-MS/MS and SIMS to obtain mass spectra of digests of 2-DE-separated proteins, and a subsequent database search, the identity of prohibitin was confirmed. Additional 30 kDa proteins that were also identified included histone ID in resting B-CLL cells, ATP synthase α and β chains, and an ADP/ATP carrier protein in the TPA-treated cells. It was concluded that an additional role of prohibitin may be to stabilize proteins in various subcellular locations during the maturation process of B-CLL cells.[238]

Cellular Proteins in Rat Basophilic Leukemia

The cellular proteins that confer a high secretory phenotype to a rat basophilic leukemia cell line (RBL-2H3.1) have been compared with those of a subclone (RBL-2H3.2) that possesses a low secretory phenotype (defined as releasing <3% of the granule content within an arbitrary period of time of observation). Of >2000 proteins resolved by 2-DE, about 40 spots exhibited differences in expression levels between the high- and low-secreting variants. These spots were excised from Coomassie-stained gels, destained, reduced, alkylated, and digested, and the resulting peptide mixtures analyzed by MALDI-TOFMS. About 20 proteins were identified with a nonreduntant protein sequence database search. Proteins preferentially expressed in the high-secreting cell variant included disulfide isomerase and elongation factor. Both proteins are associated with an increased potential for protein synthesis and metabolic activity. Another protein with elevated expression was phosphatidylethanolamine binding protein (23 kDa), a protein associated with an increase in membrane turnover during vesicle exocytosis. Several possible roles have been proposed for the increased expression of the cytosolic Cu/Zn SOD. The overexpressed α-enolase subunit and phosphoglycerate mutase (PGM) may provide increased glycolysis and energy provision to the high-secreting cells. Proteins identified in the low-secreting cell line included prohibitin and vimentin. The latter is known to be expressed by precursor mast cells but not by differentiated, mature mast cells. It was concluded that the gene expression patterns obtained by comparing the protein expression of the high- and low-secretion sublines has contributed to the development of rat basophilic leukemia cell lines as models of mast cell function.[239]

Proteome of Activated Murine B-Lymphocytes

Proteome analysis was carried out on mice B-lymphocytes that were stimulated *in vitro* by an *E. coli* lipopolysaccharide. Prior to harvesting these cells, a five-day mitogenic stimulation of the spleen was used to minimize the inclusion of cells having a non-B-cell origin. MALDI-TOFMS was used to characterize 153 spots excised from 2-DE gels. Of the identified proteins, there were 36 regulatory proteins, 2 enzymes, 15 chaperons, 15 structural proteins, 4 immunoglobulins, 1 ribosomal, and 1 histone protein. Of these, 26 were observed for the first time. The same protein species were sometimes represented by several different spots on the gels. This indicated the presence of various posttranslational modifications.[218]

6.1.4.3 Gynecologic Malignancies

Breast Cancer

Proteomic Definition of Myoepithelial Breast Cells

One baseline for the study of the proteome of malignant breast cells was provided by an evaluation of normal luminal and myoepithelial breast cells obtained in samples from reduction mammoplasties. Of the >4000 proteins detected initially with 2-DE, the quantities of approximately 170 proteins differed by >twofold between the two cell types. The molecular masses of peptides from 33 proteins, upregulated in luminal cells, and of peptides from 18 proteins, upregulated in myoepithelial cells, were determined by MALDI-TOFMS with peptide sequences being obtained by ESI-QTOFMS. The proteins were identified using the *SWISS-PROT* database with the aid of the *SEQUEST* search program. About 50% of the upregulated proteins from the luminal epithelial cells were cytoskeletal type I or II, and represented isoforms and/or posttranslationally modified species. The upregulated proteins in the myoepithelial set included tropomyosin, cytoplasmic actin, and isotypes of fructose biphosphate and pyruvate kinase M1. An additional 134 identified proteins not differentiated by cell type included proteins involved in signal transduction and metabolism. The presence of novel proteins was suggested by the acquisition of several mass spectra that could not be correlated with sequences in the databases.[240]

α-Enolase in Human Breast Carcinoma

α-Enolase was isolated from a human breast carcinoma cell line (MDA-MB231) using 2-DE, concentration with HPLC, enzymatic digestion, and analysis of 26 peptides by MALDI-TOFMS (linear mode). Sequencing was carried out using both MALDI-TOFMS (PSD mode) and LC/ESI-ITMS, the latter providing better quality data. The peptides identified represented 71% of the known amino acid sequence of α-enolase. The MS data resulted in the detection of two new, co-electrophoresing, α-enolase variants. These were identified as acetylated Asn-153Asp and Ile-152Asp/Asn-153Ile. The biological significance of these variants, with amino acid alterations at positions 152 and 153, is not known. MS/MS analyses also revealed the presence of several modifications introduced during the processing of 2-DE separation, e.g., methionine oxidation and cysteine amidoethylation, suggesting that process-induced modifications of analyte peptides can complicate the database searches and resulting interpretation.[45]

Proteins That Bind to the SH3 Domain of Mixed Linkage Kinase (MLK2) in Breast Carcinoma

MLK play a critical role in the signaling networks in breast tumor cells. The goal of one study was to identify proteins that interact with the SH3 domain of MLK2 that triggers downstream kinase cascades by linking activated cell surface receptors in the G-protein pathway. The first 100 amino acids of MLK2 (MLK2N) contain the 55 amino acid SH3 domain (residues 23-77). To confirm the integrity of the MLK2N protein, a fusion protein consisting of MLK2N and GST was constructed. MS analysis of the MLK2N polypeptide, cleaved from the construct using thrombin, revealed that while the sequence obtained matched that predicted by cDNA, there was a ~75 Da difference between measured and calculated masses. It was revealed by CID that this was an artifact of the synthesis that had resulted in the covalent modification of one cyteine residue with β-mercaptoethanol.

The construct, bound to GSH-Sepharose beads, was then used as an affinity ligand to isolate domain binding proteins from cell lysates from a cell line (MDA-MB231) that originated from a patient with intraductal carcinoma. The proteins isolated from the lysate were recovered and separated by electrophoresis. Excised protein spots were then digested with trypsin, and individual peptide fractions obtained by HPLC were used for subsequent analysis by ESIMS and ESI-MS/MS. As an illustration of the MS methodology employed,[241] the $[M+2H]^{2+}$ ion of one of the peptide fractions and the $[M+3H]^{3+}$ ion from another fraction of the same protein were subjected to CID. Their sequences were deduced using the *SEQUEST* algorithm (Figure 6.26). The first ranked sequence was accepted as correct because the differences between the normalized cross correlation parameters were >0.1 for both the first- and second-ranked sequences. The sequences were also confirmed by manual assignments of the product ions. This protein was identified as the high mobility group protein HMG1. A database consisting of 21 identified proteins has been established.[242]

Tumor-Related Proteins in Cell Lysates

A pilot study using 1-DE with electroblotting for MS analysis[243] was followed by a more advanced study using nonporous C_{18}-coated HPLC columns for the separation of proteins in three genetically related breast cells lines. The MCF-10 cell line originated from spontaneous immortalization of epithelial cells from a patient. Derived cell lines included MCF-10A cells that did not transform (used as control) and two malignant metastatic variants, MCF-10 Cala and Cald. Proteins from the soluble portions of lysed and fractionated cells were separated followed by digestion with trypsin. Peptide mass maps were obtained using MALDI-TOFMS and the proteins identified with the aid of the *MS-Fit* sequence database. A comparison of the profiles of the two malignant cell lines with that of the control revealed significant differences in the relative expression levels of several potential markers. The largest of these difference was in the expression of the tumor suppressor protein p53 (Figure 6.27, peak 7). This protein is a major regulator of normal cell growth. The concentration of p53 is often increased in malignant cells because of mutations to the protein that destroy its biological properties but increase the half-life from minutes to hours. The loss of p53 function has been associated with an unfavorable prognosis in several malignancies, including breast cancer. The form of the detected p53, i.e., wild-type or mutated, could not be determined from MS data. Another upregulated oncoprotein was the proto-oncogene kinase, c-src (Figure 6.27, peak 8). The overexpression of c-src, because of a mutation of the corresponding gene, is likely to be associated with increased phosphorylation of target proteins. Again the site of mutation could not be determined from the MS data. Other tumor-related proteins identified included c-*myc* and H-*ras*. Quantitative differences were also found between the malignant cell lines despite their common genetic background, suggesting multiple pathways were involved in the transformation of MCF-10 into the two malignant cell lines.[244] In subsequent work, a second dimension of separation was added using isoelectric focusing to separate proteins according to their pI prior to HPLC with nonporous reversed-phase columns. This is the same methodology described above in the erythroleukemia section. The approach is analogous to the classical 2-D gel methods in several aspects.[245]

Tumor-Related Proteins in Neoplastic Tissues

The MS techniques used for the study of differential protein expression in ovarian cancer (see below) have also been applied to breast cancer tissues obtained after tumor resection. Seven proteins (upregulated 2- to 7-fold) were identified in malignant tumors: two nuclear matrix proteins, heterogeneous nuclear ribonucleoproteins (hnRNP) types F and C1/C2, cytochrome BN5, phosphoprotein B23, and modified cytokeratin 6D. The last of these appeared to have been radically altered

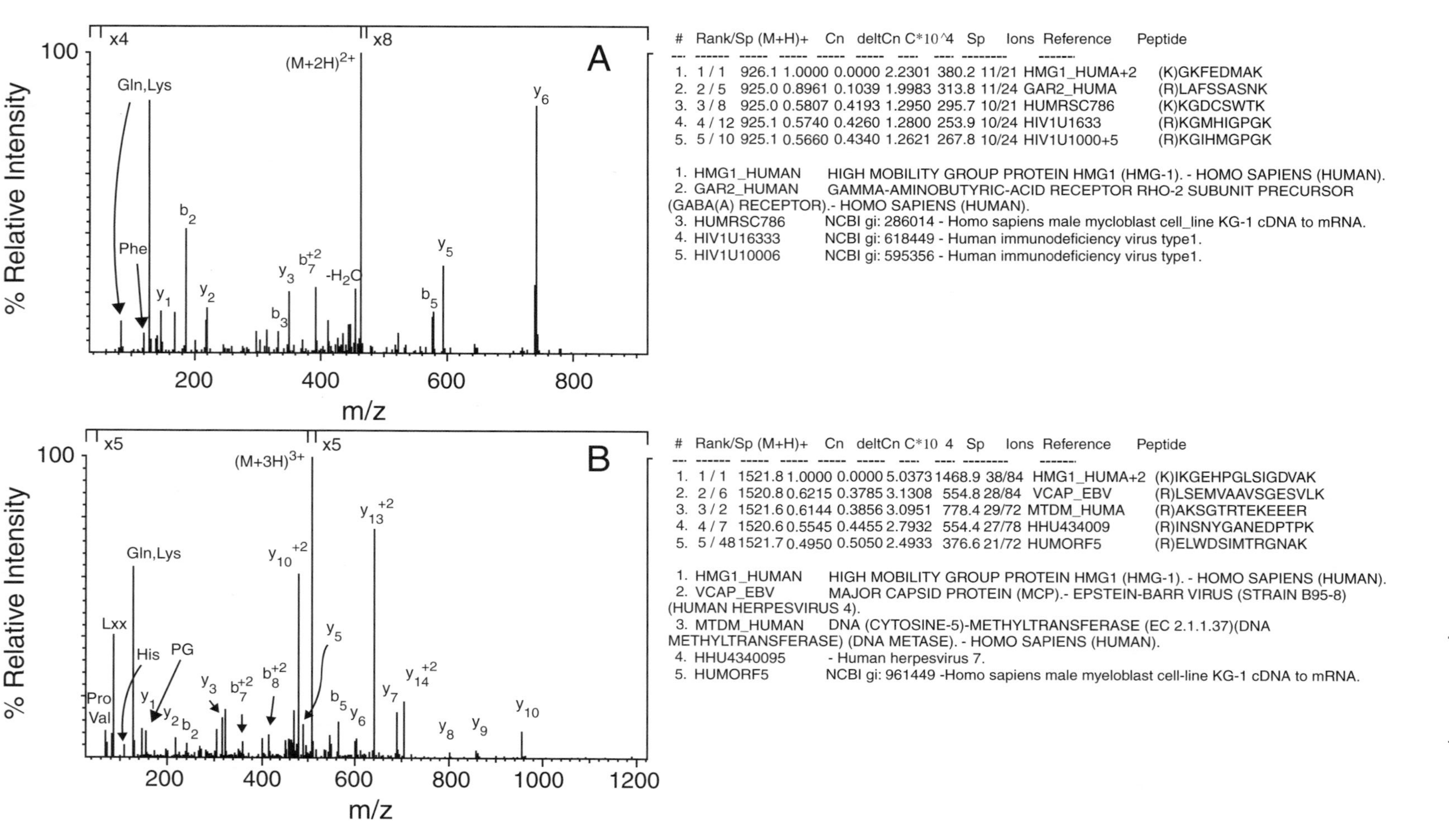

FIGURE 6.26 CID spectra of trypic peptides from the in-gel digestion of a protein spot (spot #40) of MLK2N bound protein. The major peptide ions from two fractions were subjected to CID. Spectra were analyzed using the *SEQUEST* search algorithm and the sequences identified as GKFEDMAK (panel A) and IKGEHPGLLSIGDVAK (panel B) from human high mobility group protein (HMG1). The observed product ions were assigned manually and labeled using the nomenclature proposed in References 96 and 98. Reprinted from Rasmussen, R.K. et al., *Electrophoresis,* 18, 588–598, 1997. With permission.

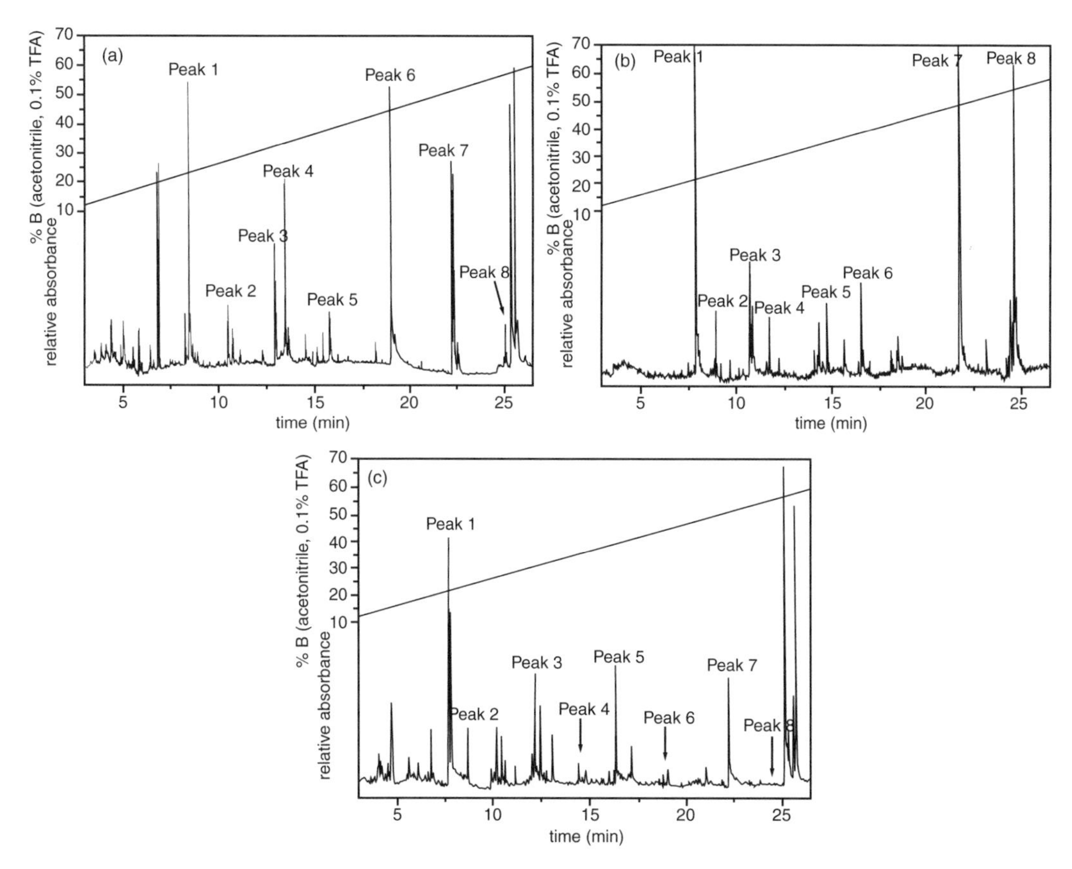

FIGURE 6.27 Nonporous reversed-phase-HPLC protein profiles of: (a) malignant human breast cell lysate (calaCL1); (b) malignant human breast cell lysate (caldCL1); (c) normal (immortalized) human breast cell lysate. Reprinted from Chong, B.E. et al., *Rapid Commun. Mass Spectrom.*, 13, 1808–1812, 1999. With permission.

through proteolytically cleavage from being 42 kDa normally into a 17 kDa fraction. The first three were identified for the first time in these tumor cells, while the others had been known from 2-DE gel-matches but without sequence determination. The methodologies were also applied to the analysis of proteins in benign tumors. Identified species included the DJ-1A protein, the cytoskeletal proteins cytokeratin 8, 18, and 19, and elongation factor 1-β. The DJ-1A protein, a protein that appears to be present only in breast fibroadenoma tissue, is an oncogene product involved in a *ras*-related signal transduction pathway.[246]

Identification of BclXL and Associated Proteins

BclXL is related to the well known Bcl-2, a protein that is amplified in breast, prostate, and lung cancers and whose functions include prevention of apoptosis, thereby promoting carcinogenesis. BclXL and associated proteins, expressed as a plasmid in 293T cells, were recovered from lysates by immunoprecipitation with anti-Flag antibody coupled to agarose beads. Ten protein bands not present in controls were digested with trypsin, analyzed by LC/ESI-MS/MS[46] and the results interpreted using databases.[247] For example, a peptide from one of the proteins was identified as KAVETHLLR (Figure 6.28) by the *SEQUEST* program using separate ion series, obtained from the N- and C-termini. Six other peptides from the same protein band were also identified by MS/MS (Table 6.8). All these peptides were matches for the human protein CLARP, a caspase-like apoptosis regulatory protein. Proteins identified in other bands included chaperone proteins, a protein kinase, a phosphatase, and others likely to be involved in the BclXL mechanism of action.[182]

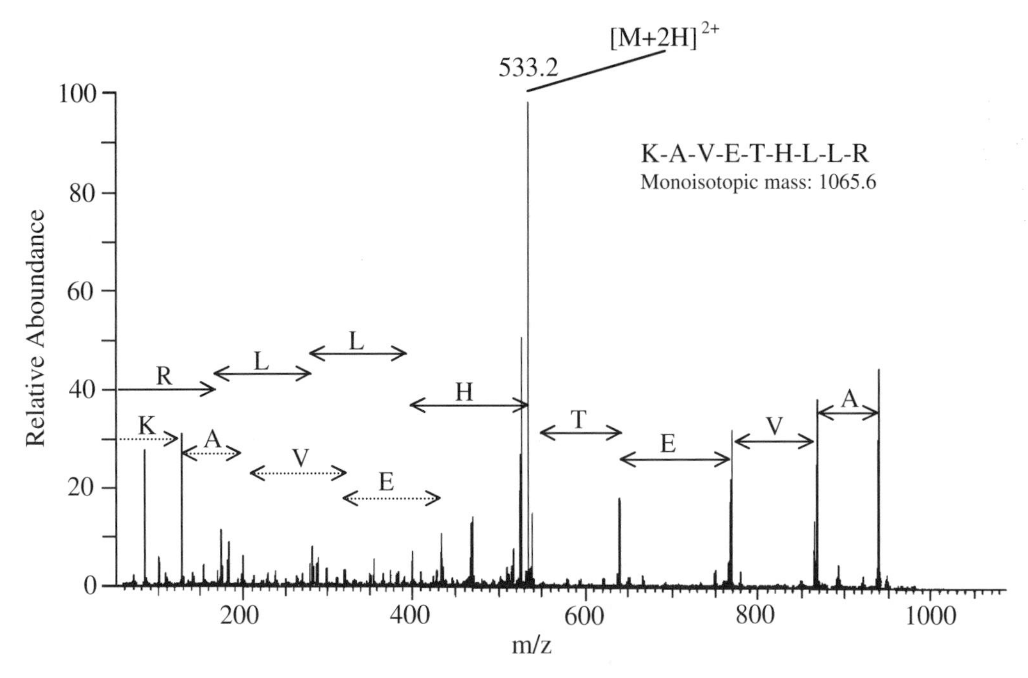

FIGURE 6.28 MS/MS spectrum of the *m/z* 533.2 ion (selected by the first quadrupole) of a single eluting peptide from one of the separated excised bands of BclXL associated proteins. The third quadrupole scanned the mass range from 50 to 1400 *m/z*. Two separate ion series were recorded by sequencing, one each from sequencing inward from the N- and C-termini. Database searching matched the sequence to KAVETHLLR from the human protein CLARP (caspase-like apoptosis regulatory protein). Six other peptides from the analysis of the selected band were matched to the same protein. Reprinted from Gygi, S.P. et al., *Electrophoresis,* 20, 310–318, 1999. With permission.

TABLE 6.8
Peptides Identified in a Single Capillary LC-MS/MS Analysis of a Separated Band, Computer Matched to CLARP[a], a Caspase-like Apoptosis Regulatory Protein

Peptide no.	Aa[b] position	Mass[c]	Charge state	Sequest X_{corr} score	Peptide sequence[d]
1	173–181	1152.6	2^+	3.18	(K) QSVQGAGTSYR
2	77–85	1065.6	2^+	2.88	(R) KAVETHLLR
3	86–94	1099.5	2^+	2.40	(R) NPHLVSDYR
4	78–85	937.5	2^+	2.17	(K) AVETHLLR
5	27–38	1292.7	2^+	4.00	(R) DVAIDVVPPNVR
6	107–117	1238.7	2^+	4.09	(K) SDVSSLIFLMK
7	50–61	1347.7	2^+	4.33	(K) LSVGDLAELLYR

[a] Genebank reference number AFOO5774.

[b] Amino acid.

[c] Monoisotopic mass.

[d] Amino acid residue prior to cleavage provided in parenthesis.

Reprinted from Gygi, S.P. et al., *Electrophoresis,* 20, 310–318, 1999.With permission.

The Molecular Chaperon 14-3-3σ in Human Breast Cancer Cells

In breast tissue cells this 23 to 30 kDA chaperon acts as a, p53-regulated, tumor suppressor that, through its interaction with cyclin-dependent kinases, controls the rate cells enter mitosis. The expression of 14-3-3σ chaperons in the prototypic breast cancer cell lines MCF-7 and MDA-MB-231, as well as in primary breast carcinomas was compared with that in normal breast epithelial cells. The various forms of 14-3-3, including α/β, δ/ξ, and σ were determined by MALDI-TOFMS peptide mass fingerprinting. The differences between measured and calculated masses were <0.2 Da and >50% of the trypsin-digested peptides matched the theoretical masses. The mass spectra obtained confirmed that the isoforms identified represented the same protein with different degrees of phosphorylation. Quantification revealed that the σ isoform is downregulated approximately 10-fold in the cancerous cells while the other isoforms are expressed at the same levels as found in normal cells. It was concluded that 14-3-3σ may be used as a marker for the non-cancerous state of breast epithelial cells.[248]

α-Tubulin Isotypes in a Human Breast Cell Line

An MS technique, developed to analyze C-terminal heterogeneity of α- and β-tubulin, has been used to study the diversity of tubulin isotypes in a human breast carcinoma cell line (MDA-MB-231). Cell extracts were resolved by 2-DE, and the proteins transferred to nitrocellulose. The region of the blot corresponding to tubulin (~50 kDa) was excised, digested with CNBr, and the released C-terminal tubulin fragments analyzed by MALDI-TOFMS in the negative ion mode. Product ion analyses were carried out using ESI-ITMS and the CID fragmentation patterns of the $[M-H]^-$ ions as well as those from PSD analyses were used to search different databases. In addition to the observation of the known tubulin isotypes, k-α1 and β1, a previously unknown α isotype (2,590 Da) was detected and its sequence determined. The amino acid sequence of the new isotype exhibited similarities with the corresponding region of the mouse α-6 isotype, however, distinct differences were also observed. Although not named at time of printing, the new isotype is likely to belong in the class 1 family of tubulin isotypes. The similarity of the mass spectra from both a taxol-stabilized microtubule preparation and a total cell extract suggested that the newly identified α-tubulin isotype is incorporated into microtubules.[249]

Search for Breast Tumor-Related Proteins in Serum

In an ongoing study evaluating the use of SELDI-TOFMS for biomarker discovery (see sections on prostate and colon cancers), a 28.3 kDa protein was detected using chips that bind proteins with an affinity for copper. During an initial evaluation of specificity, this protein was found in all patients with advanced metastatic disease (n = 21) and a smaller number of patients at other stages of breast cancer. The test was apparently negative in benign breast disease (1 positive from a total of 5 patients) and in all normal control samples (n = 23). The identity of the protein has not yet been determined.[250]

Ovarian Cancer

In a study to identify potential markers in tumor tissues obtained after resection, cell suspension, lysis, 2-DE, and in-gel proteolytic digestion were carried out using conventional techniques. The tryptic peptides were analyzed by both MALDI-TOFMS and ESI-ITMS. Seven proteins were found to be upregulated 2- to 3-fold, of which five were found in malignant tumor cell types. Of these, GST and SOD had been identified previously in ovarian tumors by gel matching but without sequence determination. Another protein identified in malignant tumors, apolipoprotein A-1, cannot be used as a marker because it also occurs in normal subjects and in treated cancer patients. Two proteins not previously reported in such tissues were retinoic acid-binding protein II, which regulates retinoic acid concentration and metabolism, and galectin 1, a carbohydrate-binding protein expressed on the surface of tumor cells and involved in cell-cell and cell-extracellular matrix adhesion and, consequently, in metastasis. A protein found in benign tumors was identified as annexin i.v. (novel identification in this tissue), a calcium and phospholipid binding protein involved in cell proliferation and differentiation. Cytokeratin 8 was identified in borderline-type tumors. An evaluation of methodology concluded that ESI-MS was appropriate for the identification of proteins <20 kDA, while MALDI-TOFMS analysis was easier and more sensitive for proteins >20 kDa when there were at least six digest masses in the 0.7 to 3.0 kDa range.[246]

6.1.4.4 Genitourinary Malignancies

Renal, bladder, and prostate cancers account for ~15% of all cancer cases. At presentation, a high percentage of cases have already developed disseminated local disease or distant metastasis. In these cases the five-year survival rates are <20%. Progress in understanding the genetic abnormalities involved in the initiation and progression of these diseases has been linked with ongoing proteomic studies intended to identify biomarkers that are both specific and indicative of the stages of the disease (reviewed in Reference 251). Aiming to investigate the potential use of SELDI-TOFMS for proteomics in uroscopy, the influence of radiocontrast medium on urinary protein profiles was investigated. In patients with impaired renal function, changes were observed in the abundance of several identified proteins such as the 11.75 kDa β_2-microglobulin.[252]

Renal Carcinoma

Several constitutive proteins have been determined in a human renal carcinoma cell line (ACHN) using Edman degradation, MALDI-TOFMS with PSD, and the *SWISS-PROT* database. The identified proteins included Cu/Zn SOD, 60S acidic ribosomal protein P2, and a heat shock protein 27 isoform. The identity of the ubiquitous and important Cu/Zn SOD was first determined using Edman-microsequencing, but three additional sequences were provided by MALDI-TOFMS, one of which was the 12-amino acid sequence GLTEGLHGFHVH (Figure 6.29). In the same study, a search was made for proteins that are regulated by the secreted cytokines γ-interferon and interleukin-4, both of which are known to modulate cellular growth and metabolism by interacting with cell surface receptors. Proteins from cytokine-stimulated cells were analyzed by 2-DE and digested with trypsin. The resulting peptides were separated by CE or HPLC for sequencing by both Edman degradation and MALDI-TOFMS. The expression of the following five proteins was increased by varying degrees in response to these cytokines: tropomyosin, heat shock protein 27, manganese SOD, GST π, and protein kinase C inhibitor I.[211]

Bladder Cancer

Bladder cancer is considered to be a “field disease,” i.e., the entire urothelium is at risk of developing tumors when exposed to tobacco metabolites or by occupational exposure to benzidine or other carcinogens. With the aim of establishing a comprehensive database of urinary proteins for patients with bladder cancer, urines were collected from approximately 50 patients the day before surgery, and a similar number of controls. After dialysis, samples were freeze-dried, resuspended in a lysis solution, and the proteins separated by 2-DE. Of 339 spots in pathological and normal samples, 21 proteins were identified by microsequencing and MALDI-TOFMS. Most of the identified proteins were plasma proteins and their variants, with the latter most probably being artifacts formed during the dialysis. Newly identified proteins associated with differential expression included a calcium binding protein, psoriasin (11 kDa), and a psoriasis-associated acid-binding protein.[253] Other identified proteins associated with the urines from cancer patients included fragments of gelsolin (49 kDa), the parent protein being a calcium-dependent actin filament of unknown role in tumors, and prostaglandin D_2 synthetase (27.4 kDa) which is involved in the formation of the arachidonic acid products observed in bladder cancer.[254] Subsequent work has extended the database to include the up- and down-regulated keratins associated with differential expression in both transitional cell carcinomas (TCC) that represent 95% of bladder cancers in the Western world,[220] and squamous cell carcinomas that are the frequent form (80%) in Africa and the Middle East.[255,256]

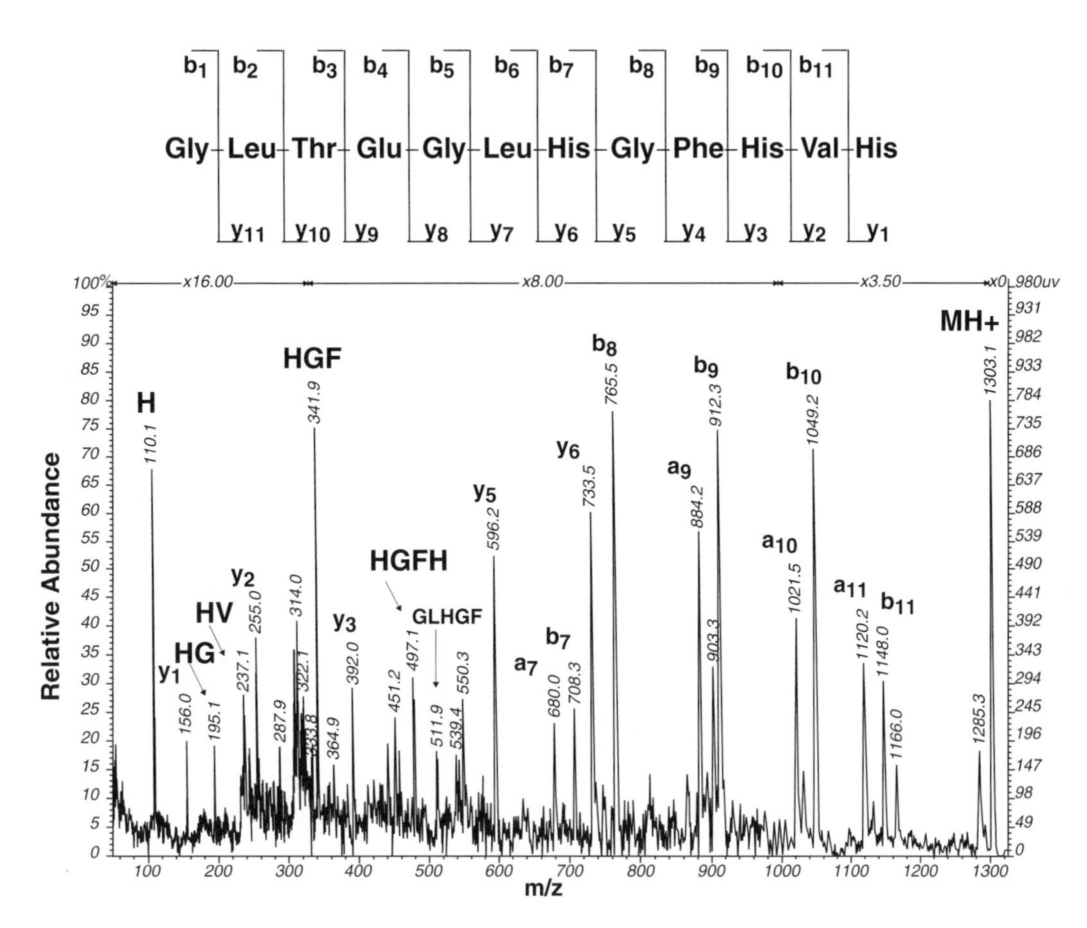

FIGURE 6.29 MALDI-TOF-PSD spectrum (reflectron mode) of an HPLC fraction of a protein (#6) from ACHN cells showing the derived 12-amino acid sequence GLTEGLHGFHVH. The protein spot was first digested (in-gel) with trypsin and the resulting peptides separated by HPLC. Reprinted from Sullivan, C.M. et al., *Cancer Res.*, 57, 1137–1143, 1997. With permission.

In an attempt to establish a diagnostic test for TCC based on specifically or differentially expressed proteins or protein clusters in urine, proteins were isolated by SELDI using chips with strong anion exchange chemistry from 94 samples including controls and from patients with TCC and other urogenital diseases. Analysis by SELDI-TOFMS resolved ~70 proteins and polypeptides in the 2 to 150 kDa mass range. The interassay reproducibility of the mass measurements for analytes with masses <10 kDa was in the 0.023 to 0.1% range. The molecular masses (std. dev.) of five individual proteins were 3353 Da (21 Da), 9495 (46.5 Da), 44.6 kDa (372.8 Da), 100.12 kDa (866.8 Da), and 133.19 kDa (772.9 Da). The 3.3 kDa protein was identified by a database search as a member of the human defensin family. This protein was also identified in bladder cells obtained by LCM but was shown not to be specific at the cellular level. The identity of the other four proteins remains unknown. Five protein clusters, within each of which there were several protein peaks, were also found to be preferentially expressed by TCC cells: 4.95 to 5.15 kDa, 5.71 to 6.00 kDa, 6.76 to 7.75 kDa, 15.00 to 16.00 kDa, and 85.00 to 92.00 kDa.

To maximize the diagnostic utility of the observed TCC-associated proteins, biomarker panels were formed by placing the five proteins and protein clusters into various combinations. If any member of the panel was observed in a sample, that combination set was considered to be positive. Of the patient urine samples analyzed, 47 to 70% were positive for one or more markers. The specificities of the individual markers ranged from 70 to 86%. When the specificities and occurrences of the individual markers and protein clusters were combined, this increased the diagnostic sensitivity for detection of TCC to 87% with a specificity of 66%. Using this combinatorial approach, even low-grade TCC could be detected with a sensitivity of 78% compared to 33% when using cytological techniques. Shortcomings of this technique are the absence of identification of the proteins and the lack of the quantification that might provide greater specificity through assessment of quantitative differences.[257]

Prostate Cancer

Although a number of novel genes are being studied as candidate prostate cancer genes, there is no known oncogene positively associated with prostate carcinoma, nor is there any proven link with *ras* mutations. Prostate specific antigen (PSA, Section 6.4.3.2), is a 22 kDa marker initially thought to be unique. However, the finding that it also occurs in patients with benign hyperplasia has led to numerous attempts, some documented in the remainder of this section, to find other biomarkers more specific for the carcinoma, differentiating it from the benign condition.

Analysis of prostate specific peptide (PSP, a member of the immunoglobulin binding factor family) in prostatic fluid by 2-DE revealed that it exists in three isoforms in this fluid, each ~15 kDa but with differing isoelectric points. The molecular masses of the isoforms were determined by ESIMS.[258] Four purported prostate cancer biomarkers, free and complexed PSA, prostate acid phosphatase (PAP, Section 6.1.3.1), PSP, and prostate specific membrane antigen (PSMA),[259] have been analyzed in their purified forms, in cell lysates, in serum, and in seminal plasma using SELDI-TOFMS. In one set of analyses, purified PSA, PSP, and PAP were exposed to three chemically defined surfaces on microchips, normal phase and two reversed phases, and PSMA was spotted on immobilized copper. The measured molecular masses agreed with known or expected values. It was possible to compare the purity of PSA from different commercial sources. The range of detection for PSA was linear in the ng/mL to μg/mL range. In other experiments, diluted serum and seminal plasma samples, that had been depleted of albumin by affinity chromatography were applied to microchip surfaces. Observed masses for the analytes were the same as for pure compounds. An additional prominent mass of ~80 kDa was confirmed to be the PSA/ACT complex. The intra-assay reproducibility of PSA mass determinations was 28,445.4 ± 31 Da with similar reproducibilities being obtained for the other analytes. When normal and cancerous cells (~2000 each) were procured from frozen tissue sections by LCM, cell lysates analyzed by SELDI-TOFMS revealed the presence of free-PSA, PAP, and PSP in both tissues. However, PSMA and several other proteins in the 1 to 10 kDa size range were found only in cells from patients with prostate cancer. An unidentified protein of 33,436 Da was also found to be upregulated in tissues from six of the prostate cancer patients.

SELDI-TOFMS can be used as an alternative to peptide map generation and sequencing for the confirmation of the identity of a protein. This is accomplished through binding specific antibodies to the chip that can then selectively recover the protein of interest. This confirmatory role is, obviously, limited to the availability and specificity of the necessary antibodies. The technique has been used for prostate tissue extracts to confirm that the masses detected for tumor-related proteins were not due to unrelated constituents of similar mass, e.g., transferrin, the mass of which is similar to that of PSA-ACT. The tumor-associated biomarkers in serum or seminal plasma were captured by binding them to antibodies on pre-activated chips for subsequent SELDI-TOFMS. The chip arrays consisted of a carbonyldiimidazole surface that reacts covalently with the amine groups of antibodies, thereby anchoring them onto the surface. After an activation procedure, anti-PSA, mouse IgG1, anti-PSMA, or anti-PSP were added to individual spots where they became covalently bound. This was followed by removal of any unbound antibodies by a series of washing and conditioning steps prior to application of serum or seminal plasma with a special bioprocessor accessory. Two antibodies could be added simultaneously to the chip arrays, allowing for multiplexed immunoassays. The presence of all the tumor-associated proteins studied, for which antibodies were available, was confirmed by either single- or multiplexed immunoassays, including both free and ACT-bound PSA. The specificity of the SELDI immunoassay resulted from the combination of discriminating power provided by the antibodies and the accuracy of measuring molecular masses with TOFMS. Quantification was also possible using standard curves constructed by plotting the log of measured peak areas against the known concentrations of purified antigens. These curves were linear and reproducible, e.g., in the 1 to 10 ng/μL range for PSA. The sensitivity of the technique for PSA approaches that of ELISA, but it is less for larger proteins.[260]

Because of the confirmed and specific upregulation of PSMA in androgen refractory and metastatic prostate cancers, baculovirus recombinant PSMA was generated, its structure and purity determined by MS, and a SELDI-based quantitative immunoassay was developed and validated.[259] In an ongoing biomarker evaluation by the same researchers, two overexpressed low molecular weight proteins of ~6.5 kDa and ~10.8 kDa size were detected by SELDI-TOFMS in the urine of a patient with confirmed prostate cancer. These compounds did not appear in the urines of a patient with benign prostatic hyperplasia or in a healthy age-matched male donor (Figure 6.30). In another ongoing study, SELDI-TOFMS has been used to compare samples, comprising ~1500 cells that were obtained by LCM, of normal, premalignant prostatic intraepithelial neoplasia (PIN) and frankly invasive prostate cancer cells. Two proteins, ~28 and ~32 kDa in size, were found to be reproducibly differentially expressed, with the PIN cells exhibiting intermediary profiles between the normal and tumor cell types. The changes in the protein profiles were consistent in 3/3 tissue sets.[219]

In a study to establish PSMA as the target of a prostate cancer-reactive monoclonal antibody (107-1A4), the target antigen was identified by SELDI-TOFMS using chips with normal phase surface chemistry. Of 19 peptides analyzed, 13 matches were found for PSMA covering 27% of the known primary sequence. In addition, ten potential N-linked glycosylation sites were identified confirming previous observations about the high carbohydrate content of PSMA.[261]

In another ongoing study on the use of SELDI-TOFMS for biomarker discovery (see the sections on breast and colon cancers), a 50.8 kDa protein was detected using chips that bind cationic proteins. When investigated in a pilot study, the protein was present in all the patients diagnosed with prostate cancer (n = 36), including all cases with a positive PSA test (n = 28) as well as those that were PSA negative (n = 8). The test was negative in all normal controls (n = 20). The identity of the protein has not yet been determined.[250]

A comparison of the 2-DE profiles of androgen-starved and androgen-stimulated versions of the highly tumorigenic M12AR prostate cancer cell line resulted in the detection of a differentially expressed protein that was significantly upregulated in the androgen-exposed cells. The protein was identified using MS/MS data (Figure 6.31, A and B) and *SEQUEST* database comparisons (Figure 6.31C), as human nucleoside diphosphate kinase A (NDKA/nm23), a gene product known for its ability to suppress metastatic activity in several human tumors, including prostate. This finding supports the observations (both *in vitro* and *in vivo*) about the survival benefit of maintaining the androgen responsiveness of prostate tumor cells.[262]

To evaluate protein expression patterns in cells, collected from benign prostatic tissues and prostate carcinomas, MALDI-TOFMS was used to analyze peptides obtained from in-gel digests of proteins separated by 2-DE. In addition to several cytoskeletal and mitochondrial proteins, three potential marker proteins were identified by a *PDQUEST* search. The expression of heat shock protein 70 and tropomyosin 1 were increased significantly in the malignant tissues, while the 40 kDa prostatic acid phosphatase protein decreased.[263]

The utility of proteome information obtainable by MS has been extended considerably by its integration into a transcriptome-based comprehensive, global gene expression approach that converges, for the tissue under consideration, the expressed sequence tag databases and transcript profiling with the proteomic data. The prostate transcriptome has been defined by assembling and annotating some 15,953 distinct transcripts from 55,000 prostate-derived expressed sequence tags and using these clusters to construct cDNA microarrays to find 20 genes which were induced by androgens.[262]

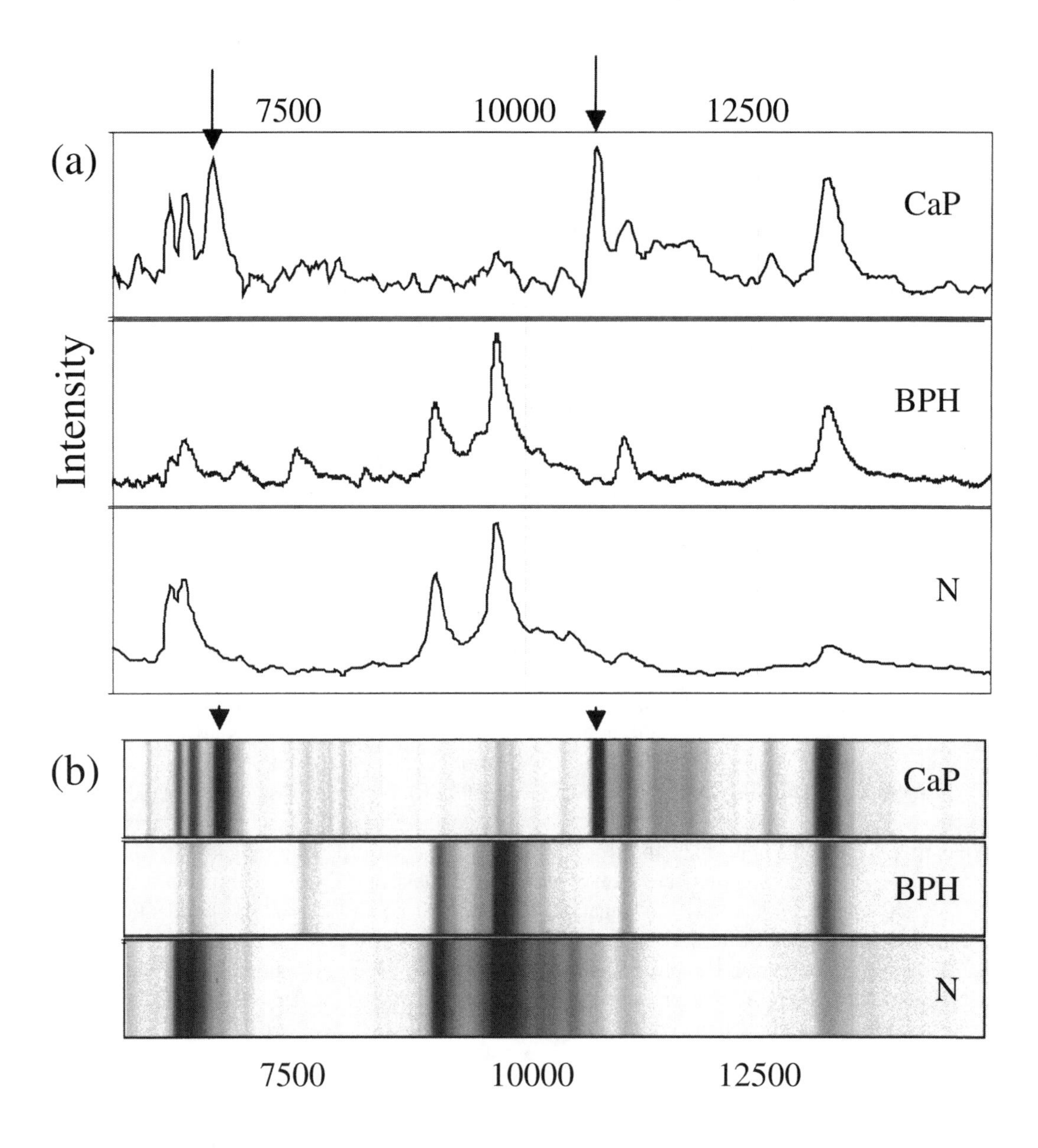

FIGURE 6.30 In an ongoing study, two overexpressed low molecular weight proteins of ~6.5 kDa and ~10.8 kDa size (arrows) were detected by SELDI-TOFMS in the urine of a patient with confirmed prostate cancer (CaP). These compounds did not appear in the urine of a patient with benign prostatic hyperplasia (BPH) or a healthy age-matched male donor (N). (a) Mass spectra in conventional format; (b) the same mass spectra in gel format view. This figure is courtesy of Dr. George L. Wright Jr. of Eastern Virginia Medical School, Norfolk, Virginia.

6.1.4.5 Gastrointestinal Malignancies

Esophageal Cancer

Although not common in the Western world, esophageal cancer occurs frequently in Asia. Normal squamous epithelium and corresponding stage II tumor cells from the same patients, ~5000 cells, obtained by LCM, were compared and ~675 distinct proteins or their isoforms (10 to 200 kDa range) visualized by 2-DE with silver staining. It was estimated that the analyses identified proteins

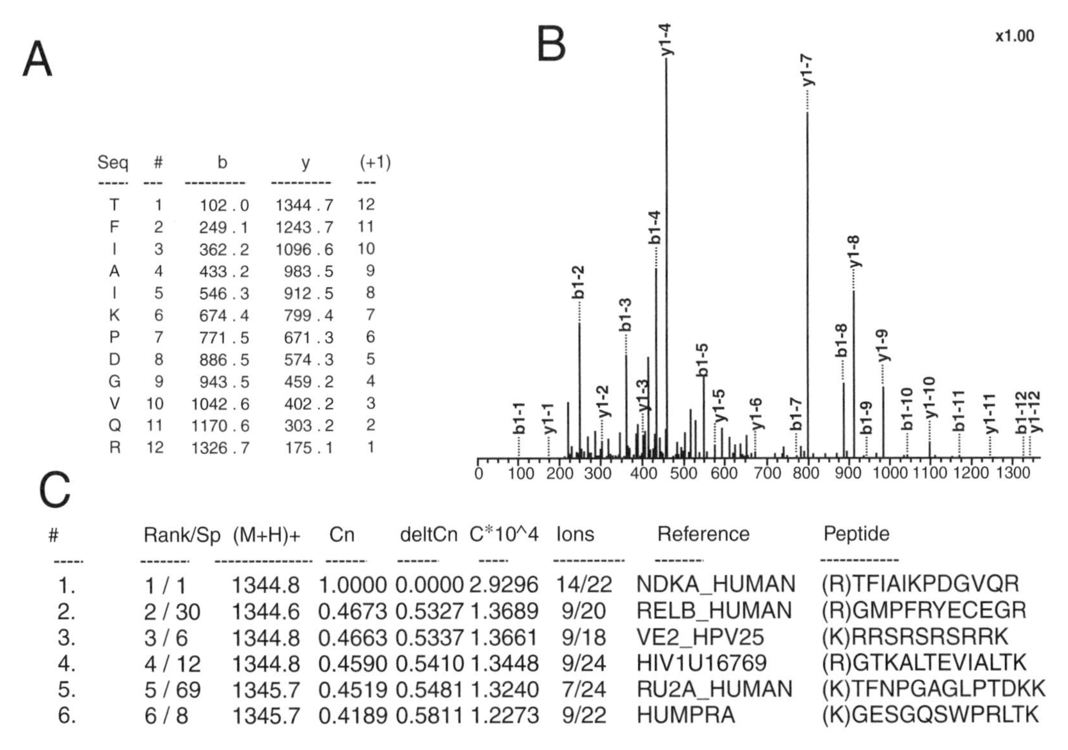

Seq	#	b	y	(+1)
T	1	102 . 0	1344 . 7	12
F	2	249 . 1	1243 . 7	11
I	3	362 . 2	1096 . 6	10
A	4	433 . 2	983 . 5	9
I	5	546 . 3	912 . 5	8
K	6	674 . 4	799 . 4	7
P	7	771 . 5	671 . 3	6
D	8	886 . 5	574 . 3	5
G	9	943 . 5	459 . 2	4
V	10	1042 . 6	402 . 2	3
Q	11	1170 . 6	303 . 2	2
R	12	1326 . 7	175 . 1	1

#	Rank/Sp	(M+H)+	Cn	deltCn	C*10^4	Ions	Reference	Peptide
1.	1 / 1	1344.8	1.0000	0.0000	2.9296	14/22	NDKA_HUMAN	(R)TFIAIKPDGVQR
2.	2 / 30	1344.6	0.4673	0.5327	1.3689	9/20	RELB_HUMAN	(R)GMPFRYECEGR
3.	3 / 6	1344.8	0.4663	0.5337	1.3661	9/18	VE2_HPV25	(K)RRSRSRSRRK
4.	4 / 12	1344.8	0.4590	0.5410	1.3448	9/24	HIV1U16769	(R)GTKALTEVIALTK
5.	5 / 69	1345.7	0.4519	0.5481	1.3240	7/24	RU2A_HUMAN	(K)TFNPGAGLPTDKK
6.	6 / 8	1345.7	0.4189	0.5811	1.2273	9/22	HUMPRA	(K)GESGQSWPRLTK

FIGURE 6.31 Identification of an androgen-regulated protein from metastatic prostate cancer cells. Protein expression profiles of starved and synthetic androgen-treated cell lysates were compared by 2-DE. Proteins demonstrating qualitative expression level differences were subjected to in-gel trypsin digestion and identified by LC-ESIMS/MS analysis. (A) and (B) show MS/MS spectrum of identified peptide, peptide sequence, and identified ion series. (C) Results from correlation of aquired peptide fragmentation spectra with database entries using SEQUEST software. The MS/MS spectrum was identified as NDKA_HUMAN (nm23). Two additional peptides were also identified from the same run (not shown). Reprinted from Nelson, P.S. et al., *Electrophoresis,* 21, 1823–1831, 2000. With permission.

in the 5×10^4 to 1×10^6 copies per cell range. Seven of the proteins were observed only in normal epithelium, i.e., they were downregulated in the tumor cells. It is likely that the patient-specific approach adopted in this study was the reason these seven downregulated proteins could be observed. Of ten proteins that appeared significantly altered in the tumor cells, two were excised, digested in-gel with protease, and the tryptic peptides analyzed by ESI-ITMS. The proteins were identified by peptide sequence matching (using four residues for each) as cytokeratin 1, that was overexpressed, and the underexpressed annexin I. This confirmed previous reports of the disregulation of these proteins in epithelial tumors.[264] In another study, lysed LCM cells from patient-matched normal esophageal epithelial and tumor cells were analyzed by SELDI-TOFMS using chips with a hydrophobic interaction (C_{18}) as the surface. Two proteins not yet identified were upregulated in a tumor-specific manner in 7/8 and 8/8 cases.[219]

Colon Cancer

To investigate the utility of LCM with respect to standard sectioning techniques for obtaining samples to be used in subsequent separation of proteins by 2-DE and analysis by MALDI-TOFMS, paired samples, obtained by standard sectioning and LCM, of both tumor and normal colon epithelial cells were obtained from colectomy patients. A comparison of proteins, separated by 2-DE and identified by peptide mass mapping, in unstained sections, with those observed in LCM samples revealed a selective reduction in the latter of nonepithelial, contaminating proteins such as hemoglobin. In addition, the LCM samples had an increased representation of intracellular epithelial structural proteins such as cytokeratins.[265]

In developing a plasma-membrane protein database for a human colon carcinoma cell line (LM1215), limitations of the 2-DE technique for the separation of hydrophobic proteins were bypassed by using only a size-based one-dimensional SDS-PAGE separation. The complex protein mixtures were separated into groups with defined molecular mass ranges by slicing the gel. After *in situ* proteolysis with trypsin,[241] the peptides were separated using HPLC[45] and analyzed by ESI-ITMS. Data-dependent switching was used to obtain CID product ion spectra. The data were compared against the *SEQUEST* database.[107] An additional advantage of this approach is the possibility of using the isotope-labeling technique to quantify normal vs. diseased sample pairs.[152]

For the tryptic digests of the gel slice containing proteins in the 160 to 190 kDa range, 319 MS/MS scans were derived. About 50% were identified by the *SEQUEST* database search, ~40% of the scans were of poor quality or represented non-peptidic constituents, and ~8% were candidates for subsequent *de novo* interpretation. Eighty-seven unique peptides were sequenced and found to have originated from 21 proteins. In all, 284 unique proteins, including 92 membrane proteins, were identified when all 16 gel slices covering the entire molecular mass range were analyzed. These included several hitherto-unidentified proteins of high molecular mass, e.g., DNA-activated protein kinase (465.3 kDa) and a number of single, seven, and twelve transmembrane helix receptors. As part of this study, a "targeted ion" technique was used to search for, and obtain, unambiguous identifications of low-abundance membrane proteins. This approach was illustrated by the identification of the human cell membrane-associated A33 antigen that is expressed in >90% of colonic cancers. After obtaining the TIC of the unfractionated products of the tryptic digestion of the gel slice covering the 40 to 45 kDa range (Figure 6.32A), only ions corresponding to the theoretical tryptic digest peptide masses of the analyte were selected for CID analysis. Thus, in the mass spectrum of the peak at the 25.2 min position on the TIC curve (Figure 6.32B), the *m/z* 1198.4, a +2 charge state ion was the fourth most intense ion and therefore not likely to be selected by data-dependent switching software; however, it was a potential ion for an A33 antigen peptide. When it was selected through the "targeted ion" method for CID analysis (Figure 6.32C) the identified sequence of YNILNQEQPLAQPASGQPVSLK was confirmatory for the A33 antigen.[37]

In another study, comparison of normal and cancerous colon mucosa by 2-DE revealed hundreds of protein spots. Among those spots only expressed in tumor tissue was a 13 kDa protein that was upregulated in 13 of 15 tumor samples.[266] The identification of the protein as S100A9 (also known as calgranulin B) was confirmed by peptide mass mapping MALDI-TOFMS. Matches were found for 13/19 peptides giving a sequence coverage of 75%.[267] The function of this protein, a member of the S100 calcium-binding protein family, is inhibition of casein kinase II, an enzyme involved in the regulation of normal cellular transcription and translation. Another calcium-binding protein, S100A8 (9 kDa), was also identified in tumors by MALDI-TOFMS; mass mapping matched in 14/21 peptides and included all the major peaks, thus providing a 97% sequence coverage. The upregulation of these two proteins, and also that of the heterodimeric complex they form, calprotectin, has been suggested as influencing the regression of colorectal carcinoma.[268] The statistically significant results of determining alterations of protein abundance in matched sets of macroscopically normal colon mucosa and colorectal carcinoma have been summarized. Downregulated in colon cancer carcinoma were liver fatty acid-binding protein, actin-binding protein/smooth muscle protein 22-alpha, and cyclooxygenase 2. Proteins that were upregulated in the same cells were several members of the S100 protein family of calcium-binding proteins, including some of their isoforms, and a novel variant of heat shock protein 70.[269]

Colonic Crypts and Polyps

Detection of cellular changes in colon epithelial crypts and polyps, caused by mutations of the tumor suppressors APC and p53, has been undertaken by comparing samples isolated from normal,

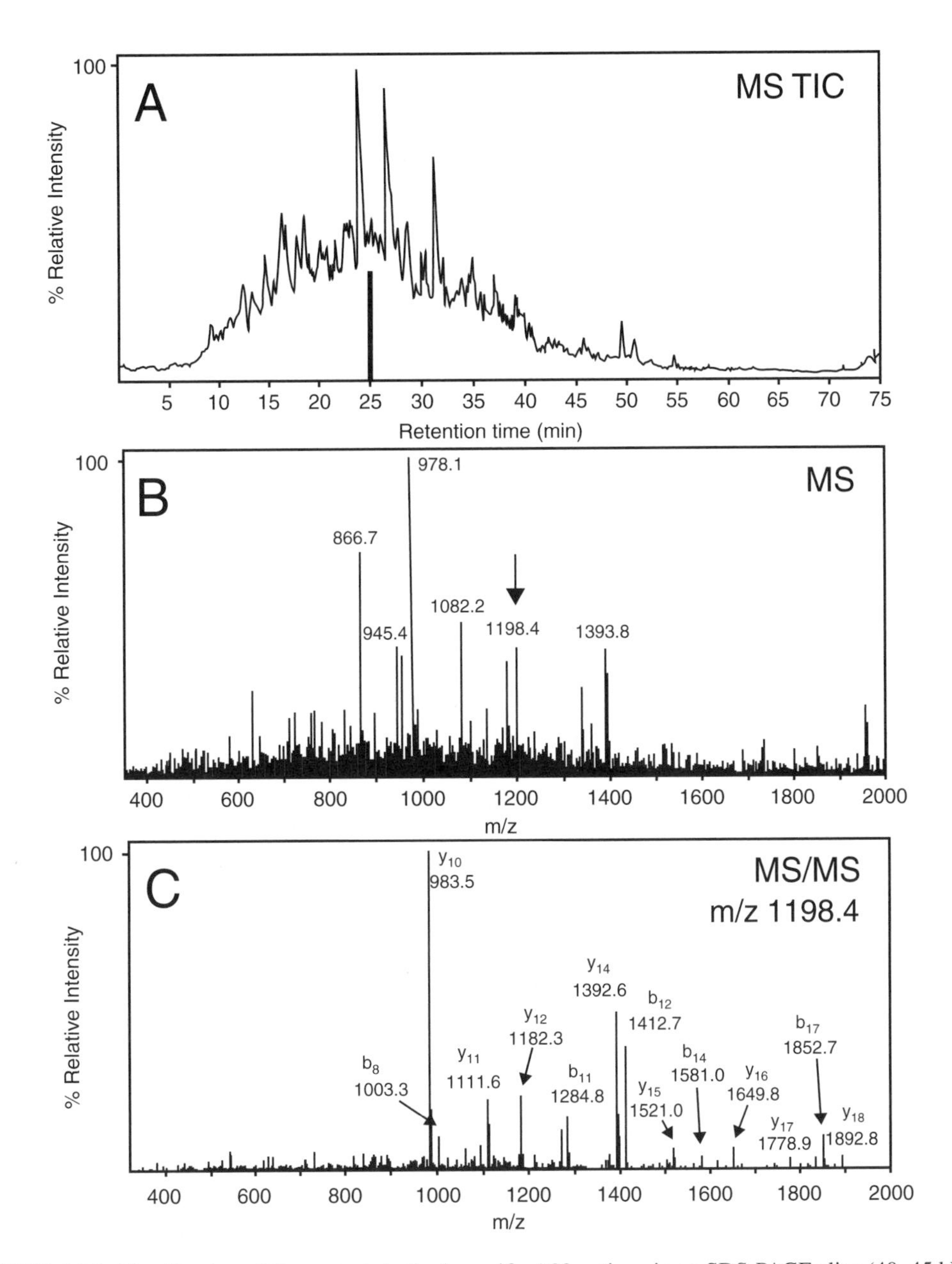

FIGURE 6.32 Identification of the gastrointestinal-specific A33 antigen in an SDS-PAGE slice (40–45 kDa). (A) Total ion current (TIC) profile of the tryptic digest. Only those ions corresponding to the theoretical tryptic digest peptide masses of the antigen were selected for MS/MS. (B) MS spectrum of the highlighted region (retention time 25.2 min) from (A). The fourth most intense ion (*m/z* 1198.4; +2 charge state) corresponding to a potential A33 antigen peptide was selected for MS/MS analysis. (C) CID MS/MS spectrum of the selected ion led to the identification of the peptide by SEQUEST as the A33 antigen peptide YNILNQEQPLAQPAS-GQPVSLK X_{corr} = 4.6282. In a separate experiment, the CID of the most intense ion in (B) (*m/z* 978.1, +2 charge state) yielded a sequence corresponding to the protein 3-ketoacyl-CoA thiolase. Reprinted from Simpson, R.J. et al., *Electrophoresis,* 21, 1707–1732, 2000. With permission.

p53-null, and multiple intestinal neoplasia (MIN) mice by using 2-DE and MS. The protein expression patterns in all epithelial crypts were comparable irrespective of the disease state. However, polyps from the animals with neoplasia contained ~60 proteins that were significantly downregulated and 4 proteins that were significantly upregulated with respect to normals. The relevant proteins were digested in-gel and the resulting peptides analyzed by LC-ESI-ITMS. Some 38 proteins were identified using the *MS-Tag* and *MS-Product* algorithms and by manual interpretation. Three isoforms of carbonic anhydrase I, that are highly expressed in normal crypts, were barely detectable in the neoplastic polyps.[270]

In another study, hundreds of protein spots were detected by 2-DE in normal epithelial colon specimens and compared with proteins recovered from human polyps. The intensity of 59 protein spots were significantly altered in the polyp specimens, and several of these were identified using MALDI-TOFMS and database searches. Upregulated proteins included numatrin (nucleophosphine/B23) and heat shock proteins 70 and 60. Downregulated proteins included the fatty acid binding protein 14-3-3σ (stratifin), cytokeratin 20, cytochrome C oxidase polypeptide Va, a Rho GDP-dissociation inhibitor, and β- and γ-actins. Most proteins with altered expression in polyps consistently exhibited differential regulation across the panel of specimens studied. It was suggested the altered expression of most of the identified proteins is a consequence of the neoplastic transformation and that some may have specific and primary activities in the process of tumorigenesis.[271]

Proteins Involved in the Development of Cellular Chemoresistance

2-DE has been used to recognize a number of protein spots that were downregulated with respect to normal cells in chemoresistant cell lines. In one protein spot from a colorectal sarcoma cell line (HT-29RNOV) that is atypically chemoresistant toward mitoxantrone, five peptides were detected by MALDI-TOFMS. This downregulated protein was identified as human adenine phosphoribosyl transferase. Two downregulated proteins were also identified in a cell line that is chemoresistant to daunorubicin (HT-29DRB). One protein yielded peptides with homologies to the protein expressed by the human breast cancer specific gene 1, while the second protein could not be identified. Similarly, interpretation of the peptide spectra, obtained by MALDI-TOFMS, led to the identification of an overexpressed protein in a resistant fibrosarcoma cell line as being the dissociation inhibitor of Rho guanine dinucleotide phosphate. A second isolated protein revealed homologies to the yeast protein yer-7; however, no human counterpart could be identified. These identified proteins with their altered expression are assumed to be involved in the modulation of the activity of protein kinase C and actin depolymerization.[272]

SELDI-TOFMS for Biomarker Discovery in Serum

In a pilot study, serum samples from 51 patients with benign, malignant, or premalignant colon diseases were compared with 20 healthy controls. Serum fractions were initially separated using ion exchange chromatography, with the fractions obtained being applied to protein chips with different surface chemistries. The best results were obtained with the SAX2-type chips that bind anionic proteins. A 13.8 kDA protein was recognized as a potential marker based on the following results: in colon cancer the protein occurred in 37/37 samples, for polyps it was 6/6, and in diverticulitis, a condition not known to be associated with increased risk of malignancy, it was present, surprisingly, in 4/4 instances. However, in other benign diseases the rate was 0/4, and for normal controls it was 0/20. The identity of the protein has not yet been determined.[273]

Liver Cancer

Hepatocellular carcinoma is one of the most common cancers in Africa, China, and Southeast Asia. Etiological risk factors include infection with the hepatitis viruses and exposure to aflatoxin B_1. During an effort to establish a protein database for liver cancer studies, approximately 400 protein spots were separated and visualized by silver staining in a human hepatocarcinoma cell line (HCC-M). Tryptic peptide masses, obtained from excised spots using MALDI-TOFMS, were searched for matches in the *SWISS-PROT* and *NCBI* nonredundant databases. Of the 272 spots which had matches returned, 191 protein species were identified (informative detailed descriptions are given in the addendum of Reference 274). In addition to several housekeeping proteins, such as α-enolase and alcohol dehydrogenase, identified proteins with expression patterns also reported as associated with other cancers or the process of carcinogenesis, included translationally controlled tumor proteins such as the epsilon isoform of the 14-3-3 protein, annexin I, II, and V, prohibitin, and thioredoxin peroxidase 1 and 2.[274] One of the 29 spots that yielded good mass spectra but was returned unmatched during the database search has been subjected to *de novo* identification by MALDI-TOFMS/MS with PSD. The sequences of 210 amino acids were used for the expressed sequence tag (EST) and assembled into a new *in silico* nuclear protein. The protein, designated as Hcc-1, was localized within the cell nucleus and its chromosome location was determined using an integrated approach incorporating a variety of molecular biology techniques. A semiquantitative analysis of Hcc-1 cDNA levels in human liver tissues, liver cell lines, and paired liver/hepatoma samples revealed that the protein was significantly overexpressed in well-differentiated hepatocellular carcinoma, but the degree of expression decreased as the carcinoma progressed to a poorly differentiated stage.[275]

Significant and reproducible quantitative and qualitative variations were obtained by image analysis of the expression levels of 99 protein spots separated by 2-DE in a comparison of the human hepatoma cell line (BEL-7404) with a normal human liver cell line (L-02). Twelve proteins were identified with the aid of *SEQUEST* database searches, using CID data obtained by LC/ESIITMS analysis of in-gel digested spots. The MS/MS analysis was performed in the data-dependent acquisition mode of the ITMS, using the ion of the highest abundance obtained in the initial, full-scan mode. Among the proteins with increased expression levels identified in the hepatoma cells were: inosine-5′-monophosphate dehydrogenase 2, a 27 kDa heat shock protein, calreticulin, calmodulin, and two homologs of reticulocalbin. A protein with a level of expression 18-fold higher in hepatoma cells than in normal liver cells was identified as GST P. Among the proteins down-regulated in hepatoma cells, tubulin beta-1, the natural killer cell enhancing factor B, and a tumor-suppressing serpin (precursor of maspin, a serine protease inhibitor) were identified. Proteins detected in liver cells but not in hepatoma cells included epidermal fatty acid-binding protein and the adipocyte-type fatty-acid-binding protein. The functional implications of these differentially expressed proteins have been discussed.[276]

Pancreatic Cancer

Human pancreatic cancer is an aggressive malignancy in which systemic metastasis occurs and for which treatment is difficult. Relapses are usually caused by development of chemoresistance. Protein expression has been studied in both the classical daunomycin resistant and atypical, mitoxantrone resistant, sublines of a human pancreatic adenocarcinoma cell culture model (EPP85-181P). The spots corresponding to three overexpressed proteins, found by 2-DE, were stained, isolated, and analyzed by MALDI-TOFMS. One of the proteins, overexpressed in both cell lines, was identified as cofilin (identifying peptide = YALYDATYESK), an actin depolymerizing protein that is a major cytoskeletal component with multiple functions. It has been suggested that non-phosphorylated cofilin has a participatory role in growth-associated processes. The other two proteins were overexpressed only in the mitoxantrone resistant subline. One of these proteins was identified as fatty acid binding protein (E-FABP, identifying peptide = ELGVGIALR). E-FABP is known to have numerous biological functions including the modulation of cell growth and proliferation, a result of its affinity for lipid ligands such as prostaglandins and leukotrienes. The third protein was stratifin (14-3-3σ protein, identifying peptide = GDYYRLAEVAT), a protein involved in a variety of cellular processes, including activation of protein kinase C. Theories have been advanced to explain the roles of these three proteins in the development of chemoresistance in this mitoxantrone-resistant subline of pancreatic adenocarcinoma.[277] The novel nuclear protein Hcc-1 that is overexpressed in well-differentiated hepatocellular carcinoma has also been shown to be significantly overexpressed in pancreatic adenocarcinoma.[275]

6.1.4.6 Other Cancers

Lung Cancer

A 35 kDa protein (TAO1) has been shown by 2-DE to be expressed in >90% of primary lung adenocarcinomas but not in other major malignancies. The pI of another 35 kDa protein (TAO2) was only slightly different from that of TAO1. Sequencing of five peptides from the excised and digested TAO1/2 spot using ESI-ITMS and CID revealed that the amino acid sequences of both proteins were identical to that of *napsin A*, an aspartic protease previously shown to be uniquely expressed in lung and kidney cells. With the aid of the peptide sequence of napsin A (TAO1/2), four EST clones were identified, with the same sequence, in a human database. The precursor of napsin A, called pronapsin A, has been determined that the processing of the precursor into the mature 35 kDa protein occurs by a cleavage between Asp-63 and Lys-64. It was concluded that napsin A (TAO1/2) may be used for the differential diagnosis of primary and metastatic lung adenocarcinoma.[278]

The MS techniques used to study differential protein expression in ovarian cancer (Section 6.1.4.3) have also been applied to lung cancer tissues obtained after tumor resection, and have shown that there are a number of upregulated proteins particular to lung adenocarcinonma. Cytokeratin 8 showed a fourfold increase; however, the protein was present as a 38 kDa variant rather than the normal 54 kDa size. Cathepsin D, an aspartic acid lysosomal endoproteinase was increased sixfold and again the protein appeared to have been proteolytically processed from 38 kDa to 27 kDa in the cancer cells. Cathepsin D is believed to be a facilitator of tumor invasion and metastasis. Two additional proteins that were uniquely expressed at high levels in primary lung adenocarcinoma could not be identified by database searches.[246]

Among other types of lung cancer, heterogeneous nuclear ribonucleoprotein (hnRNP) C1/C2 was increased threefold in small cell carcinoma[246] and a new α-tubulin isotype, expressed in a human breast cancer cell line (Section 6.1.4.3), was also detected in a human non-small cell lung cancer cell line (A549).[249]

Human lung epithelial cells are hypersensitive to low doses of ionizing radiation. However, this sensitivity is replaced with a *radioprotective response* that is induced by exposure to radiation above a threshold dose of ~0.5 Gy with obvious implications for therapy. To understand the mechanism behind this response, more than 800 proteins, with M_r in the 10 to 150 kDa range labeled with ^{35}S, were separated by 2-DE and visualized by staining from control and 0.5 Gy-irradiated human bronchial epithelial cells (L132 cells). Protein spots were excised, destained, digested with trypsin, and the masses of the resulting peptides determined using MALDI-TOFMS. Upregulated (twofold) and downregulated (two- to fivefold) proteins were identified by searching the *SWISS-PROT* protein database using tools in the *ExPASy* system. At least three of the identified proteins are known to be involved in the signaling pathway of protein kinase C, an enzyme that regulates by phosphorylation several downstream proteins in signaling cascades. The upregulated protein that predominantly affected protein kinase C, and was concomitant with the induction of radioprotection, was identified as protein kinase inhibitor 1 (13 kDa). The two other protein kinase C substrates that were upregulated in the radioprotective response were identified as translational initiation factor 2 (110 kDa) and chloride intracellular channel protein 1 (28 kDa). The former is known to enhance the expression of genes that encode for housekeeping enzymes and those proteins involved in the repair of stress-induced failure of cell cycle enzymes, while the latter is a known effector of protein kinase C activity. One of the downregulated proteins was identified as interleukin-1α (16 kDa), a protein known to be involved in the radioprotection of mouse tissues. Downregulation was also observed for another identified species, farnesyl pyrophosphate synthetase (48 kDa). This protein has been suggested as a participant in the repression of the growth of damaged cells. Another downregulated protein that is specific to lung epithelial cells was heterogeneous ribonucleoprotein A2/B1 (34 kDa). A cell-specific set of events was proposed to explain the inducible radioprotection of lung cells.[279]

Melanoma

The incidence of melanoma has been increasing at a rate faster than for any other cancer except lung cancer in women. In one study, a search was conducted for proteins in lysates of human A375 melanoma cells. Eleven proteins, isolated by 2-DE, were identified by MALDI-TOFMS and characterized by high-energy CID using a tandem magnetic MS. One of the proteins identified from the peptide sequences was stathmin. There was only 0.8 Da difference between the measured and calculated molecular masses of stathmin. The other proteins identified included α-enolase, cytokeratin, and Cu/Zn SOD.[280]

Specific populations of adjacent cryostat sections of normal epithelial and melanoma tissues were obtained by microdissection. Cells were lysed with a buffer, selected for the enrichment of the S100 protein fraction, and cell debris was removed by centrifugation. After applying 2 μL aliquots to the chip targets, several washing protocols were applied prior to adding the MALDI matrix. Reproducible protein profiles were obtained from 50 to 100 cells, while multiple analyses could be made using 500 cell samples. The protein profiles obtained by SELDI-MALDI-TOFMS were homogeneous, in contrast to those obtained in head and neck cancers (see below). Significant differences were observed between the protein profiles of tumors and surrounding epithelial regions using chips with reversed-phase or strong anionic exchange surfaces. The molecular masses of four potential markers were determined to be in the 10 to 12 kDa range; however, their identities and significance have not yet been determined.[281]

In a study aimed at comparing protein profiles in the sera of patients with malignant cutaneous melanoma with those in control subjects, sample preparation consisted of diluting the sera samples and separation into fractions with specific mass ranges using ultrafiltration with size exclusion membranes. Several proteins with molecular masses <30 kDa were detected in the sera of patients but not in the controls. The accuracy of mass determination using MALDI-TOFMS, operated in the positive ion linear mode, was in the 0.1 to 1.0% range. The concentrations of some peptide species of 2.5 to 3.5 kDa exhibited significant variations related to the clinical stages of the disease. Only measured masses of the observed proteins and peptides were reported. However, no identifications were made.[282]

A study has been conducted to explore proteins associated with the development of chemoresistance in metastatic malignant melanoma. A cell line established from the lymph node of a patient was divided into sub-lines with resistance to vindesine (microtubule poison), cisplatin (alkylator), fotemustine (alkylator), and etoposide (topoisomerase II inhibitor). Coomassie-stained spots of differentially overexpressed proteins, detected by 2-DE, were excised, digested by trypsin, and the peptides analyzed by microsequencing and MALDI-TOFMS operated in positive ion mode with delayed extraction. PSD analysis was performed in the reflectron mode. More than 1000 spots were obtained in the two pH ranges studied, and more than 40 constituents were found either up- or down-regulated, with at least seven components being absent with respect to the parental cell line. Four overexpressed proteins were identified by database searches in the resistant cell lines: translationally controlled human protein, human elongation factor 1-δ, tetratricopeptide repeat protein, and the isoform 14-3-3γ of the 14-3-3 family. None of these proteins had previously been linked to chemoresistance.[283]

Head and Neck Cancer

Protein profiles in crude tissue preparations of histopathologically characterized normal and tumor tissues were compared using SELDI-TOFMS. The tissues were disrupted manually in buffer solutions, followed by the sample and chip preparation steps described above for melanoma. Because the lysed cells were heterogeneous, the protein profiles were less reproducible than those obtained from melanoma cells by laser microdissection. A protein of 8.7 kDa was found in five of six tumor samples, but not in six normal controls. The identity of the protein has not yet been determined.[281]

Glioblastoma

To demonstrate the potential of using imaging MS (Section 6.1.4.1) for the study of human brain tumors, glioblastoma cells were implanted subcutaneously into the hind limb of a nude mouse; 12 μm thick tumor tissues were cut after the tumors grew to 1 cm in diameter. Mass spectrometric images of over 150 different proteins were detected using MALDI-TOFMS. The degree of their expression was estimated in closely located, distinct areas of tumor proliferation, as well as ischemic and necrotic areas. Mapped proteins of interest were extracted and digested with trypsin, the resulting peptides were analyzed using ESI-ITMS and identified with the aid of database searches. For example, thymosin β4 (4,964 Da), an immunoregulating protein that sequesters cytoplasmic monomeric actin, was identified as a protein with high expression levels in the proliferating areas of the glioblastoma xenograft. This protein was also identified in mouse models of prostate cancer.[66]

Neuroblastoma

Neuroblastoma and its benign counterpart, ganglioneuroma, produce different neuropeptides including the 69-residue pro-neuropeptide Y (proNPY) and its cleaved and C-terminally amidated 36-residue product, neuropeptide Y (NPY), which is biologically active. In a study of the processing of proNPY and its correlation to clinical parameters, a number of peptides with NPY-like immunoreactivity were isolated from neuroblastoma tumor tissues from children. One peptide was identified using MALDI-TOFMS as NPY3-36, an N-terminally truncated form of NPY. Identification was made by comparing the molecular and multiply charged ions of sulfoxidized (methionine at position 17) NPY to the truncated analyte. NPY3-36 was shown to have receptor selectivity different from that of intact NPY. Suppressed processing of proNPY (<50%) was correlated with advanced metastatic stage and poor clinical outcome in neuroblastomas.[284]

Neuropeptides and Pituitary Adenomas

Approximately 10% of intracranial tumors are small (<10 mm diameter), benign, and slow-growing adenomas derived from and composed of anterior cells of the pituitary gland. About 75% of pituitary adenomas are specialized tissues which excrete excessive amounts of pituitary hormones leading to endocrinologic dysfunctions, e.g., acromegaly, Cushing's disease, infertility, and menstrual irregularities. The remaining 25% are nonsecreting, large (>10 mm diameter), usually invasive macroadenomas with symptoms related to tumor size, e.g., visual loss and headaches. The discovery of a major endogenous opioid pentapeptide, methionine enkephalin (ME, YGGFM), and the 31-amino acid β-endorphin (BE, for amino acid sequence see Reference 285), together with the isolation and characterization of other neuropeptides, led to explanations of the localization and regulatory effects of neuropeptides in the anterior pituitary.[286] Mass spectrometry has been playing an increasing role in the characterization and quantification of neuropeptides.[287] The initially used EI, CI, and FD techniques have been superseded by FAB, for the characterization of fragmentation of both natural[288] and synthetic peptides.[289] The advantages of ESI have also been evaluated.[290]

The method of choice for quantification has been FAB using cesium ion guns and glycerol matrix. Synthetic deuterated peptides, added to tissues before homogenization, serve as i.s. (reviewed in Reference 291). The sophistication of techniques for the storage and processing of tissue samples has been increasing[286,292-294] leading to schemes for the quantification of neuropeptides in human pituitary tumors.[285] Both ME and BE have been quantified in nonsecreting and adenocorticotropic hormone-secreting human pituitary tumors.[295] A similar study was conducted in prolactin-secreting pituitary adenomas from the anterior lobes of postmortem pituitaries of controls and patients. ME was quantified directly, whereas BE was quantified via its corresponding tryptic peptide [BE20-24, NAIIK] obtained by trypsinolysis. MS/MS was used to maximize specificity. For example, the CID spectrum of the precursor ion of NAIIK from a pituitary was the same as that from the synthetic version, exhibiting the expected b, y, and a type fragments (Figure 6.33). BE and ME were quantified using SRM of the transition from the $[M+H]^+$ ion to the N-terminal tripeptide fragment NAI and the transition of the YGGFM ion to the N-terminal tetrapeptide

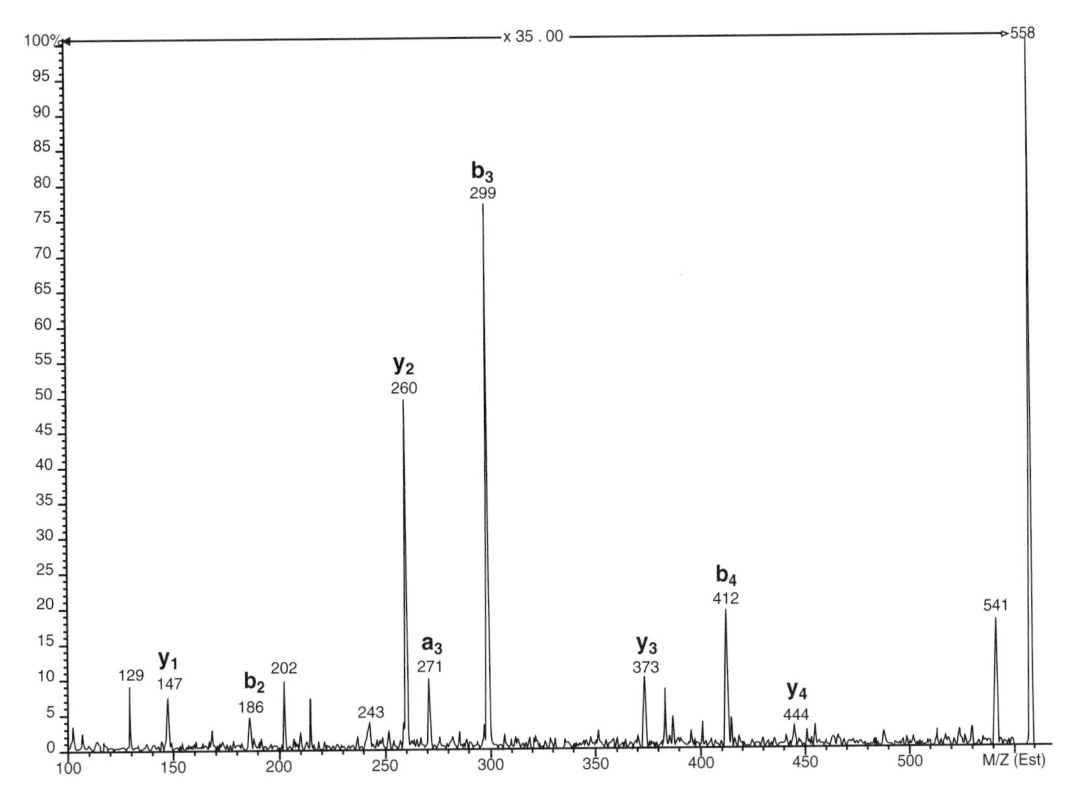

FIGURE 6.33 Product ion spectrum of NAIIK from a human pituitary tumor sample. NAIIK is a tryptic peptide (20-24) obtained from the 31-amino acid β-endorphin. Reprinted from Zhu, X. et al., *Peptides,* 16, 1097–1107, 1995. With permission.

fragment ion, YGGF, respectively (i.s.: deuterated analytes). The quantitative data from prolactin-secreting hormones revealed increased concentrations of ME but no changes in the concentrations of BE (Table 6.9). Treatment with Parlodel (bromocriptine mesylate, an ergot derivative that inhibits prolactin secretion), which is often preferred to surgery, resulted in increased ME and BE levels.[285]

The protein profile of a single postmortem pituitary from a normal subject was obtained using 2-DE. Excised protein spots were digested in-gel and the resulting peptides analyzed by MALDI-TOFMS operated in both delayed extraction and reflectron modes. Nine proteins were identified from the mass mapping data using the *SWISS-PROT* database. Of these, somatotropin (growth hormone), prolactin, and the β-chain of lutropin are hormones produced by the pituitary gland. Peptide NPP (a propiomelanocortin-derived protein) originates from the same precursor as BE. Ubiquitin thioesterase L1 is involved in the processing of several precursors in neuroendocrine tissues. Other proteins identified included the α and β-chains of hemoglobin, glutathione-S-transferase P, and glyceraldehyde-3-phosphate dehydrogenase. The molecular masses of the identified proteins were within 3 to 16% of the theoretical values except for lutropin β-chain and peptide NPP, where the observed large differences were due to glycosylation.[296]

6.1.5 Peptides and Class I Major Histocompatibility Complexes

6.1.5.1 Class I MHC Molecules

Self from nonself discrimination refers to the process by which the immune system identifies and reacts specifically against foreign entities such as infectious agents, tissues that have been transplanted from a genetically nonidentical individual, and tumors (Section 3.4.3). Immunoglobulins

TABLE 6.9
Quantitative Data of PRL-Secreting Human Pituitary Tumors

ID	SP-LI (fmol/mg)	BE (pmol/mg)	ME (pmol/mg)
Male (control)	99.4 ± 9.2	593.5 ± 121.1	76.6 ± 6.2
Male			
DM	9.6	18.7	71.1
ES	2.3	5.2	26.6
Male (on Parlodel)			
EM	20.7	665.8	2144.2
Female (on Parlodel)			
CR	89.0	1526.8	1741.7
DW	15.3	57.3	682.1
GE	9.6	97.4	747.4
SE	6.3	63.5	611.6
DA	5.8	67.4	376.5
AP	22.0	884.9	2035.5
Mean ± SEM	24.7 ± 13.1	449.6 ± 253.2	1032 ± 278.1
Female			
LV, hypothyroidism	219.6	506.9	n.d.

n.d. = not detectable

Reprinted from Zhu, X. et al., *Peptides,* 16, 1097–1107, 1995. With permission.

on the surface of B-cells sense the presence of antigenic determinants that are parts of intact antigens present as soluble foreign molecules in the circulatory system. In contrast, the receptors of T-cells do not directly recognize new antigens. Rather, specificity is engendered through surface receptors that are specific for foreign peptide antigens. For the recognition to occur, relevant *epitopes* must be displayed on the surface of host cells. These epitopes are small peptides that are noncovalently associated through high-affinity, low dissociation constant, binding with the *major histocompatibility complexes* (MHC). The MHC are special cell-surface molecules that act as hosts for the epitope. The cells that display these peptide-MHC complexes that are then recognized by T-cells, are called *antigen presenting cells* (APC). Class I MHC molecules, also known as *HLA* (*Human Leukocyte Antigen*) for human and *H-2* (*Histocompatibility*) for mouse, present peptides derived from endogenously synthesized proteins (see below). In contrast, the Class II MHC molecules display peptides derived from exogenous (ingested) antigens from the extracellular environment, e.g., soluble molecules, and internalized particles such as microbes.

T-cells bind to class I MHC molecules on target cells with the aid of *CD8* (*Cluster of Differentiation*) molecules on their surface. The response of the activated T-cells may be the direct killing of the cells that expressed the peptide-MHC complex, or the secretion of specialized molecules that, in turn, activate other cells of the immune system to attack the identified antigen. Tumor-reactive cytotoxic T lymphocytes (CTL) are known to recognize certain cancers including melanoma and renal cell, colorectal, ovarian, pancreatic, and lung carcinomas, in each case via a distinct structural motif provided by an MHC molecule (reviewed in Reference 297). Using a different method of engendering a response the Class II MHC utilize CD4 molecules to elicit T-cells, e.g., macrophages and B-cells, which release lympokines that in turn activate an immune response.

Endogenous antigen fragments that associate with Class I MHC are octa- to deca-peptides produced from both normal proteins and new or altered self-proteins expressed by cancer and virus-infected host cells. The processing cascade that generates these peptide fragments occurs in the

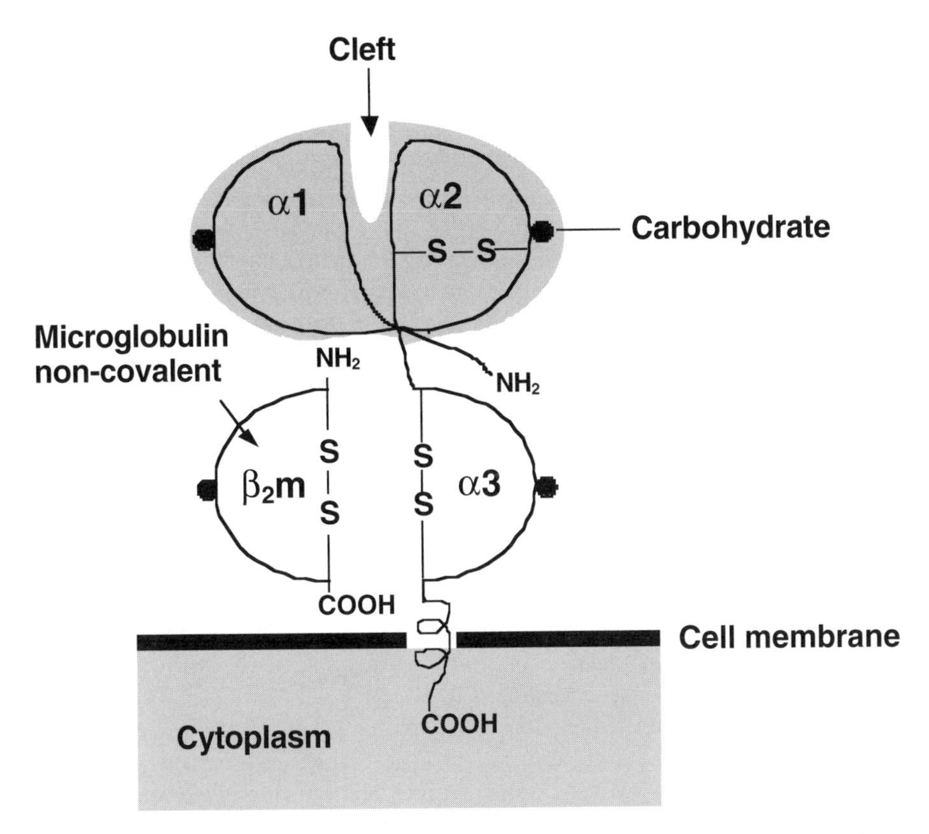

FIGURE 6.34 Relevant part of the structure of Class I MHC molecules that are heterodymeric glycoproteins. Two of the extracellular domains, α1 and α2 (each about 90 amino acids long), provide the binding side or cleft ("cradle") for the peptides derived from the processed antigens.

cytosol. It involves marking the candidate protein with the small protein ubiquitin, its degradation by multicomponent proteolytic enzymes (*proteasomes*), and the transferring of the resulting peptides into the endoplasmic reticulum with the aid of a transmembrane heterodimer polypeptide. The next steps are association of the peptide within the cleft of Class I MHC, aided by the molecular chaperon *calnexin*, and their display on the cell surface, a process that is facilitated by the Golgi apparatus and secretory vesicles.

Class I MHC molecules are heterodimeric glycoproteins. The heavy, 46 kDa polypeptide α chain, that traverses the plasma membrane, is noncovalently associated with a light, 12 kDa microglobulin chain (β2m) that is not anchored to the membrane. Two of the extracellular domains of the α chain, α1 and α2, each of which is about 90 amino acids long, provide the binding site or cleft ("cradle") for the antigenic peptides derived from the processed antigens (Figure 6.34). There are multiple types of MHC molecules, each possessing different peptide-binding specificities. Each cell expresses a pair of different HLA genes, one from each parent and encoded by multiple generic loci. Many alleles exist for each loci, e.g., some 59 alleles are known for the HLA-B locus. This results in complex and often confusing nomenclature. The 5 to 10% polymorphism is concentrated in the α1 and α2 subsections that constitute the antigenic peptide binding sites. The β2 microglobulin is not encoded by the MHC genes and is not polymorphic. Because the number of different MHC molecules per individual is limited to ~20, each type of MHC molecule can accommodate an array of peptides so as to present a large variety of antigens to the T-cells. However, there can be only one peptide per MHC molecule, and the accommodated peptides, for each type of MHC molecule, must have conserved (same or similar) amino acid residues at particular positions along their chains.

Because the peptide-MHC complexes are formed inside the cells, the T-cells actually monitor the interior of the host cells for the presence of foreign or altered antigens, thereby providing a crucial defense mechanism against microbial infection and cancer. T-cells are tolerant against the self-peptide-MHC complexes ("trash") derived from normal proteins. The display of tumor-related MHC-peptide complexes may lead to the death of the host cell (reviewed in Reference 298). Accordingly, the objectives of studying antigenic peptides include the identification of the natural and nonself peptides presented by MHC molecules and an understanding of the mechanism by which cellular proteins are processed into peptides by the multicatalytic proteasome complex for eventual presentation by the MHC. Intracellular proteins coded by mutated oncogenes are required to maintain the transformed cells but they are cycled within the cell and can therefore be expected to form epitopes. Both tumor-specific epitope peptides and self-peptides that are overexpressed by tumor cells may be candidates for T-cell-mediated tumor immunotherapy. The possibility to rationally engineer naturally processed tumor peptides that can stimulate a tumor-specific CTL response is crucial to the development of anticancer vaccines and drugs.[299-301] The potential role of mass spectrometry in vaccine development has been reviewed.[302]

6.1.5.2 Strategies and Techniques

Detection and sequencing of MHC-related peptides by MS is a daunting analytical challenge, because >10,000 distinct peptides may be separated by high-resolution HPLC or CE in a peptide extract from 10^9 cells. In addition, many peptides are present only at 10^{-15} to 10^{-18} M or even lower concentrations. Still, during the last decade, MS has become a major tool for the characterization of MHC-bound antigenic peptides, including an increasing number of tumor-associated peptides (Table 6.10). Most tumor-related antigens in Class I MHC complexes originate from known mutant proteins, e.g., gp 100, pmel117, MAGE, Melan A, p53, and tyrosinase. However, a few epitopes from melanoma that are restricted to Class II MHC have also been identified.

One of the strategies for the identification of peptide epitopes recognized by class I MHC-restricted tumor-specific CTL, is based on peptide epitopes derived from proteins whose expression is already known to be correlated with specific tumors. Another strategy, one that does not make any assumption about the nature of the candidate proteins, is based on a genetic approach, i.e., using cDNA libraries. Peptides are synthesized, based on coding sequences and consensus-binding motifs of MHC molecules, and tested for recognition by CTL reactive to various cancers. A third strategy is based on the extraction of peptides associated with MHC molecules, followed by repetitive fractionation and testing with nontumor cells that express an appropriate MHC molecule. This leads to the selection of candidate peptides for sequence determination by MS/MS. Co-elutions

TABLE 6.10
Representative Identified Human MHC Class I Epitopes

Disease	MHC molecule	Epitope	Source protein/gene	Reference
Melanoma	HLA-A2.1	YLEPGVTA	Pmel-17/gp100	318
Melanoma	HLA-A*0201	YMDGTMSQV	Tyrosinase (modified)	543
Melanoma	HLA-A3	ALLAVGATK	Pmel-17/gp 100	187
Melanoma	HLA-A*0201	ILTVLGVL	Melan 5	544
Renal cancer		WVKEKVVAL	NK4	312
Cervical cancer	HLA-1	XQFPIFLQF	HPV-18 L1	545
Cancer	H-2K (mouse)	SNFVFAGI	DEAD box p68	204

Based on deJong, A., *Mass Spectrom. Reviews,* 17, 311–335, 1998.

with synthesized peptides of known composition and CID pattern matching have been used to confirm peptide sequences obtained by CID analysis of CTL-reactive isolated fractions.[303]

Sample selection and preparation is complex and multidisciplinary (reviewed in Reference 304). A paper on the role of peptides and MHC in alloreactivity (see later) provides representative and instructive descriptions of the cell culture approach, the epitope stabilization assay, the method of peptide sensitization, and the molecular modeling employed.[305] In a typical strategy, the standard steps are acid-elution of the peptides bound to Class I MHC, isolated from tumor cells, and their fractionation with HPLC, therefore providing aliquots that can be tested for bioactive peptides. This testing is accomplished by sensitizing antigen-processing mutant cells, that predominantly express empty Class I molecules on their surface, followed by their exposure to tumor-specific CTL. Those fractions that render the surrogate target susceptible to lysis by tumor-specific T-cells are further purified prior to sequencing by one of several available MS techniques (see later). Another strategy uses concentration by immunoaffinity, release of the peptides with acetic acid, HPLC for coarse separation, loading onto a specially designed on-line membrane preconcentration-CE cartridge, and analysis by ESI-MS/MS.[306] Yet another approach to rapid purification uses affinity perfusion chromatography.[307] Regardless of the methodology, sequences deduced for native peptides should always be confirmed by comparison with synthetic peptides.

There are two other approaches for epitope identification that eliminate the need for bioassays, i.e., the generation of CTL lines for peptide identification. One technique scans genome sequences from infecting pathogens for stretches of amino acids that match a particular MHC binding motif. A limitation of this strategy is that the identity of the pathogen and the MHC binding motif must be known in advance.[308] Another strategy employs synthetic standard peptides that are co-eluted with the low abundance analytes into an LC/ESIMS/MS mass spectrometer where the sequence of the analyte is confirmed by CID. A library of 47 potential HLA-A2-binding peptides was synthesized to serve as a source for likely epitotpes. The prediction of the likelihood that one of these synthetic peptides represents an epitope was based on proteins known to be associated with gastrointestinal tumors, including p53, *ras*, carcinoembryonic antigen (CEA), Her-2, Mdm2, and Ssx2. The sensitivity of the analysis was such that as little as 68 fmol of any prospective peptide could be detected. This would correspond to 34 copies/cell in a sample of 5×10^9 cells. The method confirmed the identity of two predicted HLA-2-associated peptides, one from a p53-overexpressing B-lymphoblastoid cell line (C1R-A2) and the other from a human colon carcinoma cell line (SW1116) that contains a *ras* A12 mutation and overexpresses CEA. The technique was also applied to gastrointestinal adenocarcinomas of various origins, and histological grading. For instance, the presence of the $CEA_{694-702}$ epitope was confirmed in an intermediately differentiated adenocarcinoma of the cecum (Figure 6.35). The LOD was 400 fmol/8 $\times$ 10^8 cells, corresponding to ~300 copies/cell. No CEA-derived peptide was detected in healthy colon tissues from the same patient. An advantage of this technique is the sensitive and rapid selection of naturally processed and presented candidate peptides for subsequent lysis assay experiments with tumor-directed CTL.[309] This *predict-calibrate-detect* technique was also applied to the identification of peptides presented by HLA-A*0201 molecules on the surface of a breast carcinoma cell line (KS24.22) after induction with interferon-γ. One of the predicted epitopes, MAGE-$A1_{278-286}$, was detected by ESIMS in a peptide mixture derived from the cells. The peptide sequence was determined by CID to be KVLEYVIKV. Confirmation was provided by analysis of an identical synthetic epitope standard. Comparison of signal intensities led to an estimation of the quantity of epitope present in the tumor cells to be ~300 fmol/10^{10} cells, corresponding to ~18 copies/cell.[310]

Available MS techniques include on-line HPLC with ESIMS/MS,[304,306,311] MALDI-TOFMS with PSD,[312] and on-slide exopeptidase digestion prior to MALDI-TOFMS. Although impressive results have been obtained with ESIMS (reviewed in Reference 313), MALDI offers several advantages over ESI: (a) possibility of sequencing individual peptides in mixtures without purification; (b) considerable tolerance of detergents and salts arising from biochemical and chromatographic manipulations during sample preparation; (c) LOD of 30-100 fmol, though 1 pmol

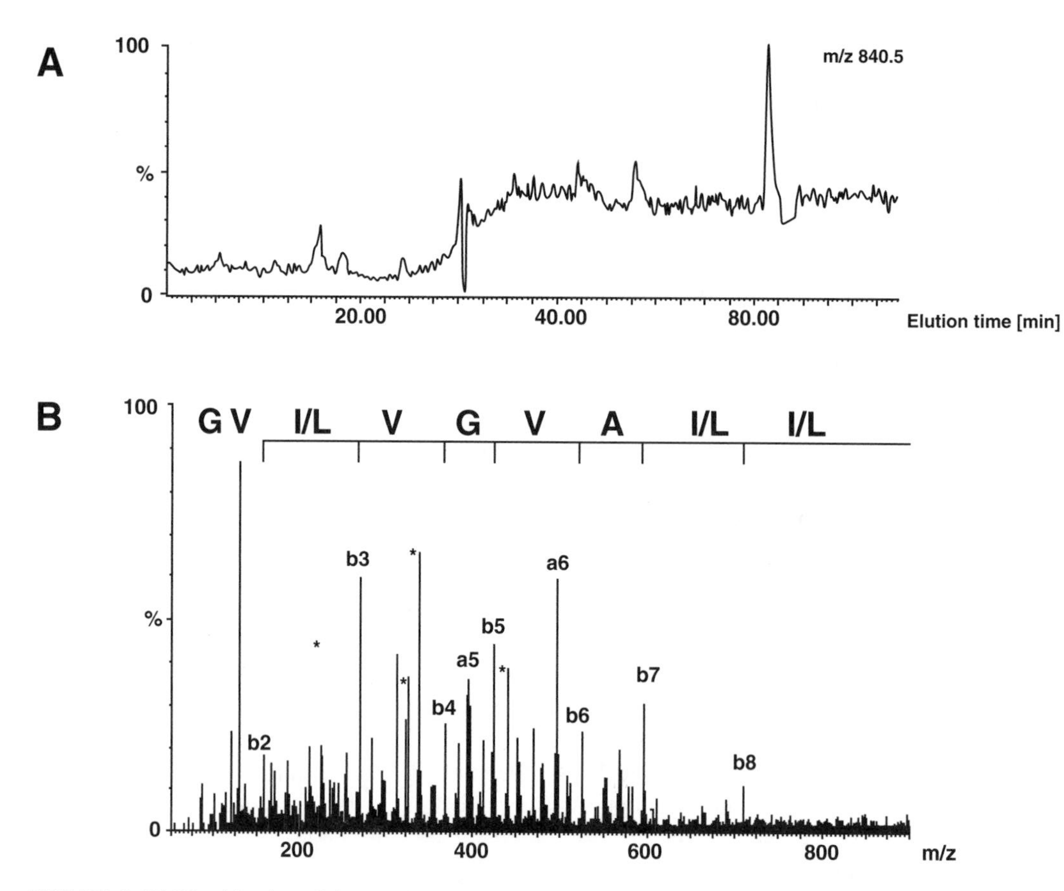

FIGURE 6.35 Identification of the predicted tumor-associated peptide $CEA_{694\text{-}702}$ in a peptide mixture eluted from SW1116 cells. (A) Mass chromatograms for *m/z* 840.5 corresponding to the $[M+H]^+$ ions of the peptide with a single peak at the same retention time as determined for the corresponding synthetic peptide. (B) CID mass spectrum recorded on $[M+H]^+$ ions at *m/z* 840.5 during a time of 62.2–63.0 min. Peaks corresponding to b and a series ions are labeled, asterisks denote internal fragments. Reprinted from Schirle, M. et al., *Euro. J. Immunol.,* 30, 2216–2225, 2000.With permission.

samples are needed for optimal results; (d) opportunity for multiple analyses of the same sample under different conditions, e.g., derivatization, hydrogen-deuterium exchange; (e) relatively easy manual or computerized interpretation because there are fewer multiply charged ions. Peptide mixtures are first analyzed by conventional, linear MALDI-TOF to obtain the molecular masses of the ions produced, followed by PSD analysis to obtain structurally significant product ions from the decay of selected precursors (reviewed in Reference 314). Evolving instrumental techniques use nanoscale LC and preconcentration CE techniques[315] combined with ITMS, ICRFTMS, and QTOFMS. For example, the multiple fragmentation capability of IT analyzers was utilized to sequence MHC-bound peptides and to investigate phosphorylation (Reference 305, see later). Current commercial ITMS include miniaturized spraying interfaces and ZoomScan, the latter providing mass accuracy of 0.01%.[316]

6.1.5.3 Peptide Antigens in Malignancies

Melanoma

Most early applications of peptide antigens were carried out on melanoma because of the availability of specific CTL and the high frequency of spontaneous regression that occurs for this tumor. In one application, Class I HLA-B7 molecules were prepared by lysing JY cells,[184] purification by immunoaffinity and fractionation by HPLC. Selected fractions were injected onto a second HPLC with direct elution into an ESIMS/MS. Some 2000 distinct peptide signals were observed in the mass range 838 to 1224 Da in quantities >30 fmol, representing an estimated 90 to 3600 copies/cell. Fifteen peptides were sequenced and the assignments confirmed by comparison with synthetic products. Figure 6.36 shows the sequence determination for one of these peptides. The CID spectra were obtained from the doubly charged ion of the free acid form (A) and from the singly charged protonated molecule after acetylation (B). The sequence was determined to be RPKSNXVXX where X denotes either Leu or Ile. Residues 1-8 were deduced from *b*-type ions while residues 8 and 9 were determined from y-type fragments.[317]

The same strategy was used for the identification of three peptides out of thousands detected that were recognized as relevant through the testing of HPLC fractions for their ability to reconstitute MHC-epitope complexes that were specific for melanoma CTL lines from two patients. The sequences were determined by CID to be SMAPGNTSV, YXEPGPVTA, and AXYDATYET. The ability of the first peptide to reconstitute the complex was two orders of magnitude larger than that of the other two peptides. The sensitivity of the method is appreciated by considering that the sequences were obtained from 15 fmol of each peptide derived from 4×10^{10} cells, or 2 copies/melanoma cell. It was determined, with the aid of synthetic peptides, that X was L rather than I. This peptide was shown to be derived from the melanoma-specific protein Pmel/gp 100 (residues 480-488). That this peptide was indeed a relevant epitope for HLA-A2.1-restricted, melanoma-specific CTL, was confirmed in cell lines derived from patients.[318]

Alloreactivity and MHC

Acute allograft rejection is the consequence of a severe T-cell response following the recognition of the presence of allogeneic MHC molecules. To identify the peptide ligand recognized by an alloreactive CTL clone raised against HLA-B27 (B*2705), 1×10^{10} cells were lysed and processed through affinity chromatography and purification. Two HPLC fractions exhibited high sensitizing activity. MALDI-TOFMS detected 15 molecular species in one fraction and 7 in the second. The sequences of two peptides appearing in both fractions were determined using both manual and automatic database searches. One sequence was RRFFPYYV, a proteasome C5-derived peptide (Figure 6.37), while the second peptide was RRLPIFSRL, a peptide that had previously been identified as a natural ligand of B*2705 and B*2709 cells.[319] Sequences were confirmed by analyzing synthetic peptides. Based on additional lysis experiments with this isolated peptide and its synthesized version, it was concluded that this proteasome-derived octamer is the epitope recognized by the human alloreactive CTL clone 27S69 that was raised against B*2705. This was the first HLA-B27 ligand shown to be the immunogenic agent in alloreactivity. Another peptide, a nonamer, found in one of the HPLC fractions also exhibited sensitizing activity and was identified as RRFFPYYVY, a natural ligand of related B* type products.[305]

Renal Carcinoma

HLA Class I molecules were recovered from a renal carcinoma cell line (A-498). The cells were lysed and immunoreactive peptides were obtained by immunoaffinity chromatography followed by acid elution and ultrafiltration. The peptides were separated by HPLC into ~30 fractions and analyzed by MALDI-TOFMS. Each collected HPLC fraction contained a few high abundance peptides and up to 100 different low abundance peptides, all with masses in the 0.7 to 1.2 kDa range. Molecular masses were determined using linear MALDI-TOFMS, and deuterium-hydrogen exchange was often used to check if initially observed ions of a given mass might, in fact, represent two components with differing numbers of exchangeable hydrogens (see below). Selected precursors were subjected to PSD analysis to obtain structurally significant product ions. Sequences were obtained using database searches and *de novo* sequencing. Sequence information was obtained for 18 peptides from 17 HPLC fractions. Comparisons with synthetic peptides confirmed the sequences of 14 of the peptides. Each was shown to have been derived from previously described cellular proteins with sequence similarities of 100% for 13 of the 14 peptides. Several of these identified

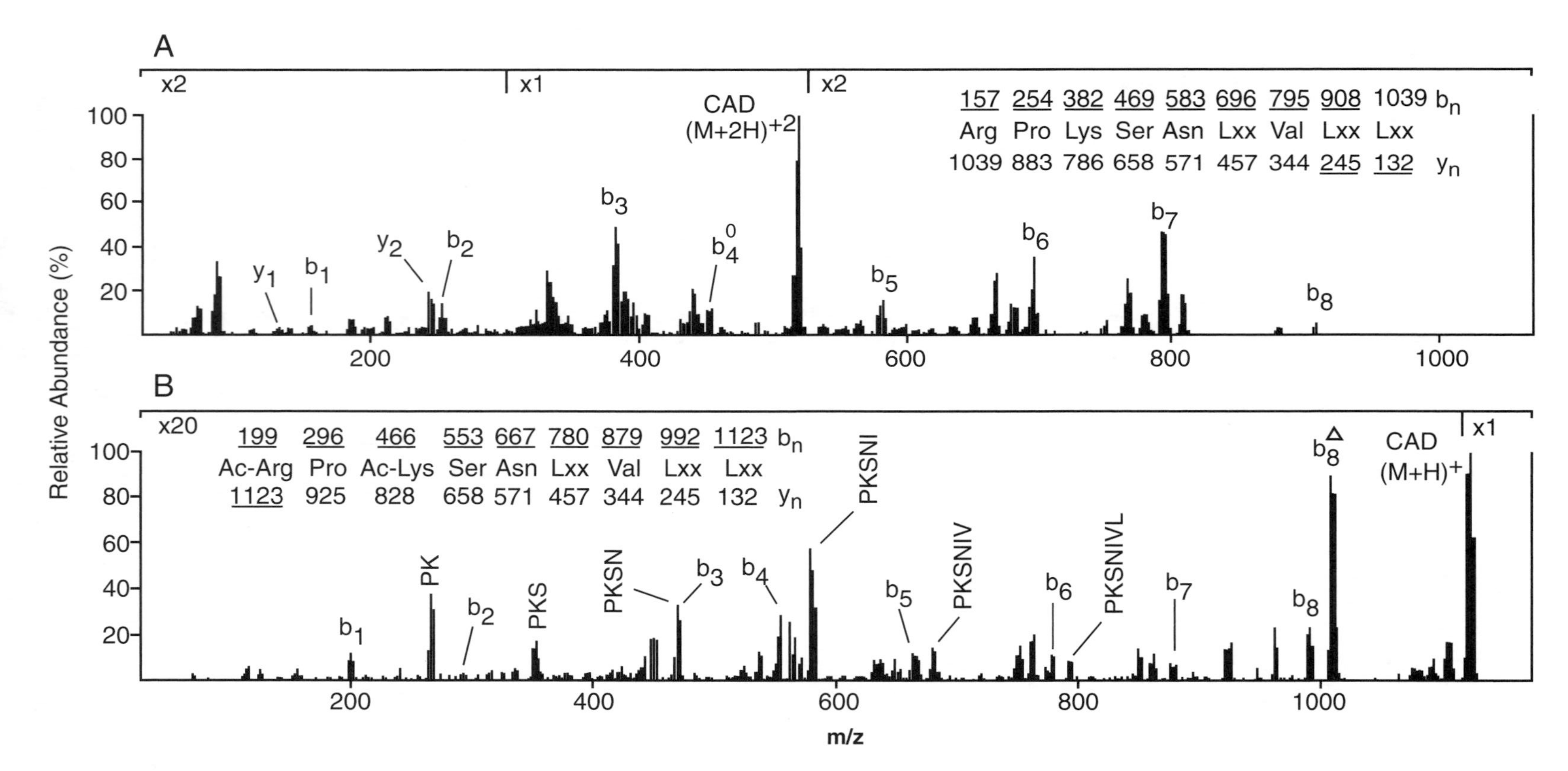

FIGURE 6.36 Mass spectrometric sequence determination of peptide 1039 from JY melanoma cells. (A) The $[M+2H]^{2+}$ (*m/z* 519) was selected to generate the spectrum of the free acid. (B) A spectrum was also obtained after acetylation (sites marked Ac) of a portion of the sample and selection of the $[M+H]^{+}$ ion (*m/z* 1123). Predicted masses for the fragments of type *b* are shown above and type *y* below the deduced sequence RPKSNXVXX (X represents Leu or Ile). Ions observed in the spectrum are underlined. Subtraction of *m/z* values for any two fragments that differ in size by a single amino acid generates a value that specifies the mass and thus the identity of the extra residue. Residues 1 to 8 were deduced from fragments of type b in the free acid spectrum, whereas residues 8 and 9 were defined by fragments of type y. The spectrum of the acetylated form allowed the assignment of residue 3 as Lys instead of Gln, which has the same mass. Reprinted from Huczko, E. et al., *J. Immunol.,* 151, 2572–2578, 1993. With permission.

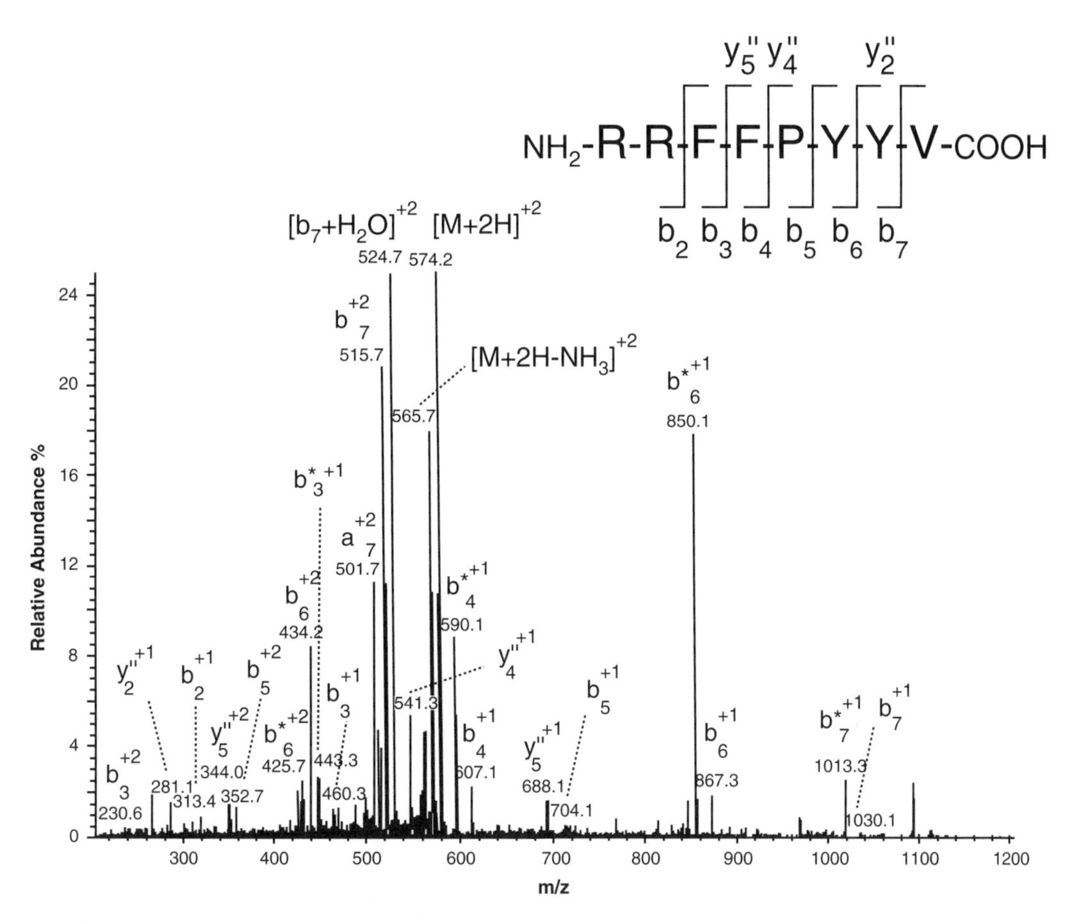

FIGURE 6.37 Sequencing of the RRFFPYYV allograft rejection epitope peptide by ESI. The MS/MS fragmentation spectrum of ion at *m/z* 574.2 is shown. The assigned peptide sequence is indicated, detailing the observed backbone fragment ions using the conventional nomenclature. Fragments labeled with asterisks originate from ions of the same series after neutral loss of ammonia (17 Da). Ions of the a series are produced by neutral loss of CO (28 Da) from ions of the b series. The precursor ion $[M+2H]^{2+}$ and a related ion resulting from loss of ammonia are indicated. The high abundance of fragments with charge 2+ is due to the presence of two basic residues at the N terminus. Reprinted from Paradela, A. et al., *J. Immunol.*, 161, 5481–5490, 1998. With permission.

HLA Class I bound peptides were shown to have been derived from housekeeping or other proteins present in normal cells.[312] Sequence determination for four analytes was also attempted by the fragment tag method[94] but they proved to be erroneous because the spectra of the analytes differed significantly from those of the respective synthetic peptides. The prominent mass signal of one of the peptides (*m/z* 1099.6) turned out to represent two components (Figure 6.38). After hydrogen-deuterium exchange, the MALDI spectra exhibited two signals, at 1117.6 Da and 1121.6 Da, revealing the presence of a major component with 18 exchangeable hydrogens, and a minor component with 22 exchangeable hydrogens. The minor component was identified as the decapeptide, PASKKTDPQK, a structure that was confirmed using a synthetic peptide. This peptide was subsequently recognized as the epitopic peptide of the renal tumor antigen, RAGE, that binds to HLA-B8, thus becoming a promising candidate target structure for CTL lysis.[312]

To map the key components of MHC Class 1 antigen processing and to characterize the constitutive and cytokine-regulated protein expression profiles in representative human carcinoma cell lines (MZ125RC and MZ1940RC), proteins separated by 2-DE were analyzed using peptide

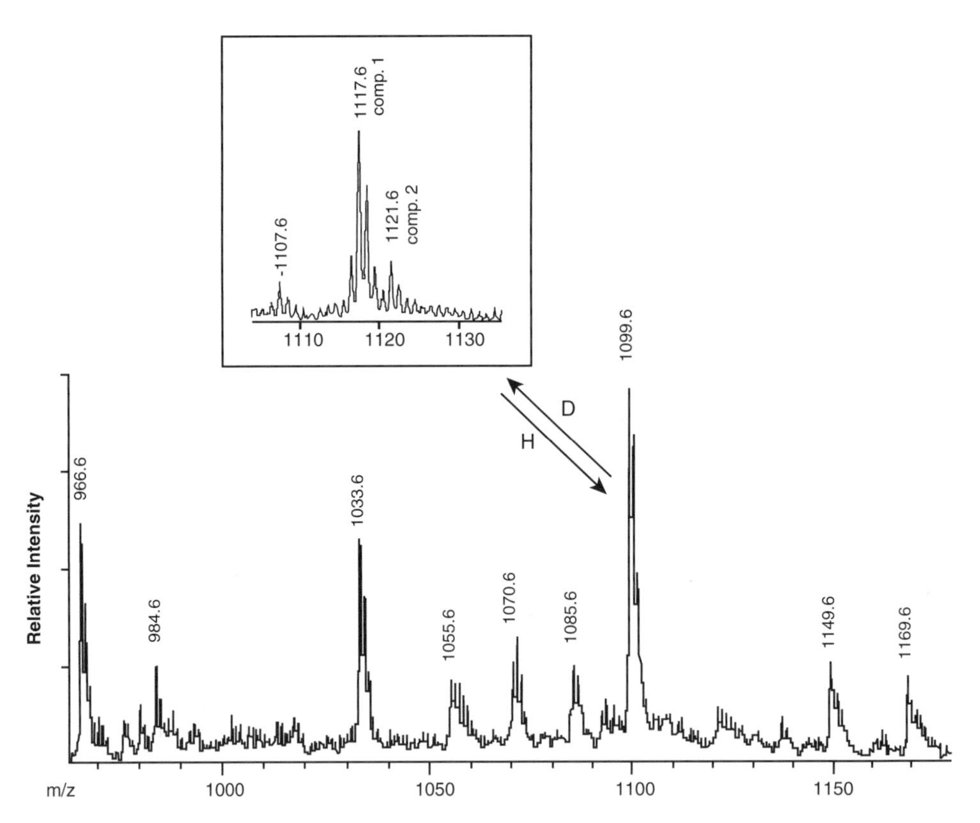

FIGURE 6.38 Typical MALDI mass spectrum recorded from a peptide peak collected from the HPLC separation of the renal carcinoma A-498 cell line peptide pool A at 27.2 min retention time. This shows a dominating mass peak at *m/z* 1099.6 which is demonstrated by comparative analysis of the deuterated sample to contain two compounds of identical precursor ion masses with differing numbers of exchangeable hydrogens, *m/z* 1117.6 and 1121.6. Reprinted from Flad, T. et al., *Cancer Res.,* 58, 5803–5811, 1998. With permission.

mass fingerprinting with MALDI-TOFMS. The following proteins were identified: β-actin, calreticulin, ER60, gp 96, grp 78/BiPm, hsp 60, hsp 70, hsp 90, and PDI. Prelimary data were presented on the comparison of protein expression using untreated and interferon-γ-treated cells.[320]

Lung Cancer

Starting with a squamous cell carcinoma cell line (VBT2), HLA-A68.2 molecules were isolated by immunoaffinity and the acid-eluted peptides purified with HPLC. Aliquots from collected fractions were added to an autologous B-lymphoblastoid cell line and tested as targets for tumor-specific CTL in an epitope reconstitution assay. The only peptide that passed this biological activity test had the sequence ETVSEQSNV, as determined by LC/ESI-MS/MS (Figure 6.39). With the exception of one amino acid, glutamic acid vs. glutamine at the sixth position, the active peptide was identical to the sequence of amino acids 581 to 589 of the elongation factor EF2. After synthesizing the $EF2_{581\text{-}589}$ sequence as well as the same sequence with a glutamine substituted for glutamic acid at position 586, activity testing led to the conclusion that a heterozygous mutation in the tumor cells accounted for the one amino acid difference.[321]

Mouse Lymphoma

A CE-ESI-MS/MS technique using membrane preconcentration was applied to detect and sequence a peptide obtained from 3×10^{-9} K^b-derived EL-4 lymphoma cells, that was present at <60 fmol level. The analyte was detected as a minor peak among several major peaks found in the TIC trace obtained during an CE/ESI-MS/MS analysis of an HPLC isolated fraction (Figure 6.40A). Examination of the

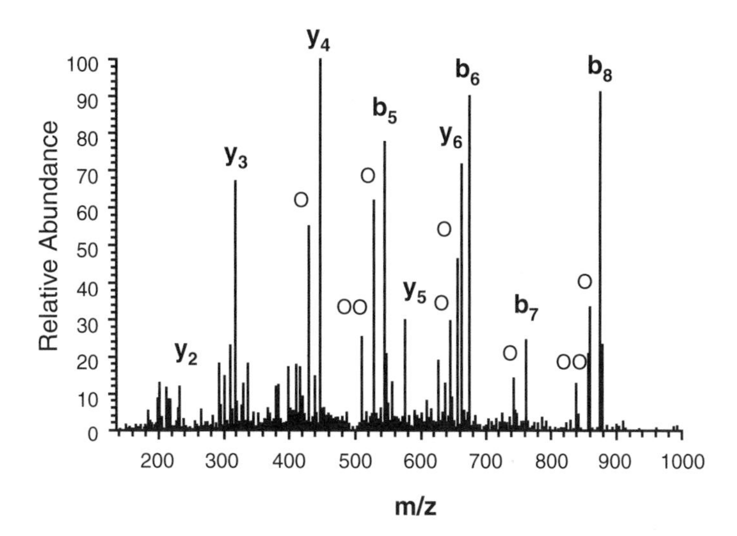

FIGURE 6.39 Mass spectrometric analysis of the *m/z* 497 candidate peptide for the squamous cell carcinoma line VBT2. The predicted masses of the type b and y fragment ions are shown above and below, respectively, as well as the deduced amino acid sequence that was obtained from CID. The loss of a single (O) or multiple (OO) water molecule is also indicated. Reprinted from Hogan, K.T. et al., *Cancer Res.*, 58, 5144–5150, 1998. With permission.

electropherogram at the appropriate retention time revealed a doubly charged species, at *m/z* 503.6 (Figure 6.40B). The CID product ions of this precursor revealed well-defined sets of y and b ions, yielding the sequence XSFKFDHX. Use of the *SEQUEST* search routine,[322] led to the peptide being identified as one derived from F-actin and thus the assignment of the Xs as I and L, respectively, for a final peptide of ISFKFDHL (Figure 6.41).[306]

Minor Histocompatibility Antigen in Bone Marrow Transplantation

The life-threatening graft vs. host disease (GvHD) that often develops after leukocyte antigen-identical bone marrow transplantation was shown to involve minor histocompatibility antigens that are recognized by T-cells. Starting with Rp cells (HA-1-expressing Epstein-Barr virus-transformed B lymphoblastoid cells), HA-A*0201 molecules were isolated and fractionated by HPLC. Two HPLC fractions that tested positive for their ability to reconstitute the HLA-1-specific lysis by T2-cells, contained more than 100 peptides. This was reduced to five peptides with initial HA-1-reconstituting activity and these were analyzed by LC/ESI-MS/MS. Based on biochemical tests, the most relevant candidate was a nonapeptide with the sequence VXHDDXXEA, where X = Leu or Ile (Figure 6.42). Comparisons with synthetic peptides confirmed the identification. Further testing showed that the only sequence recognized by three different HA-1-specific CTL clones, derived from two unrelated subjects, was VLHDDLLEA. This peptide differs by only one amino acid from VLRDDLLEA which is the HA-1-negative allelic counterpart encoded by the KIAA0223 gene from acute myelogenous leukemia KG-1. It was suggested that HA-1 allele typing of donors and recipients may reduce GvHD in bone marrow transplantation.[323]

Role of the Mutational Hotspot in p53 on Proteasomal Epitope Processing

Inactivation of the p53 tumor suppression protein, because of missense mutation at one or more mutational hotspots, preferentially renders transformed cells susceptible to lysis by CTL that are

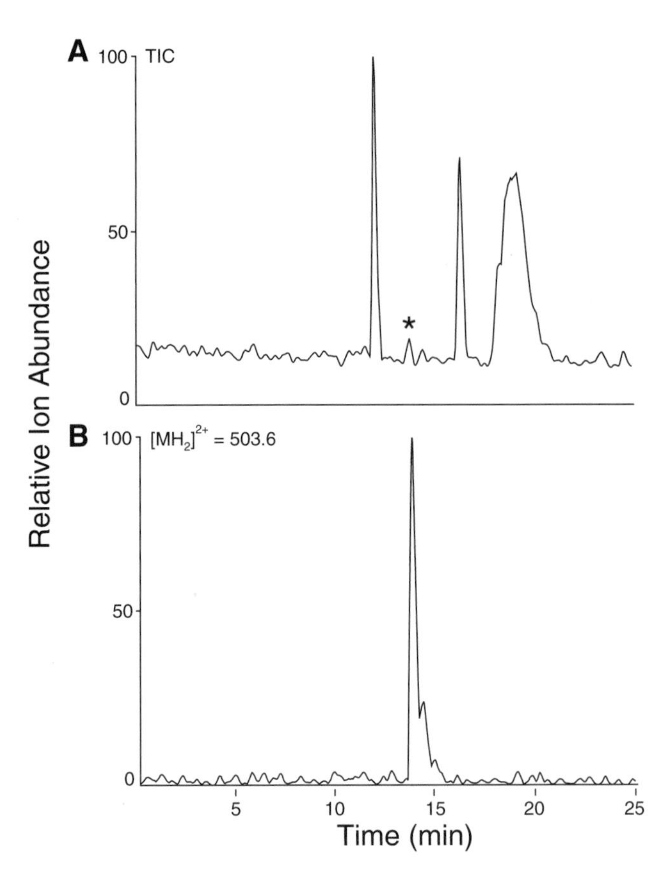

FIGURE 6.40 Membrane-preconcentration CE-MS analysis of HPLC peptide fraction from EL-4 lymphoma cells. (A) Total ion current of a 50 μL size aliquot of the diluted HPLC fraction. (B) Ion electropherogram of response marked with asterisk, shown to be a doubly charged ion at *m/z* 503.6. Reprinted from Tomlinson, A.J., Jameson, S., and Naylor, S., *J. Chromatogr. A,* 744, 273–278, 1996. With permission.

specific for the peptides presented by Class I MHC molecules originating from the proteasomal processing of the p53 proteins. Two 27-mer synthetic peptides, spanning residues 256-282 of human wild-type and mutant (R to H) p53, were incubated with 20S proteasomes purified from human T1 cells. The resulting peptides were extracted, separated, and tested for antigenic properties through a reconstitution of CTL lysis assay, by techniques similar to those described above. Identifications were made by LC/ESIMS followed by CID for sequence determination.[324] While the quality and quantity of most cleavage products generated by the 20S proteasome from the mutant and wild-type 27-mer polypeptides were comparable (Figure 6.43, A and B), the wild-type peptide digests contained >tenfold more of peptides 260-272 and 256-272 than did the digest of the mutant 27-mer peptide (Figure 6.43, C and D, panels 2 and 3). This occurred despite the equal efficiency (>95%) of the proteasome-induced degradation of both peptides (Figure 6.43, C and D, panel 4) and that comparable quantities were obtained for peptide 266-275, a peptide that did not involve cleavage between residues 272 and 273 (Figure 6.43, C and D, panel 1). It was concluded that the R to H mutation in p53 at residue 273 resulted in an abrogation of the proteasome-induced cleavage between residues 272 and 273. Experiments after additional HPLC fractionation revealed that CTL A2 264 sensitization could be achieved only with fractions from the wild-type 27-mer peptide (Figure 6.44, A and B). The identities of the observed CID fragments for these wild-type sequences were confirmed with synthetic peptides (Figure 6.44, C and D). It was concluded that the R to H mutation in the p53 hotspot at residue 273 was not only associated with malignant transformation

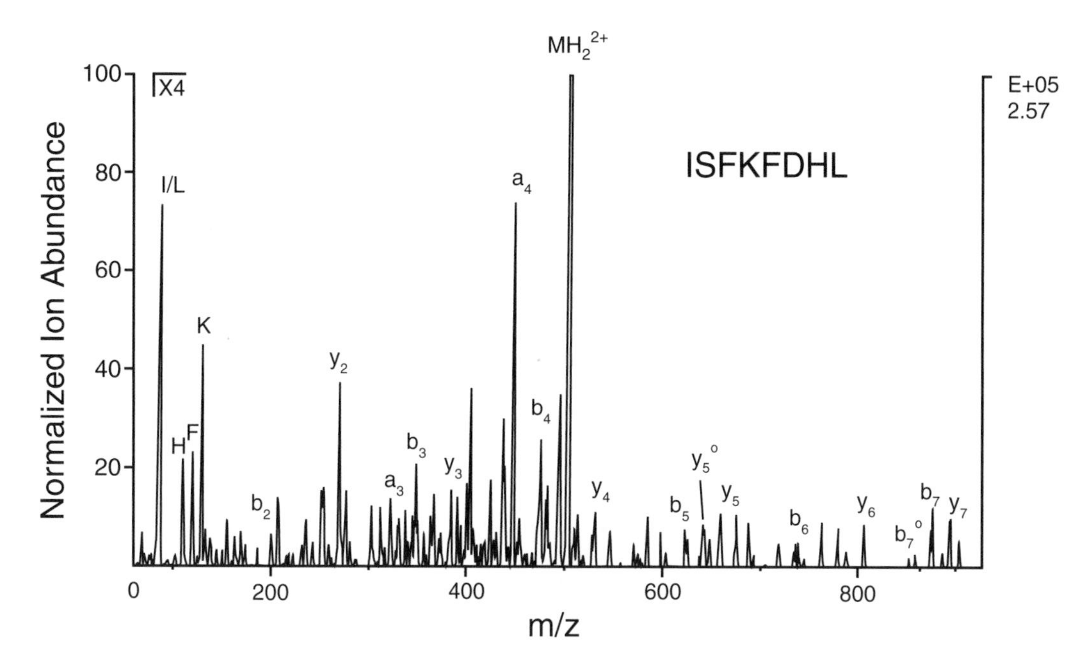

FIGURE 6.41 Product ion spectrum of the precursor ion $[M+2H]^{2+}$ shown in Figure 6.40. The peptide was identified as ISFKFDHL using the *SEQUEST* database. The peptide was derived from the protein F-actin. Reprinted from Tomlinson, A.J., Jameson, S., and Naylor, S., *J. Chromatogr. A,* 744, 273–278, 1996. With permission.

but also prevented the carboxy-terminal cleavage of the flanking peptide 264-272 by the 20S proteasome, thereby preventing the presentation of this peptide by HA*0201 molecules for CTL recognition. This finding has direct relevance to cancer vaccine design.[325]

6.1.6 Noncovalent Interactions and Protein Folding

Although the formation and structural features of noncovalent complexes in the gas phase may not accurately reflect those occurring in solution, species known to bind in solution have, in some cases, been observed as noncovalent complexes in the gas phase. Considerations that may be used to distinguish between specific and nonspecific noncovalent associations have been enumerated.[326] Most studies of this type have been carried out using ESIMS (reviewed in References 327–329).

6.1.6.1 Protein-Ligand Interactions

Ras

There are four human *ras* genes that code for a superfamily of *ras* proteins. These ~21 kDa proteins cycle between inactive GDP-bound and active GTP-bound states, and serve as molecular switches in the mitogenic signal transduction pathways that play crucial roles in regulation of cell growth and differentiation. The hydrolysis of GTP ligand to GDP is regulated by the GTPase activating protein. Unregulated activity of the *ras* proteins and/or mutations of the corresponding genes may result in situations in which the signal for growth is permanently "on" because the rate of hydrolysis of the GTP ligand to GDP is reduced to such an extent that the *ras* protein becomes locked in its active state. Mutated *ras* proteins are strongly associated with cancer, as evidenced by their presence in about 30% of tumor cell lines, including several human neoplasms.[330]

ESIMS/MS has been used to demonstrate the existence of the noncovalent complex of C-terminally truncated human *ras* (1-166), which had been expressed in *E. coli*, with GDP. Conformation

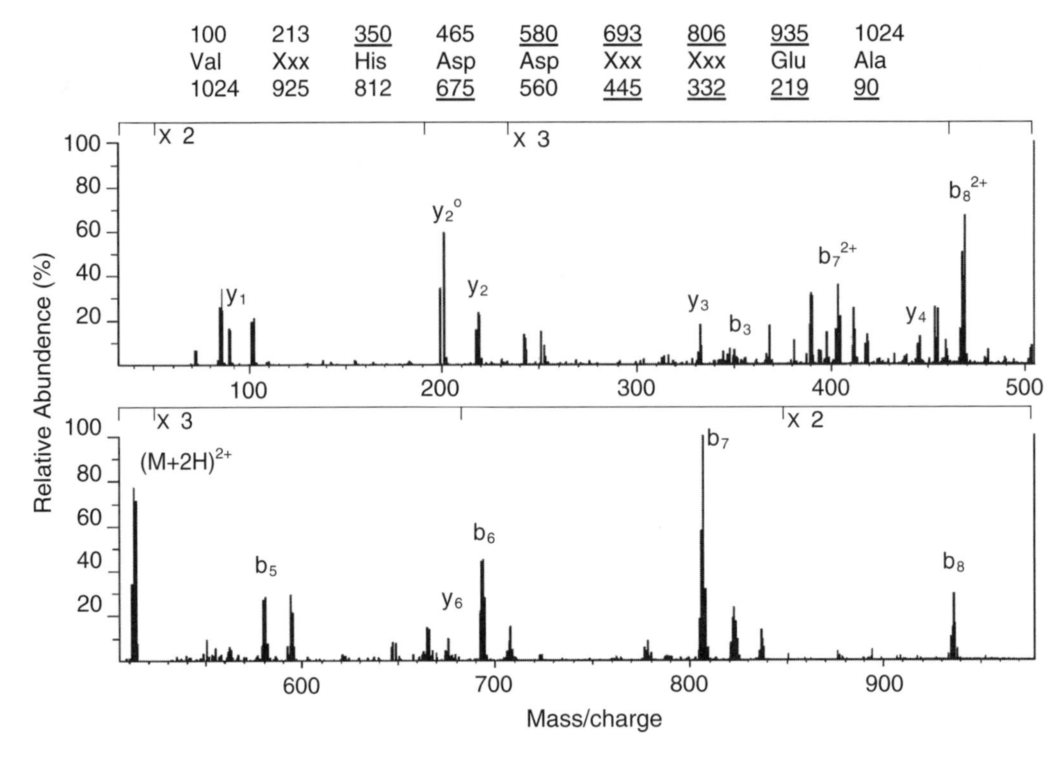

FIGURE 6.42 Sequencing a minor histocompatibility antigen HA-1 peptide by tandem mass spectrometry. CID mass spectrum of peptide candidate with *m/z* 513. The sequence was VXHDDXXEA (X = Ile or Leu). Only the synthetic peptide VLHDDLLEA coeluted with the naturally processed peptide *m/z* 513. Reprinted from den Haan, J.M. et al., *Science,* 279, 1054–1057, 1998. With permission.

of the complex strongly depended on the experimental conditions, including pH, the nature of the organic modifier, and temperature. Both the unfolded protein, 18,852±0.8 Da (theoretical MW of the apo-*ras* 18,853.3 Da), and the *ras*-GDP complex, 19,295 Da (demonstrating a 1:1 stochiometry), were detected at pH values in the 5.8 to 2.8 range. The *ras*-GTP complex is also affected by pH with only the complex being detected at pH 4.0 (Figure 6.45A). Lowering the pH to 3.4 initiated dissociation as evidenced by the appearance of free *ras* protein (Figure 6.45B), while a further decrease of the pH to 2.8 led to a nearly complete release of the ligand (Figure 6.45C). These results, combined with observations on the effects of organic modifiers on the ESI spectra,[331] led to several conclusions concerning the nature of these protein-ligand complexes.[331,332] [See also the detection and characterization of *ras* oncoprotein inhibitors by ESIMS (Section 5.3.4)].

6.1.6.2 Protein-Protein Interactions

It is now thought that a variety of cellular processes are performed and regulated by multi-protein complexes within which the proteins act cooperatively. There is, currently, considerable interest in the exploration of both the temporal and spatial aspects of protein-protein interactions. Techniques for the study of the disassembly of intact multiprotein complexes in the gas phase have been reviewed.[333] Some protein-protein complexes may also serve as therapeutic targets.

A convenient technique for identifying the proteinaceous partners with which a protein interacts starts with making a fusion protein that is immobilized on a solid support. This bound protein is then used as a "bait" around which the multi-protein complex associates. Proteins that interact nonspecifically are then washed away. The remaining protein complex is eluted, separated by one-dimensional electrophoresis, and analyzed by MS. Applications have included a study of the human

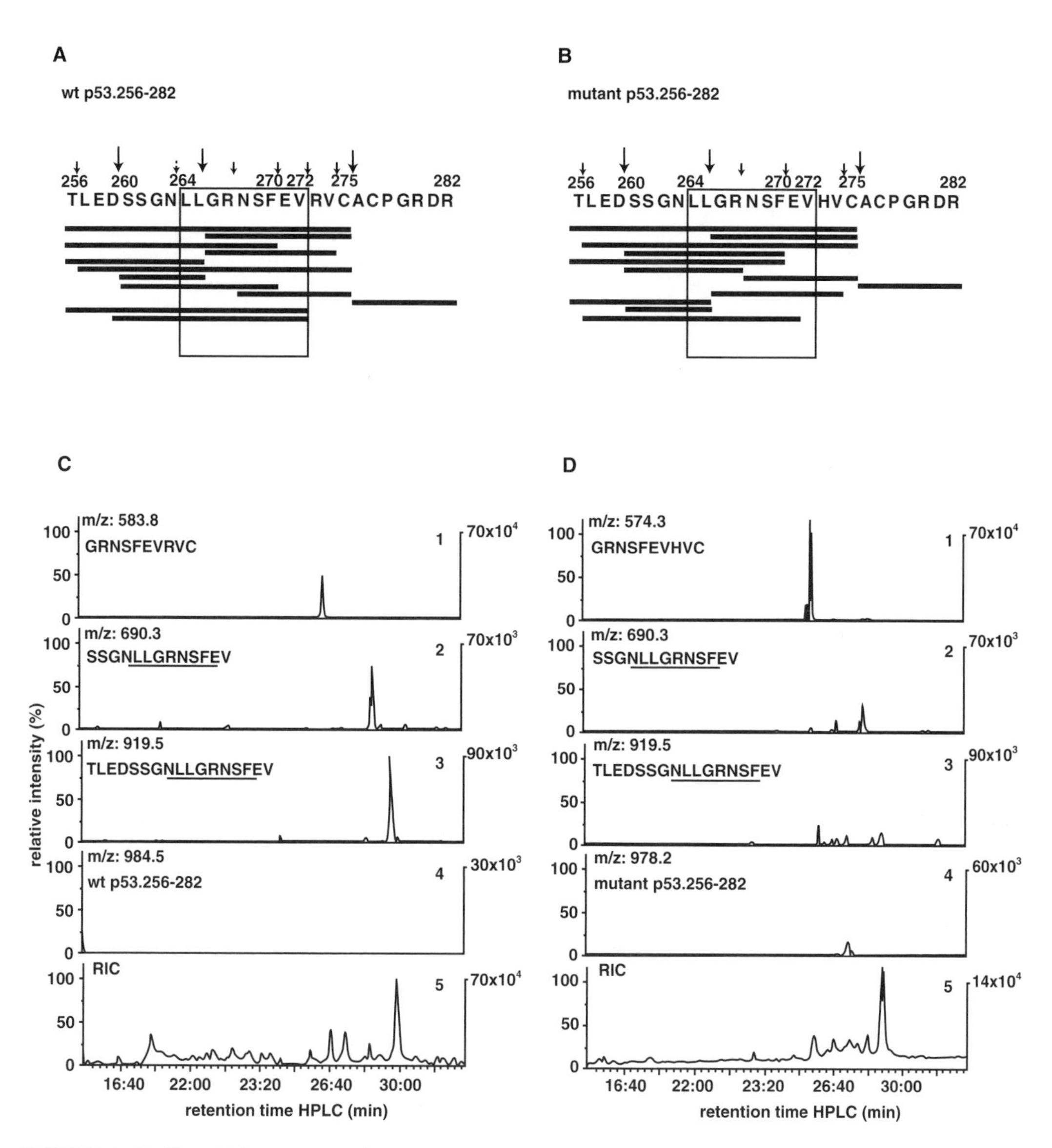

FIGURE 6.43 The p53 hotspot mutation at residue 273 (R to H) abrogates the proteasomal cleavage site between p53 residues 272 and 273. Bulk peptide products derived after 24 h from 20S proteasome-mediated degradation of synthetic WT (A and C) and mutant (273 R to H) (B and D) 27-mer polypeptides p53.256-282 covering the A*0201-restricted CTL epitope p53.264-272 (LLGRNSFEV) were separated by HPLC and abundant peptide products sequenced by MS/MS. (A and B, black bars), Cleavage products with signal intensities >threefold above background identified by MS and sequenced by MS/MS are shown in descending order. The amino acid sequence of WT (A) and mutant (B) peptide substrates with abundant (large arrows) and nonabundant (small arrows) cleavage sites are also presented. The small broken arrow (A) represent the theoretical NH_2-terminal cleavage site of the nonameric CTL epitope 264-272. The quantitative comparison of some of the relevant WT (C) and mutant (D) cleavage products is shown by the elution profiles of the doubly protonated peptide ions. Panel 1: peptide 266-275 (GRNSFEV-R/H-VC) represents a dominant product that interferes with formation of the 264-272 CTL epitope. Panels 2 and 3: the WT p53 peptides 260-272 (SSGNLLGRNSFEV) and 256-272 (TLEDSSGNLLGRNSFEV) use the C-terminal cleavage site of the minimal CTL epitope 264-272 between WT residues 272 and 273. Panel 4: uncleaved WT and mutant 27-mer substrate peptides left after proteasomal degradation and used to adjust the relative intensity scale. Panel 5: total ion current of the bulk proteasome degradation products. Reprinted from Theobald, M. et al., *J. Exp. Med.*, 188, 1017–1028, 1998. With permission.

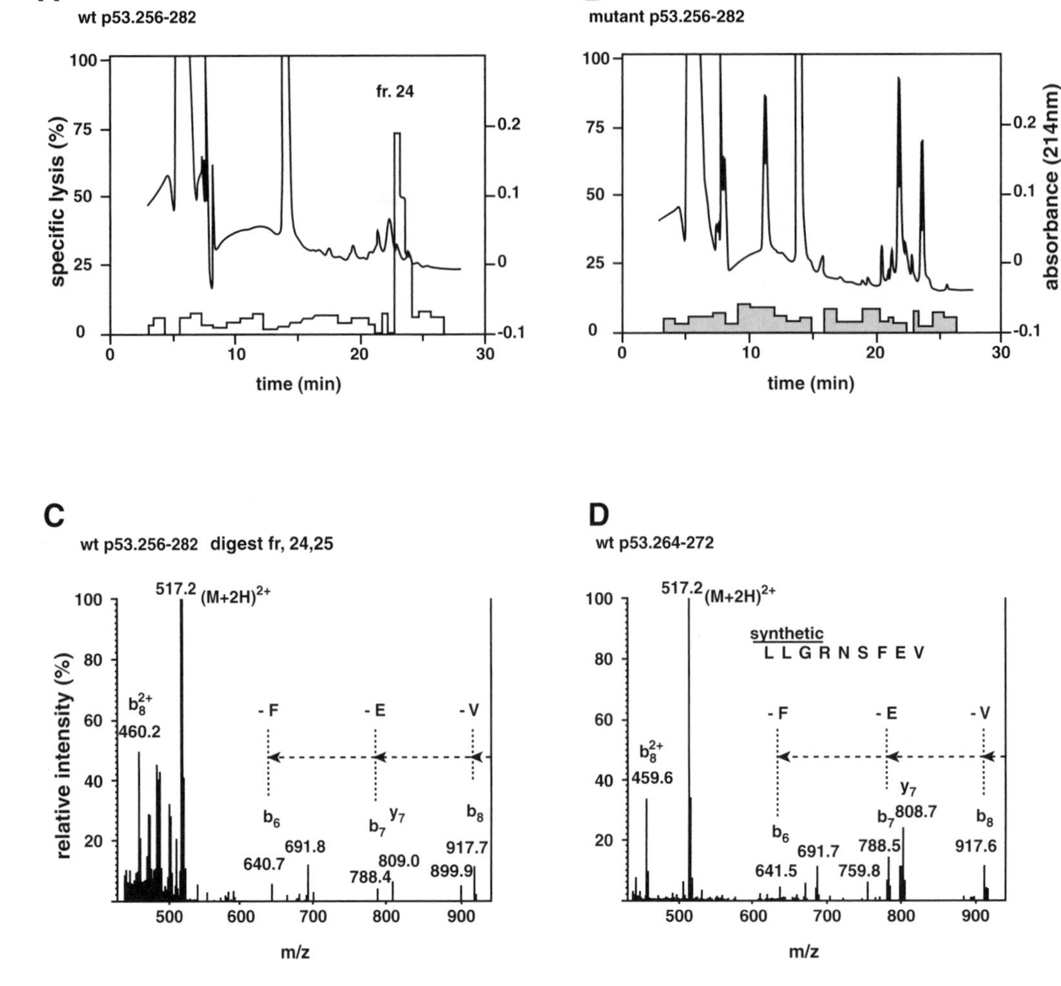

FIGURE 6.44 The p53.264-272 CTL epitope is generated by proteasomal degradation of the WT as opposed to the mutant 27-mer p53.256-282 polypeptide substrate. The HPLC profile (absorbance: –) and the specific lysis (shaded columns) of T2 targets sensitized with individual WT (A) and mutant (B) HPLC fractions are shown. CID fragments of *m/z* 517.2 (double protonated 264–272 9-mer peptide) and derived from the pooled WT fractions 24 and 25 (C) were compared with those of the synthetic 264–272 peptide (D). Fragments b8, b7, and b6, lacking the COOH-terminal residues V, E, and F, respectively, were detectable in both the pooled WT fractions (C) and the synthetic 9-mer peptide (D). Reprinted from Theobald, M. et al., *J. Exp. Med.,* 188, 1017–1028, 1998. With permission.

spliceosome in which 19 new factors were identified[130,334] and an investigation of proteins in mouse brain that would bind to immobilized profilin that resulted in the discovery of two new sets of proteins, one of which is a signaling molecule that regulates the actin cytoskeleton.[335]

A procedure called *tandem affinity purification* was recently developed for the purification of native protein-protein complexes. The method is based on preparation of a construct consisting of a tagged known protein component that is introduced into the host cell where it can interact with other cellular proteins to form a multi-protein complex. This complex is then recovered with the aid of an affinity column, designed to bind the initial protein tag. Upon recovery from the column, the components of the complex can be separated, digested, and analyzed by MALDI-TOFMS for

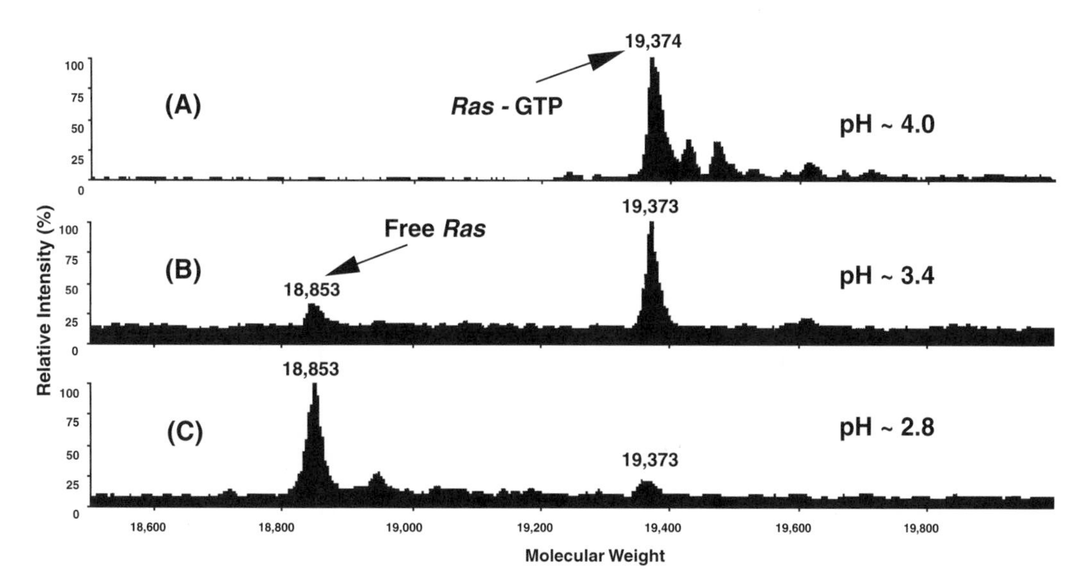

FIGURE 6.45 Deconvoluted ESI spectra of *ras*-GTP complex obtained from protein solution (no organic modifier) with (A) pH 4.0, (B) pH 3.4, (C) pH 2.8. Reprinted from Ganguly, A.K. et al., *Tetrahedron,* 49, 7985–7996, 1993. With permission.

possible identification through a database search. This technique was shown to be successful through the characterization of a hitherto unknown 3-protein complex in yeast.[8]

The genetic technique of the *in vivo yeast two-hybrid system* has emerged as a powerful tool to study protein-protein interactions.[5] A recent application that entailed analysis by MS concerned the formation of a heterocomplex between protein S100A4, a protein believed to participate in the events leading to metastasis, and another protein, S100A1. Both proteins belong to the S100 protein family, members of which are known to be multifunctional and involved in a wide variety of tumor- and other disease-related cellular processes. Direct evidence for the heterodimerization was obtained using ESI-FTICRMS. High resolution revealed the presence of at least three species, two of which represent the individual monomeric proteins. The most abundant signal corresponded to ions (carrying 11 charges) originating from the noncovalent heterodimeric complex of the two proteins. The measured mass was 24,982.00 Da vs. 24,981.61 Da calculated from the individual monomers. The biological function of the heterocomplex is not yet known.[336]

6.1.6.3 Protein-DNA Interactions

Noncovalent protein-DNA interactions are important because proteins serve as regulators of the expression of genetic information encoded in the nucleic acids. ESIMS has been used, in both the positive and negative ion modes, to provide data on the binding specificity and stochiometry of the association between polyanionic oligonucleotides and polycationic proteins.

Sequence-Specific Complexes of DNA and Transcription Factor PU.1

PU.1, a member of the ETS family of eukaryotic transciption factors, has been implicated in the regulation of gene expression during cell growth, transformation, and T-cell activation. An ESI-FTICRMS study of the binding of the 13.5 kDa DNA binding domain of transcription factor PU.1 (PU.1-DBD) to double-stranded DNA (dsDNA) oligonucleotides revealed that the formation of this protein-DNA complex and its stability in the gas phase result from sequence-specific interactions between the protein and DNA. Competitive binding studies using mixtures containing a slight excess of a mutant 19-bp dsDNA with respect to wild-type dsDNA revealed that PU.1 would form

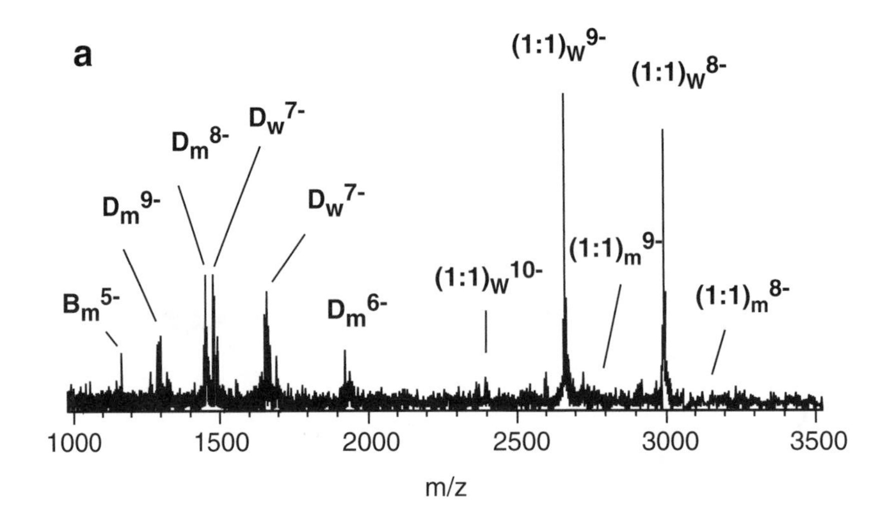

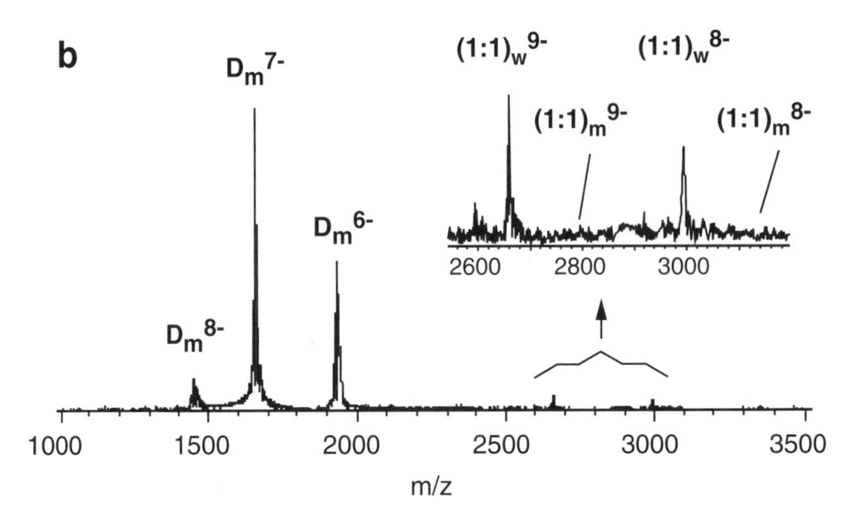

FIGURE 6.46 ESIMS spectra of PU.1-DNA binding domain (PU.1-DBD) mixtures of DNA in 10 μM NH_4OAc (pH 7.0) acquired on a FTICRMS showing specifics of wild-type DNA binding. Mixtures contain (a) PU.1-DBD (5 μM), 17-bp DNA (D_W, 15 μM), and 19-bp DNA (D_M, 20 μM); (b) PU.1-DBD (3 μM), 17-bp DNA (10 μM), and 19-bp DNA (200 μM). Reprinted from Cheng, X. et al., *Anal. Biochem.*, 239, 25–34, 1996. With permission.

a complex (with 1:1 stochiometry) only with the wild-type 17-bp DNA sequence (Figure 6.46a). There was no evidence of the protein-mutant DNA complex even for a 20-fold excess of the 19 bp dsDNA, although the excess unreacted mutant DNA was detected in the mass spectrum at lower *m/z*. The difference in binding affinity between the CGAA core of the 17-bp wild-type and the GCTA core of the 19-bp mutant dsDNA was >200-fold.[337]

Identification of Heterogeneous Nuclear Ribonucleoprotein Particle 2 (hnRNP 2) as the Product of the TLS/FUS Gene

Newly made RNA is quickly condensed into a string of closely spaced protein-containing particles (hnRNP) including at least eight different proteins. In an investigation of RNA-protein complexes formed in HeLa cell nuclear extracts, one of the separated, ATP-independent fractions contained a peak (designated as the H peak) that contained RNA-protein complexes. Four proteins, identified by nano-ESIMS, based on peptide mass and database searches, corresponded to products of previously characterized genes, HnRNP A1, B1, C, and G. Peptides from an additional protein spot were also analyzed using the sequence tag method.[94] For each peptide, the "nested set" of fragments obtained by CID allows the peptide sequence to be deduced by their respective mass differences that correspond to single amino acid residues. For example, a VSF sequence tag of one of the peptides led to the identification of its sequence as GEATVSFDDPPSAK (Figure 6.47). With the aid of two more sequences identified in a similar manner, the peptide was shown by the *GenPept* and *SWISS-PROT* databases to match the sequence of a gene known as TLS or FUS. Although the function of this gene is not known, the encoded protein (hnRNP2) has a motif sequence characteristic of RNA binding proteins and common to several hnRNP proteins. Also, it has been shown that TLS/FUS is a fusion partner in chromosomal translocations associated with human myxoid liposarcoma and a subset of myeloid leukemias.[338] Similar methodology has been used to identify the protein components of several other spliceosome-related protein complexes.[339]

6.1.6.4 Protein Conformation and Folding

Disulfide Bridges

Cysteine has a unique role among the 20 amino acids that comprise proteins. In the reduced form of cysteine, the free SH groups contribute to the biological function of the protein, e.g., serving as the active site for enzyme catalysis or as the chelating site for metal ions. In its oxidized form, cysteines help to stabilize proteins by linking two cysteines via the formation of S-S bonds. Such disulfide bonds are formed when the side-chains of two cysteine residues that are close together in the folded protein structure, react with each other, losing the hydrogen atom from each of their sulfhydryl (thiol) groups in an oxidation reaction. Disulfide bonds are relatively strong covalent bonds that reinforce the fragile (to changes in temperature and pH) tertiary structures of proteins. These structural elements of proteins are formed by the folding of polypeptides and are dependent on generation of hydrogen bonds as well as on hydrophobic interactions.

The strategy for locating disulfide bridges, and the corresponding cysteines, involves the enzymatic or chemical digestion of the protein, followed by ESIMS analysis of the resulting peptide mixture before and after the reduction of the disulfide bridges. The cysteine residues responsible for the disulfide bridges can be determined by comparing the two spectra for the disappearance of the peptide containing the disulfide bridge concurrent with the appearance of two new peptides derived from the cleavage of the S-S bond. The number of cysteines, disulfide bridges, and free-SH can be determined by measuring the molecular mass of the native protein, of the protein in which all original -SH groups are alkylated with $I-CH_2CONH_2$, and of the protein in which the S-S bonds are first reduced, followed by alkylating all -SH groups. Current mass spectrometric techniques for the characterization of disulfide bonds by various ESI and MALDI techniques, with illustrative applications, have been reviewed from different perspectives.[340–343]

The Roles of Zinc Finger 2 and Disulfide Bonds in Estrogen Receptor DNA Binding

In about one-third of estrogen receptor (ER)-positive breast cancer patients, extracted tumor ER is unable to bind to its cognate DNA estrogen response element. When not bound to DNA, the full length (67 kDa) ER and its 11 kDa DNA-binding domain (ER-DBD) suffer loss of structure and function upon oxidation by hydrogen peroxide or diamide, the latter being a reagent that oxidizes cysteine to cystine. In one set of experiments, the oxidation of ER-DBD, which contains nine cysteines and five methionines, was monitored by ESIMS/MS. Treatment with hydrogen proxide for 25 min removed four hydrogens and added zero, one, or two oxygens (Figure 6.48A). When the peroxide treatment was extended for 24 h, six hydrogens were removed and one to five oxygens were added (Figure 6.48B). When the 25-min period of oxidation was followed by addition of the thiol-reducing agent dithiothreitol (DTT), the disulfide bonds were reduced but the methionine sulfoxides survived unscathed (Figure 6.48C). Treatment with a small amount of diamide resulted in partial oxidation by the removal of four hydrogen atoms (Figure 6.48D), whereas increased

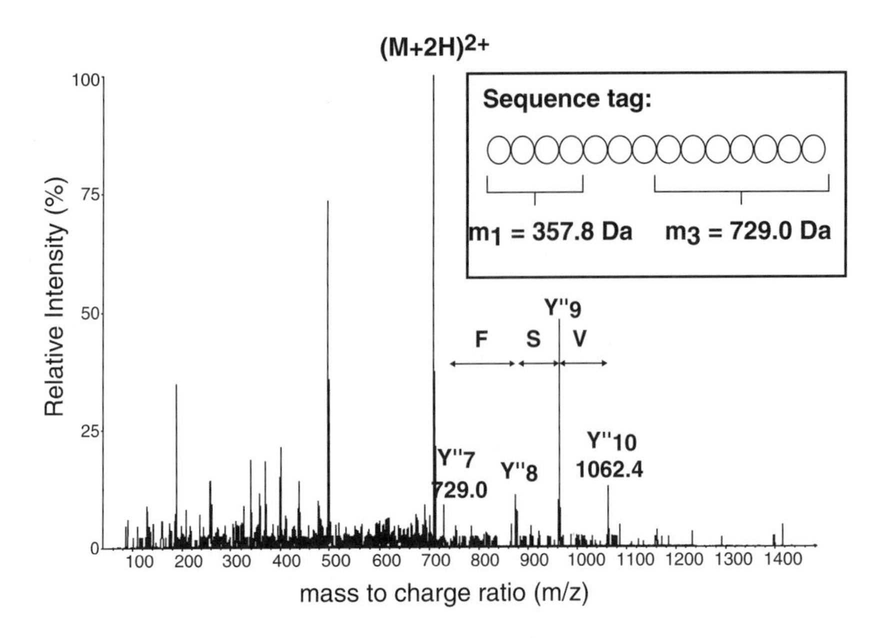

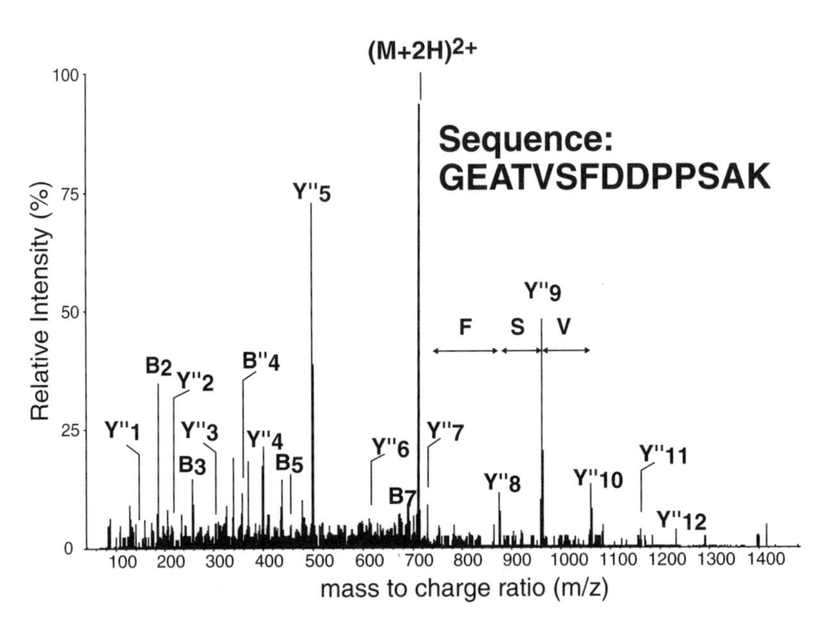

FIGURE 6.47 Identification of the protein encoded by a gene known as TLS. Upper panel, product ion spectrum of the $[M+2H]^{2+}$ ion, *m/z* 710.9, selected in the first quadrupole and then fragmented. The partial peptide sequence and its position within the peptide (peptide sequence tag) was used for database searching. m_1 is calculated from the Y″10 ion and the intact peptide mass giving the search string (729.0)VSF(106.24) and the intact peptide mass of 1,419.8 Da for the database searches. Lower panel, complete assignment of sequence ions in the fragment spectrum. The sequence of a tryptic peptide from TLS was confirmed by the fragmentation pattern after it had been retrieved by the database. Reprinted from Calvio, G. et al., *RNA,* 1, 724–733, 1995. With permission.

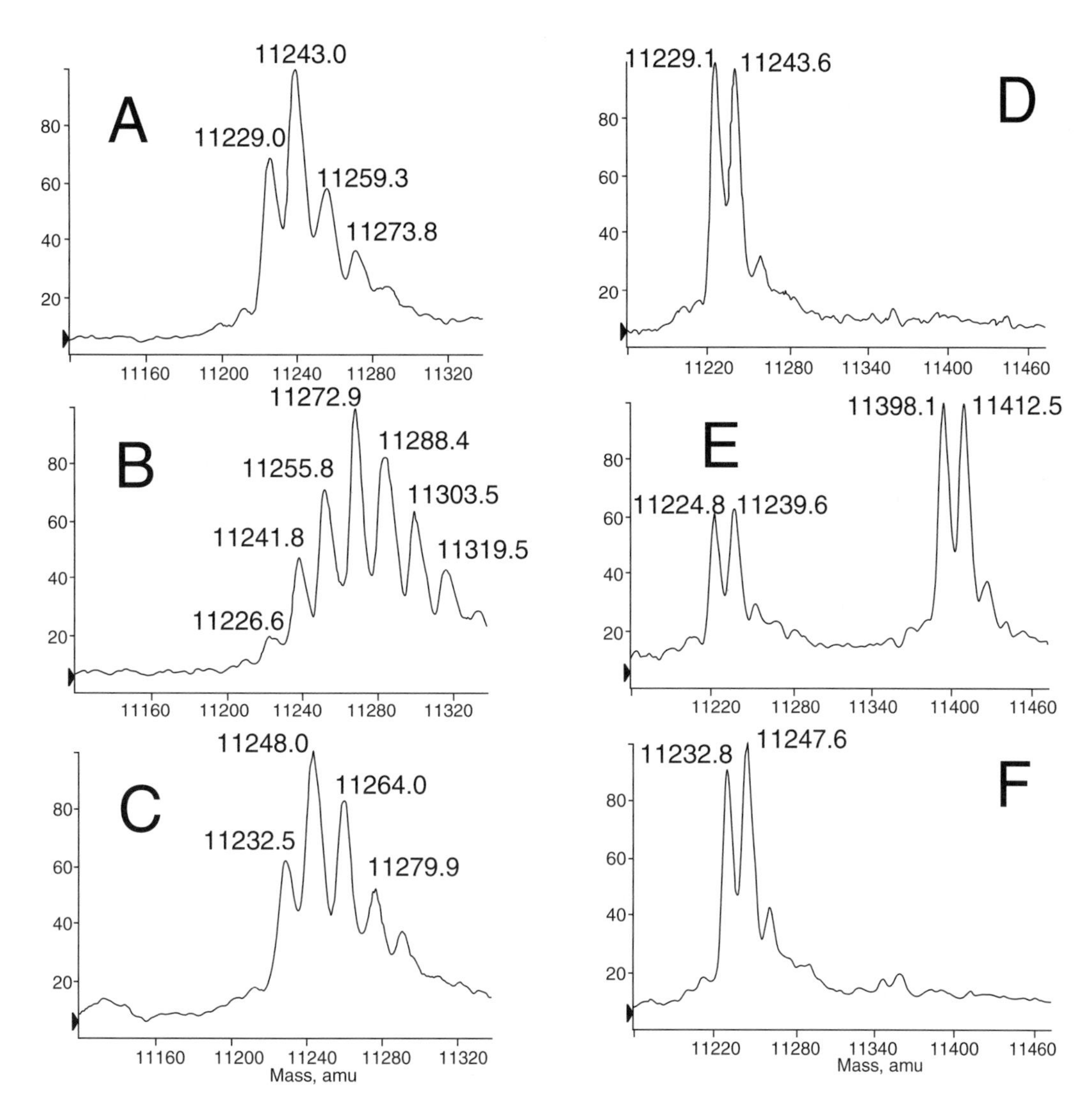

FIGURE 6.48 Deconvolution of LC/MS data from oxidation of estrogen receptor DNA binding domain (ER-DBD) by H_2O_2 or diamide. (A) After oxidation with 5 mM H_2O_2 for 25 min (loss of four Hs); (B) After oxidation with 5 mM H_2O_2 overnight (loss of six H); (C) After oxidation with 5 mM H_2O_2 for 25 min then reduction with 50 mM dithiothreitol (DTT) (fully reduced); (D) After oxidation with 5 mM diamide for 30 min (loss of four H); (E) After oxidation with 25 mM diamide for 30 min (loss of eight H plus addition of diamide); (F) After oxidation with 25 mM diamide for 30 min then reduction with 125 mM DTT (fully reduced). Reprinted from Whittal, R.M. et al., *Biochemistry,* 39, 8406–8417, 2001. With permission.

diamide concentration removed eight hydrogens and produced a new species with covalently linked diamide (Figure 6.48E). Further treatment with DTT reduced the cystines back to cysteine and removed the diamide adduct (Figure 6.48F). Further experiments with ESI-QTOFMS (and other techniques) established that while zinc inhibits thiol oxidation of the DNA-binding finger, it leaves intact the finger supporting the flexible dimerization loop. The stabilization of both a rigid DNA-binding recognition helix and a flexible helix-supported dimerization loop is accomplished cooperatively by the two $(Cys)_4$-liganded zinc finger motifs of ER-DBD. Upon oxidation, the loss of the DNA-binding function may result from the fact that the asymmetric zinc fingers become susceptible to disulfide formation which prevents the dimerization essential for DNA binding.[344]

Conformation

The concept of conformation refers to the spatial arrangement of the atoms in a molecule, e.g., the three-dimensional (3-D) shape of a protein or other macromolecule. Proteins roll off the protein-making ribosome assembly line as a linear string of amino acids. To perform their biological functions, proteins must first fold into 3-D structures that are unique to each protein. Proteins find these native states in a matter of milliseconds to seconds in bacteria or seconds to minutes for eukaryotes. These native state conformations are normally long-lived. It appears that the 3-D structures are inherently encoded in the sequence of the amino acids and that assumption of the protein's conformation is determined by energetic considerations, i.e., that conformations are aimed at minimizing free energy. A protein's conformation often changes slightly when the protein interacts with other proteins inside a cell and such changes may be critical to the normal functioning of the protein. When a protein is *denatured*, such as by interaction with solvents, the noncovalent interactions necessary to hold the folded chain together are disrupted. However, removal of the denaturing agent often allows for spontaneous *renaturing* of the protein to its native state. There is evidence that proteins can exist in intermediate forms in addition to their native folded and fully unfolded states. The *de novo* folding of nascent and translocating polypeptides *in vivo* is often controlled and/or catalyzed by proteins called molecular *chaperons* which bind to partly folded chains and help them fold along the most energetically favorable pathways as well as preventing the newly synthesized chains from associating with the wrong partners inside the crowded conditions of the cytoplasm (reviewed in Reference 345). The two major classes of ATP-dependent chaperons are the Hsp70s of the heat shock protein family and the chaperonins, such as GroEL (see later) (reviewed in Reference 346).

Strategies

The two major MS strategies for the elucidation of protein conformation are based on the charge distribution in ESI spectra and on hydrogen/deuterium (H/D) exchange. The first strategy works on the assumption that the net charge on a protein depends on its conformation. This *charge distribution* observed in the ESI spectra can be used to probe the protein's conformational state in solution. For example, heat-induced conformation changes have been detected by ESIMS.[347] In addition, experimental conditions, ranging from ESI spray parameters to the pH of the solution, influence the net charge on a protein in solution, which may be related to the conformational state of that protein.[326] Although this charge state information, a product of ESI and the conformation of the protein, is unique and easily obtainable, the all-important subtle structural alterations that result from the biological activities associated with the behavior of a protein may be difficult to monitor by this approach.

When a polypeptide is exposed to a deuterated solvent, protons attached to backbone amide functionalities begin to undergo *isotopic exchange* for deuterons (reviewed in Reference 348). The time-dependent increase in the mass of the polypeptide can be monitored by ESIMS.[349] While surface protons exposed to the deuterated solvent exchange rapidly, those protons buried in the hydrophobic core or participate in the hydrogen bonding of the α-helices and β-sheets, or are otherwise protected from exposure by the high-order structure of folded proteins, will exchange at rates that are orders of magnitude slower (reviewed in Reference 350). Monitoring H/D exchange by ESI has been applied to investigating the stability of global and regional structures[327] as well as folding pathways. Lysozyme was used to illustrate the complementary nature of the information obtained by MS and NMR.[351]

MALDI-TOFMS has also been applied to study conformational changes of proteins. Modifications of conventional conditions included adjustment of the pH of the matrix to optimize amide H/D exchange and rapid drying of chilled MALDI targets to minimize the loss of deuterons from the amide groups.[352,353] One advantage of the MALDI technique is the direct identification of sites susceptible to denaturation. This was illustrated using cytochrome C.[354]

The structural and dynamic properties of protein ions undergoing H/D exchange may also be probed with an ESI-FTICRMS (at 10^{-7} torr) by determining the accurate masses of the multiply charged ions. The approach has been illustrated by comparing the gas phase compactness of S-S cross-linked RNase A (13.7 kDa) with VP-RNase, a derivative formed by cleavage of the four disulfide bonds and alkylation of the resulting -SH groups with 4-vinylpyridine. The RNase exchanged 35 hydrogen atoms, the VP-RNase 135 hydrogens.[355]

GroEL and Protein Conformation

Although the 3-D structures of proteins appear to be inherently encoded for in their sequences, there is evidence that other factors also contribute, e.g., protein folding *in vivo* is often controlled and/or catalyzed by proteins called *chaperons*. The *E. coli* protein GroEL, required for bacteriophage assembly, is a barrel-shaped homo-tetradecamer of 57,197 Da subunits.[356] This chaperonin assists (with its co-protein GroES that participates in ATP-dependent reactions) in the "correct" folding of proteins by repeatedly binding and releasing unfolded or partially folded proteins. GroEL also prevents the formation of non-native conformations by sequestering unincorporated polypeptides during each binding interval.[357]

ESI has proved useful for probing the confirmations of GroEL-bound substrate proteins (reviewed in Reference 358). It was possible to induce the dissociation of the GroEL-substrate macromolecular complex in the collision cell of an ESI-QTOF instrument. The observed peaks at approximately *m/z* 10,000, representing charge states of 80+ to 88+ acquired during the ESI process, permitted calculation of the molecular mass of the intact GroEl complex to be 803,742 ± 616 Da (Figure 6.49). Dividing the mass of the complex by that of the monomer suggested that there were 14 subunits in this gas-phase assembly. Additional experiments at different collision energies yielded a series of ions around *m/z* 6000 with intermediate charge states, thus providing direct confirmation of the initial data that the sample consisted of 14 noncovalently bound subunits, while also indicating that these units were arranged in two heptameric rings. It was concluded that this strategy "promotes mass spectrometry from the realm of small protein and modest noncovalent complexes into the province of macromolecular assemblies with molecular masses approaching 1,000,000 Da."[359]

Folding Pathway of Recombinant Human Macrophage-Colony Stimulating Factor β (rhM-CSF)

Previous studies by FABMS explored the locations of the nine disulfide bonds and some structure-function aspects of the homodimeric rhM-CSF β.[360–362] In one study, the folding intermediates of rhM-CSF were studied with the aid of melarsen oxide [MEL, *p*-(4,6-diamino-1,3,5-triazin-2-yl)aminophenylarsenous acid], which selectively bridges spatially neighboring bis-cystenyl residues in reduced proteins and can be applied as a trapping agent in both denaturing unfolding and redox unfolding reactions. The pure dimer was fully reduced and denatured by incubation with tris-(2-carboxyethyl)phosphine (TCEP). To study redox refolding, the monomer was treated with L-γ-glutamyl-L-cysteine-glycine and bis (L-γ-glutamyl)-cysteinyl-bisglycine, and aliquots of the sample were blocked with MEL at various time intervals. After various biochemical steps, the folding intermediates were desalted by ultrafiltration, washed, and the aliquots analyzed by SDS-PAGE and ESIMS. To study reductive unfolding, the homodimer was treated with TCEP and desalted. To study denaturing unfolding, the dimer was dissolved in urea and guanidinium hydrochloride, incubated with MEL, treated with TCEP, followed by the addition of 4-vinylpyridine. The protein-containing fractions were analyzed by MS.

When the *in vitro* redox refolding of rhM-CSF β was monitored by ESIMS, abundant signals were observed at the beginning (0 h) for the completely MEL-modified rhM-CSF β monomer. The evenly distributed ion series centered around the 15+ charge state of the molecular ion (Figure 6.50A), indicating that the nine cysteine residues per analyte had added a total of four MEL groups (M·4MEL, Δm = 276 Da/MEL). After exposing the monomer to renaturation conditions

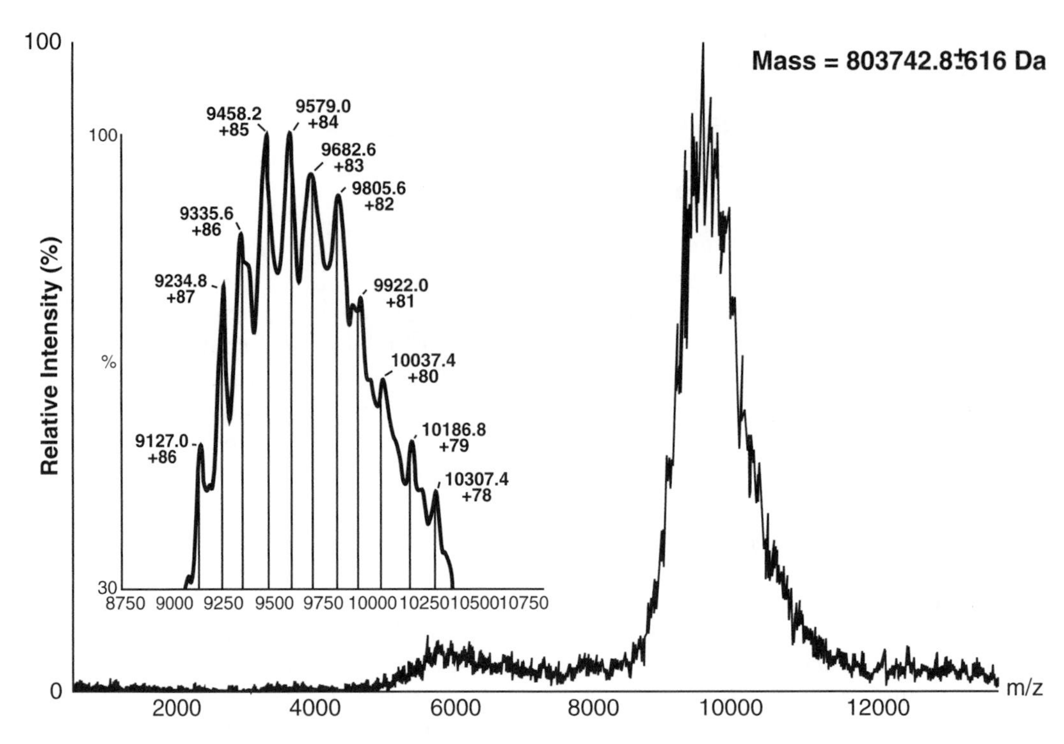

FIGURE 6.49 Nanoflow ESI mass spectrum obtained from a 1 μM solution of GroEL at pH 5.0. The spectrum represents the raw data after minimal smoothing and (insert) with higher smoothing and an expansion from *m/z* 8,750 to 10,750 with a threshold of 30% intensity. The positions and values of the centroided data used in the mass determination are shown in the expansion. The mass, 803,742.8 ± 616 Da, was calculated from the 11 charge states labeled in the spectrum. The theoretical mass of the GroEL assembly is 800,758 Da. The difference in mass between the measured and theoretical masses could be attributed either to peptide ligands bound within the cavity of GroEL or to the presence of counterions. Reprinted from Rostom, A.A. and Robinson, C.V., *J. Am. Chem. Soc.,* 121, 4718–4719, 1999. With permission.

for 6 and 42 hr), an increasingly larger number of ions representing a doubly MEL-modified species appeared (M·2MEL) with its own charge pattern, centering around the 26+ charged molecular ion (Figure 6.50, B and C). The charge distributions for the monomer remained as expected even around the 15+ charged state, suggesting the M·2MEL species was an intermediate in the refolding pathway. The major homodimeric rhM-CSF β appeared to be more compact, with the most abundant charge state at 23+ (N, Figure 6.50D).

In another set of experiments, molecular mass measurements established the addition of four carboxamidomethyl groups to the dimeric folding intermediate of rhM-CSF β. Mass spectrometric peptide mapping of the tryptic digest of this derivative established the location of the predominantly modified cysteine residues at C-157 and C-159. These residues are involved in the intermolecular disulfide bonds that hold the native rhM-CSF β dimer together.[361,363] Other experiments included ESIMS monitoring of the reductive unfolding and the denaturing unfolding of rhM-CSF β. It was concluded that the advantages of the trapping strategy included the ability to monitor the general unfolding behavior of rhM-CSF β and the uncovering of the modification sites in the folding intermediates.[364]

6.2 LIPIDS

Lipids are a heterogeneous group of compounds related more by their physical properties, such as solubility in nonpolar solvents and insolubility in water, than by chemical properties. Simple lipids

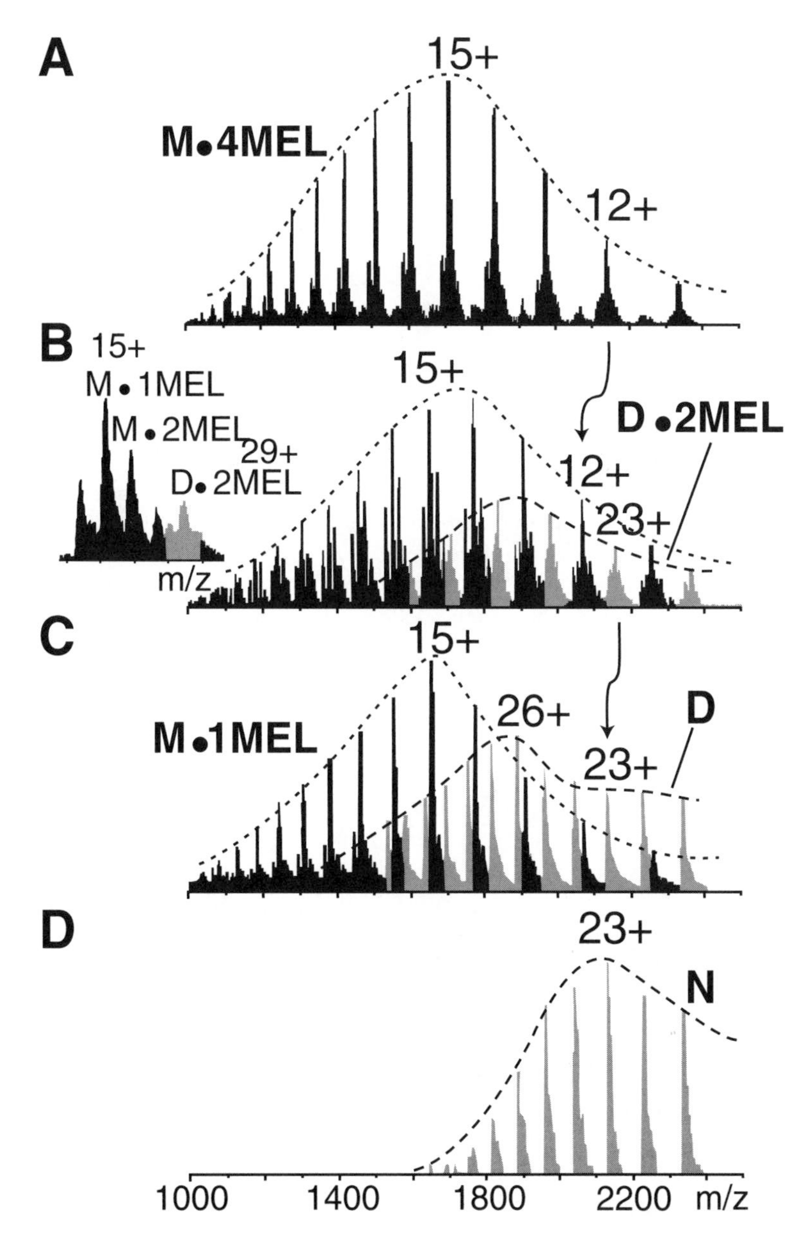

FIGURE 6.50 Monitoring of redox *in vitro* refolding of recombinant human macrophage-colony stimulating factor β (rhM-CSF β). Electrospray spectra of: A-C, melarsen oxide (MEL)-trapped folding intermediates after 0, 6, and 42 h refolding and D, mature homodimer. M•4MEL, monomeric intermedite with four MEL groups; D•2MEL, dimeric intermediate with two MEL groups; N, mature homodimer. Charge numbers denote $[M+nH]^{n+}$ ions. The different charge states of the intermediates are indicated by the dashed (dimer) and dotted (monomer) lines around the multiply charged ion series, respectively. Ion signals for the dimeric derivatives are drawn in gray. Reprinted from Happersberger, H.P. et al., *Proteins,* 2 (suppl), 50–62, 1998. With permission.

are esters of fatty acids with alcohols, e.g., fats and waxes. Precursor and derived members of the lipids family include such divergent compound classes as fatty acids, glycerol, steroids, sterols, and hormones. There are also three major classes of complex lipids that contain various other groups in addition to the constituent fatty acids and alcohols: (a) p*hospholipids* contain a phosphoric

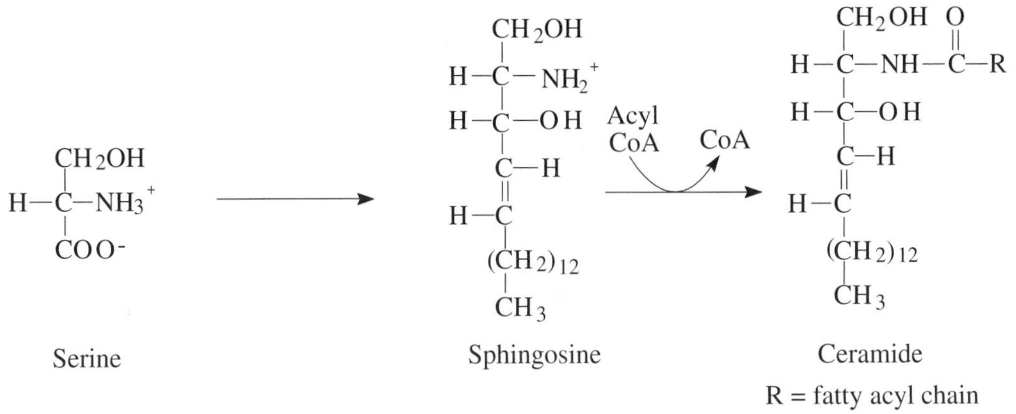

CH_2OH
H—C—NH_3^+
COO^-

Serine

CH_2OH
H—C—NH_2^+
H—C—OH
C—H
H—C
$(CH_2)_{12}$
CH_3

Sphingosine

Acyl CoA → CoA

CH_2OH O
H—C—NH—C—R
H—C—OH
C—H
H—C
$(CH_2)_{12}$
CH_3

Ceramide
R = fatty acyl chain

FIGURE 6.51 Chemical formulas of serine, sphingosine, and the ceramides.

acid residue, often with associated nitrogen-containing bases, linked to an alcohol, including sphingosine (*sphingophospholipids*) or glycerol (*glycerophospholipids);* (b) *glycolipids* contain fatty acids, sphingosine, and a carbohydrate (Section 6.4.4); and (c) *lipoproteins* in which lipids are combined with a wide variety of proteins. Other, less common complex lipids include the sulfolipids and aminolipids.

There have been several general reviews of earlier techniques for the analysis of lipids, including LC/CIMS with moving belt interface, CIMS using a direct liquid inlet, and TSPMS.[365] Other reviews have targeted the use of FABMS for the analysis of specific lipid classes including phospholipids,[366] long-chain carboxylic acids,[367] and lipoic acids.[368] A recent review concerns applications for non-volatile lipids carried out with GC/EIMS and LC/ESIMS, often with CID.[369] Although MALDIMS is well suited to the analysis of lipids, in part because of their excellent solubility in organic solvents, only a few applications have been published, e.g., using MALDI-FTICRMS for the structural characterization of phospholipids[370] and MALDI-TOFMS for the analysis of diacylglycerols, phosphatidylcholines, and (poly)phosphoinositides in neutrophilic granulocytes.[371]

6.2.1 Ceramides

Sphingosine is an 18-carbon amino alcohol, derived from serine, that has OH groups at C1 and C3, an amino group at C2, a C4,5 double bond, and a long hydrocarbon tail. When an additional hydrophobic hydrocarbon side-chain (C_{16} to C_{24}) is attached to the C2 amino group, this forms one group of the *ceramides* (Figure 6.51). The dihydroceramides are similarly derived from sphinganine (dihydrosphingosine) and both sets of ceramides are subject to modification of the sphingoid head group. The ceramides are synthesized in the endoplasmic reticulum. The main function of ceramides is to serve as part of the ubiquitous sphingolipid backbone of cell membranes. Ceramides have also been considered as *second messengers* in intracellular signaling pathways that control apoptosis, cell propagation and differentiation, cytokine synthesis, and other transduction pathways. However, this function has been questioned based on MS studies of ceramide concentrations during apoptosis.[372,373]

Mass Spectrometric Techniques

Because of the importance of ceramides, numerous techniques have been developed for their identification as well as quantification (reviewed in Reference 374). The GC/EIMS analysis of ceramides is sensitive with nM detection limits, but it requires relatively large quantities of sample along with extensive sample preparation and derivatization of the analytes.[375] However, GC/MS is still used to assess the *in vivo* synthesis of a number of volatile eicosanoids in humans[376] (Section 6.2.2).

The FAB spectra of the metal ion adducts provide structure-specific information, including the lengths of the sphingoid base, fatty acid chains and various substitutions.[377]

The analysis of ceramides and dihydroceramides by conventional ESIMS has been inadequate because of large sample requirements and insufficiently specific sample clean-up procedures that result in the generation of highly complex spectra in which most of the ions arise from interfering constituents.[378] This has led to investigation of CID and the determination that there are characteristic and dominant product ions for the sphingosine and sphinganine head groups. These ions are independent of the length of the acyl chains in the 2 to 24 carbon range, thus they can be used to detect ceramides and dihydroceramides in complex mixtures. The CID spectra also differ significantly depending on the nature of modifications to the head sphingoid group, e.g., hydroxylation and double bonds. Precursor ion scanning of ceramide standards and mixtures has led to strategies that distinguish modified ceramides from dihydroceramides. A nomenclature has been established for precursor ion cleavages, obtained by negative ion ESIMS/MS, for the identification of both the acyl and long-chain sphingoid bases.[379]

A technique developed for the quantification of ceramides in cell extracts (~10^7 cells) involved lysis followed by lipid extraction (i.s.: N-acetylsphingosine, also referred to as C2:0 ceramide, a nonnatural compound) and ESIMS analysis with CID. The LOD was ~25 nM. The viability of the technique for detecting changes in the concentrations of ceramides present in biological mixtures was established with serial dilution experiments.[380] Another ESIMS technique using CID permitted simultaneous separation and quantification of sphingolipid metabolites, including ceramides, sphingomyelin, sphingosine, sphingosylphosphorylcholine, and dimethyl-sphingosine in samples containing 0.1 to 100 ng/10^6 cells. The analytes, extracted with chloroform-methanol (nearly 100% recovery) were separated on a short HPLC column and the effluent was introduced directly into the ESI source of a QqQ instrument. Protonated molecules were the predominant ions and the CID spectra were simple with few signals. After selecting appropriate precursor and product ions for the individual analytes, the instrument was operated in SRM mode (i.s.: synthetic N10-heptadecenyl-sphyngosine and C2:0 ceramide). It was claimed that 50 samples could be analyzed daily.[381]

APCI has also been evaluated and a study of the full scan (*m/z* 100 to 1000) spectra of standards revealed that ceramides can be characterized unambiguously based on their characteristic molecular and sphingosine/sphinganine-derived ions, with appropriately selected cone (skimmer) voltages. Intact, underivatized ceramides could be identified by observing the protonated molecule and/or the Na or K adduct ions using a cone voltage set at 15 V. In-source CID at elevated cone voltages led to increased fragmentation, including loss of two molecules of water and the amidic acyl side-chain. One of the intense fragment ions (*m/z* 264) could be used to detect any sphingosine-based ceramide with the identity of the fatty acid moiety being determined from the mass difference between the protonated molecule and the *m/z* 264 ion of the sphingosine fragment.[382] In the case of the sphinganine-based ceramides, the characteristic product ion occurs, as expected, two mass units higher at *m/z* 266. The appearance of this ion has been used for the sensitive detection of the sphinganine-based class of ceramides in cell extracts.[380,381] The availability of highly characteristic product ions means that precursor ion scanning is readily accomplished and has been found to be a useful method for the identification of ceramide species in biological samples, e.g., the specific ceramide molecular species generated in T-cells.[383]

Ceramides in Leukemic HL-60 Cells

The CID technique described above[382] has been applied to a study of sphingolipid metabolites in HL-60 leukemic cells. Eleven endogenous metabolites were detected and quantified (Figure 6.52). The basal levels of the metabolites ranged from 2.5 pg/10^6 cells for dimethyl-sphyngosine to 20 μg/10^6 cells for C_{20}-sphingomyelin. By incubating cells with sphingomyelinase for differing periods of time (0 to 60 min) and quenching the enzymatic reaction by the extraction process, the concentration time-courses for the individual metabolites could be followed.[381]

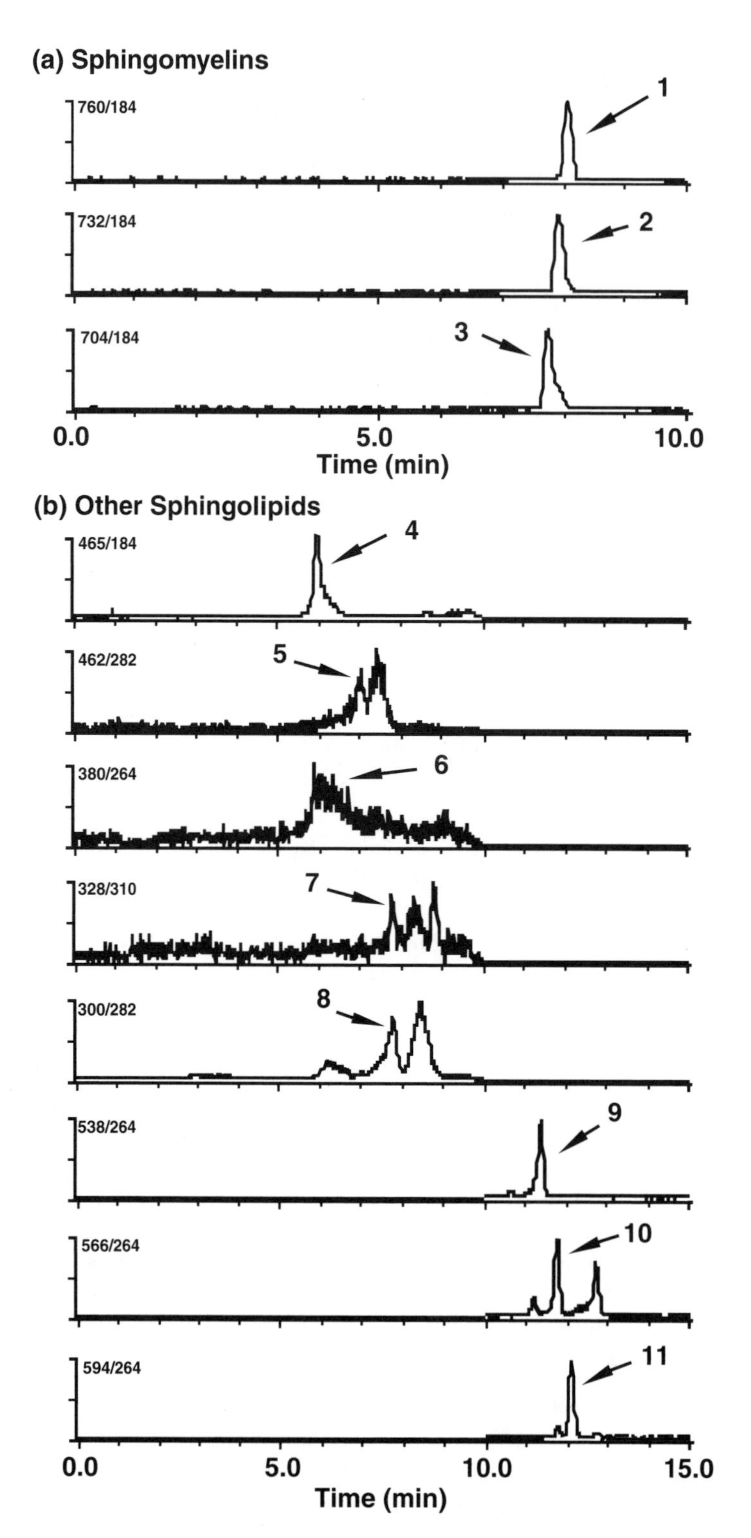

FIGURE 6.52 Multiple reaction monitoring of endogeneous sphingolipid metabolites in HL-60 cells. (a) Sphingomyelins (SM): 1, C_{20}-SM; 2, C_{18}-SM; 3, C_{16}-SM. (b) Other sphingolipids: 4, sphingosylphosphorylcholine; 5, psychosine; 6, sphingosine 1-phosphate; 7, N,N-dimethylsphingosine; 8, sphingosine; 9, C_{16}-Ceramide; 10, C_{18}-Cer; 11, C_{20}-Cer. Reprinted from Mano, N. et al., *Anal. Biochem.*, 244, 291–300, 1997. With permission.

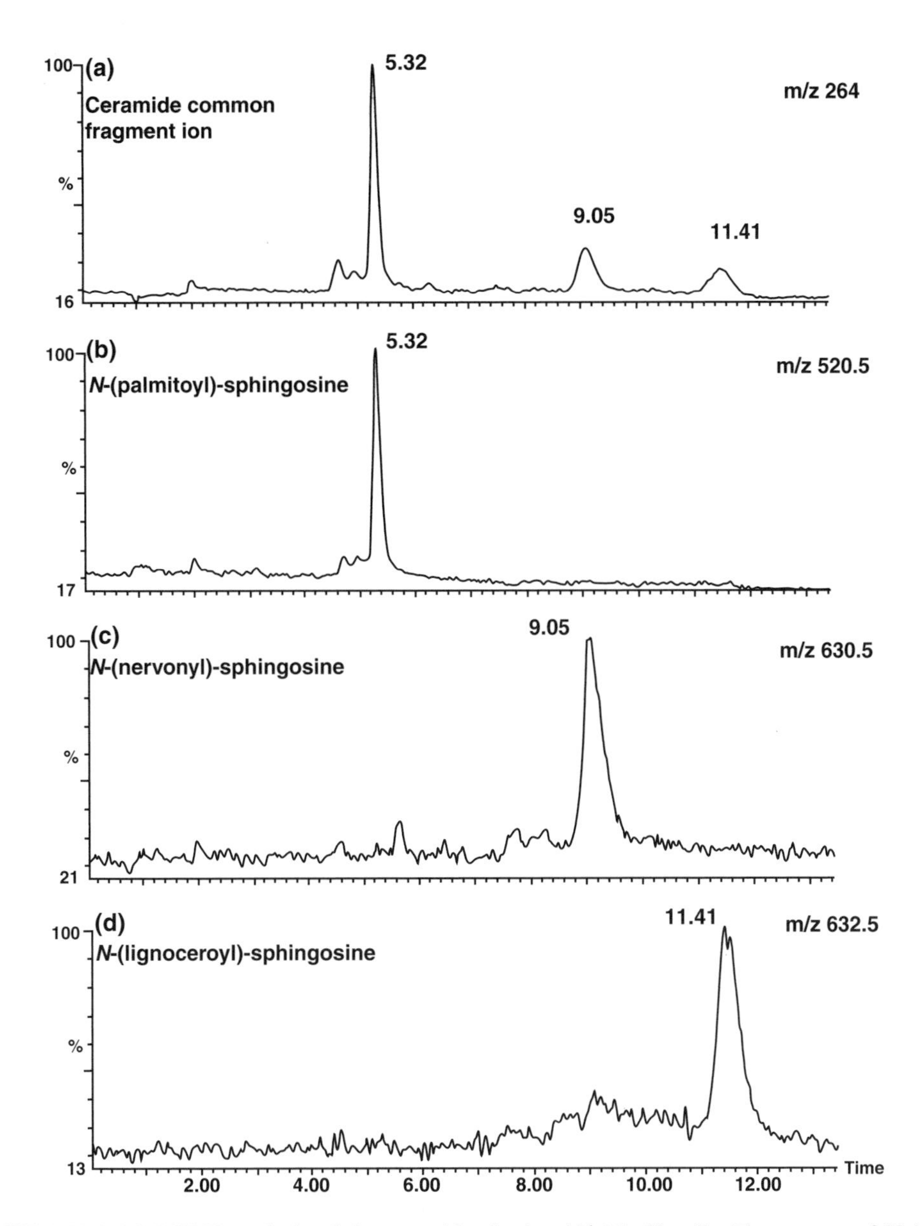

FIGURE 6.53 LC/APCIMS analysis of the ceramides in 4×10^4 HL-60 cells. The amount of N-(palmitoyl)sphingosine (50 amol/cell) and N-(nervonyl)-sphingosine (43 amol/cell) present in the organic extract was approximated by comparing responses to those from similar amounts of authentic standards. Reprinted from Couch, L.H. et al., *Rapid Commun. Mass Spectrom.*, 11, 504–512, 1997. With permission.

APCIMS has been used for the analysis of three ceramides, N-palmitoyl (C16:0), N-nervonyl (C24:1), and N-lignoceroyl-(C24:0) sphingosine in $4X10^4$ HL-60 cells.[382] Sample preparation consisted of extraction with organic solvents and separation on C_{18} LC columns. Low pmol analyte levels could be quantified using SIM (Figure 6.53).

Comparative Analysis of Ceramides in Lymphoid Cell Lines

Another of the CID-based techniques described above[380] has been applied to the comparison of ceramide profiles from three related lymphocyte cell lines, T-cell Jurkat, monocytic U937, and immature WEHI 231. Resuspended (RPMI medium) pellets, typically from 10^7 cells, were lysed with acidic methanol, the lipids extracted with chloroform, separated on silica gel columns packed

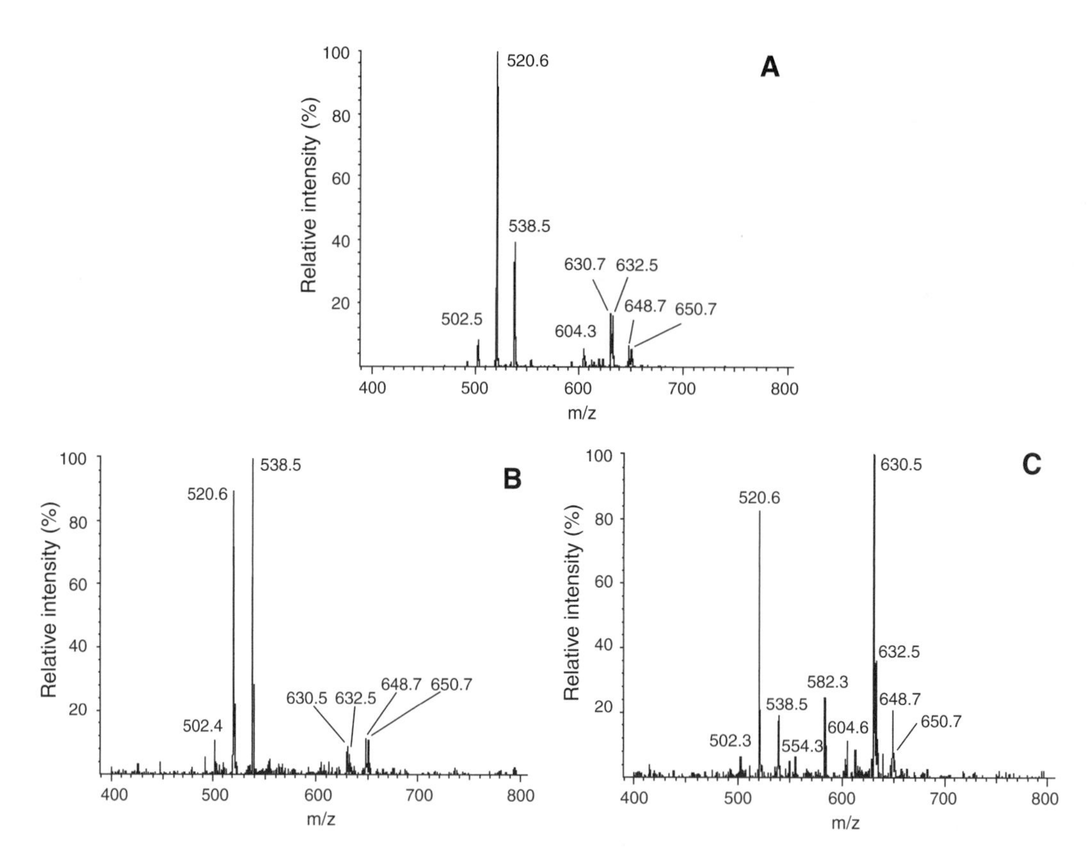

FIGURE 6.54 Precursor ion scans for the range *m/z* 400-800 are shown for lipid preparations derived from ~10^7 cells from (A) Jurkat, (B) U937, and (C) WEHI 231 cells. The major ceramide species observed were C24:0 (*m/z* 650 and 632), C24:1 (*m/z* 648 and 630). Spectra shown were averaged over a 5-min data collection window at a sample flow rate of 2 μL/min. Reprinted from Gu, M. et al., *Anal. Biochem.*, 244, 347–356, 1997. With permission.

in Pasteur pipettes, and the ceramides identified by TLC, with Coomassie staining, of aliquots from the collected fractions. The remaining fractions were pooled, dried, and reconstituted in ammonium acetate/methanol (i.s.: C2:0). The CID experiments were carried out in both product and precursor ion scan modes on a QqQ-MS. Significant differences were found between the ceramide composition of these related cell lines (Figure 6.54). Jurkat and U937 differed quantitatively with respect to C16:0 and C24:0/C24:1 ceramide. There also were marked differences between the ceramide profiles of the WEHI 231 and the other two lines, in particular with respect to the C24:1 species which was predominant in WEHI 231. It was suggested that profiling and quantifying ceramides by this ESIMS/MS technique, which requires sub-picomolar quantities of analytes, would allow for the investigation of their distribution and metabolism, even *in vivo*, leading to an elucidation of the apparently multifaceted functions of these important lipids.[380]

Ceramide Levels during Apoptosis

The method described in Reference 380 was also used to study the formation of ceramides in Jurkat T cells in which apoptosis had been induced with anti-Fas IgM. In another study, three human cell lines were investigated: T-cell Jurkat, chronic myelogeneous leukemia (K562), and melanoma (526). Lipids were extracted and purified, and 2.5 × 10^5 cell equivalents/μL of chloroform/methanol extracts were infused directly into an ESIMS/MS which was operated in negative ion mode. A significant increase in ion intensity at *m/z* 572 was detected in the Jurkat cells 24 h after initiating apoptosis using ionizing radiation. CID analyses of this ion revealed only one product ion, Cl^-,

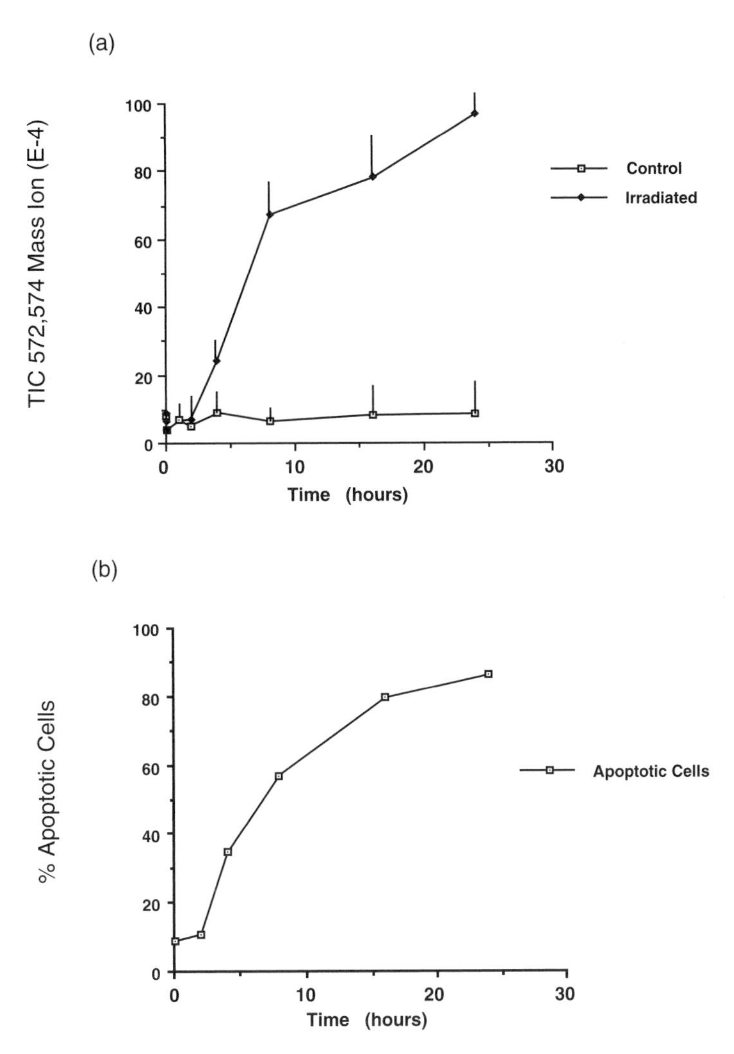

FIGURE 6.55 C16 ceramide accumulates during radiation-induced apoptosis of Jurkat cells from 0.5–24 h. Jurkat cells were exposed to ionizing radiation (20 Gy) and cultured for varying times up to 24 h. At times ranging from 0.5 to 24 h, 5×10^6 cells were harvested followed by the extraction of lipids. (a) C16:0 ceramide levels were quantified by MS, values are mean ±s.d. for three experiments; (b) The percentage of apoptoic cells, as determined by morphology. Reprinted from Thomas, R.L., Jr., et al., *J. Biol. Chem.*, 274, 30580–30588, 1999. With permission.

indicative of formation of a chlorine adduct providing the negative ion observed. The identification of the analyte as N-palmitoyl (C16:0) ceramide was confirmed by both positive- and negative-ion ESI-MS by comparison with a standard. The time course of the formation of this ceramide was established using lipid extracts of 5×10^6 cell equivalents, harvested between 0.5 and 24 h while the cells were being irradiated. There were significant differences between control and irradiated samples (Figure 6.55a). The increases in the C16:0 ceramide formation paralleled apoptosis as determined by cellular morphology (Figure 6.55b) and increases in caspase-3 activity. By comparison there was no increase in the C16:0 ceramide when the radioresistant K562 line was irradiated and the lack of apoptosis was confirmed by the cell line's inability to produce caspase-3 activity. MS analyses also confirmed long-term C16:0 ceramide accumulation in the melanoma cell line, but only in the cell fraction (~70%) that was sensitive to radiation, implying a correlation between C16:0 ceramide accumulation and radiation sensitivity. It was concluded that C16:0 ceramide has

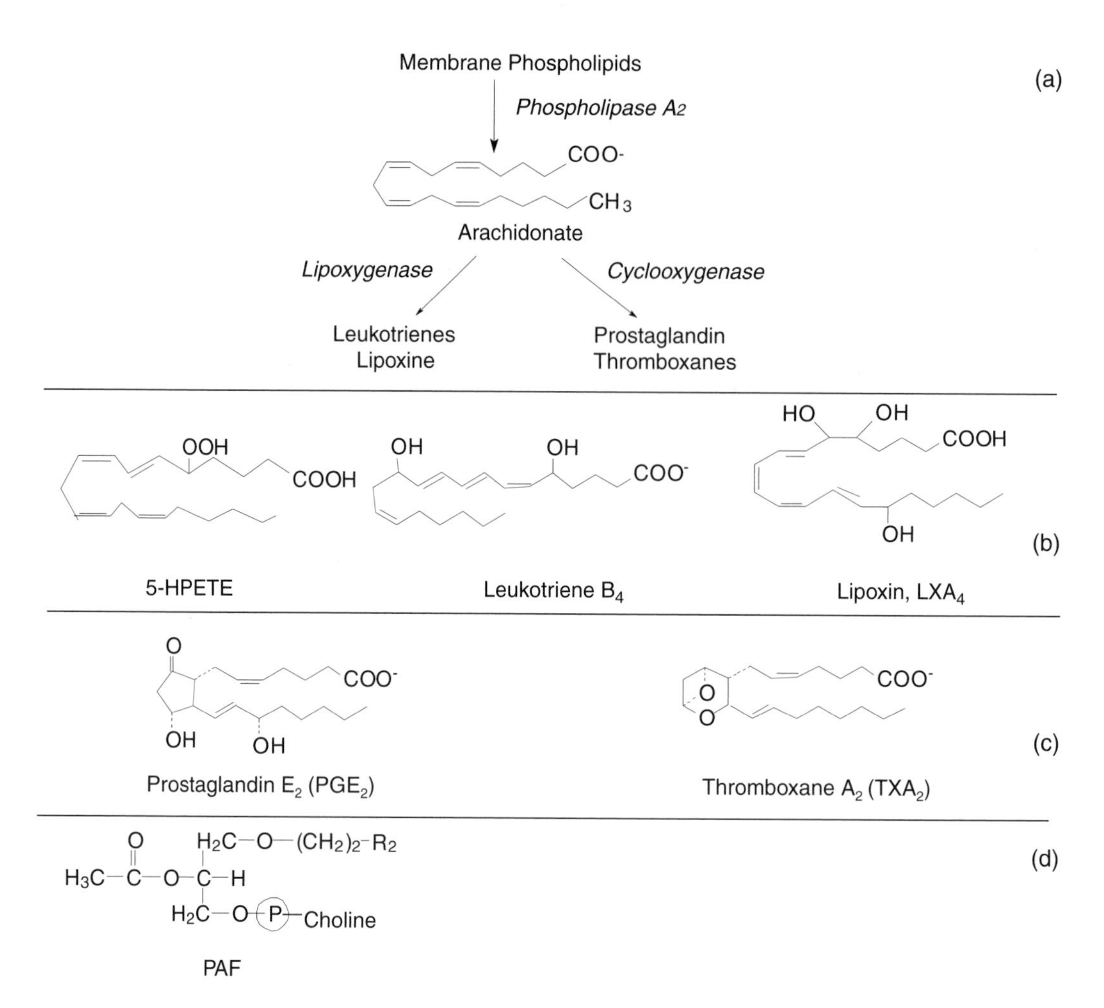

FIGURE 6.56 Chemical formulas of phospholipids. HPETE, hydroperoxyeicosatetraenoic acid, PAF, platelet activating factor.

a role in the effector (mitochondrial) phase of apoptosis, suggesting the need to investigate potential ceramide changes when apoptosis is induced by means other than radiation.[384]

6.2.2 Eicosanoids

Arachidonic Acid Metabolism

Arachidonic acid, an all cis-eicosatetraenoic acid, is a water-insoluble, 20-carbon, unsaturated ($\omega 6$, C20:4, $\Delta^{5,8,11,14}$) hydrocarbon with a carboxyl group at the end (Figure 6.56). It is stored in esterified form at the sn-2 position of the glycerol backbone of cell membrane phospholipids. In response to various stimuli, a diverse family of biologically active metabolites of arachidonic acid is synthesized within the cells and, in turn, these compounds serve as chemical communicators by activating nearby cells to initiate specific biochemical processes.

After release by phospholipase A_2, free arachidonic acid activates both lipogenase and cyclooxygenase systems of enzymes. The metabolic products resulting from the action of the lipogenase system are the hydroperoxyeicosatetraenoic acids, HPETE, e.g., 5-HPETE, 12-HPETE, and 15-HPETE. Further conversions, sometimes via intermediates, lead to the biologically active leukotrienes and lipoxins that are characterized by 3 and 4 conjugated double bonds, respectively

(Figure 6.56). The cyclooxygenase system converts arachidonate into endoperoxides that are key precursors for prostaglandins and thromboxanes, short-lived local hormones that alter the activities of the cells in which they are synthesized, as well as adjoining cells.

Lipogenase Products in Saliva of Patients with Oral Cancer

Lipogenase activity in human saliva has been attributed to the presence of epithelial cells or polymorphonuclear leukocytes. In a search for lipogenase-generated products, fatty acid-enriched fractions were prepared by solid phase extraction from mixed saliva and parotid or submandibular gland saliva fractions obtained from control subjects and patients with advanced squamous cell carcinomas (SCS) of the upper aerodigestive tract. The evaporated and reconstituted extracts were silylated and the linoleic, arachidonic, and other nonhydroxylated fatty acids quantified using GC/EIMS (i.s.: deuterated analytes). Hydroxylated fatty acids were determined by GC/EIMS (i.s.: carboxy-^{18}O-labeled analytes) after hydrogenating the double bonds.[385] In an alternative approach using GC/NCIMS in the SIM mode, hydroxyoctadecadionic acids (HODE) and hydroxyeicosatetranoic acids (HETE) were quantified as their hydrogenated pentafluorobenzylester/TMS derivatives. Using this method, normal human saliva was found to contain mainly 12-HETE and only minor quantities of 5-, 8-, 9-, or 15-HETE. Incubation of saliva with arachidonic acid followed by chiral analysis of the products revealed substantial 12 (S)-lipogenase activity in mixed saliva but not in the saliva fractions obtained directly from the salivary glands. The saliva of patients with SCS showed significantly elevated levels of free arachidonic acid and 5- and 12-HETE, as evidenced by a pair of isomer-specific carboxy-terminal and methyl-terminal fragment ions in the EI spectra (Figure 6.57). It was proven that the HETE and HODE found in mixed saliva were not excretion products of the salivary gland but originated from the SCS tissue or the macrophages or leukocytes derived from the tumor-directed inflammatory immune response.[386]

Cyclooxygenase Products of Arachidonic Acid

The cyclooxygenase mediated prostanoids include prostaglandins (PG) and thromboxanes (TXA). Prostaglandins, which exist in virtually all mammalian tissues, are synthesized *in vivo* by cyclization of the center of the 20-carbon polyunsaturated fatty acid chain (e.g., arachidonic acid) to form a cyclopentane ring. In the TXA family this cyclopentane ring is expanded by inclusion of an oxygen atom to form an oxane ring. Variations in the substituent groups attached to the rings are indicated by labels A, B etc., with the number of double bonds in the side-chains being designated by subscripts 1, 2, 3, e.g., PGE_2 (Figure 6.56).

Prostaglandins in Colorectal Mucosa

It has been suggested that the mechanism of action for the demonstrated beneficial effects of nonsteroidal antiinflammatory drugs (NSAID) on the incidence, regression, and mortality of colon cancer may be linked to the cyclooxygenase inhibitory effect of these agents. In turn this results in decreases of the concentrations of several mucosal PG that modulate tumor cell growth and facilitate tumor progression. The five prostaglandin metabolites, PGD_2, PGE_2, PGF_{2a}, TXB_2, and 6keto-PGF_{1a} were quantified in rectal mucosa biopsy specimens from patients with familial adenomatous polyposis (FAP) and controls during and after treatment with sulindac (a nonsteroidal, anti-inflammatory indene derivative). The GC/NCIMS technique used was based on SIM of the $[M-181]^-$ fragment ions of the analytes derived from their pentafluorobenzyl ester-TMS ether derivatives (i.s.: deuterated PG analogs).[387] Before sulindac treatment, PGE_2 concentrations were highest in both FAP patients and controls at a level of ~5 ng/mg protein. Only TXB_2 concentrations were statistically elevated in untreated FAP patients with respect to the controls. The concentrations of all the other analytes were not statistically different. However, after treatment with sulindac for 3 mo, the concentrations of all analytes, except PGD_2, were reduced significantly ($p < 0.05$) in the FAP patients both with respect to pretreatment FAP samples and control groups. The baseline reductions were in the 32 to 50% range but interpatient differences were very high with changes ranging from +20% to –90%. Despite the tumor regression achieved with sulindac, the drug does

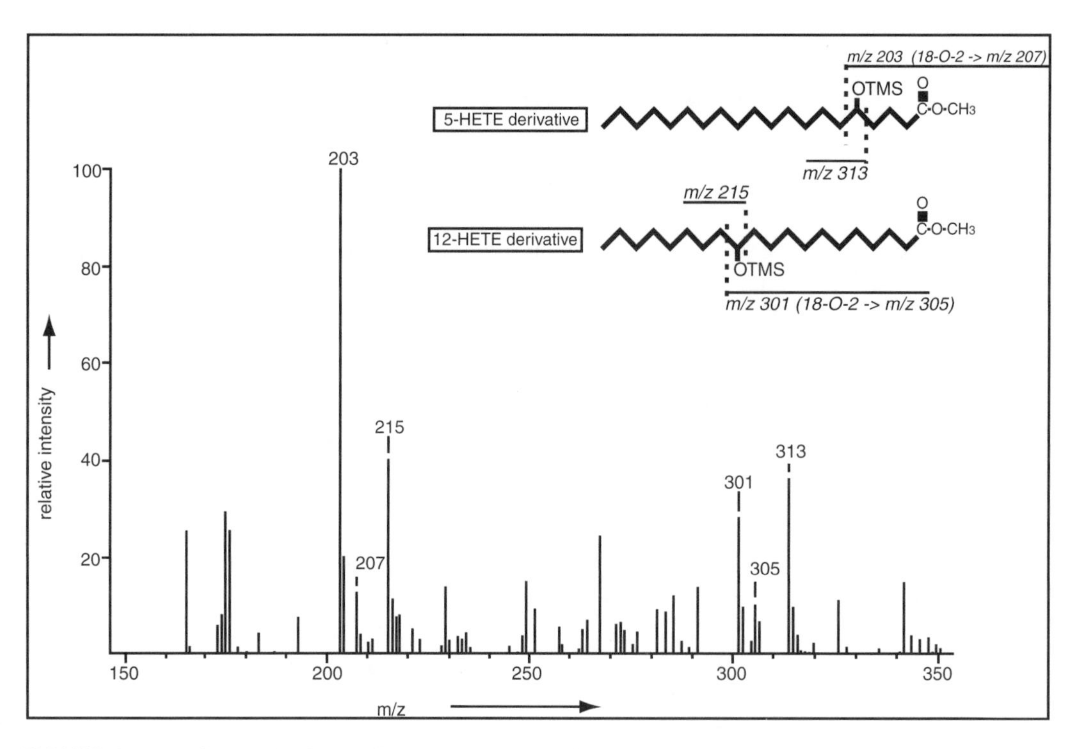

FIGURE 6.57 Background-subtracted EI mass spectrum of hydroxyeicosatetraenoic (HETE)-containing GC-peak observed in the analysis of whole saliva from a patient with squamous cell carcinoma of the hypopharynx. The presence of 5-HETE and 12-HETE is indicated by their isomer-specific fragment ions. Signals of the [carboxyl-$^{18}O^2$] labeled internal standards at *m/z* 207 and 305 are also observed. Reprinted from Metzger, K. et al., *Free Radic. Biol. Med.*, 18, 185–194, 1995. With permission.

not prevent colorectal cancer, possibly due to the eventual uncoupling of PG levels from carcinogenesis. It has been suggested that monitoring mucosal PG concentrations may serve as biomarkers of disease progression as well as indicators of patient compliance with medication regimens.[388]

In another study, an experimental selective COX-2 inhibitor (SC-58125) was used to inhibit colon cancer cell growth in two transformed human cell lines, HCA-7, which expresses high levels of COX-2 proteins and HCT-116 cells which lack the COX-2 protein. Isotope dilution GC/NCIMS was used to quantify 5 eicosanoids (PGE_2, PGD_2, $PGF_{2\alpha}$, TXB_2, and 6-keto-$PGF_{1\alpha}$) in media from cell incubations. The LOD was 4 pg/mL for each analyte. Results revealed that PGE_2 was not detectable in the HCT-116 cells. In contrast, a significant quantity of this analyte was detected in HCA-7 cells; however, metabolite production ceased when the inhibitor was added, suggesting a link between intestinal cancer growth and selective inhibiton of the COX-2 pathway.[389]

6.2.3 Lysophospholipids

Lysophospholipids (lyso-PL) are recognized as important extracellular cell signaling molecules [reviewed in Reference 390). Lysophosphatidic acids (LPA) are a group of compounds that contain a glycerol backbone with a long aliphatic chain attached to the first carbon, a hydroxy group in the second position, and a phosphate group in the third position. It was shown that the concentration of several lyso-PL in ascitic fluids is increased significantly in patients with ovarian cancer compared to patients with nonmalignant diseases. One technique, developed for the analysis of all lyso-PL with either a glycerol or a sphingosine backbone, uses solvents to extract lipids from ascites. The distibution of various lysolipids between the organic and aqueous phases in the solvent extraction protocol has been studied in connection with a quantification technique developed for plasma.[391] Parent scanning and MS/MS analyses were used to detect and confirm the structures of all extracted lyso-PL components. Subsequent quantification was carried out using SRM in the negative ion mode to monitor selected transitions from the molecular ions. Lipids with the phosphorylcholine group are positively charged, and these were analyzed in the positive ion mode by SRM. Acyl-LPA and lysophosphatidylcholine (LPC) were used as i.s.

In addition to acyl-LPA previously detected in plasma samples,[391] two new species were identified and characterized, using positive ion ESIMS, in ascites samples from patients with ovarian carcinoma: 16:0 and 18:0 alkenyl LPA7. A number of other PL were detected in the negative mode. Quantification revealed that malignant ascites from ovarian cancer patients contained significantly higher levels of lyso-PL, including total acyl-LPA, total alkyl- and alkenyl-LPA, total lysophosphatidylinositols (LPI), sphingosylphosphorylcholine (SPC), and total LPC, than nonmalignant ascites. There were no differences between malignant and benign ascites with respect to sphingosine-1-phosphate and lyso-platelet activating factor concentrations. The increased lyso-PL concentrations in the ascites of patients with ovarian cancer were not due to a general overproduction of phospholipids, since the total phospholipid contents of malignant and benign ascitic fluids were not statistically different. Accordingly, it was concluded that the members of the subclasses of alkyl-LPA, alkenyl-LPA, and, possibly, methylated lysophosphatidylethanolamines (LPE) have individual roles as extracellular cell signaling molecules in the biology of ovarian tumors.[392]

6.2.4 Platelet-Activating Factor

Platelet-activating factor (PAF) is a family of alkyl- or acyl-phospholipid autacoids consisting mainly of the alkylphospholipid, 1-O-alkyl-2-acetyl-*sn*-glycero-3-phosphorylcholine (Figure 6.56d). In addition to serving as ubiquitous and potent proinflammatory lipid mediators, PAF apparently stimulates intracellular signaling pathways linked to cellular proliferation, e.g., tyrosine phosphorylation, an action that may be counteracted by PAF receptor antagonists. The structural heterogeneity of PAF has been shown to have diverse (patho)physiological implications (reviewed in Reference 393). The molecular heterogeneity of PAF in normal human saliva was investigated by GC/EIMS using pentafluorobenzoyl acid (PFB) derivatives. Of six different PAF found and quantified, the predominant PAF (30%) was the 1-O-hexadecyl variant, present at the 0.75 pmol/mL saliva level.[394] Another method used solid-phase extraction and derivatization with PFB prior to GC/NCIMS with SIM (i.s.: D_4,$^{13}C_2$-C16:0 alkyl-PAF) for the quantification of C16:0, C18:0, and C18:1 alkyl, and the acyl PAF homologs recovered from human umbilical vein endothelial cells. The LOD was 1 fmol/injected analyte.[395]

ESI has also been evaluated for PAF. One earlier LC/ESIMS approach established that there were significant differences in LOD in glioma cells with the order being phosphatidylcholine > phosphatidylethanolamine > phosphatidylserine.[396] An HPLC-ESIMS technique, developed for PAF and PAF-related compounds, e.g., lyso-PAF, in polymorphonuclear nucleophils, used CID to obtain product ions corresponding to the phosphorylcholine group. The LOD were 0.3 ng for PAF and 3 ng for lyso-PAF.[397] An isotope dilution method using negative ion ESI provided the molecular anions of the respective analytes, which decomposed under CID to the respective carboxylate anions permitting discrimination of the isobaric PAF, 1-octadecanoyl-2-lyso-glycerophosphocholine, and 1-hexadecanoyl-2-formyl-glycerophosphocholine without chromatographic separation.[398]

Platelet-Activating Factor in Human Endometrial Cancer Cells

In a study of the production and biological activity of PAF in the HEC-1A cell line, a phospholipase C-treated material was extracted, derivatized with *t*-butyldimethylchlorosilymidazole, and analyzed by GC/EIMS (i.s.: deuterated-PAF). The mass spectrum of the analyte was the same as that of authentic C16:0 PAF. The analyte was quantified by monitoring the [M-*t*-butyl]$^+$ ion. The C16:0 PAF content of the HEC-1A cells was 40 ng PAF/17.4 mg DNA. It was also established that while PAF was capable of inducing, in a time-dependent fashion, the synthesis of DNA and the nuclear protooncogene c-*fos*, both effects could be inhibited with a PAF receptor antagonist. It was concluded that PAF participates in an autocrine mediated proliferative loop in endometrial adenocarcinoma cells.[399]

Angiogenic Role of PAF in Human Breast Cancer

Lipid extracts were obtained from primary breast carcinomas (n = 18), control breast tissues (n = 18), and T-47D and MCF-7 breast adenocarcinoma cell lines. Seven bioactive PAF components were purified by TLC and HPLC and quantified using ESIMS/MS. CID of the protonated molecules yielded product ions corresponding to the polar heads with LOD in the 0.1 to 0.3 ng/injected analyte range.[400] The different PAF species were quantified by SRM using precursor → product ion transitions, e.g., 552 → 184 for C18:0 alkyl PAF, 538 → 184 for C16:0-acyl PAF, 524 → 184 for C18:0 lysophosphatidylcholine (LPC), etc., and comparing peak areas observed with those obtained for 24 ng injections of individual standards. Breast carcinomas with high quantities of bioactive PAF (>1.5 pg individual PAF/mg dry tissue) showed CID transitions corresponding to C16:0 alkyl PAF, C18:0 LPC, C16:0 LPC, lyso-PAF, and C16:0 acyl PAF. Transitions corresponding to C16:0 alkyl PAF and C16:0 acyl PAF were absent in normal breast tissues (<0.25 pg PAF/mg tissue) and samples from breast carcinomas with low levels of PAF bioactivity. The other PAF were present in these carcinomas and the quantities of extracted bioactive PAF correlated significantly with tumor vascularization. Lipid extracts with high PAF bioactivity induced a significant angiogenic response that could be inhibited by PAF receptor antagonists. It was concluded that the lipid mediator PAF produced by cancer cells might amplify the action of polypeptide mediators, e.g., fibroblast growth factor, in tumor revascularization.[401]

6.3 OLIGONUCLEOTIDES AND NUCLEIC ACIDS

6.3.1 Mass Spectra

6.3.1.1 Nomenclature of Fragmentation

Several reviews address various aspects of the mass spectra of oligonucleotides[402-405] and the sequencing of DNA.[406] The notation for oligonucleotide fragmentation (Figure 6.58) is similar to

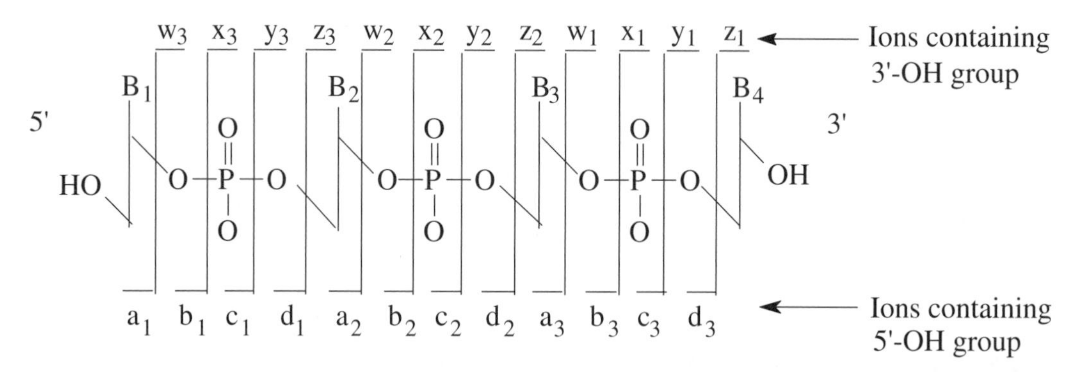

FIGURE 6.58 Notation for DNA fragmentation. The letters a, b, c, and d indicate an ion containing the 5′-OH group. The letters w, x, y, and z indicate an ion containing the 3′-OH group. Numeric subscripts indicate the point of fragmentation in number of bases from the terminal group. Based on McLuckey, S.A., Van Berkel, G.J., and Glish, G.L., *J. Am. Soc. Mass Spectrom.*, 3, 60–70, 1992.

that used for peptides (Section 6.1.1): (a) the four possible cleavages of the phosphodiester bond of the fragments containing the 5′-OH group are labeled a, b, c, and d, the corresponding fragments containing the 3′-OH group are labeled w, x, y, and z, with numerical subscripts that indicate the position of the cleavage in the nucleotide chain, i.e., the number of bases from the respective terminal group; (b) bases are notated by B, with a numerical subscript indicating the base position with respect to the 5′-end of the oligonucleotide; (c) the loss of a specific base from a fragment ion is indicated in parenthesis, e.g., a_3-B_2(A) indicates that the ion is an a-type fragment at the 3-position with an additional loss of an adenine base at the 2-position.[406] Examples of the use of this notation include the interpretation of vaious aspects of the CID spectra of oligonucleotides, such as the loss of an adenine anion[407] and the discrimination of all 3 human p53 69 base pair (bp) polymerase chain reaction (PCR) variants.[408]

Peptide nucleic acids (PNA) are a class of DNA mimics, of current interest, that consist of repeating N-(2-aminoethyl)glycine units linked by peptide bonds. The neutral peptide-like backbones contain secondary amino groups that can be attached to nucleobases. Some physical and chemical properties of PNA allow them to hybridize to complementary DNA and RNA strands. The nature of the structure of PNA confers considerably more resistance to enzymatic digestion than occurs for oligonucleotides. This and other favorable properties make PNA good candidates for investigation as antisense and antigen-type therapeutic agents. A comprehensive nomenclature has been proposed for the annotation of the CID spectra of PNA obtained by ESI-FTICRMS.[409]

6.3.1.2 Electrospray Ionization

The advents of both PCR technology (see later) and antisense chemotherapy (Section 5.3.3) have resulted in major advances in the technology and methodologies that enable the rapid synthesis of oligonucleotides that are then used as primers, probes, and DNA analogs. This has necessitated development of analytical techniques that can be used for the quality control and characterization of oligonucleotides. HPLC and CE techniques provide high separative power with resolution to the one bp level. However, MS techniques are needed to determine the confirmation of the length and base sequence, as well as other structural features of the oligonucleotides involved. Either applied directly or coupled with HPLC or CE, ESIMS has unique advantages as well as serious limitations when used as a tool for the analysis of oligonucleotides with various lengths (reviewed in Reference 405). ESI is predominantly used in the negative ion mode. This is based on the fact that there is an acidic proton at each phosphodiester bond along the oligonucleotide chain that readily permits the formation of anionic deprotonated molecules. The characteristic ESI envelope of charge states permits the use of deconvolution techniques for the determination of relatively high masses with low-resolution quadrupole mass analyzers. A serious limitation is posed by the ubiquitous formation of sodium adducts. The result of these sodium adducts is that multiple charge state envelopes are formed, thus generating a practical limit of the chain length that can be analyzed to 35-70 bases.[410] Stringent efforts can reduce the amount of sodium present. An alternative strategy is the replacement of the sodium adducts by ammonium ions which are less tightly bound to nucleotides. Both of these methods reduce the number of charge state envelopes, thus extending the mass range of the analysis and simplifying data interpretation. Practical approaches include precipitation with cold ethanol from concentrated ammonium acetate solutions and subsequent additions of triethylamine[411,412] and an on-line desalting procedure that was developed to aid in the study of the *in vivo* metabolism of phosphorothioate oligonucleotides.[413]

The advantages and limitations of on-line, ion-pair LC/ESIMS and direct infusion-ESIMS were compared using the reaction products from the solid-state synthesis of a 39-mer oligonucleotide as a model. It was concluded that while the LC method was superior for the analysis of both major and minor components, direct infusion-ESIMS (with on-line cation exchange) provided rapid and simple identification of the major components in a complex nucleic acid mixture.[414] The factors determining the performance of ESI sources coupled to QqQ, IT, and magnetic sector analyzers have been compared with respect to instrument performance characteristics in tandem operation[415] and their suitability for *de novo* sequencing of oligonucleotides.[416]

The combination of ESI with FTICRMS provides mass resolution up to 10^5. This technique has been used to analyze oligonucleotides with lengths >100 bases to an accuracy of 50 ppm[417,418] and is particularly useful for sequencing using CID.[410,419] Single-base substitutions, even the A to T polymorphism (mass difference: 9 Da), can be determined in PCR products using ESI-FTICRMS.[420] It has been concluded, based on analysis of model oligonucleotides and PCR products, that, with regard to mass accuracy and resolution, instrumental preference is ESI-FTICRMS > ESI-QMS > MALDI-TOFMS.[421]

6.3.1.3 MALDI-TOFMS

To understand the dynamics and kinetics of the ionization of DNA by MALDI in general, and in particular to explain the weak ion signals produced by poly-G when compared to poly-A, poly-C, and poly-T using both positive and negative MALDI-TOFMS, a systematic investigation of the mechanism of ionization for adenine (A), thymine (T), cytosine (C), and guanine (G) was undertaken using different MALDI matrices. It was concluded that the structure-dependent ionization of DNA by MALDI is dominated by two processes: (a) pre-protonation that occurs before laser ablation during the process of crystallization, with the acidic matrices providing $[H]^+$ ions; and (b) a matrix-independent thermal reaction that induces deprotonation of the bases. The mechanism of ionization is independent of the composition of the DNA backbone. These studies did not provide a definitive reason but did confirm that guanine is difficult to ionize. The best MALDI matrix of those tested for DNA analysis was 3-hydroxypicolinic acid (3-HPA).[422]

The combination of delayed extraction and reflectron TOF analyzers with MALDI led to significant gains in analytical performance with attributes comparable to those attained in the analysis of peptides. For example, samples as large as 50-mers may be analyzed with single settings for field strength and extraction delay time. Mass accuracy is adequate to verify sequences to 9.5 kDa, and resolution of 20,000 is available in the reflectron mode. Higher masses have been analyzed under specific conditions as noted below. Optimization of relevant instrument parameters, such as laser irradiance and wavelength, permitted detection of the oxidation of a C–S bond at 9.5 kDa, with a mass difference of only 16 Da, and the verification of the sequences of both crude synthetic oligonucleotides and purified small oligonucleotides by using timed, differential digestion and analysis of the resulting mixtures.[423]

6.3.1.4 High-Resolution MALDI-FTICRMS

Resolution higher than that available with TOF analyzers is needed to detect and explain changes due to various protonation and hydrolysis reactions that are often artifacts attributable to the action of acidic MALDI matrices of the relatively fragile oligonucleotides within the MALDI source. The effects of metastable decay may significantly limit the mass resolution attainable in FTICRMS. One methodology used an external MALDI source located 117 cm away from the FTICRMS analyzer cell with both chambers being differentially pumped using high-capacity cyropumps. The extracted ions of a mixed-base 25-mer oligonucleotide (matrix: 3-HPA) were transported into the analyzer through an rf-only quadrupole ion guide and exposed to a 0.5 s pulse of argon gas for a "cooling" effect. Under these conditions fragmentation was minimized. The delay time between ion formation and analysis was >5 s, in contrast to TOF where ions are detected a few ms after the laser pulse. The measured resolution for the analyte at *m/z* 7634 was 136,000. Sensitivity was also very high, the S/N ratio being 20:1 for a 12 pmol sample. The molecular mass of the analyte was 7635 Da, and the monoisotopic protonated molecule was clearly resolved at *m/z* 7632 (Figure 6.59). Resolution as high as 830,000 (FWHM) was achieved for smaller oligomers. To date, these are the highest resolution values reported for MS analysis of DNA.[424]

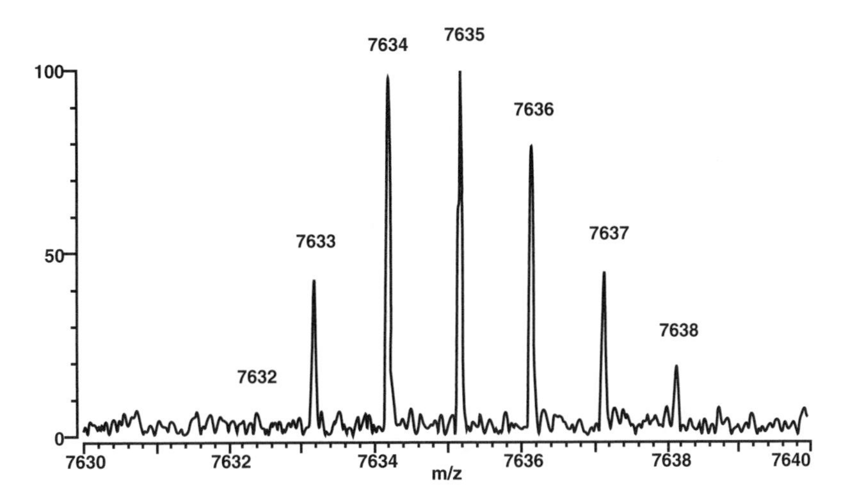

FIGURE 6.59 MALDI-FTMS spectra of mixed-base 25-mer of DNA at a mass resolution of 136,000 (FWHM). Sample size was 12 pmol. The matrix was 3-hydroxypicolinic acid. Reprinted from Li, Y. et al., *Anal. Chem.*, 68, 2090–2096, 1996. With permission.

6.3.1.5 Analysis of Large Nucleic Acids

Ion fragmentation and salt formation of the phosphate backbone have limited nucleic acid analysis by MALDI-MS with UV lasers to ~90 kDa for DNA and ~150 kDa for RNA. These mass limits have been increased significantly by using infrared lasers and a glycerol matrix, the latter providing superior reproducibility and outstanding mass precision. Molecular masses of synthetic DNA, restriction enzyme gene fragments from plasmid DNA, and RNA transcripts could be determined to ~700 kDa or 2180 nucleotides. As little as 300 amol of a test sample of 515 bp (1030 nucleotides, base peak at 318,489 Da) could be analyzed with an S/N ratio of ~20 using piezoelectric pipettes to load 1 nL size samples. Mass accuracy was <1%. Potential areas of application in genotyping and clinical diagnosis have been elaborated.[425]

In the case of ESIMS, the analysis of large nucleic acids has been limited by poor accuracy (~10%) in mass assignments and the need for extensive sample purification. Upper mass limits have been ~65 kDa (~115 bp) for PCR products and ~40 kDa (120 nucleotides) for RNA. The upper mass limits have been significantly increased by utilizing the superior mass accuracy provided by ESI-FTICRMS. After purification to remove primers, buffers, and detergents from a 500 bp PCR product (generated from the linearized bacteriophage Lambda genome), it was possible to electrospray 0.8 pmol/µL analyte. The molecular mass, 309,406 Da, was determined with a precision of 87 ppm (±27 Da). Observed negative charge states ranged from 172 to 235.[426]

6.3.2 Techniques and Strategies

6.3.2.1 Polymerase Chain Reaction (PCR)

This technique, a cornerstone of molecular biology and genetic analysis, amplifies defined DNA segments several millionfold.[427] PCR is carried out *in vitro*, does not involve cells, and provides large quantities of any desired gene from a small sample of DNA. During the first step, the double-stranded template DNA is denatured by heating. Next, specific synthetic oligonucleotide primers of ~20 bp size are annealed (hybridized) on either side of the segment and a complementary strand located between the primers is synthesized with the aid of a heat-stable DNA polymerase. The double-stranded DNA thus formed serves as the template for the next cycle. As the procedure is repeated ~30 times, the quantity of replicated DNA increases exponentially. Because the primers have to be synthesized, the beginning and ending sequences of the DNA to be cloned must be known. PCR is so sensitive that even a single copy of a DNA sequence can be detected in a sample. An important application of PCR in cancer research is the cloning of particular DNA fragments, e.g., a gene, from cells. The original template may be either DNA or RNA, thus the clone obtained is genomic DNA or transcription, cDNA, respectively.

DNA products generated by PCR amplification are usually separated according to molecular mass using gel electrophoresis with slab gels or gel-filled capillaries with detection by staining, fluorescence, or radioactive labels. Shortcomings of these techniques include resolution limitation, i.e., minimum detectable difference is one nucleotide length, need for different separation techniques for different DNA product sizes, ambiguities in size determinations, and often, the need to handle radioisotopes. The MS approach has numerous advantages. Compared to conventional techniques, resolution is vastly increased concurrent with high sensitivity (but not yet to the point of a single copy). Molecular masses up to 100 kDa level can be determined with an accuracy of 0.005%, thus differences corresponding to a single base substitution, such as 9 Da for A to T, may be detected. Accurate mass determinations mean that there is no need to use different techniques for different product sizes. Because only 100 ng DNA is needed, identifications may be made using only a few μL of reaction mixture. The mass spectrometric analyses can be carried out rapidly, often in minutes. However, this advantage is tempered by the timeconsuming sample purification required (mainly for ESI) to remove primers, salts, and other interfering materials used in the PCR process.

6.3.2.2 Differential Sequencing

The Human Genome Project will provide a reference standard (~3 billion bases of sequence data) human DNA to be used for comparisons with the DNA sequence of individual subjects for both *differential sequencing*, i.e., to characterize nondetrimental polymorphic sites, and *diagnostic sequencing*, i.e., to detect potentially harmful mutations. The expected pertinent variations are small, possibly as small as a single base, thus rapid localization of relevant regions is essential. Current detection techniques involve fluorescence tagging of one or more of the single-stranded components followed by hybridization, PCR amplification, and sequencing. Fluorescence indicates only relative hybridization efficiencies instead of unique identification of all species present, thus the determination of heterozygous samples is ambiguous. Because the m/z values depend only on base composition, differential sequencing with MS does not require fluorescence tagging. A simple comparison of measured and predicted molecular masses rapidly identifies those molecules that require further analysis. However, molecules with identical composition but differing sequences cannot be distinguished by mass spectrometry. Overlap of sample masses may be minimized by careful experimental design based on the known and predictable reference and analyte sequences of the particular situation studied.[428]

A mutation or polymorphism (see below) is almost always uniquely characterized by a change in the mass of the relevant analytes. There are at least three analytical schemes for mutation/polymorphism detection by MS (Figure 6.60): (a) distinguish between perfect and imperfect hybridization by measuring the masses of targets hybridized to fixed probes; (b) detect mutated/polymorphic sequences by determining mass differences between extended primers obtained by annealing probes to single-stranded targets upstream of the variable region followed by primer extension; and (c) confirm mutation or polymorphism by detection of only one large fragment instead of the expected two smaller fragments for normal DNA upon differential cleavage by restriction enzymes.[428]

6.3.2.3 Mass Spectrometric Methods for Genotyping

Determination of Single Nucleotide Polymorphism (SNP)

SNP are sequence variations involving a single nucleotide position. Although less polymorphic than microsatellite markers (see later), SNP occur more frequently, as often as 1 or 2 per 1000 nucleotides,

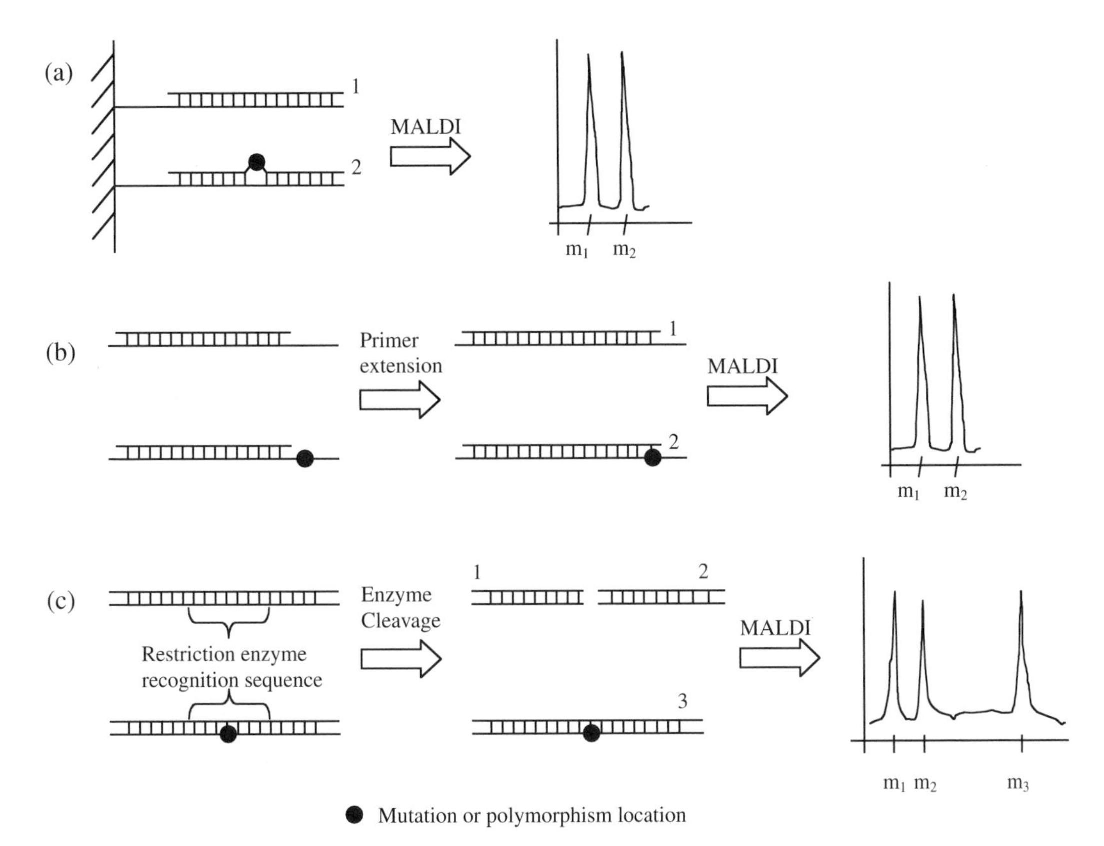

FIGURE 6.60 Examples of possible schemes for mutation/polymorphism detection via mass spectrometry, along with appearance of expected spectra. (a) Hybridization of targets to fixed probes. MS can distinguish between perfect and imperfect hybridization, based on different masses of the targets. (b) Extension of primers through known variable regions: a probe is annealed to a single-stranded target, upstream of the variable region, after which a primer extension reaction is carried out. MS distinguishes between normal and mutated/polymorphic sequence by the different mass of the extended primer. (c) Differential cleavage by restriction enzymes. In this scheme, the presence of a recognition sequence for a restriction enzyme is exploited. If no mutation is present, the enzyme cleaves the fragment into two smaller fragments (labeled 1 and 2); in the case of mutation or polymorphism, the enzyme cannot cut, therefore a single large fraction (3) is measured via MS. Reprinted from Graber, J.H., Smith, C.L., and Cantor, C.R., *Genetic Anal. Biomolec. Eng.*, 14, 215–219, 1999. With permission.

and they occur in >1% of the population. They display highly stable inheritance. This type of DNA sequence variation is usually biallelic, i.e., only two possible nucleotides are present. Maps of SNP markers can be used in genome-wide linkage and association studies of genetic traits and pharmacogenomic applications.

All the MS approaches must be able to resolve PCR products that represent closely spaced heterozygotes. The smallest mass difference between SNP alleles is 9 Da, i.e., the mass difference between A and T. Except in specific instances this level of mass resolution is usually not possible when PCR products are analyzed directly. Thus, fragments of 3 to 30 nucleotides long must be generated using hybridized probes or restriction digestion.

A number of MALDI-TOF techniques have been developed for the detection of SNP via the products of allele-discrimination reactions (reviewed in Reference 429). An example is SNP genotyping using primer extension. This involves the determination of the mass difference between an appropriate primer that is adjacent to the polymorphic site and single-base extension products of that primer. The design of the primer is critical, particularly with respect to the eventual chemical release of the extension products from the primers after purification. One approach included three sample preparation steps, PCR amplification, phosphatase digestion, and SNP primer extension with a cleavable primer. Because the final products contain both the primer (serving as i.s.) and the extension product(s), only mass differences rather than absolute masses must be determined. MALDI-TOFMS analysis was carried out automatically.[430] It was claimed that parallel sample processing with a robotic workstation and serial analysis with automatic MS make it possible to analyze thousands of samples per day.[431]

Other TOFMS-based approaches that have been developed (reviewed in Reference 432), include ligation analysis[433] and primer extension formats.[434,435] Areas of application include sequencing of PCR products,[436,437] direct mass analysis of PCR products,[438] analysis of allele-specific PCR products,[433,439] and minisequencing.[435,440-442]

ESI methods have been developed for SNP identification. In one approach, ESI-MS/MS (QqQ) was used to distinguish intact PCR products of different mass reflective of SNP, while the MS^n capabilities of IT analyzers have been explored to differentiate isobaric PCR products, i.e., products with the same mass and nucleotide composition but different sequences.[408] In another approach, flow injection with ESI-FTICRMS was employed to genotype short tandem repeats. It was concluded that this approach could provide ~300 genotypes per day.[443]

Peptide Nucleic Acid (PNA) Probes

In this approach, PNA probes complementary to the wild-type and variant sequences are hybridized to immobilized PCR products, the unbound probe is removed by washing, and the mass of the DNA probe hybrids are determined by MALDI-TOFMS. These types of hybrids have been used to determine SNP in human DNA,[444] for DNA typing of human leukocyte antigen sequences,[445] and for genetic analysis.[446] Recent developments using MALDI-TOF have included new approaches to quantification[447] and the ternary-encoding of genotypes.[448] A disadvantage of this approach is that different PNA probes must be synthesized for each variant.

The PROBE™ Technique

The Primer Oligo Base Extension technique is based on the isothermal extension of an ~20 nucleotide-long primer oligomer. The major steps of the method are: (a) the annealing of a primer to a solid-phase immobilized target template immediately upstream from the region to be investigated; (b) extension of the primer in the presence of a selected mixture of one or more deoxynucleotide-triphosphate (dNTP) and one dideoxy-nucleotide-triphosphate (ddNTP) that will terminate polymerase extension upon its incorporation; and (c) determining masses by MALDI-MS. The qualitative and quantitative composition of the mixture are based on knowledge of the local sequence and the assumption of potential mutations. The expected length and mass of the oligomer produced can be predicted from the known wild-type sequence. A major advantage of the technique is that mutations can be detected up to several bases downstream from the 3′-end of the primer. The first application was the detection of six mutations of the cystic fibrosis transmembrane regulator gene.[449] The determination of the size of microsatellite repeat regions and the detection of mutations of the receptor tyrosine kinase (RET) proto-oncogene are detailed later.[441]

MALDI on Chips

Recent trends toward miniaturization and "parallelization," aiming for speed, high throughput, and automation in diagnostic schemes, have resulted in the development of the "DNA chip." This is a small device that holds an array of a large number of DNA molecules for subsequent assessment using highly selective hybridization of fluorescence-labeled template molecules (reviewed in Reference 450). To demonstrate the technique, <μL aliquots of DNA analytes, mixed with the MALDI matrix, were dispensed with piezoelectric pipettes into 10 × 10 arrays of nL size wells that were chemically etched onto chips (12 × 12 × 0.5 mm). Because the size of these surface pits was about the same (or less) as the incident laser spot size, there was no need to make the usual multiple laser shots to search for a region that provided a strong signal (the "sweet" spot). Additional

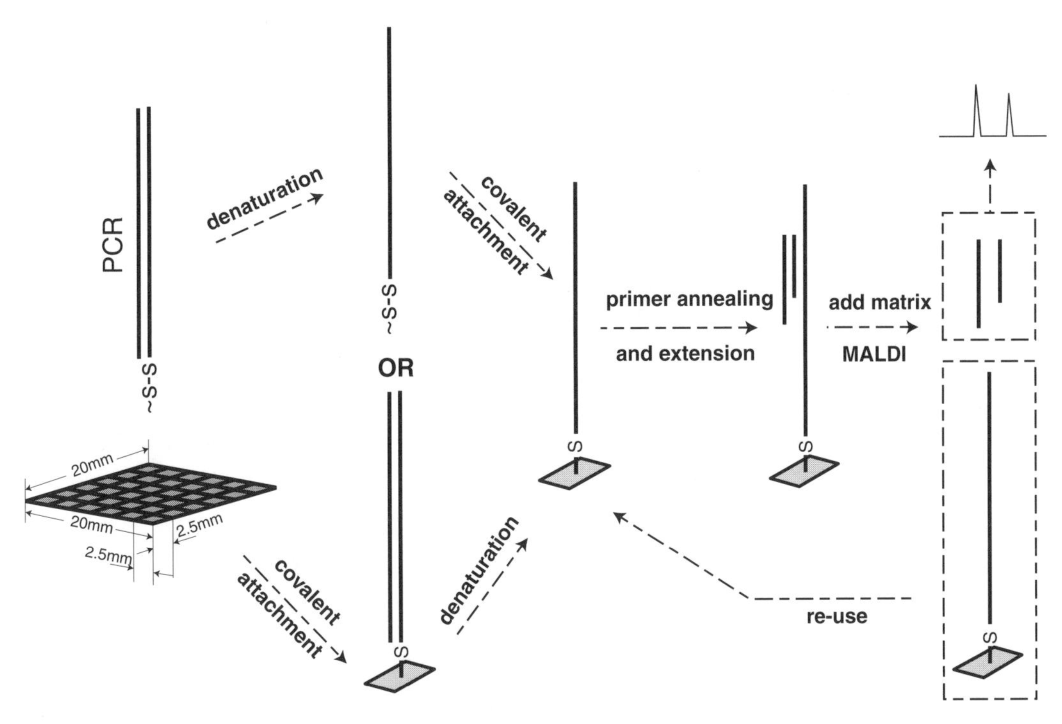

FIGURE 6.61 Schematic overview of PCR product immobilization, PROBE reaction, and MALDI-MS on a chip. Reprinted from Tang, K. et al., *Proc. Natl. Acad. Sci. USA,* 96, 10016–10020, 1999. With permission.

advantages over conventional MALDI include the small sample quantity required, the speed of analysis, reproducibility, and potential for automation.[442,451]

Other improvements in chip-based sample preparation technology have contributed to the reduction of sample requirements from μL to nL. In one approach, PCR products were prepared off-line (with a thiolated primer) and attached covalently to a silicon chip (36 wells, 6 × 6 array) using N-succinimidyl (4-iodoacetyl)aminobenzoate chemistry, a modification of previously established silicon dioxide derivatization chemistry.[452] Next, the immobilized double-stranded PCR products were denatured into single-stranded DNA in the 1 μL size wells where the PROBE reactions were carried out directly without using conventional sample tubes or microtiter plates. Alternatively, sample preparation consisted of biotinylating a nonthiolated PCR primer, denaturing the PCR products, and attachment via the biotin to the silicon surface for subsequent PROBE reactions (Figure 6.61). After adding nL quantities of MALDI matrix, the chips were transferred into the ion source of the TOFMS. The advantages of the technique were demonstrated by the unambiguous identification of the specific alleles of DNA in human platelet alloantigen models.[453]

Short Oligonucleotide Mass Analysis (SOMA)

In this approach, a type IIS restriction enzyme site is incorporated into the PCR primer that is being used to amplify a region around the SNP site. Because the cleavage sites for this enzyme are distal to the recognition site, the restriction digestion yields short fragments that can be analyzed using LC/ESIMS. Both alleles are analyzed, and both sense and antisense strands of the DNA fragments can be used to determine genotypes. An application involved detection of the polymorphism associated with a twofold increase in colon-cancer risk in Ashkenazi (eastern European) Jews. Seven variants within the APC tumor suppression gene were analyzed. A perfect correlation was obtained in a comparison with DNA sequencing.[454]

In an alternate method using ESIMS, the ends of the amplified PCR products from human mutant DNA were removed by digestion with EcoRI restriction enzyme to liberate the 3′-end products. Detection of a one-base substitution (point mutation) was possible from the mass spectra resulting from those double-stranded DNA of the PCR products that denatured in ESI. The resulting groups of ions could be assigned to sets of complementary single-stranded DNA from wild-type and mutant alleles that had been enzyme digested.[455]

6.3.2.4 Detection of Mutation

Analysis of Point Mutation of tRNA by MALDI-MS

In a two-dimensional analysis approach, native PAGE served as the first dimension to search for mobility shifts based on nucleotide substitutions in tRNA fragments. In the second dimension, the separated tRNA fragments were analyzed by MALDI-TOFMS. Because these gel-separated tRNA were not amenable to direct analysis by MALDI-TOFMS, three off-line strategies were explored for coupling the two techniques: (1) direct extraction from the gel into a buffer solution; (2) electroblotting followed by dissolution of the membrane (nitrocellulose) in the MALDI matrix; and (3) direct desorption from the membrane. Membrane dissolution was found to be the most convenient approach. Demonstrations included discrimination of a three-nucleotide deletion mutant, at the 3′ end of $tRNA^{Val}$ and 12 single-base substitutions, from the wild-type. Typical sample sizes were 40 pmol.[456]

Mutation Detection Using Tagged Synthesized Peptides

The *protein truncation test* (PTT) for the rapid detection of translation-terminating mutations is based on three steps: (1) creation of a PCR amplification product of a continuous region of coding sequence using a 5′ primer that consists of a polymerase binding site and a translation initiation site that is in-frame with the 5′ end of the test sequence; (2) generation of a radiolabeled test peptide from the PCR product in a coupled *in vitro* transcription/translation process; and (3) analysis by SDS-PAGE and audioradiography.[457] This technique was modified, using the eight-residue Flag sequence (DYKDDDDK), to yield PCR products that are tagged at their amino termini. The analytes (i.e., the test peptides) could then be separated from all other constituents of cell lysates using M2 anti-Flag monoclonal antibody coupled to agarose beads. This is a high-affinity ligand for the tag. The purified, tagged peptides were analyzed by MALDI-TOFMS in positive-ion linear mode. The results were interpreted using the *Prowl* program.[458] A relevant application is described in Section 6.3.3.5.

6.3.3 DNA Sequencing and Genetic Diagnosis

6.3.3.1 Sequencing Exons of the p53 Gene

A technique was developed in which target amplification and sequencing by MS could be carried out in a single tube. The p53 tumor suppression gene was selected as the model because of the large number of mutations in this gene that are associated with several known malignancies. Most mutations are clustered in the exons 5 to 8 area.[459] In the single tube process, exons of a commercial, pooled, genomic DNA template were amplified in a 96-well microtiter plate by PCR using flanking primers from the intron region. The amplified products were immobilized in the same wells with streptavidin-coated magnetic beads. Single-stranded DNA was obtained upon treating the beads with NaOH. The "primer walking" technique was selected to sequence 21 selected primers. The four selected termination reactions resulted in 84 sequencing reactions, all on a single PCR microtiter plate. After digestion with thermosequenase, the sequence ladders were removed from the template with NH_4OH, MALDI matrix was added, and spectra from 250 laser shots were accumulated for each sample. Using the 21 primers, 544 bases of exons 5 to 8, and 126 bases of the flanking introns were sequenced. Consuming 50 fmol of each analyte, each mass spectrum yielded ~35 bases. The accuracy of mass measurements was ~0.05%. The data obtained for all consensus sequences of exons 5 to 8 of the p53 gene were unambiguous.[436]

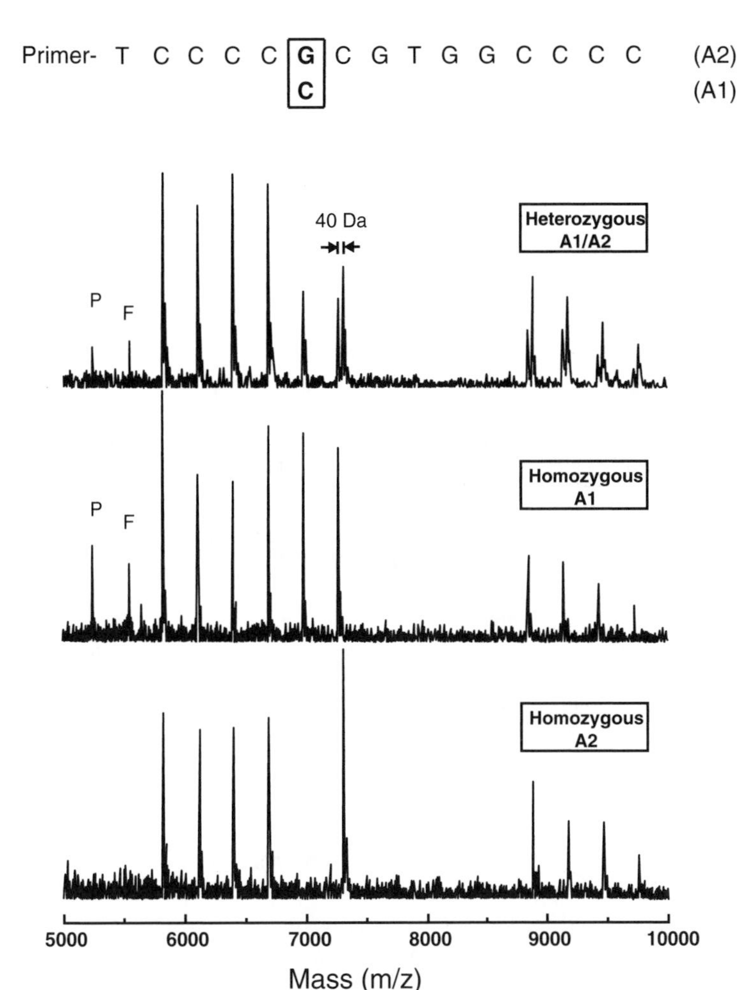

FIGURE 6.62 Sequencing of a polymorphic site of exon 4 (Arg72Pro) with primer d(AGAATGCCAGAG-GCTGC). A portion of G to C transversion was identified. P, unextended primer; F, false stop. Reprinted from Fu, D.J. et al., *Nature Biotechnol.,* 16, 381–384, 1998. With permission.

A potential application of sequencing by MS is the unambiguous detection of heterozygotes by observation of mass-shifted ladders. The common polymorphic site of p53 exon 4 was investigated in DNA from ten unrelated individuals. An identified G → C transversion permitted differentiation of homozygous and heterozygous alleles based on the 40 Da mass shift (Figure 6.62). In addition to listing the advantages of the MS approach over gel-based schemes, it was emphasized that the high throughput and parallel sample processing techniques developed are amenable to automation.[436]

6.3.3.2 Identifying Microsatellite Alleles

A "short tandem repeat" (STR) or *microsatellite* is a mono-, di-, or tetranucleotide sequence that is repeated, e.g., the commonly occurring CA dinucleotide (GT on the opposite strand) or just one base such as AAAAAAA (T on the opposite strand). These sequence polymorphisms typically repeat 5 to 30 times and are scattered throughout the entire genome. When the length of the microsatellites begins to vary as a result of the loss of mismatch repair genes, mutations occur in adjacent genes. This may eventually affect an oncogene or a tumor suppression gene. A general

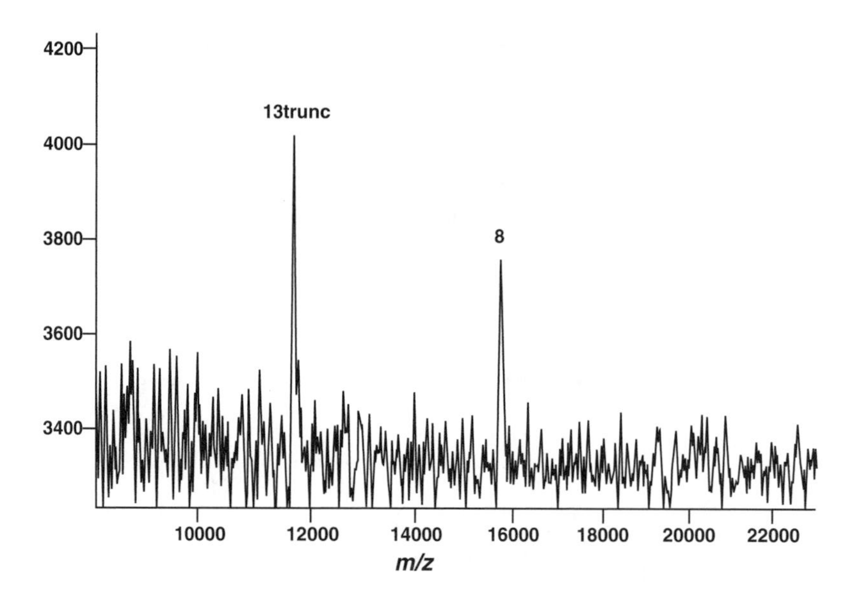

FIGURE 6.63 PROBE analysis of a sample heterozygote for an AAAT tetranucleotide repeat microsatellite located within intron 5 of the interferon-α receptor gene. 13trunc, truncated product of a 13-repeat allele (*m/z* 11,643). 8, complete tetranucleotide repeat product (*m/z* 15,718). Reprinted from Little, D.P. et al., *Anal. Chem.*, 69, 4540–4546, 1997. With permission.

perspective of STR in genomics and the current state of ESI-FTICRMS for the characterization of STR has been published.[460]

The MALDI-on-a-chip approach was applied to the identification of the AAAT repeat region in an AluVpA microsatellite within intron 5 of the interferon-α receptor located on human chromosome 21.[461] Using PROBE to process a heterozygote followed by MALDI-TOFMS analysis of the diagnostic products derived from a 50 μL PCR reaction, an unknown allele was discovered. Sequencing confirmed 13 repeats that contained a second site mutation of T → G within the *eighth* repeat. The PROBE primer was a 14-mer sense oligonucleotide that annealed adjacent to the beginning of the microsatellite on the antisense strand. The nucleotide mixture contained dATP, dTTP, dCTP, and ddGTP. MALDIMS showed that one product (M_r 11,643 Da) corresponded to the truncated tetranucleotide repeat segment that contained the mutation. A second member of a doublet, at M_r 15,718 Da, was identified as a fragment representing the complete (8-unmodified)-tetranucleotide repeat product (Figure 6.63), confirming this analyte to be heterozygote at this position.[451] Significant information on polymorphism, not possible with conventional electrophoretic sizing methods, was obtained for 28 unrelated individuals.[441]

6.3.3.3 Familial Adenomatous Polyposis (FAP)

Familial adenomatous polyposis (FAP) is an inherited, autosomal dominant disease characterized by the development of hundreds to thousands of polyps in the affected individuals during their 20s or 30s. FAP is associated with a high risk of colon cancer. Genomic DNA was isolated from 1 mL of blood from normals and patients with FAP who carried a confirmed (by mutation analysis) 5 bp AAAGA deletion at codon 1309 of the APC tumor suppression gene. This gene is mutated in some 85% of colorectal cancers. Next, a segment of the gene was amplified by PCR to produce 57 bp normal products or 52 bp (5 bp deleted) double-stranded DNA segments. The PCR products were purified, digested with EcoRI restriction endonuclease, and further purified. The ESI mass spectrum of the patient sample revealed four components (Figure 6.64) that were assigned to sense and antisense strands of the deleted and normal alleles, while the PCR products from the control subject

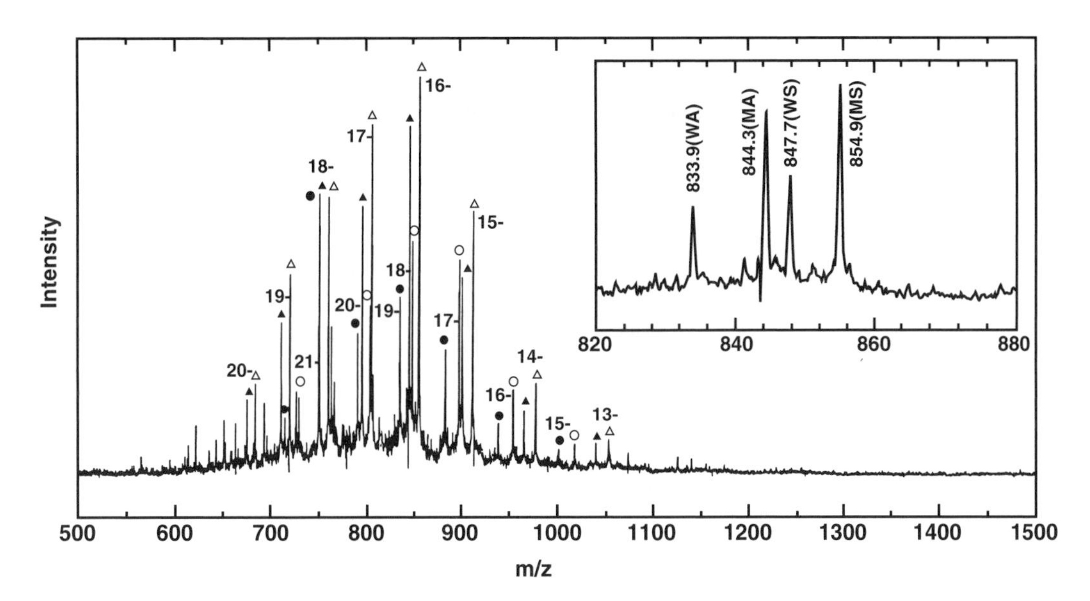

FIGURE 6.64 Mass spectrum of EcoRI digested PCR products amplified from an FAP family member. The sense and antisense strands of the digested PCR products from normal allele are indicated by open and filled circles, respectively. The sense and antisense strands from the deleted allele are indicated by open and filled triangles, respectively. The inset shows the fine structure of ion peaks assigned to the ion fragments with 18- charge from the normal (wild-type) sense strand (WS) and normal antisense strand (WA), and the ion fragments with 16- change from the deleted (mutant type) sense strand (MS) and deleted antisense strand (MA), respectively. See Table 6.11 for mass value comparisons. Reprinted from Naito, Y. et al., *J. Am. Soc. Mass Spectrom.*, 8, 737–742, 1997. With permission.

showed only two groups of peaks. The measured masses assigned to the sense (*m/z* 13,693 ± 2.1 Da) and antisense (*m/z* 13,522.5 ± 2.3 Da) strands of the deleted allele were within 0.6 to 1.3 Da of the expected values (Table 6.11). The heterozygosity revealed by the results of the ESI analysis was consistent with expectations based on the fact that the proband (the patient or member of the family that brings the family under study) inherited the mutation from an affected parent, hence only one allele included the deletion. These results confirmed only the deletions but not the sequences. It was concluded that the methodology would allow differentiation of a single substitution, i.e., A → T, that corresponds to a 9 Da mass difference.[462,463]

6.3.3.4 Multiple Endocrine Neoplasia Type 2A (MEN2)

This is a familial autosomal disorder that affects 1 in 30,000 individuals. Multiple endocrine neoplasia type 2A (MEN 2) is characterized by tumors of the thyroid C cells, often in association with hyperplasia or adenoma of the parathyroid. The predisposing gene is RET, a receptor tyrosine kinase that maps to chromosome 10q11.2. Mutations are located on the protooncogene at codon 634. Because RET mutations have been identified in virtually all families with inherited or familial MEN 2 and medullary thyroid carcinoma, identification of the specific RET proto-oncogene mutations has been an important advance.

In one approach, diagnostic products for MALDI-TOFMS analysis were obtained from extracted genomic DNA with the PROBE technique (using a 20-mer primer) to obtain exon 11 of the RET protooncogen that had been amplified by PCR using Taq polymerase. A purified pellet was mixed with the matrix and analyzed by MALDI-TOFMS. The mass spectrum of the homozygous wild-type sample revealed that the PROBE reaction with ddATP + dNTPs (N = T,C,G) caused a mass shift of 591 Da from the primer, i.e., from 6135 Da to 6726 Da (Figure 6.65a). This increase

TABLE 6.11
Expected and Measured Masses of PCR Products with Corresponding Sequences

<table>
<tr><th>Expected PCR products</th><th>Expected Mass</th><th>Measured Mass</th></tr>
<tr><td>(Normal sense strand)–</td><td></td><td></td></tr>
<tr><td>5′–AATTCCCTGCAAATAGCAGAAATAAAAGAAAAGATTGGAACTAGGTCAG–3′</td><td>15,275.8</td><td>15,277.2 ± 1.8 (a)
15,275.2 ± 1.6 (b)</td></tr>
<tr><td>|||||||||||||||||||||||||||||||||||||||||</td><td></td><td></td></tr>
<tr><td>3′–GGGACGTTTATCGTCTTTATTTTCTTTTCTAACCTTGATCCAGTCTTAA–5′</td><td>15,029.8</td><td>15,028.5 ± 4.4 (a)
15,029.0 ± 1.3 (b)</td></tr>
<tr><td>(Normal antisense strand)</td><td></td><td></td></tr>
<tr><td>(Deleted sense strand)</td><td></td><td></td></tr>
<tr><td>5′–AATTCCCTGCAAATAGCAGAAATAAAAGATTGGAACTAGGTCAG–3′</td><td>13,693.8</td><td>13,693.2 ± 2.1</td></tr>
<tr><td>|||||||||||||||||||||||||||||||||||||</td><td></td><td></td></tr>
<tr><td>3′–GGGACGTTTATCGTCTTTATTTTCTAACCTTGATCCAGTCTTAA–5′</td><td>13,523.8</td><td>13,522 ± 2.3</td></tr>
<tr><td>(Deleted antisense strand)</td><td></td><td></td></tr>
</table>

Note: Underline indicates repetitive sequence that was affected from the deletion.

Note: Sense and antisense are offset by 4 bases.

(a) Values were obtained from the normal DNA allele of the FAP patient; (b) values were reported previously for the normal volunteer.

Reprinted from Naito, Y. et al., *J. Am. Soc. Mass Spectrom.*, 8, 737–742, 1997. With permission.

in mass represented extension until the expected ddA. When the PROBE reaction used ddT instead it yielded a single peak at 8248 Da (Figure 6.65a, right side), consistent within 0.02% with the addition of the expected A_2C_3GddT. In the case of patient 1, the ddA reaction resulted in an unresolved peak at *m/z* 6731 Da, representing the expected value for the gene products heterozygous for wild-type and a C → T mutation (Figure 6.65b, left side). However, the ddT reaction yielded two clearly resolved peaks that represented the heterozygote. These were the wild-type at 8249 Da (0.04% mass error) and the C → T mutant at 6428 Da (0.08% mass error) (Figure 6.65b, right side). Similarly, data for patient 2 represented heterozygous wild-type C → A mutation: ddT reaction led to a single peak representing unresolved wild-type and C → A alleles (Figure 6.65c). The diagnosis in patient 2 might have been incorrect due to constraints of the low-resolution (~200) mass determinations, necessitating two termination reactions (ddA and ddT) for unambiguous identification of the two known mutations.[434]

As an alternate approach, allele differentiation was attempted by direct determination of the masses of the diagnostic products at high resolution, using MALDI-FTICRMS. Three synthetic 25-mer allelic sense oligonucleotides of the RET gene were prepared that included the codon at position 634 as the wild-type, C → T mutant, and C → A mutant. Positive ions were obtained from 1-10 pmol DNA in an external ion source and transferred into a 6.5 T magnetic field with the aid of an rf-only quadrupole, where they were trapped collisionally for subsequent mass analysis of the most abundant isotope peaks at high resolution, 1 to 5 × 10^4. Using this technique, the wild-type (7660.3 Da), and the G → A (7644.3 Da) and G → T (7635.3 Da) mutant alleles could be distinguished from each other, even in a mixture where masses were as little as 9.1 Da apart (Figure 6.66). It is noted that these MS techniques required neither labeling nor electrophoretic separation of the diagnostic products.[430]

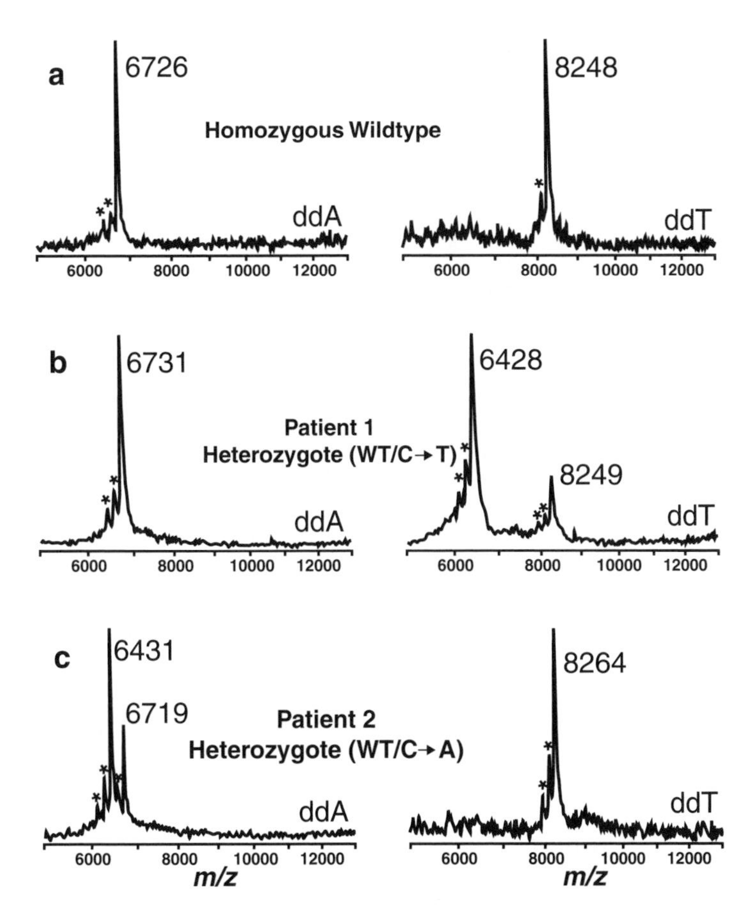

FIGURE 6.65 PROBE product spectra of the low-resolution MALDI-TOFMS of the RET oncogene for (a) wild-type; (b) Patient 1; and (c) Patient 2. Average M_r values are reported. Asterisks refer to fragmentation peaks caused by depurination of diagnostic products. Reprinted from Little, D.P. et al., *J. Mol. Med.*, 75, 745–750, 1997. With permission.

6.3.3.5 Breast Cancer Susceptibility Gene BRCA1

The MALDI-TOFMS mutation detection technique, based on tagged synthetic peptides (Section 6.3.2.4), was used to study truncations and substitutions in peptides coded for by the BRCA1 gene. In some patients with breast cancer, there is a mutation consisting of the insertion of a single adenosine in a stretch of seven adenosines at position 1129 in exon 11 of the BRCA1 gene. Two peptides were identified by MALDI-TOFMS in a sample derived from a patient (Figure 6.67B). The peptide at 3419.83 Da was also present in a control sample and represented the wild-type allele (Figure 6.67A). A second peptide, at 3114.49 Da, corresponded to the mutant peptide at the predicted mass plus one Na cation. The same technique was also used to detect another mutation that resulted in the replacement of a cysteine residue by glycine in the zinc-finger domain of the BRCA1 product. A sample derived from a subject with two wild-type alleles yielded only one peptide, at 3530.13 Da (with one Na), which was encoded by codons 51-71 of BRCA1 (Figure 6.67C). The spectrum from a carrier for the C61G mutation included a second peptide, at 3484.25 Da (including one Na). This was within one mass unit of the expected mass corresponding to the Cys to Gly substitution that yields a reduction of 45.99 Da (Figure 6.67D). It was concluded that the mass measurement accuracy of MALDI-TOFMS of 200 ppm (linear mode) would permit the detection of 366 of the 380 possible single amino acid substitutions, i.e., mass changes from 0 to 186 Da, as long as the amplicon contained <200 bases encoding for a <10 kDa

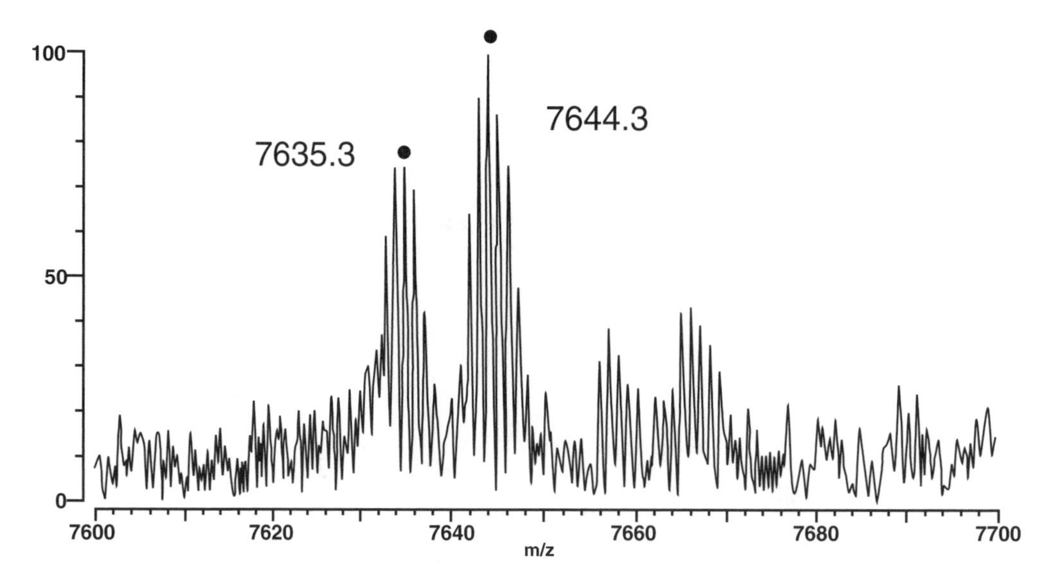

FIGURE 6.66 MALDI-FTICRMS spectrum of allelic 25-mer synthetic oligonucleotides representing the sense strand of a part of exon 11 of the RET proto-oncogene covering codon 634. The figure shows wild-type G → A/G → T heterozygote. Masses of the most abundant isotope peaks (filled circles) as expected from theoretical distributions. Spectra were also obtained for wild-type, G → A and G → T homozygotes and wild-type/G → T and G → A/G-T heterozygotes (not shown). Reprinted from Little, D.P. et al., *J. Mol. Med.,* 75, 745–750, 1997. With permission.

peptide. Compared to chip-based array strategies, this approach has the advantage that it can be used to detect unreported insertions and deletions.[458]

6.4 CARBOHYDRATES AND GLYCOCONJUGATES

6.4.1 Carbohydrates

6.4.1.1 Nomenclature and Techniques for Collision-Induced Dissociation

The nomenclature suggested for the fragmentation of carbohydrates is straightforward (Figure 6.68). Ions that retain the charge at the reducing terminus are designated X (cross-ring), Y, and Z (glycosidic). The complementary sets of ions are A, B, and C. For the X, Y, and Z ions the sugar rings are numbered from the reducing end and for A, B, and C ions they are numbered from the nonreducing end. The fragment type is shown by a subscript following the letter designation. Greek letters distinguish fragments from branched chain glycans. Superscripts show the bonds cleaved in cross-ring fragments.[464]

Ever since its development, FABMS has been used to analyze large carbohydrates and glycoconjugates (reviewed in Reference 465). More recently, MALDI-TOFMS has become the method of choice for both free carbohydrates and the carbohydrate portions of glycoconjugates. A review includes discussion of the selection of the all-important MALDI matrices as well as sample preparation and quantification techniques.[466] Post-source decay (PSD) is used to provide sequencing information[467] and an automated technique has been developed for the rapid processing and interpretation of the PSD spectra of oligosaccharides.[468]

Although ESI cannot match the sensitivity of MALDI in most cases, nanospray techniques, methods of permethylation, and reducing-terminal derivatization have received attention (reviewed in Reference 469). One approach combines permethylation and chromatographic separation with

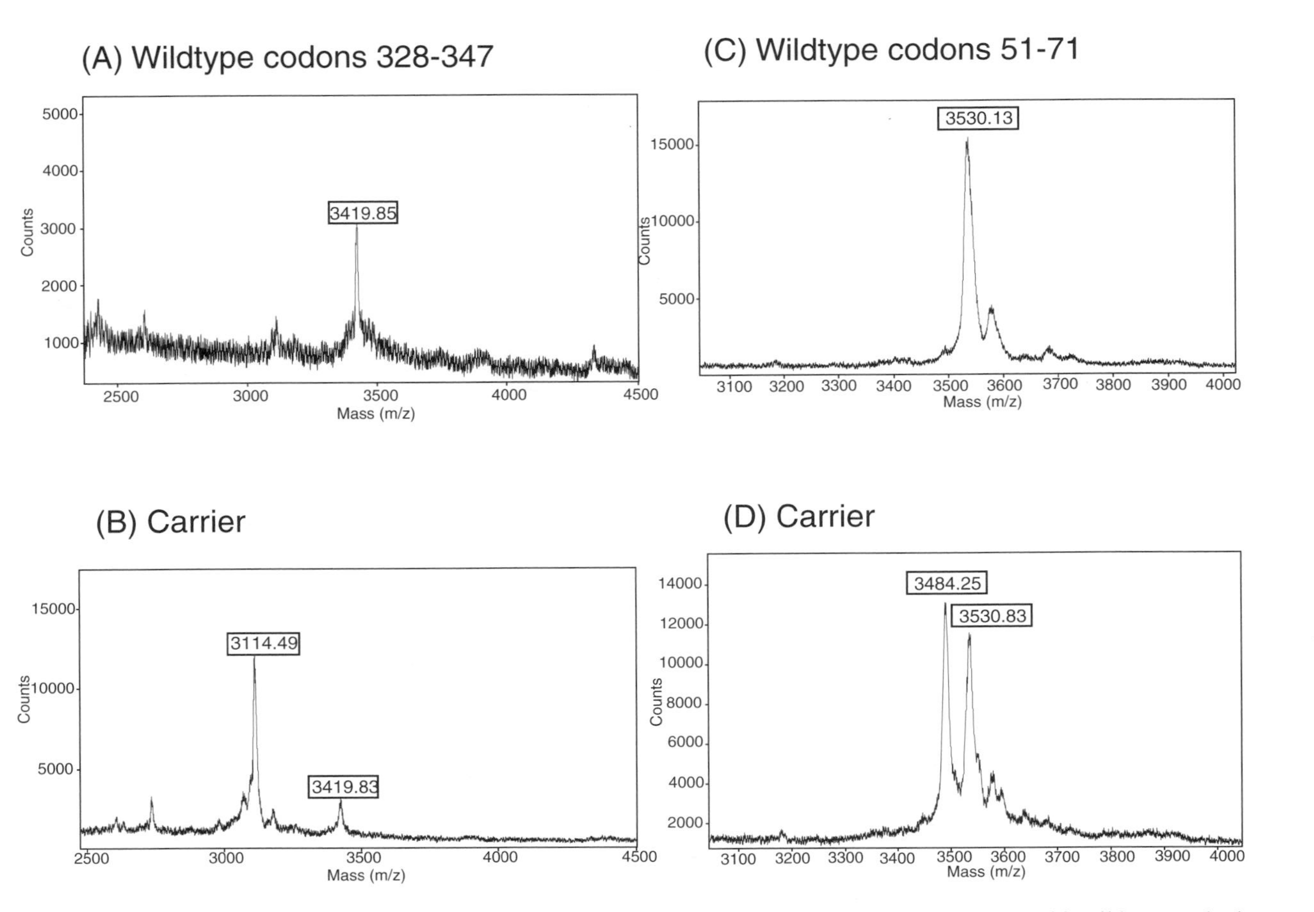

FIGURE 6.67 Spectra of peptides encoded by BRCA1. (A) Spectrum of codons 328-347 from a person with wild-type at both alleles; (B) Spectrum of a carrier heterozygous for a single base insertion at position 1,129; (C) Spectrum of codons 51-71 from a person with two wild-type alleles; (D) Spectrum of a carrier heterozygous for a T300G substitution in the BRCA1 cDNA sequence (GenBank accession no. U14680). Reprinted from Garvin, A.M., Parker, K.C., and Haff, L., *Nature Biotechnol.*, 18, 95–97, 2000. With permission.

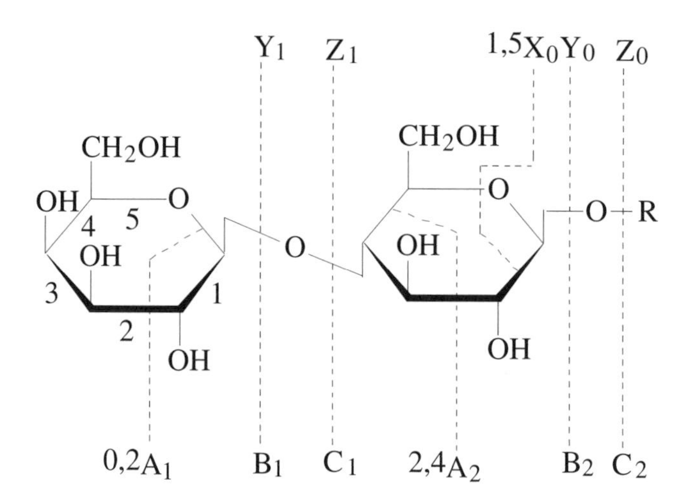

FIGURE 6.68 Scheme for fragmentation of carbohydrates. Based on Domon, B. and Costello, C.E., *Glycoconjugate J.,* 5, 397–409, 1988.

the MS^n capability of IT to characterize components of complex oligosaccharide mixtures. The technique has been tested by analysis of soluble CD4 containing various levels of sialic acid.[470]

6.4.1.2 Sialic Acids

Sialic acids are derivatives of neuraminic acid where the amino group has been modified with either an acetyl or glycolyl group. The presence of a negative charge on the carboxyl group enables sialic acids to participate in several biological processes, such as influencing glycoprotein conformations within cell membranes. The possible role of sialic acid as a tumor marker is based on observations of sialylation and sialyltransferase activity in many tumor cells (reviewed in Reference 471).

A GC/CIMS technique using SIM was developed for the determination of fully silylated (seven TMS groups) N-acetylneuraminic acid (NANA) in tumor cells (i.s.: neuraminic acid β-methylglycoside). The method was used to determine both the free and neuraminidase-susceptible NANA concentrations in tumor cells. When combined these numbers provided a total NANA concentration for the cells examined. The NANA content of neuraminidase-treated leukemic cells used in immunotherapy was ~0.3 μmol/10^9 cells with only 5×10^4 cells being required for the analysis. The total NANA content of a variety of normal, leukemic, and solid tumor cells was in the range of 0.05 to 10.0 μmol/10^9 cells.[472]

The expression of N-glycolylneuraminic acid (NeuGc) in the glycosphingolipid fractions of several types of human tumor tissues has also been measured using a GC/EIMS technique (i.s.: deuterated analog of NeuGc) and SIM. The quantity of NeuGc detected in the tumors was 0.02 to 0.5% of the total sialic acid content. No NeuGc was detected in glycosphyngolipids from normal tissues.[473]

6.4.1.3 Polysaccharides and Glycosaminoglycans

Large, linear polysaccharides that have been implicated in tumor-related processes include the chondroitins (2-amino-2-deoxy-D-Gal and D-GlcA) and hyaluronic acid (hyaluronan; the disaccharide repeat unit is glucuronic acid + 2-acetamido-2-deoxy-D-glucose). Glycosaminoglycans are highly sulfated glycans that consist of polymers of uronic acid-glucosamine repeat units in which the amino or hydroxyl groups are sulfated to various degrees, e.g., heparin. These glycans fragment extensively during MALDI ionization due to loss of sulfate. One approach to analyzing these compounds has been to form noncovalent complexes with synthetic peptides with the intention of equaling the number of sulfates with basic arginines and spacing those arginines with glycine residues.[474,475]

Decomposition of Hyaluronan (HA) by Hyaluronidase (H'ase) in Effusions from Patients with Malignant Mesothelioma

Given that hyaluronan contains ~3000 disaccharide units, 379 Da each, there is an average molecular mass of ~1.1 MDa. Negative ESI has been used for monitoring HA because it provides: (a) high sensitivity for oligosaccharides; (b) poor sensitivity for proteins, thus reducing interferences from H'ase and other protein constituents of malignant effusions and sera; and (c) the possibility of measuring very high masses with narrow mass range scanning using QqQ analyzers. HA is present at high concentrations in the pleural and peritoneal effusions of mesothelioma patients although its biological relevance is not well understood. Direct analysis of the polymer is not practical because of the high molecular weight of HA. The assessment of HA is therefore undertaken indirectly through treatment of samples with H'ase during which the rate of degradation and the nature of the products are determined. When HA in pleural fluids, sera, and an HA standard were examined the results in all cases were comparable, with no HA being found in control fluids. Spectra obtained over time revealed that H'ase acts both by cleaving disaccharides and making scissions in the middle of the chains. Thus, larger molecular mass species were encountered early in the incubation, and profiles of the multiply charged ions observed could be used for the determination of the higher mass decomposition products (Figure 6.69, a and b). Over time the products found corresponded to a range of disaccharide units of lesser molecular mass (Figure 6.69, c, right ordinate). Longer incubations created products with smaller molecular masses (Figure 6.69, c). Finally, the spectra of the end products of 24 h enzymatic incubations, consisted of singly and doubly charged ions of di-, tetra-, hexa-, octa-, and decasaccharides. The size of the decomposition products and the mechanism of H'ase action were similar in the effusions and sera despite $>10^4$-fold higher concentrations of HA in the former.

Sequencing of Specific Protein-Binding Glycosaminoglycans

Noncovalent interactions between proteins of 15 to 60 kDa size and biopolymeric saccharides have received increasing attention due to their critical influence on the stability and function of growth factors and other proteins such as endostatin. The heparin/heparan sulfate-like glycosaminoglycan family (HLGAG) appears to be especially important because of the chemical diversity that results from the differential sulfonation of individual disaccharide units. A new strategy, *surface-noncovalent affinity* mass spectrometry, has been developed for the isolation, enrichment, and sequencing of the HLGAG oligosaccharides that bind to a specific protein. The first step of this strategy is immobilization of the protein of interest on a thin hydrophobic film or metal surface followed by addition of candidate oligosaccharides in an aqueous solution. After removal of nonspecific species by washing, a synthetic basic peptide, $RG_{19}R$, in caffeic acid matrix, is added. The basicity of this peptide gives it a very high affinity for polysachharides, stronger than occurs with the native-binding proteins, therefore the complexes formed can be subsequently analyzed by MALDI-TOFMS. Because the HLGAG oligosaccharides may be depolymerized, enzymatically or chemically, before chelation, there is a possibility for obtaining complete sequence information. The technique has been demonstrated by determining the carbohydrate portions of several complexes including fibroblast growth factor-2.[476]

6.4.2 Glycosylation

The covalent glycosylation of proteins and lipids, a ubiquitous phenomenon, is considered of fundamental biological importance (reviewed in Reference 477). The carbohydrate components of glycoconjugates, called glycans, are often monosaccharides or linear or branched oligosaccharides. Glycoconjugates are major components of the outer surface of mammalian cells and the glycan components are directed into the outer environment from the membranes. The longer glycan chains of *glycoproteins* may protrude far from the membrane while the shorter glycan chains of *glycolipids* (<12 monosaccharide units) usually remain in the immediate vicinity of the cell surface. The glycan

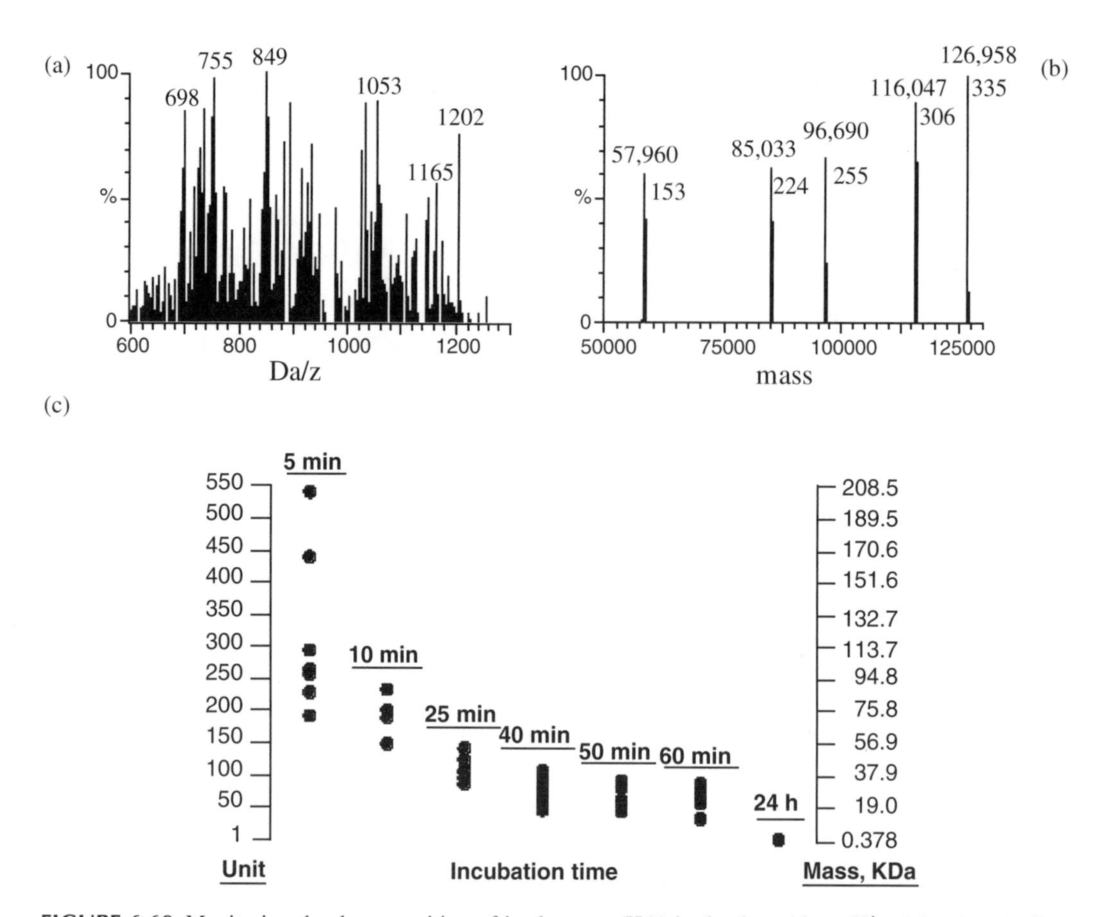

FIGURE 6.69 Monitoring the decomposition of hyaluronan (HA) by hyaluronidase (H′ase) in pleural effusions from patients with malignant mesothelioma. The disaccharide unit in HA is glucuronic acid + 2-acetamido-2-deoxy-D-glucose. There were no interferences from H'ase and the protein constituents of the fluids and sera when negative ESI was used. (a) Mass spectrum of the products obtained after a 5-min incubation with HA; (b) Transformation of (a) revealed the molecular masses of the products; (c) In addition to removing disaccharides stepwise, H'ase also made scissions in the middle of the chains, producing a number of products corresponding to a range of disaccharide units (c, left ordinate).

components of glycoconjugates have important biological roles: they protect against outside chemical or proteolytic attack; decrease immunogenicity; control glycoconjugate uptake by cells; and contribute to intercellular recognition. Alterations in the oligosaccharide sequence and other changes of the structure of cell-surface glycoconjugates have often been implicated in malignant transformation and metastatic spread. For example, proteome and glycosylation mapping have identified posttranslational modifications associated with aggressive breast cancer (reviewed in Reference 478).

The discrete and multifaceted recognition of glycoconjugates is due to the large number of permutations possible based on both the number of types of monomer and the linkages. These residues can therefore provide a huge storage of biological information, further enhanced by additional diversity due to branching. For example, 20 different monosaccharides may form ~10 million linear or branched trisaccharides in contrast to the ~8000 tripeptides that can be formed from the 20 amino acids. For hexamers the difference is seven orders of magnitude. However, only a fraction of these combinations of oligosaccharides has been used in biological systems, probably due in part to a lack of appropriate glycosyltransferases. Even beyond this level of variability there is the ubiquitous addition of phosphate, acetate, and sulfate substituents providing additional diversification.

Direct determination of O-glycosylation sites in peptides has been accomplished using high-energy CID with a magnetic instrument of EBEB design.[479] In subsequent studies, low-energy CID was successful when used with the high dynamic range of nano-ESI-QTOFMS.[480,481] O-glycosylation sites have also been determined by sequencing N-alkylated peptide fragments using ESI-QTOFMS.[482] An alternative approach has used electron capture dissociation with nano-ESI-FTICRMS to obtain c-type ions that provide direct evidence of the glycosylation sites.[483] O-GlcNAc-modified peptides have been distinguished from those not modified by exploring the lability of the modifying moiety to low-energy CID collisions.[484]

O-fucosylation is an unusual posttranslational modification suggested as being involved in cell signaling and metastasis. Earlier studies of O-fucosylation included peptide mapping by ESIMS and estimation of the status of fucosylation from the apparent loss of 146 Da in the ESI source.[485] A nano-ESI-QTOFMS technique was developed for the direct determination of the status and sites of glycosylation in O-fucosylated glycopeptides. To obtain useful fragmentation patterns, the experimental conditions used for the CID had to be modified to reflect the very labile linkage present in O-fucosylation. In conventional CID the neutral loss of sugar moieties provides straightforward evidence for the glycosylation status of the analyte peptide;[486] however, the glycosylation site cannot be determined. Accordingly, collision energy and collision gas pressure were set significantly lower than normal and the period of data acquisition was prolonged to 2 h. Under these reduced energy conditions structurally informative ions could be obtained. A disadvantage of this approach is that it cannot be coupled with HPLC in its present form. The technique has been applied to the determination, in a single analysis, of the glycosylation status, the glycan structure, and the site of glycosylation of thrombospondin-1, a large glycoprotein with several postulated tumor-related properties.[487,488]

6.4.3 Glycoproteins

Glycosylation of proteins is one of the most important posttranslational events and occurs in >50% of proteins. Glycosylation consists of a series of complex reactions catalyzed by glycosyltransferases and glycosidases in the endoplasmic reticulum and Golgi apparatus (briefly reviewed in Reference 489). N-Glycosylproteins usually contain chains of 7 to 25 monosaccharide units that are linked to asparagine residues of the protein through one specific type of N-glycosyl bond, i.e., N-acetylglucosaminyl-asparagine. The relevant asparagine residues (N) occur at the well-defined consensus sequence N-X-S/T/C, where X can be any amino acid except proline. In contrast, in O-glycosylproteins, there are four types of O-glycosidic linkages all of which are alkali-labile: (a) *mucin*-type, between N-acetyl-D-galactosamine and L-serine or L-threonine, GalNAc(α1-3)S or GalNAc(α1-3)T; (b) *proteoglycan*-type, between D-xyloses and L-serine, Xyl(β1-3)S; (c) *collagen*-type, between D-galactose and 5-hydroxy-D-lysine, Gal(β1-5)OH-L; and (d) *extensine*-type between L-arabinofuranose and 4-hydroxy-L-proline, 1-Araf(β1-4)OH-P. Unlike N-linked glycans, there is no universal consensus sequence for O-linked glycosylation, although several amino acid sequence motifs have been observed for particular glycosylation sites, e.g., mucins (see later).

Some of the general features of the glycan components of glycoproteins include: (a) structurally variable outer arms, or *antennae*, that are thought to participate in a signal recognition function; (b) *inner cores* that are nonspecific, invariant structures linked to the peptide chain; (c) structural *microheterogeneity* resulting from the partial substitution of sugar residues on a similar core structure involving the number and positions of the outermost monosaccharides; (d) the presence of *glycoforms* resulting from the fact that most glycoproteins have more than three sites that can be glycosylated; and (e) that the type and number of glycans strongly affect the 3-dimensional structure of glycoproteins, and consequently their biological function. It has been confirmed repeatedly that the structure of the glycan component of cell membrane glycoproteins is profoundly altered in neoplastic cells.

The full structural characterization of a glycoprotein involves the determination of branching points, types of linkages, and the identification of the carbohydrate isomers. Earlier MS techniques based on EI and CI involved permethylation to confer volatility and release the saccharide from the peptidic portion through β elimination. GC-EIMS is still used for composition and linkage analysis. FABMS was used in pioneering applications concerning the structure of mucins, e.g., for oncofetal carbohydrate antigens from human amniotic mucins,[490,491] for the structure of a monoclonal antibody generated against an amniotic fluid mucin-carbohydrate that recognizes a colonic tumor-associated epitope,[492] and for gangliosides (see later). More recently, both ESI and MALDI have been used in hybrid QTOF combinations (reviewed in References 489, 493, 494). Sequential digestion of particular nonreducing terminal residues followed by mass analysis after each digestion can provide information regarding the heterogeneity of the oligosaccharides at each site and the composition remaining on each glycopeptide. For example, the linkage specifically cleaved by α-mannosidase and β-mannosidase are Man(α1-2,3,6)Man and Man(β1-4)GlcNAc, respectively. With the possibility of measuring the molecular masses of peptides of up to several kDa mass with an error <10 ppm using QTOFMS and <5 ppm using FTICRMS, the information encrypted in the distribution of deltamass (mass value following the decimal point) values can be exploited for the identification and characterization of peptides. Glycosylation and phosphorylation lower the deltamass value because of the high abundance of oxygen (lipidation increases the deltamass owing to the high abundance of hydrogen). The presence of an oligosaccharide in a glycopeptide with a relatively small peptidic part results in a pronounced deviation of the deltamass value. This mass difference may be used for differentiation of peptides and carbohydrates.[495]

6.4.3.1 Mucins

Structure

Mucins are 100-1000 kDa glycoproteins with three structural characteristics: (a) they have high carbohydrate content, 50 to 90% by weight; (b) the center of the polypeptide backbones includes repeating amino acid sequences, *tandem repeats*, rich in Ser, Thr and/or Pro, where the Ser and Thr are O-linked to glycan chains that have varying structures and lengths, according to the tissue type in which they are produced; and (c) frequently occurring N-glycan chains that are appended to the peptide chain outside the tandem repeat sections. The cloning and sequencing of seven human genes (MUC1-7) have provided information on the size of the tandem repeats and the sites of N-glycosylation for these secreted mucins. The functions of four human *secretory* mucins, MUC2, MUC5A, MUC5B, and MUC7 are to lubricate and protect ductal epithelial cells in the gastrointestinal, respiratory, and reproductive tracts. Two human *membrane-bound* mucins, MUC1 and MUC4, participate in various cell-cell interactions while mucins, MUC3, MUC6, and MUC8, are not yet classified.

The MUC1 gene encodes for polymorphic epithelial mucins in which there are sequence variations in the ~20 amino acids tandem repeats. These mucins are overexpressed in tumors of the breast, lung, pancreas, gall bladder, stomach, and kidney. Tumor-associated mucins also demonstrate aberrant glycosylation with the carbohydrate chains being shorter and more highly sialylated than those from normal tissues. Biochemical evidence suggests that mucins may mask surface antigens on cancer cells, thereby protecting them from immunological surveillance. The MUC2 and MUC3 genes are also overexpressed in tumors. The proteins, from all three genes, are released into the circulation where they can be detected by antibodies. Clinical trials are in progress to evaluate whether these proteins, CA 15.3, CA 125, and DuPan-2, can be used for early detection of the recurrence of tumors as well as for monitoring of the efficacy of treatment. Mucins are also being considered as candidate structures in the development of anti-cancer vaccines that will target specific mucin and carbohydrate epitopes.

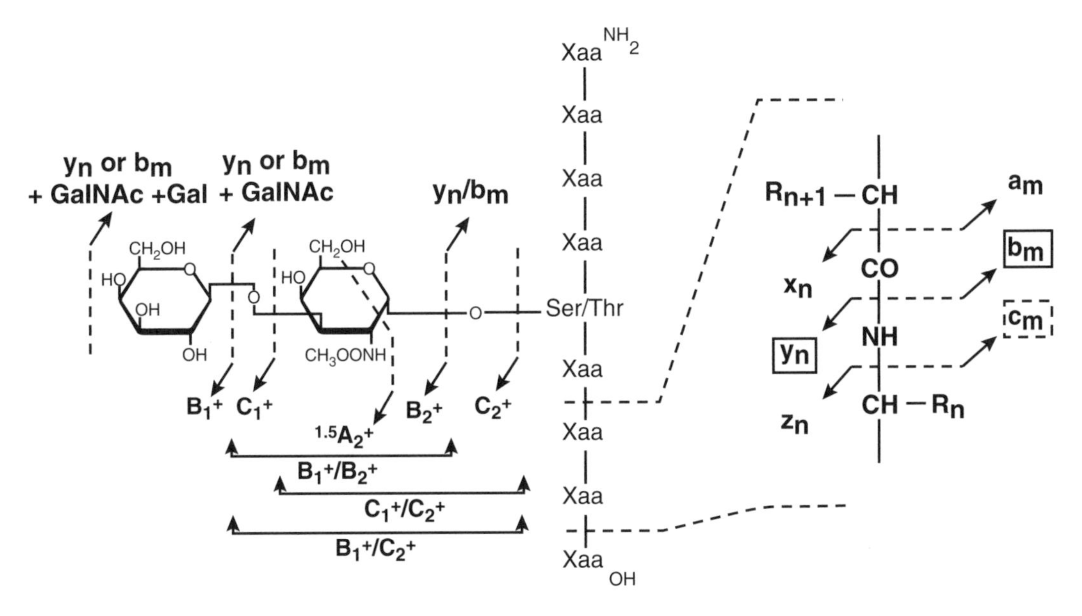

FIGURE 6.70 Nomenclature for glycopeptide fragmentation is based on the nomenclature for peptide fragmentation[96,98] in combination with the nomenclature for the fragmentation of carbohydrates.[464] Reprinted from Goletz, S. et al., *Rapid Commun. Mass Spectrom.*, 11, 1387–1398, 1997. With permission.

Mass Spectral Correlations

A nomenclature suggested for the fragmentation of glycopeptides (Figure 6.70) is a combination of the nomenclatures for peptides (Figure 6.5) and carbohydrates (Figure 6.68). This was developed in a study of the fragmentation patterns of O-linked MUC1 Tn- and TF-glycopeptides. The objective was to use MALDI-TOFMS with PSD to differentiate and compare the structures of nonacetylated and acetylated glycopeptides. Sample preparation consisted of mixing 1 pmol quantities of analytes with the matrix. The accuracy of mass determination was <0.3 Da. The fragmentation data permitted localization of the glycans on the peptides as well as the sequencing of short glycans and the peptidic backbone. Ions representing C- and N-terminal peptide fragments were presented in a parallel fashion with the carbohydrate fragment ions from the nonreducing ends. A series of relationships were revealed demonstrating both the influence and lack of influence of acetylation and the peptide backbone sequence on the type and quantity of fragment ions, including inter- and intra-ring cleavages of the glycosidic bonds and fragmentation from nonreducing ends of the glycan fractions. The advantages of this MALDI-MS technique included femtomole sensitivity, ability to differentiate between neighboring glycosylation sites, and the possibility of analyzing mixtures by determining PSD products using precursor ion selection with the aid of a pulsed field.[496]

Comparison of Instrumental Strategies for Sequencing Model MUC4 Peptides

These comparisons were made using model analytes that consisted of synthetic preparations of the decapeptide stretches from the MUC4-repeating domain, with blocked N- and C-termini. This peptide was glycosylated with only one GalNAc moiety that had been linked to one of several potential attachment sites. Mass spectra of these analytes and their nonglycosylated counterparts were obtained by nano-ESIMS coupled with QqQ and IT analyzers using low energy CID, MALDI-TOFMS with PSD, and ESI-FTICRMS with SORI-CID (SORI = Sustained Off Resonance Irradiation). The evaluation of the spectra offered instructive observations on the advantages and limitations of these significantly different instrumental approaches. High-resolution FTICRMS had unique advantages for the assignment of individual *m/z* values to mixtures of isobaric structures as well as to recognition of the positive role played by peptide chain destabilization in the detection of glycosylation sites.[497] MALDI with PSD offered advantages in detecting internal fragments for some analytes and the ESI-IT combination provided higher efficiency for the determination of glycosylation sites than did the ESI-QqQ system. There were differing needs between the instruments for setting the collision energies to obtain effective peptide sequencing. Conclusions were also presented on the performance requirements necessary for the detection of those fragments needed for determining the glycosylation sites in these monoglycosylated peptides.[498]

In a subsequent study, ESI-QTOFMS/MS was used to study the model TAP25-2 compound previously determined by MALDI/MS with PSD to be TAP25 glycosylated at Thr9 and Thr21.[497] Peptide fragments were obtained using CID of the $[M + 3H]^{3+}$ ion. The complexity of the CID spectrum was less than that encountered with PSD. The O-glycosylation site was identified by direct evidence from a series of abundant singly and doubly charged ions from the N-terminus (b ions) along with indirect evidence provided by ions from the C-terminus (y ions). It was concluded that the high level of mass accuracy and resolution of this instrument, along with straightforward ion assignments, permitted distinction of fragment ions bearing different degrees of glycosylation.[480]

Incomplete Glycan Processing of O-Linked Chains in Mucins Associated with Breast Cancer

Aberrant glycosylation in tumor cells has been shown to result in expression of carbohydrate structures with reduced complexity. This has been demonstrated by comparing MUC1 glycoforms in secretions from mammary carcinoma cells having a high metastatic phenotype (MDA-MB-231) with cell lines of a low metastatic phenotype (T47D) and using human milk as a control. Permethylated oligosaccharide alditols were analyzed by FABMS and the monosaccharide and sialic acid compositions were determined by GC-EIMS after methanolysis and TMS derivative formation. The glycoforms associated with the tumor cells were sialylated trisaccharides NeuAcα2-3(Galβ1-3)GalNAc or NeuAcα2-6(Galβ1-3)GalNAc. In addition, the cancer mucins had decreased GlcNAc/GalNAc ratios, lacked fucosylation, and the NANA was partially substituted by an N-glycosylated variant. It was concluded that the glycosylation of breast cancer-associated mucins involves incomplete glycan processing of O-linked chains and that the T47D cell line may serve as a model for carcinoma-associated alterations of O-glycosylation.[499]

O-Glycosylation During Cellular Differentiation in Colon Cancer

To investigate changes in the process of O-glycosylation through the expression of acetylgalactosaminyltransferase during differentiation of colon tumor cells (HT-29 line), microsomal preparations were made from a subpopulation that had been induced to secrete mucin by treatment with methotrexate. The degree and nature of O-glycosylation were determined before confluence when the cells were still undifferentiated (day 5), and after confluence when the mucin-secreting phenotype was expressed (day 21). A peptide substrate was used as the receptor for the acetylgalactosaminyltransferase in a time course study of enzyme activity. The peptide (TTSAPTTS) was synthesized based on the known tandem repeat peptide motif from the MUC5AC human gastric gene. It was revealed by ESI that both one and two GalNAc units could be linked to the peptide using the day 5 microsomes but only one could be attached with the day 21 microsomes. The ratios of the quantities of unreacted peptide to the glycosylated peptide, 5:1 on day 5 and 4.5:1 on day 21, suggested higher GalNAc transferase activity in confluent cells. Thus, O-glycosylation was influenced both qualitatively and quantitatively by the growth state of the cells as well as their stage of differentiation. The ESI spectra also revealed an unexplained peptidylaminotransferase activity that resulted in the addition of 2-threonine to the peptide backbone.[500]

In a related study, five oligopeptides corresponding to portions of the MUC2 tandem repeat domain were synthesized and incubated with UDP-N-acetyl-D-galactosamine and detergent-soluble microsomes prepared from a human colon carcinoma cell line (LSI74T). MALDI-TOFMS revealed that the order of the sites of the incorporation of GalNAc into the threonine residues of the MUC2 core peptides was Thr-3, Thr-6, Thr-5, Thr-2, Thr-4, and Thr-1, respectively. Oligopeptides containing alternating Thr residues were not fully glycosylated even after prolonged incubation, suggesting that the preferential order and degree of GalNac incorporation into Thr residues depends on the peptide sequence.[501]

In yet another study on the O-glycosylation of the MUC2 tandem repeat units, it was established that extracts from adenocarcinoma contained glycosylating peptides to a much greater extent than normal mucosa. It was determined by MALDI-TOFMS that glycosylation at each Thr residue occurred in a heterogeneous pattern of site-specific manner within the MUC2 tandem repeat, controlled by N-acetylgalactosyltransferase-3 in the synthesis of clustered carbohydrate antigens.[502]

O-Glycosylation of a MUC1 Tandem Repeat Model by Galactosaminyltransferase

A synthetic peptide that encompasses the known 20-amino acid tandem repeat unit of the MUC1 protein, including the immunodominant region (TAP25, $T^{1a}APPAHGVT^{9}S^{10}APDT^{14}RPAPGS^{20}T^{1b}APPA$, purity confirmed by ESIMS), was incubated with detergent-solubilized GalNAc-transferase isolated from mammary (T47D) and colon (HT29) carcinoma cells. Human skimmed milk served as the control. After isolation by ultrafiltration and purification by HPLC, four peaks were detected, collected, and identified by FABMS: unreacted TAP25 and three glycopeptides, TAP25-$(GalNAc)_1$, TAP25-$(GalNAc)_2$, and TAP25-$(GalNAc)_3$. The predominant (~80%) enzymatic product recovered from both cell types was the diglycosylated form. This was in contrast to the triglycosylated species obtained from the milk control. Because there are six possible glycosylation sites in TAP25, four Thr, and two Ser residues, the unreacted peptide and the glycosylated products were sequenced after proteolitic digestion with a papaya protease that selectively cleaves Gly-C bonds. The diglycosylated peptide gave four products; the triglycosylated peptide gave two fragments. It was determined by Edman degradation (Figure 6.71a) and FABMS (Figure 6.71b) that Thr^{9} and Thr^{1b} were the preferred sites of glycosylation, occurring within the sequences VTS or STA, and that these sites were independent of the glycosylation enzyme source. Ser^{20} was glycosylated to a lesser degree and Thr^{14} remained nonglycosylated. Experiments with antibodies revealed distortions of the peptide conformation of the binding epitope due to the glycosylation at the flanking positions.[503]

Similar experiments were conducted to explore the use of MALDI-TOFMS with PSD for the direct sequencing of O-linked glycopeptides and for the localization of glycosylation sites. In one experiment, TAP25 was glycosylated with detergent-solubilized UDP-GalNAc:polypeptide N-acetylgalactosyltransferase from the human breast carcinoma cell line T47D and a human premature milk control.[503] The di- and triglycosylated glycoforms were identical in both cases. However, the preparation from the T47D tumor line exclusively formed the monoglycosylated glycoform at Thr^{9}, while the transferase system from premature milk glycosylcosylated at Thr^{9} or Thr^{21} (Figure 6.72). The differences were explained based on enzymatic activity considerations. Methodological advantages associated with MALDI-TOFMS included the ability to (a) distinguish even neighboring glycosylation sites; (b) to localize and characterize disaccharides on MUC1-derived peptides; and (c) to distinguish six adjacent glycosylation sites in mono-, di-, and triglycosylated TAP25 products. All of this information could be obtained from a single spectrum of underivatized and uncleaved analytes, obtained straight from digestion incubates.[497]

High-Density O-Glycosylation of Tandem Repeat MUC1 Peptide from Breast Cancer Cells

The objectives of this study were to identify the sites and the degree of glycosylation within a partially deglycosylated glycopeptide fragment from the tandem repeat region of MUC1 gene products that were shed into culture medium by T47D breast cancer cells. MUC1-derived mucins were treated with exoglycosidase, resulting in a partially deglycosylated product that has lost most of the glucose and sialic acid beyond the initial, peptide-linked residues. Cleavage with clostridiopeptidase yielded a pool of glycosylated icosapeptides (PAP20) from the MUC1 tandem repeat

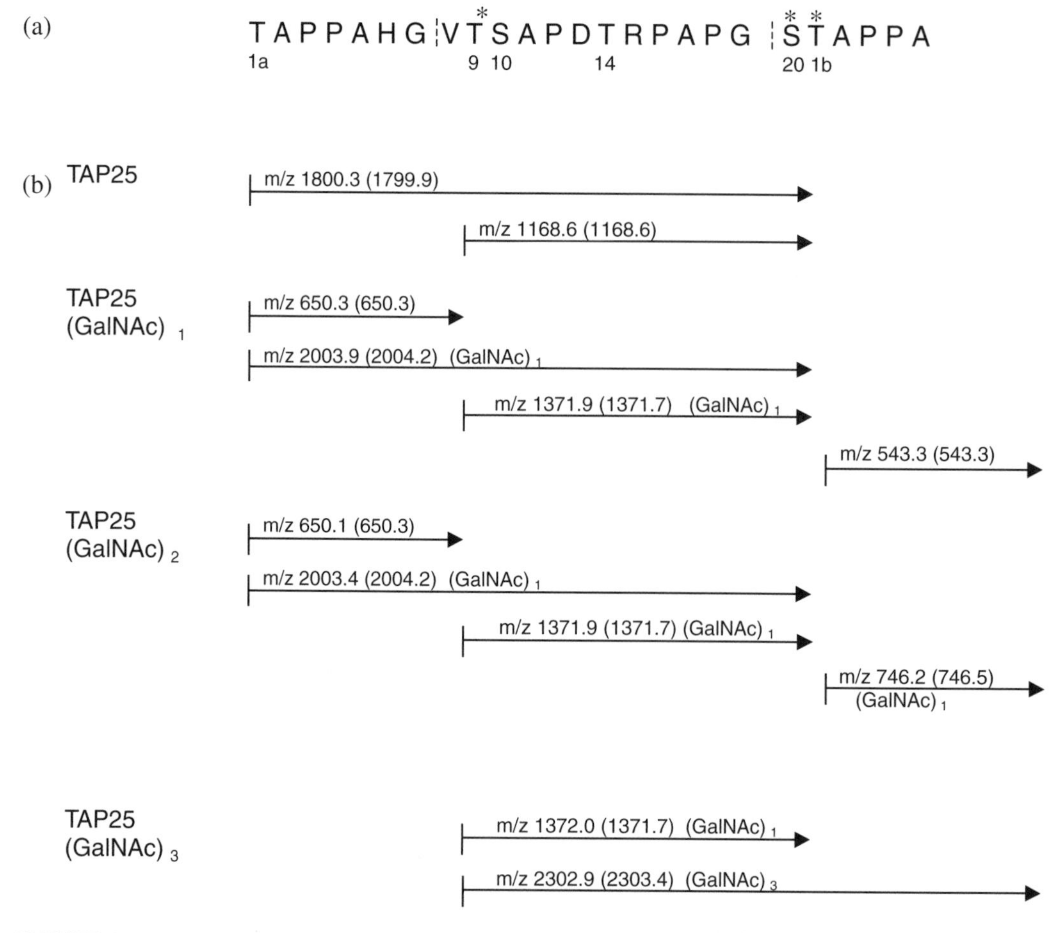

FIGURE 6.71 TAP25 is a synthetic peptide corresponding to one repeat and five overlapping amino acids of the polymorphic epithelial mucin MUC1 core protein that serves as an acceptor substrate for *in vitro* glycosylation. (a) Sequence of TAP25 with indicated positions of glycosylation (*) inferred from Edman degradation; (b) Digestion of TAP25, TAP25-$(GalNAc)_1$, TAP25-$(GalNAc)_2$, and TAP25-$(GalNAc)_3$ with the Gly-C-specific protease IV from papaya. Peptide fragments are represented by arrows, the numbers over each arrow correspond to the measured $[M+H]^+$ ions, and the values in brackets to the calculated $[M+H]^+$. Reprinted from Stadie, T.R. et al., *Eur. J. Biochem.*, 229, 140–147, 1995. With permission.

domain that, when analyzed by ESI-QTOFMS, revealed a preponderance of triply charged ions that represented a series of glycoforms of PAP20-$(HexNAc)_5$. The deconvoluted spectra of these ions were related to each other by mass increments of 162 and 291 Da, representing Hex and NeuAc moieties, respectively. The presence of this ion series confirmed that all five potential glycosylation sites in the tandem repeat region were O-glycosylated in the tumor cells.

Sequencing of the glycopeptides was next carried out to localize the O-glycosylation sites and to determine whether there had been amino acid replacements at particular points that would enable additional glycosylation to occur. For example, MS/MS analysis of a PAP20-$(HexNAc)_2$ ion revealed y type ions confirming that a substitution isomer, where Thr is inserted at the 19 position, carries a GalNAc (Figure 6.73). The degree of substitution in the breast cancer cells averaged 4.8 glycosylated sites per repeat. This appeared to be characteristic for the T47D breast cancer line when compared with the 2.6 glycosylated sites per repeat observed for mucin from lactating breast

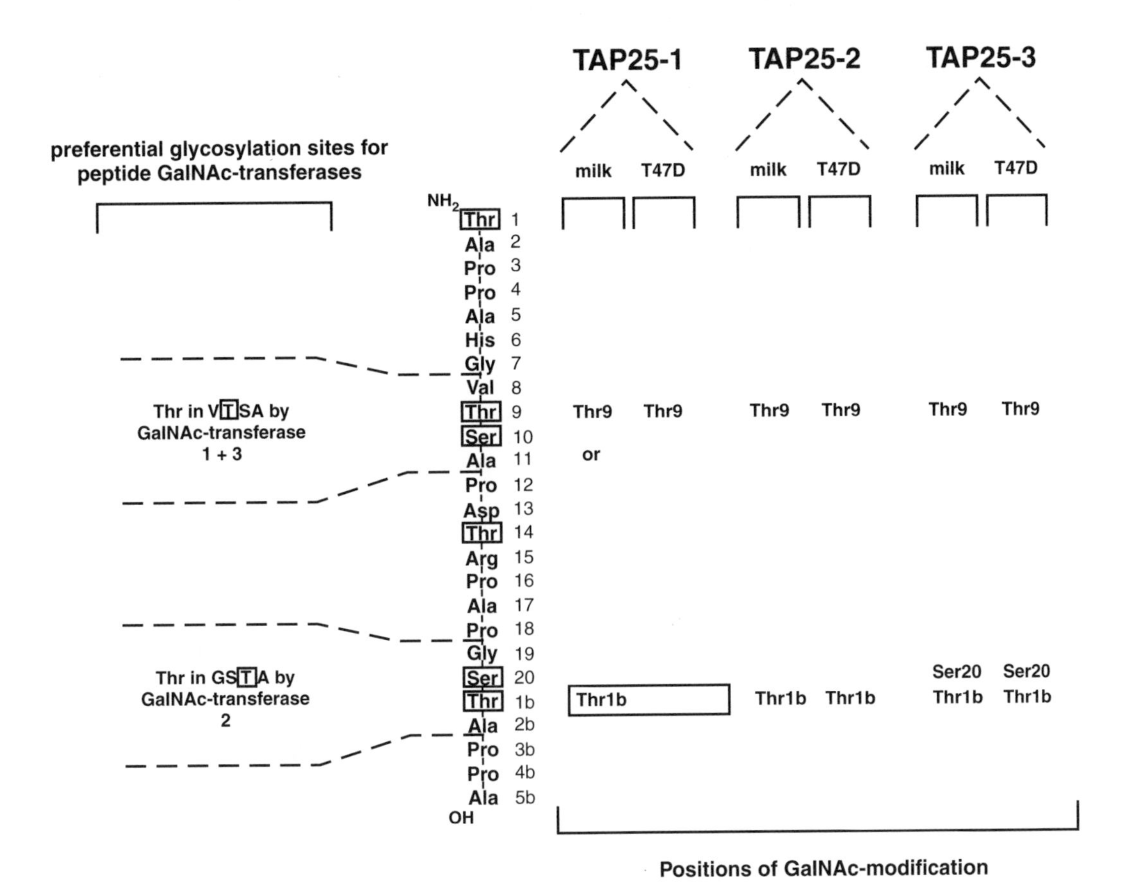

FIGURE 6.72 Summary of the glycosylation site analysis of the overlapping MUC1 repetitive sequence TAP25 (see Figure 6.71) that was glycosylated *in vitro* with detergent-solubilized UDP-GalNac:polypeptide N-acetylgalactosaminyltransferases (peptide GalNAc-transferases) from human premature skim milk or by a microsomal preparation from the human breast carcinoma cell line T47D. Reprinted from Goletz, S. et al., *Glycobiol.*, 7, 881–896, 1997. With permission.

epithelial cells.[504] Additional experimentation led to the conclusion that the repeat domain of MUC1 is polymorphic with respect to the peptide sequence.[505]

Exposed Peptide Epitopes in Underglycosylated MUC1

The MUC1 mucins, which are abundantly secreted by adenocarcinomas, are significantly underglycosylated. The presence of repeated sequences in the mucin genes suggests that peptide epitopes, such as the APDTRP core peptide sequence, are of potential importance as these epitopes may induce T-cell anergy. A monoclonal antibody (BCP8) was used to isolate and identify putative MHC Class I-associated amino acid sequences for MUC1-positive human tumor cell lines including breast adenocarcinoma (MCF-7), metastatic pancreas adenocarcinoma (CAPAN-1), and cells from a malignant pleural effusion (SKBr-3). About 3×10^9 cells were lysed and MHC Class I molecules isolated and purified.[506] Analyses of the BCP8-reactive peptide fractions by ESI-MS/MS revealed sequences representing two 3-mers (TSA and DTR) and a 6-mer (TSAPDT), all corresponding to known amino acid sequences within the MUC1 core peptide (Figure 6.74). Subsequently, a synthetic 9-mer, predicted by the MUC1 sequence, that contained all three isolated fragments (TSAPDTRPA) was shown to cause significant upregulation of MHC Class I expression. This peptide also proved to be an epitope for CTL in an *in vitro* immune response test.[507]

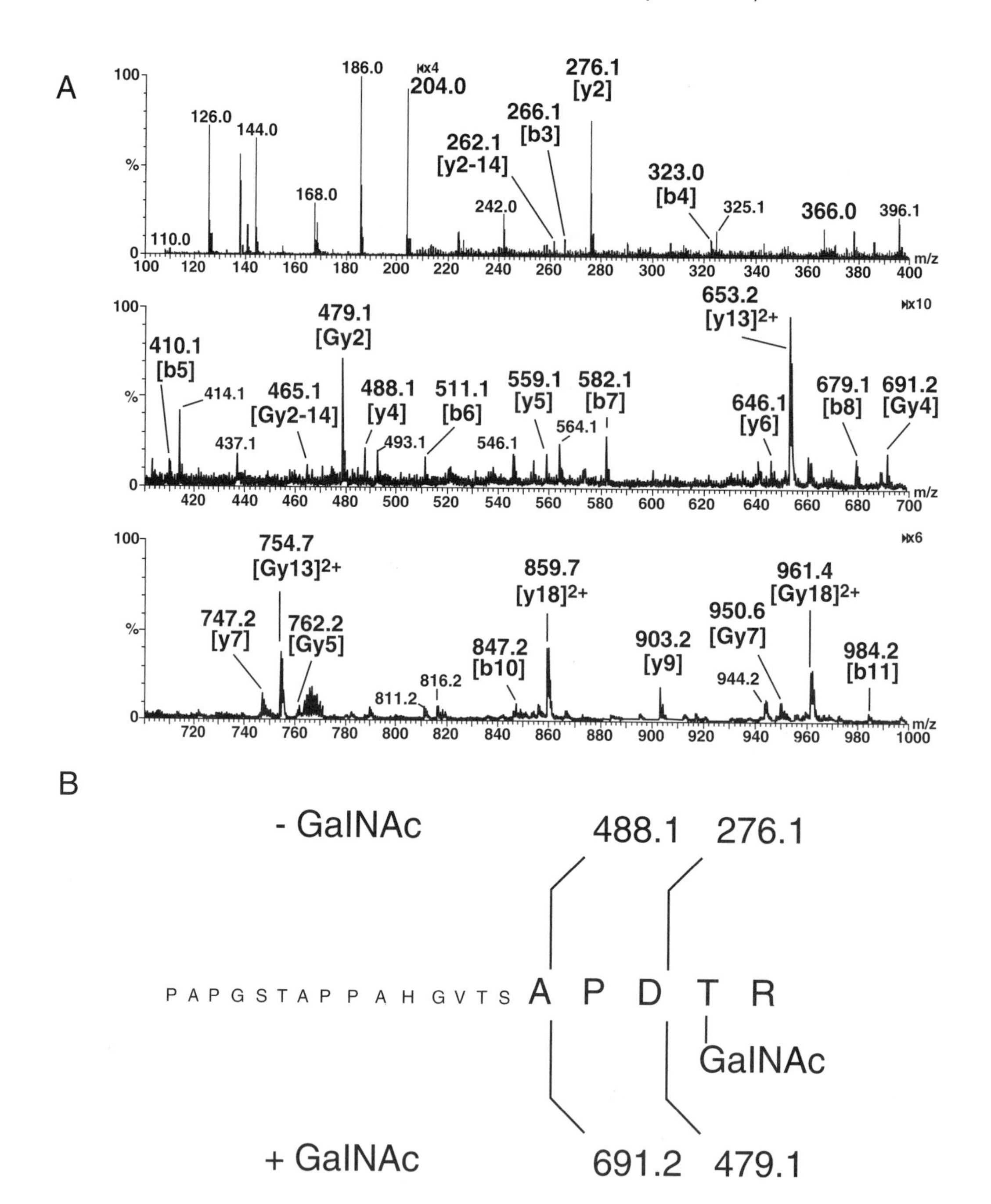

FIGURE 6.73 Secretory MUC1 was isolated for human mammary carcinoma cell line T47D, and the PAP20 glycopeptides were separated. MS/MS spectrum of the $[M+H]^+$ ion species at *m/z* 765.1 corresponding to PAP20-hexNAc$_2$. (A) A section of the MS/MS spectrum is shown that covers the mass range of singly charged fragment ions from *m/z* 100 to 1000. Two y2 ions are observed that correspond to nonglycosylated (*m/z* 276.1) and HexNAc-substituted dipeptide Thr-Arg (*m/z* 479.1). In the same way, the signals at *m/z* 488.1 and 691.2 indicate formation of a nonglycosylated (y4) or HexNAc-susbtituted tetrapeptide PDTR (Gg4), respectively. Amino acid replacements are indicated at *m/z* 262.1 (y2 -14) and 465.1 (Gg2-14), corresponding to nonglycosylated and HexNAc-substituted Ser → Arg, respectively. (B) C-terminal fragmentation scheme presenting major sequence ions of the y-series that support HexNAc substitution of the Thr-19 in PAP20 peptide. Reprinted from Muller, S. et al., *J. Biol. Chem.*, 274, 18165–18172, 1999. With permission.

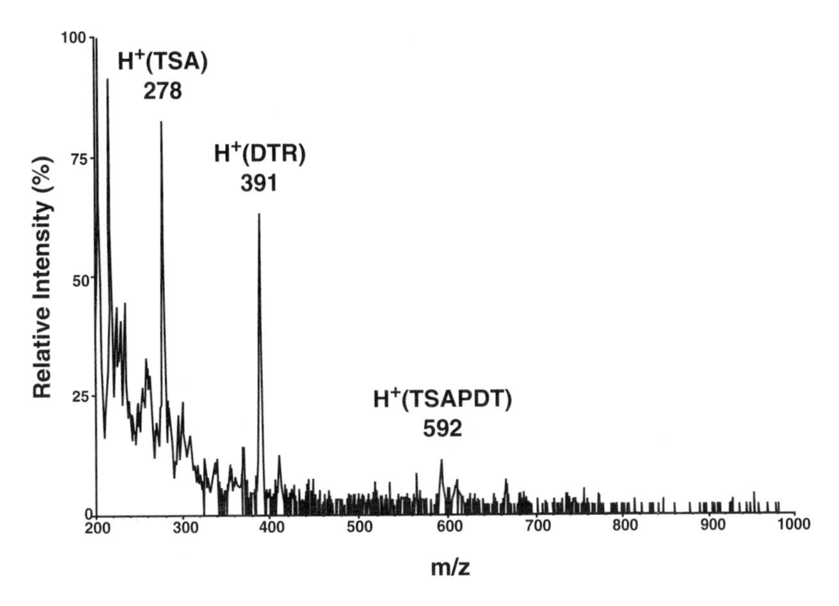

FIGURE 6.74 ESI mass spectrum showing three MUC1 peptide fragments eluted from MHC Class I proteins. The sample was obtained by affinity isolation of an HPLC fraction followed by rechromatographing. Reprinted from Agrawal, B. et al., *Cancer Res.*, 58, 5151–5156, 1998. With permission.

Glycosylation of the Tandem Repeat Unit of MUC2 Polypeptide

Following the recognition that the GalNAc-Ser/Thr cluster is essential for the antigenicity of Tn (a common blood group antigen), analysis was undertaken of the glycosylation products of the synthetic peptide, GYPKAPTTTPITTTTTVTPTPTPTGTQT, which corresponds to the repeat domain of human MUC2 and contains 14 Thr residues. Glycosylation was accomplished using microsomes from a colorectal cell line (LS180). Seven collected glycopeptide fractions were treated with trypsin. The molecular masses observed for the tryptic fragments, as determined by MALDI-MS, revealed glycosylation products that contained from 1 to 10 GalNAc residues. Those products containing 9 and 10 attached GalNAc residues (Figure 6.75A, peaks a and b) were strongly immunoreactive, while those with <6 residues (Figure 6.75B, peak c) were inactive. This provided confirmation that a cluster of GalNAc-Thr residues is required for Tn antigenicity.[508]

6.4.3.2 Antigens

Carcinoembryonic Antigen (CEA)

There are several distinct monoclonal antibodies that provide for *in vivo* diagnostic detection of the seven known immunoglobulin-like domains of CEA, an 18 kDA, highly glycosylated protein that is expressed on the surface of tumors of the breast, lung, colon, and ovary.[509] In the course of synthesizing four of the CEA domains (N, A3, B3, and A3B3), several peptides were synthesized with His_6- or His_6Met-terminating tails at the amino terminus to permit purification on Ni-NTA chelation columns (Met allows for removal of these tails by CNBr treatment after purification). The purified peptides were characterized by ESIMS (QqQ, IT, and BE analyzers) and MALDI-TOFMS. The mass of the A3 domain, as determined by ESIMS (QqQ), was 10,336 Da, instead of the expected 10,341 Da (Figure 6.76A). MALDI-TOFMS yielded 10,341 Da (Figure 6.76B), and the mass observed on the double-focusing magnetic instrument was 10,341.4 Da, within 0.23 mass units of the calculated value (Figure 6.76C). In subsequent analyses, LC/ESI-MS/MS was used to obtain peptide maps of A3 after reduction, S-alkylation, purification by HPLC, and enzymatic digestion. More than 87% of the sequence was confirmed. Similar MS/MS results were obtained for the five other domains, revealing several properties, including that the presence of two

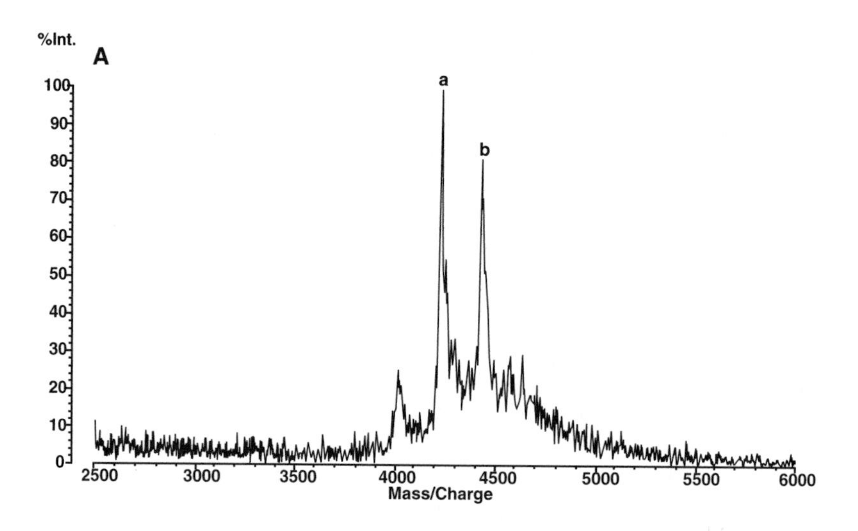

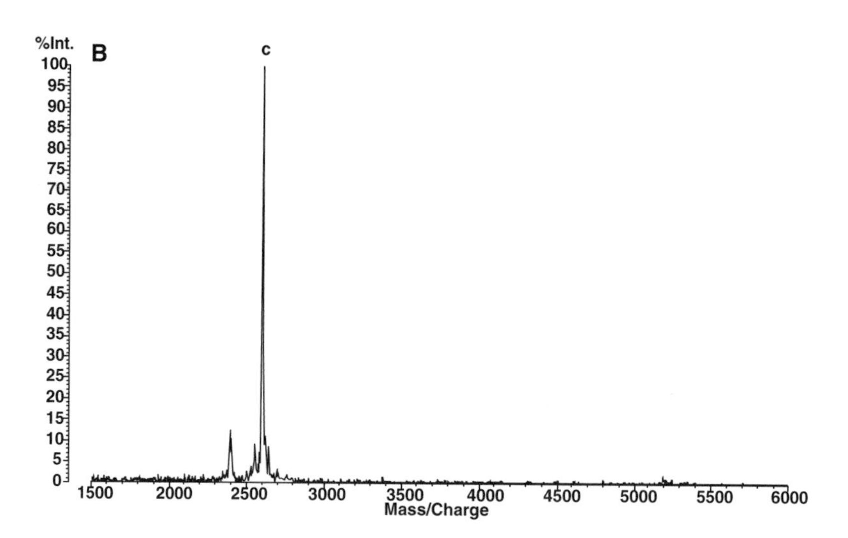

FIGURE 6.75 Glycosylation of the tandem repeat unit of the MUC2 polypeptide. MALDI-TOFMS spectra of glycopeptides in two fractions, (A) and (B), isolated by HPLC. The molecular masses of the identified peaks labeled a, b, and c were 4220, 4424, and 2592 Da, respectively. Reprinted from Inoue, M., Yamashina, I., and Nakada, H., *Biochem. Biophys. Res. Commun.*, 245, 23–27, 1998. With permission.

N-glycosylation sites in the N domain was critical to maintaining its solubility. It was also established that the techniques developed for the synthesis of long peptides (>50 residues) permitted the classification of diagnostically critical epitope groups of CEA.[510]

Prostate-Specific Antigen (PSA)

PSA is a 30 kDa single-chain serine protease produced by prostatic epithelial cells. Serum PSA is the most important clinical tumor marker currently in use for the early diagnosis and monitoring of therapy of prostate cancer (reviewed in Reference 511). There is considerable controversy about the value of the PSA test given that there are relatively large numbers of both false-positive and false-negative results.

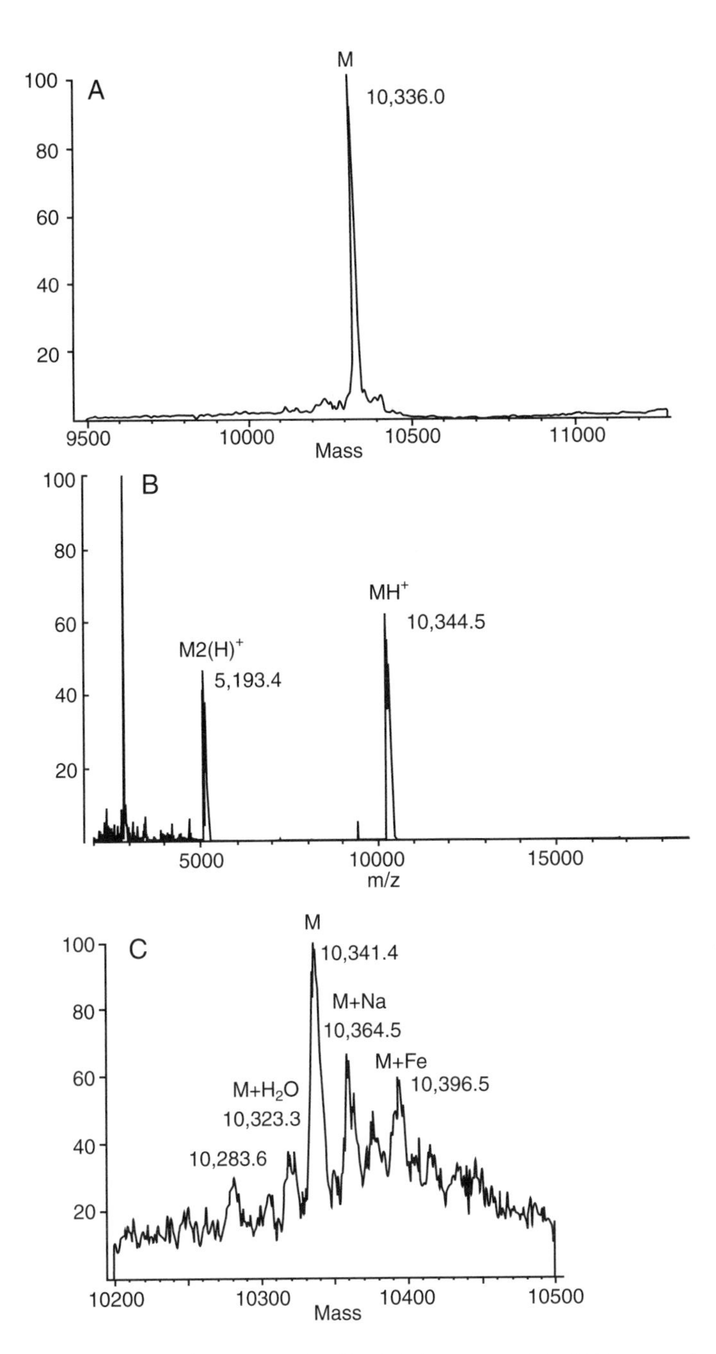

FIGURE 6.76 Mass spectral analysis of synthetic CEA A3 domain peptide. (A) Deconvoluted ESI spectrum; (B) MALDI-TOF spectrum; (C) Deconvoluted spectrum obtained on a double-focusing magnetic type mass spectrometer, providing a molecular mass within 0.23 Da of the calculated mass. Reprinted from Kaplan, B.E. et al., *J. Peptide Res.*, 52, 249–260, 1998. With permission.

The molecular weight of the antigen was reported in the 26,079-34,000 Da range using SDS-PAGE, gel filtration, or amino acid sequencing. Based on the determination of the amino acid sequence, by cDNA sequencing (237 amino acids, 5 internal disulfide bonds), the theoretical relative molecular mass, M_r, of the peptide moiety is 26,079 Da. However, the carbohydrate content and the composition and nature of the glycoforms (four side chains) have not yet been established. The

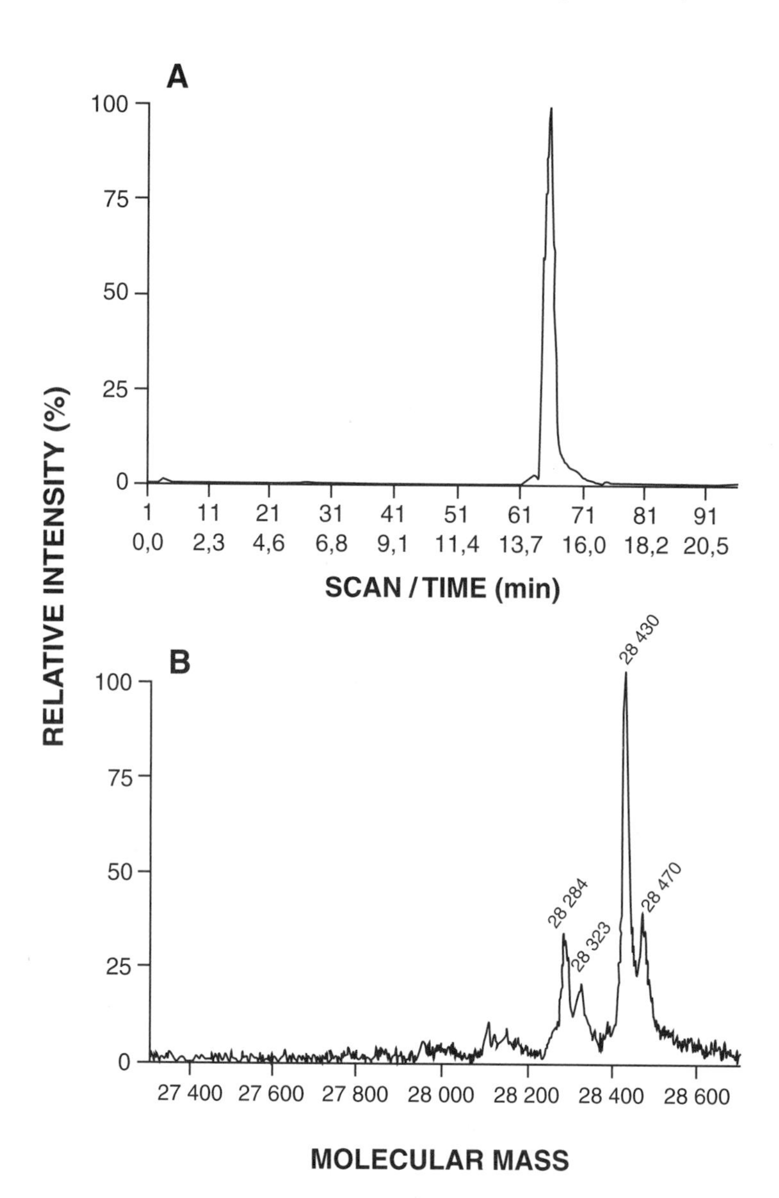

FIGURE 6.77 Prostate specific antigen (PSA). (A) Total ion current of PSA eluting from an HPLC; (B) Spectrum providing molecular masses for the PSA protein obtained by ESI. Reprinted from Belanger, A. et al., *The Prostate,* 27, 187–197, 1995. With permission.

molecular weight of intact PSA was determined using LC/ESIMS in five lots of highly purified commercial samples. The total ion current elution pattern (Figure 6.77A) showed a single peak, while the reconstructed spectrum (Figure 6.77B) revealed peaks corresponding to the major glycoforms, PSA (28,430 Da), PSA-Fuc (28,284 Da), and their potassium adducts (28,470 Da and 28,323 Da, respectively). The antigen was solubilized in a potassium buffer, hence the appearance of the potassium adducts. The M_r of PSA was 28,430 ± 1 Da for the 5 lots studied. Thus the contribution to the molecular weight of intact PSA provided by the carbohydrate moiety was 2351 Da (8.3%). The complete primary structure of the carbohydrate chain, as determined by NMR, revealed that only one N-glycosylation site is occupied in the protein, and that approximately 70% of the PSA molecules contain a fucose group in the core chitobiose moiety. It was concluded that the use of this mass spectrometric technique for the accurate and rapid characterization of PSA from various commercial lots will be a useful tool for the establishment of a reference standard to improve the currently inadequate reliability of the routine measurement of PSA in serum.[512] The molecular weight was confirmed in another study using essentially the same methodology with the main peak in the ESI spectrum envelope for PSA corresponding to the 16+ charge state that yielded 28,432 Da when deconvoluted.[513]

The precursor forms (pro-PSA) of free PSA in sera of prostate cancer patients were determined by isolation, using streptavidin-coated magnetic beads, followed by MALDI-TOFMS analysis of peptides after digestion with endoproteinase. All five serum samples analyzed contained the (-7), (-5), and (-4) pro-PSA forms, whereas the (-1) and (-2) forms were present in only three samples. The sequence of the (-7)pro-PSA was determined by ESI-QTOFMS.[514]

In human serum, the major form of PSA is as a covalent complex with a serine protease inhibitor, α_1-antichymotrypsin (ACT). The formation of this complex probably results in an inactivation of the enzymatic properties of the PSA. Free PSA was analyzed, after chemical release from the PSA-ACT complex, using MALDI-TOFMS.[515] Molecular mass analyses of the complex revealed the presence of two species, one at 78,095 ± 138 Da (67% of total protein) and the other at 82,519 ± 104 Da. It was determined by MALDI-TOFMS that ACT contained two predominant species (55,106 Da and 54,414 Da). It was concluded that MS techniques should be used for the accurate determination of the masses of both PSA and the PSA-ACT complex in serum in preparing international standards.[258] Apparently, a small portion of PSA remains in the free form in serum despite the very large excess of serine protease inhibitors. Both the free PSA and the PSA-ACT complex have been isolated and purified from a prostate cancer cell line followed by characterization using MS as well as several chromatographic and biochemical techniques.[516]

An indirect immunosorption technique was developed for the purification of PSA from human serum with the intention of eliminating the nonspecific binding and recovery of irrelevant proteins that occurs when conventional immunosorption techniques are used. The technique is based on preparation of a complex that consists of a digoxigenylated anti-PSA antibody linked to streptavidin-coated magnetic beads via a biotinylated anti-digoxigenin antibody. The specific elution of the complex consisting of PSA and digoxigenylated anti-PSA antibody is accomplished with digoxigenin-lysate. The MALDI-TOFMS mass spectrum of the eluate of a sample prepared from the serum of a patient with prostate cancer consisted of peaks corresponding to free PSA (0.19 μg/mL) and the PSA-ACT complex (1.9 μg/mL). Both the singly and doubly charged PSA-ACT ions were detected (Figure 6.78). In contrast, the mass spectrum of the eluate obtained by direct immunosorption (with propionic acid elution) was dominated by serum albumin. The PSA-ACT complex could not be observed, and only a small free PSA peak was found. Considering that the serum concentration of PSA in prostate cancer patients is in the 3 to >3000 ng/mL range, which is several orders of magnitude less than that of albumin and other serum proteins, the indirect immunosorption technique was considered to be a major improvement for serum PSA and PSA-ACT sample preparation for MS analysis. In principle, the method can be used to prepare any trace protein in a matrix where other proteins are present in overwhelming concentrations.[517]

6.4.3.3 Miscellaneous Glycoproteins

Multidrug Resistance and a 95 kDa Overexpressed Glycoprotein

There is an adriamycin-resistant human breast carcinoma cell line, MCF-7/AdrVp, that does not overexpress the P-glycoprotein but does overexpress a 95 kDa membrane glycoprotein (p95). To obtain further information about this protein, protein extracts were prepared from cell membranes, deglycosylated, isolated by preparative 2-DE, digested with trypsin *in situ* and the resulting peptides sequenced by MALDI-TOFMS. Two candidate peptides were detected (Figure 6.79). Two matches were found when these masses were searched for among the sequences of known proteins in the 30-40 kDa range (1 Da tolerance) using a nonredundant database: EVLLLAHNLPQNR and

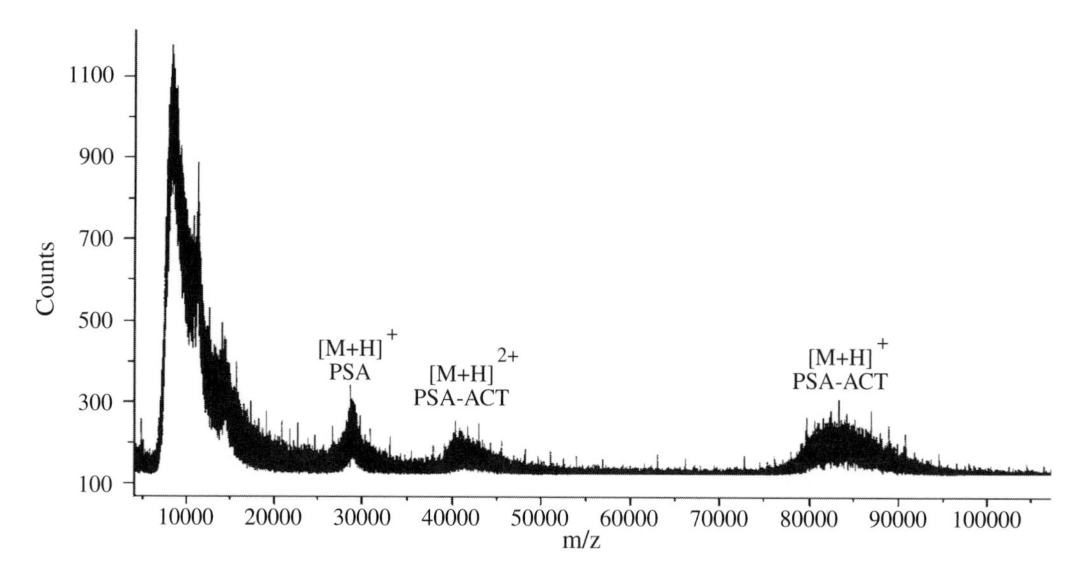

FIGURE 6.78 MALDI-TOFMS analysis of prostate-specific antigen (PSA) from human serum of a patient with prostate cancer after purification by indirect immunosorption. Final elution with digoxigenin-lysate. Free PSA, 190 ng/mL, PSA-ACT, 1.89 μg/mL (ACT, α_1-antichymotrypsin). Reprinted from Peter, J. et al., *Anal. Biochem.*, 273, 98–104, 1999. With permission.

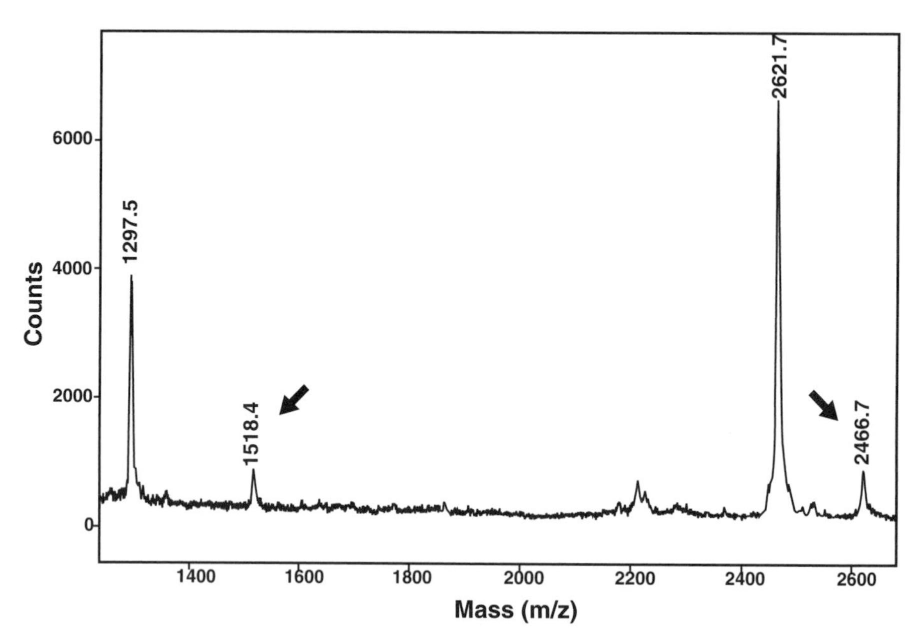

FIGURE 6.79 Mass spectral analysis of the trypsin-digested p95, the 95 kDa membrane glycoprotein over-expressed in multidrug-resistant breast cancer cells. p95 was isolated by 2-DE. Arrows: The unique trypsin fragments of p95. The detected masses of 1297.5 Da and 2466.7 Da represent, respectively, the internal standards angiotensin and adrenocorticotropic hormone. Reprinted from Ross, D.D. et al., *Cancer Res.*, 57, 5460–5464, 1997. With permission.

VDGNSLIVGYVIGTQQATPGPAYSGR. Both sequences are present in the nonspecific cross-reacting protein, NCA-90, which is an antigen related to CEA. Both peptides were also found by LC/ESIMS as well as a third peptide at *m/z* 2159. This peptide is also known to be a tryptic fragment of NCA-90. Further proof of the common identity of the fully glycosylated p95 membrane glycoprotein and NCA-90 was obtained by undertaking Northern and Western blots.[518]

Cytokines and Growth Factors

The term *cytokine*, a functional rather than structural concept, denotes a family of unrelated glycoproteins of 8-25 kDa that are secreted by cells into the extracellular fluid. Here they exert *autocrine* or *paracrine* activity by sending signals to the cell nuclei through binding to specific, high-affinity receptors on cell surfaces. Cytokines provide the means for the intercellular communication by which cellular proliferation, differentiation, and other functions, are modified. Cytokines have very high biological activities, being effective at 10^{-11} to 10^{-13} M concentrations. Most cytokines are available in high purity preparations, their primary structures and originating genes are known, and several are being used clinically as drugs with unique biological properties.

The main function of those cytokines known as *colony stimulating factors* (CSF) is to participate in the direction of the division and differentiation of bone marrow stem cells, and the precursors of blood leukocytes. The introduction of granulocyte-macrophage CSF (GMCSF) and granulocyte stimulating factor (GSF) products into clinical oncology resulted in a marked reduction in neutrophil engraftment time and the length of hospitalization required following high-dose chemotherapy with bone marrow or peripheral blood progenitor transplantation. Macrophage CSF (MCSF) is a cytokine that stimulates the survival, proliferation, differentiation, and functions of mononuclear phagocytes. The mass spectrometry of MCSF is reviewed in References 360–363, 519, 520.

The expression and glycosylation of the human fibroblast growth factor receptor subtype 4 (FGFR4) was studied using MALDI-TOFMS to identify the tryptic peptides, recognize disulfide bridges, and investigate the glycosylation sites in the extracellular domain of FGFR4. The mass spectra revealed that FGFR4 is N-glycosylated in three of the six possible sites (N-88, N-234, and N-266); however, no glycosylation was found at the other sites.[521]

Characterization of Monoclonal Antibody Glycosylation

CAMPATH-1H is a recombinant humanized murine monoclonal immunoglobulin (IgG_1) antibody that recognizes a specific antigen in human lymphocytes. The agent has been in clinical trials for treatment of non-Hodgkin's lymphoma. To investigate the glycosylation of CAMPATH-1H, oligosaccharides were released by hydrazinolysis and sequenced by ESI-MS/MS. IgG_1 derived from a murine myeloma cell line (NSO), and analyzed in this way, was found to contain five glycoforms including two that were not present in Chinese hamster ovary (CHO)-derived samples. These two had masses that were increased by 162 Da and 324 Da suggesting the addition of one and two hexose residues to the previously identified oligosaccharides. To determine if the terminal galactose residues present in these two hypergalatosylated glycoforms were α-linked, the Lys-C glycopeptide of CAMPATH-1H from NSO cells was isolated and treated with exoglycosidases that were specific for α- and β-linked nonreducing terminal galactose residues. The MALDI spectrum of the untreated NSO-derived glycopeptide revealed the five glycoforms previously observed (Figure 6.80a). Upon treatment with the α-galactosidase two glycoforms with the highest masses disappeared (Figure 6.80b). When another aliquot was treated with the β-galactosidase, the spectrum (Figure 6.80c) was consistent with expected structural intensity changes, including the formation of a compound isobaric with glycoform C (designated as C′). It was concluded that two newly discovered glycoforms of CAMPATH-1H had been identified that contain either one or two nonreducing terminal α-linked galactose residues and that these modifications may, consequently, be responsible for the compound's immunogenic properties.[522]

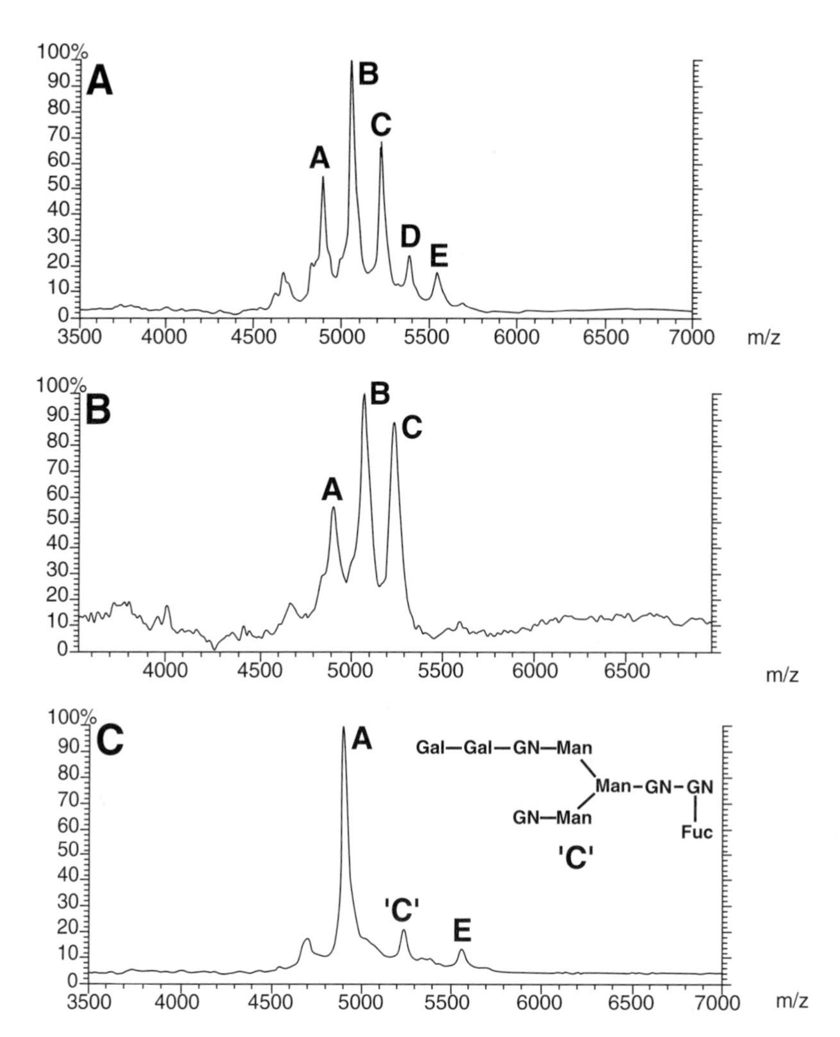

FIGURE 6.80 MALDI-TOFMS spectra of: (A) the NSO-derived CAMPATH IgG_4 glycopeptide; (B) Same glycopeptide, following α-galactosidase digestion; (C) Same glycopeptide, following β-galactosidase digestion. For each spectrum, approximately 500 fmol were spotted. The ordinate is ion abundance. The structures of A, B, C, D, and E, which are Asn-linked oligosaccharides, are given in Reference 522. Reprinted from Sheeley, D.M., Merrill, B.M., and Taylor, L.C.E., *Anal. Biochem.,* 247, 102–110, 1997. With permission.

Glycosylation of c-Myc

Mutations of c-myc, a zipper nuclear phosphoprotein that contains 439 amino acids, are associated with various types of tumors in human and other species. It has been shown previously that Thr-58 is a major *in vivo* phosphorylation site in the N-terminal transcriptional activation domain of c-myc. Thr-58 can apparently also be occupied by an O-linked GlcNAc. The fact that Thr-58 can have two posttranslational modifications has major implications for the regulation of this protein. The glycosylation site was identified in the tryptic peptide by enzymatically labeling the GlcNAc moiety with ^{3}H-galactose and analyzing the resulting peptides by MALDI-TOFMS. The peak at *m/z* 1867.3 was characterized as [FELLPTPPLSPSR + ^{3}HGalβ(1-4)GlcNAc + 2Na – H]$^+$ (Figure 6.81). Other identified peaks shown in the figure include [FELLPTPPLSPSR + Na + H_2O]$^+$ (*m/z* 1494.7) and [FELLPTPPLSPSR + GlcNAc + Na + H_2O]$^+$ (*m/z* 1697.9). All other peaks could also be assigned to structures commensurate with the process of glycosylation. The identification of Thr-58 as the glycosylation site was confirmed by PSD and Edman degradation. It was suggested that the

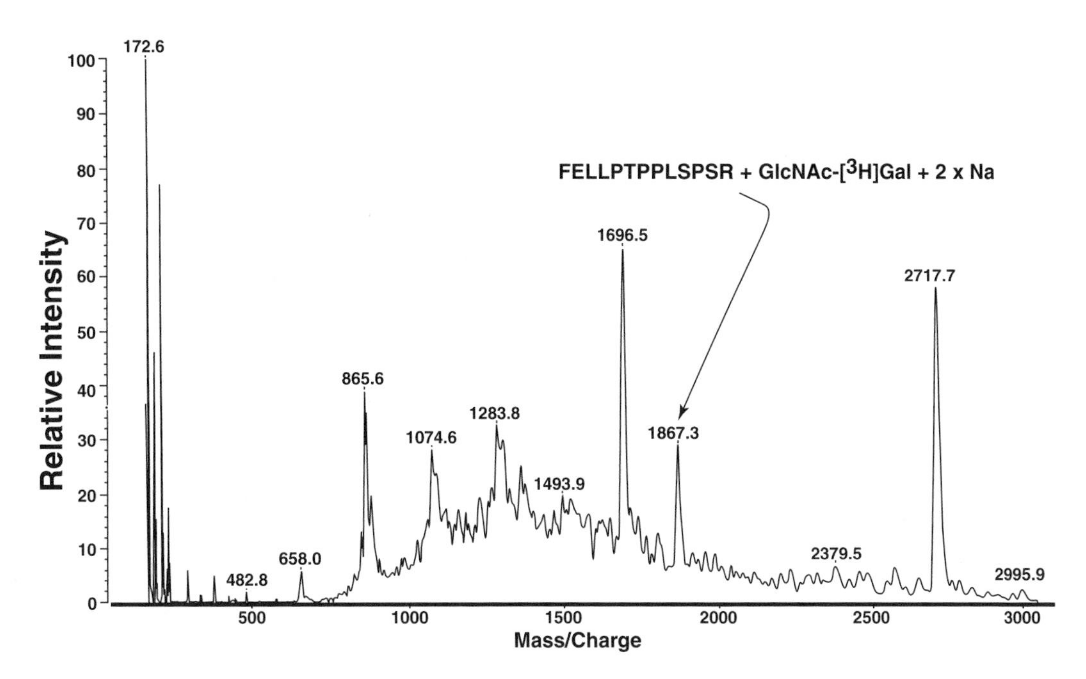

FIGURE 6.81 c-Myc is glycosylated at threonine 58, a known phosphorylation site and mutational hotspot in lymphomas. Identification of [^{3}H]galactose-labeled FELLPTPPLSPSR by MALDI-TOFMS. The *m/z* of 1867.3 represents the molecular mass of FELLPTPPLSPSR (1453.71) plus [^{3}H]Gal(β1-4)GlcNAc (367.33) plus two sodiums, -one hydrogen. For the assignment of the other peaks see the reference. Reprinted from Chou, T., Hart, G., and Dang, C., *J. Biol. Chem.*, 270, 18961–18965, 1995. With permission.

reciprocal phosphorylation/O-GlcNAc glycosylation could differentially regulate c-myc functions in different stages of the cell cycle.[523]

N-Glycan Structure of Matrix Metalloproteinase (MMP)-1 Derived from Fibrosarcoma Cells

After purifying and concentrating MMP-1 that had been derived from HT-1080 fibrosarcoma cells, the N-glycan antennae were released by enzymatic hydrolysis with N-glycosidase. Sequential glycosidase digestion of the released N-glycans was then carried out with six enzymes having different specificities. The mass spectrum of the intact N-glycosidic glycans from the fibrosarcoma cells (Figure 6.82A) was much more complex than the one obtained from MMP-1 isolated from a human fibroblast control (not shown). The mass spectrum of the N-glycans after incubation with Newcastle disease neuraminidase revealed no changes in signal intensities (Figure 6.82B). In contrast, treatment with a broad-specificity neuraminidase resulted in significant changes (Figure 6.82C). In a succeeding step the desialylated N-glycans were treated with a fucosidase. The resulting mass spectrum (Figure 6.82D) suggested that the antennae carry LewisX (Galβ(1-4)[Fucα1-3)]GlcNAc-) and LeXNAc (Gal(NAcβ1-4)[Fuc(α1-3)]GlcNAc-) structures, neither of which was present in the fibroblast samples. In addition, it was suggested that terminal GlcNAc residues were not present because no such residues were released from the desialylated and defucosylated glycans by a β-N-acetylglucosamidase (Figure 6.82E). Further structural information can be obtained from the mass spectra shown in Figures 6.82F, 6.82G, and 6.82H, which were derived from samples that were treated with other linkage specific enzymes. Because of the complexity of the spectra shown, and also because of other data that were obtained using different treatments and conditions, the reference below should be consulted for a detailed explanation of the formation of the observed peaks and the disappearance of others, as well as for the final proposed structures. The heterogeneous glycosylation pattern of the fibrosarcoma-derived MMP-1 was significantly different

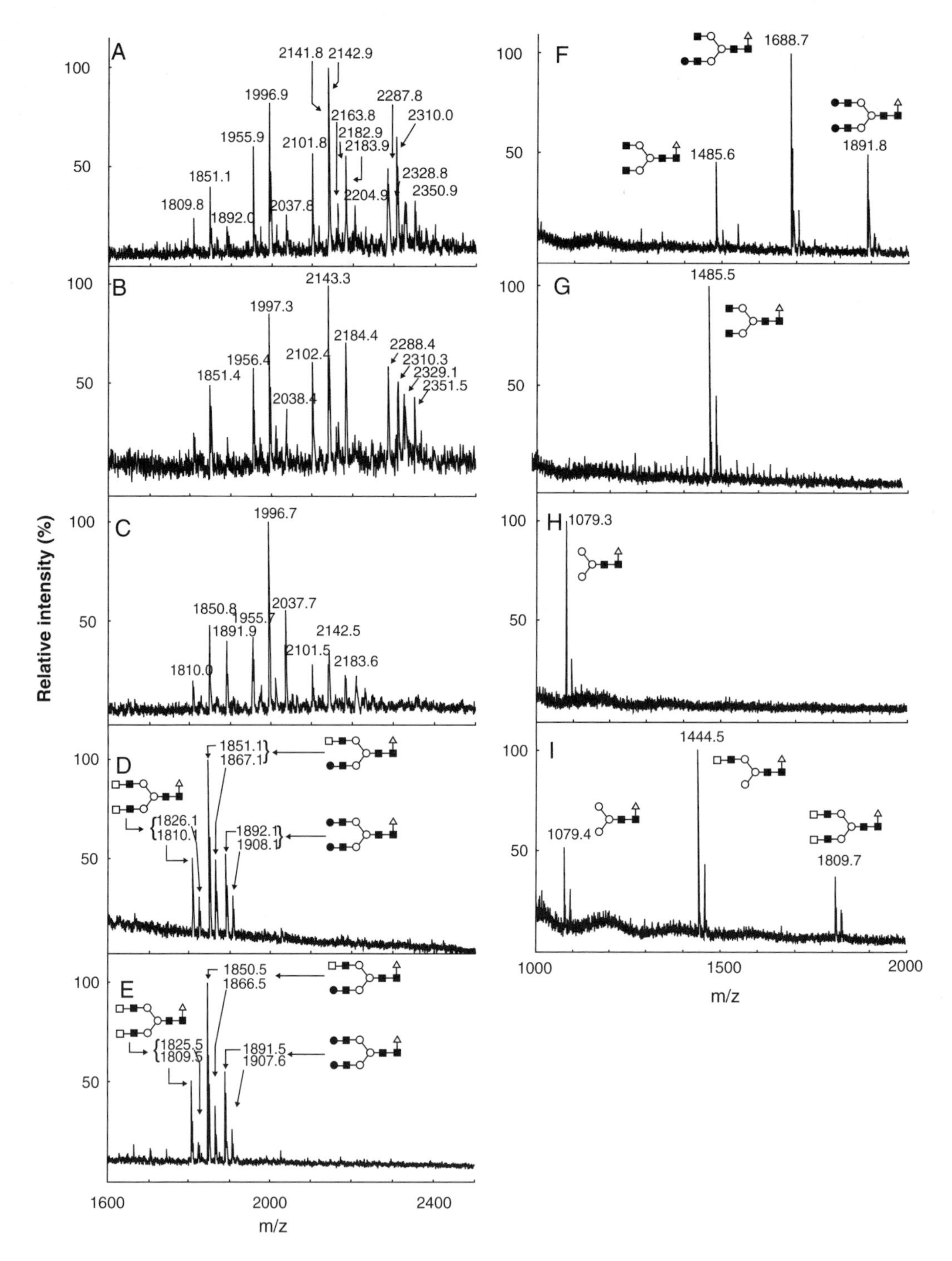

FIGURE 6.82 Analysis of matrix metalloproteinase (MMP-1) N-glycans by MALDI-TOFMS. (A) Intact N-glycans; (B) N-glycans after incubation with Newcastle disease virus neuraminidase; (C) N-glycans after incubation with *C. perfingens* neuraminidase, followed by consecutive treatments with (D) almond meal (α1-3,4)-fucosidase, (E) β-N-acetylglucosaminidase, (F) (β1-4)galactosidase, (G) β-N-acetylhexoseaminidase, and (H) β-N-acetylglucosaminidase. (I) Desialylated and defucosylated N-glycans treated with β-N-acetylhexosaminidase and β-N-acetylglucosaminidase. The symbols used for monosaccharide residues: □, galactose; ●, N-acetylglucosamine, ○, mannose, ■, N-acetylgalactosamine, Δ, fucose. Reprinted from Saarinen, J. et al., *Eur. J. Biochem.*, 259, 829–840, 1999. With permission.

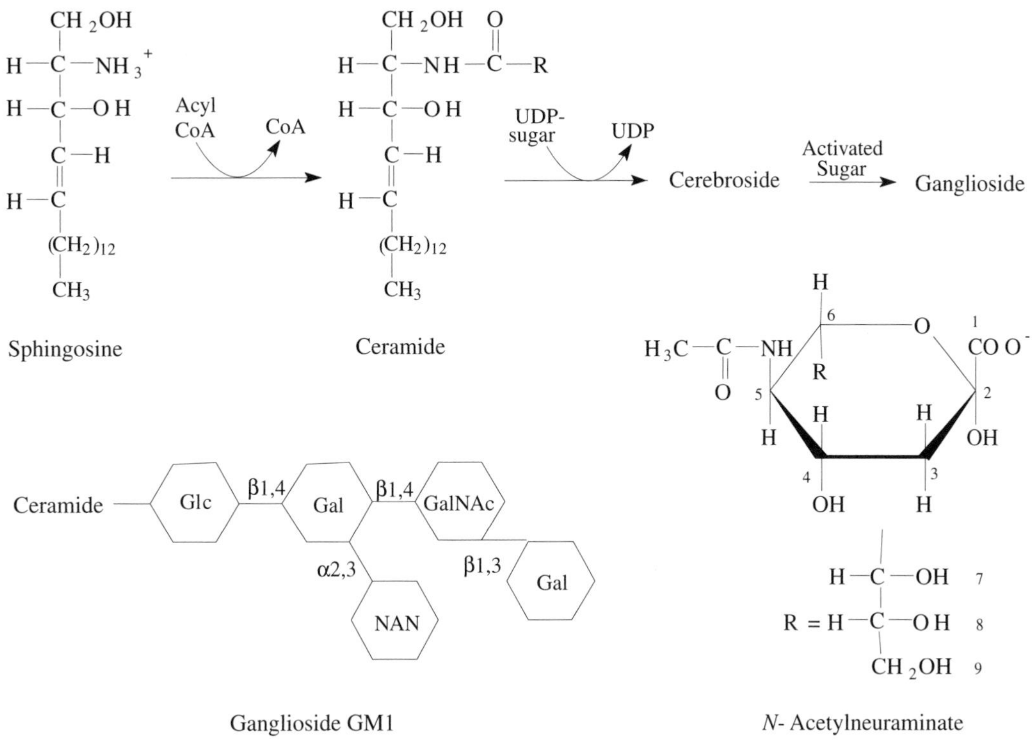

FIGURE 6.83 Chemical structures and formation of gangliosides.

from that derived from fibroblasts. It was pointed out that several glycan structures identified in the fibrosarcoma-derived MMP-1 contain structural motifs that are likely to bind to selectins. Such selectin-glycan interactions would be potentially significant in both tumor cell invasion and angiogenesis.[524]

6.4.4 Glycolipids (Gangliosides) in Various Malignancies

Sphingosines are derived from the amino acid serine in which the carbonyl group is substituted with a hydrocarbon side-chain. *Ceramides* are formed from sphingosines when an additional side-chain is attached to the amino group. *Glycosphingolipids* are derived from ceramides by the addition of a sugar or polysaccharide to the side-chain hydroxyl group of the original serine. This family of sphingolipids is ubiquitous. They are important components of the bilayers of cell plasma membranes and have major functions as cell-surface transducers and receptors for cell-cell recognition (reviewed in Reference 525). *Gangliosides* are glycosphingolipids that contain one or more *neuraminic acid* residues, usually as the N-acetyl derivative (Figure 6.83). Gangliosides are found in high concentrations in the ganglion cells of the central nervous system and, in lower concentrations, in the membranes of most other cells where the sugar residues serve as polar head groups in the lipid bilayers. Tumor-associated gangliosides are significantly different from those of normal tissues. Although most structural differences originate from aberrant glycosylation, important structural diversity has also been observed in the lipid (ceramide) portion of gangliosides.

In comparison to negative-ion FAB, negative-ion ESI of gangliosides produces significantly lower chemical background signals. A study of several native gangliosides using ESI revealed the formation of abundant $[M-H]^-$, $[M-2H]^{2-}$, and $[M-3H]^{3-}$ anions.[526] The fact that the number of negative charges on the molecular ions is correlated with the number of sialic acid residues remaining after ionization suggests that the negative charge may be located on the carboxylic group of the sialic acid. There was no adduct or cluster ion formation. Sensitivity was increased significantly when 300 pmol/μl of NH_4HCO_3 was added to the mobile phase to increase conductivity.

The synthesis of the core neolactotetraosylceramide and attachment of a sialic acid moiety to this core are considered to be of major importance during cell differentiation and tumorigenesis. The site of attachment of sialic acid to the sugar chain and the mass spectral fragmentation of four sialylated neolactotetraosyl- and neolactohexaosyl-ceramides were studied using ESI-MS/MS with a QTOF instrument. Ganglioside mixtures were obtained from human granulocytes. The ESIMS spectra obtained in positive and negative ion modes and the mechanisms of fragmentation under different MS/MS conditions were evaluated with emphasis on double cleavages.[527] The methodological aspects of analyzing gangliosides from human brain tissue samples using ESI-QTOFMS have also been explored and the respective advantages and limitations of operation in positive and negative ion modes compared to determine the substitution patterns of the sugars and other functional groups.[528,529]

Gangliosides in Pancreatic Cancer

In a study to determine the structure of the epitope to the monoclonal antibody DU-PAN-2, which is used for the diagnosis of Lewis-negative phenotype pancreatic carcinoma, the gangliosides recognized by the antibody were extracted with chloroform/methanol from the tumor of a patient exhibiting a high value in the DU-PAN-2 test. After separating the mono-, di-, and polysialoganglioside fractions using several dialysis and re-extraction steps, it was determined by electrophoresis with immunostaining that the monosialoganglioside fraction contained the DU-PAN-2-positive epitopes, and that the major tumor ganglioside was GM3. Two other ganglioside bands (named X and Y) also reacted with the DU-PAN-2 monoclonal antibody. They demonstrated faster electrophoretic mobility than GM1 but were slower than GM3. The masses of the $[M-H]^-$ ions for these compounds, as detected by negative-ion FAB, were consistent with proposed structures involving C16:0-C18 and C24:1-C18 sphingenines (Figure 6.84A) and C16:0-C18 and C24h:l-C18 sphingenines (Figure 6.84B), respectively. It can also be seen in these figures that the spectra exhibited ions that represent successive elimination of characteristic and diagnostic sugar moieties, including those corresponding to $[M-H-NeuAc]^-$, $[M-H-NeuAc-Gal]^-$, $[M-H-NeuAc-Gal-GlcNAc]^-$, $[M-H-NeuAc-Gal-GlcNAc-Gal]^-$, and $[M-H-NeuAc-Gal-GlcNAc-Gal-Glc]^-$. It was concluded from these and additional experiments that the gangliosides recognized by the DU-PAN-2 antibody are of the $IV^3\alpha$NeuAc-Lc_4Cer type, containing normal and hydroxy fatty acids. Interpretation of the spectra allowed several other possible types of structure to be eliminated from further consideration.[530]

The monoclonal antibody NS 19-9, originally developed from a colon cancer cell line, has also shown reactivity to sera of patients with pancreatic cancer. A similar approach to that given above was used to determine the nature of the ganglioside in the CA 19-9 antigen. Gangliosides extracted from patient tumors were separated into mono-, di-, and polysialoganglioside fractions. The monosialoganglioside fraction exhibited the highest activity. This fraction was further separated by HPLC and the purified antigen isolated from the most active fraction. The negative FAB mass spectrum revealed an intense $[M-H]^-$ ion as well as a series of characteristic fragment ions starting with those corresponding to $[M-deoxyHex]^-$ at *m/z* 1516 and $[M-deoxyHex-NeuAc]^-$ at *m/z* 1225, and continued down to the ceramide ion at *m/z* 536 in which the fatty acid was assumed to be palmitic acid and the long-chain base to be C18-sphingenine. Based on the carbohydrate sequence, and with confirmation by immunostaining, the ganglioside was identified as sialyl Le^a. There was a significant correlation between tumor ganglioside antigen concentrations and serum CA 19-9 levels (determined by routine RIA) but not with tumor burdens.[531]

Aberrant Fatty Acid Hydroxylation in Neuroblastoma

Neuroblastoma (a relatively common childhood tumor) sheds chemically detectable quantities of specific gangliosides with immunosuppressive activity, into the peripheral circulation. A study to compare the structures of the lipid portions of the oligosaccharide-homogeneous GM2 gangliosides

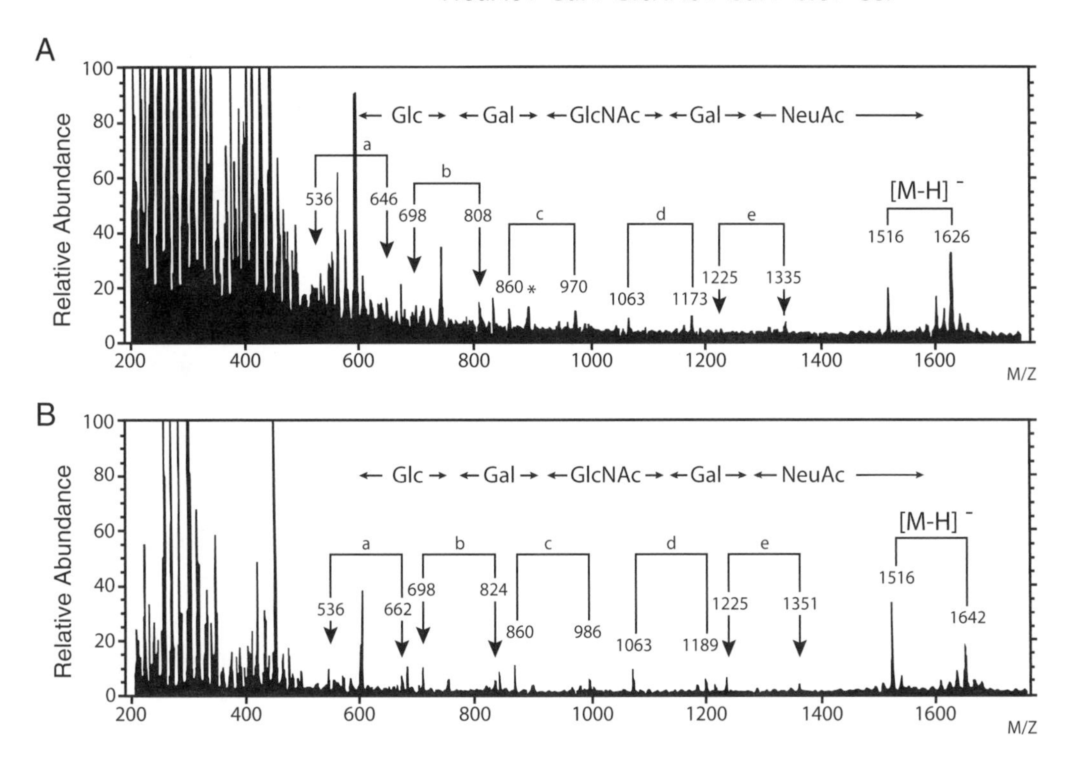

FIGURE 6.84 In a human pancreatic cancer, two gangliosides (designated as X and Y) were found in addition to the major ganglioside GM3. Negative ion FAB mass spectra of X (A) and Y (B) revealed C16:0– (*m/z* 1516) and C24:1–C18 sphingenine (*m/z* 1626) for X (A) and C16:0– (*m/z* 1516) and C24h:1–C18 sphingenine (*m/z* 1642) for Y (B). Ions originating from the successive elimination of sugar moieties were detected as shown. Matrix peaks shown with *. Reprinted from Hamanaka, Y. et al., *FEBS Lett.*, 353, 48–52, 1994. With permission.

from neuroblastoma with those from normal human brain revealed that the former exhibited significant heterogeneity associated with the presence of aberrant α-hydroxylated fatty acids.[532] Some 18 ganglioside species, extracted from neuroblastoma tissue, were initially separated using reversed-phase HPLC and then purified by normal-phase HPLC. Almost 20% of the total tumor GM2 ganglioside fraction was comprised of α-hydroxylated fatty-acid-containing species (16:0, 18:0, 20:0, and 24:1). In contrast, virtually no fatty acid hydroxylation could be detected in normal human brain gangliosides.

The separated gangliosides were analyzed by positive and negative ion FABMS in combination with CID. Negative ion FABMS of GM2 from neuroblastoma revealed the presence of at least 10 gangliosides that could be distinguished by their fatty acid composition (Figure 6.85B). This was in sharp contrast to only two distinct gangliosides in the GM2 fraction from normal human brain (Figure 6.85A). In the negative FAB spectra of neuroblastoma gangliosides, the $[M-H]^-$ ions appeared in sets of pairs with each member of the pairs being separated by 16 mass units and the sets of pairs by 28 mass units. This suggested that there were regular, periodic (C_2H_4) differences in the fatty acid chain lengths and that hydroxylation was occurring. CID of these long chain structures was undertaken using the $[Cer-O]^-$ fragment (NeuAc GM2 – 818) as the precursor ion. Prominent product ions were present and these identified the fatty acid moieties.[533] Using the mass difference between the fatty acyl group and the $[CerO]^-$ ion, the weight of the amino side-chain

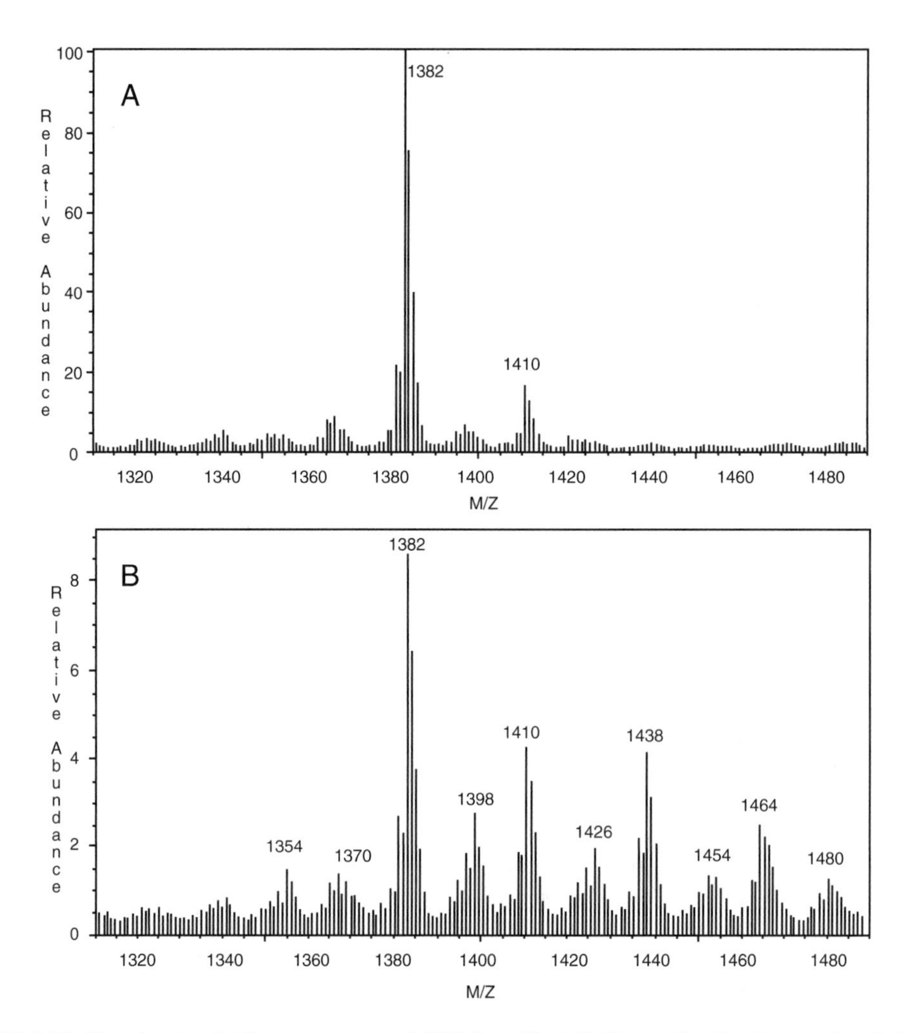

FIGURE 6.85 Negative ion FAB mass spectra of HPLC-purified GM2 gangliosides in a study of aberrant fatty acyl α-hydroxylation in human neuroblastoma. (A) Normal human brain. The pattern in neuroblastoma (B) revealed significant heterogeneity of the molecular ions. Reprinted from Ladisch, S. et al., *J. Biol. Chem.*, 264, 12097–12105, 1989. With permission.

can be calculated (Figure 6.86). The positive FABMS and negative FABMS CID fragmentation patterns provided complementary information on the structure of the ceramides fractions of this family of gangliosides. In a similar fashion, fatty acid hydroxylation was determined in tumor-associated GM3 and GD2 neuroblastoma gangliosides. It was concluded that the heterogeneity due to tumor-associated aberrant fatty acid hydroxylation of gangliosides should be considered in addition to the already-documented heterogeneity associated with the oligosaccharide portions of gangliosides.[532]

Glucosylceramides in Multidrug-Resistant Breast Cancer

Although it is believed that the major cause of multidrug resistance (MDR) is the overexpression of the membrane efflux transporter (Section 5.1.1.2), other biochemical changes are also likely to contribute. Glycosylceramides (ceramide glycoconjugates where the sugar moiety is nonspecified, i.e., glucose or galactose) are not only precursors of a large number of glycosphingolipids but are also, putatively, involved in cell proliferation, differentiation, and oncogenic transformation. They have also been considered as second messengers in signal transduction pathways that involve growth factors.

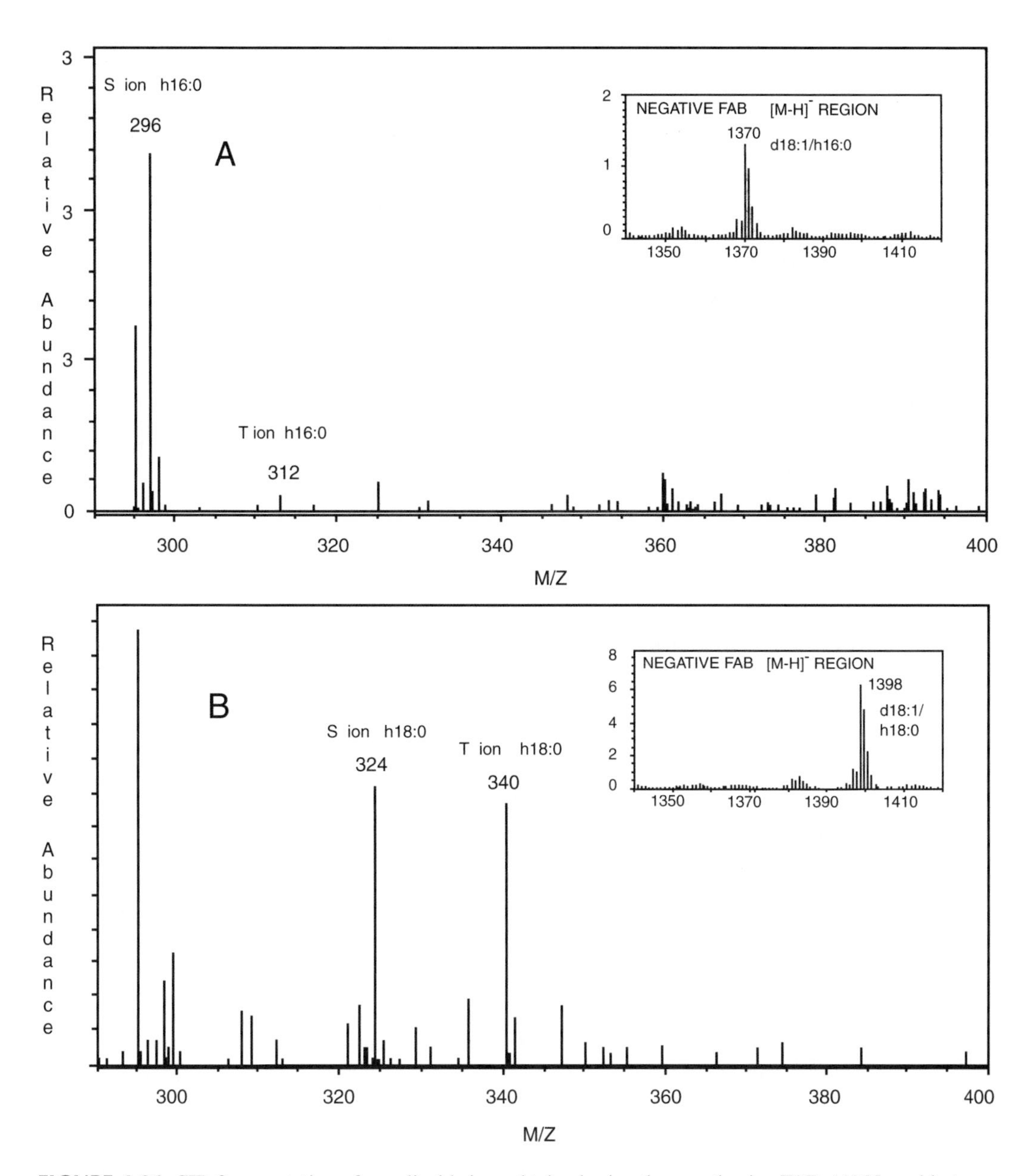

FIGURE 6.86 CID fragmentation of ganglioside ions obtained using the negative ion FAB. (A) Neuroblastoma d18:1-h16:0 GM2; (B) d18:1-h18:0 GM2. The precursor ions were $[CerO]^-$ at *m/z* 552 (A) and 580 (B). The fragment ions S and T representing the fatty acid moieties are labeled. Several unidentified ions are also present. The insets show the negative-ion FABMS spectra of the molecular ion regions. Reprinted from Ladisch, S. et al., *J. Biol. Chem.*, 264, 12097–12105, 1989. With permission.

In a study of ceramides in wild-type MCF-7 cancer cells sensitive to adriamycin and in MDR subclones, high concentrations of two unidentified lipids were found, by preparative TLC, in total lipid extracts of the latter, but not in the drug-sensitive cells. Incubation of the resistant cell line with sphingolipid and glucosylceramide biosynthesis inhibitors prevented accumulation of these lipids. The FABMS mass spectrum of one of the lipids was relatively simple (Figure 6.87B) with the $[M+H]^+$ peak at *m/z* 699 corresponding to an N-palmitoyl monoglycosylceramide. It was accompanied by an ion at *m/z* 537 that was the result of the expected loss of the hexose. Small

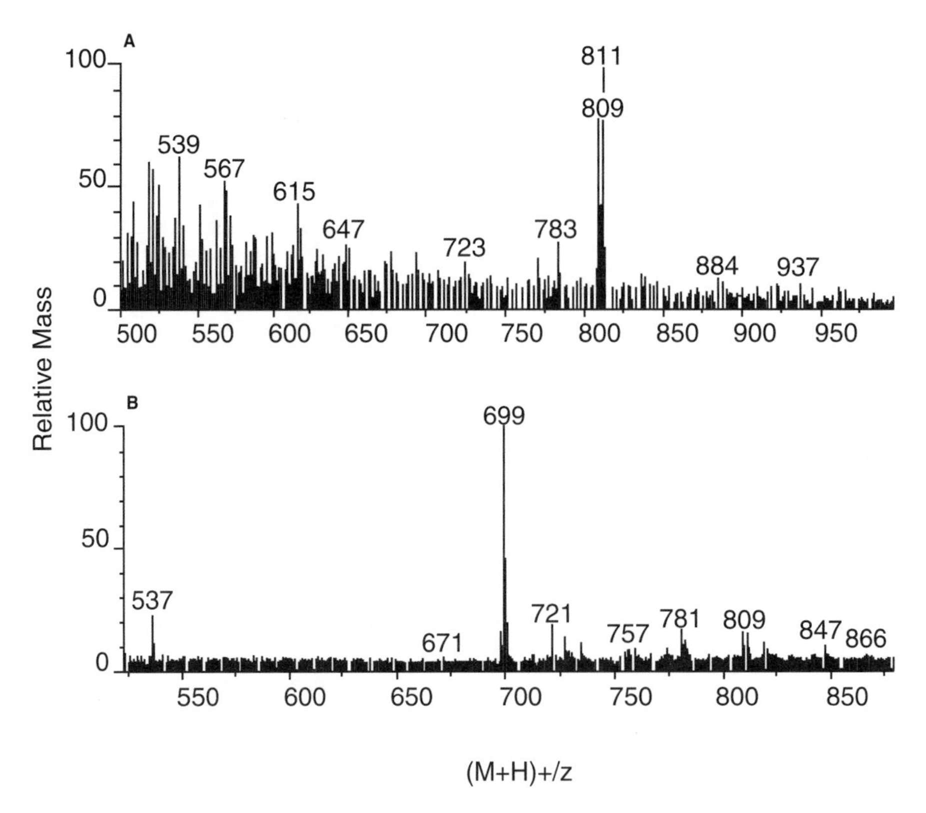

FIGURE 6.87 FAB mass spectra of TLC-separated and -purified lipids from multidrug-resistant MCF-7-AdrR cells. (A) A heterogeneous spectrum of one of the TLC bands contained predominantly N-tetracosanoyl (lignoceroyl) monoglycosylceramide at $[M+H]^+$ (*m/z* 811) and N-tetracosanoyl (nervonoyl) monoglycosylceramide at $[M+H]^+$ (*m/z* 809). (B) The single lipid in the other TLC band was identified as N-palmitoyl monoglycosylceramide. Reprinted from Lavie, Y. et al., *J. Biol. Chem.*, 271, 19530–19536, 1996. With permission.

quantities of other monoglycosylceramides with different amide side-chains were also present in the spectrum at higher masses. The FAB spectrum of the second lipid was more heterogeneous (Figure 6.87A). Interpretation of the data revealed the presence of 24:0 and 24:1 fatty acid monoglycosyl ceramides, namely N-tetracosanoyl (lignoceroyl) and N-tetracosenoyl (nervonoyl) at *m/z* 811 and *m/z* 809, respectively. The head group in both of these lipids was shown, by TLC of standards, to be glucose rather than galactose. Additional biochemical experiments established that the presence of these glycosphingolipids in the MDR cells was due to accelerated synthesis rather than hindered breakdown. Glycosylceramides have also been found to accumulate in some cell lines of drug resistant epidermal carcinoma and ovarian adenocarcinoma, suggesting the need for further studies to establish the possibility of a global nature for this association.[534]

Monosialogangliosides in Human Myelogenous Leukemia HL60 Cells

Gangliosides were extracted from HL60 cells, with human leukocytes as controls, and extensively fractionated. The structures of these gangliosides were determined using negative ion FABMS of the native forms and, after permethylation, by positive ion FABMS and direct injection ESIMS. For example, the ESIMS spectrum of permethylated gangliosides from one of the isolated HL60 fractions was a single (multiply sodiated) compound (Figure 6.88). CID of the triply charged ion provided recognizable fragment ions from the reducing side of each HexNAc residue and a full fucosylated sequence from both the reducing and nonreducing termini (Figure 6.89). The spectrum

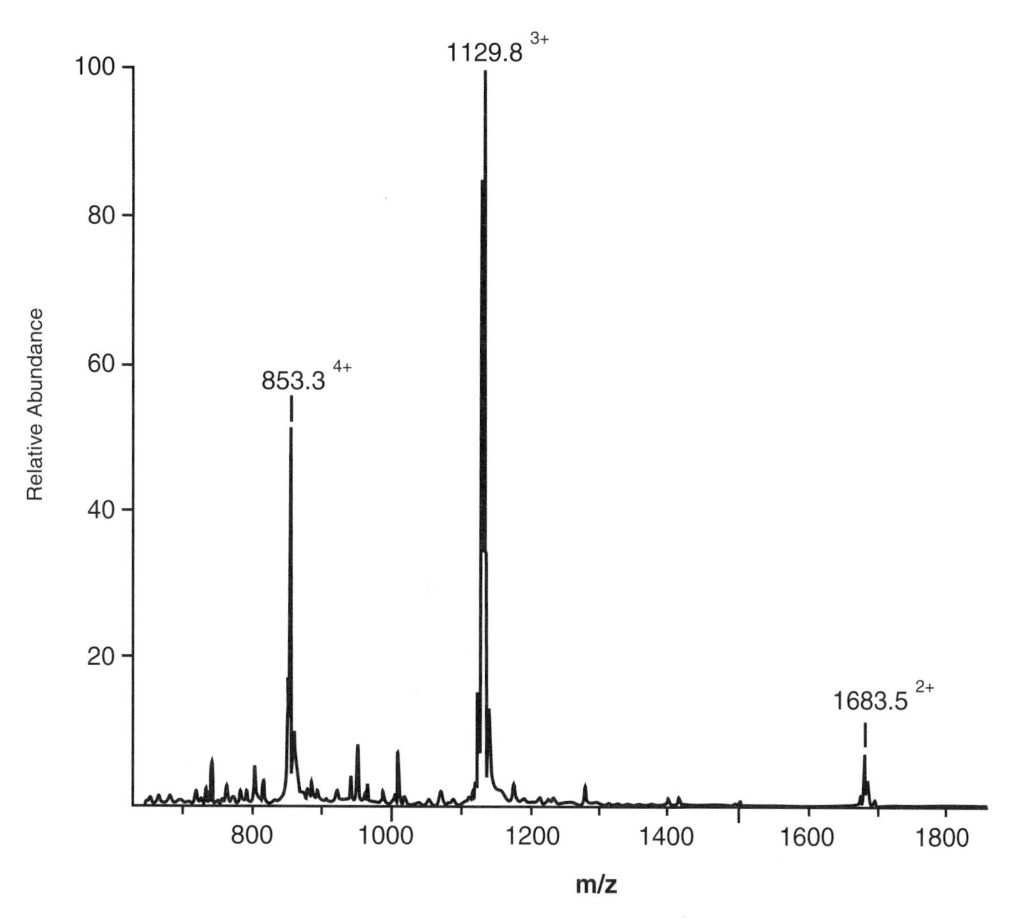

FIGURE 6.88 Electrospray mass spectrum of gangliosides in one of the separated fractions isolated from HL60 cells. See also Figure 6.89. Reprinted from Stroud, M. et al., *Biochemistry,* 35, 758–769, 1996. With permission.

also revealed that the ceramide residue was composed of a sphingenine moiety with a palmitoyl N-acyl chain. A detailed interpretation of the mass spectra of the other gangliosides from the HL60 cells revealed that in these cells there was a lack of the SLe^x gangliosides (with tetraosyl to octaosyl ceramide cores) that constitute the major class of gangliosides in epithelial tumors.[535] In a subsequent study, FABMS and ESIMS were used to determine the structure of the E-selectin-binding fractions from HL60 cells and the structural requirements for that binding.[536]

Gangliosides in Renal Cell Carcinoma

The expression level of gangliosides, particularly disialogangliosides, was correlated with the degree of metastatic potential in renal cell carcinoma (RCC). Two new disialogangliosides (G1 and G2) have been reported in a cell line (TOS-1) derived from a metastatic lesion of an RCC patient. Several techniques, including ESIMS, NMR, monoclonal antibody reactivity, and monosaccharide and fatty acid composition analysis by GC/MS, have been used to establish that these gangliosides have a hybrid structure that lies between the ganglio-series, GM2, and the lacto-series Type I. The composition of G1 has been determined to be GalNAc-disialosyl-Lc_4Cer (IV^4GalNAcAcIV^3NeuAcIII^6NeuAcLc_4). The composition of G2 is similar except it lacks the GalNAc(β1-4) substitution. The complex ESI fragmentation pattern confirmed the presence of two series of ions representing the ganglio-series and lacto-series Type 1 structures. The degree of expression of these gangliosides was correlated with the metastatic potential of the RCC.[537]

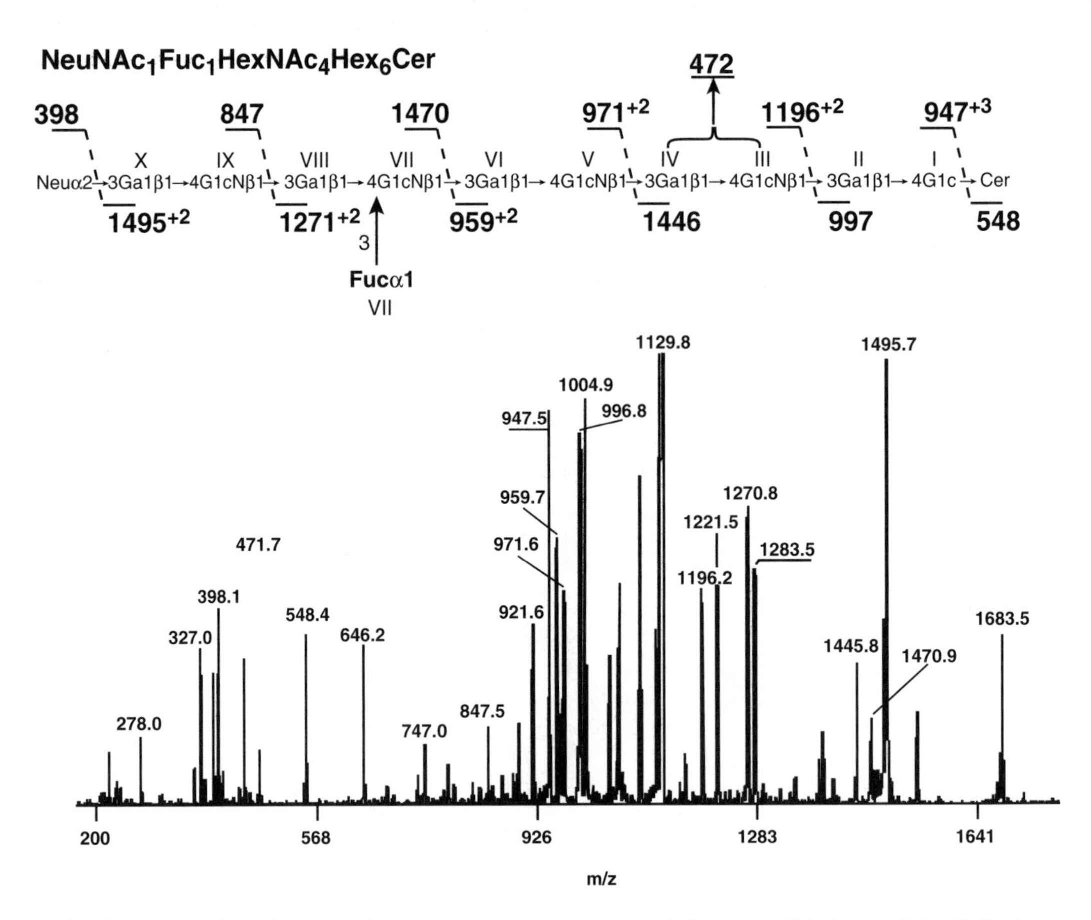

FIGURE 6.89 Collision-induced dissociation mass spectrum of the *m/z* 1129.8 ion $[M+3Na]^{3+}$ from Figure 6.88. Top: proposed structure and fragmentation scheme. Reprinted from Stroud, M. et al., *Biochemistry*, 35, 758–769, 1996. With permission.

Gangliosides in Glioma

Examination of the acid glycosphingolipid components of rat glioma isografts led to the observation of a GM3 band that migrated faster on TLC than authentic GM3. The oligosaccharide structure of the unknown was determined by GC/MS, using partially methylated alditol acetates, and by FABMS. The FAB mass spectrum of the intact analyte revealed a protonated molecule at *m/z* 1305 that corresponded to GM3 with an additional acetyl residue. It was composed of a long-chain sphingenine with C24:0 as the amido-linked fatty acid and a NANA glycosylating moiety (Figure 6.90). The identified fragment ions indicated that the acetyl group was on the ceramide fraction of this fast moving GM3 molecule and not the sugar moiety. Thus, the novel O-acetylated GM3 contained 3-O-acetyl-4-sphingosine.[538]

A new GM1-type ganglioside was detected in dissected tumors of human glioma and accounted for approximately 20% of the monosialoganglioside fraction. This ganglioside was not detectable in normal human brain tissues or in 17 other types of tumors. The structure of the ganglioside was determined by FABMS to be II^3NeuNH_2-$GgOse_4$Cer. The role of this ganglioside in human glioma growth or invasiveness is not known.[539]

Ganglioside in Metastatic Brain Tumor

To identify gangliosides associated with brain metastasis, tumor samples were analyzed from patients with a variety of primary tumors both with and without brain metastasis. Separated

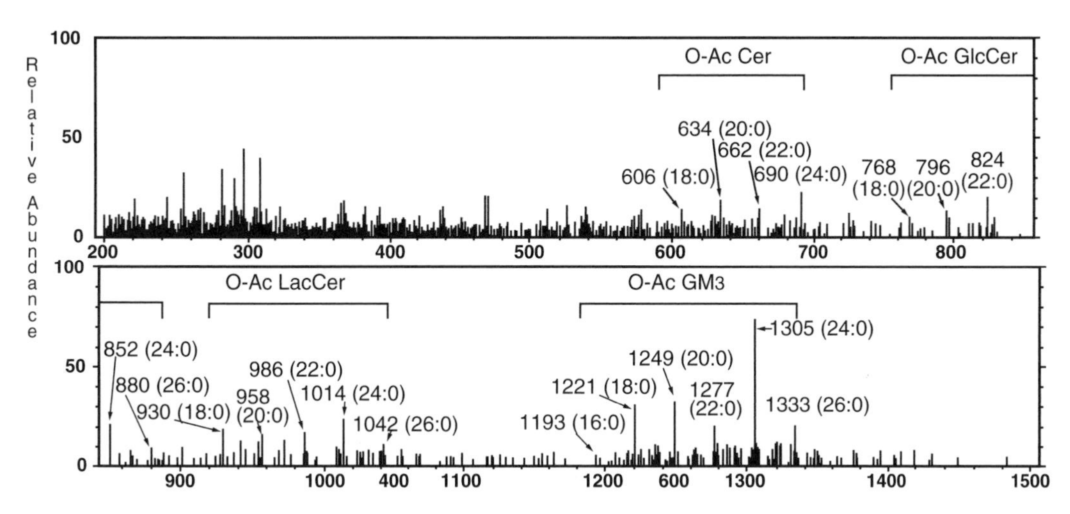

FIGURE 6.90 Negative-ion FAB mass spectrum of O-Ac-GM3 ganglioside from rat glioma. The numbers in parentheses at the fragment indicate carbon number and degree of unsaturation of the fatty acid. Reprinted from Suetake, K. et al., *FEBS Lett.,* 361, 201–205, 1995. With permission.

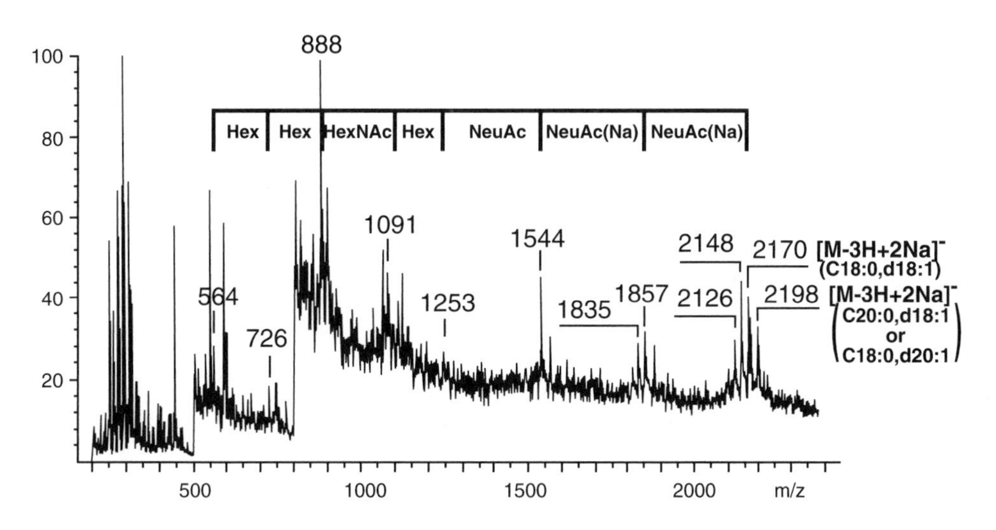

FIGURE 6.91 Secondary-ion mass spectrum of the GT1b-like ganglioside band of a brain metastatic tumor from lung adenocarcinoma. Assignments of signals derived from the ganglioside are indicated in the figure. The molecular-related ions of *m/z* 2126, 2148, and 2170 are $[M-H]^-$, $[M-2H+Na]^-$, and $[M-3H+2Na]^-$ of GT1b consisting of ceramide with C18:0 and d18:1. The molecular-related ion *m/z* 2198 correspond to $[M-3H+2Na]^-$ of GT1b with C20:0, d18:1, or C18:0, d20:1 ceramide. Reprinted from Hamasaki, H. et al., *Biochim. Biophys. Acta,* 1437, 93–99, 1999. With permission.

glycosphingolipids were analyzed using a combined far-Eastern blot-negative ion FABMS technique. Emphasis was placed on GT1b, as this has been considered to be a brain specific ganglioside. The ganglioside was identified based on the interpretation of peaks in the FAB spectrum of GT1b obtained from a brain tumor that had metastasized from a primary lung cancer (Figure 6.91) and by comparison with an authentic sample. Because GT1b was found in all samples from brain metastases but not in systemic carcinomas without brain metastasis (where GM3 was the common major component), the presence of GT1b was suggested as a marker for estimating metastatic potentials to the brain.[540]

Gangliosides and the Immune Response

It has been suggested that one mechanism by which tumor cells escape immune response to their tumor-specific transplantation antigens (TSTA) involves shedding gangliosides into the microenvironment surrounding the tumor where they can act as soluble modulators of the immune response. This was studied in a tumor model of erythroleukemia B6 origin (FBL-3) that expresses unique TSTA. The major FBL-3 ganglioside was identified as GM1b. The carbohydrate structure of this ganglioside was characterized by negative ion FABMS as consisting of NANA, GalNAc, and two galactoses. The ceramide substructure was determined by CID to contain d18:1-C16:0 and d18:1-C24:0 species indicating that GM1b is composed of two species. It was shown in a variety of biochemical and immunological experiments that these tumor-derived gangliosides inhibit syngeneic antitumor response and thus are implicated in promoting tumor formation and progression.[541]

RECOMMENDED BOOKS

Dass, C., *Principles and Practice of Biological Mass Spectrometry,* Wiley-InterScience, New York, 2000.

Kinter, M. and Sherman, N.E., *Protein Sequencing and Identification Using Tandem Mass Spectrometry,* Wiley-InterScience, New York, 2000.

Snyder, A.P., *Interpreting Protein Mass Spectra: A Comprehensive Resource*, Oxford University Press, Washington, D.C., 2000.

REFERENCES

1. Humphery-Smith, I. and Blackstock, W., Proteome analysis: genomics via the output rather than the input code, *J. Protein Chem.,* 16, 537–544, 1997.
2. Miklos, G.L.G. and Maleszka, R., Protein functions and biological contexts, *Proteomics*, 1, 169–178, 2001.
3. Jung, E., Heller, M., Sanchez, J.C., and Hochstrasser, D.F., Proteomics meets cell biology: the establishment of subcellular proteomes, *Electrophoresis*, 21, 3369–3377, 2000.
4. Chambers, G., Lawrie, L., Cash, P., and Murray, G.I., Proteomics: a new approach to the study of disease. *J. Pathology,* 192, 280–288, 2000.
5. Pandey, A. and Mann, M., Proteomics to study genes and genomes, *Nature,* 2000, 405, 837–846.
6. Wilkins, M.R., Williams, K.L., Appel, R.D., and Hochstrasser, D.F., Eds., *Proteome Research: New Frontiers in Functional Genomics*, Springer-Verlag, New York (1997).
7. Steiner, S. and Witzmann, F.A., Proteomics: applications and opportunities in preclinical drug development, *Electrophoresis,* 21, 2099–2104, 2000.
8. Rigaut, G., Shevchenko, A., Rutz, B., Wilm, M., Mann, M., and Seraphin, B., A generic protein purification method for protein complex characterization and proteome exploration, *Nature Biotechnol.,* 17, 1030–1032, 1999.
9. Yates, J.R., Mass spectrometry. From genomics to proteomics, *Trends Genet.,* 16, 5–8, 2000.
10. Andersen, J.S. and Mann, M., Functional genomics by mass spectrometry, *FEBS Lett.,* 480, 25–31, 2000.
11. Yates, J.R. Mass spectrometry and the age of proteome, *J. Mass Spectrom.,* 33, 1–19, 1998.
12. Godovac-Zimmermann, J. and Brown, L.R., Perspectives for mass spectrometry and functional proteomics, *Mass Spectrom. Reviews,* 20, 1–57, 2001.
13. Aebersold, R. and Goodlett, D.R., Mass spectrometry in proteomics, *Chem. Rev.,* 101, 269–295, 2001.
14. Harry, J.L., Wilkins, M.R., Herbert, B.R., Packer, N.H., Gooley, A.A., and Williams, K.L., Proteomics: capacity versus utility, *Electrophoresis,* 21, 1071–1081, 2000.
15. Lahm, H.W. and Langen, H., Mass spectrometry: a tool for the identification of proteins separated by gels, *Electrophoresis,* 21, 2105–2114, 2000.
16. Patterson, S.D., Proteomics: the industrialization of protein chemistry, *Curr. Opin. Biotech.,* 11, 413–418, 2000.
17. Gevaert, K. and Vandekerckhove, J., Protein identification methods in proteomics, *Electrophoresis,* 21, 1145–1154, 2000.

18. Chalmers, M.J. and Gaskell, S.J., Advances in mass spectrometry for proteome analysis, *Curr. Opin. Biotech.*, 11, 384–390, 2000.
19. Gygi, S.P. and Aebersold, R., Mass spectrometry and proteomics, *Current Opin. Chem. Biol.*, 4, 489–494, 2000.
20. Resing, K.A. and Ahn, N.G., Application of mass spectrometry to signal transduction, *Progr. Biophys. Molec. Biol.*, 71, 501–523, 1999.
21. Pandey, A., Andersen, J.S., and Mann, M., Use of mass spectrometry to study signaling pathways, *Sciences's STKE*, 1–12, 2000.
22. Corthals, G.L., Wasinger, V.C., Hochstrasser, D.F., and Sanchez, J.C., The dynamic range of protein expression: a challenge for proteomic research, *Electrophoresis*, 21, 1104–1115, 2000.
23. Kelleher, N.L., From primary structure to function: biological insights from large-molecule mass spectra, *Chem. Biol.*, 7, R37–R45, 2000.
24. Quadroni, M. and James, P., Proteomics and automation, *Electrophoresis*, 20, 664–677, 1999.
25. Kelleher, N.L., Lin, H.Y., Valaskovic, G.A., Aaerud, D.J., Fridiksson, E.K., and McLafferty, F.W., Top down versus bottom up protein characterization by tandem high-resolution mass spectrometry, *J. Am. Chem. Soc.*, 121, 806–812, 1999.
26. Hanash, S.M., Biomedical applications of two-dimensional electrophoresis using immobilized pH gradients: current status, *Electrophoresis*, 21, 1202–1209, 2000.
27. Hellman, U., Sample preparation by SDS/PAGE and in-gel digestion, *EXS*, 88, 43–54, 2000.
28. Gevaert, K., Houthaeve, T., and Vandekerckhove, J., Techniques for sample preparation including methods for concentrating peptide samples, *EXS*, 88, 29–42, 2000.
29. Michalski, W.P. and Shiell, B.J., Strategies for the analysis of electrophoretically separated proteins and peptides, *Anal. Chim. Acta*, 383, 27–46, 1999.
30. Nilsson, C.L. and Davidsson, P., New separation tools for comprehensive studies of protein expression by mass spectrometry, *Mass Spectrom. Reviews*, 19, 390–397, 2000.
31. McCormack, A.L., Schieltz, D.M., Goode, B., Yang, S., Barnes, G., Drubin, D., and Yates, J.R., Direct analysis and identification of proteins in mixtures by LC/MS/MS and database searching at the low-femtomole level, *Anal. Chem.*, 69, 767–776, 1997.
32. Opiteck, G.J., Lewis, K.C., and Jorgenson, J.W., Comprehensive on-line LC/MS of proteins, *Anal. Chem.*, 69, 1518–1524, 1997.
33. Fey, S.J. and Larsen, P.M., 2D or not 2D, *Current Opinion Chem. Biol.*, 5, 26–33, 2001.
34. Lopez, M.F., Proteome analysis: 1. Gene products are where the biological action is, *J. Chromatogr. B*, 722, 191–202, 1999.
35. Epstein, L.B., Smith, D.M., Matsui, N.M., Tran, H.M., Sullivan, C., Raineri, I., Burlingame, A.L., Clauser, K.R., Hall, S.C., and Andrews, L.E., Identification of cytokine-regulated proteins in normal and malignant cells by the combination of two-dimensional polyacrylamide gel electrophoresis, mass spectrometry, Edman degradation and immunoblotting and approaches to the analysis of their functional roles, *Electrophoresis*, 17, 1655–1670, 1996.
36. James, P., Of genomes and proteomes, *Biochem. Biophys. Res. Comm.*, 231, 1–6, 1997.
37. Simpson, R.J., Connolly, L.M., Eddes, J.S., Pereira, J.J., Moritz, R.L., and Reid, G.E., Proteomic analysis of the human colon carcinoma cell line (LIM 1215): development of a membrane protein database, *Electrophoresis*, 21, 1707–1732, 2000.
38. Sechi, S. and Chait, B.T., Modification of cysteine residues by alkylation: a tool in peptide mapping and protein identification, *Anal. Chem.*, 70, 5150–5158, 1998.
39. Shevchenko, A., Wilm, M., Vorm, O., and Mann, M., Mass spectrometric sequencing of proteins in silver-stained polyacrylamide gels, *Anal. Chem.*, 68, 850–858, 1996.
40. Erdjument Bromage, H., Lui, M., Lacomis, L., Grewal, A., Annan, R.S., McNulty, D.E., Carr, S.A., and Tempst, P., Examination of micro-tip reversed-phase liquid chromatographic extraction of peptide pools for mass spectrometric analysis, *J. Chromatogr. A*, 826, 167–181, 1998.
41. Gharahdaghi, F., Weinberg, C.R., Meagher, D.A., Imai, B.S., and Mische, S.M., Mass spectrometric identification of proteins from silver-stained polyacrylamide gel: a method for the removal of silver ions to enhance sensitivity, *Electrophoresis*, 20, 601–605, 1999.
42. Gygi, S.P., Corthals, G.L., Zhang, Y., Rochon, Y., and Aebersold, R., Evaluation of two-dimensional gel electrophoresis-based proteome analysis technology, *Proc. Natl. Acad. Sci. USA*, 97, 9390–9395, 2000.

43. Binz, P.A., Muller, M., Walther, D., Bienvenut, W.V., Gras, R., Hoogland, C., Bouchet, G., Gasteiger, E., Fabbretti, R., Gay, S., Palagi, P., Wilkins, M.R., Rouge, V., Tonella, L., Paesano, S., Rossellat, G., Karmime, A., Bairoch, A., Sanchez, J.C., Appel, R.D., and Hochstrasser, D.F., A molecular scanner to automate proteomic research and to display proteome images, *Anal. Chem.*, 71, 4981–4988, 1999.
44. Bienvenut, W.V., Sanchez, J.C., Karmime, A., Rouge, V., Rose, K., Binz, P.A., and Hochstrasser, D.F., Toward a clinical molecular scanner for proteome research: parallel protein chemical processing before and during western blot, *Anal. Chem.*, 71, 4800–4807, 1999.
45. Reid, G.E., Rasmussen, R.K., Dorow, D.S., and Simpson, R.J., Capillary column chromatography improves sample preparation for mass spectrometric analysis: complete characterization of human alpha-enolase from two-dimensional gels following in situ proteolytic digestion, *Electrophoresis*, 19, 946–955, 1998.
46. Haynes, P.A., Fripp, N., and Aebersold, R., Identification of gel-separated proteins by liquid chromatography-electrospray tandem mass spectrometry: comparison of methods and their limitations, *Electrophoresis*, 19, 939–945, 1998.
47. Martin, S.E., Shabanowitz, J., Hunt, D.E., and Marto, J.A., Subfemtomole MS and MS/MS peptide sequence analysis using nano-HPLC micro-ESI Fourier transform ion cyclotron resonance mass spectrometry, *Anal. Chem.*, 72, 4266–4274, 2000.
48. Washburn, M.P., Wolters, D., and Yates, J.R., Large-scale analysis of the yeast proteome by multidimensional protein identification technology, *Nature Biotechnol.*, 19, 242–247, 2001.
49. Wahl, J., Goodlett, D., Udseth, H., and Smith, R., Use of small-diameter capillaries for increasing peptide and protein detection sensitivity in capillary electrophoresis-mass spectrometry, *Electrophoresis*, 14, 448–457, 1993.
50. Naylor, S. and Tomlinson, A.J., Membrane preconcentration-capillary electrophoresis-mass spectrometry in the analysis of biologically derived metabolites and biopolymers, *Biomed. Chromatogr.*, 10, 325–330, 1996.
51. Emmett, M. and Caprioli, R., Micro-electrospray mass spectometry ultra-high-sensitivity analysis of peptides and proteins, *J. Am. Soc. Mass Spectrom.*, 5, 605–613, 1994.
52. Wilm, M., Shevchenko, A., Houthaeve, T., Brelt, S., Schwelgerer, L., Fotsls, T., and Mann, M., Femtomole sequencing of proteins from polyacrylamide gels by nano-electrospray mass spectrometry, *Nature*, 379, 466–469, 1996.
53. Merchant, M. and Weinberger, S.R., Recent advancements in surface-enhanced laser desorption/ionization-time of flight-mass spectrometry, *Electrophoresis*, 21, 1164–1174, 2000.
54. Fung, E.T., Thulasiraman, V., Weinberger, S.R., and Dalmasso, E.A., Protein biochips for differential profiling, *Current Opin. Biotechnol.*, 12, 65–69, 2001.
55. Rubin, R.B. and Merchant, M., A rapid protein profiling system that speeds study of cancer and other diseases, *Amer. Clin. Lab.*, 18–20, 2000.
56. Nelson, R.W., The use of bioreactive probes in protein characterization, *Mass Spectrom. Reviews*, 16, 353–376, 1997.
57. Shevchenko, A., Loboda, A., Shevchenko, A.E., Ens, W., and Standing, K.G., MALDI quadrupole time-of-flight mass spectrometry: a powerful tool for proteomic research, *Anal. Chem.*, 72, 2132–2141, 2000.
58. Baldwin, M.A., Medzihradszky, K.F., Lock, C.M., Fisher, B., Settineri, T.A., and Burlingame, A.L., Matrix-assisted laser desorption/ionization coupled with quadrupole/orthogonal acceleration time-of-flight mass spectrometry for protein discovery, identification, and structural analysis, *Anal. Chem.*, 73, 1707–1720, 2001.
59. Medzihradszky, K.F., Campbell, J.M., Baldwin, M.A., Falick, A.M., Juhasz, P., Vestal, M.L., and Burlingame, A.L., The characteristics of peptide collision-induced dissociation using a high-performance MALDI-TOF/TOF tandem mass spectrometer, *Anal. Chem.*, 72, 552–558, 2000.
60. Quenzer, T.L., Emmett, M.R., Hendrickson, C.L., Kelly, P.H., and Marshall, A.G., High sensitivity Fourier transform ion cyclotron resonance mass spectrometry for biological analysis with nano-LC and microelectrospray ionization, *Anal. Chem.*, 73, 1721–1725, 2001.
61. Shen, Y., Zhao, R., Belov, M.E., Conrads, T.P., Anderson, G.A., Tang, K., Pasa-Tolic, L., Veenstra, T.D., Lipton, M.S., Udseth, H.R., and Smith, R.D., Packed capillary reversed-phase liquid chromatography with high-performance electrospray ionization Fourier transform ion cyclotron resonance mass spectrometry for proteomics, *Anal. Chem.*, 73, 1766–1775, 2001.

62. Pasa-Tolic, L., Jensen, P.K., Anderson, G.A., Lipton, M.S., Peden, K.K., Martinovic, S., Tolic, N., Bruce, J.E., and Smith, R.D., High throughput proteome-wide precision measurements of protein expression using mass spestrometry, *J. Am. Chem. Soc.*, 121, 7949–7950, 1999.
63. Li, W., Hendrickson, C.L., Emmett, M.R., and Marshall, A.G., Identification of intact proteins in mixtures by alternated capillary liquid chromatography electrospray ionization and LC ESI infrared multiphoton dissociation Fourier transform ion cyclotron resonance mass spectrometry, *Anal. Chem.*, 71, 4397–4402, 1999.
64. Bruce, J.E., Anderson, G.A., Wen, J., Harkewicz, R., and Smith, R.D., High-mass-measurement accuracy and 100% sequence coverage of enzymatically digested bovine serum albumin from an ESI-FTICR mass spectrum, *Anal. Chem.*, 71, 2595–2599, 1999.
65. Caprioli, R.M., Farmer, T.B., and Gile, J., Molecular imaging of biological samples: localization of peptides and proteins using MALDI-TOF MS, *Anal. Chem.*, 69, 4751–4760, 1997.
66. Stoeckli, M., Chaurand, P., Hallahan, D.E., and Caprioli, R.M., Imaging mass spectrometry: a new technology for the analysis of protein expression in mammalian tissues, *Nature Med.*, 7, 493–496, 2001.
67. Traini, M., Gooley, A.A., Ou, K., Wilkins, M.R., Tonella, L., Sanchez, J.C., Hochstrasser, D.F., and Williams, K.L., Towards an automated approach for protein identification in proteome projects, *Electrophoresis*, 19, 1941–1949, 1998.
68. Jin, X., Kim, J., Parus, S., Lubman, D.M., and Zand, R., On-line capillary electrophoresis/microelectrospray ionization-tandem mass spectrometry using an ion trap storage/time-of-flight mass spectrometer with SWIFT technology, *Anal. Chem.*, 71, 3591–3597, 1999.
69. Chan, J.H., Timperman, A.T., Qin, D., and Aebersold, R., Microfabricated polymer devices for automated sample delivery of peptides for analysis by electrospray ionization tandem mass spectrometry, *Anal. Chem.*, 71, 4437–4444, 1999.
70. Figeys, D., Gygi, S.P., McKinnon, G., and Aebersold, R., An intergrated microfluidics-tandem mass spectrometry system for automated protein analysis, *Anal. Chem.*, 70, 3728–3734, 1998.
71. Figeys, D. and Aebersold, R., Nanoflow solvent gradient delivery from a microfabricated device for protein identifications by electrospray ionization mass spectrometry, *Anal. Chem.*, 70, 3721–3727, 1998.
72. Gooley, A.A., Ou, K., Russell, J., Wilkins, M.R., Sanchez, J.C., Hochstrasser, D.F., and Williams, K.L., A role for Edman degradation in proteome studies, *Electrophoresis*, 18, 1068–1072, 1997.
73. Grant, G.A., Cranksaw, M.W., and Gorka, J., Edman sequencing as tool of characterization of synthetic peptides, *Methods Enzymol.*, 289, 395–408, 1997.
74. Henzel, W.J., Tropea, J., and Dupont, D., Protein identification using 20-minute Edman cycles and sequence mixture analysis, *Anal. Biochem.*, 267, 148–160, 1999.
75. Shively, J.E., The chemistry of protein sequence analysis, *EXS*, 88, 99–118, 2000.
76. Jensen, O.N., Podtelejnikov, A., and Mann, M., Delayed extraction improves specificity in database searches by matrix-assisted laser desorption/ionization peptide maps, *Rapid Commun. Mass Spectrom.*, 10, 1371–1378, 1996.
77. Kamo, M., Kawakami, T., and Tsugita, A., Novel detection methods for modification sites using chemical cleavages and mass spectrometry, *J. Protein Chem.*, 17, 511, 1998.
78. Kamo, M. and Tsugita, A., N-terminal amino acid sequencing of 2-DE spots, *Methods Mol. Biol.*, 112, 461–466, 1999.
79. Roepstorff, P. MALDI-TOF mass spectrometry in protein chemistry, *EXS*, 88, 81–97, 2000.
80. Fenyo, D., Qin, J., and Chait, B.T., Protein identification using mass spectrometric information, *Electrophoresis*, 19, 998–1005, 1998.
81. Clauser, K.R., Baker, P., and Burlingame, A.L., Role of accurate mass measurement in protein identification strategies employing MS or MS/MS and database searching, *Anal. Chem.*, 71, 2871–2882, 1999.
82. Green, M.K., Johnston, M.V., and Larsen, B.S., Mass accuracy and sequence requirements for protein database searching, *Anal. Biochem.*, 275, 39–46, 1999.
83. Houthaeve, T., Gausepohl, H., Ashman, K., Nillson, T., and Mann, M., Automated protein preparation techniques using a digest robot, *J. Protein Chem.*, 16, 343–348, 1997.
84. Jensen, O.N., Podtelejnikov, A.V., and Mann, M., Identification of the components of simple protein mixtures by high-accuracy peptide mass mapping and database searching. *Anal. Chem.*, 69, 4741–4750, 1997.

85. Appella, E., Arnott, D., Sakaguchi, K., and Wirth, P.J., Proteome mapping by two-dimensional polyacrylamide gel electrophoresis in combination with mass spectrometric protein sequence analysis, *EXS*, 88, 1–27, 2000.
86. Wilm, M., Mass spectrometric analysis of proteins, *Adv. Protein Chem.*, 54, 1–29, 2000.
87. Bergman, T., Ladder sequencing, *EXS*, 88, 133–144, 2000.
88. Chait, B.T., Wang, R., Beavis, R., and Kent, S., Protein ladder sequencing, *Science*, 262, 89–92, 1993.
89. Hoving, S., Munchbach, M., Schmid, H., Signor, L., Lehmann, A., Staudenmann, W., Quadroni, M., and James, P., A method for the chemical generation of N-terminal peptide sequence tags for rapid protein identification, *Anal. Chem.*, 72, 1006–1014, 2000.
90. Woods, A., Huang, A., Cotter, R.J., Pasternack, G., Pardoll, D., and Jaffee, E., Simplified high-sensitivity sequencing of a major histocompatibility complex Class-I-associated immunoreactive peptide using matrix-assisted laser desorption/ionization mass spectrometry, *Anal. Biochem.*, 226, 15–25, 1995.
91. Patterson, D., Tarr, G., Regnier, F., and Martin, S., C-terminal ladder sequencing via matrix-assisted laser desorption mass spectrometry coupled with carboxypeptidase Y time-dependent and concentration-dependent digestions, *Anal. Chem.*, 67, 3971–3978, 1995.
92. Figeys, D., Corthals, G.L., Gallis, B., Goodlett, D.R., Ducret, A., Corson, M.A., and Aebersold, R., Data-dependent modulation of solid-phase extraction capillary electrophoresis for the analysis of complex peptides and phosphopeptide mixtures by tandem mass spectrometry: application to endothelial nitric oxide synthase, *Anal. Chem.*, 71, 2279–2287, 1999.
93. Link, A.J., Eng, J., Schieltz, D.M., Carmack, E., Mize, G., Morris, D.R., Garvik, B.M., and Yates, J.R., Direct analysis of protein complexes using mass spectrometry, *Nature Biotechnol.*, 17, 676–682, 1999.
94. Mann, M. and Wilm, M., Error-tolerant identification of peptides in sequence databases by peptide sequence tags, *Anal. Chem.*, 66, 4390–4399, 1994.
95. Susin, S.A., Lorenzo, H.K., Zamzami, N., Marzo, I., Snow, B.E., Brothers, G.M., Mangion, J., Jacotot, E., Costantini, P., Loeffler, M., Larochette, N., Goodlett, D.R., Aebersold, R., Siderovski, D.P., Penninger, J.M., and Kroemer, G., Molecular characterization of mitochondrial apoptosis-inducing factor, *Nature*, 397, 441–446, 1999.
96. Roepstorff, P. and Fohlman, J., Proposal for a common nomenclature for sequence ions in mass spectra of peptides, *Biomed. Mass Spectrom.*, 11, 601–604, 1984.
97. Biemann, K., Contributions of mass spectrometry to peptide and protein structure, *Biomed. Environ. Mass Spectrom.*, 16, 99–111, 1988.
98. Biemann, K., Nomenclature for peptide fragment ions (positive ions), *Methods Enzymol.*, 193, 886–887, 1990.
99. Hunt, D.F., Yates III, J.R., Shabanowitz, J., Winston, S., and Hauer, C.R., Protein sequencing by tandem mass spectrometry, *Proc. Natl. Acad. Sci. USA*, 83, 6233–6237, 1986.
100. Hall, S., Smith, D., Masiarz, F., Soo, V., Tran, H., Epstein, L.B., and Burlingame, A.L., Mass spectrometric and Edman sequencing of lipocortin I isolated by two-dimensional SDS/PAGE of human melanoma lysates, *Proc. Natl. Acad. Sci. USA*, 90, 1927–1931, 1993.
101. Spengler, B., Post-source decay analysis in matrix-assisted laser desorption/ionization mass spectrometry of biomolecules, *J. Mass Spectrom.*, 32, 1019–1036, 1997.
102. Lennon, J.J. and Walsh, K.A., Direct sequence analysis of proteins by in-source fragmentation during delayed ion extraction, *Protein Sci.*, 6, 2446–2453, 1997.
103. Katta, V., Chow, D.T., and Rohde, F., Application of in-source fragmentation of protein ions for direct sequence analysis by delayed extraction MALDI-TOF mass spectrometry, *Anal. Chem.*, 70, 4410–4416, 1998.
104. Takayama, M. and Tsugita, A., Sequence information of peptides and proteins with in-source decay in matrix assisted laser desorption/ionization-time of flight-mass spectrometry, *Electrophoresis*, 21, 1670–1677, 2000.
105. Griffiths, W.J., Nanospray mass spectrometry in protein and peptide chemistry, *EXS*, 88, 69–79, 2000.
106. Link, A.J., Hays, L.G., Carmack, E.B., and Yates, J.R., Identifying the major proteome components of *Haemophilus influenzae* type-strain NCTC 8143, *Electrophoresis*, 18, 1314–1334, 1997.
107. Eng, J., McCormack, A., and Yates, J., An approach to correlate tandem mass spectral data of peptides with amino acid sequences in a protein database, *J. Am. Soc. Mass Spectrom.*, 5, 976–989, 1994.

108. Gatlin, C.L., Kleemann, G.R., Hays, L.G., Link, A.J., and Yates, J.R., Protein identification at the low femtomole level from silver-stained gels using a new fritless electrospray interface for liquid chromatography-microspray and nanospray mass spectrometry, *Anal. Biochem.*, 263, 93–101, 1998.
109. Johnson, R.S. and Taylor, J.A., Searching sequence databases via *de novo* peptide sequencing by tandem mass spectrometry, *Methods Molec. Biol.*, 146, 41–62, 2000.
110. Roth, K.D., Huang, Z., Sadagopan, N., and Watson, J.T., Charge derivatization of peptides for analysis by mass spectrometry, *Mass Spectrom. Rev.*, 17, 255, 1998.
111. Huang, Z.H., Shen, T., Wu, J., Gage, D.A., and Watson, J.T., Protein sequencing by matrix-assisted laser desorption ionization-postsource decay-mass spectrometry analysis of the N-Tris(2,4,6-trimethoxyphenyl)phosphine-acetylated tryptic digests, *Anal. Biochem.*, 268, 305–317, 1999.
112. Lindh, I., Hjelmqvist, L., Bergman, T., Sjovall, J., and Griffiths, W.J., *De novo* sequencing of proteolytic peptides by a combination of C-terminal derivatization and nano-electrospray/collision-induced dissociation mass spectrometry, *J. Am. Soc. Mass Spectrom.*, 11, 673–686, 2000.
113. Chen, R.H., Shevchenko, A., Mann, M., and Murray, A., Spindle checkpoint protein Xmad 1 recruits Xmad 2 to unattached kinetochores, *J. Cell Biol.*, 143, 283–295, 1998.
114. Keough, T., Lacey, M.P., Fieno, A.M., Grant, R.A., Sun, Y., Bauer, M.D., and Begley, K.B., Tandem mass spectrometry methods for definitive protein identification in proteomics research, *Electrophoresis*, 21, 2252–2265, 2000.
115. Shevchenko, A., Chernushevich, I.V., Ens, W., Standing, K.G., Thomson, B., Wilm, M., and Mann, M., Rapid *'de novo'* peptide sequencing by a combination of nanoelectrospray, isotopic labeling and quadrupole/time-of-flight mass spectrometer, *Rapid Commun. Mass Spectrom.*, 11, 1015–1024, 1997.
116. Uttenweiler-Joseph, S., Neubauer, G., Christoforidis, S., Zerial, M., and Wilm, M., Automated *de novo* sequencing of proteins using the differential scanning technique, *Proteomics*, 1, 668–682, 2001.
117. Qin, J., Herring, C.J., and Zhang, X., *De novo* peptide sequencing in an ion trap mass spectrometer with ^{18}O labeling, *Rapid Commun. Mass Spectrom.*, 12, 209–216, 1998.
118. Persson, B., Bioinformatics in protein analysis, *EXS*, 88, 215–231, 2000.
119. Caporale, C., Analytical techniques and computer algorithms combined for the rapid characterization of structural peptide and protein features, *J. Peptide Res.*, 52, 421–429, 1998.
120. Appel, R.D. and Hochstrasser, D.F., Computer analysis of 2-D images, *Methods Molec. Biol.*, 112, 363–382, 1999.
121. Appel, R.D., Bairoch, A., and Hochstrasser, D.F., 2-D databases on the world wide web, *Methods Molec. Biol.*, 112, 383–392, 1999.
122. Fenyo, D., Identifying the proteome: software tools, *Curr. Opin. Biotech.*, 11, 391–395, 2000.
123. Mann, M., Hojrup, P., and Roepstorff, P., Use of mass spectrometric molecular weight information to identify proteins in sequence databases, *Biol. Mass Spectrom.*, 22, 338–345, 1993.
124. Wilkins, M.R., Gasteiger, E., Bairoch, A., Sanchez, J.C., Williams, K.L., Appel, R.D., and Hochstrasser, D.F., Protein identification and analysis tools in the ExPASy server, *Methods Mol. Biol.*, 112, 531–552, 1999.
125. Wilkins, M.R., Gasteiger, E., Wheeler, C.H., Lindskog, I., Sanchez, J.C., Bairoch, A., Appel, R.D., Dunn, M.J., and Hochstrasser, D.F., Multiple parameter cross-species protein identification using MultiIdent—a world-wide web accessible tool, *Electrophoresis*, 19, 3199–3206, 1998.
126. Perkins, D.N., Pappin, D.J.C., Creasy, D.M., and Cottrell, J.S., Probility-based protein identification by searching sequence databases using mass spectrometry data, *Electrophoresis*, 20, 3551–3567, 1999.
127. Zhang, W. and Chait, B.T., ProFound: an expert system for protein identification using mass spectrometric peptide mapping information, *Anal. Chem.*, 72, 2482–2489, 2000.
128. Mortz, E., O'Connor, P.B., Roepstorff, P., Kelleher, N.L., Wood, T.D., McLafferty, F.W., and Mann, M., Sequence tag identification of intact proteins by matching tandem mass spectral data against sequence data bases, *Proc. Natl. Acad. Sci. USA*, 93, 8264–8267, 1996.
129. Mann, M., A shortcut to interesting human genes: peptide sequence tags, expressed-sequence tags and computers, *Trends Biol. Sci.*, 212, 494–495, 1996.
130. Neubauer, G., King, A., Rappsilber, J., Calvio, C., Watson, M., Ajuh, P., Sleeman, J., Lamond, A., and Mann, M., Mass spectrometry and EST-database searching allows characterization of the multi-protein spliceosome complex, *Nature Genet.*, 20, 46–50, 1998.
131. Yates, J., Eng, J., and McCormack, A., Mining genomes: correlating tandem mass spectra of modified and unmodified peptides to sequences in nucleotide databases, *Anal. Chem.*, 67, 3202–3210, 1995.

132. Griffin, P.R., MacCoss, M.J., Eng, J.K., Blevins, R.A., Aaronson, J.S., and Yates, J.R., Direct database searching with MALDI-PSD spectra of peptides, *Rapid Commun. Mass Spectrom.*, 9, 1546–1551, 1995.
133. Eriksson, J., Chait, B.T., and Fenyo, D., A statistical basis for testing the significance of mass spectrometric protein identification results, *Anal. Chem.*, 72, 999–1005, 2000.
134. Fernandez-de-Cossio, J., Gonzalez, J., Satomi, Y., Shima, T., Okumura, N., Besada, V., Betancourt, L., Padron, G., Shimonishi, Y., and Takao, T., Automated interpretation of low-energy collision-induced dissociation spectra by SeqMS, a software aid for *de novo* sequencing by tandem mass spectrometry, *Electrophoresis*, 21, 1694–1699, 2000.
135. Wilkins, M.R., Lindskog, I., Gasteiger, E., Bairoch, A., Sanchez, J.C., Hochstrasser, D.F., and Appel, R.D., Detailed peptide characterization using PEPTIDEMASS—a World-Wide-Web-accessible tool, *Electrophoresis*, 18, 403–408, 1997.
136. Wilkins, M.R., Gasteiger, E., Tonella, L., Ou, K., Tyler, M., Sanchez, J.C., Gooley, A.A., Walsh, B.J., Bairoch, A., Appel, R.D., Williams, K.L., and Hochstrasser, D.F., Protein identification with N and C-terminal sequence tags in proteome projects, *J. Mol. Biol.*, 278, 599–608, 1998.
137. Wilkins, M.R., Ou, K., Appel, R.D., Sanchez, J.C., Yan, J.X., Golaz, O., Farnsworth, V., Cartier, P., Hochstrasser, D.F., Williams, K.L., and Gooley, A.A., Rapid protein identification using N-terminal "sequence tag" and amino acid analysis, *Biochem. Biophys. Res. Commun.*, 221, 609–613, 1996.
138. Wilkins, M.R. and Williams, K.L., Cross-species protein identification using amino acid composition, peptide mass fingerprinting, isoelectric point and molecular mass: a theoretical evaluation, *J. Theor. Biol.*, 186, 7–15, 1997.
139. Mc Lachlin, D.T. and Chait, B.T., Analysis of phosphorylated proteins and peptides by mass spectrometry, *Curr. Opin. Chem. Biol.*, 5, 591–602, 2001.
140. Loo, J.A., Brown, J., Critchley, G., Mitchell, C., Andrews, P.C., and Ogorzalek Loo, R.R., High sensitivity mass spectrometric methods for obtaining intact molecular weights from gel-separated proteins, *Electrophoresis*, 20, 743–748, 1999.
141. Eckerskorn, C., Strupat, K., Schleuder, D., Hochstrasser, D., Sanchez, J.-C., Lottspeich, F., and Hillenkamp, F., Analysis of proteins by direct-scanning infrared-Maldi mass spectrometry after 2D-Page separation and electroblotting, *Anal. Chem.*, 69, 2888–2892, 1997.
142. Schleuder, D., Hillenkamp, F., and Strupat, K., IR-MALDI-Mass analysis of electroblotted proteins directly from the membrane: comparison of different membranes, application to on-membrane digestion, and protein identification by database searching, *Anal. Chem.*, 71, 3238–3247, 1999.
143. Marshall, A.G., Hendrickson, C.L., and Jackson, G.S., Fourier transform ion cyclotron resonance mass spectrometry: a primer, *Mass Spectrom. Reviews*, 17, 1–35, 1998.
144. Goodlett, D.R., Bruce, J.E., Anderson, G.A., Rist, B., Pasa-Tolic, L., Fiehn, O., Smith, R.D., and Aebersold, R., Protein identification with a single accurate mass of a cysteine-containing peptide and constrained database searching, *Anal. Chem.*, 72, 1112–1118, 2000.
145. Marshall, A.G., Senko, M.W., Li, W., Li, M., Dillon, S., Guan, S., and Logan, T.M., Protein molecular mass to 1 Da by ^{13}C, ^{15}N double-depletion and FT-ICR mass spectrometry, *J. Am. Chem. Soc.*, 119, 433–434, 1997.
146. Shi, S.D.-H., Hendrickson, C.L., and Marshall, A.G., Counting individual sulfur atoms in a protein by ultrahigh-resolution Fourier transform ion cyclotron resonance mass spectrometry: experimental resolution of isotopic fine structure in proteins, *Proc. Natl. Acad. Sci. USA*, 95, 11532–11537, 1998.
147. Blackstock, W.P. and Weir, M.P., Proteomics: quantitative and physical mapping of cellular proteins, *Trends Biotechnol.*, 17, 121–127, 1999.
148. Mann, M., Quantitative proteomics? [news], *Nature Biotechnol.*, 17, 954–955, 1999.
149. Gygi, S.P., Rist, B., and Aebersold, R., Measuring gene expression by quantitative proteome analysis, *Curr. Opin. Biotech.*, 11, 396–401, 2000.
150. Leenheer, A. and Thienport, L., Applications of isotope dilution-mass spectrometry in clinical chemistry, pharmacokinetics, and toxicology, *Mass Spectrom. Reviews*, 11, 249–307, 1992.
151. Oda, Y., Huang, K., Cross, F.R., Cowburn, D., and Chait, B.T., Accurate quantification of protein expression and site-specific phosphorylation, *Proc. Natl. Acad. Sci. USA*, 96, 6591–6596, 1999.
152. Gygi, S.P., Rist, B., Gerber, S.A., Turecek, F., Gelb, M.H., and Aebersold, R., Quantitative analysis of complex protein mixtures using isotope coded affinity tags, *Nature Biotechnol.*, 17, 994–999, 1999.

153. Yao, X., Freas, A., Ramirez, J., Demirev, P.A., and Fenselau, C., Proteolytic ^{18}O labeling for comparative proteomics: Model studies with two serotypes of adenovirus, *Anal. Chem.*, 73, 2836–2842, 2001.
154. Krishna, R. and World, F., Identification of common post-translational modifications, *in Protein Structure: A Practical Approach*, Creighton, T.E., Ed., Oxford University, New York, (1997) 91–116.
155. Wilkins, M.R., Gasteiger, E., Gooley, A.A., Herbert, B.R., Molloy, M.P., Binz, P.A., Ou, K., Sanchez, J.C., Bairoch, A., Williams, K.L., and Hochstrasser, D.F., High-throughput mass spectrometric discovery of protein post-translational modifications, *J. Mol. Biol.*, 289, 645–657, 1999.
156. Vertommen, D., Rider, M., Ni, Y., Waelkens, E., Merlevede, W., Vandenheed, J.R., and Lint, J.V., Regulation of protein kinase D by multisite phosphorylation, *J. Biol. Chem.*, 275, 19567–19576, 2000.
157. Dass, C., Analysis of phosphorylated proteins by mass spectrometry, in *Mass Spectrometry of Biological Materials,* Larsen, B.S. and McEwen, C.N., Eds., Marcel Dekker, New York, (1998) 247–280.
158. Resing, K.A. and Ahn, N.G., Protein phosphorylation analysis by electrospray ionization-mass spectrometry, *Methods Enzymol.*, 283, 29–44, 1997.
159. Howard, S., He, S., and Withers, S.G., Identification of the active site nucleophile in jack bean alpha-mannosidase using 5-fluoro-beta-L-gulosyl fluoride, *J. Biol. Chem.*, 273, 2067–2072, 1998.
160. Quadroni, M. and James, P., Phosphopeptide analysis, *EXS*, 88, 199–214, 2000.
161. Weijland, A., Neubauer, G., Courtneidge, S.A., Mann, M., Wierenga, R.K., and Superti-Furga, G., The purification and characterization of the catalytic domain of Src expressed in *Schizosaccharomyces* pombe. Comparison of unphosphorylated and tyrosine phosphorylated species, *Eur. J. Biochem.*, 240, 756–764, 1996.
162. Verma, R., Annan, R.S., Huddleston, M.J., Carr, S.A., Reynard, G., and Deshales, R.J., Phosphorylation of $Sic1_p$ by G_1 Cdk required for its degradation and entry into S phase, *Science*, 278, 455–460, 1997.
163. Annan, R.S. and Carr, S.A., The essential role of mass spectrometry in characterizing protein structure: mapping posttranslational modifications, *J. Protein Chem.*, 16, 391–402, 1997.
164. Nuwaysir, L. and Stults, J., Electrospray ionization mass spectrometry of phosphopeptides isolated by on-line immobilized metal-ion affinity chromatography, *J. Am. Soc. Mass Spectrom.*, 4, 662–669, 1993.
165. Cao, P. and Stults, J.T., Mapping the phosphorylation sites of proteins using on-line immobilized metal affinity chromatography/capillary electrophoresis/electrospray ionization multiple stage tandem mass spectrometry, *Rapid Commun. Mass Spectrom.*, 14, 1600–1606, 2000.
166. Neubauer, G. and Mann, M., Mapping of phosphorylation sites of gel-isolated proteins by nanoelectrospray tandem mass spectrometry: potentials and limitations, *Anal. Chem.*, 71, 235–242, 1999.
167. Ogueta, S., Rogado, R., Marina, A., Moreno, F., and Vazquez, J., Identification of phosphorylation sites in proteins by nanospray quadrupole ion trap mass spectrometry, *J. Mass Spectrom.*, 35, 556–565, 2000.
168. Metzger, S. and Hoffmann, R., Studies on the dephosphorylation of phosphothyrosine-containing peptides during post-source decay in matrix-assisted laser desorption/ionization, *J. Mass Spectrom.*, 35, 1165–1177, 2000.
169. Ahn, N.G. and Resing, K.A., Toward the phosphoproteome, *Nature Biotechnol.*, 19, 317–318, 2001.
170. Zhou, H., Watts, J.D., and Aebersold, R., A systematic approach to the analysis of protein phosphorylation, *Nature Biotechnol.*, 19, 375–378, 2001.
171. Oda, Y., Nagasu, T., and Chait, B.T., Enrichment analysis of phosphorylated proteins as a tool for probing the phosphoproteome, *Nature Biotechnol.*, 19, 379–382, 2001.
172. Weckwerth, W., Willmitzer, L., and Fiehn, O., Comparative quantification and identification of phosphoproteins using stable isotope labeling and liquid chromatography/mass spectrometry, *Rapid Commun. Mass Spectrom.*, 14, 1677–1681, 2000.
173. Wind, M., Wesch, H., and Lehmann, W.D., Protein phosphorylation degree: determination by capillary liquid chromatography and inductively coupled plasma mass spectrometry, *Anal. Chem.*, 73, 3006–3010, 2001.
174. Resing, K., Mansour, S., Hermann, A., Johnson, R., Candia, J., Filasawa, K., Vande Woude, G., and Ahn, N., Determination of v-Mos-catalyzed phosphorylation sites and autophosphorylation sites in MAP kinase kinase by ESI/MS, *Biochemistry*, 34, 2610–2620, 1995.
175. Mansour, K., Resing, K., Candi, J., Hermann, A., Gloor, J., Hersklind, K., Wartmann, M., Davis, R., and Ahn, N.G., Mitogen-activated protein [MAP] kinase phosphorylation of MAP kinase kinase: determination of phosphorylation sites by mass spectrometry and site-directed mutagenesis, *J. Biochem.*, 116, 304–314, 1994.

176. Mistry, S.J. and Atweh, G.F., Stathmin expression in immortalized and oncogene transformed cells, *Anticancer Res.*, 19, 573–578, 1999.
177. Redeker, V., Lachkar, S., Siavoshians, S., Charbaut, E., Rossier, J., Sobel, A., and Curmi, P.A., Probing the native structure of stathmin and its interaction domains with tubulin, *J. Biol. Chem.,* 275, 6841–6849, 2000.
178. Zugaro, L.M., Reid, G.E., Ji, H., Eddes, J.S., Murphy, A.C., Burgess, A.W., and Simpson, R.J., Characterization of rat brain stathmin isoforms by two-dimensional gel electrophoresis-matrix assisted laser desorption/ionization and electrospray ionization-ion trap mass spectrometry, *Electrophoresis*, 19, 867–876, 1998.
179. Ainsztein, A. and Purich, D., Stimulation of tubulin polymerization by MAP-2. Control by protein kinase C-mediated phosphorylation at specific sites in the microtubule-binding region, *J. Biol. Chem.*, 269, 28465–28471, 1994.
180. Otvos, L.J., Hoffmann, R., Xiang, Z.Q., Deng, H., Wysocka, M., Pease, A.M., Rogers, M.E., Blaszczyk-Thurin, M., and Ertl, H.C.J., A monoclonal antibody to a multiphosohorylated, conformational epitope at the carboxyl-terminus of p53, *Biochim. Biophys. Acta*, 1404, 457–474, 1998.
181. Yates, J.R., Eng, J.K., McCormack, A.L., and Schieltz, D., Method to correlate tandem mass spectra of modified peptides to amino acid sequences in the protein database, *Anal. Chem.*, 67, 1426–1436, 1995.
182. Gygi, S.P., Han, D.K.M., Gingras, A.C., Sonenberg, N., and Aebersold, R., Protein analysis by mass spectrometry and sequence database searching: tools for cancer research in the post-genomic era, *Electrophoresis,* 20, 310–318, 1999.
183. Gingras, A.C., Gygi, S.P., Raught, B., Polakiewicz, R.D., Abraham, R.T., Hoekstra, M.F., Aebersold, R., and Sonenberg, N., Regulation of 4E-BP1 phosphorylation: a novel two-step mechanism, *Genes Dev.*, 13, 1422–1437, 1999.
184. Hunt, D., Henderson, R., Shabanowitz, J., Sakaguchi, K., Michel, H., Sevilir, N., Cox, A., Appella, E., and Engelhard, V., Characterization of peptides bound to Class I MHC molecule HLA-A2.1 by mass spectrometry, *Science*, 255, 1261–1263, 1992.
185. Wang, W., Meadows, L., den Haan, J., Sherman, N., Chen, Y., Blokland, E., Shabanowitz, J., Agulnik, A., Hendrickson, R., Bishop, C., Hunt, D., Goulmy, E., and Engelhard, V., Human H-Y: a male-specific histocompatibility antigen derived from the SMCY protein, *Science,* 269, 1588–1590, 1995.
186. Meadows, L., Wang, W., den Haan, J.M., Blokland, E., Reinhardus, C., Drijfhout, J.W., Shabanowitz, J., Pierce, R., Agulnik, A.I., Bishop, C.E., Hunt, D.F., Goulmy, E., and Engelhard, V.H., The HLA-A*0201-restricted H-Y antigen contains a posttranslationally modified cysteine that significantly affects T cell recognition, *Immunity,* 6, 273–281, 1997.
187. Skipper, J.C., Hendrickson, R.C., Gulden, P.H., Brichard, V., Van Pel, A., Chen, Y., Shabanowitz, J., Wolfel, T., Slingluff, C.L., Boon, T., Hunt, D.F., and Engelhard, V.H., An HLA-A2-restricted tyrosinase antigen on melanoma cells results from posttranslational modification and suggests a novel pathway for processing of membrane proteins, *J. Exp. Med.*, 183, 527–534, 1996.
188. Maltese, W.A., Posttranslational modification of proteins by isoprenoids in mammalian cells, *FASEB J.*, 4, 3319–3328, 1990.
189. Ceciliani, F., Faotto, L., Negri, A., Colombo, I., Berra, B., Bartorelli, A., and Ronchi, S., The primary structure of UK114 tumor antigen, *FEBS Lett.,* 393, 147–150, 1996.
190. Louie, D.F., Resing, K.A., Lewis, T.S., and Ahn, N.G., Mass spectrometric analysis of 40 S ribosomal proteins from rat-1 fibroblasts, *J. Biol. Chem.*, 271, 28189–28196, 1996.
190a. Abraham, J., Kelly, J., Thibault, P., and Benchimol, S., Post-translational modification of p53 protein in response to ionizing radiation analyzed by mass spectrometry, *J. Mol. Biol.,* 295, 853–864, 2000.
191. Hensel, J., Hintz, M., Karas, M., Linder, D., Stahl, B., and Geyer, R., Localization of the palmitoylation site in the transmembrane protein p12E of Friend murine leukemia virus, *Eur. J. Biochem.*, 232, 373–380, 1995.
192. Vinh, J., Loyaux, D., Redeker, V., and Rossier, J., Sequencing branched peptides with CID/PSD MALDI-TOF in the low-picomole range: application to the structural study of the posttranslational polyglycylation of tubulin, *Anal. Chem.*, 69, 3979–3985, 1997.
193. Vinh, J., Langridge, J.L., Bre, M.H., Levilliers, N., Redeker, V., Loyaux, D., and Rossier, J., Structural characterization by tandem mass spectrometry of the posttranslational polyglycylation of tubulin, *Biochem.*, 38, 3133–3139, 1999.

194. van Etten, R., Davidson, R., Stevis, P., MacArthur, H., and Moore, L., Covalent structure, disulfide bonding, and identification of reactive surface and active site residues of human prostatic acid phosphatase, *J. Biol. Chem.*, 26, 2313–2319, 1991.
195. Dahse, R., Fiedler, W., and Ernst, G., Telomeres and telomerase: biological and clinical importance, *Clin. Chem.*, 43, 708–714, 1997.
196. Shay, J.W., Zou, Y., Hiyama, E., and Wright, W.E., Telomerase and cancer, *Hum. Mol. Genet.*, 10, 677–685, 2001.
197. Bodnar, A.G., Quellette, M., Frolkis, M., Holt, S.E., Chiu, C., Morin, G.G., Harley, C.B., Shay, J.W., Lichtsteiner, S., and Wright, W.E., Extension of life-span by introduction of telomerase into normal human cells, *Science*, 279, 349–352, 1998.
198. Lingner, J. and Cech, T.R., Purification of telomerase from *Euplotes aediculatus:* requirement of primer 3′ overhang, *Proc. Natl. Acad. Sci. USA*, 93, 10712–10717, 1999.
199. Lingner, J., Hughes, T.R., Shevchenko, A., Mann, M., Lundblad, V., and Cech, T.R., Reverse transcriptase motifs in the catalytic subunit of telomerase, *Science*, 276, 561–567, 1997.
200. Villa, R., Folini, M., Lualdi, S., Veronese, S., Daidone, M.G., and Zaffaroni, N., Inhibition of telomerase activity by a cell-penetrating peptide nucleic acid construct in human melanoma cells, *FEBS Lett.*, 473, 241–248, 2000.
201. Muzio, M., Chinnaiyan, A.M., Kischkel, F.C., O'Rourke, K., Shevchenko, A., Ni, J., Scaffidi, C., Bretz, J.D., Zhang, M., Gentz, R., Mann, M., Krammer, P.H., Peter, M.E., and Dixit, V.M., FLICE, a novel FADD-homologues ICE/CED-3-like protease, is recruited to the CD95 (Fas/APO-1) death-inducing signaling complex, *Cell*, 85, 817–827, 1996.
202. Jones, M.D., Patterson, S.D., and Lu, H.S., Determination of disulfide bonds in highly bridged disulfide-linked peptides by matrix-assisted laser desorption/ionization mass spectrometry with post-source decay, *Anal. Chem.*, 70, 136–143, 1998.
203. Huang, H.S., Chen, C.J., Lu, H.S., and Chang, W.C., Identification of a lipoxygenase inhibitor in A431 cells as a phospholipid hydroperoxide glutathione peroxidase, *FEBS Lett.*, 424, 22–26, 1998.
204. Dubey, P., Hendrickson, R.C., Meredith, S.C., Siegel, C.T., Shabanowits, J., Skipper, J.C.A., Engelhard, V.H., Hunt, D.F., and Schreiber, H., The immunodominant antigen of an ultraviolet-induced regressor tumor is generated by a somatic point mutation in the DEAD box helicase p68, *J. Exp. Med.*, 185 No. 4, 695–705, 1997.
205. Heller, M., Godlett, D.R., Watts, J.D., and Aebersold, R., A comprehensive characterization of the T-cell antigen receptor complex composition by microcapillary liquid chromatography-tandem mass spectrometry, *Electrophoresis,* 21, 2180–2195, 2000.
206. Roos, M., Soskic, V., Poznanovic, S., and Godovac-Zimmermann, J., Post-translational modifications of endothelin receptor B from bovine lungs analyzed by mass spectrometry, *J. Biol. Chem.*, 273, 924–931, 1998.
207. Soskic, V., Gorlach, M., Poznanovic, S., Boehmer, F.D., and Godovac-Zimmermann, J., Functional proteomics analysis of signal transduction pathways of the platelet-derived growth factor β receptor, *Biochem.*, 38, 1757–1764, 1999.
208. Nakanishi, T., Kadomatsu, K., Okamoto, T., Tomoda, Y., and Muramatsu, T., Expression of midkine and pleiotropin in ovarian tumors, *Obstet. Gynecol.,* 90, 285–290, 1997.
209. Fabri, L., Maruta, H., Muramatsu, H., Muramatsu, T., Simpson, R.J., Burgess, A.W., and Nice, E.C., Structural characterisation of native and recombinant forms of the neurotrophic cytokine MK, *J. Chromatogr.*, 646, 213–225, 1993.
210. Stam, K., Stewart, A., Qu, G., Iwata, K., Fenyo, D., Chait, B.T., Marshak, D., and Haley, J.D., Physical and biological characterization of a growth-inhibitory activity purified from the neuroepithelioma cell line A673, *Biochem. J.*, 305, 87–92, 1995.
211. Sullivan, C.M., Smith, D.M., Matsui, N.M., Andrews, L.E., Clauser, K.R., Chapeauronge, A., Burlingame, A.L., and Epstein, L.B., Identification of constitutive and gamma-interferon- and interleukin 4-regulated proteins in the human renal carcinoma cell line ACHN, *Cancer Res.*, 57, 1137–1143, 1997.
212. Vercoutter-Edouart, A.S., Czeszak, X., Crepin, M., Lemoine, J., Boilly, B., Le Bourhis, X., Peyrat, J.P., and Hondermarck, H., Proteomic detection of changes in protein synthesis induced by fibroblast growth factor-2 in MCF-7 human breast cancer cells, *Exp. Cell Res.*, 262, 59–68, 2001.
213. Hotta, T., Asai, K., Takeda, N., Tatematsu, A., Nakanishi, K., Eksioglu, Y.Z., Isobe, I., and Kato, T., Neuroblastoma growth factors derived from neurofibroma (NF1): participation of uridine in a neuroblastoma growth, *J. Neurochem.*, 60, 312–319, 1993.

214. Bichsel, V.E., Liotta, L.A., and Petricoin, E.F., Cancer proteomics: from biomarker discovery to signal pathway profiling, *Cancer J.*, 7, 69–78, 2001.
215. Alaiya, A.A., Franzen, B., Auer, G., and Linder, S., Cancer proteomics: from identification of novel markers to creation of artificial learning models for tumor classification, *Electrophoresis*, 21, 1210–1217, 2000.
216. Pacholski, M.L. and Winograd, N., Imaging with mass spectrometry, *Chem. Rev.*, 99, 2977–3005, 1999.
217. Stoeckli, M., Farmer, T.B., and Caprioli, R.M., Automated mass spectrometry imaging with a matrix-assisted laser desorption ionization time-of-flight instrument, *J. Am. Soc. Mass Spectrom.*, 10, 67–71, 1999.
218. Frey, J.R., Fountoulakis, M., and Lefkovits, I., Proteome analysis of activated murine B-lymphocytes, *Electrophoresis*, 21, 3730–3739, 2000.
219. Paweletz, C.P., Gillespie, J.W., Ornstein, D.K., Simone, N.L., Brown, M.R., Cole, K.A., Wang, Q.H., Huang, J., Hu, N., Yip, T.T., Rich, W.E., Kohn, E.C., Linehan, W.M., Weber, T., Taylor, P., Emmert-Buck, R., Liotta, L.A., and Petricoin, E.F., Rapid protein display profiling of cancer progression directly from human tissue using a protein biochip, *Drug Develop. Res.*, 49, 34–42, 2000.
220. Celis, A., Rasmussen, H.H., Celis, P., Basse, B., Lauridsen, J.B., Ratz, G., Hein, B., Ostergaard, M., Orntoft, T., and Celis, J.E., Short-term culturing of low-grade superficial bladder transitional cell carcinomas leads to changes in the expression levels of several proteins involved in key cellular activities, *Electrophoresis*, 20, 355–361, 1999.
221. Emmert-Buck, M.R., Bonner, R.F., Smith, P.D., Chuaqui, R.F., Zhuang, Z., Goldstein, S.R., Weiss, R.A., and Liotta, L.A., Laser capture microdissection, *Science*, 274, 998–1001, 1996.
222. Gillespie, J.W., Ahram, M., Best, C.J., Swalwell, J.I., Krizman, D.B., Petricoin, E.F., Liotta, L.A., and Emmert-Buck, M.R., The role of tissue microdissection in cancer research, *Cancer J.*, 7, 32–39, 2001.
223. Spahr, C.S., Davis, M.T., McGinley, M.D., Robinson, J.H., Bures, E.J., Beierle, J., Mort, J., Courchesne, P.L., Chen, K., Wahl, R.C., Yu, W., Luethy, R., and Patterson, S.D., Towards defining the urinary proteome using liquid charomatograph-tandem mass spectrometry. I. Profiling an unfractionated tryptic digest, *Proteomics*, 1, 93–107, 2001.
224. Davis, M.T., Spahr, C.S., McGinlet, M.D., Robinson, J.H., Bures, E.J., Beierle, J., Mort, J., Yu, W., Luethy, R., and Patterson, S.D., Towards defining the urinary proteome using liquid chromatography-tandem mass spectrometry. II. Limitations of complex mixture analyses, *Proteomics,* 1, 108–117, 2001.
225. Muller, E.C., Schumann, M., Rickers, A., Bommert, K., Wittmann-Liebold, B., and Otto, A., Study of Burkitt lymphoma cell line proteins by high resolution two-dimensional gel electrophoresis and nanoelectrospray mass spectrometry, *Electrophoresis*, 20, 320–330, 1999.
226. Brockstedt, E., Rickers, A., Kostka, S., Laubersheimer, A., Dorken, B., Wittmann-Liebold, B., Bommert, K., and Otto, A., Identification of apoptosis-associated proteins in a human Burkitt lymphoma cell line, *J. Biol. Chem.*, 273, 28057–28064, 1998.
227. Rickers, A., Brockstedt, E., Mapara, M.Y., Otto, A., Dorken, B., and Bommert, K., Inhibition of CPP32 blocks surface IgM-mediated apoptosis and D4-GDI cleavage in human BL60 Burkitt lymphoma cells, *Eur. J. Immunol.*, 28, 296–304, 1998.
228. Brockstedt, E., Otto, A., Rickers, A., Bommert, K., and Wittmann-Liebold, B., Preparative high-resolution two-dimensional electrophoresis enables the identification of RNA polymerase B transcription factor 3 as an apoptosis-associated protein in the human BL60-2 Burkitt lymphoma cell line, *J. Protein Chem.*, 18, 225–231, 1999.
229. Essmann, F., Wieder, T., Otto, A., Muller, E.C., Dorken, B., and Daniel, P.T., GDP dissociation inhibitor D4-GDI (Rho-GDI 2), but not the homologous Rho-GDI 1, is cleaved by caspase-3 during drug-induced apoptosis, *Biochem. J.*, 346, 777–783, 2000.
230. Toda, T., Sugimoto, M., Omori, A., Matsuzaki, T., Furuichi, Y., and Kimura, N., Proteomic analysis of Epstein-Barr virus-transformed human B-lymphoblastiod cell lines before and after immortalization, *Electrophoresis*, 21, 1814–1822, 2000.
231. Joubert-Caron, R., Le Caer, J.P., Montandon, F., Poirier, F., Pontet, M., Imam, N., Feuillard, J., Bladier, D., Rossier, J., and Caron, M., Protein analysis by mass spectrometry and sequence database searching: a proteomic approach to identify human lymphoblastoid cell line proteins, *Electrophoresis*, 21, 2566–2575, 2000.

232. Chen, Y., Jin, X., Misek, D., Hinderer, R., Hanash, S.M., and Lubman, D.M., Identification of proteins from two-dimensional gel electrophoresis of human erythroleukemia cells using capillary high performance liquid chromatography/electrospray-ion trap-reflectron time-of-flight mass spectrometry with two-dimensional topographic map analysis of in-gel tryptic digest products, *Rapid Commun. Mass Spectrom.*, 13, 1907–1916, 1999.
233. Banks, J.F. and Gulcicek, E.E., Rapid peptide mapping by reversed-phase liquid chromatography on nonporous silica with on-line electrospray time-of-flight mass spectrometry, *Anal. Chem.*, 69, 3973–3978, 1997.
234. Chen, Y., Wall, D., and Lubman, D.M., Rapid identification and screening of proteins from whole cell lysates of human erythroleukemia cells in the liquid phase, using non-porous reversed phase high-performance liquid chromatography separations of proteins followed by matrix-assisted laser desorption/ionization mass spectrometry analysis and sequence database searching, *Rapid Commun. Mass Spectrom.*, 12, 1994–2003, 1998.
235. Wall, D.B., Kachman, M., Gong, S., Hinderer, R., Parus, S., Misek, D.E., Hanash, S.M., and Lubman, D.M., Isoelectric focusing nonporous RP HPLC: a two-dimensional liquid-phase separation method for mapping of cellular proteins with identification using MALDI-TOF mass spectrometry, *Anal. Chem.*, 72, 1099–1111, 2000.
236. Keim, D., Hailat, N., Hodge, D., and Hanash, S.M., Proliferating cell nuclear antigen expression in childhood acute leukemia, *Blood*, 76, 985–990, 1990.
237. Voss, T., Ahorn, H., Haberl, P., Dohner, H., and Wilgenbus, K., Correlation of clinical data with proteomics profiles in 24 patients with B-cell chronic lymphocytic leukemia, *Int. J. Cancer*, 91, 180–186, 2001.
238. Woodlock, T.J., Bethlendy, G., and Segel, G.B., Prohibitin expression is increased in phorbol ester-treated chronic leukemic b-lymphocytes, *Blood Cells Mol. Dis.*, 27, 27–34, 2001.
239. Carroll, K., Ray, K., Helm, B., and Carey, E., Two-dimensional electrophoresis reveals differential protein expression in high- and low-secreting variants of the rat basophilic leukemia cell line, *Electrophoresis*, 21, 2476–2486, 2000.
240. Page, M.J., Amess, B., Townsend, R.R., Parekh, R., Herath, A., Brusten, L., Zvelebil, M.J., Stein, R.C., Waterfield, M.D., Davies, S.C., and O'Hare, M.J., Proteomic definition of normal luminal and myoepithelial breast cells purified from reduction mammoplasties, *Proc. Natl. Acad. Sci. USA*, 96, 12589–12594, 1999.
241. Moritz, R.L., Eddes, J.S., Reid, G.E., and Simpson, R.J., S-pyridylethylation of intact polyacrylamide gels and in situ digestion of electrophoretically separated proteins: a rapid mass spectrometric method for identifying cysteine-containing peptides, *Electrophoresis*, 17, 907–917, 1996.
242. Rasmussen, R.K., Ji, H., Eddes, J.S., Moritz, R.L., Reid, G.E., Simpson, R.J., and Dorow, D.S., Two-dimensional electrophoretic analysis of human breast carcinoma proteins: mapping of proteins that bind to the SH3 domain of mixed linkage kinase MLK2, *Electrophoresis*, 18, 588–598, 1997.
243. Chong, B.E., Lubman, D.M., Rosenspire, A., and Miller, F., Protein profiles an identification of high performance liquid chromatography isolated proteins of cancer cell lines using matrix-assisted laser desorption/ionization time-of-flight mass spectrometry, *Rapid Commun. Mass Spectrom.*, 12, 1986–1993, 1998.
244. Chong, B.E., Lubman, D.M., Miller, F.R., and Rosenspire, A.J., Rapid screening of protein profiles of human breast cancer cell lines using non-porous reversed-phase high performance liquid chromatography separation with matrix-assisted laser desorption/ionization time-of-flight mass spectral analysis, *Rapid Commun. Mass Spectrom.*, 13, 1808–1812, 1999.
245. Chong, B.E., Yan, F., Lubman, D.M., and Miller, F.R., Chromatofocusing nonporous reversed-phase high-performance liquid chromatography/electrospray ionization time-of-flight mass spectrometry of proteins from human breast cancer whole lysates: a novel two-dimensional liquid chromatography/mass spectrometry method, *Rapid Commun. Mass Spectrom.*, 15, 291–296, 2001.
246. Bergman, A.C., Benjamin, T., Alaiya, A.A., Waltham, M., Sakaguchi, K., Franzen, B., Linder, S., Bergman, T., Auer, G., Appella, E., Wirth, P.J., and Jornvall, H., Identification of gel-separated tumor marker proteins by mass spectrometry, *Electrophoresis*, 21, 679–686, 2000.
247. Ducret, A., Van Oosrveen, I., Eng, J.K., Yates, J.R., and Aebersold, R., High throughput protein characterization by automated reverse-phase chromatography/electrospray tandem mass spectrometry, *Protein Sci.*, 7, 706–719, 1998.

248. Vercoutter-Edouart, A.S., Lemoine, J., Le Bourhis, X., Louis, H., Boilly, B., Nurcombe, V., Revillion, F., Peyrat, J.P., and Hondermarck, H., Proteomic analysis reveals that 14-3-3 sigma is down-regulated in human breast cancer cells, *Cancer Res.*, 61, 76–80, 2001.
249. Rao, S., Aberg, F., Nieves, E., Horwitz, S.B., and Orr, G.A., Identification by mass spectrometry of a new α-tubulin isotype expressed in human breast and lung carcinoma cell lines, *Biochemistry*, 40, 2096–2103, 2001.
250. Watkins, B., Szaro, R., Ball, S., Knubovets, T., Briggman, J., Hlavaty, J.J., Kusinitz, F., Stieg, A., and Wu, Y., Detection of early-stage cancer by serum protein analysis, *Amer. Lab.*, 32–36, 2001.
251. Unwin, R.D., Knowles, M.A., Selby, P.J., and Banks, R.E., Urological malignancies and the proteomic-genomic interface, *Electrophoresis*, 20, 3629–3637, 1999.
252. Hampel, D.J., Sansome, C., Sha, M., Brodsky, S., Lawson, W.E., and Goligorsky, M.S., Toward proteomics in uroscopy: urinary protein profiles after radiocontrast medium administration, *J. Amer. Soc. Nephrol.*, 12, 1026–1935, 2001.
253. Ostergaard, M., Wolf, H., Orntoft, T.F., and Cells, J.E., Psoriasin (S100A7): a putative urinary marker for the follow-up of patients with bladder squamous cell carcinomas, *Electrophoresis*, 20, 349–354, 1999.
254. Rasmussen, H.H., Orntoft, T.F., Wolf, H., and Celis, J.E., Towards a comprehensive database of proteins from the urine of patients with bladder cancer, *J. Urology*, 155, 2113–2119, 1996.
255. Celis, J.E., Ostergaard, M., Rasmussen, H.H., Gromov, P., Gromova, I., Varmark, H., Palsdottir, H., Magnusson, N., Andersen, I., Basse, B., Lauridsen, J.B., Ratz, G., Wolf, H., Orntoft, T.F., Celis, P., and Celis, A., A comprehensive protein resource for the study of bladder cancer: http://biobase.dk/cgi-bin/celis, *Electrophoresis*, 20, 300–309, 1999.
256. Celis, J.E., Wolf, H., and Ostergaard, M., Bladder squamous cell carcinoma biomarkers derived from proteomics, *Electrophoresis*, 21, 2115–2121, 2000.
257. Vlahou, A., Schellhammer, P.F., Mendrinos, S., Patel, K., Kondylis, F.I., Gong, L., Nasim, S., and Wright, G.L.J., Development of a novel proteomic approach for the detection of transitional cell carcinoma of the bladder in urine, *Am. J. Pathology*, 158, 1491–1502, 2001.
258. Bedzyk, W.D., Larsen, B., Gutteridge, S., and Ballas, R.A., Molecular mass determination for prostate-specific antigen and alpha 1-antichymotrypsin complexed *in vitro*, *Biotechnol. Appl. Biochem.*, 27, 249–257, 1998.
259. Xiao, Z., Jiang, X., Beckett, M.L., and Wright, G.L.J., Generation of a baculovirus recombinant prostate-specific membrane antigen and its use in the development of a novel protein biochip quantitative immunoassay, *Protein Exper. Purif.*, 19, 12–21, 2000.
260. Wright, G.L., Cazares, L.H., Leung, S., Nasim, S., Adam, B., Yip, T., Schellhammer, P.F., Gong, L., and Vlahou, A., Proteinchip surface enhanced laser desorption/ionization (SELDI) mass spectrometry: a novel biochip technology for detection of prostate cancer biomarkers in complex protein mixtures, *Prostate Cancer Prostatic Diseases*, 2, 264–276, 1999.
261. Wang, S., Diamond, D.L., Hass, G.M., Sokoloff, R., and Vessella, R.L. Identification of prostate specific membrane antigen (PSMA) as the target of monoclonal antibody 107-1A4 by proteinchip; array, surface-enhanced laser desorption/ionization (SELDI) technology, *Int. J. Cancer*, 92, 871–876, 2001.
262. Nelson, P.S., Han, D., Rochon, Y., Corthals, G.L., Lin, B., Monson, A., Nguyen, V., Franza, B.R., Plymate, S.R., Aebersold, R., and Hood, L., Comprehensive analyses of prostate gene expression: convergence of expressed sequence tag databases, transcript profiling and proteomics, *Electrophoresis*, 21, 1823–1831, 2000.
263. Alaiya, A.A., Oppermann, M., Langridge, J., Roblick, U., Egevad, L., Brindstedt, S., Hellstrom, M., Linder, S., Bergman, T., Jornvall, H., and Auer, G., Identification of proteins in human prostate tumor material by two-dimensional gel electrophoresis and mass spectrometry, *Cell Mol. Life Sci.*, 58, 307–311, 2001.
264. Emmert-Buck, M.R., Gillespie, J.W., Paweletz, C.P., Ornstein, D.K., Basrur, V., Appella, E., Wang, Q.H., Huang, J., Hu, N., Taylor, P., and Petricoin, E.F., 3rd., An approach to proteomic analysis of human tumors, *Mol. Carcinog.*, 27, 158–165, 2000.
265. Lawrie, L.C., Curran, S., McLeod, H.L., Fothergill, J.E., and Murray, G.I., Application of laser capture microdissection and proteomics in colon cancer, *Mol. Pathol.*, 54, 253–258, 2001.
266. Jungblut, P.R., Zimny Arndt, U., Zeindl Eberhart, E., Stulik, J., Koupilova, K., Pleissner, K.P., Otto, A., Muller, E.C., Sokolowska Kohler, W., Grabher, G., and Stoffler, G., Proteomics in human disease: cancer, heart and infectious diseases. *Electrophoresis*, 20, 2100–2110, 1999.

267. Stulik, J., Kovarova, H., Macela, A., Bures, J., Jandik, P., Langr, F., Otto, A., Thiede, B., and Jungblut, P., Overexpression of calcium-binding protein calgranulin B in colonic mucosal diseases, *Clin. Chim. Acta*, 265, 41–55, 1997.
268. Stulik, J., Österreicher, J., Koupilová, K., Knizek, J., Macela, A., Bures, J., Jandik, P., Langr, F., Dedic, K., and Jungblut, P.R., The analysis of S100A9 and S100A8 expression in matched sets of macroscopically normal colon mucosa and colorectal carcinoma: the S100A9 and S100A8 positive cells underlie and invade tumor mass, *Electrophoresis*, 20, 1047–1054, 1999.
269. Stulik, J., Koupilová, K., Österreicher, J., Knizek, J., Macela, A., Bures, J., Jandik, P., Langr, F., and Dedic, K., Protein abundance alterations in matched sets of macroscopically normal colon mucosa and colorectal carcinoma, *Electrophoresis*, 20, 3638–3646, 1999.
270. Cole, A.R., Ji, H., and Simpson, R.J., Proteomic analysis of colonic crypts from normal, multiple intestinal neoplasia and p53-null mice: a comparison with colonic polyps, *Electrophoresis*, 21, 1772–1781, 2000.
271. Melis, R. and White, R., Characterization of colonic polyps by two-dimensional gel electrophoresis, *Electrophoresis*, 20, 1055–1064, 1999.
272. Sinha, P., Hutter, G., Kottgen, E., Dietel, M., Schadendorf, D., and Lage, H., Search for novel proteins involved in the development of chemoresistance in colorectal cancer and fibrosarcoma cells *in vitro* using two-dimensional electrophoresis, mass spectrometry and microsequencing, *Electrophoresis*, 20, 2961–2969, 1999.
273. Liotta, L. and Petricoin, E.F., Molecular profiling of human cancer, *Nature Rev. Genetics*, 1, 48–56, 2000.
274. Seow, T.K., Ong, S.E., Liang, R.C.M., Ren, E.C., Chan, L., Ou, K., and Chung, M.C.M., Two-dimensional electrophoresis map of the human heptatocellular carcinoma cell line, HCC-M, and identification of the separated proteins by mass spectrometry, *Electrophoresis*, 21, 1787–1813, 2000.
275. Choong, M.L., Tan, L.K., Lo, S.L., Ren, E.C., Ou, K., Ong, S.E., Liang, R.C., Seow, T.K., and Chung, M.C., An integrated approach in the discovery and characterization of a novel nuclear protein over-expressed in liver and pancreatic tumors, *FEBS Lett.*, 496, 109–116, 2001.
276. Yu, L.R., Zeng, R., Shao, X.X., Wang, N., Xu, Y.H., and Xia, Q.C., Identification of differentially expressed proteins between human hepatoma and normal liver cell lines by two-dimensional electrophoresis and liquid chromatography-ion trap mass spectrometry, *Electrophoresis*, 21, 3058–3068, 2000.
277. Sinha, P., Hutter, G., Kottgen, E., Dietel, M., Schadendorf, D., and Lage, H., Increased expression of epidermal fatty acid binding protein, cofilin, and 14-3-3-σ (stratifin) detected by two-dimensional gel electrophoresis, mass spectrometry and microsequencing of drug-resistant human adenocarcinoma of the pancreas, *Electrophoresis*, 20, 2952–2960, 1999.
278. Chuman, Y., Bergman, A.-C., Ueno, T., Saito, S., Sakaguchi, K., Alaiya, A.A., Franzen, B., Bergman, T., Arnott, D., Auer, G., Appella, E., Jornvall, H., and Linder, S., Napsin A, a member of the aspertic protease family, is abundantly expressed in normal lung and kidney tissue and is expressed in lung adenocarcinomas, *FEBS Lett.,* 462, 129–134, 2000.
279. Gamble, S.C., Dunn, M.J., Wheeler, C.H., Joiner, M.C., Adu-Poku, A., and Arrand, J.E., Expression of proteins coincident with inducible radioprotection in human lung epithelial cells, *Cancer Res.*, 60, 2146–2151, 2000.
280. Clauser, K., Hall, S.C., Smith, D., Webb, J., Andrews, L., Tram, H., Epstein, L.B., and Burlingame, A.L., Rapid mass spectrometric peptide sequencing and mass matching for characterization of human melanoma proteins isolated by two-dimensional PAGE, *Proc. Natl. Acad. Sci. USA*, 92, 5072–5076, 1995.
281. von Eggeling, F., Davies, H., Lomas, L., Fiedler, W., Junker, K., Claussen, U., and Ernst, G., Tissue-specific microdissection coupled with Protein Chip array technologies: applications in cancer research, *Biotechniques*, 29, 1066–1069, 2000.
282. Ferrari, L., Seraglia, R., Rossi, C.R., Bertazzo, A., Lise, M., Allegri, G., and Traldi, P., Protein profiles in sera of patients with malignant cutaneous melanoma, *Rapid Commun. Mass Spectrom.*, 14, 1149–1154, 2000.
283. Sinha, P., Kohl, S., Fischer, J., Hutter, G., Kern, M., Kottgen, E., Dietel, M., Lage, H., Schnolzer, M., and Schadendorf, D., Identification of novel proteins associated with the development of chemoresistance in malignant melanoma using two-dimensional electrophoresis, *Electrophoresis*, 21, 3048–3057, 2000.

284. Bjellerup, P., Theodorsson, E., Jornvall, H., and Kogner, P., Limited neuropeptide Y precursor processing in unfavourable metastic neuroblastoma tumours, *British J. Cancer*, 83, 171–176, 2000.
285. Zhu, X., Robertson, J.T., Sacks, H.S., Dohan, F.C., Tseng, J., and Desiderio, D.M., Opioid and tachykinin neuropeptides in prolactin-secreting human pituitary adenomas, *Peptides*, 16, 1097–1107, 1995.
286. Zhu, X. and Desiderio, D.M., Peptide quantification by tandem mass spectrometry, *Mass Spectrom. Reviews*, 15, 213–240, 1996.
287. Desidero, D.M., Mass spectrometry, high performance liquid chromatography, and brain peptides, *Biopolymers*, 40, 257–264, 1996.
288. Dass, C., Fridland, G.H., Tinsley, P.W., Killmar, J.T., and Desiderio, D.M., Characterization of β-endorphin in human pituitary by fast atom bombardment mass spectrometry of trypsin-generated fragments, *J. Peptide Protein Res.*, 34, 81–87, 1989.
289. Giorgianni, S.B. and Desiderio, D.M., Fast atom bombardment mass spectrometry of synthetic peptides, *Methods Enzymol.*, 289, 478–499, 1997.
290. Dass, C., Kusmierz, J.J., Desiderio, D.M., Jarvis, S.A., and Green, B.N., Electrospray mass spectrometry for the analysis of opioid peptides and the quantification of endogenous methionine enkephalin and β-endorphin, *J. Am. Soc. Mass Spectrom.*, 2, 149–156, 1991.
291. Desiderio, D.M., Mass spectrometric quantification of neuropeptides, in *Protein and Peptide Analysis by Mass Spectrometry*, Chapman, J.R., Ed., Humana Press, Totowa, NY, (1996) 57–65.
292. Lovelace, J.L., Kusmierz, J.J., and Desiderio, D.M., Analysis of methionine enkephalin in human pituitary by multi-dimensional reversed-phase high-performance liquid chromatography, radioreceptor assay, radioimmunoassay, fast atom bombardment mass spectrometry, and mass spectrometry-mass spectrometry, *J. Chromatogr.*, 562, 573–584, 1991.
293. Kusmierz, J., Dass, C., Robertson, J.T., and Desiderio, D.M., Mass spectrometric measurement of β-endorphine and methionine enkephalin in human pituitaries: tumors and post-mortem controls, *Int. J. Mass Spectrom. Ion Proc.*, 111, 247–262, 1991.
294. Desiderio, D.M. and Zhu, X., Quantitative analysis of methionine enkephalin and β-endorphin in the pituitary by liquid secondary ion mass spectrometry and tandem mass spectrometry, *J. Chromatogr. A*, 794, 85–96, 1998.
295. Desiderio, D., Kusmierz, J., Zhu, X., Dass, C., Hilton, D., and Robertson, J., Mass spectrometric analysis of opioid and tachykinin neuropeptides in non-secreting and ACTH-secreting human pituitary adenomas, *Biol. Mass Spectrom.*, 22, 89–97, 1993.
296. Beranova-Giorgianni, S. and Desiderio, D.M., Mass spectrometry of the human pituitary proteome: identification of selected proteins, *Rapid Commun. Mass Spectrom.*, 14, 161–167, 2000.
297. Lehner, P.J. and Cresswell, P., Processing and delivery of peptides presented by MHC class I molecules, *Curr. Opin. Immunol.*, 8, 59–67, 1996.
298. Shresta, S., Pham, C., Thomas, D.A., Graubert, T.A., and Ley, T.J., How do cytotoxic lymphocytes kill their target?, *Curr. Opin. Immunol.*, 10, 581–587, 1998.
299. Rosenberg, S.A., Yang, J.C., Schwartzentruber, D.J., Hwu, P., Marincola, F.M., Topalian, S.L., Restifo, N.P., Dudley, M.E., Schwarz, S.L., Spiess, P.J., Wunderlich, J.R., Parkhurst, M.R., Kawakami, Y., Seipp, C.A., Einhorn, J.H., and White, D.E., Immunologic and therapeutic evaluation of a synthetic peptide vaccine for the treatment of patients with metastatic melanoma, *Nature Med.*, 4, 321–327, 1998.
300. Ayyoub, M., Mazarguil, H., Monsarrat, B., Van den Eynde, B., and Gairin, J.E., A structure-based approach to designing non-natural peptides that can activate anti-melanoma cytotoxic T cells, *J. Biol. Chem.*, 274, 10227–10234, 1999.
301. Rosenberg, S.A., The identification of cancer antigens: impact on the development of cancer vaccines, *Cancer J.* 6(suppl. 2), S142–S149, 2000.
302. Poland, G.A., Ovsyannikova, I.G., Johnson, K.L., and Naylor, S., The role of mass spectrometry in vaccine development, *Vaccine*, 19, 2692–2700, 2001.
303. Wang, W., Gulden, P.H., Pierce, R.A., Shabanowitz, J.A., Man, S.T., Hunt, D.F., and Engelhard, V.H., A naturally processed peptide presented by HLA-A*0201 is expressed at low abundance and recognized by an alloreactive $CD8^+$ cytotoxic T cell with apparent high affinity, *J. Immunol.*, 158, 5794–5804, 1997.
304. Walker, C.R., Sherman, N.E., Shabanowitz, J., and Hunt, D.F., Mass spectrometry in immunology: identification of a minor histocompatibility antigen, in *Mass Spectrometry of Biological Materials*, Larsen, B.S. and McEwen, C.N., Eds., Marcel Dekker, New York, (1998) 115–136.

305. Paradela, A., Garcia-Peydro, M., Vazquez, J., Rognan, D., and de Castro, J.A., The same natural ligand is involved in allorecognition of multiple HLA-B27 subtypes by a single T cell clone: role of peptide and the MHC molecule in alloreactivity, *J. Immunol.*, 161, 5481–5490, 1998.
306. Tomlinson, A.J., Jameson, S., and Naylor, S., Strategy for isolating and sequencing biologically derived MHC class I peptides, *J. Chromatogr. A*, 744, 273–278, 1996.
307. Malik, P. and Strominger, J.L., Perfusion chromatography for very rapid purification of class I and II MHC proteins, *J. Immunol. Methods*, 234, 83–88, 2000.
308. Brockman, A.H., Orlando, R., and Tarleton, R.L., A new liquid chromatography/tandem mass spectrometric approach for the identification of class I major histocompatibility complex associated peptides that eliminates the need for bioassays, *Rapid Commun. Mass Spectrom.*, 13, 1024–1030, 1999.
309. Schirle, M., Keilholz, W., Weber, B., Gouttefangeas, C., Dumrese, T., Becker, H.D., Stevanovic, S., and Rammensee, H.G., Identification of tumor-associated MHC class 1 ligands by a novel T cell-independent approach, *Euro. J. Immunol.*, 30, 2216–2225, 2000.
310. Pascolo, S., Schirle, M., Guckel, B., Dumrese, T., Stumm, S., Kayser, S., Moris, A., Wallwiener, D., Rammensee, H.G., and Stevanovic, S., A MAGE-A1 HLA-A A*0201 epitope identified by mass spectrometry, *Cancer Res.*, 61, 4072–4077, 2001.
311. van der Heeft, E., ten Hove, J.G., Herberts, C.A., Meiring, H.D., van Els, C.A., and de Jong, A., A microcapillary column switching HPLC-electrospray ionization MS system for the direct identification of peptides presented by a major histocompatibility complex class 1 molecule, *Anal. Chem.*, 70, 3742–3751, 1998.
312. Flad, T., Spengler, B., Kalbacher, H., Brossart, P., Baier, D., Kaufmann, R., Bold, P., Metzger, S., Bluggel, M., Meyer, H.E., Kurz, B., and Muller, C.A., Direct identification of major histocompatibility complex class I-bound tumor-associated peptide antigens of a renal carcinoma cell line by a novel mass spectrometric method, *Cancer Res.*, 58, 5803–5811, 1998.
313. de Jong, A., Contribution of mass spectrometry to contemporary immunology, *Mass Spectrom. Reviews*, 17, 311–335, 1998.
314. Purcell, A.W. and Gorman, J.J., The use of post-source decay in matrix-assisted laser desorption/ionisation mass spectrometry to delineate T cell determinants, *J. Immunol. Methods,* 249, 17–31, 2001.
315. Russo, R.E., Shabanowitz, J., and Hunt, D.F., A novel ESI source for coupling capillary electrophoresis and mass spectrometry: sequence determination of tumor peptides at the attomole level, *J. Microcolumn Separt.*, 10, 281–285, 1998.
316. Marina, A., Garcia, M.A., Albar, J.P., Yague, J., de Castro, J.A.L., and Vazquez, J., High-sensitivity analysis and sequencing of peptides and proteins by quadrupole ion trap mass spectrometry, *J. Mass Spectrom.*, 34, 17–27, 1999.
317. Huczko, E., Bodnar, W., Benjamin, D., Sakaguchi, K., Zhu, N., Shabanowitz, J., Henderson, R., Appella, E., Hunt, D., and Engelhard, V., Characteristics of endogenous peptides eluted from the Class 1 MHC molecule HLA-B7 determined by mass spectrometry and computer modeling, *J. Immunol.*, 151, 2572–2578, 1993.
318. Cox, A., Skipper, J., Chen, Y., Henderson, R., Darrow, T., Shabanowitz, J., Engelhard, V., Hunt, D., and Slingluff, C.L., Identification of peptide recognized by five melanoma-specific human cytotoxic T cell lines, *Science*, 264, 716–719, 1994.
319. Fiorillo, M.T., Meadows, L., D'Amato, M., Shabanowitz, J., Hunt, D.F., Appella, E., and Sorrentino, R., Susceptibility to ankylosing spondylitis correlates with the C-terminal residue of peptides presented by various HLA-B27 subtypes, *Eur. J. Immunol.*, 27, 368–373, 1997.
320. Lichtenfels, R., Ackermann, A., Kellner, R., and Seliger, B., Mapping and expression pattern analysis of key components of the major histocompatibility complex class I antigen processing and presentation pathway in a representative human renal carcinoma cell line, *Electrophoresis*, 22, 1801–1809, 2001.
321. Hogan, K.T., Eisinger, D.P., Cupp, S.B., Lekstrom, K.J., Deacon, D.D., Shabanowitz, J., Hunt, D.F., Engelhard, V.H., Slingluff, C.L., and Ross, M.M., The peptide recognized by HLA-A68.2 restricted, squamous cell carcinoma of the lung-specific cytotoxic T lymphocytes is derived from a mutated elongation factor 2 gene, *Cancer Res.*, 58, 5144–5150, 1998.
322. Pemberton, L.A., Kerr, S.J., Smythe, G., and Brew, B.J., Quinolinic acid production by macrophages stimulated with IFN-gamma, TNF-alpha, and IFN-alpha, *J. Interferon. Cytokine. Res.*, 17, 589–595, 1997.

323. den Haan, J.M., Meadows, L.M., Wang, W., Pool, J., Blokland, E., Bishop, T.L., Reinhardus, C., Shabanowitz, J., Offringa, R., Hunt, D.F., Engelhard, V.H., and Goulmy, E., The minor histocompatibility antigen HA-1: a diallelic gene with a single amino acid polymorphism, *Science*, 279, 1054–1057, 1998.
324. Eggers, M., Boes Fabian, B., Ruppert, T., Kloetzel, P.M., and Koszinowski, U.H., The cleavage preference of the proteasome governs the yield of antigenic peptides, *J. Exp. Med.*, 182, 1865–1870, 1995.
325. Theobald, M., Ruppert, T., Kuckelkorn, U., Hernandez, J., Haussler, A., Ferreira, E.A., Liewer, U., Biggs, J., Levine, A.J., Huber, C., Koszinowski, U.H., Kloetzel, P.M., and Sherman, L.A., The sequence alteration associated with a mutational hotspot in p53 protects cells from lysis by cytotoxic T lymphocytes specific for a flanking peptide epitope, *J. Exp. Med.*, 188, 1017–1028, 1998.
326. David, L. and Zhang, Z., Probing noncovalent structural features of proteins by mass spectrometry, *Mass Spectrom. Reviews*, 13, 341–356, 1994.
327. Winston, R.L. and Fitzgerald, M.C., Mass spectrometry as a readout of protein structure and function, *Mass Spectrom. Reviews*, 16, 165–179, 1997.
328. Loo, J.A., Studying noncovalent protein complexes by electrospray ionization mass spectrometry, *Mass Spectrom. Reviews*, 16, 1–23, 1997.
329. Pramanik, B.N., Bartner, P.K., Mirza, U.A., Liu, Y.H., and Ganguly, A.K., Electrospray ionization mass spectrometry for the study of non-covalent complexes: an emerging technology, *J. Mass Spectrom.*, 33, 911–920, 1998.
330. Pruitt, K. and Der, C.J., Ras and Rho regulation of the cell and oncogenesis, *Cancer Lett.*, 171, 1–10, 2001.
331. Ganguly, A.K., Pramanik, B.N., Huang, E.C., Tsarbopoulos, A., Girijavallabhan, V.M., and Liberles, S., Studies of the *Ras*-GDP and *Ras*-GTP noncovalent complexes by electrospray mass spectrometry, *Tetrahedron*, 49, 7985–7996, 1993.
332. Ganguly, A.K., Pramanik, B.N., Tsarbopoulos, A., Covey, T., Huang, E., and Fuhrman, S., Mass spectrometric detection of the noncovalent GDP-bond conformational state of the human H-*ras* protein, *J. Am. Chem. Soc.*, 114, 6559–6560, 1992.
333. Rostom, A.A. and Robinson, C.V., Disassembly of intact multiprotein complexes in the gas phase, *Current Opin. Struct. Biol.*, 9, 135–141, 1999.
334. Gottschalk, A., Neubauer, G., Banroques, J., Mann, M., Luhrmann, R., and Fabrizio, P., Identification by mass spectrometry and functional analysis of novel proteins of the yeast (U4/U6.U5) tri-snRNP, *EMBO J.*, 18, 4535–4540, 1999.
335. Witke, W., Podtelejnikov, A.V., Di Nardo, A., Sutherland, J.D., Gurniak, C.B., Dotti, C., and Mann, M., In mouse brain profilin I and profilin II associate with regulators of the endocytic pathway and actin assembly, *EMBO J.*, 17, 967–976, 1998.
336. Tarabykina, S., Kriajevska, M., Scott, D.J., Hill, T.J., Lafitte, D., Derrick, P.J., Dodson, G.G., Lukanidin, E., and Bronstein, I., Heterocomplex formation between metastasis-related protein S100A4 (Mts1) and S100A1 as revealed by the yeast two-hybrid system, *FEBS Lett.*, 475, 187–191, 2000.
337. Cheng, X., Morin, P.E., Harms, A.C., Bruce, J.E., Ben-David, Y., and Smith, R.D., Mass spectrometric characterization of sequence-specific complexes of DNA and transcription factor PU.1 DNA binding domain, *Anal. Biochem.*, 239, 25–34, 1996.
338. Calvio, G., Neubauer, G., Mann, M., and Limond, A., Identification of hnRNP P2 as TLS/FUS using electrospray mass spectrometry, *RNA*, 1, 724–733, 1995.
339. Ajuh, P., Küster, B., Panov, K., Zomerdijk, J., Mann, M., and Lamond, A.I., Functional analysis of the human CDC5L complex and identification of its components by mass spectrometry, *EMBO J.*, 19, 6569–6581, 2000.
340. Przybylski, M. and Glocker, M.O., Electrospray mass spectrometry of biomacromolecular complexes with non-covalent interactions—new analytical perspectives for supramolecular chemistry and molecular recognition processes, *Angew. Chem. Int. Ed. Engl.*, 35, 806–826, 1996.
341. Happersberger, H.P., Przybylski, M., and Glocker, M.O., Selective bridging of bis-cysteinyl residues by arsonous acid derivatives as an approach to the characterization of protein tertiary structures and folding pathways by mass spectrometry, *Anal. Biochem.*, 264, 237–250, 1998.
342. Barbirz, S., Jakob, U., and Glocker, M.O., Mass spectrometry unravels disulfide bond formation as the mechanism that activates molecular chaperone, *J. Biol. Chem.*, 275, 18759–18766, 2000.
343. Happersberger, H.P., Bantscheff, M., Barbirz, S., and Glocker, M.O., Multiple and subsequent MALDI-MS on-target chemical reactions for the characterization of disulfide bonds and primary structures of proteins, *Methods Molec. Biol.*, 146, 167–184, 2000.

344. Whittal, R.M., Benz, C.C., Scott, G., Semyonov, J., Burlingame, A.L., and Baldwin, M.A., Preferential oxidation of zinc finger 2 in strogen receptor DNA-binding domain prevents dimerization and, hence, DNA binding, *Biochemistry*, 39, 8406–8417, 2001.
345. Ranson, N.A., White, H.E., and Saibil, H.R., Chaperonins, *Biochem. J.*, 333, 233–242, 1998.
346. Feldman, D.E. and Frydman, J., Protein folding *in vivo*: the importance of molecular chaperons, *Current Opin. Struct. Biol.*, 10, 26–33, 2000.
347. Mirza, U., Cohen, S., and Chait, B.T., Heat-induced conformational changes is proteins studied by electrospray ionization mass spectrometry, *Anal. Chem.*, 65, 1–6, 1993.
348. Wade, D., Deuterium isotope effects on noncovalent interactions between molecules. *Chem. Biol. Interact.*, 117, 191–217, 1999.
349. Miranker, A., Robinson, C., Radford, S., and Dobson, C., Investigation of protein folding by mass spectrometry, *FASEB J.*, 10, 93–101, 1996.
350. Last, A.M. and Robinson, C.V., Protein folding and interactions revealed by mass spectrometry, *Curr. Opin. Chem. Biol.*, 3, 564–570, 1999.
351. Miranker, A., Robinson, C.V., Radford, S.E., Aplin, R.T., and Dobson, C.M., Detection of transient protein folding populations by mass spectrometry, *Science*, 262, 896–900, 1993.
352. Mandell, J.G., Falick, A.M., and Komives, E.A., Measurement of amide hydrogen exchange by MALDI-TOF mass spectrometry, *Anal. Chem.*, 70, 3987–3995, 1998.
353. Mandell, J.G., Falick, A.M., and Komives, E.A., Identification of protein-protein interfaces by decreased amide proton solvent accessibility, *Proc. Natl. Acad. Sci. USA*, 95, 14705–14710, 1998.
354. Yang, H.H., Li, X.C., Amft, M., and Grotemeyer, J., Protein conformational changes determined by matrix-assisted laser desorption mass spectrometry, *Anal. Biochem.*, 258, 118–126, 1998.
355. Suckau, D., Shi, Y., Beu, S., Senko, M., Quinn, J., Wampler III, F., and McLafferty, F., Coexisting stable conformations of gaseous protein ions, *Proc. Natl. Acad. Sci. USA*, 90, 790–793, 1993.
356. Sigler, P.B., Xu, Z., Rye, H.S., Burston, S.G., Fenton, W.A., and Horwich, A.L., Structure and function in GroEL-mediated protein folding, *Annu. Rev. Biochem.*, 67, 581–608, 1998.
357. Coyle, J.E., Texter, F.L., Ashcroft, A.E., Masselos, D., Robinson, C.V., and Radford, S.E., GroEL accelerates the refolding of hen lysozyme without changing its folding mechanism. *Nature Struct. Biol.*, 6, 683–690, 1999.
358. Robinson, C.V., Gross, M., and Radford, S.E., Probing conformations of GroEL-bound substrate proteins by mass spectrometry, *Methods Enzymol.*, 290, 296–313, 1998.
359. Rostom, A.A. and Robinson, C.V., Detection of the intact GroEL chaperonin assembly by mass spectrometry, *J. Am. Chem. Soc.*, 121, 4718–4719, 1999.
360. Glocker, M.O., Arbogast, B., Milley, R., Cowgill, C., and Deinzer, M.L., Disulfide linkages in the *in vitro* refolded intermediates of recombinant human macrophage-colony-stimulating factor: analysis of the sulfhydryl alkylation of free cysteine residues by fast-atom bombardment mass spectrometry, *Proc. Natl. Acad. Sci. USA*, 91, 5868–5872, 1994.
361. Glocker, M., Arbogast, B., Schreurs, J., and Deinzer, M., Assignment of the inter- and intramolecular disulfide linkages in recombinant human macrophage colony stimulating factor using fast atom bombardment mass spectrometry, *Biochemistry*, 32, 482–488, 1993.
362. Zhang, Y.H., Yan, X., Maier, C.S., Schimerlik, M.I., and Deinzer, M.L., Structural comparison of recombinant human macrophage colony stimulating factor beta and a partially reduced derivative using hydrogen deuterium exchange and electrospray ionization mass spectrometry, *Protein Sci.*, 10, 2336–2346, 2001.
363. Glocker, M., Arbogast, B., and Deinzer, M., Characterization of disulfide linkages and disulfide bond scrambling in recombinant human macrophage colony stimulating factor by fast-atom bombardment mass spectrometry of enzymatic digests, *J. Am. Soc. Mass Spectrom.*, 6, 638–643, 1995.
364. Happersberger, H.P., Stapleton, J., Cowgill, C., and Glocker, M.O., Characterization of the folding pathway of recombinant human macrophage-colony stimulating-factor β (rhM-CSFβ) by bis-cysteinyl modification and mass spectrometry, *Proteins*, 2 (Suppl), 50–62, 1998.
365. Kim, H.Y. and Salem, Jr., N., Liquid chromatography-mass spectrometry of lipids, *Prog. Lipid Res.*, 32 No. 3, 221–245, 1993.
366. Murphy, R.C. and Harrison, K.A., Fast atom bombardment mass spectrometry of phospholipids, *Mass Spectrom. Reviews*, 13, 57–75, 1994.
367. Kuksis, A. and Myher, J., Application of tandem mass spectrometry for the analysis of long-chain carboxylic acids, *J. Chromatogr. B*, 671, 35–70, 1995.

368. Kataoka, H., Chromatographic analysis of lipoic acid and related compounds, *J. Chromatogr. B*, 717, 247–262, 1998.
369. Murphy, R.C., Fielder, J., and Hevko, J., Analysis of nonvolatile lipids by mass spectrometry, *Chem. Rev.*, 101, 479–526, 2001.
370. White, F., Seldomridge, S., and Marshall, A., Structural characterization of phospholipids by matrix-assisted laser desorption/ionization Fourier transform ion cyclotron resonance mass spectrometry, *Anal. Chem.*, 67, 3979–3984, 1995.
371. Schiller, J., Arnhold, J., Muller, B.M., Reichl, S., and Arnold, K., Lipid analysis by matrix-assisted laser desorption ionization mass spectrometry: a methodological approach, *Anal. Biochem.*, 267, 46–56, 1999.
372. Watts, J.D., Gu, M., Patterson, S.D., Aebersold, R., and Polverino, A.J., On the complexities of ceramide changes in cells undergoing apoptosis: lack of evidence for a second messenger function in apoptotic induction, *Cell Death Different.*, 6, 105–114, 1999.
373. Watts, J.D., Aebersold, R., Polverino, A.J., Patterson, S.D., and Gu, M., Ceramide second messengers and ceramide assays, *Trends Biochem. Sci.*, 24, 228–229, 1999.
374. Cremesti, A.E. and Fischl, A.S., Current methods for the identification and quantitation of ceramides: an overview, *Lipids*, 35, 937–945, 2000.
375. Hansson, G., Li, Y., and Karlsson, H., Characterization of glycosphingolipid mixtures with up to ten sugars by gas chromatography and gas chromatography-mass spectrometry as permethylated oligosaccarides and ceramides released by ceramide glycanase, *Biochemistry*, 28, 6672–6678, 1989.
376. Tsikas, D., Application of gas chromatography-mass spectrometry and gas chromatography-tandem mass spectrometry to assess *in vivo* synthesis of postaglandins, thromboxane, leukotrienes, isoprostanes and related compounds in humans, *J. Chromatogr. B*, 717, 201–245, 1998.
377. Ann, Q. and Adams, J., Structure-specific collision-induced fragmentations of ceramides cationized with alkali-metal ions, *Anal. Chem.*, 65, 7–13, 1993.
378. Caldas, E., Jones, A., Winter, C., Ward, B., and Gilchrist, D., Electrospray ionization mass spectrometry of sphinganine analog mycotoxins, *Anal. Chem.*, 67, 196–207, 1995.
379. Raith, K. and Neubert, R.H., Structural studies on ceramides by electrospray tandem mass spectrometry, *Rapid Commun. Mass Spectrom.*, 12, 935–938, 1998.
380. Gu, M., Kerwin, J.L., Watts, J.D., and Aebersold, R., Ceramide profiling of complex lipid mixtures by electrospray ionization mass spectrometry, *Anal. Biochem.*, 244, 347–356, 1997.
381. Mano, N., Oda, Y., Yamada, K., Asawaka, N., and Katayama, K., Simultaneous quantitative determination method for sphingolipid metabolites by liquid chromatography/ionspray ionization tandem mass spectrometry, *Anal. Biochem.*, 244, 291–300, 1997.
382. Couch, L.H., Churchwell, M.I., Doerge, D.R., Tolleson, W.H., and Howard, P.C., Identification of ceramides in human cells using liquid chromatography with detection by atmospheric pressure chemical ionization, *Rapid Commun. Mass Spectrom.*, 11, 504–512, 1997.
383. Watts, J.D., Gu, M., Polverino, A.J., Patterson, S.D., and Aebersold, R., Fas-induced apoptosis of T cells occurs independently of ceramide generation, *Proc. Natl. Acad. Sci. USA*, 94, 7292–7296, 1997.
384. Thomas, Jr., R.L., Matsko, C.M., Lotze, M.T., and Amoscato, A.A., Mass spectrometric identification of increased C16 ceramide levels during apoptosis, *J. Biol. Chem.*, 274, 30580–30588, 1999.
385. Lehmann, W., Metzger, K., Stephan, M., Wittig, U., Zalan, I., Habenicht, A., and Furstenberger, G., Quantitative lipoxygenase product profiling by gas chromatography negative-ion chemical ionization mass spectrometry, *Anal. Biochem.*, 224, 227–234, 1995.
386. Metzger, K., Angres, G., Maier, H., and Lehmann, W., Lipoxygenase products in human saliva: patients with oral cancer compared to controls, *Free Radic. Biol. Med.*, 18, 185–194, 1995.
387. Hubbard, W.C., Litterst, C.L., Lin, M.C., Bleecker, E.R., Eggleston, J.C., McLemore, T.L., and Boyd, M.R., Profiling of prostaglandin biosynthesis in biopsy fragments of human lung carcinomas and normal human lung by capillary gas chromatography-negative ion chemical ionization mass spectrometry, *Prostaglandins*, 32, 889–906, 1986.
388. Giardiello, F.M., Spannhake, E.W., DuBois, R.N., Hylind, L.M., Robinson, C.R., Hubbard, W.C., Hamilton, S.R., and Yang, V.W., Prostaglandin levels in human colorectal mucosa: effects of sulindac in patients with familial adenomatous polyposis, *Dig. Dis. Sci.*, 43, 311–316, 1998.
389. Sheng, H., Shao, J., Kirkland, S.C., Isakson, P., Coffey, R.C., Morrow, J., Beauchamp, R.D., and DuBois, R.N., Inhibition of human colon cancer cell growth by selective inhibition of cyclooxygenase-2, *J. Clin. Invest.*, 99, 2254–2259, 1997.

390. Moolenaar, W., Kranenburg, O., Postma, F., and Zondag, G., Lysophosphatidic acid: G-protein signaling and cellular responses, *Curr. Opin. Cell Biol.*, 9, 168–173, 1997.
391. Xiao, Y., Chen, Y., Kennedy, A.W., Belinson, J., and Xu, Y., Evaluation of plasma lysophospholipids for diagnostic significance using electrospray ionization mass spectrometry (ESI-MS) analyses, *Ann. N. Y. Acad. Sci.*, 905, 242–259, 2000.
392. Xiao, Y.J., Schwartz, B., Washington, M., Kennedy, A., Webster, K., Belinson, J., and Xu, Y., Electrospray ionization mass spectrometry analysis of lysophospholipids in human ascitic fluids: comparison of the lysophospholipid contents in malignant vs nonmalignant ascitic fluids, *Anal. Biochem.*, 290, 302–313, 2001.
393. Pinckard, R.N., Woodard, D.S., Showell, H.J., Conklyn, M.J., Novak, M.J., and McManus, L.M., Structural and (patho)physiological diversity of PAF, *Clin. Rev. Allergy*, 12, 329–359, 1994.
394. Woodard, D.S., Mealey, B.L., Lear, C.S., Satsangi, R.K., Prihoda, T.J., Weintraub, S.T., Pinckard, R.N., and McManus, L.M., Molecular heterogeneity of PAF in normal human mixed saliva: quantitative mass spectral analysis after direct derivatization of PAF with pentafluorobenzoic anhydride, *Biochim. Biophys. Acta*, 1259, 137–147, 1995.
395. Weintraub, S.T., Satsangi, R.K., Sprague, E.A., Prihoda, T.J., and Pinckard, R.N., Mass spectrometric analysis of platelet-activating factor after isolation by solid-phase extraction and direct derivatization with pentafluorobenzoic anhydride, *J. Am. Soc. Mass Spectrom.*, 11, 170–181, 2000.
396. Kim, H., Wang, T., and Ma, Y., Liquid chromatograph/mass spectrometry of phospholipids using electrospray ionization, *Anal. Chem.*, 66, 3977–3982, 1994.
397. Silvestro, L., Da Col, R., Scappaticci, E., Libertucci, D., Biancone, L., and Camussi, G., Development of a high-performance liquid chromatographic-mass spectrometric technique, with an ionspray interface, for the determination of platelet-activating factor (PAF) and lyso-PAF in biological samples, *J. Chromatogr.*, 647, 261–269, 1993.
398. Harrison, K.A., Clay, K.L., and Murphy, R.C., Negative ion electrospray and tandem mass spectrometric analysis of platelet activating factor (PAF) (1-hexadecyl-2-acetyl-glycerophosphocholine), *J. Mass Spectrom.*, 34, 330–335, 1999.
399. Maggi, M., Bonaccorsi, L., Finetti, G., Carloni, V., Muratori, M., Laffi, G., Forti, G., Serio, M., and Baldi, E., Platelet-activating factor mediates an autocrine proliferative loop in the endometrial adenocarcinoma cell line HEC-1A[1], *Cancer Res.*, 54, 4777–4784, 1994.
400. Savu, S.R., Silvestro, L., Sorgel, F., Montrucchio, G., Lupia, E., and Camussi, G., Determination of 1-O-acyl-2-acetyl-sn-glyceryl-3-phosphorylcholine, platelet-activating factor and related phospholipids in biological samples by high-performance liquid chromatography–tandem mass spectrometry, *J. Chromatogr. B*, 682, 35–45, 1996.
401. Montrucchio, G., Sapino, A., Bussolati, B., Ghisolfi, G., Rizea-Savu, S., Silvestro, L., Lupia, E., and Camussi, G., Potential angiogenic role of platelet-activating factor in human breast cancer, *Am. J. Pathol.*, 153, 1589–1596, 1998.
402. Kong, Y., Zhu, Y., and Zhang, J., Ionization mechanism of oligonucleotides in matrix-assisted laser desorption/ionization time-of-flight mass spectrometry, *Rapid Commun. Mass Spectrom.*, 15, 57–64, 2001.
403. Nordhoff, E., Kirpekar, F., and Roepstorff, P., Mass spectrometry of nucleic acids, *Mass Spectrom. Reviews,* 15, 67–138, 1996.
404. Deforce, D.L. and Van den Eeckout, E.G., Analysis of oligonucleotides by ESI-MS, *Adv. Chromatogr.*, 40, 539–566, 2000.
405. Deforce, D.L. and Van den Eeckout, E.G., Analysis of oligonucleotides by ESI-MS, *Adv. Chromatogr.*, 40, 539–566, 2000.
406. Murray, K.M., DNA sequencing by mass spectrometry, *J. Mass Spectrom.*, 31, 1203–1215, 1996.
407. McLuckey, S.A., Van Berkel, G.J., and Glish, G.L., Tandem mass spectrometry of small, multiply charged oligonucleotides, *J. Am. Soc. Mass Spectrom.*, 3, 60–70, 1992.
408. Krahmer, M.T., Walters, J.J., Fox, K.F., Fox, A., Creek, K.E., Piris, L., Wunschel, D.S., Smith, R.D., Tabb, D.L., and Yates, J.R.I., MS for identification of single nucleotide polymorphisms and MS/MS for discrimination of isomeric PCR products, *Anal. Chem.*, 72, 4033–4040, 2000.
409. Flora, J.W. and Muddiman, D.C., Comprehensive nomenclature for the fragment ions produced from collisional activation of peptide nucleic acids, *Rapid Commun. Mass Spectrom.*, 12, 759–762, 1998.
410. Little, D.P., Aaserud, D.J., Valaskovic, G.A., and McLafferty, F.W., Sequence information from 42-108-mer DNAs (complete for a 50-mer) by tandem mass spectrometry, *J. Am. Chem. Soc.*, 118, 9352–9359, 1996.

411. Potier, N., VanDorsselaer, A., Cordier, Y., Roch, O., and Bischoff, R., Negative electrospray ionization mass spectrometry of synthetic and chemically modified oligonucleotides, *Nucleic Acids Res.*, 22, 3895–3903, 1994.
412. Limbach, P., Crain, P., and McCloskey, J., Molecular mass measurement of intact ribonucleic aids via electrospray ionization quadrupole mass spectrometry, *J. Am. Soc. Mass Spectrom.*, 6, 27–39, 1995.
413. Gaus, H.J., Owens, S.R., Winniman, M., Cooper, S., and Cummins, L.L., On-line HPLC electrospray mass spectrometry of phosphorothioate oligonucleotide metabolites, *Anal. Chem.*, 69, 313–319, 1997.
414. Huber, C.G. and Krajete, A., Comparison of direct infusion and on-line liquid chromatography/electrospay ionization mass spectrometry for the analysis of nucleic acids, *J. Mass Spectrom.*, 35, 870–877, 2000.
415. Premstaller, A., Ongania, K.H., and Huber, C.G., Factors determining the performance of triple quadrupole, quadrupole ion trap and sector field mass spectrometers in electrospray ionization tandem mass spectrometry of oligonucleotides. 1. Comparison of performance characteristics, *Rapid Commun. Mass Spectrom.*, 15, 1045–1052, 2001.
416. Premstaller, A. and Huber, C.G., Factors determining the performance of triple quadrupole, quadrupole ion trap and sector field mass spectrometer in electrospray ionization mass spectrometry. 2. Suitability for *de novo* sequencing, *Rapid Commun. Mass Spectrom.*, 15, 1053–1060, 2001.
417. Muddiman, D.C., Anderson, G.A., Hofstadler, S.A., and Smith, R.D., Length and base composition of PCR-amplified nucleic acids using mass measurements from electrospray ionization mass spectrometry, *Anal. Chem.*, 69, 1543–1549, 1997.
418. Muddiman, D.C., Null, A.P., and Hannis, J.C., Precise mass measurement of a double-stranded 500 base-pair (309 kDa) polymerase chain reaction product by negative ion electrospray ionization Fourier transform ion cyclotron resonance mass spectrometry, *Rapid Commun. Mass Spectrom.*, 13, 1201–1204, 1999.
419. Medzihradszky, K.F. and Burlingame, A.L., The advantages and versatility of a high-energy collision-induced dissociation-based strategy for the sequence and structural determination of proteins, *Methods*, 6, 284–303, 1994.
420. Tsuneyoshi, T., Ishikawa, K., Koga, Y., Naito, Y., Baba, S., Terunuma, H., Arakawa, R., and Prockop, D.J., Mass spectrometric gene diagnosis of one-base substitution from polymerase chain reaction amplified human DNA. *Rapid Commun. Mass Spectrom.*, 11, 719–722, 1997.
421. Krahmer, M.T., Johnson, Y.A., Walters, J.J., Fox, K.F., and Nagpal, M., Electrospray quadrupole mass spectrometry analysis of model oligonucleotides and polymerase chain reaction products: determination of base substitutions, nucleotide additions/deletions, and chemical modifications, *Anal. Chem.*, 71, 2893–2900, 1999.
422. Kong, Y., Zhu, Y., and Zhang, J., Ionization mechanism of oligonucleotides in matrix-assisted laser desorption/ionization time-of-flight mass spectrometry, *Rapid Commun. Mass Spectrom.*, 15, 57–64, 2001.
423. Juhasz, P., Roskey, M.T., Smirnov, I.P., Haff, L.A., Vestal, M.L., and Martin, S.A., Applications of delayed extraction matrix-assisted laser desorption ionization time-of-flight mass spectrometry to oligonucleotide analysis, *Anal. Chem.*, 68, 941–946, 1996.
424. Li, Y., Tang, K., Little, D.P., Köster, H., Hunter, R.L., and McIver, R.T.J., High resolution MALDI Fourier transform mass spectrometry of oligonucleotides, *Anal. Chem.*, 68, 2090–2096, 1996.
425. Berkenkamp, S. and Hillenkamp, F., Infrared MALDI mass spectrometry of large nucleic acids, *Science*, 281, 260–262, 1998.
426. Muddiman, D.C., Null, A.P., and Hannis, J.C., Precise mass measurement of a double-stranded 500 base-pair (309 kDa) polymerase chain reaction product by negative ion electrospray ionization Fourier transform ion cyclotron resonance mass spectrometry, *Rapid Commun. Mass Spectrom.*, 13, 1201–1204, 1999.
427. Taylor, G.R. and Robinson, P., The polymerase chain reaction: from functional genomics to high-school practical classes, *Current Opin. Biotechnol.*, 9, 35–42, 1998.
428. Graber, J.H., Smith, C.L., and Cantor, C.R., Differential sequencing with mass spectrometry, *Genetic Anal. Biomolec. Eng.*, 14, 215–219, 1999.
429. Jackson, P.E., Scholl, P.F., and Groopman, J.D., Mass spectrometry for genotyping: an emerging tool for molecular medicine, *Molec. Medicine Today*, 6, 271–276, 2000.
430. Wu, K., Shaler, T., and Becker, C., Time-of-flight mass spectrometry of underivatized single-standed DNA oligomers by matrix-assisted laser desorption, *Anal. Chem.*, 66, 1637–1645, 1994.

431. Li, J., Butler, J.M., Tan, Y., Lin, H., Royer, S., Ohler, L., Shaler, T.A., Hunter, J.A., Pollart, D.J., Monforte, J.A., and Becker, C.H., Single nucleotide polymorphism determination using primer extension and time-of-flight mass spectrometry, *Electrophoresis*, 20, 1258–1265, 1999.
432. Griffin, T.J. and Smith, L.M., Single-nucleotide polymorphism analysis by MALDI-TOF mass spectrometry, *Trends Biotechnol.*, 18, 77–84, 2000.
433. Jurinke, C., van den Boom, D., Jacob, A., Tang, K., Worl, R., and Köster, H., Analysis of ligase chain reaction products via matrix-assisted laser desorption/ionization time-of-flight mass spectrometry, *Anal. Biochem.*, 237, 174–181, 1996.
434. Little, D.P., Braun, A., Darnhofer-Demar, B., Frilling, A., Li-Y, McIver, R.T. and Köster, H., Detection of RET proto-oncogene codon 634 mutations using mass spectrometry, *J. Mol. Med.*, 75, 745–750, 1997.
435. Fei, Z., Ono, T., and Smith, L.M., MALDI-TOF mass spectrometric typing of single nucleotide polymorphisms with mass-tagged ddNTPs, *Nucleic Acids Res.*, 26, 2827–2828, 1998.
436. Fu, D.J., Tang, K., Braun, A., Reuter, D., Darnhofer-Demar, B., Little, D.P., O'Donnell, M.J., Cantor, C.R., and Köster, H., Sequencing exons 5 to 8 of the p53 gene by MALDI-TOF mass spectrometry, *Nature Biotechnol.*, 16, 381–384, 1998.
437. Kirpekar, F., Nordhoff, E., Larsen, L.K., Kristiansen, K., Roepstorff, P., and Hillenkamp, F., DNA sequence analysis by MALDI mass spectrometry, *Nucleic Acids Res.*, 26, 2554–2559, 1998.
438. Ross, P.L., Davis, P.A., and Belgrader, P., Analysis of DNA fragments from conventional and microfabricated PCR devices using delayed extraction MALDI-TOF mass spectrometry, *Anal. Chem.*, 70, 2067–2073, 1998.
439. Ross, P., Hall, L., and Haff, L.A., Quantitative approach to single-nucleotide polymorphism analysis using MALDI-TOF mass spectrometry, *Biotechniques*, 29, 620–629, 2000.
440. Higgins, G.S., Little, D.P., and Köster, H., Competitive oligonucleotide single-base extension combined with mass spectrometric detection for mutation screening, *BioTech.*, 23, 710–714, 1997.
441. Braun, A., Little, D.P., Reuter, D., Muller-Mysok, B., and Köster, H., Improved analysis of microsatellites using mass spectrometry, *Genomics*, 46, 18–23, 1997.
442. Little, D.P., Braun, A., O'Donnell, M.J., and Köster, H., Mass spectrometry from miniaturized arrays for full comparative DNA analysis, *Nature Med.*, 3, 1413–1416, 1997.
443. Hannis, J.C. and Muddiman, D.C., Genotyping short tandem repeats using flow injection and electrospray ionization Fourier transform ion cyclotron resonance mass spectrometry, *Rapid Commun. Mass Spectrom.*, 15, 348–350, 2001.
444. Ross, P.L., Lee, K., and Belgrader, P., Discrimination of single-nucleotide polymorphisms in human DNA using peptide nucleic acid probes detected by MALDI-TOF mass spectrometry, *Anal. Chem.*, 69, 4197–4202, 1997.
445. Jiang-Baucom, P. and Girard, J.E., DNA typing of human leukocyte antigen sequence polymorphism by peptide nucleic acid probes and MALDI-TOF mass spectrometry, *Anal. Chem.*, 69, 4894–4898, 1997.
446. Griffin, T.J., Tang, W., and Smith, L.M., Genetic analysis by peptide nucleic acid affinity MALDI-TOF mass spectrometry, *Nature Biotechnol.*, 15, 1368–1372, 1997.
447. Ross, P., Hall, L., and Haff, L.A., Quantitative approach to single-nucleotide polymorphism analysis using MALDI-TOF mass spectrometry, *Biotechniques*, 29, 620–629, 2000.
448. Griffin, T.J. and Smith, L.M., Genetic identification by mass spectrometric analysis of single-nucleotide polymorphism: ternary encoding of genotypes, *Anal. Chem.*, 72, 3298–3302, 2000.
449. Braun, A., Little, D.P., and Köster, H., Detecting CFTR gene mutations by using primer oligo base extension and mass spectrometry, *Clin. Chem.*, 43, 1151–1158, 1997.
450. Graber, J.H., O'Donnell, M.J., Smith, C.L., and Cantor, C.R., Advances in DNA diagnostics, *Current Opin. Biotechnol.*, 9, 14–18, 1998.
451. Little, D.P., Cornish, T.J., O'Donnell, M.J., Braun, A., Cotter, R.J., and Köster, H., MALDI on a chip: Analysis of arrays of low-femtomole to subfemtomole quantities of synthetic oligonucleotides and DNA diagnostic products dispensed by piezoelectric pipet, *Anal. Chem.*, 69, 4540–4546, 1997.
452. O'Donnell, M.J., Tang, K., Köster, H., Smith, C.L., and Cantor, C.R., High-density, covalent attachment of DNA to silicon wafers for analysis by MALDI-TOF mass spectrometry, *Anal. Chem.*, 69, 2438–2443, 1997.
453. Tang, K., Fu, D., Julien, D., Braun, A., Cantor, C., and Köster, H., Chip-based genotyping by mass spectrometry, *Proc. Natl. Acad. Sci. USA*, 96, 10016–10020, 1999.

454. Laken, S.J., Jackson, P.E., Kinzler, K.W., Vogelstein, B., Strickland, P.T., Groopman, J.D., and Friesen, M.D., Genotyping by mass spectrometric analysis of short DNA fragments, *Nature Biotechnol.*, 16, 1352–1356, 1998.
455. Ember, L.R., The nicotine connection, *C&EN*, 8–18, 1994.
456. Wei, J. and Lee, C.S., Polyacrylamide gel electrophoresis coupled with matrix-assisted laser desorption/ionization mass spectrometry for tRNA mutant analysis, *Anal. Chem.*, 69, 4899–4904, 1997.
457. Roest, P.A., Roberts, R.G., Sugino, S., van Ommen, G.J., and den Dunnen, J.T., Protein truncation test (PTT) for rapid detection of translation-terminating mutations, *Hum. Mol. Genet.*, 2, 1719–1721, 1993.
458. Garvin, A.M., Parker, K.C., and Haff, L., MALDI-TOF based mutation detection using tagged *in vitro* synthesized peptides, *Nature Biotechnol.*, 18, 95–97, 2000.
459. Hainaut, P., Soussi, T., Shomer, B., Hollstein, M., Greenblatt, M., Hovig, E., Harris, C.C., and Montesano, R., Database of p53 gene somatic mutations in human tumors and cell lines: updated compilation and future prospects, *Nucleic. Acids. Res.*, 25, 151–157, 1997.
460. Null, A.P. and Muddiman, D.C., Perspectives on the use of electrospray ionization Fourier transform ion cyclotron resonance mass spectrometry for short tandem repeat genotyping in the post-genome era, *J. Mass Spectrom.*, 36, 589–606, 2001.
461. McInnis, M.G., Lutfalla, G., Slaugenhaupt, S., Petersen, M.B., Uze, G., Chakravarti, A., and Antonarakis, S.E., Linkage mapping of highly informative DNA polymorphisms within the human interferon-alpha receptor gene on chromosome 21, *Genomics*, 11, 573–576, 1991.
462. Naito, Y., Ishikawa, K., Koga, Y., Tsuneyoshi, T., Terunuma, H., and Arakawa, R., Molecular mass measurement of polymerase chain reaction products amplified from human blood DNA by electrospray ionization mass spectrometry, *Rapid Commun. Mass Spectrom.*, 9, 1484–1486, 1995.
463. Naito, Y., Ishikawa, K., Koga, Y., Tsuneyoshi, T., Terunuma, H., and Arakawa, R. Genetic diagnosis by polymerase chain reaction and electrospray ionization mass spectrometry: detection of five base deletion from blood DNA of a familial adenomatous polyposis patient, *J. Am. Soc. Mass Spectrom.*, 8, 737–742, 1997.
464. Domon, B. and Costello, C.E., A systematic nomenclature for carbohydrate fragmentation in FAB-MS/MS spectra of glycoconjugates, *Glycoconjugate J.*, 5, 397–409, 1988.
465. Peter-Katalinic, J., Analysis of glycoconjugates by fast atom bombardment mass spectrometry and related MS techniques, *Mass Spectrom. Reviews*, 13, 77–98, 1994.
466. Harvey, D.J., Matrix-assisted laser desorption/ionization mass spectrometry of carbohydrates, *Mass Spectrom. Reviews*, 18, 349–450, 1999.
467. Spina, E., Cozzolino, R., Ryan, E., and Garozzo, D., Sequencing of oligosaccharides by collision-induced dissociation matrix-assisted laser desorption/ionization mass spectrometry, *J. Mass Spectrom.*, 35, 1042–1048, 2000.
468. Mizuno, M., Sasagawa, T., Dohmae, N., and Takio, K., An automated interpretation of MALDI/TOF postsource decay spectra of oligosaccharides. 1. Automatic peak assignment, *Anal. Chem.*, 71, 4764–4771, 1999.
469. Harvey, D.J., Identification of protein-bound carbohydrates by mass spectrometry, *Proteomics*, 1, 311–328, 2001.
470. Delaney, J. and Vouros, P., Liquid chromatography ion trap mass spectrometric analysis of oligosaccharides using permethylated derivatives, *Rapid Commun. Mass Spectrom.*, 15, 325–334, 2001.
471. Narayanan, S., Sialic acid as a tumor marker, *Ann. Clin. Lab. Sci.*, 24, 376–384, 1994.
472. Roboz, J., Suzuki, R., and Bekesi, G., Determination of total and neuraminidase-susceptible N-acetylneuraminic acid in normal and neoplastic cells by selected ion monitoring, *Anal. Biochem.*, 87, 195–205, 1978.
473. Kawai, T., Kato, A., Higashi, H., Kato, S., and Naiki, M., Quantitative determination of N-glycolylneuraminic acid expression in human cancerous tissues and avian lymphoma cell lines as a tumor-associated sialic acid by gas chromatography-mass spectrometry, *Cancer Res.*, 51, 1242–1246, 1991.
474. Juhasz, P. and Biemann, K., Mass spectrometric molecular-weight determination of highly acidic compounds of biological significance via their complexes with basic polypeptides, *Proc. Natl. Acad. Sci. USA*, 91, 4333–4337, 1994.
475. Juhasz, P. and Biemann, K., Utility of non-covalent complexes in the matrix-assisted laser desorption ionization mass spectrometry of heparin-derived oligosaccharides, *Carbohydr. Res.*, 270, 131–147, 1995.

476. Keiser, N., Venkataraman, G., Shriver, Z., and Sasisekharan, R., Direct isolation and sequencing of specific protein-binding glycosaminoglycans, *Nature Med.*, 7, 123–128, 2001.
477. Reuter, G. and Gabius, H.J., Eukaryotic glycosylation: whim of nature or mulipurpose tool?, *Cell Mol. Life Sci.*, 55, 368–422, 1999.
478. Dwek, M.V., Ross, H.A., and Leathem, A.J., Proteome and glycosylation mapping identifies post-translational modifications associated with aggressive breast cancer, *Proteomics*, 1, 756–762, 2001.
479. Medzihradszky, K.F., Gillece-Castro, B., Towsend, R., Burlingame, A.L., and Hardy, M., Structural elucidation of O-linked glycopeptides by high energy collision-induced dissociation, *J. Am. Soc. Mass Spectrom.*, 7, 319–328, 1996.
480. Hanisch, F.-G., Green, B.N., Bateman, R., and Peter-Katalinic, J., Localization of O-glycosylation sites of MUCI tandem repeats by QTOF ESI mass spectrometry, *J. Mass Spectrom.*, 33, 358–362, 1998.
481. Alving, K., Paulsen, H., and Peter-Katalinic, J., Characterization of O-glycosylation sites in MUC2 glycopeptides by nanoelectrospray QTOF mass spectrometry, *J. Mass Spectrom.*, 34, 395–407, 1999.
482. Hanish, F.G., Jovanovic, M., and Peter-Katalinic, J., Glycoprotein identification and localization of O-glycosylation sites by mass spectrometric analysis of deglycosylated/alkylaminylated peptide fragments, *Anal. Biochem.*, 290, 47–59, 2001.
483. Mirgorodskaya, E., Roepstorff, P., and Zubarev, R.A., Localization of O-glycosylation sites in peptides by electron capture dissociation in a Fourier transform mass spectrometer, *Anal. Chem.*, 71, 4431–4436, 1999.
484. Haynes, P.A. and Aebersold, R., Simultaneous detection and identification of O-GlcNAc-modified glycoproteins using liquid chromatography-tandem mass spectrometry, *Anal. Chem.*, 72, 5402–5410, 2000.
485. Harris, R.J., Leonard, C.K., Guzzetta, A.W., and Spellman, M.W., Tissue plasminogen activator has an O-linked fucose attached to threonine-61 in the epidermal growth factor domain, *Biochemistry*, 30, 2311–2314, 1991.
486. Medzihradszky, K.F., Gillece-Castro, B.L., Settineri, C.A., Townsend, R.R., Masiarz, F.R., and Burlingame, A.L., Structure determination of O-linked glycopeptides by tandem mass spectrometry, *Biomed. Environ. Mass Spectrom.*, 19, 777–781, 1990.
487. Macek, B., Hofsteenge, J., and Peter-Katalinic, J., Direct determination of glycosylation sites in O-fucosylated glycopeptides using nano-electrospray quadrupole time-of-flight mass spectrometry, *Rapid Commun. Mass Spectrom.*, 15, 771–777, 2001.
488. Fosteenge, J., Huwiler, J.G., Macek, B., Hess, D., Lawler, J., Mosher, D.F., and Peter-Katalinic, J., C-mannosylation and O-fucosylation of the thrombospondin type 1 module, *J. Biol. Chem.*, 276, 6485–6498, 2001.
489. Dell, A. and Morris, H.R., Glycoprotein structure determination by mass spectrometry, *Science*, 291, 2351–2356, 2001.
490. Hanisch, F., Uhlenbruck, G., Peter-Katalinic, J., and Egge, H., Structural studies on oncofetal carbohydrate antigens (CA 19-9, CA 50, and CA 125) carried by O-link sialyloligosaccharides on human amniotic mucins, *Carbohydr. Res.*, 178, 29–47, 1988.
491. Hanisch, F. and Reter-Katalinic, J., Structural studies on fetal mucins from human amniotic fluid. Core typing of short-chain O-linked glycans, *Eur. J. Biochem.*, 205, 527–535, 1992.
492. Schwonzen, M., Schmits, R., Baldus, S., Vierbuchen, M., Hanisch, F., Pfreundschuh, M., Diehl, V., Bara, J., and Uhlenbruck, G., Monoclonal antibody FW6 generated against a mucin-carbohydrate of human amniotic fluid recognises a colonic tumour-associated epitope, *Brit. J. Cancer*, 65, 559–565, 1992.
493. Burlingame, A.L., Characterization of protein glycosylation by mass spectrometry, *Current Opin. Biotechnol.*, 7, 4–10, 1996.
494. Mirgorodskaya, E., Krogh, T.N., and Roepstorff, P., Characterization of protein glycosylation by MALDI-TOFMS, *Methods Molec. Biol.*, 146, 273–292, 2000.
495. Lehmann, W.D., Bohne, A., and von der Lieth, C.W., The information encrypted in accurate peptide masses: improved protein identification and assistance in glycopeptide identification and characterization, *J. Mass Spectrom.*, 35, 1335–1341, 2000.
496. Goletz, S., Leuck, M., Franke, P., and Karsten, U., Structure analysis of acetylated and non-acetylated O-linked MUC1-glycopeptides by post-source decay matrix-assisted laser desorption/ionization mass spectrometry, *Rapid Commun. Mass Spectrom.*, 11, 1387–1398, 1997.
497. Goletz, S., Thiede, B., Hanisch, F.G., Schultz, M., Peter-Katalinic, J., Muller, S., Seitz, O., and Karsten, U., A sequencing strategy for the localization of O-glycosylation sites of MUC1 tandem repeats by PSD-MALDI mass spectrometry, *Glycobiol.*, 7, 881–896, 1997.

498. Alving, K., Korner, R., Paulsen, H., and Peter-Katalinic, J., Nanospray-ESI low-energy CID and MALDI post-source decay for determination of O-glycosylation sites in MUC4 peptides, *J. Mass Spectrom.*, 33, 1124–1133, 1998.
499. Hanisch, F., Stadie, T.R., Deutzmann, F., and Peter-Katalinic, J., MUC1 glycoforms in breast cancer cell line T47D as a model for carcinoma-associated alterations of O-glycosylation, *Eur. J. Biochem.*, 236, 318–327, 1996.
500. Hennebicq-Reig, S., Tetaert, D., Soudan, B., Kim, I., Huet, G., Briand, G., Richet, C., Demeyer, D., and Degand, P., O-glycosation and cellular differentiation in a subpopulation of mucin-secreting HT-29 cell line, *Exp. Cell Res.*, 235, 100–107, 1997.
501. Iida, S., Takeuchi, H., Kato, K., Yamamoto, K., and Irimura, T., Order and maximum incorporation of N-acetyl-D-galactosamine into threonine residues of MUC2 core peptide with microsome fraction of human-colon-carcinoma LS174T cells, *Biochem. J.*, 347, 535–542, 2000.
502. Inoue, M., Takahashi, S., Yamashina, I., Kaibori, M., Okumura, T., Kamiyama, Y., Vichier-Guerre, S., Cantacuzene, D., and Nakada, H., High density O-glycosylation of the MUC2 tandem repeat unit by N-acetylgalactosaminyltransferase-3 in colonic adenocarcinoma extracts, *Cancer Res.,* 61, 950–956, 2001.
503. Stadie, T.R., Chai, W., Lawson, A.M., Byfield, P.G., and Hanisch, F., Studies on the order and site specificity of GalNAc transfer to MUCI tandem repeats by UDP-GalNAc: polypeptide N-acetylgalactosaminyltransferase from milk or mammary carcinoma cells, *Eur. J. Biochem.*, 229, 140–147, 1995.
504. Muller, S., Goletz, S., Packer, N., Gooley, A., Lawson, A.M., and Hanisch, F.G., Localization of O-glycosylation sites on glycopeptide fragments from lactation-associated MUC1. All putative sites within the tandem repeat are glycosylation targets *in vivo*, *J. Biol. Chem.*, 272, 24780–24793, 1997.
505. Muller, S., Alving, K., Peter-Katalinic, J., Zachara, N., Gooley, A.A., and Hanisch, F.G., High density O-glycosylation on tandem repeat peptide from secretory MUC1 of T47D breast cancer cells, *J. Biol. Chem.*, 274, 18165–18172, 1999.
506. Henderson, R.A., Cox, A.L., Sakaguchi, K., Appella, E., Shabanowitz, J., and Hunt, D.F., Direct identification of an endogenous peptide recognized by multiple HLA-A2.1-specific cytotoxic T cells, *Proc. Natl. Acad. Sci. USA*, 90, 10275–10279, 1993.
507. Agrawal, B., Reddish, M.A., Christian, B., VanHeele, A., Tang, L., Koganty, R.R., and Longenecker, B.M., The anti-MUC1 monoclonal antibody BCP8 can be used to isolate and identify putative histocompatibility complex Class I associated amino acid sequences, *Cancer Res.*, 58, 5151–5156, 1998.
508. Inoue, M., Yamashina, I., and Nakada, H., Glycosylation of the tandem repeat unit of the MUC2 polypeptide leading to the synthesis of the Tn antigen, *Biochem. Biophys. Res. Commun.*, 245, 23–27, 1998.
509. Hefta, L., Chen, F., Ronk, M., Sauter, S., Sarin, V., Oikawa, S., Nakazato, H., Hefta, S., and Shively, J., Expression of carcinoembryionic antigen and its predicted immunoglobulin-like domains in HeLa cells for epitope analysis, *Cancer Res.*, 52, 5647–5655, 1992.
510. Kaplan, B.E., Hefta, L.J., Blake, R.C., Swiderek, K.M., and Shively, J.E., Solid-phase synthesis and characterization of carcinoembryonic antigen (CEA) domains, *J. Peptide Res.*, 52, 249–260, 1998.
511. Brawer, M.K., Prostate-specific antigen: current status, *CA Cancer J. Clin.*, 49, 264–281, 1999.
512. Belanger, A., van Halbeek, H., Graves, H., Grandbois, K., Stamey, T., Huang, L., Poppe, I., and Labrie, F., Molecular mass and carbohydrate structure of prostate specific antigen: studies for establishment of an international PSA standard, *The Prostate*, 27, 187–197, 1995.
513. Nagasaki, H., Watanabe, M., Komatsu, N., Kaneko, T., Dube, J.Y., Kajita, T., Saitoh, Y., and Ohta, Y., Epitope analysis of prostate-specific antigen (PSA) C-terminal-specific monoclonal antibody and new aspects for the discrepancy between equimolar and skewed PSA assays, *Clin. Chem.*, 45, 486–496, 1999.
514. Peter, J., Unverzagt, C., Krogh, T.N., Vorm, O., and Hoesel, W., Identification of precursor forms of free prostate-specific antigen in serum of prostate cancer patients by immunosorption and mass spectrometry, *Cancer Res.,* 61, 957–962, 2001.
515. Peter, J., Unverzagt, C., and Hoesel, W., Analysis of free prostate-specific antigen (PSA) after chemical release from the complex with a1-antichymotrypsin (PSA-ACT), *Clin. Chem.,* 46, 474–482, 2000.
516. Vaisanen, V., Lovgren, L., Hellman, J., Piironen, T., Lilja, H., and Pettersson, K., Characterization and processing of prostate specific antigen (hk3) and human glandular kallikrein (hK2) secreted by LNCaP cells, *Prostate Cancer Prostatic Diseases*, 2, 91–97, 1999.

517. Peter, J., Unverzagt, C., Lenz, H., and Hoesel, W., Purification of prostate-specific antigen from human serum by indirect imunosorption and elution with a hapten, *Anal. Biochem.*, 273, 98–104, 1999.
518. Ross, D.D., Gao, Y., Yang, W., Leszyk, J., Shively, J., and Doyle, L.A., The 95-kilodalton membrane glycoprotein overexpressed in novel multidrug-resistant breast cancer cells is NCA, the nonspecific cross-reacting antigen of carcinoembryonic antigen, *Cancer Res.*, 57, 5460–5464, 1997.
519. Glocker, M.O., Kalkum, M., Yamamoto, R., and Schreurs, J., Selective biochemical modification of functional residues in recombinant human macrophage colony-stimulating factor beta (rhM-CSF beta): identification by mass spectrometry, *Biochemistry*, 35, 14625–14633, 1996.
520. Wilkins, J.A., Cone, J., Randhawa, Z.I., Wood, D., Warren, M.K., and Witkowska, H.E., A study of intermediates involved in the folding pathway for recombinant human macrophage colony-stimulating factor (M-CSF): evidence for two distinct folding pathways, *Protein Sci.*, 2, 244–254, 1993.
521. Tuominen, H., Heikinheimo, P., Loo, B.M., Kataja, K., Oker-Blom, C., Uutela, M., Jalkanen, M., and Goldman, A., Expression and glycosylation studies of human FGF receptor 4, *Protein Expr. Purif.*, 21, 275–285, 2001.
522. Sheeley, D.M., Merrill, B.M., and Taylor, L.C.E., Characterization of monoclonal antibody glycosylation: comparison of expression systems and identification of terminal α-linked galactose, *Anal. Biochem.*, 247, 102–110, 1997.
523. Chou, T., Hart, G., and Dang, C., c-Myc is glycosylated at threonine 58, a known phosphorylation site and a mutational hot spot in lymphomas, *J. Biol. Chem.*, 270, 18961–18965, 1995.
524. Saarinen, J., Welgus, H.G., Flizar, C.A., Kalkkinen, N., and Helin, J., N-Glycan structures of matrix metalloproteinase-1 derived from human fibroblasts and from HT-1080 fibrosarcoma cells, *Eur. J. Biochem.*, 259, 829–840, 1999.
525. Huwiler, A., Kolter, T., Pfeilschifter, J., and Sandhoff, K., Physiology and pathophysiology of sphingolipid metabolism and signaling, *Biochim. Biophys. Acta*, 1485, 63–99, 2000.
526. Ghardashkhani, S., Gustavsson, M., Breimer, M., Larson, G., and Samuelsson, B., Negative electrospray ionization mass spectrometry analysis of gangliosides, sulphatides and cholesterol 3-sulphate, *Rapid Commun. Mass Spectrom.*, 9, 491–494, 1995.
527. Metelmann, W., Muthing, J., and Peter-Katalinic, J., Nano-electrospray ionization quadrupole time-of-flight tandem mass spectrometric analysis of a ganglioside mixture from human granulocytes, *Rapid Commun. Mass Spectrom.*, 14, 543–550, 2000.
528. Metelmann, W., Vukelic, Z., and Peter-Katalinic, J., Nano-electrospray ionization time-of-flight mass spectrometry of gangliosides from human brain tissue, *J. Mass Spectrom.*, 36, 21–29, 2001.
529. Metelmann, W., Peter-Katalinic, J., and Muthing, J., Gangliosides from human granulocytes: a nano-ESI QTOF mass spectrometry fucosylation study of low abundance species in complex mixtures, *J. Am. Soc. Mass Spectrom.*, 12, 964–973, 2001.
530. Hamanaka, Y., Hamanaka, S., Shinagawa, Y., Suzuki, T., Inagaki, F., Suzuki, M., and Suzuki, A., Ganglioside antigen of DU-PAN-2 in a human pancreatic cancer, *FEBS Lett.*, 353, 48–52, 1994.
531. Hamanaka, Y., Hamanaka, S., and Suzuki, M., Sialyl Lewis(a) ganglioside in pancreatic cancer tissue correlates with the serum CA 19-9 level, *Pancreas*, 13, 160–165, 1996.
532. Ladisch, S., Sweeley, C., Becker, H., and Gage, D., Aberrant fatty acyl alpha-hydroxylation in human neuroblastoma tumor gangliosides, *J. Biol. Chem.*, 264, 12097–12105, 1989.
533. Domon, B. and Costello, C.E., Structure elucidation of glycosphingolipids and gangliosides using high-performance tandem mass spectrometry, *Biochemistry*, 27, 1534–1542, 1988.
534. Lavie, Y., Cao, H., Bursten, S.L., Giuliano, A.E., and Cabot, M.C., Accumulation of glucosylceramides in multidrug-resistant cancer cells, *J. Biol. Chem.*, 271, 19530–19536, 1996.
535. Stroud, M., Handa, K., Salyan, M., Ito, K., Levery, S., Hakomori, S., Reinhold, B.B., and Reinhold, V.N., Monosialogangliosides of human myelogenous leukemia HL60 cells and normal human leukocytes. 1. Separation of E-selectin binding from nonbinding gangliosides,and absence of sialosyl-Le(x) having tetraosyl to octaosyl core, *Biochemistry*, 35, 758–769, 1996.
536. Stroud, M., Handa, K., Salyan, M., Ito, K., Levery, S., Hakomori, S., Reinhold, B.B., and Reinhold, V.N., Monosialogangliosides of human myelogenous leukemia HL60 cells and normal human leukocytes. 2. Characterization of E-selectin binding fractions, and structural requirements for physiological binding to E-selectin, *Biochemistry*, 35, 770–778, 1996.
537. Ito, A., Levery, S.B., Saito, S., Satoh, M., and Hakomori Si, S., A novel ganglioside isolated from renal cell carcinoma, *J. Biol. Chem.*, 276, 16695–16703, 2001.

538. Suetake, K., Tsuchihashi, K., Inaba, K., Chiba, M., Ibayashi, Y., Hashi, K., and Gasa, S., Novel modification of ceramide: rat glioma ganglioside GM3 having 3-O-acetylated sphingenine, *FEBS Lett.,* 361, 201–205, 1995.
539. Fredman, P., Mansson, J.E., Dellheden, B., Bostrom, K., and von Holst, H., Expression of the GM1-species, [NeuN]-GM1, in a case of human glioma, *Neurochem. Res.*, 24, 275–279, 1999.
540. Hamasaki, H., Aoyagi, M., Kasama, T., Handa, S., Hirakawa, K., and Taki, T., GT1b in human metastatic tumors: GT1b as a brain metastasis-associated ganglioside, *Biochim. Biophys. Acta*, 1437, 93–99, 1999.
541. Mckallip, R., Ruixiang, L., and Ladisch, S., Tumor gangliosides inhibit the tumor-specific immune response, *J. Immunol.*, 163, 3718–3726, 1999.
542. Roepstorff, P., Mass spectrometry in protein studies from genome to function, *Curr. Opin. Biotech.*, 8, 6–13, 1997.
543. Mosse, C.A., Meadows, L., Luckey, C.J., Kittlesen, D.J., Huczko, E.L., Slingluff, C.L., Shabanowitz, J., Hunt, D.F., and Engelhard, V.H., The class I antigen-processing pathway for the membrane protein tyrosinase involves translation in the endoplasmic reticulum and processing in the cytosol, *J. Exp. Med.*, 187, 37–48, 1998.
544. Castelli, C., Storkus, W., Maeurer, M., Martin, D., Huang, E., Pramanik, N., Nagabhushan, T., Parmiani, G., and Lotze, M., Mass spectrometric identification of a naturally processed melanoma peptide recognized by CD8 cytotoxic T lymphocytes, *J. Exp. Med.*, 181, 363–368, 1995.
545. Garcia, A.M., Ortiz-Navarrete, V.F., Mora-Garcia, M.L., Flores-Borja, F., Diaz-Quinonez, A., Isibasi-Araujo, A., Trejo-Becerril, C., Chacon-Salinas, R., Hernandez-Montes, J., Granados-Arreola, J., de Leo, C., and Weiss-Steider, B., Identification of peptides presented by HLA class I molecules on cervical cancer cells with HPV-18 infection, *Immunol. Lett.*, 67, 167–177, 1999.
546. Topalian, S.L., Gonzales, M.I., Parkhurst, M., Li, Y.F., Southwood, S., Sette, A., Rosenberg, S.A., and Robbins, P.F., Melanoma-specific CD4+ T cells recognize nonmutated HLA-DR-restricted tyrosinase epitopes, *J. Exp. Med.*, 183, 1965–1971, 1996.
547. Elder, J., Schnolzer, M., Hasselkus-Light, C., Henson, M., Lerner, D., Phillips, T., Wagaman, P., and Kent, S., Identification of proteolytic processing sites within the gag and pol polyproteins of feline immunodeficiency virus, *J. Virol.*, 67, 1869–1876, 1993.
548. Papac, D., Thornburg, K., Bullesbach, E., Crouch, R., and Knapp, D., Palmitylation of a G-protein coupled receptor. Direct analysis by tandem mass spectrometry, *J. Biol. Chem.*, 267, 16889–16894, 1992.
549. Jedrzejewski, P.T. and Lehmann, W.D., Detection of modified peptides in enzymatic digests by capillary liquid chromatography/electrospray mass spectrometry and a programmable skimmer CID acquisition routine, *Anal. Chem.*, 69, 294–301, 1997.
550. Rudd, P.M., Guile, G.R., Küster, B., Harvey, D.J., Opdenakker, C., and Dwek, R.A., Oligosaccharide sequencing technology, *Nature*, 388, 205–207, 1997.

Index

A

Accelerator mass spectrometry, 117
  food, 168
  HAA, 163
Accurate mass, 7, 64
Acetylation, 398
Acetylator phenotype, 144
N-Acetylcysteine, 330
Acquired immunity, 97
Acrylonitrile, 128, 129
Acute allograft rejection, 443
Adjuvant chemotherapy, 205
Adjuvant therapy, 100
Aflatoxins, 158
Air
  diesel fuels in, 152
  gasoline in, 152
Alkylating agents, mechanism of action, 211
Alkyl sulfonates, 224
Allele differentiation, 484
Allicin, 330
Ames test, 112
Amino acid residue masses, 374
4-Aminobiphenyl, hemoglobin adduct, 143
  bladder carcinoma, 145
  fetuses, 145
Aminoglutethimide, 273
Amsacrine, 245
Analyte volatility, 9
Analyzer(s)
  electrostatic, 43
  high-resolution, 18
  laser microprobe, 33
  low-resolution, 18
  magnetic, 46
  magnetic sector, 42
  quadrupole, 39, 40, 45
  tandem, 45
  TOF, 36
Anaphase, 87
Anaplasia, 85
Anchorage dependency, 86
Androgens, 277
Angiogenesis inhibitors, 290
Anthracyclines, 246, 247
Antibodies, 96
Antibody-directed enzyme prodrug therapy, 302
Antiestrogens, 278
Antigen presenting cells, 438
Antigens, 96, 402, 499
Antimetabolites, 265
Antimicrotubule agents, 254
Antineoplastic peptides, 377
Antisense oligonucleotides, 299
APCI, *see* Atmospheric pressure chemical ionization
Apoptosis, 86
D-Arabinitol, 271
Arachidonic acid
  cyclooxygenase products, 469
  metabolism, 468
Aromatase inhibitors, 273
Aromatic amines, 125
Aromatic hydrocarbons, 117
Aromatic pollutants in waste, 156
Array detector, 52
Arsenic, 177
  analysis, ICPMS, 178
  trioxide, 304
  urine, 178
Atmospheric pressure chemical ionization (APCI), 25
Autophosphorylation, 89

B

Base peak, 5
Basophilic leukemia, 417
BCNU, 213
Bc1XL, breast cancer and, 421
Benign tumors, 81
Benzene, 117
  binding, accelerator MS, 117
  emission, tailpipes, 153
  metabolism, 117, 118
  metabolites
    hemoglobin adducts, 120
    quantification, 119
  oxide, 119
  smoking and, 142
Benzidine, 125, 126
Benzidine-N-glucuronide, 126
Benzoquinones, 121
Bioinformatics, 63
Biological response modifiers (BRMs), 100
Biomarker(s), 113
  discovery, 407, 432
  DNA adducts, 113
  hemoglobin adducts, 116
Biotherapy, 100
Bladder cancer, 424
Bone marrow transplantation, minor histocompatibility antigen and, 447
Bottom up approach, 364
Brain stathmin isoforms, 392
Breast cancer, 418

Breast cancer susceptibility gene BRCA1, 485
BRMs, *see* Biological response modifiers
Broccoli, 328
Burkitt's lymphoma, 94, 409
Busulfan, 226
Butadiene
  macromolecular adducts, 133
  metabolism, 132

C

CAD, *see* Collisionally-activated decomposition
Calgranulin B, 430
Calicheamins, 290
Californium-252 plasma desorption, 35
Call kill hypothesis, 201
Camptothecin, 238
Cancer, *see also* specific type
  clinical presentation, 98
  complications, 101
  diagnosis, 98
  diagnostic workup, 98
  distribution, 82
  epidemiology, incidence, 82
  genetic abnormalities, 90
  as genetic disorder, 90
  incidence, 82
  risk factors, 83
  treatment, 98
Capillary electrochromatography (CEC), 56
Capillary electrophoresis, stacking in, 115
Capillary isotachophoresis (CITP), 56
Capillary zone electrophoresis (CZE), 56
Carbohydrates, nomenclature of CID of, 486
Carboplatin, 233
Carcinoembryonic antigen, 499
Carcinogen(s)
  absorption, 109
  biomarkers, 107
  in cigarette smoke, 138
  classification, 107
  distribution, 109
  identification, well water, 172
  known human, 108
  mechanics of action, 107
  possible human, 109
  probable human, 109
  risk assessment, 107
  storage, 109
Carcinogenesis, accumulation of mutations, 94
Carcinogenicity, evidence, 107
Carmustine, 213
β-Carotene, 325
Carotenoids, 325
Carrier effect, 67
Carrier-mediated transport, 111
Catechins, 321
CCNU, 213, 228
CEC, *see* Capillary electrochromatography
Cell
  cultures, 103
  cycle, 86
    checkpoints, 88
    cyclins, 88
    mitosis, 88
    phase-specific drugs, 202
  growth, abnormalities of, 85
  -mediated immunity, 97
Cellular chemoresistance, 432
Cellular communication, 85, 89
Cellular proliferation, 85
Ceramide(s), 462
  leukemic HL-60 cells, 463
  levels during apoptosis, 466
  lymphoid cell lines, 465
Channeltrons, 52
Channel type multipliers, 50
Chaperon 14-3-3σ, 423
Checkpoints, 88
Chemical carcinogenesis, key steps, 111
Chemical ionization (CI), 24
  gases, 25
  reagents, 25
Chemical messengers, 89
Chemical prevention, 305
Chemopreventives, classifications of, 306
Chemotherapy
  regrowth of cells during, 202
  types of, 205
Chlorambucil, 223
Chromium, 179
Chronic lymphocytic leukemia, 415
CI *see* Chemical ionization
CID
  constant neutral loss mode, 21
  precursor (parent) ion mode, 21
  precursor-product mode, 21
Cigarette smoke, carcinogens in, 138
Cisplatin, 229, 233
CITP, *see* Capillary isotachophoresis
CLARP, breast cancer and, 421
Class I MHC molecules, 437
Clinical trials, 206, 210
Clonality, 102
c-myc glycosylation, 506
Coal fly ash, 156
Coke plant workers, PAH profiles of, 123
Collisionally-activated decomposition (CAD), 60
Collision-induced dissociation, 60
Colon cancer, 429
Colonic crypts, 430
Colonic polyps, 430
Colony stimulating factors, 505
Combination chemotherapy, 204, 205
Combinatorial chemistry, 207
Complete response, 210
Complex mixtures, monitoring, 156
Compost-amended soil, 176
Cone voltage, 61
Constant neutral loss mode, 21
Constant neutral loss scanning, 62

Contact inhibition, 86
Continuous flow FAB, 30
Conversion dynode, 50
Coomassie staining, 367
Cotinine, 138
  formation, from nicotine, 139
  in hair, quantification, 141
  in saliva, quantification, 141
  in serum, quantification, 139
  in urine, quantification, 140
C-terminal degradation, 372
Cu/Zn superoxide dismutase, 417, 435
Cyclins, 88
Cyclophosphamide, 135, 211
  DNA adducts, 218
  quantification, 215
  in urine, 136
Cyclotron resonance analyzers, 14
Cysteinylation, 395
Cytogenetics, 99
Cytokine-regulated proteins, 406
Cytokines, 505
Cytostatic agents, 202, 203
Cytotoxic drugs
  exposure by health care personnel to, 134
  resistance, 204
  toxicity, 204
Cytotoxic therapy, 201
CZE, *see* Capillary zone electrophoresis

D

Dacarbazine, 229
Dactinomycin, 254
Daidzein, 316
DALPC, *see* Large protein complexes
Data systems, 15
Daunomycin, 246
Daunorubicin, 245
DE, *see* Delayed extraction
1-DE, 367
2-DE, 367
Debulking surgery, 100
Delayed extraction (DE), 39
De novo sequencing, 376
N-Desmethyltamoxifen, 287
Diagnostic cytology, 98
Diagnostic histology, 98
Diesel exhaust, 125
Diesel fuel, 154
Differential-display proteomics, 407
Differential sequencing, 476
Differentiating agents, 304
Diffuse tumors, 82
Dioxins, 170
Direct laser desorption ionization, 33
Disulfide
  bridges, 455
  isomerase, 408
DNA
  adducts, 91
    ethylene oxide, 130
    potential sites, 113
  analysis, 101
  damage
    oxidation, 179
    radiation, 179
  packaging, 94
  sequencing, 480
Dolastatin, 263
Double-focusing, 36, 46
Doxorubicin, 245
Drug(s)
  conjugates to proteins, 289
  development, 206, 209
  discovery, 206
Dwell time, 66

E

Edman degradation, 369
Eicosanoids, 468
Electroblotting, 367
Electron
  -capture dissociation, 60
  ionization, 10, 23
Electrospray ionization (ESI), 26
  mechanism, 27
  microfabricated, 29
  multiplexed, 29
  nanospray, 29
  silicone chip-based, 29
Electrostatic analyzer, 43
Elements, 177
ELISA, *see* Enzyme-linked immunosorbent assay
Ellipticine, 245
Endocrine
  ablation, 272
  therapy, 271
Endostatin, 295
Eniluracil, 267
α-Enolase, 408, 418
Environmental carcinogens, 137
Enzyme-linked immunosorbent assay (ELISA), 115
Epicatechin, 321
Epigallocatechin, 321
Epigenetic carcinogens, 107
1,2-Epoxitamoxifen, 286
Erythrocytes, 95
Erythroleukemia, 412
ESI, *see* Electrospray ionization
Esophageal cancer, 428
Ethylene oxide, 129
Etiologic factors, carcinogenic risk and, 84
Etoposide, 243
Evidence of carcinogenicity, 107
Exemestane, 276
ExPASy server, 380

F

FAB, *see* Fast atom bombardment
Facilitated transport, 111
Familial adenomatous polyposis, 482
Farnesylation, 396
Fast atom bombardment (FEB), 30
FD, *see* Field desorption
FI, *see* Field ionization
Fibroblast growth factor, 406
Field
  carcinogenesis, 305
  desorption (FD), 34
  ionization (FI), 34
Field disease, 424
FindMod, 386
Fine isotopic structure of proteins, 382
Fingerprinting, 370
Flat top peaks, 68
Flavanols, 321
Flavonoids, 307, 312
FLICE, 399
5-Fluorouracil (5-FU), 265
Food, 158, 308
Formestane, 278
Fotemustine, 228
Fourier transformation, 14
Fourier transform ion-cyclotron resonance (FTICR), 44
  advantages, 45
  steps of operation, 45
Frame-shift mutation, 91
FTICR, *see* Fourier transform ion-cyclotron resonance
5-FU, *see* 5-Fluorouracil
Fumagillol, 295
Functional status, 101

G

Gangliosides, 509
  glioma, 516
  immune response, 518
  lipids in, 510
  metastatic brain tumor, 516
  pancreatic cancer, 510
  renal cell carcinoma, 515
Garlic, 329
Gasoline, 125, 152
Gastrointestinal cancers, 428
Gene
  knockout, 104
  targeting, 101, 104
Genetic diagnosis, 480
Genistein, 297, 316, 318
Genitourinary malignancies, 424
Genotoxic carcinogens, 107
Genotyping, 476
GF, *see* Growth fraction
Ginsenosides, 322, 325
Glioblastoma, 436
Glioma, gangliosides, 516
Glow discharge, 35
Glucosylceramides, multidrug-resistant breast cancer, 512
Glutathione, 113
Glycans, 489
Glycoconjugates, 486
Glycoforms, 491
Glycolipids, 509
Glycoproteins, 491, 503
Glycosaminoglycans, 488, 489
Glycosphingolipids, 509
Glycosylation, 489, 491, 499
O-Glycosylation, 494
Gompertzian growth, 86
Grading, tumor, 99
Granulocytes, 95
Green tea, 321
GroEL, 459
Growth
  factors, 505
  fraction (GF), 86
  patterns, 85
Gynecologic malignancies, 418

H

HAA, *see* Heterocyclic aromatic amines
Head and neck cancer, 435
Hedamycin, 252
Helicobacter pylori, 176
Hematologic malignancies, 82, 409
Hematopoiesis, 95
Hemoglobin adducts
  advantages, 116
  ethylene, 129
  GC methods, 116
  propylene, 129
Hereditary cancers, 83
HERP index, *see* Human exposure/rodent potency index
Heterocyclic aromatic amines (HAA), 161
  accelerator MS, 163
  Danube River, 174
  DNA adducts, 166
  food, 163
  metabolic activation, 166
  quantification, 163
  structures, 164
  urine, 165
Hexamethylene bisacetamide, 304
Hexapoles, 41
High-energy collisions, 60
Human exposure/rodent potency (HERP) index, 108
Human myelogenous leukemia, monosialogangliosides, 514
Hyalurinan, 489
Hyaluronidase, 489
4-Hydroxyandrost-4-ene-3,17-dione, 275
α-Hydroxytamoxifen, 285

I

ICAT, *see* Isotope-coded affinity tags
ICP, arsenic, 178
IDEnT peptides, 382
Ifosfamide, 219
Imaging mass spectrometry, 407
Immobilized metal affinity chromatography, 388
Immune response, 97, 518
Immunity
  acquired 97
  cell-mediated, 97
Immunoassays, 114
Indirect immunosorption, 503
Induction chemotherapy, 205
Inductively coupled plasma, 34
Inlets
  batch, 53
  direct, 53
  gases, 53
  high throughput, 58
  membrane, 54
  reservoir, 53
  static direct, 54
  volatile liquids, 53
In-source fragmentation, 61
Instrumentation, for peptide mapping, 369
Intact proteins, mass, 381
Intercalators, 246
Interfaces
  capillary electrophoresis systems, 56
  electrophoresis, 57
  gas chromatograph, 54
  liquid chromatograph, 55
Internal standards, 67
Ion(s)
  activation methods, 59
  adduct, 6
  detectors, 14
  dimeric, 6
  fragment, 6
  guides, quadrupoles as, 41
  isobaric, 18
  isotopic, 6
  molecular, 6
  multiply charged, 5, 6
  negative, 6
  positive, 6
  precursor, 6
  product, 6
  progeny fragment, 6
  -sensitive photoplate detector, 51
  singly charged, 5, 6
  sources, 9
    chemical ionization, 24
    electron ionization, 10, 23
    electrospray ionization, 10
  traps (IT), 41
  types of, 6
Ionization
  atmospheric pressure chemical, 25
  electrospray, 26
  fast atom bombardment, 30
  field, 34
  field desorption, 34
  inductively coupled plasma, 34
  matrix-assisted laser desorption, 11, 31
  mechanisms of, 6
  photoionization, 33
  REMPI, 33
  SELDI, 33
  techniques, 9
  thermal, 36
Irinotecan, 238
Isobaric ions, 18
Isoflavones, 316, 318
  quantification, 314
  soy, 317
Isoflavonoids, 310, 314
Isothiocyanates, 327, 328
Isotope
  -coded affinity tags (ICAT), 385, 388
  dilution, 388
  patterns, 67
Isotopic clusters, 20
Isotopic double-depletion, 383
IT, *see* Ion traps

K

Knockout mouse, 105

L

Lab-on-a-chip, 59
Ladder sequencing, 371
Large nucleic acids, 475
Large protein complexes, direct analysis of, 375
Laser capture microdissection, 409
Laser microprobe analyzers, 33
Leaf burning, 157
Leukemias, 82
Lignans, 307, 310, 314
Limit of detection, 17
d-Limonene, 322
Lipids, 460, 510
Lipogenase products, 469
Liver cancer, 432
Lomustine, 213
Low-energy collisions, 60
Lung cancer, 434, 446
Lymphoblastoid cells, protein expression, 412
Lymphocytes, 96
Lymphomas, 82
Lysophospholipids, 470

M

t,t-MA, *see trans,trans*-Muconic acid
Macrophage-colony stimulating factor, 459, 505
Macrophages, 96
Magnetic analyzers, 42, 46
Magnetic sector analyzers, 14, 42
Mainstream smoke, 137
Major histocompatibility complexes (MHC), 437, 438
MALDI, *see* Matrix-assisted laser desorption ionization
Malignant mesothelioma, 489
Malignant neoplasms, 81
Malignant tumors, 81
Mascot, 379
Mass, 5
  accuracy, 16
  analyzer(s), 12, 36
    comparison, 13
    magnetic sector, 14
    transmission-quadrupole, 12
  -to-charge ratio (*m/z*), 5
  isotopomers, 68
  mapping, 370
  range, 16
  scale, linearity, 64
  spectra, 5
  spectral libraries, 63
  spectrometer systems, 8
Mass spectrometry
  global scope of, 18
  journals, 69
  /mass spectrometry (MS/MS), 59
  range of applications, 22
  resources, 69
  scope of, 22
  secondary-ion, 30
Matrix
  -assisted laser desorption ionization (MALDI), 11, 31
    advantages, 32
    on chips, 478
    matrices, 32
    mechanism, 32
  metalloproteinase, 507
Mattauch-Herzog design, 46
Matthieu diagram, 39
MCF-10, 419
Medroxyprogesterone acetate, 19
Melanoma, 435, 442
Melphalan, 222
Membrane inlet MS, 171
6-Mercaptopurine, 265
Metallothionein, 214, 224
Metaphase, 87
Metastable ions, 60
Metastatic brain tumor, ganglioside, 516
Methotrexate (MTX), 268
4-(Methylnitrosamino)-1-(3-pyridyl)-1-butanone (NNK), 147
  afferent, smoking cessation, 151
  cervical mucus, 148
  hemoglobin adducts of as dosimeters, 149
  newborns, 151
  nonsmokers, 150
  urinary metabolites, 149
MHC, *see* Major histocompatibility complexes
MHC-related peptides
  detection, 440
  sequencing, 440
Microarray systems, 59
Microchannel plate, 52
Microfabricated fluidic systems, 58
Microheterogeneity, 491
Microsatellite, 481
Microsomal mixed-function oxidases, 112
Microtubule-associated protein, phosphorylation, 392
Microtubules, 87, 254
Midkine, 404
Mitogen-activated protein kinase kinase, phosphorylation, 389
Mitosis, 87
Mitotic spindles, 256
Mitoxantrone, 248
Mixed linkage kinase, 418
MoAbs, *see* Monoclonal antibodies
Molecular dosimetry, 109
Molecular epidemiology, 82, 109
Molecular mass, 7
Molecular probes, 102
Molecular separator, 55
Molecular switches, 89
Monoclonal antibodies (MoAbs), 96, 505
Monoisotopic mass, 7
Mouse
  lymphoma, peptide antigens and, 446
  models, 104
MS-Fit, 379, 413
MS/MS, *see* Mass spectrometry/mass spectrometry
MS-Tag, 380
MTX, *see* Methotrexate
MUC1
  tandem repeat, 495
  underglycosylated, 497
Mucin, 492
  breast cancer and, 494
  colon cancer and, 494
(a)Mucin, 491
Muconic acid, children, 154
trans,trans-Muconic acid (*t,t*-MA), 118, 142
MUC4 peptides, 493
Multichannel detector, 52
Multidrug resistance, 205, 222, 503
Multiple endocrine neoplasia, 483
Multiple intestinal neoplasia mice, 432
Multiple myeloma, 82
Multipoint detectors, 52
Municipal sludge, 176
Murine B-lymphocytes, 418
Mutation(s), 90
  accumulation of, 94
  analysis, 480
Mutational hotspot in p53, 447
m/z, *see* Mass-to-charge ratio

N

Nano-electrospray, 29
Necrosis, 86
Neoplasm, 81
Neoplastic transformation, 103
Neovascularization, 290
Neuroblastoma, 436
Neuropeptides, 436
Nicotine
  cotinine formation from, 139
  metabolites, quantification, 137
  quantification, 137
6-Nitrochrysene, 128
Nitrofluoranthrenes, 155
Nitrogen mustards, 211
Nitropyrenes, 125, 155
Nitrosamines, 168
N-Nitrosodiethanolamine, 169
Nitrosoureas, 227
NNK, *see* 4-(Methylnitrosamino)-1-(3-pyridyl)-1-butanone
Nomenclature of CID
  of carbohydrates, 486
  of oligonucleotides, 472
  of peptides, 374
Nominal mass, 7
Noncovalent interactions, 449
Nonspecific immune response, 97
Nonsteroidal inhibitors, 273
Nonvolatile analytes, 11
Normalization, 5
Northern blotting, 102
N-terminal degradation, 372
Nucleic acids, 472
Nutritional prevention, 305

O

Occupational carcinogens, 117
Oligonucleotides, 472
  ESI spectra, 473
  MALDI-TOF, 474
  nomenclature of CID of, 472
Oltipraz, 306, 330
Oncogene(s), 91
  characteristics, 92
  expression, examples, 93
$^{16}O/^{18}O$ isotopic doublets, 377
Opportunistic fungal infections, 271
Orthogonal acceleration, 38
Orthogonal spraying, 28
Ovarian cancer, 423
Oxaliplatin, 235
Oxidation, DNA damage, 179

P

Paclitaxel, 258
  metabolism, 260
  quantification, 260
PAF, *see* Platelet-activating factor
PAH, *see* Polycyclic aromatic hydrocarbons
Palliative surgery, 100
Palmitoylation, 398
Pancreatic cancer, 433
Partial response, 210
Passive diffusion, 110
Passive smoking, 137
PCR, *see* Polymerase chain reaction
PepFrag, 380
PepIndent/Multi/Indent, 379
PepSea, 379
Peptide(s), 361
  antigens
    lung cancer and, 446
    in malignancies, 442
    mouse lymphoma and, 446
    renal carcinoma and, 443
  fragmentation, nomenclature, 373
  fragments, proteolytic, 368
  mapping, 370
  MUC4, 493
  nomenclature of CID of, 374
  nucleic acids, 399, 473, 478
  preconcentration for ES, 368
  sequence tags, 380
Peptide Search, 380
Pesticides, 170, 171
p53 gene, 480
Pharmacodynamics, 210
Pharmacogenomics, 204
Pharmacokinetics, 209
Phase I trials, 210
Phase II
  biotransformations, 112
  conjugating enzymes, 112
  trials, 210
Phase III trials, 210
Phase IV studies, 210
Phosphatases, 387
Phospholipid hydroperoxide glutathione peroxidase, 401
Phosphoproteins, 387
Phosphorylated residues, identification, 387
Phosphorylation, 387
Photodissociation, 60
Photodynamic therapy, 302
Photofrin, 302
Photoionization, 33
Phytoestrogens, 307, 313
Pituitary adenomas, 436
Pituitary proteins, 437
Plastic monomers, 128
Platelet-activating factor (PAF), 471, 472
Platelet-derived growth factor β receptor, 403
Platelets, 95
Platinum compounds, 229
  adducts, 236
  biotransformation, 235
  distribution, 230
  inductively-coupled, 233

JM216, 235
quantification, 231
Pleiotropin, 404
Point mutations, 91
Polycyclic aromatic hydrocarbons (PAH), 122, 142
air, 155
atmospheric mutagens, 155
food, 170
smoking and, 142
soil, 174
Polyglycylation, 398
Polymerase chain reaction (PCR), 102, 475
Polymorphism, 439
Polysaccharides, 488
Positional isotopomers, 68
Possible human carcinogens, 109
Postlabeling, 114
Post-source decay (PSD), 62
Posttranslational modifications, 385, 395
^{32}P-postlabeling, 114
Precursor
(parent) ion mode, 21
-product mode, 21
scanning, 62
Prenylation, 396
Primary chemotherapy, 205
Primer Oligo Base Extension, 478
Probable human carcinogens, 109
PROBE™, 478, 483
Procarbazine, 229
Procarcinogens, 112
Product-ion scanning, 62
ProFound, 379
Prohibitin, 417
Propylene oxide, 129, 130
Prostaglandins, 469
Prostate
acid phosphatase, 398, 426
cancer, 426
-specific antigen (PSA), 426, 500, 503
specific membrane antigen, 426
specific peptide, 426
Protein(s), 361
chip arrays, 368
conformation, 455, 458, 459
cytokine-regulated, 406
delivery into ion source, 365
digestion, 365
-DNA interactions, 453
drugs conjugated to, 289
fine isotopic structure of, 382
folding, 449, 455
identification
database searches, 377
by MS alone, 366
software for, 379
strategies, 363
induced by apoptosis, 410
intact, 381
isolation, 365
kinases, 387
-ligand interactions, 449
microtubule-associated, 392
p53, 419
pituitary, 437
profiles, 408
-protein interactions, 450
sequencing, Edman degradation, 369
translational repression, 394
truncation test, 480
tumor-associated, 407, 419
Proteoglycan, 491
Proteolytic ^{18}O labeling, 385
Proteome, 361
analysis, 361
technology, 361
urinary, 409
Proteomics, 361
differential-display, 407
global, 361
quantification, 383
targeted, 361
tumor, 407
Protonated molecule, 6
Proto-oncogenes, 91
Proximate carcinogens, 112
PSA, *see* Prostate-specific antigen
PSD, *see* Post-source decay
Purine analogs, 265
Pyrimidine analogs, 265

Q

QTOF operation, 47
Quadrupole analyzer, 39, 40, 45
Quadrupole ion traps, 13, 41
Quercetin, 313, 320
o-Quinone, 287
Quinone methide, 287

R

Radiation
DNA damage, 179
therapy, 100
Radioprotective response, 434
Raney nickel, 134
Ras
oncoprotein inhibitors, 301
protein-ligand interactions and, 449
Rayleigh instability limit, 26
Reactive ultimate carcinogen, generation, 110
Receptors, 89
Recombinant human macrophage-colony stimulating factor β, 461
Reflectrons, 38
Relative molecular mass, 7
REMPI, *see* Resonance-enhanced multiphoton ionization
REMPI-TOF, 156
Renal carcinoma, 424, 443

Renal cell carcinoma, gangliosides, 515
Resolution, 15
Resolving power, 15
Resonance-enhanced multiphoton ionization (REMPI), 33
13-*cis*-Retinoic acid, 331
Retinoids, 331
Reverse geometry, 46
RNA analysis, 101
Robotic liquid handling, 58

S

Sample introduction
  interfaces for, 52
  systems, 7
Sarcolysin-containing peptide, 378
Scan speed, 16
Scope of applications, mass spectra and, 17
Scoring algorithms, 379
Secondary ion mass spectrometry (SIMS), 30, 34
Second-hand smoking, 137
Second messengers, 89, 462
Secretory mucins, 492
SELDI, *see* Surface-enhanced laser desorption ionization
Selected ion monitoring (SIM), 65
  profiles, 66
  quantification, 67
Selected reaction monitoring (SRM), 62, 67
Selenium, 178, 331, 332
Selenoproteins, 333
Sensitivity, 17
Sequencing
  antineoplastic peptides, 377
  de novo, 376
  Edman degradation, 369
  exons, 480
  ladder, 371
  nano-ESI-MS/MS, 374
  tandem mass spectrometry, 372
SEQUEST, 380, 419, 421, 430
Sheathless interfaces, 56
Short oligonucleotide mass analysis, 479
Short tandem repeat, 481
Shotgun identification, 374
Sialic acids, 488
Sidestream smoke, 137
Signal transduction, 89
SIM, *see* Selected ion monitoring
SIMS, *see* Secondary ion mass spectrometry
Single cysteine-containing peptide, 382
Single nucleotide polymorphism, 476
Smoke exposure, 143
SNP genotyping, 477
SO, *see* Styrene-7,8-oxide
Soft ionization, 10
Software
  packages, 377
  for protein identification, 379
Soil, 171, 174
Solid tumors, 81
Southern blot analysis, 102
Soyfoods, 309
Soy isoflavone, 309, 317
Spark source, 35
Specificity, 17
Sphingosine, 462, 509
Sporadic cancers, 83
SRM, *see* Selected reaction monitoring
Stable isotope dilution, 67
Staging, tumor, 99
Stathmin, 392, 412, 413
Steroidal inhibitors, 275
Steroid-receptor complex, 272
Stratified randomization, 210
Styrene, 133, 134
Styrene-7,8-oxide (SO), 133
Sugar cane soot, 157
Sulforaphane, 328
Sulfur-containing compounds, 327
Suramin, 249
Surface-enhanced laser desorption ionization (SEDLI), 33, 368
  breast cancer, 423
  prostate specific antigen, 426
  -TOFMS for biomarker discovery, 432
Surface-induced dissociation, 60
Surface-noncovalent affinity, 489
SWISS-PROT, 381, 432
Synchronization, 201

T

TAAs, *see* Tumor-associated antigens
Tamoxifen
  mechanism of action, 278
  metabolism, 279, 282
  metabolites, 278
  quantification, 284
Tandem affinity purification, 452
Tandem analyzers, 45
Tandem mass spectrometry, 59
Tandem repeat(s), 492
  MUC1 peptide, breast cancer, 495
  unit, 499
Tandem-in-space techniques, 61
Tandem-in-time techniques, 63
Target dose, 111
Targeted drug delivery, 289
Targeted drug therapy, 289
Targretin, 332
Taxanes, 258
Taxol, 258
Taylor cone, 26
T-cell antigen receptor complex, 403
Tea, 321
Telomerase, 86, 399
Telophase, 87
Temoporfin, 304
Teniposide, 243
Terpenes, 322

Therapeutic drug monitoring, 210
Thermal ionization, 36
Thiotepa, 224
Time-of-flight (TOF)
  advantages, 37
  analyzers, 36
  orthogonal acceleration, 38
Time-lag focusing, 39
TNP-470 fumagillol, 295
Tobacco, 137
Tobacco-specific N-nitrosamines
  carcinogenic action, 146
  metabolism, 146
TOF, *see* Time-of-flight
Toluenediamines, 127
Top down approach, 364
Topoisomerase(s), 237
  inhibitors, 237, 239
  I inhibitors, 238
  II inhibitors, 243
Topoisomers, 237
Topotecan, 242
Toremifene, 288
Transcription, 362
Transcription factor PU.1, 453
Transcriptomes, 361
Transgenic animals, 104
Translational repression protein, 394
Transmission-quadrupole mass analyzer, 12
Trapped-ion techniques, 63
Treatment modalities, 100
Trichloroethylene, 157
Trimetrexate, 269
α-Tubulin isotypes, 423
Tubulins, 254
Tumor, 81
  -associated antigens (TAAs), 97
  -associated proteins, 407
  grading, 99
  immunology, 95
  markers, 99
  proteomics, 407
  -related proteins, 419
  staging, 99
  suppressor genes, 91, 94
    characteristics, 92
    examples, 93

U

Underglycosylated MUC1, 497
Urban air, 156
Uridine as growth factor, 406
Urinary metabolites, 313
Urinary proteome, 409

V

Vacuum systems, 15
Velbamine, 256
Velocity-focusing, 36
Vimentin, 408, 417
Vinblastine, 256
Vinca alkaloids, 256
Vincristine, 256
Vindoline, 256
Vinorelbine, 256
Vinyl chloride
  DNA adducts, 131
  monomer, 173
Vitamin E, 332
Vitamins, 331

W

Water, 171
  HAA, 174
  phenolic xenoestrogens, 174
  vinyl chloride, 173
Watson-Biemann separator, 55
Wine, 321

Y

Yeast two-hybrid system, 453

Z

Zinc finger 2, 455
Z-spray, 28